Pathways to Astronomy

Second Edition

Stephen E. Schneider
Professor of Astronomy
University of Massachusetts, Amherst

Thomas T. Arny
Professor of Astronomy, Emeritus
University of Massachusetts, Amherst

Boston Burr Ridge, IL Dubuque, IA New York San Francisco St. Louis
Bangkok Bogotá Caracas Kuala Lumpur Lisbon London Madrid Mexico City
Milan Montreal New Delhi Santiago Seoul Singapore Sydney Taipei Toronto

PATHWAYS TO ASTRONOMY, SECOND EDITION

This book is printed on acid-free paper.

1 2 3 4 5 6 7 8 9 0 QPD/QPD 0 9 8

ISBN 978-0-07-340445-5
MHID 0-07-340445-4

Publisher: *Thomas Timp*
Sponsoring Editor: *Debra B. Hash*
Director of Development: *Kristine Tibbetts*
Senior Developmental Editor: *Mary E. Hurley*
Senior Marketing Manager: *Lisa Nicks*
Project Manager: *April R. Southwood*
Senior Production Supervisor: *Sherry L. Kane*
Senior Media Project Manager: *Sandra M. Schnee*
Designer: *Tara McDermott*
Cover/Interior Designer: *Rokusek Design*
Lead Photo Research Coordinator: *Carrie K. Burger*
Photo Research: *Mary Reeg*
Compositor: *ICC Macmillan Inc.*
Typeface: *10.5/12 Minion*
Printer: *Quebecor World Dubuque, IA*

Library of Congress Cataloging-in-Publication Data

Schneider, Stephen E. (Stephen Ewing), 1957–
Pathways to astronomy / Stephen E. Schneider, Thomas T. Arny. —2nd ed.
p. cm.
Includes index.
ISBN 978-0-07-340445-5 — ISBN 0-07-340445-4 (hard copy : alk. paper) 1. Astronomy—Textbooks. I. Arny, Thomas. II. Title.
QB43.3.S36 2009
520–dc22

2008019729

www.mhhe.com

To my father, who taught me the night sky when I was little.

—Steve

ABOUT THE AUTHORS

Stephen Schneider is a professor and Thomas Arny is an emeritus professor in the Astronomy Department at the University of Massachusetts at Amherst, which is part of the Five College Astronomy Department (comprising faculty from UMass, Amherst, Hampshire, Mt. Holyoke, and Smith Colleges). Both are recipients of their college's Outstanding Teacher Award, and they have collectively taught introductory astronomy for over 50 years to students with a wide variety of backgrounds.

Professor Schneider's interest in astronomy began at the amateur level when he was a child. He pursued an astronomy degree from Harvard as an undergraduate and obtained his Ph.D. from Cornell. His dissertation work received the Trumpler Award of the Astronomical Society of the Pacific, and he was named a Presidential Young Investigator. He works closely with science teachers presenting many workshops and special courses each year. He also loves to draw and paint.

Thomas T. Arny received his undergraduate degree from Haverford College and his Ph.D. in astronomy from the University of Arizona. In addition to his interest in astronomy, he has a long-standing fascination with the natural world: weather (especially atmospheric optics such as rainbows), birds, wildflowers, and butterflies.

BRIEF CONTENTS

PART IV

STARS AND STELLAR EVOLUTION 395

PART V

GALAXIES AND THE UNIVERSE 565

CONTENTS

PREFACE

APPROACH

There are many astronomy textbooks available today, but *Pathways to Astronomy* offers something different . . .

Created by two veteran teachers of astronomy, both recipients of outstanding teaching awards, *Pathways* breaks down introductory astronomy into its component parts. The huge and fascinating field of astronomy is divided into 84 units from which you can selectively choose topics according to your interests, while maintaining a natural flow of presentation.

One of the frustrations created by other current astronomy textbooks is that each chapter covers such a wide array of topics and is wed to such a specific order of presentation that it is difficult for students to absorb the large amount of material, and it is difficult for the professor to link the chapter readings and review questions to his or her own particular approach to teaching the subject. Whether you are learning astronomy for the first time or teaching it for the tenth, *Pathways* offers greater flexibility for exploring astronomy in the way you want.

The unit structure allows the new learner and the veteran professor to relate the text more clearly to college lectures. Each unit is small enough to be easily tackled on its own or read as an adjunct to the classroom lecture. For the faculty member who is designing a course to relate to current events in astronomy or a particular theme, the structure of *Pathways* makes it easier to assign reading and worked problems that are relevant to each topic. For the student of astronomy, *Pathways* makes it easier to digest each topic and to clearly relate each unit to lecture material.

Each unit of *Pathways to Astronomy* is like a mini-lecture on a single topic or closely related set of ideas. The same material covered in other introductory astronomy texts is included, but it is broken up into smaller self-contained units. This gives greater flexibility in selecting topics than is possible with the wide-ranging chapter in a traditional text that covers the same material as four or five *Pathways* units.

Even though the units are written to be as independent as possible, they still flow naturally from one to the next or even in alternative orders—different *Pathways*—through the book. Professors can select units to fit their course needs and cover the units in the order they prefer. They can choose individual units that will be explored in lecture while assigning other units for self-study. Or they can cover all the units in full depth in a content-rich course. With the short length of units, students can more easily digest the material covered in an individual unit before moving on to the next unit. And because the questions and problems are focused on a single topic, it is much easier to determine mastery of each topic.

The unit format also provides an opportunity to take some extra steps beyond the ordinary text. The authors have included some material of special interest that introduces topics most introductory texts do not offer—for example, units on calendar systems and special relativity. More advanced material within a particular unit topic is also organized toward the end of the unit so that the essentials are covered first—also providing flexibility for assigning readings.

Pathways to Astronomy makes it easy to tailor readings and exercises so they fit best within a course's structure. It also provides opportunities to travel down some fascinating paths to enhance a course or to provide additional reading for advanced students.

NEW TO THIS EDITION

In every Unit

- **Added and revised figures:** Figures were updated throughout the book for clarity and to include some of the most interesting new images available. Nearly every unit has new or revised figures.
- **Revised text:** While updating recent results and findings, we have kept the order and content of presentation largely the same as the first edition. We have modified the text for greater clarity wherever we saw an opportunity. Revisions include but are not limited to:
  - **Unit 1:** Revised Pluto's status and introduced "dwarf planets."
  - **Unit 9:** Added information about the Maya long-count cycle.
  - **Unit 12:** Added discussion of "eccentricity" of elliptical orbit.
  - **Unit 13:** Moved explanation of altitude and azimuth coordinates from Unit 31. The analemma-shaped motion of the Sun is now introduced.
  - **Unit 18:** New section and figures illustrate different orbit shapes.
  - **Unit 19:** New section on calculating tidal effects.
  - **Unit 20:** Section added to specifically discuss conservation of mass.
  - **Unit 31:** Much-expanded discussion and figures about small telescopes for amateur use.
  - **Unit 32:** New planet definition and recent discoveries about trans-Neptunian objects now discussed.
  - **Unit 34:** Updated information and figures on latest findings about exoplanets.
  - **Unit 35:** Added discussion and illustration of magnetic reversals.
  - **Unit 36:** More thorough discussion of global warming and ozone depletion.
  - **Unit 40:** Recent findings about quantity of water in polar caps on Mars discussed, including *Reconnaissance Orbiter* image of polar cap canyon.
  - **Unit 46:** Further discussion of Pluto and Eris.
  - **Unit 48:** Expanded discussion and pictures of meteorite classes. Recent evidence of meteorite fragments in mammoth tusks also discussed.
  - **Unit 51:** Added discussion of proton storms and effect on satellites and power grids. Discussion of the Sun and global warming based on new sun-spot and sea surface temperature data also added.
  - **Unit 52:** Added explanations of transverse velocity and space velocity.
  - **Unit 60:** Text greatly revised in light of recent modeling and observations of protostars. Many new images from the Hubble Space Telescope added.
  - **Unit 64:** Used recent models of post-main-sequence evolution to update figures and text.
  - **Unit 66:** New section on hypernovae and gamma-ray bursts.
  - **Unit 71:** New section on evolution of the Galaxy through mergers.
  - **Unit 73:** Added discussion of motions perpendicular to the Galactic disk and dark matter.
  - **Unit 75:** "Red sequence" and "blue sequence" of galaxies now described.
  - **Unit 78:** New image and discussion of "Bullet Cluster" and implications for dark matter.
  - **Unit 81:** Expanded discussion and new illustrations of early universe.
  - **Unit 82:** New figures compare recent Supernovae Ia data and cosmological models, and a new section discusses the ultimate fate of the universe, heat death, and proton decay.
  - **Unit 83:** Substantial updates to discussion and figures describing origin of life on Earth.

- **Unit 84:** Updated information about current SETI efforts, including Allen array.
- **Revised and enhanced review questions at the end of every unit:** There are now about twice as many review questions, designed for a thorough review of each unit and organized to reflect the unit content.
- **Revised and enhanced "Problems" section at the end of every unit:** With input from five different researchers and educators, we more than doubled the number of quantitative questions, addressing a wider range of topics and levels in every unit.

Mathematical Symbols

- To help students work with the math and avoid confusion, throughout the text we have chosen unique symbols for each physical quantity and used them consistently.
- Where tradition dictates that the same letter be used for different quantities (such as acceleration and semimajor axis), we have used distinctive fonts.
- A new symbol glossary has been created as part of the glossary to help students keep track of the meaning of symbols in formulas.

Looking Up A new looking up piece, Looking Up #9: Southern Circumpolar Constellations, has been added at the beginning of the text.

Cosmic Periodic Table We have produced a periodic table geared toward astronomy. It is introduced in Unit 4 as part of the discussion of elements. This table is designed to convey information of interest to astronomers, such as:

- Relative abundances.
- Condensation temperatures.
- Mass excess (relative to iron-56).
- Origins of elements and radioactive lifetimes of unstable elements.

Star Charts We have added seasonal star charts, and to help students of astronomy learn about the night sky, we have added a month-by-month guide to finding the Moon and planets.

You may contact your McGraw-Hill sales representative for an even more detailed list of revision changes.

FEATURES

Looking Up Illustrations It can be challenging to link introductory astronomy to the sky around us. The nine "Looking Up" full-page art pieces provide another pathway to astronomy, connecting what we actually see when "looking up" at the night sky with the more academic side of astronomy. Each illustration displays a large-scale photograph of one or more constellations in the night sky. Each also contains close-up photographs and illustrations of some of the most interesting telescopic objects with cross-references to the text. Details are also given regarding the objects' distances from Earth, along with three-dimensional illustrations of some of the stars or other objects within the field of view. The Looking Up Illustrations begin on page xxv, following the Acknowledgments.

Star Chart and Planetarium Activities A good star chart and a planetarium program also help to link the study of astronomy to the night sky. *Pathways to Astronomy* offers a foldout star chart and seasonal star charts for northern hemisphere observers. We also provide the remarkable *Starry Night* software. These will help students to take that next step beyond the book—exploring the night sky.

Starry Night

See the Pathways *2nd edition ARIS site for online Starry Night™ planetarium exercises.*

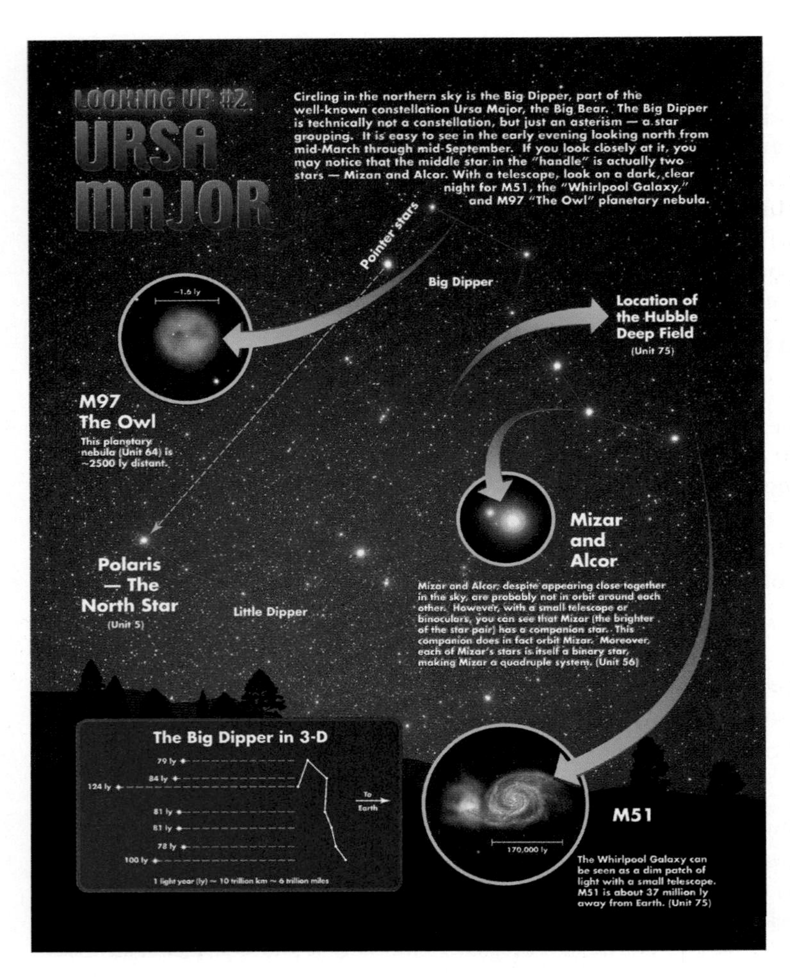

Detailed Art *Pathways to Astronomy* has taken each illustration a step further than the norm. Many figures are annotated to describe the processes that are actually happening within the illustration. Photos are often inserted next to the illustration for comparison so students can see the process in reality.

Writing Style *Pathways to Astronomy* provides coverage of technically complex ideas without confusing students. The authors give the students a reason to read every sentence. They relate many astronomical concepts to common, familiar experiences, and they convey the drama inherent in the processes at work in our universe. This engaging style helps students to learn from what they read, allowing the professor more class time to explore concepts in depth.

End of Unit Material The end-of-unit sections include hundreds of review questions, mathematical problems, and test yourself questions to help students master the unit material. These elements allow students to apply what they have learned before moving on to another unit. The answers for all the test yourself questions are provided at the back of the text.

Thought Questions Dozens of thought questions are scattered throughout the margins of the units. These questions are designed to invite students to think beyond the text, and to think about questions that have no easy answer.

Detailed Math Steps For some mathematical material appearing in the text, additional steps and explanations have been provided in the margins rather than including these within the text. This gives instructors flexibility in the depth of mathematical emphasis they wish to include.

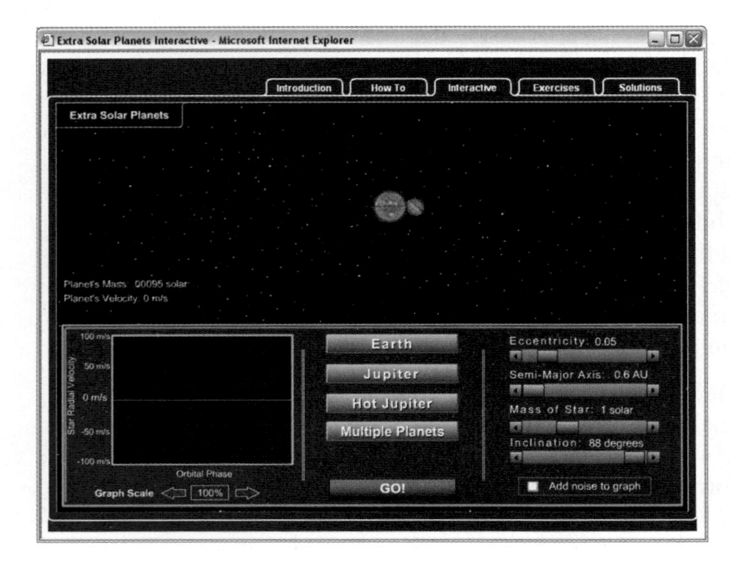

Electronic Media Integration with the Text Interactive and Animation icons have been placed where additional understanding can be gained through an animation or interactive experience on ARIS (see the description of the ARIS supplement).

Interactives McGraw-Hill is proud to bring you an assortment of 23 outstanding Interactives like no other. Each Interactive is programmed in Flash for a stronger visual appeal. These Interactives offer a fresh, dynamic method to teach astronomy basics. Each Interactive allows users to manipulate parameters and gain a better understanding of topics such as blackbody radiation, the Bohr model, a "solar system builder," retrograde motion, cosmology, and the H-R diagram by watching the effect of these manipulations. Each Interactive includes an analysis tool (interactive model), a tutorial describing its function, content describing its principal themes, related exercises, and solutions to the exercises. Users can jump between these exercises and analysis tools with just a mouse click.

SUPPLEMENTS

McGraw-Hill offers various tools and technology products to support *Pathways to Astronomy*. Instructors can obtain teaching aids by calling the Customer Service Department at 800-338-3987 or contacting their local McGraw-Hill sales representative.

Starry Night This planetarium software is available with every text. This software lets users investigate the inner-workings and features of the universe at varying levels of detail using a variety of tools, including the ability to examine the night sky on a date they enter from a location they enter.

www.mhhe.com/schneider

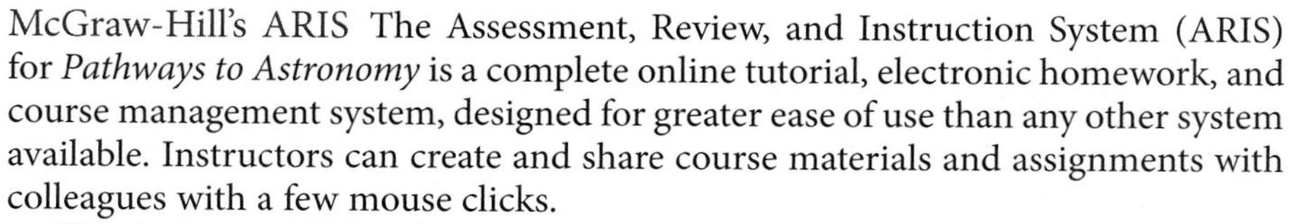

McGraw-Hill's ARIS The Assessment, Review, and Instruction System (ARIS) for *Pathways to Astronomy* is a complete online tutorial, electronic homework, and course management system, designed for greater ease of use than any other system available. Instructors can create and share course materials and assignments with colleagues with a few mouse clicks.

The design of ARIS makes it easy for students to take full advantage of the following tools:

- **Interactive student technology:** Includes 23 outstanding Astronomy Interactives, Animations, and Online Quizzes.
- **Text-specific features:** Includes Multiple-Choice Quizzes and Test Yourself Questions.
- **General astronomy features:** Includes Astronomy Time Line, Astronomy Picture of the Day, and Constellation Quizzes.
- **Additional instructor resources:** Includes Instructor's Manual, PowerPoint Lecture Outlines, introductory materials, sample syllabi, and clicker questions.

All assignments, quizzes, tutorials, and Interactives are directly tied to text-specific materials in *Pathways to Astronomy;* but instructors can also edit questions, import their own content, and create announcements and due dates for assignments. ARIS features automatic grading and reporting of easy-to-assign homework, quizzing, and testing. All student activity within McGraw-Hill's ARIS

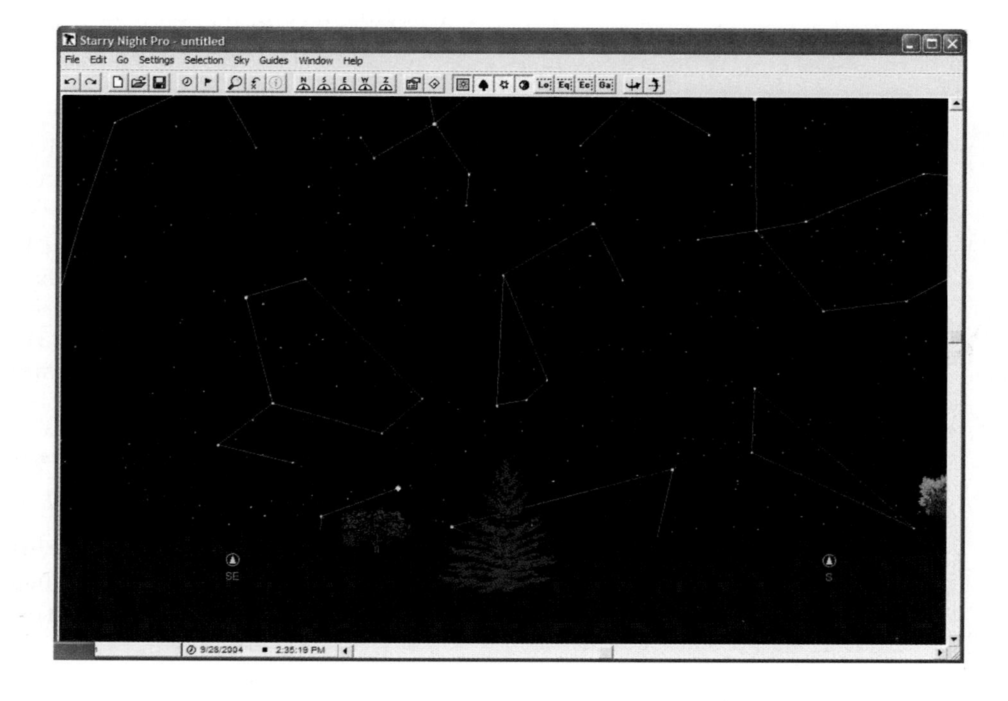

is automatically recorded and available to the instructor through a fully integrated grade book that can be downloaded to Excel.

COMPUTERIZED TEST BANK ONLINE

A comprehensive bank of test questions is provided within a computerized test bank powered by McGraw-Hill's flexible electronic testing program EZ Test Online (www.eztestonline.com). EZ Test Online allows you to create paper and online tests or quizzes in this easy-to-use program!

Imagine being able to create and access your test or quiz anywhere, at any time, without installing the testing software. Now, with EZ Test Online, instructors can select questions from multiple McGraw-Hill test banks or author their own, and then either print the test for paper distribution or give it online.

Test Creation

- Author/edit questions online using the 14 different question-type templates.
- Create printed tests or deliver online to get instant scoring and feedback.
- Create question pools to offer multiple versions online—great for practice.
- Export your tests for use in WebCT, Blackboard, PageOut, and Apple's iQuiz
- Compatible with EZ Test Desktop tests you've already created
- Sharing tests with colleagues, adjuncts, and TAs is easy

Online Test Management

- Set availability dates and time limits for your quiz or test
- Control how your test will be presented
- Assign points by question or question type with drop-down menu
- Provide immediate feedback to students or delay until all finish the test
- Create practice tests online to enable student mastery
- Your roster can be uploaded to enable student self-registration

Online Scoring and Reporting

- Automated scoring for most of EZ Test's numerous question types
- Allows manual scoring for essay and other open-response questions
- Manual re-scoring and feedback is also available
- EZ Test's grade book is designed to easily export to your grade book
- View basic statistical reports

Support and Help

- User's Guide and built-in page-specific help
- Flash tutorials for getting started on the support site
- Support Website: www.mhhe.com/eztest
- Product specialist available at 1-800-331-5094
- Online Training: http://auth.mhhe.com/mpss/workshops/

Presentation Center

Electronic book images and assets for instructors

Build instructional materials wherever, whenever, and however you want!

Accessed from *Pathways* ARIS website, **Presentation Center** is an online digital library containing photos, artwork, animations, and other media types that can be used to create customized lectures, visually enhanced tests and quizzes, compelling course websites, or attractive printed support materials. Assets are copyrighted by McGraw-Hill Higher Education, but can be used by instructors for classroom purposes. The visual resources in this collection include:

- **Art** Full-color digital files of all illustrations in the book can be readily incorporated into lecture presentations, exams, or custom-made classroom materials. In addition, all files are pre-inserted into PowerPoint slides for ease of lecture preparation.
- **Photos** The photos collection contains digital files of photographs from the text, which can be reproduced for multiple classroom uses.
- **Animations and interactives** Numerous full-color animations and the astronomy interactives, illustrating important processes, are also provided. Harness the visual impact of concepts in motion by importing these files into classroom presentations or online course materials.

Also residing on your textbook's ARIS website are:

- **PowerPoint Lecture Outlines** Ready-made presentations that combine art, and lecture notes are provided for each chapter of the text.
- **PowerPoint Slides** For instructors who prefer to create their lectures from scratch, all illustrations and photos are pre-inserted by chapter into blank PowerPoint slides.

Instructor's Manual The Instructor's Manual is found on ARIS and can be accessed only by instructors. This manual includes hints for teaching with this text, additional thought and discussion questions, answers to end-of-unit material, and sample syllabi.

Personal Response Systems and Questions Personal response systems "clickers" can bring interactivity into the classroom. A wireless response system gives the instructor and students immediate feedback from the entire class. The wireless response pads are essentially remotes that are easy to use and engage students. Clickers allow the instructor to motivate student preparation, interactivity, and active learning to receive immediate feedback and know what students understand. A text-specific set of questions, formatted in Powerpoint, is available via download from the *Pathways* ARIS site.

Electronic Books If you or your students are ready for an alternative version of the traditional textbook, McGraw-Hill brings you innovative and inexpensive electronic textbooks. By purchasing e-books from McGraw-Hill, students can save as much as 50 percent on selected titles delivered on the most advanced e-book platform available.

E-books from McGraw-Hill are smart, interactive, searchable, and portable, with such powerful tools as detailed searching, highlighting, note taking, and student-to-student or instructor-to-student note sharing. E-books from McGraw-Hill will help students to study smarter and quickly find the information they need. Students will also save money. Contact your McGraw-Hills sales representative to discuss e-book packaging options.

ACKNOWLEDGMENTS

Writing and revising a text such as *Pathways* is a collaboration with everyone who reads or uses it. We are deeply grateful to everyone who offered a suggestion, or pointed out a mistake, or found a place where we might improve the first edition. Our sincere thanks to the following for their constructive reviews, new ideas and suggestions:

B. N. Achar *University of Memphis*
Nadine Barlow *Northern Arizona University*
Lloyd Black *Rowan University*
Giovanni Bonvicini *Wayne State University*
James Brau *University of Oregon*
Steve Brown *Oklahoma State University, Oklahoma City*
Charles Burkhardt *St. Louis Community College, Flors Valley*
Jim Caffey *Drury University*
Michael Carini *Western Kentucky University*
Thomas Corwin *University of North Carolina, Charlotte*
John Cowan *University of Oklahoma, Norman*
Orville Day *East Carolina University*
Gregory Dolise *Harrisburg Area Community College*
Mark Edwards *Hofstra University*
Robert Egler *North Carolina State University*
Paul Eskridge *Minnesota State University, Mankato*
Juhan Frank *Louisiana State University*
Marc Gagne *West Chester University of Pennsylvania*
Richard Gelderman *Western Kentucky University*
Timothy Giblin *College of Charleston*
J. E. Hasbun *University of West Georgia*
Wayne Hayes *Greenville Technical College*
Melissa Hayes-Gehrke *University of Maryland, College Park*
David Hedin *Northern Illinois University*
Jeffrey Hopkins *Midlands Technical College*
James Imamura *University of Oregon*
Mike Inglis *Suffolk Community College, Selden*
Asaad Istephan *Madonna University*
William Keel *University of Alabama,Tuscaloosa*
William Koch *Johnson County Community College*
Thomas Krause *Towson University*
J. Patrick Lestrade *Mississippi State University*
Dave Ludwikoski *Community College of Baltimore County, Catonsville*
Norman Markworth *Stephen F. Austin State University*
Matt Marone *Mercer University*
Stephan Martin *Santa Rosa Junior College*
Mark McGovern *Antelope Valley College*
Janet Mclarty-Shroeder *Cerritos College*
Scott Miller, *Pennsylvania State University, University Park*
Andrea J. Pepper *Georgia Perimeter College*
Bradley Peterson *Ohio State University, Columbus*
Cynthia Peterson *University of Connecticut, Storrs*
Raymond Pfeiffer *The College of New Jersey*
Panos Photinos *Southern Oregon University*
Slawomir Piatek *Rutgers University*
Paul Sipiera *William Rainey Harper College*

John Stanford *Georgia Perimeter College*
Christopher Taylor *California State University, Sacramento*
Rusty Towell *Abilene Christian University*
Bruce Twarog *University of Kansas*
Paul Wiita *Georgia State University*
Lisa Will *Mesa Community College*

We are particularly indebted to the following, who provided detailed reviews for the second edition and helped us develop additional questions for each unit:

Lyle Ford *University of Wisconsin, Eau Claire*
Karl Haisch Jr. *Utah Valley State College*
Thomas Hockey *University of Northern lowa*
Ana M. Larson *University of Washington*
Henry Leckenby *St. Mary's University of Minnesota*
Craig Lincoln *St. Louis Community College*
Dwight Russell *Baylor University*
James Schombert *University of Oregon*
Paul J. Thomas *University of Wisconsin, Eau Claire*
Glenn P. Tiede *Bowling Green University*

Finally, we want to thank Wendy Nelson for careful editing of the revised manuscript, Elizabeth Berry for great help in revising the glossary, and the entire publishing team at McGraw-Hill for their hard work and encouragement (April Southwood, Liz Recker, and many others).In this edition, we were able to work much more directly with the electronic images and page proofs, providing new opportunities to adjust and improve the text and figures. Thanks to the K4 team at Mcgraw-Hill for helping to make this work. SES would also like to thank Cynthia Beth Rubin (*Rhode Island School of Design*) for teaching him some basics of working with digital media.

We also want to thank again those who reviewed the first edition through several stages of the manuscript, offering many suggestions and helping us to develop the design of *Pathways*.

Jennifer E. Bachman *Whatcom Community CoIlege*
William G. Bagnuolo, Jr. *Georgia State University*
Nadine Barlow *Northern Arizona University*
Cecilia Barnbaum *Valdosta State University*
Peter A. Becker *George Mason University*
Debra L. Burris *Oklahoma City Community College*
Eugene R. Capriotti *The Ohio State University*
George L. Cassiday *University of Utah*
Stan Celestian *Glendale Community College*
Thomas M. Corwin *UNC Charlotte*
John I. Cowan *University of Oklahoma*
Christopher]. Crow *Indiana/Purdue University, Fort Wayne*
Deborah M. Dann *Corning Community College*
Robert N. DeWitt *Penn State Altoona*
Ryan E. Droste *Trident Technical College, Charleston*
Robert J. Dukes, Jr. *College of Charleston*
Robert A. Egler *North Carolina State University*
Paul B. Eskridge *Minnesota State University*
Rica Sirbaugh French *MiraCosta College*
Robert B. Friedfeld *Stephen F. Austin State University*
Donna H. Gifford *Pima Community College*
Terry L. Goforth *Southwestern Oklahoma State University*

Alec Habig *University of Minnesota, Duluth*
Harold M. Hastings *Hofstra University*
Scott Hildreth *Chabot College*
Paul Hintzen *California State University, Long Beach*
N. Brian Hopkins *Rio Salado College*
William Hussong *College of DuPage*
James N. Imamura *University of Oregon*
Douglas R. Ingram *Texas Christian University*
Charles Kerton *Iowa State University*
Dave Kriegler *University of Nebraska*
Claud Lacy *University of Arkansas*
Henry J. Leckenby *University of Wisconsin, Oshkosh*
Loris Magnani *University of Georgia*
Phil Matheson *Utah Valley State College*
Roy C. McCord *Irvine Valley College*
José Mena-Werth *University of Nebraska at Kearney*
Zdzislaw E. Musielak *The University of Texas at Arlington*
Arnold L. O'Brien *University of Massachusetts, Lowell*
Cynthia Peterson *University of Connecticut*
Bob Powell *State University of West Georgia*
Mike Reed *Missouri State University*
Richard Rees *Westfield State College*
R. S. Rubins *University of Texas at Arlington*
M. Alper Sahiner *Seton Hall University*
Ann Schmiedekamp *Penn State University, Abington College*
Larry C. Sessions *Metropolitan State College of Denver*
James R. Sowell *Georgia Tech*
Michael E. Summers *George Mason University*
Sharon A. Swihart *University of Massachusetts/Amherst*
Lisa M. Will *Mesa Community College*
J. Wayne Wooten *Pensacola Junior College/University of West Florida*

Special thanks and appreciation also are due to those who contributed to the production of the ancillaries that accompany *Pathways to Astronomy:*

Cecilia Barnbaum *Valdosta State University*
Debra L. Burris *Oklahoma City Community College*
John J. Cowan *University of Oklahoma*
Patrick Koehn *Eastern Michigan State University*
David Kriegler *University of Nebraska*
Henry J. Leckenby *University of Wisconsin, Oshkosh*
Roy S. Rubins *University of Texas at Arlington*
Larry C. Sessions *Metropolitan State College of Denver*
Paul J. Thomas *University of Wisconsin, Eau Claire*
J. Wayne Wooten *Pensacola Junior College/University of West Florida*

Acknowledgments for Interactives

McGraw-Hill would like to thank Adam Frank, professor at the University of Rochester and president of Truth-N-Beauty, as well as the other employees of Truth-N-Beauty, especially Ted Pawlicki and Carol Latta.

LOOKING UP #1

NORTHERN CIRCUMPOLAR CONSTELLATIONS

For observers over most of the northern hemisphere, there are 5 constellations that are circumpolar (Unit 5), remaining visible all night long: Ursa Major (the Big Bear), Ursa Minor (the Little Bear), Cepheus (the King), Cassiopeia (the Queen), and Draco (the Dragon). The brightest stars in Ursa Major and Ursa Minor form two well-known asterisms: the Big and Little Dippers.

Delta Cepheus

A pulsating variable star (Unit 63) at a distance of ~300 ly.

Cassiopeia

Cepheus

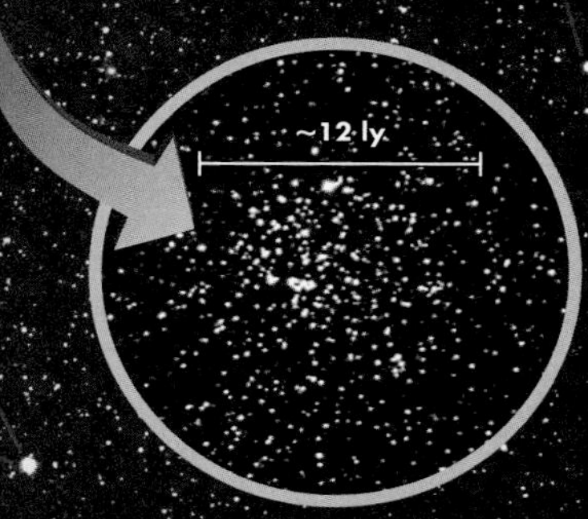

M52

This is an open star cluster (Unit 69). Its distance is uncertain — perhaps 3000–5000 ly.

Draco

Polaris (The North Star)

Little Dipper

M101

This spiral galaxy is ~27 million light years from us (Unit 75).

Thuban

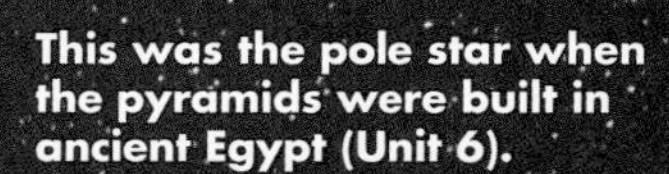

This was the pole star when the pyramids were built in ancient Egypt (Unit 6).

M81 and M82

Gravitational interactions between M81 and M82 have triggered star formation in both galaxies (Unit 75).

Big Dipper

Cassiopeia in 3-D

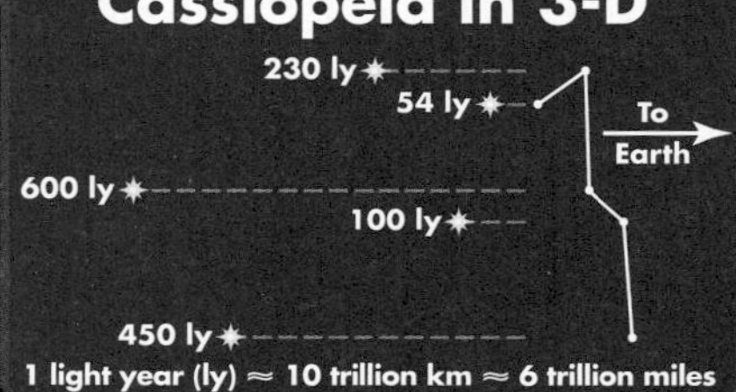

1 light year (ly) ≈ 10 trillion km ≈ 6 trillion miles

NORTH

LOOKING UP #2

URSA MAJOR

Circling in the northern sky is the Big Dipper, part of the well-known constellation Ursa Major, the Big Bear. The Big Dipper is technically not a constellation, but just an asterism — a star grouping. It is easy to see in the early evening looking north from mid-March through mid-September. If you look closely at it, you may notice that the middle star in the "handle" is actually two stars — Mizan and Alcor. With a telescope, look on a dark, clear night for M51, the "Whirlpool Galaxy," and M97 "The Owl" planetary nebula.

Pointer stars

Big Dipper

~1.6 ly

Location of the Hubble Deep Field

(Unit 75)

M97 The Owl

This planetary nebula (Unit 64) is ~2500 ly distant.

Mizar and Alcor

Mizar and Alcor, despite appearing close together in the sky, are probably not in orbit around each other. However, with a small telescope or binoculars, you can see that Mizar (the brighter of the star pair) has a companion star. This companion does in fact orbit Mizar. Moreover, each of Mizar's stars is itself a binary star, making Mizar a quadruple system. (Unit 56)

Polaris — The North Star

(Unit 5)

Little Dipper

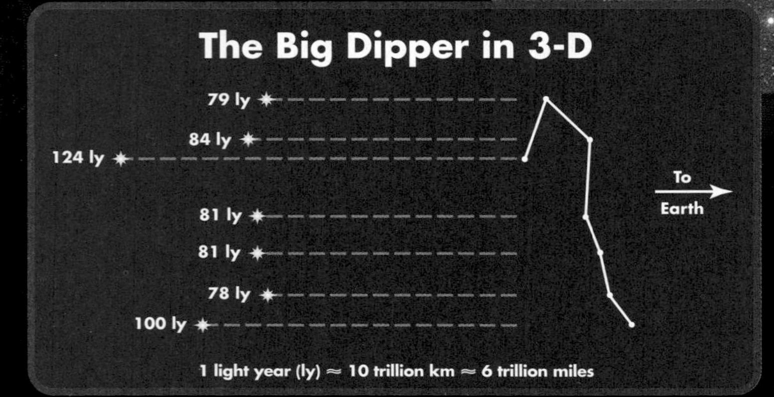

M51

The Whirlpool Galaxy can be seen as a dim patch of light with a small telescope. M51 is about 37 million ly away from Earth. (Unit 75)

LOOKING UP #3
M31 & PERSEUS

The galaxy M31 lies in the constellation Andromeda, near the constellations Perseus and Cassiopeia. Northern hemisphere viewers can see M31 dimly with the naked eye, but easily with binoculars. It is about 2.9 million ly from us. The best times to see it are August through December.

The Double Cluster

If you scan with binoculars from M31 toward the space between Perseus and Cassiopeia, you will see the Double Cluster — two groups of massive, luminous but very distant stars. The Double Cluster is best seen with binoculars. The two clusters are about 7000 ly away and a few hundred light years apart. (Unit 69)

~150,000 ly

M31 Andromeda Galaxy (Unit 76)

Perseus

Perseus is easy to identify: it looks a little like the Eiffel tower.

California Nebula

An emission nebula (Unit 72).

M45 Pleiades

(Unit 69)

Capella

The brightest star in the constellation Auriga, the Charioteer. A binary star (Unit 56)

Auriga

Perseus in 3-D

1 light year (ly) ≈ 10 trillion km ≈ 6 trillion miles

1300 ly
250 ly
230 ly
34 ly
110 ly
500 ly
100 ly
500 ly
560 ly
700 ly

To Earth

LOOKING UP #4
SUMMER TRIANGLE

The Summer Triangle consists of the three bright stars Deneb, Vega, and Altair. (Unit 58) These stars, the brightest to our eyes in the constellations Cygnus, Lyra, and Aquila, respectively, rise in the east shortly after sunset in late June and are visible throughout the northern summer and into late October (when they set in the west in the early evening). Deneb is intrinsically the brightest of the three, although Vega looks the brightest to us. Deneb looks dim only because it is so much farther from us than Vega and Altair.

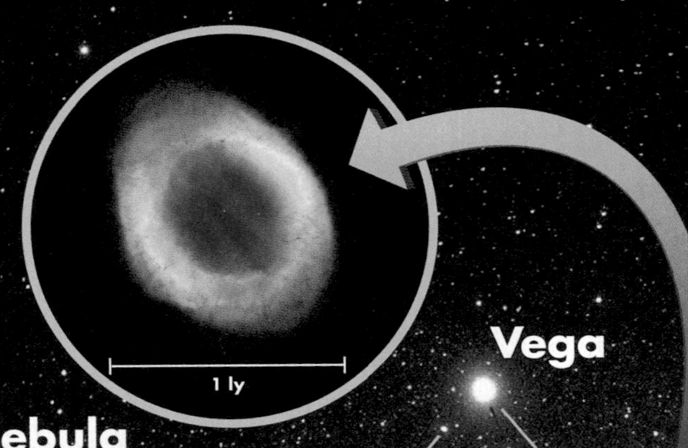

M57
ing Nebula

This planetary nebula (Unit 64) is about 2000 ly away from us. It is 7000 years old and has a white dwarf at its center.

Vega

Lyra

Cygnus

Epsilon Lyra

A double, double star (Unit 56)

Deneb

Deneb is a Blue Supergiant (Unit 66), one of the most luminous stars we can see, Deneb emits about 250,000 times more light than the Sun.

Altair

Alberio

This star pair (easily seen in a small telescope) shows a strong color contrast (orange and blue). Astronomers disagree about whether they orbit each other or just happen to lie in the same direction in the sky. (Unit 56)

M27 Dumbbell Nebula

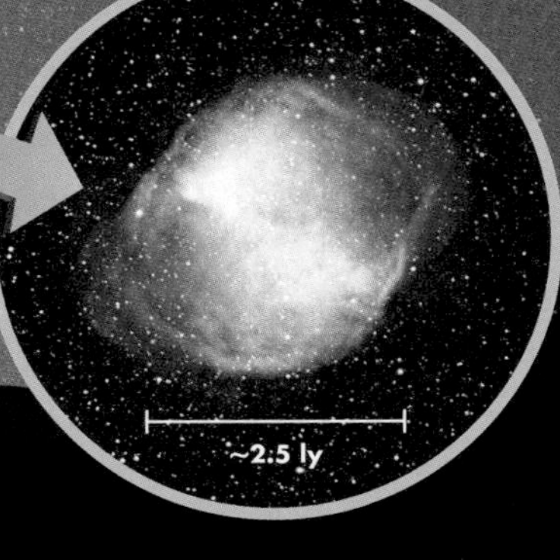

Another planetary nebula (Unit 64), The Dumbbell is about 900 ly distant and is about 2.5 ly in diameter.

The Summer Triangle in 3-D

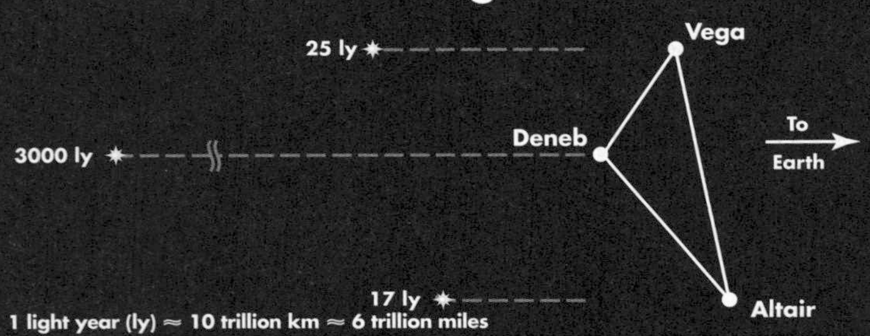

1 light year (ly) ≈ 10 trillion km ≈ 6 trillion miles

LOOKING UP #5

TAURUS

Taurus — The Bull
One of the constellations of the zodiac and one of the creatures hunted by Orion. Taurus is visible in the evening sky from November through March. The brightest star in Taurus is Aldebaran, the eye of the bull.

M1 The Crab Nebula

~7 ly

The Crab Nebula is the remnant of a star that blew up in the year AD 1054 as a supernova. It is about 5000 ly away from us. (Units 26 and 66)

Zeta Tauri

This open star cluster is easy to see with the naked eye and looks like a tiny dipper. It is about 400 ly from Earth. (Unit 69)

M45 Pleiades

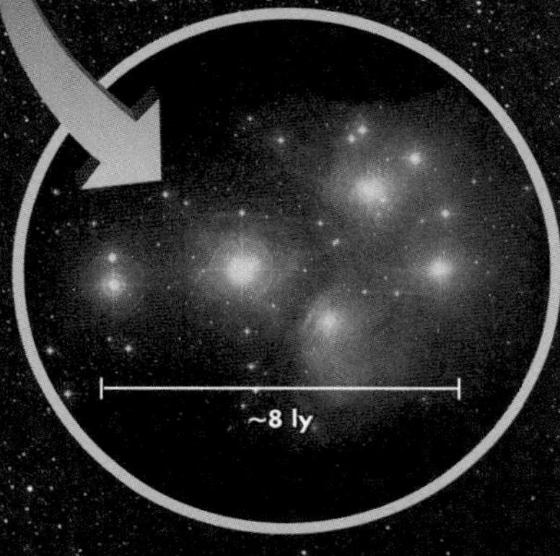

Aldebaran

Aldebaran is a red giant star. (Unit 62) It is about 65 ly from us and its diameter is about 45 times the diameter of the Sun. It lies between us and the Hyades.

Hyades

The "V" in Taurus is a nearby star cluster, about 137 ly away. It is easy to see its many stars with binoculars. (Unit 69)

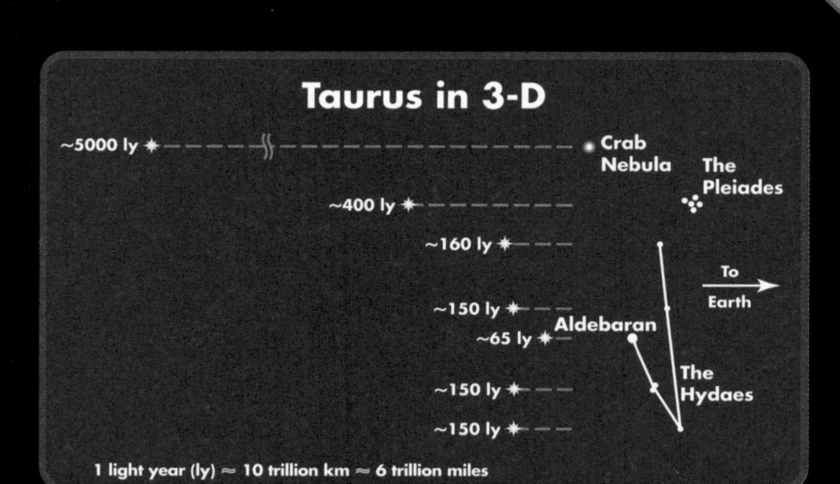

LOOKING UP #6

ORION

Orion is easy to identify because of the three bright stars of his "belt." You can see Orion in the evening sky from November to April, and before dawn from August through September.

Betelgeuse

Betelgeuse is a Red Supergiant star (Unit 62) that has swelled to a size that is as large as our Solar System out to Mars. Its red color indicates that it is relatively cool for a star, about 3000 Kelvin.

Horsehead Nebula

The horsehead shape is caused by dust in an interstellar cloud blocking background light. (Unit 72)

1.8 ly

3 ly

M42 Orion Nebula

The Orion Nebula is an active star-forming region rich with dust and gas. (Units 60, 72)

1 ly

1 ly

Rigel

Rigel is a Blue Supergiant star (Unit 66). Its blue color indicates a surface temperature of about 10,000 Kelvin.

Orion in 3-D

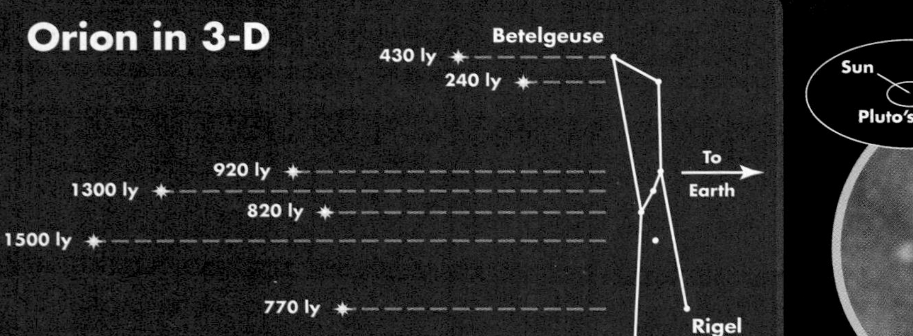

Sun

Pluto's orbit

A protoplanetary disk

This is the beginning of a star; our early Solar System may have looked like this! (Unit 34)

LOOKING UP #7

SAGITTARIUS

Sagittarius marks the direction to the center of the Milky Way. It is best located by its "teapot" shape, with the Milky Way seeming to rise like steam from the spout. For northern latitude viewers, the constellation is best seen in the evening, July to September. For such viewers, it is low on the southern horizon. Many star-forming nebulae are visible in this region. (Units 60,72)

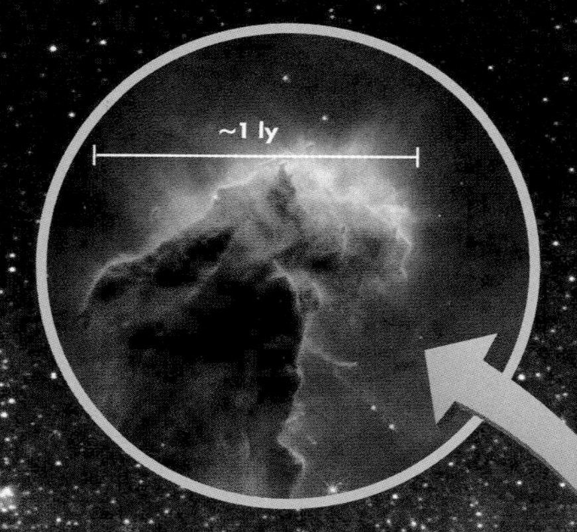

M16 Eagle Nebula

This young star cluster and the hot gas around it lie about 7000 ly from Earth.

~70 ly

M17 Swan Nebula

M20 Trifid Nebula

The distance to the Trifid Nebula (so named because of the black streaks that divide it into thirds) is very uncertain. It lies between 2000 and 8000 ly away. This makes its size very uncertain, too.

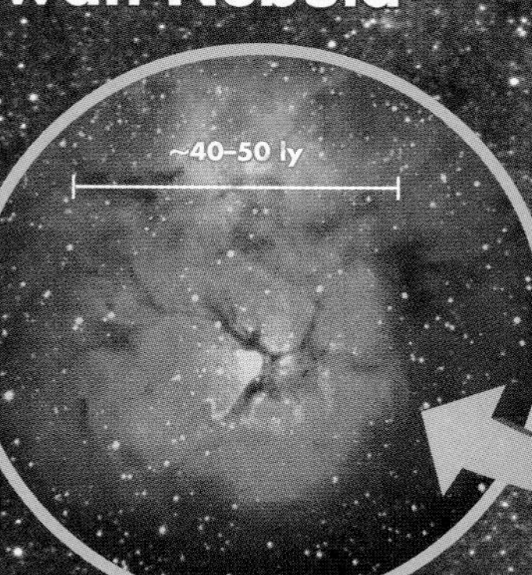

M22

~100 ly

M22 is one of the many globular clusters that are concentrated toward the core of our galaxy. Easy to see with binoculars, it is just barely visible to the naked eye. It is about 11,000 ly away from us. (Unit 69)

Center of the Milky Way

(Unit 73)

M8 Lagoon Nebula

The "teapot" of Sagittarius

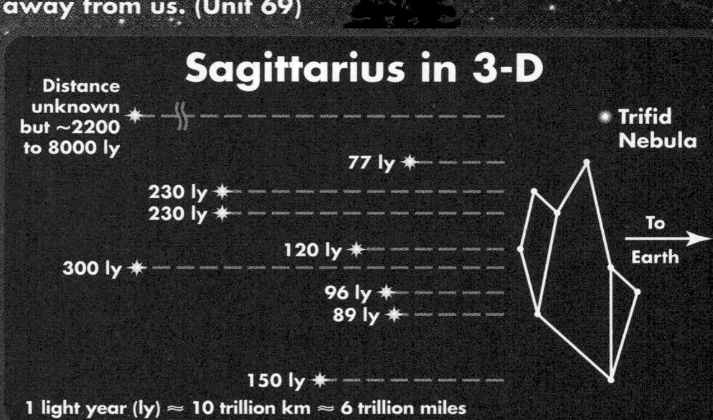

LOOKING UP #8

CENTAURUS AND CRUX, THE SOUTHERN CROSS

These constellations are best observed from the southern hemisphere. Northern hemisphere viewers can see Centaurus low in the southern sky in May – July. Crux may be seen just above the southern horizon in May and June from the extreme southern US (Key West and South Texas).

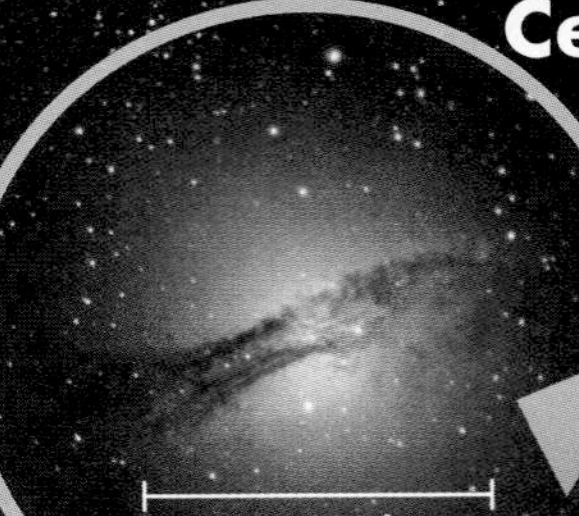

Centaurus A

This unusual galaxy, ~11 million ly distant, is one of the brightest radio sources in the sky (Unit 77)

Alpha Centauri

Proxima Centauri

This dim star is the nearest star to the Sun, 4.22 ly distant (Unit 52)

~50 ly

The Jewel Box

NGC 4755, an open star cluster (Unit 69) ~500 ly from us.

Crux
The Southern Cross

The Coal Sack

An interstellar dust cloud (Unit 72)

Eta Carinae

A very high-mass star doomed to die young (Unit 60) ~8000 ly distant

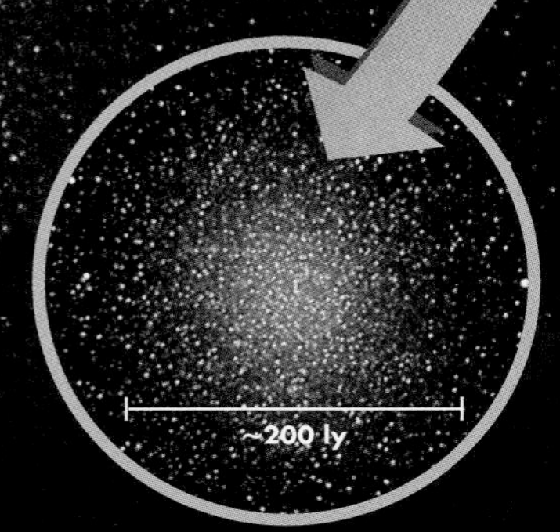

Omega Centauri

The largest globular cluster in the Milky Way (Unit 69), ~17,000 ly distant and containing millions of stars

Southern Cross in 3-D

352 ly
321 ly
228 ly
363 ly
88 ly

To Earth

1 light year (ly) ≈ 10 trillion km ≈ 6 trillion miles

LOOKING UP #9
SOUTHERN CIRCUMPOLAR CONSTELLATIONS

Most of the constellations in this part of the sky are dim, but observers in much of the southern hemisphere can see the Magellanic Clouds circling the south celestial pole throughout the night.

Crux
The Southern Cross

Musca

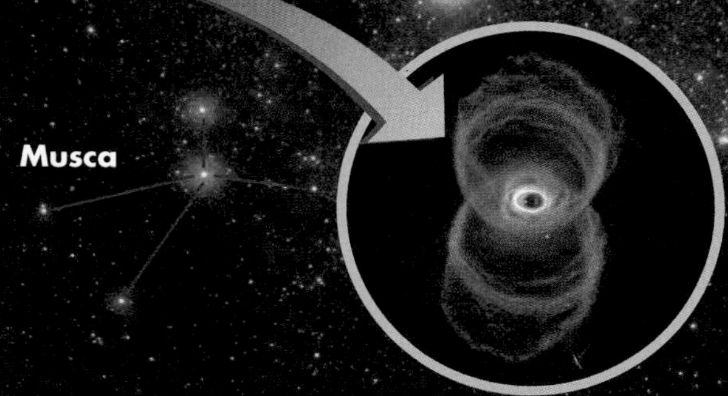

Hourglass Nebula

A planetary nebula (Unit 64) ~8000 ly distant

Apus

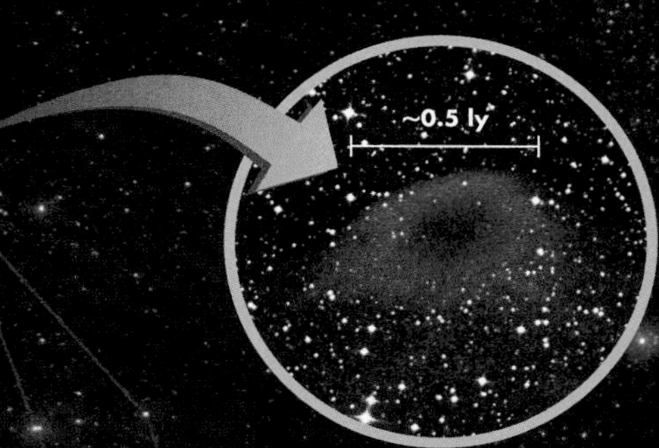

Thumbprint Nebula

A Bok globule (Unit 60) about 600 ly distant

Octans
The constellation closest to the south celestial pole is named after a navigational instrument, the octant.

Chamaeleon

The South Celestial Pole

No bright stars lie near the south celestial pole, but the southern cross points toward it.

Volans

Mensa

Hydrus

Small Magellanic Cloud

A dwarf galaxy orbiting the Milky Way at a distance of about 200,000 ly (Unit 76)

Large Magellanic Cloud

A small galaxy orbiting the Milky Way at a distance of about 160,000 ly (Unit 76)

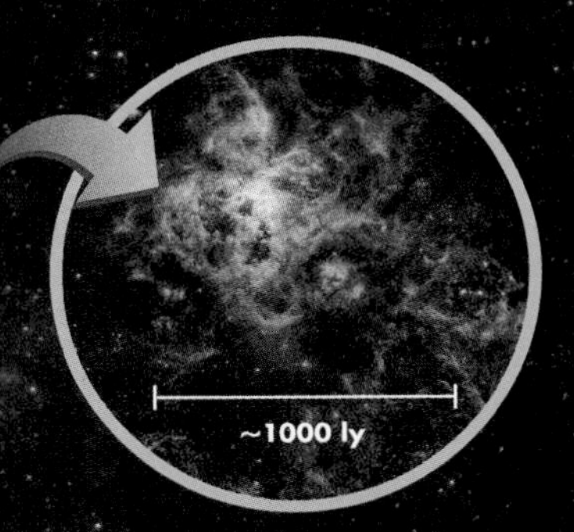

Tarantula Nebula

A star-formation region (Unit 60) in the Large Magellanic Cloud larger than any known in the Milky Way.

Pathways to Astronomy

Part One

The content of some Units will be enhanced if you have previously studied some earlier Units in this textbook. These Background Pathways are listed below each Unit image.

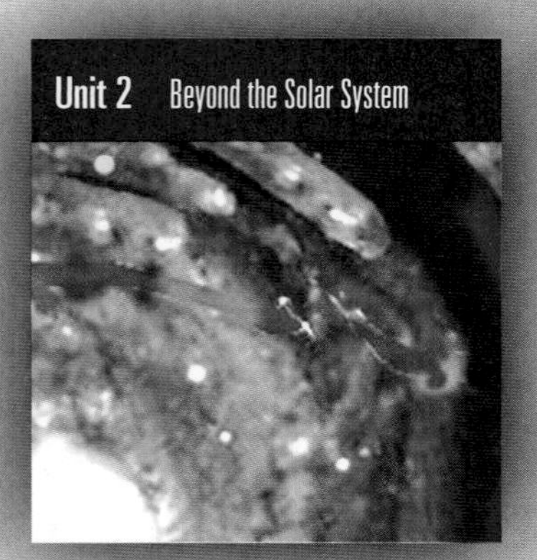

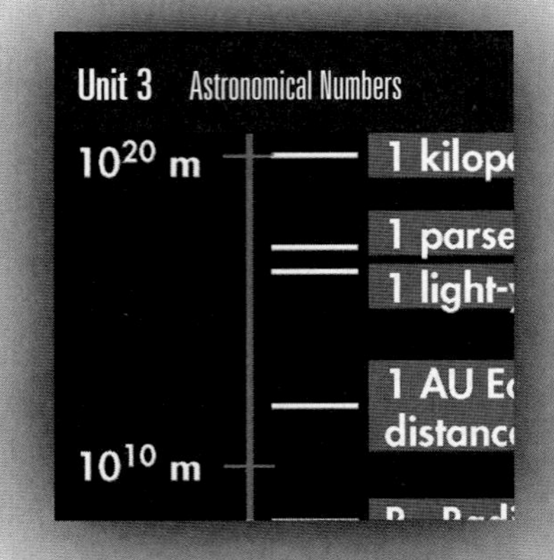

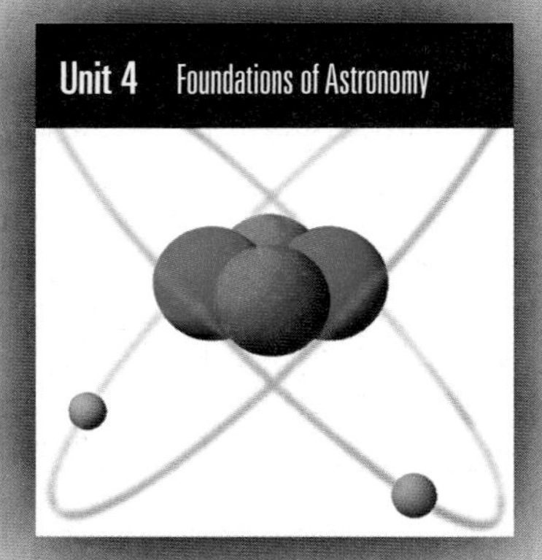

Background Pathways: Units 1 and 2

Background Pathways: Unit 5

Background Pathways: Unit 6

The Cosmic Landscape

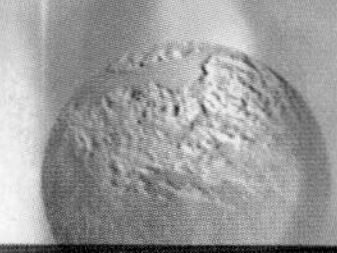

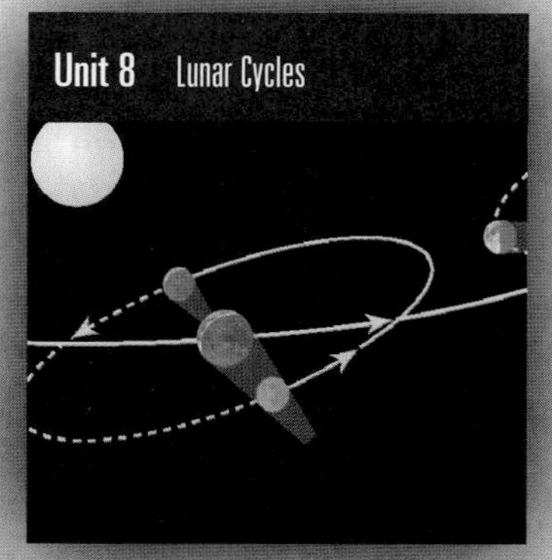

Background Pathways: Unit 6

Background Pathways: Units 6 and 8

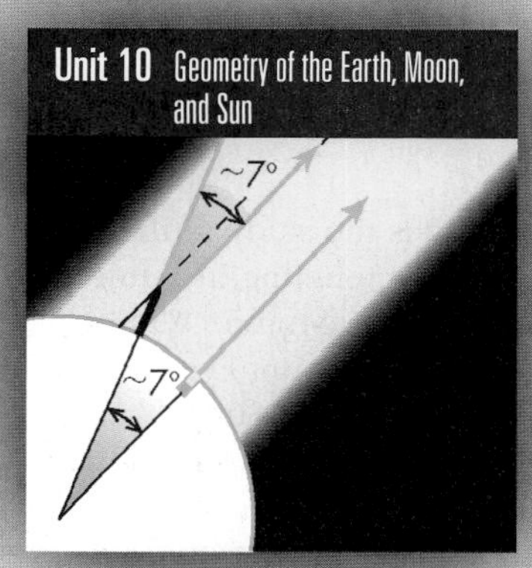

Background Pathways: Unit 8

Background Pathways: Units 1 and 6

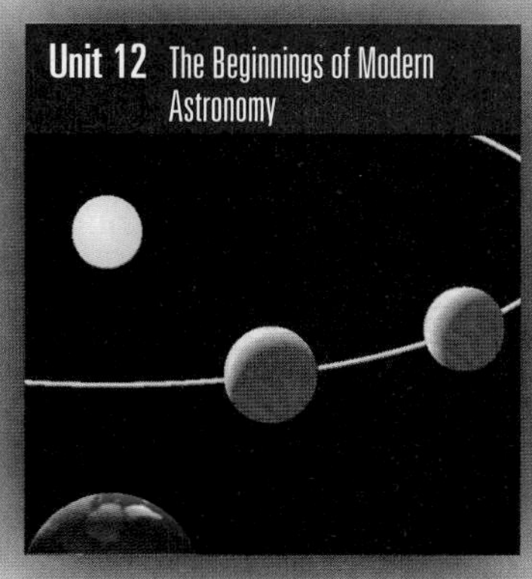

Background Pathways: Unit 11

Background Pathways: Units7,8,and11

unit

1

Our Planetary Neighborhood

Astronomy is the study of the universe: from the Earth itself to the most distant galaxy, from deciphering its nature at the beginning of time to predicting its eventual fate in the remote future. Within its vast space lie planets with dead volcanoes whose summits dwarf Mount Everest. There are stars a thousand times the size of the Sun—so large that in the Sun's place, they would swallow up the Earth. And there are galaxies—slowly whirling systems of billions of stars—so vast that the Earth is smaller by comparison to them than a single grain of sand is to the Earth itself. On this small planet within this immense cosmic landscape, life has evolved over billions of years, giving rise to creatures who seek to understand the nature of the universe.

Through astronomy we gain the ability to study places so remote that there is no possibility of our ever visiting them. We gain insights into alien environments unlike anything found on Earth. And our insights provide new perspectives on our home planet and about ourselves.

The vast variety of planets, their distances, and their sizes are almost unimaginable. We will explore the Earth and its neighboring planets in detail in Part III; but to gain some sense of scale, we begin with a brief look at the Earth and the Earth's neighborhood—the Solar System.

1.1 THE EARTH

We begin with the Earth, our home **planet** (Figure 1.1). This spinning ball of rock and metal, coated with a thin layer of gas and liquid, is huge by human standards, but it is one of the smaller bodies in the cosmic landscape. Nevertheless, it is an appropriate place to start because, as the base from which we view the universe, it determines what we can see. We cannot travel to any but the nearest objects in our quest to understand the universe. Instead, we are like children who know their neighborhood well but for whom the larger world is still a mystery, known only from books and television.

But just as children use knowledge of their neighborhood to build their image of the world, so astronomers use their knowledge of Earth as a guide to distant worlds. The size of the Earth and features on it, for example, are useful reference points for appreciating the sizes of other objects. We will often refer to other planets in terms of their radii relative to the Earth's own radius of about 6371 kilometers (3909 miles). Similarly, it is convenient to refer to other planets' masses in terms of the Earth's own mass, which is itself so large that it is difficult to imagine: 5,970,000,000,000,000,000,000,000 kilograms, or about 6 billion trillion tons.

Although few people realize it, we use the size of the Earth for defining the fundamental unit of the **metric system** (Unit 3): the meter. The meter was originally defined to be one 10-millionth of the distance from the equator to the

The calculations made in the original determination of the meter were slightly off, so that the equator-to-pole distance is today found to be 10,002 kilometers. More significantly, though, the Earth is not a perfect sphere, so the circumference around the equator is larger than around the poles—about 40,075 kilometers. The definition of the meter today is based on the wavelength of light (Unit 22) produced by a particular kind of laser.

FIGURE 1.1
The planet Earth, our home, with blue oceans, white clouds, and multihued continents. An image taken by Apollo 17 astronauts on their way to the Moon in 1972.

The internal heat of planets appears to come from two sources—heat left over from their formation and the decay of radioactive elements in their interior (Unit 35). Their ability to retain this heat depends mostly on their size.

North Pole—so that the Earth's circumference equals 40,000 kilometers (about 25,000 miles). When this system was introduced, it was a convenient way to measure the distances ships traveled on the Earth's oceans, and it offered a less arbitrary standard than the many other measurement systems used at the time.

The geological processes that occur on Earth provide another kind of measure—one for interpreting the processes that shape the other planets. When a volcano spews molten lava, it provides a hint that below the surface our planet is extremely hot. During the last century geologists discovered that this internal heat drives slow but powerful currents that shake our planet's crust, move continents, build mountains, and heave up volcanoes. We can carry over our understanding of such geological processes here on Earth to help us make hypotheses about the processes that create similar features on other planets. And when we discover different features on other worlds, we can use them to help us think about the Earth in new ways.

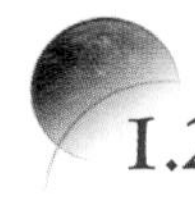

1.2 THE MOON

Our nearest neighbor, the Moon, is a world profoundly different from the Earth. The Moon is our **satellite,** orbiting the Earth nearly 400,000 kilometers (about a quarter million miles) away. A string stretched from the Earth to the Moon could wrap around the Earth almost 10 times. The Moon is much smaller, only about one-quarter our planet's diameter. The Moon also has symbolic significance for us—it marks the present limit of direct human exploration of space.

With the naked eye, the Moon appears to be a quiet glowing orb (Figure 1.2A); but with a small telescope or binoculars, we can see that the Moon is a rocky

FIGURE 1.2
(A) The Moon as we see it with unaided eyes. (B) The Moon's surface as seen through a small telescope.

Q What impressions of the character of the Moon do you have based on observing it with the naked eye? How does your impression change after viewing it through a telescope or examining high-resolution photographs? How do you imagine people's perceptions changed after the invention of the telescope in the early 1600s?

world—somewhat like the Earth, yet utterly unlike the Earth as well (Figure 1.2B). Instead of white whirling clouds, green-covered hills, and blue oceans, we see an airless, pitted ball of rock. Instead of crumpled mountain ranges and volcanoes, the Moon's surface is peppered with craters blasted into the surface when bodies crashed into its surface. We see evidence of a history of steady pounding by objects ranging from the microscopic to the size of mountains, impacting at speeds more than 10 times faster than any rifle bullet. Some of the larger collisions carved out craters more than 100 kilometers (60 miles) in diameter, while the innumerable smaller impacts pulverized the surface rock to rubble and dust.

The Earth, so near to the Moon in space, must have suffered a similar pounding, yet looks utterly different. Why do these two worlds bear so little resemblance? Much of the explanation lies in their greatly different mass. The Moon's mass is only about 1/80th the Earth's, and its smaller bulk made the Moon less able to retain internal heat or an atmosphere after its formation. With less internal heat, the Moon's interior is quiet. Heat-driven motion, so important in shaping and changing the Earth's surface over the eons, is nearly absent in the Moon. Without an atmosphere, the Moon is not protected from small impacting objects, which on Earth are vaporized through the heat of friction before reaching the ground. And without erosion caused by an atmosphere or modification of the surface caused by geological activity, the Moon's surface exhibits all of its old scars. Much of the record of Earth's past has been erased, but the Moon can help us reconstruct our planet's ancient history.

1.3 THE PLANETS

Beyond the Moon, circling the Sun as the Earth does, are seven other planets, sister bodies of Earth. To the unaided eye, the other planets are mere points of light whose positions shift slowly from night to night. But by observing them, first with Earth-based telescopes, then ultimately by remotely piloted spacecraft, we have learned that they are truly other worlds.

In order of increasing distance from the Sun, the eight planets are Mercury, Venus, Earth, Mars, Jupiter, Saturn, Uranus, and Neptune. These worlds have dramatically different sizes and landscapes. For example:

- Craters scar the airless surface of Mercury.
- Dense clouds of sulfuric acid droplets completely shroud Venus.

FIGURE 1.3
Images of the eight planets along with the dwarf planets Ceres, Pluto, and Eris. The correct relative sizes are shown at bottom.

- White clouds, blue oceans, green jungles, and red deserts tint Earth.
- Huge canyons and deserts spread across the ruddy face of Mars.
- Immense atmospheric storms with lightning sweep across Jupiter.
- Trillions of icy fragments orbit Saturn, forming its bright rings.
- The spin axis is tipped almost sideways for pale blue Uranus.
- Icy methane clouds whirl in the deep-blue atmosphere of Neptune.

Figure 1.3 shows pictures of these eight distinctive bodies and reveals something of their relative size and appearance. We can indicate the sizes of the planets by comparing them to Earth's radius, which we abbreviate as $R_{\oplus}$ using the international symbol $\oplus$ for the Earth. The planets range in size from Jupiter, with a radius of 11 Earth radii (11 $R_{\oplus}$), down to Mercury with a radius of slightly over one-third Earth's radius (0.38 $R_{\oplus}$).

Our two nearest neighboring planets, Venus and Mars, provide deeper insights into our own planet because they are so similar to, and yet so different from, the Earth. They are the closest to Earth in size, with radii of 0.95 and 0.53 $R_{\oplus}$, respectively. Their landscapes include features that resemble many seen on Earth—for example, they both have volcanoes and mountain ranges—so we conclude that geologic processes like those that shaped the Earth must have occurred on both.

Despite their similarities in size and in distance from the Sun, Venus and Mars have dramatically different atmospheres. On Venus we would be crushed and cooked by its intensely hot, dense atmosphere, whereas on Mars we would suffocate and freeze. And the other worlds astronomers have studied are even more alien. Every other planet and moon that we have found would also be hostile to human life. By studying the factors that make other planets so different from Earth, we may also begin to understand the potential consequences of our own impact on the Earth.

Pluto is no longer regarded as one of the major planets in the Solar System (Unit 46). It is now classified as a "dwarf planet" or "plutoid" by the International Astronomical Union, which is the organization responsible for astronomical nomenclature. In this new definition, the asteroid Ceres (Unit 41) has also been reclassified as a dwarf planet. In fact, Pluto is exceeded in size by the recently discovered Eris, another icy dwarf planet that orbits even farther from the Sun.

1.4 THE SUN

All of the planets of our Solar System are dwarfed by the Sun, whose immense gravity holds them in orbit. The Sun is a **star,** a huge ball of gas over 100 times the diameter of the Earth and over 300,000 times more massive. The differences in size between the Sun, Jupiter, and Earth are illustrated in Figure 1.4.

The Sun, of course, differs from the planets in more than just size: It generates energy in its core by nuclear reactions that convert hydrogen into helium. The Sun is producing energy at a furious rate—more in just one second than all of the bombs

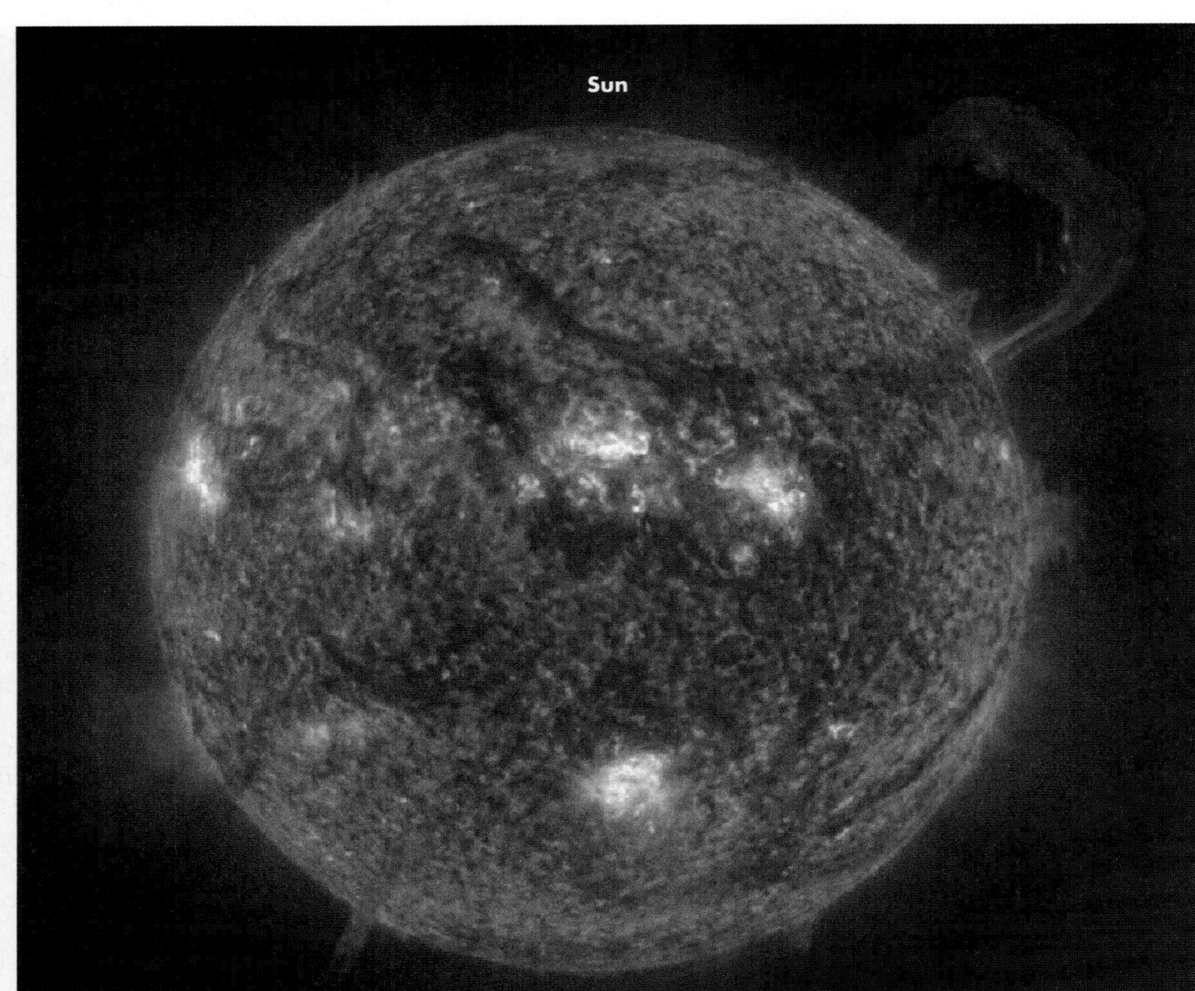

FIGURE 1.4
The Sun as viewed through a filter that allows its hot outer gases to be seen. The Earth and Jupiter are shown to scale beside it for comparison. (The filter allows astronomers to see very hot helium gas.)

Q The Sun and all the planets are hotter in their interiors than on their surfaces. What different sources of energy might produce their internal heat?

and the energy ever generated on Earth during human history. The energy from the intensely hot core flows to the Sun's surface, which is more than a thousand times cooler than the core but is still hot enough to vaporize iron. From there the Sun's energy streams into space, illuminating and warming the planets' surfaces.

The energy the Sun can produce is enormous but nonetheless limited. The Sun's stream of energy has already lasted more than 4 billion years—enough time for life to form and evolve on Earth and for intelligent creatures who can marvel at these things to have come into being. However, much evidence indicates that the Sun will run out of fuel eventually, in about another 5 or 6 billion years. It will then fade away like a cooling ember. Thus astronomy helps us to look deep into the past and far into the future as we consider phenomena of an enormous range in sizes and distances.

1.5 THE SOLAR SYSTEM

The Sun and the bodies orbiting it form the **Solar System,** bound together by the enormous gravity of the Sun. In addition to the eight planets, the Solar System is filled with a vast number of smaller bodies—satellites (moons) orbiting the planets and asteroids and comets orbiting the Sun. However, the mass of all of the planets and other objects in the Solar System does not add up to even 1% of the Sun's mass.

The paths that the planets follow around the Sun (Figure 1.5) form a set of huge approximately circular rings, one inside the other, with the Sun at their center. The orbits of the planets lie in nearly the same plane, so that the whole system is

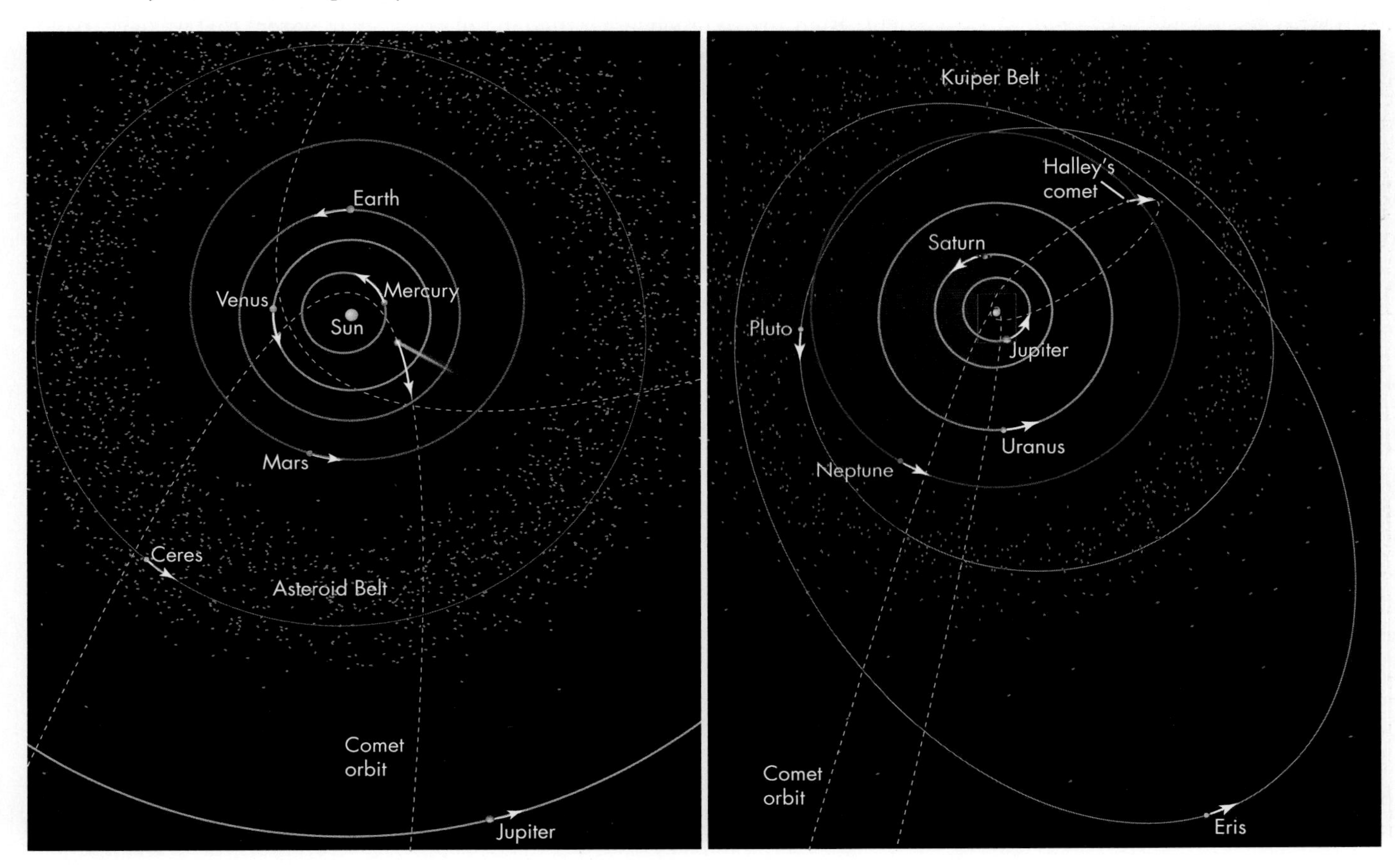

FIGURE 1.5
Sketch of the positions and orbits of the planets and dwarf planets in our Solar System on March 20, 2011. The orbit of Halley's comet along with another typical comet orbit are also shown. The approximate location of small bodies in the asteroid belt and Kuiper belt are indicated. To show the orbits to scale, the inner and outer Solar System are shown separately.

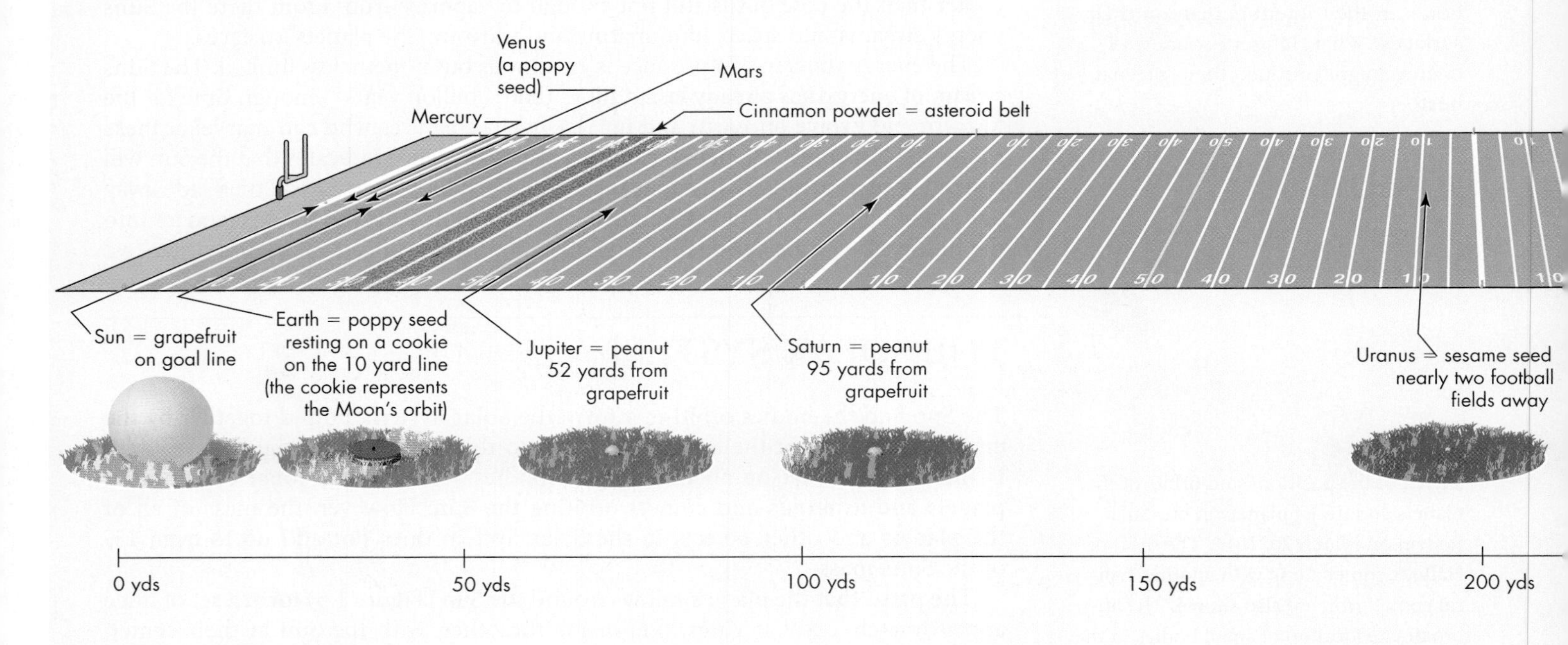

FIGURE 1.6
A scale model of the Solar System with the Sun the size of a grapefruit. At this scale all of the planets that are visible without a telescope fit within a single football field. Uranus and Neptune each require another whole field, while Pluto, Eris, and many other icy bodies are even farther away.

disk-shaped. The outermost planet—Neptune—lies about 4.5 billion kilometers (about 2.8 billion miles) from the Sun. The distances between planets in the outer Solar System are so much larger than in the inner Solar System that we have to plot these two regions separately to see them clearly.

Beyond Neptune's orbit are innumerable small objects orbiting in the "Kuiper belt" much as beyond Mars there are vast numbers of small bodies orbiting in the asteroid belt. The orbits of these objects are generally less circular and more tilted than the planets' orbits. This is especially true of comets (Unit 47) whose orbits carry them between the inner and outer parts of the Solar System.

If we constructed a scale model of the Solar System with the Sun represented by a grapefruit, Earth would be the size of a grain of sand or a poppy seed, orbiting about 9 meters (about 30 feet) away. This scale is illustrated in Figure 1.6, where the sizes and distances of the planets are illustrated on a string of U.S. football fields. In this scale model, the dwarf planets Pluto and Eris would be specks of dust orbiting more than 0.35 kilometer (0.22 mile) away.

1.6 THE ASTRONOMICAL UNIT

It is as difficult to comprehend distances in the Solar System reported in kilometers as it is to comprehend the distance between New York and Tokyo in millimeters. Whenever possible, we try to use units of measurement—such as millimeters or kilometers—appropriate to the size of what we seek to measure.

In earlier times people used units that were quite literally at hand, such as finger widths or the spread of a hand to measure a piece of cloth, and paces to measure the size of a field. In the same tradition, although on a different scale, astronomers have adopted **units**—quantities used for measurement—related to familiar objects, such as the Earth. For example, the Earth's distance from the Sun makes a good unit for measuring the size of the Solar System.

As the Earth orbits the Sun, its distance varies from 147.1 to 152.1 million kilometers (91.4 to 94.5 million miles). The **astronomical unit,** abbreviated **AU,**

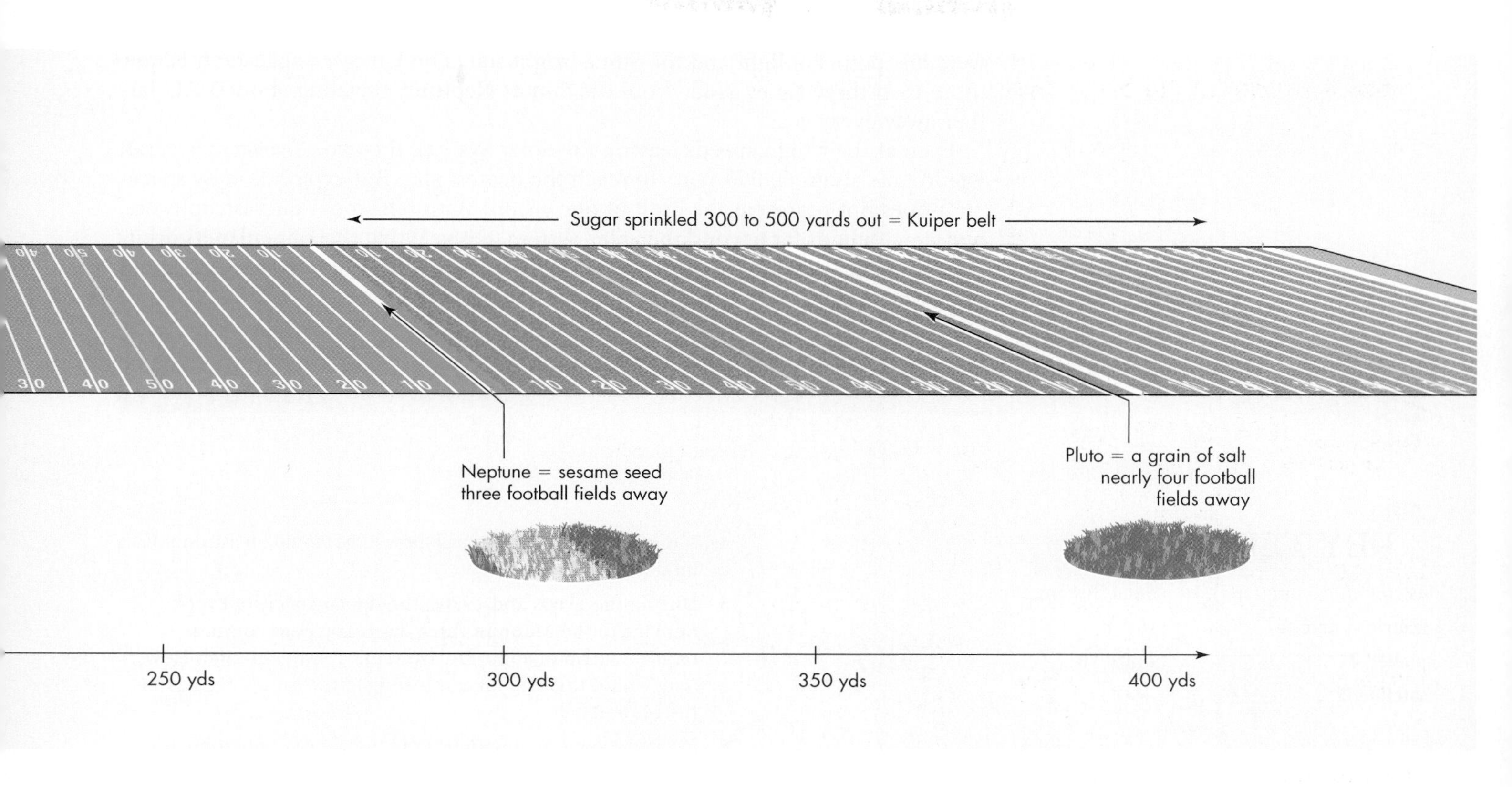

is approximately the average of these extremes, or about 150 million kilometers (93 million miles). If we use the AU to measure the scale of the Solar System, Mercury turns out to be 0.4 AU from the Sun, while Neptune is about 30 AU.

The Solar System remains the limit to our exploration of the universe with spacecraft. Most of our probes have explored only the inner part of the Solar System, although we have sent a few to explore the outer planets. The *Voyager I* and *II* spacecrafts launched in 1977 are our most distant explorers. In 1990 *Voyager I,* having passed Pluto's orbit at more than 40 AU, looked back and made the image of the Solar System shown in Figure 1.7. From this distance the planets

Q What do you suppose are the effects of the varying distance of the Earth from the Sun? How could we observe that?

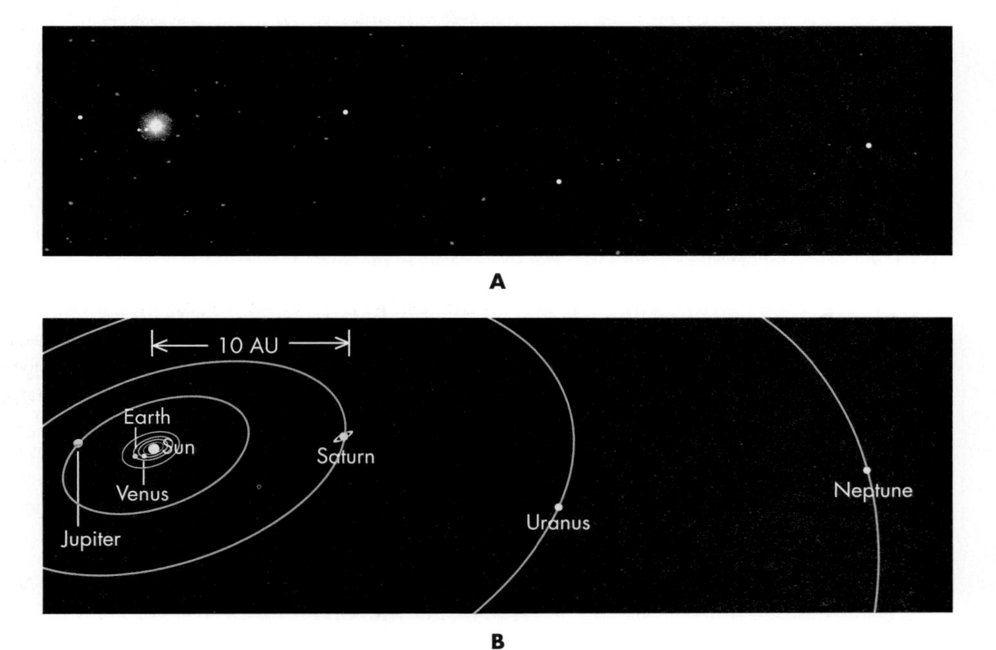

FIGURE 1.7
(A) This view of the Solar System is based on a series of real images made by the *Voyager I* spacecraft. The craft was about 40 AU from the Sun and about 20 AU above Neptune's orbit. The images of the planets (mere dots because of their immense distance) and the Sun have been made a little bigger and brighter in this view for clarity. Mercury is lost in the Sun's glare; Mars happened to lie nearly in front of the Sun at the time the image was made, so it too is invisible. (B) A sketch of the orbits of the planets, showing approximately where each object was located when the image was made (February 1990). Notice the immense, empty spaces between the planets and the nearly concentric rings of the planets' orbits, which make a nearly flat system.

were just points of light and the Sun a bright star. The *Voyager I* spacecraft is now more than three times as far from the Sun as Neptune, traveling about 3 AU farther away every year.

Even at their high speeds leaving the Solar System, the two *Voyager* spacecraft would take about 60,000 years to reach the nearest star. But exploration by spacecraft does not represent the limit of our vision. With telescopes and astrophysics, our view extends far beyond the Solar System to reveal that there are planets orbiting other stars, and there are billions of other stars with much to teach us about our own star.

KEY TERMS

astronomical unit (AU), 10
metric system, 4
planet, 4
satellite, 5
Solar System, 9
star, 8
units, 10

QUESTIONS FOR REVIEW

1. What are the eight planets in order of distance from the Sun?
2. What is a dwarf planet? Can you name the three objects currently in this category?
3. What planets are most similar to Earth?
4. About how many times bigger in radius is the Sun than the Earth? How many times bigger in mass?
5. Besides the Sun and planets, what other kinds of objects are members of the Solar System?
6. What is an astronomical unit?

PROBLEMS

1. If you use a volleyball as a model of the Earth, how big would 1 kilometer be on it? The volleyball has a circumference of 68 cm.
2. What would be the circumference and diameter (circumference $= \pi \times$ diameter) of a ball that would represent the Moon if the Earth were a volleyball? What kind of ball or object matches this size?
3. If the Earth were a volleyball, what would be the diameter of the Sun? What object matches this size?
4. If the Earth were a volleyball, how large would an astronomical unit be?
5. During the 1960s and 1970s, the *Apollo* spacecraft took humans to the Moon in three days. Traveling to Mars requires a trip of about 2 astronomical units in total. How long would this trip take, traveling at the same speed as to the Moon?
6. Using the same assumptions as in the previous question, how long would it take to travel to Pluto, about 40 astronomical units away?

TEST YOURSELF

1. Which of the following lists gives the sizes of the objects from smallest to largest?
   a. Moon, Earth, Pluto, Mars, Jupiter
   b. Pluto, Mars, Moon, Earth, Jupiter
   c. Jupiter, Pluto, Mars, Moon, Earth
   d. Moon, Mars, Jupiter, Earth, Pluto
   e. Pluto, Moon, Mars, Earth, Jupiter
2. How many times larger is the Sun's diameter compared to Earth's?
   a. 2 times
   b. 5 times
   c. 10 times
   d. 25 times
   e. 100 times
3. Venus and Mars have features that resemble the sand dunes seen in Earth deserts, but the Moon does not. What feature is the Moon *lacking* that most likely explains this difference?
   a. Stronger gravity
   b. Life
   c. Liquid water
   d. An atmosphere
   e. Sand (or other fine particles)

unit

2 Beyond the Solar System

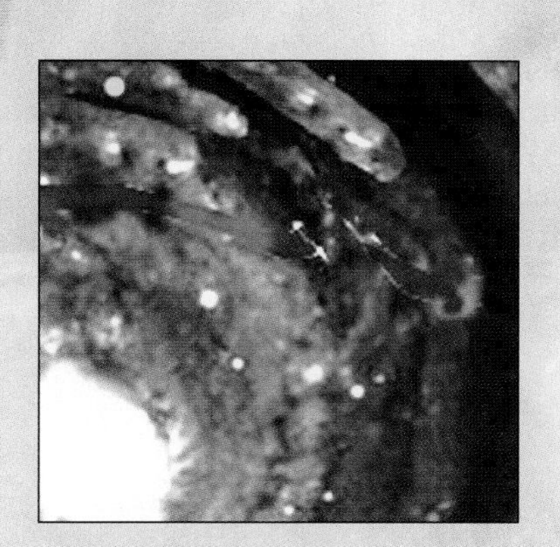

Beyond the Solar System lie billions of other stars. On a clear night, even with just your eyes, you can see that the stars are not scattered randomly across the sky but are often clumped into small groups. That clumping is the work of **gravity**, a force that attracts every object toward every other object across the farthest reaches of space. You can see that attraction in everyday life—if you let go of a book, the Earth's gravitational force makes the book fall. The same force holds the whole Earth together and gives the planets and the Sun their round shapes.

Gravity holds a star together against the enormous forces generated by the fusion of matter in the star's core. Indeed, without that furious outpouring of energy, the gravitational attraction pulling the matter together in a star is so powerful that the star must necessarily collapse in on itself. Moreover, the force of gravity spans the empty space between objects and can link separate objects together into gravitationally bound systems. It keeps the Moon in its orbit around the Earth and holds our planet in its orbit around the Sun. The Solar System is an example of one of the *smaller* systems of objects bound together by their mutual gravitational attraction.

The pull of gravity creates enormous structures—clusters of thousands of stars, galaxies of hundreds of billions of stars, or even clusters of galaxies pulled together and orbiting each other at millions of kilometers per hour. The intense gravity around exotic objects like black holes and quasars can pull in matter at immensely greater speeds—close to the speed of light—leading to titanic collisions and enormous energy bursts.

2.1 STELLAR EVOLUTION

The night sky is filled with myriad stars, some much like the Sun, but others thousands of times larger or smaller. The brightest stars produce over a billion times more light than the dimmest stars. Some stars are much hotter than the Sun and shine a dazzling blue-white, whereas others are cooler and glow a deep red.

The Sun is a fairly typical star and its properties provide a convenient comparison for understanding other stars. For example, astronomers often describe stars' masses as ranging from 0.1 to 100 $M_\odot$(0.1 to 100 "solar masses"). The $\odot$ symbol is the international symbol for the Sun. The mass of a star appears to be the most critical factor in determining its difference from other stars. The stronger gravity of a more massive star drives up the rate of nuclear fusion dramatically, while smaller stars consume their fuel at a more leisurely pace.

In Part IV of this book, we will see how astronomers have pieced together evidence that allows us to understand how stars are born, how they change as they age, and the dire fates they face when they run out of fuel. These changes, called **stellar evolution** by astronomers, are driven by the inexorable pull of gravity. Discovering the story of stars' lives has been one of the great triumphs of astrophysics and the scientific method during the last few centuries.

Making sense of the story of how stars are born and die has led to a greatly expanded vision of the kinds of objects that exist in the universe. We now know of a wide variety of objects that are beyond anything imagined a century ago. There are huge numbers of "failed" stars—objects that weren't quite massive enough to start the fusion of hydrogen to helium of the stars that shine in the night sky. And when stars die, they leave behind bizarre corpses—huge amounts of matter compacted into tiny objects such as white dwarfs, neutron stars, and black holes.

The life stories of other stars tell us that when the Sun runs out of hydrogen fuel, it will undergo drastic changes as it restructures itself to use helium fuel instead. We can predict that the Sun will go through a phase in which it will expand and nearly swallow up the Earth. The enormous energy output during this phase will

A

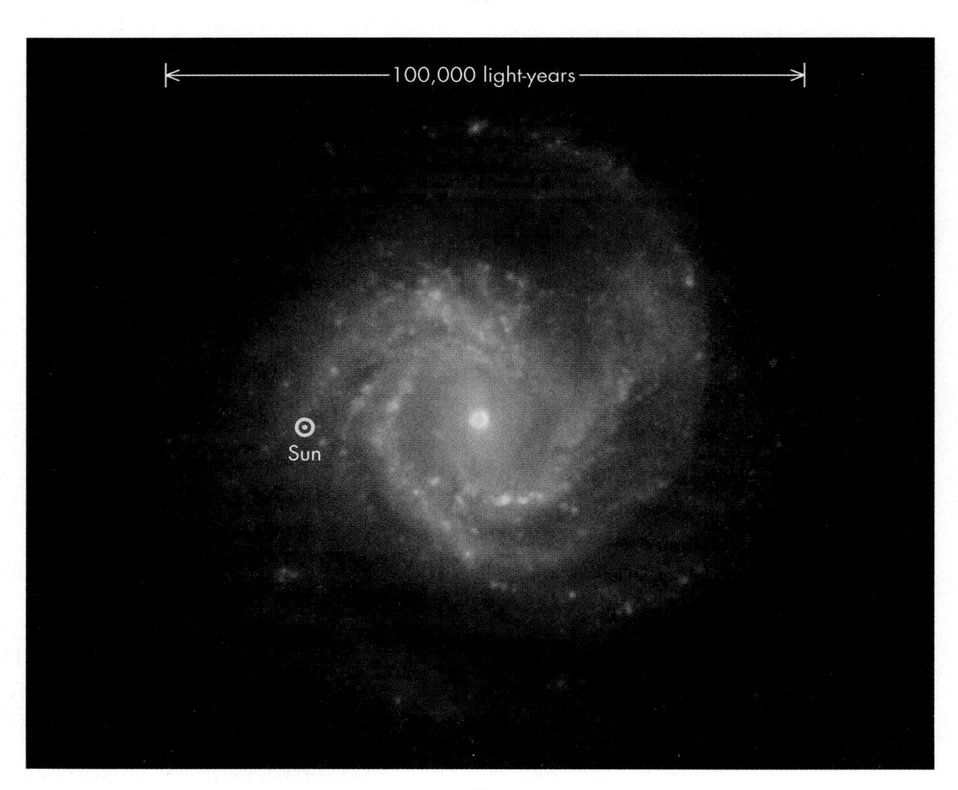

B

FIGURE 2.1
The Milky Way Galaxy. (A) A side view made by plotting stars in the 2MASS star catalog. (B) An image of a galaxy thought to look very similar to the Milky Way if seen from above.

melt our planet entirely. The Sun will end this luminous phase by blowing away its own atmosphere and then slowly fading like a cooling ember after a campfire has burned out.

2.2 THE MILKY WAY GALAXY

Q Both the Solar System and the Milky Way Galaxy are disk shaped. Why do you think this might be?

Our Sun and the stars we see at night are part of an immense system of stars called the Milky Way Galaxy. The **Milky Way Galaxy** is a cloud of several hundred billion stars with a flattened disklike shape somewhat like that of the Solar System (Figure 2.1), but roughly 100 million times the diameter. The Sun and other stars orbit around the center of the Milky Way, but our Galaxy is so vast that it takes the Sun several hundred million years to complete one trip around this immense disk.

In the Milky Way, as in many other **galaxies,** stars intermingle with immense clouds of gas and dust, as we shall explore in Part V of this book. These mark the sites of stellar birth and death. Within cold, dark clouds, gravity may draw the gas into dense clumps that eventually turn into new stars, lighting the gas and dust around them (Figure 2.2). Stars eventually burn themselves out, but they do not disappear quietly. Stars like the Sun last billions of years before their final phases when they tear themselves apart before fading away. The most massive stars die after just a few million years in titanic explosions, spraying matter outward to mix with the vast clouds of gas lying between the stars. This matter from exploded stars is ultimately recycled into new stars.

In this huge swarm of stars and clouds, the Solar System is all but lost—a single grain of sand on a vast beach—forcing us again to grapple with the problem of scale. Stars are almost unimaginably remote: even the nearest one to the Sun is about 40 trillion kilometers (25 trillion miles) away—about 10,000 times farther than Neptune. Such distances are so immense that analogy is often the only way to grasp them. For example, if we think of the Sun as a pinhead, the nearest star would be another pinhead about 60 kilometers (35 miles) away, and the space

FIGURE 2.2
Photographs of interstellar clouds in the Milky Way taken with an earthbound telescope. On the scale of these pictures, the orbit of Neptune is about 1000 times smaller than the period ending this sentence. (A) A cold, dark cloud and a star cluster beside it. Dust in the cloud blocks our view of the stars behind it. (B) A group of clouds heated by young stars. Glowing hydrogen in the clouds creates their red color.

A

B

Q Is it likely that there are many planets in our Galaxy?

between them would be nearly empty. On this scale, the Sun is to the size of the Milky Way as a pinhead is to the size of the Sun itself.

2.3 THE LIGHT-YEAR

While the astronomical unit works well for describing distances in our Solar System, distances between stars are so immense that the AU is an inappropriately small unit. The second nearest star (after the Sun!) is hundreds of thousands of AU away. A convenient way of describing such distances is the light-year.

Measuring a distance in terms of a unit of time may at first sound peculiar, but we do it all the time in everyday life. For example, we say that our town is a two-hour drive from the city, or our dorm is a five-minute walk from the library. In making such statements, we imply that we are traveling at a standard speed: freeway driving speed or a walking pace.

Astronomers have a superb speed standard: the speed of light in empty space. This is a constant of nature equal to 299,792,458 meters per second. We usually round this off to 300,000 kilometers per second (about 186,000 miles per second). Moving at this constant, universal speed, light in one year travels a distance defined to be 1 **light-year,** abbreviated as 1 ly. This works out to be about 10 trillion kilometers (6 trillion miles). As an example of the use of light-years, the star nearest our Sun is 4.2 light-years away. Although we achieve a major convenience in adopting such a huge distance for our scale unit, we should not lose sight of how truly immense such distances are. For example, if we were to count off the kilometers in a light-year, one every second, it would take us about 300,000 years!

We can use the light-year for setting the scale of the Milky Way Galaxy. In light-years, the visible disk is over 100,000 light-years across, with the Sun orbiting roughly 25,000 light-years from the center. In the Sun's vicinity, stars are separated by typically a few light-years, but they are much more crowded toward the center and gradually thin out in the outer regions, so that there is no precisely defined outer edge. Throughout the disk of the Milky Way are scattered gas clouds with sizes of up to hundreds of light-years. Some of these clouds contain more than a million times the Sun's mass, but they are so diffuse that their overall density is less than a billionth-trillionth of the density of the air we breathe.

2.4 GALAXY CLUSTERS AND BEYOND

Having gained some sense of scale for the Milky Way, we now resume our exploration of the cosmic landscape, pushing out to the realm of other galaxies. Here we find that just as stars are gravitationally bound into star clusters and galaxies, so galaxies are themselves bound into **galaxy groups** and **galaxy clusters.**

The Milky Way is part of a collection of galaxies called the **Local Group.** It is "local," of course, because it is the one we inhabit. The Local Group is a relatively small concentration of galaxies, containing only about four dozen galaxies, but it is still about 3 million light-years in diameter. The Milky Way is the second largest galaxy in the Local Group after the Andromeda Galaxy, also known as

FIGURE 2.3
Photograph of M31, the Andromeda Galaxy. This is the largest galaxy in our Local Group.

M31 is visible to the naked eye in the constellation Andromeda. See Looking Up #3: M31 & Perseus.

M31, the catalog number of a galaxy about 2.5 million light-years away from us (Figure 2.3).

To gain some perspective on the size of the Milky Way and Local Group, imagine that we built a model in which a light-year was scaled down to just 1 meter. At this scale Neptune's orbit would fit inside the head of a pin, and the Sun would be the size of a virus. The Milky Way Galaxy would be the size of a large city—about 100 km (60 miles) across, while M31 would be another city about 2500 km away (Figure 2.4). Given the enormous scale of this model, light crawls along about as fast as a plant might grow—just a meter per year; so that even at light speed, a trip from the Milky Way to M31 would take 2.5 million years!

The Milky Way is currently moving at about 90 kilometers per second (about 300,000 kilometers per hour, or 200,000 miles per hour) toward M31. Originally these two galaxies were moving away from each other, but the gravitational pull between them caused them to slow down, stop, and ultimately fall back toward each other. In about 5 billion years the Milky Way and M31 will crash into each other, and the stars and gas in the Milky Way's disk will be flung into wild new orbits, probably sending our Solar System flying off into intergalactic space.

We have to adjust the scale of our thinking again to imagine even larger scales. The Local Group is one of dozens of galaxy groups surrounding a much larger assemblage of galaxies called the **Virgo Cluster,** like suburbs surrounding a major city. The Virgo Cluster is centered about 50 million light-years away and itself contains thousands of galaxies. The central region of the Virgo Cluster is shown in Figure 2.5.

The term *Local Supercluster* has sometimes been used to describe the enormous "metropolitan area" of the Virgo Cluster with its surrounding galaxy groups, but this collection of groups and clusters is just a modest example of an even larger scale of clustering. The term **supercluster** is now generally reserved for collections of many galaxy clusters (and their associated surrounding regions), gravitationally bound to one another in structures that span hundreds of millions of light-years.

The mind boggles at these enormous gravitational structures, and perhaps you have begun to wonder whether this hierarchy of structures extends ever upward. But structures of such vast size are about the largest objects we can see before we take the final jump in scale to the **universe** itself. Although for centuries our knowledge of the visible universe was confined by the limits of our telescopes and instruments, today we are reaching a fundamental limit: We can see only as far away as light has had time to travel in the age of the universe. The best evidence today indicates that the universe is 13.7 billion years old, and therefore we cannot see any farther than light can have traveled in that time. The largest superclusters we see span nearly 1% of the visible universe, but as best we can determine the universe is relatively uniform over larger spans than this.

Within the 13.7 billion light-years we can see, there is no hint of an edge or change in the nature of the universe. This suggests that the universe must be far larger than what we can see. Current ideas of how the universe began predict that the universe is immensely larger, perhaps extending limitlessly. Regardless of our uncertainty about the known universe's overall size, we can observe that it has a well-ordered hierarchy of smaller structures. Small objects are clustered into larger systems, which are themselves clustered: satellites around planets, planets around stars, stars in galaxies, galaxies in clusters, clusters in superclusters, as illustrated in Figure 2.6 and Table 2.1. This orderly structure originated as a result of the pull of gravity and the ways that matter interacts on different scales.

Q Suppose we extended the scale model in Figure 2.4. How far away would the Virgo Cluster be? How large would the observable universe (out to 13.7 billion light-years) be?

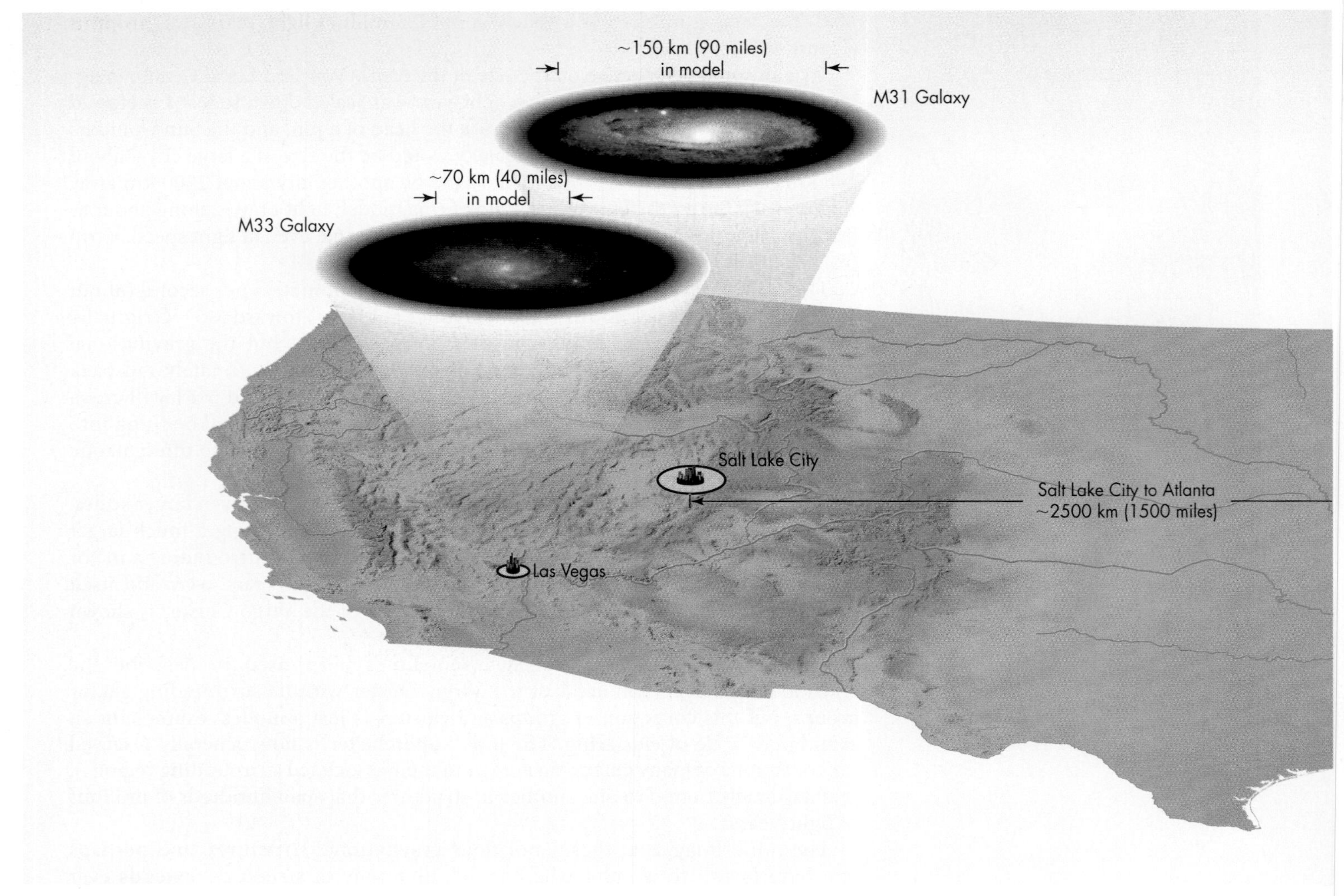

FIGURE 2.4
A scale model of the Local Group in which 1 meter represents 1 light-year. At this scale the Local Group is nearly the size of the United States, and the three largest galaxies in the group are the size of metropolitan areas including their surrounding suburbs. The entire Solar System is the size of a pinhead on this scale.

2.5 THE STILL-UNKNOWN UNIVERSE

Over the last century, astronomers have uncovered evidence that the entire universe behaves in ways that were utterly unexpected a hundred years ago. Measurements show that galaxies are flying away from each other as a result of an all-encompassing explosion, called the **Big Bang.** Albert Einstein's general theory of relativity showed that this is the wrong way to interpret what we see, though—space itself is expanding, carrying the galaxies along with it. Einstein's theories, now cornerstones of modern physics, have forced us to reevaluate our most basic notions of space and time, and we are discovering that there is much more to the universe than we can directly see through our telescopes.

Q The idea of dark matter may seem quite unusual, but can you think of any situations in which you have been able to find evidence that something was present without actually seeing it? Were you later proved correct?

Far more matter is found to fill the universe than what we can see in planets, stars, gas clouds, galaxies, and so on. Even after we correct for the amount of matter that is hard to detect, there is strong evidence that there is about 10 times more utterly invisible **dark matter** that is made of strange substances that *cannot* form

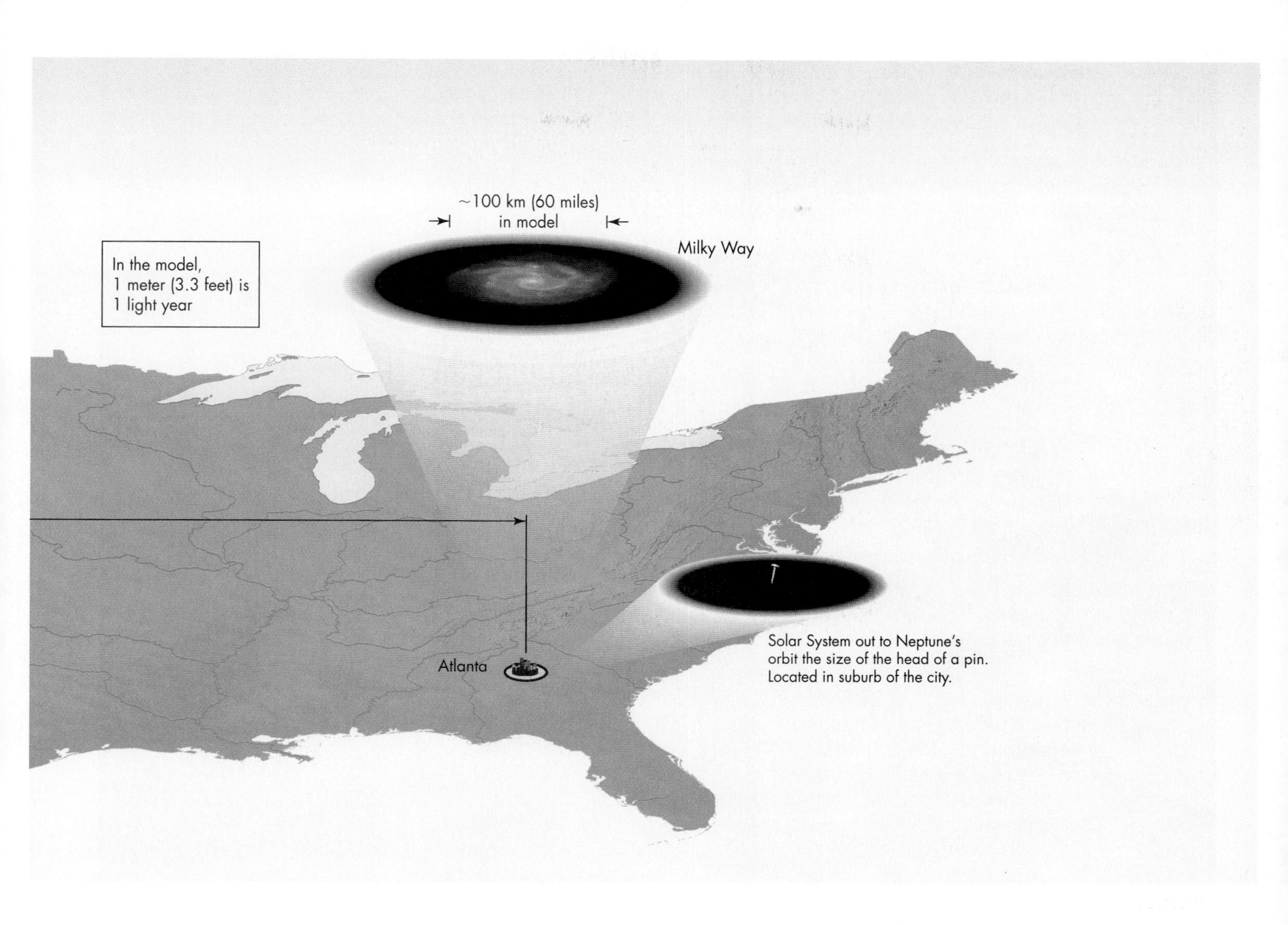

FIGURE 2.5
Photograph of the central region of the Virgo Cluster. The Milky Way and entire Local Group have been slowed in their motion away from this galaxy cluster by its strong gravitational pull.

stars or planets. Yet this matter appears to be the main component of every system the size of a galaxy or larger. The evidence that such matter exists comes from a variety of sources. The motion of stars in the outer parts of galaxies is so fast that they would fly off into intergalactic space unless much more mass is present than we can see. The same has been found true for galaxies in galaxy clusters. That is, the gravitational force that would be exerted by the observable stars and galaxies is far too small to hold the systems together. This is the subject of intense ongoing investigations, and it may be that particle physicists, exploring the submicroscopic world at scales far smaller than atoms, will uncover the nature of this strange kind of matter.

Over the last decade, astronomers have also accumulated evidence for an even stranger substance that has been dubbed **dark energy.** This is an energy that pervades every corner of space—even the emptiest vacuum. Einstein predicted the possibility of such an all-pervasive energy, but based on the evidence of his day, he abandoned the idea. New evidence suggests that not only does dark energy exist, it "outweighs" all of the visible and dark matter in the universe. If the latest measurements are correct, dark energy will drive the expansion of the universe faster and faster, perhaps in the extremely remote future causing space to expand so rapidly that it will tear apart the visible universe!

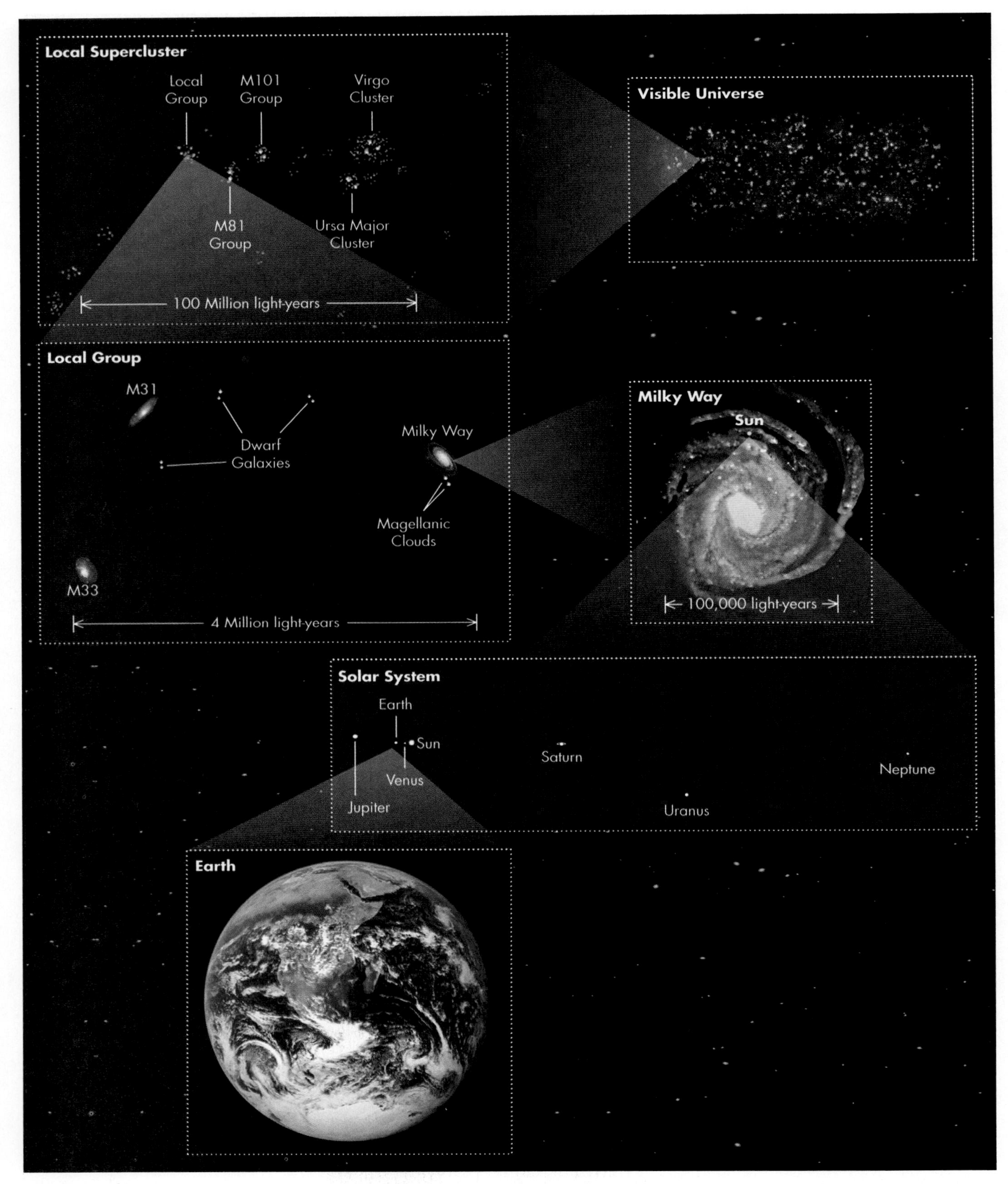

FIGURE 2.6
How the Earth fits into the universe.

TABLE 2.1 Scales of the Universe	
Object	**Approximate Radius**
Earth	6400 km (4000 miles) = $R_\oplus$
Moon's orbit	380,000 km $\approx$ 60 $\times$ $R_\oplus$
Sun	700,000 km = $R_\odot \approx$ 109 $\times$ $R_\oplus$
Earth's orbit	150 million km = 1 AU $\approx$ 214 $\times$ $R_\odot$
Solar System to Neptune	4.5 billion km $\approx$ 30 AU
Nearest star	270,000 AU $\approx$ 4.2 light-years (ly)
Milky Way Galaxy	50,000 ly
Local Group	1.5 million ly
Local Supercluster	50 million ly
Visible Universe	13.7 billion ly

KEY TERMS

Big Bang, 18
dark energy, 19
dark matter, 18
galaxy, 15
galaxy cluster, 16
galaxy group, 16
gravity, 13
light-year, 16
Local Group, 16
Milky Way Galaxy, 15
stellar evolution, 14
supercluster, 17
universe, 17
Virgo Cluster, 17

QUESTIONS FOR REVIEW

1. What do astronomers mean by stellar "evolution"?
2. What is a galaxy?
3. Roughly how big across is the Milky Way Galaxy?
4. How is a light-year defined?
5. To what systems (in increasing order of size) does the Earth belong?
6. What do the terms *dark matter* and *dark energy* refer to?

PROBLEMS

1. If the Milky Way were the size of a nickel (about 2 centimeters in diameter):
   a. How big would the Local Group be?
   b. How big would the Local Supercluster be?
   c. How big would the visible universe be? (The data in Table 2.1 may help you here.)
2. If we detected radio signals of intelligent origin from the Andromeda Galaxy (M31) and immediately responded with a radio message of our own, how long would we have to wait for a reply?
3. Are galaxies father apart (compared to their diameters) than solar systems?
4. If the Milky Way is moving away from the Virgo Cluster at 1000 kilometers per second, how long does it take for the distance between them to increase by 1 light-year?
5. The Milky Way is moving toward the larger galaxy M31 at about 100 kilometers per second. M31 is about 2 million light-years away. How long will it take before the Milky Way collides with M31 if it continues at this speed?
6. The most distant galaxies seen by the Hubble Space Telescope are 12 billion light-years away. Using the scale in Section 2.4 (where the Milky Way is about the size of a large city), how far would that be in the model?

TEST YOURSELF

1. You write your home address in the order of street, town, state, and so on. Suppose you were writing your cosmic address in a similar manner. Which of the following is the correct order?
   a. Earth, Milky Way, Solar System, Local Group
   b. Earth, Solar System, Local Group, Milky Way
   c. Earth, Solar System, Milky Way, Local Group
   d. Solar System, Earth, Local Group, Milky Way
   e. Solar System, Local Group, Milky Way, Earth
2. Which of the following astronomical systems is/are held together by gravity?
   a. The Sun
   b. The Solar System
   c. The Milky Way
   d. The Local Group
   e. All of them
3. The light-year is a unit of
   a. time
   b. distance
   c. speed
   d. age
   e. All of the above

unit 3

Astronomical Numbers

Astronomy deals with a greater range of numbers than any other science. It is challenging to grasp, even figuratively, the vast range of the measurements needed to study the universe (Figure 3.1). To understand a planet that has a diameter of 13,000,000 meters, we also need to understand the interactions between atoms that are only 0.0000000001 meters in diameter. And as we discuss other aspects of the universe, we will be considering sizes, masses, brightness, energy, and times that span even greater ranges.

To deal with this vast array of numbers, astronomers use four different strategies: (1) the metric system, (2) scientific notation, (3) special units, and (4) approximation. We will look at each of these approaches in turn, but we can summarize these procedures briefly as follows:

The *metric system,* as opposed to the English system, allows easy conversions between larger and smaller **units** of measure, such as between meters and kilometers. To carry out many calculations, we must express measurements in fundamental units such as meters, seconds, and kilograms. Faced with numbers like 13,000,000 and 0.0000000001, astronomers use a way of abbreviating these numbers called *scientific notation*. This notation keeps track of the order of magnitude of a number separately from its precise value.

Metric units do not always prove convenient for interpreting the sizes of things, so sometimes we create special units with a clear physical meaning, such as light-years (Unit 2.3). Often the essential information we need about a number is just its approximate size, so astronomers round off values. This avoids focusing on a level of precision that is not important. We say, for example, that the astronomical unit, which stands for the Earth's distance from the Sun, is 150 million kilometers even though it is known to be 149,597,870 kilometers.

This mathematical background is important for understanding much of the material in astronomy, but it does not necessarily have to be mastered all at once. Return to this unit whenever the numbers seem overwhelming.

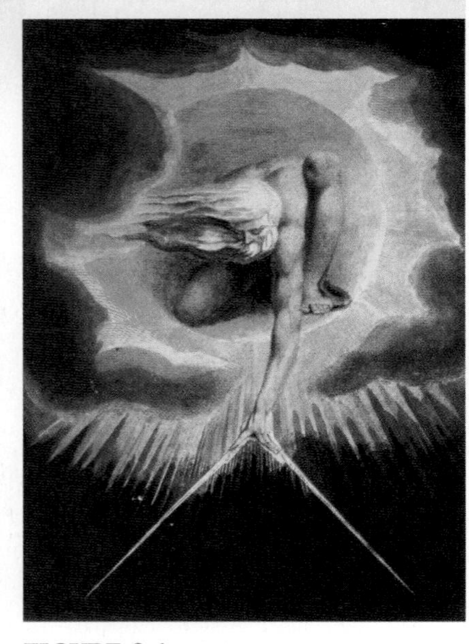

FIGURE 3.1
The "Ancient of Days" taking the measure of the universe. Etching by William Blake (c. 1794).

3.1 THE METRIC SYSTEM

Before the metric system was introduced in 1791, widely varying systems of measurement were used in different countries, or even in different regions within a single country. For example, the Paris foot was 6.63% longer than the English foot, while the Spanish *vara* was 8.67% shorter than the English yard. Similarly, the French pound (*livre*) was 10.2% larger than the English pound, and many other weight measures were in common use throughout even just Europe. Confusion about measurements was widespread and required frequent conversion between different units.

Napoleon Bonaparte is often said to have been short, but contemporaries described him as being of average or slightly above average height for his time. This mistake was apparently made because the height reported in Paris feet was misinterpreted.

That same confusion caused the loss of a Mars *Orbiter* in 1999, when the company building navigation jets provided its thrust data in English units while NASA mission controllers believed the numbers were in metric units. Firing the thrusters sent the craft too close to the planet, where it burned up in the atmosphere. Today the metric system has been widely adopted; the complex system of English units is still the primary choice in only the United States, Liberia, and Myanmar. In the end, the units used should not interfere with appreciating the important ideas of astronomy, but it is helpful to be familiar with the metric system for the calculations we will carry out.

The great advantage of the **metric system** over the English system is that different units are related by factors of 10, making it much simpler to convert between different units—1 kilometer is 1000 meters, whereas 1 mile is 5280 feet. We can usually convert between metric units by just "moving the decimal point" instead of multiplying or dividing by a conversion factor like 5280 feet per mile.

The trick to converting from one unit to another is to "multiply by 1". *One* can be written in many ways. If a and b are equal, then $a/b = 1$. In the conversion carried out here, because $1 \text{ km} = 10^3 \text{ m}$, we have multiplied by $\frac{10^3 \text{ m}}{1 \text{ km}}$. This lets us cancel out the kilometer units because they appear in both the numerator and the denominator.

Each time we move the decimal point, we are performing the equivalent of multiplying or dividing by factors of 10. Mathematically, we can use **exponents** to indicate the *number* of times we multiply by 10. An exponent indicates how many times we multiply a quantity by itself. For example, there are 1000 meters in a kilometer. We can write 1000 as "10 to the third power" or "10 to the exponent 3" as follows:

$$1000 = 10 \times 10 \times 10 = 10^3$$

When we convert meters to kilometers, we also move the decimal point over three places:

$$1 \text{ km} = 1. \text{ km} \times \frac{10^3 \text{ m}}{\text{km}} = 1000.\text{m}$$

where we have used the abbreviations m for meters and km for kilometers, and we have explicitly included the decimal point for clarity.

When we multiply by 10^3, we move the decimal point to the right by 3 places. If we were to divide by 10^3, we would move the decimal point to the *left* by 3 places. This second case can also be written as multiplying by 10^{-3}. In other words "10 to the power negative 3" is the equivalent of one thousandth (1/1000). When we take 10 (or any number) to the "zeroth power" the result is always one: $10^0 = 1$.

Similarly, we can write 1 billion as

$$1{,}000{,}000{,}000. = 10 \times 10 \times 10 \times 10 \times 10 \times 10 \times 10 \times 10 \times 10 = 10^9$$

or one millionth as

$$0.000001 = \frac{1}{1{,}000{,}000.} = \frac{1}{10 \times 10 \times 10 \times 10 \times 10 \times 10} = \frac{1}{10^6} = 10^{-6}.$$

In brief, rather than writing out all the zeros, we use the exponent to tell us the number of zeros.

It is often forgotten today that the English system also offers many intermediate size scales, but their relationships are complex. The inch is divided into 12 "lines," the foot into 12 inches, the yard into 3 feet, the "rod" into 5.5 yards, the "chain" into 4 rods, the "furlong" into 10 chains, the mile into 8 furlongs, and the "league" into 3 miles. The acre, still commonly used as a measure of area, is one furlong long by one chain wide. As if this is not enough to give one a headache, weights and volumes were divided in completely different ways!

In the metric system, **metric prefixes** identify various possible **powers of 10**. The prefix *kilo-*, for example, indicates 1000, while *milli-* indicates one thousandth, and *mega-* indicates 1 million. These prefixes can be added to any unit of measure to create a new unit that is closer to sizes we are interested in discussing: millimeter, kilogram, or megabyte, for example. The last term, *megabyte,* indicates a million pieces of information (such as letters or digits) stored in a computer and demonstrates how we can use metric prefixes in front of any unit of measure we like.

Table 3.1 shows the standard metric prefixes along with their meanings in words and exponential notation. This is a complete listing of metric prefixes; in this book we use only the seven shown in boldface in the table. A nanometer, for example, is a unit we will use when discussing light waves. It is a billionth of a meter, or 10^{-9} m, and can be abbreviated *nm.* To convert 5 nanometers to meters, we would move the decimal point to the left by 9 places: $5. \text{ nm} = 0.000000005 \text{ m}$. Thus, starting from a fundamental unit of measure like the meter, the metric prefixes give us a wide range of units that are appropriate in different contexts.

TABLE 3.1 Metric Prefixes

Power of 10	Exponential Notation	Metric Prefix	Abbreviation
septillion (trillion trillion)	10^{24}	yotta	Y
sextillion (billion trillion)	10^{21}	zetta	Z
quintillion (million trillion)	10^{18}	exa	E
quadrillion (thousand trillion)	10^{15}	peta	P
trillion	10^{12}	tera	T
billion	$\mathbf{10^{9}}$	**giga**	G
million	$\mathbf{10^{6}}$	**mega**	M
thousand	$\mathbf{10^{3}}$	**kilo**	k
hundred	10^{2}	hecto	h
ten	10^{1}	deca	da
tenth	10^{-1}	deci	d
hundredth	$\mathbf{10^{-2}}$	**centi**	c
thousandth	$\mathbf{10^{-3}}$	**milli**	m
millionth	$\mathbf{10^{-6}}$	**micro**	μ
billionth	$\mathbf{10^{-9}}$	**nano**	n
trillionth	10^{-12}	pico	p
quadrillionth	10^{-15}	femto	f
quintillionth	10^{-18}	atto	a
sextillionth	10^{-21}	zepto	z
septillionth	10^{-24}	yocto	y

TABLE 3.2 MKS Units

Quantity	Metric Unit	MKS Equivalent	English Equivalent
Length	meter	m	3.28 feet
Mass	kilogram	kg	2.2 pounds (of mass)
Time	second	sec	(same)
Area	square meter	m^2	10.76 square feet
Volume	liter (L)	10^{-3} m^3	1.06 U.S. quarts
Speed	meters per second	m/sec	2.24 miles per hour
Acceleration	meters per square second	m/sec^2	3.28 feet per square second
Density	kilograms per liter	10^3 kg/m^3	0.036 pounds per cubic inch
Force	newton (N)	$kg \cdot m/sec^2$	0.225 pounds (of force)
Energy	joule (J)	$kg \cdot m^2/sec^2$	0.000948 BTUs
Power	watt (W)	$kg \cdot m^2/sec^3$	0.00134 horsepower

Any system of physical measurements requires three fundamental units—those describing length, mass, and time. In the metric system these are the meter (which is about 10% longer than the yard), the kilogram (which is about 2.2 pounds), and the second. This set of units defines the **MKS system,** which stands for meter, kilogram, and second. Units for measuring other quantities, such as force, energy, and power, can all be written in terms of these fundamental units. Some of the more common kinds of metric units we use in this book, and their nearest equivalent in the English system, are listed in Table 3.2.

Because of difficulties in reproducing a precisely consistent mass from a liter of water, the kilogram has been redefined today in terms of a platinum-alloy cylinder stored at the International Bureau of Weights and Measures in France.

One unit commonly used in this book, the liter—a unit of volume—is not quite as simple a combination of MKS units as the others. Because we will often be comparing the **densities** (mass per volume) of materials, the density in kilograms per liter is a convenient reference. The liter is equivalent to one thousandth of a cubic meter. This in turn leads to the original definition of the kilogram, which is the mass of one liter of water. Water has a density of 1 kilogram per liter, but we will encounter centers of dying stars where the density is enormously larger, and regions of interstellar space where the density is a minuscule fraction of this.

The conversion here is accomplished first by converting kilowatts to watts, multiplying by, $1 = \frac{1000\ \text{W}}{1\ \text{kW}}$, and then again multiplying by $1 = \frac{0.00134\ \text{horsepower}}{1\ \text{W}}$.

The units in Table 3.2 can all employ metric prefixes, so the specifications for a car's power would be likely listed as, say, 200 kilowatts. To convert this to the English unit of horsepower, we could carry out the following calculation:

$$200\ \text{kW} = 200{,}000\ \text{W} \times 0.00134\ \text{horsepower/W} = 268\ \text{horsepower}$$

Also be aware that in the English system, the term *pound* is used interchangeably for a mass (a measure of the amount of matter) as well as for the gravitational force with which the Earth pulls on that mass (see Unit 14.1 for details). As we will discover, the same mass weighs a different amount on different planets.

3.2 SCIENTIFIC NOTATION

When we make a calculation in scientific notation, we usually begin by converting all of the values into MKS units. For example, the mass of the Sun is 1,989,000,000,000,000,000,000,000,000,000 kilograms. Partly because there is no metric prefix close to the size of such an enormous number, we use a more concise way to express such numbers, called *scientific notation.*

Scientific notation combines the powers-of-10 notation just described with the particular value of the number. We divide the number into a value between 1 and 10 and a power of 10 that when multiplied together yield the original number. Thus we can write 600 (six hundred) as

$$600 = 6 \times 10 \times 10 = 6 \times 10^2.$$

We write 543,000 (five hundred forty-three thousand) as

$$543{,}000 = 5.43 \times 10 \times 10 \times 10 \times 10 \times 10 = 5.43 \times 10^5$$

and 21 millionths becomes

$$\frac{21}{1{,}000{,}000} = 0.000021 = \frac{2.1}{100{,}000} = \frac{2.1}{10^5} = 2.1 \times 10^{-5}.$$

Once again the exponent indicates how we have moved the decimal point. The Sun's mass expressed this way is 1.989×10^{30} kg.

With scientific notation, multiplying and dividing very large numbers becomes easier. This is because when we multiply two powers of 10 we just add the exponents, whereas to divide we subtract them. For example,

$$10^2 \times 10^5 = (10 \times 10) \times (10 \times 10 \times 10 \times 10 \times 10) = 10^{2+5} = 10^7$$

or as an example of division,

$$\frac{10^8}{10^5} = \frac{10 \times 10 \times 10 \times 10 \times 10 \times 10 \times 10 \times 10}{10 \times 10 \times 10 \times 10 \times 10} = 10 \times 10 \times 10 = 10^3$$

We can write this as a pair of general rules:

$$10^A \times 10^B = 10^{A+B} \quad \text{and} \quad 10^A/10^B = 10^{A-B}.$$

An important thing to remember is that $10^A + 10^B$ does *not* equal 10^{A+B}. To add or subtract two numbers expressed in scientific notation, *first* convert both to the *same* power of 10, then add or subtract the values "out front."

As an illustration of the use of scientific notation, we can calculate the number of kilometers in a light-year. To do this, we multiply light's speed by the number of seconds in a year. A year is 365¼ days, each day having 24 hours, each hour 60 minutes, and each minute 60 seconds, so the total number of seconds in a year is

The conversion here is accomplished by multiplying by 1 written in three different ways. Make sure you can identify them.

$$365.25\ \text{days} \times \frac{24\ \text{hours}}{1\ \text{day}} \times \frac{60\ \text{minutes}}{1\ \text{hour}} \times \frac{60\ \text{seconds}}{1\ \text{minute}}$$

$$= 31{,}557{,}600\ \text{seconds} \approx 3.156 \times 10^7\ \text{sec}.$$

The speed of light is 299,793 km/sec = 2.998×10^5 km/sec. Multiplying the speed by the time gives us the distance:

$$\begin{aligned} 1 \text{ ly} &= \text{speed of light} \times 1 \text{ year} \\ &= 2.998 \times 10^5 \text{ km/sec} \times 3.156 \times 10^7 \text{ sec} \\ &= 2.998 \times 3.156 \times 10^{5+7} \text{ km} \\ &\approx 9.46 \times 10^{12} \text{ km} \end{aligned}$$

or nearly 10 trillion kilometers.

3.3 SPECIAL UNITS

For good or bad, astronomers have invented several new units to describe objects and phenomena that are far more immense than what we encounter on Earth. Usually these units make it easier to gain physical intuition, but they add to the number of units to learn, and they are not as easy to convert as metric units.

The **light-year,** the distance light travels in a year, is a good example. In principle we could use metric prefixes and a unit like "petameters" (Table 3.1), and write a light-year as 9.46 petameters or 9.46 Pm. However, the light-year has such a useful interpretation that we prefer to introduce it as another unit. Specifically, when we see an object 10 million light-years away, the light has taken 10 million years to reach us. That means that we are seeing the object *as it was 10 million years ago.* This becomes even more interesting as we look out billions of light-years, when the universe was just a fraction of its current age. We are literally able to look back toward the beginning of time, and the light-year unit helps us understand what, or rather "when," we are seeing.

A list of the special units we will use in this book is given in Table 3.3. Many of these units are used to relate other objects to the more familiar Sun and Earth,

TABLE 3.3 Special Units

Quantity	Special Unit	Abbreviation	Metric Equivalent
Length	Earth's radius	$R_\oplus$	6.37×10^6 m
	Sun's radius	$R_\odot$	6.97×10^8 m
	Astronomical unit (Earth–Sun distance)	AU	1.50×10^{11} m
	Light-year (distance light travels in one year)	ly	9.46×10^{15} m
	parsec (distance calculated by special geometric technique)	pc	3.09×10^{16} m
Mass	Earth's mass	$M_\oplus$	5.97×10^{24} kg
	Sun's mass	$M_\odot$	1.99×10^{30} kg
Time	Year (orbital period of Earth)	yr	3.16×10^7 sec
	Sun's lifetime (estimated total time Sun will generate energy)	$t_\odot$	3.16×10^{17} sec
Speed	Speed of light (through empty space)	c	3.00×10^8 m/sec
Acceleration	Earth's surface gravity (the rate at which falling objects accelerate)	g	9.81 m/sec^2
Energy	Kiloton of TNT (energy released by a standard 1000-ton bomb)	kt	4.18×10^{12} J
Power	Sun's luminosity (energy the Sun generates per second)	$L_\odot$	3.86×10^{26} W

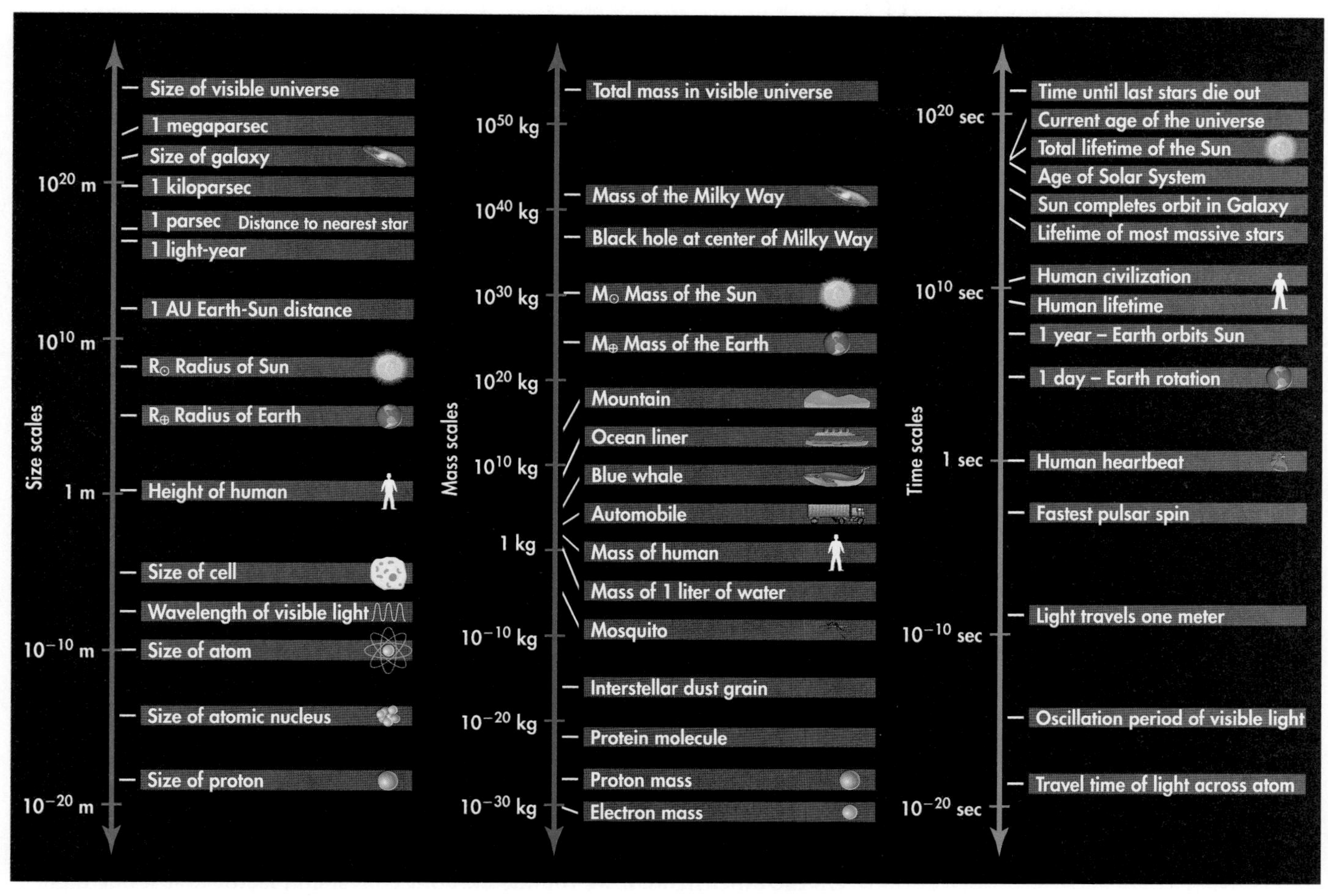

FIGURE 3.2
Scales showing relative sizes, masses, and times of special units and various objects and processes in the universe. Each is shown on a scale of powers of 10—ranging from the submicroscopic to the astronomical.

such as the Sun's power output or the length of Earth's year. These special units add a level of complication to a calculation whenever we need to convert to the MKS system, but some of the units were invented to make our calculations easier. For example, the astronomical unit simplifies the calculation of orbital parameters of objects in the Solar System (Unit 12) as well as of stars that orbit each other (Unit 17). The **parsec** was invented as a unit to simplify the formula for the primary method we have for finding stars' distances (Unit 52). The relative sizes of the parsec and other special units for describing sizes, masses, and times are illustrated in Figure 3.2. These additional units are intended to clarify rather than confuse.

3.4 APPROXIMATION

Astronomy is a science in which we often have to make uncertain estimates. Even though we can measure some numbers precisely, many others are subject to such uncertainty that there is no point in keeping track of more than the first couple of digits of a number.

For example, we know that there are 3.261633 light-years in a parsec. However, we rarely know the distance of a star more precisely than within 10%. If a star's

In the technical literature, astronomers often report numbers with a "plus or minus" range to indicate the uncertainty in a value, like 71 ± 6. Then if we multiplied the number by 2, we would find 142 ± 12, with both the value and the uncertainty doubling. This notation becomes cumbersome, especially when we start multiplying a set of numbers each of which has its own precision.

distance is listed as 10 parsecs, it is *not* correct to say the star is at a distance of 32.61633 light-years. The latter number gives the appearance of precision that does not exist. It would be more appropriate to convert 10 parsecs to 33 light-years or even 30 light-years—these better reflect the level of precision with which such distances are known.

Generally, measured values are reported showing only the number of digits that are accurately known. Values written as 374,000 or 0.00512 or 1.90×10^7, for example, are each reported to three **significant digits.** Having three significant digits means that the first two digits (37, 51, 19 in these three examples) are well determined, and the last one is fairly accurate, although it may have some uncertainty. Scientific notation here is useful for indicating when a trailing 0 is significant. This would be unclear in the example if 1.90×10^7 were written 19,000,000.

Any result from multiplying or dividing numbers in a formula can be only as accurate as the *least* accurate measurement used. For example, the nearest star, Proxima Centauri, is measured to be 1.3 parsecs away—to two digits of precision. We can multiply this by the number of light-years per parsec to get the distance in light-years:

$$1.3 \text{ pc} \times \frac{3.261633 \text{ ly}}{\text{pc}} = 4.2401229 \text{ ly} \approx 4.2 \text{ ly}$$

We **round off** the result to 4.2 ly, with two digits to reflect the level of precision for this calculation.

Notice that because we are limited by the least precise number, there was really no need to use all of the digits 3.261633 in our calculation. If we rounded off our ly-to-pc conversion factor to 3.26, we would have gotten the same result:

$$1.3 \text{ pc} \times \frac{3.26 \text{ ly}}{\text{pc}} = 4.23 \text{ ly} \approx 4.2 \text{ ly}$$

On the other hand, if we rounded off our conversion factor to just one digit, 3, we get this:

$$1.3 \text{ pc} \times \frac{3 \text{ ly}}{\text{pc}} = 3.9 \text{ ly} \approx 4 \text{ ly}$$

The result should be reported to only one digit because one of the numbers we used had only one significant digit.

Finally, note that if we round the conversion factor to 3.3, the result is

$$1.3 \text{ pc} \times \frac{3.3 \text{ ly}}{\text{pc}} = 4.29 \text{ ly} \approx 4.3 \text{ ly}$$

which is rounded *up* because the digit (9) after the last significant digit, 2, is greater than or equal to 5. Unfortunately, when we carry out a calculation with several numbers that are approximate, we multiply the error, and so the result has a *greater* uncertainty than the least accurate of the numbers.

Good use of approximation is an art as well as science. For example, someone might ask whether the distance of 4.2 light-years to Proxima Centauri is its distance from the Earth or the Sun. We know the Earth–Sun separation very accurately: 1 AU is 149,597,900 km. Should we worry about this when working out the correct distance? A little reflection will show that as big as an AU is, it is utterly insignificant compared to 4.2 light-years. The first time you think about this, you may require some calculations to convince yourself. Those calculations show that light traverses 1 AU in just 8½ minutes—so just as you would not worry about 8½ minutes of the time it takes to complete an undergraduate degree, you need not worry about it for the distance to Proxima Centauri.

KEY TERMS

QUESTIONS FOR REVIEW

1. What are the advantages of the metric system?
2. What is meant by a positive exponent? a negative exponent? an exponent of zero?
3. How is a number expressed in scientific notation?
4. What nonmetric units do astronomers frequently use?
5. What are significant digits?
6. When calculations are carried out with several numbers of different precision, what is the precision of the result of the calculation?

PROBLEMS

1. The radius of the Sun is 7×10^5 kilometers, and that of the Earth is about 6.4×10^3 kilometers. Show that the Sun's radius is approximately 100 times the Earth's radius.
2. A typical bacterium has a diameter of about 10^{-6} meters. An atom has a diameter of about 10^{-10} meters. How many times smaller than a bacterium is an atom?
3. Using scientific notation, show that it takes sunlight about 8½ minutes to reach Earth from the Sun.
4. How many 1-kiloton bombs would need to be exploded to produce 3.86×10^{26} Joules, the amount of energy emitted by the Sun in 1 second?
5. If I want to work out the distance to the Andromeda Galaxy, do I need to be concerned about whether the distance is from the center of the Milky Way or from the Sun's location? Why or why not?
6. Suppose two galaxies move away from each other at 6000 km/sec and are 300 million (3×10^8) light-years apart. If their speed has remained constant, how long has it taken them to move that far apart? Express your answer in years.
7. Using scientific notation, evaluate $(4 \times 10^8)^3 / (5 \times 10^{-6})^2$.
8. Using scientific notation, evaluate $(3 \times 10^4)^2 / \sqrt{4 \times 10^{-6}}$.
9. Napoleon Bonaparte was reported to be about 5 feet 2 inches tall as reported by contemporaries and at his autopsy. If these were in Paris units, 6.63% longer than English inches and feet, how tall was he actually in (a) English units, and (b) metric units?

TEST YOURSELF

1. An orbiting spacecraft travels about 4×10^4 kilometers in about 5×10^3 seconds. What is its speed?
   a. 8×10^0 kilometers/second
   b. 2×10^8 kilometers/second
   c. 8×10^6 kilometers/second
   d. 9×10^7 kilometers/second
   e. 1×10^1 kilometers/second
2. Which of the following is *not* equivalent to 30 kilometers?
   a. 30,000,000 millimeters
   b. 3×10^6 cm
   c. 30,000 meters
   d. 3×10^3 meters
   e. 0.03 megameters
3. From the following, choose the best approximation for the sum of 3.14×10^{-1} and 6.86×10^4.
   a. 10.00×10^3
   b. 6.86×10^4
   c. 3.14×10^{-1}
   d. 1.00×10^3
   e. 3.72×10^3

unit 4

Foundations of Astronomy

It seems that whenever we try to understand *larger* objects, we need to know more about ever smaller elements of the *submicroscopic* world. To study planets we must understand atoms and molecules; to study stars we have to understand the nuclei of atoms; to study the universe we need to understand subatomic particles.

The discoveries we have made about the submicroscopic properties of matter in laboratories here on Earth appear to apply to the rest of the universe. This is remarkable in itself, given how tiny a piece of the universe the Earth represents. In studying the cosmos, we encounter extremes of temperature, size, gravity, and energy that no one could dream of testing in Earth-based laboratories.

Our understanding of the universe has not come easily. Astronomy builds on many ideas from physics, chemistry, geology, mathematics, and biology. Learning how to extend our understanding from our experiments here on Earth to the far reaches of the universe is part of the science of astronomy. And astronomers sometimes encounter surprises when comparing what we have learned from our Earth-based experiments to more distant realms. These surprises may lead to new ideas for revising or extending our theories. This understanding represents the intellectual work of thousands of men and women over thousands of years, through the testing of ideas. That testing is part of what we generally call the *scientific method*.

4.1 THE SCIENTIFIC METHOD

In its simplest form, the **scientific method** is the procedure by which scientists construct their ideas about the universe and its contents, regardless of whether those ideas concern stars, planets, living things, or matter itself. In the scientific method, a scientist proposes an idea—a **hypothesis**—about some property of the universe and then tests the hypothesis by experiment. Ideally the experiment's results either support the hypothesis or refute it. Astronomers face a special challenge in applying the scientific method because they cannot experiment with their subject matter directly. Astronomy is primarily an observational science, like geology or the study of human behavior; so the strength of astronomers' theories is tied to their ability to predict phenomena or circumstances not yet observed.

One of the requirements for a hypothesis to be regarded as scientific is that it must be clear how it can be disproved. In experimental sciences this may be the prediction of the outcome of a laboratory experiment. For an observational science like astronomy, it might be a prediction that, for example, whenever we find a star with two particular properties, we will find that it also has a third property. To disprove the hypothesis, we need only demonstrate that the prediction is incorrect.

By requiring hypotheses to be disprovable, science not only provides a way of introducing new ideas; it also encourages scientists to test and retest older ideas from new perspectives. This openness to retesting should not be interpreted as vagueness or uncertainty—far from it. For an idea to stand up to this constant testing, it must be robust, providing us with much of the most rigorously tested information we possess.

Once a hypothesis has been thoroughly tested and verified, it may be termed a *theory* or *law.* When we use the word *theory* here, we do *not* mean that the idea is unproved or tentative, and the word *law* should not be interpreted as meaning the idea is beyond testing. Rather, theories and laws have achieved wide acceptance by successful testing over a long time. **Laws** are generally mathematical statements, whereas **theories** are generally expressed in words.

Sometimes more complex ideas are described as **models.** Models express relationships between different quantities and how they affect each other. Usually astronomical models are expressed mathematically or geometrically, and often they oversimplify a complex set of relationships to try to identify the most important elements. For example, contrary models were developed to explain the motions of planets, one assuming that the Earth was the center of their motion, another that the Sun was. Both models can accurately predict the motions we see, but today we recognize the Sun-centered model as the more nearly correct of the two models. However, even it is not quite right, because in fact the bodies in the Solar System all interact with each other through the force of gravity in a more complex way than simply circling the Sun. And in some situations the Earth-centered model is more convenient for describing the motions we see—as when we speak of the "Sun setting" as opposed to saying that the Earth rotated so that the Sun became hidden below our local horizon.

When an idea has achieved the status of a theory or law, its ability to make accurate predictions does not end even when a new idea overtakes it. For example, in the late 1600s Isaac Newton proposed several mathematical relationships describing motion and gravity, known as Newton's laws. They are extremely successful mathematical models that are still used today despite the fact that they are now known to make small errors in the prediction of motions under extreme conditions beyond our common experience, as shown in the early 1900s by Albert Einstein. He proposed revisions to these laws, known as the theories of special and general relativity, which explain how space and time are altered by motion and gravity. Scientists have subjected Einstein's theories to an enormous number of tests, and they have passed all such tests with the highest precision. Einstein's theories may someday also need revision; but it should be understood that, as with Newton's laws, their validity will not suddenly vanish. The many tests of the theories' predictions show that they work with a high degree of precision over a great range of circumstances.

Application of the scientific method is no guarantee that its results will be believed. For example, even before 300 B.C.E., Greek philosophers pursued several lines of inquiry demonstrating that the Earth is a sphere (Unit 10). Yet despite the evidence supporting this hypothesis, for thousands of years many people continued to believe the Earth to be flat. Today many people believe that it is possible to make a spaceship move faster than the speed of light (this belief is depicted in many science fiction movies), even though Einstein's theory of relativity has provided compelling evidence that this cannot happen.

We can be quite confident of the longest-standing theories because so many scientists have attempted to overturn them. Nevertheless, an exciting aspect of science is that it invites skepticism. At the forefront of the sciences, debate is very lively, with new hypotheses being proposed to explain new measurements, and challenges to accepted theories always encouraged. Individual scientists often dispute particular pieces of evidence. One astronomer may find the evidence supporting a hypothesis convincing, while another astronomer may think that the experiment was done incorrectly or the data were analyzed improperly. This is all part of the scientific process.

We need therefore to keep in mind that when we discuss ideas, they are never "proved," and sometimes they are not even widely accepted. This is especially true of ideas at the frontiers of our knowledge—for example, those dealing with the origin and structure of the universe or those dealing with black holes. However, the tentativeness of such ideas does not always stop astronomers from being positive about them, leading the Russian physicist Lev Landau to joke that astrophysicists are "often in error, but never in doubt." Therefore, keep in mind that some of the ideas we discuss in this book will be vastly improved on, or perhaps proved wrong, in the future. That is not a failing of science, however. It is its strength.

4.2 THE NATURE OF MATTER

One of the most powerful ideas in astronomy is that we can apply to the universe at large what we learn about nature here on Earth. This is sometimes called the hypothesis of **universality.**

Starting from this hypothesis, the step-by-step application of physical laws to the nature of matter drives us to many extraordinary conclusions. We will conclude that there are exotic things in the universe such as black holes, quasars, neutron stars, and dark matter, as discussed in later units. Yet the volume of space we have explored directly with space probes is just a thousandth-trillionth-trillionth-trillionth (10^{-39}) of the volume of the visible universe. How can we be sure that the properties of matter and physical laws do not change elsewhere in the universe?

The elements were created by several astronomical processes indicated in the table and discussed later in the text. If all radioactive elements are excluded, only 79 remain.

Our studies of matter on Earth show that it is made up of tiny particles called **atoms,** and 92 kinds of these occur naturally. The simplest type, a hydrogen atom, is about one 10-billionth of a meter (10^{-10} m) in diameter, so that 10 million (10^7) hydrogen atoms could be put in a line across the diameter of the period at the end of this sentence. Despite this tiny size, atoms themselves have structure. Every atom has a small (about 1/10,000th the atom's diameter) central core called the **nucleus** that is surrounded by one or more even smaller lightweight particles called **electrons** (Figure 4.1A). The nucleus is composed of two kinds of heavy particles called **protons** and **neutrons.**

The different kinds of atoms are displayed in the **periodic table** of the elements. The periodic table organizes the atoms in order of the number of protons, or **atomic number.** The table is organized so that atoms in the same column have similar chemical properties. The first part of the periodic table is shown in Figure 4.2, and the full table is shown inside the back cover. A wealth of additional information is shown for each atom, including the **atomic mass,** which is essentially the sum of the number of protons and neutrons in the atom's nucleus.

Superficially, an atom resembles a miniature solar system, but different physical forces and behaviors are at play than the **gravitational force** between planets and the Sun. At subatomic scales, particles do not behave in ways familiar to us—their motions are more wavelike (Unit 21). It is impossible to predict their precise position at any moment, so the location of the electron is described in terms of an "orbital cloud" (Figure 4.1B).

Experiments show that electrons repel each other, and protons repel each other, but electrons and protons *attract* each other. This situation is different from gravitational force, where all kinds of matter exert only an attraction on each other. You can get some sense of the effects of electrical forces by rubbing a balloon on your hair in dry weather. The balloon collects electrons from your hair, while your hair is left with an excess of protons. The electric charge on the balloon attracts strands of your hair, while your hair stands on end because the strands of like-charged hair repel each other. This is the same force at play on an atomic level.

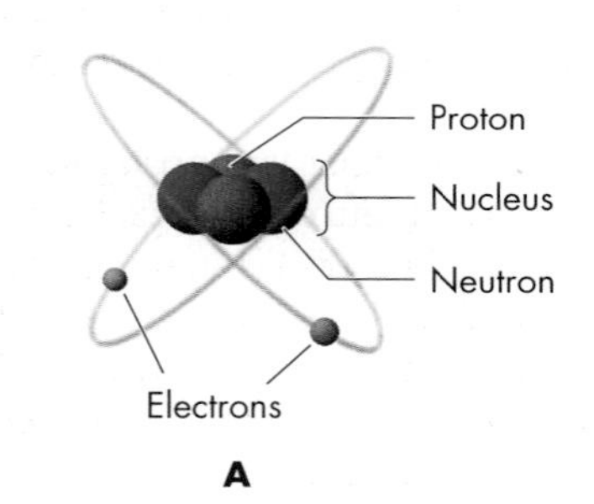

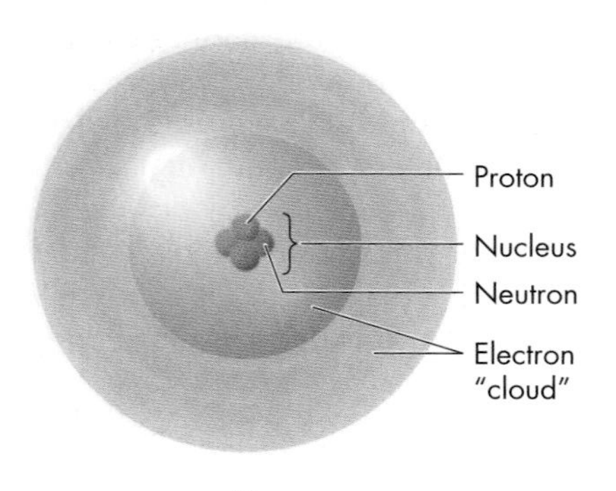

FIGURE 4.1
An atom consists of a nucleus around which electrons orbit. The nucleus is itself composed of particles called protons and neutrons. (A) A classical view of an atom. (B) A more modern view with the electrons in "orbital clouds."

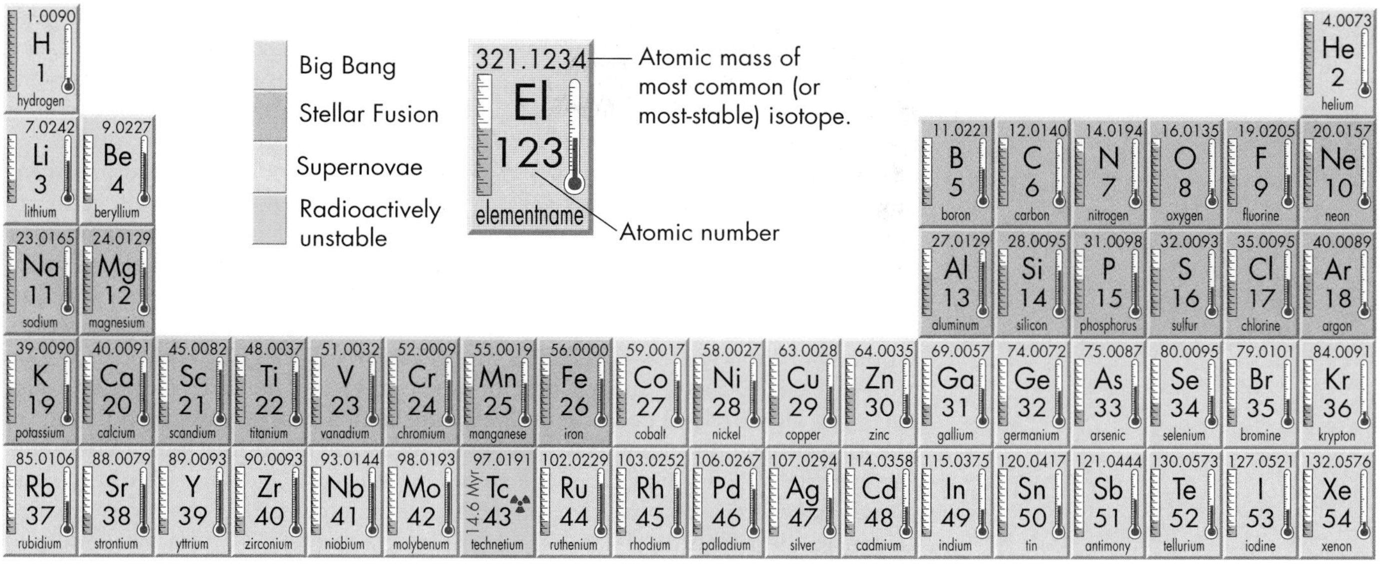

Big Bang
Stellar Fusion
Supernovae
Radioactively unstable

321.1234 — Atomic mass of most common (or most-stable) isotope.
El
123 — Atomic number
elementname

1	2	3	4	5	6	7	8	9	10	11	12	13	14	15	16	17	18
1.0090 H 1 hydrogen																	4.0073 He 2 helium
7.0242 Li 3 lithium	9.0227 Be 4 beryllium											11.0221 B 5 boron	12.0140 C 6 carbon	14.0194 N 7 nitrogen	16.0135 O 8 oxygen	19.0205 F 9 fluorine	20.0157 Ne 10 neon
23.0165 Na 11 sodium	24.0129 Mg 12 magnesium											27.0129 Al 13 aluminum	28.0095 Si 14 silicon	31.0098 P 15 phosphorus	32.0093 S 16 sulfur	35.0095 Cl 17 chlorine	40.0089 Ar 18 argon
39.0090 K 19 potassium	40.0091 Ca 20 calcium	45.0082 Sc 21 scandium	48.0037 Ti 22 titanium	51.0032 V 23 vanadium	52.0009 Cr 24 chromium	55.0019 Mn 25 manganese	56.0000 Fe 26 iron	59.0017 Co 27 cobalt	58.0027 Ni 28 nickel	63.0028 Cu 29 copper	64.0035 Zn 30 zinc	69.0057 Ga 31 gallium	74.0072 Ge 32 germanium	75.0087 As 33 arsenic	80.0095 Se 34 selenium	79.0101 Br 35 bromine	84.0091 Kr 36 krypton
85.0106 Rb 37 rubidium	88.0079 Sr 38 strontium	89.0093 Y 39 yttrium	90.0093 Zr 40 zirconium	93.0144 Nb 41 niobium	98.0193 Mo 42 molybenum	97.0191 14.6 Myr Tc 43 technetium	102.0229 Ru 44 ruthenium	103.0252 Rh 45 rhodium	106.0267 Pd 46 palladium	107.0294 Ag 47 silver	114.0358 Cd 48 cadmium	115.0375 In 49 indium	120.0417 Sn 50 tin	121.0444 Sb 51 antimony	130.0573 Te 52 tellurium	127.0521 I 53 iodine	132.0576 Xe 54 xenon

FIGURE 4.2
The periodic table of the elements displays properties of each type of atom. The full periodic table is shown on the inside back cover and displays astronomical properties of each kind of atom in addition to the basic chemical properties.

To model this idea, the proton is designated as having a "positive" **electric charge,** and an electron as having a "negative" electric charge. A neutron, as its name suggests, is neutral and has no charge. With this description, we can say that electric charges of the same sign repel each other (+ repels +− ; − repels −) and opposites attract (+ attracts −). That attraction is what holds the electrons in their orbits around the nucleus of an atom.

The light we see is also made up of elementary particles, called **photons,** which can be generated or affected by electrically charged particles such as electrons and protons. The characteristics of the light emitted by each kind of atom are distinct, and a careful analysis of photon energies allows us to determine the kind of matter that generated the light. Astronomers have used this fact to test the hypothesis of universality. In fact, by examining the characteristics of light emitted by very distant objects, we are also testing whether the universe was the same in the past when the light left those objects. We find that everywhere we look in the universe, light has the same characteristics that we find in matter on Earth. The same kinds of atoms made of the same kinds of electrons, protons, and neutrons undergo the same kinds of chemical reactions we observe here on Earth as they did "a long time ago in a galaxy far, far away."

4.3 THE FOUR FUNDAMENTAL FORCES

The world of matter at submicroscopic scale is a whole universe unto itself, but we need to understand the universe at this scale to understand the larger universe. For example, reactions occurring in the nuclei of atoms are the source of power for stars, and the characteristics of the fundamental forces in nature are linked to how the universe itself formed.

At this most basic level, there are only a few kinds of elementary particles interacting through just four forces. Two of the four forces are relatively common to our everyday experience: gravitational force and electric force. Actually, the electric force that holds the atom together is also fundamental to the magnetic force that makes a compass work or holds magnets on the door of your refrigerator. Moving electric charges generate fields of magnetic force, and moving magnets can generate electric fields. Scientists refer to them jointly as **electromagnetic force.** For subatomic particles, the electromagnetic force is far more important than the gravitational force. The strength of the electromagnetic force between two protons is 10^{36} (a trillion trillion trillion) times stronger than the gravitational force.

Two other forces play critical roles inside atoms. One of these, called the strong nuclear force or simply the **strong force,** binds protons and neutrons together to form an atom's nucleus. It is strong in the sense that, for two protons in a nucleus, it is about 100 times stronger than the electric repulsion they have for each other. Although the effects of the strong force cannot be seen directly in everyday life, without it the nuclei of atoms would fly apart, and so would our familiar world. The last of the four forces is called the **weak force** because it is about a billion times weaker than the strong force between protons. The weak force can cause one kind of particle to decay into another through a process described further in the following section. The strong and weak forces are both short-range forces that have virtually no influence outside an atomic nucleus. The strength of the weak force can become more nearly comparable to the other forces at very, very small distances—less than 10^{-18} (a millionth-trillionth of a meter—only a thousandth of the size of a proton.

The strong and weak forces each have their own associated "charges." These are not electric charges, but different properties of the subatomic particles that respond to the particular force. The strong force has three kinds of charge that must always occur in balanced combinations, analogous to the way red, blue, and green light can be combined to make white light. This analogy has led the charge associated with the strong force to be called "color charge." Associated with the weak force is a "weak charge," which has a complex representation that depends on a pair of properties in different combinations.

If the different kinds of charges are neutralized in a particle, then it will not respond to the associated force. This is why gravity is the dominant force at astronomical size scales despite being so extremely weak compared to the other forces. The great strength of the electromagnetic force keeps pulling electrons into an atom until their negative charges cancel out the positive charge exerted by the protons of the nucleus. As a result, the net electromagnetic force of the Sun on our planet is negligible compared to the gravitational force. The same occurs with the strong force and the weak force. Gravity remains because its "charge"—mass—does not come in opposites that can cancel each other's effects.

In the language of particle physics, each of the four forces is "mediated," or carried, by a particle. We just discussed the photon's interaction with electric charges; it is the carrier of the electromagnetic force. A particle called a **gluon** (the name invented to suggest its strong gluelike properties) similarly carries the strong

TABLE 4.1 Fundamental Forces

Force	Associated Property	Effect	Range	Carrier Particle	Relative Strength
Gravitational	Mass	All masses attract each other.	Infinite but weakens with distance	Gravitation	10^{-36}
Electromagnetic	Electric charge	Opposites attract, likes repel.	Infinite but weakens with distance	Photon	1
Strong	Color charge	Three colors combine to make neutral combinations.	$\approx 10^{-15}$ meters (distance between protons in atomic nucleus)	Gluon	10^{2}
Weak	Weak charge	Massive particles decay to lower-mass particles.	$\approx 10^{-18}$ meters (1/1000th proton diameter)	W and Z	10^{-7}

force, while particles called W and Z carry the weak force. Physicists hypothesize that another particle—the **graviton**—carries the gravitational force, although this particle has not been detected. Some of the basic properties of the forces are summarized in Table 4.1.

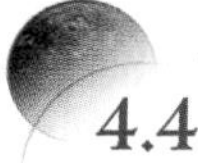

4.4 THE ELEMENTARY PARTICLES

The inner space of an atom is a universe that is very foreign to our everyday experience. Many kinds of subatomic particles populate this universe and respond to the four forces in unfamiliar ways. In interactions, a number of properties, like charge, are always conserved; but some particles can be converted from one type to another, and mass and energy can sometimes be converted into each other.

Protons and neutrons are built of elementary particles called **quarks.** Two types of quarks, called the *up quark* (*u*) and *down quark* (*d*), combine in sets of three to make the proton (*uud*) and neutron (*ddu*). The strong force acting between quarks grows stronger with distance, making it impossible to remove a quark from nuclear particles to study in isolation. However, particle physicists can "see" the internal lumpy structure of protons or neutrons when they bombard them with high-energy electrons, and from this kind of experiment they can determine much about the characteristics of quarks.

The weak force can convert quarks from one type into another. In fact, the weak force, despite its name, is important because it is the only force that can change one type of particle into another. It can cause a particle to spontaneously decay into a less massive particle—for example, causing a neutron to decay into a proton. In this process a down quark turns into an up quark and emits two other particles, an electron and a **neutrino,** which balance the electric and weak charges in the interaction.

The discovery of the neutrino is a good example of the scientific method in action. Physicists studying the decay of the neutron in the 1930s could see that the proton and electron produced in the decay did not have as much energy as predicted by existing theories of energy and mass conservation. Various hypotheses were put forward, including the idea that there was a new particle that interacted so weakly that it was difficult to detect. This new particle had to have no electric charge and extremely little mass compared to any other known particle, and many scientists doubted the possibility of such a "ghost" particle. However, the neutrino was finally detected more than 20 years later, confirming the hypothesis and reinforcing the earlier conservation theories.

Neutrinos remain one of the most mysterious of the particles known today. For many years they were thought to have no mass, but it is now suspected that they do have a tiny (but still unmeasured) mass. Besides gravity, they interact only through the weak force, whereas electrons interact through the weak and electromagnetic forces and quarks interact through all the forces. Because the neutrino's strongest interaction is through such a weak force, it truly can seem "ghostly." The weak force requires that neutrinos must be *extremely* close to interact with another elementary particle. They might have to pass within 10^{-24} m of a quark—a millionth the diameter of a proton—for an interaction to be likely, so neutrinos can travel through an atomic nucleus without much chance of interacting with any of the particles. As a result, normal matter looks nearly transparent to them. Neutrinos pass through the Earth more readily than photons pass through a window! On the rare occasion when they interact with other particles, though, they have the ability to change their character, turning a down quark into an up quark, for example (Figure 4.3).

Each particle also has a corresponding "antiparticle" with opposite properties. When an antiparticle meets its partner, they annihilate each other, leaving nothing but energy. Antiparticles, and their role at the beginning of the universe, are discussed further in Unit 81.

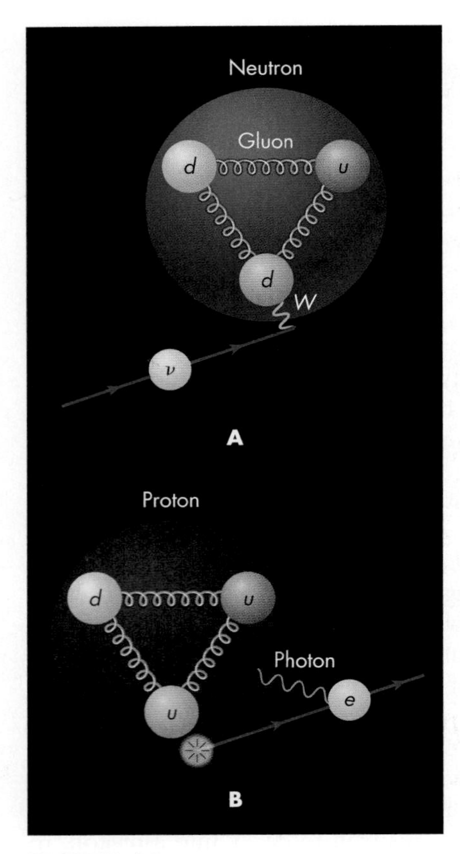

FIGURE 4.3
Illustration of the interaction of a neutrino (ν) with the quarks in a neutron. (A) The neutrino passes very close to a down (d) quark, and they exchange a weak force (W) particle. (B) The down quark becomes an up (u) quark and an electron is emitted. The strong force between quarks is illustrated by gluon exchange, whereas the electron responds to electromagnetic force through photons.

Up quarks, down quarks, electrons, and neutrinos make up nearly all of the matter in the known universe. There are two sets of more massive particles with properties that parallel those of these four particles, making "generations" of particles in the terminology of physicists, as illustrated in Figure 4.4. These alternate generations of matter are created at high temperatures and during high-energy collisions, but the weak force causes them to decay down to the first generation of particles. The other generations of quarks have been given fanciful names—*strange, charm, bottom,* and *top*—because they exhibit properties that we do not see in normal matter. Corresponding to the electron, the two other generations of particles are called the *muon* and *tauon,* and they each have an associated neutrino, called the *muon neutrino* and the *tauon neutrino.*

These particles, along with the carrier particles of the forces, are the basis of the **standard model** of particle physics. This model was developed before all of the particles were discovered and it predicted their properties before they were detected, so today it is regarded as a robust theory of the fundamental nature of matter and the forces or interactions that occur between its particles.

In principle, then, the story of the universe has just a few unique actors, and if we understood physics and chemistry perfectly, we should be able to predict the stunning array of phenomena that we find throughout the visible universe. The complexity that grows out of the interactions of these elementary particles produces many surprises, however. It seems impossible that any theory of these particles could have predicted that 13.7 billion years into its history the universe would create beings who would contemplate these questions!

The science of astronomy uses what we have learned from particle physics and other sciences to interpret what we see. Occasionally astronomers have had to hypothesize some piece of "missing physics," or a previously unknown force or particle, because of contradictions between what is observed and what was predicted. For example, our understanding of the neutrino's mass has been changed by studies of the neutrinos emitted by the Sun. In this way astronomy can sometimes drive forward new understandings in particle physics—or chemistry, geology, and biology. Astronomy is detective work in which we piece together what is known and guess at what must be missing to build a more complete, consistent explanation of the phenomena we observe.

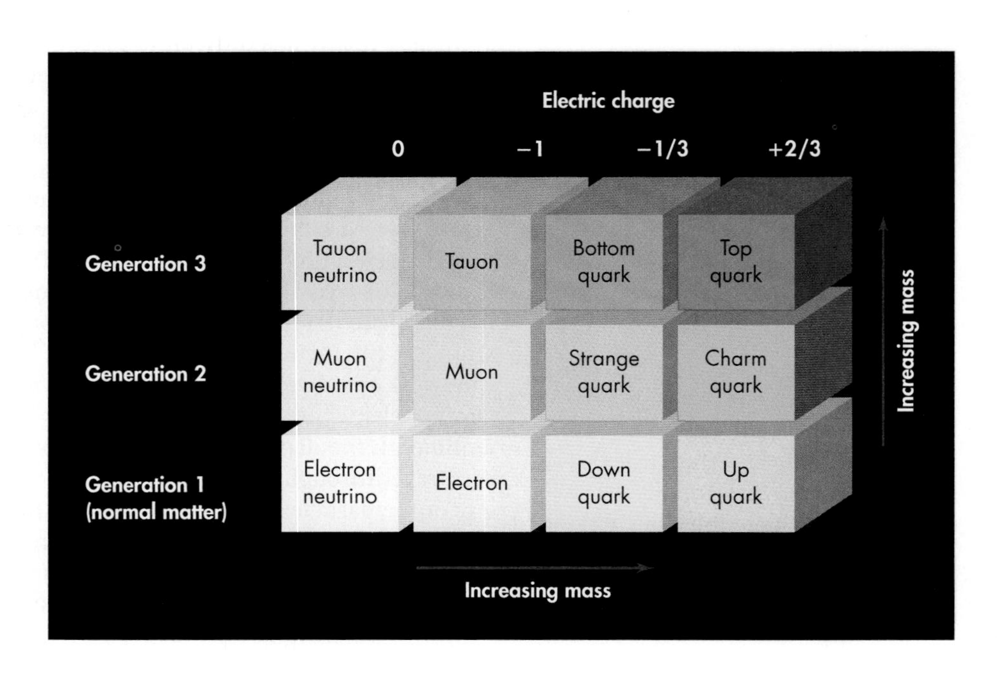

FIGURE 4.4
The elementary particles of the standard model. The general direction of increasing mass is shown, although the down quark appears to be more massive than the up quark. The particles of generations 2 and 3 tend to decay into the first-generation particles except when energies are very high, as in the very early moments of the universe.

KEY TERMS

QUESTIONS FOR REVIEW

1. What is meant by the scientific method?
2. Can a theory ever be proved to be true beyond any doubt?
3. What force(s) hold the Sun together?
4. What particles make up an atom? Which are used to determine its atomic number?
5. What forces(s) hold an atom together?
6. What is a neutrino? What is a quark?

PROBLEMS

1. Make a hypothesis about the cause of some weather phenomenon you have observed. For example, what produces lightning or hail? Describe the different observations you are familiar with that support your hypothesis. What experiment could be conducted that could disprove your hypothesis?
2. If an atom were the size of a city (10 km across), how large would a proton be?
3. Imagine that the Earth and the Moon were both positively charged by an amount proportional to their masses, so that their electrical repulsion canceled out their gravitational attraction. Using the relative strengths of gravity and the electromagnetic force, how much mass in the Earth would have to be positively charged?
4. Explain at a fundamental level what is happening when you take clothes out of a dryer and find them clinging to each other. What do you think is happening when you hear a crackling sound as you pull a sock and shirt apart?
5. If the strong force grows larger with distance, explain why the force is limited in its range to acting within the atomic nucleus.
6. It is estimated that a neutrino has to pass within a distance of about 10^{-24} meters of a quark for a high likelihood of interaction. If an atom has a radius of 10^{-10} meters and 12 nuclear particles in its nucleus, what is the probability that a neutrino passing through an atom will undergo an interaction? (Hint: Think of the atom as a target and the area around each quark as a bull's-eye.)

TEST YOURSELF

1. According to the standard model, which of the following is not an elementary particle?
   a. Photon
   b. Proton
   c. Electron
   d. Neutrino
   e. Up quark
2. The best measure of the validity of an astronomical hypothesis is
   a. Its level of acceptance by scientists.
   b. The ease with which it can be understood by nonscientists.
   c. The sophistication of the mathematics it requires.
   d. Its ability to make predictions about future observations.
   e. How well it describes what has been previously observed.
3. Of the following, the best model of the physical structure of an atom is
   a. electrons orbiting a dense nucleus of protons and neutrons.
   b. electrons and positrons orbiting a microscopic black hole.
   c. protons and electrons orbiting around a nucleus of neutrons.
   d. protons orbiting a low-density nucleus of electrons and neutrons.
   e. neutrons, protons, and electrons all in orbit around each other in an "atomic cloud."

unit 5

The Night Sky

One of nature's spectacles is the night sky seen from a clear, dark location with the stars scattered across the vault of the heavens. From ancient records we know that the pattern of stars has changed little over the last several thousand years. Thus, the night sky affords us a direct link with our remote ancestors as they tried to understand the nature of the heavens. When you look up at the stars, you might be a shepherd in ancient Egypt, a hunter-gatherer on the African plains, a trader sailing along the coast of Persia, or even an airplane navigator in the early twentieth century.

Astronomical observations are part of virtually every culture and include the obvious events that anyone who watches the sky can see without any equipment, such as the setting of the Sun in the western sky and the moving pattern of stars in the night sky. Sadly, many of the astronomical phenomena well known to ancient people are not familiar to us today because the bright lights of cities make it hard to see the sky and its rhythms. Therefore, if we are to appreciate the context of astronomical ideas, we need to first understand what our distant ancestors knew and what we ourselves can learn by watching the sky throughout a night.

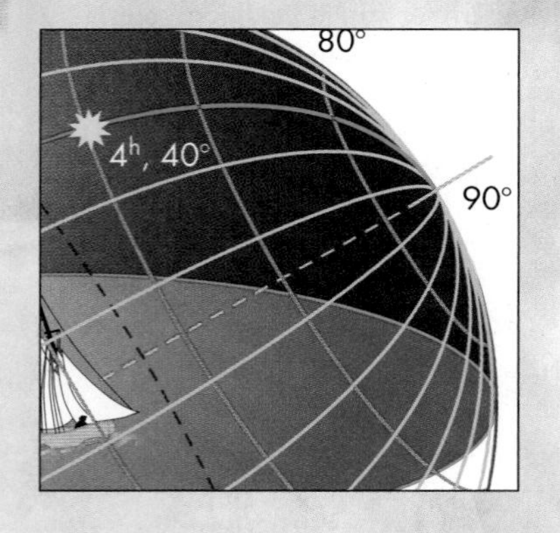

Background Pathways

Unit 1: Our Planetary Neighborhood 4

Unit 2: Beyond the Solar System 13

5.1 THE CELESTIAL SPHERE

From a dark location on a clear night, thousands of stars are visible to the naked eye. The stars appear to be sprinkled on the inside of a huge dome that arches overhead. At any moment as you look at the stars, another half of this dome lies below the **horizon,** hidden by the solid Earth beneath your feet. If we took a spaceship far enough away from the Earth, we would see the whole panorama of stars surrounding us.

Although the stars look as if they form a dome over us, the stars are at vastly different distances from us. For example, the nearest star is about 4 light-years away, while others we can see are thousands of times more distant. And even the planets, the Sun, and the Moon, which is 100 million times closer than the nearest star, are still so distant that we are unable to get a direct sense of their true three-dimensional arrangement in space without precise measurements and careful deduction.

Nevertheless, for the purposes of studying the patterns of the night sky with our naked eyes, we can treat all stars as if they are at the same distance from us—a distant starscape on the interior of a giant **celestial sphere,** with the Earth at its center, as depicted in Figure 5.1. The celestial sphere is a model (Unit 4.1). It does not represent physical reality, but it serves as a useful way to visualize the arrangement and motions of celestial bodies.

For clarity, in Figure 5.1 and many other figures throughout the book, the sizes of astronomical bodies are exaggerated compared to the distances between them.

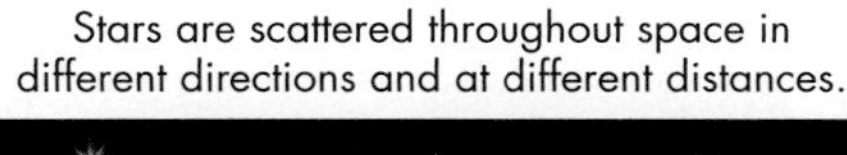

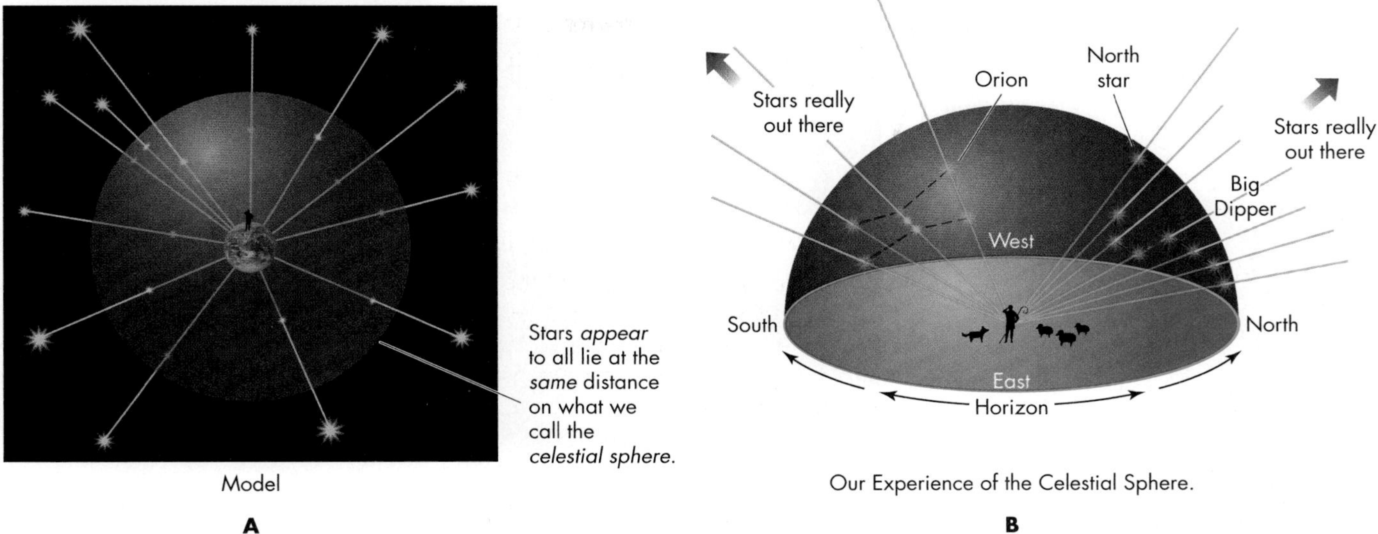

FIGURE 5.1
(A) Although the stars are scattered through space at very different distances, they appear to lie at the same distance from us on what we call the *celestial sphere*. Note: Sizes and distances are drastically exaggerated. For Earth at the size shown, the nearest star would really be 10,000 km (6000 miles) away and would be about 100 times larger than the Earth. (B) The celestial sphere meets the ground at the horizon.

5.2 CONSTELLATIONS

The patterns formed by brighter stars sometimes suggest the shapes of animals, personages, or objects of cultural relevance. For example, the pattern of stars shown in Figure 5.2A looks a little like a lion, although other cultures saw a dragon or a sphinx in this pattern of stars. Today we call this **constellation**—a recognized grouping of stars—Leo. This name has been carried down through the centuries from the Latin word for "lion," and the connection with a lion dates back millennia earlier.

Today the International Astronomical Union recognizes 88 official constellations that cover the entire celestial sphere. About half of these constellations, particularly the larger ones, have ancient roots. Some smaller constellations were named more recently by cartographers making maps of the skies and filling in areas of the sky between the better-known constellations. Some groupings of stars remain in common use even though they are not part of the 88 recognized constellations. These "unofficial" groupings of stars are called **asterisms.** For example, the pattern of stars that makes up the head of Leo is sometimes called the "Sickle." Similarly, the "Big Dipper" is an asterism within the constellation Ursa Major (Latin for "large bear").

Many constellations bear no obvious resemblance to their namesakes. It takes some imagination to see a swan in the constellation Cygnus (Figure 5.2B), which is one of the more recognizable constellations. In some cases, factors other than their shape may have played a role. The location of constellations like Ursa Major and Ursa Minor (large bear and small bear), always in the northern part of the sky, may have made early sky watchers think of real or legendary bears in northern lands. In some cases the names are connected with legends regarding other constellations, such as Canis Major and Canis Minor, the large and small dogs who were hunting

FIGURE 5.2
The two constellations Leo (A) and Cygnus (B) with figures sketched in to help you visualize the animals they resemble.

companions of Orion—one of the more recognizable constellations (see Looking Up #6: Orion). Some areas of the southern sky that are not visible from Europe were given names by early European navigators and explorers. This yielded names like Telescopium, Microscopium, and Antlia, named respectively for the telescope, microscope, and the air pump!

Although most of the constellation names come to us from the classical world of ancient Greece and Rome, many other astronomical terms show the influence of Islam and other Middle Eastern cultures. For example, the names **zenith** (for the point of the sky that is straight overhead) and **nadir** (for the point on the celestial sphere directly below you and opposite the zenith) come from Arabic words. So do the names of nearly all the bright stars—Aldebaran, Betelgeuse, and Zubeneschamali, to name a few. Many of these star names describe the parts of a constellation; for example, the second brightest star in Leo is called Denebola, from Arabic words meaning "the lion's tail."

Q What point on the Earth is directly opposite where you live? Where is the zenith for someone there?

Keep in mind that stars in a constellation generally have no physical relation to one another. They simply happen to be in more or less the same direction in the sky. Also, all stars move through space; but as seen from Earth, their positions change very slowly, usually taking tens of thousands of years to make any noticeable shift. Thus we see today virtually the same pattern of stars that was seen by ancient peoples.

5.3 DAILY MOTION

If you watch a star near the horizon, in as little as 10 minutes you will notice a shift in its position. If you locate a star high in the sky and then come back to look at it after few hours, you will find it has changed position dramatically. Everyone has noticed these phenomena for one star—the Sun. Our cycle of day and night and the motion of the Sun across the sky occur because, as the Earth spins, we face different parts of the celestial sphere. The same kinds of apparent motions occur with stars as well.

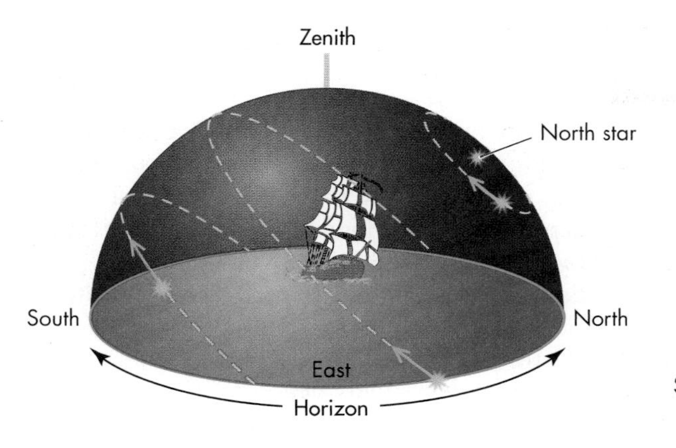

FIGURE 5.3
Stars appear to rise and set as the celestial sphere rotates overhead. Also shown are the celestial equator and celestial poles and where they lie on the celestial sphere with respect to the Earth's equator and poles.

Star rise and star set

From our perspective standing on Earth, it seems as if the celestial sphere rotates around a stationary Earth (Figure 5.3) once each day. Objects within the Solar System, including the Sun, shift relative to the stars on the celestial sphere, but these changes are generally only perceptible over the course of days or weeks. These motions are gradual enough that in 24 hours the objects remain at nearly the same positions on the celestial sphere.

As with the Sun, the stars, the planets, and the Moon all rise in the east, move across the sky, and set in the west. Ancient peoples had no compelling reasons to believe that the Earth spun, so they attributed all daily motion—that of the Sun, Moon, stars, and planets—to the turning of the vast celestial sphere overhead. Today, of course, we know that it is not the celestial sphere that spins but the Earth; however, we still speak of the Sun "rising" and "setting."

As the celestial sphere turns overhead, two points on it do not move, as you can see in Figure 5.3. These points are defined as the north and south **celestial poles.** The celestial poles lie exactly above the North and South Poles of the Earth, and just as our planet turns about a line running from its North Pole to its South Pole, so the celestial sphere rotates around the celestial poles from our perspective on Earth. The star **Polaris** in the constellation Ursa Minor happens to lie close to the north celestial pole. This makes Polaris an important navigation aid, frequently called the "North Star," because it lies close to the direction of true north. There is no comparably bright star anywhere near the south celestial pole, so finding true south from the stars usually requires some triangulation between brighter constellations fairly far from the pole (see Looking Up #9: Southern Circumpolar Constellations).

Another useful sky marker frequently used by astronomers is the **celestial equator.** The celestial equator lies directly above the Earth's equator, just as the celestial poles lie above the Earth's poles, as Figure 5.3 shows. Stars on the celestial equator rise due east and set due west.

5.4 LATITUDE AND LONGITUDE

The part of the sky we can see depends on where we are located on the Earth. Our location is determined by a **longitude,** defining our east–west position, and a **latitude,** defining our north–south position. Some parts of the celestial sphere rotate into view at different times depending on our longitude, but some parts may remain forever hidden unless we move to a different latitude.

We can divide a map of the Earth into a grid by drawing lines of latitude, each at a fixed distance from the equator, and lines of longitude running from pole to pole. The positions of these lines are measured not in a linear unit like kilometers, but instead by an angle in degrees. Ninety degrees (90°) makes up a right angle, and 360° makes up a full circle. Latitudes are measured north and south relative to the equator; for example, Boston, Massachusetts, and Rome, Italy, are both about 42°N, while Buenos Aires, Argentina, and Sydney, Australia, are both about 34°S.

The position east or west has no such natural reference line as the equator. By international agreement, the north–south line that runs through the Royal Observatory in Greenwich, England, is used to mark 0° longitude. Using our sample cities above, Boston is 71°W, Rome is 12°E, Buenos Aires is 58°W, and Sydney is 151°E of this line.

Any two locations at the same latitude will see the same parts of the celestial sphere during a night, so Boston's sky is the same as Rome's. The difference in longitude means they do not see it at the same time. For anyone observing the

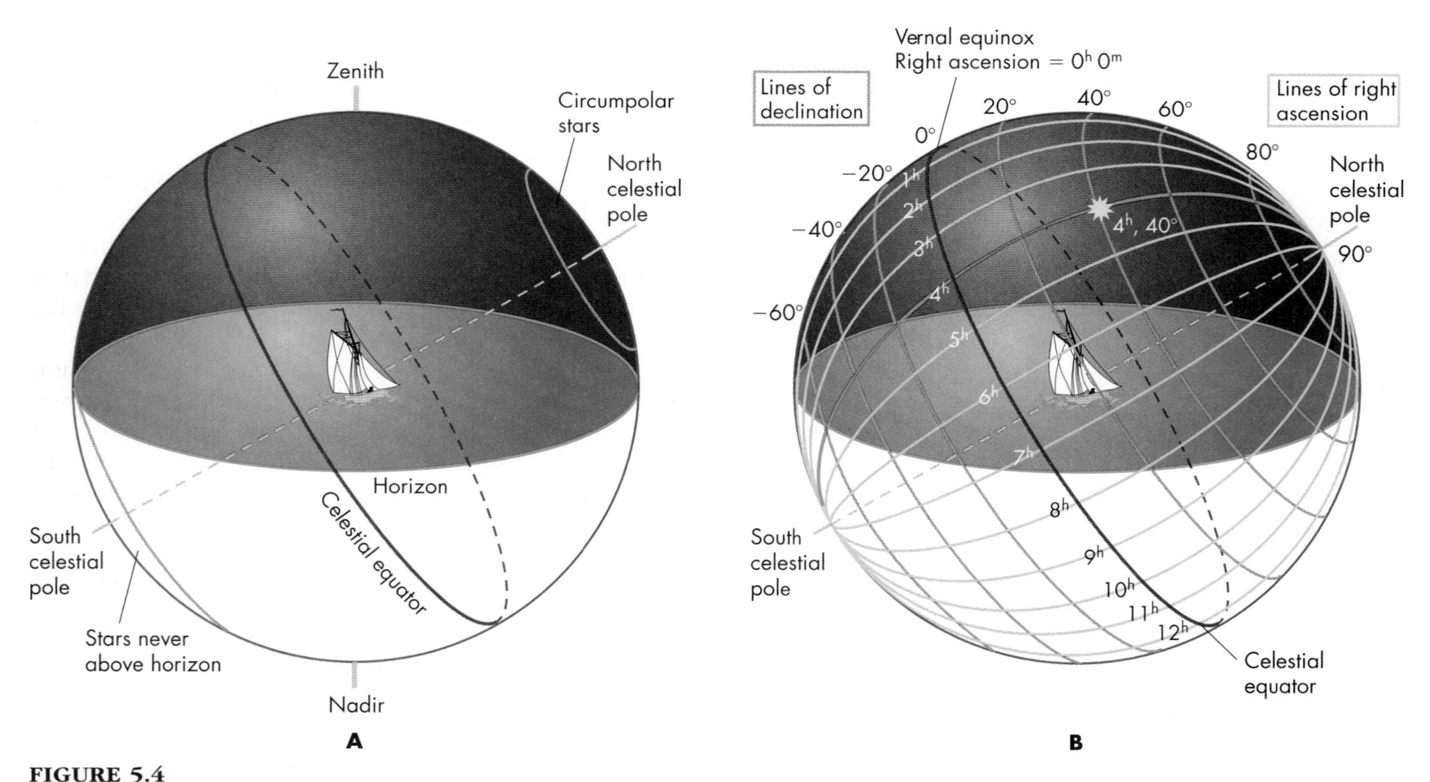

FIGURE 5.4

(A) Some stars rise and set, whereas circumpolar stars near the north celestial pole remain above the horizon continuously. Others near the south celestial pole never rise for a northern observer. (B) Locating a star according to right ascension and declination.

sky north of the equator, there will be some stars near the north celestial pole that always remain above the horizon, said to be **circumpolar,** and some stars near the south celestial pole that always remain below the horizon (Figure 5.4A). The reverse is true south of the equator.

The most extreme case is at the poles. If you were standing at the North Pole, Polaris, the North Star, would remain always straight overhead. The Earth under your feet would permanently block the view of the southern half of the sky, while all the stars north of the celestial equator would be circumpolar, appearing to move around you parallel to the horizon. Again, the reverse would be true if you were at the South Pole, but there happens to be no stellar equivalent of a "South Star" located almost directly overhead.

The location of the celestial poles is particularly important for navigation by the stars because the celestial pole is at an angle above the horizon equal to your latitude. Observing from the Earth's North Pole, 90°N, the north celestial pole is 90° above the horizon. From the equator, the north celestial pole is just on the horizon—we could say 0° above it. Thus, if you can find the celestial pole, you can determine your latitude.

Q If you were standing on the Earth's equator, where would you look to see the north celestial pole? Could you see this pole from Australia?

Q Can you think of a way to build a simple device to measure the angle of a star above the horizon?

5.5 CELESTIAL COORDINATES

The coordinate grid used by astronomers is similar to the longitude and latitude system used to describe positions on Earth. A star's location in the sky is described by a **right ascension,** defining its east–west position, and a **declination,** defining its north–south position. Right ascension (or **RA** for short) plays the same role as longitude, while declination (or **dec**) plays the same role as latitude.

The celestial sphere can thus be divided into a grid consisting of east–west lines parallel to the celestial equator, and north–south lines connecting one celestial pole to the other (Figure 5.4B). Declination values run from +90° to −90° (from the north to the south celestial poles), with 0° at the celestial equator. Right ascension values can likewise be recorded in degrees, but they are commonly listed in "hours." Just as we can divide a circle into 360°, we can divide it into 24 "hours." Each hour of RA equals 15°; that is, $360° \div 24 = 15°$. The convenience of this system is that if a star at RA = 2^h is overhead now, a star at RA = 5^h will be overhead three hours from now. The right ascension of an object can be further refined to minutes (m) and seconds (s), just as we divide time intervals.

With a set of coordinate lines established, we can now locate astronomical objects in the sky the same way we can locate places on the Earth. Astronomers use star charts for this purpose, much as navigators use maps to find places on Earth. Part of a detailed star chart is shown in Figure 5.5. It shows the location of stars and other objects. It also gives some indication of the relative brightness of the stars by marking their positions with larger or smaller dots. Many charts also have information about the season and time of night at which the stars are visible. The foldout star chart in the back of this book labels the dates when different stars will be overhead at 8 P.M. in the evening—a common time to be outside viewing the stars.

In Figure 5.5 an oval-shaped object is located at right ascension 0 hours 42.7 minutes (0^h 42.7^m), declination +41°. This is a nearby galaxy called M31, located in the Local Group. Because its declination very nearly matches the latitude of Boston and Rome, M31 passes nearly through the zenith once each day as observed from both cities. From any location, if a star's declination matches your latitude, it will pass through your zenith.

The location of the chart in Figure 5.5 can be found on the foldout chart at the back of this book. Compare these charts to the photograph of the sky in Looking Up #3: Perseus & M31.

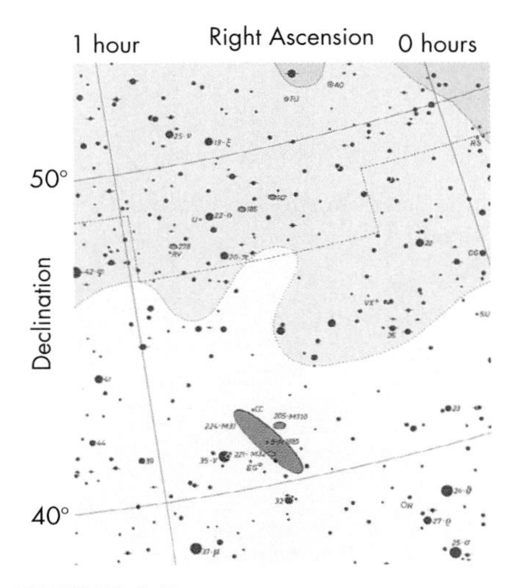

FIGURE 5.5
Small portion of a detailed star chart. Black circles are stars. Their size indicates their brightness—larger circles are brighter stars. Red ellipses are galaxies. The shaded blue area is the Milky Way.

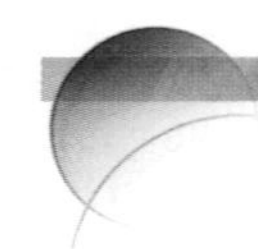

KEY TERMS

asterism, 39
celestial equator, 41
celestial pole, 41
celestial sphere, 38
circumpolar, 43
constellation, 39
declination (dec), 43
horizon, 38
latitude, 42
longitude, 42
nadir, 40
Polaris, 41
right ascension (RA), 43
zenith, 40

QUESTIONS FOR REVIEW

1. What is the celestial sphere?
2. What is a constellation? How does it differ from an asterism?
3. How are latitudes and longitudes defined on the Earth?
4. How are the celestial poles and celestial equator related to the Earth?
5. What is right ascension? declination?
6. Why is right ascension measured in hours, rather than degrees?

PROBLEMS

1. What are your latitude and longitude? What are the latitude and longitude at the point on the Earth exactly opposite you (in the direction of your nadir)? What geographic features are located there?
2. Your observatory lies at a latitude of 45°N and a longitude of 90°W. You wish to observe three celestial objects that have declinations of +87°, −40°, and −67°. Which of these are objects you can observe?
3. The constellations Ursa Major (the Great Bear), in the Northern Hemisphere, and Crux (the Southern Cross), in the Southern Hemisphere, can be used to locate the north and south celestial poles, respectively. Using the star charts at the end of the book, can you show how?
4. Boston and Rome are at the same latitude, but Boston is at a longitude of 71°W, while Rome is at 12°E. If M31 passes through the zenith at a particular time in Rome, how many hours later will it pass through the zenith in Boston?
5. How far is the celestial equator from your zenith if your latitude is 42°N? 34°S?
6. Draw a celestial sphere, as shown in Figure 5.3, as it would appear if you were standing at the North Pole. What celestial object lies directly above you? What celestial object lies on the horizon? Now draw a celestial sphere as it would appear if you were standing at the equator. What celestial object lies directly above you? What celestial object lies on the horizon?
7. Using a protractor, draw a diagram of the Earth. Mark the poles, the equator, and your own latitude. With a ruler, find the line that is "tangent" to your location—that is, the line should touch the Earth's surface at your location while being parallel to the surface at this point. This tangent line represents your horizon. Draw a line extending from pole to pole to indicate the Earth's axis, and extend both this line and your horizon line until they cross. Show that your latitude is the same as the angle of the celestial pole with respect to your horizon.

TEST YOURSELF

1. What makes the star Polaris special?
   a. It is the brightest star in the sky.
   b. The Earth's axis points nearly at it.
   c. It sits a few degrees above the northern horizon from any place on Earth.
   d. It is part of one of the most easily recognized constellations in the sky.
   e. All of the above
2. Where must you be located if a star with a declination of 10° passes through your zenith?
   a. The North Pole
   b. A longitude of 10°W
   c. A latitude of 10°N
   d. A longitude of 10°E
   e. A latitude of 10°S
3. If you are standing at the Earth's North Pole, which of the following will be at the zenith?
   a. The celestial equator
   b. The Moon
   c. M31
   d. The north celestial pole
   e. The Sun

Starry Night

See the Pathways *2nd edition ARIS site for online Starry Night™ planetarium exercises.*

unit 6

The Year

Background Pathways

For people long ago, observations of the heavens had more than just curiosity value. Because so many astronomical phenomena are cyclic—that is, they repeat at a regular interval—they can serve as timekeepers. The most basic of these cycles is the rhythm of day and night, as the celestial sphere appears to rotate about the Earth (Unit 5). This cycle is not completely uniform—days and nights alternately lengthen and shorten over the year. This slow rhythm is tied to a gradual shift of the Sun's apparent position relative to the "fixed stars" on the celestial sphere. The shifting position of the Sun also leads to seasonal changes in the weather and temperature.

The motion of the Sun against the celestial sphere provides a means for tracking these changes predictably. For example, when is it time to plant crops? Or move to the next location to ensure a ready supply of water? Or prepare for winter? Some of the impetus for studying the heavens probably came from the desire to plan for future events. Many ancient peoples built monumental structures to mark the changing position of the Sun. An example is the Mayan pyramid at Chichén Itzá. The pyramid is designed so that on the first day of spring or fall, shadow play creates the appearance of a serpent sliding down the staircase (Figure 6.1).

6.1 ANNUAL MOTION OF THE SUN

As the Earth orbits the Sun, the stars that are visible each night change. The shift is so slow that it is difficult to appreciate from one night to the next, but in the span of a month the changes become obvious. Because these movements repeat yearly, they are called *annual motions.*

FIGURE 6.1
On the equinoxes, the Sun casts a shadow that resembles a serpent slithering down the steps of the Mayan pyramid (left side in photograph) at Chichén Itzá. The head of the serpent is depicted in a sculpture at the base of the stairs.

If you watch the sky each evening over several months, you will discover that new constellations appear in the eastern sky and old ones disappear from the western sky. Across most of North America, Europe, and Asia on an early July evening, the constellation Scorpius will be visible in the southern half of the sky. However, on December evenings, the brilliant constellation Orion, the hunter, is visible instead.

The realization that different stars are visible at different times of the year was extremely important to early peoples because it provided a way to predict the changing of the seasons. A farmer might be tricked into planting too early by a short spell of warmer-than-normal weather in the late winter, but experience would teach that each year the stars could reliably predict when spring was arriving. For example, if you live in the Northern Hemisphere and the early evening sky shows Leo in the south instead of Scorpius or Orion, it will soon be time to plant. Even in semitropical climates, where seasonal temperature differences are much smaller, planting with accuracy is also necessary to avoid crop damage due to annual flooding or dry periods.

Constellations by season

The changing of constellations throughout the year is caused by the Earth's motion around the Sun. As the Earth **revolves** (orbits) around the Sun, the Sun's glare blocks our view of the part of the celestial sphere that lies in the direction of the Sun, making the stars that lie beyond the Sun invisible, as Figure 6.2 shows. For example, in early June, a line from the Earth to the Sun points toward the

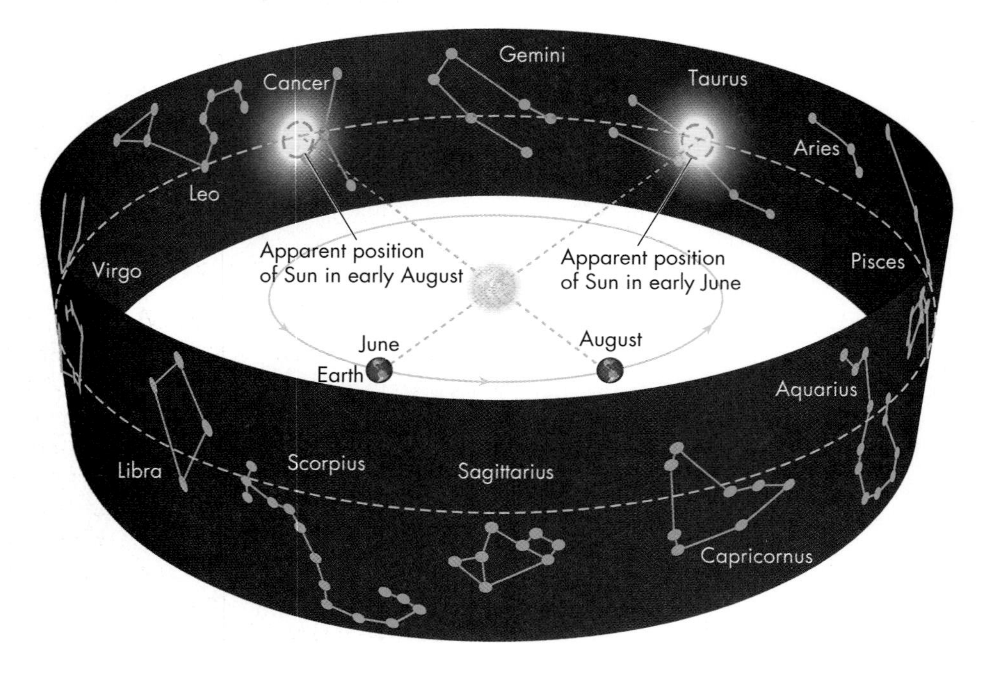

FIGURE 6.2
As the Earth orbits the Sun, the Sun appears to move around the celestial sphere through the background stars. The Sun's path is called the *ecliptic*. The Sun appears to lie in Taurus in June, in Cancer during August, in Virgo during October, and so forth. Therefore the constellations we see after sunset change with the seasons. Note that the ecliptic is the extension of the Earth's orbital plane out to the celestial sphere. (Sizes and distances of objects are not to scale.)

constellation Taurus, and its stars are completely lost in the Sun's glare. In the dusk after sunset, however, it is possible to see the neighboring constellation, Gemini, just above the western horizon. A month later, from the Earth's new position, the Sun lies in the direction of Gemini, causing this constellation to disappear in the Sun's glare. Looking to the west just after sunset, it is possible to see, just barely, the dim stars of the constellation Cancer above the horizon. A month after that, the Sun is in Cancer, and the constellation Leo is visible just above the horizon after sunset.

Month by month, the Sun hides one constellation after another. It is like sitting around a campfire and not being able to see the faces of the people on the far side. But if we get up and walk around the fire, we can see faces that were previously hidden. Similarly, the Earth's motion allows us to see stars previously hidden in the Sun's glare.

6.2 THE ECLIPTIC AND THE ZODIAC

The name *ecliptic* arises because only when the new or full moon crosses this line can an eclipse occur. (See Unit 8.)

If we could mark on the celestial sphere the path traced by the Sun as it moves through the constellations, we would see that it moves around the celestial sphere, as illustrated in the lower part of Figure 6.2. Astronomers call the path that the Sun traces across the celestial sphere the **ecliptic.** You can see in Figure 6.2 that the ecliptic is the extension of the Earth's orbit onto the celestial sphere, just as the celestial equator is the extension of the Earth's equator onto the celestial sphere.

The ecliptic passes through a dozen constellations, which are collectively called the **zodiac.** The word *zodiac* comes from Greek roots meaning "animals" (as in **zo**ology) and "circle" (as in **dia**meter). That is, *zodiac* refers to a circle of animals, which for the most part its constellations represent: Aries (ram), Taurus (bull), Gemini (twins), Cancer (crab), Leo (lion), Virgo (virgin), Libra (scale), Scorpius (scorpion), Sagittarius (archer), Capricornus (goat), Aquarius (water bearer), and Pisces (fish). Actually, the ecliptic passes through a thirteenth constellation, Ophiuchus (serpent holder), during the first half of December (between Scorpius and Sagittarius); but this constellation was not included in the zodiac, probably because of some uncertainty in ancient times about the precise path of the Sun and some vagueness about the boundaries of the constellations.

Q Can you think of an astronomical reason why the zodiac may have been divided into 12 signs rather than 13—or some other number entirely?

The names of some of the constellations of the zodiac may have originated in the seasons when the Sun passed through them. For example, rainy weather in much of Europe during winter was foretold by the Sun's appearance in the constellation Aquarius (the water bearer). Likewise, the harvest time was indicated by the Sun's appearance in Virgo (the virgin), a constellation often depicted as the goddess Proserpine, holding a sheaf of grain.

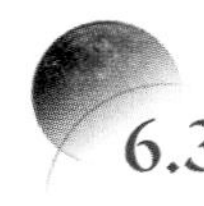

6.3 THE SEASONS

INTERACTIVE

Seasons

Many people believe that we have seasons because the Earth's distance from the Sun changes. They assume that summer occurs when we are closest to the Sun and winter when we are farthest away. It turns out, however, that the Earth is several million kilometers closer to the Sun in early January, when the Northern Hemisphere is coldest, than it is in July. Clearly then, seasons must have some other cause.

Q When it is summer in the Northern Hemisphere, what is the season in the Southern Hemisphere? What does this demonstrate about possible causes of the seasons?

To see what causes our seasons, we need to look at how our planet is oriented in space. As the Earth orbits the Sun, our planet also spins or **rotates.** That spin is around a line—the **rotation axis**—which we might imagine running through the Earth from its North Pole to its South Pole. The Earth's rotation axis is not perpendicular to its orbit around the Sun. Rather, it is tipped by 23.5° from the vertical, as shown in Figure 6.3A.

FIGURE 6.3
(A) The Earth's rotation axis is tilted 23.5° to its orbit around the Sun. (B) The Earth's rotation axis keeps nearly the same tilt and direction as it revolves (orbits) around the Sun. As a result, sunlight falls more directly on the Northern Hemisphere during half of the year and on the Southern Hemisphere during the other half of the year. (Sizes and distances are not to scale.)

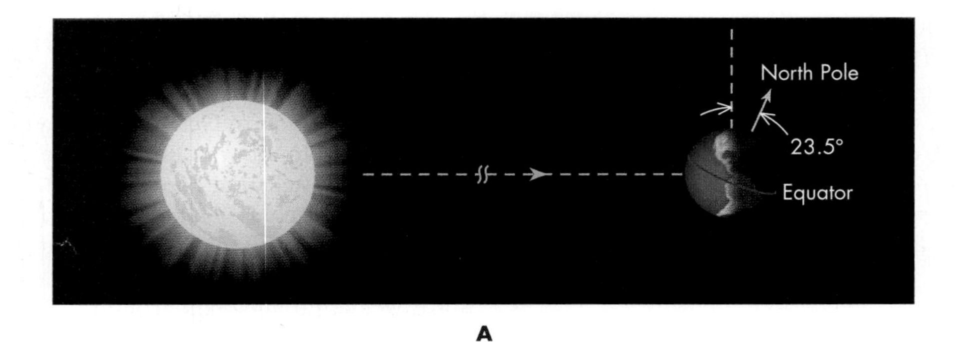

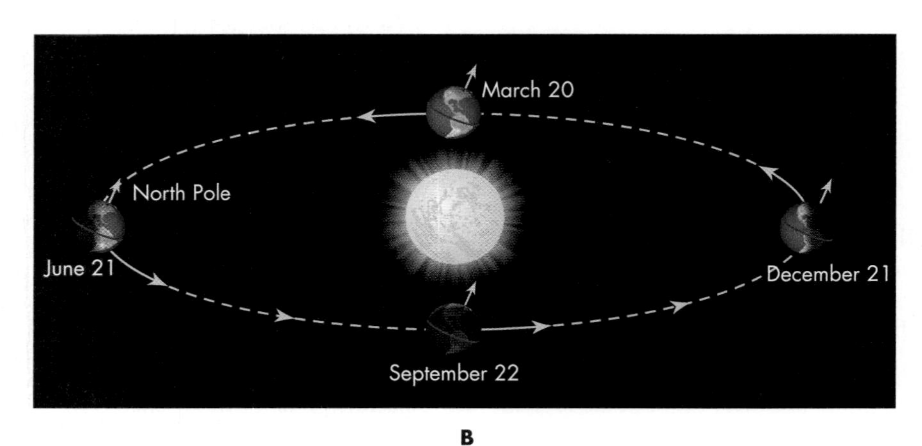

ANIMATIONS
Earth's rotation axis

The constancy of Earth's tilt is a consequence of the *conservation of angular momentum.* (See Unit 20.)

As our planet moves along its orbit, its rotation axis maintains nearly exactly the same tilt and direction, as Figure 6.3B shows. That is, the Earth behaves much like a giant spinning top. In fact, this tendency to maintain its orientation is shown by every spinning object. It is what keeps a rolling coin from tipping over and it is why a quarterback puts "spin" on a football.

The constancy of our planet's tilt as we move around the Sun causes sunlight to fall more directly on the Northern Hemisphere for half of the year and more directly on the Southern Hemisphere for the other half of the year, as illustrated in Figure 6.3B. This in turn changes the amount of heat each hemisphere receives from the Sun.

A surface directly facing the Sun receives the most concentrated sunlight (Figure 6.4). If the surface receives the sunlight at an angle, the light is spread out over a larger area and therefore is less concentrated. An astronomer might express this in terms of the energy received per square meter. A portion of the Earth directly facing the Sun receives about 1300 watts on every square meter. Where the surface is tilted at an angle to the Sun's light, the same 1300 watts are spread out over a larger area on the ground, and each square meter of the Earth's surface receives only a fraction as much energy. You take advantage of this effect instinctively when you warm your hands at a fire by holding your palms flat toward the fire. You also may have experienced the high temperature of pavement or a beach around noon, when the Sun is shining most directly on it, whereas the same surface will be cooler in the late afternoon, even though it is not shaded, when sunlight strikes it more obliquely.

As the Earth orbits the Sun, the same region will face the Sun either more or less directly depending on the time of year. For example, a region in the Northern Hemisphere receives sunlight most directly in late June and most obliquely in late December, as illustrated in Figure 6.5. The direct sunlight produces the strongest heating, whereas the large angle in December produces the least. This heating difference is enhanced because the Earth's tilt also leads to more hours of daylight in

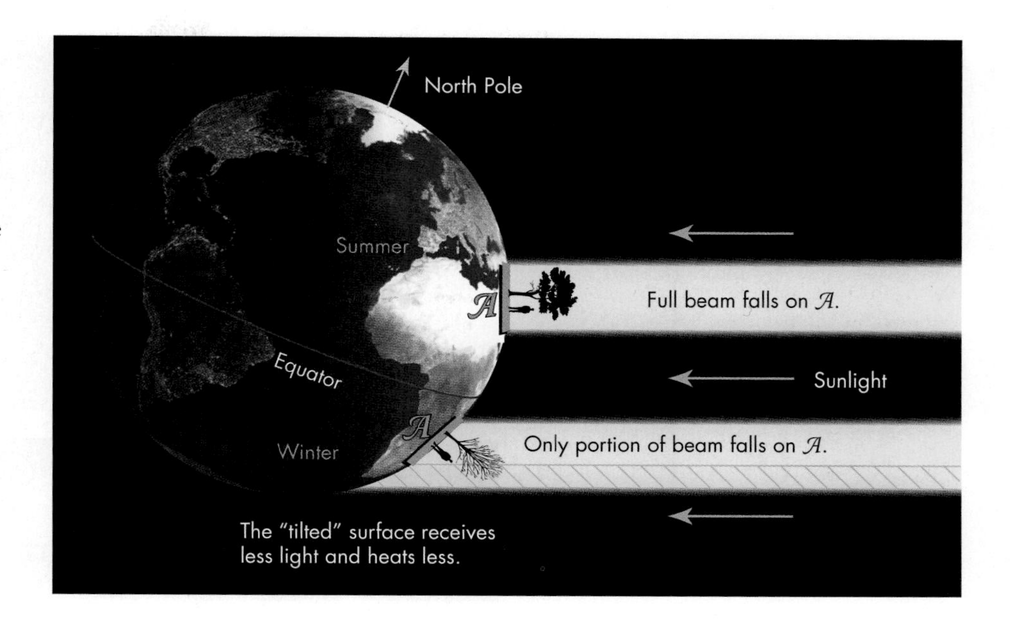

FIGURE 6.4
The portion of the Earth's surface directly facing the Sun receives more concentrated light (and thus more heat) than other parts of the Earth's surface. The same size "beam" of sunlight (carrying the same amount of energy) spreads out over a larger area where the surface is "tilted."

the spring and summer than in the fall and winter (Unit 7.2). As a result, we not only receive the Sun's light more directly, we receive it for a longer time. Therefore, the seasons are caused by the tilt of the Earth's rotation axis. From Figure 6.5 you can also see that this makes the seasons reversed between the Northern and Southern Hemispheres; when it is summer in one, it is winter in the other.

An important point here is not to confuse the "directness" of the Sun's light with one hemisphere being closer to the Sun. It is true that the Northern Hemisphere of the Earth is a few thousand kilometers closer to the Sun than the Southern Hemisphere during the northern summer. However, the effect of this difference in distance is tiny. Compared to the millions of kilometers of distance to the Sun, this difference in distance between the two hemispheres does not produce a change of even one-tenth of a degree in the temperature. By contrast, the differing angle at which the Sun shines on higher latitudes during the year changes the solar energy absorbed by the ground by a factor of two or more. The same size bundle of sunlight is spread out over a much larger area in winter, as shown in the last panel of Figure 6.5; this is the source of seasonal heating changes.

6.4 THE ECLIPTIC'S TILT

The Sun's seasonal motion

The tilt of the Earth's rotation axis not only causes seasons, it makes the ecliptic tilted with respect to the celestial equator. The Sun lies north of the celestial equator for half of the year and south of the celestial equator for the other half of the year. As you can see in Figure 6.3, in June the Northern Hemisphere is tipped toward the Sun. As a result, the Sun lies north of the celestial equator. But in December, when the Earth is on the other side of the Sun, the Sun lies south of the celestial equator.

The consequence of such north–south motion is that the Sun's path—the ecliptic—must cross the celestial equator, and therefore the ecliptic must be tilted with respect to that line, as the sequence of sketches in Figure 6.6 shows. This motion of the Sun north and south in the sky during the year is also why at noon the Sun is so high in the summer sky and so low in the winter sky. For example, at a latitude of 40°, the noon Sun is about 73.5° above the horizon on the first day of summer, but it is only about 26.5° above the horizon on the first day of winter.

FIGURE 6.5
Why the Sun at noon is high in the sky in summer and low in the sky in winter.

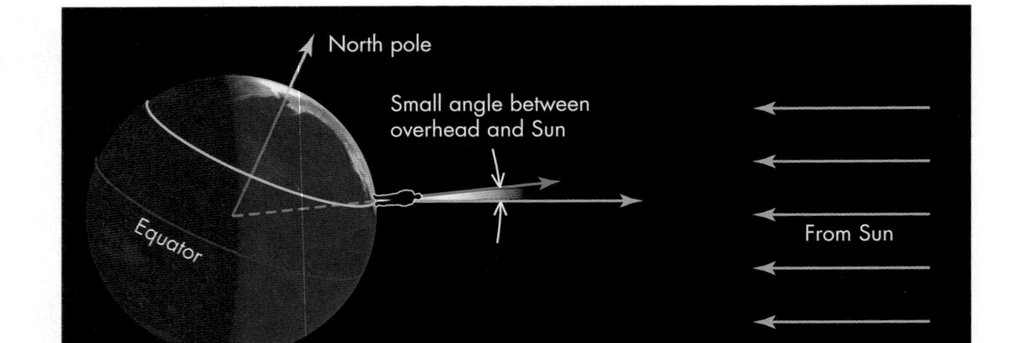

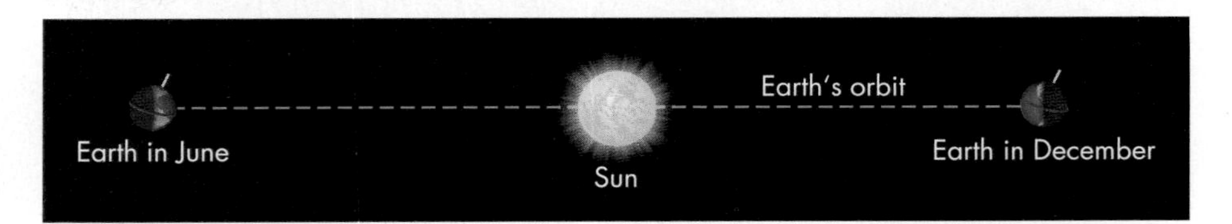

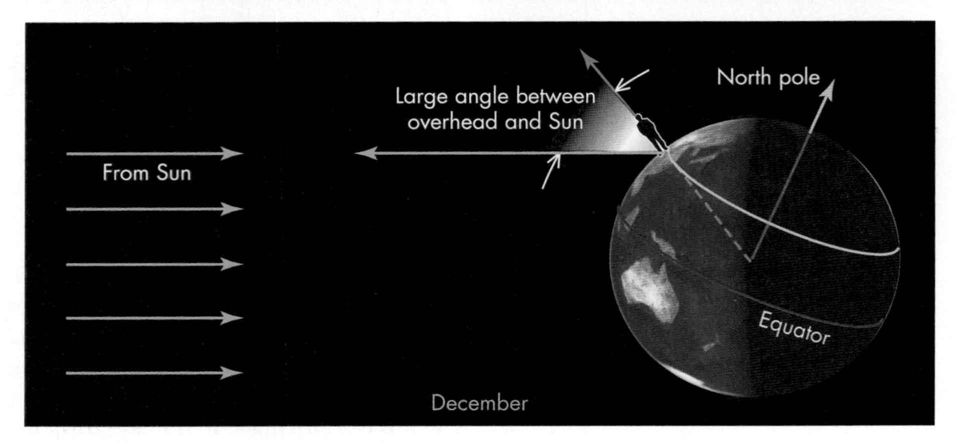

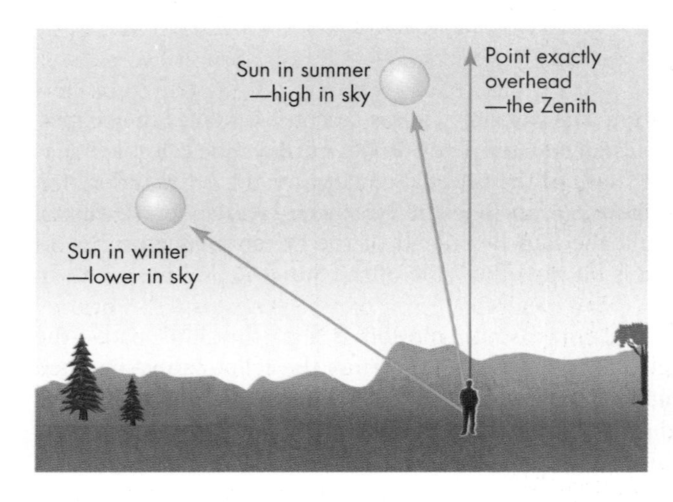

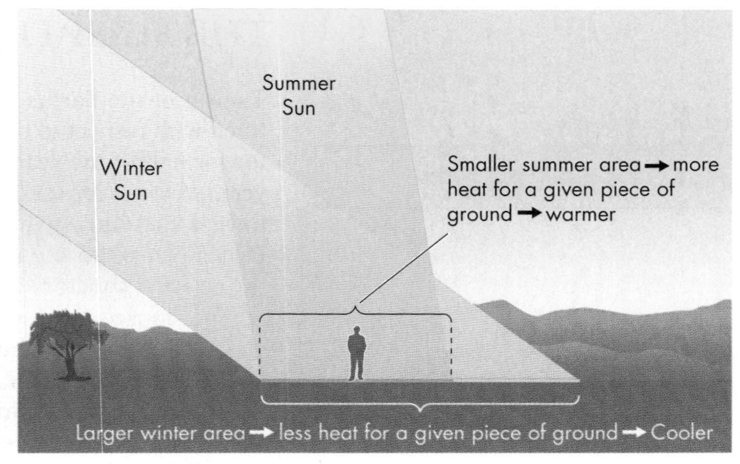

The summer and winter beams carry the same amount of energy, but spread that energy over very different amounts of ground.

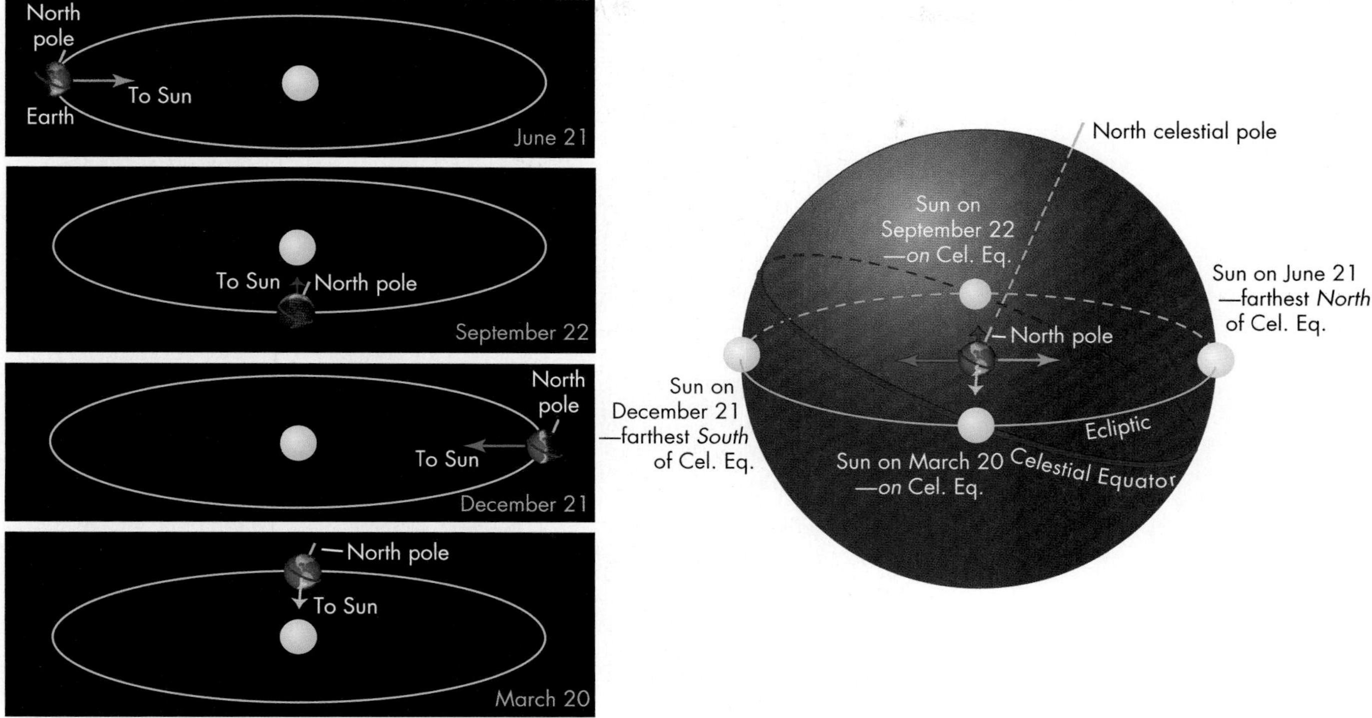

FIGURE 6.6
As the Earth orbits the Sun, the Sun's position with respect to the celestial equator changes. The Sun reaches 23.5° *north* of the celestial equator on June 21 but 23.5° *south* of the celestial equator on December 21. The Sun crosses the celestial equator on about March 20 and September 22 each year. The times when the Sun reaches its extremes are known as the solstices; the times when it crosses the celestial equator are the equinoxes. (The dates can sometimes vary because of the extra day inserted in leap years.)

6.5 SOLSTICES AND EQUINOXES

The exact dates of the equinoxes and solstices vary slightly from year to year, mainly because of differences in the calendar due to leap years, but also because of slight variations in the Earth's orbit.

The tilt of the ecliptic with respect to the celestial equator means that during the year, the points on the horizon where we see the Sun rise and set are not due east and west, except when the Sun is crossing the celestial equator. This occurs on two days of the year called the **equinoxes,** from the Latin for "equal night," so named because the length of the night is approximately equal to the length of the day on those dates. The **vernal equinox** occurs near March 20, when the Sun is moving from the Southern Hemisphere of the celestial sphere into the Northern Hemisphere. Six months later the **autumnal equinox** occurs near September 22 as the Sun crosses the celestial equator on its way south. In the Northern Hemisphere, these dates mark the first day of spring and fall, respectively, but in the Southern Hemisphere this is reversed.

On every other day of the year the Sun rises either to the north or south of due east in a regular, predictable fashion as illustrated in Figure 6.7A. From the vernal equinox to the autumnal equinox (during the Northern Hemisphere spring and summer, and the Southern Hemisphere fall and winter), the Sun rises in the northeast and sets in the northwest. During the rest of the year, the Sun rises in the southeast and sets in the southwest.

Midway between the equinoxes, the Sun reaches its farthest point north or south on the celestial sphere, 23.5° from the celestial equator. At these times of year,

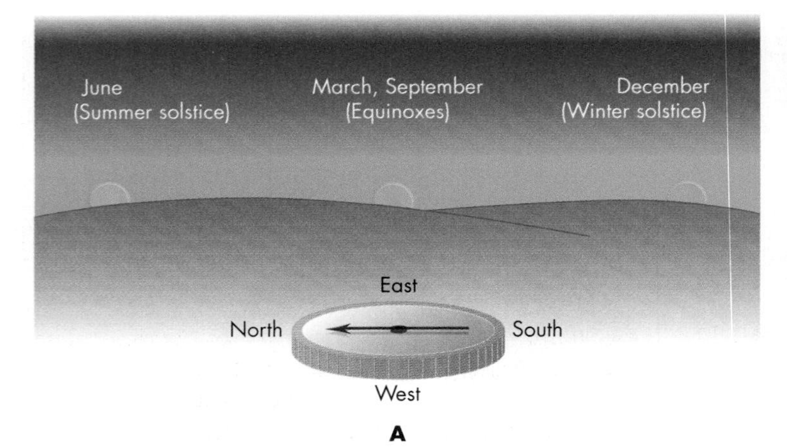

FIGURE 6.7
(A) The direction of the rising and setting Sun changes throughout the year. At the equinoxes, the rising and setting points are due east and due west. The sunrise direction shifts slowly north from March until the summer solstice, after which it shifts back, reaching due east at the autumn equinox. The sunrise direction continues moving south until the winter solstice, then reverses direction again back to the north. The sunset point similarly shifts north from approximately December 21 until June 21 and south from June 21 until December 21. (B) Sketch showing the alignment of the stones with respect to the solstice at Stonehenge, a stone monument built by the ancient Britons on Salisbury Plain, England. The orientation of the stones marks the seasonal rising and setting points of the Sun. (C) A photograph of sunrise on the summer solstice at Stonehenge.

the Sun pauses in its north–south motion and changes direction. Accordingly, these times are called the **solstices,** meaning the Sun (*sol*) stops its northward or southward motion and begins to reverse direction. The solstices occur close to June 21 and December 21. In terms of celestial coordinates (Unit 5.5), the Sun is at a declination of 0° on the equinoxes, while it is at plus or minus 23.5° on June 21 and December 21, respectively. The right ascension of the Sun is, by definition, zero on the vernal equinox, around March 20. The Sun's position at the moment it crosses the celestial equator on this date is used to define 0 hours (or 0°) for the right ascension system. On June 21 the Sun moves to a right ascension of about 6 hours, then 12 hours on September 23, then 18 hours on December 21, before returning to 0 hours of right ascension a year later (see Figure 6.6).

The Sun's position in the celestial sphere is shown in star charts by the curving line of the ecliptic—see the foldout star chart in the back of the book.

The seasonal shifts of the Sun also define three kinds of regions on the Earth: the polar regions, the tropics, and, lying in between these, the temperate latitudes. The polar regions mark the latitudes where the Sun does not rise during some portion of the year. This occurs within 23.5° of the poles—north of 66.5°N, the **Arctic Circle,** and south of 66.5°S, the **Antarctic Circle.** The tropics lie between latitudes 23.5°S and 23.5°N, where the Sun passes directly overhead at some time during the year. The northern limit of tropical latitudes, 23.5°N, is called the **Tropic of Cancer**

TABLE 6.1 Monthly Average Temperatures in Four Cities (in Degrees Celsius)

City	Latitude	Jan.	Feb.	Mar.	April	May	June	July	Aug.	Sept.	Oct.	Nov.	Dec.
Buenos Aires	34°S	23.5	22.7	20.6	16.7	13.3	10.4	10.0	11.1	13.2	16.0	19.3	22.0
Boston	42°N	−2.2	−1.6	2.5	8.2	14.1	19.4	22.5	21.5	17.3	11.5	5.5	0.0
Rome	42°N	7.1	8.2	10.5	13.7	17.8	21.7	24.4	24.1	20.9	16.5	11.7	8.3
Sydney	34°S	22.1	22.1	21.0	18.4	15.3	12.9	12.0	13.2	15.3	17.7	19.5	21.2

because the Sun was in the constellation Cancer on the summer solstice when this term was defined. The southern limit is the called the **Tropic of Capricorn** after the constellation where the Sun was at the winter solstice. The Sun in no longer in these constellations at the solstices because of *precession* (next section).

The seasons "officially" begin on the solstices and equinoxes, with northern spring running from the vernal equinox to the solstice in June. Even though the longest day is on the first day of summer, the hottest period of the year occurs roughly six weeks later, as shown for four cities in Table 6.1. The delay, known as the *lag of the seasons,* results from the oceans and land being slow to warm up in summer. Similarly, there is about a six-week lag after the shortest day of the year until the coldest period of the year.

Just as the changing position of the Sun against the constellations could be used as an indicator of the seasons to ancient peoples, so too could the northward and southward journeys of the Sun. One well-known example of a structure to mark these journeys is Stonehenge, the ancient stone circle in England (Figure 6.7B). Although its exact use in ancient times is lost to us, it appears that it was laid out so that seasonal changes in the Sun's position could be observed by noting which stone arches the Sun rose or set through. For example, on the summer solstice an observer standing at the center of this immense circle of vertical stones would see the rising Sun framed by an arch (Figure 6.7C). A variety of structures were built by cultures all over the world to detect the extreme limits of the Sun's motion in the sky because of the importance of knowing when the changes of seasons occur.

Q During the course of the year, the sunset (and sunset) position shifts. How does the amount of the shift depend on latitude?

6.6 PRECESSION

If you watch a spinning top, you will see that it "wobbles," often more extremely as it slows down. That it wobbles is another way of saying that its rotation axis slowly shifts direction (Figure 6.8). The spinning Earth wobbles too, in a motion called **precession.** Precession occurs very slowly for the Earth. A single "wobble" takes about 26,000 years, but it has both interesting and important consequences.

Currently the northern end of our planet's rotation axis points at the star Polaris. But this is only temporary. When the Egyptian pyramids were built 4000 years ago, the "North Star" was Thuban (meaning "the star") in the constellation Draco (see Looking Up #1: Northern Circumpolar Constellations). In the future the axis will continue shifting direction past Polaris, and it will not point close to any bright stars for thousands of years. In about 7000 years the south celestial pole will be very close to a star slightly brighter than Polaris in the constellation Vela. In that future time there will be a "South Star." In 12,000 years the rotation axis will have shifted so that the north celestial pole points fairly close to the bright star Vega. Then we will have a new, much brighter "North Star."

Precession also slowly alters Earth's climate. At this time we are closest to the Sun during the northern winter. In about 13,000 years we will be farthest from the Sun during the northern winter. This will make seasons in the Northern Hemisphere more severe at that time. Precession is suspected to be one of the components that affect long-term changes in climate, which may have triggered past ice ages.

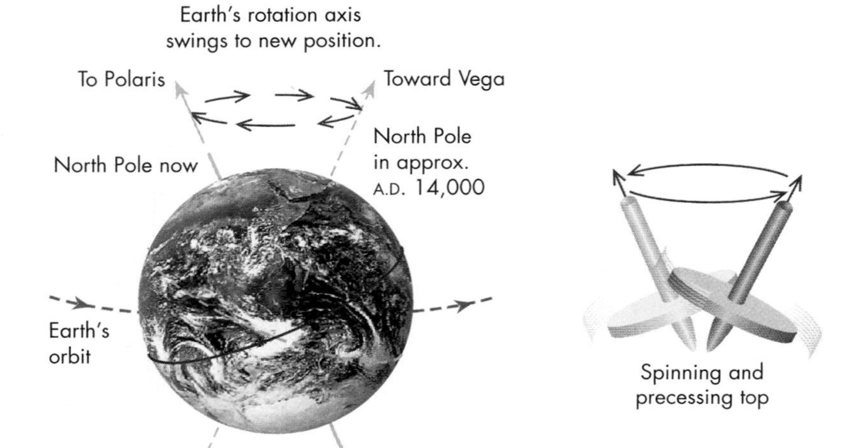

FIGURE 6.8
Precession makes the Earth's rotation axis swing around, slowly tracing out a circle in the sky.

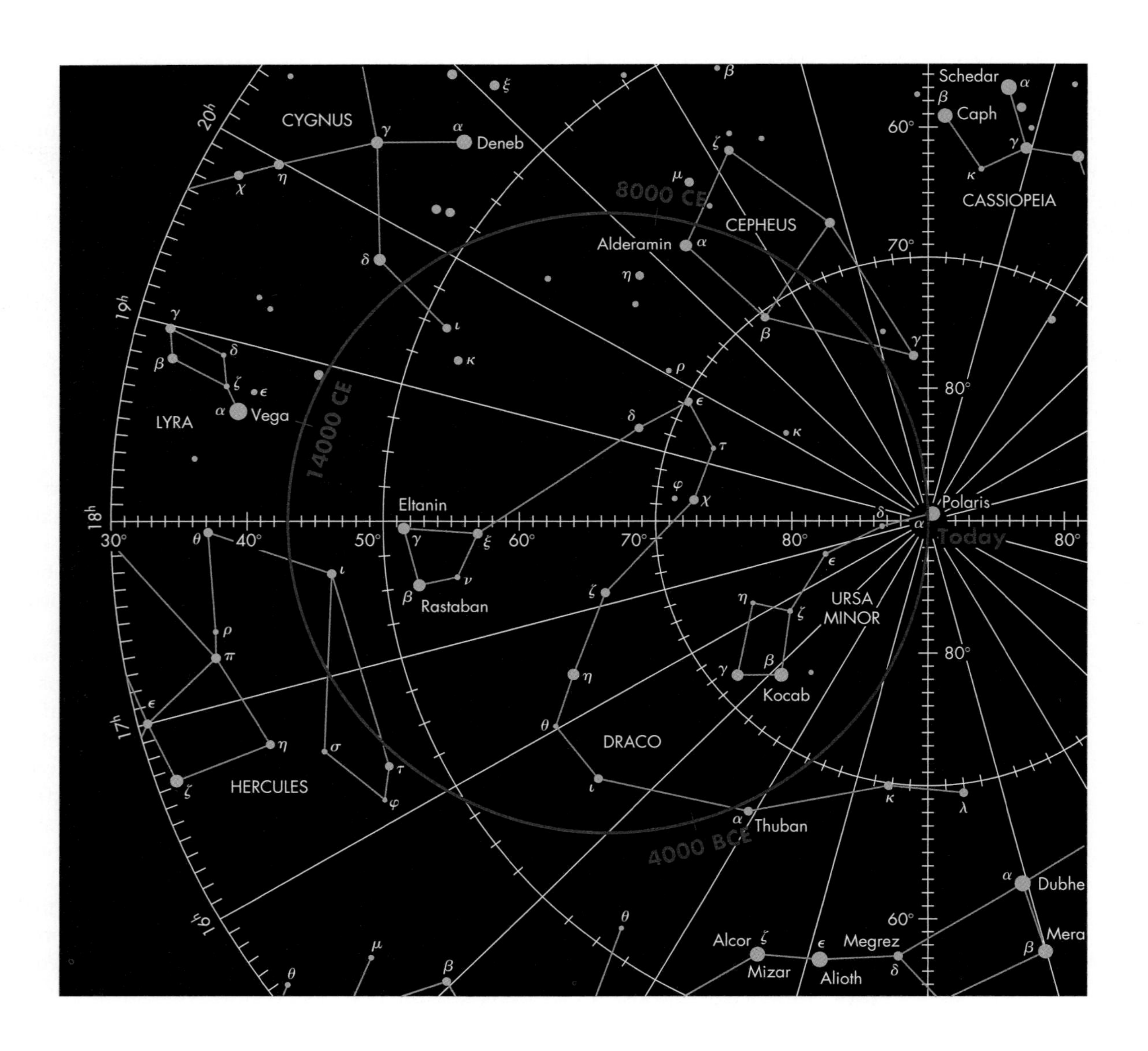

KEY TERMS

QUESTIONS FOR REVIEW

1. What is the ecliptic? What is the zodiac?
2. What causes the seasons?
3. When it is winter in Australia, what season is it in the United States?
4. Where is the Sun located on the celestial sphere during the equinoxes and solstices?
5. Why is the summer solstice *not* the hottest day of the year?
6. How does the Sun's position on the horizon at sunset change through the course of the year?
7. What effect does precession of the Earth's rotation axis have on the Sun's location in the zodiac?

PROBLEMS

1. If the shape of the Earth's orbit were unaltered but its rotation axis were shifted so that it had no tilt with respect to the orbit, how would seasons be affected?
2. Suppose the Earth's axis were tilted by 10° instead of 23.5°. Where would the tropics and arctic regions be? How would seasons be different?
3. Suppose the Earth's axis were tilted by 50° instead of 23.5°. Where would the tropics and arctic regions be? How would seasons be different?
4. Suppose the Earth's axis were tilted by 90° instead of 23.5°. Where would the tropics and arctic regions be? How would the seasons be different?
5. Describe the motion you would see on the solstices and the equinoxes if you were observing the Sun from the Arctic Circle, at a latitude of 66.5°N.
6. If you wished to observe a star with a right ascension of 12^h, what would be the best time of year to observe it? What would be the best time to observe a star with a right ascension of 6^h? (Also see Unit 5.5.)

TEST YOURSELF

1. From what location on Earth will the Sun always rise due east and set due west?
   a. The North Pole
   b. The South Pole
   c. The equator
   d. A latitude of 23.5°N
   e. Nowhere
2. On what day(s) of the year are nights longest at the equator?
   a. They are the same length throughout the year there.
   b. The solstices
   c. The equinoxes
   d. Approximately June 21
   e. Approximately December 21
3. During winter in either hemisphere the temperature is lower because the Sun
   a. stops moving.
   b. is farthest south.
   c. doesn't rise as high in the sky.
   d. has a lower temperature.
   e. is farther away due to the Earth's eccentric orbit.

Starry Night

See the Pathways *2nd edition ARIS site for online Starry Night™ planetarium exercises.*

unit 7

The Time of Day

From before recorded history, people have used events in the heavens to mark the passage of time. The day was the time interval from sunrise to sunrise, and the time of day could be determined from how high the Sun was in the sky. As our ability to independently measure time has become more accurate, we have found that the apparent motions of the Sun across the sky are not as uniform as we once thought, and in this age of rapid travel and communications it no longer makes sense for each town to set its own time according to the Sun. With high-precision modern clocks we have even detected a gradual slowing of the Earth's spin!

Each of these adjustments to our understanding of how to keep time provides an insight into the workings of astronomy.

Background Pathways

Unit 6: The Year 45

7.1 THE DAY

The length of the day is set by the Earth's rotation speed on its axis. One day is defined to be one rotation. However, we must be careful how we measure our planet's rotation. For example, we might use the time from one sunrise to the next to define a day. That, after all, is what sets the day–night cycle around which we structure our activities. However, we would soon discover that the time from sunrise to sunrise changes steadily throughout the year as a result of the seasonal change in the number of daylight hours. A better time marker is the time it takes the Sun to move from its highest point in the sky on one day (what we technically call **apparent noon**) to its highest point in the sky on the next day—a time interval that we call the **solar day.**

We often divide a day into "**A.M.**" and "**P.M.,**" which stand for **ante meridian** and **post meridian,** respectively. The **meridian** is a line that divides the eastern and western halves of the sky. The meridian extends from the point on the horizon due north to the point due south and passes directly through the zenith, the point exactly overhead. As the Sun moves across the sky (Figure 7.1), it crosses the meridian at the time called *apparent noon*. Before (*ante*) noon is thus A.M., while after (*post*) noon is P.M.

Apparent solar time is what a sundial measures, and during the year this time may be ahead or behind clock time by up to about a quarter of an hour. This variation arises from the Earth's orbital characteristics (Unit 13). Although the Sun's position determines the day–night cycle, it does not make a stable reference for measuring Earth's spin.

We can avoid most of the variation in the day's length if, instead of using the Sun, we use a star as our reference. For example, if we pick a star that crosses our meridian at a given moment and measure the time it takes for that same star to return to the meridian again, we will find that this time interval repeats quite precisely. However, this interval is not 24 hours, but about 23 hours, 56 minutes, and 4.0905 seconds. This day length, measured with respect to the stars, is called a **sidereal day,** which is divided into correspondingly shorter hours, minutes, and seconds of **sidereal time.**

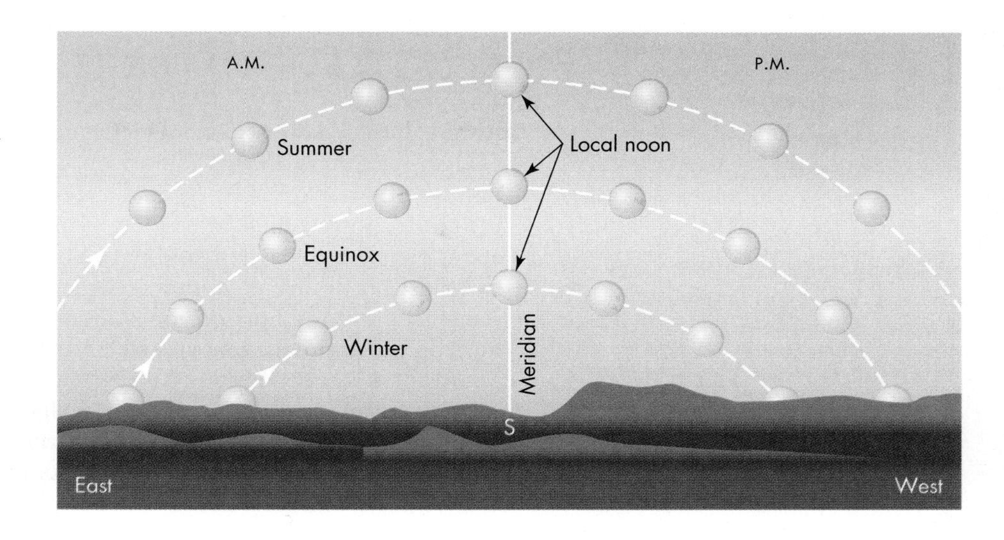

FIGURE 7.1
The Sun rises in the east, crosses the meridian at local noon, then sets in the west. This figure depicts the path of the Sun seen from the Northern Hemisphere at the equinoxes and summer and winter solstices.

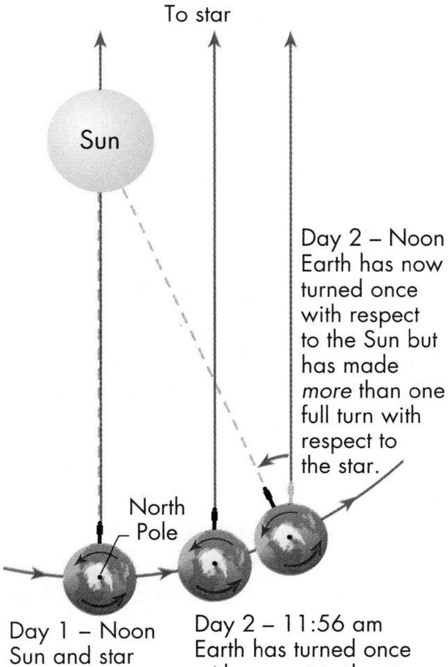

Astronomers find that a clock set to run at this speed is extremely useful. It is not just that the sidereal day is much more stable. Another reason is that, at a given location, any particular star will always rise at the same sidereal time. To avoid the nuisance of A.M. and P.M., sidereal time is measured on a 24-hour basis. For example, the bright star Procyon in the constellation Canis Minor (small dog) rises at about 10 P.M. in November but at about 8 P.M. in December and 6 P.M. in January by solar time. However, on a clock keeping sidereal time, it always rises at the same time at a given location: about 01:30 by the sidereal clock.

Why is the sidereal day shorter than the solar day? We can see the reason by looking at Figure 7.2, where we measure the interval between successive apparent noons—a solar day. Let us imagine that at the same time we are watching the Sun, we can also watch a star, and that we measure the time interval between its passages across the meridian—a sidereal day.

As we wait for the Sun and star to move back across the meridian, the Earth moves along its orbit. The distance the Earth moves in one day is so small compared with the star's distance that we see the star in essentially the same direction as on the previous day. However, we see the Sun in a measurably different direction, as Figure 7.2 shows. The Earth must therefore rotate a bit more before the Sun is again on the meridian. That extra rotation, needed to compensate for the Earth's orbital motion, makes the solar day slightly longer than the sidereal day.

It is easy to figure out how much longer, on the average, the solar day must be. Because it takes us 365¼ days to orbit the Sun and because there are 360° in a circle, the Earth moves approximately 1° per day in its orbit around the Sun. That means that for the Sun to reach its noon position, the Earth must rotate approximately 1° past its position on the previous day. Another way of thinking about this is that the Sun is slowly moving eastward across the sky through the stars at the same time as the Earth is rotating. Therefore, in a given "day," the Earth must rotate a bit more to keep pace with the Sun than it would to keep pace with the stars.

In 24 hours there are 1440 (24 × 60) minutes, and during this time the Earth rotates 360°. Therefore, for the Earth to rotate about 1° extra so that the same side is facing the Sun again, takes about 1440/360 = 4 minutes. The solar day is therefore about 4 minutes longer than the sidereal day.

To star

Sun

Day 2 – Noon
Earth has now turned once with respect to the Sun but has made *more* than one full turn with respect to the star.

North Pole

Day 1 – Noon
Sun and star are both overhead.

Day 2 – 11:56 am
Earth has turned once with respect to the star. Star is back overhead, but the Sun isn't.

FIGURE 7.2
The length of the day measured with respect to the stars is not the same as the length measured with respect to the Sun. The Earth's orbital motion around the Sun makes it necessary for the Earth to rotate slightly more before the Sun will be back overhead. (Motion is exaggerated for clarity.)

7.2 LENGTH OF DAYLIGHT HOURS

Hours of daylight

Although each day lasts 24 hours, the number of hours of daylight, or the amount of time the Sun is above the horizon, changes dramatically throughout the year unless you are close to the equator. For example, in northern middle latitudes, including most of the United States, southern Canada, and Europe, summer has about 15 hours of daylight and only 9 hours of night. In the winter, the reverse is true. Just north of the **Arctic Circle** at a latitude of 66.5° (90° − 23.5°), the Sun remains above the horizon for 24 hours on the summer solstice and below the horizon the entire day on the winter solstice. On the equator, day and night are each 12 hours every day.

This variation in the number of daylight hours is caused by the Earth's tilted rotation axis. Remember that as the Earth moves around the Sun, its rotation axis points in very nearly a fixed direction in space. As a result, the Sun shines more directly on the Northern Hemisphere during its summer and at a more oblique angle during its winter. The result (as you can see in Figure 7.3) is that a large fraction of the Northern Hemisphere is illuminated by sunlight at any time in the summer, but a small fraction is illuminated in the winter. So as rotation carries us around the Earth's axis, only a relatively few hours of a summer day are unlit, but a relatively large number of winter hours are dark. On the first days of spring and autumn (the equinoxes), the hemispheres are equally lit, so that day and night are of equal length everywhere on Earth.

If we change our perspective and look out from the Earth, we see that during the summer the Sun's path is high in the sky, so that the Sun spends a larger portion of the day above the horizon (Figure 7.1). This gives us not only more heat but also more hours of daylight. On the other hand, in winter the Sun's path across the sky is much shorter, giving us less heat and fewer hours of light.

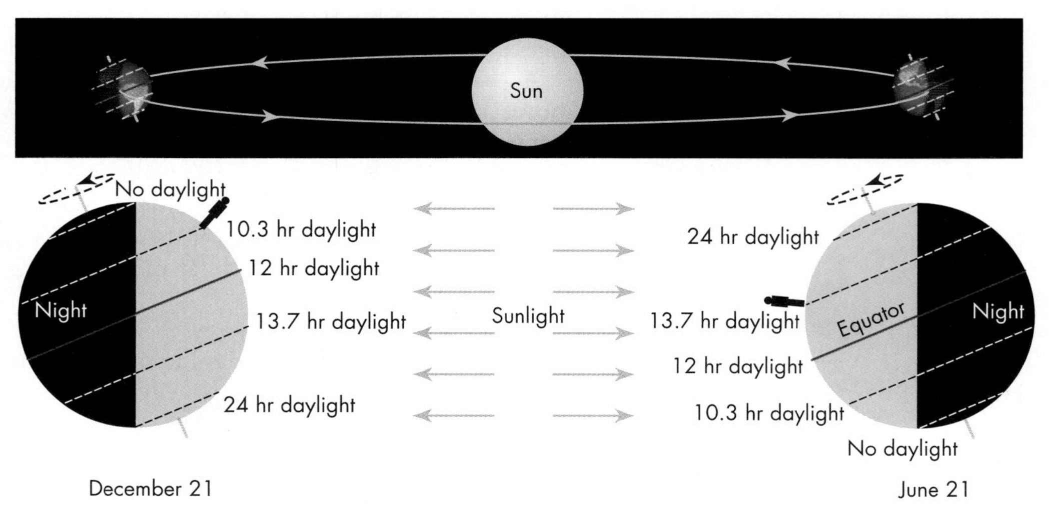

FIGURE 7.3
The tilt of the Earth affects the number of daylight hours. Locations near the equator always receive about 12 hours of daylight, but locations toward the poles have more hours of darkness in winter than in summer. In fact, above latitudes 66.5°, the Sun never sets for part of the year (the *midnight sun* phenomenon) and never rises for another part of the year. At the equinoxes, all parts of the Earth receive the same number of hours of light and dark. (Sizes and separation of the Earth and Sun are not to scale.)

7.3 TIME ZONES

Because the Sun is our basic timekeeping reference, most people like to measure time so that the Sun is highest in the sky at about noon. Technically this is unnecessary now that we have good electronic clocks that can keep time independent of the Sun. Nevertheless, it is a tradition that is hard to break, and as a result, clocks in different parts of the world are set to different times so that the local clock time approximately reflects the position of the Sun in the sky. Because the Earth is round, the Sun can't be "overhead" everywhere at the same time, so it can't be noon everywhere at the same time.

By the late 1800s, with the increasing speed of travel and communications, it became confusing for each city to maintain its own time according to the position of the Sun in the sky. By international agreement, the Earth was therefore divided into 24 major **time zones,** centered every 15° of longitude, in which the time differs by one hour from one zone to the next. With this system, clocks in a time zone all read the same, and they are at most a half hour ahead of or behind what they would be if the time were measured locally. Many regions use local geographic features or political borders to define the boundaries between time zones rather than strictly following the longitude limits (see Figure 7.4). Authorities in a few

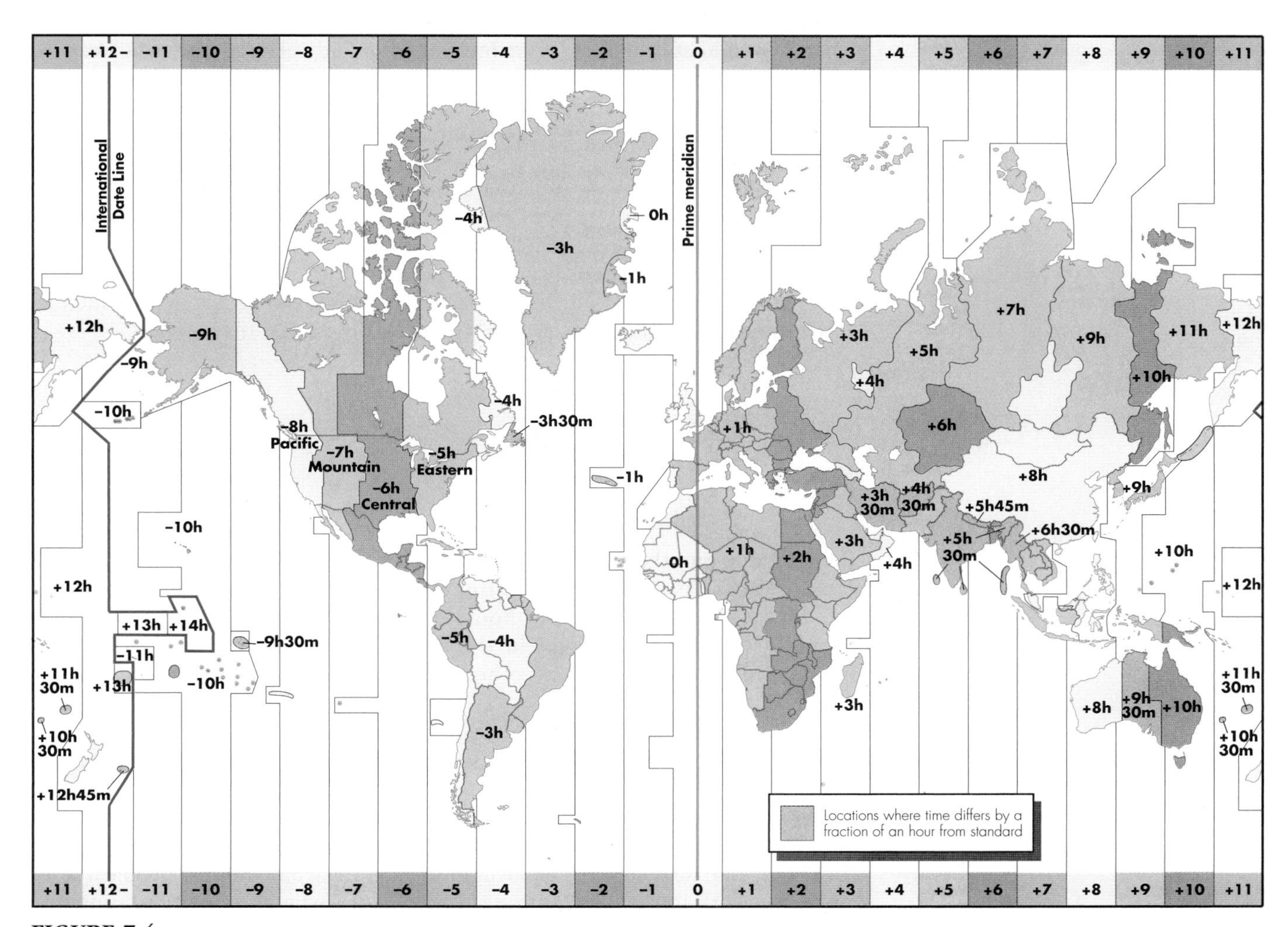

FIGURE 7.4
Time zones of the world and the international date line. Local time = universal time + numbers on top or bottom of chart.

countries and regions did not adopt this agreement, choosing instead to maintain a time standard that was closer to local time. For example, Newfoundland, India, Nepal, and portions of Australia are offset by 30- or 15-minute differences from the international standard.

Across the lower 48 United States, the time zones are, from east to west, Eastern, Central, Mountain, and Pacific. Within each zone the time is the same everywhere and is called **standard time.** In the eastern zone the time is denoted Eastern Standard Time (EST), whereas in the central zone it is denoted Central Standard Time (CST). If you traveled across the United States, you would reset your watch as you crossed from one time zone to another, adding one hour for each time zone as you moved from west to east and subtracting one hour when you moved from east to west.

Q In Jules Verne's story *Around the World in 80 Days,* the travelers discover that although they have experienced 80 days and nights, they still have one more day before 80 days have passed in London. How can this be?

If you travel through many time zones, you may need to make such a large time correction that you shift your watch past midnight. For example, if you could travel westward quickly enough that little time elapsed, setting your watch back each time you crossed a time zone, you could end up at your starting point with your watch turned backward by 24 hours. But you would not have traveled back in time!

A traveler does not actually "gain" a day: when you cross longitude 180° (roughly down the middle of the Pacific Ocean), you add a day to the calendar if you are traveling west and subtract a day if you are traveling east. For example, you could celebrate the New Year in Japan and take a flight after midnight to Hawaii, where it would still be the day before, so you could celebrate the New Year that night too! The precise location where the day shifts is called the **international date line** (Figure 7.4). It generally follows 180° longitude but bends around extreme eastern Siberia and some island groups to ensure that they keep the same calendar time as their neighbors.

The nuisance of having different times at different locations can be avoided by using **universal time,** abbreviated as UT. Universal time is the time kept in the time zone containing the longitude zero, which passes through Greenwich, England. By using UT, which is based on a 24-hour system to avoid confusion between A.M. and P.M., two people at remote locations can decide to do something at the same time without worrying about what time zones they are in.

7.4 DAYLIGHT SAVING TIME

In many parts of the world, people set clocks ahead of standard time during the summer months and then back again to standard time during the winter months. This has the effect of shifting sunrise and sunset to later hours during the day, thereby creating more hours of daylight during the time when most people are awake. Time kept in this fashion is called **daylight saving time** in the United States. In some other parts of the world it is called "Summer Time."

Q If daylight saving time saves energy, why not just have people come to work earlier when the Sun rises earlier and later when it rises later?

Daylight saving time was originally established during World War I as a way to save energy. By setting clocks ahead, less artificial light was needed in the evening hours, In effect, it is a method to get everyone to wake up and go to work an hour earlier than they would normally and take advantage of the earlier rising of the Sun. Clocks are not set ahead permanently because in some parts of the country the winter sunrise is so late that daylight saving time would require people to wake and go to work in the dark. In the United States daylight saving time runs from the second Sunday in March to the first Sunday in November. In Europe it runs from the last Sunday in March to the last Sunday in October, whereas in Australia it runs in exact reverse. Many other countries, and even some states within the United States, follow different rules, while most tropical countries keep their clocks fixed on standard time.

7.5 LEAP SECONDS

Highly accurate time measurements have now allowed us to detect a gradual slowing of the Earth's spin. The second was defined as 1 part in 86,400 ($24 \times 60 \times 60$) of a day, based on astronomical measurements made more than a century ago. With modern atomic clocks it has been determined that the length of the mean solar day is now about 86,400.002 seconds, and it is increasing by about 0.0014 second each century. Over a year the 0.002 seconds add up to most of a second: $365 \times 0.002 \approx 0.7$ second each year. This means that an accurate clock set at midnight on New Year's Eve would signal the beginning of the next new year almost 1 second too early. By the year 2100, an accurate clock will be off by 1.2 seconds after a year.

This time change might sound insignificant, but over long periods its effects accumulate. Fossil records from corals that put down visible layers each day indicate that 400 million years ago the Earth's year had about 400 days. The Earth took the same amount of time to orbit the Sun, but the Earth took only about 22 hours to spin on its axis.

Note that atomic clocks are *not* radioactive. They use an internal vibration frequency of cesium atoms to determine precise time intervals.

With the precise timing needed today for technological uses such as the Global Positioning System (GPS), even millisecond errors are critical. To keep our clocks in agreement with the Sun and stars, we now adjust atomic clocks with a **leap second** every year or two. This is coordinated worldwide so that clocks everywhere remain in agreement.

The slowing of the Earth is a consequence of the interaction of the spinning Earth with ocean tides, which are held relatively stationary by the Moon's gravity. In effect, the eastern coasts of the continents "run into" high tide as the solid Earth spins beneath the tide, and this creates a **tidal braking** force that is slowing the Earth. Tides are discussed further in Unit 19.

KEY TERMS

A.M. (ante meridian), 56
apparent noon, 56
Arctic Circle, 58
daylight saving time, 60
international date line, 60
leap second, 61
meridian, 56
P.M. (post meridian), 56
sidereal day, 56
sidereal time, 56
solar day, 56
standard time, 60
tidal braking, 61
time zone, 59
universal time (UT), 60

QUESTIONS FOR REVIEW

1. Where is the meridian located?
2. How is the sidereal day defined? Why do the sidereal and solar days differ in length?
3. What is universal time?
4. What are the advantages and disadvantages of time zones?
5. How does time shift across the international dateline?
6. Why are "leap seconds" added to our clocks every few years?

PROBLEMS

1. Your friend lives in a town at a longitude 5° to the east of you. Both of you define "noon" as when the Sun reaches its highest point in the sky. How do your clocks differ from each other?
2. If Russia had all clocks set to Moscow time (time zone +3 in Figure 7.4), instead of using many time zones, what time would the Sun rise at the easternmost tip of Siberia on the equinoxes? Examining Figure 7.4, what places in the world have the Sun crossing the meridian earliest according to local time? What places have it latest?
3. One city is at a longitude of 78°E while a second on is at 95°W. How many hours will pass between a star crossing the meridian in the first city and in the second? If the time zone of each city is based on the closest longitude that is a multiple of 15° and the star passes overhead in the first city exactly at midnight, what time (by the clock) will it pass over the second city?
4. Suppose the Earth's spin slowed down until there were just 180 days in a year. Compare the length of a sidereal and a solar day in this new situation. (Do not redefine units of time—just express them in terms of our current hours, minutes, and seconds.)
5. It is thought that the angle of the Earth's axis varies by a few degrees over tens of thousands of years. If the angle were 20° instead of 23.5°, what would be the angle of the noontime Sun from the horizon at your own latitude on the solstices?
6. What is your longitude, and what is the longitude of the center of the nearest time zone? Calculate what time the Sun should, on average, cross your own meridian, assuming that the Sun crosses the meridian at exactly 12:00 at the center of your time zone.

TEST YOURSELF

1. In which of the following locations can the length of daylight range from zero to 24 hours?
   a. Only on the equator
   b. At latitudes closer than 23.5° to the equator
   c. At latitudes between 23.5° and 66.5° north or south
   d. At latitudes greater than 66.5° north or south
   e. Nowhere on Earth
2. In which of the following locations is the length of daylight 12 hours throughout the year?
   a. Only on the equator
   b. At latitudes closer than 23.5° to the equator
   c. At latitudes between 23.5° and 66.5° north or south
   d. At latitudes greater than 66.5° north or south
   e. Nowhere on Earth
3. Daylight saving time
   a. reduces the amount of daylight in summer and shifts it to winter.
   b. corrects clocks for errors caused by the Earth's tilted axis.
   c. results in the Sun crossing the meridian around 1 P.M.
   d. keeps clocks in closer agreement with the position of the Sun in the sky.
   e. All of the above

See the Pathways *2nd edition ARIS site for online Starry Night™ planetarium exercises.*

unit

8

Lunar Cycles

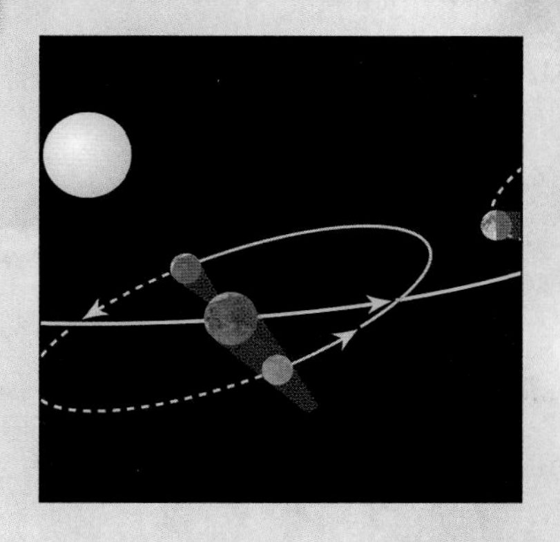

Background Pathways

Unit 6: The Year 45

The Moon is one of the loveliest of astronomical sights; its beauty is extolled in literature, poetry, song, and art. The Moon is forever changing shape, brightness, position, and the times when it is visible. Sometimes it is so bright that it can illuminate the night well enough to hike or even read by its light, whereas at other times it provides little light or is altogether absent. The pattern of these changes may not be obvious unless you spend many successive nights studying the Moon. The Moon's changes follow a regular, clocklike 29½-day cycle. This is the origin of our time period of a **month**—the term coming from the word **Moon**.

The Moon rises in the east and sets in the west. Also, like the Sun, the Moon shifts its position across the background stars from west to east, but about 12 times more rapidly. You can easily detect this motion if the Moon happens to lie close to a bright star. In as little as 10 minutes you can see the Moon shift eastward with respect to the star. This rapid motion makes the Moon the most quickly changing of the astronomical bodies that we can regularly see with the naked eye.

8.1 PHASES OF THE MOON

If you spot the Moon in the west shortly after sunset one evening, on subsequent nights it will have shifted progressively farther east, and in about two weeks it will be in the eastern sky as the Sun sets. These changes occur as the Moon orbits the Earth, causing the Moon to rise and set about 50 minutes later each day. If the Earth did not spin, there would be about two weeks between the Moon's rising in the west and its setting in the east.

One of the most striking features of the Moon is that its shape seems to change from night to night in what is called the cycle of lunar **phases.** That is, starting from when the Moon appears as a thin **crescent** just after sunset, more of it becomes illuminated each night—it **waxes**—until after about two weeks it appears as a fully illuminated disk. Then it steadily decreases—it **wanes**—back to a thin crescent again, and finally disappears for a day or two before beginning the cycle again. In addition to *crescent,* several descriptive names are used for the phases seen during the period of approximately 29½ days of the **lunar month.** When we see more than half of the Moon lit, it is **gibbous;** when completely illuminated, it is **full;** and when completely dark, it is **new.**

The half-lit Moon is also given a special name. Because this phase occurs one-quarter or three-quarters of the way through the lunar cycle, it is called a **first-quarter moon** or **third-quarter moon.** The **quarter,** new, and full phases refer to particular moments in the orbit of the Moon around the Earth—although we often refer to the Moon being full, for example, if it is within about a day of that phase. A new moon occurs when the Moon lies as near as possible to the line between us and the Sun. A full moon occurs when the Moon is on the other side of the Earth from the Sun, opposite it in the sky. The quarter moon occurs when the Moon is 90° away from the Sun in the sky. All of these terms can become confusing, so they are summarized in Table 8.1 and illustrated in Figure 8.1.

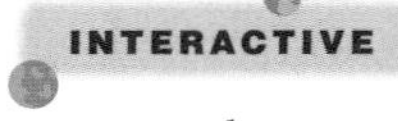

Lunar phases

TABLE 8.1 The Lunar Cycle

Position in Orbit (Figure 8.1)	Day	Phase	Location and Time for Viewing
1	0.0	New	In the same direction as the Sun
2	0.0–7.4	Waxing crescent	Near the Sun in the evening sky
3	7.4	First quarter	90° from the Sun in the evening sky
4	7.4–14.7	Waxing gibbous	Far from the Sun in the evening sky
5	14.7	Full	Opposite the Sun, rising at sunset
6	14.7–22.1	Waning gibbous	Far from the Sun in the morning sky
7	22.1	Third quarter	90° from the Sun in the morning sky
8	22.1–29.5	Waning crescent	Near the Sun in the morning sky
1	29.5	New	In the same direction as the Sun

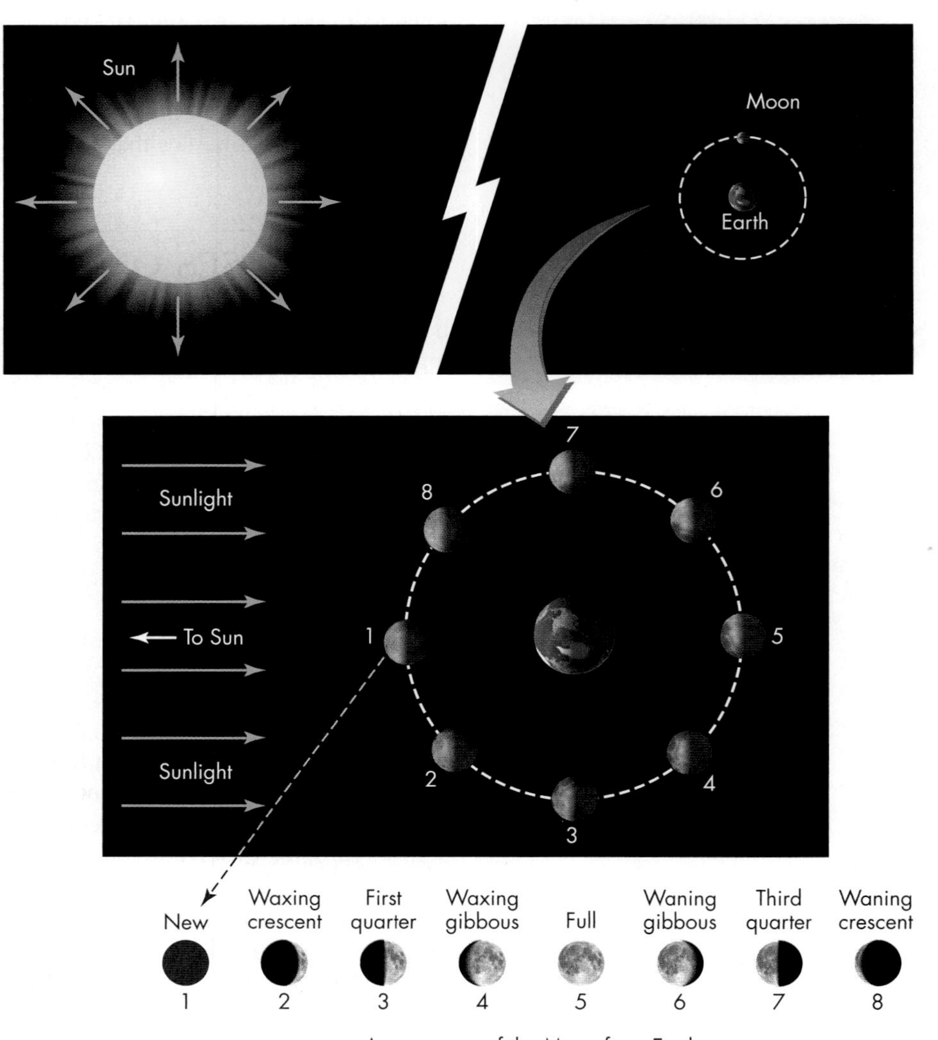

FIGURE 8.1
The cycle of the phases of the Moon from new to full and back again. The phases are caused by our seeing different amounts of the Moon's illuminated half. (Sizes and distances of objects are not to scale.)

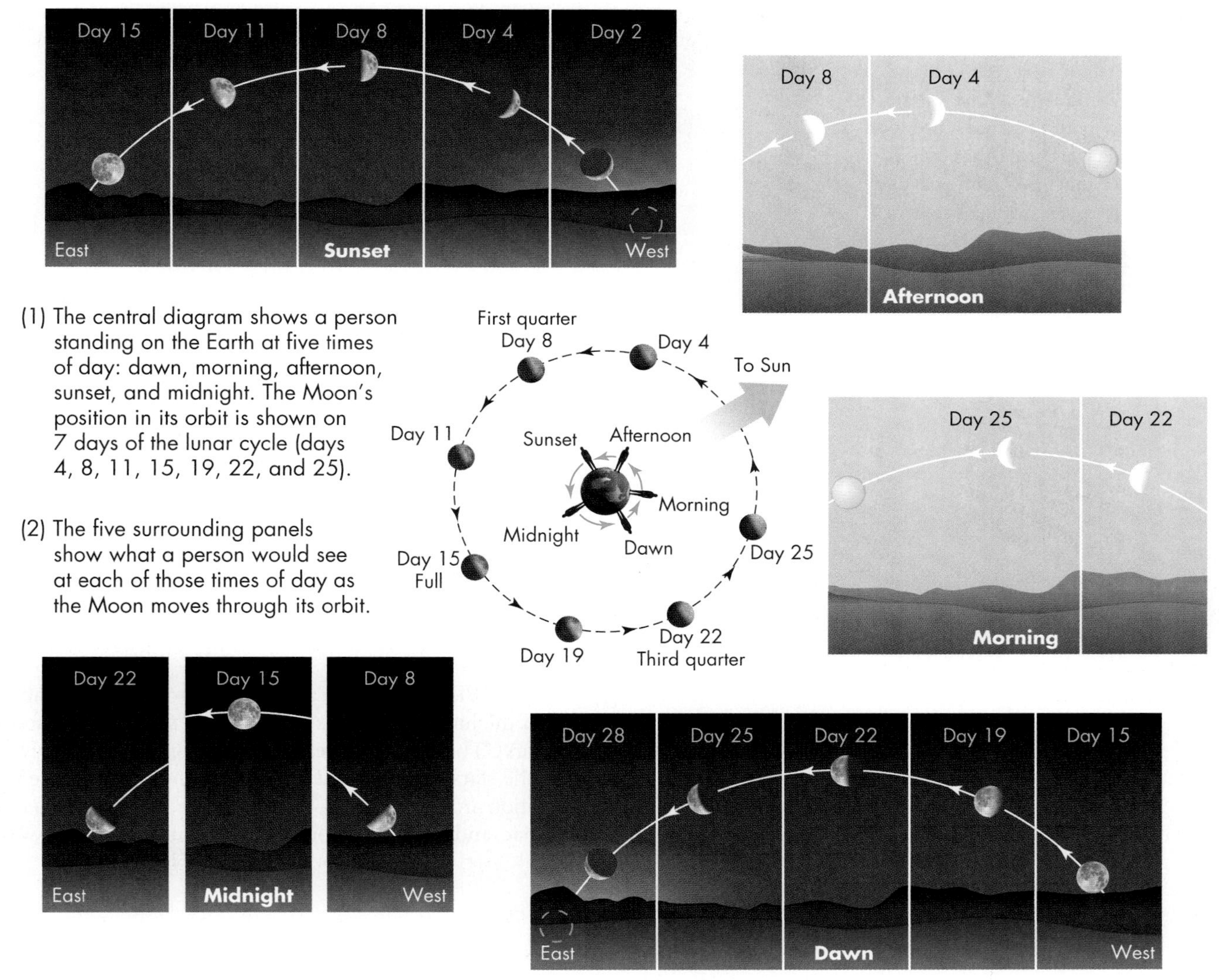

FIGURE 8.2
When the Moon is visible. This figure shows where to look for the Moon and how it appears at different times of day as it goes through its monthly cycle of phases. The central sketch shows the Earth and Moon and where the Moon is in its orbit at different phases, indicated here by the number of days since the new moon. The yellow arrow points toward the Sun.

The times of day and night when you can see the Moon also change throughout the lunar cycle. This is indicated in Table 8.1 and shown graphically in Figure 8.2.

Many people mistakenly assume that the changes in shape are caused by the Earth's shadow falling on the Moon. However, you can deduce that this cannot be the explanation because crescent phases occur when the Moon and Sun lie approximately in the same direction in the sky, and the Earth's shadow must therefore point away from the Moon. In fact, we see the Moon's shape change because as it moves around us, we see different amounts of its illuminated half. For example, when the Moon lies approximately opposite the Sun in the sky, the side of the Moon toward the Earth is fully lit. On the other hand, when the Moon lies approximately between us and the Sun, its fully lit side is turned nearly completely away from us, and therefore we glimpse at most a sliver of its illuminated side, as illustrated in Figure 8.1.

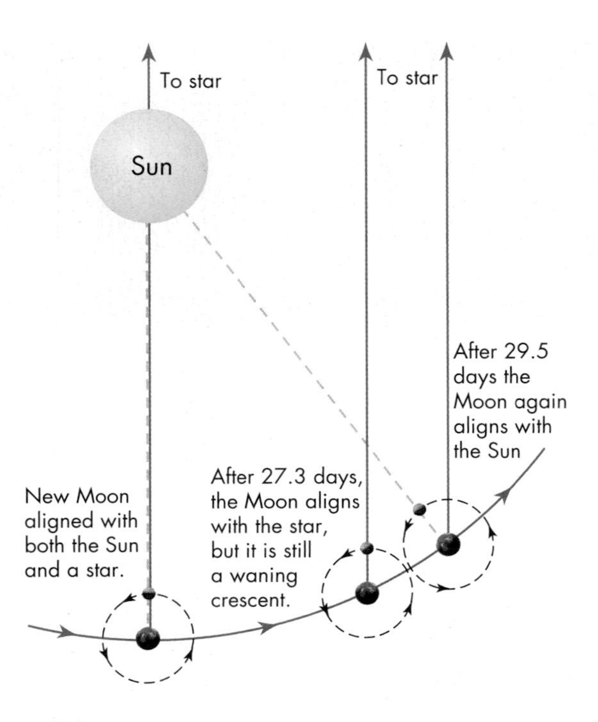

FIGURE 8.3
The *sidereal* month is the time the Moon takes to complete an orbit relative to the distant stars. This is about 27.3 days, less than the lunar month because as the Moon is orbiting the Earth, the Earth is orbiting the Sun. It takes about two additional days for the Moon to come back in alignment with the Sun.

Another confusing point is that the Moon's orbit around the Earth is actually about 27.3 days, not the 29.5 days of the lunar cycle. This is also called the **sidereal month** (from the Latin *siderea* for "starry") because it is the time the Moon takes to return to the same position relative to the stars, as illustrated in Figure 8.3. The difference is a result of the Earth's orbital motion around the Sun—after a month, the Sun shifts into a new constellation of the zodiac, and so the position of the new moon shifts as well.

8.2 ECLIPSES

Eclipse

An **eclipse** occurs when the Moon lies exactly between the Earth and the Sun, or when the Earth lies exactly between the Sun and the Moon, so that all three bodies lie on a straight line. When this happens the Moon will cast a shadow on the Earth or vice versa. A **solar eclipse** occurs whenever the Moon passes directly between the Sun and the Earth and blocks our view of the Sun, as depicted in Figure 8.4. A **lunar eclipse** occurs when the Moon passes through the Earth's shadow as it moves through the portion of its orbit opposite the Sun, as shown in Figure 8.5. Thus solar eclipses can occur *only* when the Moon is new, and lunar eclipses *only* when the Moon is full.

Total eclipses are amazing sights. During a total *solar* eclipse, the Moon completely covers the Sun. The midday sky may become dark as night for several minutes. Birds and other animals may react as if it were nightfall, and the glow from the outer atmospheric layers of the Sun becomes visible. During a total *lunar* eclipse, the Moon usually turns dark orange or red for over an hour. It does not usually become completely black in the Earth's shadow because some sunlight is bent by the Earth's atmosphere into the shadow (Figure 8.6). During a lunar eclipse, if you were looking at the Earth from the Moon, the Earth would appear as a fiery ring of red—in effect you would be seeing sunsets from all over the Earth simultaneously!

The name *ecliptic* for the Earth's orbital plane around the Sun arises because only when the new or full moon crosses this line can an eclipse occur.

Given the remarkable nature of these events, it is not surprising that early people recorded them and sought (successfully) to predict them. Ancient Babylonian and

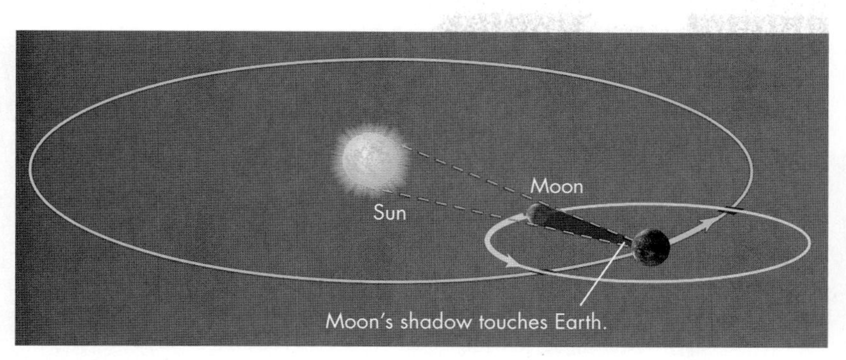

FIGURE 8.4
A solar eclipse occurs when the Moon passes between the Sun and the Earth so that the Moon's shadow touches the Earth. The photo inset shows what the solar eclipse looks like from Earth, from the center of the Moon's shadow, where the Sun is completely covered.

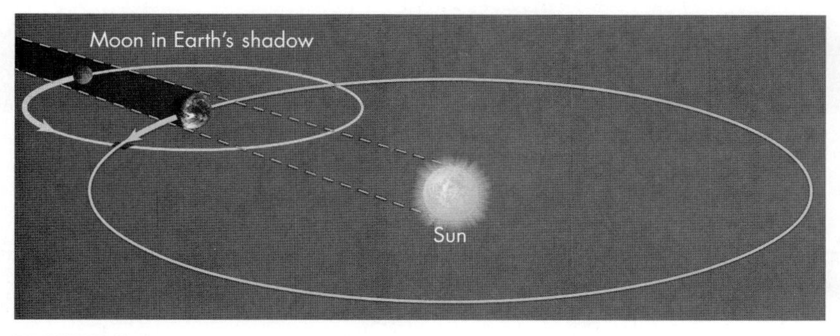

FIGURE 8.5
A lunar eclipse occurs when the Earth passes between the Sun and the Moon, causing the Earth's shadow to fall on the Moon. Some sunlight leaks through the Earth's atmosphere, casting a deep reddish light on the Moon. The photo inset shows what the eclipse looks like from Earth.

FIGURE 8.6
As sunlight falls on the Earth, some passes through the Earth's atmosphere and is slightly bent so that it ends up in the Earth's shadow. In its passage through our atmosphere, most of the blue light is removed, leaving only the red. That red light then falls on the Moon, giving it its ruddy color at totality.

Chinese astronomers kept records stretching back centuries and, based on the patterns they found, were able to predict future eclipses. In more recent times astronomers have used total solar eclipses to study the outer layers of the Sun, which are normally hidden from sight by the enormously brighter central region of the Sun.

Because the Moon is so small compared with the Earth, its shadow is small, and therefore you can see a total solar eclipse only from within a narrow band on the Earth, as illustrated in Figure 8.7. On the other hand, when the Moon

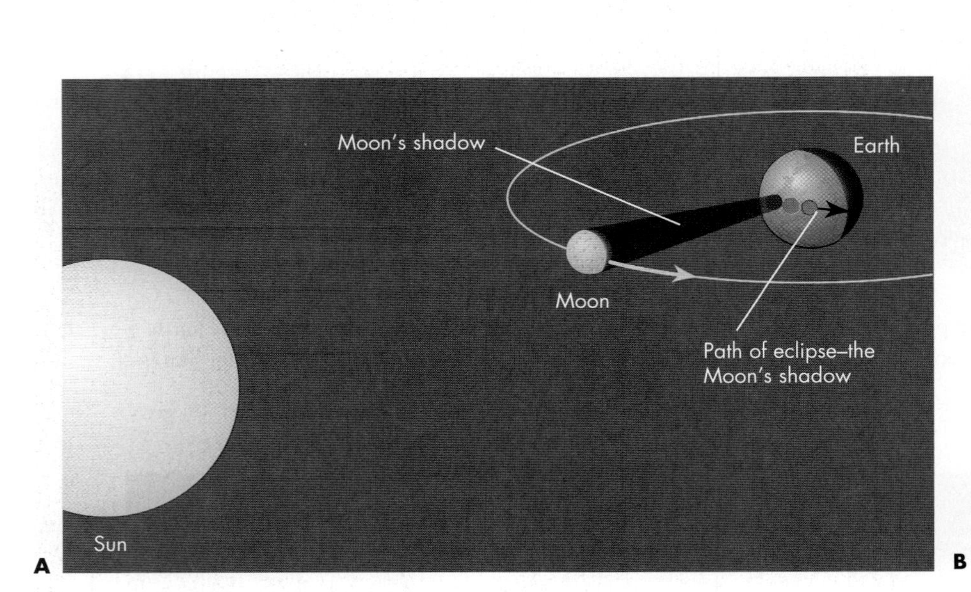

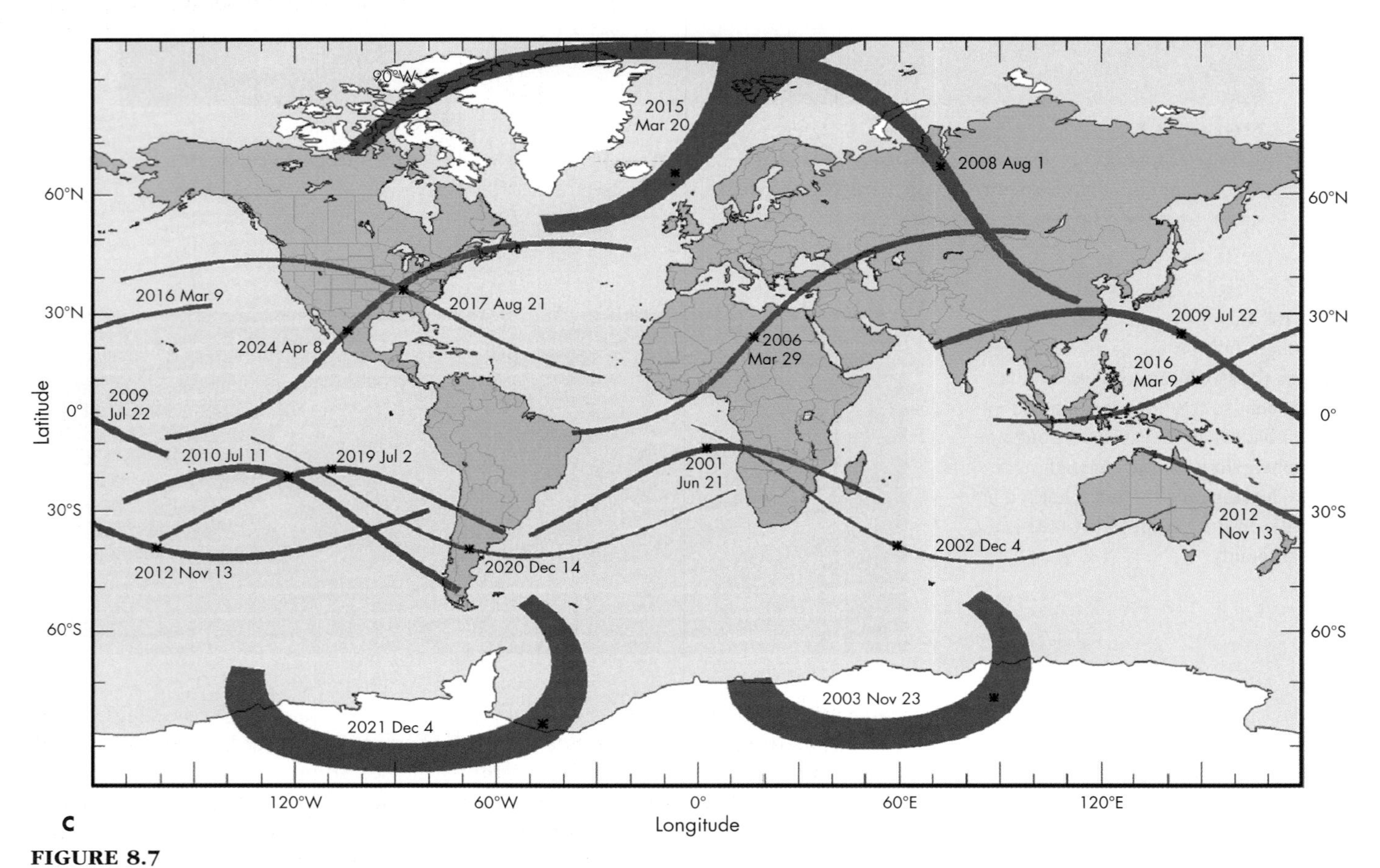

FIGURE 8.7
(A) Sketch of how the Moon's shadow travels across the Earth. (B) Series of images from a location in the path of the Moon's shadow showing the Moon moving across the face of the Sun. (C) Location of recent and upcoming total solar eclipses (through 2025).

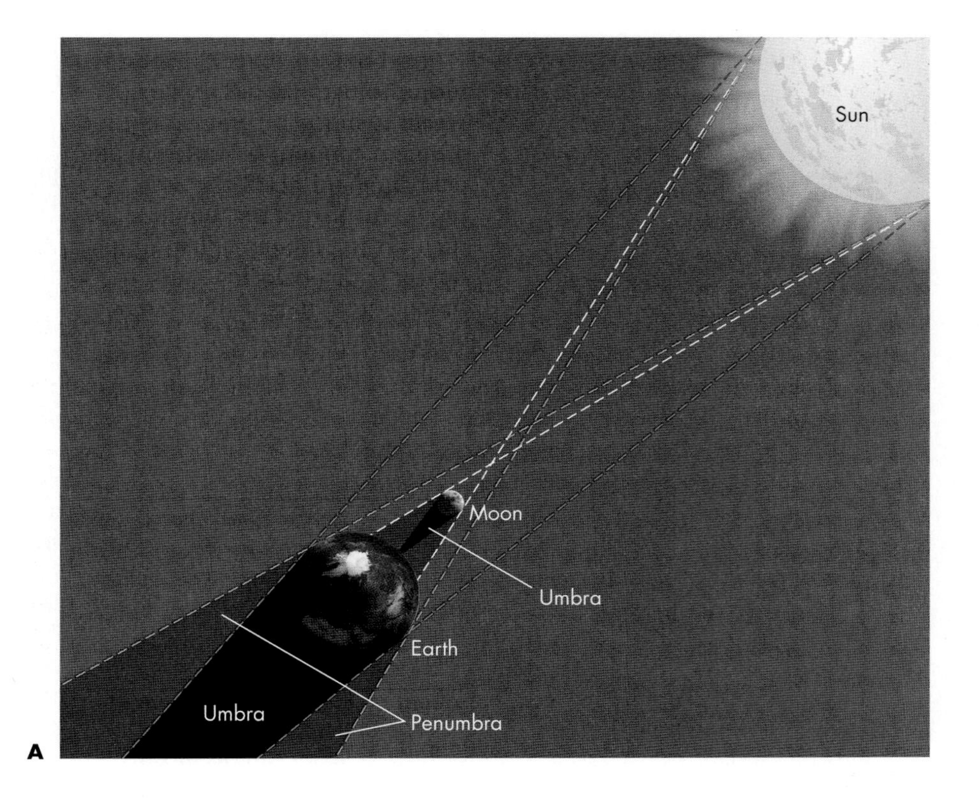

FIGURE 8.8
(A) The shadow of the Earth contains regions where the Sun is completely blocked (the umbra) and regions where sunlight is only partially blocked (the penumbra).
(B) A photograph of the Earth from space during a solar eclipse shows the Moon's shadow on the Earth. The umbra is almost black, while the surrounding region of the penumbra goes from black to fully lit, depending on how much of the Sun is covered by the Moon.

Astronomers can calculate when and what types of eclipses occurred in the distant past and will occur into the distant future. Counting all kinds of partial and total eclipses visible from somewhere on Earth, there are always at least two solar and two lunar eclipses each year, and there can be as many as five of either. Total eclipses are less common, although some years have as many as two total solar eclipses or three total lunar eclipses.

Q The Moon's distance from the Earth is measured to be gradually increasing. When the Moon is much farther away than it is now, what will happen to the frequency of the various kinds of eclipses?

passes through the Earth's shadow in a lunar eclipse, the eclipse is visible from half of the Earth—anywhere that the Moon is above the horizon at the time of the eclipse.

Most solar eclipses pass almost unnoticed because the Moon only partially blocks the Sun's light. The shadow cast by the Moon or by the Earth contains regions where the Sun's light is blocked completely, called the **umbra,** and an outer region where the Sun's light is blocked only partially, called the **penumbra** (Figure 8.8). From the Earth, when we are in the Moon's penumbra, we

FIGURE 8.9
An annular eclipse of the Sun. The Moon is at a distant point in its orbit, so it appears smaller and cannot block the Sun entirely.

call it a **partial eclipse.** These occur about twice as often as total eclipses, and they are visible from a much larger portion of Earth's surface than total solar eclipses.

The Moon's orbit around the Earth is not perfectly circular, so sometimes it is far enough that the umbra does not quite reach the Earth. When this happens, the Moon will appear smaller than the Sun (Figure 8.9), so even if we are in a position where the Sun and Moon are exactly aligned, we can still see the Sun's surface in a ring or *annulus* around the Moon. This is a type of partial eclipse is called an **annular eclipse**.

Similarly, some portion of the Moon passes through the penumbra of the Earth's shadow about twice as often as it completely enters the umbra, for a total lunar eclipse in which the Earth completely blocks the light of the Sun from directly striking any part of the Moon. In partial lunar eclipses, only part of the Moon enters the umbra. The weakest types of partial eclipses are called **penumbral eclipses** because no part of the Moon enters the umbra. They are often so minor that they are not marked on astronomical calendars, although someone living on the Moon would observe the Earth partially eclipsing the Sun.

You may wonder why we do not have eclipses every month. The answer is that the Moon's orbit is tipped by about 5° with respect to the Earth's orbit around the Sun (Figure 8.10). Because of this tilt, even if the Moon is new its shadow may pass above or below Earth. As a result, no eclipse occurs. Similarly, when the Moon is full, it may pass above or below the Earth's shadow so that again no eclipse occurs. Eclipses can occur only when the Moon is within about 1° of crossing the ecliptic—the plane of the Earth's orbit around the Sun. This usually occurs only twice during the year (Figure 8.11).

For example, if a solar eclipse occurs on a given day in May, a lunar eclipse is likely to occur roughly two weeks later or earlier (the interval between full and new moons), and it is even possible to have two partial solar eclipses on two new moons in a row. Half a year later, in November, the Moon will again be close to the ecliptic when it is new or full, and another set of eclipses may occur.

The times of year at which eclipses can occur gradually shift from year to year because the plane of the Moon's orbit "wobbles." This wobble (called *precession;* see also Unit 6.6) swings the Moon's orbital plane back to the same orientation every 18.6 years, so the paths of eclipses tend to be similar. This can be seen in Figure 8.7, where similarly-shaped shadow paths occur about 18 years apart.

The 18 year, 11 day, 8 hour period, called the *saros,* between similar eclipses was discovered by ancient astronomers. For a solar eclipse, the eight additional hours shift the location where the eclipse can be seen about one-third of the way around the globe. However after three of these periods (about 54 years, 31 days), a similar eclipse occurs near the original location.

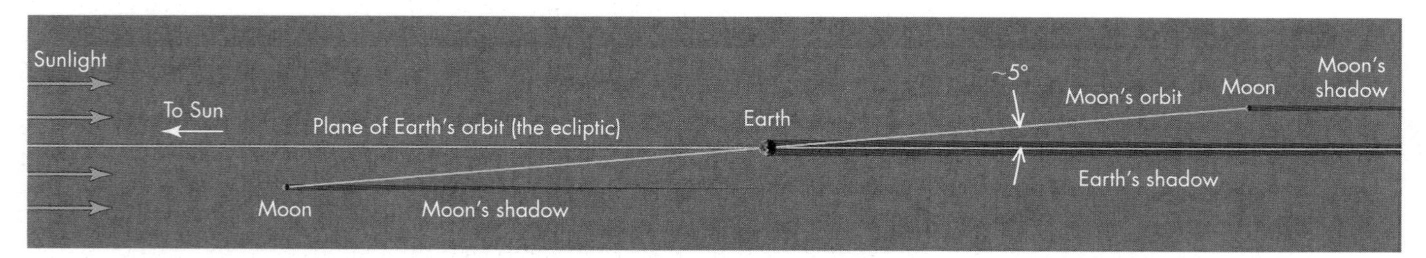

FIGURE 8.10
The Moon's orbit (seen here edge-on) is tilted with respect to the Earth's orbit by about 5°. The Earth and Moon are drawn to correct relative size and separation. Note how thin their shadows are.

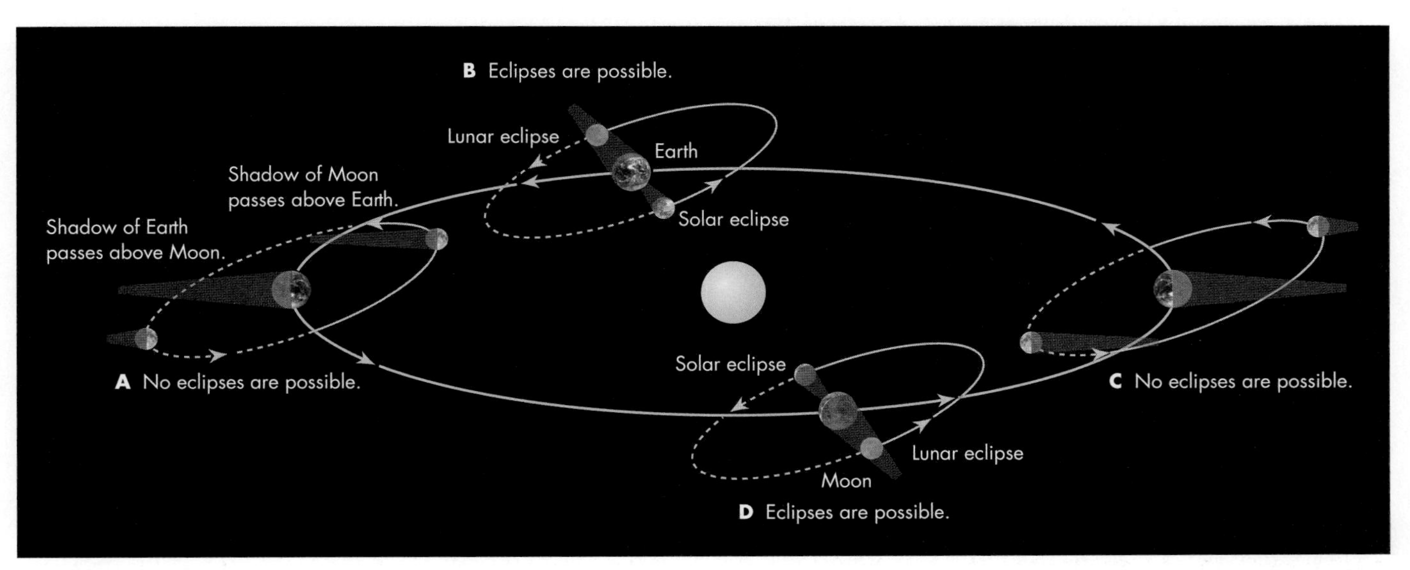

FIGURE 8.11
The Moon's orbit keeps approximately the same orientation as the Earth orbits the Sun. Because of its orbital tilt, the Moon generally is either above or below the Earth's orbit. Thus the Moon's shadow rarely hits the Earth, and the Earth's shadow rarely hits the Moon, as you can see in (A) and (C). Eclipse seasons occur when the Moon's orbital plane, if extended, intersects the Sun. A solar eclipse will then occur at new moon and a lunar eclipse at full moon, as you can see in (B) and (D).

8.3 MOON LORE

The Moon figures prominently in folklore around the world. Anyone who spends time far from the glow of city lights will have noted how different a moonlit night is from the pitch dark when the Moon is below the horizon. This clearly affects animal behavior—predators and prey using the extra light or the cover of darkness to their own advantage. However, most stories concerning the Moon's powers are false.

For example, people often claim that the full moon triggers antisocial behavior—hence the term *lunacy.* Such supposed correlations with lunar phase do not stand up under closer scrutiny. Careful examination of the data show that automobile accidents, murders, admissions to clinics, and so forth do not correlate with lunar phase. It is true that the extra light at night around the time of the full moon makes it easier to see and do things outside, so we may simply be more aware of what happens then.

Q What natural phenomena that occur on Earth do you think might be linked to the lunar cycle? How could you test your idea?

The position of the full moon in the sky changes during the year. The full moon is almost opposite the Sun, so in winter the full moon is located nearly where the Sun is on the celestial sphere in summer, and vice versa. Thus the full moon traces a path through the sky similar to that of the Sun six months earlier or later, so the full moon reaches a much higher point in the sky in the winter than in the summer (compare Figure 8.12 to Figure 7.1).

Special names are sometimes given to the full moon at different times of the year. Some examples from North American folklore are listed in Table 8.2. The full moon nearest the time of the autumn equinox is often called the harvest moon because, as it rises in the east at sunset, the light from the harvest moon helps farmers see to get the crops in. Full moons in other months also have special names, but only the names "harvest moon" and "hunter's moon" (for a full moon in the October hunting season) are widely used.

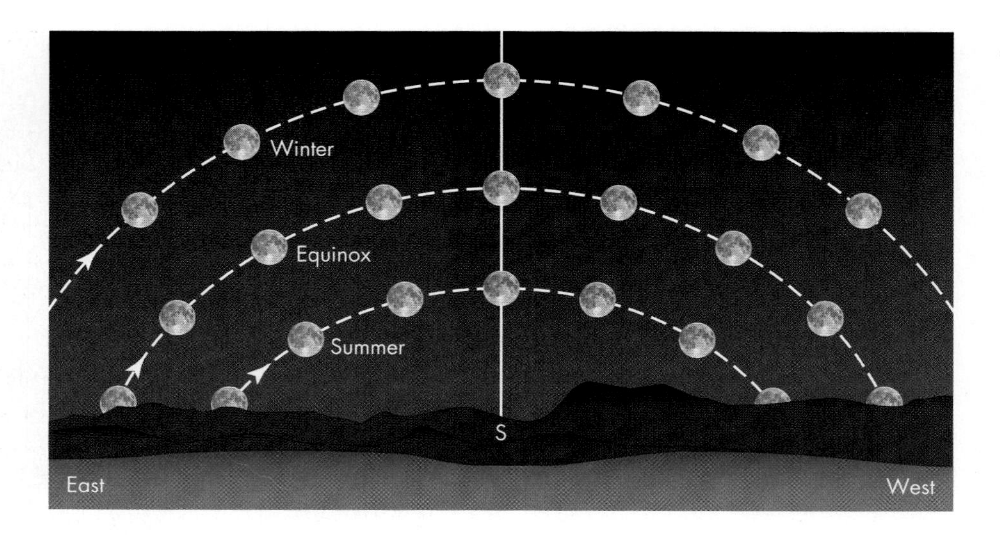

FIGURE 8.12
Path of the full moon through the sky during different seasons. It reaches its highest point in the sky at midnight, but this is lower in the sky in summer than in winter.

TABLE 8.2 Names Used for Full Moons

Month	Name of Full Moon	Month	Name of Full Moon
January	Old moon	July	Thunder or hay moon
February	Hunger moon	August	Grain or green corn moon
March	Sap or crow moon	September	Harvest moon
April	Egg or grass moon	October	Hunter's moon
May	Planting moon	November	Frost or beaver moon
June	Rose or flower moon	December	Long night moon

It is unclear what the origin is of the phrase *once in a blue moon,* indicating a rare event. One suggestion is that this phrase refers to months with two full moons. It is unusual to have two full moons in one month because the cycle of phases is 29.5 days; therefore, unless the Moon is full on the first day of the month, the next full moon will fall in the following month. The odds of the Moon being full on the first of the month are about 1 in 30, so two full moons in a given month happen about every 2½ years. On some calendars the symbol for a second full moon in a month is printed in blue ink. This printing practice appears to be more recent than the earliest uses of the phrase, however, so a "blue moon" must have meant something else originally.

It is possible, in fact, that on rare occasions the Moon may look blue. This odd coloration can be caused by particles in the Earth's atmosphere. Normally our atmosphere filters out blue colors and allows red to pass through, which is why the Sun looks red when it is low in the sky. However, if the atmosphere contains particles whose size falls within a narrow range, the reverse may occur. Dust from volcanic eruptions or smoke from forest fires occasionally has just the right size particles to filter out the red light, allowing mainly the blue colors of the spectrum to pass through. Under these unusual circumstances, we may therefore see a "blue moon."

8.4 THE MOON ILLUSION

To most people the Moon appears to be much larger when they see it rising or setting than when it is high in the sky. But if you measure the Moon's apparent diameter carefully, you will find it to be slightly *smaller* when it is near the horizon than when it is nearly overhead. In fact, when the Moon is nearest its zenith, we are slightly closer to it. The difference in distance occurs because, as the Earth spins, we are carried closer to and farther from the Moon by a distance equal to the Earth's radius, and when we see the Moon overhead, we are on the part of the Earth closest to the Moon.

This misperception of sizes, known as the **Moon illusion,** is still not fully understood. We know that it is an optical illusion caused, at least in part, by the observer's comparing the Moon with objects seen near it on the horizon, such as distant hills and buildings. You know those objects are big even though their distance makes them appear small. Therefore, you unconsciously magnify both them and the Moon, making the Moon seem larger. You can verify this sense of illusory magnification by looking at the Moon through a narrow tube that blocks out objects near it on the skyline. You can also hold up a small coin at arm's length and compare its size to the Moon's. Observed these ways, you will find that the Moon appears roughly the same size regardless of where it is on the sky.

Figure 8.13 shows a similar effect. Because you know that the rails are really parallel, your brain ignores the apparent convergence of the railroad tracks and mentally spreads the rails apart. That is, your brain provides the same kind of enlargement to the circle near the rails' convergence point as it does to the rails, causing you to perceive the middle circle as larger than the lower one, even though they are the same size.

FIGURE 8.13
Circles beside converging rails illustrate how your perception may be fooled. The bottom circle looks smaller than the circle on the horizon but is in fact the same size. Similarly, the circle high in the sky looks smaller than the circle on the horizon.

KEY TERMS

annular eclipse, 70
crescent moon, 63
eclipse, 66
first-quarter moon, 63
full moon, 63
gibbous, 63
lunar eclipse, 66
lunar month, 63
month, 63
moon illusion, 73
new moon, 63
partial eclipse, 70
penumbra, 69
penumbral eclipse, 70
phase, 63
quarter moon, 63
sidereal month, 66
solar eclipse, 66
third-quarter moon, 63
total eclipse, 66
umbra, 69
wane, 63
wax, 63

QUESTIONS FOR REVIEW

1. How long does it take the Moon to go through a cycle of phases?
2. What are the names of the different phases, and when do they occur?
3. How is phase related to the angle between the Moon and Sun?
4. What is the phase of the Moon during a total solar eclipse? during a total lunar eclipse?
5. Why don't we have a lunar and a solar eclipse every month?
6. Why does the Moon look red during a total lunar eclipse?
7. What parts of the Sun are visible during a total solar eclipse?
8. Why does the Moon look larger when it is near the horizon?

PROBLEMS

1. What time would a first-quarter moon rise? What time would it set? Draw a diagram of the Earth, Moon, and Sun to explain your answers.
2. You see a half-full moon visible in the morning sky. How long do you have to wait for the next full moon?
3. The Moon takes about 27.3 days to complete a trip around the ecliptic and return to the same position relative to the stars. This is called its sidereal period. Show that the extra 2.2 days needed to complete a cycle of lunar phases requires the Moon to move through the same angle that the Earth has moved around the Sun in 29.5 days.
4. Show that the ratio of the Moon's diameter to the Sun's diameter is very similar to the ratio of the Moon's distance from the Earth to the Sun's distance from the Earth.
5. A total solar eclipse occurred over Europe on August 11, 1999. Based on the 18 year, 11 day, 8 hour period between repetitions of similar eclipses, a similar total eclipse will occur somewhere on Earth at a similar latitude. When and where will it be?
6. Describe the sequence of events that an observer on the Moon would see when we are seeing a total lunar eclipse on Earth.

TEST YOURSELF

1. Suppose you observe a solar eclipse shortly before sunset. The phase of the Moon must be
   a. full.
   b. new.
   c. first quarter.
   d. third quarter.
2. During a solar eclipse,
   a. the Earth's shadow falls on the Sun.
   b. the Moon's shadow falls on the Earth.
   c. the Sun's shadow falls on the Moon.
   d. the Earth's shadow falls on the Moon.
   e. the Earth stops turning.
3. Each day, the Moon rises about
   a. the same time.
   b. an hour later.
   c. It depends on the year.
   d. an hour earlier.
   e. It depends on the season.

Starry Night

See the Pathways *2nd edition ARIS site for online Starry Night^{TM} planetarium exercises.*

unit 9

Calendars

Different cultures record the passage of time in many ways. The array of approaches may seem bewildering at first, but they are based on just a few astronomical cycles.

We group days into weeks, months, and years to track time periods, which reflect natural cycles that shape our lives. A problem arises, though. These various time periods are not simple multiples of one another. The lunar cycle is about 29.53 days, and the year (from vernal equinox to vernal equinox) is about 365.24 days; so the number of days in the month or year is not a whole number, and the number of weeks in a month, or months in a year, is not a whole number. As a result, there is no obvious or best way to record the passage of time.

For example, is it more important that spring arrives on the same date, that the Moon is new on the first day of the month, or that the method of counting days is simple? This question has been answered in different ways by different cultures. Some developed complex systems for keeping their calendars connected with the astronomical cycles, while others ignored the incompatibilities and let their calendars lose that connection.

Background Pathways

Unit 6: The Year 45

Unit 8: Lunar Cycles 63

9.1 THE WEEK

The origin of the seven-day week is uncertain. It is approximately a quarter of the lunar cycle, but there is no simple interval of days that will keep alignment with the 29.53 day lunar month. The seven-day week was used by the Babylonians and Sumerians, although other cultures used anywhere from six to ten days. In ancient Rome an eight-day week (marking the interval between market days) was used (Figure 9.1); in more recent history, after the French Revolution, a ten-day week was briefly adopted in France.

It is possible that there are seven days in the week to recognize the seven visible objects that move across the sky with respect to the stars: the Sun, the Moon, and the planets Mercury, Venus, Mars, Jupiter, and Saturn. We can see the names of some of these bodies in our English day names (Sunday, Monday, and Saturday). The influence is even clearer in the romance languages, which have their roots in Latin. An example is Spanish, with the day names *lunes, martes, miércoles, jueves*, and *viernes* relating to the Moon, Mars, Mercury, Jupiter, and Venus.

Some English day names come to us through the names of Germanic gods, many of whom have a direct parallel with the Greco-Roman gods after whom the planets are named. For example, Tuesday is named for Tiw, a god of war like Mars (giving Spanish *martes*). Wednesday is named for Woden, the chief god of Germanic peoples, identified with Mercury (*miércoles*). Thursday is named for Thor, the thunder god. He had powers like those of Jupiter, who was also called Jove (*jueves*). Friday is named for Freya, a love goddess like Venus (*viernes*).

Despite the astronomical connections to days' names, the week may simply reflect a human cycle: reasonable periods for work and rest, for purchasing goods and social gatherings. The seven-day week became dominant worldwide with the

Q: Where else do the names of ancient gods show up in everyday life? What qualities do the names convey?

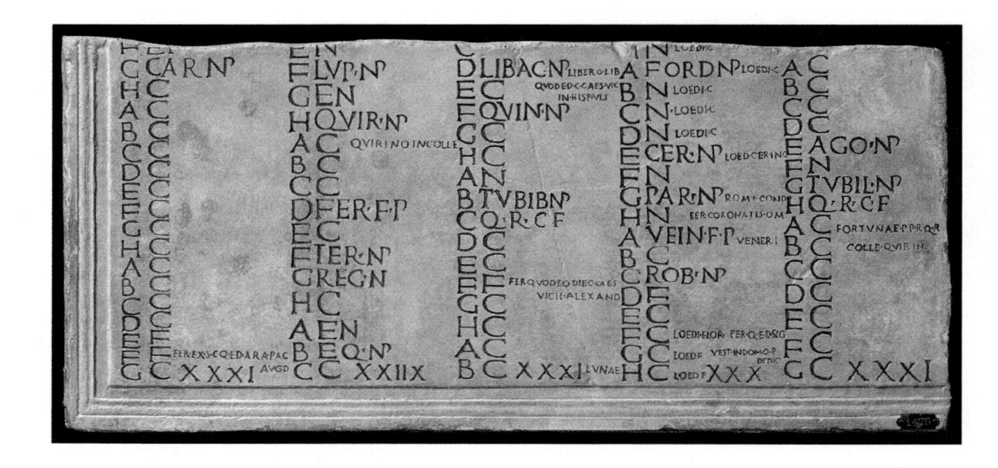

FIGURE 9.1
A portion of an ancient Roman calendar showing the first five months of the year. (Note the Roman numerals indicating the number of days in each month at the bottom of each column.) The letters A through H indicate the 8 days of the week.

spread of Jewish, Christian, and Islamic cultures, which all base their week on the biblical account of the creation of the world in seven days, as well as several other cultures that also used seven days without a biblical origin. Longer time periods such as the month and year have a clearer connection to astronomical cycles.

9.2 THE MONTH

The month is the next largest unit that is used by nearly all cultures. This time interval, and its name, derives from the Moon's cycle of phases. The interval between new moons is 29.53059… days, and because the year has about 365¼ days in it, there are about 12 lunar cycles per year.

However, 12 lunar cycles end up 11 days short of a full year, so there is no way to build a simple calendar that reflects both cycles. This has led to several different calendar systems used by different cultures and religions.

The commonly used calendar system maintains the "month," but it has no relationship to the phases of the Moon. It is an interval of 28 to 31 days, 12 of which add up to a year. This system is really just a solar calendar, maintaining the shorter period we call a month only as a timekeeping convenience. The history of this system is quite complicated, and we will return to it in Section 9.3.

Other calendar systems have retained a much closer connection to the Moon. For example, the Islamic calendar system is purely lunar, defining a year as 12 lunar cycles (Figure 9.2). However, this means dates shift relative to the seasonal position of the Sun by almost 11 days each year. For example, the holy month of Ramadan began on September 24 in 2006 but on September 13 in 2007 and September 2 in 2008. This system is a logical development for a culture that arose in the Middle East, where seasonal differences are not especially pronounced and where night travel by moonlight through the desert was commonplace.

The Chinese and Jewish calendars are called **lunisolar calendars** because they maintain a connection to the cycles of both the Sun and the Moon by adding a 13th month about every three years. The extra month is added whenever the calendar gets too far out of alignment with the seasonal year, so some years have about 354 ($\approx 12 \times 29.53$) days while others have about 384 ($\approx 13 \times 29.53$) days. For example, the Jewish New Year, Rosh Hashanah, occurs on September 30, 2008, then shifts to September 19 and September 9 in the subsequent two years. It shifts back to September 29 in 2011 after a year in which a 13th month is added between the sixth and seventh named months.

1426 A.H.

1 MUHARRAM مُحَرَّم Feb-Mar 2005

SAT	SUN	MON	TUE	WED	THU	FRI
					1 (10)	2 (11)
3 (12)	4 (13)	5 (14)	6 (15)	7 (16)	8 (17)	9 (18)
10 (19)	11 (20)	12 (21)	13 (22)	14 (23)	15 (24)	16 (25)
17 (26)	18 (27)	19 (28)	20 (1)	21 (2)	22 (3)	23 (4)
24 (5)	25 (6)	26 (7)	27 (8)	28 (9)	29 (10)	30 (11)

2 SAFAR صَفَر Mar-Apr 2005

SAT	SUN	MON	TUE	WED	THU	FRI
1 (12)	2 (13)	3 (14)	4 (15)	5 (16)	6 (17)	7 (18)
8 (19)	9 (20)	10 (21)	11 (22)	12 (23)	13 (24)	14 (25)
15 (26)	16 (27)	17 (28)	18 (29)	19 (30)	20 (31)	21 (1)
22 (2)	23 (3)	24 (4)	25 (5)	26 (6)	27 (7)	28 (8)
29 (9)						

⋮

11 DHUL QA'DAH ذُو الْقَعْدَة Dec-Jan 2005-06

SAT	SUN	MON	TUE	WED	THU	FRI
	1 (4)	2 (5)	3 (6)	4 (7)	5 (8)	6 (9)
7 (10)	8 (11)	9 (12)	10 (13)	11 (14)	12 (15)	13 (16)
14 (17)	15 (18)	16 (19)	17 (20)	18 (21)	19 (22)	20 (23)
21 (24)	22 (25)	23 (26)	24 (27)	25 (28)	26 (29)	27 (30)
28 (31)	29 (1)					

12 DHUL HIJJAH ذُو الْحِجَّة Jan 2006

SAT	SUN	MON	TUE	WED	THU	FRI
		1 (2)	2 (3)	3 (4)	4 (5)	5 (6)
6 (7)	7 (8)	8 (9)	9 (10)	10 (11)	11 (12)	12 (13)
13 (14)	14 (15)	15 (16)	16 (17)	17 (18)	18 (19)	19 (20)
20 (21)	21 (22)	22 (23)	23 (24)	24 (25)	25 (26)	26 (27)
27 (28)	28 (29)	29 (30)				

FIGURE 9.2
An Islamic calendar, showing the first two and last two months of the year 1426. Corresponding dates in the common calendar are shown in green (in 2005–2006 C.E.).

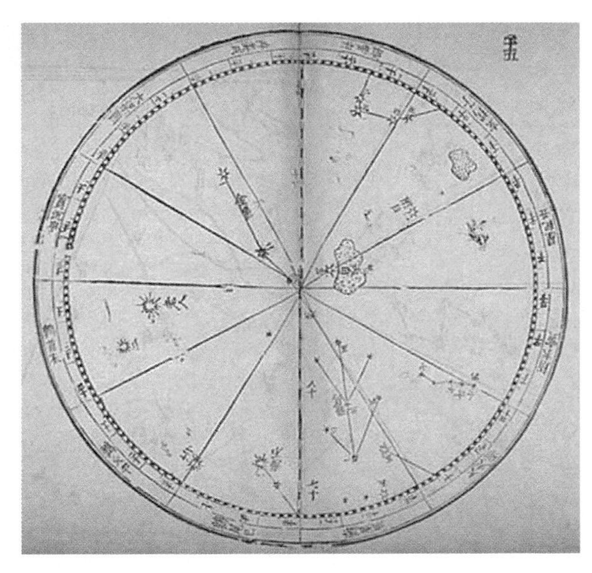

FIGURE 9.3
A Chinese calendar dividing up the year by days and position of the Sun along its path through the stars.

The Chinese calendar is similar, but the month names are based on the segment of the ecliptic where the Sun is located when the Moon is new (Figure 9.3). This is somewhat similar to naming the months according to the constellation of the zodiac that the Sun is in at each new moon. When two new moons occur while the Sun is in the same segment of the ecliptic, there can be two of the same named month in a row. This can happen, for example, when a new moon occurs just as the Sun enters the segment of the ecliptic designated for "August," and then another new moon occurs just before the Sun leaves that segment. In such a case, there will be a second August that year. Using this procedure adds an extra month once every three or so years at varying times during the year. This shifts the date of the Chinese New Year in a pattern similar to that for Rosh Hashanah.

9.3 THE ROMAN CALENDAR

Our commonly used calendar is based on one developed by the Romans about 200 B.C.E. (before the Common Era—see Section 9.5). In fact, the word *calendar* is itself of Roman origin. There is some controversy about how the original Roman calendar was organized. It may have had only 10 months, and it probably began on the first day of spring (the vernal equinox) rather than in January.

Many of the names of our months reflect that early calendar system. For example, if the year began in March, then September, October, November, and December would be the 7th (Sept.), 8th (Oct.), 9th (Nov.), and 10th (Dec.) months, each word's root reflecting the Latin name for the number of the month. Origins of the names of the months, some of which are less certain, are listed in Table 9.1.

Because the Roman calendar was not required to conform to the cycles of astronomical phenomena, it became a form of political patronage. The priests who regulated the calendar would add days and even months to please one group, and take days off to punish another. Think about the possibilities when rent and bills are due! Such confusion resulted from these abuses that in 46 B.C.E. Julius Caesar

TABLE 9.1 Origin of the Names of the Months

Month	Origin of the Name
January	Janus (gate), the two-faced god looking to the past and future; hence, beginnings
February	*Februa* ("expiatory offerings")
March	The god Mars
April	Etruscan *apru* (April), probably shortened from the Greek Aphrodite, goddess of love and earlier of the underworld
May	Maia's month; Maia ("she who is great"), the eldest of the Pleiades and the mother of Hermes by Zeus
June	Junius, an old Roman noble family (from Juno, wife and sister of Jupiter, equal to Greek Hera)
July	Julius Caesar (*Julius* means "descended from Jupiter"; the *Ju-* of June and July are the same: Jupiter, "Sky-father")
August	Augustus Caesar (*augustus* means "sacred" or "grand")
September–December	"Seventh month" to "tenth month" (the *-ember* may come from the same root as month)

asked the astronomer Sosigenes to design a calendar that would fit the astronomical events better and give less room for the priests and politicians to manipulate it. The resulting calendar, known as the **Julian calendar,** consisted of 12 months, alternating in order between 31 or 30 days, except February, which is discussed further in the next section.

The Julian calendar barely survived Caesar before the politicians were at it again. First the name of the month *Quintilis* (originally the fifth month of the Roman year) was changed to *Julio* to honor Julius Caesar—hence our July. Next, on the death of Julius Caesar's successor, Augustus Caesar, an able and highly respected leader, the following month, *Sextilis*, was renamed in his honor—hence the name August. However, the order of months originally made August 30 days long, and it would have been impolitic to have Augustus's month a day shorter than Julius's; so August was changed to have 31 days, and all the following months had the number of their days changed to reestablish the 30/31 alternation. Unfortunately, this used up one more day than the year allowed. Poor February, already one day short, was trimmed by a second day, leaving it with only 28 days. With only minor modifications, this is the calendar we use today.

9.4 THE LEAP YEAR

The ancient Egyptians knew that the year is not exactly 365 days long. It takes about 365¼ days for the Earth to complete an orbit around the Sun. Because we can't have fractions of a day in the calendar, a calendar based on a year of 365 days will come up 1 day short every 4 years.

Does a quarter of a day really matter? Consider that the seasons are set by the orientation of the Earth's rotation axis with respect to the Sun, not by how many days have elapsed. We therefore want to make sure that we start each year when the Earth has the same orientation. Otherwise the seasons get out of step with the calendar. For example, because in 4 years you will lose 1 day, in 120 years you will lose a month, and in 360 years you will lose an entire season. With a 365-day year, in a little over three centuries April would begin in what is now January.

This problem is corrected by the **leap year,** a device introduced with the Julian calendar to keep the calendar in step with the seasons. The leap year corrects for the quarter day by adding a day to the calendar every fourth year. The extra day is added to February, which alternates between 28 and 29 days. The tradition of making adjustments to February may date back to when February was the last month, and year-end adjustments were made to keep the calendar in agreement with the vernal equinox. The civil calendar of India today has a similar system of leap years. It has a set of 12 months of 30 or 31 days that begins at the vernal equinox, and the first month of the year increases from 30 to 31 days during a leap year.

Unfortunately, the year is actually a little shorter than 365¼ days, so the leap year corrects by a tiny bit too much. To address this problem, the Julian calendar was modified in 1582 at the direction of Pope Gregory XIII. The common calendar we use today is therefore known as the **Gregorian calendar.** The change this calendar system introduced was to eliminate leap years in centuries that are not divisible by 400. Thus, 1700, 1800, and 1900 were not leap years, but 1600 and 2000 were.

The inauguration of the Gregorian calendar in 1582 was not a smooth transition. Because it was adopted roughly 1600 years after the Julian calendar, the error in the relationship between the calendar and the seasons had accumulated to about 10 days. To bring the calendar back into synchrony with the seasons, Pope Gregory XIII simply eliminated 10 days from the year 1582 so that the day after October 4 became October 15.

Q: If you were to create a calendar completely on your own, what would it look like?

Although the changeover was accepted in much of Europe, non-Catholic countries such as Protestant England refused to abide by the Pope's edict. The change was not made in England until 1752. In Russia the change was not made until the revolution in the early 1900s. Other countries (Greece and Turkey, for example) changed in the 1920s.

9.5 THE CHRONICLING OF YEARS

The common scheme for numbering years began over a millennium ago in the year 532. But why "532"? The numbering of years by different cultures often relates to important religious dates, or the estimated starting date of a nation, or the years since an important ruler came to power. Sometimes these numbering systems were developed decades or even centuries after the event whose anniversary they are intended to mark, so the numbering is somewhat uncertain.

For example, until 532, throughout Europe the year was still commonly counted from the supposed founding of Rome (in 753 B.C.E., according to the common calendar), which made 532 the year 1285. Others, though, marked 532 as the year 248, reckoning from the year that Diocletian became emperor of Rome. Dionysius Exiguus, a monk of this era, carried out calculations for determining the date of the Christian holiday of Easter, which is related to the date of the full moon after the vernal equinox. The monk published a table of Easter dates beginning in the year 532 "*anno domini*" (or **A.D.**, meaning "in the year of the Lord," *not* "after death") to honor Jesus's life instead of Rome or its emperors. However, the monk was a better astronomer than historian. Christian scholars even at the time pointed out that this could not mark the years since the birth of Jesus because of obvious inconsistencies with other historical dates. But the table enjoyed wide circulation, and the idea of marking years this way became popular in Christian countries. So despite its historical inaccuracy, it became widely used and has become the common numbering system used today.

For the years that preceded A.D. 1, the convention became to write them as 1, 2, 3 . . . **B.C.**, standing for "before Christ." There is no year 0, which complicates mathematical treatment of years and introduces confusion about when centuries and millennia begin. In fact, the twenty-first century and the third millennium began on January 1, *2001*, because a decade, century, or millennium is complete only after its 10th, 100th, or 1000th year is completed. December 31, 2000, marked such a completion. Hence, starting from A.D. 1, the transition from A.D. 1999 to A.D. 2000 marked the completion of only 1999 years, not two full millennia.

Recently two different abbreviations have begun to replace A.D. and B.C. They are **C.E.** and **B.C.E.**, which stand, respectively, for "Common Era" and "before the Common Era." "Common Era" refers to our present calendar, which is used nearly worldwide for most business purposes and thereby avoids reference to a particular religion. Yet another abbreviation—**B.P.**, which stands for "before the present (era)"—is used, especially in archaeological, paleontological, and geological works. B.P., used for dates determined by analyzing radioactive carbon, takes 1950 C.E. as its base year.

The Jewish calendar is dated from the biblical creation of the world, so Rosh Hashanah in 2000 began the year 5761. The Islamic calendar is dated from Mohammed's emigration to Medina in 622 C.E. Because the Islamic year is shorter than the solar year, its numbering increases by one year relative to the common system every 33 or 34 years. The Islamic year 1421 began April 6, 2000.

The Chinese year-numbering system is based on cycles of 60 years, broken into five sequences of 12 years, each year of which is associated with a particular animal—Rat, Ox, Tiger, Rabbit, Dragon, Snake, Horse, Sheep, Monkey, Chicken, Dog, Pig. Every two years are associated with a particular element—Wood, Fire, Earth, Metal, Water—and the combination of animal and element characteristics repeats after 60 years. We are currently in the 79th of these 60-year cycles; year 1 of this cycle was in 1984, associated with the Rat and Wood. According to tradition this system originated in 2697 B.C.E. Continuous chronologies have been maintained since about the ninth century B.C.E.

Only a few calendar systems have survived in relatively common use. Other cultures had systems that used additional astronomical cycles. For example, the Maya included the motions of Venus in their time reckoning, making their calendar quite complex (Figure 9.4). The Mayan calendar involved 20-day "weeks" and a complex series of multiples of these "weeks" making up time periods of approximately 1 year, 20 years, 400 years, and 5000 years. One of these "long-count" periods of about 5000 years ends in 2012. Although Mayan scholars do not think this had a significant meaning in Mayan religion, some popular claims have been made that suggest otherwise.

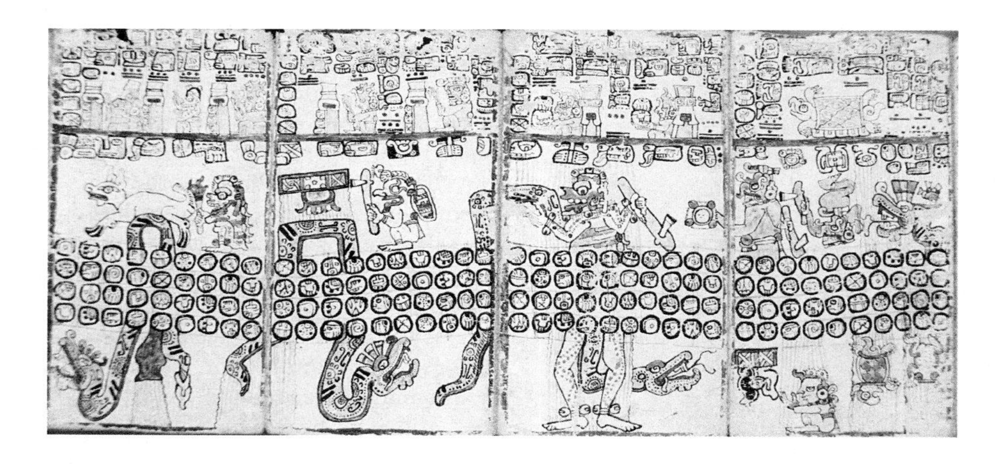

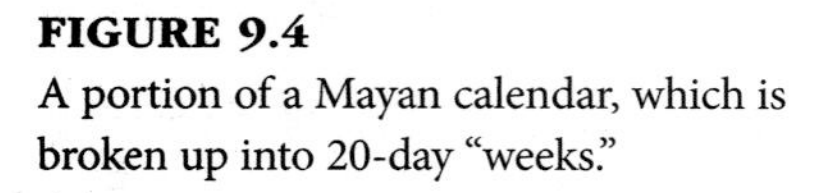

FIGURE 9.4
A portion of a Mayan calendar, which is broken up into 20-day "weeks."

KEY TERMS

A.D. (*anno domini*), 79
B.C. (before Christ), 79
B.C.E. (before the Common Era), 80
B.P. (before the present), 80
C.E. (Common Era), 80
Gregorian calendar, 79
Julian calendar, 78
leap year, 78
lunisolar calendar, 76

QUESTIONS FOR REVIEW

1. What connection does the seven-day week have to astronomical phenomena?
2. What is the difference between lunar, lunisolar, and solar calendars?
3. What is the purpose of having leap years?
4. What was the purpose of the change to the Gregorian calendar system in recent centuries?
5. What establishes the initial year in the various calendar systems?
6. When and why did the year-numbering of the common calendar system begin to be used?

PROBLEMS

1. If the same year-numbering systems were used today as were used in Europe before 532 C.E., what year would we currently report?
2. What year will 2050 C.E. be in the Chinese, Islamic, and Jewish calendar systems?
3. Suppose a 12-month lunar calendar year (such as in the Islamic calendar) began on January 1. How many solar years would pass before the lunar calendar year would again begin within 1 week of January 1? How many lunar calendar years would have elapsed?
4. The first day of the month in the Islamic calendar begins when the crescent moon is first spotted. Usually the earliest this is possible is about 16 hours after the new moon. What is the angular distance between the Moon and the Sun at this time? Approximately how long after the Sun sets will the Moon set if it is just 16 hours past new?
5. Under the Julian calendar system, the year averages to 365.25 days after 4 years. In the Gregorian system, determine the average length of the year after 400 years. If a country needed to convert from the Julian to the Gregorian calendar now, how many days would it need to remove from the calendar?
6. Suppose the year were 365.170 days long. How would you design a pattern of leap years to make the average year length exactly match this value within a span of 1000 years?

TEST YOURSELF

1. Suppose that the length of the year were 365.2 days instead of 365.25 days. How often would we have leap year? Every
   a. 2 years
   b. 5 years
   c. 10 years
   d. 20 years
   e. 50 years
2. Why does the Chinese New Year occur on different dates in the Western calendar each year?
   a. The Chinese year always starts when the Moon is new.
   b. The precession of the Earth's axis shifts it a little each year.
   c. The date is based on when Mars undergoes retrograde motion.
   d. The date is chosen each year for when a positive horoscope is cast.
   e. The Chinese calendar slips by a quarter-day each year because it has no leap days.
3. Why was the Roman calendar system of a leap year every 4 years changed in the 1500s?
   a. The year is not exactly 365.25 days long.
   b. The Earth had slowed down its spin since ancient Roman times.
   c. Precession of the Earth had changed the length of the year.
   d. The lunar phases had gone out of phase with our months.

unit

10

Geometry of the Earth, Moon, and Sun

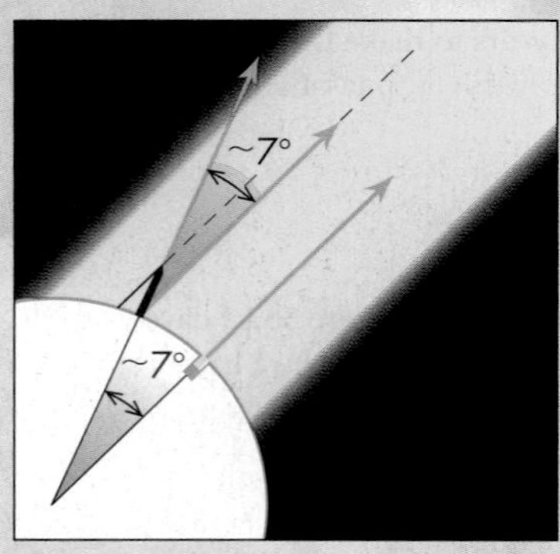

Background Pathways

It is possible to draw several important conclusions about the sizes of and distances between the Earth, Moon, and Sun based on careful but simple measurements. As far as we know, the ancient Greek astronomers of classical times were the first to do this. The values they determined were not always highly accurate, but they were surprisingly good. They demonstrated that you do not need telescopes, high-speed communications, computers, and satellites to determine, for example, that the Earth is a sphere with a circumference of about 40,000 kilometers.

We study ancient ideas of the heavens not so much for what they tell us about the heavens but to learn how observation, geometry, and careful reasoning can lead us to a deeper understanding of the universe. Moreover, these same kinds of geometric ideas are still widely used to measure the sizes of astronomical objects.

10.1 THE SHAPE OF THE EARTH

The ancient Greeks knew that the Earth was round. As long ago as about 500 B.C.E., the mathematician **Pythagoras** (about 560–480 B.C.E.) was teaching that the Earth was spherical, but the reason for his support of this idea was as much mystical as rational. Like many of the ancient philosophers, he believed that the sphere was the perfect shape and that the gods would therefore have utilized that perfect form in creating the Earth.

By 300 B.C.E., however, **Aristotle** (384–322 B.C.E.) was presenting arguments for the Earth's spherical shape that were based on simple naked-eye observations that anyone could make. Such reliance on direct observation was the first step toward generating scientifically testable hypotheses about the contents and workings of the universe. For instance, Aristotle noted that if you look at a lunar eclipse when the Earth's shadow falls on the Moon, the shadow can be clearly seen as curved, as Figure 10.1A shows. As he wrote in his treatise "On the Heavens,"

> *The shapes that the Moon itself each month shows are of every kind—straight, gibbous, and concave—but in eclipses the outline is always curved; and, since it is the interposition of the Earth that makes the eclipse, the form of this line will be caused by the form of the Earth's surface, which is therefore spherical.*

Another of Aristotle's arguments that the Earth is spherical was based on the observation that a traveler who moves south will see stars that were previously hidden below the southern horizon, as illustrated in Figure 10.1C. For example,

Q Some people still believe the Earth is flat. What "proof" would you offer them that it is round? What argument do you think would be most persuasive?

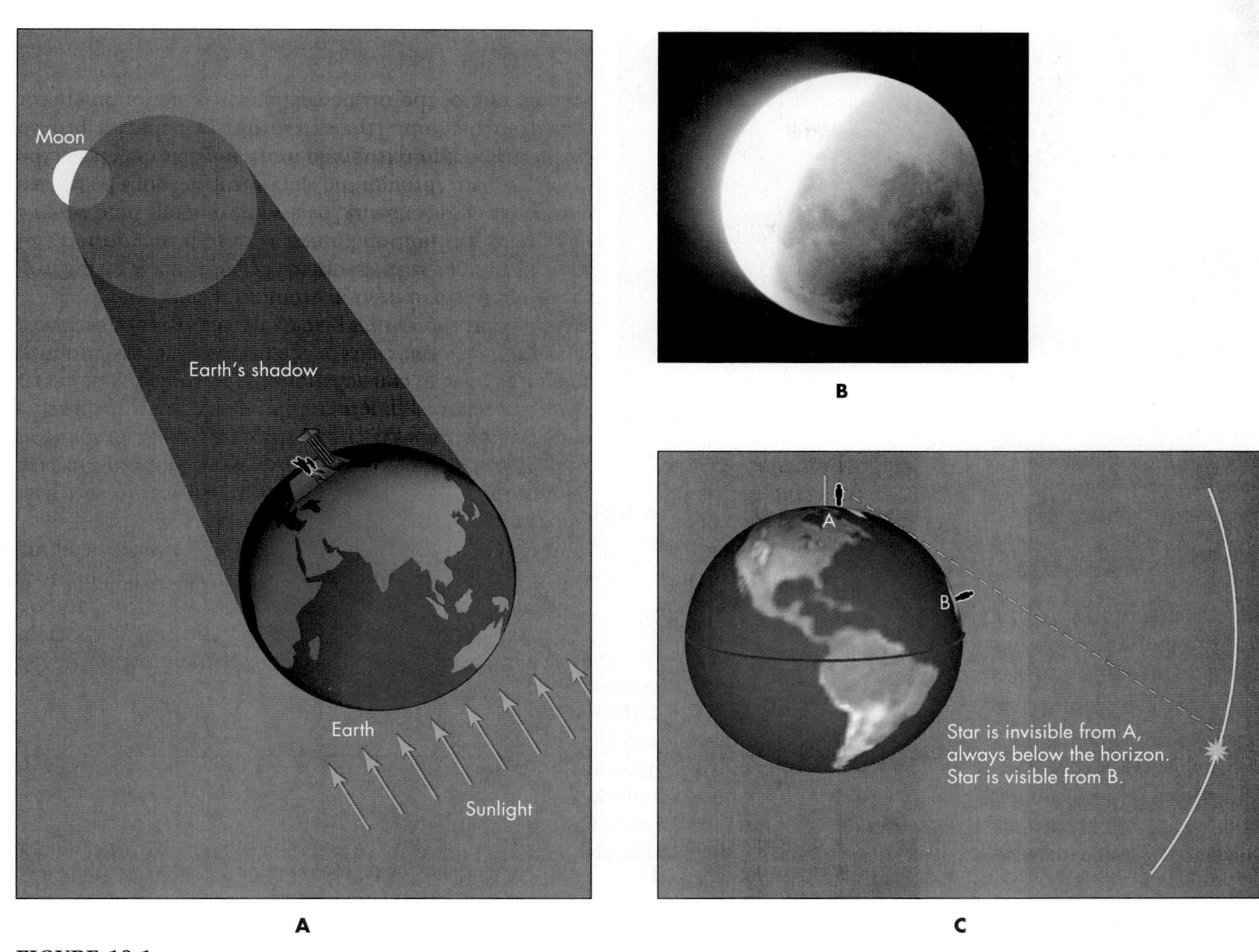

FIGURE 10.1
(A) During a lunar eclipse, we see that the Earth's shadow on the Moon is curved. Thus the Earth must be round. (B) Photograph of a partial lunar eclipse. (C) As a traveler moves from north to south on the Earth, different stars become visible. Some disappear below the northern horizon, whereas others, previously hidden, become visible above the southern horizon. This variation would not occur on a flat Earth.

the bright star Canopus is easily seen in Miami but is below the horizon in Boston. This could not happen on a flat Earth.

10.2 DISTANCE AND SIZE OF THE SUN AND MOON

One of the most remarkable ancient Greek astronomers was **Aristarchus** of Samos (about 310–230 B.C.E). He was able to estimate the size and distance of the Moon and Sun. Even though his values were not highly accurate, they gave at least the correct sense of which was larger and which was smaller than the Earth, as well as approximate indications of their distances from Earth. Few of his writings survive intact, but references to his works by later astronomers allow us to reconstruct many of his findings. For example, by comparing the size of the Earth's shadow on the Moon during a lunar eclipse to the size of the Moon's disk, as can be estimated

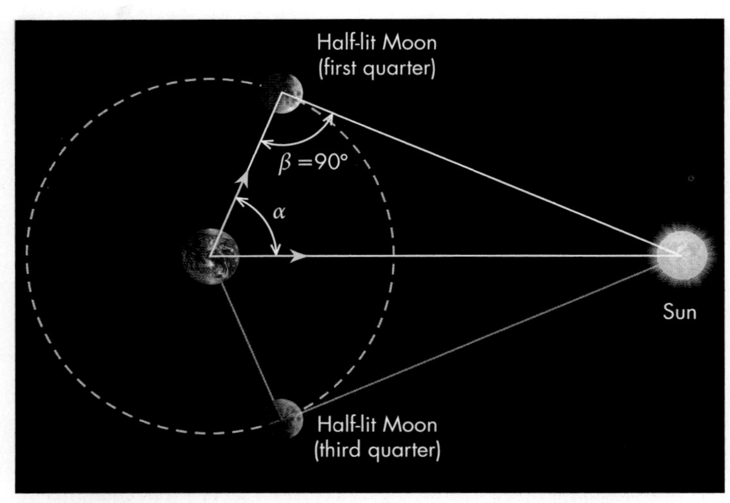

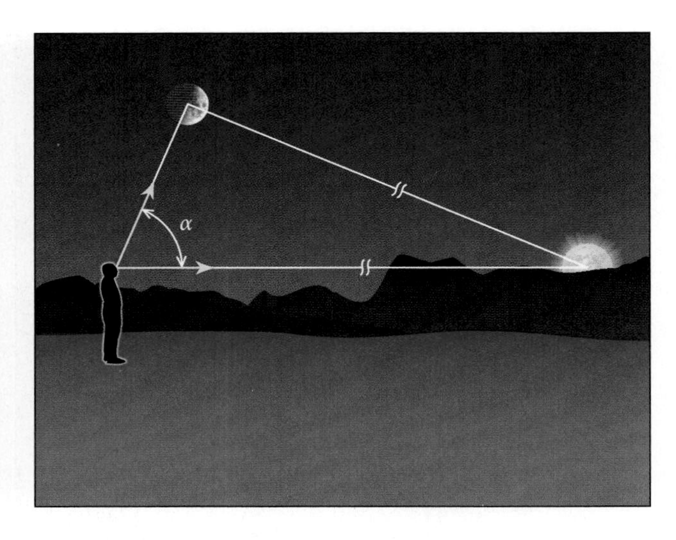

FIGURE 10.2
Aristarchus estimated the relative distance of the Sun and Moon by observing the angle α between the Sun and the Moon when the Moon is exactly half lit. Angle β must be 90° for the Moon to be half lit. Knowing the angle α, he could then set the scale of the triangle and thus the relative lengths of the sides. (Sizes and distances are not to scale.)

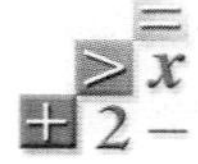

If the circumference of the Moon's orbit is 220 times Earth's diameter, it is $440 \times R_{\oplus}$. The circumference of a circle is 2π times its radius, so the Moon's orbital radius is $440 \times R_{\oplus}/2\pi \approx 70 \times R_{\oplus}$.

Even with modern instruments, it is impossible to measure the angle between the quarter Moon and the Sun well enough to obtain the Sun's distance accurately. If the Sun were as close as about 10 times the Moon's distance, though, there would be about a day less between new Moon and first quarter than between first quarter and full Moon, which would be fairly easy to detect.

in Figure 10.1A, Aristarchus calculated that the Moon's diameter was about one-third of the Earth's. His estimate that the Moon's diameter is about 0.33 that of the Earth's is not far from the correct value of about 0.27.

Aristarchus then estimated the distance to the Moon by noting that when the Moon passes through the middle of the Earth's shadow during a lunar eclipse, it takes about three hours to complete its passage through the shadow. On the other hand, the Moon takes about 660 hours to complete its passage around the zodiac. If the Earth's shadow is about the same size as the Earth itself, then the entire circumference of the Moon's orbit is about 220 (= 660/3) times bigger than the Earth's diameter, and its distance is about 70 times the Earth's radius. This is not far from the modern estimate that the Moon is about 60 Earth radii away from the planet.

Aristarchus also calculated the Sun to be about 20 times farther away from the Earth than the Moon is. He carried out this calculation by measuring the angle between the Sun and the Moon when the Moon was exactly half lit at the first- and third-quarter phases (Figure 10.2). From the apparent size in the sky of the Sun and the Moon and their relative distances, and his calculation of the Moon's diameter relative to the Earth's, Aristarchus deduced that the Sun's diameter was about seven times that of the Earth (see Section 10.4). This is far smaller than we know it is today—the Sun is more than 100 times larger than the Earth. Nevertheless, it was the first demonstration that the Sun is much bigger than the Earth and much more distant than the Moon.

It was perhaps his recognition of the large size of the Sun that led Aristarchus to propose the revolutionary idea that the Earth revolved around the Sun rather than the reverse. Aristarchus was, of course, correct; but his idea was too radical, and another 2000 years passed before scientists became convinced of its correctness. However, this was not mere stubbornness. There was a good reason for not concluding that the Earth moves around the Sun. If it did, the positions of stars should change during the year as the Earth moved from one side to the other within the celestial sphere. Looking at Figure 10.3, you can see that the nearby star should appear to lie at a different angle from, and in a different position with respect to, each of the more distant stars in January and in July.

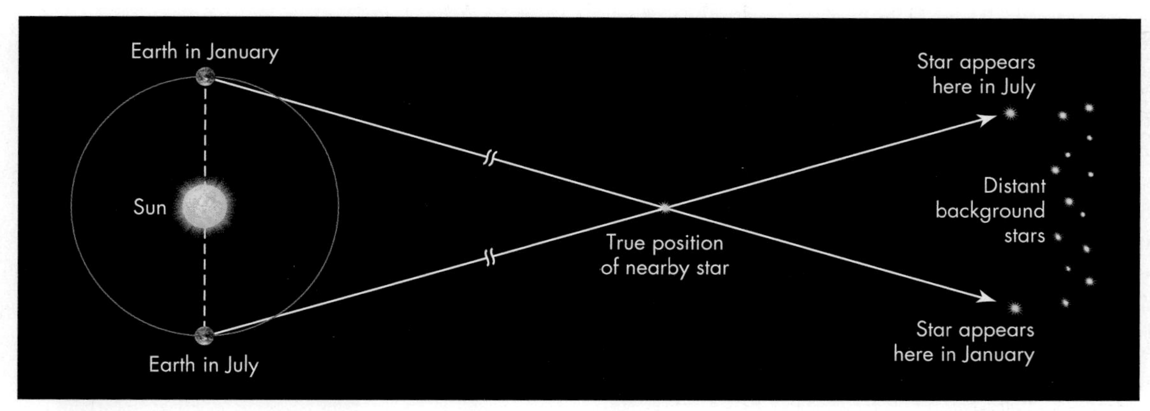

FIGURE 10.3
Motion of the Earth around the Sun causes *stellar parallax*—a shift in the position of nearby stars. Because the stars are so remote, this shift is too small to be seen by the naked eye. This lack of observable parallax led the ancient Greeks to deduce incorrectly that the Sun could not be the center of the Solar System. (Sizes and distances are not to scale.)

This shift in a star's apparent position resulting from the Earth's motion around the Sun is called **parallax,** and Aristarchus's critics were right in supposing that it should occur. Actually, many astronomers of the time assumed the stars were all at the same distance, but even so there would be a parallax effect. If the Earth moved around inside this hypothesized celestial sphere, we would see changes in the stars' apparent separation—just as we can tell how we are moving in the dark by looking at nearby lights. Because no parallax was seen, astronomers rejected Aristarchus's idea. What they failed to realize was that because of the immense distance to the stars, the parallax is so tiny that we need a powerful telescope and very precise equipment to measure it. In fact, parallax of stars was not successfully measured until 1838 (see Unit 52). Thus an idea may be rejected for reasons that are logically correct but are flawed because of limited data accuracy.

10.3 THE SIZE OF THE EARTH

Even though few astronomers accepted Aristarchus's idea that the Earth moves around the Sun, they did accept his estimates of the sizes and distances of the Sun and Moon. However, these measurements were all scaled to the size of the Earth, whose size was not known. The Moon might have one-third the diameter of the Earth, but was that 1000 km or 100,000 km?

Eratosthenes (276–195 B.C.E), head of the famous library at Alexandria in Egypt, made the first measurement of the Earth's size. He obtained a value for its circumference that very nearly matches the presently measured value. Eratosthenes's demonstration is one of the most elegant ever performed because it so superbly illustrates how scientific inference links observation and logic.

Eratosthenes, a geographer as well as an astronomer, heard that to the south, in the Egyptian town of Syene (the present city of Aswân), the Sun would be directly overhead at noon on the summer solstice and thus it cast no shadow. Proof of this was that on the solstice the Sun lit the bottom of a deep well there. However, on that same day, the Sun did cast a shadow in Alexandria, where he lived. Knowing the distance between Alexandria and Syene and appreciating the power of geometry, Eratosthenes realized he could deduce the size (circumference) of the Earth. He analyzed the problem as follows: Because the Sun is far away from the Earth, its light travels in parallel

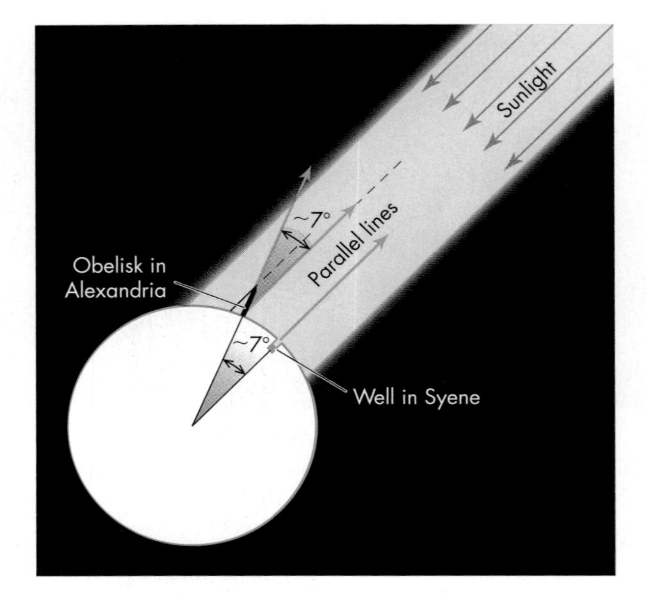

FIGURE 10.4
Eratosthenes's calculation of the circumference of the Earth. The Sun is directly overhead on the summer solstice at Syene in southern Egypt. On that same day, Eratosthenes found the Sun to be 1/50th of a circle (about 7°) from the vertical in Alexandria, in northern Egypt. Eratosthenes deduced that the angle between two verticals placed in northern and southern Egypt must be 1/50th of the circumference of the Earth.

rays toward the Earth. Thus two rays of sunlight, one hitting Alexandria and the other shining down the well, are parallel lines, as depicted in Figure 10.4.

With our modern perspective, we would say that this occurred because Syene was at a latitude close to 23.5°N, whereas Alexandria lies at about 31°N where the Sun never passes through the zenith.

At noon on the summer solstice, the Sun's rays passing through Syene are aimed straight at the center of the Earth. However, because of the curved surface of the Earth, parallel rays passing through Alexandria would miss the Earth's center. We can calculate the angle between the two cities with a geometric construction. Draw a straight line from the center of the Earth outward so that it passes vertically through the Earth's surface in Alexandria. The line makes the same angle with all of the Sun's parallel rays, so it also indicates the angle of latitude between the two cities (see Figure 10.4). The reason is that the corresponding angles formed where a single line crosses two parallel lines are equal (a geometric theorem).

This angle can be measured with sticks and a protractor (or their ancient equivalent) and is the angle between the direction to the Sun and a straight stick pointing vertically upward (see Figure 10.4). Eratosthenes found this angle to be about 1/50th of a circle, or roughly 7°. Therefore the angle formed by a line from Alexandria to the Earth's center and a line from the well to the Earth's center must also be 1/50th of a circle.

The "stadium" is an ancient unit of measure used by many cultures, and with different lengths, so the exact value used by Eratosthenes is uncertain and may have ranged from about 0.15 to about 0.21 km. The modern use of the word *stadium* comes from the use of the term to refer to a racetrack one stadium long.

Eratosthenes reasoned that if the angle between Alexandria and Syene was 1/50th of a circle, the distance between these cities must be 1/50th the circumference of the Earth. Because he knew the distance between Alexandria and the well to be 5000 stadia (where a stadium is about 0.16 kilometer), the distance around the entire Earth must be 5000 stadia × 50, or 250,000, stadia. When expressed in modern units, this is roughly 40,000 kilometers, or a diameter of about 13,000 kilometers, which is approximately the size of the Earth as we know it today.

By modern standards, there is absolutely nothing wrong with this technique. You can use it yourself to measure the size of the Earth. Eratosthenes's success was a triumph of logic and the scientific technique. The method required that he assume the Sun was so far away that its light reached Earth along parallel lines. That assumption, however, was supported by the earlier set of measurements made by Aristarchus that the Sun was much more distant than the Moon (Section 10.2).

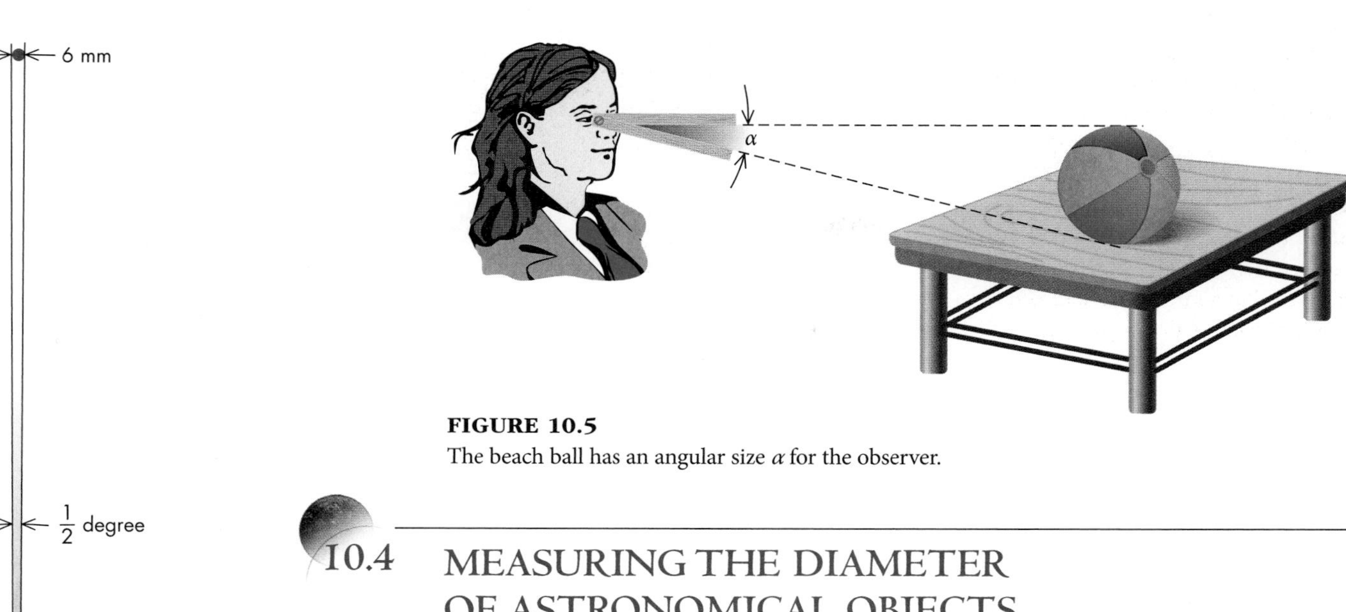

FIGURE 10.5
The beach ball has an angular size α for the observer.

10.4 MEASURING THE DIAMETER OF ASTRONOMICAL OBJECTS

One of the most basic properties of an astronomical object is its size. Because we cannot stretch a tape measure across the disk of the Sun or Moon, how do we know the size of such heavenly bodies?

The basic method for measuring the size of a distant object was worked out long ago and is still used today. This method involves first measuring how big an object *looks,* a quantity called its **angular size.** We can measure an object's angular size by drawing imaginary lines to each side of it, as shown in Figure 10.5, and then measuring the angle between the lines using a protractor.

As an astronomical example, we can measure the angular size of the Moon, or equivalently, because it is a round object, its *angular diameter*. The Moon's angular size is small, and the Moon moves on the sky, so it is not practical to aim lines on each side. Instead, we can determine its angular size by the following procedure: Hold up a ruler at arm's length in front of the Moon. If you do this, you might find, for example, that the Moon is about 6 mm across on the ruler. Next measure the distance from your eye to the ruler; about 70 cm in this example. Draw this carefully to scale on a sheet of paper (divide all of the sizes by 3 to fit it on a standard sheet of paper). This yields a picture something like Figure 10.6. Measuring the angle between the lines with a protractor, we find that the Moon's angular size is about a half degree (0.5°).

Notice that an object's angular size depends on its distance (Figure 10.7). For example, if you move closer to an object, its angular size increases. Therefore,

The angular size of an object depends inversely on its distance.

From its angular size and distance, we can find an object's *linear size*—that is, its size in units of length such as meters. To work out a mathematical formula relating angular size and linear size, imagine that you are at the center of a circle whose radius is d, the distance to the body. The circle passes through the object, as illustrated in Figure 10.8. Let ℓ be the linear size of the body. Draw lines from the center to each end of ℓ letting the angle between the lines be α (the Greek letter alpha), the object's angular size.

We now determine the object's linear size, ℓ by forming the following proportion: ℓ is to the circumference of the circle as α is to the total number of degrees around the circle, which we know is 360. Thus

$$\frac{\ell}{\text{circumference}} = \frac{\alpha}{360^\circ}$$

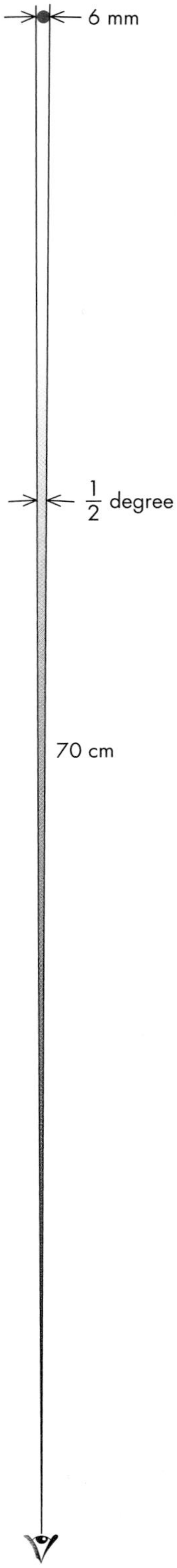

$\frac{1}{2}$ degree

70 cm

FIGURE 10.6
Scale drawing to find the angular diameter of the Moon. It is equivalent in size to a 6 mm object held 70 cm from your eye, or about 0.5°.

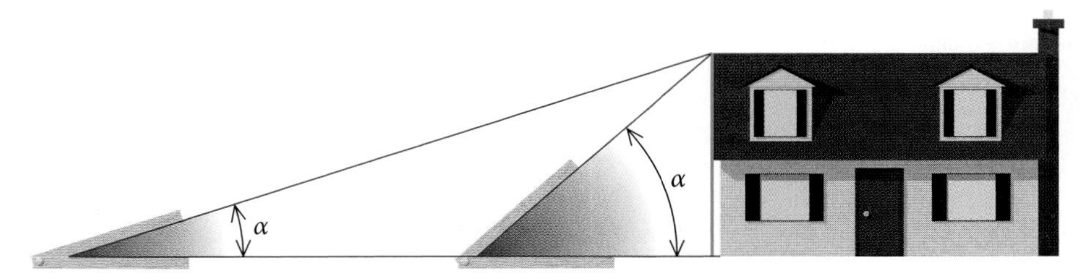

FIGURE 10.7
Angular size grows smaller in inverse proportion as the distance becomes larger.

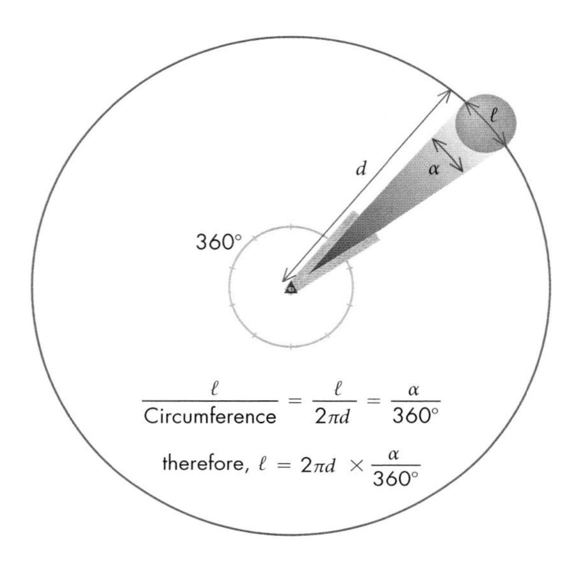

FIGURE 10.8
How to determine linear size from angular size.

However, we know from geometry that the circle's circumference is $2\pi \times d$. Thus we can solve for ℓ by multiplying both sides by the circumference:

$$\ell = \text{circumference} \times \frac{\alpha}{360°} = 2\pi \times d \times \frac{\alpha}{360°}.$$

Dividing 360° by 2π gives an important formula relating angular size, linear size, and distance:

$$\ell = d \times \frac{\alpha}{57.3°}.$$

ℓ = linear size

α = angular size

d = distance

This is known as the *angular size formula,* or sometimes the *small-angle formula,* because it is most accurate for small angles. As an example of how we might apply this formula, we use it to measure the Moon's diameter. We stated previously that the Moon's angular size is about 0.5°. Its distance is about 384,000 kilometers. Thus its diameter is

$$\ell = 384{,}000 \text{ km} \times \frac{0.5°}{57.3°} = 3400 \text{ km}.$$

So the Moon's diameter is a little more than one-quarter of the Earth's.

The angular size formula has an error of only about 1% for objects 28° across, and less than 0.1% at angles smaller than about 9°.

KEY TERMS

QUESTIONS FOR REVIEW

1. What evidence did ancient astronomers have that the Earth is round?
2. How was the circumference of the Earth determined by Eratosthenes?
3. How was the size of the Moon estimated by ancient observations?
4. Why was Aristarchus's estimate of the distance to the Sun uncertain?
5. What is the difference between *angular size* and *linear size*?
6. If you triple your distance from an object, what happens to its angular size?

PROBLEMS

1. On average, Mercury is 0.387 times Earth's distance from the Sun, and Pluto is 39.53 times Earth's distance from the Sun. What is the Sun's angular diameter as seen from Mercury? from Pluto?
2. Suppose you are an alien living on the fictitious warlike planet Myrmidon and you want to measure its size. The Sun is shining directly down a missile silo 1000 miles to your south, while at your location the Sun is 36° from straight overhead. What is the circumference of Myrmidon? What is its radius?
3. A cloud directly above you is about 10° across. From the weather report you know that the cloud is 2,200 m high. How wide is the cloud?
4. The great galaxy in Andromeda has an angular diameter along its long axis of about 5°. Its distance is about 2.2 million light-years. What is its linear diameter?
5. Suppose you see a person whom you know is 1.8 m (6.0 ft) tall, standing at a distance where he appears to have an angular height of 2°. How far away is he?
6. Aristarchus's critics argued that parallax was not observed and therefore the Earth could not be in motion. In fact we know now that the stars are very far away and parallax motion is very small. Let's examine this by estimating the change in angular size of the angular separation between two stars 10 light-years away from the Sun. The two stars are 3 light-years apart. Calculate the angular separation of the two stars when the Earth is closest to them in its orbit and when it is farthest away, on the opposite point on its orbit. If human vision (without the aid of a telescope) can at best distinguish a difference of about 0.02°, would the change in angular separation be detectable to the naked eye?

TEST YOURSELF

1. The circular shape of the Earth's shadow on the Moon led early astronomers to conclude that
   a. the Earth is a sphere.
   b. the Earth is at the center of the Solar System.
   c. the Earth must be at rest.
   d. the Moon must orbit the Sun.
   e. the Moon is a sphere.
2. The ancient Greeks are *not* credited with
   a. measuring the size of the Earth.
   b. finding the relative sizes of the Earth and Moon.
   c. determining that the Earth is round.
   d. detecting the parallax of stars.
   e. suggesting a heliocentric model of the Solar System.
3. If the Moon were half as far away from the Earth, its angular size would be
   a. twice as big.
   b. half as big.
   c. four times as big.
   d. one-quarter as big.
   e. the same as it is now.

unit

11

Planets: The Wandering Stars

Five of the brightest "stars" in the sky do not remain in the same position on the celestial sphere, but instead gradually shift from night to night. The motions of these objects are quite complex and difficult to predict. The Greeks gave them the name ***planetai,*** meaning "wanderers," from which our word ***planet*** comes.

Until the invention of telescopes, **planets** were thought of as special stars, free to move in the sky, unlike the thousands of other stars visible on the celestial sphere. This freedom of movement suggested they had special powers, and perhaps this inspired their association with the gods Mercury, Venus, Mars, Jupiter, and Saturn. When it was eventually proposed that the motions of these objects could be more easily understood if Earth was a "wandering star" too, is it any wonder that this was a difficult idea for most to accept?

Background Pathways

11.1 MOTIONS OF THE PLANETS

Planets move against the background of stars because of a combination of the Earth's and their own orbital motion around the Sun. One of the more important features of this motion is that the planets always remain within a narrow band on either side of the ecliptic, within the constellations of the zodiac, as the Sun does (Unit 6). The planets remain close to the ecliptic because their orbits, including that of the Earth, all lie in nearly the same plane (Figure 11.1).

Like the Sun and the Moon, the planets usually move from west to east past the stars. This does not mean that the planets rise in the west and set in the east. As seen from a spinning Earth, planets always rise in the east and set in the west because their movements on the celestial sphere are far slower than the Earth's rotation. The motion of a planet relative to constellations of the zodiac can be seen by marking off the planet's position on a star chart over a period of several months. Figure 11.2 illustrates such a plot and shows that planets normally move eastward through the stars as a result of their orbital motion around the Sun.

Pluto, Eris, and a number of other small Solar System bodies sometimes move outside the zodiac.

Sometimes, however, this movement is interrupted. Periodically a planet will move west with respect to the stars, a condition known as **retrograde motion** and shown in Figure 11.3. The word *retrograde* means "backward moving," and when a planet is in retrograde motion, its path through the stars turns or loops backward for a month or more. All planets undergo retrograde motion when they lie in the same direction as the Earth from the Sun. This extraordinary behavior proved difficult to explain and became one of the major challenges of astronomers for almost 2000 years.

Q If you see a bright "star" in the sky, how can you tell whether it is a star or a planet (such as Venus) without using a telescope?

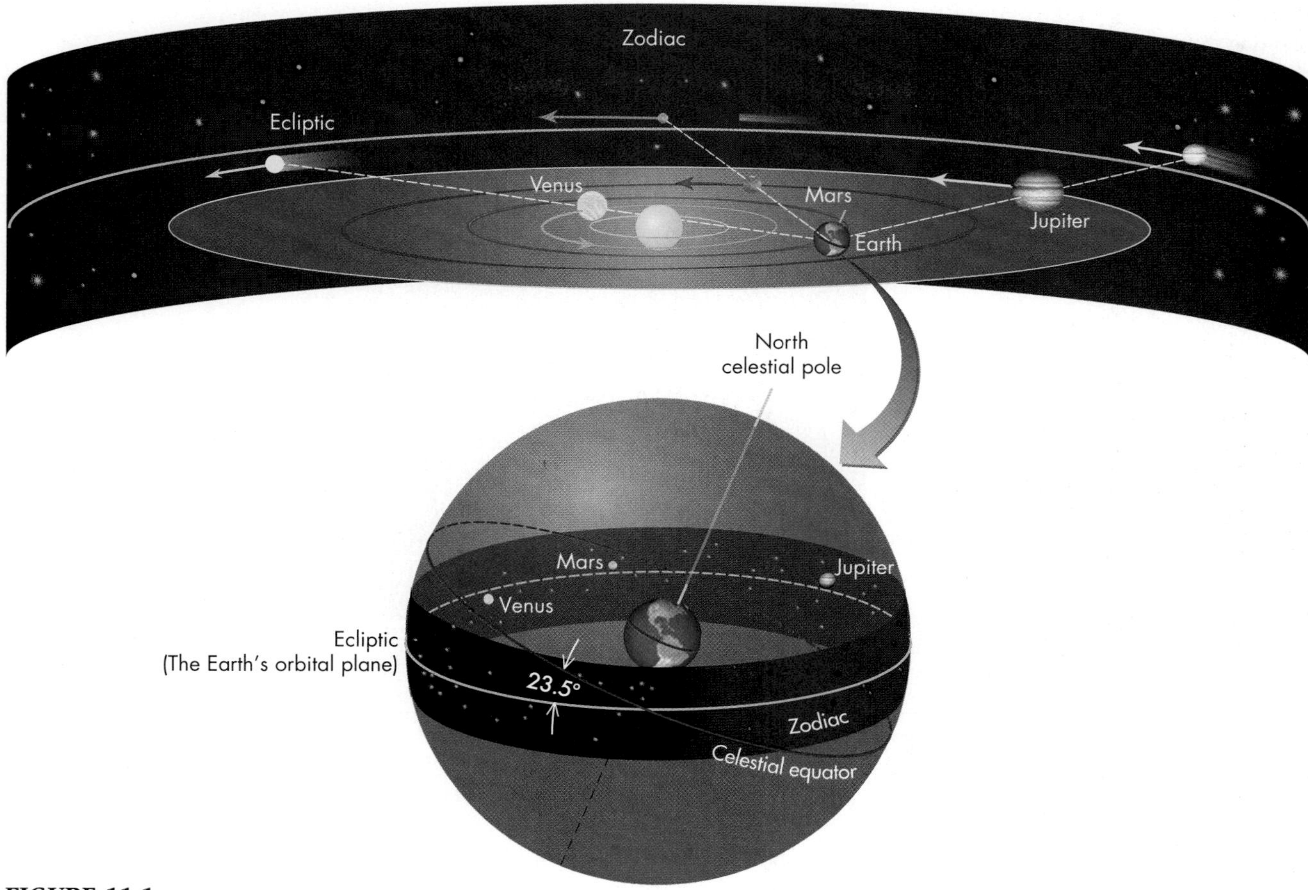

FIGURE 11.1
The planets move through the zodiac, which is tilted 23.5° with respect to the celestial equator because the Earth's rotation axis is tilted by that amount with respect to the plane in which the planets orbit.

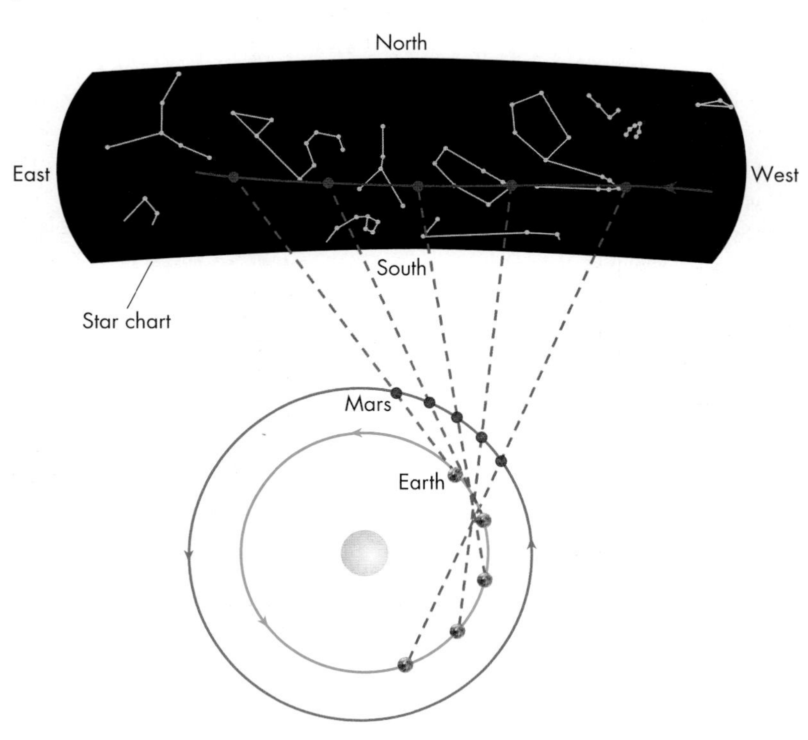

FIGURE 11.2
A planet's eastward drift against the background stars plotted on the celestial sphere. Note: Star maps usually have east on the left and west on the right, so that they depict the sky when looking south.

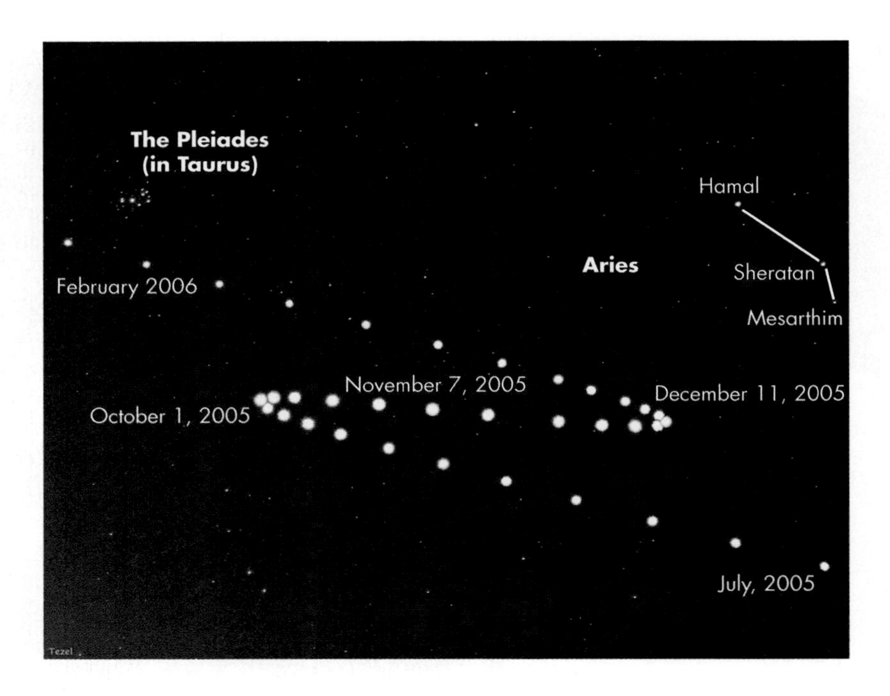

FIGURE 11.3
Images showing the retrograde motion of Mars in late 2005. Pictures were taken over an eight-month period, approximately one week apart as Mars moved through the constellations Aries and Taurus. Note how Mars became brightest during its retrograde motion when it was closest to Earth.

11.2 EARLY THEORIES OF RETROGRADE MOTION

As one observes the sky, the most basic of observations is that everything seems to move around the Earth from east to west. From earliest Greek times, this led to the model that the Earth was at the center of the universe and the planets and stars moved around it. Models of the universe of this type are called Earth-centered or **geocentric models.**

Figure 11.4 shows a typical geocentric model based on the work of the Greek astronomer Eudoxus, who lived about 400–347 B.C.E. The Sun, Moon, and

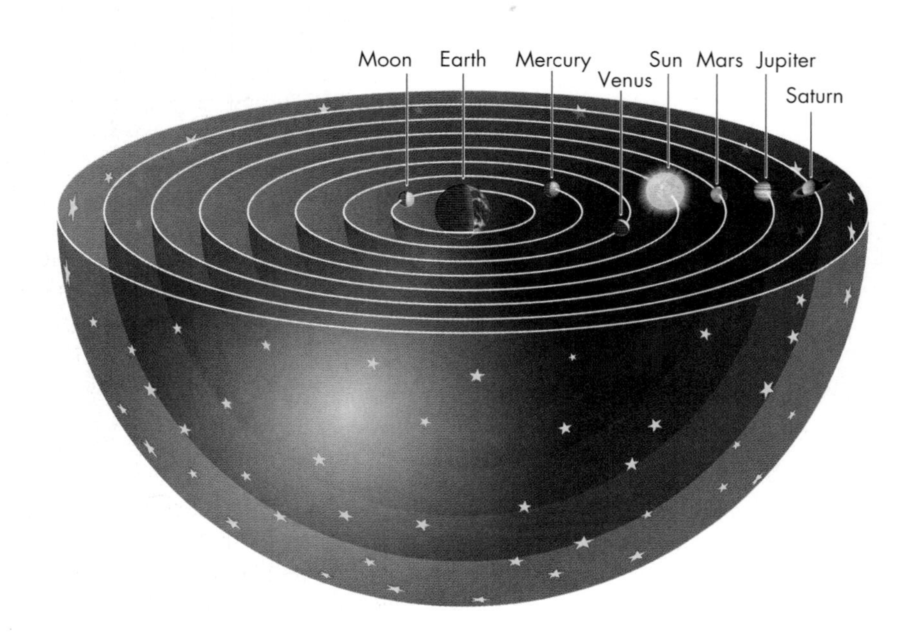

FIGURE 11.4
A cutaway view of the geocentric model of the Solar System according to Eudoxus. The celestial bodies were pictured as being carried around the Earth on transparent spheres. The spheres were imagined to spin around Earth once each day, but at slightly different rates so that the Sun, Moon, and planets moved at slightly different rates relative to the stars.

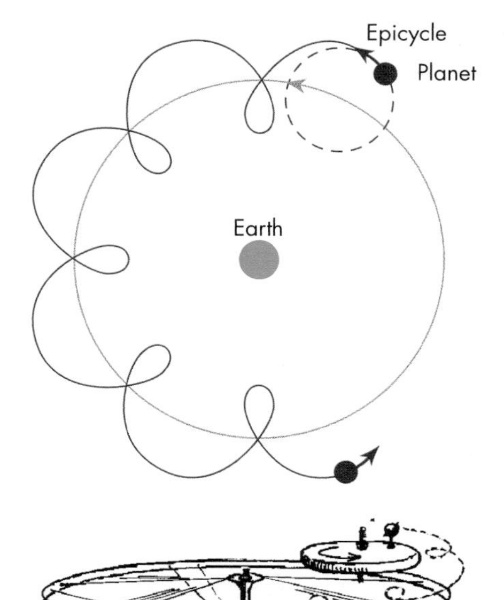

FIGURE 11.5
To explain the retrograde motion of the planets in the geocentric model, Ptolemy employed epicycles. The planets were pictured as being attached to their transparent sphere by a smaller sphere that rotated at a different rate, producing a looping motion. Epicycles are a bit like a bicycle wheel with a spinning Frisbee attached to its rim.

Ptolemy's model

planets all revolve around the Earth. The bodies that move fastest across the sky are those that are nearest to the Earth. Thus the Moon completes its orbit most rapidly and is nearest, whereas Saturn, whose path through the stars takes roughly 29 years to complete, is located the farthest out (of the planets then known). By assuming that each body was mounted on its own revolving transparent (crystalline) sphere and by tipping the rotation axis of each sphere slightly, Eudoxus was able to give a satisfactory explanation of the normal motions of the heavenly bodies. However, this model does not explain retrograde motion, unless one believes that the crystalline spheres sometimes stop turning, reverse direction, stop again, and then resume their original motion. This idea is clumsy and unappealing.

Eudoxus was able to develop a more complex model to explain retrograde motion by requiring that each planet's crystalline sphere should itself be attached to another crystalline sphere that rotated at a different angle. By combining steady rotations of both spheres, he was able to make zigzag paths that roughly resembled retrograde motion. Over the next five centuries, astronomers proposed variants of this complex model, leading to a model that could predict the planets' positions with good accuracy, developed by Claudius **Ptolemy,** the great astronomer of Roman times, who lived about 100 to 170 C.E.

Ptolemy lived in Alexandria, Egypt, which at that time was one of the intellectual centers of the world, in part because of its magnificent library. Ptolemy fashioned a model of planetary motions in which each planet moved on one small circle, which in turn moved on a larger one (Figure 11.5). The small circle, called an **epicycle,** was supposed to be carried along on the large circle like a Frisbee spinning on the rim of a bicycle wheel or the children's toy called a "Spirograph."

According to Ptolemy's model, the motion of the planet from east to west across the night sky is caused by the rotation of the large circle (the bicycle wheel, in our analogy). Retrograde motion occurs when the epicycle carries the planet in a reverse direction (caused by the rotation of the Frisbee, in our analogy). Thus, with epicycles it is possible to account for retrograde motion, and Ptolemy's model was able to predict planetary motions with fair precision.

However, discrepancies remained between the predicted and the true positions of the planets. This led to further modifications of the model, each of which led to slightly better agreement but at the cost of adding further complexity. Ptolemy's model survived until the 1500s thanks to its success at predicting the positions of the planets. In the end it was rejected because it was too complex to be plausible: Simplicity is an important element of scientific theory. As the medieval British philosopher William of Occam wrote in the 1300s, "Entities must not be unnecessarily multiplied." This expresses the idea that a scientific model that requires many parts or parameters to explain a phenomenon is less desirable than a simpler model that achieves the same result, a principle known as **Occam's razor.** It expresses a common situation encountered in science: If we begin with incorrect assumptions, as new evidence accumulates, we will often have to add more and more complexity to our models to explain discrepancies between the evidence and the model.

Although we call our numerals "Arabic," they were borrowed by Arab traders from the civilizations farther east in what is now India.

Ptolemy's era was one of decay and general political instability for Roman civilization, and much of what we know of his work (and of Greek and Roman civilization in general) we owe to Islamic scholars around the southern edge of the Mediterranean, who preserved and expanded on his work from about 700 to 1200 C.E. In addition, Islamic scholars revolutionized mathematical techniques through innovations such as algebra (which is an Arabic word) and Arabic numerals, as well as making many detailed measurements of the motions of the Sun, Moon, and planets. These greatly aided in the development of a mathematical understanding of the universe.

11.3 THE HELIOCENTRIC MODEL

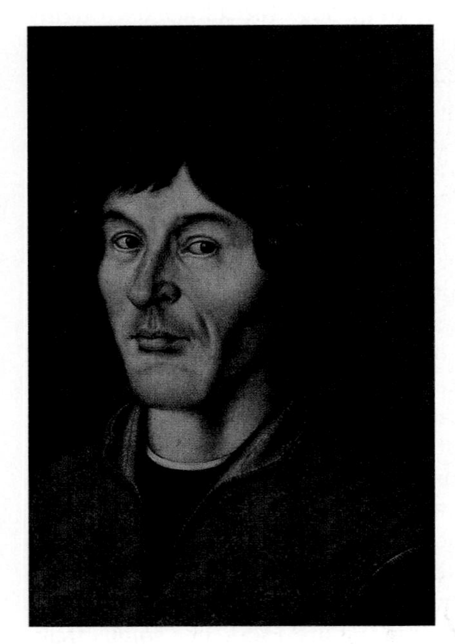

FIGURE 11.6
Nicolaus Copernicus (1473–1543).

The man who began the demolition of the geocentric model and started a revolution in astronomical thinking was a Polish physician and lawyer by the name of Nicolaus **Copernicus** (1473–1543). Copernicus (Figure 11.6) tried many modifications of the geocentric model to explain the centuries of data on planetary positions that had been collected since Ptolemy's geocentric model was developed. All of his attempts failed. He finally decided to reconsider the possibility that the Earth moves around the Sun.

A **heliocentric model** places the Sun (*helios* in Greek) at the center of a system of planets, and the Earth is just one of the planets. Such a model had been proposed nearly 2000 years earlier by Aristarchus (Unit 10.2), but had been rejected at the time. Nevertheless, Copernicus showed that such a model offered a far simpler explanation of retrograde motion. In fact, if the planets orbit the Sun, retrograde motion becomes a simple consequence of one planet on a smaller, faster orbit overtaking and passing another on a larger, slower orbit.

To see why retrograde motion occurs, look at Figure 11.7. Here we see the Earth and Mars moving around the Sun. The Earth completes its orbit, circling the Sun, in 1 year, whereas Mars takes 1.88 years to complete an orbit, with the Earth overtaking and passing Mars every 780 days. Lines drawn from the Earth to Mars show that Mars appears to change its direction of motion against the background stars each time the Earth overtakes it. A similar phenomenon occurs when you drive on a highway and pass a slower car. Both cars are, of course, moving in the same direction. However, as you pass the slower car, it looks as if it is moving backward against the stationary objects behind it.

INTERACTIVE
Retrograde motion

Retrograde motion occurs around the time when Mars is directly opposite the Sun in the sky, which we call **opposition** (Figure 11.8). When Mars is at opposition, it is also nearest Earth, as well as being brightest and easiest to see. Jupiter and

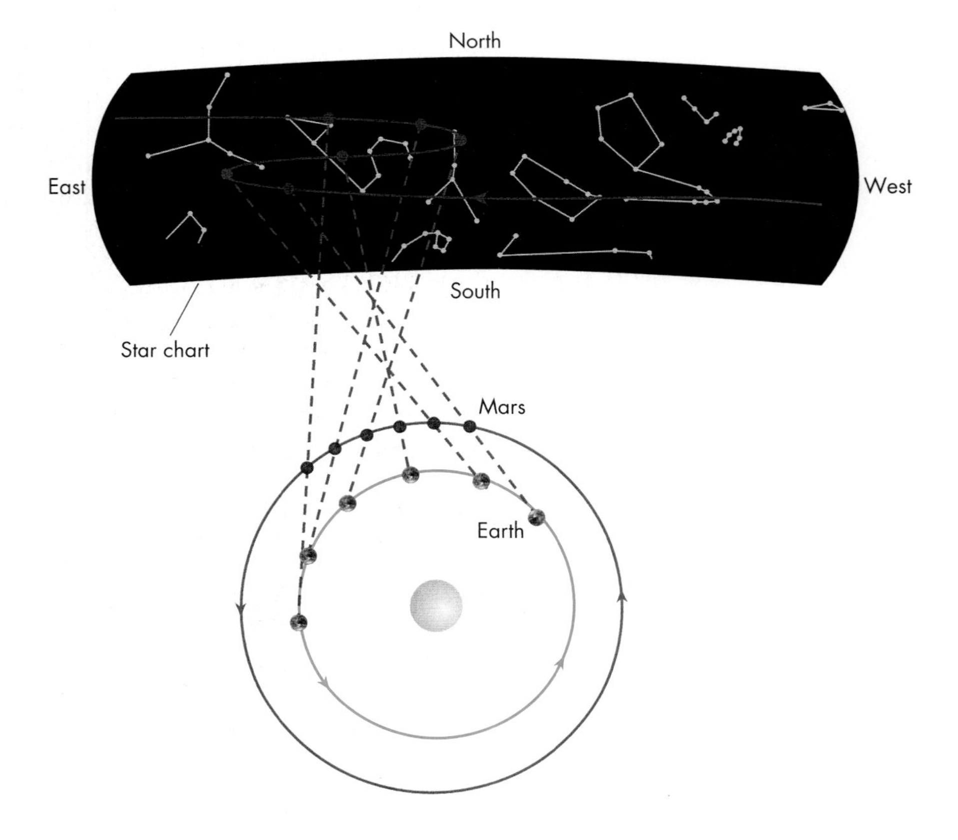

FIGURE 11.7
Why we see retrograde motion. As Earth approaches and passes Mars, Mars appears to move backward against the background stars.

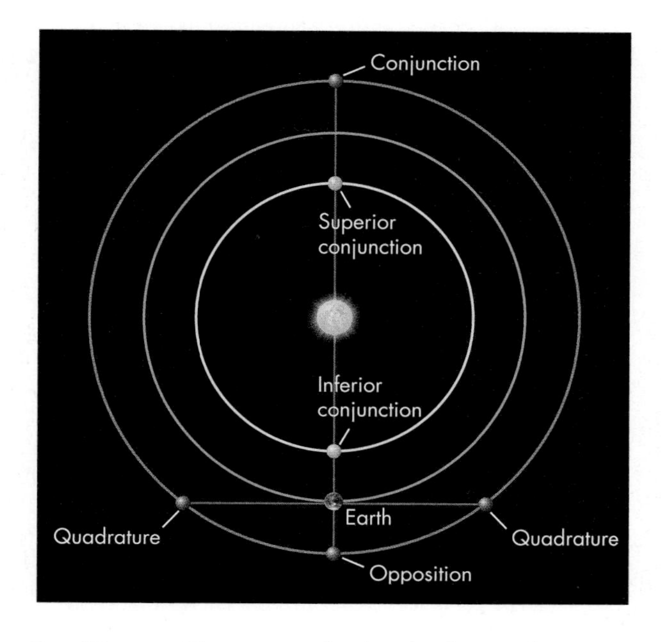

FIGURE 11.8
Planetary configurations: opposition, superior conjunction, inferior conjunction, and quadratures.

Saturn (as well as Uranus, Neptune, and other bodies in the outer Solar System) exhibit retrograde motion around the time of opposition, but Mercury and Venus are different. Because they are nearer the Sun than the Earth, they can never be in opposition. Instead they appear to reverse direction against the stars as they pass between the Earth and the Sun. Astronomers use the term **conjunction** to refer to when a planet lies in the same direction as the Sun. All of the planets come into conjunction when they are on the far side of the Sun (often called **superior conjunction**) as illustrated in Figure 11.8. Venus and Mercury undergo retrograde motion around the time when they are passing the Earth in **inferior conjunction** between the Earth and the Sun.

Retrograde motion of Mars

Copernicus found that he could determine the relative sizes of the planets' orbits from information about the planets' configurations. For example, Venus and Mercury never get very far from the Sun, as shown in Figure 11.9. Mercury can never be more than 28° from the Sun, and Venus never more than 47° from the

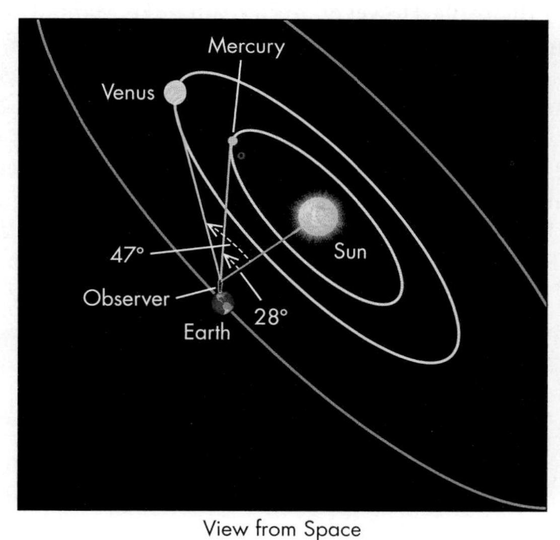

View from Space

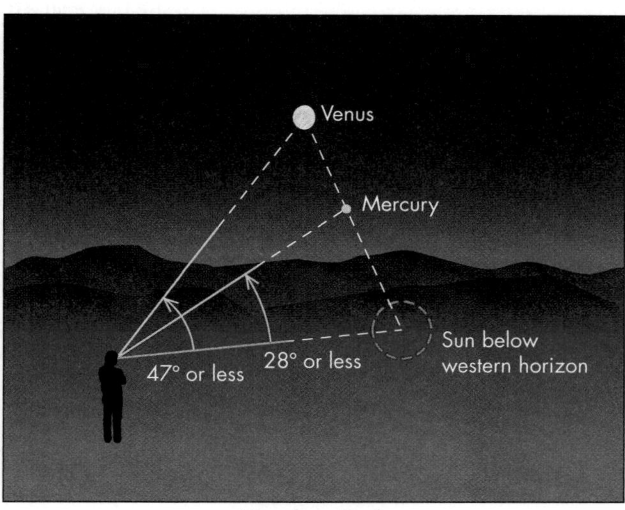

View from Earth

FIGURE 11.9
The greatest elongations of Mercury and Venus, and the Evening Star phenomenon. The left-hand diagram also shows that Mercury and Venus can *never* appear more than 28° and 47°, respectively, from the Sun.

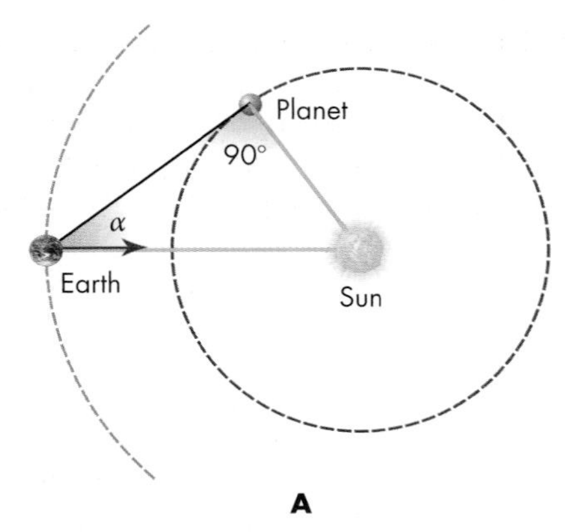

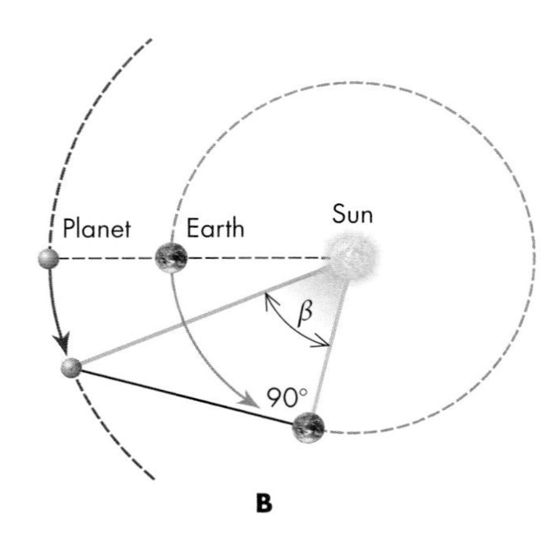

FIGURE 11.10
How Copernicus calculated the distance to the planets. (A) When an inner planet appears in the sky at its farthest point from the Sun, the planet's angle on the sky away from the Sun, α, can be measured. You can see from the figure that at the same time the angle from our line of site to the Sun is 90°. The planet's distance from the Sun can then be calculated with geometry, knowing that the Earth–Sun distance is 1 AU. (B) Finding the distance to an outer planet requires determining how long it takes the planet to move from being opposite the Sun in the sky (the planet rises at sunset) to when the Sun–Earth–planet angle is 90° (the planet rises or sets at midnight). Knowing that time interval, one then calculates what fraction of their orbits the Earth and planet moved in that time. Multiplying those fractions by 360° gives the angles the planet and Earth moved. The difference between those angles gives angle β. Finally, using geometry and the value of angle β as just determined, the planet's distance from the Sun can be calculated.

Sun, as seen on the sky. The size of this largest angular separation from the Sun, or **greatest elongation,** is determined by the sizes of these planets' orbits relative to the Earth's orbit. By assuming that each planet's orbit is a concentric circle around the Sun, we can make a geometric construction (or use trigonometry) to find their sizes (Figure 11.10A).

Determining the sizes of the other planets' orbits is a little more complicated. For this, Copernicus noted how long the planets took to move from opposition (180° from the Sun) to **quadrature** (90° from the Sun), as shown in Figure 11.10B. For a distant planet like Saturn, the time between opposition and quadrature is nearly a quarter of the orbit. However, for a planet like Mars, whose orbit is not much bigger than the Earth's, the portion of the orbit between opposition and quadrature is a fairly small fraction of the total orbit. Again, a geometric construction can be used to establish the relative size of the planets' orbits.

Remarkably, then, by using the heliocentric model, Copernicus not only could provide a simpler explanation of retrograde motion, he could even calculate the relative sizes of all of the planets' orbits (Table 11.1). He could not determine how far away the Earth was from the Sun, but he could calculate that Jupiter was about five times farther from the Sun than the Earth and that Mercury was about

TABLE 11.1 Planetary Distances from the Sun According to Copernicus

Planet	Distance in AU According to Copernicus	Actual Distance in AU
Mercury	0.38	0.39
Venus	0.72	0.72
Earth	1.00	1.00
Mars	1.52	1.52
Jupiter	5.22	5.20
Saturn	9.17	9.54

one-third as far. The distances found in this manner were expressed in terms of the Earth's distance from the Sun—the astronomical unit—whose value was not accurately known until several hundred years later. Table 11.1 illustrates that Copernicus's calculations agree remarkably well with modern values. These distances became an important starting place for astronomers who came later.

11.4 THE COPERNICAN REVOLUTION

Copernicus described his model of a Sun-centered universe in one of the most influential scientific books of all time, *De revolutionibus orbium coelestium* (On the Revolutions of the Celestial Orbs). Because his ideas were counter to the beliefs of the day, they were met with hostility. Copernicus quietly circulated a manuscript of his ideas for about 30 years before allowing the book to be published in 1543. Reportedly he saw the first copy while on his deathbed.

Some criticism of Copernicus's work was justified because his model could not predict the positions of the planets any better than Ptolemy's. This lack of success arose in part because Copernicus assumed that the planetary orbits were circles, an idea that was disproved only in the next century. To make his model more accurate, Copernicus even resorted to the use of epicycles to adjust his predictions of the planets' positions. Furthermore, his model again raised the question of why no stellar parallax (Unit 10) could be seen.

Q Imagine a planet far beyond Saturn. What would its retrograde motion look like? How does the idea of parallax compare to the idea of retrograde motion?

There were counterarguments to the objections, however. The lack of parallax could be explained if the stars were very distant. On a technical level, Copernicus's model did not require large epicycles to correct his positions—because the circular orbits were not bad approximations to the planets' orbits. Thus it could be said that Copernicus's model was no more successful than Ptolemy's at predicting the positions of planets as observed from Earth, but it offered a simpler explanation of retrograde motion.

The main objection to Copernicus's views of planetary motion was that they ran counter to the widely accepted teachings of the ancient Greek philosopher Aristotle. These views were supported both by "common sense" and by religious belief at that time. After all, when we observe the sky, it looks as if it moves around us. Moreover, we do not detect any sensations caused by the Earth's motion—it feels at rest. And if the Earth is just one of the planets, it is not the center of the universe. This mixture of rational and irrational objections made even many scientists slow to accept the Copernican view.

However, conditions were favorable for such new ideas. The cultural Renaissance in Europe was at its height; the Protestant Reformation had just begun; Europeans were sailing to the new world; and there was a growing acceptance of the immensity of the universe. In such an intellectually stimulating environment, new ideas flourished and found a more receptive climate than in earlier times. In the late 1500s the English astronomer Thomas Digges argued in favor of the Copernican model, proposing that the stars were other suns at enormous and various distances. The Italian philosopher and monk Giordano Bruno went so far as to claim that the planets around these other stars were inhabited.

The aesthetic appeal of the heliocentric system led to a growing acceptance of the model despite church efforts to suppress it. Bruno was burned at the stake in 1600 for his heretical ideas, but others continued their quest for a deeper understanding of the workings of the cosmos. Are the stars other suns? How can we reconcile motion of the Earth with our everyday experience that it seems stationary? What keeps the planets in their orbits? The steady work of science was on the verge of answering these and many more challenging questions.

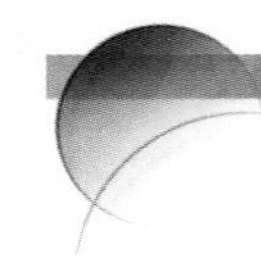

KEY TERMS

conjunction, 95
Copernicus, 94
epicycle, 93
geocentric model, 92
greatest elongation, 96
heliocentric model, 94
inferior conjunction, 95
Occam's razor, 93
opposition, 94
planet, 90
Ptolemy, 93
quadrature, 96
retrograde motion, 90
superior conjunction, 95

QUESTIONS FOR REVIEW

1. What is retrograde motion?
2. List the arguments in favor of and against Copernicus's model of a Sun-centered universe.
3. What is the best location for Mercury to appear in the sky for us to observe it?
4. What is the best location for Mars to appear in the sky for us to observe it?
5. Why is the time it takes Earth to overtake and pass Mars different from Mars's orbital period?
6. How did Copernicus estimate the planets' distances from the Sun?

PROBLEMS

1. If we discover a new planet in the Solar System and observe that it takes about one Earth year between oppositions, what could we say about it?
2. If you lived on Mars, what would the greatest elongation of the Earth be?
3. When Mars is in conjunction, within an angle of about 5° of the Sun, we lose radio contact with our spacecraft there. Based on the synodic period of Mars, estimate approximately how long this period of communication problems lasts.
4. Make a geometric construction (or use trigonometry) to show that Venus's orbit is about 0.72 times the size of Earth's, based on the fact that Venus's greatest elongation is 47°.
5. Redraw Figure 11.7 to show that retrograde motion can also be seen for a planet closer to the Sun than the Earth. Near what position would we expect to see retrograde motion for such a planet? What would an observer on this planet see of the Earth's motion?
6. Mars spends, on average, about 11.5% of its orbit moving from opposition to quadrature. Make a geometric construction (or use trigonometry) to show that Mars's orbit is about 1.52 times the size of Earth's.

TEST YOURSELF

1. A planet in retrograde motion
   a. rises in the west and sets in the east.
   b. shifts westward with respect to the stars.
   c. shifts eastward with respect to the stars.
   d. will be at the north celestial pole.
   e. will be exactly overhead no matter where you are on Earth.
2. Which are the only planets that can be seen in opposition?
   a. All planets with orbits larger than the Earth's
   b. All planets that undergo retrograde motion
   c. All planets with epicycles
   d. All planets that are larger than the Earth
   e. All of the above
3. Occam's razor is used to
   a. discriminate between models based on their simplicity.
   b. measure the angle between the Sun and planets.
   c. explain why planets change the direction of their orbits.
   d. describe the path that the planets follow across the ecliptic.
   e. execute heretics.

Starry Night

See the Pathways *2nd edition ARIS site for online Starry Night™ planetarium exercises.*

unit

12

The Beginnings of Modern Astronomy

Background Pathways

Unit 11: Planets: The Wandering Stars 90

A story is told that in medieval Europe, a gathering of scholars debated how many teeth a horse has—according to Aristotle versus the Bible. The debate went on for hours with no apparent resolution. A young novitiate proposed going out to the stable to look in a horse's mouth and count. After being castigated for his unscholarly proposal, he was ejected from the assembly!

This story may be no more than legend, but it illustrates a difference in scientific approach before about 1600. Scholars used to believe that one should look for answers in the writings of great authorities of the past. Scientists today take it for granted that if they have questions or disagreements, they should conduct experiments and make new observations to settle the issue. In the century following Copernicus's proposal that the Earth orbited the Sun, new observations finally settled the issue.

12.1 PRECISION ASTRONOMICAL MEASUREMENTS

Early first steps in solving the problem of astronomical motion were made by the sixteenth-century astronomer Tycho **Brahe** (1546–1601), pronounced *TEE-koh BRAH-hee.* Born into the Danish nobility, Brahe (Figure 12.1) used his position and wealth to indulge his passion for study of the heavens, a passion based in part on his professed belief that God placed the planets in the heavens to be used as signs to mankind of events on Earth. Driven by this interest in the skies, he designed, had built, and used instruments that permitted the most accurate pretelescopic measurements ever made. The instruments he used were similar to giant protractors that he could turn to measure positions in the sky with high precision (Figure 12.2). His instruments allowed him to measure the positions of planets to about 1 **arc minute,** or 1/60th of a degree. This is 30 times smaller than the angular size of the Moon, and about the limit of human vision.

Q Because the sky moves overhead, Tycho Brahe's time measurements had to be precise within a few seconds to achieve an accuracy of 1 arc minute for positions relative to the horizon. Can you think of ways to measure positions on the celestial sphere that might not require such accurate timing?

Brahe did more than just record planetary positions. In 1572, when he saw an exploding star (what we now call a supernova—Unit 64), he showed that this bright new object maintained a fixed position in the celestial sphere and therefore had to be far beyond the supposed spheres that move the planets. When a bright comet appeared in 1577, he showed that it lay far beyond the Moon, not within the Earth's atmosphere as taught by the ancients. These observations suggested that the heavens were more changeable and more complex than previously accepted.

Although Brahe could see the virtues of the simplicity offered by the Copernican model, he was unconvinced of its validity because even his instruments were unable to detect stellar parallax (Unit 10)—the apparent shift in star positions expected if the Earth moves around the Sun. Therefore, he offered a compromise

FIGURE 12.1
Tycho Brahe (1546–1601).

FIGURE 12.2
Picture of Tycho Brahe's astronomical instruments at Uraniborg.

model in which all of the planets except the Earth went around the Sun, while the Sun, as in earlier models, circled the Earth. Brahe was the last of the great astronomers to hold that the Earth was at the center of the universe. Brahe's model won favor with those who sought to maintain the idea of a geocentric universe. It was difficult to disprove because it was essentially the same as the Copernican model, except for the change of perspective about whether the Earth moved around the Sun or vice versa. Others, though, began to find additional reasons for doubting that the Earth was so central to the rest of the universe.

12.2 THE NATURE OF PLANETARY ORBITS

FIGURE 12.3
Johannes Kepler (1571–1630).

A few years before he died, Tycho Brahe hired a younger assistant, Johannes **Kepler** (1571–1630). They did not get along well, and Kepler (Figure 12.3) was fired and rehired during their brief association.

Kepler was a clever and hardworking man with a strong grasp of geometry and unusual ideas. For example, he searched for a fundamental relationship between the spacing between the planets and how tightly various geometrical figures could nest inside each other—such as a sphere inside a cube. Kepler did not, however, think highly of Brahe's planetary model, which he described as a "pretzel."

When Brahe died, his observational data passed into Kepler's hands. Brahe's family accused Kepler of stealing them, but their real value was apparent only after years of Kepler's painstaking work. He was able to derive from this huge set of precise data a detailed picture of the path of the planet Mars. Whereas all previous investigators had struggled to fit the planetary paths to circles, Kepler showed—with Brahe's superb data—that Mars moved not along a circular path but rather along an **ellipse.**

An ellipse can be drawn with a pencil inserted in a loop of string that is hooked around two thumbtacks. If you move the pencil while keeping it tight against the string, as shown in Figure 12.4A , you will draw an ellipse. Each point marked by a tack is called a **focus** of the ellipse. The shape of an ellipse is described by its

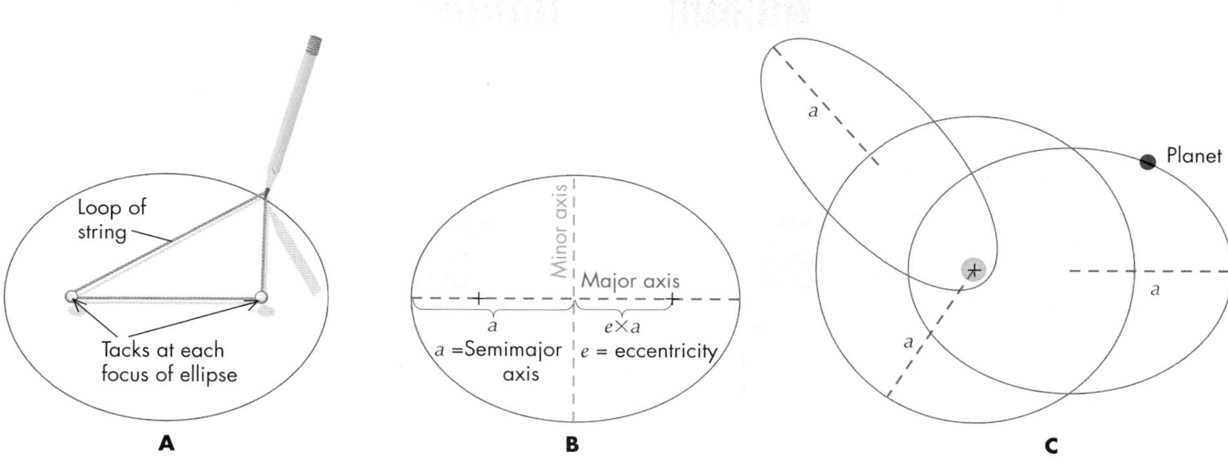

FIGURE 12.4
(A) Drawing an ellipse. (B) The major and minor axes. The semimajor axis, *a*, is half of the major axis. The amount by which the Sun is off-center in the ellipse determines the eccentricity, *e*, of the ellipse. (C) The Sun lies at one focus of the ellipse. Three different orbits are shown that have the same size semimajor axis but differing eccentricities.

long and short dimensions, called its *major axis* and *minor axis* (Figure 12.4B) respectively. For most calculations the most important value is half the major axis length, or the **semimajor axis.**

Kepler's careful measurements of Mars's orbit showed that the Sun is located *not* at the center of the ellipse but off-center, at one focus of the ellipse (Figure 12.4C). Astronomers use the term **eccentricity** to describe how round or "stretched out" an ellipse is. Mathematically, the eccentricity *e* indicates how far the Sun is from the center of the ellipse as a fraction of the semimajor axis (Figure 12.4C). If the Sun is exactly at the center, the two foci are on top of each other so the eccentricity is zero. In this case the orbit is a circle and the semimajor axis is the same as the radius of the circle. As the eccentricity of the orbit increases toward 1, the position of the Sun becomes increasingly offset from the center. Figure 12.4C illustrates ellipses with several different eccentricities but the same value for the semimajor axis.

Kepler found further that Mars moves faster in its orbit when it is closer to the Sun. He found a way to describe this geometrically by imagining the motion of a line joining the planet to the Sun: The planet's speed changes so that the line sweeps out equal areas in equal time intervals, as illustrated by the shaded areas in Figure 12.5B. For the areas to be equal, the distance traveled along the orbit in a given time must grow larger the closer the planet is to the Sun. With the elliptical shape and speed variations of the orbit now established, Kepler was able to obtain excellent agreement between the calculated and the observed positions not only of Mars but also of the other planets.

Kepler also discovered a relationship between the size of a planet's orbit and how long it takes to orbit the Sun—its orbital **period.** Kepler determined that the period depends on the semimajor axis of the orbit according to a mathematical formula. The period increases as the semimajor axis increases, but not as a simple proportionality. He found that the square of the orbital period measured in years equals the cube of the semimajor axis measured in astronomical units (Unit 1.6). This is illustrated in Table 12.1.

INTERACTIVE
Kepler's third law

Kepler's discoveries of the nature of planetary motions are expressed in what are known today as **Kepler's three laws:**

I. **Planets move in elliptical orbits with the Sun at one focus of the ellipse** (Figure 12.5A).

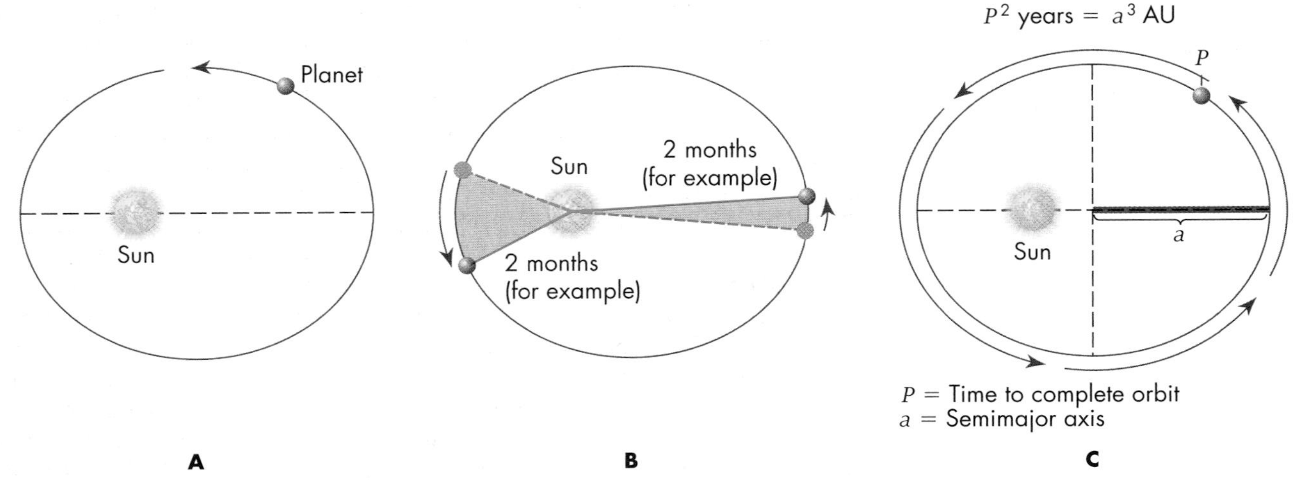

FIGURE 12.5
Kepler's three laws. (A) A planet moves in an elliptical orbit with the Sun at one focus. (B) A planet moves so that a line from it to the Sun sweeps out equal areas in equal times. Thus the planet moves fastest when nearest the Sun. For the drawing a two-month interval is chosen. (C) The square of a planet's orbital period (in years) equals the cube of the semimajor axis of its orbit (in AU).

TABLE 12.1 Table Illustrating Kepler's Third Law

Planet	Distance from Sun (a) in Astronomical Units	Orbital Period (P) in Years	a^3	P^2
Mercury	0.387	0.241	0.058	0.058
Venus	0.723	0.615	0.378	0.378
Earth	1.0	1.0	1.0	1.0
Mars	1.524	1.881	3.54	3.54
Jupiter	5.20	11.86	141	141
Saturn	9.54	29.46	868	868

II. **The orbital speed of a planet varies so that a line joining the Sun and the planet will sweep over equal areas in equal time intervals** (Figure 12.5B).

III. **The amount of time a planet takes to orbit the Sun, P, is related to its orbit's size, a** (Figure 12.5C):

$$P^2 = a^3$$

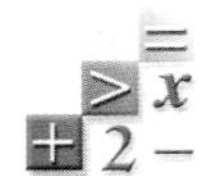

P = orbital period, measured in years
a = semimajor axis, measured in AU

Kepler's three laws describe the essential features of planetary motion around the Sun. They describe not only the shape of a planet's path but also its speed and distance from the Sun.

Kepler's laws are *empirical* findings—that is, they are determined directly from measurements rather than from a theoretical idea about what causes the planets to move. They remain valid today, even though we no longer agree with many of Kepler's own interpretations of their causes. Later scientists discovered that these laws can all be derived from more fundamental laws about the nature of motion and gravity. Describing the heavens with mathematical laws revolutionized our way of thinking about the universe. Without such mathematical formulations of physical laws, much of our technological society would be impossible. Kepler's laws were a major breakthrough in our quest to understand the universe.

Kepler's second law is a consequence of the conservation of angular momentum, which we will discuss more fully in Unit 20.

Kepler's laws

Q Kepler spent much of his life searching for connections between pure geometry and nature. Where have you encountered geometric patterns in nature?

It is perhaps ironic that such mathematical laws should come from Kepler, because so much of his work is tinged with mysticism. For example, Kepler's third law evolved from his attempts to link planetary motion to music, using the mathematical relations known to exist between different notes of the musical scale. Kepler even attempted to compose "music of the spheres" based on such a supposed link. Nevertheless, despite such excursions into these nonastronomical matters, Kepler's discoveries remain the foundation for our understanding of how planets move.

The third law has implications for the relative speeds of planets whose orbits are at different distances from the Sun. Because the third law states that $P^2 = a^3$, a planet far from the Sun (larger a) has a longer orbital period (P) than one near the Sun. If the third law had stated that $P = a$, then the speeds of the planets would all have been the same because the distance traveled in the orbit is proportional to the semimajor axis. However, the third law implies that planets nearer the Sun move faster than outer planets. For example, the Earth takes 1 year to complete its orbit; but Jupiter, whose orbit is about 5 times larger than the Earth's, takes about 12 years. Thus Jupiter moves about 0.42 (= 5/12) times as fast as Earth. This difference in speed is the source of retrograde motion—as Earth overtakes and passes Jupiter, we perceive Jupiter as moving backward, as discussed in Unit 11.

Kepler's third law allows us to predict the period of any body orbiting the Sun if we know its semimajor axis. For example, Pluto was discovered in 1930 at a distance of about 40 AU from the Sun. Assuming that this was approximately its semimajor axis, a, we can determine the period of its orbit from Kepler's third law as follows. The cube of the semimajor axis (40^3) must equal the square of the period P in years, so

$$a^3 = 40^3 = 64{,}000 = P^2.$$

Taking the square root of 64,000 gives us $P \approx 250$ years. That is, we can predict Pluto will take about 250 years to complete an orbit, which is longer than we have known of its existence. Modern detailed measurements of its orbit show that this initial estimate was quite good: Pluto's orbital period is 248.1 years, and its semimajor axis is 39.48 AU.

It is also possible to find the semimajor axis of a body orbiting the Sun from its orbital period. For example, in 1705 the English astronomer Sir Edmond Halley realized that a bright comet had been reported every 76 years in historical records. Known today as Comet Halley, we now understand that this object becomes visible only in the part of its orbit when it is close to the Sun. Its semimajor axis can be calculated by taking the square of its period (76^2) and equating that to its semimajor axis cubed:

$$P^2 = 76^2 = 5800 = a^3.$$

Taking the cube root of 5800, we find that $a \approx 18$ AU. In this case, the comet must have a highly elliptical orbit, because it is observed to come within 0.6 AU of the Sun. Therefore, for the major axis of the ellipse to be 36 AU (twice a), Comet Halley must travel out to beyond 35 AU—nearly as far as Pluto's orbit.

12.3 THE FIRST TELESCOPIC OBSERVATIONS

At about the same time that Tycho Brahe and Johannes Kepler were striving to understand the motion of heavenly bodies, the Italian scientist **Galileo** Galilei (1564–1642) was likewise trying to understand the heavens. However, his approach was dramatically different.

FIGURE 12.6
Galileo Galilei (1564–1642).

By longstanding tradition, Galilei is usually referred to by his first name, Galileo. The same used to be true of Tycho Brahe, but the more standard reference by last name is growing more common.

Kepler introduced the word *satellite*. When he saw the moons of Jupiter with a small telescope, their motion around the planet made him think of attendants or bodyguards—*satelles* in Latin.

Phases of Venus

Q Brahe argued that the Sun orbited the Earth but that the other planets orbited the Sun. Could Brahe's model explain the phases of Venus as observed by Galileo? Why or why not?

Galileo (Figure 12.6) was interested not just in celestial motion but in all aspects of motion. He studied falling bodies, swinging weights hung on strings, and so on. In addition, he used the newly invented telescope to study the heavens and interpret his findings, and what he found was astonishing.

In looking at the Moon, Galileo saw that its surface had mountains and was in that sense similar to the surface of the Earth. Therefore, he concluded that the Moon was not some mysterious ethereal body but a ball of rock. He discovered that the Sun had dark spots (now known as **sunspots**) on its surface. He noticed that the position of the spots changed from day to day. This showed that the Sun not only had blemishes and was not a perfect celestial orb, but also that it changed. These observations conflicted with previously held conceptions of the heavens as perfect and unchanging. In fact, by observing the changing position of the spots from day to day, Galileo deduced that the Sun rotated.

These observations gave evidence that the Sun and Moon are physical bodies. This was a problem for geocentric theories. To explain the rapid motion of all the objects in the sky around the Earth (once every 24 hours or so), early astronomers supposed the heavenly bodies were made of some light substance that could move at extremely high speeds—perhaps like the beam of a searchlight. It was harder to conceive of mountains of matter swinging around the Earth at such speeds.

Galileo looked at Jupiter and saw four smaller objects orbiting it, which he concluded were moons of the planet. These bodies, known today in his honor as the **Galilean satellites** (Unit 46), proved that at least some bodies in the heavens do not orbit the Earth. Seeing these satellites remain in orbit around Jupiter also refuted the notion that if the Earth moved, the Moon would fall behind it. But perhaps even more important, the motion of these bodies raised the crucial problem of what held them in orbit.

Galileo also looked at Saturn and discovered that it did not appear as a perfectly round disk but that it appeared to have "satellites" that remained stationary on either side of the planet—these were actually Saturn's rings, but his telescope was too small and too crudely made (inferior to today's inexpensive binoculars) to show them clearly. When Galileo examined the faint band of light on the celestial sphere that we call the Milky Way, he saw that it was populated with an uncountable number of stars. This single observation, demonstrating that there were far more stars than previously thought, shook the complacency of those who believed in a simple Earth-centered universe.

Galileo's observations of Venus were perhaps the most decisive in ruling out the geocentric model. Galileo observed that Venus went through a cycle of phases like the Moon, as illustrated in Figure 12.7. Consider what the geocentric model predicts: Venus should be on a crystalline sphere inside the Sun's own crystalline sphere. Because it never reaches an angle larger than 47° from the Sun, sunlight shining on Venus would always illuminate it from behind, like a crescent moon.

By contrast, the heliocentric model predicted that Venus would cross both in front of and behind the Sun from our perspective. When Venus is on the far side of the Sun, its angular size is small (because of the large distance), but we see the side of the planet illuminated by the Sun. Therefore its phase is nearly full. On the other hand, when it is passing between the Earth and the Sun, it will be angularly large, but we are facing mostly the dark side of the planet, with only a thin crescent illuminated. This behavior was exactly what Galileo found, leaving no doubt that Venus orbited the Sun.

Galileo's probings into the nature of the universe led him into trouble with religious "law." He was a vocal supporter of the Copernican view of a Sun-centered universe and wrote and circulated his views widely and somewhat tactlessly. He presented his arguments as a dialogue between a wise teacher and a nonbeliever

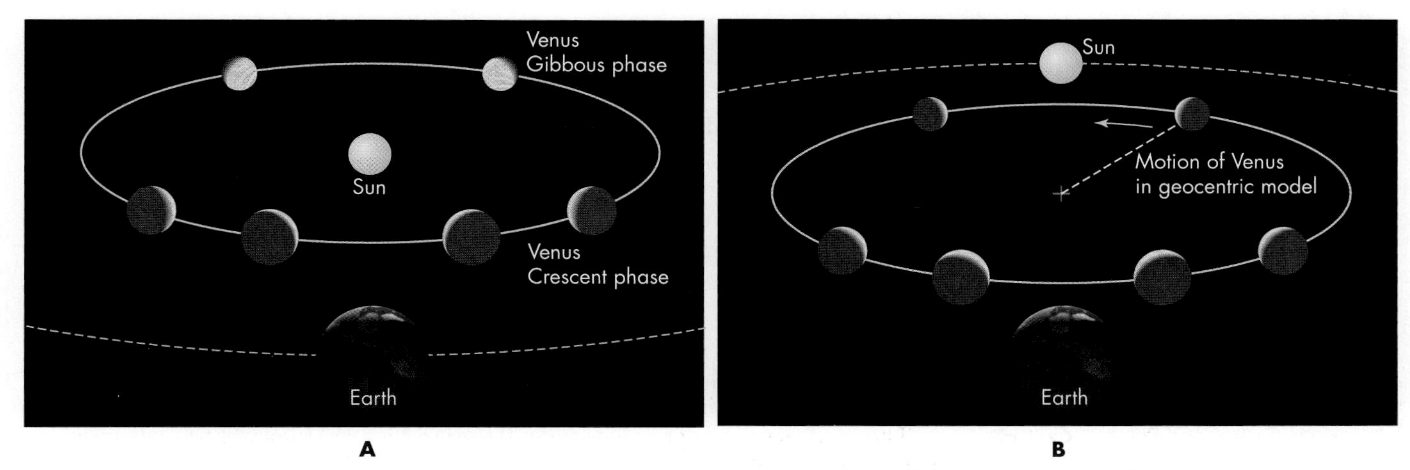

FIGURE 12.7
As Venus orbits the Sun, it goes through a cycle of phases (A). The phase and its position with respect to the Sun show conclusively that Venus cannot be orbiting the Earth. The gibbous phases Galileo observed occur for the heliocentric model but cannot happen in the Earth-centered Ptolemaic model, as illustrated in (B).

in the Copernican system whom he called *Simplicio* and who, according to his detractors, was patterned after the pope. Although the pope was actually a friend of Galileo, conservative churchmen urged that Galileo should be brought before the Inquisition because his views that the Earth moved were counter to the teachings of the Catholic Church. Considering that his trial took place when the papacy was attempting to stamp out heresy, Galileo escaped relatively lightly. He was forced to recant his "heresy" and was put under house arrest for the remainder of his life. Only in 1992 did the Catholic Church admit it had erred in condemning Galileo for his ideas.

Galileo's contributions to science would be honored even had he not made so many important observational discoveries. Beyond his revolutionary work in astronomy, he carried out fundamental experiments in physics that provided the beginnings of our understanding of the nature of motion and gravity. He was one of the principal founders of the experimental method for studying scientific problems, and his methods provided the direction for a new approach toward studying the cosmos.

KEY TERMS

arc minute, 99
Brahe, 99
eccentricity, 101
ellipse, 100
focus, 100
Galilean satellite, 104
Galileo, 103
Kepler, 100
Kepler's three laws, 101
period, 101
semimajor axis, 101
sunspot, 104

QUESTIONS FOR REVIEW

1. How were Tycho Brahe's observations critical to the development of astronomy?
2. What is the shape of a planet's orbit? Where is the Sun located relative to this shape?
3. How does a planet's speed change as it orbits the Sun?
4. What is the relationship between the size of a planet's orbit and the length of time it takes to complete an orbit?
5. How did Galileo's observations help to rule out a geocentric model of the Solar System?
6. Which planets can we see through a telescope as a crescent? When does this happen?

PROBLEMS

1. All objects orbiting the Sun obey Kepler's laws. Suppose a comet is found with an orbital period of 64 years. If its orbit is circular, what is its distance from the Sun?
2. In 2003, astronomers discovered Sedna, an object in the outer Solar System with a semimajor axis of 526 AU. What is its orbital period?
3. Suppose an asteroid is discovered with an orbit that brings it as close as 0.5 AU from the Sun and as far away as 1.5 AU. What is the semimajor axis of its orbit? What is its orbital period? What is the eccentricity of its orbit?
4. If a comet orbits the Sun and reaches 1 AU at its closest approach to the Sun, and its orbital period is 27 years, what is its maximum distance from the Sun?
5. For an asteroid in a circular orbit at 4 AU, calculate its orbital speed. Compare this to the speed of the Earth in its orbit of 29.8 km/sec.
6. A comet is discovered with an orbit that extends from Venus's orbit to Neptune's orbit. Use Kepler's second law to estimate how many times faster the comet is moving when it is closest to the Sun compared to when it is farthest from the Sun. (Hint: The area swept out over a short period of time when the planet is at these two extremes can be approximated by a triangle, and the area of a triangle equals one-half of its base times its height.)

TEST YOURSELF

1. Galileo used his observations of the changing phases of Venus to demonstrate that
   a. the Sun moves around the Earth.
   b. the universe is infinite in size.
   c. the Earth is a sphere.
   d. the Moon orbits the Earth.
   e. Venus follows an orbit around the Sun rather than around the Earth.
2. Kepler's third law
   a. relates a planet's orbital period to the size of its orbit around the Sun.
   b. relates a body's mass to its gravitational attraction.
   c. allowed him to predict when eclipses occur.
   d. allowed him to measure the distance to nearby stars.
   e. showed that the Sun is much farther away than the Moon.
3. Which of the following did the models of Copernicus and Kepler have in common?
   a. Planets move in elliptical orbits.
   b. The inner planets move faster than the outer planets.
   c. The motions of the planets are uniform.
   d. All of the above

unit 13

Observing the Sky

Experiencing the astronomical changes that occur each night, or over several days or months, will give you a much deeper appreciation of the cosmos. This unit offers a number of projects for exploring the sky and connecting your observations to the models and theories discussed in earlier units.

There are many important aspects of the universe you can learn by simply watching the sky. You do not need fancy equipment. Just a star atlas or a computer with planetarium software can help you get oriented as you look at the sky. Best of all, stargazing is fun and connects you to cultural traditions reaching into prehistory as well as current scientific research. If you find astronomical observation rewarding, there are opportunities to engage in more serious pursuits, such as searching for new comets or monitoring variable stars. These possibilities are discussed further in Unit 31.

Background Pathways

13.1 LEARNING THE CONSTELLATIONS

One of the best ways to get started in your studies of the sky is to learn the constellations. All you need is a star chart, a dim flashlight, and a place that is dark and has an unobstructed view of the night sky. A foldout star chart is located in the back of this book. The star chart gives directions for how to hold it so that you can begin matching it to the sky for the date and time at which you are observing. The planetarium software program included with this book can provide star charts tailored to your location and time.

Start by determining which way is north, using a compass if necessary. Then try to locate a few of the brighter stars, matching them up with the chart. For example, a large asterism called the Summer Triangle spans three constellations. It consists of the three bright stars conspicuous in the summer evening: Deneb (in Cygnus, the Swan), Altair (in Aquila, the Eagle), and Vega (in Lyra, the Harp), shown in Figure 13.1. This will give you some sense of how big a piece of the sky the chart corresponds to. A photograph of this region of the sky is shown in Looking Up #4: Summer Triangle.

Next, try to identify a few star patterns. Focus at first on just some brighter ones. For example, if you live at midlatitude in the Northern Hemisphere (as in most of the United States, Canada, Europe, and Asia), the Big Dipper—the asterism that is part of the constellation Ursa Major—is a good group to start with because it is circumpolar for anyone living at a latitude north of about 35°N (see Looking Up #2: Ursa Major).

As you attempt to find and identify stars, your spread hand held at arm's length makes a useful scale. For most people, a fully spread hand at arm's length covers an

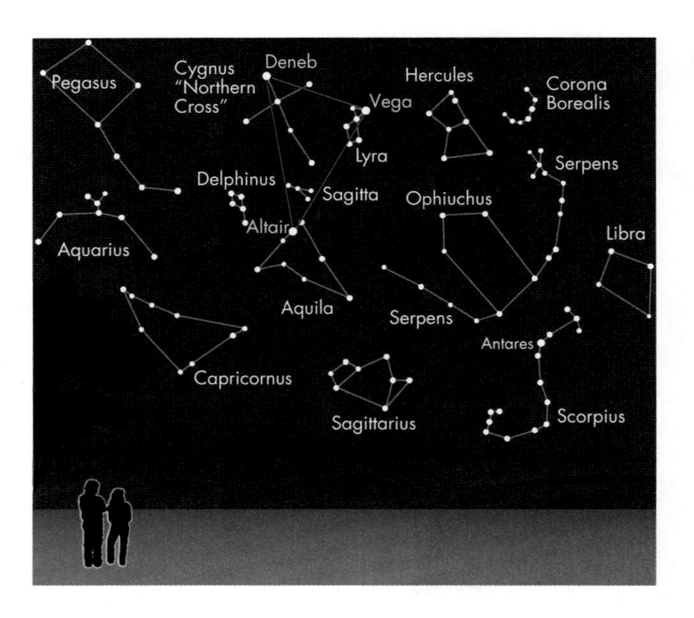

FIGURE 13.1
Dominating the night sky in July, August, and September are the three bright stars Vega, Altair, and Deneb, which form the Summer Triangle. This sketch shows almost half of the whole sky looking south (from mid-northern latitudes) at about 8 P.M. in early September.

angular size of about 20° of sky, or about the length of the Big Dipper from tip of handle to bowl, as shown in Figure 13.2. For smaller distances, you can use finger widths: held at arm's length, your thumbnail is about 2° wide, and the tip of your little finger is about 1° wide.

The Big Dipper not only is easy to spot, but is also an excellent signpost to other asterisms and stars. For example, the two stars at the end of its "bowl" away from the "handle" (see Figure 13.2) are called the "pointers" because they point, roughly, to the North Star, Polaris, about 30° or 1½ handspreads away. If you extend the arc formed by the stars in the handle of the Big Dipper, you will find a path that

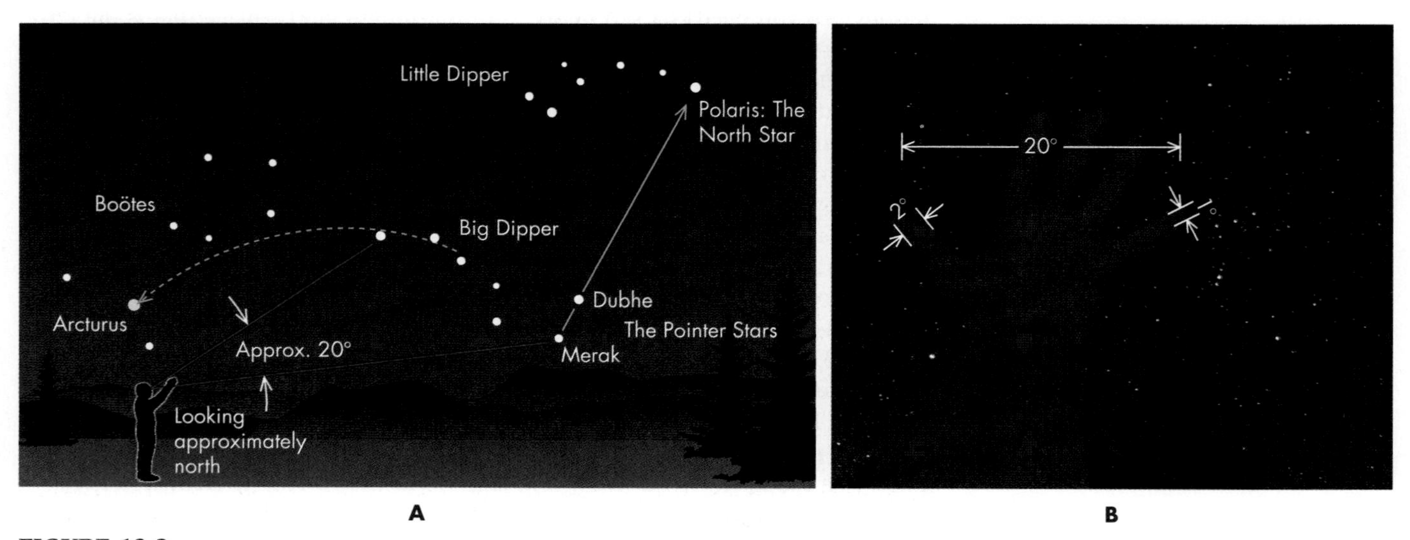

FIGURE 13.2
(A) The Big Dipper, part of the constellation Ursa Major, the Great Bear. A line through the two pointer stars points toward Polaris. The Big Dipper spans about 20° of the sky. The sky is shown approximately as it looks in mid-September at about 8 P.M. from mid-northern latitudes. (B) You can estimate angular separations on the sky using your hand stretched out at arm's length in front of you. Your handspread is about 20°, your thumb is about 2° wide, and the tip of your little finger is about 1° wide.

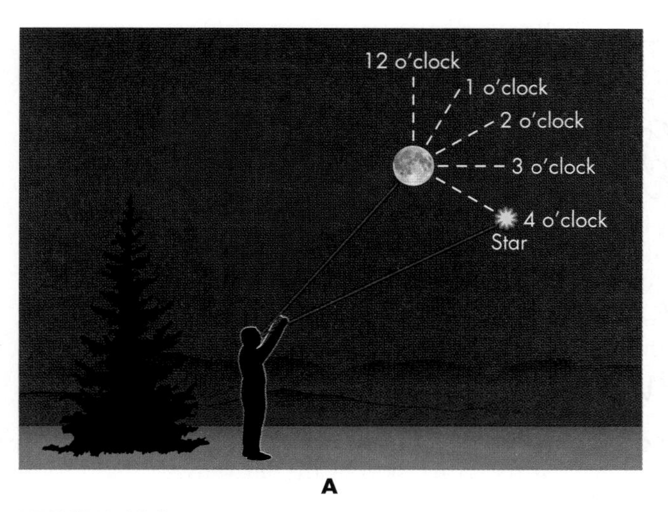

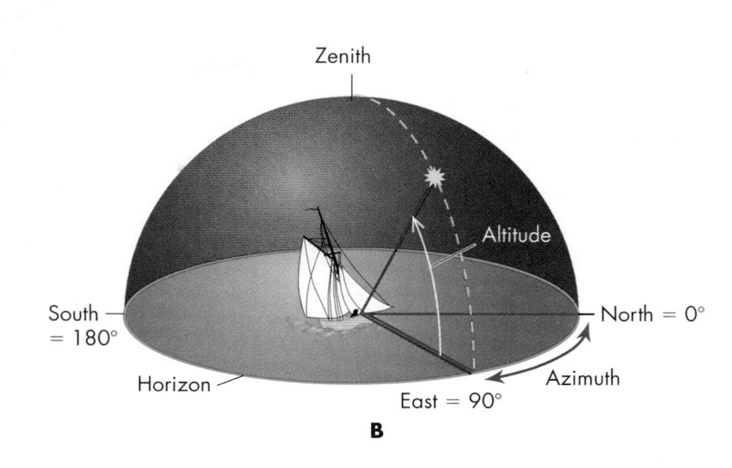

FIGURE 13.3
(A) Describing the location of stars by clock position. The star is half a handspread from the Moon and at the 4 o'clock position. (B) A star's position can be indicated by its altitude above the horizon and its azimuth measured eastward around the horizon from true north.

curves to the bright star Arcturus ("follow the arc to Arcturus"). Arcturus is also about 1½ handspreads away from the Big Dipper in the constellation Boötes (*boh-OH-teez*).

This scaling of sky distances with your hand makes it easy to point out stars to other people. For example, you can say that a star is half a handspread away from the Moon and at the 4 o'clock position, as illustrated in Figure 13.3A. A more general method for locating a star is to measure its altitude and azimuth, as shown in Figure 13.3B. The star's **altitude** is its angle above the horizon, while its **azimuth** is defined as the angle measured eastward from true north to the point on the horizon below the star.

Once you recognize a few constellations, you may find that learning the stories behind them will help you find and remember their shapes and locations. Star lore is part of virtually all cultures. The ancient Greeks, the Pawnee tribes of the American Midwest, and the Australian aborigines, for example, all created stories about the star groupings they saw in the sky. Because the star groupings do not alter their appearance except on time scales of tens of thousands of years, the night sky we see is essentially the same night sky that ancient peoples saw. Star lore can therefore link us to our ancestors in the remote past.

It has been suggested that many such stories were created as aids to memory, especially important when familiarity with the stars could be literally a matter of life or death to a farmer or a navigator. Scientists have even shown that baby birds learn to recognize star patterns and movements and use them to navigate safely—unguided by their parents—across thousands of miles of ocean to their winter homes. Perhaps we too have such instinctive faculties that help us learn the stars.

The native inhabitants of North America had a story about the Big Dipper. Its bowl represented a huge bear, and the handle represented three warriors in pursuit of the bear. They had wounded it, and it was bleeding. The red color of autumn leaves was said to be caused by the bear's blood dripping on them when the constellation lies low in the sky during the evening hours of the autumn months.

Stories are also told that connect many constellations. For example, in the winter sky you can see the Hunter, Orion (see Looking Up #6: Orion), and the maiden who, according to the myth, refused to fall in love with him. The story also involves Orion's hunting dogs (Canis Major and Canis Minor), a bull (Taurus), a rabbit (Lepus), the maiden's sisters (the Pleiades—a cluster of stars in the constellation Taurus—see Looking Up #5: Taurus), and a scorpion (Scorpius).

According to the myth, the king of the island Chios had a lovely daughter, Merope. Chios was filled with savage beasts, and to rid his kingdom of these dangerous animals, the king called on Orion to kill the beasts and make his kingdom safe. When the task was done, Orion met Merope and made unwelcome advances. In punishment, he was blinded by the king, but after doing penance, he had his sight restored. After reaching an old age, however, Orion one day stepped on a scorpion, which stung and killed him. On his death, the gods placed him in the sky with his faithful dogs (one of whom chases Lepus, the rabbit), forever attacking the wild bull, Taurus. Beyond the bull, Merope and her sisters (the Pleiades) run from the hunter, who pursues them each night across the sky. The scorpion was also placed in the sky, but on the other side of the heavens so that Orion would never again be threatened by it. (Orion is visible in the evening only in the winter, whereas Scorpius is visible in the evening only in the summer.) Like many sky myths, the Orion story has several versions, but the one described here fits together many of the astronomical references.

There are many other stories about constellations, but the one just described may give you some sense of those that have been handed down over thousands of years of written and oral history. Explore these stories as you learn the constellations: They will help you remember the relative locations in the sky of the various constellations.

13.2 MOTIONS OF THE STARS

Many people are surprised when they are told that the stars rise and set and move across the sky in much the same way that the Sun does. However, it is easy to show that stars move.

Use a tripod or a stick that you can poke into the ground so it will stand upright. Get a second straight stick that you can tape or affix to the upright in some manner. A ruler taped to a camera tripod would be ideal. Find a bright star and sight along the top stick toward the star, as sketched in Figure 13.4. If you now

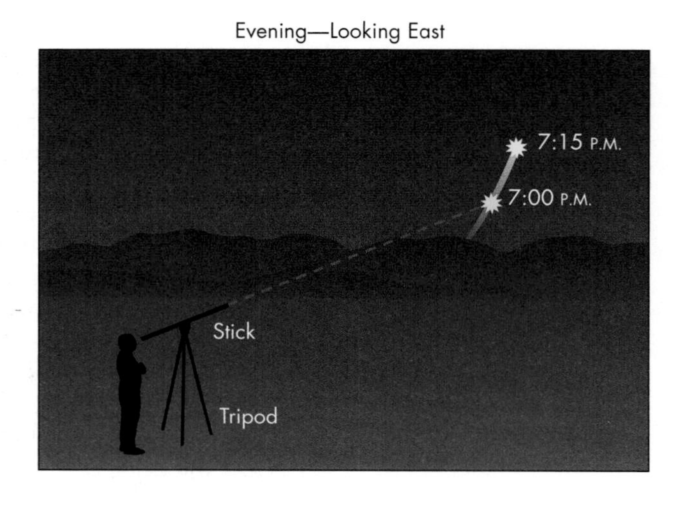

FIGURE 13.4
A sketch illustrating how to observe the motion of the stars across the sky sighting along a stick.

FIGURE 13.5
A time exposure showing how Polaris remains essentially fixed while the sky appears to pivot around it.

wait about fifteen minutes and again sight along the stick, you will see that it no longer points to the star.

You can do this experiment indoors if you can see a star through a window. Set up a tripod or other object close to the window to help you establish a sightline toward the star. Sighting over the top of your tripod, place a mark on the glass where the star appears to be with a grease pencil or piece of tape. Again, after about 15 minutes, the star's motion will be clearly visible.

The same motion can be demonstrated in just a few moments with any small telescope mounted on a tripod. If the telescope is set to point at a star, the star will steadily drift across the field of view. The image seen through a telescope may be reversed, but you can determine what direction the star is moving by seeing which way the telescope has to be adjusted to bring the star back to the center of the field of view.

This experiment will not work if you look for motion of the North Star! The North Star, Polaris, lies less than 1° from the north celestial pole. Because of its position there, it is the only moderately bright star in the northern sky that shows no obvious motion during the night. Its relatively fixed position is illustrated by the time exposure in Figure 13.5, showing the stars' apparent rotation around the north celestial pole.

Because Polaris lies nearly above the Earth's North Pole, it is useful in orienting yourself to compass directions. Polaris marks the end of the handle of the Little Dipper, an asterism that is part of the constellation Ursa Minor, the Little Bear. It can be located easily if you can find the Big Dipper, as illustrated in Figure 13.2 and Looking Up #1: Northern Circumpolar Constellations. The counterclockwise motion of the stars around the celestial pole can be observed by noting the "o'clock" position of one of the bright stars around Polaris (Figure 13.3). If you come back in about 2 hours, the position will be about an hour "earlier" on the imaginary clock.

If you observe the motion of a star rising in the east, how does the angle between its path and the horizon depend on your latitude?

In all of these experiments the stars are moving to the west of their original position. What is meant by "west" on the sky often appears to be quite different from what we might think of as west on the ground. A star's westward motion may have it moving up, down, or at various angles to the horizon. This is particularly true in Figure 13.5, where westward motion is to the left for stars above Polaris and to the right for stars below it.

13.3 MOTION OF THE SUN

Just as for other stars, the rising and setting of the Sun teaches us about the Earth's daily rotation. However, by observing the Sun for about a week we can also detect the effects of the Earth's tilted axis of rotation. You can observe the Sun's shift north and south on the celestial sphere based on its location at sunset. If you observe the Sun setting from the same location for even a few nights, you can begin to trace the patterns that led ancient peoples to build remarkable structures like Stonehenge.

Find a spot, perhaps on a hill or out a window facing west, where you can see the western horizon in the evening. Take a photograph or make a sketch of the western horizon, noting hills, buildings, or trees that might serve as reference marks. Use your hand and fingers to estimate the angular size of features on the horizon, as discussed in Section 13.1. From your viewing spot, watch the sunset and mark on your sketch where the Sun goes down, noting the date and time.

Although the Sun is often dimmed enough at sunset for safe viewing, it is not always so. The only way to be certain it is safe photographing the Sun at sunset is to use a camera that shows you the image of the Sun indirectly on a display screen, such as most digital cameras.

Make observations for as many consecutive days as you can. You will discover that the Sun's position changes by an obvious amount in a single day near the equinoxes, but much more slowly near the solstices (Unit 6). Observe how the times of sunset change as its position changes. If you enjoy photography, you might try taking photographs of the sunset such as those shown in Figure 13.6. It is extremely important, however, never to look directly at the Sun, especially through any kind of magnifying lenses, because doing so can damage your eyes.

FIGURE 13.6
A pair of photographs taken eight days apart from the same location. The pictures were taken just before sunset, close to the date of the autumnal equinox. The sunset position changed by more than 4° during this time—just about the size of the Sun's apparent diameter each day. The width of the outstretched thumb in the bottom picture helps to indicate a scale of about 2°.

13.4 MOTIONS OF THE MOON AND PLANETS

Determine from Table 13.1 when the Moon is a few days past new so it will be visible in the early evening. Go outside shortly after sunset and look for the Moon in the west near where the Sun went down. If you have a clear, dark sky, you may even be able to see the dark side of the Moon, lit up by reflected light from the Earth as shown in Figure 13.7.

To keep your chart clean for future use, you may want to apply removable stickers or write lightly with a pencil.

On the foldout star chart in the back of the book, you will find a list of dates along the top. At 8 P.M. the stars below this date lie along the meridian—the north–south line running overhead. The stars below the next date to the left are overhead an hour later, and to the right an hour earlier. Once the sky is dark, you will be able to see most of the stars six hours to the right and six hours to the left of the ones that are overhead. As the sky darkens, locate the brighter stars near the Moon, and then mark on the chart where the Moon is with respect to those stars. Finally, sketch the Moon's shape.

Q When the Moon is in its crescent phase, what phase would the Earth appear to be in for someone on the Moon? How bright would "earthlight" be on the Moon, compared to moonlight on Earth?

Repeat this process for the next four or five nights. The Moon will set about 50 minutes later each evening; note these times and adjust your observing time accordingly. After watching for a few nights, mark out the Moon's path on the star chart. Ideally, you might want to follow the Moon's track for about two weeks, although as the Moon reaches its third quarter (three weeks after new moon), you will have to stay up late because the third-quarter moon does not rise until about midnight. Early risers who get up before dawn can watch the Moon become a waning crescent as it approaches the new phase.

Morning and Evening Stars

You can also use a star chart to study the motion of the planets. If it is visible, Venus is a good choice because it is bright and moves rapidly across the sky. It is often called the **Evening Star** because it stands out as the brightest "star" in the evening sky. However, Venus spends half its time as the **Morning Star** and is sometimes too close to the Sun to be seen easily, so it is not always a convenient target for observation.

The locations where Venus and other planets are visible each month are given in Table 13.1. The table indicates whether the planets are more easily observed in the evening sky (blue) or morning sky (green), and when they are close to the Sun. Mercury is always quite challenging to see, requiring a clear view when it is just beginning to get dark, near to where the Sun has set or is about to rise. The dates when it reaches its greatest angle from the Sun are indicated in the table, and you will have the best chance of finding it within a week of those dates.

FIGURE 13.7
This photograph shows the waxing crescent Moon a few days after the New Moon. The bright crescent portion is lit by the Sun, but the dark side is also visible, illuminated by light reflected off the Earth. The light reflected from the Earth is similar to the dim illumination we receive on Earth from moonlight.

If you watch the planets long enough, you can see how closely they follow the ecliptic—shown in the star chart by a curving line that crosses both sides of the equator. Because the outer planets move relatively slowly across the sky, you should space out your observations, perhaps marking positions once a week rather than every night. You will also be able to see whether they are making direct or retrograde motion against the background stars. Retrograde motion occurs for a month or more around the time the planets are closest to Earth in their orbits. This occurs for the outer planets for the months where the constellation is marked in black in Table 13.1.

The interval between these periods of retrograde motion (or any other successive planetary configurations such as opposition or conjunction) is called the **synodic period.** The synodic period differs from the planet's orbital period because both the Earth and the other planets move around the Sun. Therefore the interval between oppositions is neither an Earth year nor the other planet's orbital period. For example, the Earth takes over two years to catch up to and overtake Mars after an opposition. The Earth overtakes the slower-moving, more distant planets more quickly, and the interval between oppositions is closer to one year. For example, the Martian synodic period is about 780 days, whereas the Saturnian synodic period is 378 days.

TABLE 13.1 Moon and Planet Finder

The dates of the New Moon and the locations of the five bright planets are given for each month. The table is continued on the foldout star chart in the back of the book. The abbreviation for the constellation where the planet can be found is given, listed in green when the planet is primarily in the morning sky and blue for the evening sky. The constellation name is in black when the planet is in opposition to the Sun, indicating that it rises at sunset and sets at sunrise, and "Sun" is listed when the planet is nearly in line with the Sun (conjunction). For Venus, the month when it is at its greatest elongation from the Sun is marked by an asterisk. Mercury is always fairly close to the Sun and can only be seen shortly after sunset or shortly before sunrise. The date of its greatest elongation is given, and the best opportunity for seeing it is generally within about one week of this date. The Moon will lie close to the ecliptic at the Sun's position on the given date, then shifts about 13° to the east each subsequent day.

		New Moon	Mercury	Venus	Mars	Jupiter	Saturn
2008	Jan	8	22	Sgr	Tau	Sgr	Leo
	Feb	7	Sun	Cap	Tau	Sgr	Leo
	Mar	7	3	Aqr	Gem	Sgr	Leo
	Apr	6	Sun	Psc	Gem	Sgr	Leo
	May	5	14	Tau	Cnc	Sgr	Leo
	Jun	3	Sun	Sun	Leo	Sgr	Leo
	Jul	3	1	Cnc	Leo	Sgr	Leo
	Aug	1,30	Sun	Vir	Vir	Sgr	Leo
	Sep	29	11	Vir	Vir	Sgr	Sun
	Oct	28	22	Sco	Vir	Sgr	Leo
	Nov	27	Sun	Sgr	Sco	Sgr	Leo
	Dec	27	Sun	Cap	Sun	Sgr	Leo
2009	Jan	26	4	Sgr*	Sgr	Sun	Leo
	Feb	25	13	Cap	Cap	Cap	Leo
	Mar	26	Sun	Sun	Aqr	Cap	Leo
	Apr	25	26	Psc	Psc	Cap	Leo
	May	24	Sun	Tau	Psc	Cap	Leo
	Jun	22	13	Gem*	Ari	Cap	Leo
	Jul	22	Sun	Cnc	Tau	Cap	Leo
	Aug	20	24	Vir	Tau	Cap	Leo
	Sep	18	Sun	Vir	Gem	Cap	Sun
	Oct	18	6	Sco	Cnc	Cap	Vir
	Nov	16	Sun	Sgr	Cnc	Cap	Vir
	Dec	16	18	Cap	Leo	Cap	Vir
2010	Jan	15	27	Sun	Cnc	Aqr	Vir

13.5 A SUNDIAL: ORBITAL EFFECTS ON THE DAY

The preceding projects reveal many basic features of the sky and planetary motion. A surprising amount of even more sophisticated information about the nature of Earth's orbit can be learned from careful observations of a sundial over the course of a year.

A sundial can be as simple as a flagpole or any other fixed tall pole where you can mark the shadow cast by the top of the pole. If we were to measure the length of the solar day from noon to noon with a stopwatch, we would discover that it is in general *not* exactly 24 hours. We can do this with a flagpole by marking when the shadow lies exactly along a north–south line. The time between successive noons varies by as much as half a minute at different times during the year.

A

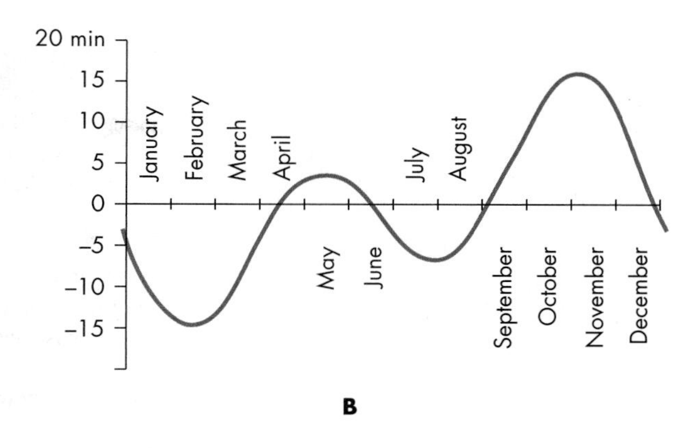

FIGURE 13.8
(A) A series of photographs of the Sun at noon (clock time) throughout the year show that the Sun is sometimes a little east and sometimes a little west of the meridian. The shape it traces on the sky is called an "analemma." (B) The equation of time is the correction that must be applied to sundial time to determine mean solar time.

Our clocks do not change speed during the year, of course, but instead use the average day length during a year. That average day length is called the **mean solar day,** which has, by definition, 24 hours of clock time. Therefore, even if you lived on the central longitude of your time zone (Unit 7), the Sun would not normally lie along the meridian when your clock indicates noon. This is illustrated in Figure 13.8A, which shows photographs that were taken at the same time on many days over the course of a year. The Sun moves north and south because of the Earth's tilted axis, but it also makes a figure-eight pattern, called an **analemma.**

The difference in length between the mean solar day and the true solar day accumulates to a difference of over 16 minutes between clock time and time based on the position of the Sun at different times of year. This difference is described by the **equation of time,** which is shown graphically in Figure 13.8B. The equation of time gives the correction needed on a sundial if it is to give the same time as your watch. The difference may seem just a curiosity today; but for a navigator using the Sun to determine a ship's longitude, it could cause an error of more than 300 km (200 miles) if no corrections were made.

The variation in the solar day arises because of two effects: The Earth's orbit is not circular, and the Earth's axis is tilted with respect to the orbit. Both of these have similarly sized effects on the length of the day; but they follow different patterns that are offset in time, which makes the equation of time quite complicated.

> **Q** What differences would you see between a sundial in the Northern Hemisphere and one in the Southern Hemisphere?

The effect of the Earth's elliptical orbit on the day is illustrated in Figure 13.9. In January, when the Earth is closest to the Sun, the Earth sweeps through a larger angle in its orbit (Kepler's second law; Unit 12.2), and the Sun appears to shift farther eastward on the celestial sphere than it does on average. This in turn means the Earth has to rotate a bit farther to face the Sun again, with the interval between successive noons becoming about 10 seconds longer than average. This effect adds in the same direction for about half the year, and consequently solar time falls further and further behind clock time. The reverse applies in the months of April through September, when the Earth is farther from the Sun. During this half of the year, solar time gets ahead of clock time.

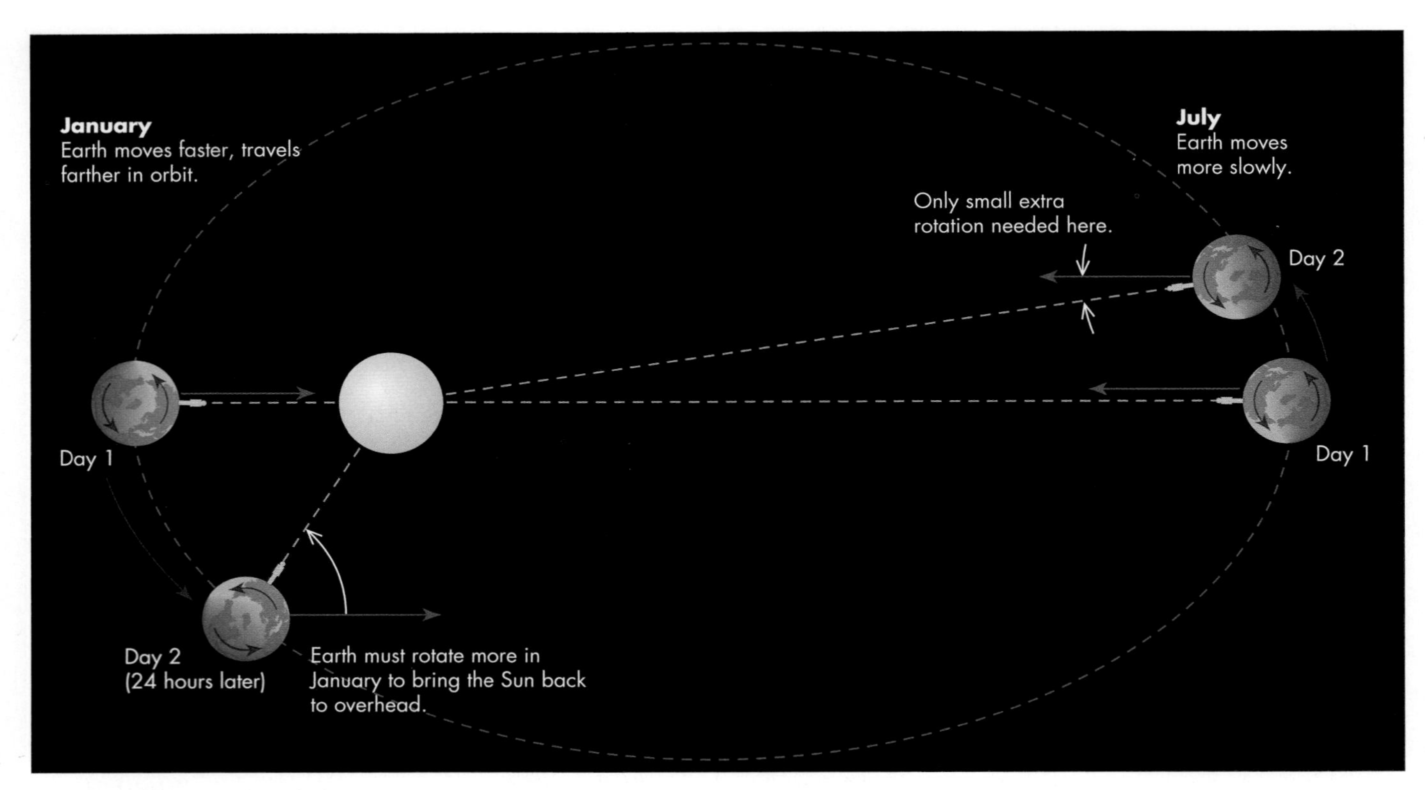

FIGURE 13.9
As the Earth moves around the Sun, its orbital speed changes as a result of Kepler's second law of motion. For example, the Earth moves faster in January when it is near the Sun than in July when it is far from the Sun. Therefore in 24 hours the Earth moves farther along its orbit in January than in July. As a result, the Earth must turn slightly more in January to bring the Sun back to overhead. This makes the interval between successive noons longer in January than in July and means they are not exactly 24 hours. For that reason, time is kept using a "mean Sun" that moves across the sky at the real Sun's average rate. (The extremes of the Earth's distance from the Sun have been greatly exaggerated in this figure.)

The effect of the Earth's tilted axis on the day is easiest to understand if we think of how the Sun shifts position on the celestial sphere. Suppose the Earth's orbit were perfectly circular and the Sun shifted by the same distance along the ecliptic each day. In this case the Sun would make its most rapid progress in an eastward direction among the stars when it was moving due east rather than when part of its motion was northward or southward. (This may seem more obvious when we speak of a car driving at a fixed speed; clearly the more directly eastward it is driven, the more progress to the east it will make in one day.) Because the ecliptic is tilted, the Sun moves due east among the stars, parallel to the celestial equator, only during the two solstices. It makes the least eastward progress when its motion north or south is largest at the equinoxes. Thus, the Earth has to turn farther when the Sun's apparent position has shifted farther eastward, and this effect can lengthen the day by up to about 20 seconds at the solstices, shortening it by the same amount near the equinoxes.

Because both effects reach maxima near the end of the year (winter solstice and closest approach to the Sun), the longest day of the year is actually a couple of days after the winter solstice—about 30 seconds longer than average. This is not the same as the length of daylight hours, which are shortest (in the Northern Hemisphere) at this time of year as discussed in Unit 7.

The elliptical orbit of the Earth and its changing speed also explains why the times between equinoxes and solstices are not the same throughout the year. The average dates for the start of each season are March 20, June 21, September 22 and December 21. This gives 93 days each for northern spring and summer, but just 90 and 89 days for autumn and winter. This difference was a puzzle to ancient astronomers. It was finally understood when Kepler (Unit 12) discovered that planets travel on elliptical orbits with a varying speed.

KEY TERMS

altitude, 109
analemma, 115
azimuth, 109
equation of time, 115
Evening Star, 113
mean solar day, 115
Morning Star, 113
synodic period, 113

QUESTIONS FOR REVIEW

1. How can you use your hand to measure angles between objects in the sky?
2. What methods can you use for describing positions in the sky?
3. How can you demonstrate that stars change their position throughout the night?
4. Of the Sun, Moon, and planets, which move most quickly, and which move most slowly, relative to the stars?
5. Which planets appear as "Morning" or "Evening" stars?
6. Why does sundial time differ from clock time?

PROBLEMS

1. Use your hand and fingers to estimate the angular size of at least three constellations. Sketch the constellations to scale with your measurements.
2. Sailors have "handy" rules for estimating the time until sunset. Approximately how many minutes before sunset is the Sun "one finger" above the horizon at the equator? At 45° N or S latitude?
3. It is said in the text that the Moon sets about 50 minutes later each evening. Given that the lunar cycle of phases is 29.53 days, calculate a more accurate value for this time.
4. Calculate a precise value for how many degrees the Moon moves each day, keeping in mind that to complete one lunar month of 29.53 days, the Moon must once again align with the Sun, which also has shifted along the ecliptic during that time.
5. Even at greatest elongation, Mercury will not always be visible high above the horizon after sunset or before sunrise. This is primarily because the angle between the ecliptic and the horizon is shallow sometimes during the year. In that case, Mercury is far from the Sun, but still close to the horizon. When the ecliptic is more nearly perpendicular to the horizon in your viewing location, Mercury will have a higher altitude. By examining the foldout star chart, estimate what times of year the ecliptic will be at the best angle for you to observe Mercury in the evening sky. Is there an upcoming greatest elongation that is favorable for you?
6. If you mark the position of the shadow cast by the top of a flagpole at the same clock time every day throughout the year, you will find that the marks trace out a figure-eight pattern. Explain why this happens by referring to the equation of time.

TEST YOURSELF

1. Shortly after sunset, you see a bright star above the eastern horizon. Which of the following statements is true of where you might see this star at a later time?
   a. Within a few hours the star sets below the eastern horizon.
   b. Within twelve hours, the star will have set in the west.
   c. The star remains about the same place all night.
   d. The next evening, the star will be in the west.
2. Suppose you see Venus at greatest elongation (farthest from the Sun) in the evening tonight. About how long will you have to wait to see Venus at greatest elongation in the evening again?
   a. Venus's orbital period (225 days)
   b. Venus's rotation period (243 days)
   c. Earth's orbital period (365 days)
   d. Venus's synodic period (584 days)
3. The equation of time describes differences between clock time and sundial time caused by
   a. wobbling of the Earth's axis.
   b. the tilt of the Earth's axis relative to its orbit.
   c. tugs on Earth's orbit by other planets.
   d. changes in Earth's speed as it orbits the Sun.
   e. Both (b) and (d) are true.

Part Two

The content of some Units will be enhanced if you have previously studied some earlier Units in this textbook. These Background Pathways are listed below each Unit image.

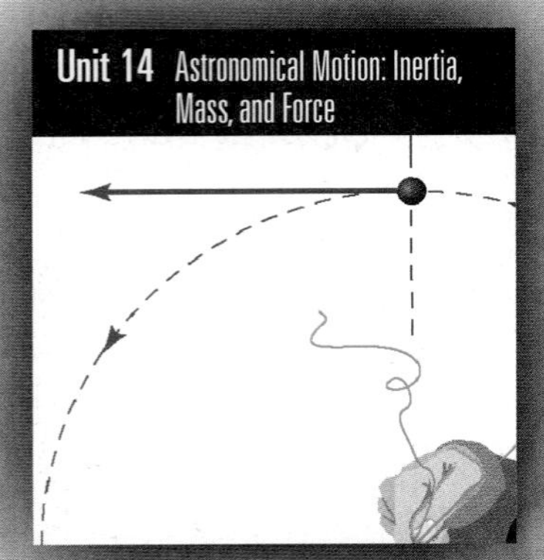

Background Pathways: Unit 12

Background Pathways: Unit 14

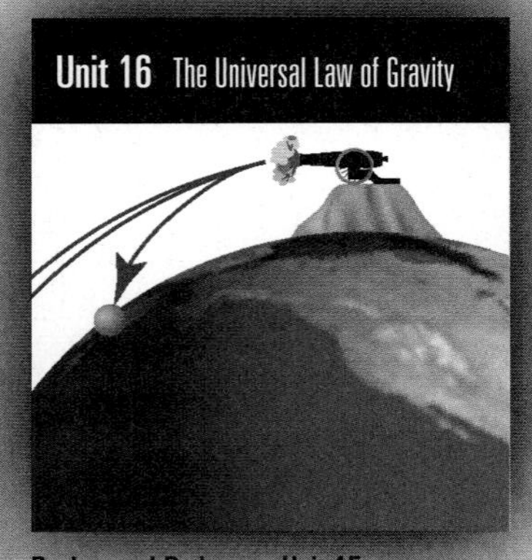

Background Pathways: Unit 15

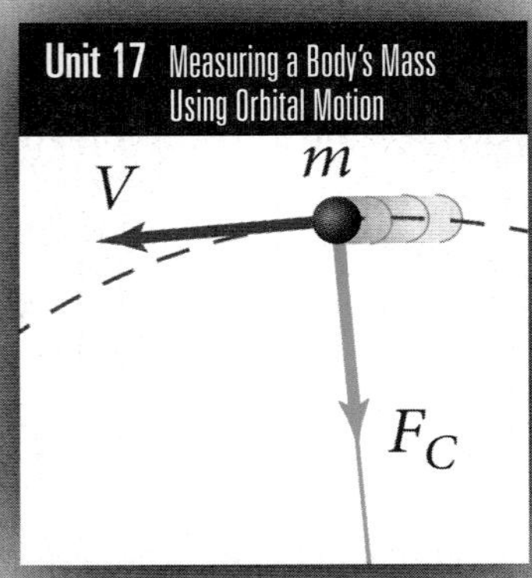

Background Pathways: Unit 16

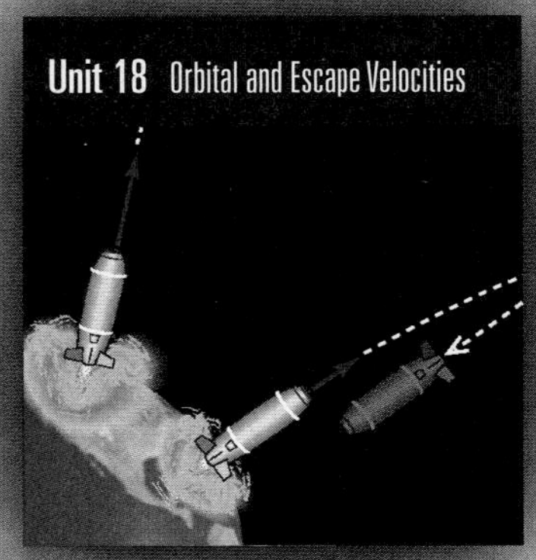

Background Pathways: Unit 16

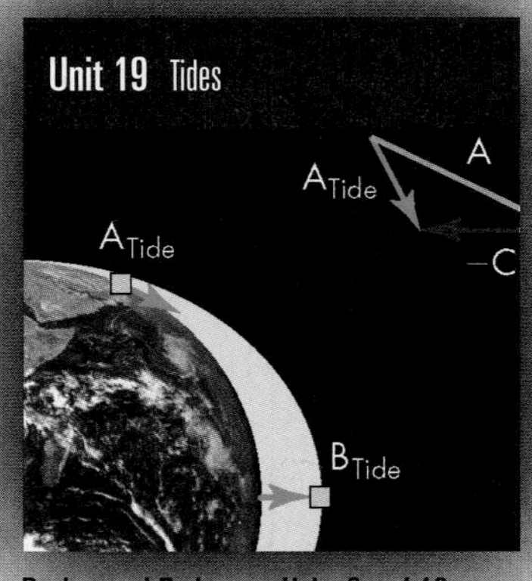

Background Pathways: Units 8 and 16

Background Pathways: Unit 16

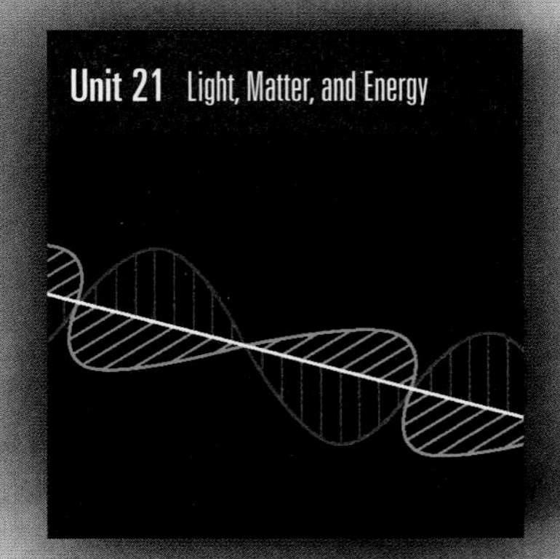

Background Pathways: Unit 4

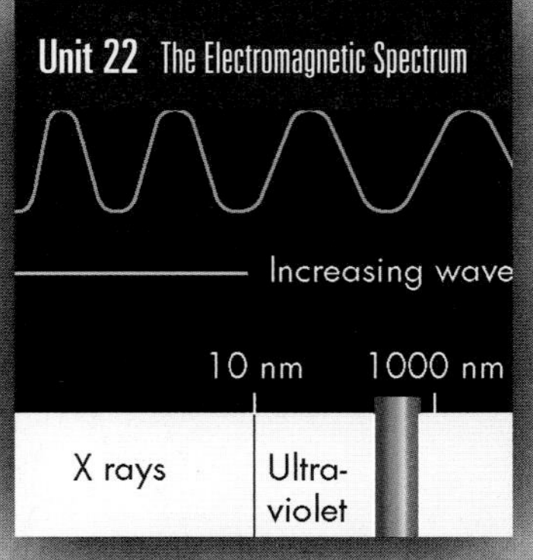

Background Pathways: Unit 21

Probing Matter, Light, and Their Interactions

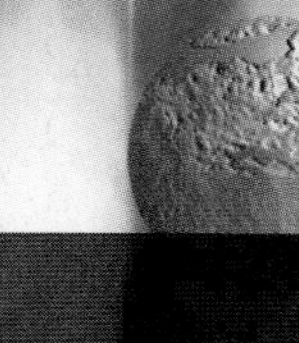

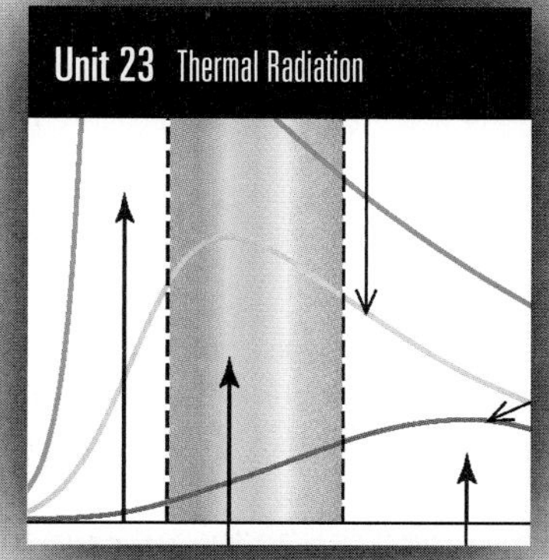

Background Pathways: Unit 22

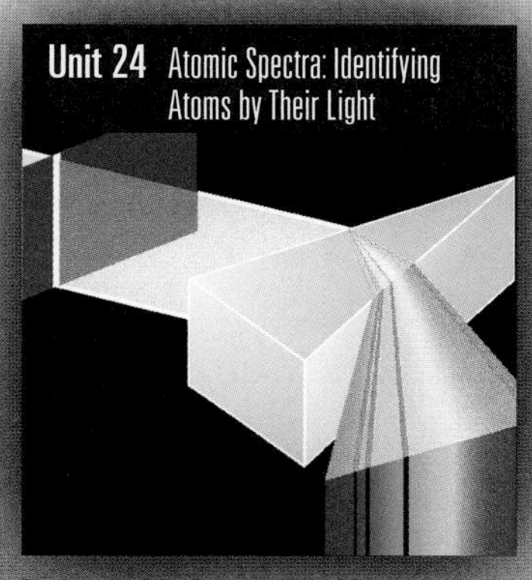

Background Pathways: Unit 23

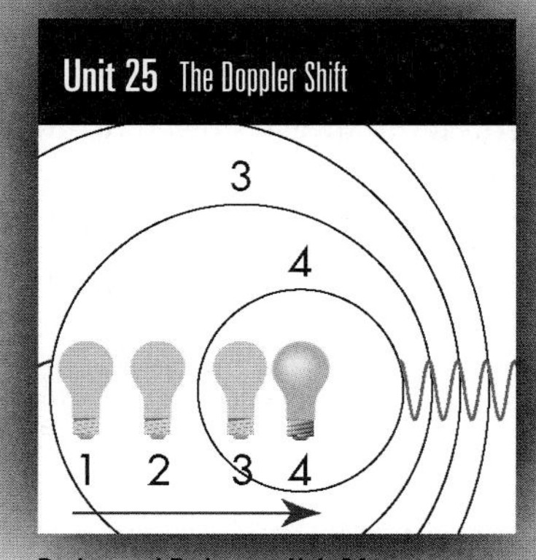

Background Pathways: Unit 24

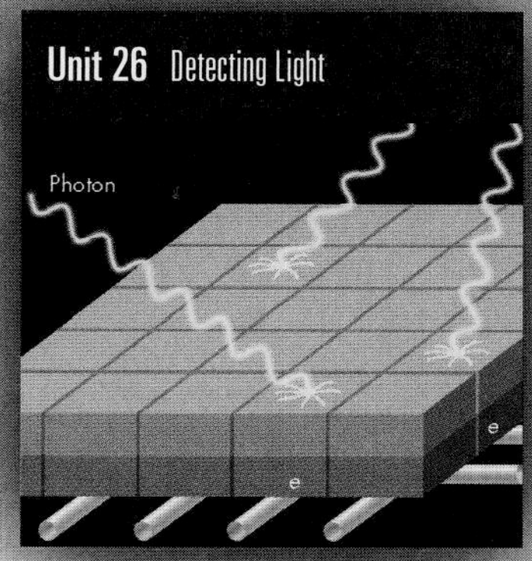

Background Pathways: Unit 22

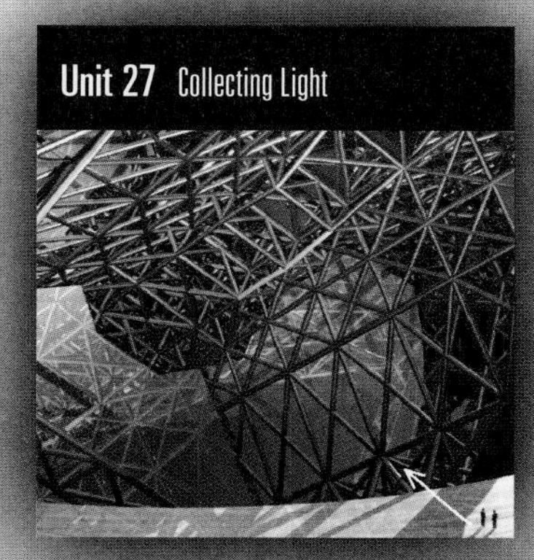

Background Pathways: Unit 26

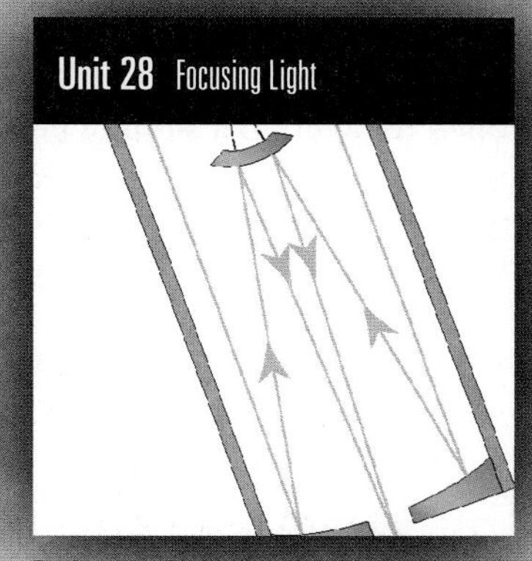

Background Pathways: Unit 26

Background Pathways: Unit 26

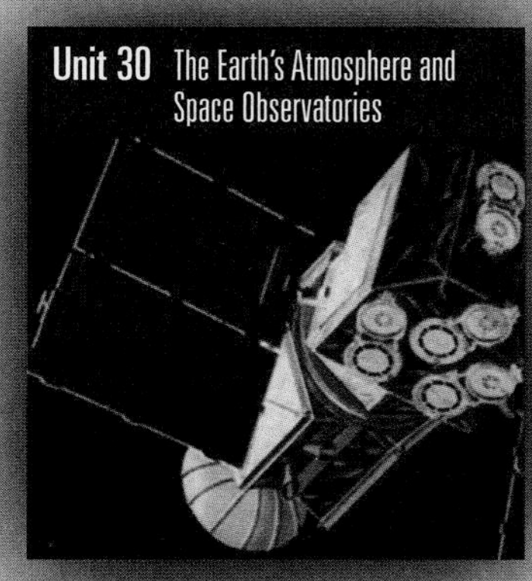

Background Pathways: Unit 22

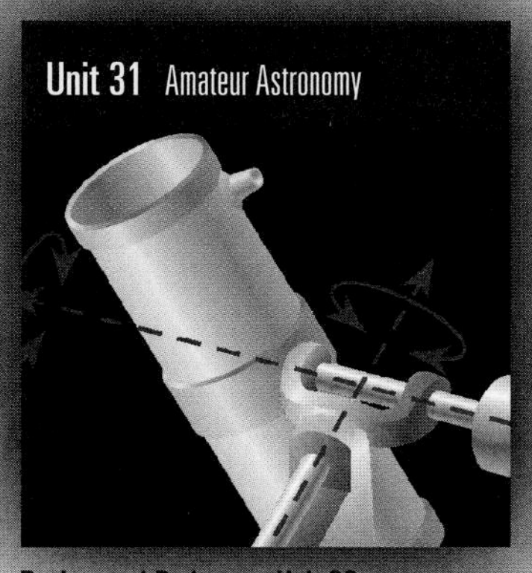

Background Pathways: Unit 28

unit

14

Astronomical Motion: Inertia, Mass, and Force

Background Pathways
Unit 12: The Beginnings of Modern Astronomy 99

Astronomers of antiquity did not make the connection between gravity and astronomical motion that we recognize today. People puzzled over why, if the Earth moved, they did not simply fly off into space; and they were also mystified about what kept the planets moving in their orbits.

The solutions to these mysteries began with a series of careful experiments conducted by Galileo Galilei in the early 1600s. Apart from his famous—and perhaps fictitious—demonstration of weights dropped from the Leaning Tower of Pisa, Galileo experimented with projectiles and with balls rolling down planks. The behavior of balls rolling down planks sounds far removed from the behavior of planets, but these basic experiments led him to recognize several properties of motion. A new understanding of forces and motion was essential to make Copernicus's heliocentric model of the Solar System plausible.

14.1 INERTIA AND MASS

Central to Galileo's laws of motion is the concept of **inertia.** Inertia is the tendency of a body at rest to remain at rest and of a body in motion to keep moving in a straight line at a constant speed. Galileo's contemporary Johannes Kepler introduced the term, but Galileo demonstrated it by real experiment.

In one such experiment, Galileo rolled a ball down a sloping board repeatedly and noticed that it always sped up as it rolled down the slope (Figure 14.1). He next rolled the ball up a sloping board and noticed that it always slowed down as it approached the top. He hypothesized that if a ball rolled on a flat surface and no forces—such as friction—acted on it, its speed would neither increase nor decrease but remain constant. That is, in the absence of forces, inertia keeps an object already in motion moving at a fixed speed.

Inertia is familiar to all of us in everyday life. Apply the brakes of your car suddenly, and the inertia of the bag of groceries beside you keeps the bag moving forward at its previous speed until it hits the dashboard or spills onto the floor. We commonly think about the amount of inertia an object has in terms of how heavy it is, but our senses can be fooled—your own body or an object you are carrying may feel lighter underwater or heavier on an amusement park ride.

In scientific terms we measure inertia by an object's **mass.** Mass can be described as the amount of matter an object contains. It is generally measured in kilograms.

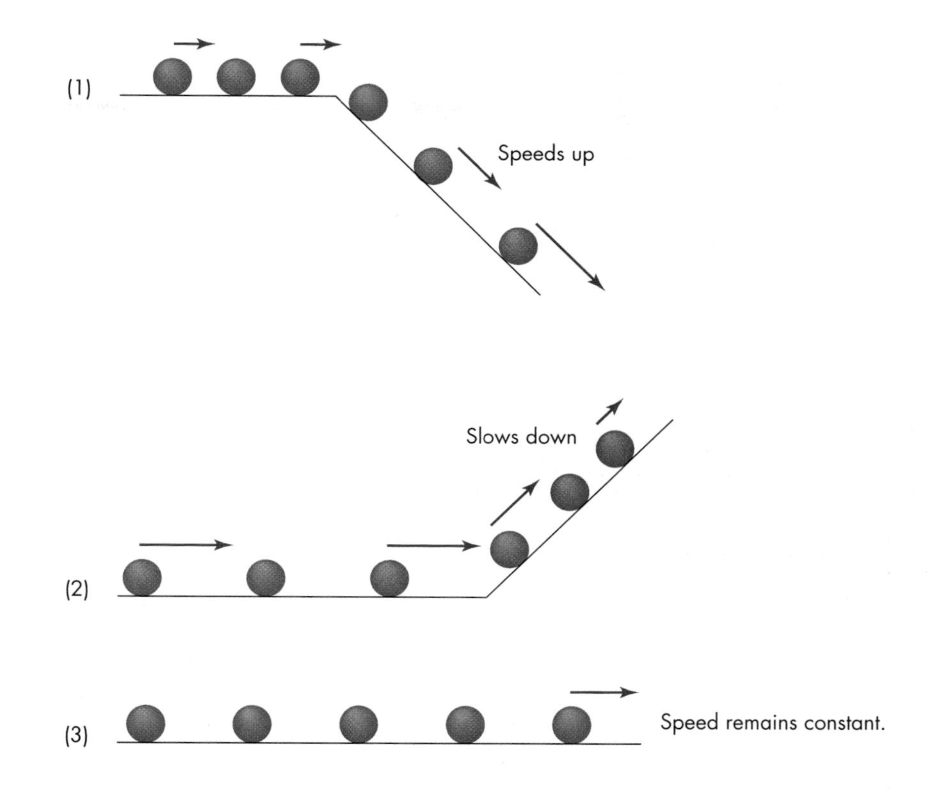

FIGURE 14.1
A ball rolling down a slope speeds up. A ball rolling up a slope slows down. A ball rolling on a flat surface rolls at a constant speed if no forces (including friction) act on it.

Q Suppose you were in an elevator standing on a bathroom scale, and the cable broke, allowing the elevator to fall. What would your weight register on the scale? Why?

One kilogram—abbreviated 1 kg—equals 1000 grams. For example, a liter (1.1 quarts) of water has a mass of 1 kg. Different substances may have the same mass in a larger or smaller volume—for example, about 1/3 liter of rock has a mass of 1 kg, and about 800 liters of air have this mass. The kilogram standard is a cylinder of an alloy of platinum, copies of which are kept by governments around the world. All other measurements of mass are made relative to these official standards.

It is important to remember that mass is not the same as **weight.** Because an object's mass is the amount of matter in the object, its mass in kilograms is a fixed quantity. An object's weight, however, is a measure of the net force on it. Your weight on the Moon will be different from your weight on Earth, because the strength of the Moon's gravitational pull on you is different. Your weight can also vary on Earth because of other forces acting on you. For example, the buoyancy of water may make you feel lighter or even weightless. In an elevator, as it first starts rising, you may feel heavier—and in fact if you took a bathroom scale into the elevator, you would discover that your weight does increase momentarily. On the other hand, when an elevator starts downward, you will feel momentarily lighter, and in an orbiting space capsule, astronauts feel weightless. We experience changes in weight as a result of the gravitational force on us, other forces acting on us, or the way our surroundings move; but no matter where we are, we have the same mass.

Some scientists prefer to say that weight refers only to the gravitational pull on you, and other effects (like buoyancy or orbiting in a spacecraft) change only your apparent weight.

14.2 THE LAW OF INERTIA

From his experiments on the manner in which bodies move and fall, Galileo deduced the first correct "laws of motion." But a more complete understanding was achieved by another scientist, arguably the greatest of all time, who was born the

FIGURE 14.2
Isaac Newton (1642–1727).

year Galileo died. Isaac **Newton** (1642–1727) made astounding contributions to mathematics, physics, and astronomy. Moreover, Newton (Figure 14.2) pioneered the modern studies of motion, optics, and gravity. Many of these ideas were conceived when he was 23, forced to stay home from college because the plague was ravaging England. Newton was a fascinating individual. He was a deeply religious man and wrote prolifically on theological matters as well as science.

In his attempts to understand the motion of the Moon, Newton not only deduced the law of gravity; realizing that the mathematical methods he needed did not exist, he invented calculus! What is especially remarkable about Newton's work is that the discoveries he made in the seventeenth century form the basis for calculating the trajectories of spacecraft today.

Newton recognized the special importance of inertia and helped clarify various aspects of it. He described it in what is now called **Newton's first law of motion** (sometimes referred to simply as the *law of inertia*). The law can be stated as follows:

I. A body continues in a state of rest, or in uniform motion in a straight line at a constant speed, unless made to change that state by forces acting on it.

An important point here is that inertia causes an object to resist changes in either speed or direction. This is again exemplified by groceries on a car seat. If the car turns a corner at a constant speed, the grocery bag will slide on the seat and tip over. Its inertia keeps it going in the same direction as before unless you apply a force to it. To prevent the bag from continuing with its former speed and direction, you need to apply a force on it. Whether you are stopping or turning, you must reach over and hold the bag to keep it from falling over.

Q How is the term *inertia* used in everyday conversation? How similar is that meaning to the physical meaning?

Because the speed and direction of motion are both important, scientists use a quantity that incorporates both: **velocity**. A velocity might be written as 100 kilometers per hour (kph), or 60 mph, to the northeast. In space we have to define a velocity in three dimensions. A body's velocity changes if either its speed or direction changes. This lets us simplify Newton's first law:

I. A body maintains the same velocity unless forces act on it.

14.3 FORCES

FIGURE 14.3
Balanced forces lead to no change in velocity. (Note that gravity is also applying a downward force on the box, but this is balanced by the table's upward force.)

In effect, Newton's first law defines what a **force** is—anything that can cause a body to change velocity. In some situations forces may be applied to an object, yet there is no *net* force—that is, the combination of all forces acting on the object cancel out. For example, if a box is at rest and pushed equally by two opposing forces, the forces are balanced. Therefore the box experiences no net force and does not move (Figure 14.3).

Newton's first law may not sound impressive at first, but it carries an idea that is crucial in astronomy: If a body is changing speed or moving along a curved path, some net force must be acting on it.

Actually, Newton was preceded in stating this law by the seventeenth-century Dutch scientist Christian Huygens. However, Newton went on to develop additional physical laws and—more important for astronomy—showed how to apply them to the universe. For example, if we tie a mass to a string and swing it in a circle, Newton's first law tells us that the mass's inertia will carry it in a straight line if no forces are acting on it. What force, then, is acting on the circling mass? The force is the one exerted by the string, preventing the mass from moving in a straight line and keeping it moving in a circle. We can feel that force

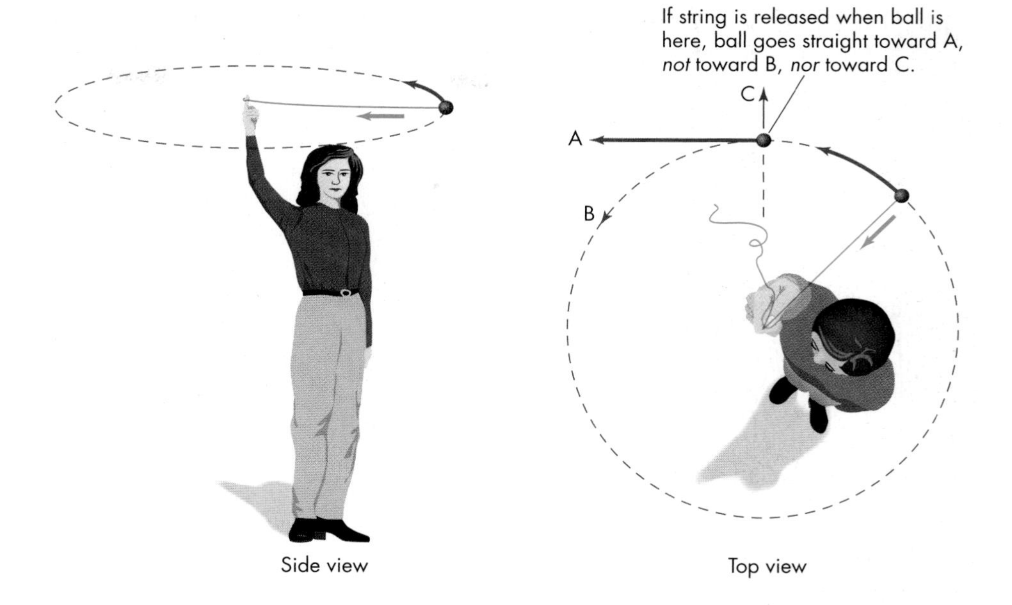

FIGURE 14.4
For a mass on a string to travel in a circle, a force (green arrow) must act along the string to overcome inertia. Without that force, inertia makes the mass move in a straight line.

as a tug on our hand from the string, and we can see its importance if we suddenly let go of the string. With the force no longer acting on it, the mass flies off in a straight line, demonstrating the first law, as illustrated in Figure 14.4.

If we watch the mass in the last example a little longer, though, we will notice that its path is not completely straight even after we let go of the string. If we release it parallel to the ground, we will notice that the trajectory of the mass begins curving down toward the ground. Newton's great insight was that the Earth must therefore be exerting a force on the object—the force of gravity.

We can translate this example to an astronomical setting and apply it to the orbit of the Moon around the Earth, the orbit of the Earth around the Sun, or the path of the Sun around the center of the Milky Way. Each of these paths is curved. Therefore a force must be acting in each case, a force that tugs on objects through space like an invisible string.

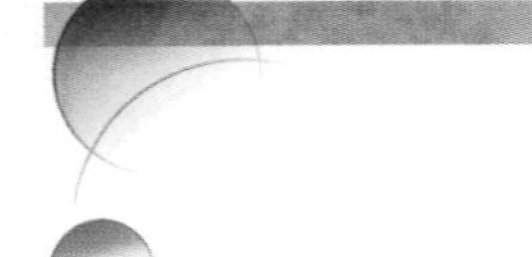

KEY TERMS

force, 122
inertia, 120
mass, 120
Newton, Isaac, 122
Newton's first law of motion, 122
velocity, 122
weight, 121

QUESTIONS FOR REVIEW

1. What is meant by *inertia?*
2. What is the difference between mass and weight?
3. Why does an object sliding on the floor come to a stop? Does this violate the law of inertia?
4. What does Newton's first law of motion tell you about the difference between motion in a straight line and motion along a curve?
5. When a ball slows down while it is rolling up a slope, what force is changing the ball's velocity?
6. Why is it necessary to specify direction when speaking of velocity or force, but not when considering other quantities such as mass?

PROBLEMS

1. Which do you think has more inertia—a small inflated balloon or a bowling ball? If each were moving toward you at 1 meter per second (1 m/sec), which would be easier to catch? Why?
2. Which do you think has more momentum, a bowling ball moving at 1 m/sec on the Earth, or a bowling ball moving at 1 m/sec on the Moon? Is there any difference?
3. In some amusement park rides, you are spun in a cylinder and pressed against the wall as a result of the spin. People sometimes describe that effect as being due to "centrifugal force." What is really holding you against the wall of the spinning cylinder? Drawing a sketch and using Newton's first law may help you answer this question.
4. A cinder block can be weightless in space. Would you be hurt if you kicked it with your bare foot while it was weightless in space? Explain your answer using Newton's first law.
5. Consider the situation shown in Figure 14.4. If the speed of the ball were higher, would this require a greater or lesser force applied by the string?
6. Forces in the MKS system are given in units called "newtons," where 1 newton acting for 1 second on a 1-kilogram-mass object will change the object's velocity by 1 m/sec. For example, suppose an object with a mass of 20 kg is moving at a rate of 6 m/sec in a straight line. A force of 30 newtons acting opposite the direction of motion of the object for 4 sec will bring the object to a halt.
   a. What do you think the resulting motion of the object would be if the force acted in the opposite direction?
   b. How do you think the object would behave if the force continued after 4 sec?

TEST YOURSELF

1. Which of the following demonstrate(s) the property of inertia?
   a. A car skidding on a slippery road
   b. The oil tanker Exxon Valdez running aground
   c. A brick sitting on a tabletop
   d. Whipping a tablecloth out from under the dishes on a table
   e. All of the above
2. If an object moves along a curved path at a constant speed, you can infer that
   a. a force is acting on it.
   b. it is accelerating.
   c. it is in uniform motion.
   d. Both (a) and (b) are true.
   e. Neither (a) nor (b) is true.
3. The mass of a 5 kg bowling ball would be _______ if it were located in deep space, far from any star or planet.
   a. zero
   b. much smaller
   c. slightly smaller
   d. the same

unit 15

Force, Acceleration, and Interaction

Background Pathways

Unit 14: Astronomical Motion: Inertia, Mass, and Force 120

Suppose a force acts on an object. How much deviation from straight-line or, as it is sometimes called, uniform motion will the force produce? To answer that question, we need to define carefully what we mean by *motion*.

Motion of an object is a change in its position, which we can characterize in two ways: by the direction of the object and by its speed. For example, a car is moving east at 40 miles per hour. If the car's speed and direction remain constant, we say it has a constant velocity. If the car changes either its speed or direction, it is no longer moving uniformly, as depicted in Figure 15.1. Any change in velocity is defined as an **acceleration**.

15.1 ACCELERATION

We are all familiar with acceleration as a change in speed. For example, when we step on the accelerator in a car and it speeds up, we say the car is accelerating. Although in everyday usage *acceleration* implies an increase in speed, scientifically *any* change in speed is an acceleration. So in scientific terms, a car "accelerates" when we apply the brakes and it slows down.

In the previous example we produced acceleration by changing the car's speed. We can also produce acceleration by changing the car's direction of motion. For example, suppose we drive a car around a circular track at a steady speed. At each moment, the car's direction of travel is changing, and therefore its velocity is

Uniform motion
(Same speed, same direction)

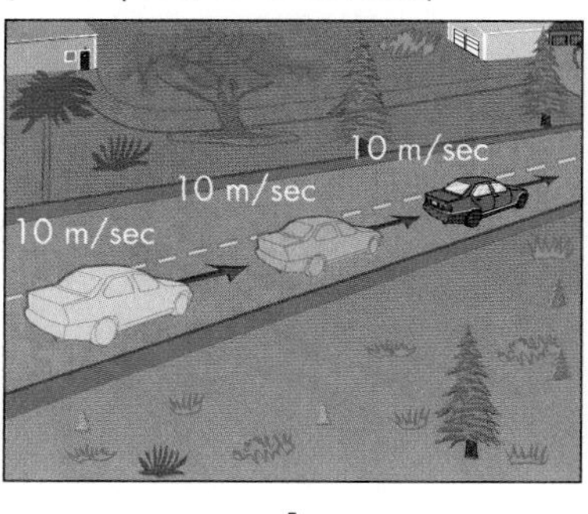

A

Acceleration
(A change in speed)

B

Acceleration
(A change in direction)

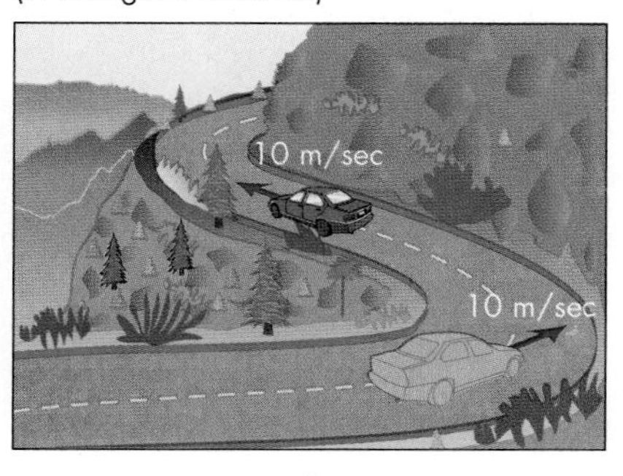

C

FIGURE 15.1
Views of a car in uniform motion and accelerating. (A) Uniform motion implies no change in speed or direction. The car moves in a straight line at a constant speed. If an object's speed (B) or direction (C) changes, the object undergoes an acceleration.

changing. Similarly, a mass swung on a string or a planet orbiting the Sun is experiencing a change in velocity and is therefore accelerating. In fact, a body moving in a circular orbit constantly accelerates, even if its speed is not changing.

The acceleration of an object is defined as its change in velocity divided by the time taken to change it. This can be written mathematically as

$$\text{acceleration} = \frac{\text{change in velocity}}{\text{change in time}}$$

or, using symbols,

$$a = \frac{\Delta V}{\Delta t}.$$

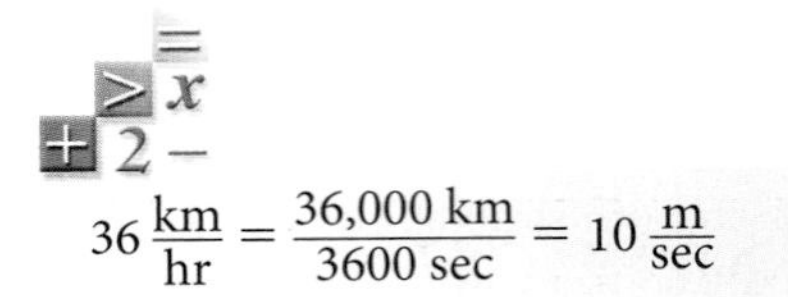

$$36\,\frac{\text{km}}{\text{hr}} = \frac{36{,}000\text{ km}}{3600\text{ sec}} = 10\,\frac{\text{m}}{\text{sec}}$$

Suppose a car is traveling at 10 meters per second (this is the same as 36 km per hour, or about 22 mph) along a straight road. If the car increases its speed to 30 m/sec over 5 seconds, we would say that its change of velocity is $\Delta V = 30$ m/sec $-$ 10 m/sec $=$ 20 m/sec, while the time change is $\Delta t = 5$ sec. The acceleration is therefore

$$a = \frac{20\text{ m/sec}}{5\text{ sec}} = 4\,\frac{\text{m}}{\text{sec}^2}.$$

Acceleration is usually written in units of "m/sec^2," and we might say that the car's acceleration during this 5-second interval is "four meters per second squared" or "four meters per second per second." What this means is that the speed changes, on average, by 4 m/sec during each second. After 1 second of this acceleration, the car sped up by 4 m/sec and was traveling at 14 m/sec. After 2 seconds, it was traveling at 18 m/sec. After 3 seconds, it was traveling at 22 m/sec.

To describe velocities in two dimensions, we can use "vector addition." The speed eastward could be represented by an arrow to the right 10 units long: ⟶. The speed northward could be represented by a vertical arrow 1 unit long: ↑. The length of the two added together can be found from the Pythagorean formula: $\sqrt{10^2 + 1^2} = \sqrt{101} = 10.05$

Note that the overall speed of the car barely changes, even though we have changed the component of velocity northward by 1 km/sec.

Suppose the driver turns the car slightly to avoid an obstacle in the right lane. The driver maintains the same speed eastward (10 m/sec), but at the end of the turn the car now has a component of motion 1 m/sec northward. Depending on how quickly the turn is completed, the acceleration we would feel inside the car might be large or small. If the direction change is done gradually over 2 seconds, the acceleration is small: $a = (1\text{ m/sec})/2\text{ sec} = 0.5\text{ m/sec}^2$—a gentle push to the north. But if it is made quickly, in just 0.2 seconds, the acceleration is much larger: $a = (1\text{ m/sec})/0.2\text{ sec} = 5\text{ m/sec}^2$—an acceleration that might topple a bag of groceries.

How do we produce acceleration? Newton realized that for a body to accelerate, a force must act on it. By pressing the accelerator pedal, we make the engine run faster and transmit a force to the car by turning the tires faster. By turning the steering wheel, the tires transmit a sideways force to the car that changes its

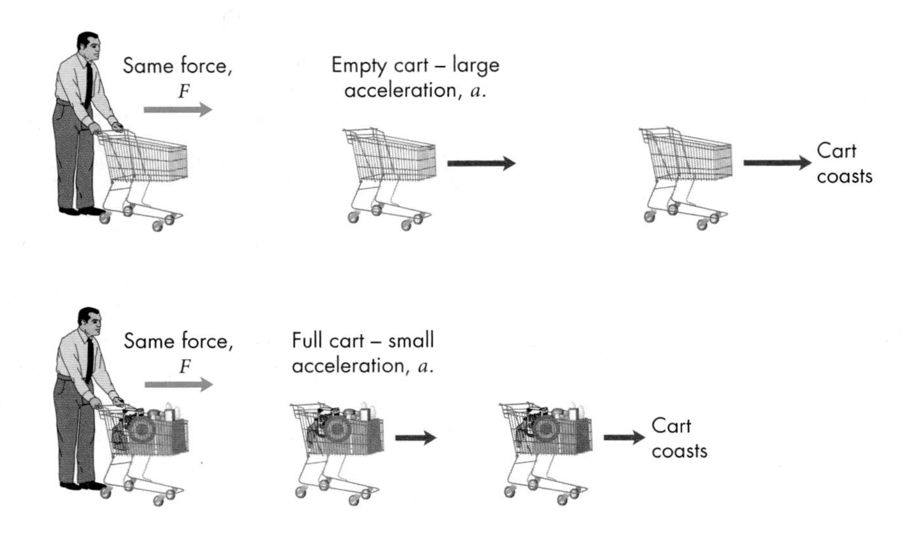

FIGURE 15.2
A loaded cart will not accelerate as easily as an empty cart.

direction of motion. For example, to accelerate—change the direction of—a mass whirling on a string, we must constantly exert a pull on the string. Similarly, to accelerate a shopping cart, we must exert a force on it. In addition, experiments show that the acceleration we get is proportional to the force we apply. That is, greater force produces a larger acceleration. For example, if we push a shopping cart gently, its acceleration is slight. If we push harder, its acceleration is greater. But experience shows us that more than just force is at work here. For a given push, the amount of acceleration also depends on how full the cart is. A lightly loaded cart may scoot away under a slight push, but a heavily loaded cart hardly budges given the same push, as illustrated in Figure 15.2. Thus the acceleration produced by a given force also depends on the amount of matter being accelerated.

15.2 NEWTON'S SECOND LAW OF MOTION

In MKS units (Unit 3), the unit of force is appropriately enough called the newton, which is the force needed to accelerate a 1-kg mass to a speed of 1m/sec in 1 second (1 newton = 1kg · m · sec²). This is about one-tenth the downward force of a 1-kg mass in Earth's gravitational field, and about a quarter pound of force in English units.

With an understanding of how to measure the acceleration of an object, we are now prepared to write an equation that can describe how forces affect the motions of any object in the universe. This is **Newton's second law of motion,** and it is surprisingly simple. Mathematically, in its most familiar form, the law states this:

$$F = m \times a$$

In words,

I. The force (F) acting on an object equals the product of its acceleration (a) and its mass (m).

The way the equation is commonly written, as it is here, would be useful if we wanted to determine what force must be acting on an object of known mass that is undergoing a measured acceleration. Forces are expressed in the MKS unit **newtons.**

Much more often we are interested in predicting how we will change the velocity of an object (accelerate it) when applying a known force. Thus a useful way of thinking about Newton's formula is this:

a = acceleration (in m/sec²)

F = force (in newtons or kg · m/sec²)

m = mass (in kg)

$$a = F/m.$$

In words,

II. The amount of acceleration (a) that a body will experience is equal to the force (F) applied divided by the body's mass (m).

Another way of saying this is that the acceleration is proportional to the force and inversely proportional to the mass of the object.

Incidentally, we can also write Newton's second law as $m = F/a$. This form is useful for measuring the mass of an object, independent of the gravitational pull it is experiencing—or even if it is floating, weightless, in space. This is the way astronauts measure their mass while in orbit. A known force is applied, and the resulting acceleration is measured; these numbers are "plugged into" the equation, and the mass is calculated.

Q: In what situations in everyday life have you experienced Newton's second law?

Astonishingly, this simple equation allows scientists to predict virtually all features of a body's motion. With $a = F/m$ and with knowledge of the masses and the forces in action, engineers and scientists can, for example, target a spacecraft safely between Saturn and its rings by setting off its thrusters to produce forces on the spacecraft to change its velocity (speed and direction) by a predictable amount even though it is millions of miles away from its target and hundreds of millions of miles away from the Earth.

15.3 ACTION AND REACTION: NEWTON'S THIRD LAW OF MOTION

FIGURE 15.3
Skateboarders illustrate Newton's third law of motion. (A) When X pushes on Y, an equal push is given to X by Y. (B) When X pushes on a much heavier person Z, an equal and opposite push is given to each, but Z moves off at a slower speed.

Newton's studies of motion led him to yet another critical law, which relates the forces that bodies exert on each other. This additional relation, **Newton's third law of motion,** is sometimes called the *law of action–reaction:*

> **III. When two bodies interact, they create equal and opposite forces on each other.**

This law is sometimes counterintuitive because we usually think of one object supplying a force and another receiving it; but the force is felt by both with equal intensity. Two skateboarders side by side may serve as a simple example of the third law (Figure 15.3). If X pushes on Y, both move. According to Newton's law, when X exerts a force on Y, Y exerts a force on X, so that both accelerate.

Suppose, though, that one of the skateboarders has a much larger mass than the other. No matter which of the skateboarders pushes on the other, the one with the smaller mass will move away faster. The force F acting on each is the same, but because of Newton's second law the resulting accelerations differ: $a = F/m$. The skateboarder with the smaller mass m will have the larger acceleration because m appears in the denominator. A larger force would have to be applied to the heavier skateboarder to make him or her accelerate as much as the lighter skateboarder.

The gravitational force between the Earth and the Sun affords an astronomical example of Newton's second and third laws and at the same time leads us a step closer to understanding gravity. According to Newton's third law, the gravitational force of the Earth on the Sun must be exactly the same as the gravitational force of the Sun on the Earth. Why, then, does the Earth orbit the Sun and not the other way around? The answer is the same as for the skateboarders: Because of Newton's second law, $a = F/m$. Thus, even though the Earth and Sun exert precisely equal forces upon each other, the Sun accelerates 300,000 times less because it is 300,000 times more massive than the Earth. Because the Earth's acceleration is so much larger than the Sun's, the Earth does most of the moving. In fact, however, the Sun does move a little bit as the Earth (and every other planet) orbits it, much as you must lean back and move in a circle yourself if you swing a heavy weight around you (Figure 15.4). Because the Sun is so much more massive than all of the planets, it wobbles by only a fraction of its own radius. Astronomers have been able to detect other stars wobbling in this way, giving us evidence of planets orbiting hundreds of other stars (Unit 34).

FIGURE 15.4
A hammer thrower feels an equal and opposite force matching the force he exerts on the swinging hammer.

Q Suppose the skateboarder pushed off a wall instead of another skateboarder. How does Newton's third law apply?

KEY TERMS

acceleration, 125
newton, 127
Newton's second law of motion, 127
Newton's third law of motion, 128

QUESTIONS FOR REVIEW

1. How does acceleration differ from velocity? From force?
2. What is happening when an object has a velocity in the positive X direction and an acceleration in the negative X direction at the same time?
3. What is Newton's second law?
4. Under what circumstances can an object experience a force yet maintain an unchanging speed?
5. How can objects move if every force is associated with an equal and opposite force?
6. If you exert on the Earth a gravitational force equal to the force the Earth applies on you, why is it you, and not the Earth, that falls?

PROBLEMS

1. A car driving at 90 kph brakes and comes to a stop in 3 seconds. What was its average acceleration during this time in units of meters/second^2?
2. In metric units, forces are measured in "newtons," which can also be written as kilograms × meters/second^2. What is the acceleration experienced by a 100-kg mass if a 100-newton force is applied to it? What is the acceleration experienced by a 200-kg mass if the same force is applied to it?
3. A person with a weight of 800 newtons (about 180 pounds) and a person with a weight of 500 newtons (about 110 pounds) are both standing on the same layer of ice over a pond. Is the force of the ice on each person the same (because it is the same surface) or different (for some other reason)? Are the two individuals equally likely to fall through the ice? Please explain your answers.
4. If a 10-newton (10 kg m/sec^2) force were applied to an initially stationary 1-kg mass, how fast would it be moving after 1 second? after 2 seconds? Assume that no other forces are acting on the mass.
5. If the same force in the previous problem were applied to a 2-kg mass, how fast would it moving after 1 second? What force would be needed to accelerate it at the same rate as the 1-kg mass?
6. Suppose that a 1000-kg rocket's thrusters give it an acceleration of 20 m/sec^2 when it is launched from Earth's surface. If Earth's gravity applies a downward force of approximately 10,000 newtons on the rocket, what is the force of the thrusters on the rocket? If the same rocket were launched from the surface of another planet, would the acceleration of the rocket be greater or less than the acceleration experienced by the rocket taking off from the Earth? Explain your answer.

TEST YOURSELF

1. A rocket blasts propellant out of its thrusters and "lifts off," heading into space. What provided the force to lift the rocket?
   a. The propellant pushing against air molecules in the atmosphere
   b. The propellant heating and expanding the air beneath the rocket, and so pushing the rocket up
   c. The action of the propellant accelerating down, giving a reaction force to the rocket
   d. The propellant reversing direction as it strikes the ground below the rocket, then bouncing back and pushing the rocket up
2. Which of the following cases does *not* describe an acceleration?
   a. A car rounding a curve at a steady 50 kph
   b. A car changing its speed from 100 kph to 90 kph
   c. A car falling off a cliff
   d. A race car driving at 200 kph on a straight highway
3. If you apply the same force to two carts, the first with a mass of 100 kg, the second with a mass of 10 kg, the acceleration of the 100-kg cart will be ____ the acceleration of the 10-kg cart.
   a. 10 times larger than
   b. 10 times smaller than
   c. the same as

unit

16

The Universal Law of Gravity

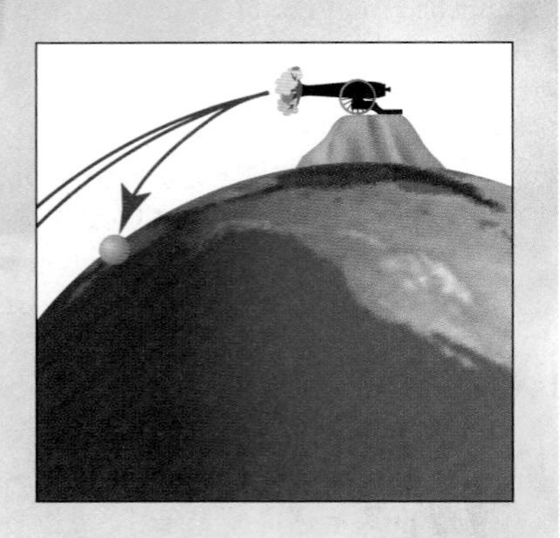

Background Pathways

Unit 15: Force, Acceleration, and Interaction 125

According to one story, Newton realized gravity's role when he saw an apple falling from a tree. The apple falling down to the Earth's surface made him speculate whether Earth's gravity might extend to the Moon. Newton realized that if the Earth's gravitational pull reached all the way to the Moon, it could provide the force that keeps the Moon circling the Earth—like a string pulling on a twirling mass.

16.1 ORBITAL MOTION AND GRAVITY

Much of Newton's work is highly mathematical, but as part of his discussion of orbital motion he described a thought experiment to demonstrate how a body can move in orbit. Thought experiments are not actually performed; rather, they serve as a way to think about problems. In Newton's thought experiment, we imagine a cannon on a mountain peak firing a projectile (Figure 16.1A). From our everyday experience, we know that whenever a body is thrown horizontally, gravity pulls it downward so that its path is an arc. Moreover, the faster we throw the body, the farther it travels before striking the ground.

Now let us imagine increasing the projectile's speed more and more, allowing it to travel ever farther. However, as the distance traveled by the ball becomes very large, we see that the Earth's surface curves away below the projectile (Figure 16.1B). Therefore, if the projectile moves at the right speed, its curvature downward will match the curvature of the Earth's surface, and the projectile will never hit the ground. Such is the nature of orbital motion and how the Moon orbits the Earth. The Moon is "falling," but because of its sideways motion it always misses the Earth. This does not answer the question of *how* the Moon got its sideways motion (presumably it was not fired out of a cannon!); but once set up with the right velocity, it can continue "falling" forever.

If we continue increasing the speed, the cannonball will begin to swing farther and farther out from the Earth in an elliptical path. If its speed is great enough, however, it will escape from the Earth and never return (Unit 18).

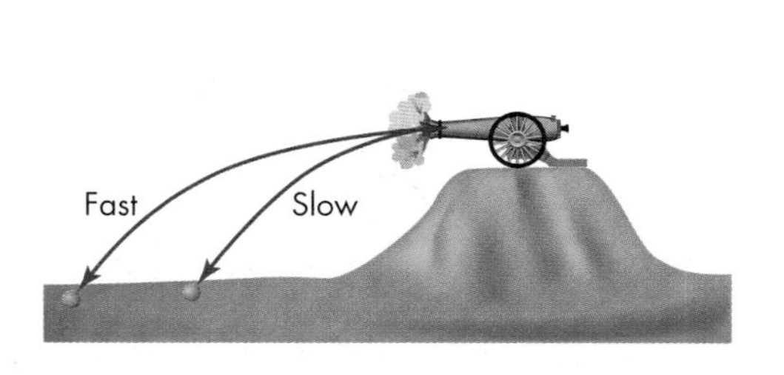

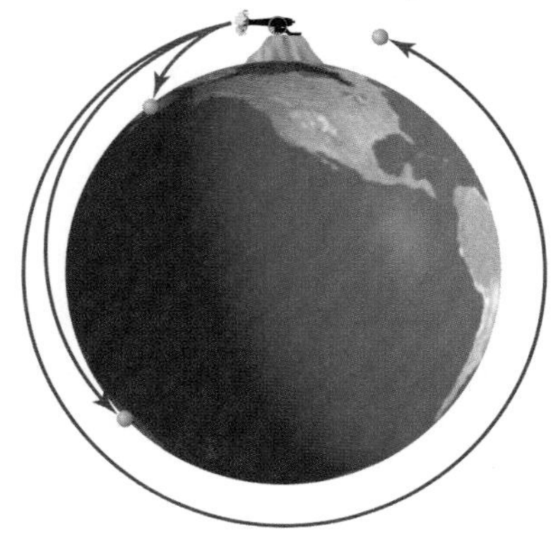

FIGURE 16.1
(A) A cannon on a mountain peak fires a projectile. The faster the projectile is fired, the farther it travels before hitting the ground. (B) At a sufficiently high speed, the projectile travels so far that the Earth's surface curves out from under it as fast as it falls, and the projectile is in orbit.

We can phrase this thought experiment more specifically using Newton's first law of motion. According to that law, in the absence of forces, the projectile would travel in a straight line at constant speed. But because a force—gravity—is acting on the projectile, its path is not straight but curved. Moreover, the law helps us understand that the projectile does not stop because it has inertia.

Newton's cannon

Notice that in this discussion we used no formulas. All we needed was Newton's first law and the idea that gravity supplies the force. To make further progress—for example, to determine how rapidly the projectile must move to be in orbit—we need laws that have a mathematical formulation.

16.2 NEWTON'S UNIVERSAL LAW OF GRAVITATION

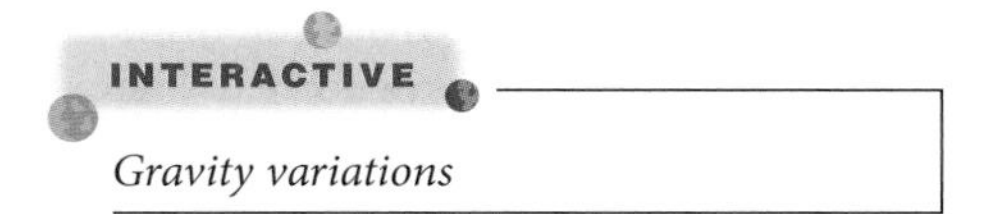

Gravity variations

Newton was not the first person to attempt to discover and define the force that holds planets in orbit around the Sun. Nearly 100 years before Newton, Kepler recognized that some force must hold the planets in their orbits and proposed that something similar to magnetism might be responsible. Newton was not even the first person to suggest that gravity is responsible. Other members of the Royal Society in England speculated about gravity's role, but none was able to present a convincing case. Years after developing his initial ideas, Newton published his law of gravity in 1687 in one of the milestones in the history of science, his *Philosophiae Naturalis Principia Mathematica* (Mathematical Principles of Natural Philosophy). He demonstrated the properties that gravity must have if it is to control planetary motion. Moreover, Newton went on to derive the **law of gravity** in mathematical form, allowing astronomers to predict the position and motion of the planets and other astronomical bodies.

Q Why do you suppose most spacecraft are launched to the east? Why are they generally launched from near the equator?

The law of gravity must describe the force that acts in many circumstances—a falling apple, the Moon, or the planets—with a single equation. We can begin to see what form this equation must have if we consider all of these different situations. First, it seems clear that gravity must depend on mass, because larger bodies, like the Sun, produce a stronger force than smaller bodies like the Earth or an apple. And the pull an object exerts must also depend on the mass of the object being pulled—for example, the Earth must pull on the Moon with a far larger force than it exerts on the apple, in order to overcome the Moon's larger inertia. Newton determined that the only way to explain these diverse situations and make the law of gravity consistent with his laws of motion was that the gravitational force between two bodies must depend on the product of their masses. Finally, Newton determined that the force must grow weaker with distance to explain the slower speeds of the outer planets in their orbits about the Sun (Kepler's third law, Unit 12.2).

Newton analyzed these issues and concluded the following:

Every mass exerts a force of attraction on every other mass. The strength of the force is directly proportional to the product of the masses divided by the square of the distance between them.

An important note about the distances used in this calculation: it is the distance between the centers, or technically the **centers of mass,** of the two objects. For a spherical object the center of mass is at the center, but for more complicated shapes it is essentially a balance point for the mass distribution—for your body, this is roughly at the center of your pelvis. Thus, if you are standing on the Earth, the distance used to calculate the gravitational force exerted on your body is not zero but approximately the radius of the Earth, or about 6380 km.

ANIMATIONS

Force of gravity

We can write this extremely important result in a shorthand mathematical manner by defining several algebraic variables. Let M and m be the masses of the

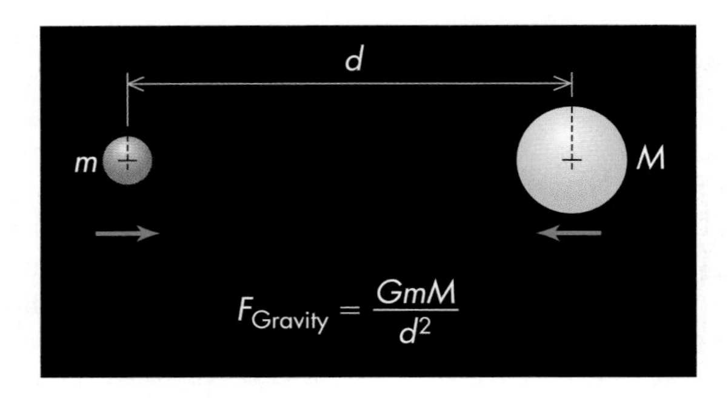

FIGURE 16.2
Gravity produces a force of attraction (green arrows) between bodies. The strength of the force depends on the product of their masses, m and M, and the square of their separation, d^2. G is the universal gravitational constant.

F_G = Force of gravity
G = Newton's gravitational constant
M, m = Masses of objects attracting each other
d = Distance between their centers

Because the unit of force, the newton, is defined as 1 kg · m/sec^2, we can also write this as $G = 6.67 \times 10^{11}$ m^3/sec^2 · kg.

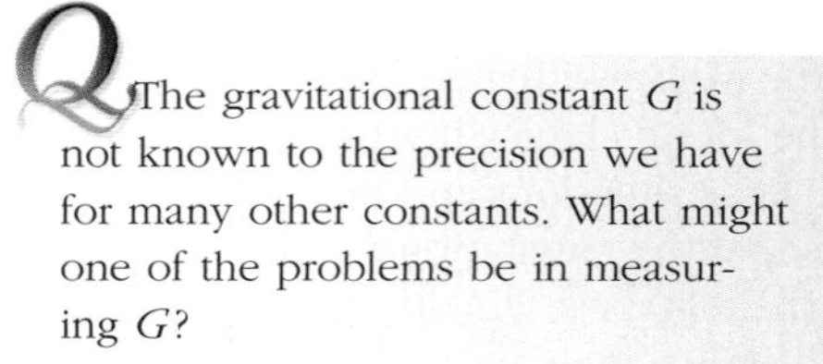

The gravitational constant G is not known to the precision we have for many other constants. What might one of the problems be in measuring G?

two bodies (Figure 16.2), and let the separation between their centers be d. Then the strength of the gravitational force between them, F_G, is

$$F_G = G\frac{M \times m}{d^2}$$

The factor **G** is a constant, a conversion factor, that lets us translate from the units of mass and distance on the right side of the equation to units of force for F_G. The value of G is found by measuring the force between two bodies of known mass and separation—for example, two large lead masses in a laboratory. The resulting value for G depends on the units chosen to measure M, m, d, and F_G. For example, if M and m are measured in kilograms, d in meters, and F_G in newtons, then

$$G = 6.67 \times 10^{-11} \text{ newtons} \cdot \text{meters}^2/\text{kilogram}^2.$$

As long as we make measurements using the same units, G is the same whether we are dealing with stars, planets, or apples.

Writing the law of gravity as an equation helps us see several important points. If either M or m increases, and the other factors remain the same, the force increases by the same amount. We call this a *direct proportionality*. On the other hand, if d (the distance between the objects) increases, the force gets weaker. In fact, it weakens as the square of the distance. That is, if the distance between two masses is doubled, the gravitational force between them decreases by a factor of four, not two. We call this an *inverse-square proportionality*.

Finally, the law of gravity shows us that even though one body's gravitational force on another weakens with increasing distance, the gravitational force never completely disappears. The gravitational attraction of a body reaches across the entire universe. The Earth's gravity not only holds you onto its surface but also extends to the Moon and exerts the force that holds the Moon in orbit around the Earth. Earth's pull extends even to distant stars and galaxies, although its pull is minuscule and just one among the forces from countless other objects.

16.3 SURFACE GRAVITY AND WEIGHT

Recall Galileo's observation that balls with different weights dropped together from the same height all strike the ground simultaneously. This seems counterintuitive at first because we feel a greater force downward when we heft a large mass than a light one. In fact a feather does fall more slowly than a cannonball if there is air resistance, but in a vacuum they fall at the same speed.

The Earth *does* pull with a larger force on a massive object, but this is counterbalanced by the object's greater inertia. It requires a greater force to accelerate an object with larger inertia, and gravity provides just the right force to accelerate all objects at the same rate on our planet's surface. We can show this mathematically using Newton's laws.

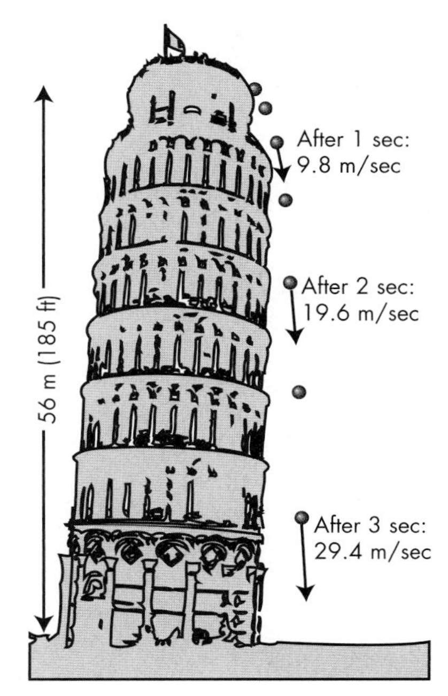

FIGURE 16.3
A ball dropped from the Leaning Tower of Pisa will take more than 3 seconds to reach the ground. At 1 g acceleration, after each second it will be falling 9.8 m/sec faster, striking the ground at a speed of about 33 m/sec.

Consider the gravitational acceleration on an object with mass m dropped near the surface of Earth. The Earth has a mass $M_\oplus$ and a radius $R_\oplus$. The distance between the center of the object and the center of the Earth is approximately $R_\oplus$, so we can use $R_\oplus$ as a close approximation to the actual distance. We can use Newton's second law, $a = F/m$, to calculate the acceleration due to the Earth's gravity. This acceleration is usually written as g. Using Newton's universal law of gravity, we find

$$g = \frac{F_G}{m} = G\frac{m \times M_\oplus}{m \times R_\oplus^2} = G\frac{M_\oplus}{R_\oplus^2}$$

This is often simply called the **surface gravity.**

Note that the mass of the falling object does not matter. The surface gravity, g, depends only on the mass and radius of the Earth and the gravitational constant G. If we "plug in" the values for these quantities, we can find the acceleration that all objects experience at the Earth's surface:

$$g = 6.67 \times 10^{-11}\,\frac{\text{newton} \cdot \text{m}^2}{\text{kg}^2} \times \frac{5.97 \times 10^{24}\,\text{kg}}{(6.37 \times 10^6\,\text{m})^2}$$
$$= 9.81\,\frac{\text{newton}}{\text{kg}}$$
$$= 9.81\,\frac{\text{m}}{\text{sec}^2}$$

In the last step we have used the definition of a **newton:** 1 newton = 1 kg · m/sec^2.

Therefore all objects at the Earth's surface accelerate at this same rate. After falling for one second, they have a downward velocity of 9.81 m/sec (about 35 kph, or 22 mph); after two seconds, 19.62 m/sec (70 kph, 44 mph); after three seconds, 29.43 m/sec (105 kph, 66 mph)—increasing by 9.81 m/sec each second. This acceleration is illustrated in Figure 16.3 for a ball dropped from the Leaning Tower of Pisa. The ball would take over 3 seconds to reach the ground 56 m below, by which time it would be falling at about 33 m/sec (120 kph, or 75 mph).

The letter "g" is often used in describing other rates of acceleration. For example, you may experience up to about 5 g's on an amusement park ride or race car, and automobile airbags are triggered by a negative acceleration (*deceleration*) of about 10 g's.

We can also compare the rates of acceleration experienced at the surface of other astronomical bodies. The Moon's surface gravity is about 0.17 g—that is, on the Moon you would weigh about 17% of what you weigh on the Earth. Figure 16.4 shows astronaut John Young jumping about a meter in the air as he salutes while wearing a spacesuit and life support system with a combined mass of about 90 kg (about 200 pounds). On the Moon your mass remains the same; but your weight depends on the forces you are experiencing, and the Moon's force on you at its surface is only 17% of the Earth's force.

Q Does the saying "The bigger they are, the harder they fall" make sense given Newton's laws and the law of gravity?

Q In Figure 16.4, why do you suppose the flag appears to be waving? Can you locate the astronaut's shadow? How might you use the shadow to determine how high he is jumping?

FIGURE 16.4
Astronaut John Young making a jumping salute at the *Apollo 16* lunar landing site near Descartes Crater. Despite a total mass of over 170 kg (370 pounds) in his space suit, he easily jumps up because of the Moon's weak gravity.

KEY TERMS

center of mass, 131

g (acceleration due to gravity at Earth's surface), 133

G (a constant), 132

law of gravity, 131

newton (unit of force), 133

surface gravity, 133

QUESTIONS FOR REVIEW

1. Describe the variables that determine the force of gravity between two objects.
2. In what sense is gravitation an "inverse-square law"?
3. What is the function of the gravitational constant *G*?
4. The force of gravity holds the Moon in orbit, while the same force causes an apple released from a tree to fall to the ground. Why does the same force have different effects on these two objects?
5. What is surface gravity?
6. How is it possible that an object on a planet more massive than the Earth could have the same weight on that planet's surface as on the Earth?

PROBLEMS

1. Suppose you were standing at the top of a *very* tall tower, 6370 km tall. What would be the gravitational force of the Earth on you at the top of this tower compared to the force you feel standing on Earth's surface? (Hint: You do not have to do as much arithmetic if you work this out with proportions.)
2. What would be the weight of an object that weighs 180 newtons on the Earth if it were moved to a planet with twice the mass of the Earth and a radius twice the radius of the Earth?
3. The Moon's radius is 1.74×10^6 m, while its mass is 7.35×10^{22} kg. Find the surface gravity on the Moon from these values.
4. The Sun's radius is 6.97×10^8 m, while its mass is 1.99×10^{30} kg. Find the surface gravity on the Sun from these values. How much would you weigh if you could stand on the Sun's surface?
5. The mass of the planet Jupiter is 1.90×10^{27} kg, and the minimum distance from us to Jupiter is 6.30×10^{11} m. The mass of a loaded 18-wheeler is 35,000 kg.
   a. What is the gravitational force of Jupiter on you?
   b. What is the force of the 18-wheeler on you if you are standing 2 meters away from its center of mass?
6. Using the data in Appendix Table 5, find the surface gravity at the cloud-tops of Uranus.

TEST YOURSELF

1. The Earth's mass is about 80 times larger than the Moon's. What is the ratio of the gravitational force the Earth exerts on the Moon to the gravitational force the Moon exerts on the Earth?
   a. 80 to 1
   b. 1 to 80
   c. 1 to 1
   d. 80^2 to 1
   e. 1 to 80^2
2. If the distance between two bodies is quadrupled, the gravitational force between them is
   a. increased by a factor of 4.
   b. decreased by a factor of 4.
   c. decreased by a factor of 8.
   d. decreased by a factor of 16.
   e. decreased by a factor of 64.
3. Gravity
   a. is the result of the pressure of the atmosphere on us.
   b. occurs between objects that are touching each other (or that are both touching the atmosphere).
   c. is the force larger objects exert on smaller ones.
   d. is the attraction between all objects that have mass.
   e. is caused only by planets and the Sun.

unit

17

Measuring a Body's Mass Using Orbital Motion

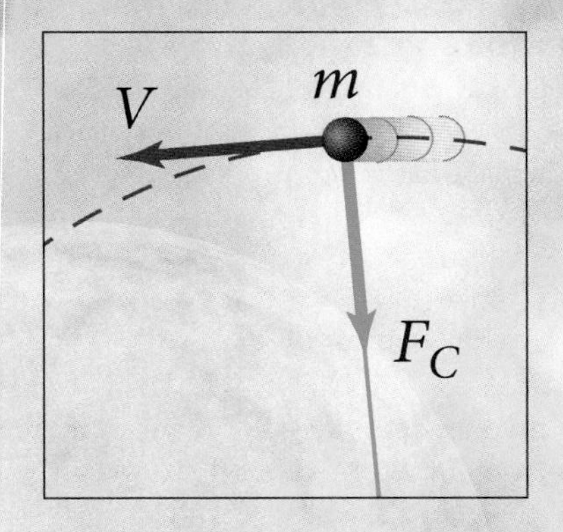

Background Pathways

Unit 16: The Universal Law of Gravity 130

Knowledge of orbital motion is important for more than simply understanding the paths of astronomical objects. From properties such as the size and period of an orbit, astronomers can deduce the masses of one or both of the orbiting objects.

The method for determining an astronomical object's mass was first worked out by Newton using his laws of motion and gravity. The underlying idea is simple: The masses of the orbiting bodies determine the gravitational force between them. The gravitational force in turn sets the properties of the orbit. Thus, from knowledge of the orbit, astronomers can work backward to find the masses of the objects.

17.1 MASSES FROM ORBITAL SPEEDS

Consider the case of a small body in a circular orbit around a large body. To simplify our calculation we will suppose that the small body has such a small mass that it only negligibly shifts the position of the more massive body. We can therefore treat the more massive object as essentially stationary. These restrictions are met to high precision in many astronomical systems, such as the Earth's motion around the Sun and the Sun's motion around the center of the Milky Way. These assumptions simplify the mathematics, but the results turn out to be essentially the same as if we were to consider more complex cases.

From Newton's first law, we know that there must be a force acting on a body that moves along a circular path. This **centripetal force** must be applied to any body moving in a circle, whether it is a car rounding a curve, a mass swung on a string, or the Earth orbiting the Sun.

Newton used calculus (which he invented to help solve this type of problem) to determine the acceleration a body undergoes when it travels in a circle, as illustrated in Figure 17.1. He derived the following equation for the centripetal force, F_C, on a mass m moving with a velocity V at a distance d from the center of the circle:

F_C = Centripetal force

m = Mass moving on circular path

V = Velocity of circular motion

d = Distance from center of circular motion

$$F_C = \frac{m \times V^2}{d}$$

Without going into the details of the derivation of this formula, we can still understand why it has these dependencies. The *size* of the change in velocity at any moment and the *rate* at which it changes are both proportional to the velocity. Hence the result depends on V^2. The turns become tighter and the acceleration is

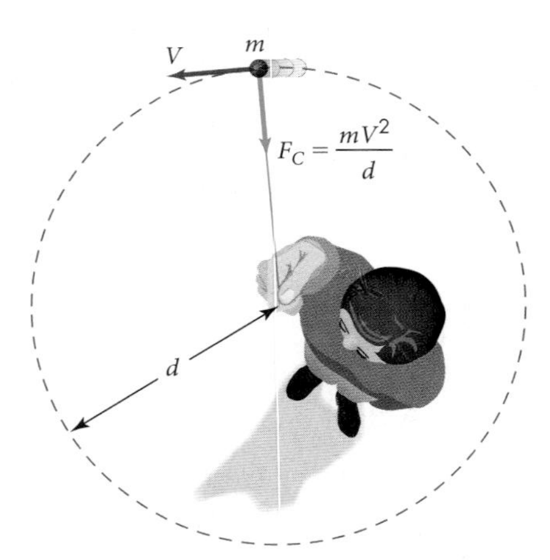

FIGURE 17.1
The centripetal force, F_C, depends on the mass and speed at which an object swings in a circle as well as the object's distance from the center.

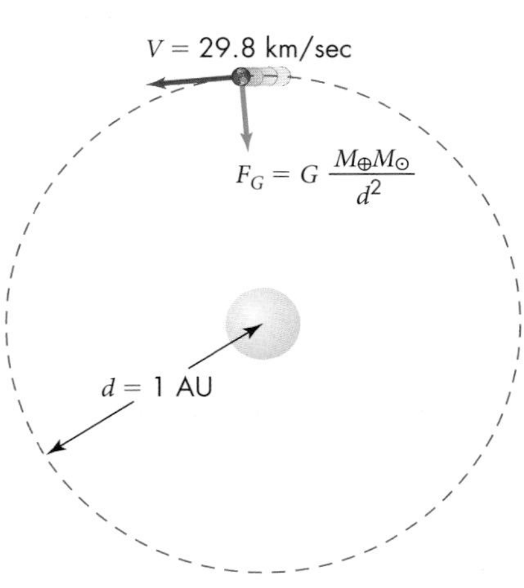

FIGURE 17.2
The gravitational force between the Sun and the Earth holds the Earth in its nearly circular orbit.

Q If you are rounding a corner in a car, what does the centripetal force equation tell you about what conditions might make the car skid?

greater if the circle is smaller, which gives an inverse dependence on the distance from the center. Finally, the mass term comes from Newton's second law—a larger mass takes a greater force to turn.

Using this equation, we can find the mass of a star if we know the speed and radius of a planet's orbit around it. Let the star's mass be M and the planet's mass be m, with the planet's mass assumed to be much smaller than M. Assume that the planet moves in a circular orbit at a distance d from the Sun with a velocity V. The gravitational attraction between the star and the planet provides the force that deflects the planet from its tendency to move in a straight line, creating the force needed to produce the observed centripetal acceleration.

Step-by-step derivation:

$$F_G = F_C$$

$$G\frac{m \times M}{d^2} = \frac{m \times V^2}{d}$$

$$G\frac{M}{d} = V^2$$

$$\frac{d}{G} \times G\frac{M}{d} = \frac{d}{G} \times V^2$$

$$M = \frac{d \times V^2}{G}$$

For an object in a circular orbit, the gravitational force F_G must equal the centripetal force F_C. If we set $F_G = F_C$ and carry out the algebra, we find that the mass is related to the size and speed of the orbit as follows:

$$M = \frac{d \times V^2}{G}$$

Therefore, we can determine the mass of an object if we know the speed and distance of another body orbiting it. And we do not even need to know the mass of the orbiting body!

For example, the orbit of the Earth around the Sun (Figure 17.2) allows us to determine the Sun's mass. The Earth's orbital velocity V is 29.8 km/sec, or 2.98×10^4 m/sec. Using this value together with the Earth–Sun distance (1 AU = 1.50×10^{11} m), we find that the Sun's mass must be:

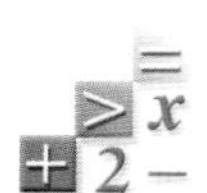

Step-by-step derivation of the Earth's orbital speed: $V = 2\pi \times d/P = 6.28 \times 1.50 \times 10^{11}$ m/(3.16×10^7 sec) = 2.98×10^4 m/sec = 29.8 km/sec.

$$M_{\odot} = \frac{1.50 \times 10^{11}\text{ m} \times (2.98 \times 10^4\text{ m/sec})^2}{6.67 \times 10^{-11}\text{ m}^3/\text{sec}^2 \cdot \text{kg}} = 2.0 \times 10^{30}\text{ kg}.$$

This same method can be used to calculate, for example, the mass of the Earth from an orbiting satellite, or the mass of a galaxy from an orbiting star. This is an especially convenient way of finding masses, because astronomers have methods for determining the speed of an orbiting object from the way an object's motions affect the light it emits (Unit 25).

17.2 KEPLER'S THIRD LAW REVISITED

In the previous section we could have written the expression for M in a slightly different way: by expressing the velocity, V, in terms of the orbital circumference ($2\pi d$) and the period, P. If we were to do that, we would end up with:

$$M = \frac{4\pi^2}{G} \times \frac{d^3}{P^2}$$

Astronomers and other scientists often use subscripts as reminders about the units to use or the meaning of symbols in an equation. The subscripts do not otherwise affect how we carry out the arithmetic.

This expression bears a certain resemblance to Kepler's third law that $P^2 = a^3$, where P is measured in years and a in astronomical units. As a reminder that these values need to be measured in these units, we sometimes write P_{yr} and a_{AU}.

The resemblance to Kepler's third law is a little clearer if we rewrite it as:

$$1 = \frac{a^3_{AU}}{P^2_{yr}}$$

The difference from Newton's formula is that instead of measuring distances in meters, periods in seconds, and masses in kilograms, Kepler could just use the Earth's orbital parameters as a reference, ignoring the Sun's mass and the constants (4, π, and G). Kepler's version also applies to any elliptical orbit for which we can measure the semimajor axis a, not just the distance in a circular orbit.

Newton's version of Kepler's third law is a modified form of the equation that applies not just to objects orbiting the Sun but to any two objects orbiting each other (Figure 17.3). It applies even if the two bodies have similar masses, and even if the orbit is elliptical. Newton showed that the sum of the two bodies' masses obeys the following law:

$$M_A + M_B = \frac{a^3_{AU}}{P^2_{yr}}$$

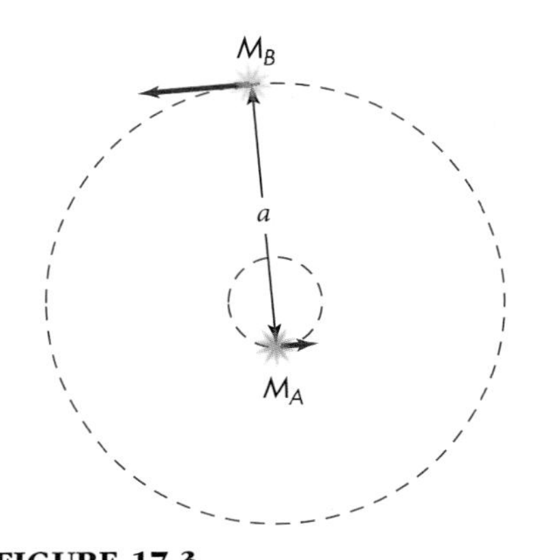

FIGURE 17.3
A pair of stars in orbit around each other. The sum of their masses can be determined by the modified form of Kepler's third law.

where the masses of the two orbiting bodies, M_A and M_B, are in units of the Sun's mass $M_\odot$.

Newton's version of the law matches Kepler's version within the Solar System because the Sun's mass is so much larger than anything else. As a result, the sum of the Sun's mass and any other object's mass is not much different from the Sun's mass alone. For example, the Earth's mass is just 0.000003 $M_\odot$, so when it is added to the Sun's mass, the left side of the equation becomes 1.000003. It requires very precise measurements to detect the small difference of this value from 1. The only planet for which it is not too difficult to detect the difference is Jupiter, which has a mass of about 0.001 $M_\odot$.

The modified form of Kepler's law is especially important because we can use it to measure the masses of stars that are orbiting each other (Unit 56). Whenever we can measure the orbital period, P, of two stars in years, along with the semimajor axis of their orbit, a, in AU, we can determine the sum of the two stars' masses.

With these equations, gravity becomes a tool for determining the mass of astronomical bodies, and we shall use this method many times throughout our study of the universe.

KEY TERMS

centripetal force, 135

Newton's version of Kepler's third law, 137

QUESTIONS FOR REVIEW

1. What is a centripetal force?
2. In what situations have you experienced a centripetal force?
3. How does a centripetal force depend on the speed and radius of the circular motion?
4. Why is it generally unnecessary to know the mass of a planet if we want to measure the mass of the star it orbits?
5. What are the differences between Kepler's third law and Newton's version of this law?
6. Might alien astronomers living in a different system of planets orbiting a star have derived Kepler's third law too?

PROBLEMS

1. Given that Jupiter is about five times farther from the Sun than the Earth, calculate its orbital velocity. How many years does it take Jupiter to complete an orbit around the Sun?
2. Derive the mass of the Earth from the Moon's orbital period of 27.3 days and orbital distance of 3.84×10^8 m.
3. Suppose that we wanted to create a version of Kepler's third law for objects orbiting the Earth instead of the Sun. If we measure the distance to satellites in terms of the distance from the Earth to the Moon (lunar units, perhaps, or LU), then what should our units of time be to make the relationship the simple

$$1 = \frac{a_{LU}^3}{P_?^2}\ ?$$

4. A pair of identical stars is separated by 25.0 AU, and the stars orbit each other with a period of 62.0 years. What are the masses of the stars?
5. The Sun orbits the center of the Milky Way Galaxy at about 220 km/sec at a distance from the center of about 2.6×10^{20} meters (about 28,000 light-years). What is the mass of the Milky Way inside the Sun's orbit? (Note: In this problem, we assume that the Galaxy can be treated as a sphere of matter. Strictly speaking, this is not precisely correct, but the more elaborate math needed to calculate the problem properly ends up giving almost the same answer.)
6. Jupiter's orbital period is 11.8622 years and its semimajor axis is 5.203 AU. Show that this implies a mass for Jupiter of approximately 1/1000th the Sun's mass. If Jupiter had a completely negligible mass, but orbited at the same distance from the Sun, what would its orbital period be?

TEST YOURSELF

1. If you are riding a merry-go-round and experience a centripetal acceleration of 0.1*g*, and the merry-go-round starts spinning twice as fast, how big will your acceleration be?
   a. 0.05 *g*
   b. 0.1 *g*
   c. 0.2 *g*
   d. 0.3 *g*
   e. 0.4 *g*
2. You determine the mass of a galaxy from the observation that it is rotating at 200 km/sec at a distance 100,000 ly from its center. Later you make a measurement and find that it is rotating at the same speed two times farther from the center. How many times bigger will the new mass you calculate be compared to the old mass you calculated?
   a. 100,000 times bigger
   b. 200 times bigger
   c. 2 times bigger
   d. 4 times bigger
   e. 200,000 times bigger
3. In the distant past, it was thought that the Moon orbited four times closer to the Earth. What would its speed have been in its orbit at this distance compared to its currently measured speed?
   a. 4 times slower
   b. 4 times faster
   c. 2 times slower
   d. 2 times faster
   e. The same

unit

18

Orbital and Escape Velocities

Background Pathways

Unit 16: The Universal Law of Gravity 130

V_{circ} = Velocity of a circular orbit

G = Newton's gravitational constant

M = Mass of body being orbited

R = Radius of orbit (from center of body being orbited)

What goes up does not always come down. Without friction to slow it down, a satellite around a planet, or a planet around the Sun, can remain in orbit essentially forever. Newton's image of a cannon ball circling the Earth (Unit 16) is not far removed from the satellites of today.

If we were to continue Newton's thought experiment with a cannon beyond the point where the ball circled the Earth, the ball would begin to travel farther and farther out. It would follow an elliptical path, as found by Kepler, and return to its starting point. At a high enough speed, however, the ball would continue to travel outward forever.

18.1 CIRCULAR ORBITS

Most spacecraft orbit the Earth only as far up as necessary to avoid friction from the Earth's atmosphere. The international space station, for example, orbits at about 360 km (220 miles) above the Earth's surface. Even at that height, drag from the very tenuous atmosphere there causes it to drop about 50 m closer to Earth each day. Periodic visits by the space shuttle bring not only food and supplies, but an opportunity to boost the station back to higher altitudes.

Compared to the Earth's radius of about 6400 km, the space station's, and most other satellites', orbits are less than 10% farther from the Earth's center than the Earth's own surface. If we calculate the balance between gravity and centripetal force (Unit 17), we find that an object in a circular orbit must have an **orbital velocity** of

$$V_{circ} = \sqrt{\frac{GM}{R}}$$

where M is the mass of the planet (or other object) being orbited and R is the orbital distance from the *center* of the planet. Without deriving this equation, we can see that it makes sense because the orbital speed is larger for a more massive planet, and it grows slower at larger radii, where gravity is weaker.

For an orbit just above the Earth's surface (Figure 18.1) the circular orbital speed is

$$\begin{aligned} V_{circ} &= \sqrt{\frac{GM_{\oplus}}{R_{\oplus}}} \\ &= \sqrt{\frac{6.67 \times 10^{-11}\ \text{m}^3/\text{sec}^2 \cdot \text{kg} \times 5.97 \times 10^{24}\ \text{kg}}{6.37 \times 10^{6}\ \text{m}}} \\ &= \sqrt{6.25 \times 10^{7}\ \text{m}^2/\text{sec}^2} \\ &= 7900\ \text{m/sec} \end{aligned}$$

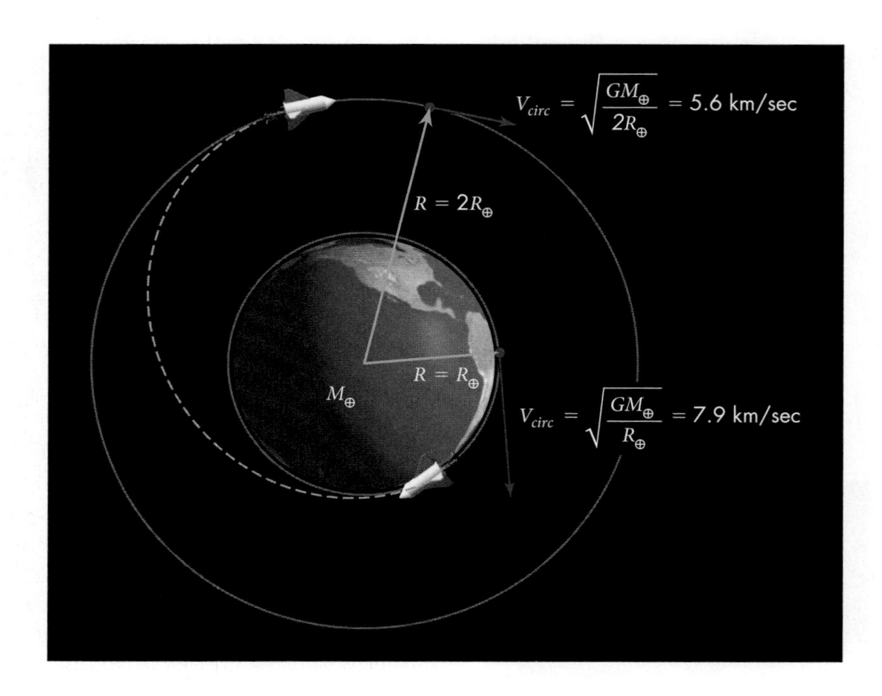

FIGURE 18.1
A circular orbit is slower at larger distances from the Earth. A rocket has to fire its thrusters to speed up to travel out to the slower orbit. At twice the Earth's radius, the circular velocity is only 5.6 km/sec, but the rocket is traveling slower than this. It will complete its elliptical orbit, "falling" back close to the Earth unless it fires its thrusters again to match the circular orbit velocity.

Orbital velocity

or a little under 8 km/sec. Notice that for a satellite orbiting at larger radii, the circular velocity becomes slower.

The slower speed in higher orbits leads to an oddity—to "slow down" a rocket's orbit, it must "speed up." For example, to travel out to the larger, slower orbit in Figure 18.1, the rocket must fire its thrusters to speed itself up. That puts it into an elliptical orbit that carries it out to the larger radius. When it reaches the larger orbital radius, it is traveling more slowly than the outer circular orbit speed. It will fall back along its elliptical orbit unless it again fires its rockets to speed up a second time and match the circular orbit speed at this radius.

Q If you were commanding the space shuttle and wanted to reach the space station several thousand kilometers ahead of you, how would you maneuver to reach it?

18.2 ESCAPE VELOCITY

To overcome a planet's gravitational force and escape into space, a rocket must achieve a critical speed known as the **escape velocity.** Escape velocity is the speed at which an object needs to move away from a body so as not to be drawn back by its gravitational attraction. We can understand how such a speed might exist if we think about throwing an object into the air. The faster the object is tossed upward, the higher it goes and the longer it takes to fall back. Escape velocity is the speed an object needs so that it will never fall back, as depicted in Figure 18.2. Thus escape velocity is of great importance in space travel if craft are to move away from one body and not be drawn back to it. However, escape velocity is also important for understanding many astronomical phenomena, such as whether a planet has an atmosphere and the nature of black holes.

The escape velocity, V_{esc}, for a spherical body such as a planet can be found from the law of gravity and Newton's laws of motion. It is given by the following formula:

$$V_{esc} = \sqrt{\frac{2GM}{R}}.$$

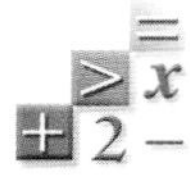

V_{esc} = Velocity needed to escape from gravitational pull of a body
G = Newton's gravitational constant
M = Mass of body
R = Starting distance from center of body

Note how similar this formula is to the circular orbit speed formula. The escape velocity is only $\sqrt{2}$ (≈ 1.414) times larger than the circular orbit speed at any radius. Thus if we know one, we can multiply or divide by $\sqrt{2}$ to find the other.

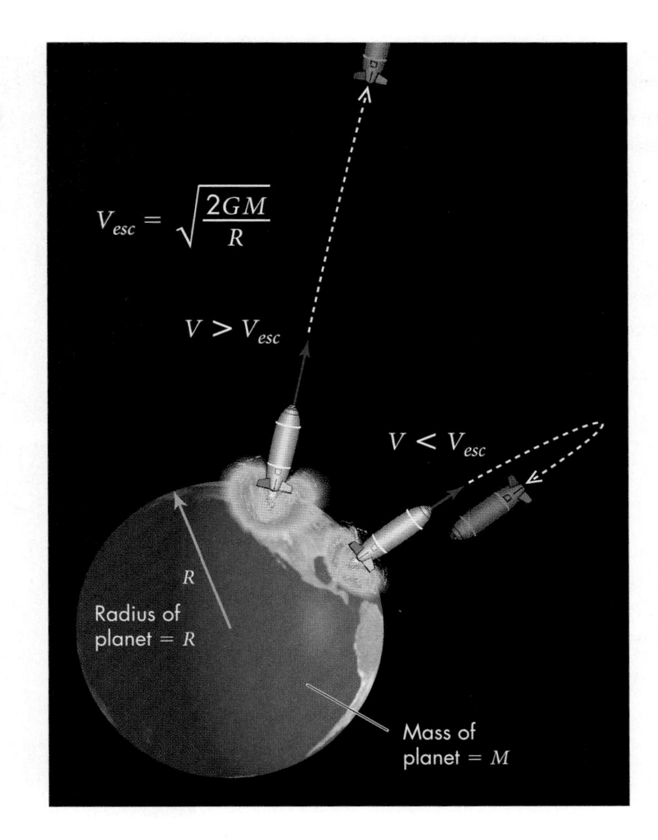

FIGURE 18.2
Escape velocity is the speed an object must have to overcome the gravitational force of a planet or star. The escape velocity from the surface of a planet depends on the mass and radius of the planet.

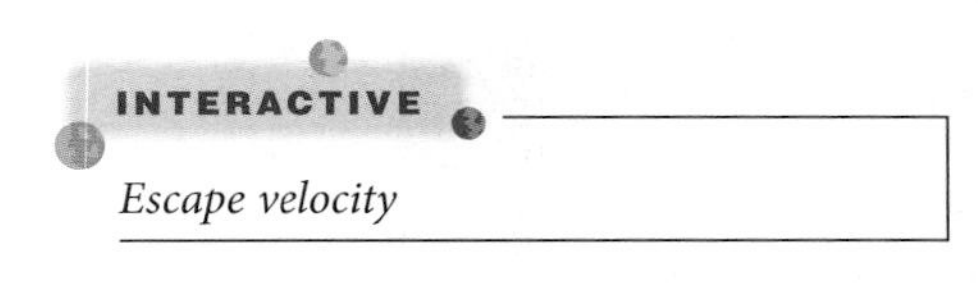

Escape velocity

ANIMATIONS
Escape velocity

Multiplying $\sqrt{2}$ times the speed of 7.9 km/sec calculated in the previous section gives the escape velocity from Earth's surface: 11.2 km/sec.

Notice in the equation for V_{esc} that if two planets of the same radius are compared, the one with the larger mass will have the larger escape velocity. On the other hand, if two planets of the same mass are compared, the one with the smaller radius will have the greater escape velocity.

Notice also that the escape velocity does *not* depend on the mass of the spacecraft, and neither does the orbital velocity. This is a common point of confusion. These speeds are independent of the mass of the spacecraft for essentially the same reason that bodies of different masses fall at the same rate, as Galileo demonstrated. It does, however, require a greater force to accelerate a larger mass to the same speed, so more power is needed to launch a heavier spacecraft, as common sense would suggest.

To illustrate the use of the formula, we calculate the escape velocity from the Moon. The Moon's mass is 7.35×10^{22} kilograms. Its radius is 1.74×10^{6} meters. We insert these values in the formula for escape velocity and find:

$$
\begin{aligned}
V_{esc}(Moon) &= \sqrt{\frac{2 \times 6.67 \times 10^{-11}\ \mathrm{m^3/sec^2 \cdot kg} \times 7.35 \times 10^{22}\ \mathrm{kg}}{1.74 \times 10^{6}\ \mathrm{m}}} \\
&= 2.4 \times 10^{3}\ \mathrm{m/sec} \\
&= 2.4\ \mathrm{km/sec.}
\end{aligned}
$$

This low escape velocity compared to the Earth's 11.2 km/sec means it is much easier to blast a rocket off the Moon than off the Earth.

Q How would the escape velocity change if we were starting not from the surface of a planet but on a launch platform twice as far from the center?

The low escape velocity from the Moon is also one of the primary reasons why it lacks an atmosphere (Figure 18.3). The average speed of molecules in a gas at the Earth's and Moon's temperatures is typically about 0.5 km/sec. Individual molecules may randomly travel faster, though. It has been found that if the escape

FIGURE 18.3
A low escape velocity is one of the factors that leads to the absence of an atmosphere on small planets and satellites such as the Moon.

velocity is not at least 10 times larger than the average molecular speed of a gas, a planet does not generally retain the gas. By contrast, the Sun's escape velocity is 617 km/sec; thus it is able to hang on to the gas in its atmosphere even though it is heated to a high temperature.

Q Can two bodies have the same escape velocity but different densities?

The escape velocity formula applies not only to planets, but to any massive body. For example, a black hole (Unit 68) is a body, such as a collapsed star, whose radius is so small that the escape velocity exceeds the speed of light. We can even consider whether the entire universe is expanding at a speed greater than its "escape velocity" and will expand forever or collapse back on itself (Unit 82).

18.3 THE SHAPES OF ORBITS

Circular and escape velocities represent just two possibilities out of an array of possible motions that bodies follow in a gravitational field. At speeds slower than the escape velocity, objects follow elliptical orbits, but at higher velocities, the orbits are described by other geometric shapes.

The many possible trajectories of a spacecraft are illustrated in Figure 18.4. This figure illustrates what might happen to a spacecraft in circular orbit around the Earth if it fired its engines to make its speed slower or faster. If the spacecraft fires

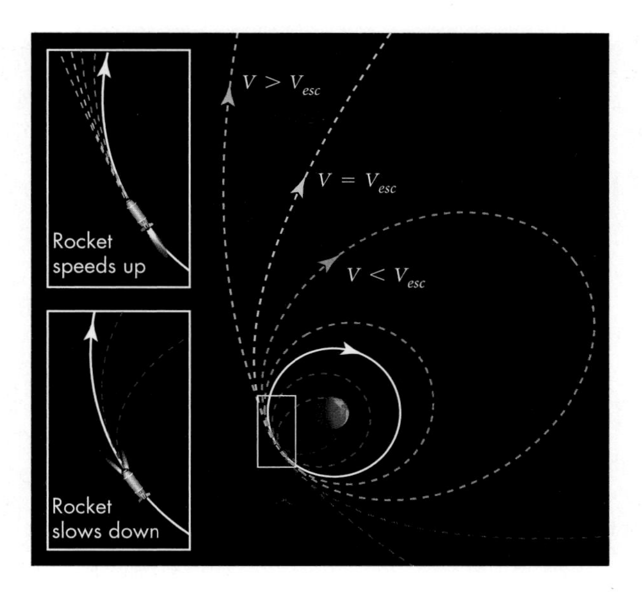

FIGURE 18.4
From a circular orbit around the Earth (shown in white), if a spacecraft fires its engines, it can end up in a variety of different trajectories. If it slows down, it will end up in an elliptical orbit (red dashed lines), possibly even colliding with Earth. The more it speeds up, the more stretched out its elliptical orbit will become (orange). If it speeds up to the escape velocity, it will end up on a "parabolic" trajectory (yellow) that never returns to Earth. Faster still and it follows a "hyperbolic" path (blue).

thrusters to slow itself down, it will now follow an elliptical orbit with the center of the Earth at the farther focus of the ellipse (see Unit 12.2). If it slows down too much, its elliptical orbit may cross the surface of the Earth, and the spacecraft will crash unless further maneuvers are made.

If the spacecraft fires its engines to speed up, it will also enter an elliptical orbit—up to a limit. This time the orbit follows an ellipse with the Earth at the nearer focus of the ellipse. The more the speed is increased, the greater the "eccentricity" of the ellipse (Unit 12.2). As the speed gets close to the escape velocity, the elliptical orbits will become more elongated, carrying the spacecraft far from the Earth before it "falls back" along its elliptical trajectory.

At the escape velocity, the trajectory follows a mathematical curve called a **parabola.** At even higher speeds the trajectory will follow a flatter curve called a **hyperbola.** These bowl-shaped curves are illustrated in Figure 18.4 in yellow and blue. The spacecraft would never return once it began to travel on this trajectory unless it slowed itself down again relative to the Earth. Parabolic and hyperbolic trajectories are followed by some other objects in nature, such as some comets that enter the Solar System at such high speeds that they pass through and then travel outward never to be seen again.

Aerospace engineers have considered building a launching platform in high Earth orbit, similar to the idea illustrated in Figure 18.4. If spacecraft were to be launched from such a platform, the escape velocity would be smaller because the distance from the center of the Earth, R, would be greater. Moreover, because they would already be traveling at orbital velocity, they would need to be boosted only by the difference between escape and circular velocities at that distance.

Q What would be the best way to orient the orbit of a launching platform around Earth for travel to other parts of the Solar System? Where would you want to be in the orbit to get the maximum boost from the Earth's own orbit?

Even though launching from an orbiting space platform would make the escape from Earth's pull easier, that is only half the challenge. As a spacecraft moves beyond Earth's immediate vicinity, the effects of the Sun's gravity becomes more apparent. Even at Earth's orbital distance, the Sun still exerts a substantial force. At 1 AU, the escape velocity from the Sun is about 42 km/sec. For probes to reach the outer parts of the Solar System therefore requires accelerating spacecraft to about 12 km/sec faster than Earth's orbital speed of 30 km/sec. It is almost as difficult to send a space probe to Mercury, even though it is closer to the Sun. The spacecraft must slow down by about 8 km/sec relative to Earth's orbit in order to make its way so close to the Sun.

KEY TERMS

escape velocity, 140
hyperbola, 143
orbital velocity, 139
parabola, 143

QUESTIONS FOR REVIEW

1. How does the speed for a circular orbit depend on the distance from a planet? How does it depend on the mass of the planet?
2. What must be done to switch from a smaller circular orbit to a larger one? from a larger to a smaller one?
3. What is meant by *escape velocity?*
4. What is the relationship between escape velocity and circular velocity?
5. If you are orbiting at a fixed distance from an object, does the orbital or escape velocity depend on the actual radius of the object?
6. What shape trajectories do objects follow if they are traveling at the escape velocity or higher?

PROBLEMS

1. Calculate the orbital speed for a satellite 1000 km above the Earth's surface, using the fact that $M_{\oplus} = 5.97 \times 10^{24}$ kg and $R_{\oplus} = 6.37 \times 10^{6}$ m.
2. At what distance would a satellite orbiting the Earth be *geosynchronous* (orbiting the Earth once every 24 hours)?
3. The escape velocity of the Sun can be considered to be the escape velocity of the solar system.
   a. Calculate the escape velocity of the Sun at its surface.
   b. Calculate the escape velocity of the Sun at the orbit of the Earth.
4. What is the escape velocity from the surface of an asteroid with a radius of 50 km and a mass of 1.0×10^{18} kg? (These are the approximate values for the asteroid Hekate.) If very good pitchers can throw a fast ball with a speed of 162 kph (or 101 mph), could they throw the ball off the asteroid?
5. What is the escape velocity from a galaxy at a radius of 50,000 ly if the mass of the galaxy is 10^{12} times the Sun's mass?
6. Which body has a larger escape velocity, Mars or Saturn?

   $M_{\text{Mars}} = 0.1\ M_{\oplus}$ $\quad R_{\text{Mars}} = 0.5\ R_{\oplus}$

   $M_{\text{Saturn}} = 95\ M_{\oplus}$ $\quad R_{\text{Saturn}} = 9.4\ R_{\oplus}$

TEST YOURSELF

1. Why is the orbital speed of a satellite orbiting 100 km above the Moon's surface smaller than the orbital speed of a satellite orbiting 100 km above the Earth's surface?
   a. The Moon's radius is smaller.
   b. The Moon's mass is smaller.
   c. The Moon is more than 100 km from the Earth.
   d. The Moon has no atmosphere.
   e. All of the above
2. Suppose the Sun suddenly collapsed in on itself, dropping to half its current radius. How big would the escape velocity from the surface be compared to its current value?
   a. ½ as big
   b. 2 times bigger
   c. $\sqrt{2}$ times bigger
   d. 4 times bigger
   e. The same
3. Suppose the Sun suddenly collapsed in on itself, dropping to half its current radius. How big would the escape velocity be at Earth's orbital distance of 1 AU compared to its current value?
   a. ½ as big
   b. 2 times bigger
   c. $\sqrt{2}$ times bigger
   d. 4 times bigger
   e. The same

unit

19 Tides

Anyone who has spent even a few hours by the ocean knows that the water's level rises and falls during the day. A blanket set on the sand 10 feet from the water's edge may be under water an hour later, or an anchored boat may be left high and dry. This regular change in the height of the ocean is called the **tides**, which are caused primarily by the Moon.

In fact, tides occur everywhere bodies interact gravitationally: between planets and their satellites; between stars that orbit each other; and between neighboring galaxies. Tidal interactions are a consequence of the way gravitational forces vary with distance. In this unit we explore Earth–Moon tidal interaction specifically, but the same analysis might be applied to a pair of stars or a pair of galaxies.

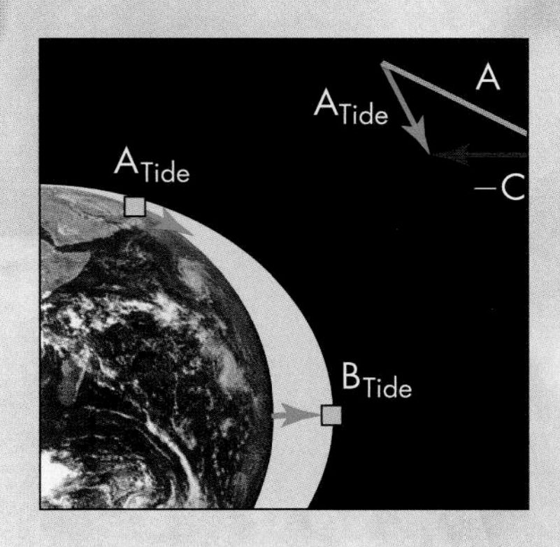

Background Pathways

Unit 8: Lunar Cycles 63

Unit 16: The Universal Law of Gravity 130

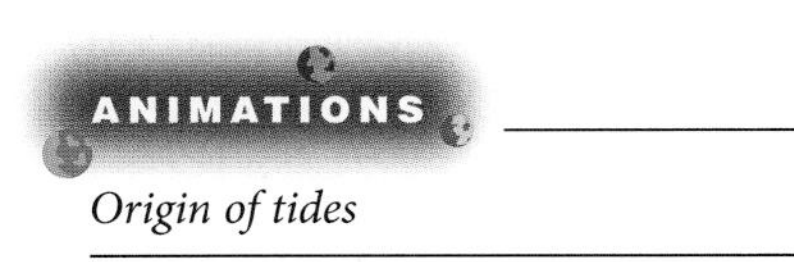

Origin of tides

19.1 CAUSE OF TIDES

Just as the Earth exerts a gravitational pull on the Moon, so too the Moon exerts a gravitational attraction on the Earth. Until now we have considered the gravitational pull between interacting bodies as if each body were a discrete whole, but in fact gravity pulls on every atom within each body. Recall, too, that the force of gravity weakens with distance because of the $1/d^2$ dependence of gravity. Hence the Moon pulls on different parts of the Earth with different strengths. For example, the attraction is stronger on the side of the Earth nearer the Moon and weaker on the far side (see Figure 19.1). Having forces of different strength pulling on opposite sides of a body produces a stretching effect. Astronomers call this stretching effect a "differential gravitational force" or simply a **tidal force.**

The tidal force draws water in the oceans into **tidal bulges** on the side of the Earth facing the Moon as well as on the Earth's far side, as shown in Figure 19.2. The tidal bulge on the far side of the Earth may seem counterintuitive. Consider the following analogy: Imagine that you are being lifted up by hundreds of strings tied to all parts of you and your clothing. Suppose the strings tied to the hair on your head are being pulled a bit harder than the rest—your hair will end up being pulled upward until it stands on end. And suppose that the strings tied to your feet and shoelaces are pulled a bit less strongly—your feet and shoelaces will dangle downward, away from the direction in which you are being pulled. This stretching effect is produced by the difference of forces across your body.

The oceans' tidal bulges are drawn toward the points facing and opposite the Moon, but the Earth spins. The Earth's rotation therefore carries us first into one bulge and then the next. As we enter the bulge, if we are on the ocean, the water level rises, and as we leave the bulge, the level falls. Because there are two bulges, we are carried into high water twice a day; these are the twice-daily high tides.

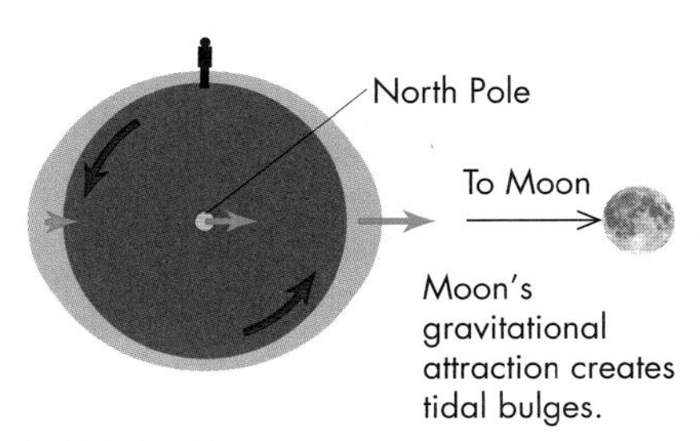

FIGURE 19.1
The difference in the Moon's gravitational pull between the near and far sides of the Earth creates tidal bulges on the Earth.

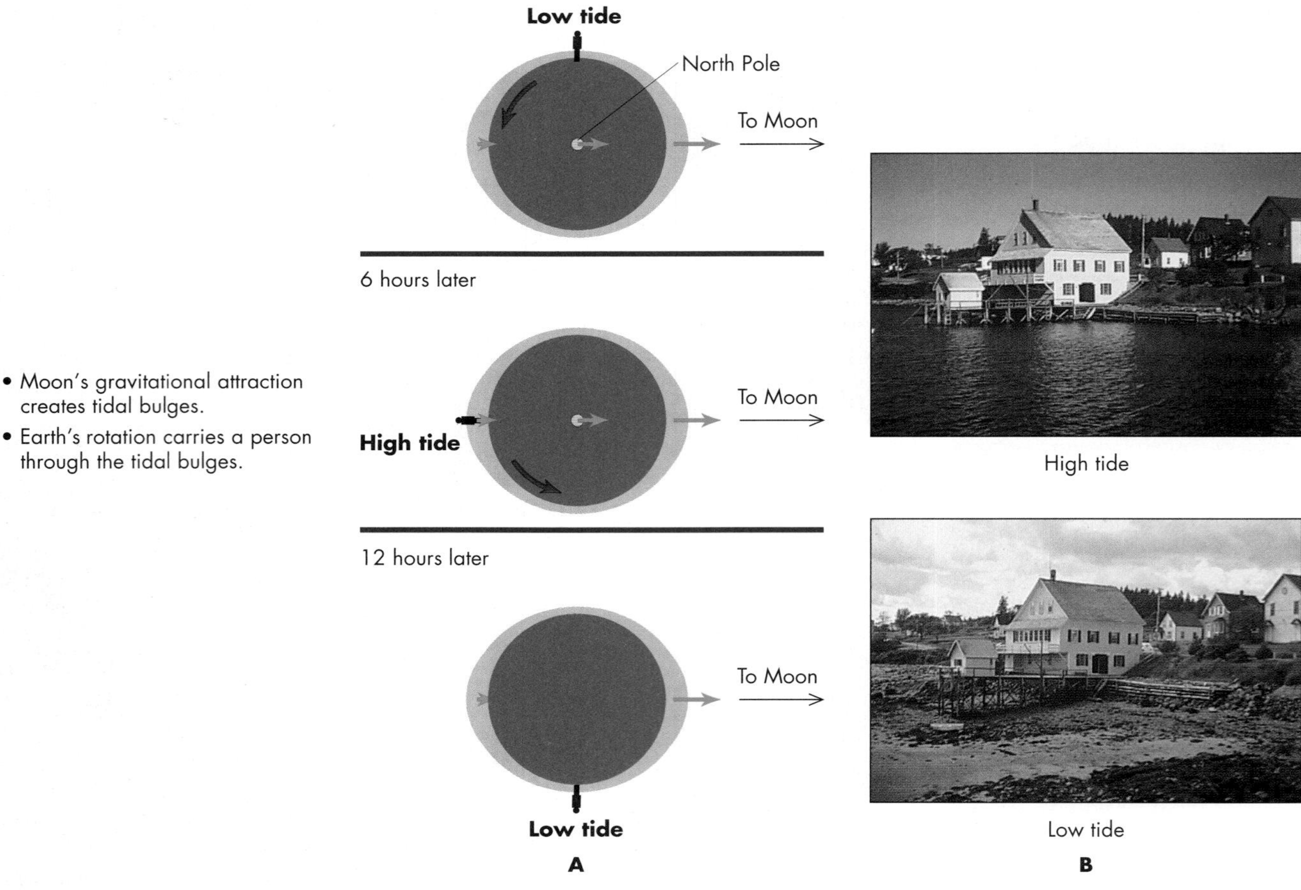

FIGURE 19.2
As the Earth rotates, it carries points along the coast through the tidal bulges. Because there are two bulges where the water is high and two regions where the water is low, we get two high tides and two low tides each day at most coastal locations.

Q The Moon does not generally lie directly above the Earth's equator, but may be as much as about 30° north or south of the equator. How can this cause the two high tides to have different strengths?

Between the times of high water, as we move out of the bulge, the water level drops, making two low tides each day (Figure 19.3).

This simple picture becomes more complicated when the tidal bulge reaches shore. In most locations the tidal bulge has a depth of about 2 meters (6 feet), but it may reach 10 meters (30 feet) or more in some long narrow bays (as you can see in the photographs of high and low tides along the coast of Maine—Figure 19.2) and may even rush upriver as a *tidal bore*—a cresting wave that flows upstream. On some rivers, surfers ride the bore upstream on the rising tide. The rising tide water in these regions is funneled into a narrower region, "piling" up the water to greater depth.

The motion of the Moon in its orbit makes the tidal bulge shift slightly from day to day. Thus high tides come about 50 minutes later each day, the same delay as in moonrise (see Unit 7).

19.2 THE SIZE OF THE TIDAL FORCE

How big is this effect? It turns out to be quite small. When the Moon is overhead (or on the far side of the Earth below your feet), you weigh about one ten-millionth less than when the Moon is near the horizon.

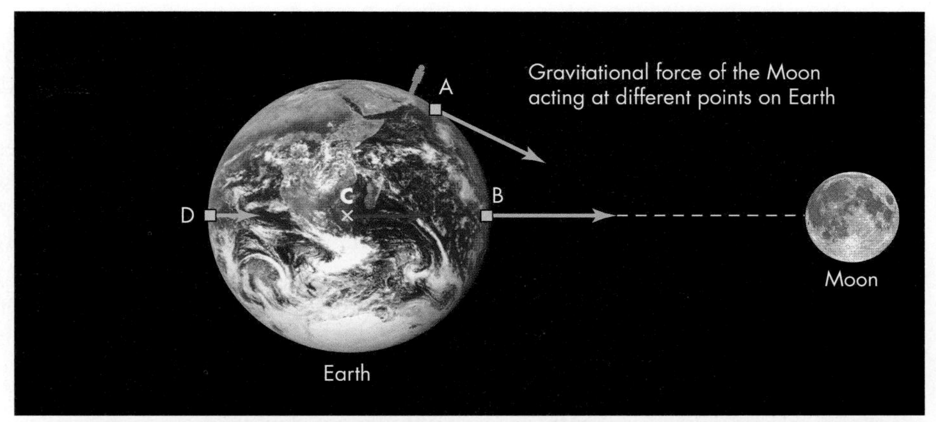

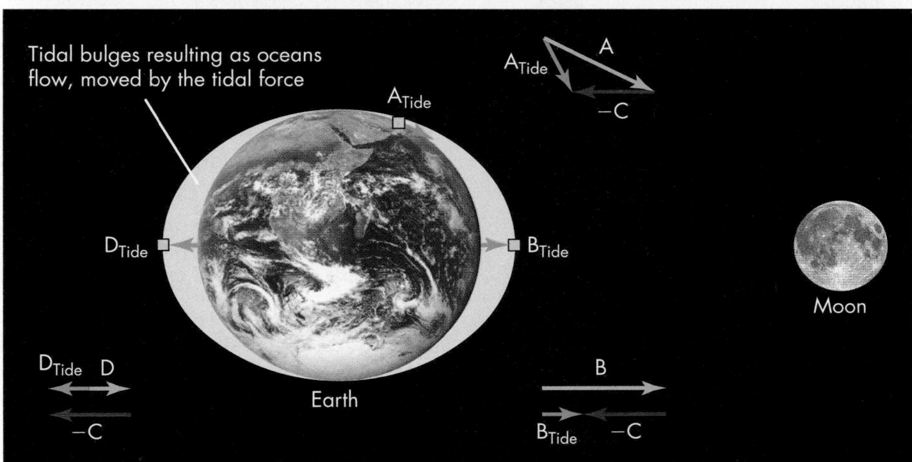

FIGURE 19.3
(Top) Arrows schematically show the Moon's gravitational force at different points on the Earth. (Bottom) Tidal forces from the point of view of an observer on the Earth. These arrows represent the difference between the Moon's gravitational force at a given point and its force at the Earth's center (C). Graphically, you can find the tidal force by "adding" the arrows. The figure shows schematically how to do this, but details are omitted.

To calculate the strength and direction of tidal forces, we must add and subtract vectors, shown as force arrows in Figure 19.3. This is done by putting each vector's tail to the head of the previous vector being summed up—or if subtracting a vector, doing this with a vector in the opposite direction. The final vector is from the tail of the first vector to the head of the last vector added together.

We can calculate the size of the tidal force by comparing the forces felt on the side of the Earth nearest the Moon relative to the Earth itself. For example, consider 1 kg of water at the average distance between the Earth's and Moon's centers: 3.84×10^8 m. Because the Moon's mass is 7.35×10^{22} kg, the gravitational force of the Moon on this kilogram of water will be:

$$F_M = G\frac{M \times m}{d^2} = 6.67 \times 10^{-11} \text{ newton} \cdot \text{m}^2/\text{kg}^2 \, \frac{7.35 \times 10^{22} \text{ kg} \times 1 \text{ kg}}{(3.84 \times 10^8 \text{ m})^2}$$
$$= 3.33 \times 10^{-5} \text{ newton}$$

This is the force felt on average by each kilogram of water all over the Earth's surface.

A kilogram of water on the side of the Earth nearest the Moon is 6370 km (the Earth's radius, $R_\oplus$) closer to the Moon, so it feels a slightly larger force:

$$F_{M,near} = G\frac{M \times m}{(d - R_\oplus)^2}$$
$$= 6.67 \times 10^{-11} \text{ newton} \cdot \text{m}^2/\text{kg}^2 \, \frac{7.35 \times 10^{22} \text{ kg} \times 1 \text{ kg}}{(3.84 \times 10^8 \text{ m} - 6.37 \times 10^6 \text{ m})^2}$$
$$= 3.44 \times 10^{-5} \text{ newton.}$$

The *difference* of these two forces, 1.1×10^{-6} newton, is the tidal force. This is the force that the kilogram of water feels relative to the solid Earth, pulling it in a direction away from the surface. The same kilogram of water feels a force of 9.8 newtons "downward" due to the Earth's gravity. This is almost ten million times stronger, so there is no chance that the Moon will pull the water off the surface!

Such a tiny force is barely measurable, so why do the oceans change in height so dramatically? The answer is that water can flow, and the tidal force can create forces that pull the water along the surface. For example, in Figure 19.3, water at point A feels a net force that causes it to flow toward the point directly beneath the Moon. On the far side of the Earth from the Moon, there are similar net tidal forces along the surface causing water to flow toward the point farthest from the Moon. Here again, the flow is not a strong current of water traveling thousands of kilometers, but mostly small shifts in the position of water in oceans all over the world creating the excess we see in the tidal bulges.

19.3 SOLAR TIDES

The Sun also creates tides on the Earth. But although the Sun is much more massive than the Moon and exerts a larger gravitational force on the Earth, it is also much farther away; so the *differential* force from one side of the Earth to the other is smaller. The result is that the Sun's tidal effect on the Earth is only about one-half the Moon's. Nevertheless, it is easy to see the effect of their tidal cooperation in **spring tides,** which are unusually large tides that occur at new and full moons. At these times the lunar and solar tidal forces work together, adding their separate tidal bulges, as illustrated in Figure 19.4A. Notice that spring tides have nothing to do with the seasons; rather they refer to the "springing up" of the water at new and full moons.

It may seem odd that spring tides occur at both new and full moons because the Moon and Sun pull together when the Moon is new but in opposite directions when it is full. However, the Sun and the Moon each create two tidal bulges, and the bulges combine regardless of whether the Sun and Moon are on the same or opposite sides of the Earth. On the other hand, at first and third quarters, the Sun's and the Moon's tidal forces work at cross-purposes, creating tidal bulges at right angles to one another, as shown in Figure 19.4B. The **neap tides** that result are therefore not as extreme as average high and low tides.

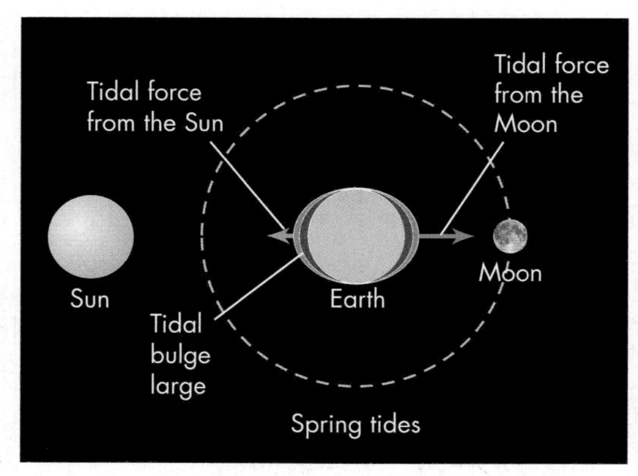

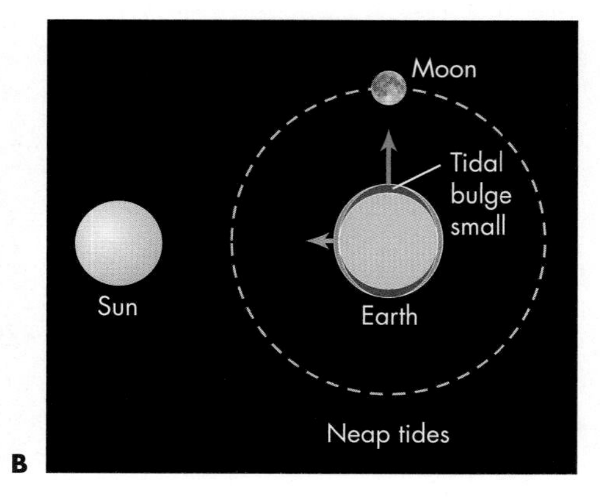

FIGURE 19.4
The Sun's gravity creates tides too, though its effect is only about half that of the Moon. (A) The Sun and Moon each create tidal bulges on the Earth. When the Sun and Moon are in line, their tidal forces add together to make larger-than-average tides. (B) When the Sun and Moon are at 90° as seen from Earth, their tidal bulges are at right angles and partially nullify each other, creating smaller-than-average tidal changes.

19.4 TIDAL BRAKING

Over a long period of time the effects of tides tend to slow down an object's rotation, a phenomenon known as **tidal braking.** Figure 19.5 shows how the Moon is currently tidally braking the Earth. As the Earth spins, friction between the ocean and the solid Earth drags the tidal bulge ahead of the imaginary line joining the Earth and Moon, as depicted in Figure 19.5. The Moon's gravity pulls on the bulge, as shown by the arrow in the figure, and holds it back. The resulting drag is transmitted through the ocean to the Earth, slowing its rotation the way your hand placed on a spinning bicycle wheel slows the wheel. The result of this slowing of our planet's spin is that the day lengthens by about 0.002 seconds each century. At this rate, after about 100 billion years, the Earth's spin would slow to the point where it was locked into synchrony with the Moon's orbital period, although the Sun will not live long enough for this to happen.

Similar tidal effects have slowed the rotation of most of the moons in the Solar System to the point where one side of the moon always faces the planet. The Sun has probably caused tidal braking of Mercury and Venus, which experience stronger solar tides than the Earth because they are closer to the Sun. These effects are explored further in Part Three of this book. The effects of tidal braking may have far-reaching consequences. For example, some scientists speculate that life may be unlikely to form on a planet that always has the same face to its star.

Q: One kind of alternative energy source is tidal power. How could you extract energy out of the tides?

The tidal braking that the Moon applies to the Earth leads to a reaction force on the Moon itself. Because the Earth's spin pulls the tidal bulge a little bit ahead of the Moon, the bulge exerts a pull that is a little off-center from the Moon's orbit. This produces a small torque on the Moon, as illustrated by the purple arrows in Figure 19.5. Even though tidal bulges are produced on both sides of the Earth, the one on the side closer to the Moon exerts a larger force because of gravity's weakening strength at larger distances. This is not just a hypothetical result. Apollo astronauts placed special reflectors on the Moon, and by bouncing laser light from them and recording the round-trip travel time, we have determined that the Moon is moving away from the Earth at 4 cm (about 1.5 inches) per year.

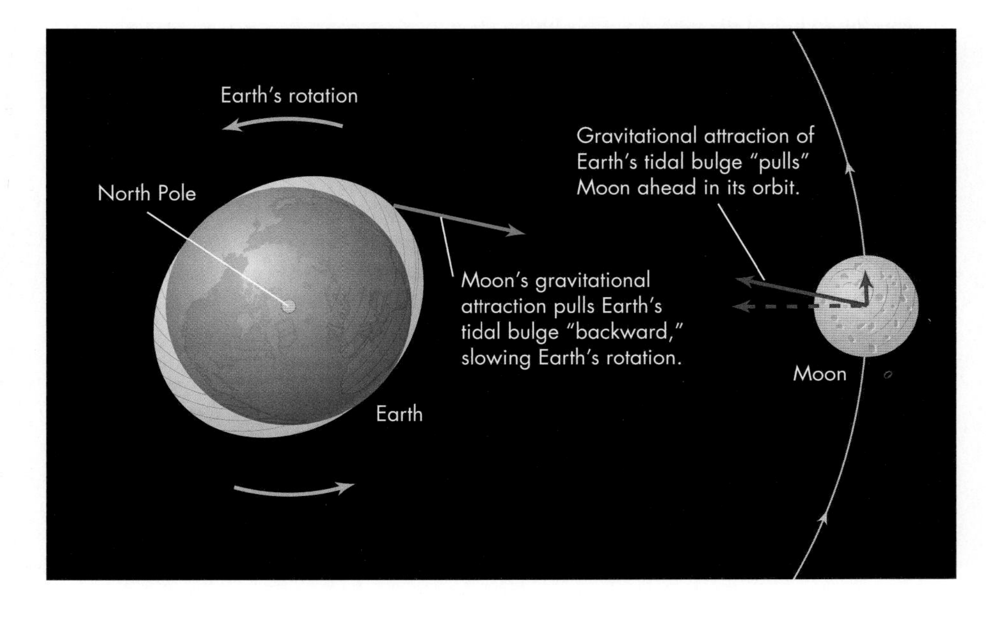

FIGURE 19.5
Tidal braking slows the Earth's rotation and speeds up the Moon's motion in its orbit. Friction between the oceans and Earth's solid crust gradually slows the Earth's rotation. At the same time, the friction "drags" the bulges of water "ahead" of the Earth–Moon line, producing a reaction force on the Moon that causes it to orbit farther from the Earth.

KEY TERMS

neap tide, 148
spring tide, 148
tidal braking 149
tidal bulge, 145
tidal force, 145
tides, 145

QUESTIONS FOR REVIEW

1. What is a tidal force?
2. Why is there a tidal bulge on the side of the Earth opposite the Moon?
3. If the same side of the Earth faced the Moon at all times, how would this affect the tides?
4. If there were no coastal effects, would you expect the tides to be larger on the coast of Maine or on the equator?
5. Why are the tides caused by the Moon larger in effect than the tides caused by the Sun?
6. What are spring and neap tides?

PROBLEMS

1. At what times would you expect high and low tides when the Moon is full?
2. At what times would you expect high and low tides when the Moon is at first quarter?
3. Carry out the same calculation as in Section 19.2 for the tidal force on the far side of the Earth. Show that this tidal force has approximately the same strength as the tidal force on the near side but points in the opposite direction.
4. If the Moon were massive enough, it could pull a kilogram of water off of the Earth's surface with its tidal forces. How massive would it have to be for this to happen?
5. Calculate the Sun's tidal force on 1 kilogram of water. The Sun's mass is 1.99×10^{30} kg, and the distance between the Earth and Sun is 1.50×10^{11} m.
6. Using the law of gravity, calculate the tidal force of Mars on a kilogram of water. The mass of Mars is 6.42×10^{23} kg, and the minimum distance between the Earth and Mars is 7.78×10^{10} m.

TEST YOURSELF

1. As a result of the Moon's gravitational pull, when would you weigh less?
   a. Whenever it is high tide locally
   b. Whenever it is low tide locally
   c. Only when the Moon is overhead
   d. Only if I were near one of the Earth's poles
   e. My weight never changes as a result of the Moon's gravity.
2. Low tide during the new moon occurs at about
   a. midnight.
   b. 6 P.M.
   c. midnight and noon.
   d. noon.
   e. 6 A.M. and 6 P.M.
3. Ocean tides are caused primarily by
   a. seismic pressure waves beneath the surface.
   b. sunlight reflecting off waves.
   c. the Moon's gravitational pull.
   d. tectonic motion of the spreading ocean floor.
   e. all of these.

unit

20 Conservation Laws

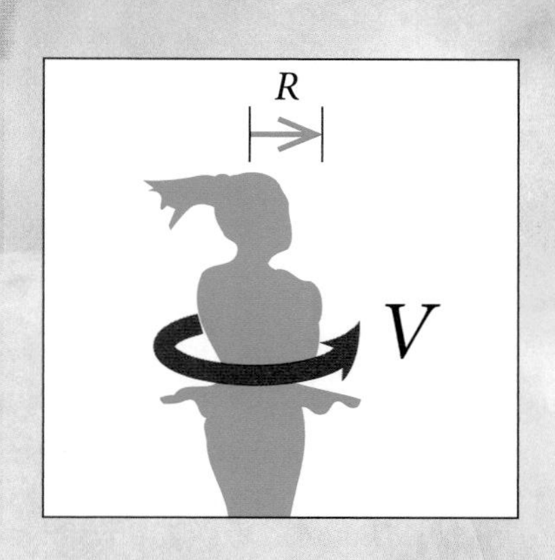

Background Pathways

Unit 16: The Universal Law of Gravity 130

Newton's laws of motion are but one set of laws important for understanding the universe. Another set of laws, the conservation laws, are also very powerful. Some physical properties of matter or a system remain the same—are conserved—in almost all circumstances. For example, in cooking (and other chemical processes) the numbers of each type of atom are conserved; the atoms may combine with each other in different kinds of molecules, but no atoms are created or destroyed. Knowing this allows us to make a number of predictions about what is possible when we cook a certain combination of ingredients. It may not teach us how to bake a cake, but it can help us predict the consequences of leaving out certain ingredients!

Conservation laws identify fundamental properties that do not change under almost any condition. We will discuss three such laws here: conservation of energy, conservation of mass, and conservation of angular momentum. The power of these conservation laws is that if we can measure the energy, the mass, or the angular momentum of a system at one time, we know that all subsequent changes in the system can occur only in ways that conserve all three quantities. With these laws we can predict how fast a collapsing star will spin or the explosive energy an asteroid will have if it strikes the Earth.

20.1 CONSERVATION OF ENERGY

Energy is a familiar term from everyday conversation, but that usage is not always the same as is meant in the sciences. **Energy** can come in many forms, but in general it can be described as the ability to generate motion. Thus a moving ball may be able to hit a stationary ball and set it in motion. The moving ball therefore has energy. Energy can also be present where there is no motion. When you pick up a ball off the ground and hold it at rest in your hand, the ball is stationary, but if you let go of the ball, it will be pulled downward by gravity and set into motion. Thus, picking it up gave it a different form of energy that would allow it to generate motion.

Energy is neither created nor destroyed—it just changes forms. This can be described by the law of conservation of energy, which states the following:

> **The energy in a closed system may change form, but the total amount of energy does not change as a result of any process.**

The idea of a *closed system* is important: It means that energy is not exchanged with anything outside the system. For example, the Earth is not a closed system because it receives energy from the Sun and radiates heat energy into space. Very few systems remain isolated at all times, but often we can examine an individual process—energy conversion during a roller coaster ride, for example—as if it occurred in isolation.

To appreciate the power of this law, we need to look more closely at how we measure energy as well as some of the forms it may take. Energy is interesting in

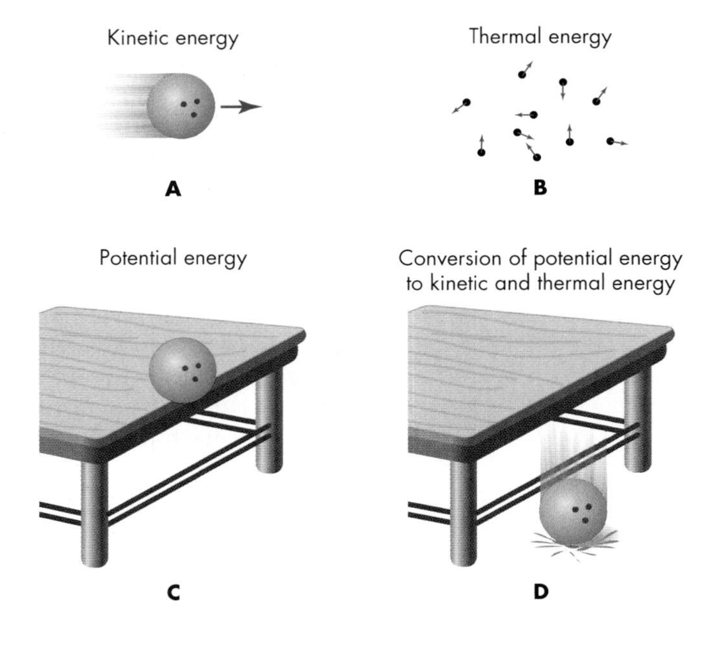

FIGURE 20.1
Energy can be found in many forms. A moving body has kinetic energy (A), while the energy we associate with heat is produced by the motions of particles within the substance (B). Energy can also be found in a "potential" form, such as a bowling ball at the edge of a table (C). If it falls off the edge of the table, it produces kinetic energy as it falls to the floor (D).

part because it can have so many forms. The most obvious form is the energy of an object that is already in motion, or **kinetic energy.** A simple example of kinetic energy is the energy of a moving car or a thrown ball (Figure 20.1A). Physicists measure energy in **joules,** the unit of energy in the MKS system (see Unit 3), named after the nineteenth-century British physicist James Joule, who helped verify the law of conservation of energy. The kinetic energy of a body of mass m moving with a speed V can be expressed mathematically as

E_K = Kinetic energy
m = Mass of moving object
V = Speed of moving object

$$E_K = \frac{1}{2} m \times V^2$$

If m is measured in kilograms and V in meters per second, the calculation will give the energy in joules.

Another form of energy is heat, or **thermal energy.** Heat is also energy of motion, but at the atomic or molecular level. That is, the hotter something is, the more rapidly its atoms and molecules move within it. Sometimes that motion is a vibration. Other times it is a random careening of particles (Figure 20.1B).

Yet another common form of energy is **potential energy.** Potential energy is energy that has not been liberated but is available. For example, water stored behind a dam, and a bowling ball poised on the edge of a table, both have potential energy (Figure 20.1C). If the dam bursts or the ball rolls off the table, the potential energy can be released and converted into motion. These examples of potential energy result from the gravitational attraction between the Earth and the water or the bowling ball. To lift an object up against Earth's gravity takes energy, and that energy is stored until the object falls back to the Earth. The **gravitational potential energy** can be written mathematically as

E_G = Gravitational potential energy
m, M = Masses of objects
d = Distances between objects' centers
G = Newton's gravitational constant

$$E_G = -G\frac{m \times M}{d}$$

where G, m, M, and d are all defined as in the law of gravity (Unit 16). Again, if all the quantities are measured in MKS units, the resulting energy will be in joules. The form of this equation is what you might expect—potential energy has a bigger magnitude for larger masses, and it grows weaker with distance as gravity grows weaker. A curious aspect of this equation, though, is the minus sign. We discuss this further a little later.

> **Q** What kind of energy is nuclear energy?

The usefulness of the law of conservation of energy is that when you add together all the forms of energy at the beginning of a process, their sum must equal the total amount of energy in all its forms at the end. As an example, let's consider the bowling ball on the table edge. If it is just sitting there, it initially has only potential energy. If it falls from the table, that potential energy is converted to energy of motion as the ball falls. Upon hitting the ground it may look as if the energy has disappeared: The ball is at rest again so that it no longer has energy of motion. Likewise, it has given up its potential energy. However, its energy has *not* disappeared. On impact, the ball shakes the ground and gives its energy of motion to the atoms in the ground. Their motion produces heat and sound (vibrations of atoms in the air), so that the ball's energy has changed form yet again (Figure 20.1D). If you pick up the bowling ball and set it back on the table, in lifting the ball you will expend the same amount of **chemical energy**—potential energy in the electrical bonds of molecules such as fat and sugar stored in your body—as the potential energy that the bowling ball gains.

> **Q** How will the temperature of a gas cloud change as gravity causes it to collapse?

An astronomical example that illustrates the conservation of energy and its conversion to different forms is an asteroid approaching Earth. Initially it has energy of motion. During its passage through the atmosphere, a small amount of the energy of motion is converted to heat and light as it tears through the atmosphere. The asteroid's kinetic energy drives the asteroid deep into the solid crust of the Earth; as it slows, it transfers its kinetic energy by compressing and heating the rock at the point of impact. This violent compression is transferred to surrounding rock, expanding outward in all directions. In a fraction of a second, so much heat is generated that the asteroid and surrounding rock melt or vaporize. Some rock is blasted up and out, giving it kinetic energy, and seismic waves (Unit 35) travel through the Earth, shaking the ground thousands of kilometers away. Most of the asteroid's kinetic energy ends up as thermal energy, which is eventually radiated away into space (Unit 23). Some of the energy remains as potential energy—stored in chemical bonds created by reactions in the heat of the blast or stored as gravitational potential energy in rock pushed upward in the region surrounding the blast crater. Thus the energy changes form, but the total energy at the beginning equals the total energy at the end.

Galileo's experiments showing that a ball slows down as it goes up a ramp or speeds up as it goes down (Unit 14) can also be understood in terms of conservation of energy. As the ball rolls up a ramp (Figure 20.2) its kinetic energy decreases as its gravitational potential energy increases. Extending this idea, we can phrase the question of how fast a rocket must travel to leave the Earth (its *escape velocity,* Unit 18) in terms of the kinetic energy it must have to completely overcome the Earth's gravitational potential energy. The rocket slows, and as it reaches an infinite distance where the gravitational pull approaches zero, its kinetic energy also

> **Q** What forms of energy have you used today?

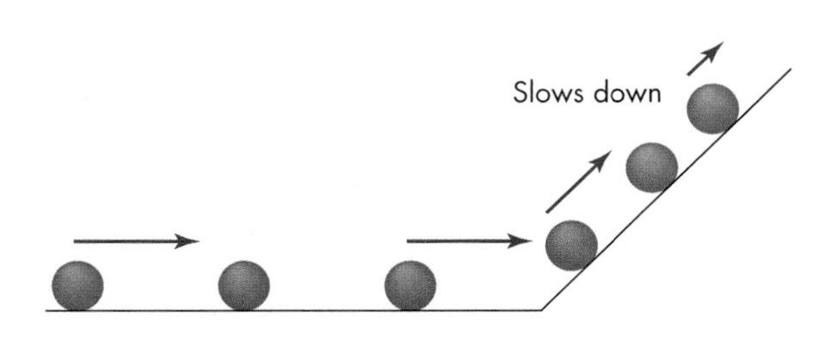

FIGURE 20.2
A ball rolling up a ramp loses kinetic energy as its gravitational potential energy increases. When it reaches its highest point and momentarily stops before rolling back down, all of its kinetic energy has been converted to potential energy.

Assigning a negative value to gravitational potential energy may seem strange, but the alternatives are perhaps stranger. If we assigned "zero energy" to when two objects were put close together, we would have to say that the potential energy was very large when the objects were barely feeling each other's gravity—and the gravitational potential energy would still become negative if the objects were put even closer together.

drops to zero. If a rocket has enough (positive) kinetic energy to overcome the (negative) gravitational potential energy, it will escape from the Earth's surface. In fact, we can recover the formula for escape velocity by setting $E_K + E_G = 0$.

This helps explain why gravitational potential energy (or potential energy based on any attractive force) has a negative value. When two objects are very far apart, the force of gravity becomes negligible, so it is natural to describe the gravitational potential energy as near zero. If the two objects fall together, their kinetic energy increases—subtracting energy from the gravitational potential energy. If we start with nearly zero joules of gravitational potential energy and subtract thousands of joules, E_G must become more negative as gravity pulls objects closer together.

Curiously, on astronomical scales many objects in the universe, including perhaps the entire universe itself, exhibit a near balance between the gravitational potential and kinetic energies. Both kinds of energy may be individually large, but their sum is usually close to zero.

20.2 THE CONSERVATION OF MASS (ALMOST)

Even before the discovery of the conservation of energy, scientists had proposed another fundamental conservation law: the conservation of mass. This was thought to be precisely true, until a very interesting exception was discovered by Albert Einstein in the 1900s. In fact, it is so close to being true in everyday circumstances that for almost all practical purposes it can be treated as a conservation law.

In the 1700s, chemical conversions and interactions that occurred between substances were not fully understood. However, a number of scientists had noted that the total mass of the material at the end of any chemical experiment was the same as at the beginning. The more carefully all of the products of the experiment were collected and measured, the more precisely this rule was found to be obeyed.

By the early 1800s, the English chemist and physicist James Dalton realized that this behavior of matter could be explained if matter was made up of atoms—submicroscopic particles that had specific unchanging properties and masses (also see Unit 4). The different kinds of atoms are listed in the Periodic Table of the Elements, shown in the back inside cover of this book. These atoms can be arranged in different combinations or molecules, but they cannot be destroyed or altered. Even though the matter might appear to change form drastically, in fact the same number of each type of atom is present at the end of an experiment as at the beginning. As a result, the mass is also conserved.

The notion of the conservation of matter began to be altered in the 1900s. It was discovered that atoms themselves are made of smaller subatomic particles, such as electrons, protons, and neutrons (Unit 4). In some circumstances those particles can be pulled apart—allowing one type of atom to be converted into another. The interactions that cause these kinds of conversions generally require extremely high energies and are quite rare at Earthly temperatures. However, they are not so rare at the temperatures found inside stars.

At about the same time Albert Einstein showed that mass and energy can be converted into one another. For example, the subatomic particles that make up atoms of helium have a lower mass when combined to form atoms of carbon. The loss in mass is balanced by a release of energy. The conversion of atoms from one type to another proves to be the source of energy of the Sun and other stars (Unit 50).

One way of understanding this relationship between energy and mass is to think of mass as a kind of extremely compact form of potential energy. In fact, Einstein showed that mass and energy need to be treated similarly. He showed, for example, that gravity depends not just on the mass of particles but on their energy as well. We tend to think of mass and energy as being very different quantities, partly because the energies we deal with in everyday life are minuscule compared to the potential energy bundled up in matter.

To be exact, physicists today refer to conservation of "mass-energy." In most circumstances, however, mass and energy are each separately conserved to a very high degree of precision.

20.3 CONSERVATION OF ANGULAR MOMENTUM

You have perhaps seen water going down a drain and noticed that as the water moves inward toward the drain hole, it spirals faster and faster. Likewise, ice skaters who go into a spin with outstretched arms and then fold their arms inward spin even faster. These are both examples of the conservation of angular momentum.

Angular momentum is a measure of the tendency of a spinning object to keep spinning with its rotation axis pointing in the same direction unless acted on by an external "rotational force," called a **torque,** that works to alter the rotation. The distinction between a force and a torque is that some forces act in directions that do not affect rotation, while the strength of a torque is measured by its ability to increase or decrease rotation. If a mass m is moving around a rotation axis at a distance R with a velocity V, then mathematically its angular momentum is $m \times V \times R$. The law of conservation of angular momentum states this:

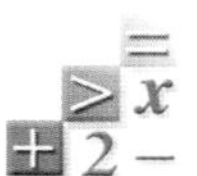

m = Mass of rotating object

V = Speed of rotation around axis

R = Radius from axis of rotation

If no external force acts to change the spin rate, then $m \times V \times R$ remains constant.

How does this explain the faster spin of the ice skater whose arms are pulled in? If $m \times V \times R$ is constant, and if the skater makes R (the length of the extended arms) shorter, then V must get larger (Figure 20.3). Notice that when skaters are pulling their own arms in, no *external* forces are involved. Likewise, as the water flowing toward a drain circles nearer the drain hole, its R decreases; therefore its V of rotation must increase.

Q Suppose you were in a spinning spacecraft. How could you use a jet thruster to stop your spin? If you can slow your spin, in what sense is angular momentum conserved?

We have already encountered an astronomical example of this conservation law in Kepler's second law of planetary motion. As a planet moves closer to the Sun, its orbital velocity must increase to keep $m \times V \times R$ constant. Because nearly all objects have some intrinsic amount of rotation, if any process causes them to change size, they will change their spin rate in response.

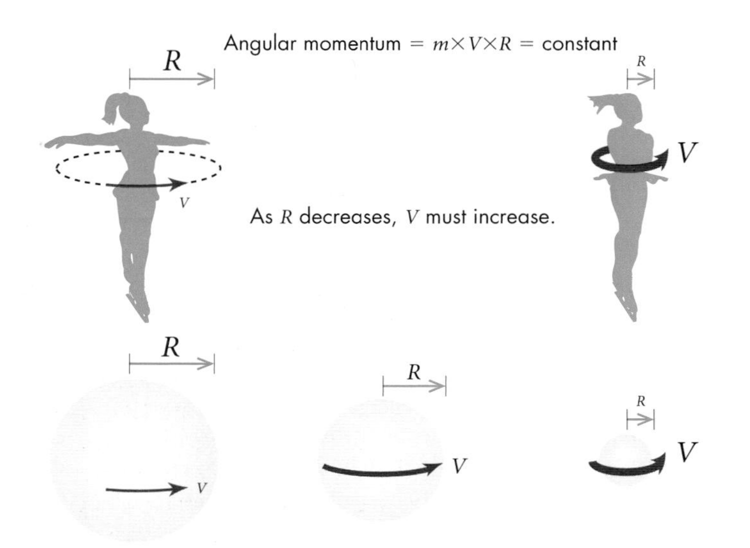

FIGURE 20.3
Conservation of angular momentum spins up a skater and a spinning star. Angular momentum is proportional to a body's mass, m, times its rotation speed, V, times its radius, R. Conservation of angular momentum requires that mathematically the product $m \times V \times R$ remains constant. Thus if a given mass changes its R, its V must also change. In particular, if R decreases, V must increase, as the ice skater demonstrates by spinning faster as she draws her arms in. Similarly, a rotating star spins faster if its radius shrinks.

Another example of the conservation of angular momentum is tidal braking (Unit 19.4). As the Earth's spin slows due to its interaction with ocean tides, its angular momentum in decreasing. However, the lost angular momentum is transferred to the Moon, whose angular momentum increases as its orbital radius grows larger. Using the law of conservation of angular momentum, we can deduce the rate at which the Moon must be moving away based on the rate of slowing of the Earth's spin, and this is confirmed by experiment.

KEY TERMS

angular momentum, 155
chemical energy, 153
conservation law, 151
energy, 151
gravitational potential energy, 152
joules, 152
kinetic energy, 152
potential energy, 152
thermal energy, 152
torque, 155

QUESTIONS FOR REVIEW

1. What is a conservation law?
2. An object slows to a halt as it rolls uphill. What happened to the kinetic energy?
3. What are the different forms of energy?
4. In what situations is mass conserved, and what happens to it when it is not conserved?
5. How does the conservation of angular momentum explain why a planet moves faster when it is closer to the Sun in its orbit?
6. What is a torque?

PROBLEMS

1. A basketball has a mass of about 0.5 kg. How fast would it have to be thrown to have 1 joule of kinetic energy? Convert your answer to kph (or mph if that is more familiar to you). Is that a hard throw or a light throw?
2. An object with a mass of 1.00×10^5 kg is at a height of 3.00×10^4 m, headed directly away from the Earth. What energy will this object have when it comes back and strikes the Earth?
3. Show that if the gravitational potential energy and kinetic energy add up to zero for an object on the Earth's surface, you can derive the escape velocity formula given in Unit 18.
4. Suppose that a comet with a mass of 10^9 kg strikes the Earth at a velocity of 40 km/sec. How much energy would be released by the impact? Convert this energy into the equivalent of "kilotons of TNT" (kt), where 1 kt $\approx 4.2 \times 10^{12}$ joules.
5. Astronomers estimate that about 100 metric tons (1×10^5 kg) of meteorites strike the Earth every day.
   a. If the meteorites increased the mass of the Earth by 1% without adding any angular momentum, how long would the day become?
   b. The Earth's mass is 5.97×10^{24} kg. How many years would it take to increase the Earth's mass by 1% at this rate?
6. Suppose that a star collapses inward until it is reduced to one-tenth of its original size while losing no mass.
   a. How many times faster will its surface be moving?
   b. If initially it was spinning around once every 100 hours, how long will it take to spin around after it collapses?

TEST YOURSELF

1. A volleyball has about half the mass of a basketball. If they are both moving at the same speed, the volleyball's kinetic energy would be________the basketball's.
   a. 2 times more than
   b. 1/2 as much as
   c. 4 times more than
   d. 1/4 as much as
   e. the same as
2. Suppose a comet has an elliptical orbit that brings it twice as close to the Sun at its closest approach as when it is at its largest separation. If its speed is 10 km/sec at its largest separation, what will its speed be when it is closest?
   a. 5 km/sec
   b. 10 km/sec
   c. 20 km/sec
   d. 40 km/sec
3. How many times more chemical energy must a car expend to reach 100 kph compared to 50 kph?
   a. 50
   b. 4
   c. 100
   d. 2
   e. 200

unit

21

Light, Matter, and Energy

Background Pathways

Unit 4: Foundations of Astronomy 30

Our home planet is so far from other astronomical bodies that, with few exceptions, we cannot bring samples from them to study in our laboratories. And even for objects within the range of our spacecraft, physical conditions may make it impossible to send probes to make direct measurements. For example, if we want to know how hot the Sun is, we cannot stick a thermometer into it. Nor can we directly sample the atmosphere of a distant star. However, we can sample such remote bodies indirectly by analyzing their light.

At first this might appear to be an enormous limitation—allowing us to say something about only the direction to an object and the quantity of light it produces. However, if we can understand the energetics of light and the way it interacts with matter, we can also learn about what the body is made of and its temperature. Light, therefore, is our key to studying the universe; but to use the key, we need to understand the nature of light and matter and how they interact.

21.1 THE NATURE OF LIGHT

Light is radiant energy; that is, it is energy that can travel through space from one point to another without the need for a direct physical link. Light is, therefore, very different from, for example, sound. Sound can reach us only if it is carried by a medium such as air or water. Light can reach us even across empty space. In empty space we can see the burst of light of an explosion, but we will hear no sound from it at all.

Light's capacity to travel through the vacuum of space is paralleled by another special property: its high speed. In fact, the speed of light is an upper limit to all motion (Unit 53). In empty space, light travels at the incredible speed of 299,792,458 meters per second—a speed that would take an object seven times around the Earth in one second! The speed of light in empty space is a universal constant and is denoted by c, and although it is very precisely measured, for most purposes it can be rounded off to 300,000 km/sec.

The speed of light is slower when it travels through matter. The speed at which light travels through glass is about 60% of its speed in a vacuum; through water, about 75%; through air, about 99.97%.

Observation and experimentation on light throughout the last few centuries have produced two different models of what light is and how it works. According to one model, light is a wave that is a mix of electric and magnetic energy, changing together, as depicted in Figure 21.1. The ability of such a wave to travel through empty space comes from the interrelatedness of electricity and magnetism.

You can see this relationship between electricity and magnetism in everyday life. For example, when you start your car, turning the ignition key sends an electric current from the battery to the starter. There the current generates a magnetic force that turns over the engine. Similarly, when you pull the cord on a lawn

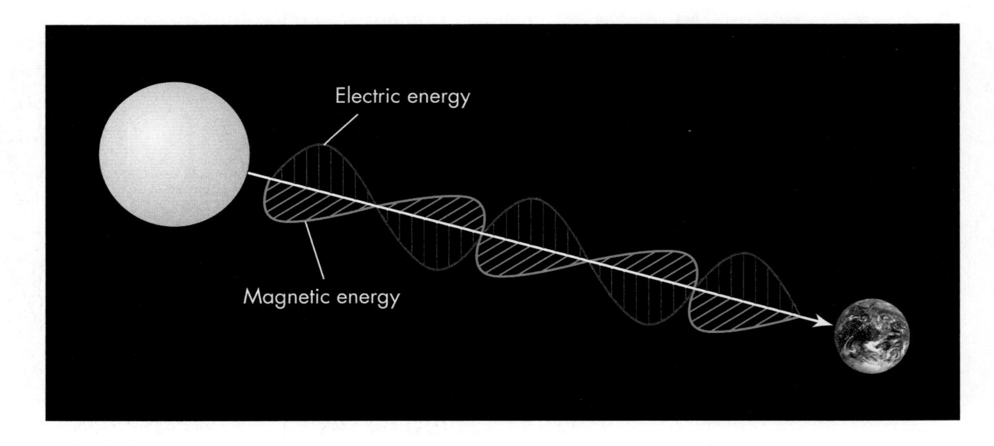

FIGURE 21.1
A wave of electromagnetic energy moves through empty space at the speed of light, 299,792.5 kilometers per second. The wave carries itself along by continually changing its electric energy into magnetic energy, and vice versa.

mower, you spin a magnet, generating an electric current that creates the spark to ignite the gas in the engine.

Light can travel through empty space because a small disturbance of an electric field creates a magnetic disturbance in the adjacent space, which in turn creates a new electric disturbance in the space adjacent to it, and so on. In this fashion, light can move wavelike through empty space, "carrying itself by its own bootstraps." Thus we say that light is an **electromagnetic wave.** When such a wave encounters the electrons in matter, it may interact with them, a little bit like the way a water wave makes a boat rock. It is from such tiny disturbances in our eyes or in an antenna that we can detect electromagnetic waves.

Many of the early experiments on light demonstrated this distinctly wavelike behavior. For example, sources of light can **interfere** with each other like the interacting waves in an ocean, reinforcing and canceling each other's effects in complex patterns. And like ocean waves passing through a narrow inlet into a bay, light passing through an opening does not travel in a straight line, but spreads out. This is an important consideration in the design of telescopes because it limits the sharpness of images a telescope can produce (Unit 29).

The wave model of light works well to explain many phenomena—for example, the length between waves determines the color of light (Unit 22). However, the wave model fails to explain other properties of light. If light were like a water wave, for example, we could deliver any amount of electromagnetic energy by varying the intensity of the electromagnetic wave. However, we find something quite different for light. For example, the red light we see coming from a stoplight is made up of little parcels containing 3×10^{-19} joules of energy. We cannot get half this much energy or 1.7 times this much energy from red light, just whole multiples of this much energy. We can vary the energy in a water wave by making the waves taller or shorter, but light behaves as if, in the analogy to water waves, we can make waves in a bathtub that are 3, 6, or 9 centimeters tall, but no size in between. To describe this property, we say that the energy of light is **quantized;** that is, it comes in discrete packets.

Joules are the MKS unit of energy (Unit 3).

Based on this property, light is better described as being made of particles. These particles are called **photons** (Figure 21.2). You can have 1 or 2 or 3 photons, but not a half a photon or 1.7 photons. In this model the photons are packets of energy. When they enter your eye, they produce the sensation of light by striking your retina like microscopic bullets, releasing their energy and causing a chemical change in your photoreceptor cells (Unit 31). In empty space, photons move in a straight line at the speed of light.

Q If you start a water wave from one side of a bathtub (or a sink or pool), does any water travel from one side of the tub to the other? How can you test your idea?

For centuries, scientists argued over whether light consisted of waves or particles, pointing out experiments in which the opposing model could not give an adequate explanation. By the early twentieth century, however, it became clear that

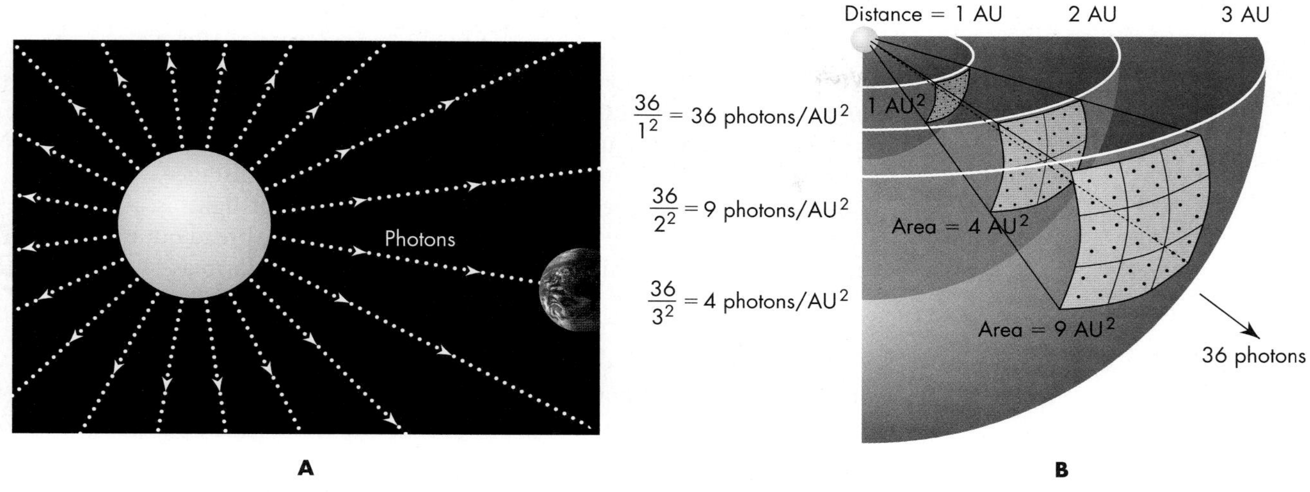

FIGURE 21.2
(A) Photons—particles of electromagnetic energy—stream away from a light source at the speed of light. (B) Photons become spread over successively larger areas as the distance from the source of light increases. The number of photons in the same area decreases as a function of the inverse square of the distance.

Photons from a light source

neither model completely describes light. Moreover, at the submicroscopic scales in which it operates, light has a more complex dual nature that we can describe only incompletely with our wave and particle models. So today scientists describe light as having a **wave–particle duality,** and they use whichever model—wave or particle—best describes or explains a particular phenomenon. For example, the way light reflects off a mirror is easily understood if you imagine photons striking the mirror and bouncing back just the way a ball rebounds when thrown at a wall. On the other hand, the focusing of light by a lens is best explained by the wave model.

21.2 THE EFFECT OF DISTANCE ON LIGHT

An individual photon traveling through space maintains its speed and energy; however, we also know that objects appear dimmer when they are farther away. This dimming is the result of the light's spreading out over a larger area as the photons move away from the source of light—much as the height of a wave produced by a rock tossed into a pond grows shorter as the wave ring expands.

This spreading out of photons explains why the planets in the inner Solar System are warmer than the outer planets (Unit 32), even though they receive photons that have the same energy. For example, photons traveling away from the Sun become spread farther apart, as illustrated in Figure 21.2. Suppose we traced the path of 36 photons that were spread out over a spherical surface area of 1 AU^2 at Earth's distance. At 2 AU they would be spread out over a surface area of 4 AU^2, at 3 AU they would be spread out over a surface area of 9 AU^2, and so on. The number of photons reaching each square AU of surface area is decreasing as a function of the distance *squared.*

We can make this more mathematically precise by noting that at the distance of the Earth from the Sun, the photons are spread over the area of a sphere that has a radius of 1 astronomical unit (the Earth–Sun distance). The surface area of this sphere, given by the geometric formula $4\pi \times (1\ \text{AU})^2$, is about 12 square AU. At Jupiter's distance from the Sun—about 5 AU—the light is spread out over a sphere with a surface area of $4\pi \times (5\ \text{AU})^2$,which is an area 25 times larger than at the Earth's distance. Thus each square meter of the Earth's surface receives 25 times more light than each square meter of Jupiter's surface. As a result, the Sun looks 25 times dimmer at Jupiter's distance than at the Earth's distance.

Astronomers describe this change in the amount of light as a change in its **brightness.** At a distance d from the source, the light is spread out over an area of $4\pi\ d^2$, so we can write this relationship in mathematical form as

$$\text{brightness} = \frac{\text{total light output}}{4\pi d^2}$$

Inverse-square law

This relationship is known as the **inverse-square law** because the light grows dimmer in proportion to the square of the distance. The inverse-square law is important for determining how much energy is received at different distances from a source of light. The relationship also allows us to calculate the distances to many stars (Unit 54).

21.3 THE NATURE OF MATTER

Just as scientists have puzzled over the nature of light, they have also puzzled over the nature of matter. Like so many of our ideas about the nature of the universe, our ideas about matter date back to the ancient Greeks. For example, Leucippus and his student Democritus, who lived around the fifth century B.C.E. in Greece, taught that matter is composed of tiny indivisible particles. They called these particles *atoms,* which means "uncuttable" in Greek. They argued, for example, that the water in a tub could be subdivided into smaller and smaller pieces—droplets—only down to some finite size for the smallest particle of water.

Our current model for the nature of atoms dates back to the early 1900s and the work of the New Zealand physicist Ernest Rutherford. He showed that an atom can be thought of as a dense core called a *nucleus* around which smaller particles called *electrons* orbit (Figure 21.3). As described in Unit 4, the force that holds the electrons in orbit is the electrical attraction between the positively charged protons in the nucleus of the atom and the negative electric charge of the electrons. The orbits of electrons are extremely small. For example, the diameter of the smallest electron orbit in a hydrogen atom is only about 10^{-10} (1 ten-billionth) meter. The nucleus of the atom, where the protons and neutrons are packed together, is less than one ten-thousandth as big as the electron's orbital.

The structure of an atom with electrons orbiting an atomic nucleus superficially resembles planets orbiting the Sun. The nucleus of the atom is similar to the Sun's position in the Solar System in that it contains almost all of the mass and provides the primary source of attraction for the orbiting bodies. However, the resemblance ends there.

Unusual effects operate at an atomic level that have no counterpart in larger systems. Our model of electrons as particles is incomplete because, like light, they too have a wave–particle duality. Their wave nature is not another kind of electromagnetic wave, and it is not related to the fact that an electron has an electric charge; rather, it concerns fluctuations of "electron-ness." In the early twentieth century it was discovered that all matter behaves like a wave in certain respects, although usually this is apparent only for the individual subatomic particles. The existence and position of such particles can fluctuate according to rules that are described by the area of physics called *quantum mechanics.*

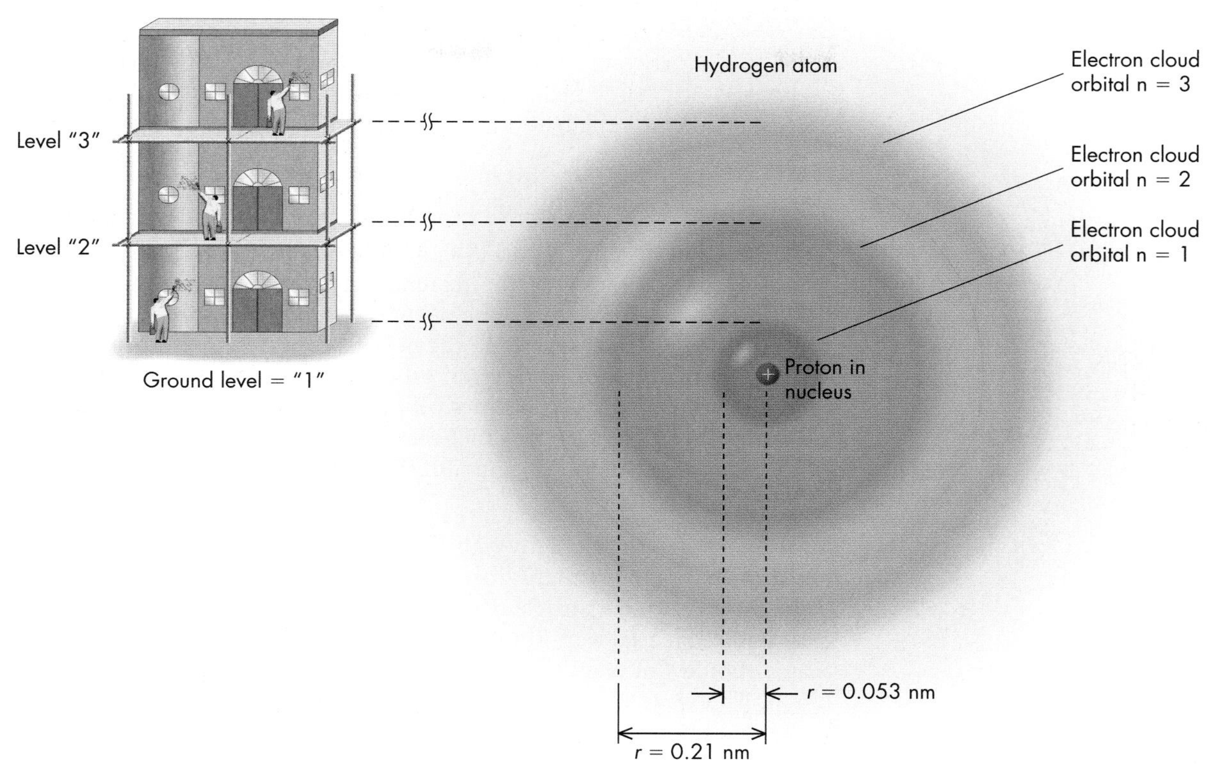

FIGURE 21.3
Just as the painters can be only at levels 1, 2, 3, . . . of the scaffold (and cannot "float in between"), so too an electron must be in orbital 1, 2, 3, . . .

In quantum mechanics the location of a particle is described in terms of a wave function that describes the probability of finding it at any position.

One of the most important consequences of electrons' duality is that their orbits may have only certain quantized sizes. Although a planet may orbit a star at any distance, an electron may orbit an atomic nucleus only at certain distances—much as when you climb a set of stairs, you can gain only certain discrete heights. This reason for this peculiar steplike character to electrons' orbits was worked out by the French physicist Louis de Broglie (pronounced *de-BROY*) in the early 1900s. De Broglie showed that the electron's wave nature forces the electron to move only in orbits whose circumference is a whole number of waves. If it moved in an orbit whose circumference was ½ or 1½ times the length of a wave, the crest of a subsequent wave would land where the trough of the wave had been previously as it circled the nucleus, and the electron wave would "cancel" itself out. This property prevents the electron from falling into the nucleus despite the electrical attraction because the smallest orbit, or **ground state,** is just one electron wave long.

The wave nature of the electron has another important effect. It "smears out" the position of the electrons. Just as a water wave does not exist at a single point, the electron wave cannot be defined as a single point. As a result, although we have described the electrons as orbiting like tiny particles around the nucleus, a better description is that they exist as a three-dimensional electron wave or **orbital.** These discrete, yet smeared out, orbitals are illustrated as green "clouds" in Figure 21.3.

Electrons in orbitals have another property totally unlike those of planets in orbit: The electrons can shift from one orbital to another as they interact with light. This shifting changes their energy, as can be understood by a simple analogy.

The electrical attraction between the nucleus and the electron creates a force between them like a spring. If the electron increases its distance from the nucleus, the spring must stretch. The atom requires energy to accomplish this "stretch." Similarly, if the electron moves closer to the nucleus, the spring contracts and the atom must give up, or emit, energy. We perceive that emitted energy as electromagnetic waves.

We can try to describe this interaction in two ways corresponding to the wave–particle duality of nature at submicroscopic scales:

Particles: An electron is orbiting an atomic nucleus at a particular distance. It is struck by a photon that contains the exact amount of energy needed to knock the electron into the next larger possible orbit. The electron absorbs the photon and jumps into the new, larger orbit.

Waves: An atom has an electron orbital of a particular wave configuration. An electromagnetic wave comes by with electric and magnetic fields oscillating at just the exact rate so that the orbital interacts with the wave, absorbing energy from it, and begins to vibrate in a new, more energetic mode in a larger orbital.

Neither description is exactly right because matter at these scales is both a particle *and* a wave—a duality that is foreign to our everyday experience.

In some ways the particle description is clearer. Discrete particles interact in a clear and sequential way. But it may lead one to wonder how the electron "knows" that it should not react when a photon with slightly less or slightly more energy hits it. The wave picture is more challenging in other ways, but the interaction between a photon and an electron may be clearer. The electron orbital responds to the electromagnetic oscillations, much as the string of a musical instrument may begin vibrating in response to a sound wave of the right pitch, but not to all pitches.

For our purposes, we can simplify the discussion of electron behavior by focusing on just the energies that are involved. When an electron moves between orbitals—an **electronic transition**—the energy change depends on the particular orbital and on the kind of atom. To understand why, we need to say a bit more about atomic structure.

21.4 THE CHEMICAL ELEMENTS

The structure of atoms determines both their chemical properties and their light-emitting and light-absorbing properties. For example, iron and hydrogen not only have very different atomic structures, but they also emit photons with different energies. From such differences astronomers can deduce whether an astronomical body—a star or a planet—contains iron, hydrogen, or whatever chemicals happen to be present. Therefore, understanding the structure of atoms ultimately helps us understand the nature of stars.

Q In the periodic table, why do you suppose the number of neutrons increases relative to the number of protons at high atomic numbers?

Iron and hydrogen are examples of what are called chemical **elements.** The main characteristics of a chemical element are determined by the number of protons in its nucleus. For example, hydrogen atoms contain 1 proton; helium atoms contain 2 protons; carbon, 6; oxygen, 8; and iron, 26. A complete list of the elements is given in the Periodic Table of the Elements shown on the inside back cover of the book (see also Unit 4). The **atomic number** of each element indicates the number of protons in its nucleus.

Although the identity of an element is determined by the number of protons in its nucleus, the chemical properties of each element are determined by the number of electrons orbiting its nucleus. The number of electrons normally equals the

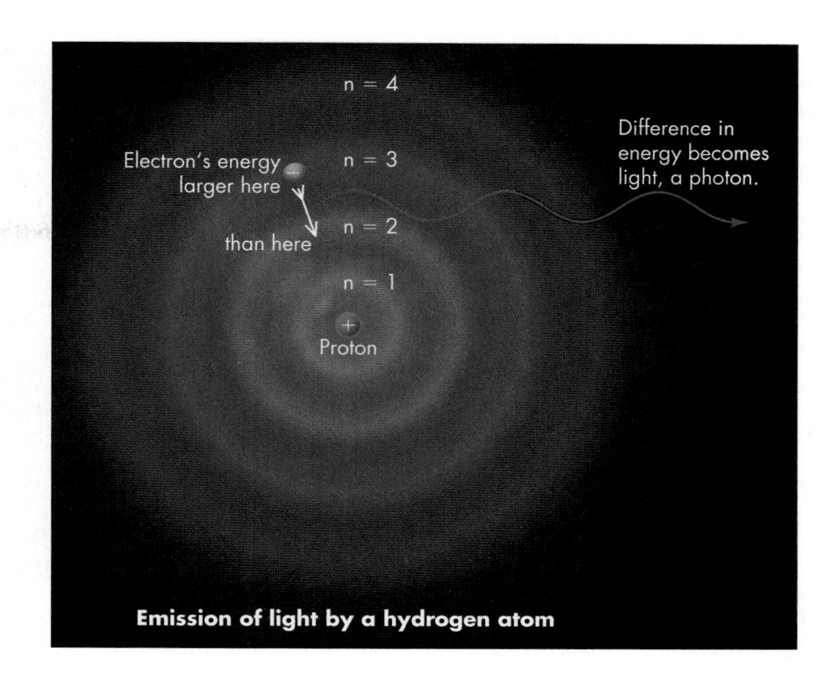

FIGURE 21.4
Energy is released when an electron drops from an upper to a lower orbital, causing the atom to emit electromagnetic radiation.

number of protons. This means that each atom normally has an equal number of positive and negative electrical charges and is therefore electrically neutral.

Unfortunately, an atom with six protons does not simply have electron orbitals that interact with photons six times more energetic than those that interact with hydrogen, or any other such simple relationship. The electron orbital clouds interact with each other in complex ways, giving each element a rich variety of unique atomic properties. We study these in detail in Unit 24, but here we provide an overview of the principles involved.

We have seen that when an electron moves from one orbital to another, the energy of the atom changes. If the electron is in its innermost orbital, closest to the nucleus, the atom's energy is at its lowest possible energy, or ground state. If one or more electrons is shifted outward into a larger orbital, the atom's energy is increased. Such an atom is said to be in an **excited state**.

Although the energy of an atom may change, the energy does not just appear or disappear; rather, it changes form (see Unit 20 on the conservation of energy). According to this principle, if an atom loses energy, exactly the same amount of energy must reappear in some other form, such as an electromagnetic wave.

How is electromagnetic radiation created? When an electron drops from one orbital to another, it alters the electric energy of the atom. As we described earlier, such an electrical disturbance generates a magnetic disturbance, which in turn generates a new electrical disturbance. Thus the energy released when an electron drops from a higher to a lower orbital becomes an electromagnetic wave, a process called **emission** (Figure 21.4).

Emission plays an important role in many astronomical phenomena. The aurora (northern lights) is an example of emission by atoms in the Earth's upper atmosphere. Gas in the Sun's outer atmosphere and clouds of gas in deep space also emit light as electrons drop to lower energy orbitals. The light from the flame of a candle is another example of emission, and when different elements are burned, they produce different colors because of their different electron energy levels (Figure 21.5).

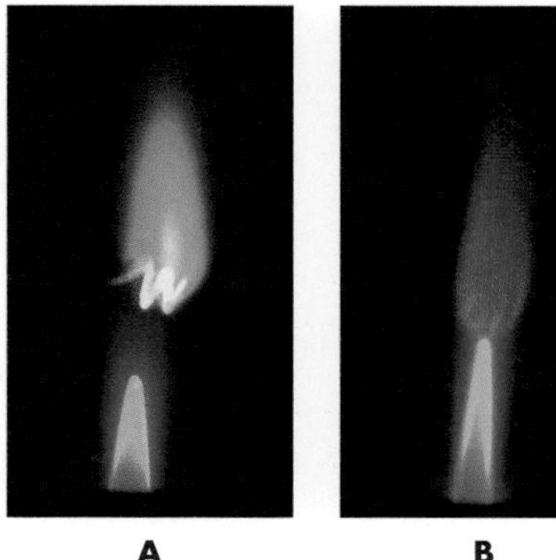
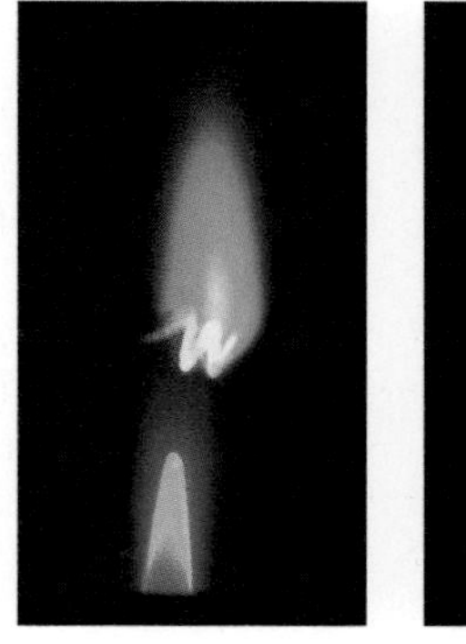

FIGURE 21.5
When different elements are heated in a flame, they emit different wavelengths of light. When placed in the burner flame, copper (A) appears green, while strontium (B) appears red.

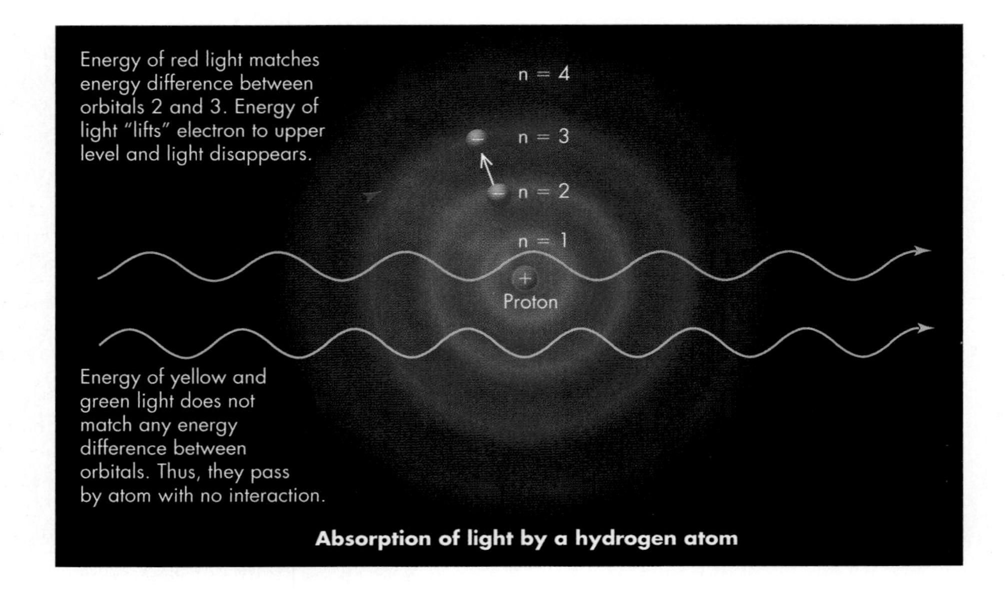

FIGURE 21.6
An atom can absorb a photon if the photon's energy matches the energy required for one of the atom's electrons to jump to a higher orbital.

Emission of electromagnetic waves

Absorption of electromagnetic waves

The reverse process, in which light's energy is stored in an atom as energy, is called **absorption** (Figure 21.6). Absorption lifts an electron from a lower to a higher orbital and excites the atom by increasing the electron's energy. Absorption may even **ionize** the atom—this occurs if the electron gains so much energy that it leaves orbit completely. Absorption is important in understanding such diverse phenomena as the temperature of a planet and the identification of star types, as we will discover in later units.

Whether you picture emission and absorption as photons and electrons knocking about or as waves exciting vibrations, the important thing to keep track of is how much energy has been transferred. You may find it helpful in understanding emission and absorption if you think of an analogy. Absorption is a bit like drawing an arrow back, preparing to shoot it from a bow. Emission is like the arrow being shot. In one case, energy of your muscles is transferred to and stored in the flexed bow. In the other, the stored energy is released as the arrow takes flight.

KEY TERMS

absorption, 164
atomic number, 162
brightness, 160
electromagnetic wave, 158
electronic transition, 162
element, 162
emission, 163
excited state, 163
ground state, 161
interfere, 158
inverse-square law, 160
ionize, 164
light, 157
orbital, 161
photon, 158
quantized, 158
wave–particle duality, 159

QUESTIONS FOR REVIEW

1. Why is light called an *electromagnetic wave?*
2. What is a photon? What is meant by duality?
3. How is the brightness of an object different from its luminosity (the total light output of the object)?
4. What is the inverse-square law?
5. What determines whether a particular photon can interact with an atom?
6. What is the difference between emission and absorption in terms of what happens to an electron in an atom?

PROBLEMS

1. Given that the Sun is 150 million kilometers from the Earth, calculate how long it takes light to travel from the Earth to the Sun.
2. Suppose you are operating a remote-controlled spacecraft on Mars from a station here on Earth. How long will it take the craft to respond to your command if Mars is at its nearest point to Earth? (You will need to look up a few numbers in the Appendix.)
3. a. At Mars's distance of 1.52 AU from the Sun, how many times less bright is the light from the Sun than at Earth's distance?
   b. What about at Mercury's distance of 0.39 AU?
4. The Sun is about 10 billion times brighter than the next brightest star, Sirius, (as seen from Earth). How far would we have to be from the Sun for the Sun to be as bright as Sirius? (The distance from the Earth to the Sun is 1.00 AU).
5. Suppose that the electron in a hydrogen atom is in the fourth energy level (the third excited state). How many different-energy photons can be released as the electron drops back to the ground state?
6. Describe, using basic principles, the story of a green photon that was generated in the Sun, was reflected off some object on Earth, and then entered your eye. What happened at the atomic level along each step of its journey? For example, what does it mean that you "see" the photon?

TEST YOURSELF

1. Light is an electromagnetic wave in that it
   a. is made of magnetized electrons moving back and forth.
   b. can be generated by the motions of electrical charges.
   c. travels through water in microscopic oscillations.
   d. carries no energy.
2. When a photon interacts with an atom, what changes occur in the atom?
   a. The atomic number increases.
   b. The nucleus begins to glow.
   c. An electron changes its orbital.
   d. The photon becomes trapped in orbit.
3. Describing an atom's orbitals as being *quantized* refers to the fact that
   a. the atom is made of individual electrons, protons, and neutrons.
   b. the electron can oscillate only with certain discrete patterns.
   c. there is the same number of electrons as protons.
   d. there is a whole number of electrons in an atom.
   e. the electron's electric charge has a single fixed value.

unit

22

The Electromagnetic Spectrum

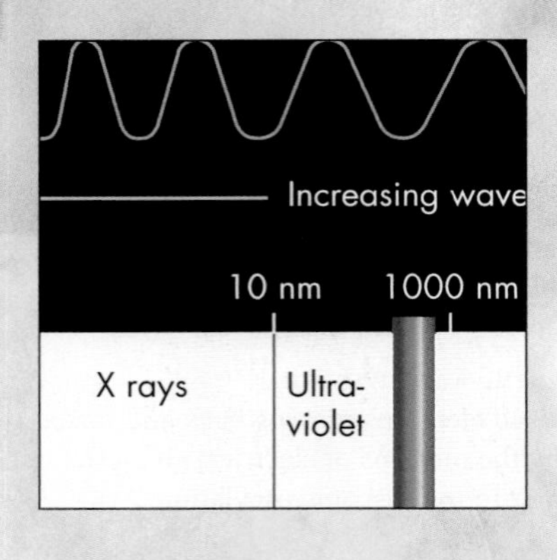

Background Pathways

Unit 21: Light, Matter, and Energy 157

There is more to light than what meets the eye. Our eyes can detect electromagnetic waves—known as *electromagnetic radiation*—with certain energies, but there are waves we cannot see. You are already familiar with many other forms of electromagnetic waves from many different sources. For example, radio waves, X-rays, and ultraviolet light are also electromagnetic radiation.

Other kinds of electromagnetic radiation differ from visible light only in the length of their electromagnetic waves. We can alternatively describe this in terms of the energies their photons contain. X-rays are rapidly oscillating waves; equivalently, we might call them highly energetic photons. Radio waves are slowly oscillating waves—or low-energy photons. This entire range of waves is called the *electromagnetic spectrum.*

22.1 THE COLORS OF VISIBLE LIGHT

Regardless of whether we consider light to be a wave or a stream of photon particles, our eyes perceive one of its most fundamental properties as color. Human beings can see colors ranging from deep red through orange and yellow into green, blue, and violet. The colors to which the human eye is sensitive define what is called the **visible spectrum.** But what property of photons or electromagnetic waves corresponds to light's different colors?

According to the wave interpretation of light, light's color is determined by the light's **wavelength,** which is the spacing between wave crests (Figure 22.1). That is, instead of describing a quality, light's *color,* we can specify a quantity, its *wavelength,* denoted by the Greek letter lambda, λ. The wavelengths of visible light are very small—roughly the size of a bacterium. For example, the wavelength of deep-red light is about 7×10^{-7} meter, or 700 nanometers. The wavelength of violet light is about 4×10^{-7} meter, or 400 nanometers. Intermediate colors have intermediate wavelengths.

Sometimes it is useful to describe electromagnetic waves by their frequency rather than their wavelength. **Frequency** is the number of wave crests that pass a given point in 1 second. The number of waves or "cycles" per second is indicated with the MKS unit **hertz** (abbreviated as Hz), named after Heinrich Hertz, who pioneered the broadcast and reception of radio waves in the late 1800s. In equations the frequency is usually denoted by the Greek letter nu: ν.

The frequency and wavelength of a wave have a simple relationship to each other. When the wavelength is longer, the frequency is lower (slower); when the wavelength is shorter, the frequency is higher (faster). If you float in ocean waves, as illustrated in Figure 22.2, you can experience this in terms of the rate at which you bob up and down (frequency) as waves pass by.

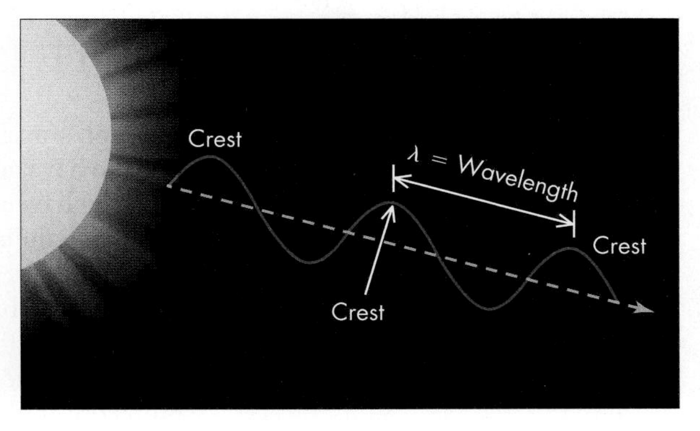

FIGURE 22.1
The distance between crests defines the wavelength, λ, for any kind of wave, be it water or electromagnetic.

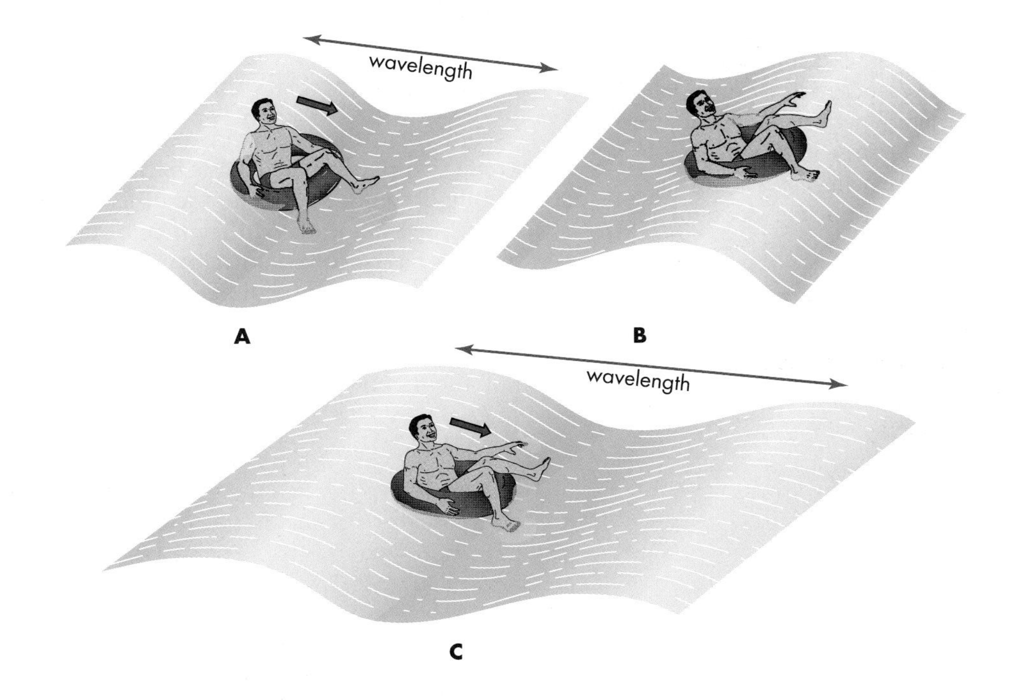

FIGURE 22.2
The frequency at which you bob up and down while floating in the water depends on the wavelength of the wave. Short-wavelength waves move past you quickly, so you bob up and down at a high frequency (A–B). A long-wavelength wave (C) takes longer to pass by, even though the speed of the wave crest is the same, so the frequency is slower.

The speed of light is reduced when light travels through materials that appear transparent to us, such as glass, water, and gases. Furthermore, different colors—wavelengths—of light are slowed differently. For example, in nearly all materials, blue light travels slightly more slowly than red light.

The relationship between the frequency and wavelength of a wave is determined by the wave speed, because in one vibration a wave must travel a distance equal to one wavelength. Thus if we multiply the wavelength λ by the number of waves per second, ν, the product is the speed of light:

$$\lambda \times \nu = c.$$

Because all light waves travel at the same speed, c, the wavelength determines the frequency and vice versa. We will generally use λ to characterize electromagnetic waves, but ν is just as good. For example, yellow light vibrates about 500 trillion times a second (5×10^{14} Hz); red light vibrates a little slower; blue light a little faster.

Be sure to remember that *longer wavelengths of visible light correspond to redder colors.* Equivalently, *lower frequencies correspond to redder colors.* We will use this information many times as we interpret the significance of different colors of light coming from astronomical objects.

22.2 WHITE LIGHT

Although wavelength is an excellent way to specify most colors of light, some light seems to have no color. For example, the Sun high in the sky and an ordinary lightbulb appear to have no dominant color. Light from such sources is called **white light.**

White light is not a special color of light; rather it is a mixture of all colors. That is, the sunlight we see is made up of all the wavelengths of visible light—a blend of red, yellow, green, blue, and so on—and our eyes perceive the combination of all these as white. Newton demonstrated this property of sunlight by a simple but elegant experiment. He passed sunlight through a prism (Figure 22.3) so that the light was spread out into the visible spectrum (which we see as a rainbow of colors). He then recombined the separated colors with a lens and reformed the beam of white light.

The adding of light of different colors is very different from the way that pigments of paint mix, because the pigments reflect the color we see and absorb other colors. When you mix red, green, and blue paint, you will generally get a dark gray or brown color because most wavelengths of light are being absorbed. This is called "subtractive" color mixing.

You can see how colors of light mix if you look at a color television screen with a magnifying glass. You will notice that the screen is covered with tiny red, green, and blue dots. In a red object, only the red spots are lit. In a blue object, only the blue spots are illuminated. In a white object, all three color spots are lit, and your brain mixes these three colors to form white. Other colors are made by appropriate blending of red, green, and blue.

You may have noticed that photos taken in artificial light look much more yellow than the scene you remember, or outdoor photographs may look bluer than you remember. That is because most lightbulbs are much more yellow than sunlight, but we adjust to the difference. We generally see a white piece of paper as white even under different color lighting conditions. Presumably because our senses have evolved to make us aware of differences in our surroundings, we ignore the ambient "color" of sunlight just as in time we learn to ignore a steady background sound or smell.

White light and the enormous variety of colors we can see represent just some of the richness possible for electromagnetic waves ranging between 400 and

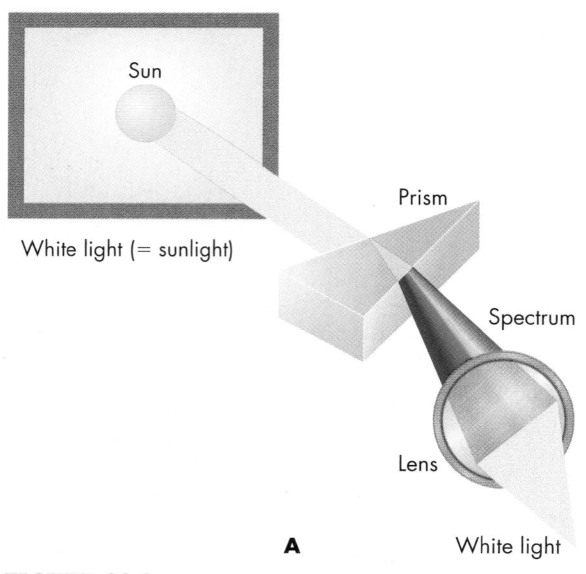

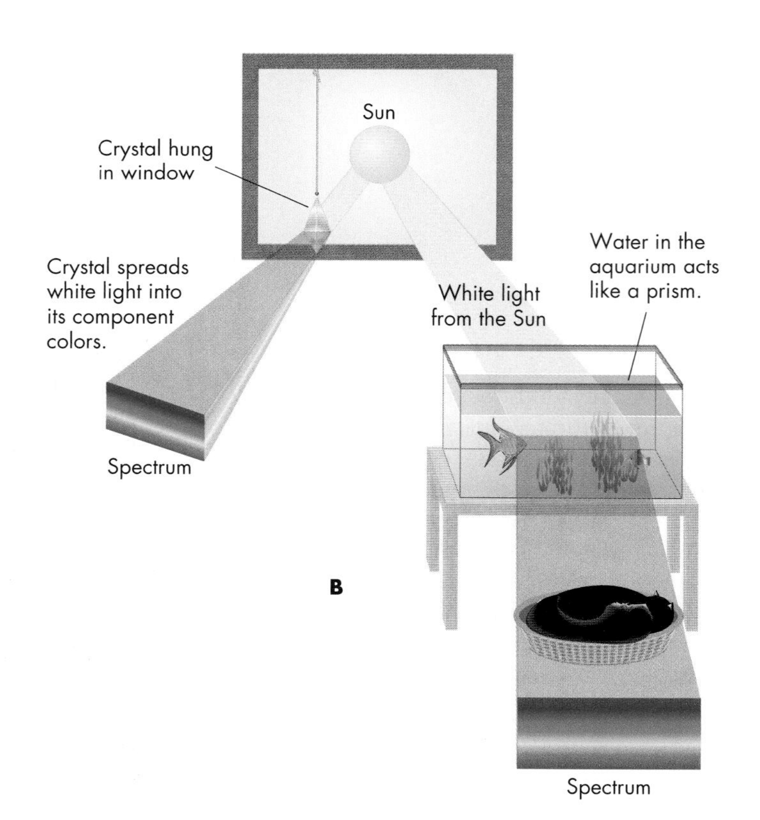

FIGURE 22.3
(A) A prism spreads "white" light into its component colors (a spectrum). Combining the colors again with a lens makes the light "white" again. (B) Spectra in everyday life.

700 nm. Yet today there are instruments capable of detecting electromagnetic waves with wavelengths anywhere from thousands of kilometers long down to 10^{-18} meter or less. Ordinary visible light falls in a narrow section in the middle of a broad spectral range (see Figure 22.4 and Table 22.1). Objects may emit and absorb light in every part of the electromagnetic spectrum, creating "colors" completely outside the range of our eyes. Using our eyes alone, an object may appear to be dark; but it may be the source of an assortment of different-wavelength photons when viewed with a camera that is sensitive to a different part of the electromagnetic spectrum.

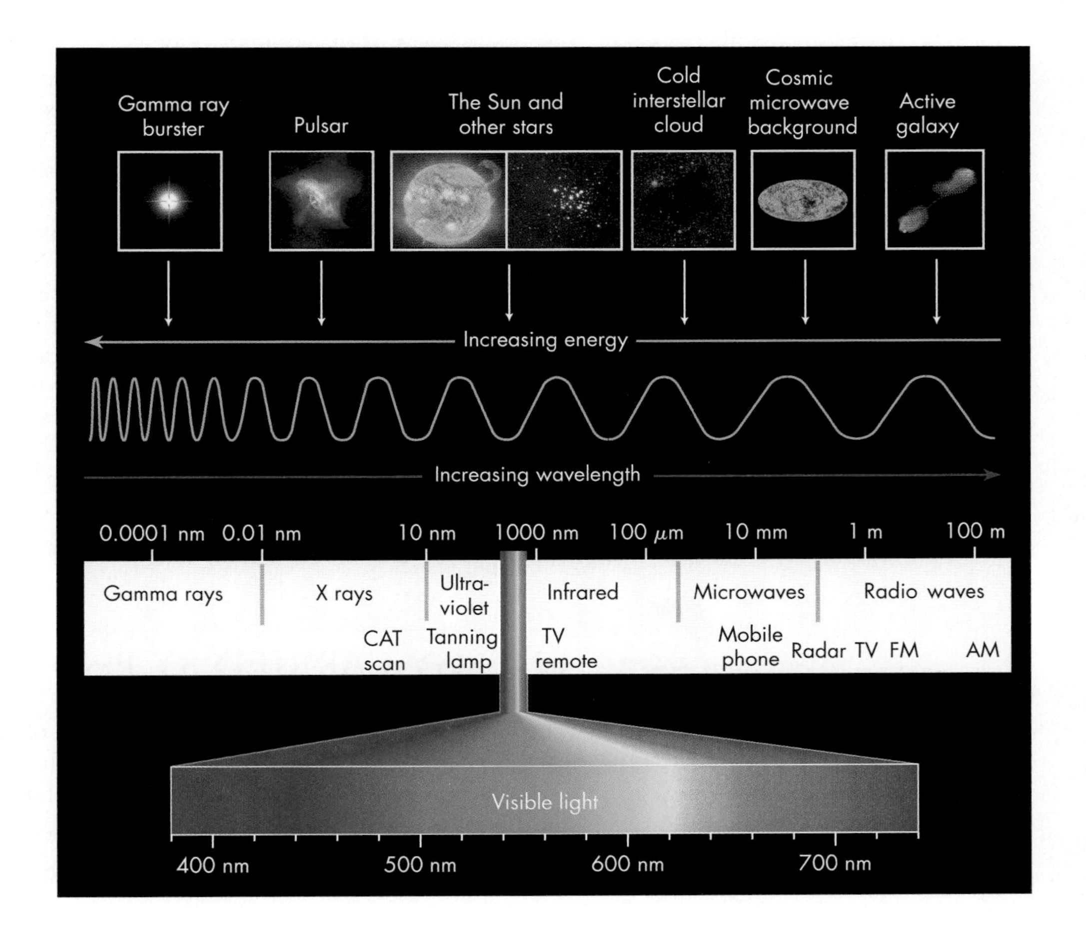

FIGURE 22.4
The electromagnetic spectrum.

TABLE 22.1 Electromagnetic Spectrum

Wavelength	Kind of Radiation	Astronomical Sources
100–500 meters	Radio (AM broadcast)	Pulsars (remnants of exploded stars)
10–100 meters	Short-wave radio	Active galaxies
1–10 meters	TV, FM radio	Solar radio outbursts, interstellar gas
10–100 centimeters	Radar, cell phones	Planets, active galaxies
1–100 millimeters	Microwaves	Interstellar clouds, cosmic background radiation
1–10^3 micrometers	Infrared	Young stars, planets, interstellar dust
400–700 nanometers	Visible light	Stars, Sun
10–300 nanometers	Ultraviolet	Massive stars
0.01–10 nanometers	X-rays	Collapsed stars, hot gas in galaxy clusters
10^{-7}–0.01 nanometer	Gamma rays	Active galaxies and gamma-ray bursters

22.3 INFRARED AND ULTRAVIOLET RADIATION

The exploration of the electromagnetic spectrum beyond visible light began in 1800, when Sir William Herschel (discoverer of the planet Uranus) showed that heat radiation, such as you feel from the Sun or from a warm radiator, though invisible, is related to visible light.

Although humans cannot see infrared radiation, several kinds of snakes, including the rattlesnake, have special infrared sensors located just below their eyes. These allow the snake to "see" in total darkness, helping it to find warm-blooded prey such as rats.

Herschel was trying to measure heat radiated by astronomical sources. He projected the Sun's spectrum onto a tabletop and placed a thermometer in each color to measure its energy. He was surprised that when he put a thermometer just off the red end of the visible spectrum, the thermometer registered an elevated temperature there just as it did in the red part of the spectrum. He concluded that some form of invisible energy detectable as heat existed beyond the red end of the spectrum, and he therefore called it **infrared.** Even though your eyes cannot see infrared light, when it strikes your skin it deposits energy that warms the skin, and the nerves in your skin can feel the heat.

Just a year later, in 1801, the German scientist Johann Ritter was studying the light sensitivity of chemicals that were later to become the basis for photography. When he shone light through a prism onto one of these chemicals, he discovered that there was an intense photochemical reaction beyond the last visible portion of the spectrum at its violet end—**ultraviolet** radiation.

Infrared and ultraviolet radiation differ in no physical way from visible light except in their wavelength. Infrared has longer wavelengths and ultraviolet shorter wavelengths than visible light (see Table 22.1). Both can be described as waves or photons of electromagnetic energy, like visible light. They just cannot be seen by the human eye.

22.4 ENERGY CARRIED BY PHOTONS

The warmth we feel on our face from a beam of sunlight demonstrates that light carries energy, but not all wavelengths carry the same amount of energy. It turns out that the amount of energy, E, carried by a photon depends on its wavelength, λ. Each photon of wavelength λ carries an energy, E, given by

$$E = \frac{h \times c}{\lambda}$$

In MKS units, E is measured in joules and λ is measured in meters, and c is the speed of light. Laboratory measurements of the constant h have determined its value: $h = 6.63 \times 10^{-34}$ joule-second, and $h \times c = 1.99 \times 10^{-25}$ joule-meter. So for a red photon with a wavelength of 670 nm,

$$E = \frac{h \times c}{\lambda} = \frac{1.99 \times 10^{-25}\,\text{joule} \cdot \text{m}}{670\ \text{nm}}$$

$$= \frac{1.99 \times 10^{-25}\,\text{joule} \cdot \text{m}}{6.7 \times 10^{-7}\,\text{m}}$$

$$= 3.0 \times 10^{-19}\,\text{joule}$$

The quantity h is called *Planck's constant,* a constant of nature that describes the fundamental relationship between energies and waves in the submicroscopic realm of quantum mechanics. This equation can also be written in terms of the frequency of a photon. From the equation relating wavelength and frequency, we find that $\nu = c/\lambda$, so the energy of a photon is equivalently $E = h \times \nu$.

Thus the wavelength of a photon determines how much energy it carries. Long-wavelength, low-frequency photons contain less energy. In other words,

> **A photon carries an amount of energy proportional to its frequency and inversely proportional to its wavelength.**

For example, an ultraviolet photon with a wavelength of 300 nm carries twice as much energy as a 600-nm yellow photon and three times as much energy as a 900-nm infrared photon. In fact, ultraviolet photons of sufficiently short wavelength carry so much energy that they can break apart atomic and molecular bonds, which is why ultraviolet light gives you a sunburn but an infrared "heat lamp" does not.

22.5 RADIO WAVES

Q It might help you remember the connection between energy, wavelength, and frequency to make waves on a rope (or a telephone cord). Tie one end to a doorknob and shake the loose end of it. To make short-wavelength waves along the rope, how rapidly (frequently) do you shake the rope? Does it take more of your energy to make short waves or long waves?

Q Why do you suppose that the centers of some filled pastries grow hot in a microwave oven while their outer portions do not grow as hot?

James Clerk Maxwell, a Scottish physicist, predicted the existence of **radio waves** in the mid-1800s. It was some 20 years later, however, before Heinrich Hertz produced them experimentally in 1888, and another 50 years had to pass before Karl Jansky discovered naturally occurring radio waves coming from cosmic sources.

Radio waves range in length from millimeters to hundreds of meters and longer, making them much longer than visible and infrared waves. An AM "radio" is a receiver for electromagnetic waves that are hundreds of meters long; FM radio corresponds to wavelengths of a few meters. Actually radio wavelengths are more commonly listed by their frequency. You can see this on a radio dial, where you tune in a station by its frequency in kilohertz or megahertz (thousands or millions of waves per second) rather than its wavelength.

Today we can generate radio waves and use them in many ways, ranging from communication to radar to cooking. Given the enormous span of wavelengths and uses, the wide range of radio waves is sometimes subdivided, with the shorter wavelengths being called **microwaves.** These are generally described as waves of about 10 cm or less, although there is no definite standard on what the boundaries are between infrared, microwave, and radio. Most microwave ovens, for example, generate electromagnetic waves with wavelengths of about 12 cm.

The strong connection between radio waves and their use for communication systems confuses many people into thinking that radio waves are a kind of sound wave. Not so! Radio waves are a kind of light, traveling at the same enormous speed as all forms of electromagnetic radiation. If radio waves traveled anywhere near as slowly as sound, you would have to wait tens of seconds for every response when calling on a cell phone to a friend just a few kilometers away, and hours when calling across the country!

Many kinds of astronomical objects such as forming stars, exploding stars, active galaxies, and interstellar gas clouds generate these low-energy radio waves by natural processes. Understanding this radiation reveals information about the temperature of the gas, the molecules present, magnetic fields in the region, and a wide variety of other factors. Astronomers detect this radiation with radio telescopes to study the kinds of astrophysical processes and conditions that are different from those that generate visible light.

22.6 HIGH-ENERGY RADIATION

Over the last several decades, astronomers have also probed very short-wavelength regions called the **X-ray** and **gamma ray** parts of the spectrum. X-ray wavelengths are even shorter than those of visible and ultraviolet light, typically between 0.01 and 10 nanometers, and gamma-ray wavelengths are shorter yet (see Table 22.1).

The fact that X-rays and gamma rays have such short wavelengths implies that they carry a great deal of energy and are generated only in regions with extremely high temperatures or high-energy reactions taking place. These wavelengths therefore reveal to astronomers some of the universe's most exotic objects and processes, such as hot gas falling into black holes.

The high energy of these photons allows them to penetrate many objects that are opaque to visible light. One familiar example is the ability of X-rays to penetrate soft tissue, which allows us to make diagnostic medical X-ray images. The X-ray photons that are not blocked by bone or other dense material travel through and expose a photographic negative—much like Ritter's 1801 experiment with ultraviolet light.

Q Our eyes are also insensitive to infrared light, but there are "infrared night-vision goggles." How do you suppose we can see infrared light with these?

Note that because X-rays are outside the range that our eyes can see, there is no such thing as a pair of "X-ray glasses" that lets you see through objects. If you looked through a pair of glasses that transmitted only X-rays, you would see . . . *nothing!*

Despite the enormous variety of electromagnetic waves, they are all the same physical phenomenon: the vibration of electric and magnetic energy traveling at the speed of light. The essential difference between kinds of electromagnetic radiation is merely their wavelength—or frequency—or energy.

KEY TERMS

electromagnetic radiation, 166
electromagnetic spectrum, 166
frequency, 166
gamma ray, 171
hertz, 166
infrared, 170
microwaves, 171
radio waves, 171
ultraviolet, 170
visible spectrum, 166
wavelength, 166
white light, 168
X-ray, 171

QUESTIONS FOR REVIEW

1. How are color and wavelength related?
2. What is meant by the *electromagnetic spectrum?*
3. Why is it more dangerous to be exposed to a weak X-ray source than to a strong infrared source?
4. Does all electromagnetic radiation travel at the same speed?
5. What are the bands of the electromagnetic spectrum from short to long wavelengths?
6. In what ways is it possible and impossible for people to personally detect energy outside the visible spectrum?

PROBLEMS

1. The frequency of a radio station transmitter is 90 megahertz. What is its wavelength?
2. Microwave ovens operate at 12-cm wavelength by making water molecules oscillate at the frequency of the electromagnetic radiation. What frequency is this? Explain at the atomic level why some substances heat up and others do not. How does this differ from what happens in conventional ovens at the atomic level?
3. A yellow photon has wavelength of 500 nm. What is its frequency? How much energy does it have?
4. What is the wavelength of a photon that has 1000 times as much energy as a photon with a wavelength of 500 nm? Would it be more dangerous to be exposed to 1 million photons with a wavelength of 500 nm, or 1000 photons with 1000 times the energy of a 500-nm photon? Why?
5. The energies of electrons in their atomic orbitals are typically about 10^{-18} joule. What wavelength of light would have photons with this energy? Which band of electromagnetic radiation is this in?
6. a. What is the energy (in joules) of an extremely powerful gamma ray with a wavelength of 10^{-7} nanometer?
   b. How many of these gamma-ray photons would add up to the same amount of energy as a housefly in flight, which has a kinetic energy of about one millionth of a joule as it flies along?

TEST YOURSELF

1. Which type of electromagnetic radiation has the longest wavelength?
   a. Ultraviolet
   b. Visible
   c. X-ray
   d. Infrared
   e. Radio
2. Which kind of photon has the highest energy?
   a. Ultraviolet
   b. Visible
   c. X-ray
   d. Infrared
   e. Radio
3. A photon may be Doppler-shifted to a lower frequency if the source of light is moving away from us at high speed. If the frequency drops to half its former value, the energy of the photons will
   a. drop by half too.
   b. double.
   c. remain the same.

unit

23 Thermal Radiation

When a body becomes hotter, the atoms and molecules inside it vibrate more rapidly, undergo more collisions, and generally interact more energetically. It is not surprising, then, that a hot object generally emits more electromagnetic radiation and that the photons emitted tend to have higher energies. For example, when a chunk of charcoal becomes hot, it may glow with an orange color. This kind of electromagnetic emission is called **thermal radiation**.

Thermal radiation is produced by many objects that you might not think of as "glowing." Your body or an animal's body produces infrared thermal radiation, for example (Figure 23.1). On the other hand, some objects, like the collapsed remnants of stars, are so hot that their thermal radiation is mostly at X-ray wavelengths. Not all kinds of materials produce thermal radiation—in particular, low-density gas produces emission at the individual wavelengths associated with electrons changing orbitals. A wide variety of denser materials produce thermal emission that looks similar no matter what the chemical composition is of the emitting materials.

Wherever thermal radiation occurs, it follows two simple rules: (1) The photons have higher energy when the material is hotter, and (2) the total amount of radiation produced increases rapidly as the temperature climbs. These two rules have mathematical formulations that allow us to learn about the physical conditions in material from the thermal radiation it emits. These provide important tools for studying objects throughout the universe.

Background Pathways
Unit 22: The Electromagnetic Spectrum 166

23.1 BLACKBODIES

A **blackbody** is an object that absorbs all the radiation falling on it. Because such a body reflects no light, it looks black to us when it is cold, hence its name. For an object to appear black, it must be capable of absorbing all wavelengths of electromagnetic radiation. This implies that the substance is capable of undergoing energy transitions corresponding to all wavelengths. This is *not* true of a low-density gas, but it is true or nearly true for a wide range of other substances.

When blackbodies are heated, they can also radiate at all wavelengths. Once again, this is because transitions of all energies (or equivalently, all wavelengths) are possible. Thus they are both excellent absorbers and excellent emitters. Moreover, the intensity of their radiation changes smoothly from one wavelength to the next with no gaps or narrow peaks of brightness. Very few objects are perfect blackbodies, but many of the objects astronomers study are near enough to being blackbodies that their radiation looks very similar. For example, a piece of charcoal, an electric stove burner, the Sun, and the Earth all produce thermal emission that closely resembles the idealized blackbody.

FIGURE 23.1
A picture of the infrared thermal radiation from a dog. The dog's eyes are warmest and produce the most light, while its nose is cold and emits little light.

On the other hand, gases (unless compressed to a high density) are generally not blackbodies and do not produce thermal emission. The wavelengths emitted by a gas are determined by its composition and the electron energy-level transitions that are possible for the kinds of atom and molecules present in the gas.

What makes a dense material different? Within it, the orbitals of each atom's electrons are disturbed by neighboring atoms. The possible energies of the

electrons are no longer precisely defined as they are for isolated atoms, so an incoming photon is likely to encounter an atom that can absorb its energy. Solid materials often retain some characteristics of their component atoms and molecules, which is why the things we see are not all black. For example, a leaf appears green because it absorbs most wavelengths *except* those with green colors.

As a solid or dense gas becomes hotter, its similarity to a blackbody increases. This is because the vibrations and energetic collisions associated with higher temperatures cause the electrons' orbitals to be disturbed even more, further "smearing out" the range of energies—and wavelengths—they can absorb or emit.

23.2 COLOR, LUMINOSITY, AND TEMPERATURE

One important property we can infer about an object from the light it emits is its temperature. Temperature gives rise to subtle differences in the colors of stars at night. And by observing beyond the wavelengths of visible radiation, we can detect objects both much cooler and much hotter than stars.

A hot object emits light, as you can easily see if you turn on the burner of an electric stove. You can also see from such a burner that the color of the light is related to the temperature of the burner. In particular, as the burner grows hotter, at first it emits infrared light that you might sense as heat; then it shifts to red, then orange (Figure 23.2A). This relation between color and temperature allows astronomers to measure the temperature of stars and other astronomical objects from the wavelengths of light they emit. They can make this measurement using a relation called **Wien's law,** named for the German physicist Wilhelm Wien (pronounced *Veen*), who discovered it in the 1890s.

Wien's law states that the wavelength at which an object radiates most strongly is inversely proportional to the object's temperature:

Hotter bodies radiate more strongly at shorter wavelengths.

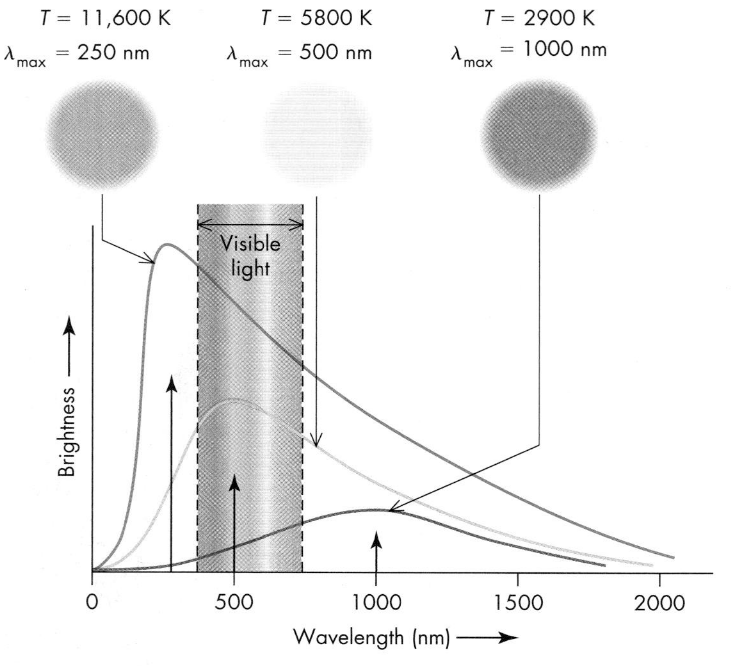

FIGURE 23.2
(A) The hotter burner glows more orange than the cooler burner. (B) As an object is heated, the wavelength at which it radiates most strongly, λ_{max}, shifts to shorter wavelengths, a relation known as *Wien's law.* Note also that as the object's temperature rises, the amount of energy radiated increases at *all* wavelengths.

We can see this effect illustrated in Figure 23.2B, where we plot the amount of energy radiated at each wavelength (color) for three bodies of different temperatures. Notice that the hotter body has its most intense emission (highest point) at a shorter wavelength than the cooler body. This is what gives it a different color.

Note also that for objects of the same size, the hotter object emits more energy at every wavelength than the cooler one does. This is obvious looking at a stove burner. As it grows hotter, going from dull red to orange, it also grows much brighter. This is also the basis for the dimmer switches on lights: Changing the amount of electricity going to the lamp changes the temperature of the filament in the lightbulb and thereby its brightness. The light output of an object is known as its **luminosity.** What is observed is that even if the filament rises only a little bit in temperature, its luminosity increases substantially. Thus,

The luminosity of a hot body rises rapidly with temperature.

A mathematical form of this principle is formulated as the *Stefan-Boltzmann law.*

Before we can write out these mathematical laws in a form that allows us to carry out specific calculations, we must decide what temperature scale we wish to use. To make this choice, it will be helpful if we briefly discuss what temperature measures.

23.3 MEASURING TEMPERATURE

Fahrenheit temperatures, °F, can be calculated by using the formula $T_{°F} = 9/5\ T_K - 459.4$, where T_K is the temperature in Kelvin. The Fahrenheit scale was designed to have 0°F at the temperature at which concentrated saltwater freezes and 100°F at body temperature.

An object's temperature is directly related to its energy content and to the speed of its molecular motion. That is, the hotter the object, the faster its atoms move and the more energy it possesses. Similarly, if the body is cooled sufficiently, molecular motion within it slows to a virtual halt and its energy approaches zero. In the late 1800s the British scientist Lord Kelvin devised a temperature scale based on these properties (Figure 23.3). At zero on the **Kelvin** scale, molecular motion essentially ceases and objects have no heat energy. This is known as **absolute zero**—the lowest possible temperature. Notice that negative temperatures therefore have no meaning on the Kelvin scale because there cannot be less motion than none.

It is simple to convert from Kelvin to the more familiar Celsius scale. The temperature in Kelvin equals 273.15 plus the Celsius temperature. For example, the

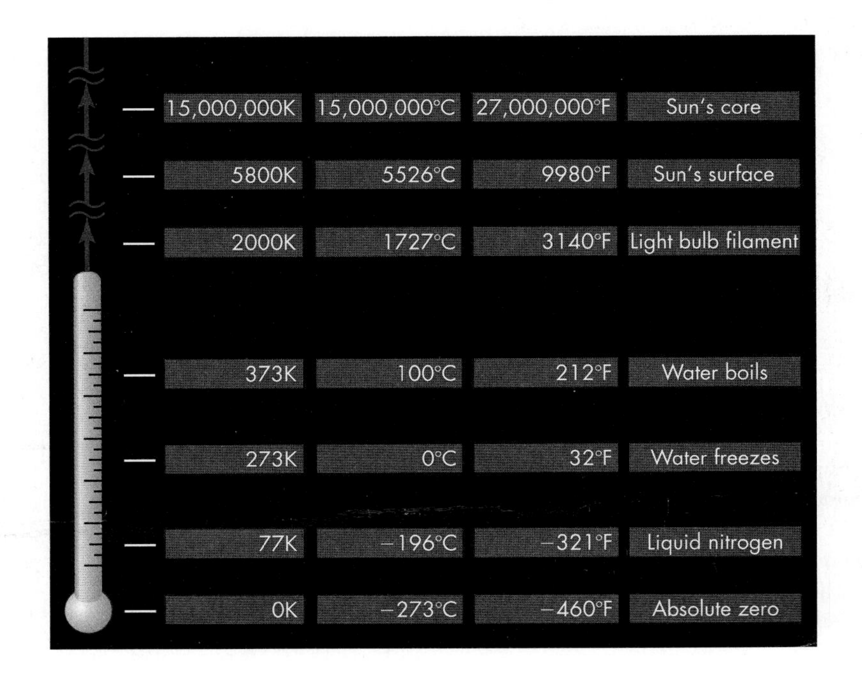

FIGURE 23.3
Temperatures in Kelvin (K) and on the Celsius and Fahrenheit scales.

Note that temperatures on the Kelvin scale are not given in "degrees" but are simply called "Kelvin."

freezing and boiling points of water, 0°C and 100°C, are approximately 273 K and 373 K, respectively. Room temperature is about 300 Kelvin.

The Kelvin scale is useful in astronomy because the temperature in Kelvin is directly proportional to an object's **thermal energy** (the energy due to the internal vibrations and motions of the atoms within the object). Thus when the thermal energy doubles, the value of the temperature in Kelvin also doubles. By contrast, when the temperature of an object doubles in the Celsius and Fahrenheit scales, it is *not* true that its thermal energy doubles.

Because of its direct relation to so many physical processes, we will use the Kelvin scale in most of the remainder of this book. With a temperature scale chosen, we can now learn how we might measure a star's temperature.

23.4 TAKING THE TEMPERATURE OF ASTRONOMICAL OBJECTS

Wien's law is important because we can use it to measure how hot something is simply from the color of light it radiates most strongly. To measure a distant body's temperature using Wien's law, we proceed as follows. First we measure the body's brightness at many different wavelengths to find the particular wavelength at which it is brightest (that is, its wavelength of maximum emission). Then we use the law to calculate the body's temperature. To see how this is done, however, we need a mathematical expression for the law.

If we let T be the body's temperature measured in Kelvin, and we let λ_{max} be the wavelength in nanometers at which the body radiates most strongly (Figure 23.2B), Wien's law can be written in this form:

$$T = \frac{2.9 \times 10^6 \text{ K} \cdot \text{nm}}{\lambda_{max.}}.$$

The subscript *max* on λ is to remind us that it is the wavelength of maximum emission.

As an example, we can measure the Sun's temperature. The Sun turns out to radiate most strongly at a wavelength of about 500 nanometers. Substituting $\lambda_{max} = 500$ nanometers, we find

$$T_{\odot} = \frac{2.9 \times 10^6 \text{ K} \cdot \text{nm}}{500 \text{ nm}} = 5800 \text{ K}.$$

The Sun is not a perfect blackbody; nevertheless, this result is within about 20 K of the value found by much more sophisticated analyses.

You might note that the wavelength at which the Sun radiates most strongly corresponds to a blue-green color, yet the Sun looks yellow-white to us. The reason we see it as whitish is related to how our eyes perceive color. Physiologists have found that the human eye interprets sunlight (and light from all extremely hot bodies) as whitish, with only tints of color. Such hot bodies emit a significant amount of light at all visible wavelengths, not just at the color corresponding to the wavelength of maximum emission. Thus cool stars look white tinged with red, whereas very hot stars look white tinged with blue.

Wien's law works quite accurately for most stars and planets or almost any material that has atoms packed tightly together. One important thing to remember, though, is that the law tells us only about thermal radiation *emitted* by an object. An object may simultaneously be emitting thermal radiation and *reflecting* light from another source. For example, the red color of an apple and the green color of a lime have nothing to do with their temperatures. They both do emit thermal radiation, but if they are at normal room temperature, the radiation they emit will be primarily in the infrared range (Figure 23.4).

Q Suppose you want to design a paint to keep a building cool in bright sunlight. At what wavelengths would you want it to absorb light well, and at what wavelengths would you want it to reflect light well?

FIGURE 23.4
Side-by-side infrared and visible photographs of the same scene. Most of the visible light we see on Earth is light reflected off a surface, not thermal radiation.

23.5 THE STEFAN-BOLTZMANN LAW

In the late 1800s two Austrian scientists, Josef Stefan and Ludwig Boltzmann, showed how the total amount of light emitted by an object increases as its temperature rises, as indicated in Figure 23.2B. The **Stefan-Boltzmann law**, as their discovery is now called, states that a body of temperature T (measured in Kelvin) radiates an amount of energy each second equal to σT^4 per square meter. The quantity σ is called the **Stefan-Boltzmann constant**, and its value is 5.67×10^{-8} watt per meter^2 per Kelvin^4. Like Wien's law, this law is precisely true for blackbodies, but it is also quite accurate for many different kinds of materials, both solids and gases, as long as the material is relatively dense. It works well for embers of coal, a lightbulb filament, or the surface of a star.

The Stefan-Boltzmann law shows mathematically what we observe as rapid brightening as an object gets hotter. If the temperature doubles, for example, the light output increases by a factor of 16. We can show this mathematically for material heated to one temperature that's twice another. For example, suppose a lightbulb filament is heated to 1000 K. The amount of light it generates is

$$\begin{aligned}\sigma \times (1000\text{ K})^4 &= (5.67 \times 10^{-8}\text{ watt/m}^2/\text{K}^4) \times (10^3)^4 \times \text{K}^4\\ &= 5.67 \times 10^{-8} \times 10^{12}\text{ watt/m}^2\\ &= 5.67 \times 10^4\text{ watt/m}^2.\end{aligned}$$

On the other hand, if its temperature is raised to 2000 K, the luminosity is

$$\begin{aligned}\sigma \times (2000\text{ K})^4 &= (5.67 \times 10^{-8}\text{ watt/m}^2/\text{K}^4) \times (2 \times 10^3)^4 \times \text{K}^4\\ &= 5.67 \times 10^{-8} \times 16 \times 10^{12}\text{ watt/m}^2\\ &= 90.72 \times 10^4\text{ watt/m}^2.\end{aligned}$$

Comparing the second line in each of these calculations, you can see that the only difference is the factor of 16, which comes from taking 2 to the fourth power: $2^4 = 2 \times 2 \times 2 \times 2 = 16$. It would not matter what pair of temperatures we put into the equation: As long as one is *two times hotter* than the other, the amount of light generated by each square meter of the hotter object will always be *16 times more luminous* than the cooler one.

The other thing to notice from these calculations is the enormous wattages generated by material at these high temperatures. A square meter (about 10 square feet)

Q Why do you suppose a lightbulb filament is usually a thin winding wire instead of simply being thicker and straight? How might this thinking apply to the thermal radiation from a rocky asteroid versus the same amount of rock broken up into small pieces?

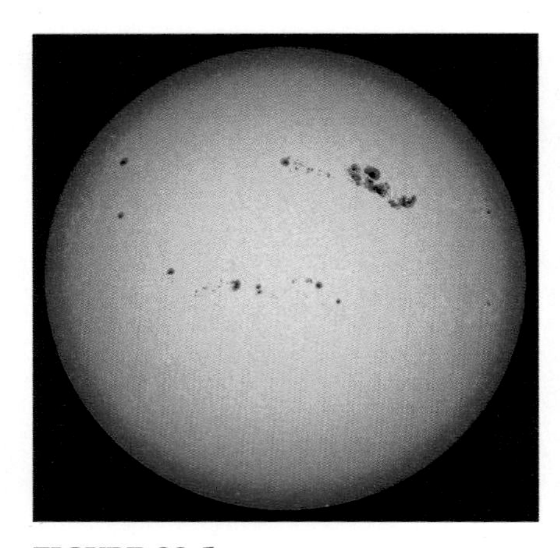

FIGURE 23.5
A portion of the Sun's surface. The darker spots in the picture are cooler than surrounding regions, so they look dark by contrast. However, they are generating light.

of material heated to 2000 K generates over 900,000 watts of luminous power! A lightbulb filament is typically about 2000 to 2500 K. If the total surface area of the filament is one square centimeter, that's equivalent to 10^{-4} m^2. Such a filament would produce about 91×10^4 watt/$m^2 \times 10^{-4}$ $m^2 = 91$ watts of thermal energy.

For a 2000-K filament, Wien's law tells us that most of the emission of lightbulbs is at infrared wavelengths we cannot see. Typically, less than 10% of the power is produced as visible light. By raising the square-centimeter filament's temperature to 2500 K, we would now produce 221 watts. In addition, Wien's law tells us that if we raised the temperature, we would shift the peak of the emission closer to optical wavelengths, so the amount of visible light would increase even more.

The Sun's surface is about 5800 K, so compared to the 1000-K thermal emission we just calculated, the Sun's surface must generate $5.8^4 = 5.8 \times 5.8 \times 5.8 \times 5.8 =$ 1130 times more power. This is over 64 million watts per square meter emitted by every square meter over the entire surface of this enormous body! When you look at photographs of the Sun's surface (Figure 23.5), you will see some regions that are darker and some lighter. These are regions with slightly different temperatures. The strong dependence of luminosity on temperature means that a region that is at 5000 K will look dimmer than the surrounding regions.

KEY TERMS

absolute zero, 175
blackbody, 173
Kelvin, 175
luminosity, 175
Stefan-Boltzmann constant, 177
Stefan-Boltzmann law, 177
thermal energy, 176
thermal radiation, 173
Wien's law, 174

QUESTIONS FOR REVIEW

1. How are color and temperature related?
2. Under what circumstances can an object that looks yellow be hotter than one that looks red?
3. What are the advantages of the Kelvin temperature scale?
4. What is Wien's law?
5. What is the Stefan-Boltzmann law?
6. What could explain how you might have two lightbulbs of equal brightness, both producing thermal radiation, but one bluer than the other?

PROBLEMS

1. What is the temperature range for objects whose wavelength at maximum falls within the visible spectrum?
2. A lightbulb radiates most strongly at a wavelength of about 3000 nanometers. How hot is its filament?
3. How hot would a source need to be to emit its maximum radiation in the FM range of the radio dial (100 MHz)? Why don't common objects interfere with our radio reception?
4. The Earth's temperature averaged over the year is about 300 Kelvin. At what wavelength does it radiate most strongly? What part of the electromagnetic spectrum does this wavelength lie in? Can you see it?
5. If a lightbulb filament could be heated to 10,000 K, how much more thermal radiation would it produce than if it were at 2500 K?
6. The total light an object emits depends on its temperature and its surface area. How large would a star with a surface temperature of 3000 K need to be to have the same luminosity as a star with a surface temperature of 6000 K? (Please answer in terms of the size of the 6000-K star.)

TEST YOURSELF

1. Suppose we doubled the thermal energy of a rock that had a temperature of 7°C = 45°F = 280 K. What would its new temperature be?
   a. 14°C
   b. 90°F
   c. 560 K
   d. All of the above
2. If the temperature of an object doubles, the wavelength where it emits the most amount of light will be
   a. 4 times longer.
   b. 2 times longer.
   c. the same.
   d. 2 times shorter.
   e. 4 times shorter.
3. If the temperature of an object doubles, the total amount of its thermal radiation will be
   a. the same.
   b. 2 times larger.
   c. 4 times larger.
   d. 8 times larger.
   e. 16 times larger.

unit

24 Atomic Spectra: Identifying Atoms by Their Light

Background Pathways

Unit 23: Thermal Radiation 173

The keys to determining the composition and conditions of an astronomical body are the wavelengths of light it absorbs and emits. The technique used to capture and analyze the light from an astronomical body is called **spectroscopy.** In spectroscopy, electromagnetic radiation that is emitted or reflected by the object being studied is collected with a telescope and spread into its component colors to form a spectrum. For visible wavelengths this is done by passing the light through a prism or through a grating consisting of numerous tiny, parallel lines (Figure 24.1). Because photons of particular wavelengths are emitted and absorbed by atoms as electrons shift between orbitals, the spectrum of light will bear an imprint of the kinds of atoms that the light interacted with. This allows astronomers to search for various atoms' "signatures" by measuring how much light is present at each wavelength.

Spectroscopy is such an important tool for astronomers that we will examine it at a fairly high level of detail. Specifically, why does an atom produce a unique spectral signature? To understand that, we need to recall how light is produced.

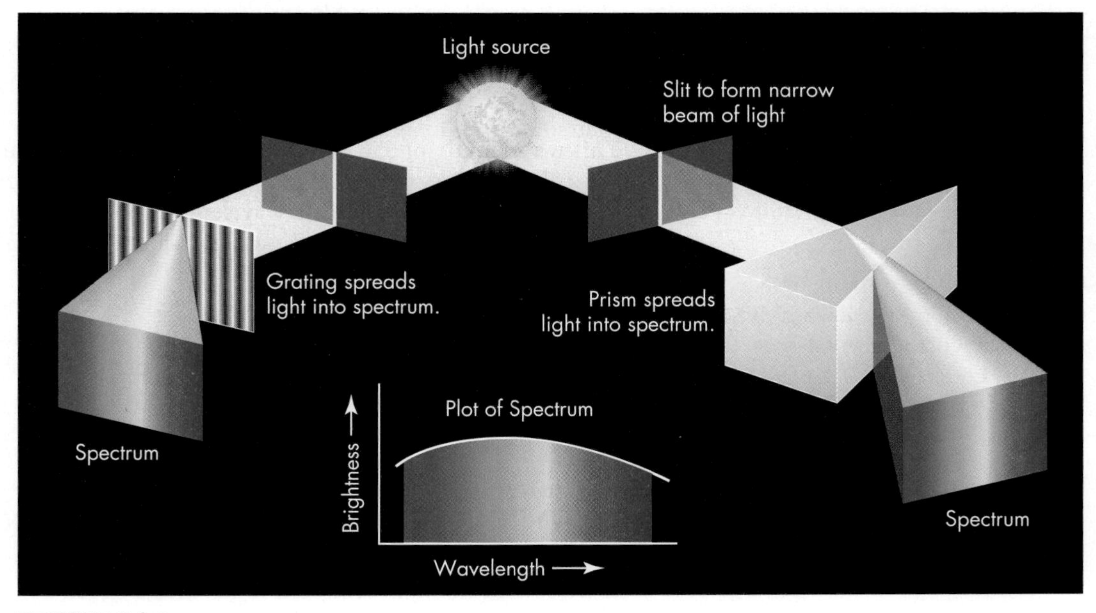

FIGURE 24.1
Sketch of a spectroscope and how it forms a spectrum. Either a prism or a grating may be used to spread the light into its component colors.

24.1 HOW A SPECTRUM IS FORMED

When an electron moves from one orbital to another, the atom's energy changes by an amount equal to the difference in the energy between the two orbitals. Because the atom's energy is determined by what orbitals its electrons move in, orbitals are often referred to as **energy levels.**

Suppose we heat up a gas of atoms. Heating speeds up the atoms, causing more forceful and frequent collisions that can knock an atom's electrons into more energetic orbitals. At the same time, the electrical attraction between the nucleus and the electron draws the electron back down to lower energy levels. As the electron shifts downward, the atom's energy decreases, and the energy that is lost appears as a photon.

As we saw in Unit 21, the orbitals of electrons cannot have just any energy. As a result of the wavelike nature of matter they can have only discrete energy levels. The allowed energy levels are very difficult to calculate for all but the simplest atom, hydrogen, which consists of a single electron orbiting a proton.

In the early 1900s, Danish physicist Niels Bohr developed a model of the hydrogen atom that gives its possible energy levels. The **Bohr model** of hydrogen describes the electron as orbiting at a set of discrete radii. If we number the energy levels starting from the *ground state* as $n = 1, 2, 3$, and so on, then the electron's orbital radius is $0.053 \times n^2$ nanometers. (A nanometer is 10^{-9} meter; see Unit 3.) The energy of level n can be described by the formula:

E_n = energy of atom in state n

n = level starting at ground state ($n = 1$)

$$E_n \text{ (hydrogen)} = -2.18 \times 10^{-18} \text{ joule}/n^2.$$

As n gets larger, the energies approach zero, which reflects the weaker electrical attraction that the electron feels far away from the nucleus.

Suppose we look at an electron shifting from orbital 3 to orbital 2, as shown in Figure 24.2A. The wavelength of the emitted light can be calculated from the energy difference ΔE of the levels and the relation between energy and wavelength: $\Delta E = hc/\lambda$ (Unit 22). Using the energy level formula we find:

$$\lambda = \frac{h \times c}{\Delta E} = \frac{1.99 \times 10^{-25} \text{ joule} \cdot \text{m}}{2.18 \times 10^{-18} \text{ joule} \left(\frac{1}{2^2} - \frac{1}{3^2}\right)}$$

$$= \frac{1.99 \times 10^{-25}}{3.03 \times 10^{-19}} \text{ m} = 6.56 \times 10^{-7} \text{m}$$

This wavelength, of 656 nanometers, is in the visible part of the spectrum, a bright red color. An electron dropping from orbital 3 to orbital 2 in a hydrogen atom always produces light of this wavelength. In fact, in astronomical images where there is glowing red-colored gas, it is almost certainly produced by hydrogen atoms undergoing this transition.

Q In what ways is the hydrogen energy level formula similar to the formula describing gravitational potential energies (Unit 20)?

If, instead, the electron moves between orbital 4 and orbital 2 (Figure 24.2B), there will be a different change in energy because orbital 4 has higher energy than orbital 3. The larger energy change corresponds to a shorter-wavelength photon. A calculation of its energy change leads in this case to a wavelength of 486 nanometers, a turquoise blue color.

24.2 IDENTIFYING ATOMS BY THEIR LIGHT

INTERACTIVE

The Bohr atom

If we spread the light from a hot gas into a spectrum, we will see that in general the spectrum contains light at only certain wavelengths. In a gas made up of hydrogen atoms, we will see light from atoms undergoing changes between many different energy levels simultaneously, as depicted in Figure 24.3. Thus hydrogen gas produces a spectrum containing the red 656-nanometer and turquoise

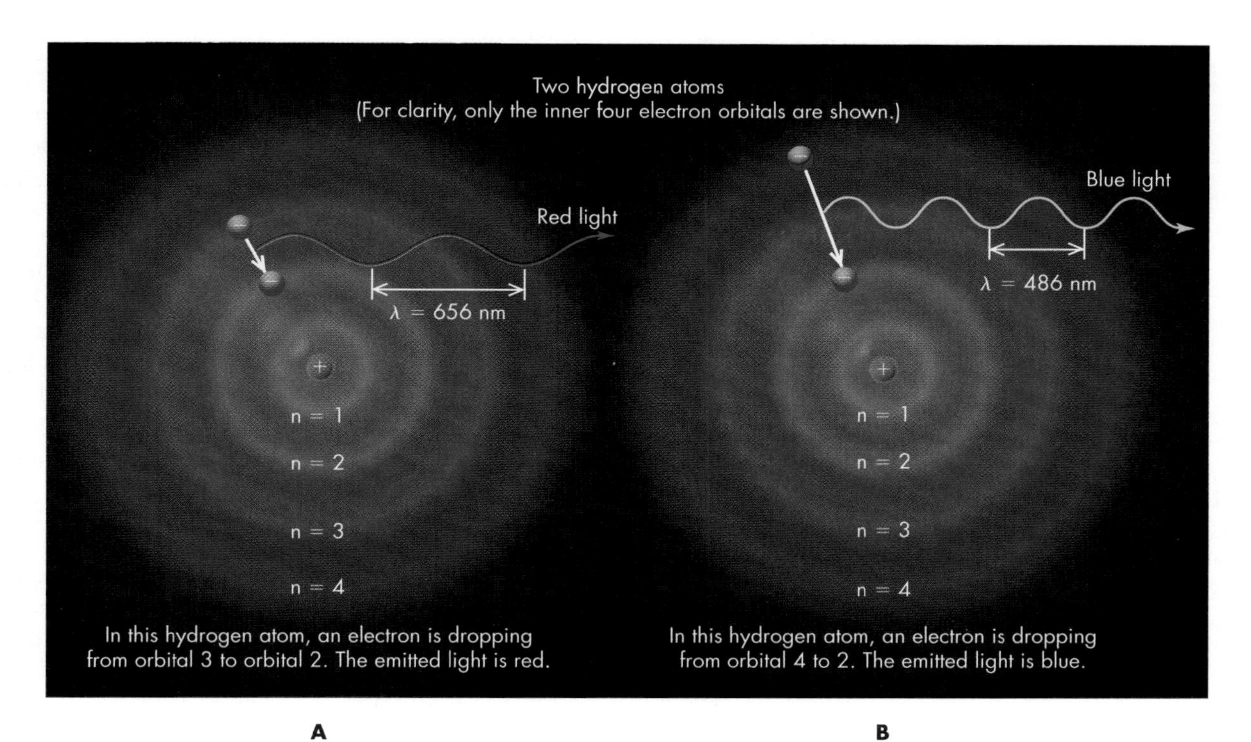

FIGURE 24.2
Emission of light from a hydrogen atom. The energy of an electron dropping from an upper to a lower orbital is converted to light. The light's color depends on the orbitals involved.

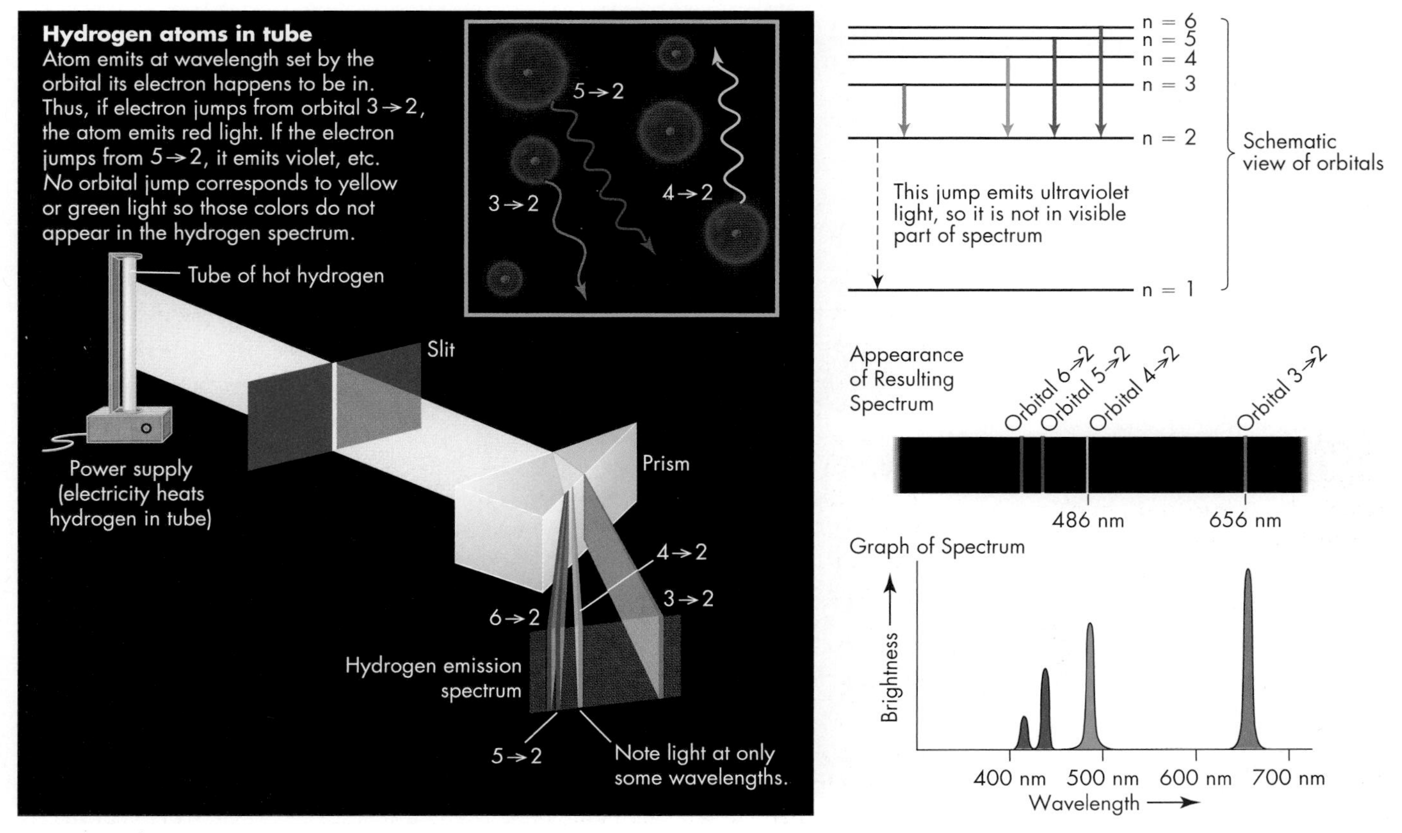

FIGURE 24.3
The spectrum of hydrogen in the visible wavelength range.

486-nanometer lines, as well as violet light corresponding to electrons dropping from orbitals 5 to 2, and other lines that lie outside the wavelength range that our eyes can detect. This is known as an **emission-line spectrum.** This set of lines is the "signature" of hydrogen, unique to this element, and it allows astronomers to determine whether an astronomical object contains hydrogen.

If instead of hydrogen we looked at heated helium, we would see a very different spectrum. The reason is simple. Helium has two protons instead of one, and the electrons interact with each other, which leads to a different set of electron energy levels and thus a different set of spectral lines, as you can see in Figure 24.4. The spectrum thus becomes a means of identifying what atoms are present in a gas.

In fact helium was first discovered in 1868 by spectroscopic observations of the Sun during an eclipse. Astronomers recognized that the wavelength of the strong yellow line in its spectrum had no known counterpart, and they gave the name *helium* to the unknown element after the Greek word for the Sun, *helios*. Helium was not discovered by chemists on Earth until the 1890s.

Q If an atom had just five different possible energy levels for its electrons, how many different wavelengths of light might it be possible for the atom to emit?

For each element present in a gas we will see specific sets of spectral lines. We will see no light at most other colors because there are no electron orbital transitions corresponding to those energies. Therefore, the spectrum shows a set of brightly colored lines separated by wide, dark gaps. Unlike hydrogen, there is no simple mathematical formula we can write down to describe the energy levels, but the patterns of spectral lines have been determined from laboratory studies.

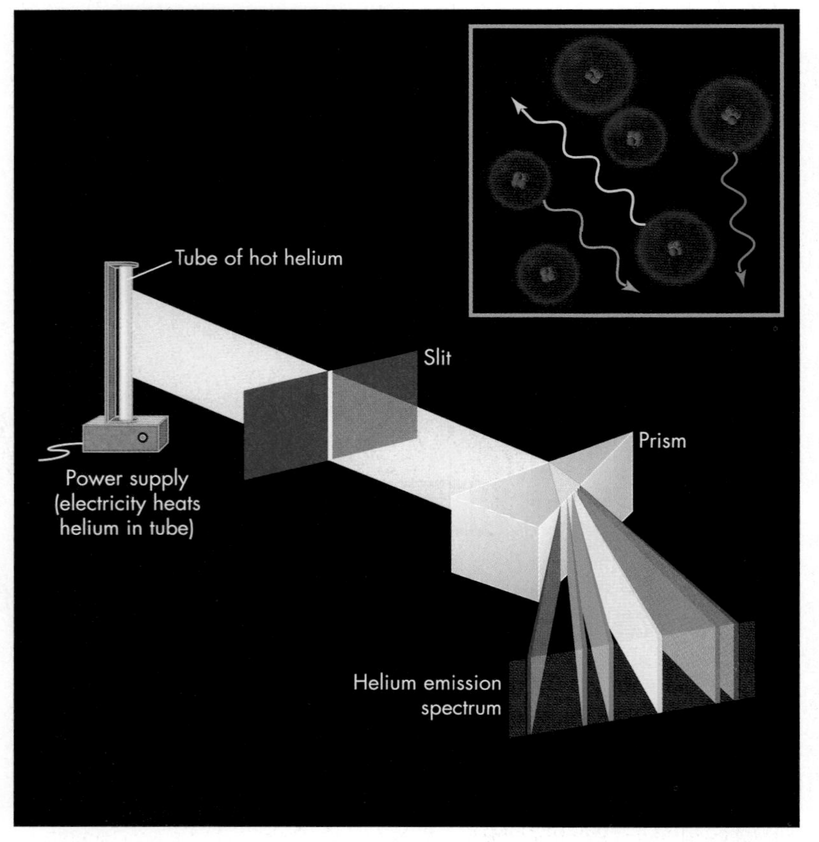

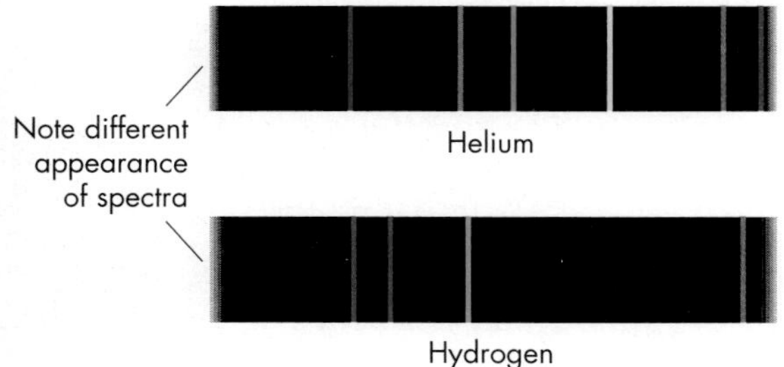

FIGURE 24.4
The spectrum of helium in the visible wavelength range.

It is also possible to identify atoms in a gas from the wavelengths at which they absorb light. Light can be absorbed by an atom only if the energy of its wavelength corresponds to the energy difference between two energy levels in the atom. If the wavelength does not match, or the atom is not in the right starting state, the photon cannot be absorbed. The photon will simply move past the atom, leaving itself and the atom unaffected.

Q If hydrogen atoms must be in the second energy level to absorb 656-nm photons, how might hydrogen's ability to absorb these photons be affected by temperature?

For example, suppose we shine a beam of light that initially contains all the colors of the visible spectrum through a box full of hydrogen atoms. If we examine the spectrum of the light after it has passed through the box, we will find that certain wavelengths of the light are missing from the spectrum (Figure 24.5). In particular, the spectrum will contain gaps that appear as dark lines at 656 nanometers and 486 nanometers, precisely the wavelengths at which the hydrogen atoms emit. The **absorption-line spectrum** is, in effect, the opposite of the emission-line spectrum, because an atom's set of possible absorption lines have exactly the same wavelengths as its possible emission lines.

The gaps in the spectrum are created when photons at 656 nanometers and 486 nanometers interact with hydrogen atoms' electrons, lifting them from orbital 2 to 3 or orbital 2 to 4, respectively. Light at other wavelengths in this range has no effect on the atoms. Thus we can tell that hydrogen is present from either its emission or its absorption spectral lines.

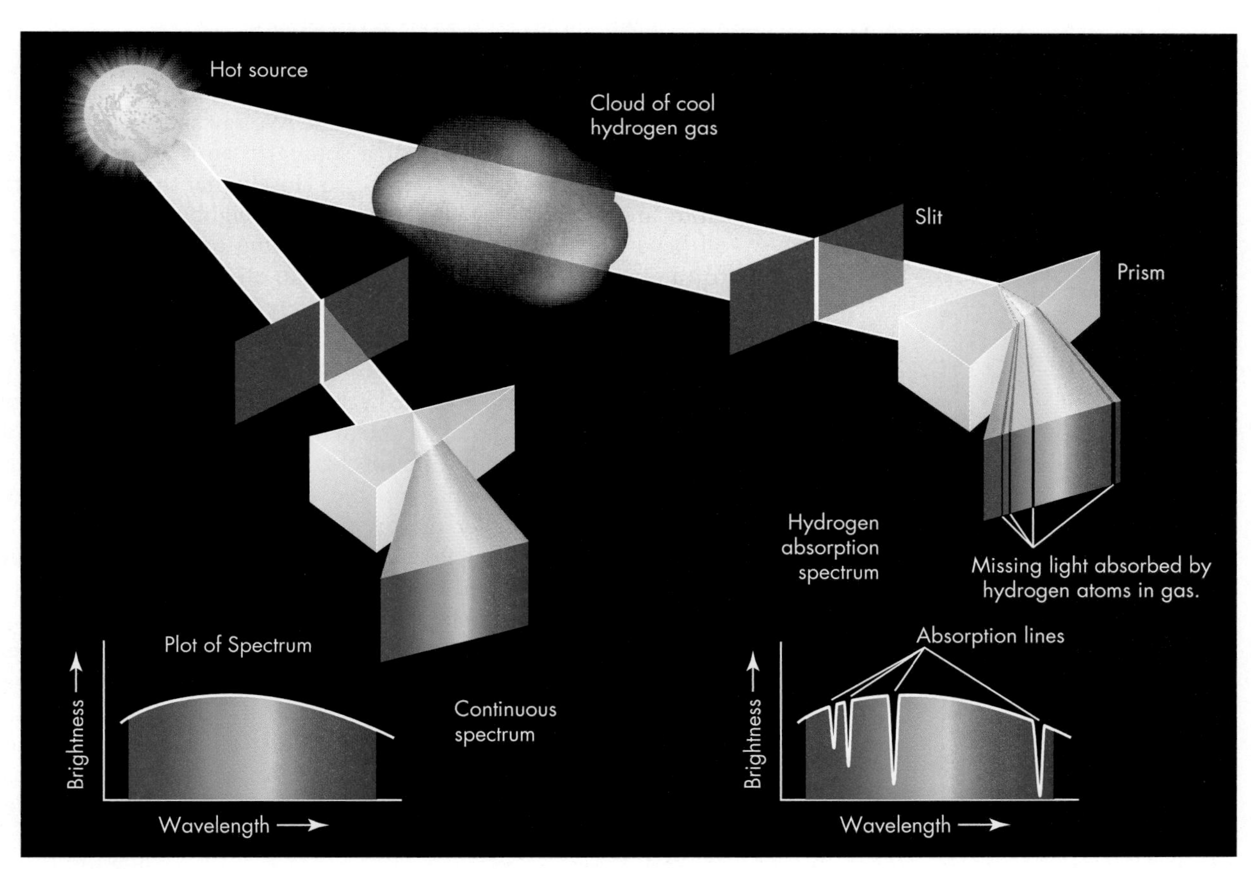

FIGURE 24.5
A hot, dense substance produces a continuous spectrum. Atoms in an intervening gas cloud absorb only those wavelengths whose energy equals the energy difference between their electron orbitals. The absorbed energy lifts the electrons to upper orbitals. The lost light makes the spectrum darker at the wavelengths where it is absorbed.

In our discussion we have considered light emitted and absorbed by individual atoms in a gas. If the atoms are linked to one another to form molecules, such as water or carbon dioxide, the new configuration of the electron orbitals results in an entirely different set of emission and absorption lines, so the molecules can also be identified from spectroscopy. In fact, even solid objects may imprint spectral lines on light that reflects off them. For example, when light from the Sun reflects from an asteroid, spectral features appear that were not present in the original sunlight. This gives astronomers information about the surface composition of bodies too cool to emit significant light of their own.

We conclude that in general we can identify the kinds of atoms or molecules that are present by examining either the bright or the dark spectral lines. Gaps in the spectrum at 656 nanometers and 486 nanometers imply that hydrogen is present. Similar gaps at other wavelengths would show that other elements are present. By matching the observed gaps to a directory of absorption lines, we can identify the atoms and molecules that are present. This is the fundamental way astronomers determine the chemical composition of astronomical bodies.

24.3 TYPES OF SPECTRA

The basic forms of spectra we have discussed can be categorized into three basic classes, and each implies a set of physical conditions. This classification system was first formulated by the German physicist Gustav Kirchhoff in the mid-1800s, so it is usually known as **Kirchhoff's laws:**

1. Some sources emit light in such a way that the intensity changes smoothly with wavelength and all colors are present. We say such a light has a **continuous spectrum** (Figure 24.6A). For a source to emit a continuous spectrum, its atoms must in general be packed so closely that the electron orbitals of each atom are distorted by the presence of neighboring atoms. Isolated atoms all behave essentially identically, absorbing or emitting frequencies specific to the type of atom. However, when an atom is bumping into other atoms, the electron energy levels become altered by various amounts so that each atom can interact with photons of different energies, allowing photons of all wavelengths to be emitted or absorbed. Such conditions are typical of solid or dense materials such as the heated filament of an incandescent lightbulb, a glowing piece of charcoal, or the denser gas in the atmosphere of a star.
2. Conversely, an emission-line spectrum (Figure 24.6B) implies that the atoms or molecules are well separated and the atoms are hot enough to have excited the electrons into higher energy levels so they can emit the light at the specific frequencies associated with a drop to a lower energy level. Emission-line spectra are usually produced by hot, low-density gas, such as that in a fluorescent tube, the aurora, and many interstellar gas clouds.
3. An absorption-line spectrum (Figure 24.6C) is produced when light from a hot, dense body (as in law 1 above) passes through cooler gas between it and the observer. In this case nearly all the colors are present, but light is either missing or much dimmer at the specific wavelengths corresponding to energy transitions in the gas atoms. Absorption-line spectra are seen, for example, in most star's spectra produced by cooler low-density gas above the hotter layers (that produce a continuous spectrum) beneath. Cool interstellar gas clouds between a star and us can also produce absorption lines in a star's spectrum.

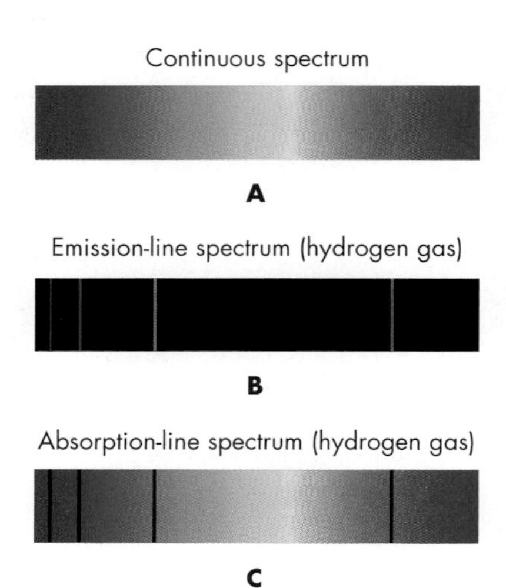

FIGURE 24.6
Types of spectra: (A) continuous, (B) emission-line, and (C) absorption-line.

Kirchhoff's laws give us a starting point for interpreting the physical conditions that need to be present to produce each kind of spectrum we have discussed.

24.4 ASTRONOMICAL SPECTRA

The spectrum of the Sun is shown in Figure 24.7. The spectrum is shown in two different styles. Figure 24.7A is an image of the light after it has been spread out by a prism or grating. Astronomers often find it more useful to plot a line graph of the brightness of the light at each wavelength as in Figure 24.7B.

The Sun's spectrum is typical of most stars. It shows a continuous spectrum interrupted by narrow absorption lines. In the graph (Figure 24.7B), the absorption lines are seen as sharp dips in the brightness. Astronomers can often identify which elements produce these absorption lines from large directories of spectral lines. By matching the wavelength of the line of interest to a line in the table, astronomers can determine what kind of atom or molecule created the line.

The relative number of atoms of each type is also indicated in the periodic table inside the back cover of this book.

Some of the lines are faint and hard to see, whereas other lines are obvious and strong. The strength or weakness of a given line depends on the number of atoms or molecules absorbing (or emitting, if we are looking at an emission line) at that wavelength. Complicating the interpretation, the number of atoms or molecules that can absorb or emit depends not just on how many of them are present but also on their temperature. Knowing the temperature, however, astronomers can deduce from the strength of emission and absorption lines the relative quantity of each atom.

Because the 10% of atoms in the Sun besides hydrogen are more massive than hydrogen, hydrogen contributes "only" about 71% of the Sun's mass.

We can see from the spectral lines that the Sun contains hydrogen from the features at 486 nm and 656 nm discussed in Section 24.1. In fact, when a detailed calculation is made of the strength of the lines, it turns out that roughly 90% of the atoms in the Sun are hydrogen. When the Sun's spectrum is examined in detail, thousands of absorption lines from nearly every element are found. Table 24.1 shows a partial list of the Sun's composition based on an analysis of the absorption lines. Analyses of other stars and other astronomical objects show that the composition of the Sun is quite typical. The first eight elements listed in the table are the eight most abundant in the Sun and throughout the universe, accounting for over 99.8% of all atomic matter.

Table 24.1 shows one other interesting piece of data for each of these elements—the energy needed to ionize the element. The energy listed is the amount needed to remove an electron when the atom is in its ground state. For a star such as the Sun, many of the atoms are already in an excited state, so less energy is required to ionize the atom. For example, hydrogen atoms in the $n = 2$ level require only one-fourth as much energy to be ionized as is listed in the table. This is the energy of a 365-nm photon. Ultraviolet photons can ionize many kinds of atoms in a star's atmosphere.

When an atom can be ionized by photons, the effect on a spectrum is different from the narrow absorption lines we have seen so far. *Any* photon that exceeds the **ionization energy** threshold can be absorbed by the atom. As a result, we do

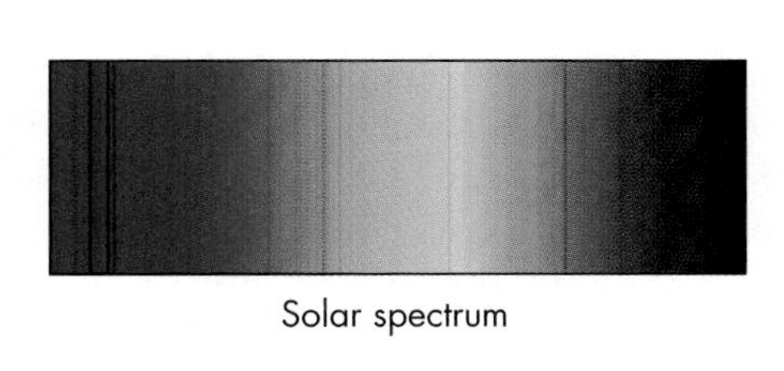

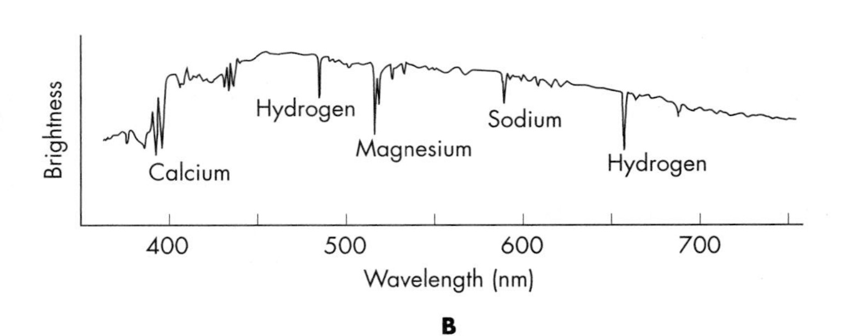

FIGURE 24.7
(A) The spectrum of the Sun. Note the narrow dark absorption lines. (B) A graphical representation of the Sun's spectrum. Several of the absorption lines are identified according to the element in the Sun's atmosphere that is causing the absorption.

TABLE 24.1 Composition of a Typical Star, Our Sun

Element	Number of Protons (Atomic Number)	Relative Number of Atoms	Percentage by Mass	Energy to Remove Outermost Electron
Hydrogen	1	10^{12}	71.1%	2.179×10^{-18} J
Helium	2	9.64×10^{10}	27.4%	3.940×10^{-18} J
Carbon	6	2.88×10^{8}	0.25%	1.804×10^{-18} J
Nitrogen	7	7.94×10^{7}	0.08%	2.328×10^{-18} J
Oxygen	8	5.75×10^{8}	0.65%	2.182×10^{-18} J
Neon	10	8.91×10^{7}	0.13%	3.454×10^{-18} J
Silicon	14	4.07×10^{7}	0.06%	1.306×10^{-18} J
Iron	26	3.47×10^{7}	0.14%	1.266×10^{-18} J
Gold	79	8	0.00000011%	1.478×10^{-18} J
Uranium	92	0.4	<0.000000007%	0.992×10^{-18} J

Note: The table lists eight of the most common elements along with gold and uranium to illustrate how extremely rare they are. Data on relative number of atoms drawn from Lodders (2003) *The Astrophysical Journal*, vol. 591, pp. 1220–1247.

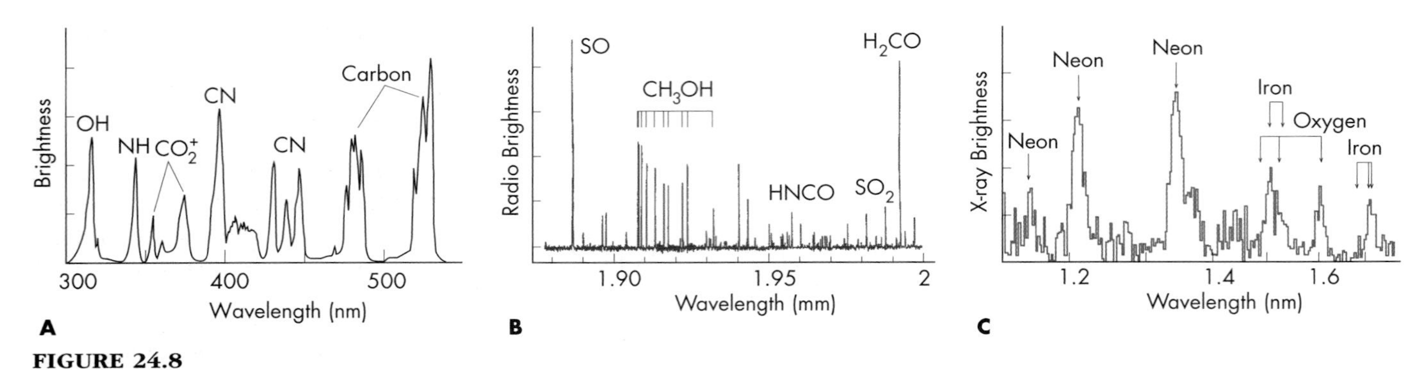

FIGURE 24.8
Emission-line spectra at a variety of different wavelengths. (A) A spectrum of a comet at visible and ultraviolet wavelengths. (B) A microwave spectrum of a cold interstellar cloud. (C) An X-ray spectrum of hot gas from an exploding star. Many of the atoms and molecules responsible for the emission lines are identified.

not see a narrow absorption line; instead, all photons at wavelengths shorter than the threshold can be absorbed. Stars' spectra generally show a sharp drop in light output at ultraviolet wavelengths because of the photons' ability to ionize atoms.

Spectroscopy is not limited to stars or to visible wavelengths of light. Figure 24.8 shows spectra of a comet, an interstellar gas cloud, and an exploding star. All three spectra show emission lines, but in very different parts of the electromagnetic spectrum. The best region to examine for spectral lines often depends on the temperature of the gas. A radio spectrum is well suited to a cold interstellar cloud, while X-ray observations are appropriate for a hot exploding star.

In both the comet and the interstellar cloud, the temperatures are low enough for molecules—some quite complex—to have formed. By contrast, the wavelengths of the emission lines seen in the X-ray spectrum tell us that these atoms are highly ionized. In fact, the gas is so hot that 7 electrons have been stripped from the neon atoms, 9 from the oxygen, and 16 from the iron atoms! Regardless of the wavelength region we use, the spectrum allows us to determine what kinds of materials are present and their physical conditions.

KEY TERMS

absorption-line spectrum, 183
Bohr model, 180
continuous spectrum, 184
emission-line spectrum, 182
energy level, 180
ionization energy, 185
Kirchhoff's laws, 184
spectroscopy, 179

QUESTIONS FOR REVIEW

1. What can astronomers learn about astronomical objects from their spectra?
2. Why don't atoms emit a continuous spectrum?
3. Some atoms of different elements produce spectral lines that are very close to the same wavelength. How can we deduce which kind of atom they come from?
4. How do the wavelengths of an element's emission lines relate to the wavelengths of its absorption lines?
5. What elements does spectroscopy indicate are the most common in the Sun's atmosphere?
6. How it is possible for the same cloud of gas to provide an emission spectrum to one observer, and an absorption spectrum to a different observer?

PROBLEMS

1. How could you use a spectrum produced from sunlight reflected off the atmosphere of Venus to determine something about Venus's atmospheric composition?
2. If you were to look at the spectrum of the gas flame of a stove or the blue part of a Bunsen burner flame, what sort of spectrum would you expect to see: absorption, emission, or continuous? Why?
3. In the formula for hydrogen's energy levels, find the energy difference between levels $n = 1$ and $n = 2$, and determine the wavelength of a photon with this energy. What part of the electromagnetic spectrum is this?
4. Using the formula for hydrogen's energy levels, calculate the energy of a photon emitted when hydrogen goes from level 143 to level 142 (a transition that has been observed by astronomers). What is the wavelength of this spectral line? What part of the electromagnet spectrum is this?
5. Based on the formula for hydrogen's energy levels, what is the energy of an electron if n goes to infinity? Explain. Would a beam of photons with energy $= 1.1 \times 10^{-18}$ joules affect a hydrogen atom that is in its ground state? What if it was in the $n = 2$ state?
6. From the data in Table 24.1, what wavelength photon can ionize helium? What wavelength can ionize uranium?

TEST YOURSELF

1. An astronomer finds that the visible spectrum of a mysterious object shows bright emission lines. What can she conclude about the source?
   a. It contains cold gas.
   b. It is an incandescent solid body.
   c. It is rotating very fast.
   d. It contains hot, relatively tenuous gas.
   e. It is moving toward Earth at high speed.
2. Most stars have spectra showing dark lines against a continuous background of color. This observation indicates that these stars
   a. are made almost entirely of hot, low-density gas.
   b. are made almost entirely of cool, low-density gas.
   c. have a warm interior that shines through hotter, high-density gas.
   d. have a hot interior that shines through cooler, low-density gas.
3. Suppose we detect the 656-nm wavelength of the $n = 3$ to $n = 2$ transition of hydrogen from a 5000-K gas cloud. If the cloud was heated to 10,000 K, what would the wavelength of the $n = 3$ to $n = 2$ transition now be?
   a. 278 nm
   b. 1312 nm
   c. 5656 nm
   d. 658 nm
   e. 656 nm

unit 25

The Doppler Shift

If we observe light from a source that is moving toward or away from us, we will find that the wavelengths we receive from it are altered by its motion. If it moves toward us, its wavelengths will be shorter. If it moves away from us, its wavelengths will be longer, as illustrated in Figure 25.1. Furthermore, the faster the source moves, the greater those changes in wavelength will be. This change in wavelength caused by motion is called the **Doppler shift,** and it is a powerful tool for measuring the speed of approach or recession of astronomical objects. Motion perpendicular to the line of sight creates no Doppler shift because the source is neither approaching nor moving away from us.

The Doppler shift occurs for all kinds of waves. You have probably heard the Doppler shift of sound waves from the siren of an emergency vehicle: The siren's noise drops from a high pitch to a low pitch (corresponding to a shift from short to longer wavelengths of the sound) as the approaching vehicle passes you and then moves away (Figure 25.1A). The same thing happens with electromagnetic waves, as when, for example, the Doppler shift of a radar beam that bounces off your car reveals to a law enforcement officer how fast your car is moving (Figure 25.1B).

Background Pathways

Unit 24: Atomic Spectra: Identifying Atoms by Their Light 179

The Doppler effect

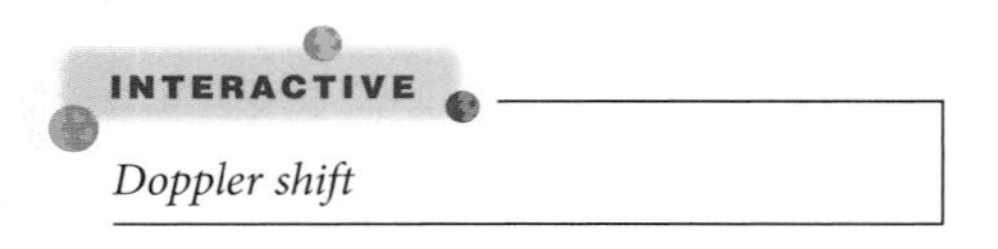

Doppler shift

25.1 CALCULATING THE DOPPLER SHIFT

We can understand the Doppler shift by thinking about the spacing between wave crests. If a source is generating electromagnetic waves at a particular frequency (like those hydrogen atoms might emit, for example) and is stationary, the wave crests reach us with the normal "rest" wavelength λ. If the object is moving toward us, the object still generates the waves at the same frequency, but in the time between the generation of one wave crest and the next, the object has moved closer, so the wavelength we observe, λ', is shorter. This is illustrated in Figure 25.2, where the position of the light source is shown at the time it emitted the crest of each wave. The wave crest expands out spherically around the position from which it was emitted, but the spacing between wave crests is tighter in the direction in which the source is moving.

The difference between λ' and λ is usually written $\Delta\lambda$ (called "delta lambda" or "the change in λ"). For example, suppose a spectral line is generated at 656 nm (λ), but the line instead appears to be at 658 nm (λ'). The change in wavelength is

$$\Delta\lambda = \lambda' - \lambda = 658 \text{ nm} - 656 \text{ nm} = 2 \text{ nm}.$$

This is the distance the source has moved in the time it took to generate the wave.

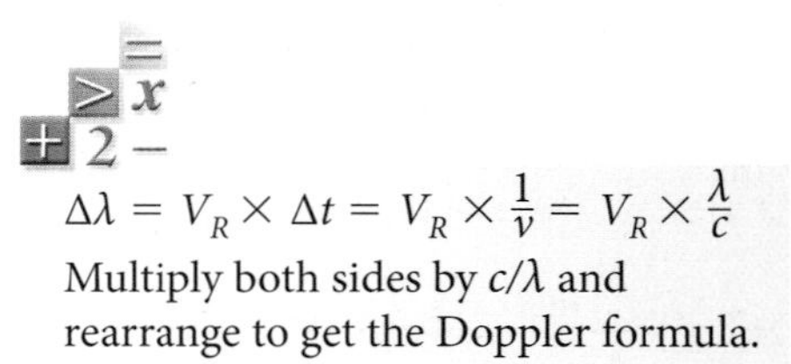

$\Delta\lambda = V_R \times \Delta t = V_R \times \frac{1}{\nu} = V_R \times \frac{\lambda}{c}$

Multiply both sides by c/λ and rearrange to get the Doppler formula.

The time between the generation of one wave crest and the next is, by definition, the inverse of the frequency, $1/\nu$. With the aid of the relationship between wavelength and frequency ($\lambda \times \nu = c$, Unit 22), the time between wave crests can

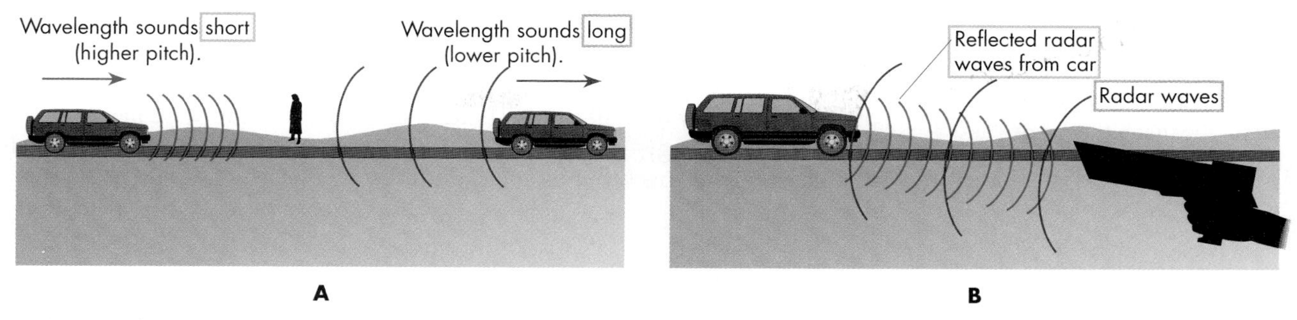

FIGURE 25.1
The Doppler shift: Waves shorten as a source approaches and lengthen as it recedes. (A) The Doppler shift of sound waves from a passing car. (B) The Doppler shift of radar waves in a speed trap.

FIGURE 25.2
A source of light moving toward you will shorten the distance the next wave must travel to reach you, so the time between wave crests grows shorter. If the light source moves away, the distance grows larger and the wavelength longer.

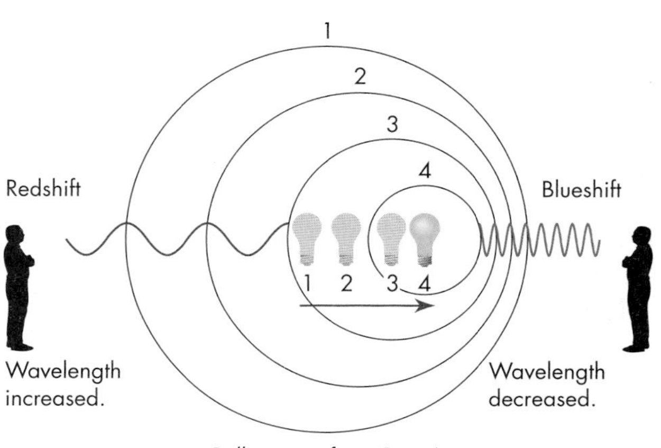

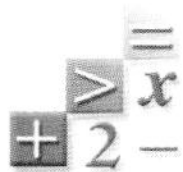

V_R = Radial velocity (toward or away from observer)
$\Delta\lambda$ = Change from normal wavelength
λ = Normal wavelength
c = Speed of light

Q Wien's law indicates that the wavelength of light is longer from a cooler object. How could we tell the difference between this and an object that is moving away from us?

be written as c/λ. So the speed of the object along our line of sight, called its **radial velocity,** V_R, is

$$V_R = \frac{\Delta\lambda}{\lambda} \times c.$$

This formula relates the change in wavelength to the source's speed, allowing astronomers to find out how fast a source is moving and whether the motion is toward or away from us. By convention, $\Delta\lambda$ and V_R are negative if the object is approaching us (so that the observed wavelength, λ', is shorter than the normal rest wavelength) and positive if it is moving away from us. In the previous example, the speed is

$$V_R = \frac{2 \text{ nm}}{656 \text{ nm}} \times c = 0.003 \times 300{,}000 \text{ km/sec} = 900 \text{ km/sec}$$

and the object must be moving away from us because the wavelength has grown longer.

Doppler shift measurements can be made at any wavelength of the electromagnetic spectrum. But regardless of the wavelength region observed, astronomers refer to shifts that *increase* the measured wavelength of the radiation as **redshifts** and those that *reduce* the wavelength as **blueshifts.** This harkens back to the early days of spectroscopy when only the visible portion of the electromagnetic spectrum was observable.

Incidentally, we can also understand the Doppler shift in the particle model of light by thinking of the motion of a source toward or away from us as adding to or

The constancy of the speed of light was one of the great mysteries resolved by Einstein's theory of special relativity (see Unit 53).

subtracting from the energy of the photon. This would be like throwing a ball to a receiver while running. If you are running toward the receiver, the ball will arrive with a greater impact; but if you are running away, the impact will be weaker, even though you throw the ball equally hard in each case. However, we have to be careful with this analogy—in the case of the photon, experiments conclusively show that its energy changes, but its speed does *not*.

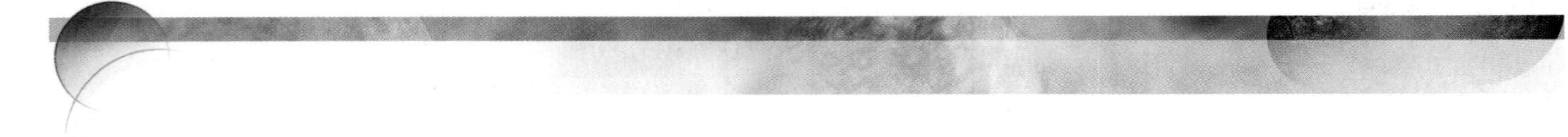

KEY TERMS

blueshift, 189
Doppler shift, 188
radial velocity, 189
redshift, 189

QUESTIONS FOR REVIEW

1. What is the Doppler shift?
2. What is the difference between an object's velocity and its radial velocity?
3. Do we always see a Doppler shift when an object is moving?
4. How could two objects be moving at different velocities, but show the same redshift?
5. If you made a spectrum of a star and it was redshifted by 1 nm at the red end of the spectrum, how much would it be redshifted at the violet end?
6. What is the difference between a redshift and a blueshift?

PROBLEMS

1. About how fast would you have to be driving toward a red stoplight for it to appear green?
2. The $n = 3$ to $n = 2$ transition of hydrogen is normally at 656.3 nm. If it is detected from a star at 655.5 nm, what is the radial velocity of the star?
3. Suppose that a star is rotating at a speed of 50 km/sec at its equator. Its spectral lines will be spread out because one side of the star will be redshifted and the other side will be blueshifted. For the 656.3-nm line of hydrogen, over how wide a range in wavelengths will the spectral line be spread?
4. Hydrogen produces a spectral line at 656.3 nm. Oxygen produces a spectral line at 645.5 nm. If you detected a galaxy with a spectral line at 655.0 nm, what Doppler shift would you derive if you assumed it was produced by hydrogen? by oxygen? How could you determine which was correct?
5. Astronomers study planets in the Solar System with radar. Suppose you were observing Venus, which orbits the Sun at 35 km/sec, from the Earth, which orbits at 30 km/sec. Draw a diagram of the two planets' orbits (Venus orbits at about 0.7 AU) and use your diagram and vectors representing the velocities to estimate the radial velocity we would see when Venus is at its greatest elongations and when it is in conjunction with the Sun.
6. Hydrogen also produces spectral lines at radio wavelengths, notably at 21.1 cm. If a galaxy is moving away from us at 10% of the speed of light, at what wavelength will we detect this line? Convert this into a frequency.

TEST YOURSELF

1. If an object's spectral lines are shifted to longer wavelengths, the object is
   a. moving away from us.
   b. moving toward us.
   c. very hot.
   d. very cold.
   e. emitting X-rays.
2. What properties of stars can be determined from their spectra?
   a. Chemical composition
   b. Surface temperature
   c. Radial velocity
   d. All of the above
   e. None of the above
3. The wavelength of a radio emission line from a galaxy normally seen at 100 cm is detected at 101 cm. This implies that the galaxy is moving
   a. toward us at 1% of the speed of light.
   b. away from us at 1% of the speed of light.
   c. toward us at 1 centimeter per second.
   d. away from us at 1 centimeter per second.

unit

26 Detecting Light

Like all scientists, astronomers rely heavily on observations to guide them in proposing hypotheses and in testing theories already developed. Unlike most scientists, however, astronomers usually cannot actively probe the objects they study. Instead, they must perform their observations passively, collecting whatever light and other forms of radiation have been emitted by the bodies they seek to study.

The quest to understand the cosmos ultimately depends on the quality and detail of the data we can collect. Sometimes that quality and detail are limited by the available technology. Sometimes they are constrained by the physical properties of the light we are attempting to detect. In this unit we present an overview of the instruments used for astronomical observation. In subsequent units we will explore in greater detail several of the limitations and challenges posed by these instruments.

Background Pathways

Unit 22: The Electromagnetic Spectrum 166

26.1 TECHNOLOGICAL FRONTIERS

Astronomy is the most ancient of the sciences, but many of its most exciting discoveries are very recent. For example, in the long history of astronomy, only in the last 200 years have astronomers known that "light"—electromagnetic radiation—extends beyond the visible spectrum; and only in the last half century have instruments been developed to study these other bands of electromagnetic radiation. And at visual wavelengths, telescopes have progressed to achieve remarkable precision and size (Figure 26.1).

The design and construction of instruments to collect and detect electromagnetic radiation of all wavelengths is a branch of astronomy called **instrumentation.**

FIGURE 26.1
One of the twin Keck telescopes on the summit of Mauna Kea, Hawaii, during a total eclipse of the Sun. These became the second and third largest individual optical telescopes in the world when a slightly larger telescope of similar design was completed in 2008.

The amazing images and data that have changed many of our ideas about the nature of the cosmos in recent decades are due to the remarkable strides made in instrumentation.

The most recognizable tool of astronomical technology is the **telescope.** Telescopes allow astronomers to observe things that are not visible to the naked eye either because they are too dim or because they emit radiation outside the visible range of the electromagnetic spectrum. Telescopes are basically devices for collecting a large amount of electromagnetic radiation, focusing this light onto detectors, and magnifying the images to resolve fine detail. Astronomers strive to build bigger telescopes because the amount of light that can be collected depends on the size of the telescope's **aperture,** or collecting area (Unit 27). Collecting more light allows us to see dimmer and more distant objects. If we collect enough light, it is possible to examine the spectrum of radiation from a source, which reveals details of its physical and chemical nature (Unit 24).

A variety of telescope designs use different techniques to take the light that enters the telescope aperture and concentrate or **focus** it (Unit 28). Many of the principles involved in focusing light are similar whether we are dealing with radio waves or X-rays; but the construction details differ enormously because the precision of the construction is determined by the wavelength of the light being observed. This allows radio telescopes to be built with rougher surfaces; however, they have to be built much larger to achieve a comparable **resolution**—a telescope's ability to detect fine detail. This is because resolution is also a function of the size of the telescope relative to the wavelength of the radiation it is detecting (Unit 29).

The Earth's own atmosphere presents challenges for observing electromagnetic radiation. The atmosphere absorbs some wavelengths and alters the path of light in ways that badly degrade the quality of images. For these reasons, a number of space telescopes have been developed in recent years and placed in orbit above the atmosphere (Unit 30). However, putting telescopes in space presents significant challenges—and expense—and one of the goals of modern instrumentation is to find ways to compensate for the atmosphere for ground-based telescopes and to improve the resolution of telescopes with new designs.

26.2 DETECTING VISIBLE LIGHT

Modern telescopes bear little resemblance to the long tubes depicted in cartoons (Figure 26.2). Moreover, professional astronomers almost never sit at the eyepiece of a telescope. And few astronomers ever wear lab coats! Not only are astronomers not needed close to a telescope, but the heat from their bodies can ruin the image quality. With the major research telescopes of today, an astronomer is more likely to be dressed in jeans and a T-shirt, operating the telescope from a computer terminal that may be thousands of miles away. She or he will examine the data being collected, along with a wide array of data about the telescope's performance and weather conditions, and attempt to gauge whether enough data of sufficient quality have been collected to move on to the next set of observations. To make the most efficient use of the instruments, it is becoming more common for telescopes to be operated, and the data checked, by computer programs. A team of astronomers develops computer programs to observe the targeted sources efficiently, then waits for the requested data but has no immediate control over the telescope at all.

"It's somewhere between a nova and a supernova -- probably a pretty good nova."

FIGURE 26.2
Astronomers discussing the nature of their discovery? (Copyright © 2005 by Sidney Harris)

Before astronomical photography became widespread in the late 1800s, astronomers did stare through telescope eyepieces, and they wrote down data or made sketches of the objects being observed. However, despite its marvelous abilities, the human eye has difficulty detecting very faint light (Unit 31). Many astronomical objects are too dim for their few photons to create a sensible effect on the eye. For example, even if you were to look at a galaxy through the largest telescope available today, the light is so spread out that it would appear merely

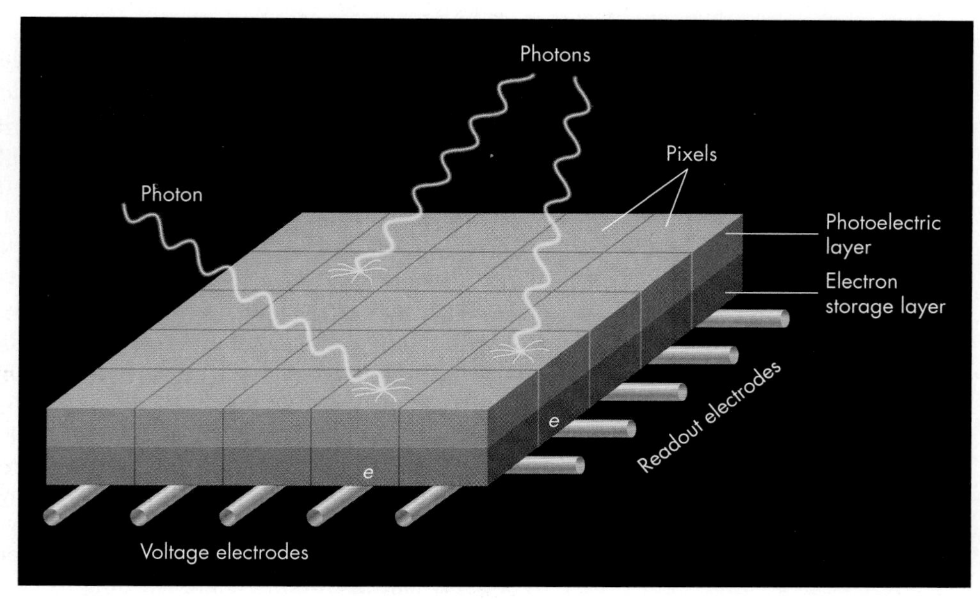

FIGURE 26.3
Simplified diagram of a CCD. Photons striking the photoelectric layer free an electron (e). A positive voltage applied to one set of electrodes attracts the electrons and holds them in place under each pixel. During the readout phase, the voltages are changed to "push" the collected electrons along the readout electrodes.

as dim smudges to your eye. Only by storing galaxies' light, sometimes for many hours, can we discern the detailed structure of the galaxies. Thus, to see very faint objects, astronomers must use detectors that can store light in some manner.

Albert Einstein explained the photoelectric effect in 1905, providing one of the foundations for understanding the particle-like nature of light. His work showed that light is composed of discrete photons with energies that depend on their wavelength. His Nobel Prize was awarded for this work, though he is even better known for his development of relativity theory, explaining the nature of space and time (Unit 53).

Until the 1980s, astronomers generally used photographic film to record the light from the objects they were studying. Astronomers today, however, almost always use an electronic detector similar to those in digital cameras, called a **CCD,** which stands for "charge-coupled device." These devices use the **photoelectric effect,** in which a photon that has sufficient energy causes an electron to become unbound from the atom it is orbiting. Thus when the incoming light strikes the "semiconductor" surface of the CCD, it frees electrons to move within the material. The surface is divided into thousands of little squares, or **pixels** (Figure 26.3). Electrical voltages on the surface pull the freed electrons into the nearest pixel. Here they are temporarily stored before being "read out" by the electronics. Then the whole chip is cleared and returned to its original state. The number of freed electrons in each pixel is proportional to the number of photons hitting it (that is, proportional to the intensity of the light). An electronic device coupled to a computer then scans the CCD, counting the number of electrons in each pixel and generating a picture.

Q Many CCDs are refrigerated to very low temperatures. How do you suppose keeping the temperature low might make the image quality better? (Consider thermal effects on atoms within the CCD.)

Such cameras are extremely efficient, generally recording more than 75% of the photons striking them. This allows astronomers to record images much faster than with film, which might detect only 10% of the photons. Furthermore, unlike film, CCDs are reusable, and the digital images can be processed rapidly by a computer. Astronomers work with the digital images to correct them for instrumental effects, remove extraneous light, and determine the exact amount of light striking different parts of the image.

26.3 OBSERVING AT NONVISIBLE WAVELENGTHS

Many astronomical objects radiate electromagnetic energy at wavelengths outside the range of visible light, and so astronomers have devised new ways to observe such objects. For example, cold clouds of gas in interstellar space emit little visible light but large amounts of radio energy. To observe them, astronomers use radio telescopes, which employ technologies like those used for detecting television or radio station transmissions, but which are far more sensitive.

Long-wavelength radio waves cause electrons in an antenna to oscillate in response, and the small electric currents of these moving electrons are detected and amplified.

At infrared and shorter wavelengths astronomers generally use CCDs or similar technologies to detect photons. For example, dust clouds in space are too cold to emit detectable visible light, but they do radiate infrared photons. Normal CCDs are much more sensitive to infrared photons than photographic film (with wavelengths up to about twice the wavelength of optical photons). But longer-wavelength photons do not have enough energy to free electrons in the photoelectric material. New photoelectric materials have been developed for this purpose. These detectors must be kept very cold, however, because heat from the instrument itself or any surrounding equipment will generate photons of these energies, which will overwhelm the photons from the astronomical source of interest.

Astronomers use X-ray and gamma-ray telescopes to observe, for example, the hot gas that accumulates around black holes. A single X-ray photon has sufficient energy to free many electrons within the CCD material used, and thus the number of electrons freed can be used as a measure of the energy of the incoming X-ray photon.

These technologies improved dramatically in the last few decades. An astronomer spending several years using cutting-edge infrared detectors around 1980 to collect data for a Ph.D. dissertation could with today's instruments collect those same data in five minutes! Astronomy at other wavelengths is likewise advancing rapidly, and astronomers are learning to interpret an ever-growing array of data as these other areas begin to "catch up" with the much better established field of visible-wavelength astronomy.

One of the challenges in interpreting data collected at nonvisible wavelengths is how to display it. Because our eyes cannot see these other wavelengths, astronomers must devise ways to depict what such instruments record. One way is with **false-color images,** as shown in Figure 26.4. In a false-color image, colors are used to represent information from other wavelengths in a way that takes advantage of our eyes' ability to discern fine differences in color. In a common type of false-color image, colors are used to represent the intensity of radiation that we cannot see—often using colors as they are shown in a geographic map, with reds and oranges representing the high levels (like mountains) and blues and purples representing low areas (like oceans). For example, in Figure 26.4A (a radio image of a galaxy and the jet of hot gas spurting from its core), astronomers color the regions emitting the most intense radio energy red; they color areas emitting somewhat less energy yellow and the faintest areas blue.

Figure 26.4B shows a different kind of false-color image (of the gas shell ejected by an exploding star). It is an X-ray image where different energy X-ray photons have been assigned colors analogously to how we see colors in the visible spectrum—blue for high energies, green for medium, and red for lower energies.

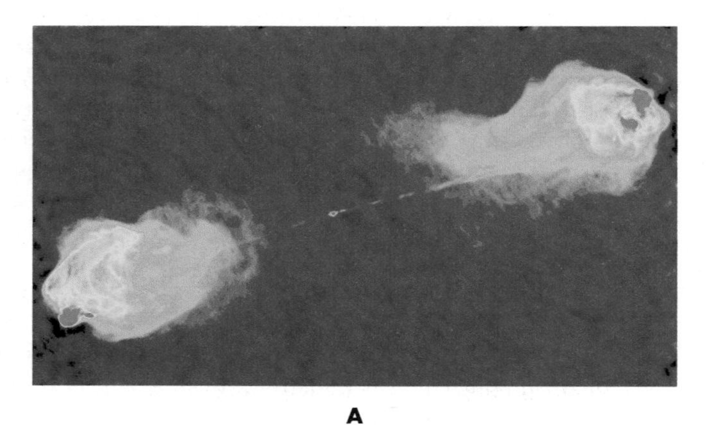

A

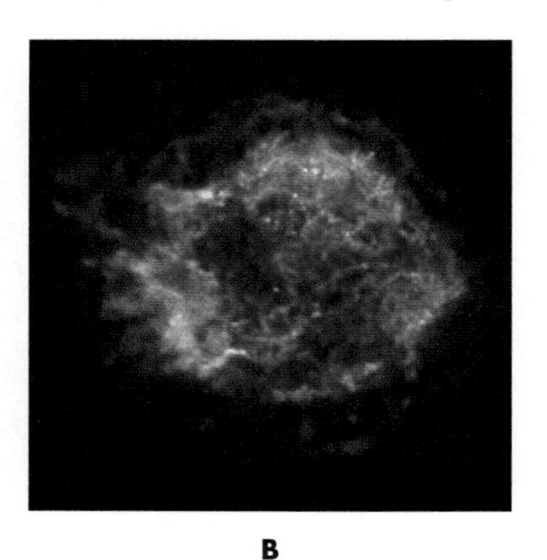

B

FIGURE 26.4

(A) False-color picture of a radio galaxy. We can't see radio waves, so colors are used to represent their brightness—red brightest, blue dimmest. (B) False-color X-ray picture of Cas A, an exploding star.

A false-color image may be coded to depict data in many different ways. Colors may be used to represent different wavelength ranges—for example, red representing radio, yellow representing infrared, and blue representing X-ray emission. Or the colors may be coded to indicate derived quantities like the temperature of gas, the strength of magnetic fields, or Doppler shift values. These techniques of visualizing data help astronomers to "see" connections between processes going on within an astronomical object.

26.4 THE CRAB NEBULA: A CASE HISTORY

The history of observational astronomy is illustrated by the story of the discovery and gradual piecing together of the astronomical mystery of an exploding star, first seen nearly 1000 years ago. Although the story began with naked-eye observations, it continues with observations made with telescopes on the ground and in space. Moreover, the story illustrates how astronomers have come to rely on observing radiation at many wavelengths, not just visible light.

On July 4, 1054 C.E., just after sunset, astronomers in ancient China and other east Asian countries noticed a brilliant star near the crescent moon in a part of the sky where no bright star had previously been seen. They wrote of this event, "In the last year of the period *Chih-ho,* a guest star appeared. After more than a year it became invisible." The star was so bright that for several weeks it was visible during the day; but it gradually faded away, and after more than a year it became invisible and was forgotten.

Curiously, no records of this spectacular event have been found in European archives. It is difficult to imagine that such a spectacular event, visible for over a year from most of the Earth, went unnoticed by anyone in Europe. Some scholars speculate that written records of this event might have been suppressed because they ran counter to religious beliefs of the medieval church.

Nearly 700 years later, in 1731, John Bevis, a British physician and amateur astronomer, noticed with his telescope a faint patch of light in the constellation Taurus. In the late 1700s the French astronomer Charles Messier, who was hunting for comets, developed a catalog of similar dim patches of light that might be confused with comets. This object was first on his list, and so it became known as "Messier 1" or M1. Messier 1 can be seen with a small telescope—Looking Up #5 shows its location in Taurus.

In 1844 William Parsons, an Irish astronomer and telescope builder (also known as Lord Rosse), noticed that M1 contained filaments that to his eye resembled a crab (Figure 26.5A). He therefore named it the "Crab Nebula." Modern photographs (Figure 26.5B) look quite different from Rosse's sketch, but perhaps still resemble a crab—for someone with a generous imagination!

Messier's catalog contains more than 100 of the brightest of these nebulae (Latin for "clouds"). Their positions are given in the Appendix Table 13. A number of them are identified on the Looking Up illustrations. The complete Messier catalog is given in Appendix Table 13.

In 1921 John Duncan, an American astronomer, compared two photographs of the nebula taken 12 years apart and noticed that it had increased slightly in diameter. He therefore deduced that the nebula was expanding. At the same time, several other astronomers came across the ancient Chinese records and noticed the coincidence in position of the nebula with the report of the exploding star. Then, seven years later, Edwin Hubble, at Mount Wilson Observatory in California, measured the increase of size more accurately and calculated from the rate of expansion that the nebula was about 900 years old—roughly matching the date recorded by the Chinese astronomers. The speed of the nebula's expansion calculated from these observations was astonishingly large—thousands of kilometers per second. Astronomers realized that the Chinese astronomers had witnessed an explosion that had sent gas flying violently outward at almost unimaginable speeds.

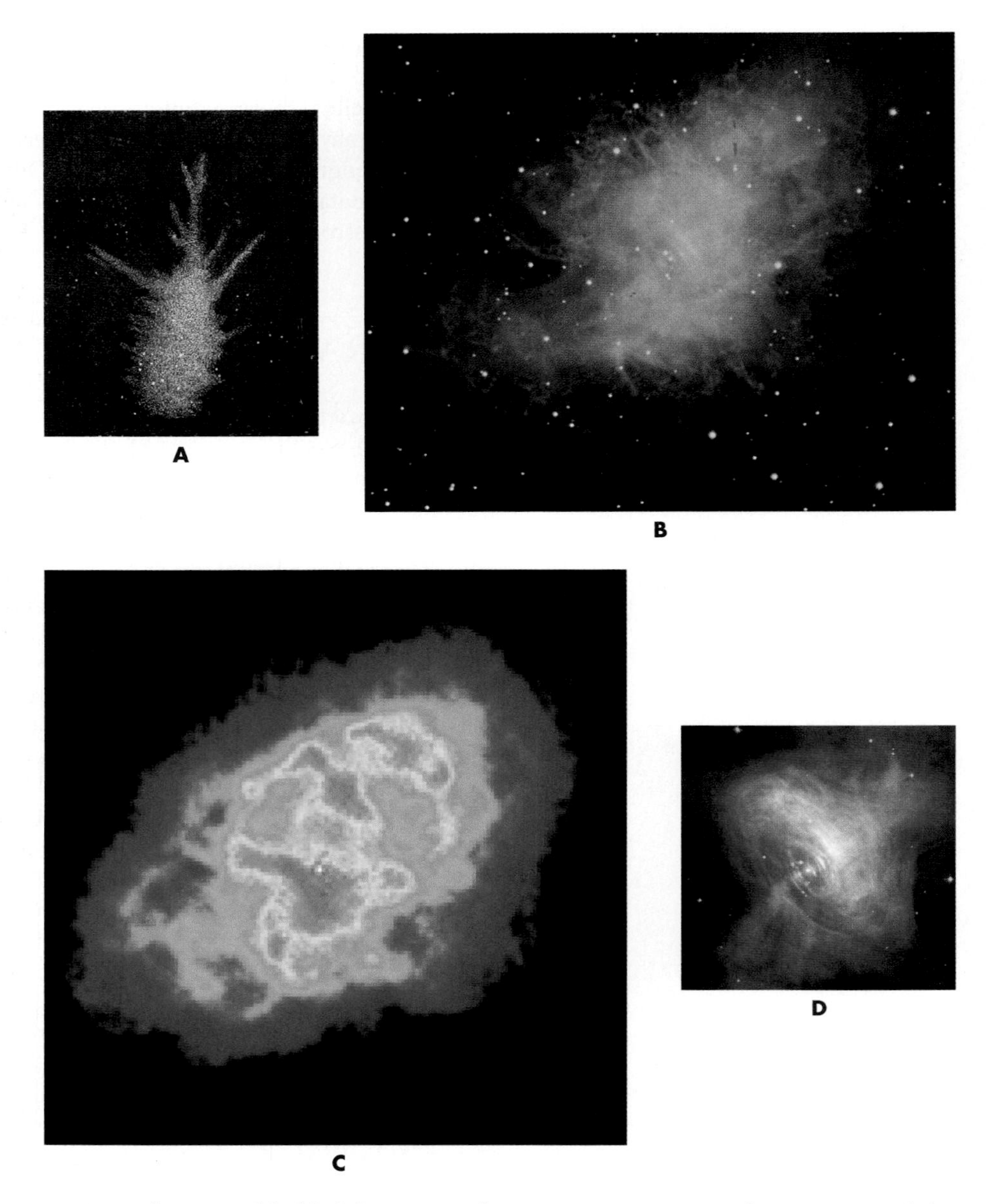

FIGURE 26.5
(A) Lord Rosse's 1844 drawing of the "Crab Nebula." (B) Visible-light photograph of the Crab Nebula. (C) Radio image of the Crab Nebula. (D) X-ray image in false color of the Crab Nebula. X-rays are shown as blue/white while optical light is shown as red. The flattened round shape of the glowing gas suggests a spinning disk.

Since the second half of the twentieth century, astronomers have examined the Crab Nebula with telescopes using virtually all wavelength bands and, in doing so, have added yet more to their understanding of a star's demise. For example, the nebula is a powerful source of both radio waves (Figure 26.5C) and X-ray radiation (Figure 26.5D). These observations have led astronomers to the conclusion that this type of *supernova* explosion occurs after massive stars use up their fuel—teaching us that stars may die violent deaths (Units 65, 66).

The assortment of modern observations at many wavelengths has given astronomers one of their most detailed understandings of the death throes of a star. From radio-wavelength observations, they deduce that the "remains" of the original star are spinning incredibly fast—about 30 revolutions per second. This in turn provides an essential clue that the star has collapsed to such a small size that it is as dense as the nucleus of an atom—an object called a *neutron star* (Unit 67). The X-ray observations reveal details of high-energy particles being ejected at near the speed of light from the dead star and trapped in rings around the star by its strong magnetic fields. Thus, by observing at a variety of wavelengths, astronomers have deduced far more than they could have from observations at one wavelength alone.

Q Based on what you know about the interactions of light and matter, what kinds of information might you hope to learn by studying radiation in the different electromagnetic bands?

KEY TERMS

aperture, 192
CCD, 193
false-color image, 194
focus, 192
instrumentation, 191
photoelectric effect, 193
pixel, 193
resolution, 192
telescope, 192

QUESTIONS FOR REVIEW

1. What advantages does a larger telescope have over a smaller one?
2. What are the advantages of having a telescope in space? Is this true of all wavelengths?
3. How does a CCD work?
4. What is a false-color image?
5. How have astronomical observing techniques changed over the last century?
6. How were modern astronomers able to determine that the "Crab Nebula" was produced by an explosion almost a thousand years ago?

PROBLEMS

1. Suppose that you are using an older CCD camera that collects only 40% of the light that would otherwise be gathered. How would this affect the magnification of the image you are taking? How would this affect the brightness of the image?
2. If a visible-wavelength photon striking a CCD can liberate one electron, approximately how many electrons could be liberated by an X-ray photon with a wavelength of 1 nm? with a wavelength of 0.1 nm? How might it be possible to distinguish between one photon of higher energy versus many photons of lower energy?
3. One of the 10-m diameter Keck telescopes can observe 500-nm visible-wavelength photons. If a radio telescope observing 10-cm photons was built to be proportionally as large relative to the photons it was observing, what would its diameter be? What about a gamma-ray telescope observing 0.01-nm photons?
4. The diameter of your eye's pupil is about 8 mm, while the Arecibo radio telescope has a diameter of about 300 m. For what wavelengths of light does the Arecibo radio telescope have the same relative size to the wavelength as your eye does to visible light?
5. To provide clear images, the surface of a telescope mirror should be smooth to within 1/10 of a wavelength of the light being collected. What size is this for a visible-wavelength telescope? for a radio telescope? for a gamma-ray telescope? Compare the sizes of these largest imperfections to physical objects.
6. Small knots of gas in the Crab Nebula are observed to be moving away from the position of the dead star at its center. Measurements of the position from photographs taken in 1940 and 1990 show that knots originally located 150 arc seconds from the center had moved 8 to 9 arc seconds farther out. From this information, estimate the range of dates when the explosion might have occurred.

TEST YOURSELF

1. Which of the following is *not* a reason astronomers seek to build larger telescopes?
   a. Larger telescopes can resolve greater detail.
   b. Larger telescope collect more photons.
   c. Larger telescopes can detect fainter objects.
   d. Larger telescopes are less affected by the Earth's atmosphere.
2. Which of the following is an *advantage* of a CCD over photographic film?
   a. CCDs can collect light for a long time.
   b. CCDs record a greater percentage of the photons striking them.
   c. CCDs are not affected by blurring of the Earth's atmosphere.
   d. CCDs do not need to be corrected for instrumental effects.
   e. All of the above
3. Which of the following is an important difference between radio waves and other wavelengths of electromagnetic radiation for the design of astronomical instrumentation?
   a. Radio waves are detected as sounds, not light.
   b. Radio waves travel much more slowly than other electromagnetic radiation.
   c. Radio photons have very small energies.
   d. Radio waves cannot be focused.
   e. All of the above

unit

27 Collecting Light

Because most astronomical objects are so remote, their radiation is extremely faint by the time it reaches Earth. For example, to collect enough light to study the most remote galaxies, astronomers use telescopes with mirrors the size of swimming pools, and it may be necessary to point a telescope for hours at a spot so small that it would require thousands of years to look at every object in the sky this way. Therefore astronomers have to be selective in their approach—choosing objects to study that may help to confirm or reject a detailed hypothesis or to refine a theory.

Moreover, special instruments must often be used to extract the information desired from the radiation—instruments that can measure the brightness, the spectrum, and the position of objects to high precision. These instruments generally require much more light than is needed simply to detect that an object is present at a particular position in the sky, and this in turn requires larger telescopes to keep the required observing time within practical limits.

Background Pathways

Unit 26: Detecting Light 191

27.1 MODERN OBSERVATORIES

Telescopes are essentially large "photon buckets" used to collect light over a large area and then concentrate the light for use by a high-sensitivity detector. The challenge is to build a telescope as large as possible but with every surface and separation built to a very exacting standard—to within a fraction of a wavelength of the electromagnetic radiation being observed. This high precision is required to keep the light waves **coherent:** The electromagnetic waves passing through different parts of lenses or bouncing off different parts of mirrors must all arrive at the detector with their wave crests in unison to achieve the maximum effect. If the waves arrive out of unison, they will partially cancel each other—somewhat like how the rowers in a large racing shell must row in unison for maximum effect.

The immense telescopes and the associated equipment that astronomers use are expensive. Therefore, the largest telescopes are often national or international facilities. For example, the United States operates the National Optical Astronomical Observatory and the National Radio Astronomy Observatory, Europe operates the European Southern Observatory, and dozens of other countries each operate many different telescopes. They are open to use by astronomers from around the world through a competitive review of observing proposals. Many colleges and universities have their own research telescopes (in addition to smaller ones near campus for instructional purposes). Altogether, several thousand observatories exist around the world, and observatories can be found on every continent. There are even telescopes at the South Pole in Antarctica. These take advantage of the long night and extreme dryness of the bitterly cold antarctic air to view wavelengths of light that are absorbed by water vapor.

The individual optical telescope with the largest "collecting area" is currently the Gran Telescopio Canarias (GTC) in the Canary Islands, Spain (Figure 27.1). The telescope "array" with the greatest collecting area in the world is the VLT (Very Large Telescope), which consists of four 8-meter telescopes that can work as

FIGURE 27.1
Photograph of the Gran Telescopio Canarias (GTC) in the Canary Islands, Spain. This became the largest visible-wavelength telescope in the world in 2008. The telescope's huge mirror is shown here under construction—36 mirror segments combine to make an overall mirror 10.4 meters in diameter. As the telescope tracks a star or galaxy being observed, all 36 mirrors can be measured with a laser and then adjusted to keep them in precise alignment.

FIGURE 27.2
Conceptual design for a 100-m diameter optical telescope, nicknamed the "Overwhelmingly Large" (OWL) telescope.

a unit. The VLT is operated by a consortium of European countries and Chile and is located in the extremely dry, high-altitude mountains in northern Chile.

To observe the tiny trickle of light from the faintest objects, bigger telescopes are needed. Several designs for 30- to 50-meter diameter telescopes are currently being considered, and even larger telescopes appear to be feasible (Figure 27.2). Such large telescopes might allow us to detect reflected light from an Earth-size planet around another star, or to study the very first stages of star and galaxy formation shortly after the universe began. Building such a large structure precise to about 50 nanometers is extremely challenging, but it appears possible by making the mirror from hundreds of independently-controlled segments.

Radio telescopes are generally the largest of all. Radio astronomers must collect very large numbers of these lowest-energy photons to produce a signal powerful enough to be detected. Fortunately, because the wavelengths being observed are of the order of centimeters or longer, the telescopes need be built only to a precision of millimeters, which is well within the capability of current manufacturing techniques.

Q How thick is a piece of paper in nanometers? Determine this by measuring the thickness of a stack of 100 sheets. How does this compare to the precision that an optical telescope must maintain?

FIGURE 27.3
Photograph of the 300-meter diameter Arecibo radio telescope in Puerto Rico.

One such huge radio telescope is located in Arecibo, Puerto Rico (Figure 27.3). The Arecibo radio telescope is more than 300 meters (1000 feet) in diameter. The huge "dish" is fixed in place in a natural basin and consequently has limited ability to look in different directions. It depends in large part on the rotation of the Earth to observe different parts of the sky. Fully steerable radio telescopes of 100 meters in diameter are located in West Virginia and in Bonn, Germany. Radio astronomers are also thinking about ways to collect signals from very faint sources, and plans for a "Square Kilometer Array," which would have a collecting area of 1 kilometer square, are currently being tested. In the more distant future, astronomers would like to build a radio telescope on the far side of the Moon, where it would be shielded from most human-made radio-wavelength transmissions.

Q The "dish" of the Arecibo telescope is fastened to the ground. How do you suppose it is possible to look in any direction other than straight overhead with this telescope?

27.2 COLLECTING POWER

For us to see an object, light or photons from it must enter our eyes. How bright the object appears to us depends on the number of its photons that enter our eye per second. Astronomers can increase this amount by collecting photons with a telescope, which then funnels the photons to our eye. The bigger the telescope's **collecting area,** the more photons it gathers, as shown in Figure 27.4. The result is a brighter image, which allows us to see dim stars that are invisible in telescopes with smaller collecting areas. The area of a circular collector of diameter D is

$$\text{Collecting area} = \frac{\pi}{4} \times D^2$$

(or equivalently, $\pi \times \text{radius}^2$). As a result, a relatively small increase in the radius of the collecting area gives a larger increase in the number of photons caught. For example, doubling the radius of a lens or mirror increases its collecting area by a factor of 4. Because the collecting area is so important to a telescope's performance, astronomers usually describe a telescope by the diameter of its **aperture**—the area over which it collects photons. Thus the 10-meter Keck Telescope in Hawaii has mirrors spanning 10 meters (roughly 30 feet) in diameter.

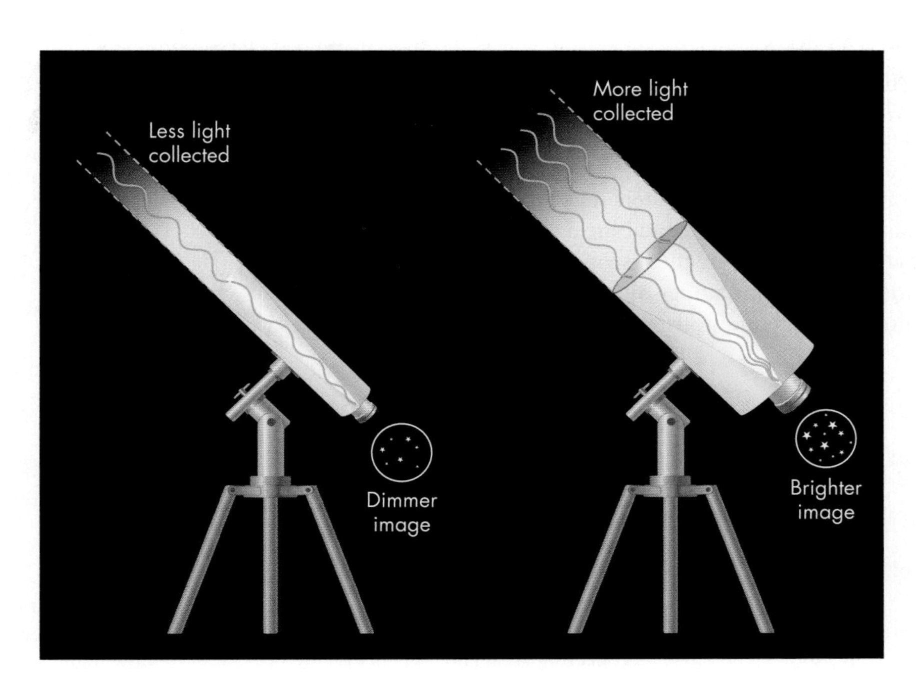

FIGURE 27.4
A large lens collects more light (photons) than a small one, leading to a brighter image.

When your eye's pupil is completely dilated, it has an aperture with a diameter of about 8 mm. The Hubble Space Telescope has an aperture with a diameter of 2.4 m. Comparing the size of the collection area for the HST with that of your eye's pupil, we get:

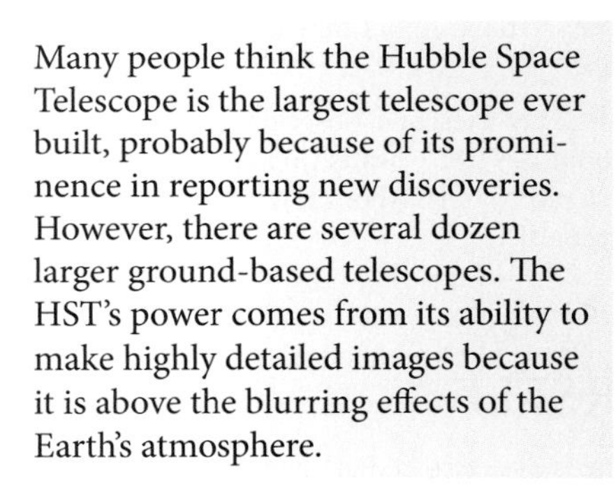
Many people think the Hubble Space Telescope is the largest telescope ever built, probably because of its prominence in reporting new discoveries. However, there are several dozen larger ground-based telescopes. The HST's power comes from its ability to make highly detailed images because it is above the blurring effects of the Earth's atmosphere.

$$\frac{\frac{\pi}{4}(2.4\text{ m})^2}{\frac{\pi}{4}(8\text{ mm})^2} = \left(\frac{2400\text{ mm}}{8\text{ mm}}\right)^2 = 300^2 = 90{,}000.$$

Thus the Hubble Space Telescope collects 90,000 times more light per second than your eye. Carrying out the same calculation for the Keck telescope shows that it collects more than 1.5 million times more light than your eye, and 17 times more light than the Hubble Space Telescope. The advantage of the Hubble, of course, is that it is free of the atmospheric distortions that Earth-based collected light is subject to (Unit 30).

27.3 FILTERING LIGHT

Telescopes generally collect a wide range of wavelengths. For example, most visible-wavelength telescopes also work into the infrared and ultraviolet. However, astronomers rarely collect all of the incoming light, but instead **filter** out all but a narrow range of wavelengths, selected to learn about the particular objects they are studying.

Filters are often thin pieces of colored glass. The filter glass is designed to reject all wavelengths but those in the range of interest and to be highly transparent within that range. At visual wavelengths astronomers use an assortment of filters:

B, or blue, transparent primarily at 390–480 nm
V, or "visual," transparent at 500–590 nm
R, or red, transparent at 570–710 nm.

Some specialized filters select narrow ranges of wavelength to examine the light from a single spectral line, whereas other filters extend into the ultraviolet and infrared regions.

Different kinds of film, CCDs, and lenses may behave differently depending on the wavelength of the light. By narrowing the range of wavelengths observed, filters make it easier to compare and combine data gathered with different telescopes and detectors and under different atmospheric conditions.

By observing stars and other objects through filters, astronomers can characterize more precisely how much red light or blue light an object emits, which can help to determine its temperature (Unit 23). In addition, filters help to reject unwanted light—when we observe an object known to emit most strongly at one wavelength, a filter can be chosen to admit light at that wavelength and no others.

Note that a filter does not "color" the light passing through it. When you look through a blue filter, you are seeing just the blue photons from each object, not changing the wavelength of other color photons. Most objects produce or reflect some photons of every wavelength, so a yellow object will not necessarily appear black through a blue filter.

KEY TERMS

aperture, 200
coherent, 198
collecting area, 200
filter, 201

QUESTIONS FOR REVIEW

1. What is meant by the term *collecting area?*
2. When the diameter of one telescope is twice that of another, how does its collecting area compare?
3. How does a larger telescope make it easier to see faint objects?
4. Why do astronomers use filters instead of collecting all wavelengths of light?
5. Why would you be unable to detect spectral lines of hydrogen if you observed them through a V filter?

PROBLEMS

1. If you look at stars with binoculars with lenses about 8 cm (3 inches) in diameter, how many times fainter a star should you be able to see compared to what you can see with your unaided, fully dilated eye, with the pupil open to about 8 mm?
2. Compare the collecting area of one of the 36 segments of the 10.4-meter GTC to the collecting area of the 2.4-meter Hubble Space Telescope. Which is larger? By what factor?
3. A group of stars is observed through several different filters, and some stars are seen to be brightest when observed through either the V (500–590 nm) or R (570–710 nm) filter.
   a. Assuming a star is brightest in the middle of the R filter, use Wien's law (Unit 23.4) to estimate its temperature.
   b. If a star is brightest through the V filter, use Wien's law to estimate the range of temperatures it might have.
4. The energy of the light reaching Earth from a "first magnitude" star is approximately 10^{-6} watts (joules per second) over each square meter of a telescope's collecting area. Assuming that this light is composed of photons of wavelength 500 nm, about how many photons are arriving in a square meter each second? How many photons enter your eye each second if your pupil is 8 mm in diameter?
5. At a radio wavelength around 30 cm, the Arecibo radio telescope can collect about 3×10^{-14} watts (joules per second) of radiation from the Crab Nebula. About how many photons is it collecting each second?
6. The orbiting Chandra X-ray telescope, observing at about 0.1 nm, can collect about 2×10^{-13} watts (joules per second) of radiation from the Crab Nebula. About how many photons is it collecting each second?

TEST YOURSELF

1. The Palomar telescope is 5 meters in diameter, whereas the GTC telescope is approximately 10 meters in diameter. How many times larger is the GTC telescope's collecting area?
   a. 2 times
   b. 4 times
   c. 5 times
   d. 10 times
   e. 25 times
2. Which of the following is a good reason for building larger telescopes?
   a. They permit us to detect dimmer stars.
   b. They make stars look much bigger.
   c. They let us see much dimmer surface brightness levels.
   d. All of the above
3. The purpose of a blue filter is to
   a. alter the wavelength of light we observe to make it easier to detect.
   b. eliminate the color of the sky from photographs.
   c. let through only the blue wavelengths of light.
   d. allow a telescope to collect more light.

unit

28 Focusing Light

Background Pathways

In addition to a telescope's role as a large "photon bucket," telescopes are designed to distinguish the direction from which light arrives and to **focus** the light. Focusing light is the process of bringing the electromagnetic waves together to form the strongest and most detailed signal possible.

Galileo Galilei is sometimes credited with the invention of the telescope, but the invention seems to have been the work of the Dutch spectacle maker Hans Lippershey in 1608. News of the invention spread rapidly, and by the summer of 1609, Thomas Harriott, an English mathematician and scientist, had used a telescope to study the Moon. Just a few months later, Galileo, quickly grasping the idea behind the design of the telescope, built a telescope of better quality than any previously available, capable of magnifying images by about a factor of 20. With this telescope he not only observed the Moon but discovered Jupiter's moons, Venus's phases, and the vast numbers of stars in the Milky Way. Most important, he published and interpreted his findings (Unit 12.3).

The competition to build better telescopes has continued nonstop to this day, and now includes specialized telescopes for every wavelength range. Astronomers have not only sought to build bigger telescopes that can detect dimmer objects (Unit 27), they have developed improved designs for focusing the light to provide more detailed images. The methods used for focusing light fall into two broad categories, employing either a lens or a mirror to concentrate the light that enters the main aperture of the telescope. Each of these methods has advantages and disadvantages, as we discuss next.

28.1 REFRACTING TELESCOPES

The first telescopes used **lenses** to focus the light. Transparent substances like the glass of a lens can alter the path of electromagnetic waves. This effect is called **refraction.** Eyeglasses and the lenses inside our eyes use refraction to focus light.

Telescopes are called **refractors** if they use a lens as the primary means to focus the light that enters the aperture. The lens of a refractor focuses the light by bending the rays, as shown in Figure 28.1. A lens that is just the right shape can collect light over the entire aperture and bend it so that it arrives at a focus behind the lens, creating an image. The Figure depicts the light from two stars, a yellow star straight in front of the telescope and a red star above it. The light from each star is focused to a point on a flat region called the **focal plane.** This is where photographic film or a CCD might be placed, although often there are additional lenses in front of the detector to provide extra magnification or to correct for distortions produced by a single lens.

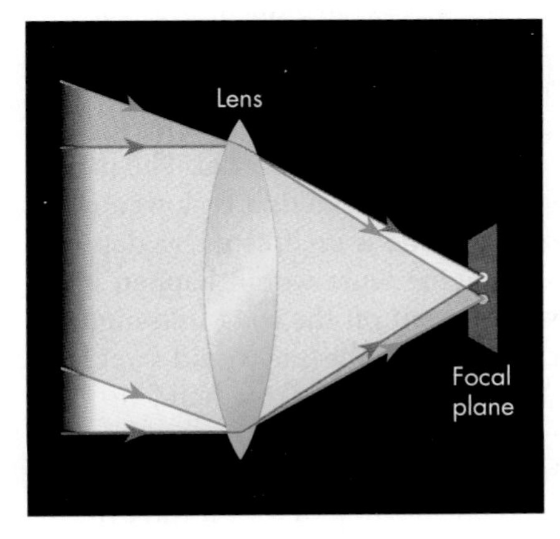

FIGURE 28.1
Light is shown arriving from a yellow star straight in front of the lens and from a red star above it. Each star's light is brought to a focus at the focal plane.

Lenses can bend light because when light moves from one transparent material into another at an angle, the direction in which the light travels is bent. Bending occurs when light from a distant star enters the Earth's atmosphere, or when that light travels from air into water or glass. The bending is generally stronger with a greater difference in density of the materials. For example, glass is slightly more dense than water, and light entering glass from air is bent slightly more than

Ocean waves travel more slowly where the water becomes less deep, so a change in wave direction can sometimes indicate where the water grows shallow.

FIGURE 28.2
Refraction of light in a glass of water. Notice how the pencil appears to be bent.

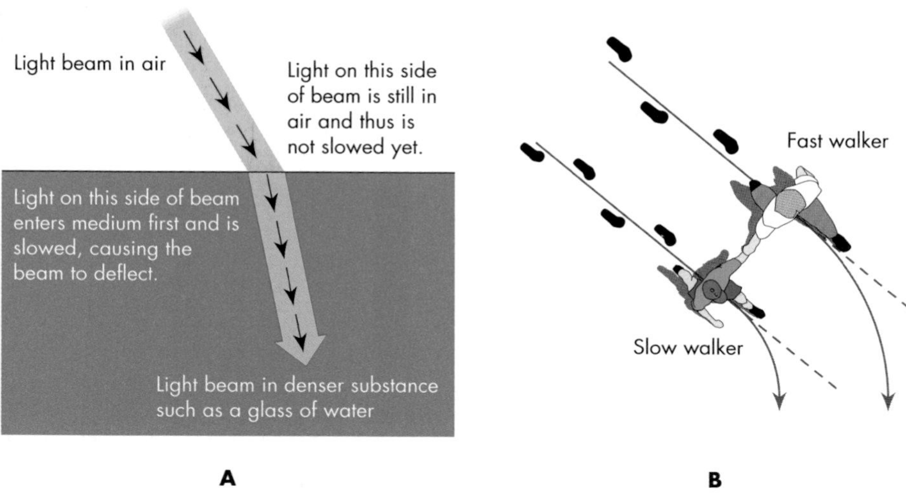

FIGURE 28.3
Cause of refraction. (A) Light entering the dense medium is slowed, while the portion still in the less dense medium proceeds at its original speed. (B) A similar effect occurs when you walk hand-in-hand with someone who walks more slowly than you do.

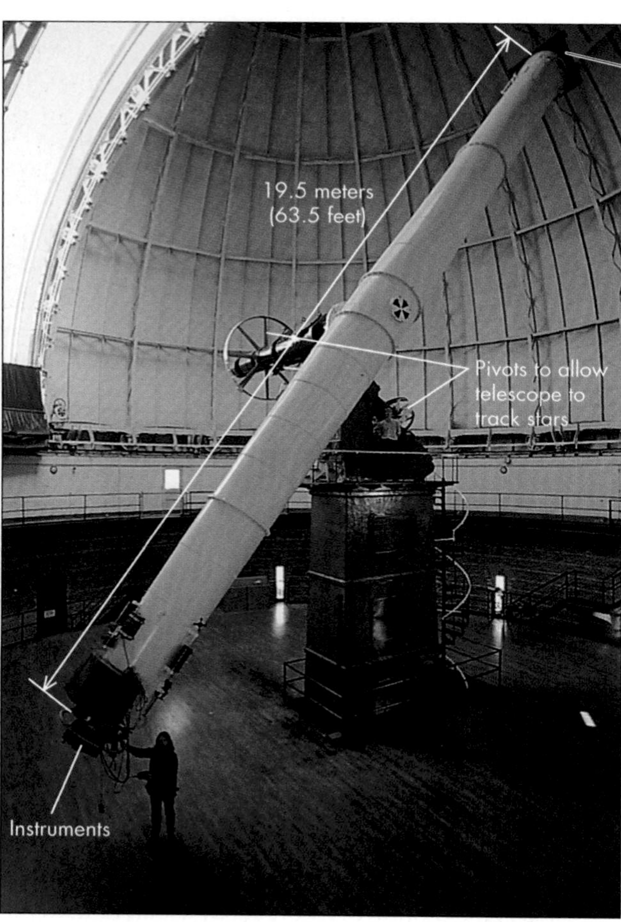

FIGURE 28.4
A refracting telescope. Completed in 1897 for the University of Chicago's Yerkes Observatory in Williams Bay, Wisconsin, this refractor has a lens approximately 1 meter (40 inches) in diameter, making it the world's largest refracting telescope.

when it enters water. However, a clear piece of glass underwater may be hard to see because of the small difference in refracting properties.

You can easily see the effects of refraction by placing a pencil in a glass of water and noticing that the pencil appears bent, as shown in Figure 28.2. The pencil in water also illustrates an important property of refraction. If you look along a pencil placed partly in the water and change the pencil's tilt, you will see that the amount of bending (refraction) changes. Exactly vertical rays are not bent at all, nearly vertical rays are bent only a little, and rays entering at a grazing angle are bent most.

Light is bent because its speed changes as it enters matter, generally slowing in denser material. This decrease in the speed of light arises from its interaction with the atoms through which the electromagnetic waves move. To understand how this reduction in light's speed makes it bend, imagine a light wave approaching a slab of material. The part of the wave that enters the material first is slowed while the part remaining outside is unaffected, as depicted in Figure 28.3A. Imagine what would happen if the right side of your car went off the road into mud. As that side of the car slowed, the car would swerve to the right. Or if two people are walking hand-in-hand (as sketched in Figure 28.3B), if the one on the right slows down, the direction of their motion turns to the right. So, too, if one portion of a light wave moves more slowly than another, the light's path will bend.

Figure 28.4 shows a photograph of the world's largest refractor, the 1-meter (40-inch) diameter Yerkes telescope of the University of Chicago. This telescope, completed in 1897, was one of the last

Q If you are scuba diving and you look up at the surface of the water overhead, you will see light from the sky in a region only out to about 45° away from the zenith. How might refraction explain this?

large-lens telescopes ever built. Building a telescope with such a large lens presents serious structural problems. The Yerkes lens has a mass of over 200 kilograms (over 450 pounds). This massive piece of glass has to be supported by its edges, where the glass is thinnest, so the lens flexes slightly, causing image distortions. In addition, the large mass of glass is located at the end of a long telescope tube, which must be even more massive—about 20 tons—to keep it from flexing. Building larger refractors would require even more massive telescopes with even greater problems of structural flexing. To build larger-aperture telescopes, an alternative approach was needed.

28.2 REFLECTING TELESCOPES

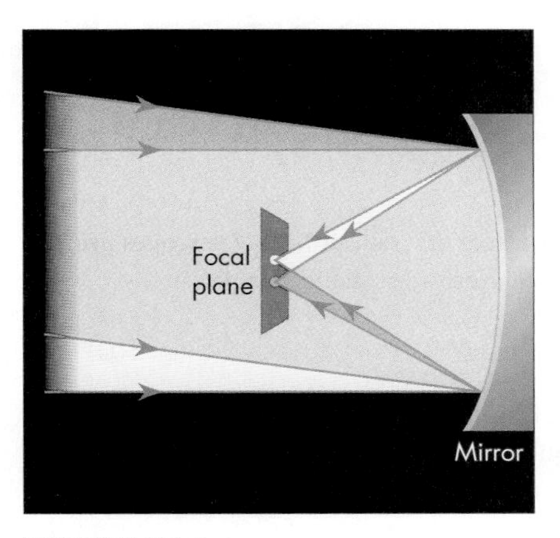

FIGURE 28.5
A curved mirror focuses light from a yellow star straight in front of it and a red star above that. The focal plane is in front of the mirror and blocks some of the starlight.

Not long after the invention of telescopes, people realized that mirrors could be used to bring light to a focus; but a practical design to do this was not worked out until 1670 by Isaac Newton. Technology at the time did not make it easy to build *reflecting telescopes*—**reflectors** for short—so it was not until the late 1800s that reflectors began to replace refractors. Today almost all research telescopes, whether used for visible, radio, or X-ray wavelengths, employ mirrors to collect and focus light.

As Figure 28.5 shows, a curved mirror can focus light rays reflected from it. The Figure depicts the light from two stars, a yellow star straight in front of the telescope and a red star above the yellow star. Each star's light is focused at a point on the focal plane, but this is now in front of the mirror, so the observer or detector blocks some of the light from reaching the mirror. If the mirror is big enough, the fraction of the mirror blocked by a detector (or even an astronomer!) may be only a small fraction of the aperture. Some large telescopes contain a "cage" in which an astronomer may ride inside the telescope at the focal plane.

A variety of other designs have been developed to move the focus outside the path of the incoming light. Newton's solution was to use a **secondary mirror** to deflect the light out to the side (Figure 28.6). This is today known as a **Newtonian telescope,** and it remains a popular design for smaller reflectors. Most modern research telescopes, whether optical or radio, use a secondary mirror that reflects the light back through a hole cut in the middle of the main mirror. This is called a **Cassegrain telescope,** named for the French sculptor Guillaume Cassegrain, who designed it in 1672. The secondary mirror in all of these designs blocks part of the light, but only a small fraction of the total, from reaching the **primary mirror.**

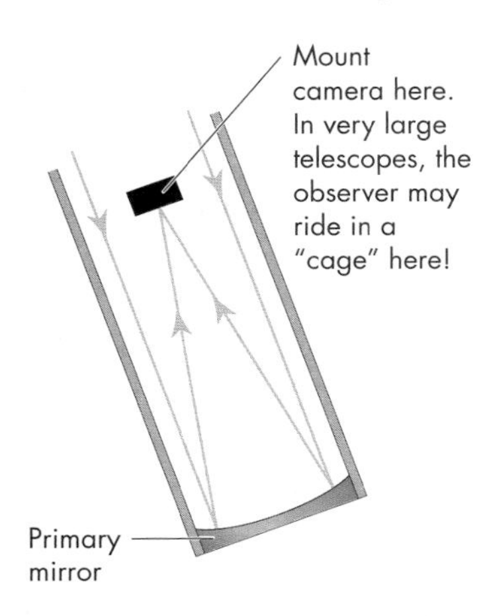

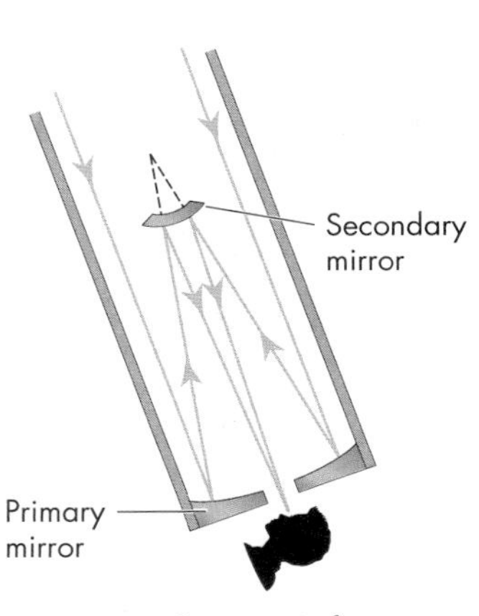

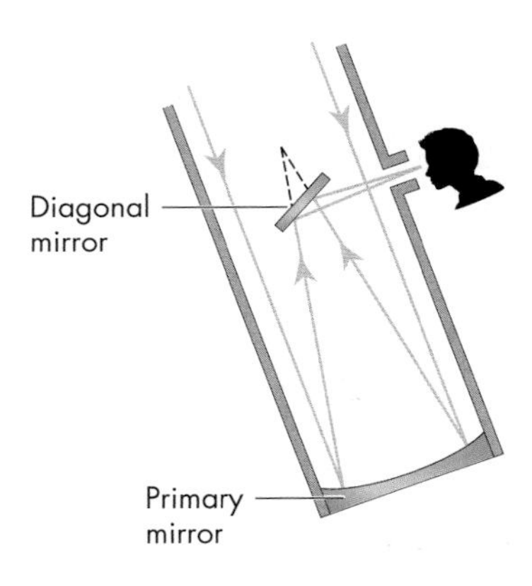

FIGURE 28.6
Sketches of different focus arrangements for reflectors.

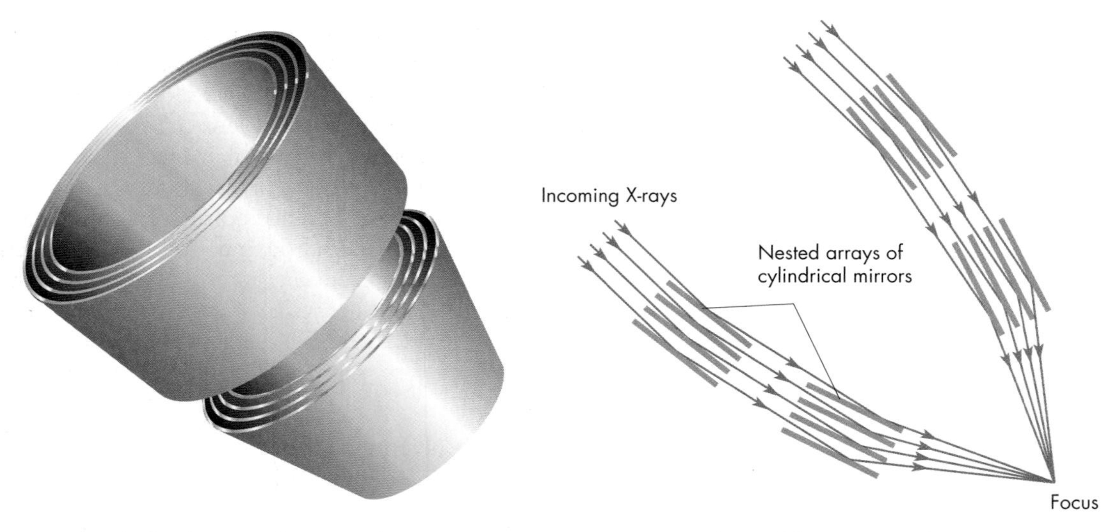

A The cylindrical mirrors of an X-ray telescope

B Cross-sectional view of the grazing-incidence mirrors showing how they focus X-rays

FIGURE 28.7
Grazing incidence optics in an X-ray telescope. X-rays pass through normal mirrors, so X-ray telescopes bring X-rays to a focus by letting them graze the surfaces of nested cylindrical mirrors at shallow angles.

Newton's first telescope mirror was made from polished metal because the technique of laying a thin reflective coat of metal on polished glass had not yet been developed.

In optical telescopes the mirrors are usually made of glass that has been shaped to a smooth curve, polished, and then coated with a thin layer of aluminum or some other highly reflective material. The reflecting surface must be smooth to about one-tenth of the wavelength of the photons bouncing off it; otherwise the light will scatter in many directions. Glass can be polished to such a smoothness, and the electromagnetic waves therefore can be brought together in sharp focus.

A radio telescope designed to study 10-centimeter wavelength light, by contrast, needs to be smooth only at the level of 1 centimeter, so it can be built of metal plates bolted together. At the other extreme, X-ray (and gamma-ray) photons have wavelengths that may be smaller than the size of an atom. X-ray photons encountering an optical telescope mirror would be scattered in many directions or absorbed by the mirror. However, they will reflect off a surface if they encounter it at a very shallow angle. (You can demonstrate this principle to yourself with visible light by tilting a rough surface so light hits it at a glancing angle.) The mirror designs for X-ray telescopes therefore look completely different from optical designs, but they achieve the same result of funneling the photons toward the detector (Figure 28.7).

When the Hubble Space Telescope was first put into service, it was discovered that the surface had been polished to the wrong shape because of a manufacturing error. Fortunately, it was possible to insert a lens to correct for the shape error, much as you can wear glasses to correct for focusing problems caused by the shape of your eye.

It is important that the electromagnetic waves arrive at the detector in unison, so the shape of the reflector is critical in addition to its smoothness. The mirror needs to be designed so that after electromagnetic waves bounce off the mirror, the crests of the waves come together at the focus all at the same time. By achieving this simultaneous arrival, the strength of the electromagnetic vibration at the focus will be the concentrated strength of the entire wave collected over the whole aperture. If the mirror shape is not correct, not all portions of the crest of the wave will arrive at the same time, and they will not combine to the maximum strength. As a result, the signal will be weaker at the detector, and the images will be blurry.

Moreover, as a telescope moves, its mirrors and lenses must keep their same precise shapes and relative positions if the images are to be sharp. This is one of the most technically demanding parts of building a large telescope, because large pieces of glass or metal bend slightly when their positions shift. Unlike lenses, however, a mirror can be supported from behind, thereby helping to hold the glass in shape.

Development of large reflectors proceeded rapidly during the early 1900s, greatly surpassing the apertures of the largest refractors. A 5-meter (17-foot) diameter mirror was completed in 1948 at the Palomar observatory in California—the mirror alone weighing nearly 15 tons, the whole telescope nearly 500 tons. Building larger telescopes along similar lines was found impractical because the weight of glass and supporting structures grew too rapidly with size.

Recent innovations have worked around the problem of building extremely massive, stiff "back structures." One way is to collect and focus the light with several smaller mirrors and then align each one individually. So instead of a telescope having a single large mirror, it may have many small ones. Mirrors designed this way are called *segmented mirrors*. Currently two of the largest such segmented mirrors are in the twin 10-meter (33-foot) Keck telescopes (Figure 28.8A), operated by the California Institute of Technology and the University of California and located on the 4200-meter (14,000-foot) volcanic peak Mauna Kea in Hawaii. The total weight of the glass in this design is about the same as in the Palomar 5-meter telescope, but the total telescope weight is only about half as much. Each Keck telescope consists of 36 separate mirrors (visible in Figure 28.8A) that are kept aligned by lasers that measure precisely the tilt and position of each mirror. If any misalignment is detected, tiny motors shift the offending mirror segment to keep the image sharply in focus. The same principle is being used to build large new radio telescopes.

Another approach is to build mirrors much thinner and lighter than those in earlier, smaller telescopes. For many years telescope mirrors were made thicker as they were made larger, to keep them stiff. By making a back structure that uses sensors and tiny motors to compensate for bending as the mirror is tilted, much as described for the Keck telescopes, a thin mirror can be kept precisely shaped. A number of thin 8-meter (about 26-foot) glass mirrors have been built in recent years, achieving overall weights of glass and telescope that are similar to those of the Palomar 5-meter telescope. Four are used in unison at the VLT (Very Large Telescope) of the European Southern Observatory in Chile, giving the combination of the four the largest total collecting area of any current visible-wavelength observatory. A photograph of the 8-meter Gemini telescope on Mauna Kea is shown in Figure 28.8B.

FIGURE 28.8
(A) Photograph of one of the twin Keck telescopes. The 36 mirrors cover an area 10 meters (about 33 feet) in diameter, making them the world's largest single optical telescopes. (B) One of the two Gemini telescopes. One of this telescope pair is on Mauna Kea, Hawaii; the other one is in Chile. The mirror of each telescope is about 8 meters (26 feet) in diameter. The yellow lines show the light path through the telescope.

28.3 REFLECTORS VERSUS REFRACTORS

FIGURE 28.9
Photograph of the 100-meter diameter radio telescope in Green Bank, West Virginia. The main reflector is designed to reflect the light at an angle so the secondary mirror does not block the telescope's aperture.

After 1900, almost all telescopes built for research were reflectors. This was primarily because of the expense of building larger lenses and the technological problems of supporting such massive pieces of glass, as we have discussed. Note that lenses are still frequently used as part of the optics of many research telescopes, but these telescopes continue to be called *reflectors* if a mirror does the primary work of focusing the light that enters the aperture.

Refracting telescopes have the advantage that the focus is behind the lens. This simplifies the design, and there is no need for a secondary mirror or other equipment that partially blocks the telescope aperture. Image quality is degraded by anything that blocks part of the aperture. For example, the "spikes" seen on stars in most astronomical images are caused by the support structures that hold the secondary mirror in place in a reflector.

One way to eliminate the problem of having a secondary mirror blocking the primary is to construct a reflector with an *off-axis* focus. The idea is to design the primary mirror so that it reflects the light at an angle, bringing it to focus at a point outside the field of view of the aperture. The German/English astronomer Sir William Herschel built a telescope of this design in the 1780s. However, because the primary mirror needs to be asymmetric, such telescopes are substantially more difficult to build and are relatively uncommon. The 100-meter (330 foot) diameter Green Bank Telescope (GBT), built using this idea, is shown in Figure 28.9.

Even though refractors do not have any blockage in their aperture, they have some significant disadvantages because of the properties of glass. The lenses may be as opaque as a chunk of concrete to shorter-wavelength ultraviolet photons. Mirrors, by contrast, are generally much less selective in the wavelengths of electromagnetic radiation they reflect, allowing much broader wavelength coverage with a single telescope. Mirrors also reflect light at the same angle, independent of wavelength, but the same is not true of lenses. The angle by which glass bends light depends on the light's wavelength. So, for example, the red wavelengths of light may be out of focus when the blue wavelengths are sharp, creating images fringed with color (Figure 28.10A). This property of refracting materials is a problem for making a sharp image.

This effect on image quality is called *chromatic aberration* and is a problem for cameras as well as telescopes.

28.4 COLOR DISPERSION

Different wavelengths of light travel at slightly different speeds in most materials and are therefore refracted by different amounts. Shorter wavelengths of light are generally refracted more strongly. Thus when white light—which is a mix of all colors—passes through a refracting material at an angle, it is spread into a spectrum, or rainbow, in a process called **dispersion.** This happens to light passing through the glass of a **prism** (Figure 28.10B) or through drops of water in a rainstorm.

Because of this effect of lenses, most modern research instruments avoid lenses if mirrors can achieve the same result. The dispersion caused by glass is one reason astronomical mirrors have reflective aluminum coating on top of the glass instead of underneath it (as in a conventional mirror). However, this has the disadvantage that telescope mirrors can be easily marred and must be recoated over time as the metal becomes tarnished by exposure to air.

As we saw in Unit 24 and elsewhere, dispersion is enormously useful for carrying out spectroscopy on the light from astronomical sources. Astronomers can

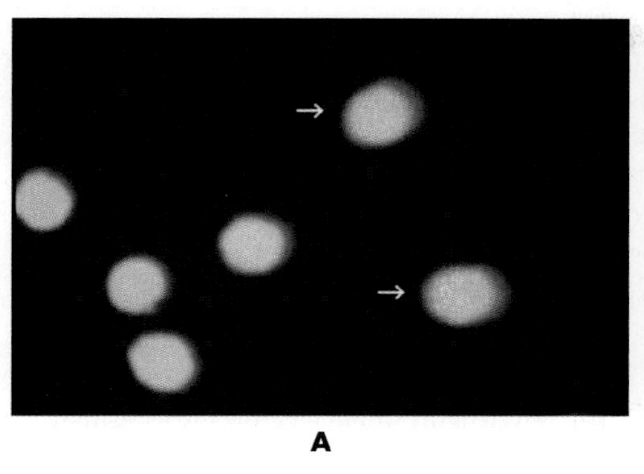

A

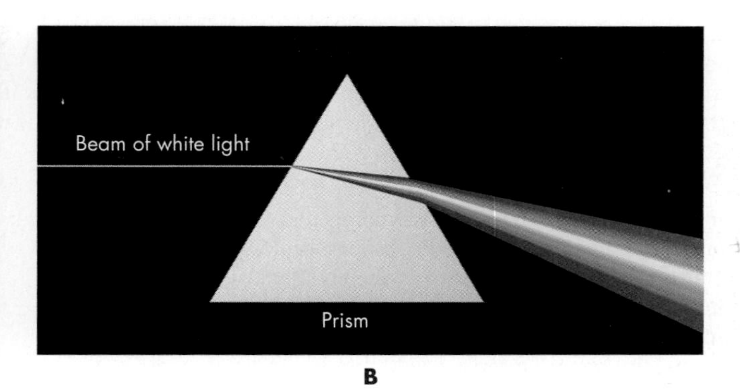

B

FIGURE 28.10
(A) The dispersion of light caused by glass can result in different wavelengths focusing differently. The result can appear in images as fringes of color. (B) Glass refracts blue light more than red light, producing a spectrum.

use a large prism to split the light of all of the astronomical sources within the telescope's field of view into spectra all at once. Stars and other objects may display slight or large differences in the characteristics of their spectra even though their color and appearance otherwise look the same. This gives astronomers an important tool for classifying objects and discovering new kinds of sources.

KEY TERMS

Cassegrain telescope, 205
dispersion, 208
focal plane, 203
focus, 203
lens, 203
Newtonian telescope, 205
primary mirror, 205
prism, 208
reflector, 205
refraction, 203
refractor, 203
secondary mirror, 205

QUESTIONS FOR REVIEW

1. What do we mean by the phrase *to focus light?*
2. What is the difference between reflecting and refracting telescopes?
3. Why is it necessary for reflecting telescopes to have a secondary mirror?
4. Why have the large telescopes built in the last century all been reflectors, instead of refractors?
5. What technologies are needed to allow the giant reflector mirrors of today to maintain a sharp focus as they are aimed in different directions?
6. How is dispersion of light useful in astronomy?

PROBLEMS

1. An image will appear upside down when we look through a simple refracting telescope, as can be seen from Figure 28.1. Trace the light through a similar diagram to show what orientation objects would have through a Cassegrain telescope. Assume the secondary mirror in the Cassegrain telescope is flat.
2. Suppose a reflector has a primary mirror with a diameter of 100 cm, and a secondary mirror that blocks a central region with a diameter of 25 cm. What fraction of the collecting area does the secondary mirror block? What would be the diameter of a telescope with the same collecting area if no part of its area were blocked?
3. The front surface of your eye (the cornea) does most of the focusing of light entering your eye; the back wall of the eye is like a focal plane. Draw a diagram similar to Figure 28.1 to show how your eye focuses the light rays from a distant star straight in front of you. Now draw light rays for a nearer point of light (a few times farther away from your cornea than the cornea is from the back of the eye). The light rays will be bent by the cornea by about the same angle as the distant starlight, so do they now focus closer or farther from the cornea? A nearsighted person is able to focus on nearby objects, but not distant ones. What does this tell you about the shape of the cornea for a nearsighted person?

4. Most refracting telescopes have long telescope tubes. It should be possible to build a lens that focuses light much closer, without such a long tube. Sketch how the shape of the lens would need to be different for a lens that focuses closer. Use your diagram to show why such a close-focusing lens would probably have more severe problems with chromatic aberration.
5. After the Palomar 5-m telescope was completed, astronomers estimated that the mass of larger telescopes would increase as the (mirror diameter)^3 or (mirror diameter)^4 because of the greater mass of glass and support needed to keep the entire telescope stiff. If the 10-m Keck telescopes followed this rule, how many times more massive would they have been than the Palomar telescope?
6. The *index of refraction* measures the factor by which light slows down in a substance. The index of refraction of water is 1.33. If you sent a pulse of light through water and a pulse of light through air (index of refraction of 1.00), how much longer would it take the light to travel 1 km through the water? (The speed of light in air is 299,800 km/sec.)

TEST YOURSELF

1. Suppose that a reflector and a refractor have the same diameter aperture. Which of the following would be a disadvantage of using the refractor?
   a. Refractors bend the light at too large an angle.
   b. Refractors do not focus all colors at the same place.
   c. Refractors have less magnification.
   d. Refractors allow in less light.
2. Suppose you were examining a pulse of radio signals from a distant civilization far away across interstellar space. Knowing that the ionized gas in interstellar space causes dispersion of radio waves, what effect would you expect this to have on the signal?
   a. The signal would be slowed down—stretched out to fill a much longer time.
   b. The path would be bent so the signal would come from a different direction than it started from.
   c. The time when the pulse arrived would be different for different wavelengths.
   d. The wavelengths would all grow longer as they ran out of energy.
3. How does the speed of light in glass compare to the speed of light in empty space?
   a. It is faster in glass.
   b. It is faster in space.
   c. It is the same in both.

unit

29

Telescope Resolution

Background Pathways

People often ask how high a magnification a telescope can produce. In reality, any telescope can deliver as high a magnification as we want with the right eyepiece. What is more important to understand is the **resolution** of a telescope—the level of detail it is capable of revealing. A highly magnified blurry image is of little use.

Resolution is determined in part by the quality of the optics; but even with perfectly made lenses and mirrors, there is a fundamental limit on resolution imposed by the aperture size of a telescope. The larger the aperture, the finer the detail it is capable of detecting. This is the other reason astronomers strive to build larger telescopes—so they can resolve increasingly finer structural details of the objects they observe.

29.1 RESOLUTION AND DIFFRACTION

If you mark two black dots close together on a piece of paper and look at them from the other side of the room, at a great enough distance your eyes will no longer be able to see them as separate spots. Similarly, stars that lie very close together, or markings on a planet, may not be clearly distinguishable. The smallest separation between features that a telescope is able to distinguish is a measure of the telescope's resolution.

Resolution is fundamentally limited by the wave nature of light. Whenever waves pass through an opening, they are bent at the edges of the opening by a phenomenon called **diffraction.** Figure 29.1A shows how water waves, all initially parallel to each other, are diffracted after they pass through a narrow opening. The central portion of the set of waves travels straight through the opening, but weaker waves radiate away from the edges of the opening.

Light waves are similarly diffracted at the edge of a telescope's aperture. The result is that the light waves from a star are shifted into slightly different directions, smearing out the light. Figure 29.1B shows an image of a star from the Hubble Space Telescope. The image has been enhanced to bring out the very faint **diffraction pattern** produced when a star is imaged. Diffraction spreads the star's image into a blur, even though the light is coming from a single point within this image. A small percentage of the light is diffracted at larger angles and produces the complex outer parts of the diffraction pattern.

The complex set of rings and spikes seen in the diffraction pattern is caused by interference between waves coming through different parts of the aperture or diffracting from the edges of any other structures within the telescope's aperture. **Interference** occurs when light waves add to or subtract from each other. The diffracted light travels through the telescope along slightly different paths, offset from its original direction. Light diffracted by different edges of the aperture and

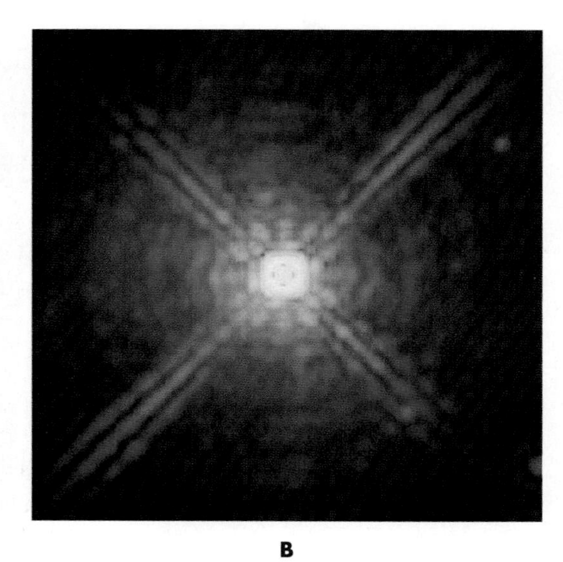

FIGURE 29.1
(A) Water ripples diffracted as they pass (moving down in image) through a narrow opening. (B) Image of a star made with the Hubble Space Telescope. Light coming from a single point is spread out by diffraction. Diffraction at the edges of the aperture and at internal support structures produces a very faint, complex diffraction pattern.

following slightly longer or shorter paths may arrive at the detector with the wave crests adding together or canceling each other out. Where they add together, we see a brighter spot in the diffraction pattern. Where the interference causes the waves to cancel one another, we see a darker spot in the diffraction pattern. The diffraction rings are produced by interactions between light waves diffracted from different parts of the edges of the aperture. The X-shaped pattern surrounding the star is caused by light diffracted by the support structure holding a secondary mirror in the telescope.

One way of describing the resolution of a telescope is to find the closest pair of stars that can still be discerned from each other. If the stars are separated by too small an angle, diffraction by the telescope aperture (or poor optics) will overlap the light from the stars to such an extent that they become indistinct from each other.

The diffraction pattern can make brighter stars appear larger in an astronomical image even though the light from all stars is essentially coming from a single point. This is especially true when images are optimized to examine fainter objects. When the image is brightened, the inner portion of the diffraction pattern around a bright star may become completely "burned out" (completely white or bright). For the brightest stars, the diffraction pattern will be "burned out" even farther from the star. Detecting a faint star near a bright star can be difficult because a dim object may become lost within the bright star's diffraction pattern even if the separation is relatively large. This makes it extremely difficult to image such objects as planets around other stars.

29.2 CALCULATING THE RESOLUTION OF A TELESCOPE

The best possible resolution of a telescope depends on both the wavelength of the radiation being observed and the diameter of the lens or mirror that collects the light. Resolution can be improved by using a larger mirror or lens, and it can also be improved by observing at shorter wavelengths.

Detailed calculations show that if two points of light are separated by an angle α (measured in arc seconds) and are observed at a wavelength λ (measured in

An arc second is a unit of angle and is equal to 1/3600 of a degree.

nanometers), they cannot be seen as separate sources unless they are observed through a telescope whose diameter D (in centimeters) is larger than

$$D_{cm} > 0.02\ \lambda_{nm}/\alpha_{arcsec},$$

where the subscripts cm, nm, and arcsec have been added as reminders of the units that must be used for measuring each term. Notice that the diameter needed to resolve two sources increases as the sources get closer together.

Alternatively, we can express the smallest separation angle a telescope is capable of resolving by rearranging the equation terms:

$$\alpha_{arcsec} > 0.02\ \lambda_{nm}/D_{cm}.$$

α_{arcsec} = smallest resolvable angle (in arc seconds)

λ_{nm} = wavelength of observation (in nanometers)

D_{cm} = diameter of telescope aperture (in centimeters)

For example, a telescope of diameter 100 centimeters (about 40 inches) observing visible light ($\lambda \approx 500$ nanometers) will be able to resolve two stars as little as 0.1 arc seconds apart.

Radio telescopes, despite their enormous size, observe very long wavelengths of light. In consequence, they usually have much poorer resolution than optical telescopes. The Arecibo radio telescope has a diameter of 300 m ($D = 30{,}000$ cm). Observing a radio wavelength of 10 cm ($\lambda = 10^8$ nm), the smallest resolvable angle at Arecibo is

$$0.02 \times 10^8/30{,}000 \approx 70 \text{ arcsec.}$$

The unaided human eye (pupil diameter 0.8 cm) observing visible wavelengths (500 nm) can resolve

$$0.02 \times 500/0.8 \approx 13 \text{ arcsec.}$$

Thus the human eye can resolve finer detail than the largest single radio telescope in the world! Of course, if our eyes were sensitive to radio wavelengths, they would have extremely poor resolution.

Q: Most people would say that they can resolve greater detail in bright light, when their pupils are very small, than in dim light. What other factors might account for this impression?

29.3 INTERFEROMETERS

The limitation on resolution caused by diffraction is a major impediment to studying distant objects, where the angular size becomes very small. There are structural and financial limits to building larger and larger telescopes, so how can we achieve better resolution?

An answer to this problem begins to become apparent when we realize that a telescope aperture does not need to be continuous. If you take binoculars or a small telescope and put narrow strips of black tape in an X across the front of them—leaving gaps where light can pass through—the image you see through the binoculars does not have an X through it. The separate parts of the lens where light can still pass will focus the light perfectly well. Similarly, if you blacked out portions of the mirror on a reflecting telescope, the telescope would still function. The interesting thing is that as long as portions near the edges of the mirror remain uncovered, the resolution remains about the same. Less light may be collected, and the diffraction pattern would grow more complex, but the resolution would be little affected.

If you try putting black tape over a lens, do not stick the tape to the glass of the lens! Some lenses have special coatings or might be difficult to clean.

The pieces of the aperture actually do not need to be connected at all. As long as we can determine the relative positions of the waves precisely enough, we can add the light waves together so that they are in sync. More remarkable still, it is possible to record the electromagnetic wave from the separate "pieces" and later join together the recorded signals to form the equivalent of a single extended telescope aperture. This process is now regularly carried out at radio

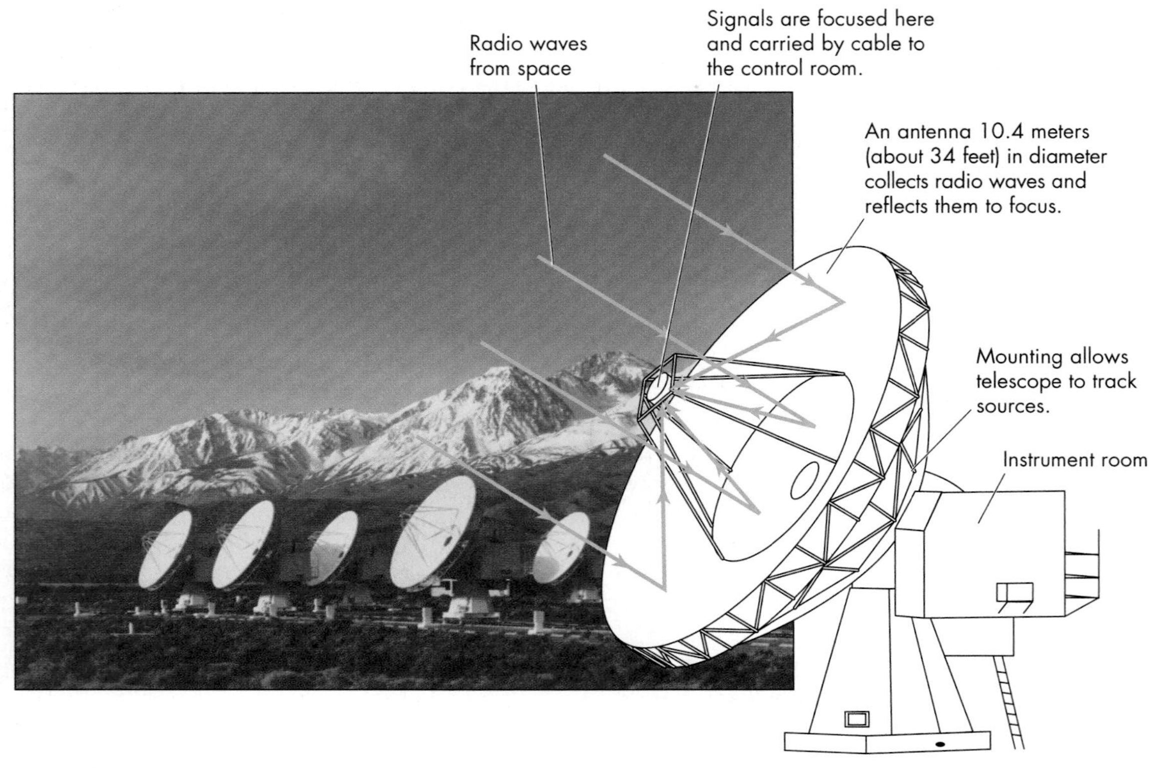

FIGURE 29.2
Photograph of the Owens Valley Millimeter Wavelength radio telescope, operated by the California Institute of Technology. In the background you can see the Sierra Nevada mountain range. The "telescope" is an array of six separate dishes that collect the radio waves. The captured radiation is then combined by computer to increase the resolution of the instrument.

wavelengths (Figure 29.2) to join the signals from many separate telescopes. Major radio telescope arrays are located in many countries, the largest in steady use being the VLBA (Very Long Baseline Array) that spans the United States from the Caribbean to Hawaii. Radio astronomers sometimes arrange to make observations with even larger arrays by linking observations from radio telescopes around the world, and even space-borne radio telescopes, to achieve a resolution the equivalent of a single telescope that has an aperture more than 10,000 kilometers across!

Q Interferometers can provide superb resolution, but most observations are still made with individual telescopes. What sorts of trade-offs do you think are likely to be made when using an interferometer?

Astronomers call a combined set of telescopes like this an **interferometer,** named after the principle of interference discussed earlier. To achieve the equivalent of a single large aperture, the same wave crest must be combined from the individual telescopes so that the interference between waves does not cause them to subtract from each other. Because the telescopes may be widely separated, the same wave crest will arrive at the different telescopes at different times. The electromagnetic waves from each telescope must be delayed by corresponding amounts of time so that the combined wave generates the strongest possible coherent signal. This indicates that the waves are all being brought together in unison—similar to the function of properly shaped lenses and mirrors in an individual telescope.

The result of this process is the ability to produce images in which the resolution is set not by the size of the individual telescopes but rather by their

FIGURE 29.3
Photograph of an infrared and optical wavelength interferometer (IOTA). Light from the object of interest is collected by the three telescopes and sent to a control room. Computers there combine the light and reconstruct an image of the object.

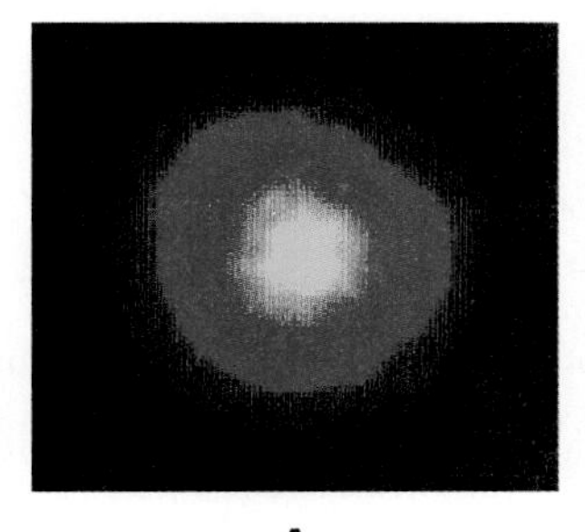

A

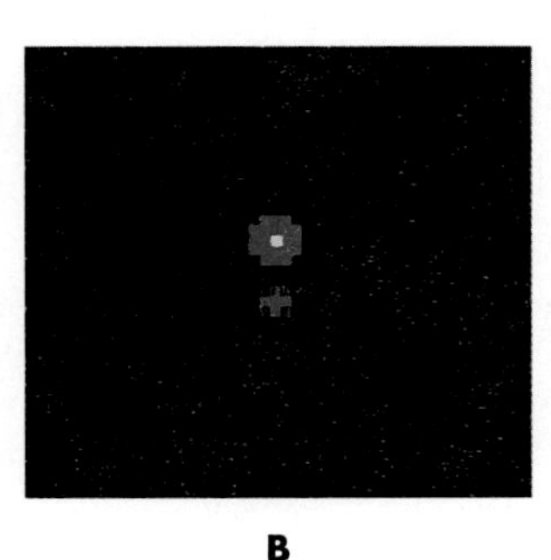

B

FIGURE 29.4
(A) A young star observed with an ordinary telescope. (B) The same star observed with an interferometer. The higher resolution of the interferometer reveals that the "star" is actually two stars in orbit around one another.

separation. The twin Keck telescopes and the four telescopes of the VLT are designed to be used in this way. A photograph of an interferometer using three telescopes is shown in Figure 29.3. For example, if the mirrors are 100 meters apart, the interferometer has the same resolution as a telescope 100 meters in diameter.

The high resolution of interferometers is far beyond what can be obtained in other ways. Figure 29.4A shows a view of two closely spaced stars as observed with a single telescope. Their images are blended as a result of diffraction. Figure 29.4B shows the same stars observed with an interferometer after the image has been processed by a computer. The two stars can now be easily distinguished: The two separate mirrors produce the resolution of a single mirror with a diameter equal to the spacing between them.

KEY TERMS

diffraction, 211
diffraction pattern, 211
interference, 211
interferometer, 214
resolution, 211

QUESTIONS FOR REVIEW

1. What is meant by the *resolution* of a telescope?
2. If the optics could be made perfectly, why would there still be a limit on a telescope's resolution?
3. For telescopes operating at different wavelengths, how big must their apertures be to achieve the same resolution?
4. What effects do blockages within a telescope's aperture have on the quality of the image it produces?
5. What is an interferometer?
6. What advantages do we gain using an interferometer?

PROBLEMS

1. If I am trying to resolve a double star with a 10-cm diameter telescope, should I observe the star using a blue filter or a red filter? What resolution do you expect with each?
2. What is the smallest resolvable angle of the 2.4-m Hubble Space Telescope when it is observing 240-nm ultraviolet light? What is its resolution when it is observing 1.2-micron infrared light?
3. Find the smallest resolvable angle of the 100-m diameter Green Bank Telescope when it is observing radio waves at 20-cm wavelength. What is its resolution when observing a 3-mm wavelength?
4. Suppose that for an interferometer to work properly, the electromagnetic waves must arrive "in sync" to 1% of the period of the wave. What time difference is this for a 20-cm wavelength radio wave? What time difference is this for a 500-nm visible-wavelength wave? (Recall that the period of the wave is the inverse of its frequency ν.)
5. What is the smallest resolvable angle of a radio interferometer that extends 10,000 km when it is observing at 20-cm wavelength? at 3-mm wavelength?
6. Suppose you wanted to build a telescope capable of resolving a planet the size of Earth that is 10 light-years distant. (A) Calculate the angular diameter of Earth if it were that far away (see Unit 10.4). To see any detail on the surface, we would want to see an angular size about ten times smaller than this. (B) Calculate the diameter of a telescope that would be needed to resolve this angular size if observing at 500 nm. Is this feasible?

TEST YOURSELF

1. A telescope's resolution measures its ability to see
   a. fainter sources.
   b. more distant sources.
   c. finer details in sources.
   d. larger sources.
   e. more rapidly moving sources.
2. Astronomers use interferometers to
   a. observe extremely dim sources.
   b. measure the speed of remote objects.
   c. detect radiation that otherwise cannot pass through our atmosphere.
   d. improve the ability to see fine details in sources.
   e. measure accurately the composition of sources.
3. Suppose we made observations with a telescope, but found that we needed better resolution. Which of the following would allow us to make higher-resolution observations?
   a. Observe with a telescope that has a bigger mirror.
   b. Use a telescope with a lens instead of a mirror.
   c. Use a telescope with a mirror made of gold.
   d. Observe with the same telescope but at longer wavelengths.
   e. All of the above

unit 30

The Earth's Atmosphere and Space Observatories

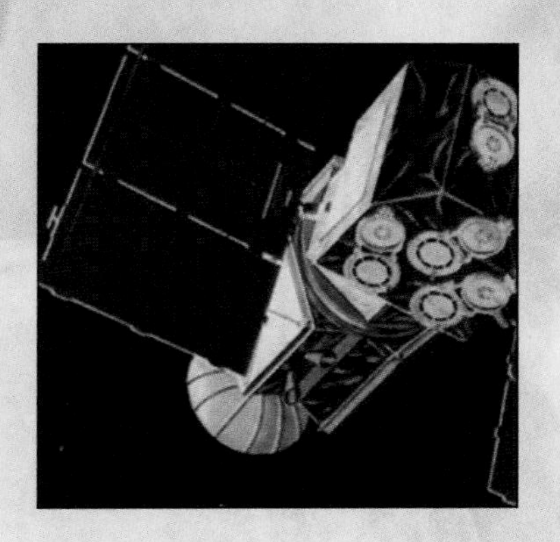

Background Pathways

Unit 22: The Electromagnetic Spectrum 166

The Earth's atmosphere presents challenges for carrying out astronomical observations. Although our atmosphere is transparent at visible wavelengths, it is not transparent at all wavelengths. Moreover, the atmosphere distorts the radiation that does pass through it. For example, sometimes it acts like an imperfect lens, distorting light as it moves from space into progressively denser layers of air. To complicate matters further, these effects vary with wavelength and weather conditions.

The atmosphere makes it challenging to build telescopes that work well. Some wavelengths cannot penetrate the Earth's atmosphere, so they cannot be observed except from space. For other wavelengths, conditions may be better in space, but observations from the ground are possible too. In these cases there are trade-offs to consider. Space telescopes are generally much more expensive to build, launch, and maintain than ground-based telescopes. Also, limitations on the weight that can be launched into space set limits on the size of a space telescope. For such reasons astronomers have explored many technological innovations, which by compensating for the distorting effects of the Earth's atmosphere will improve observations made from the ground.

30.1 ATMOSPHERIC ABSORPTION

Telescopes operating at many wavelengths face a major obstacle: Most of the radiation they seek to measure cannot penetrate the Earth's atmosphere. Gases in the Earth's atmosphere absorb electromagnetic radiation, and the amount of this absorption varies greatly with wavelength. For example, atmospheric gases affect visible light relatively little, so our atmosphere is nearly completely transparent to the wavelengths we see with our eyes. On the other hand, some of the gases strongly absorb infrared radiation while others absorb ultraviolet radiation.

The transparency of the atmosphere to visible light compared with its non-transparency to infrared and ultraviolet radiation creates what is called an **atmospheric window.** An atmospheric window is a wavelength region in which electromagnetic energy comes through our atmosphere easily compared with other nearby wavelengths (Figure 30.1). For example, there is a large window at radio wavelengths that makes most ground-based radio observations practical.

Atmospheric absorption

Other wavelength ranges present greater difficulties. Water (H_2O) and carbon dioxide (CO_2) absorb most infrared wavelengths, although there are some narrow windows within the range. This blockage by molecules in the Earth's

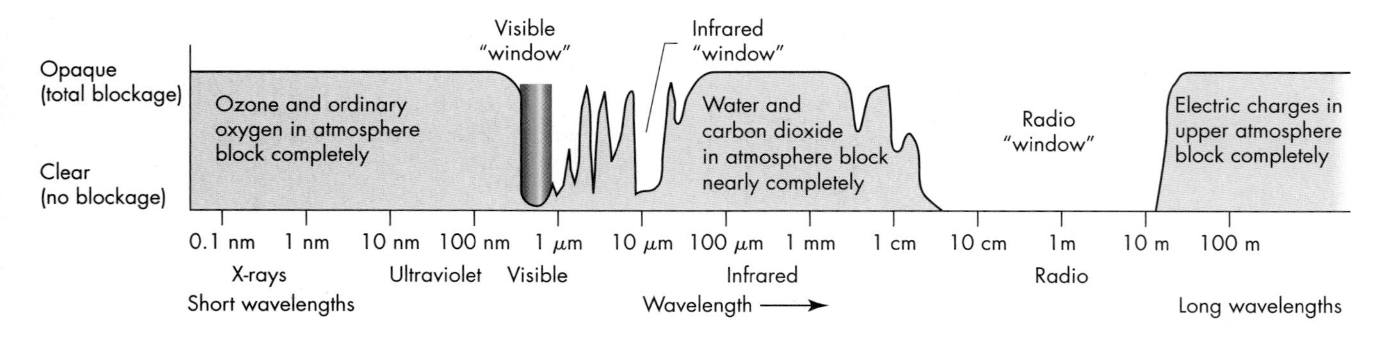

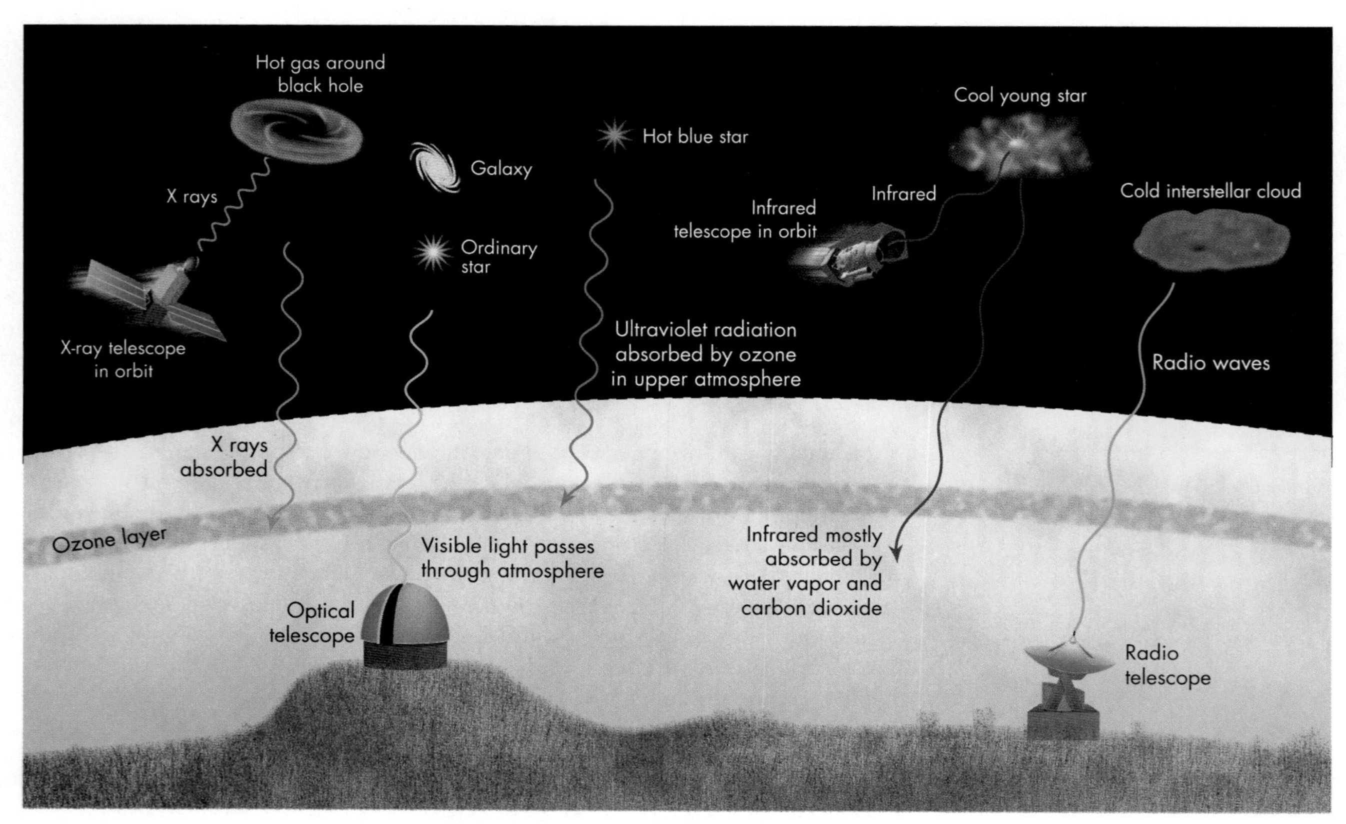

FIGURE 30.1
Atmospheric absorption. Wavelength regions where the atmosphere is essentially transparent, such as the visible spectrum, are called *atmospheric windows.* (Wavelengths and atmosphere are not drawn to scale.)

atmosphere makes it difficult to detect emission from water or carbon dioxide located elsewhere in the cosmos. At shorter wavelengths than visible light, ozone (O_3) and ordinary oxygen (O_2) strongly absorb ultraviolet radiation, while oxygen and nitrogen absorb X-rays and gamma radiation. As a result, only from space can we observe the high-energy photons from some of the most violent processes occurring in the cosmos.

Because the atmospheric window at visible wavelengths permits us to observe the Sun and other stars, we can do much astronomy from the ground. But the blockage at other wavelengths is beneficial, too. For example, because our atmosphere blocks wavelengths shorter than visible light, it protects us from high-energy photons that are dangerous to living organisms. Similarly, because our atmosphere is not transparent to infrared wavelengths, it traps infrared radiation

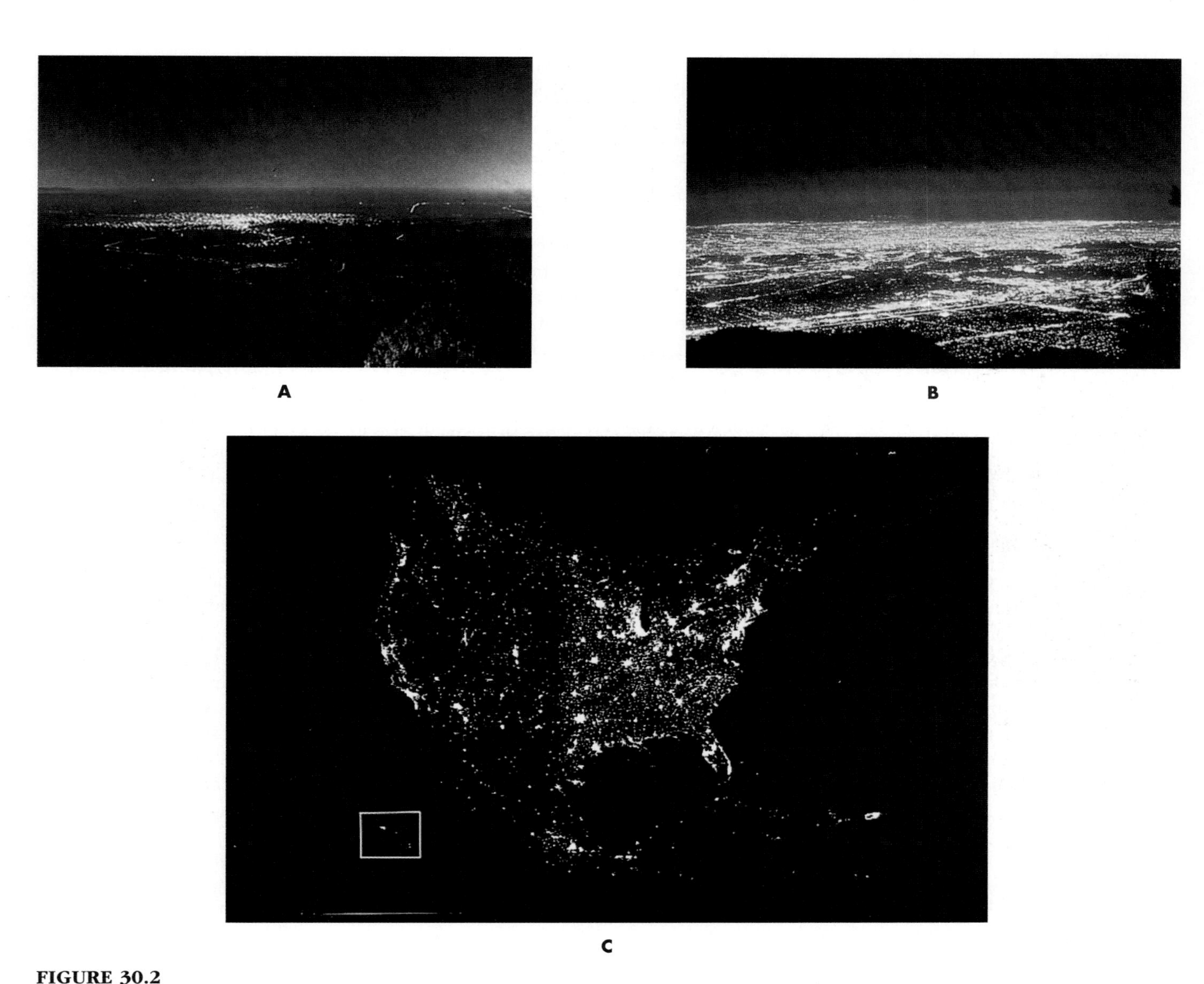

FIGURE 30.2
Photographs illustrating light pollution. (A) Los Angeles basin viewed from Mount Wilson Observatory in 1908. (B) Los Angeles at night in 1988. (C) Notice the pattern of the interstate highway system visible in the satellite picture of North America at night.

emitted by the warmed ground at night, retaining much of its heat and preventing the oceans and us from freezing.

In recent decades astronomers have had to contend with another factor that limits their observations from the ground: **light pollution.** Most inhabited areas are peppered with nighttime lighting such as streetlights and outdoor advertising displays (Figure 30.2A and B). Although some such lighting may increase safety, much of it is wasted energy, illuminating unessential areas and spilling light upward into the sky, where it serves no purpose. Figure 30.2C shows a satellite view of North America at night and illustrates the wasted energy created by light pollution. Such stray light can interfere with astronomical observations. Many early observatories were built in cities to make them more accessible, but they have become essentially useless for research because of light pollution. In some places astronomers have persuaded regional planning bodies to develop lighting codes to minimize light pollution. Light pollution not only wastes energy and interferes

Get involved in your community by letting people know that they can save money by using light fixtures that have reflectors to keep the light shining downward. They are paying for the electricity that a poorly designed fixture uses to light up the night sky!

with astronomy; it also destroys a part of our heritage—the ability to see stars at night. The night sky is a beautiful sight, and it is shameful to deprive people of it.

30.2 ATMOSPHERIC SCINTILLATION

ANIMATIONS

Why stars twinkle

Even from a location where there is no light pollution and the air is clear, the atmosphere still causes significant problems for observations at visible wavelengths. Light's path bends as it travels through materials of different density. This is the same principle of refraction that allows lenses to work (Unit 28), but even pockets of air of different density can alter the path of light.

If you have ever noticed a star twinkling in the night sky, then you have observed atmospheric **scintillation**. This is caused by various pockets of warmer and cooler air that have slightly different densities. Each acts as a lens on the light traveling through it. As hot air rises or as winds and turbulence stir the air, these pockets will cross in front of a star, magnifying and demagnifying, and bending the path of the light first one way and then another. The color dispersion of these "lenses" will even make the star's color appear to vary. These effects are particularly apparent when we view a star near the horizon, where the light travels a longer path through the atmosphere. Depending on atmospheric conditions, a star may flicker and shift more or less, conditions astronomers call, respectively, bad or good **seeing**.

Q One way you can often tell a planet from a bright star without using a telescope is by how much they twinkle. Planets tend to shine more steadily than a star in the same position. Why do you suppose there is a difference?

With bad seeing, the starlight you see at any instant is a blend of light from many slightly different directions, which smears the star's image and makes it dance (Figure 30.3). You can see a similar effect if you look down through the water at something on the bottom of a swimming pool. If the surface of the water has even slight disturbances, a pebble or coin on the bottom seems to dance around. Scintillation in our atmosphere limits the ability of Earth-bound observers to see fine details in astronomical objects. The dancing image of a star or planet distorts its picture when recorded by a camera or other device.

Until recently, ground-based astronomers had to submit to the distortions of seeing, but now they can partially compensate for such distortions in several ways. One technique involves observing a bright star simultaneously with the object of interest. Extremely rapid measurements are made of the bright star's shifting position, then tiny motors make compensating adjustments to the tilt of a secondary mirror to keep the bright star's position the same. Because the pockets of air in

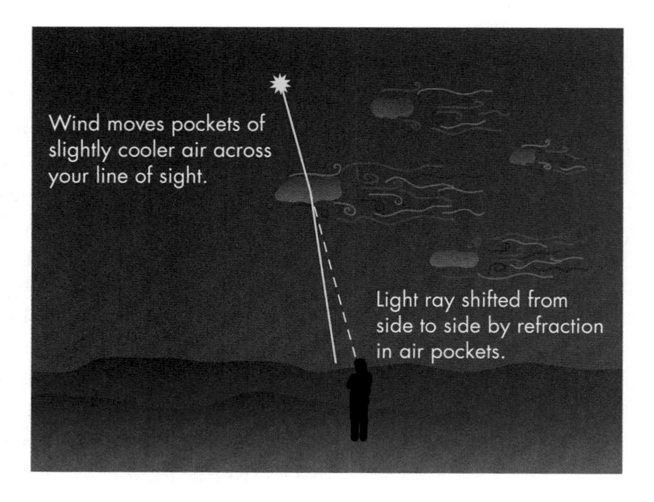

FIGURE 30.3
Twinkling of stars (seeing) is caused by moving atmospheric irregularities that refract light in random directions.

FIGURE 30.4
A laser beam creates an artificial star whose image serves as a reference to eliminate the atmosphere's distortion of real stars. This photograph was taken at the Starfire Optical Range of the Phillips Laboratory at Kirtland Air Force Base in New Mexico.

the atmosphere are fairly large, faint objects in the vicinity of the bright star are also kept stationary, and the image remains sharp. Astronomers are learning to make even more rapid adjustments for atmospheric effects, eliminating many of the distortions of atmospheric seeing. This technique, called **adaptive optics,** has already given astronomers dramatically improved views through the turbulence of our atmosphere. Because bright stars are not present next to all of the objects we might want to observe, astronomers have developed a technique using a powerful laser beam to create an artificial star high in the Earth's atmosphere, as shown in Figure 30.4.

30.3 ATMOSPHERIC REFRACTION

Refraction by the atmosphere causes a larger, yet less obvious, effect. As light enters the atmosphere from space, its overall path is bent (Figure 30.5A). The bending is so large near the horizon that we can see stars that are actually below the horizon! Just as for the refraction of light by a lens (Unit 28), light coming nearly straight down into the atmosphere is little affected. Light entering the atmosphere at a steeper angle is bent by a larger amount (Figure 30.5B).

At the horizon, the effect of the bending is about half a degree. As a result, when the Sun or Moon appears to be just touching the horizon, it is in fact below the horizon—if the Earth had no atmosphere, it would already be out of view. Thus atmospheric refraction alters the time at which the Sun seems to rise or set. It also

FIGURE 30.5
(A) Atmospheric refraction makes the Sun or a star look higher in the sky than it really is. (B) Refraction is stronger for objects nearer the horizon. (C) The Sun looks flattened because refraction "lifts" its lower edge more than its upper edge. The Sun's reflection in a band below it on the water is called the *glitter path*.

Q How do the refraction and absorption of light by the atmosphere relate to the color of the Moon during a total lunar eclipse?

distorts the shape of the Sun because of differences in the amount of bending from the bottom to the top of the Sun's disk (Figure 30.5C).

By "lifting" the Sun's image above the horizon, even though the Sun has set, refraction slightly extends the length of daylight hours. As a result, the day of the year when the Sun appears to be above the horizon for exactly 12 hours is not the equinox, but rather a few days before the first day of spring and a few days after the first day of autumn. This depends on latitude and atmospheric conditions, but it turns out that near latitude 40°N, on St. Patrick's Day (March 17) the day usually has 12 hours between sunrise and sunset, whereas the actual equinox occurs on about March 21.

30.4 OBSERVATORIES IN SPACE

Figure 30.6 shows several of the major telescopes that have been launched. Some, like the Hubble Space Telescope, which was launched in 1990, have operated for many years thanks to their low orbit around Earth that allows servicing missions with the space shuttle to repair problems and upgrade equipment. Most other space telescopes, like the Spitzer Infrared Space Telescope, have a shorter lifetime because the detectors require coolants such as liquid helium to keep them close to a temperature of absolute zero. When the supply of liquid helium is used up after several years, it is unlikely that it can be replaced because the telescope was placed in deep

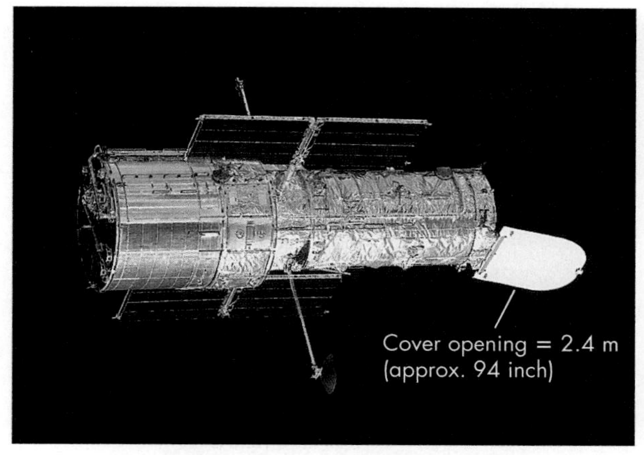

Hubble Space Telescope (13.6 m long) – HST

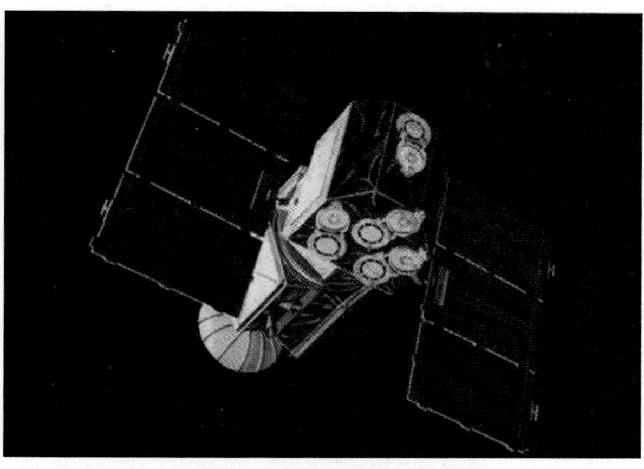

Extreme Ultraviolet Explorer – EUVE

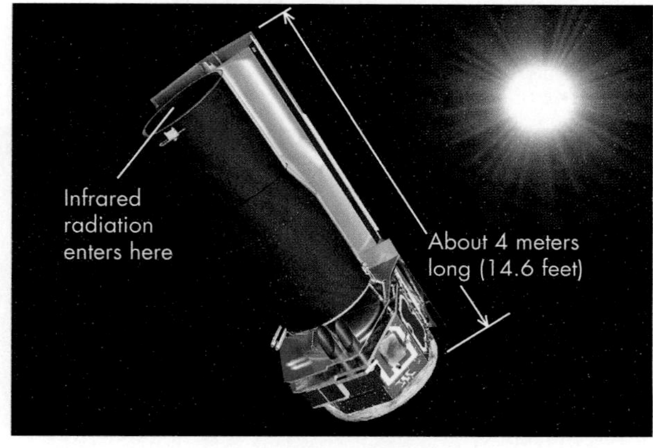

Spitzer Infrared Space Telescope

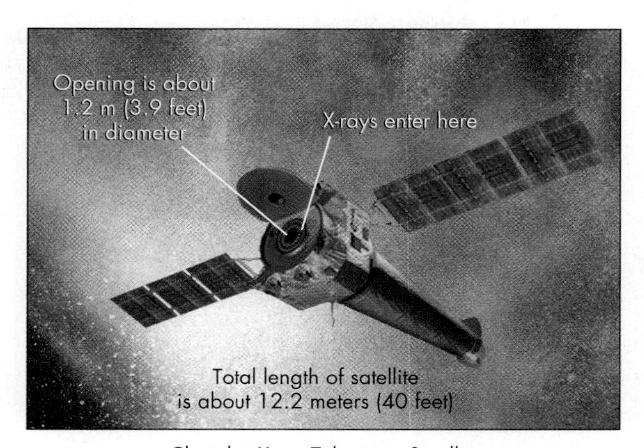

Chandra X-ray Telescope Satellite

FIGURE 30.6
Photograph of the Hubble Space Telescope and drawings of some other space observatories: Spitzer, EUVE, and Chandra.

space millions of kilometers from Earth. It was located far from Earth so that the Earth's own heat would not interfere with observations. By contrast, the Hubble Space Telescope orbits about 500 km (300 miles) above the Earth's surface.

More than 50 space telescopes have been launched by countries around the world to examine all different parts of the electromagnetic spectrum. Many of these missions were brief and were designed to give us our first clues about cosmic radiation outside of the Earth's atmospheric windows. With each mission we have developed a clearer idea of the characteristics of electromagnetic emission, leading to improved designs. Recent telescopes like the Chandra X-ray telescope, launched in 1998, have provided some of the first images of X-ray emissions, with resolutions comparable to visible-wavelength images.

Of the many orbiting telescopes used by astronomers, the Hubble Space Telescope (HST) is the most ambitious to date. The HST is designed to observe at visible, infrared, and ultraviolet wavelengths and has a mirror 2.4 meters (about 8 feet) in diameter. Its instruments allow it to take both pictures and spectra of astronomical objects. Although the HST initially had a number of problems, astronauts have repaired the major defects; and astronomers are now delighted with the clarity of its images (Figure 30.7). These images reveal details difficult to see with telescopes on the ground because such telescopes must peer through the blurring effects of our atmosphere.

Sombrero Galaxy
A system of billions of stars and dark dusty interstellar matter.

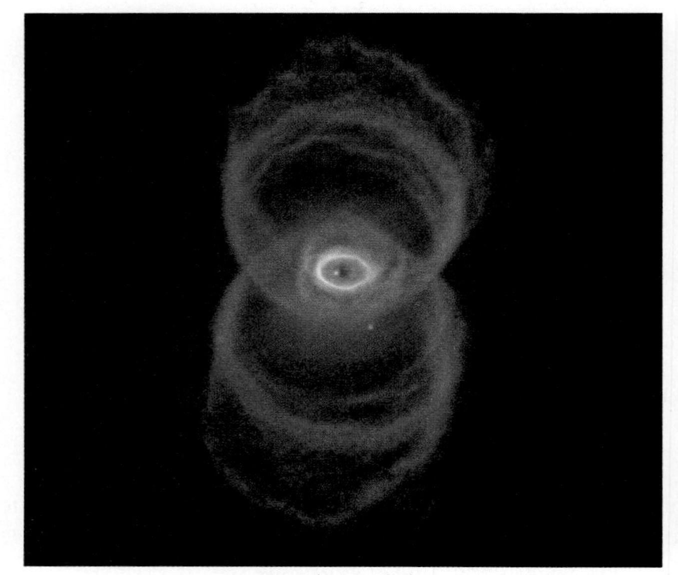

Hourglass Nebula
A dying star

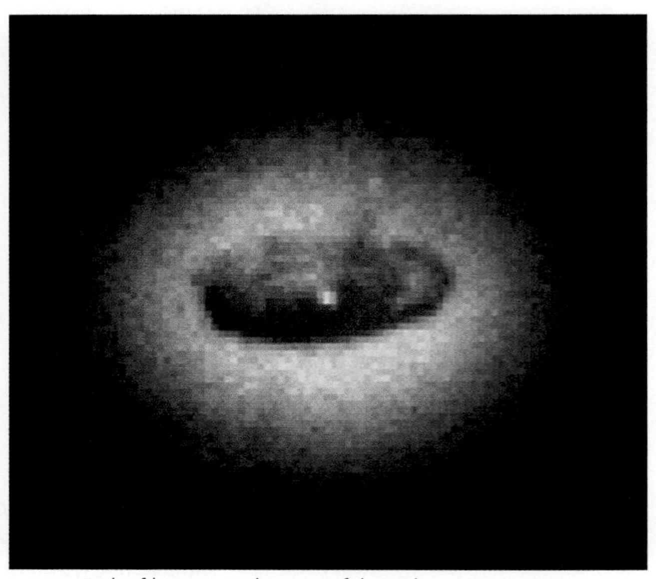

Disk of hot gas in the core of the galaxy NGC 4261;
a black hole may lie in the disk's center.

FIGURE 30.7
Some images obtained with the Hubble Space Telescope.

Despite the freedom from atmospheric blurring and absorption that space observatories enjoy, much astronomical work will be done from the ground for the foreseeable future. Ground-based telescopes can be built much larger than orbiting telescopes. Moreover, equipment problems on the ground can be corrected without the expense, delay, danger, and complexity of a space shuttle launch.

Because huge telescopes in space or even on the Moon will remain only dreams for years to come, astronomers choose with care the location of ground-based observatories. Sites are picked to minimize clouds and the inevitable distortions and absorption of even clear air. Thus nearly all observatories are in dry, relatively cloud-free regions of the world, such as the American Southwest, the Chilean desert, Australia, and a few islands, such as Hawaii and the Canary Islands off the coast of Africa. Moreover, astronomers try to locate observatories on mountain peaks to get them above the haze that often develops close to the ground in such dry locales and to improve the seeing.

KEY TERMS

adaptive optics, 221
atmospheric window, 217
light pollution, 219
scintillation, 220
seeing, 220

QUESTIONS FOR REVIEW

1. What is an atmospheric window?
2. What do astronomers mean by *seeing?*
3. How can light pollution be reduced?
4. Why does the Sun look flattened when it is close to the horizon?
5. What are the advantages and disadvantages of putting telescopes in space? on a mountaintop?
6. How does adaptive optics make ground-based images appear sharper?

PROBLEMS

1. Why do stars twinkle more when they are near the horizon than when they are near the zenith? Draw a diagram to explain your answer.
2. Astronomers have built telescopes that operate from high-flying airplanes. What wavelength ranges would this be most useful for accessing? What potential challenges would arise when observing from an airplane?
3. In the polar regions, there can be entire days when the Sun remains below the horizon. Given the refraction of the atmosphere, what is the latitude farthest from the poles where you might experience a 24-hour night?
4. Suppose that in the atmosphere there are pockets of air of different density that are 10 meters across. If winds move these pockets of air at a speed of 30 kph, for how long is an image through a telescope likely to remain stable? How would having many of these pockets along your line of sight change your time estimate?
5. A telescope in low orbit around the Earth takes about 90 minutes to circle the Earth. If the telescope remained oriented in the same way relative to the Earth, by how many arc seconds would its direction change relative to the stars during a 1-second exposure? (Actually, space telescopes are stabilized by gyroscopes which maintain their orientation relative to the stars.)
6. If the refraction of the atmosphere causes the Sun's position to shift by 0.5° at the horizon, calculate how long the day would be on the equinox for someone living at the equator. Would that same day be longer or shorter for someone living at latitude 45°? Explain using a diagram.

TEST YOURSELF

1. Which of the following telescopes would be most suitable for ground-based observations?
   a. Radio telescope
   b. X-ray telescope
   c. Ultraviolet telescope
   d. Gamma-ray telescope
   e. Infrared telescope
2. Which of the following explains the advantage of the Hubble Space Telescope compared with ground-based telescopes?
   a. It is the largest reflector ever made.
   b. It can detect X-rays from space.
   c. It is not affected by atmospheric scintillation.
   d. It is closer to the objects it is imaging.
   e. All of the above
3. What is the purpose of adaptive optics?
   a. To make telescope designs more flexible so they can fit in smaller buildings
   b. To allow telescopes to look in several directions without having to turn the primary mirror
   c. To increase the effective collecting area of a telescope by capturing more photons
   d. To rapidly adjust for the changing distortions caused by the Earth's atmosphere
   e. All of the above

unit

31 Amateur Astronomy

Anyone with access to a small telescope has better equipment than Galileo ever had. With such equipment and access to a dark sky, that person can become an amateur astronomer. The pleasures of the hobby can range from the aesthetic satisfaction of taking a lovely photograph to the thrill of discovering a new comet or an exploding star. Much of the enjoyment of the night sky comes from learning the constellations and studying the changing positions of the Sun, Moon, and planets (Unit 13). Binoculars will let you begin to appreciate many of the more remote objects of the night sky, and a small telescope will let you see much more. Here we offer a few suggestions for taking your interest in astronomy a step further.

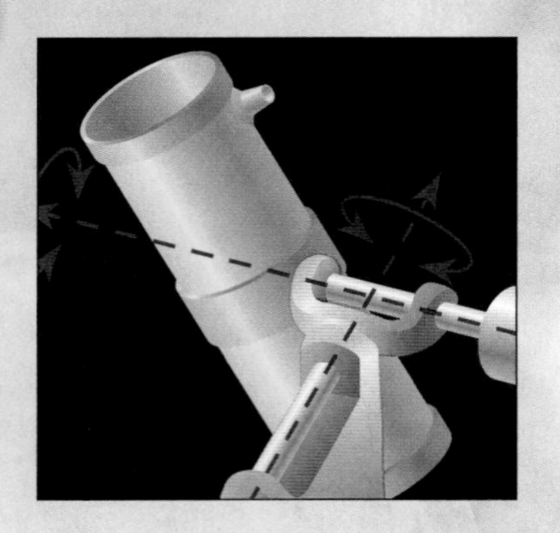

Background Pathways

Unit 28: Focusing Light 203

31.1 THE HUMAN EYE

The human eye is a remarkable device. It not only detects visible-wavelength photons, but can discriminate between different wavelengths. It has a self-adjusting aperture and can adjust the sensitivity of its receptor cells by nearly a factor of a million. Some parts of the eye allow for high resolution and color discrimination, while other portions permit us to detect extremely low light levels. Understanding how the eye works will help you get the most from your observations of the night sky.

Your eye (Figure 31.1) contains many of the same elements as a telescope. The eye's **iris** can expand or contract to adjust the aperture, or **pupil,** to allow in more or less light. Light passing through the pupil is focused on the back surface called the **retina.** The shape of a transparent layer of your eye called the **cornea** does most of the focusing of light entering your eye, but an internal lens can be adjusted to allow the eye to focus at different distances.

The light focused on the retina can trigger a response in light-sensitive nerve cells called **rods** and **cones,** so named because of their shapes. There are more than 100 million of these "detectors"—akin to the pixels of a CCD—that send signals to your brain, providing your sense of vision. The rods and cones all use **photopigments**—proteins that change form when they absorb light, causing a

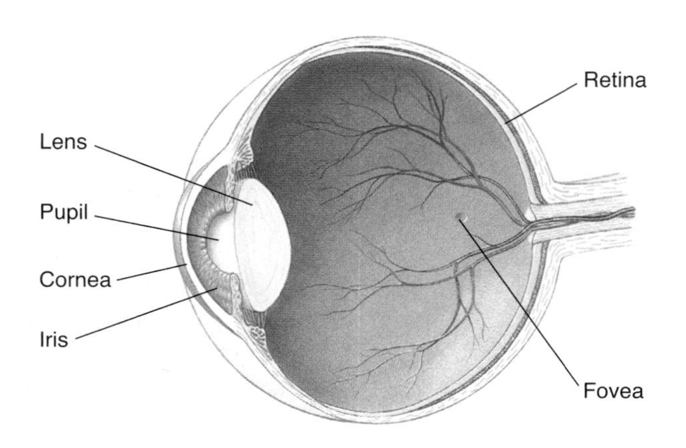

FIGURE 31.1
Structure of the human eye. The transparent cornea and lens focus light onto the retina at the back of the eye, which contains the rods and cones. The center of each eye's visual field is focused on the fovea. Focusing is accomplished by contraction and relaxation of the muscles that adjust the curvature of the lens.

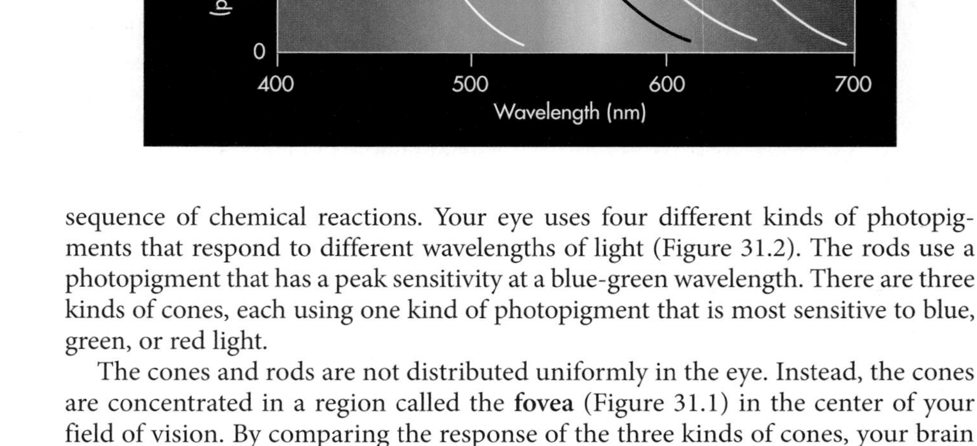

FIGURE 31.2
The spectral sensitivities of the photopigments in rods and three types of cones.

sequence of chemical reactions. Your eye uses four different kinds of photopigments that respond to different wavelengths of light (Figure 31.2). The rods use a photopigment that has a peak sensitivity at a blue-green wavelength. There are three kinds of cones, each using one kind of photopigment that is most sensitive to blue, green, or red light.

The cones and rods are not distributed uniformly in the eye. Instead, the cones are concentrated in a region called the **fovea** (Figure 31.1) in the center of your field of vision. By comparing the response of the three kinds of cones, your brain assigns a color. Note that a 500-nm photon can excite any of the photopigments, so your brain assigns a perceived color based on the relative strength of the reaction by different cones. Digital cameras employ a similar strategy to produce color images. Alternate pixels are behind one of three different color *filters* (Unit 27.3), mimicking the way the eye works.

Your color perception is strongly concentrated in the middle of your field of vision, so your color discrimination is best when you look straight at an object, and it drops sharply away from there. There are few cones beyond 10° from the center of your vision, and almost none beyond 40°. You may think you see colors in your peripheral vision, but that is because your eye constantly glances in different directions to fill in detail, and your brain remembers the colors from when you saw an object earlier. Your brain also compares the images from both eyes to estimate the distances of objects. In addition, it tries to fill in missing information from parts of the eye where there are no cones and rods—such as the blind spot about 15° to the outside of the center of your field of vision in each eye, where the nerves come together to carry the signals to your brain.

Q To "see" your blind spot, put a large black dot near one end of a piece of paper, then from the dot draw a straight line that is about 20 cm (8 inches) long. Close one eye and hold the paper in front of you with the line horizontal and the dot at the right side for your right eye (or vice versa). Look first at the dot and then slowly "slide" your center of vision along the line. You will find that at a particular angle the dot seems to disappear. Why do you suppose we are unaware of our own blind spots?

31.2 YOUR EYES AT NIGHT

You will discover that the longer you stay outside in dim light, the more sensitive your eyes will become and the fainter the stars you will be able to see. This is the result of physiological changes in your eye referred to as **dark adaption.**

The simplest change in your eye occurring in dim light is that the pupil opens wider. This is easy to verify by looking at yourself in a mirror in a dimly lit room. In bright sunlight your pupil can shrink to a diameter of as little as 1 millimeter, but in total darkness its diameter may expand to about 7 or 8 millimeters, which allows more light to enter your eye. A change from 1 to 8 millimeters is a factor of 8 in diameter and therefore a factor of 64 in the amount of light admitted to the eye. The maximum size of the pupil tends to decrease with age to about 5 or 6 mm.

Your eyes undergo another change in the dark. The rods can build up the level of photopigments within them until they are about 1 million times more sensitive to light than they are in full daylight. The process takes about 20 minutes to get fully established but is undone by even a few seconds' exposure to bright light. Thus, once you are dark-adapted, you should stay away from bright lights for as long as you intend to observe. Smoking, alcohol, and other drugs also reduce your ability to adapt to the dark by reducing the iris's ability to expand and by interfering with the chemistry of the rods, cones, and other nerve cells. Getting enough vitamin A is important too, because this is an essential chemical for the photopigments in your eye.

The cones have much more limited ability to increase their sensitivity than the rods; so as light levels decrease, your vision relies more on the rods. This has several effects: Your ability to discriminate between colors declines or disappears completely; the wavelengths of light you are sensitive to shifts to blue wavelengths, a phenomenon known as the *Purkinje effect;* and your best sensitivity to light shifts outside of the perceived center of your field of vision in the fovea.

The Purkinje effect can change your perception of the relative brightness of different-colored objects. In full daylight, the eye's overall response is strongest to yellow-green colors. At low light levels, you become most sensitive to the blue-green colors that the rods' photopigment responds to. This is possibly the result of natural selection because the average color of starlight is bluer than sunlight, so eyes responsive to blue will therefore aid survival. As a result of this effect, your perception of the relative brightness of two differently colored stars will change as your eyes adapt to the dark—an important consideration if you are trying to study the changes in brightness of **variable stars** (Unit 63) by comparing them to neighboring stars. It is certainly the case that night-flying insects see blue light better than yellow. That is why bug zappers use a blue light to attract insects, and also why a yellow lightbulb is often used for outdoor night lighting to be less attractive to insects.

Within a circle about 1° across in the center of your field of vision, the fovea contains no rods at all. The number of rods relative to cones climbs steadily in the surrounding area of the retina. You can apply this information practically with the following observational technique for viewing dim objects. Look at a point about 5° away from the dim object you are trying to see (about the length of your outstretched thumb). By doing so you gain sensitivity to faint objects. This technique, called **averted vision,** places the focused image of the object you are trying to see on a part of the iris with a high density of rods. This is not easy to do at first, but with practice you will be able to see some extremely faint objects that are "invisible" when you look straight at them.

31.3 CHOOSING A TELESCOPE

Although your eyes can see remarkably faint stars and even a few galaxies, a small telescope will greatly increase the number of objects you can observe and will offer you far better and more interesting views of the Moon, planets, and even more remote objects. The basics discussed in previous units about the different kinds of telescopes and how they work (Unit 28) apply to small telescopes as well as research telescopes. However, there is an even wider range of telescope designs available for amateur astronomy, so selecting the best one for your needs can be confusing.

A good starting size is a 10- to 15-cm (4- to 6-inch) aperture telescope. When a telescope is referred to by size, as in "a 10-cm telescope," the size refers to the diameter of the primary mirror or lens (Unit 27). With a 10-cm telescope, you can easily study our Moon, the moons of Jupiter, the rings of Saturn, and most of the star clusters, nebulae, and galaxies in the Messier Catalog (Appendix Table 13). A relatively inexpensive design is a Newtonian reflector (Unit 28.2), although a wide variety of other designs are available with various advantages and disadvantages.

A well-manufactured 10-cm telescope can provide just about as sharp a view as you can get. The resolution of a telescope is limited by diffraction, and for a 10-cm aperture this limit is about 1 arc second (Unit 29.2). This is about the same as the smearing typically caused by atmospheric turbulence or "seeing" (Unit 30.2) on a good night from most locations. A larger telescope will offer relatively little improvement in the detail it is possible to see, unless you are observing under exceptional atmospheric conditions.

Refractors are also an excellent choice, but they tend to be more expensive for the same size aperture. Because glass causes color dispersion (Unit 28.4), you should avoid inexpensive refractors that have a single-element primary lens. All better-quality refractors use two or three lenses designed to correct each other's color dispersion. Good-quality refractors have what are called "achromatic" lenses and the best have "apochromatic" lenses—indicating that the color problems of the lens have been corrected to different degrees. Refractors are particularly favored for bright objects such as the Moon and planets, because they can give the crispest and highest contrast images. This is because they have a clear aperture without secondary mirrors or internal support structures that would scatter light inside the telescope and make the diffraction pattern worse.

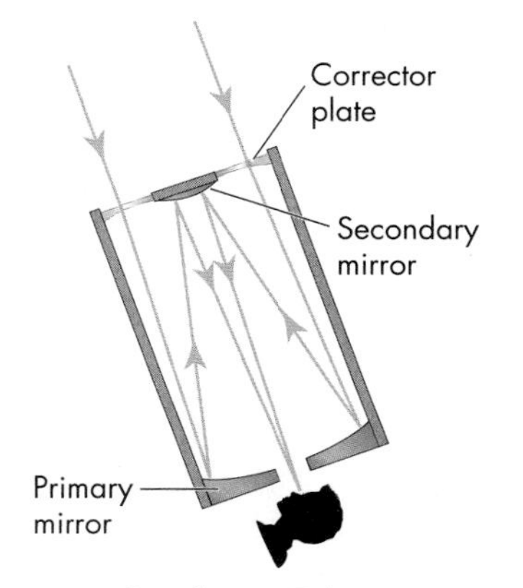

FIGURE 31.3
Diagram of a catadioptric telescope. This type of telescope has a corrector plate—a large lens over the main aperture—that works in combination with the primary mirror to focus the light. A Schmidt design with a Cassegrain focus is illustrated. A Maksutov design, another popular type, uses a different shape for the corrector plate.

If you plan to take your telescope on the road, perhaps to enjoy dark skies in the wilderness, you will probably want it to be compact. A popular portable design is called a **catadioptric telescope,** which is a hybrid between a reflector and a refractor. This type of telescope uses both a large mirror and a large lens, called a corrector plate, to focus the light entering the aperture, as illustrated in Figure 31.3. Catadioptric telescopes are generally built using the Cassegrain design (Unit 28.2). This folds the path of the light back on itself, sending the light through a hole in the primary mirror, making the telescope more compact. The idea of a catadioptric telescope is that the corrector lens and mirror are relatively simple shapes to manufacture, keeping the cost lower than a refractor of the same aperture. Similar to a refractor, the telescope body is a solid sealed unit, so dust and misalignment are less problematic than for a reflector. The secondary mirror blocks part of the aperture, but the corrector plate is used to hold it in place, so no additional support structures are needed.

Selecting a telescope with a larger aperture of 20 cm or more will allow you to view faint objects more easily. The larger aperture will also expand the number of faint objects you can explore, but at increasing expense and decreasing portability. Keep in mind that using your eyes well can have more effect on what you will be able to see than increasing your telescope's size. Keep your eyes shielded from bright lights for at least half an hour, and learn to use averted vision when looking through an eyepiece. With these steps you will be able to see things through the telescope that would otherwise be invisible to you.

31.4 COMPLETING THE TELESCOPE

Many other aspects than just the telescope design contribute to having a successful experience observing the night sky. The telescope gathers the light and brings it to focus, but viewing the focused image requires additional optics called "eyepieces." The telescope must be held on a steady "mount," but that mount also has to be able to move smoothly and easily so you can follow the motion of the celestial sphere. These too offer a wide variety of choices.

Notice that we have said nothing about magnification to this point. Some telescopes advertise their "magnifying power," but this is an unimportant number because any telescope can magnify by any amount with different eyepieces. You can determine the magnification of a telescope–eyepiece combination by dividing the **focal length** of the telescope by the focal length of the eyepiece:

$$\text{Magnification} = \frac{\text{Telescope focal length}}{\text{Eyepiece focal length}}$$

The focal length of a refractor is usually the length of the telescope tube—the distance over which it focuses the light entering the aperture down to a point. With the multiple bounces and shaped mirrors in many reflectors, the focal length may not be obvious. However, you can calculate it by multiplying the telescope's aperture size by its *f-ratio:*

$$\text{Telescope focal length} = (\text{Aperture diameter}) \times (\text{f-ratio}).$$

For example, a 10-cm f/8 telescope has a focal length of 10 cm × 8 = 80 cm.

Eyepieces almost universally have their focal length listed on them in millimeters. Thus for a telescope with an 80-cm focal length, an 8-mm eyepiece will give a magnification of

$$\text{Magnification} = \frac{80\text{ cm}}{8\text{ mm}} = \frac{800\text{ mm}}{8\text{ mm}} = 100.$$

Likewise, a 4-mm eyepiece on this telescope will give a magnification of 200, whereas a 16-mm eyepiece will give a magnification of 50. Distortions caused by the atmosphere make a magnification of about 100 to 200 the useful limit.

Binoculars are normally labeled with a pair of numbers such as "8 × 50." These two numbers refer to the magnification (8×) and the diameter of the aperture in millimeters (50 mm). They are often also labeled with a third number, such as 5°, which indicates the angle across the "field of view" you can see within the binoculars.

Some telescopes are sold with low-quality eyepieces, and you can improve your viewing substantially by upgrading them. Eyepieces come in a wide range of designs, some working better for people with glasses, others offering a wider field of view, and not all will work well with every telescope. Your best way to find a good match may be to visit an astronomy club convention or "star party," where you may get the chance to try different members' eyepieces with your own telescope.

Finally, it is important to decide how your telescope will be supported. Most important is that the mount be sturdy. At a magnification of a factor of 100, tiny vibrations of the telescope caused by wind or the touch of your hand will make the image jiggle, hopelessly blurring it. There are two main designs for mounting a telescope to choose between.

An **alt–az mount** (Figure 31.4A) is the simplest and least expensive. It allows the telescope to swing up and down in altitude from the horizon to the zenith, and pivot in azimuth to point toward different compass directions (Unit 13). This lets

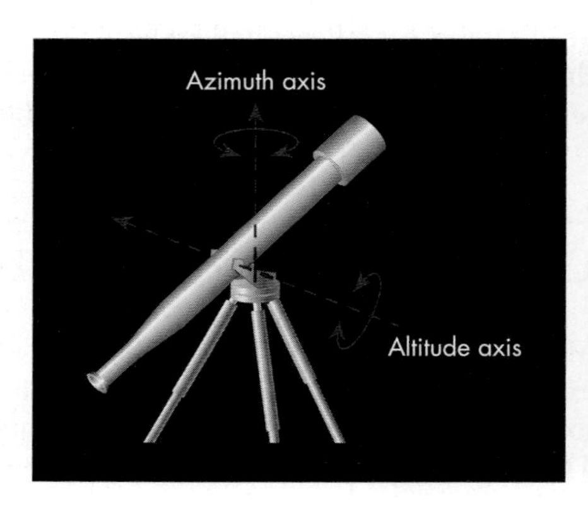

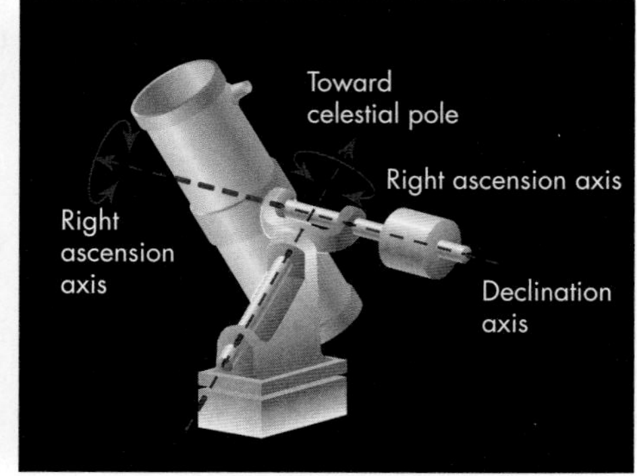

FIGURE 31.4
Telescope mount designs. (A) An alt–az mount allows the telescope to pivot around the horizon (azimuth) and angle up and down (altitude). (B) An equatorial mount pivots around an axis pointing toward the celestial pole (right ascension) and angle north and south (declination).

you see all parts of the sky, but unless you are observing from the North or South Pole, it will not be easy to follow objects on the celestial sphere as the Earth rotates. Because of the Earth's rotation, objects will move out of your field of view within a minute or two, or even seconds under high magnification.

To compensate for the Earth's rotation, an **equatorial mount** (Figure 31.4B) has a pivot axis that can be pointed at the celestial pole. With an equatorial mount, you can swing the telescope from the polar axis until you match an object's declination, and then pivot the telescope to match its right ascension (Unit 5.5). By rotating the telescope about the pivot axis, you can exactly counteract the effects of the Earth's rotation and keep the telescope pointed at the same celestial coordinate. Telescopes on equatorial mounts are more difficult to balance and generally must be heavier than alt–az mounts. However, they allow you to follow objects smoothly because they are oriented to rotate parallel to the Earth's axis—and with a motor drive, they can remain pointing at the same object for long periods.

If you exactly align your telescope (no easy task), you can find faint astronomical objects by their right ascension and declination. Generally, though, it is easier to locate an object by pointing to the approximate position relative to bright stars as determined from a sky chart or planetarium software program. Then progressively close in on the position of the object by "star hopping" to fainter stars using the small "finder scope" mounted on the side of the telescope. Some newer telescopes offer computer-guided "GoTo" systems to point the telescope. GoTo systems can also guide alt–az mounts to track celestial objects, and as their price has declined this has become a popular combination.

Before you actually buy a telescope, you might want to talk with a local amateur astronomer or astronomy instructor. If you can find an astronomy club, some of the members may have secondhand telescopes they are willing to sell at reduced prices. For information about new telescopes, browse through magazines like *Astronomy* or *Sky & Telescope*—publications widely read by both amateur and professional astronomers—which contain many advertisements for small telescopes.

31.5 ASTRONOMICAL PHOTOGRAPHY

Most astronomical pictures in books look little like what you see through a telescope, because your eyes cannot store up light as a camera can. You do not need expensive equipment to take photographs like the one shown in Figure 31.5. What you need for astronomical photography is a camera with as much manual control as possible. Many modern cameras have automatic focus and exposure features, which may frustrate your attempts to use them in low-light conditions.

Many interstellar clouds and galaxies are large enough that we could easily see them but for the limited sensitivity of our eyes. To photograph them, you do not need a telescope, but you do need to make a long exposure while keeping your camera steady. The camera focus should be set to infinity and its aperture opened fully (at the lowest f-stop number). Start with an exposure of 10 or so seconds, and then experiment with different exposure times. If you are using film, stick to one of the high-speed color films (ASA 400 or higher). For many 35-millimeter cameras, you can make long exposures by putting the camera on the B (for "flashbulb") setting and holding the shutter release down for the desired time. For long exposures you should mount the camera on a tripod or other stable platform, then use a cable release to open the shutter without shaking the camera. Some digital cameras have settings for long exposure times or will sense the low light levels and stay open automatically for a longer time.

If you expose for more than about 15 seconds, the Earth's rotation will smear the star image into a streak. Deliberately allowing the smearing to occur can

FIGURE 31.5
Sunset view of four planets strung along the zodiac on March 1, 1999. Their straight-line arrangement results from the flatness of the Solar System. From top to bottom, you can see Saturn, Venus, Jupiter, and Mercury (nearly lost in the twilight).

FIGURE 31.6
A picture made with a fixed camera showing star trails. The constellation Orion is rising, along with the planet Saturn (the bright trail to the left).

Q In the star-trail picture, how long was the exposure? In what direction was the observer looking? Were the stars rising or setting?

produce dramatic pictures of what are called "star trails" (see Figure 31.6). To make star-trail pictures, leave the shutter open for 20 minutes or so. Longer exposures are possible, but if the Moon is out or if there is much light pollution (Unit 30), the image may become foggy.

To take untrailed long exposures or to use a telephoto lens, you will need a way to compensate for the Earth's rotation. Many telescopes allow you to attach a

FIGURE 31.7
(A) Photograph of the constellation Orion made with an ordinary camera. (B) Picture of the Orion Nebula taken with a small backyard telescope.

Q If you did not have a telescope to compensate for the Earth's rotation, you might be able to attach your camera to a board, itself attached by hinges to a base. How would you need to orient the hinge to compensate for the Earth's rotation? What could you build to slowly tilt the board at the same rate the sky is "turning"?

Visit NASA's Astronomy Picture of the Day (APOD) website for hundreds of examples of amateur photographs in addition to pictures made at major observatories and space observatories.

camera to the body of the telescope. If the telescope has an equatorial mount, then you can point the camera in any direction (it need not be the same direction the telescope is looking) while you follow a star with the telescope. This will keep the telescope (and everything attached to it) at the same orientation with respect to the celestial sphere. Figure 31.7A shows a picture of the constellation Orion made in this way. You can also attach a camera to a telescope with an alt–az mount, but here you need to point the camera close to the same direction that the telescope is pointing, and even so there will be some twisting of the image in long exposures because the telescope tube rotates relative to the celestial sphere as it tracks the star.

Photography is easier with a motor drive on the telescope, but it is important to keep in mind that the telescope must remain well balanced or you may strain the motor. The weight of the camera and any lenses should be counterbalanced by weights elsewhere on the telescope so there is no net force pulling the telescope to turn in one direction or another. The highest-magnification pictures can, of course, be taken with special adaptors that allow your camera to look through the telescope (Figure 31.7B).

With the declining price of CCD detectors (Unit 26.2), amateur astronomers are producing spectacular images that were not possible even with professional equipment a decade ago. However, try the simpler photographic techniques first—you will be amazed at some of the faint star clusters and nebulae that you can image without investing in expensive equipment.

As you continue your explorations of the night sky, you should look for local astronomy clubs and visit planetariums. There are also organizations such as the American Association of Variable Star Observers (AAVSO), which will help you become involved in international efforts to monitor stars that have exhibited unusual behavior in the past. Through such monitoring projects, amateur astronomers make important contributions to the field of astronomy by alerting the astronomical community to unexpected events, such as outbursts from stars or the appearance of new comets.

31.6 SURFACE BRIGHTNESS

Always look well away from the Sun. It can damage your eyes permanently. It should never be looked at directly except with specialized equipment designed specifically for the purpose.

Look at a star with your naked eyes. Next, look at the same star through binoculars; you will notice it appears brighter. If, however, during the day you look at birds or trees through the same binoculars, they will appear larger, but not brighter. Now, with one eye, try looking at the blue sky through binoculars or a telescope, and look at the sky with the other eye unaided. The sky will appear the same or perhaps even brighter with the unaided eye. What these exercises demonstrate is that a telescope does not increase the "surface brightness" of an object.

We can express the **surface brightness** of the sky, a planet, or a distant galaxy as the number of photons received per square degree over the surface of the object. With a telescope you will collect more photons from each square degree than with your naked-eye observation; but the telescope also magnifies the image, spreading its photons out by a compensating amount. Hence, if we use a telescope with an aperture big enough to collect 100 times more photons than your naked eye can collect, the magnification of the image will spread the image's light over an area at least 100 times larger. As a result, the amount of light reaching one square millimeter at the back of your eye, or a CCD detector at the back of a telescope, will be no greater however big a telescope is used. This is why the sky does not appear any brighter through a telescope.

But why do stars appear brighter when magnified? Stars are so far away that they look like points of light to your eye. Even when they are magnified by a telescope to cover an area 100,000 times larger, and we are collecting 100,000 times more photons from the star, the magnified image of the star still remains imperceptibly small. The image size is smaller than the pixel size of a CCD or of a cell in your eye, either of which is then further limited by telescope diffraction (Unit 29) and by the blurring of the Earth's atmosphere (Unit 30). As a result, the increased numbers of photons appear still to be coming from one point, making the star look brighter, but not discernibly bigger.

When you look through a telescope at a dim astronomical object, such as a nebula or galaxy, it almost never matches its appearance in photographs. Photographs exhibit low surface brightness details you could never see with your eyes. If magnification cannot make an object appear brighter, how can photographs look so bright? The answer is that these photographs collect more photons in each square degree by making long exposures. Unlike our eyes, a CCD or photographic film can accumulate the photons over a long period of time. Even if very few photons arrive each second from a low surface brightness region, by training our CCD or open camera shutter on the region for hours, we will gather many photons. At the end of our collection period, their cumulative numbers will cause the area to appear brighter.

KEY TERMS

QUESTIONS FOR REVIEW

1. What part of a telescope or observatory corresponds most closely to the parts of your eye: iris, cornea, pupil, lens, retina, rods, fovea, brain?
2. What is meant by *dark adaption?*
3. Why does *averted vision* allow you to see dimmer objects?
4. What determines the magnification of a telescope?
5. What are the advantages of different types of mounts for telescopes?
6. Why does the image of a galaxy or other extended object look so much dimmer than a photograph made through the same telescope?

PROBLEMS

1. How can two objects have the same brightness in the sky, but appear to have different levels of visibility in a telescope? Explain which would be easier to observe on a night with a high amount of background light, a star cluster or a galaxy.
2. Suppose you are using a 10-cm (4-inch) diameter telescope with a focal length of 1000 mm (40 inches).
   a. What is the f-ratio?
   b. What focal length eyepiece will give you a magnification factor of 100?
3. Suppose you are using a 20-cm (8-inch) f/10 Schmidt-Cassegrain telescope. You have a set of eyepieces with labeled focal lengths of 6, 12, 20, and 40 mm. What magnification will each eyepiece give you? Why will images probably look a bit blurry with the 6-mm eyepiece?
4. Suppose you have a pair of 8×60 binoculars, meaning they have a magnification of 8× and an aperture 60 mm in diameter.
   a. Compared to your eyes with your pupils dilated to 8 mm, how many times more light do the binoculars collect?
   b. If the binoculars have an f-ratio of 4, what is their focal length? What is the focal length of the eyepieces?
5. Similar to an eye, cameras vary f-ratio by changing the aperture. Suppose you have a camera lens labeled as having a focal length of 50 mm with f-ratios varying from f/1.4 to f/30.
   a. What is the range of aperture diameters for this camera?
   b. When the f-ratio is small, the "depth of field" (range of distances over which objects remain in focus) becomes smaller. Why would the focus be likely to remain sharper when the aperture is smaller?
6. a. Your eye has an aperture that can vary between 1 and 8 mm in diameter. Estimate approximately what the range of f-ratios is for your eye.
   b. People often report that objects look sharper in brighter light, which is the opposite of what you might expect based on the diffraction limit when the pupil is smaller. Discuss how both the focusing and the detection of light in your eye may lead to this perception.

TEST YOURSELF

1. In very dim light,
   a. your pupils are smallest.
   b. your rods and cones grow much more sensitive to light.
   c. your color vision is at its most sensitive.
   d. you can see things more clearly if you stare straight at them.
   e. All of the above
2. As a star rises and moves across the sky, which of the following change(s)?
   a. Its right ascension
   b. Its declination
   c. Its azimuth
   d. Both (a) and (b)
   e. None of the above
3. What is the primary advantage of a 40-cm reflector over a 20-cm refractor?
   a. The reflector will collect about 4 times more light.
   b. The reflector will let you see astronomical features that are twice as small.
   c. Reflectors transmit more light because they do not use any lenses.
   d. The reflector will provide twice the magnification.

Part Three

The content of some Units will be enhanced if you have previously studied some earlier Units in this textbook. These Background Pathways are listed below each Unit image.

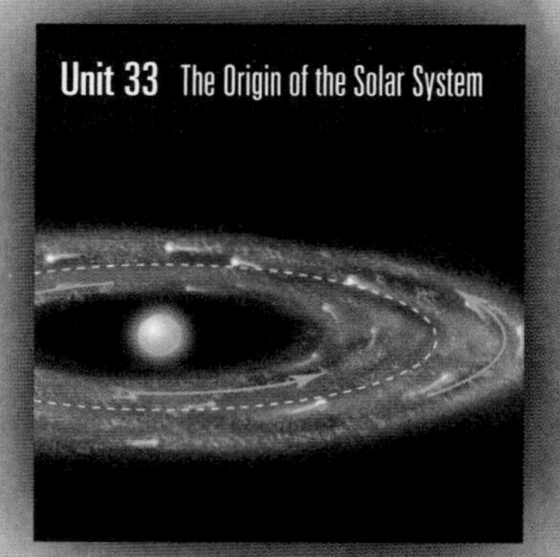

Background Pathways: Unit 32

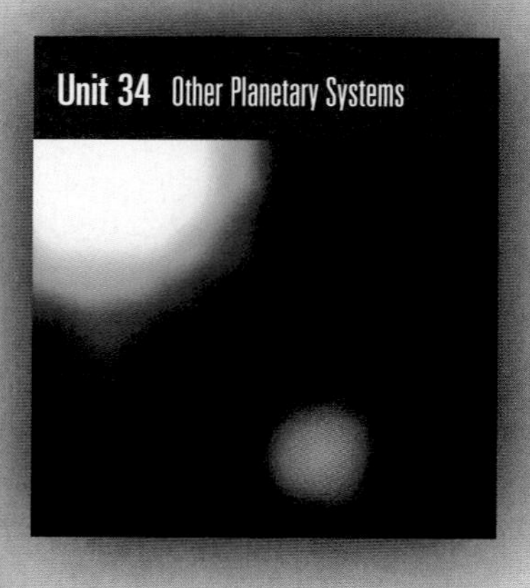

Background Pathways: Unit 32

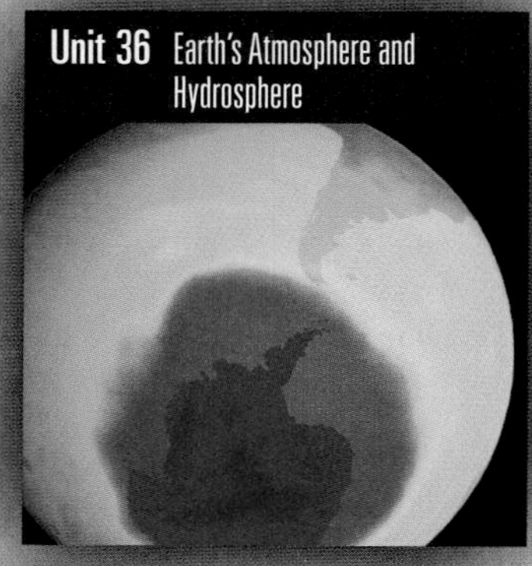

Background Pathways: Unit 35

Background Pathways: Unit 35

Background Pathways: Unit 37

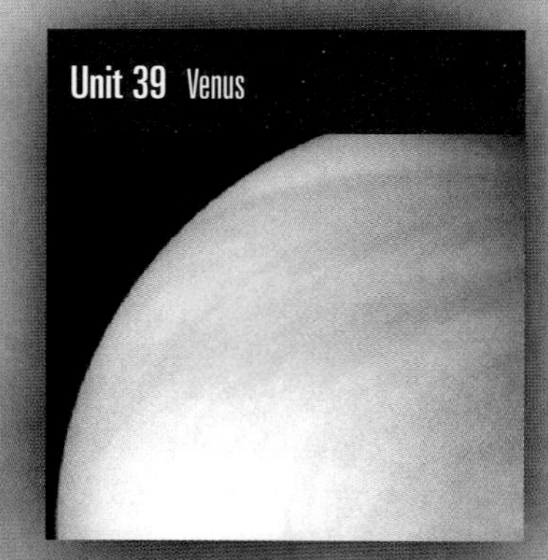

Background Pathways: Unit 36

Background Pathways: Unit 36

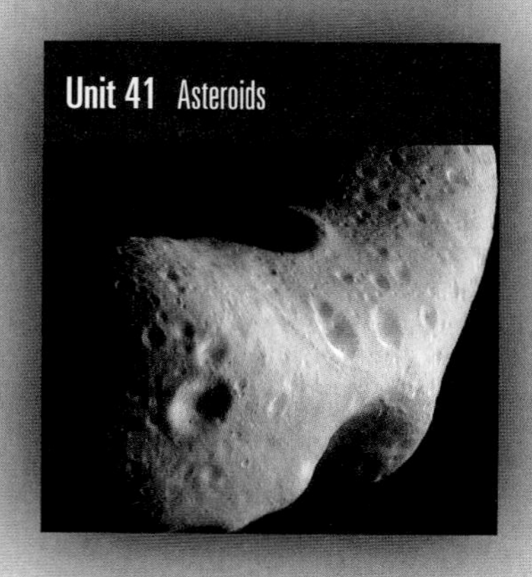

The Solar System

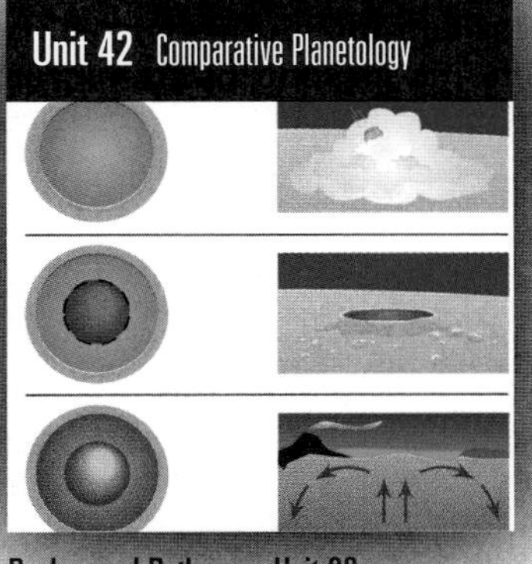

Background Pathways: Unit 32

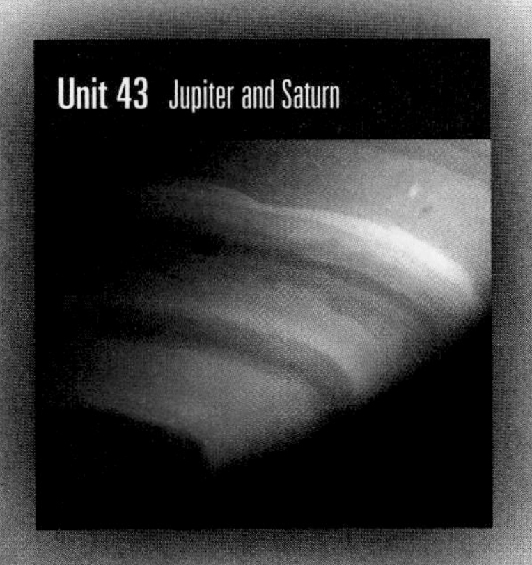

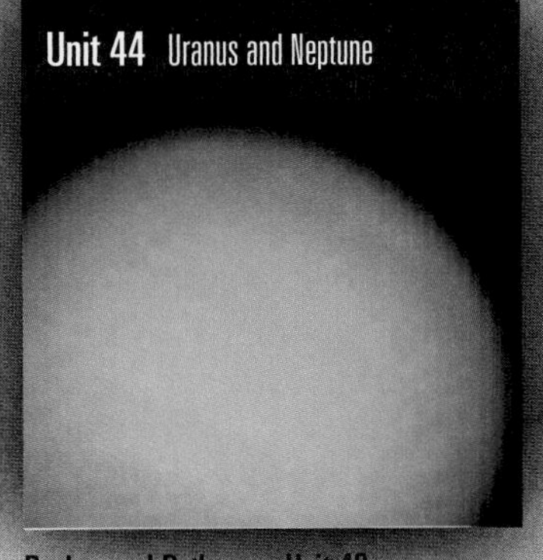

Background Pathways: Unit 43

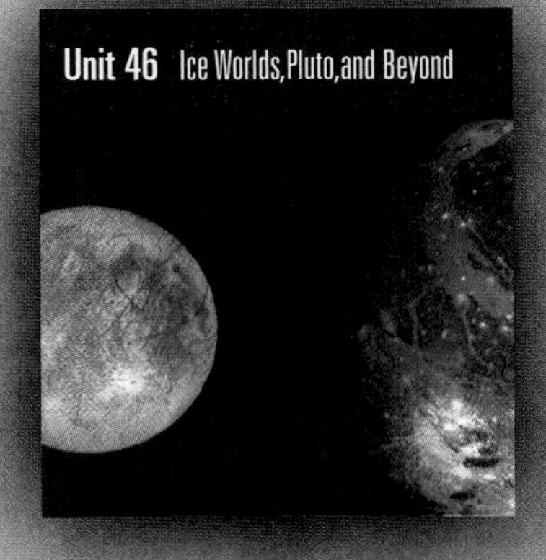

unit

32 The Structure of the Solar System

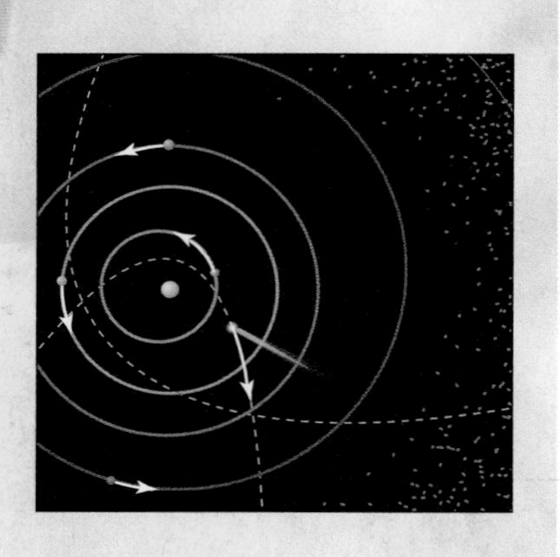

The **Solar System** contains the Sun, the eight planets, their moons, and swarms of asteroids and comets. The various objects in the Solar System differ enormously in size, composition, and temperature, but the Solar System as a whole possesses an underlying order. This order is based on physical conditions today as well as when the Solar System formed; and from the patterns they see, astronomers attempt to read the story of how our Solar System came to be. This unit examines the overall structure of the Solar System and its components, and discusses the patterns in the properties we find. Each of these elements is explored in greater detail in subsequent units.

32.1 COMPONENTS OF THE SOLAR SYSTEM

The Solar System is defined as the Sun and the objects that are held in orbit around it by the Sun's gravity. These other objects are far smaller than the Sun, in total contributing less than one seven-hundredth of the Sun's **mass**—the equivalent in mass to the weight of a car's steering wheel versus the rest of the vehicle. Within the tiny fraction of the Solar System's mass outside the Sun there are probably more than a million celestial objects larger than mountains, with the largest being the planet Jupiter. These objects emit no visible light of their own but shine by reflected sunlight, and astronomers have cataloged only a fraction of them. Even more objects smaller than mountains orbit unseen and uncounted.

On the scale of humans, every one of these objects is big; the fact that altogether they amount to so little compared to the Sun is simply an indication of how enormous the Sun is. The Sun is a star, a huge ball of extremely hot gas. Its enormous gravity crushes and heats matter in its core so intensely that nuclear reactions take place (Unit 49). The Sun's gravity, heat, and light are largely responsible for the nature of everything else within the Solar System. The Sun is the center of the Solar System, but it is very different from the objects that orbit it. A more complete understanding of the Sun is best gained in the context of studying other stars, so Solar System astronomy focuses on the objects in orbit about the Sun.

Planets are a big step down in size from stars. Even Jupiter has only about one one-thousandth the mass of the Sun. And Jupiter itself contains more mass than everything else in the Solar System combined. Jupiter is the fourth of the eight objects recognized as planets. In order of increasing orbital distance from the Sun, the officially recognized planets are Mercury, Venus, Earth, Mars, Jupiter, Saturn, Uranus, and Neptune. The differences in sizes of the planets can be seen in Figure 32.1, which also shows a portion of the Sun's edge to illustrate how the Sun dwarfs even the large planets.

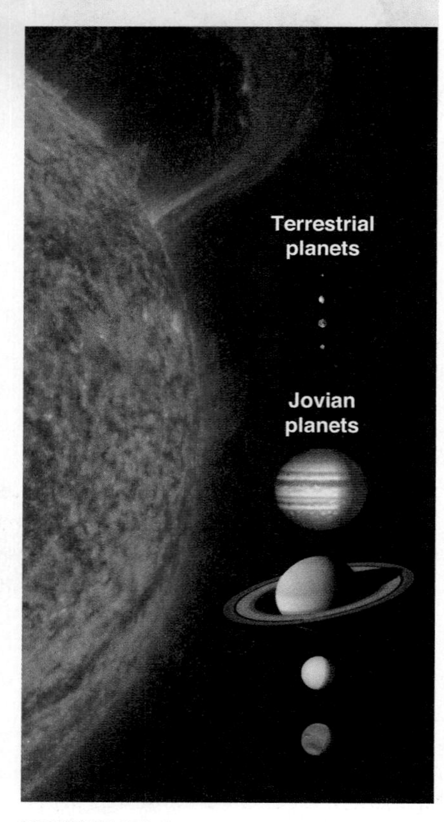

FIGURE 32.1
The planets and Sun to scale.

The **inner planets**—Mercury, Venus, Earth, and Mars—are small rocky planets with relatively thin or no atmospheres. This type of planet is often called an

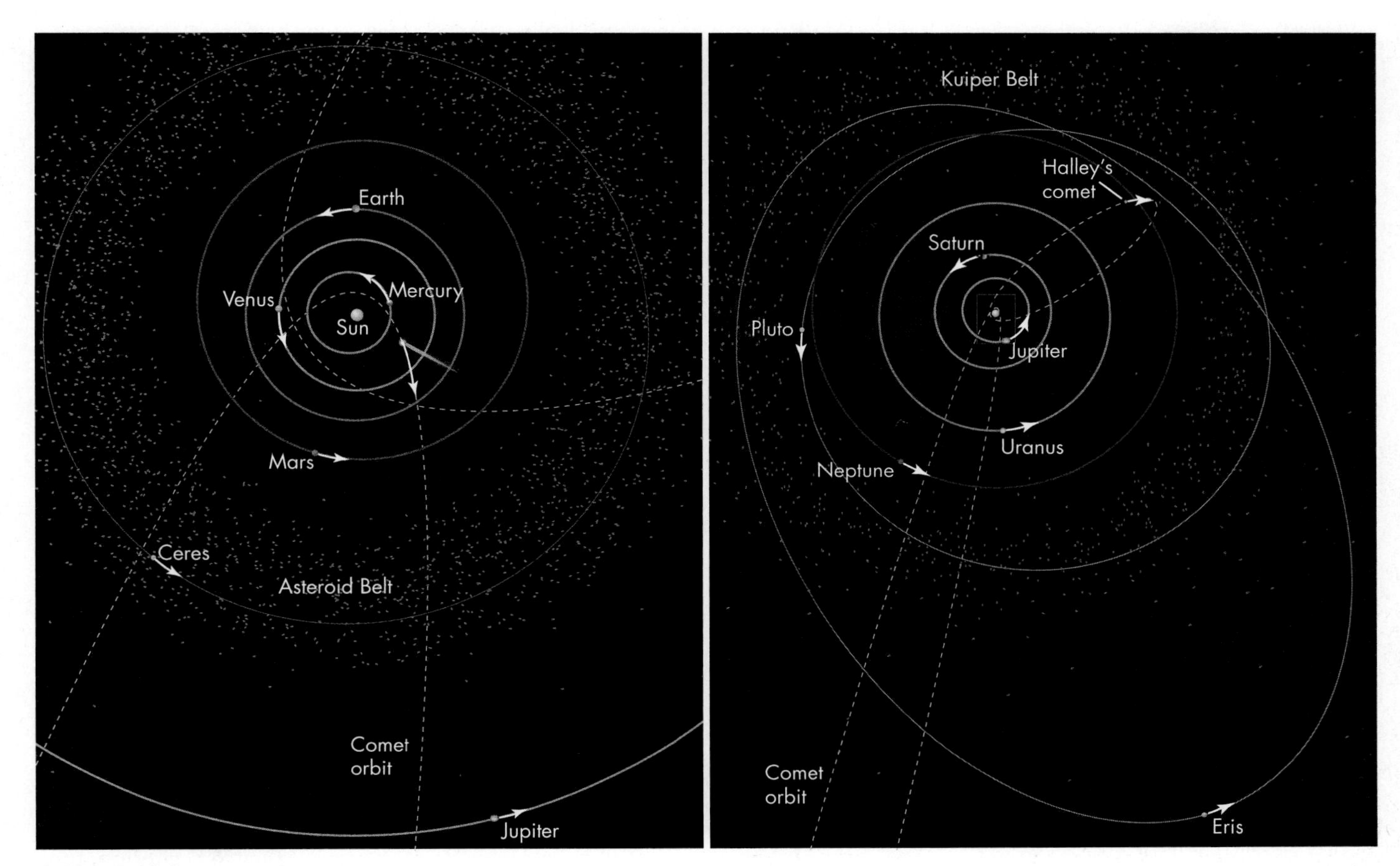

FIGURE 32.2
Diagrams of the inner and outer Solar System seen from above. The planets' orbits are shown in the correct relative scale in the two drawings.

Earth-like planet, or a **terrestrial planet.** The Earth is the largest terrestrial planet, although Venus is nearly as big. Mars has about half of Earth's diameter and only about a tenth of Earth's mass. Mercury has 38% of Earth's diameter and only 6% of Earth's mass. From examining the differences and similarities between these planets, we can better understand the processes that shape the Earth.

The planets' orbits are illustrated in Figure 32.2. In this view we are "looking down" on the orbits of the Earth and other planets from far above the Earth's and Sun's north poles. Viewed from this direction, the planets' orbits are roughly circular. Outside of Mars's orbit is a region called the **asteroid belt.** It gets its name because a majority of the small objects known as **asteroids** (Unit 41) orbit there. Asteroids are rocky or metallic bodies with diameters ranging from meters up to hundreds of kilometers. The asteroids are irregular in shape, except for the largest, Ceres, which has a diameter about 7% of the Earth's. Because its gravity has pulled it into a rounded shape, it is termed a **dwarf planet.**

Astronomers sometimes speak of the **superior planets.** This refers to planets having an orbit larger than Earth's and therefore includes Mars.

The **outer planets**—Jupiter, Saturn, Uranus, and Neptune (Figure 32.2)—are also called **Jovian** (Jupiter-like) **planets** or **gas giants** because of their thick atmospheres. They are huge compared with the terrestrial planets: 15 to 320 times more massive than Earth, and 4 to 11 times its diameter. The Jovian planets have no distinct surface. Instead, their atmospheres thicken with depth and eventually liquefy under intense pressure. Deeper still, the liquid material gradually compresses to the same density as rock on Earth, but remains gaseous or liquid because of intense heat there. Deep in the center there is probably a solid or molten rock and iron **core.** However, we could never "land" on Jupiter because we would simply sink ever deeper into its interior until our spaceship was crushed by pressures great enough to compress solid rock.

The outer Solar System has a layout similar to that of the inner Solar System, but on a far grander scale (Figure 32.2). Outside of Neptune's orbit is a region called the **Kuiper** (*KY-per*) **belt** named after a Dutch-born American astronomer. Analogous to the asteroid belt beyond Mars's orbit, the Kuiper belt contains vast numbers of objects that orbit the Sun out to about twice Neptune's orbit. These **trans-Neptunian objects** (or **TNOs**) are composed primarily of ices in this remote, cold region of the Solar System.

Until 2006, one of the TNOs, Pluto, was called a planet. However, over the preceding decade it had become clear that Pluto was just one of the largest among thousands of objects orbiting beyond Neptune. Dozens of the TNOs were found with sizes not much smaller than Pluto. Then in 2005, Eris—a body 25% more massive than Pluto—was discovered. Many questions arose. Was Eris a planet too? What about all of the objects that are just a little smaller than Pluto? And what about potentially huge numbers of large objects that might be discovered in future years as our telescopes and detection techniques improve?

It may seem surprising, but there was no officially accepted definition of "planet" before 2006. The term comes from ancient times and means a "wanderer" because of the changing position of these objects on the celestial sphere (Unit 11). In both popular and astronomical usage, a planet is a large body orbiting the Sun, but how big is big enough? Pluto has only one twenty-fifth the mass of Mercury, the smallest body called a planet since ancient times. Astronomers convened at the *International Astronomical Union* assembly in 2006, and developed a definition with two requirements. A planet must be (1) so massive that its gravity pulls it into a roughly spherical shape, and (2) the dominant mass in the neighborhood of its orbit. The second part of this definition means the objects lying in the asteroid belt and Kuiper belt are not planets. However, if they meet the first criterion, they can still be labeled dwarf planets. As a result, Pluto and Eris are both labeled dwarf planets—and several dozen more TNOs may soon be added to that category.

A dwarf planet orbiting beyond Neptune was given the official designation of "plutoid" in 2008.

Actually, there are seven **satellites** (moons) of the planets that are both larger and more massive than either Pluto or Eris, including our own Moon. Astronomers are steadily finding more small satellites of the Jovian planets with new satellite missions and improving telescopes (Unit 45). Astronomers have so far identified 63 satellites of Jupiter, 60 of Saturn, 27 of Uranus, and 13 of Neptune. Among the terrestrial planets, Mars has 2, Earth has 1, while Mercury and Venus have no satellites. Pluto has 3 satellites and Eris has 1, and some smaller TNOs and even some asteroids are known to have small satellites. The satellites generally share some basic characteristics with their host planets, being made of similar materials—rocky materials in the case of our Moon and Mars's two tiny satellites, and ices of some compounds related to the gaseous elements of the Jovian planets.

The number of known satellites is constantly increasing. The numbers quoted are current as of mid 2008.

The outer Solar System continues beyond the Kuiper belt out to perhaps a few thousand times farther than Neptune's orbit—about one-quarter of the distance to the next nearest star. Sunlight is so dim this far from the Sun that little light is reflected back to us, and we know relatively little about this region. Our primary clue about the objects that populate this region comes from **comets** (Unit 47). These are icy bodies that spend most of their time in the outer Solar System; but some have orbits that carry them into the inner Solar System, perhaps no more frequently than once in a million years. A few approach so close to the Sun that their frozen outer layers are vaporized by the Sun's heat, then stream away from the Sun in long "tails" that are sometimes visible to the naked eye. Astronomers hypothesize that a large population of objects orbit in what is called the **Oort cloud,** named after the Dutch astronomer who proposed its existence. The Oort cloud is thought to surround the Solar System

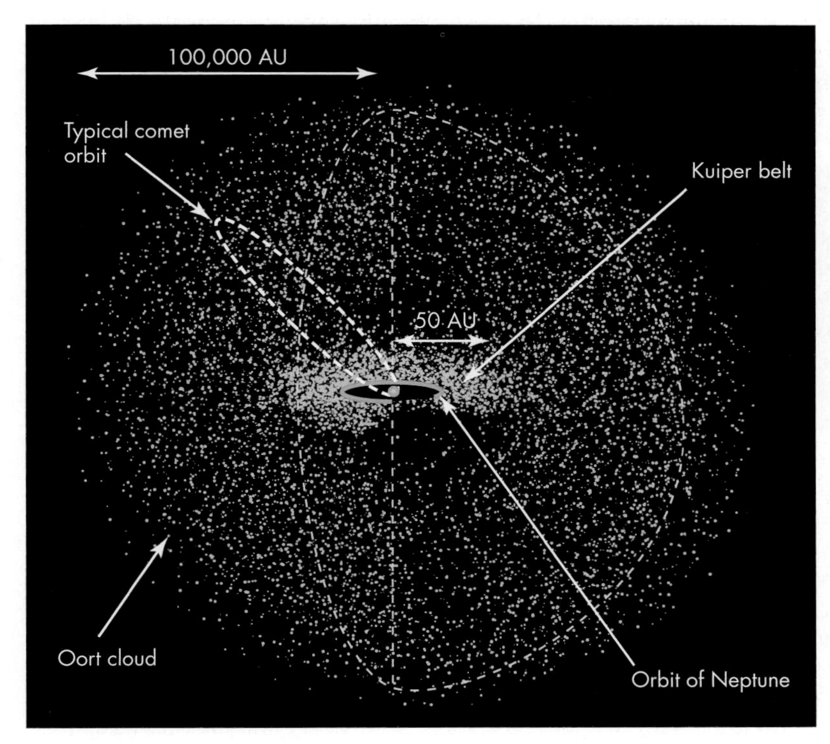

FIGURE 32.3
Sketch of the Oort cloud and the Kuiper belt. The scale shown is only approximate. Orbits and bodies are not to scale.

(Figure 32.3), with most of its objects remaining there in very large orbits that may never enter the inner Solar System. Hypotheses about the total mass contained in the Oort cloud vary greatly, from a small fraction of the Earth's mass to as much as Jupiter's mass. Such great uncertainty arises because of the difficulty in determining the likelihood that an Oort cloud object will come close enough to the inner Solar System to be detected by our telescopes. Therefore, we cannot say precisely what percentage of the cloud's population we *do* observe.

Astronomers suspect that for most of the comets that come into the inner Solar System (and that therefore can be well studied), their orbits have been disturbed by the gravitational effects of a distant star or a close encounter with another object. The comets entering the inner Solar System are generally quite small—diameters of 50 km or less. Until recently it was supposed that all of the Oort cloud objects were this small, but recent discoveries have revealed the presence of objects comparable in size to Pluto beyond the Kuiper belt. One of these, named Sedna, has an orbit that carries it out to 30 times Neptune's distance, but currently it is only about three times farther than Neptune. Sedna would not have been detected with current instrumentation if it had been in a more distant part of its orbit. It seems likely that there are many more even larger worlds awaiting discovery in these most remote portions of the Solar System.

32.2 ORBITAL PATTERNS IN THE SOLAR SYSTEM

The orbits of Pluto and many other TNOs are sufficiently elliptical that they come closer to the Sun than Neptune during part of their orbit. Although these orbits may appear to cross in diagrams, the orbits are generally quite tilted relative to the rest of the Solar System, so collisions are highly improbable.

The eight planets move around the Sun in nearly circular orbits, all lying in almost the same plane, as shown in the top view in Figure 32.2 and the side view in Figure 32.4. The planets all travel around the Sun in the same direction: counterclockwise as seen from above the Earth's North Pole. Their orbits could be contained in a thin disk, a spinning "pancake," whose thickness relative to its diameter would be about the same as three CDs stacked together.

As the planets orbit the Sun, each also spins on its rotation axis. The spin is again generally in the same direction as the direction in which planets orbit the Sun, and the Sun's own rotation is in the same direction as well—counterclockwise as seen from above the Earth's North Pole. The planets' rotation axes are tilted relative to the plane of planetary orbits, but generally not far from the orientation of the Sun and the planets' orbits. There are exceptions, however. Venus spins slowly but backward (clockwise), while Uranus's axis lies nearly in the **orbital plane** (Figure 32.5) of the planet. The flattened structure of the Solar System and the generally consistent orbital and spin properties of the bodies in it are two of its most fundamental features, and any theory of the Solar System must explain them.

The satellites of planets also generally move in a regular pattern along approximately circular paths that are roughly in the same plane as each planet's equator. Thus each planet and its moons resemble a miniature Solar System—an important clue to the origin of the satellites. There are a few exceptions, such as Neptune's moon Triton, which orbits in a backward direction—leading astronomers to hypothesize that it was once an independent object in the Kuiper belt that was captured by Neptune.

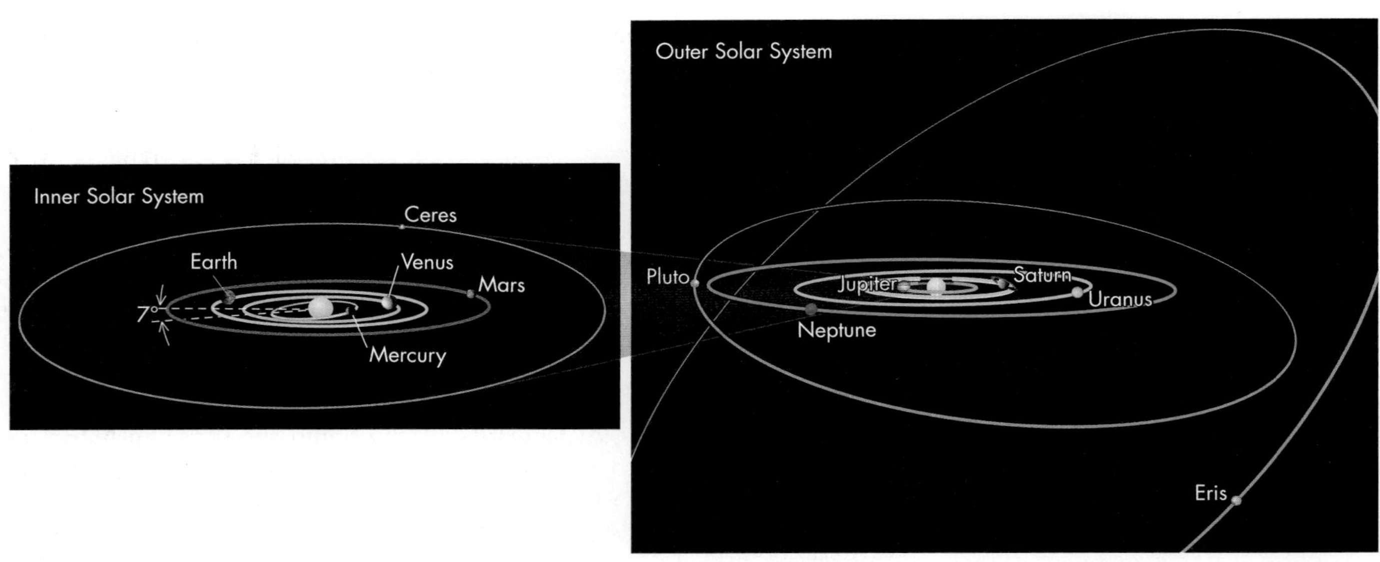

FIGURE 32.4
Planets and their orbits from the side. The left-hand portion illustrates the inner Solar System, including the orbit of the dwarf planet Ceres. The lower portion shows the outer Solar System, showing the large angle of the orbits of the dwarf planets Eris and Pluto relative to the major planets. Orbits and bodies are not to the same scale.

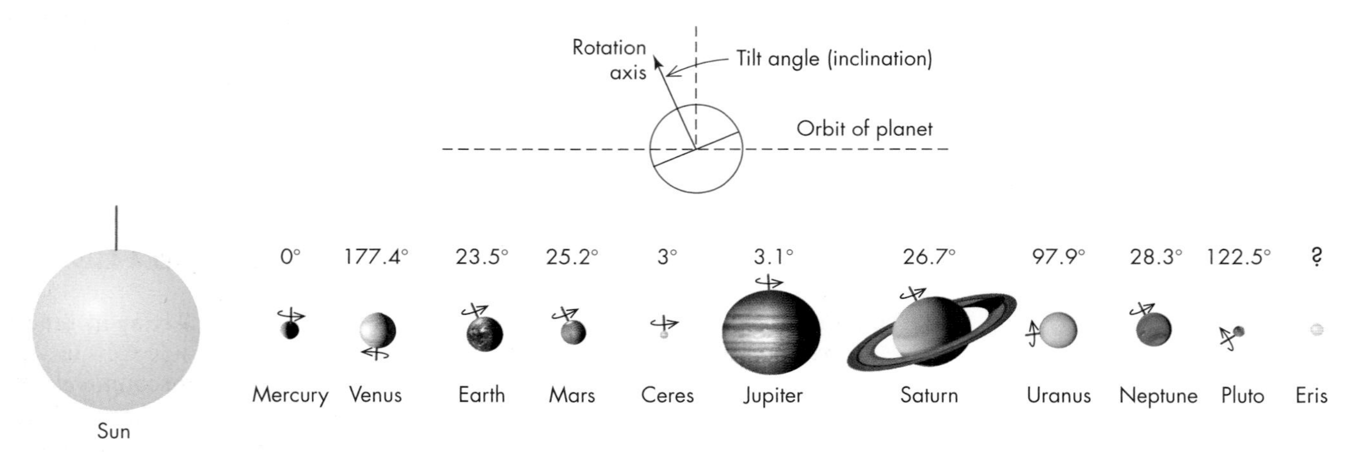

FIGURE 32.5
The tilt of the rotation axes of the planets and dwarf planets. The diagram illustrates the orientation of the rotation axes of the planets relative to their orbit around the Sun. Most planets spin around their axis in approximately the same counterclockwise sense as they orbit the Sun; however, Venus spins backward and Uranus's axis almost lies in its orbital plane.

One feature that is not fully understood about the orbits of the planets is their spacing. The separations between the inner planets are relatively small—between 0.3 and 0.4 AU—while the distances between the orbits of the outer planets are much larger—between 4 and 11 AU. The entire inner Solar System is smaller than the separation between neighboring planets in the outer Solar System. This makes it difficult to show the inner and outer planets' orbits in the same figure when we wish to illustrate their orbits to scale. What is known is that not all configurations of planets are stable for long periods of time because of gravitational interactions between the planets. The Solar System was either formed this way or gradually evolved to a very stable set of orbits. We are just beginning to uncover the characteristics of other planetary systems. There are not very many where we have detected more than two planets, but the other systems detected so far look quite different from our own, and it appears that some of them are currently changing (Unit 34).

32.3 COMPOSITIONS IN THE SOLAR SYSTEM

By studying the spectrum of light that the Sun emits (Units 24 and 55), we find that the Sun is made mostly of hydrogen (about 71% of its mass) and helium (about 27%), but it also contains small percentages of nearly all the other chemical elements (oxygen, carbon, iron, uranium, and so forth) in vaporized form. The Jovian planets appear to be similar to the Sun in composition, and their masses are likewise dominated by the element hydrogen.

That the inner planets are rocky and the outer planets are hydrogen-rich or made of ice is important to our understanding of the history of the Solar System. By *rock* we mean material composed of substances like silicon and oxygen (SiO_2) with an admixture of other heavy elements such as aluminum (Al), magnesium (Mg), sulfur (S), and iron (Fe). By *ice* we mean **volatile** substances that would be liquids and gases at typical Earth temperatures, such as the hydrogen-rich compounds water (H_2O), frozen ammonia (NH_3), and frozen methane (CH_4). At the low temperatures found far from the Sun, these hydrogen-rich compounds form solids but have much lower densities than rock.

From spectroscopy we can learn a great deal about the composition of an object's surface. However, spectra give no clue about what lies deep inside a planet where light cannot penetrate. To learn about the interior, astronomers must therefore use alternative methods. One such technique uses the planet's density.

32.4 DENSITY AND COMPOSITION

The average **density** of a planet is its mass divided by its volume. Both mass and volume can be measured relatively easily. For example, Newton's law of gravitation (Unit 16) allows us to determine a body's mass from its gravitational attraction on a second body orbiting it (Unit 17). From this law, we can calculate a planet's mass by observing the orbital motion of one of its moons or a passing spacecraft. We can determine a planet's volume ($\mathcal{V}$) from the geometric formula for a sphere: $\mathcal{V} = 4\pi R^3/3$, where R is the planet's radius. We can measure R in several ways—for example, from its angular size and distance (Unit 10).

With the planet's mass, M, and volume, $\mathcal{V}$, known, calculating its average density (usually represented by the Greek letter rho, ρ) is straightforward: We divide M by $\mathcal{V}$ (Figure 32.6). For example, the Earth's radius is 6.37×10^6 m and its mass is 5.97×10^{24} kg, so its density is

$$\begin{aligned}\rho_\oplus &= \frac{M_\oplus}{\mathcal{V}_\oplus} \\ &= \frac{5.97 \times 10^{24}\ \text{kg}}{4\pi(6.37 \times 10^6\ \text{m})^3/3} \\ &= \frac{5.97 \times 10^{24}\ \text{kg}}{1.08 \times 10^{21}\ \text{m}^3} \\ &= 5530\ \text{kg/m}^3 \\ &= 5.53\ \text{kg/liter}\end{aligned}$$

The last step in this calculation was made by noting that a cubic meter contains exactly 1000 liters. A liter bottle of soda is a familiar volume, and it makes a useful

FIGURE 32.6
Measuring a planet's mass, radius, and average density. Mass can be determined from the orbit of a satellite (Unit 17), and volume can be measured after we determine the radius of the planet from its angular size (Unit 10).

Volume

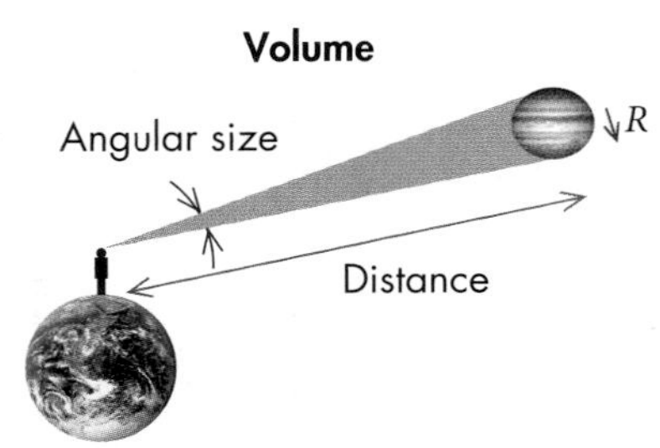

Measure angular size of planet, and use relation between angular size and distance to solve for planet's radius, R. Calculate volume, $\mathcal{V}$, of planet:

$$\mathcal{V} = \frac{4\pi R^3}{3}$$

for a spherical body of radius R.

Mass

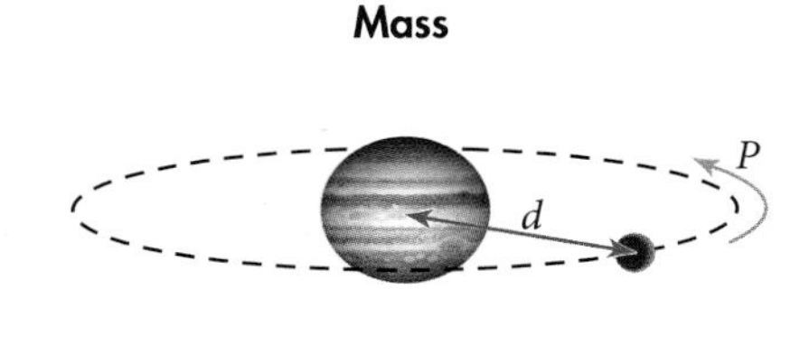

Observe motion of a satellite orbiting planet. Determine satellite's distance, d, from planet and orbital period, P. Use Newton's form of Kepler's third law:

$$M = \frac{4\pi^2 d^3}{GP^2}$$

Insert measured values of d and P, and value of constant G. Solve for M.

Average Density

Average density, ρ, equals mass, M, divided by volume, $\mathcal{V}$:

$$\rho = \frac{M}{\mathcal{V}}$$

Densities are often reported in a variety of units. However, many of them are equivalent. For example, 1 kilogram per liter is the same density as 1 gram per cubic centimeter or 1 metric ton per cubic meter. The number of kilograms per liter is also approximately the same as the number of pounds per pint in British units.

comparison for discussing densities. Keep in mind that water and ice have a density of about 1 kilogram per liter.

Density gives an important clue to an object's composition. For example, it would be easy to tell if a closed box contained a block of iron or a block of wood. This is the basis of the famous story about the ancient Greek scientist Archimedes leaping from his bath and running down the street shouting "Eureka!" He had been given the task of determining whether the king's crown was solid gold, but he was not allowed to damage it. He realized from the way that his own body displaced water in a bathtub that a mass of gold with the same mass as his body would be much smaller—because of its high density—and would displace much less water. Thus if the crown did not displace the same amount of water as an equal weight of gold, it must contain another metal, like lead or silver, which each have different densities. Likewise, we can use the density of the Earth or any planet to estimate its composition.

One way to make such an estimate of composition is to compare a planet's density with the densities of candidate materials to find a likely match. We can narrow the range of likely candidate materials by considering the most common elements that make up the Sun or the Earth's surface rocks. Studies of their compositions show that some elements are more common than others—for example, iron is much more abundant than other dense substances such as gold, silver, or lead. Knowing that the average density of the Earth (5.53 kilograms per liter) is intermediate between silicate rock (about 3.0 kilograms per liter) and iron (about 7.9 kilograms per liter), we can infer that the Earth has an iron core beneath its rocky, silicate-rich crust, a supposition that has been supported by more complex techniques (see Unit 35).

All the terrestrial planets have an average density similar to the Earth's (3.9 to 5.5 kilograms per liter). On the other hand, all the Jovian planets have a much smaller average density (0.7 to 1.7 kilograms per liter), which is similar to that of ice. There are a number of asteroids for which we have accurate mass and size measurements, and their average densities are in the range of 2.0 to 3.5 kilograms per liter. The solid bodies in the outer Solar System, including Pluto, are generally

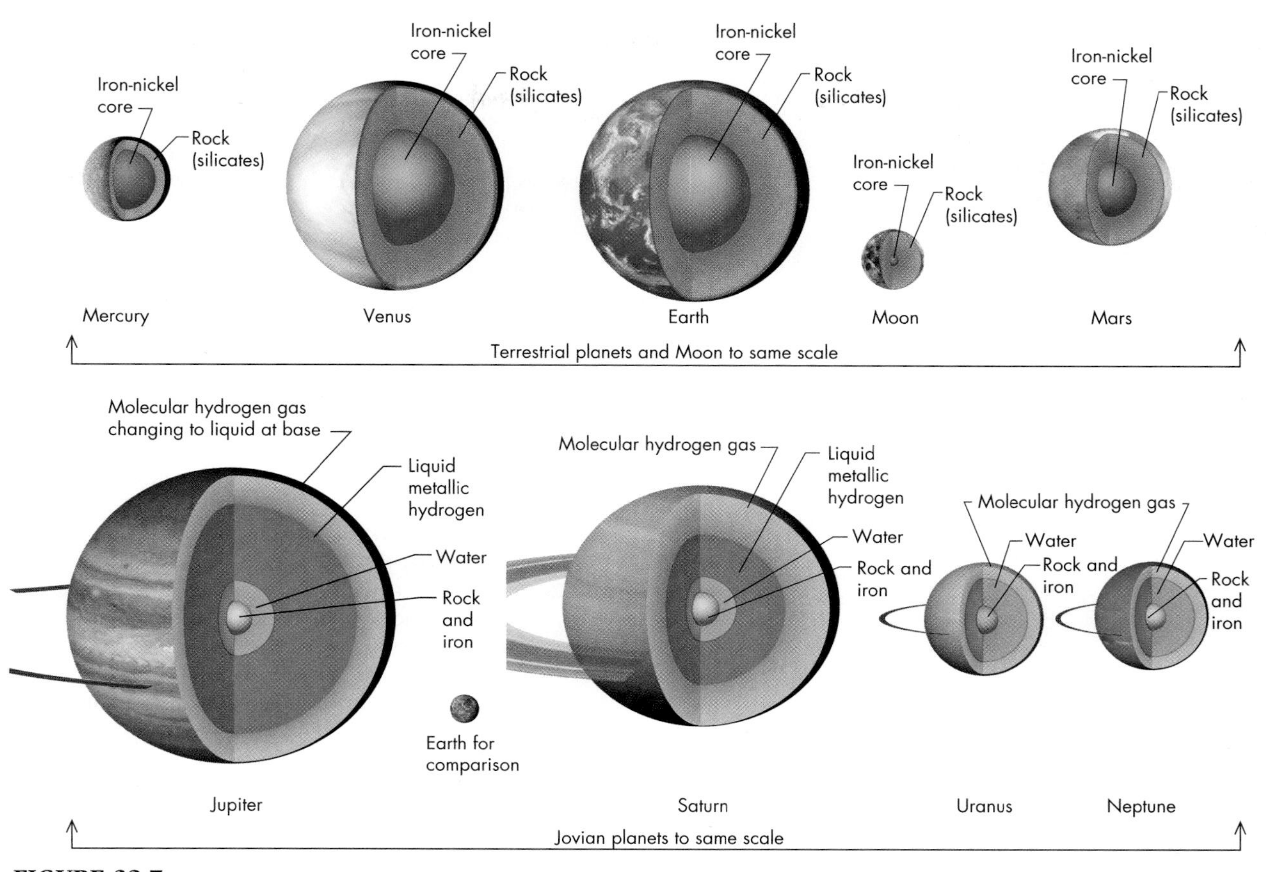

FIGURE 32.7

Sketches of the interiors of the planets. Details of sizes and composition of inner regions are uncertain for many of the planets. (The Earth's Moon is shown for comparison.)

between 1.0 and 2.0 kilograms per liter. Further analyses indicate that the objects in the inner Solar System are rocky with a decreasing fraction of iron the farther they are from the Sun. In the larger bodies of the inner Solar System, the iron has sunk to the core, as shown in Figure 32.7. In the outer Solar System, the densities of the gas giants and the solid bodies indicate that they must contain mainly low-density materials in their interiors.

Beneath their deep atmospheres the gas giants have cores of iron and rock about the size of the terrestrial planets, as illustrated in Figure 32.7. In addition to density determinations, astronomers deduce the existence of these cores in two other ways. First, if the outer planets have the same relative amount of heavy elements as the Sun, they should contain several Earth masses of iron and silicates. Because these substances are much denser than hydrogen, they must sink to the planet's core. Secondly, mathematical analyses of these planets indicate that the distortion to their shape caused by their rotation, along with details of their gravitational fields sensed by spacecraft, can be best explained if they have small, dense cores.

Our discussion of the composition of the planets not only underlines the differences between the two regions of the Solar System but also furnishes another clue to their origin: The planets and Sun were all made from the same material. Astronomers think this because Jupiter and Saturn have a composition almost identical to that of the Sun. On the other hand, the inner planets have a similar composition except for the missing volatile compounds, presumably lost because of the higher temperatures of the inner Solar System.

KEY TERMS

asteroid, 239
asteroid belt, 239
comet, 240
core, 239
density, 243
dwarf planet, 239
gas giant, 239
inner planet, 238
Jovian planet, 239
Kuiper belt, 240
mass, 238
Oort cloud, 240
orbital plane, 241
outer planet, 239
satellite, 240
Solar System, 238
superior planet, 239
terrestrial planet, 239
trans-Neptunian object (TNO), 240
volatile, 243

QUESTIONS FOR REVIEW

1. What properties distinguish the terrestrial from the Jovian planets?
2. What is a dwarf planet?
3. How are asteroids and TNOs similar? How do they differ?
4. What is meant by a volatile substance, and why would we expect to find objects made out of volatile substances far from the Sun?
5. Why is the density of an object an important clue to its composition?
6. How can we determine the density of a distant planet?
7. What is the Oort cloud? Where is it located, and what kind of objects come from it?

PROBLEMS

1. In Appendix Tables 5 and 6, look up the sizes and distances from the Sun of Jupiter, Neptune, Earth, Mercury, Pluto, and Ceres. Imagine the Sun is scaled down to the size of a large beach ball, 1 meter in diameter.
   a. What diameter would these six objects have if scaled down by the same factor?
   b. What would their masses be if their density remained the same?
2. What would be the distance from the Sun for each of the bodies listed in problem 1, scaled down by the same factor that would make the Sun 1 meter in diameter?
3. To arrive at the average density of Mercury ($\rho = 5.43$ kg/L), we could imagine different combinations of materials than given in the text. What would be the problems with the following models?
   a. A rocky interior ($\rho \approx 3$ kg/L) surrounded by an iron crust ($\rho \approx 8$ kg/L)?
   b. An iron interior surrounded by a thick layer of ice ($\rho \approx 1$ kg/L)?
4. Calculate the densities of Venus and Jupiter from the following data: The mass and radius of Venus are 4.87×10^{24} kilograms and 6051 kilometers, respectively. The mass and radius of Jupiter are about 1.9×10^{27} kilograms and 71,492 kilometers, respectively. How do these numbers compare with the density of rock (about 3 kg/L) and water (1 kg/L)?
5. Use the data in Appendix Table 5, " Physical Properties of the Planets," to list the planets in order according to several different parameters:
   a. smallest to largest radius.
   b. smallest to largest mass.
   c. most to least dense. Note that the order of some planets is changed in each case. Can you draw any approximate conclusions from these results? For example, are smaller planets always more dense? What other factors might explain why some planets are out of order with the overall trends you find?
6. You can calculate the acceleration due to gravity at the surface of a planet using the formula from Unit 16.3: $g = {}^{GM}\!/_{R^2}$ where M and R are the mass and radius of the planet, and G is the gravitational constant (see Appendix Table 1). How much do you weigh on Earth? Determine how much you would weigh on Jupiter, Mars, and Ceres. (Hint: Because your mass would not change, you can simply compare the values of g for the other planets with that of Earth, 9.8 m/sec^2.)

TEST YOURSELF

1. Which of the following planets are primarily rocky with iron cores?
   a. Venus, Jupiter, and Neptune
   b. Mercury, Venus, and Pluto
   c. Mercury, Venus, and Earth
   d. Jupiter, Uranus, and Neptune
   e. Mercury, Saturn, and Pluto
2. Why is Pluto not considered a Jovian planet?
   a. Its mass and radius are so small.
   b. It is so far out in the Solar System.
   c. Its interior is mostly rock and iron.
   d. Its atmosphere is rich in oxygen, making it more like the Earth.
   e. It is not really orbiting the Sun but is simply drifting through the outer edge of the Solar System.
3. More than 99% of the mass of the Solar System is contained in
   a. the Earth.
   b. the Sun.
   c. the Jovian planets.
   d. Jupiter.
   e. asteroids.

unit

33

The Origin of the Solar System

Background Pathways

Unit 32: The Structure of the Solar System 238

How did the Solar System form? What processes gave it its main features? Because we were not around to witness its birth, our explanation of its origin must be a reconstruction based on observations that we make now, long after the event. The general features of the Solar System, such as its flatness and two main families of planets, provide powerful clues that have helped astronomers reconstruct how the Solar System must have formed. Some of the basic aspects of this theory were developed more than 200 years ago, but recent findings and the ability to carry out detailed physical simulations on computers allow us to refine our understanding of what happened in our home's remote past.

Any theory of the Solar System's origin must explain a number of facts:

- The Solar System is flat, with the planets orbiting in the same direction.
- The composition of the outer planets is similar to the Sun's; the inner planets lack the gases that condense at low temperatures.
- There are rocky planets near the Sun and massive, gaseous planets farther out.

We have listed only the most important observed features that our theory must explain. Astronomers have many additional clues that must also be explained by a successful theory. These clues include the properties of asteroids, the number of craters on planetary and satellite surfaces, and the detailed chemical composition of surface rocks and atmospheres.

33.1 THE AGE OF THE SOLAR SYSTEM

We can add one more clue about the formation of the Solar System to the set of its features. Several pieces of information indicate when the Solar System formed. All existing evidence is consistent with the various bodies in the Solar System having formed at one time—about 4.5 billion years ago. Geological features on Earth point to a history of many billions of years for our planet. Likewise, we find a similar age for the Sun, based on its current brightness and temperature and its rate of nuclear fuel consumption (Unit 61). However, the most precise age information comes from a clock that ticks inside the rocks of the terrestrial planets.

A number of naturally occurring atoms undergo **radioactive decay.** For many of the heaviest elements, all forms of the atom are radioactive, as illustrated by the elements marked in green in the periodic table in the inside back cover of this book. Even the lighter elements have some **isotopes**—nuclei with different numbers of neutrons—that are radioactive.

Processes inside the nucleus of a radioactive atom, such as the weak force (Unit 4), cause the nucleus to split apart spontaneously—breaking apart into lower-mass atoms. At any given moment there is a small chance of a radioactive atom splitting. The atom may last for billions of years, or it may decay in the next

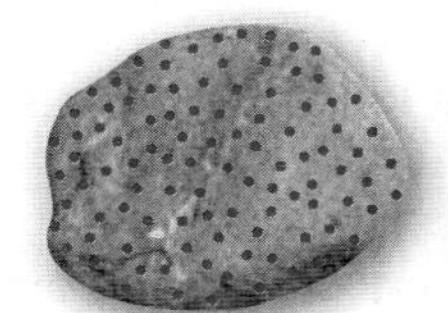
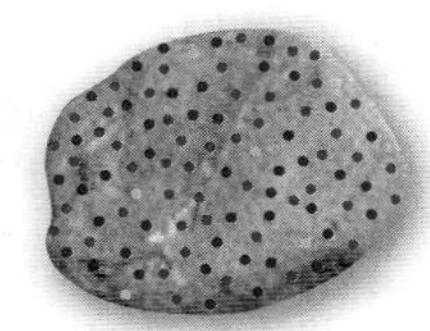
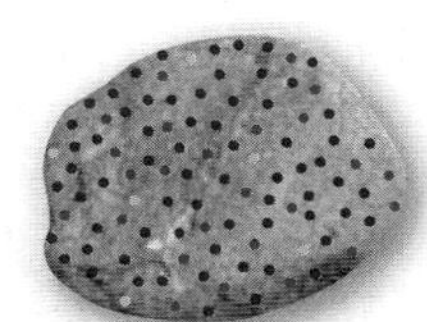
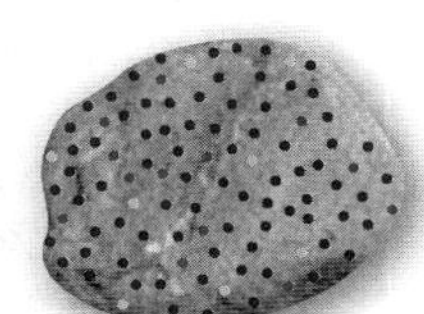
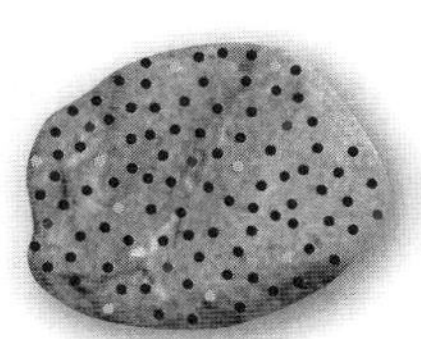

FIGURE 33.1
The amount of argon compared to potassium in a sample of rock gives information about the rock's age. Potassium decays radioactively into calcium about 90% of the time and into argon about 10% of the time.

second. This is like a lottery. The atom may "win" (and split apart) on the next tick of the clock or it may "play the lottery" over and over again without winning.

The average time for an atom to decay is called its **half-life.** For example, the isotope potassium-40 has a half-life of 1.28 billion years. If you were to start with 100 kg of these atoms, after 1.28 billion years, 50 kg (on average) of the "parent" atoms would have decayed into "daughter" atoms, which in this case are atoms of calcium and the gas argon (Figure 33.1). After another 1.28 billion years, half of the *remaining* potassium decays, leaving 25 kg of the original potassium. Note that there is no increase in the chances for the remaining potassium to decay—just as your odds for winning the lottery do not improve just because you have lost before.

After each half-life passes, another half of the remaining atoms decay. So after one half-life, one-half remains; after two half-lives, one-quarter remains; after three half-lives, one-eighth remains; and so forth. These fractions can also be written like this:

$$\frac{1}{2} = \frac{1}{2^1} \quad \frac{1}{4} = \frac{1}{2^2} \quad \frac{1}{8} = \frac{1}{2^3}.$$

With the passing of each half-life, the number of remaining parent atoms declines by another power of 2. This can be expressed by a mathematical formula. If the half-life is τ (the Greek letter tau), after a time t the fraction remaining will be

t = Time since last molten
τ = Half-life of element

$$Fraction = \left(\frac{1}{2}\right)^{t/\tau}.$$

For example, after three half-lives, $t = 3 \times \tau$, so the fraction is $(1/2)^3 = 1/8$, matching what we just found.

This equation also lets you calculate how much of the original isotope remains, for any portion of a half-life; so after 1.5 half-lives ($t = 1.5\tau$), for example, the fraction remaining is $(½)^{1.5} = 0.35$.

We can use this property to determine the age of a mineral that incorporates potassium within its structure. As the potassium-40 decays inside the mineral, an increasing amount of calcium and argon will be bound inside the mineral even though these atoms would not normally be included when the mineral formed. Therefore, the more daughter atoms a rock contains relative to the original radioactive atoms, the older the rock is. The more argon relative to potassium within a rock sample, the older the rock must be, as shown in Figure 33.1.

This method reveals the length of time since the rock was last molten. When the rock is liquid, the decay products (like calcium and argon), which are not normally part of the chemical structure of the rock, escape. This resets the "clock." By this method geologists find that the oldest rocks on Earth are about 4 billion years old. Such ancient rocks are found in many places—northern Canada, southern Africa, and Australia. Thus portions of the Earth's crust solidified for a final time 4 billion years ago, and the Earth itself must have formed sometime earlier. Small mineral inclusions in some rocks have been dated to 4.4 billion years. These are thought to be bits of rock that cooled even closer to the beginning of the Earth's history and became imbedded in other rocks without having been melted.

Even older rocks have been found among samples from the Moon and **meteorites,** which are small pieces of asteroids that have struck the Earth (Unit 48). These have ages up to about 4.5 billion years. Because the Moon and asteroids are small,

Q Why would a smaller body tend to cool more rapidly than a large body? Can you think of any examples of this in your own experience?

they are expected to have cooled off rapidly (Unit 35), and hence their ages are probably closer to the age of the Solar System. In addition, a variety of radioactive elements with different half-lives are incorporated in different kinds of minerals. All of them give consistent values for the age of the Solar System.

From such evidence, we assume the Earth also formed about 4.5 billion years ago. Moreover, we can deduce that most of its crust has been molten at one time or another since then. Thus we can add the following to our clues about the Solar System's formation:

For all bodies in the Solar System whose ages have so far been determined, the evidence indicates that they formed about 4.5 billion years ago.

Four and one-half billion years is an immense age. To illustrate, if those billions of years were compressed into a single year, all of human recorded history would have happened during just the last minute of the year. The brevity of human life compared to the vast age of the Earth prevents us from observing how truly dynamic our planet is. Mountains and seas appear to us to be permanent and unchanging, but in fact they change dramatically over the vast epochs of the Earth's existence. Such changes ultimately are caused by the heat in the Earth's interior, which creates motion in the Earth's interior and crust (Unit 35).

33.2 BIRTH OF THE SOLAR SYSTEM

Solar nebula theory

The strongest theory for the origin of the Solar System derives from a hypothesis proposed in the eighteenth century by Immanuel Kant, the great German philosopher, and Pierre-Simon Laplace, a French mathematician. Kant and Laplace independently proposed what is now called the **solar nebula theory:** The Solar System originated from a rotating, disk-shaped cloud (*nebula* is Latin for "cloud") of gas and dust, with the outer part of the disk becoming the planets and the center becoming the Sun. This theory offers a natural explanation for the flattened shape of the system and the common direction of motion of the planets around the Sun.

The modern form of the theory proposes that the Solar System was born 4.5 billion years ago from an **interstellar cloud,** an enormous rotating aggregate of gas and dust like the one shown in Figure 33.2. Such clouds are common between the stars in our Galaxy even today, and there is strong evidence today that all stars have formed from them. Thus, although our main concern here is with the birth of the Solar System, we should remember that our theory applies more broadly and implies that most stars could have planets, or at least surrounding disks of dust and gas from which planets might form (Unit 34).

FIGURE 33.2
Photograph of an interstellar cloud (the dark region at top), which may be similar to the one from which the Solar System formed. The dark cloud is known as Barnard 86. The star cluster is NGC 6520.

Interstellar clouds are the raw material of the Solar System, so we need to describe them more fully. Although such clouds are found in many shapes and sizes, the one that became our Sun and planets was probably a few light-years in diameter and contained about twice the present mass of the Sun. The cloud probably began with a composition similar to what we estimate from spectroscopy in the Sun's atmosphere today (Unit 24). About 71% of its mass was hydrogen gas and about 27% was helium gas, with the remaining 2% including all the other chemical elements. Some elements, such as iron, silicon, and oxygen, form small solid "dust" particles called **interstellar grains.**

Interstellar grains range in size from large molecules to micrometers or larger. They are thought to be made of a mixture of silicates (crystals of silicon and oxygen), iron compounds, carbon compounds, and water frozen into ice. Astronomers deduce the presence of these substances from their spectral lines, which are seen in starlight that has passed through dense dust clouds. Moreover, a few hardy interstellar dust grains, including tiny diamonds, have been found in ancient meteorites. This direct evidence from grains and the data from spectral lines shows that these elements occur in proportions similar to those we observe

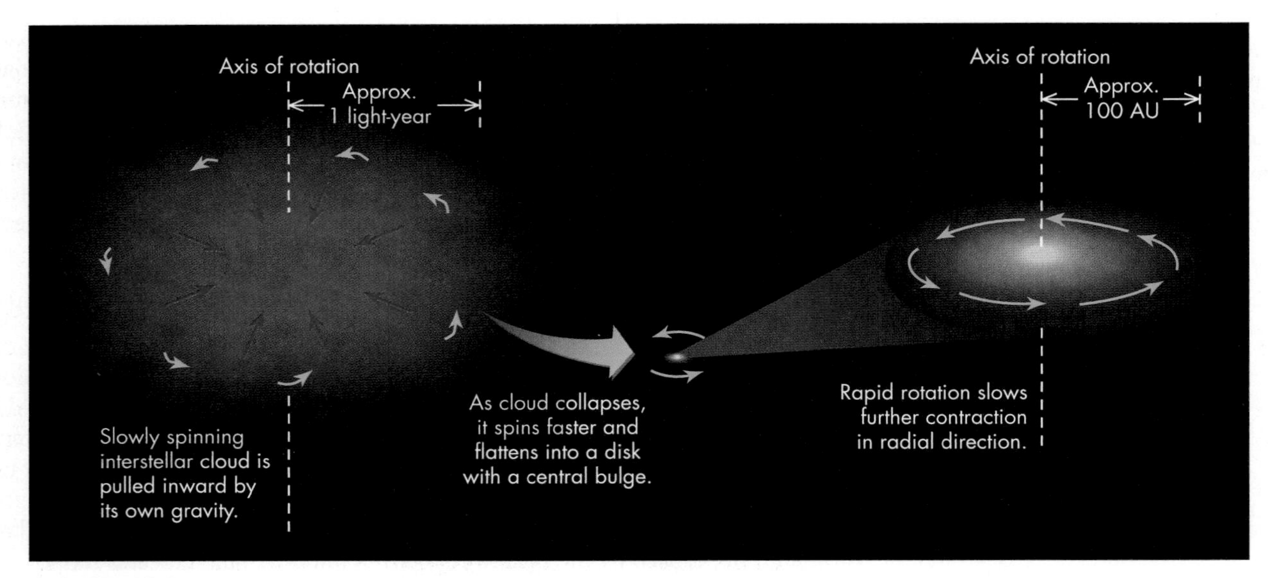

FIGURE 33.3
A sketch illustrating the collapse of an interstellar cloud to form a rapidly spinning disk. Note that the final size of the disk is not shown to scale—in actuality it would be about 60,000 times smaller than the cloud from which it formed!

in the Sun. This compositional similarity provides additional support for the view that the Sun and its planets formed from an interstellar cloud.

The cloud began its transformation into the Sun and planets when the gravitational attraction between the matter in the cloud caused it to collapse inward, as shown in Figure 33.3. Astronomers hypothesize that the collapse may have been triggered by a star exploding nearby or by a collision with another interstellar cloud. But regardless of its initial cause, the infall was not uniform around the cloud's center. Instead, because the cloud was rotating, it flattened into a spinning disk.

Flattening occurred because rotation retards the inward motion in the plane of rotation, perpendicular to the cloud's rotation axis. This is a result of a basic physical process—the conservation of angular momentum—which causes an object to spin faster if its radius becomes smaller (see Unit 20.2). The flattening caused by spinning happens in a pizza parlor, where the chef flattens the dough by tossing it into the air with a spin. Unlike the pizza dough, however, the inward pull of gravity made the cloud smaller even as it spun faster. It took a few million years for the cloud to collapse and become the **solar nebula.** Most of the matter collected in the center of the nebula, eventually forming the Sun. A fraction of the material was revolving too fast to merge into the center and formed the surrounding disk.

The whole solar nebula was probably about 200 AU in diameter and perhaps 10 AU thick. During the collapse phase, gravitational potential energy was converted into thermal energy, providing the initial heating even before the Sun became a star (see Unit 20.1). The inner parts of the disk were hot, heated by radiation from the young Sun and by the impact of gas falling onto the disk during its collapse. The outer parts of the nebula were cold, both because the radiation from the early Sun was weak and because collisions were slower in this region of weaker gravitational attraction. This set the stage for the formation of the planets.

33.3 FROM DUST GRAINS TO PLANETESIMALS

Condensation occurs when a gas cools and its molecules stick together to form liquid or solid particles. For condensation to happen, the gas must cool below a critical temperature, the value of which depends on the substance condensing and the surrounding pressure. For example, suppose we start with a cloud of vaporized iron at a temperature of 2000 K. If we cool the iron vapor to about 1400 K, tiny flakes

Technically, *condensation* is the change from gas to liquid, and *deposition* is the change from gas to solid; but it is common practice to use the term *condensation* to refer to both.

FIGURE 33.4
Water vapor cools as it leaves the kettle. The cooling makes the vapor condense into tiny liquid water droplets, which we see as the "steam."

Q Imagine that silicates condensed at 200 K and that the less dense water condensed at 1000 K. How would the structure of the Solar System and composition of the planets differ from what they are today?

Q What objects might have survived from the planetesimal stage? How would they be different today, compared to 4.5 billion years ago?

of iron will condense from it. Silicates condense at about the same temperature, so flakes of rocky material will condense if the hot gas contains silicon and oxygen.

At lower temperatures other substances will condense. For example, water can condense at room temperature, which you can see as steam escapes from a boiling kettle (Figure 33.4). Molecules in the hot water vapor come into contact with the cooler air of the room. As the vaporized water cools, its molecules move more slowly, so that when they collide, electrical forces can bind them together, first into pairs, then into small clumps, and eventually into the tiny droplets that make up the cloud of steam we see above the spout.

An important feature of condensation is that when a mixture of vaporized materials cools, the materials with the highest vaporization temperatures condense first. Thus as a mixture of gaseous iron, silicate, and water cools, it will make iron grit when its temperature reaches 1400 K; silicate grains form when the temperature drops below 1300 K; and when the temperature drops below about 180 K, water crystals form. This is a bit like putting a bowl of hot chicken soup in the freezer. First the fat condenses and freezes, then the broth.

However, the condensation process will not occur if the temperature never drops sufficiently. Therefore, in a region of the solar nebula where the temperature never cools below 500 K, water will not condense but iron and silicates will. By studying how different atoms and molecules would likely condense in the solar nebula, we can estimate an average condensation temperature for each element. These are given in Table 33.1 for the ten most common elements in the solar nebula. They are also indicated by small thermometer symbols for each element in the periodic table shown in the inside back cover of this book.

A sequence of condensation would have occurred in the solar nebula because the young Sun heated the inner part of the disk more than the outer parts (Figure 33.5). Iron and silicates, which condense even at high temperatures, could condense almost everywhere within the disk. However, out to nearly the orbit of Jupiter it was too hot for water to condense. Water, ammonia, methane, and other easily vaporized substances were present as gases in the inner solar nebula, but they could not form solid particles there. A small portion of these substances could still combine chemically with silicate grains, so that the rocky material from which the inner planets formed contained small quantities of water and other gases, as shown by the overall composition of the Earth given in Table 33.1.

The tiny particles that condensed from the nebula stuck together into bigger pieces in a process called **accretion.** The process of accretion is a bit like building a snowman. You begin with a handful of loose snowflakes and squeeze them together to make a snowball. Then you add more snow by rolling the ball on the ground. As the ball gets bigger, the larger surface area makes it easier for more snow to stick, and the ball rapidly grows.

Similarly, in the solar nebula tiny grains stuck together and formed bigger grains that grew into clumps. The details of this process are not entirely certain, but dust grains can often become electrically charged and then stick to each other because of electrical forces. This can happen as the grains collide with each other and with gas molecules—this is similar to the buildup of electrical charges on clothes as they tumble in a dryer, making socks stick to shirts. Subsequent collisions, if not too violent, allowed these smaller particles to grow into objects ranging in size from millimeters to kilometers. These larger objects are called **planetesimals**—small, planetlike bodies.

Because the planetesimals near the Sun formed primarily from silicate and iron particles, while those farther out were cold enough that they could incorporate ice and frozen gases, there were two main types of planetesimals: rocky iron ones near the Sun and icy rocky iron ones farther out (Figure 33.6). This, then, explains the second observation we described at the beginning of this unit—the change in composition between the inner and outer Solar System.

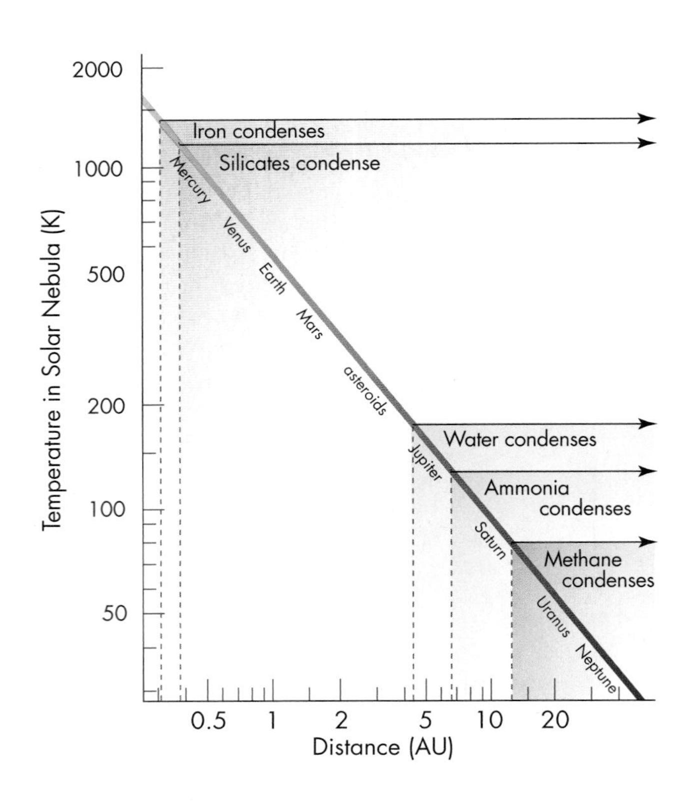

FIGURE 33.5
Graph of the how the temperature in the solar nebula approximately decreased as a function of distance from the Sun. The temperature and radius at which several major atoms and molecules condense is indicated in the plot. These substances remained gaseous at smaller distances from the Sun because the temperature was too high for them to condense.

TABLE 33.1 Condensation Temperatures of Major Elements

Element	Condensation Temperature*	Percentage by Mass in Sun	Percentage by Mass in Earth
Hydrogen	180 K (H_2O)	71.1%	0.0033%
Helium	3 K	27.4%	0.00000002%
Carbon	80 K (CH_4)	0.25%	0.045%
Nitrogen	130 K (NH_3)	0.08%	0.0004%
Oxygen	1300 K (silicates), 180 K (H_2O)	0.65%	30.1%
Neon	9 K	0.12%	0.0000000004%
Magnesium	1300 K (silicates)	0.07%	13.9%
Silicon	1300 K (silicates)	0.06%	15.1%
Sulfur	700 K (FeS)	0.04%	2.9%
Iron	1400 K	0.14%	32.1%

*Condensation temperatures depend on what other elements are present. For example, oxygen will combine with silicon to form silicates that condense at high temperature. However, oxygen is so much more abundant than silicon, that a large amount remains after the silicon is all condensed. Similarly hydrogen will condense out in water and other molecules, but most of the hydrogen remains as a gas after it combines with all the other elements that it can.

33.4 FORMATION OF THE PLANETS

As planetesimals grew, their masses became large enough that they developed significant gravitational attraction, pulling matter onto them more rapidly and pulling them into collisions with other planetesimals. Computer simulations indicate that some collisions would shatter both bodies, but less violent collisions led to merging into larger bodies.

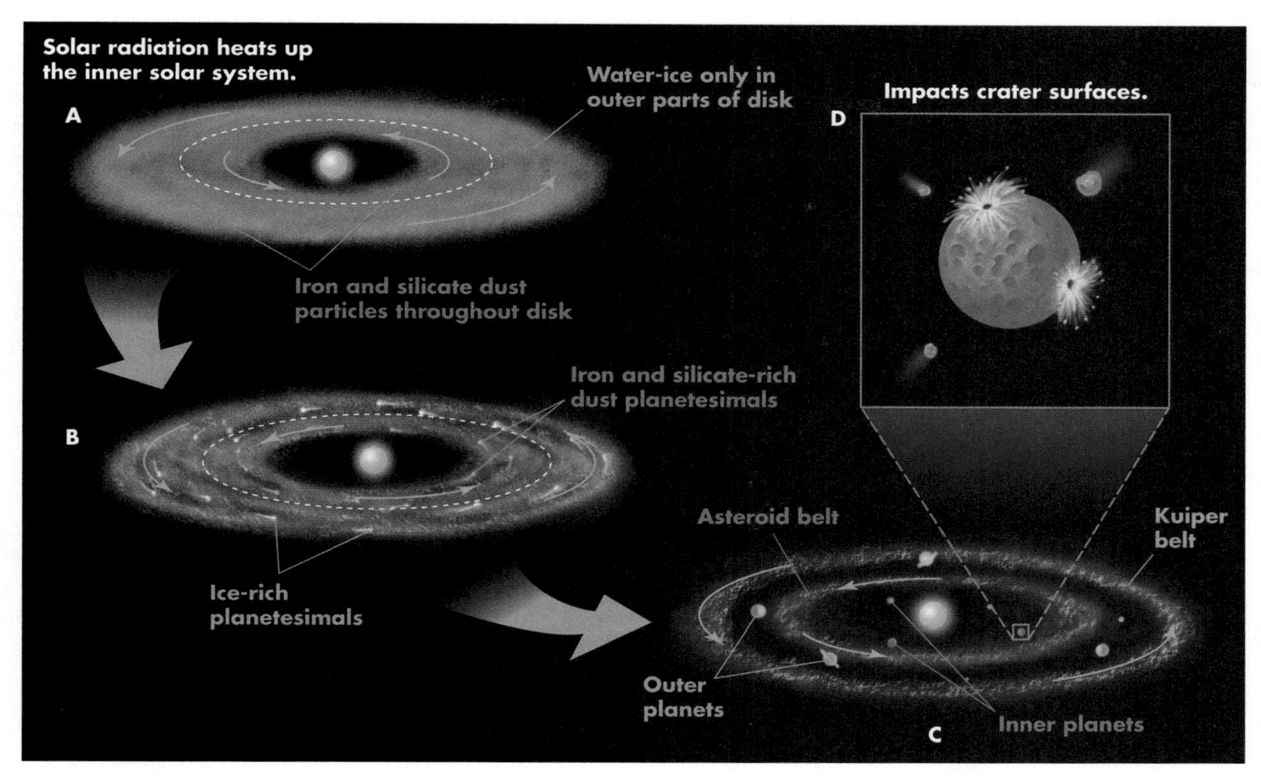

FIGURE 33.6
An artist's depiction of how the planets may have formed in the solar nebula.

Merging of the planetesimals increased their mass and thus their gravitational attraction. That, in turn, helped them grow even more massive by drawing planetesimals into clumps or rings around the Sun. Within these clumps, growth proceeded even faster, so that over about 100,000 years, larger and larger objects formed, as depicted in Figure 33.6 B and C. At the same time, their orbits would have steadily grown more nearly circular—averaging out the more eccentric orbits of the individual planetesimals that combined to make them.

Planet growth may have been especially rapid in the outer parts of the solar nebula. Planetesimals there had more material from which to grow, because ice was about 10 times more abundant than silicate and iron compounds. Additionally, once a planetesimal grew to several times the mass of the Earth, it was able to attract and retain gas by its own gravity. Because hydrogen and helium gas were overwhelmingly the most abundant materials in the solar nebula, planets large enough to tap that reservoir could grow vastly larger than those that formed from solid material only. Thus Jupiter, Saturn, Uranus, and Neptune may have begun as bodies of ice and rock, but their gravitational attraction resulted in their becoming surrounded by the huge envelopes of hydrogen-rich gases that we see today. The smaller, warmer bodies of the inner Solar System could not capture hydrogen and therefore remained small and lack that gas. This explains the third observation we mentioned at the beginning of this unit—that there are rocky planets near the Sun and massive, gaseous ones farther out.

It is not certain that planetesimals were a necessary stage in the formation of the Jovian planets. Using computer simulations to study that distant time, astronomers try to solve Newton's laws of motion for the complex mix of dust and gas that we think made up the solar nebula. The solutions reveal what might have happened in these early stages of planet formation. One of the more interesting findings of such calculations is that giant planets, particularly planets as large as Jupiter or Saturn, may have formed directly from slightly denser regions of gas in the disk. Far from the Sun, where the gas is cold, gravity can more easily overcome the resistance of warmer gas to being squeezed into a smaller region. This may have allowed gravity to pull gas together to make the giant planets without the need to first form cores from planetesimals. Both hypotheses are consistent with the large abundance of hydrogen in the Jovian planets and their dense cores. In this direct formation model, heavy elements captured from the nebular gas and dust would have sunk to the core later in the process of formation.

The moons of the Jovian planets probably were formed from planetesimals orbiting the growing planets. Once the body that would eventually become Jupiter grew massive enough, its gravitational force drew in additional material, and it would have become ringed with debris spinning in a disk in a miniature version of the solar nebula itself. Thus moon formation was a scaled-down version of planet formation, and so the satellites of the outer planets have the same regularities as the planets around the Sun. All four giant planets have flattened satellite systems in which the satellites (with few exceptions) orbit in the same direction (Unit 45) as the planet spins. This is true even of Uranus, with its oddly tipped axis, so its moons orbit almost perpendicular to the plane of the Solar System.

Many of the satellites of the gas giants are huge themselves. Two exceed Mercury's diameter and would certainly be considered full-fledged planets if they were orbiting the Sun on their own. A few of the larger moons even have atmospheres. However, their small size made their gravity too weak to draw in hydrogen and helium gas, so these moons are composed mainly of rock and ice. This gives them solid surfaces. These distant moons might in the future be ideal bases for studying the planets that have no surface to land on.

33.5 LATE-STAGE BOMBARDMENT

As planetesimals struck the growing planets, their impact heated them. You can demonstrate how impacts generate heat by hitting a small piece of metal repeatedly with a hammer and then feeling the metal. It will be warmer than it was before you began hitting it. Technically, therefore, impact heating is release of gravitational potential energy. Another way to think of this is as an application of the conservation of energy: Energy of motion becomes energy of heat.

Differentiation

For the forming Earth, the impacting objects acted as the hammer, and gravity was the force that drove them onto the young Earth. Imagine the vast heating created as mountain-size masses of material slammed into a planet at speeds far greater than the speed at which a bullet is shot from a gun (Figure 33.6D). In addition, many elements in the solar nebula were radioactive and would release heat as they decayed. The heat melted the planets and allowed matter with high density (such as iron) to sink to their cores, while matter with lower density (such as silicate rock) "floated" to their surfaces in a process called **differentiation** (Figure 33.7). You have seen differentiation at work if you have ever had the misfortune to melt a container of chocolate chip ice cream. When you open the box, you find that all the chips have sunk to the bottom and the air in the ice cream has risen to the top as foam. Besides redistributing material within a planet according to its density, differentiation contributed to the heating. As denser materials are pulled downward through less dense material, heat is generated by their friction.

Eventually, as the bombardment slowed, the surfaces of the solid bodies cooled and formed a crust; but the continuing fall of planetesimals blasted out huge craters, such as those we see on the Moon and on all other bodies in the Solar System with solid surfaces. Occasionally an impacting body was so large that it did more than simply leave a crater. For example, one hypothesis proposes that Earth's Moon was created when the young Earth was struck by a Mars-size planetesimal.

A

B

C

FIGURE 33.8
Sources of our atmosphere and oceans. (A) Volcanic gas vent today. Gas from ancient eruptions built some of our atmosphere. (B) Planetesimals collide with young Earth, melting its surface, which then releases gases. (C) Comets strike young Earth and vaporize.

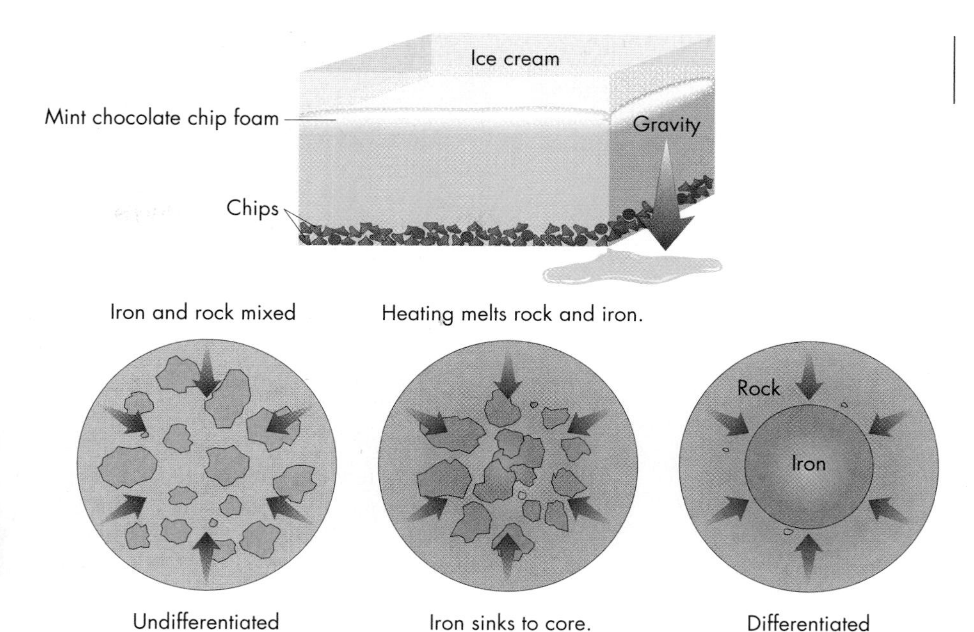

FIGURE 33.7
If a box of chocolate chip ice cream warms up, it will "differentiate" as the dense chocolate chips sink to the bottom of the box. Similarly, melting lets much of a young planet's iron sink to its core.

The extreme tilt of the rotation axis of Uranus may also have arisen from a planetesimal collision. In short, planets and satellites were brutally battered as the large planetesimals collided during the late stages of planet building.

Although planet building consumed most of the planetesimals, some survived to form small moons, the asteroids, and trans-Neptunian objects. It is thought that the asteroid belt consists of planetesimals and their fragments that were unable to assemble into a planet because of constant "stirring" by Jupiter's gravitational force. Jupiter's gravity (and that of the other giant planets) also disturbed the orbits of icy planetesimals in the outer Solar System, tossing some in toward the Sun and others outward in elongated orbits to form the swarm of comet nuclei of the Oort cloud. Those that remain in the disk out to about twice Neptune's orbit form the Kuiper belt (Unit 32).

Atmospheres were the last part of the planet-forming process for the terrestrial planets. Unlike the Jovian planets, the inner planets were not massive enough to capture gas from the solar nebula. As a result, they ended up deficient in hydrogen and helium and hydrogen compounds. Moreover, shortly after formation, their surfaces were so hot that they would probably have driven off most gases that might have been present.

Astronomers have proposed several hypotheses to explain how Earth's oceans and atmosphere formed. According to one hypothesis, the gases and liquids on Earth's surface were originally trapped inside the solid material that eventually became the Earth. When that material was heated—in association with volcanic activity (Figure 33.8A) or by the violent impact of asteroids hitting the surface of the young Earth (Figure 33.8B)—the gases escaped and formed our atmosphere.

According to another hypothesis, the gases and liquids were not originally part of the Earth but were brought here by comets after the Earth's surface cooled. Comets are made mostly of a mixture of frozen water and gases. When a comet strikes the Earth, the impact melts the ices and vaporizes the frozen gas. With enough impacts, comets could have delivered sufficient water and gas to form the oceans and atmosphere (Figure 33.8C).

ANIMATIONS

Origin of the atmosphere

A theory about the origin of atmospheres must also explain why Mercury and the Moon have essentially no atmosphere. This can be explained by their low masses, which has two important consequences. First, a small terrestrial planet has a lower escape velocity than a large one (Unit 18), so it cannot hold on to an atmosphere as effectively. Second, smaller planets have less volcanic activity, so there is less volcanic outgassing to start with. If there was no source of replenishment, all terrestrial planets would eventually lose their atmospheres. This is because the molecules in an atmosphere have a wide spread of speeds, and there are always a few moving fast enough in the upper atmosphere to escape. For the Earth, the rate of this loss is so small that it would require far longer than the age of the Earth to lose its atmosphere. For small worlds the loss is much quicker, unless the temperatures are very low, which makes the thermal motions slower. This explains why some moons in the outer Solar System have atmospheres even though their escape velocity is smaller than our Moon's.

Only a few hundred thousand to a million years were needed to assemble the planets from the solar nebula. This is long in the human time frame but only a fraction of a percent of the Solar System's 4.5 billion years. All of the objects within the Solar System are therefore about the same age—the fourth property of the Solar System, mentioned at the beginning of this unit. Of course, the rain of infalling planetesimals tapered off gradually over hundreds of millions of years, and continues to some degree to this day as comets and asteroids continue to collide with and become part of the planets. However, the major period of Solar System building was completed relatively quickly.

One process still had to occur before the Solar System became what we see today: The residual gas and dust between the planets were removed. Just as a finished house is swept clean of the debris of construction, so too was the Solar System. In the sweeping process, the Sun was probably the cosmic broom, with its intense heat driving a flow of tenuous gas outward from its atmosphere—a flow that continues today. As that flow impinged on the remnant gas and dust around the Sun, this debris was pushed away from the Sun to the fringes of the Solar System. Such gas flows are seen in most young stars, and the Sun was probably no exception.

KEY TERMS

accretion, 251
condensation, 250
differentiation, 254
half-life, 248
interstellar cloud, 249
interstellar grain, 249
isotope, 247
meteorite, 248
planetesimal, 251
radioactive decay, 247
solar nebula, 250
solar nebula theory, 249

QUESTIONS FOR REVIEW

1. How are radioactive elements in rocks used to estimate the age of the Solar System? Why aren't all rocks found to have the same age?
2. What is the solar nebula? What shape does it have, and why?
3. How do the different types of planets relate to the temperature at which different elements condense?
4. What is the difference between condensation and accretion? What are planetesimals?
5. What is differentiation? Why is it more likely to have occurred in the larger bodies in the Solar System?
6. What different hypotheses could explain why some terrestrial planets have atmospheres and others do not?

PROBLEMS

1. Use the data from Figure 33.1 to draw a graph showing the ratio of argon atoms to potassium-40 atoms that a rock will contain over a 5.12-billion year history. Draw a curve through the points. From your curve, estimate the age of a rock that

contains half as many daughter argon atoms as the potassium isotope from which they formed. Estimate the age of a rock that contains an equal amount of the two.

2. The common form of uranium (^{238}U) has a half-life of 4.5 billion years. It decays through a sequence of unstable isotopes to lead (^{206}Pb), releasing a number of helium nuclei in the process. (Radioactive decay is the source of most of the helium in Earth's atmosphere today.) Find the ratio of the number of ^{238}U to ^{206}Pb atoms you would expect to find for rocks that last melted 1, 2, 3, and 4 billion years ago.
3. Table 33.1 indicates that the element iron contributes 32.1% of the Earth's mass, oxygen 30.1%, and silicon 15.1%.
   a. Assume that all of the iron in the solar nebula in the Earth's vicinity was incorporated into the Earth. Use the data in Table 33.1 to calculate what percentage of the available silicon and oxygen were incorporated.
   b. What percentage of the other major elements were included in the Earth?
   c. Nearly all of the oxygen in the Earth is locked up in silicates. On average, how many oxygen atoms are there per silicon atom? (Note that silicon has an atomic mass of 28, while oxygen is 16.)
4. Suppose that the Jovian planets have the same overall composition as the Sun.
   a. From Table 33.1, 0.14% of a Jovian planet's mass would be iron. What fraction of their mass would be silicates, assuming that oxygen atoms combined with silicon in the ratio 3-to-1?
   b. What would be the mass of rock (silicates) and iron in Jupiter compared to Earth's mass?
   c. What would be the mass of rock and iron in Uranus compared to the Earth's mass? What terrestrial planet's mass is this closest to?
5. The conservation of angular momentum (Unit 20) says that a contracting gas cloud must spin faster as it gets smaller.
   a. Suppose the material orbiting today at 1 AU began 1 ly away. By what factor has its orbital radius changed?
   b. Material orbiting at 1 AU today (like the Earth) is moving at 30 km/sec. According to the conservation of angular momentum, how fast was its rotation speed when it was at a radius of 1 ly?
6. A "rule of thumb" in planetary science is that a planet can hold on to a gas for the age of the Solar System if the average speed of molecules in the gas is less than one-sixth the escape velocity of the planet (this takes into account the fact that molecules high in an atmosphere are influenced by a smaller pull from gravity). The average speed of a molecule depends on the gas temperature and the mass of the molecule according to the following equation:

$$V_{gas} = 157 \frac{\text{m}}{\text{sec}} \sqrt{\frac{\text{temperature}}{\text{molecule's mass}}}$$

   The molecular mass used in this formula is determined by adding up the atomic masses of its constituents, so hydrogen molecules (H_2) have mass 2, nitrogen molecules (N_2) have mass 28.
   a. What is the average speed of a hydrogen molecule in Earth's atmosphere if the gas temperature is 300 K? The Earth's escape velocity is 11,200 m/sec. Can the Earth hold on to a hydrogen atmosphere?
   b. What is the average speed of a nitrogen molecule in Earth's atmosphere? Can the Earth maintain a nitrogen atmosphere?
   c. What is the average speed of a hydrogen molecule on Jupiter, where the temperature is 160 K? Jupiter's escape velocity is 59,500 m/sec. Can Jupiter hold on to a hydrogen atmosphere?
   d. If Jupiter were placed at Mercury's distance from the Sun, where the temperature would rise to about 450 K, would it be able to hold on to its hydrogen atmosphere?

TEST YOURSELF

1. One explanation of why the planets near the Sun are composed mainly of rock and iron is that
   a. the Sun's magnetic field attracted all the iron in the young Solar System into the region around the Sun.
   b. the Sun is made mostly of iron. The gas ejected from its surface is therefore iron, so that when it cooled and condensed, it formed iron-rich planets near the Sun.
   c. the Sun's heat made it difficult for other substances such as ices and gases to condense near it.
   d. the statement is false. The planets nearest the Sun contain large amounts of hydrogen gas and subsurface water.
   e. the Sun's gravitational attraction pulled iron and other heavy material inward and allowed the lighter material to float outward.
2. Which of the following features of the Solar System does the solar nebula hypothesis explain?
   a. All the planets orbit the Sun in the same direction.
   b. All the planets move in orbits that lie in nearly the same plane.
   c. The planets nearest the Sun contain only small amounts of substances that condense at low temperatures.
   d. All the planets and the Sun, to the extent that we know, are the same age.
   e. All of the above.
3. The numerous craters we see on the solid surfaces of so many Solar System bodies are evidence that
   a. they were so hot in their youth that volcanoes were widespread.
   b. the Sun was so hot that it melted all these bodies and made them boil.
   c. these bodies were originally a mix of water and rock. As the young Sun heated up, the water boiled, creating hollow pockets in the rock.
   d. they were bombarded in their youth by many solid objects.
   e. all the planets were once part of a single, very large and volcanically active mass that subsequently broke into many smaller pieces.

unit

34 Other Planetary Systems

The solar nebula theory (Unit 33) explains how many of the features of the Solar System probably arose out of its formation process, and the theory predicts that planet formation is a normal part of star formation. Astronomers have long searched for planets orbiting stars other than the Sun. Their interest in **exoplanets,** as these distant worlds are called, is motivated not merely by the wish to detect other planets. Astronomers also hope that the study of such systems will help us better understand the Solar System.

Astronomers had long speculated about the likelihood of planets existing around other stars, but it was not until the 1990s that new technologies provided a wealth of data confirming their existence. By 2002, 100 exoplanets had been discovered, by 2008 more than 300 were known, and the number is increasing rapidly. Astronomers are also discovering young star systems that appear to be in their planet-forming phase. Our understanding of the nature of planetary systems is growing rapidly, and although evidence now suggests that planetary systems are common, it may be that ones like our own Solar System are uncommon.

34.1 YOUNG PLANETARY SYSTEMS

If we could find a star just beginning to form, then according to the solar nebula theory it should be surrounded by a disk of dust and gas. The gas and dust would extend out into interstellar space, but in the inner regions—tens of astronomical units across—the disk would be warmed by the young star and heated by collisions, making it easier to detect.

We can hunt for such systems in interstellar clouds in which we can identify young stars. This has become possible in recent years with high-resolution observations using the Hubble Space Telescope as well as ground-based telescopes employing adaptive optics to correct for the blurring effects of Earth's atmosphere (Unit 30). In addition, a variety of observations at longer wavelengths are able to detect infrared emission from the warm gas and dust. For example, Figure 34.1 shows pictures made with the Hubble Space Telescope of gas and dust disks near the Orion Nebula (see also Looking Up #6: Orion). The pictures show **protoplanetary disks:** disks of dark, dusty material orbiting young stars. The stars at the centers of these disks are less than 1 million years old—so young that they have not yet become hot enough to emit much visible light. The orbiting material has become concentrated enough that it blocks light from the Orion Nebula behind it.

As the gas and dust in a disk continue to condense, they may eventually form planets. Figure 34.2A, shows a ring of dusty material around a young star that is estimated to be about 10 million years old. The dust is heated by light from the star, and this false-color infrared image shows emission from the warmed particles at a

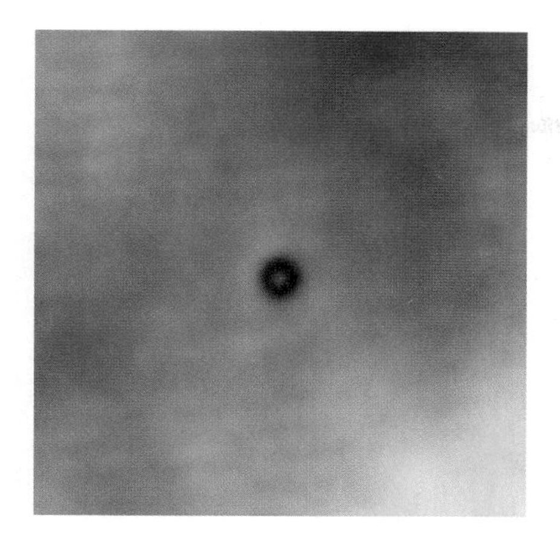
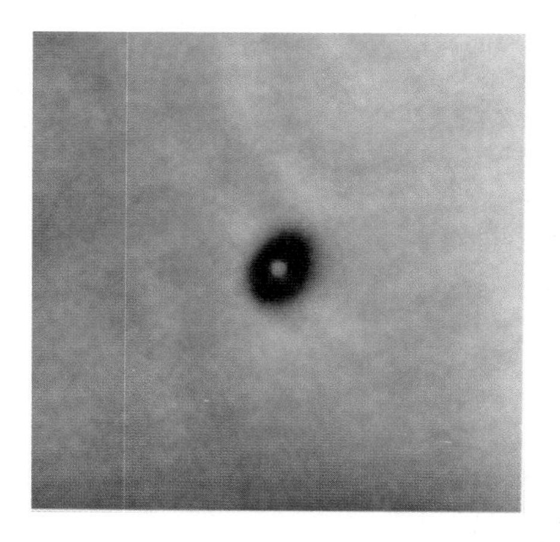

FIGURE 34.1
Two young stars in the Orion Nebula. The glowing red spot at the center of each image is a newly formed star, less than 1 million years old. The dark regions surrounding each star are disks of gas and dust. The dust in the disks is dense enough to block the background light coming from this large star-formation region. The two disks are about 150 AU and 290 AU in diameter.

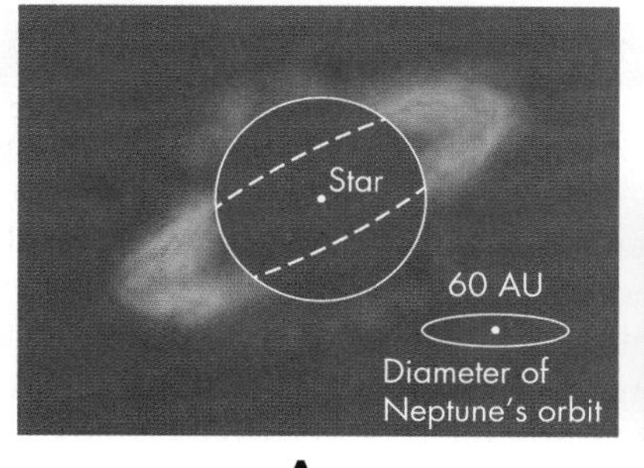

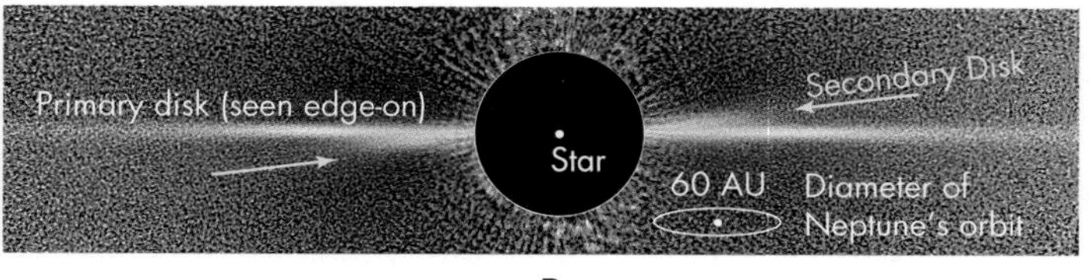

FIGURE 34.2
Disks around two young stars, both estimated to be about 10 million years old. The central region in both images was masked off to block out bright starlight from the central star. (A) Dust disk around the star HR 4796A is concentrated in a ring about the size of the Solar System's Kuiper belt. (B) The disk around Beta Pictoris is oriented almost precisely edge-on to us. A secondary disk seen in the image probably indicates the presence of a planet orbiting at a slight angle to the primary disk.

distance similar to that of the trans-Neptunian objects in the Solar System. Little or no dust remains nearer to the star than this ring. Indeed, observations of most young stars that are more than a few million years old rarely show any dust near the stars. This may indicate that planets have formed in the inner part of the system and have swept up most of the remaining dust. Figure 34.2B shows another protoplanetary disk around the star Beta Pictoris, which is also approximately 10 million years old. This disk is seen almost exactly edge-on, so it is not possible to directly see a hole in its center. However, the Hubble Space Telescope image shows that some of the dust is tilted, making a secondary disk. Models suggest that this tilted disk is probably caused by a planet orbiting at a slight angle to the rest of the disk—much as the planets in the Solar System do not all orbit in precisely the same plane.

Q What aspect of the image in Figure 34.2B indicates that the dust is in a ring? Is it possible that closer to the star there is dusty material that is too cold to emit much infrared light?

These images confirm that processes occur around young stars that are similar to the hypothesized solar nebula; but detecting the planets themselves requires a different approach. Once the scattered material is locked up into a body as small as a planet, it is difficult to image it directly.

34.2 DETECTING EXOPLANETS

Astronomers also refer to exoplanets as extrasolar planets.

Directly observing exoplanets is extremely difficult because the planets are so small compared with the stars they orbit, and they shine primarily by light reflected from their stars. Moreover, they are so far away that from our perspective they are separated from their stars by a very small angle, so they are drowned

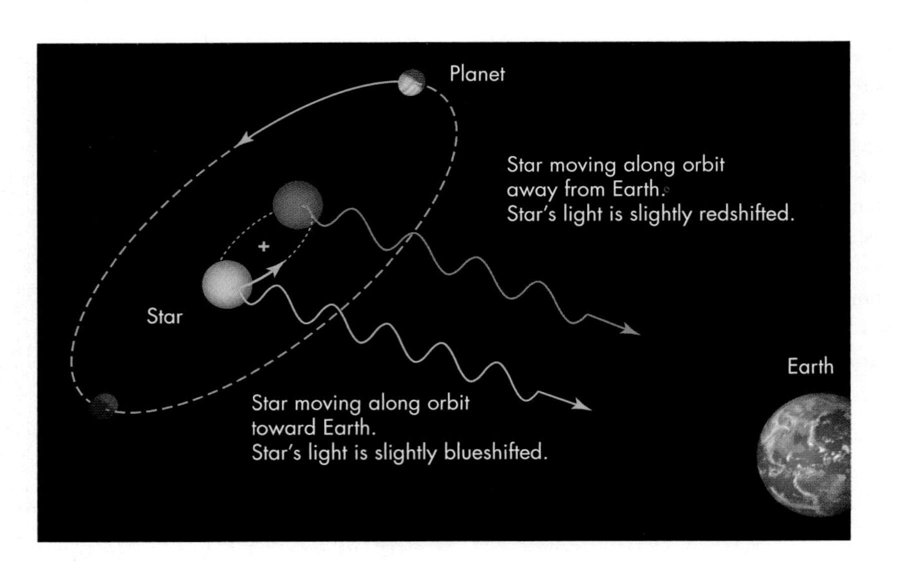

FIGURE 34.3
As an unseen planet orbits a star, the star's position "wobbles." This produces a changing Doppler shift. From the size and period of the wobble, the planet's mass and distance can be determined.

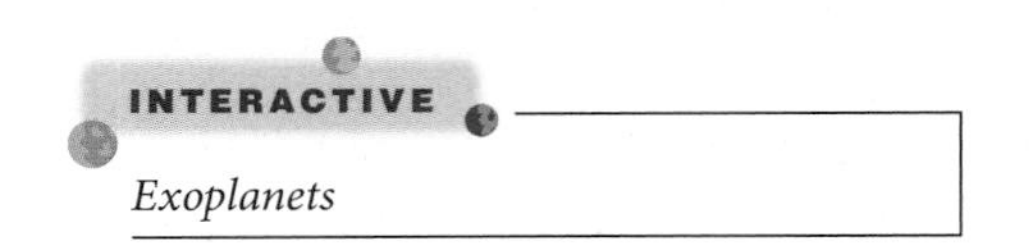

Exoplanets

out by the light of their stars. Therefore, astronomers have detected these planets mostly by their effects on the stars they orbit.

As a planet orbits its star, the planet exerts a gravitational force on the star as a result of Newton's law of action–reaction (Unit 15.3). That force makes the star's position wobble slightly, just as you wobble a little if you swing a heavy weight around you. This allows planets to be detected by the **Doppler shift method,** so named because the wobble creates a changing Doppler shift (Unit 25) in the star's light (Figure 34.3). From that shift, and its change over time, astronomers can deduce the planet's orbital period, mass, and distance from the star. Astronomers have discovered more than 300 exoplanets (as of late 2008) by this technique.

Figure 34.4 illustrates one of the more complex systems yet found, containing five planets orbiting the star 55 Cancri. As each planet orbits the star, it tugs the star with a regular repeating pattern; and after all of the planets have made several orbits, it is possible to reconstruct their masses and distances. This star is only a few percentage points more massive than the Sun, but clearly its planetary system is very unlike our own. Three of the planets—all giants by Solar System standards—orbit closer to their star than Mercury orbits the Sun. The closest takes only 2.8 days to orbit the star! The fourth planet out from the star has an orbit at roughly 1 AU, but its mass is more than twice Neptune's. At this distance, its surface temperature may not be far different from Earth's, but its orbit is so elliptical that its distance from 55 Cancri ranges from about Earth's distance from the Sun to less than Venus's distance from the Sun.

> **Q** How would our Solar System have differed if Jupiter were a brown dwarf instead of a gaseous planet?

Astronomers have detected exoplanets orbiting stars with masses ranging from less than 0.1 $M_{\odot}$ to more than 4 $M_{\odot}$. Most of the searches have focused on stars with masses about equal to the Sun's. Figure 34.5 shows the layouts of most of the planetary systems in which two or more exoplanets have been discovered so far. The figure indicates the estimated mass of each of these planets and compares them to the inner planets of the Solar System.

The most massive exoplanet discovered so far is about 17 times Jupiter's mass. More-massive objects have been detected, but these are not considered planets. Instead they are called **brown dwarfs**—objects that are too low in mass to fuse hydrogen like a true star but are so massive that they undergo nuclear fusion of some isotopes during their formation.

> **Q** Given its mass and orbit, what do you suppose the small exoplanet orbiting Gliese 581 might be like?

At the low-mass limit of present detection, all of the exoplanets found to date are much more massive than the Earth. The smallest object yet found (discovered in 2007) has about 1/60th of Jupiter's mass—about five times the Earth's mass. It

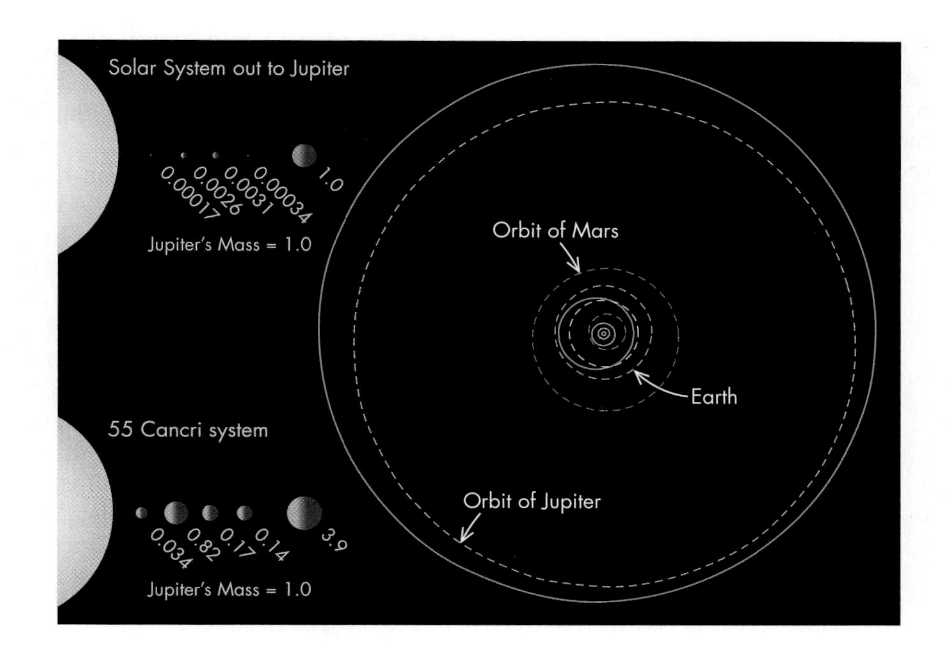

FIGURE 34.4
The 55 Cancri system contains five known planets around a star that is very similar to the Sun. The estimated masses (compared to Jupiter) for these planets and their orbits are compared with the five innermost planets in the Solar System. The figure also shows the approximate relative sizes of the planets. The fourth planet out in the 55 Cancri system orbits its star at about the same distance as the Earth from the Sun, but its mass is more than twice that of Neptune.

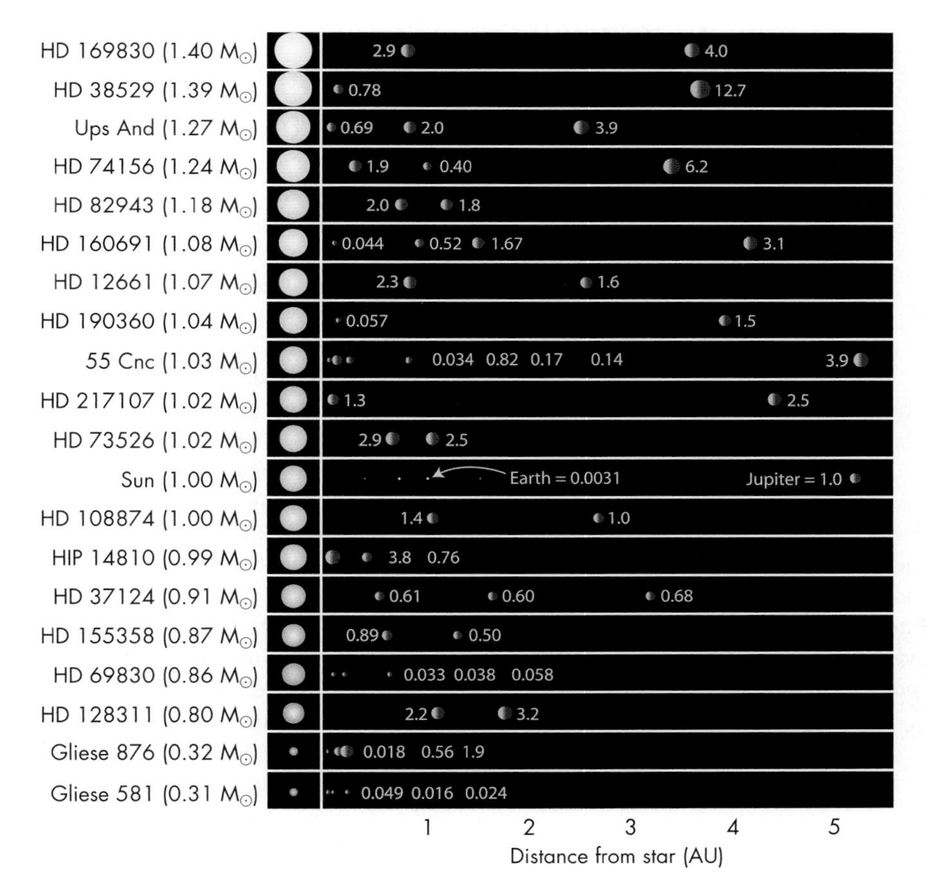

FIGURE 34.5
Comparison of the orbital radii and relative sizes of exoplanets with the Solar System. Most of the systems with two or more known exoplanets are shown, organized according to the mass of the star that they orbit. The sizes of the dots are based on the mass of each planet, and approximately indicate their true relative size. The numbers indicate the mass of each planet in units of Jupiter's mass.

is shown as the small dot second closest to the star Gliese 581 in Figure 34.5. It is not clear whether this object is a terrestrial planet or more like Uranus (which has a mass 14 times Earth's). The innermost exoplanet in the Gliese 581 system orbits only 0.059 AU from its star, taking only 5.4 days to complete a "year." Gliese 581 is much less luminous than the Sun, so even though the planet is so close to the star, it would not become as hot as you might expect. Astronomers estimate that it might be possible for liquid water to be present at this orbital distance. On the other hand, this planet has a mass similar to Uranus's, so it is not clear that it would be similar to terrestrial planets.

The detection of so many Jovian-mass planets in other star systems tells us that planet formation is common; but looking at Figure 34.5, you can see that none of these planetary systems look much like our own. A large majority of the planets found so far have masses comparable to Jupiter's or larger. However, unlike Jupiter, most of these objects orbit extremely close to their stars. In fact, over 40% of the exoplanets orbit closer to their stars than Mercury does to the Sun! Based on our understanding of planet formation in the Solar System (Unit 33), we would not expect gas giants to form so close to a star.

Q How do you suppose the spectral lines of the gases in a planet might be altered when the gases are at a high temperature?

Does this imply that our model for planet formation is wrong? Perhaps, but what we are seeing may be a result of how we search. Our current technology allows us to detect only a relatively large wobble in a star's position. Planets that exert a tug strong enough to be detected have to be massive and near their star. It is not yet possible to detect exoplanets as small as Earth, and it is very difficult to detect exoplanets as far away from their star as Jupiter is from the Sun. Therefore we would not have been able to detect other systems that *are* like the Solar System. As our technology improves, so will our ability to observe smaller wobbles, perhaps revealing many more kinds of planetary system configurations.

34.3 MIGRATING PLANETS

Finding gas giant planets so near to stars raises many questions. How can a gas giant form so close to a star? How can it survive there? Astronomers are pleased that the discovery of extrasolar planets confirms the general planet formation hypothesis, they are puzzled by how unlike our own planetary system most of these remote planetary families are. These new findings have led astronomers to consider whether processes might cause a planet's orbit to "migrate" to a new distance from its star.

Until recently, most astronomers had assumed that planets would remain in orbits close to where they formed. Computer simulations of the early stages of our Solar System suggest, however, that planets' orbits may shift as they interact with the material in the protoplanetary disk. These calculations show that frictional and gravitational interactions between the forming planets and leftover material in the disk of dust and gas can shift the planets' orbits either inward or outward. The young planet "eats" into the material like a cosmic Pac-man, being drawn in the direction where more material is present.

The interactions between a young planet and material in the protoplanetary disk mostly take place at a distance, and do not result in a merger. Material orbiting at a larger radius has a higher orbital angular momentum (Unit 20.3), so any interactions with it will result, on average, in an increase in the angular momentum of the planet's orbit.

According to the simulations, Neptune might have formed about 20 AU from the Sun and moved out to its present distance of about 30 AU as it consumed material in the outer Solar System. Protoplanetary material that did not merge with the forming planet may have undergone gravitational interactions that sent it sailing into the outer Solar System, forming the Kuiper belt. Other material was redirected into the inner Solar System, where perhaps it added water and other volatiles to the young terrestrial planets. Some of the simulations even suggest that there were other Uranus-mass objects in the Solar System that were ejected in close encounters with the migrating Neptune.

Migration of planets could have important consequences for planetary systems. For example, if a giant planet migrates inward toward its star, it would probably swallow smaller, Earth-size planets as it passes them, or send them careening into chaotic orbits. Thus small planets, suitable for life as we know it, may form but fail to survive in such systems.

Many of the massive exoplanets detected so far also have very elliptical orbits, rather than the nearly circular ones in our own system. Having a massive planet on a very elliptical orbit also does not bode well for the survival of Earth-like planets in these systems. As a massive planet sweeps into the inner portion of a star system, its strongly varying gravitational influence is likely to disturb the orbits of other planets, either ejecting them from the system or causing them to fall into their star.

Some evidence suggests this fate may have befallen planets in a few of these remote systems. A number of the stars with extrasolar planets are appreciably richer in iron than our Sun, which may be because they have swallowed Earth-like planets and vaporized them. The iron from the vaporized planet's core then enriches the star's atmosphere. This is not the only interpretation, however. It has been found that stars with large amounts of iron in their atmospheres are more likely to have planets than stars with iron levels similar to the Sun's. It may be that exoplanets form more easily from an interstellar cloud in which there is a greater fraction of heavy elements.

Of the star systems that have been investigated for planets, planets have been detected orbiting about 25% of the stars that have levels of iron in their atmosphere 3× higher than the Sun; planets have so far been detected orbiting only about 3% of the stars that have about the same level of iron in their atmosphere as the Sun.

All of the unusual features of other planetary systems suggest that planetary system formation can take a variety of paths. So far, the Solar System appears to be exceptional in its layout and in the stability of its planets' orbits. However, the biases of the Doppler effect technique for detecting planets need to be kept in mind. There may be many planetary systems more similar to our own, waiting to be discovered.

34.4 OTHER METHODS OF EXOPLANET DETECTION

Several other techniques for detecting exoplanets have been implemented in recent years, and these may help broaden our understanding of the planet formation process. One of the most exciting techniques—direct imaging—has succeeded in detecting four or five possible planets. This method has only worked so far for young, hot planets located far from their stars. For example, the first found is about 50 AU from its star (Figure 34.6), detected from its own infrared emission.

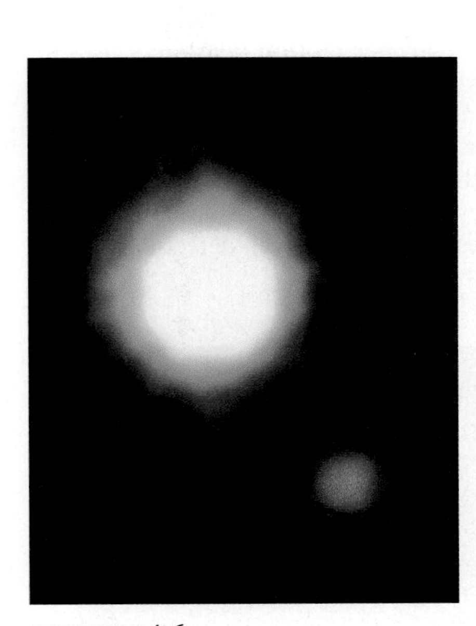

FIGURE 34.6
The first image of an exoplanet, made at infrared wavelengths with the 8-meter Very Large Telescope (VLT) in Chile using adaptive optics. The exoplanet, which is seen glowing red from its infrared emission, is 55 AU from "2MASSWJ1207334-393254," which is itself a brown dwarf.

Not all astronomers agree that the objects detected by direct imaging are planets. Because their orbits are so large, astronomers have not detected their orbital motion and therefore cannot directly measure their masses, so they may be brown dwarfs. In fact, the object shown in Figure 34.6 is itself orbiting a brown dwarf.

Another way of detecting exoplanets is called the **transit method,** in which a planet is detected as it passes in front of its star, temporarily dimming the star's light. This method also favors the detection of large planets close to the star because small planets block little light, and it is quite unlikely that a distant planet's orbit will line up precisely to pass in front of the star. This method has detected only about 10% as many exoplanets as the Doppler method to date, but it reveals much more information about the exoplanet.

One problem with the Doppler method is that it cannot give us the exact mass of the planet. This is because we do not know the angle of the exoplanets' orbits relative to our line of sight. For example, if a planet orbited perpendicular to our point of view, the star would wobble neither toward nor away from us, so we would not see a Doppler shift. As the angle becomes more nearly edge-on, the Doppler shift wobble becomes larger. We use the size of the wobble to estimate a planet's mass, so we may be underestimating the planet's mass. If an exoplanet transits in front of the star, however, we know for certain that the orbit is edge-on from our point of view.

Exoplanets that transit in front of the star also allow us determine their diameters. Because most are so near their stars, we might ask whether these are gas giants or some odd type of extremely massive terrestrial planet. An exoplanet orbiting the Sun-like star HD 209458 moves between us and the star every 3.5 days. At such times, the planet blocks a fraction of the star's light. From the amount of dimming, astronomers can deduce that the diameter of the planet is about 1.3 times the diameter of Jupiter. The planet's mass is estimated to be slightly less than Jupiter's mass, so its density implies that the planet is in fact a gas giant.

When a planet passes in front of its star, astronomers also have an opportunity to study the planet in even greater detail. Some of the star's light passes through the planet's atmosphere, so that gas in the planet's atmosphere imprints detectable absorption lines on the star's spectrum. In 2001 the Hubble Space Telescope detected the element sodium in the atmosphere of HD 209458 at a level consistent with Jupiter's atmospheric composition.

Astronomers have also detected oxygen and water vapor in HD 209458. Substances important for life as we know it. However, remember the extremely short orbital period of 3.5 days for this exoplanet. Using Newton's version of Kepler's third law (Unit 17), astronomers deduce that the planet orbits a mere 0.045 AU from its star—roughly one-eighth the distance at which Mercury orbits our Sun. Analysis of the spectral lines indicates that the atmosphere is very hot—about 10,000 K (18,000°F). The gases seen are "boiling" off a planet that is almost certainly uninhabitable.

Another technique for detecting exoplanets uses an unusual property of gravity: Gravity can bend the path of light. If we observe a distant star, and a nearer star passes in front of it, the light from the background star can be focused in our direction, making the background star appear to brighten (Figure 34.7). If a planet orbits the nearer star, the planet can produce a smaller brightening event. The gravitational force acts as a lens to focus the light of the distant star, so this is known as the **gravitational lensing method.** So far this method has detected only a few exoplanets, but it has the potential for detecting very low mass objects.

A future method for detecting exoplanets involves detecting the sideways "wobble" of the central star. This is called the **proper motion method;** it uses the same effect studied with the Doppler shift method, but it observes the sideways shift in the sky. The sideways shift is larger, the farther away a planet is from the star, so more-distant planets produce larger shifts. Because orbits this far out take a long time to complete, though, the method will require many years of high-precision position measurements to yield results. Spacecraft missions are currently being planned that could make measurements that might even detect Earth-mass planets at Earth-like distances around nearby stars.

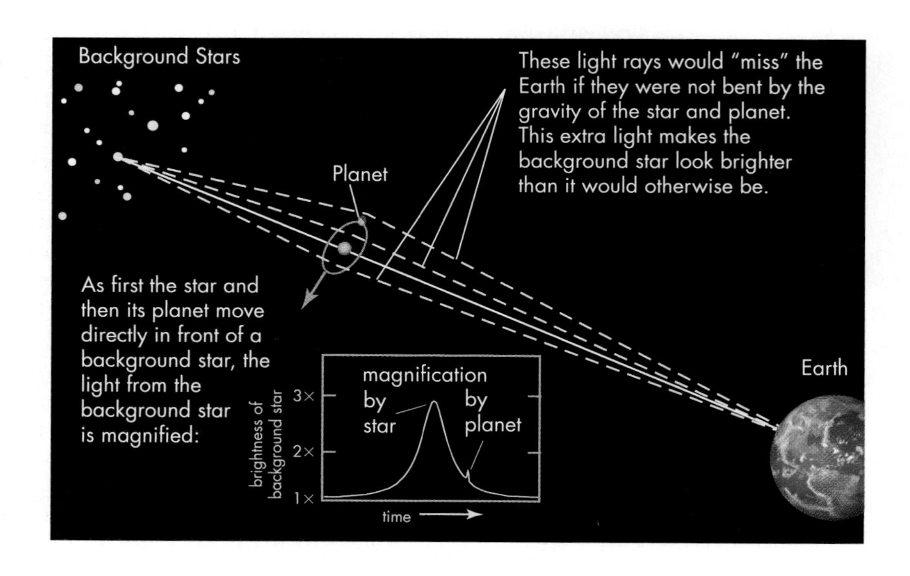

FIGURE 34.7
Detecting an exoplanet by gravitational lensing. When one star passes precisely in front of another, the nearer star can bend the light from the background star. This can focus the light of the background star, making it look brighter. A planet in orbit around the nearer star can cause a smaller brightening event.

KEY TERMS

brown dwarf, 260
Doppler shift method, 260
exoplanet, 258
gravitational lensing method, 264
proper motion method, 264
protoplanetary disk, 258
transit method, 263

QUESTIONS FOR REVIEW

1. What observations indicate that planets are currently forming around other stars?
2. Why is it so difficult to directly "see" a planet orbiting another star? Can you think of an analogy using objects here on Earth?
3. What different methods have astronomers used to detect planets outside the Solar System?
4. How does Newton's third law apply to our detection of exoplanets? Why is it easier to detect Jupiter-like planets than Earth-like planets?
5. What aspects of the exoplanets discovered so far cause us to question whether our Solar System is "typical" or not?
6. How can a planet "migrate" in its orbit?

PROBLEMS

1. Suppose a Jupiter-size exoplanet (radius 71,500 km) passed in front of a Sun-size star (radius 696,000 km). What percentage of the star's light would be blocked by the exoplanet?
2. The Doppler method for finding planets can be used to see how fast the Sun moves in response to a planet orbiting it.
   a. Using Kepler's third law ($P^2 = a^3$, for P in years and a in AU), find the period of the orbit and its circumference if the semimajor axis is 0.05 AU, 0.5 AU, or 5 AU.
   b. Calculate how fast a planet would be orbiting the Sun in meters per second at each of these three distances.
   c. The Sun also "orbits" the point of balance between the Sun and the planet (although this point may be inside the Sun). According to Newton's third law, the Sun's speed around this point must be smaller than an orbiting planet's in proportion to the ratio of their masses. What is the Sun's speed in each case above if the orbiting planet has Jupiter's mass? the Earth's mass?
3. When planets orbit a star, both orbit their common "center of mass" (Unit 17), which lies at a point between them closer to the center of the more massive object in proportion to the ratio of their masses. (If the planet is $1/10^{th}$ as massive as the star, the point will be 10× farther from the planet than from the star.)
   a. By how many AU does the Sun shift in each of the cases in the previous problem?
   b. If viewed from 30 light-years away, how large will its angular shift be? (Refer to the angular size formula in Unit 10.)
4. Gliese 581 is a star with a mass 31% of the Sun's mass, and it generates only about 1.3% as much light as the Sun.
   a. Use Newton's version of Kepler's third law to show that the nearest planet, which takes 5.37 days to orbit the star, must orbit at a distance of 0.041 AU.
   b. Compare the brightness of light received from Gliese 581 at the distance of 0.041 AU to the brightness of light received at Earth from the Sun. (Use the inverse-square law—Unit 21.2.)
5. Astronomers observe planets orbiting stars with a variety of masses. If the mass of the star is smaller, its force on a planet will be smaller, so the planet orbits more slowly. On the other hand, the star will be farther from the center of mass because the ratio of masses is more nearly equal, so its orbital speed will be faster. It is not immediately apparent which will have the greater effect. The goal of this problem is to show that it is easier to detect the same mass planet around a less massive star using the Doppler method.
   a. Use the technique of problem 1 to find the speed of a 1 solar-mass star in response to a Jupiter-mass planet in orbit at 1 AU.
   b. Use Newton's version of Kepler's third law (Unit 17.2) to find the period of a Jupiter-mass planet orbiting at 1 AU around a 0.25-solar-mass star. (Note that the planet's mass is still very small by comparison to the star.)
   c. What is the speed of the planet and 0.5-solar-mass star above?

TEST YOURSELF

1. How have astronomers detected most planets orbiting around other stars?
   a. Using the latest generation of telescopes, the images have high enough resolution to see the planets.
   b. The planets cause the stars to wobble, which can be detected from shifts in the wavelengths of light from the star.
   c. The planets generate X-rays that can be seen from telescopes in space.
   d. As the planets spin, they generate radio waves that can be detected with radio telescopes.
   e. All of the above.
2. Given the characteristics of exoplanets discovered so far, what are the most likely conclusions we might draw about how the solar nebula theory should be modified for the formation of other planetary systems?
   a. Terrestrial planets need not be composed of rock and iron.
   b. Massive planets usually are not composed of light elements.
   c. Not all interstellar clouds spin fast enough to form planets.
   d. Planets' orbits may change size within the nebula.
   e. Small planets do not form in most planetary systems.
3. The Doppler shift method for detecting the presence of exoplanets is best able to detect
   a. massive planets near the star.
   b. massive planets far from the star.
   c. low-mass planets near the star.
   d. low-mass planets far from the star.

unit

35

The Earth as a Terrestrial Planet

In the simplest terms, Earth (Figure 35.1) is a huge ball of rock and metal. With a diameter of 12,742 kilometers (7918 miles) and a mass of 5.97×10^{24} kilograms, it is the largest of the "rocky" bodies that inhabit the inner part of the Solar System. It is followed in size by Venus, Mars, Mercury, the Moon, the asteroid Ceres, and then the other asteroids.

Observations of the terrestrial planets and the Moon have been carried out for centuries, revealing many similarities and many differences between terrestrial worlds of different sizes and at different distances from the Sun. Of course, the Earth is the best-studied planet, so we begin our exploration of the planets with an astronomer's view of our home planet.

Background Pathways

Unit 32: The Structure of the Solar System 238

35.1 COMPOSITION OF THE EARTH

Although we have called the Earth a ball of rock and metal, we should be more precise. Rocks are composed of minerals, and minerals in turn are composed of chemical elements. Analysis of the surface rocks of the Earth shows that the two most common elements in them are oxygen and silicon, which usually combine to make types of rock called **silicates.** For example, ordinary sand (particles of the mineral quartz) is nearly pure silicon dioxide (SiO_2). Table 35.1 lists the most abundant elements in the Earth's crust. Many of these other elements combine with silicon and oxygen to form other types of silicates.

Although we can directly sample the surface composition of our planet, the composition of its interior is much more difficult to determine. We can deduce what lies inside Earth in two ways: by studying earthquake waves and by analyzing the Earth's density. The density of the Earth is 5.5 kilograms per liter—an average over the whole planet. This is greater than the density of the crust, which is composed primarily of silicates with an average density of about 3 kilograms per liter. We can conclude that the interior parts of the Earth must have a density greater than 5.5 kilograms per liter, which, averaged together with the crust, gives the Earth's overall measured density. That by itself does not tell us what lies inside the Earth; but if we examine the common crustal elements in Table 35.1, the only one with a high density is iron, which has a density of 7.9 kilograms per liter, making it a likely candidate. More detailed analysis requires that we examine the Earth's interior more directly.

If we ask how the Earth's core can be studied, your first reaction might be to say, "Why not drill a very deep hole and take a look or pull out samples?" However, the deepest hole yet drilled into the Earth penetrates only 12 kilometers, a mere scratch when compared to the Earth's 6380-kilometer radius. If the Earth were an apple, the deepest drilled holes would not have broken the apple's skin. Thus to study the Earth's core, we must rely on indirect means such as earthquakes.

FIGURE 35.1
The solid Earth. A reconstructed image, with the atmosphere and oceans removed.

The mass and number columns are different because different atoms have different masses. A low-mass element like hydrogen contributes many atoms but little overall mass to crustal rocks.

Oxygen and hydrogen in crustal rocks are not gases, but are chemically bound to other elements there to form minerals.

TABLE 35.1 Composition of the Earth's Crust

Chemical Element (Symbol)	Percentage of Element in Crust, by Mass	Percentage of Element in Crust, by Number
Oxygen (O)	46%	60%
Silicon (Si)	28%	21%
Aluminum (Al)	8%	6%
Iron (Fe)	6%	2%
Calcium (Ca)	4%	2%
Magnesium (Mg)	2%	2%
Sodium (Na)	2%	2%
Potassium (K)	2%	1%
Titanium (Ti)	0.6%	0.2%
Hydrogen (H)	0.1%	3%
Others (combined)	<1%	<1%

When earthquakes shake and shatter rock, they generate **seismic waves** that travel outward from the location of the quake through the body of the Earth. The speed of the waves depends on the properties of the material through which they move. We can measure that speed by carefully timing the arrival of seismic waves at places far from the quake that created them. Analysis of the speed along different paths then reveals to scientists a picture of the Earth's interior. Thus seismic waves allow us to "see" inside the Earth much as doctors use ultrasound waves to "see" inside our bodies (Figure 35.2). You can use a similar, though obviously much cruder, technique to locate wall studs by thumping areas of a plaster wall with your knuckle and listening for differences in the character of the sound waves.

Seismic waves in the Earth are of two main types: **P waves** and **S waves.** P waves form as matter in one place vibrates against the adjacent matter ahead of it, producing compression and decompression in an oscillating fashion (Figure 35.3A). P waves are like sound waves, and they travel easily through both solids and liquids. By contrast, S waves form as matter is shaken from side to side, like a wriggle in a shaken rope or a long spring (Figure 35.3B). In a liquid, material easily slips past adjacent matter, and S waves do not propagate. Thus S waves can travel only through solids.

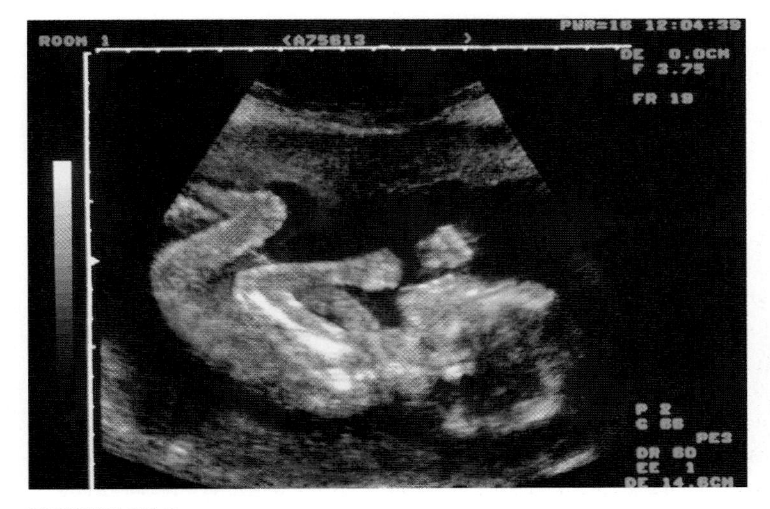

FIGURE 35.2
A sonogram allows a doctor to "see" inside a patient. Here an unborn fetus can be "seen" by reconstructing how sound waves reflect off surfaces within the mother's abdomen.

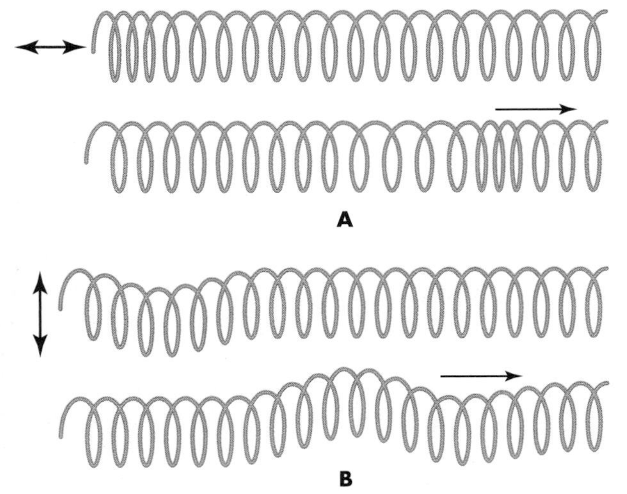

FIGURE 35.3
(A) A P wave can be made when you push and pull a spring, compressing and stretching it. (B) An S wave can be made in a spring when you shake it side to side.

P waves are also known as primary or pressure waves. S waves are also known as secondary or shear waves.

Seismic waves

Because of this difference in wave behavior, if a laboratory on the far side of the Earth from an earthquake detects P waves but no S waves, the seismic waves must have encountered a region of liquid on their way from the earthquake to the detecting station (Figure 35.4). Observations show precisely this effect, from which we infer that the Earth has a liquid core. From the speed of the waves we can also estimate the density and composition of the materials that the waves pass through.

Analysis of the seismic waves from many quakes indicates that the Earth's interior has four distinct regions. Figure 35.5 illustrates these different layers and their relative sizes. The surface layer is a solid, low-density **crust** about 20 to 70 kilometers (12 to 43 miles) thick and composed of rocks that are mainly silicates. Beneath the crust is a region of hot, but not quite molten, rock called the **mantle.** This region displays properties indicating that it may be composed largely of the mineral olivine, an iron-magnesium silicate. This mineral gets its name from its olive color. Pieces of it are sometimes carried to the surface in the lava that erupts from volcanoes (Figure 35.6). The mantle extends roughly halfway to the Earth's center and, despite not being liquid, is capable of slow motion when stressed, much as a wax candle can be bent by steady pressure when it is warm. Beneath the mantle is a region of dense liquid material, probably a mixture of iron,

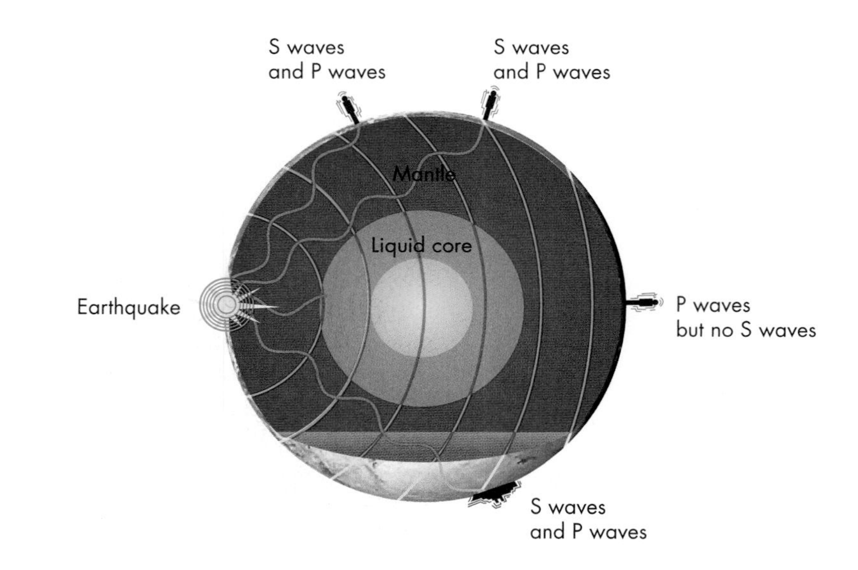

FIGURE 35.4
P and S waves move through the Earth, but the S waves cannot travel through the liquid core.

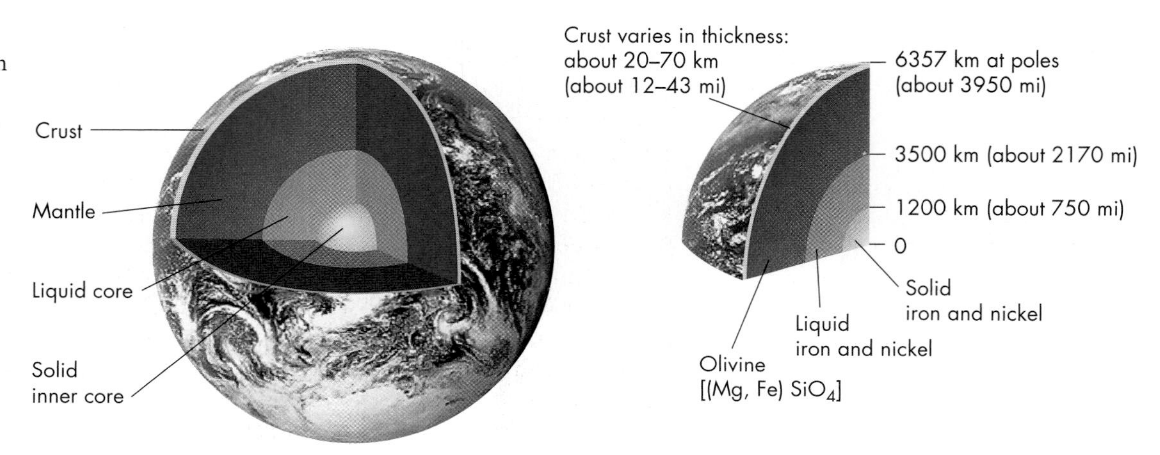

FIGURE 35.5
An artist's reconstruction of the Earth's interior.

FIGURE 35.6
Olivine (the greenish crystals) in a lava sample.

nickel, and perhaps sulfur, called the liquid or **outer core.** At the very center is a solid or **inner core,** probably also composed of iron and nickel.

You can see from this discussion that the Earth's interior structure is layered so that the densest material (the iron and nickel) is at the center and the lower-density, lighter materials (silicates) are near the surface in the crust and mantle. This separation of materials by density probably occurred while the Earth was forming and still in a molten state—high-density materials sinking while low-density materials floated to the surface—in a process of differentiation (Unit 33.5).

You may be puzzled over why there is a solid core inside a liquid core at the Earth's center. If it is hot enough to melt part of the interior of the Earth, why is the very center not liquid as well? The solid core is not cooler; rather, it has a higher melting temperature because it is under greater pressure. At very high pressures, a previously melted material may solidify. You can understand why this happens in the following way. For a solid to form, the atoms composing it must be able to link up to their neighbors to form rigid bonds. Heating makes the atoms move faster, breaking their bonds, so the material becomes liquid. However, if the material is highly compressed, the atoms may be forced so close together that, despite the high temperature, bonds to neighbors may hold and keep the substance solid.

The compression needed to solidify the Earth's inner core comes from the weight of the overlying material. The thousands of kilometers of rock above the Earth's deep interior generate enormous pressure there, which squeezes what would otherwise be molten iron into a solid.

35.2 HEATING OF THE EARTH'S CORE

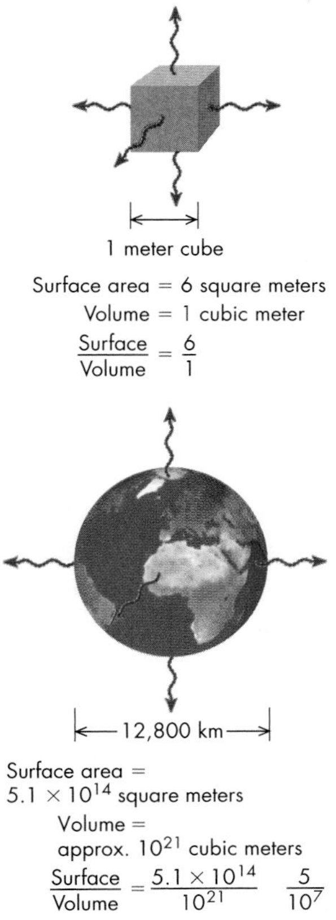

1 meter cube

Surface area = 6 square meters
Volume = 1 cubic meter
$\frac{\text{Surface}}{\text{Volume}} = \frac{6}{1}$

12,800 km

Surface area = 5.1×10^{14} square meters
Volume = approx. 10^{21} cubic meters
$\frac{\text{Surface}}{\text{Volume}} = \frac{5.1 \times 10^{14}}{10^{21}} \quad \frac{5}{10^7}$

FIGURE 35.7
Heat readily escapes from small rocks but is retained in larger bodies.

The Earth's interior is much hotter than its surface (a fact that figures in folklore and theology). Anyone who has seen a volcano erupt can hardly doubt this. As you descend from the surface, the temperature rises about 2 K with every 100 meters in depth. If this temperature increase continued at the same rate all the way to the center, the Earth's core would be 120,000 K! However, by measuring the amount of heat escaping from the deep interior, and from laboratory studies of the properties of heated rock, geologists estimate that the temperature in the core of the Earth is "only" about 6500 K—easily able to melt rock and iron.

What makes our planet's heart so hot? Most scientists think that the Earth formed hot. As we discussed in Unit 33, the Earth formed from many smaller bodies drawn together by their mutual gravity. The impact of these colliding and accreting bodies generated heat. After the bombardment stopped, the Earth's surface cooled, but its interior has remained hot. The Earth has behaved much like a baked potato taken from the oven, cooling on the outside but remaining hot inside because heat leaks only slowly from its interior to its surface. The rate of this heat loss depends on the total surface area exposed to the cold of space. Compared to its total volume of rock, a smaller object has proportionally more surface area through which it can lose heat, as Figure 35.7 illustrates. The result is that larger bodies cool more slowly than smaller bodies. *Thus we expect that larger objects (such as the Earth) will have a hotter interior than smaller objects (such as the Moon).* Another way of thinking about this is that the core of a smaller object has a thinner outer layer of rock to blanket it from the cold of space, so it cools faster.

The Earth has also been heated since its birth by natural radioactivity in the Earth's interior. Thus the Earth generates heat much as a nuclear reactor does. All rock contains trace amounts of naturally occurring radioactive elements such as uranium. A **radioactive element** is one that naturally breaks down into lighter elements in a process called **radioactive decay.** For example, the element uranium can spontaneously split apart into thorium and helium, releasing energy that generates heat. The thorium then splits, releasing more heat, and a sequence of such decays continues releasing heat until a stable isotope of lead is reached. The heat so created in a small piece of rock at the Earth's surface escapes into space, and so the rock's temperature does not increase appreciably. However, in the Earth's deep interior, the heat is trapped by the outer rocky layers, slowing its escape. With the heat trapped, the temperature of the Earth's interior gradually rises, and the rock eventually melts. Scientists are unsure whether radioactivity or impacts during formation had the greater effect, but both are thought to have made important contributions to the Earth's internal temperature.

35.3 THE EARTH'S DYNAMIC SURFACE

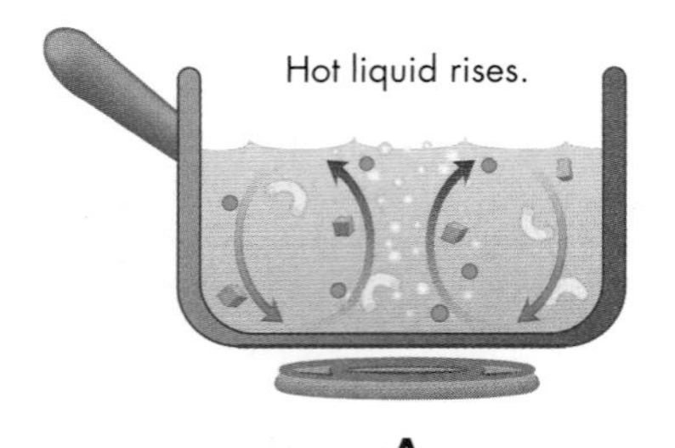

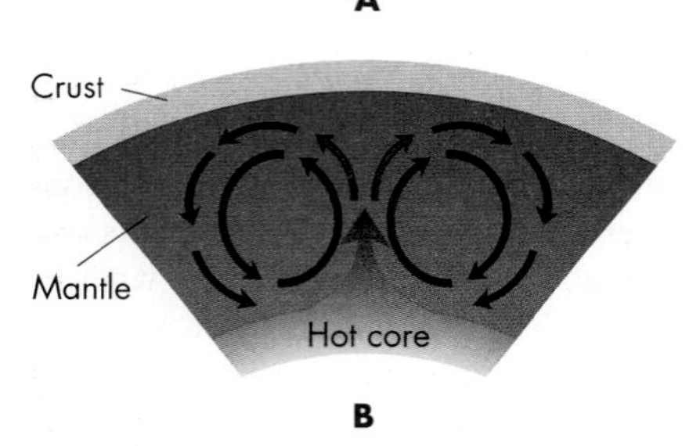

FIGURE 35.8
(A) The soup in a pan on the stove undergoes convection. Hot soup near the burner rises to the surface, where it cools and then sinks. (B) Heat from the Earth's hot core drives slow convection currents in the mantle.

As heat works its way out from the Earth's core, it drives movement of the rock in the Earth's interior. Heating often causes motion, as you can see by watching a pan of soup on a hot stove. If you look into the pan as it heats, you will see some of the soup, usually right over the burner, slowly rising from the bottom to the top, while some will be sinking again (Figure 35.8A). Such circulating movement of a heated liquid or gas is called **convection.** Convection occurs because heated matter expands and becomes slightly less dense than the cooler material around it. Being less dense ("lighter"), it rises—the principle on which a hot-air balloon operates. As the hotter material flows upward, it carries heat along with it. In this way convection not only causes motion but also carries heat.

Convective motions in a pot of soup are easy to see, but our planet's crust and mantle are not moving so rapidly, and the motions are hidden below many kilometers of rock. Nevertheless, when the rock in the Earth's interior is heated, it too may develop convective motions, but they are very, very slow. It takes hundreds of millions of years for rock to circulate through the mantle, and therefore it appears nearly motionless over the history of human observations.

Deep in the Earth's interior, rock at the base of the mantle is heated and molten. Just like heated soup, the heated rock expands, becoming less dense and slightly buoyant. This pushes it upward relative to the surrounding denser rock in great, slow plumes. When such a plume nears the crust, it turns and flows parallel to the surface below the crust (Figure 35.8B). Gradually, the hot material loses its heat to the outside and then cools, grows more dense, and sinks downward again to be rewarmed and rise once more. As the mantle material circulates below the crust, it drags the surface layers with it, shaping the Earth's surface in a process called **plate tectonics.**

The term *plate* is used because the pieces of the Earth's crust that move are very thin (about 50 kilometers deep) but many thousands of kilometers across. Plate motion is a little like the dry crust that forms on the top of a pan of boiling oatmeal if you do not stir it regularly. As heat generates convection in the liquid underneath, pieces of the crusty surface are pushed around by the motion underneath. Similarly, the Earth's crustal plates move in response to the motion of the mantle.

What evidence do we have that the plates are moving? Plate motion is only centimeters per year—about the rate at which fingernails grow. However, by using the global positioning system (GPS), scientists can measure the distance between continents to an accuracy of millimeters and can "see" the separation change from one year to the next. For example, Europe and North America are moving apart at about 2.5 centimeters per year. At this rate, it would have taken about 150 million years for the Atlantic to grow as wide as it is now—which is consistent with the age of the oldest rocks on the Atlantic's floor.

Rifting

Continental plate

Rifting makes oceans widen.

20mm/year

Subduction

Subduction builds coastal mountains.

Sinking material

A

B

FIGURE 35.9
(A) Rifting occurs where rising material in the mantle pulls crustal plates apart at the planet's surface. (B) Subduction builds mountain chains where plates collide and material sinks back toward the interior of the Earth.

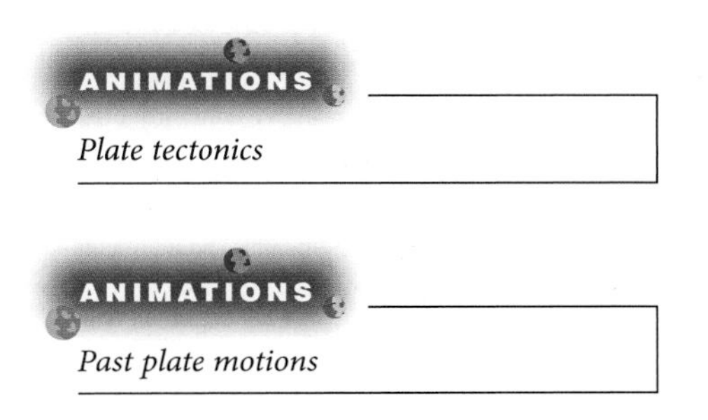

When the mantle stretches and spreads the crust, it breaks apart in a phenomenon called **rifting** (Figure 35.9A). Rifting has reshaped the Earth's surface dramatically. The Atlantic Ocean has "opened" over the past 250 million years by this process, separating Europe and Africa from North and South America. Running roughly north–south through the middle of the Atlantic is a ridge where molten rock rises to the surface at the edge between plates that are being carried away from each other. Other ocean ridges can also be seen in the map of the ocean floor shown in Figure 35.10A.

The circulation of material in the mantle also drags pieces of crustal material toward each other, driving them into massive, low-speed collisions. Along these converging plate edges, the surface buckles to form **mountain ranges,** such as the Rockies and Andes along the western coast of the Americas. Deep trenches are found in the ocean floor where one plate is driven back down into the mantle in a process called **subduction** (Figure 35.9B). These trenches can be seen around much of the rim of the Pacific Ocean in Figure 35.10A.

Q Why do you suppose that such slow motion of the plates can lead to violent earthquakes?

The slow motions of plates on Earth's surface have more drastic effects, such as earthquakes and volcanoes. These tend to occur at plate boundaries where pieces of the crust are pushing and pulling against each other (Figure 35.10B). For example, friction may temporarily stop two plates from sliding past each other, causing them to stick. Pressure will then build until the rock breaks, freeing the stuck plates and generating a sudden lurch in the crust—an earthquake. These same shifts may move crustal rock to different depths, changing the pressure on it. Hot rock moved upward can melt because of the decreased pressure. This is the principle behind the Earth's core being liquid at shallower depth and solid farther down.

A volcano might originate within crustal rock at a depth of about 50 kilometers and a temperature of about 1300 K, which can become molten when the pressure on it is reduced. The molten rock is less dense that its surroundings, so it begins rising upward through fissures in the surrounding rock. If it rises quickly enough that it does not have time to cool and resolidify, it will eventually get to within a few kilometers of the surface, where the pressure becomes so low that gases begin to boil out of the molten rock, forming bubbles within it. This further lowers the molten rock's density, helping it to rise even more rapidly, finally erupting through

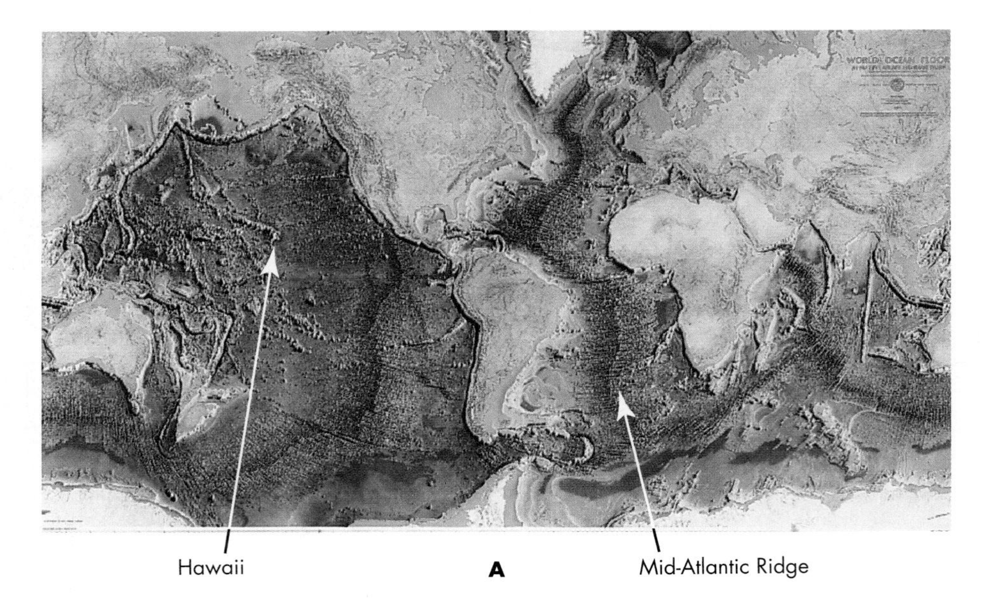

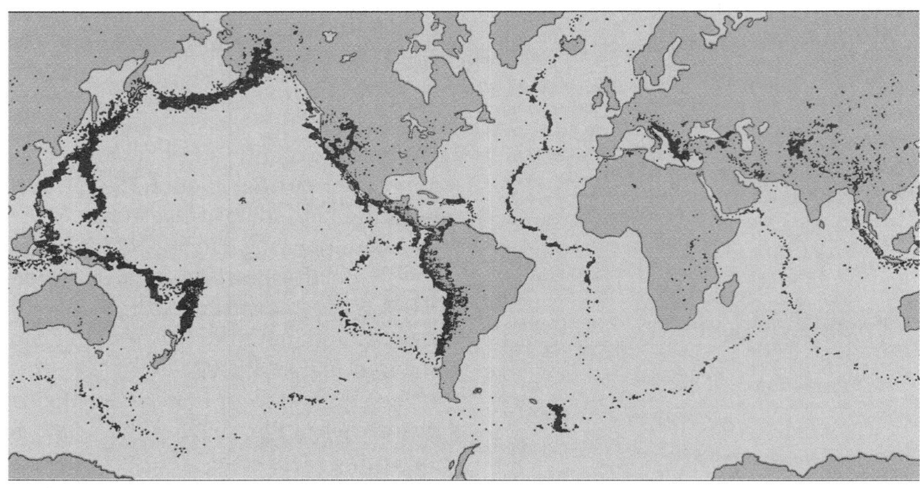

FIGURE 35.10
(A) A map of the Earth's ocean floors. Rifting occurs along long ridges along the ocean floor where plates are moving apart, such as the Mid-Atlantic Ridge. Deep trenches occur where one plate is forced down under another, which can be seen around much of the rim of the Pacific Ocean. (B) Locations of earthquakes (red dots) identify many plate boundaries.

Source: Copyright by Marie Tharp 1977/2003. Reproduced by permission of Marie Tharp Oceanographic Cartographer, One Washington Ave., South Nyack, New York 10960

the surface. This last step is a little like releasing the pressure on a can of warm soda—with an ensuing eruption!

When a plate moves over a region where there is a rising plume of heat from the Earth's interior, a volcano may also form. The Hawaiian island chain represents just such a situation. As the plate moves to the northwest, it carries the former active volcanoes away from the rising plume. New volcanoes form over the plume, trailed by a string of successively older extinct volcanic islands to the northwest. The older of these former volcanoes have subsided back below the surface of the ocean, but they can be seen as a long chain of undersea mountains on the seafloor (Figure 35.10A).

From a wide variety of geological evidence, it has been possible to reconstruct the history of plate motion for more than half a billion years. Figure 35.11 illustrates how this motion has altered the location and appearance of Earth's continents. Plate tectonics has sculpted the landscape of our planet, but it is ultimately driven by the heat within Earth. We therefore expect that tectonic motions will be a factor only on terrestrial planets that are large enough to have maintained a hot interior. Small bodies such as the Moon and Mercury have cooled off much more, and indeed they show no signs of current tectonic motion.

Q Imagine you are living a few hundred million years from now. How might plate tectonics affect your transcontinental travel plans?

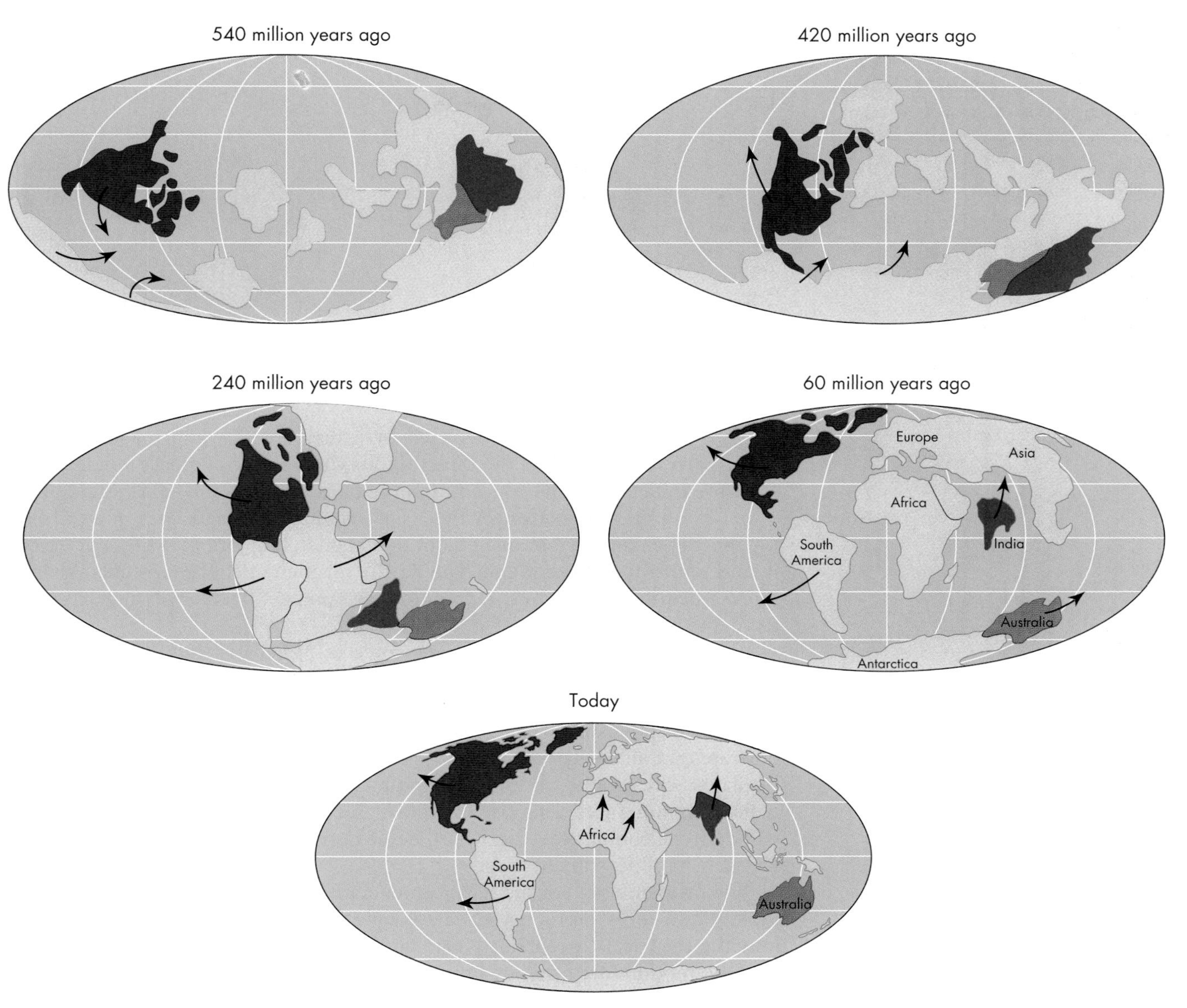

FIGURE 35.11
Over the past 540 million years, the appearance of Earth's surface has changed dramatically as the result of plate tectonics. Two hundred and forty million years ago there was essentially one giant continent. Before that the land areas were split apart in a completely different configuration.

35.4 THE EARTH'S MAGNETIC FIELD

Many astronomical bodies have magnetic properties, and the Earth is no exception. Magnetic forces are "carried" by what is called a **magnetic field.** They cause a compass needle to turn so it points in the direction of the field. Magnetism resembles gravity in that it exerts a force on a distant object and requires no direct physical contact. Unlike gravity, magnetism can be either attractive or repulsive—more like electric forces—but a magnet always has one side that is attractive and another side that is repulsive to another magnet.

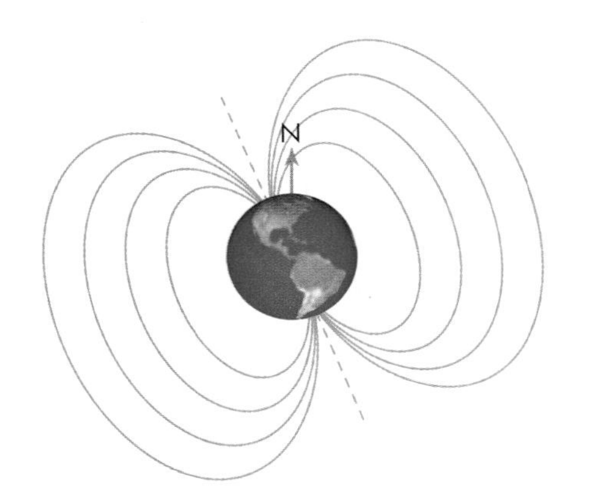

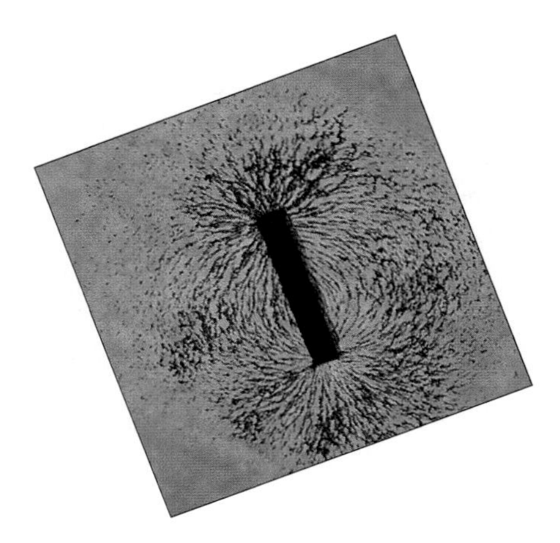

FIGURE 35.12
A schematic view of Earth's magnetic field lines and a photograph of iron filings sprinkled around a toy magnet, revealing its magnetic field lines.

Magnetic fields are often depicted by a diagram showing magnetic lines of force. Each line represents the direction in which a tiny compass would point in response to the field. The concentration of lines indicates the field's strength, with more lines implying a stronger field. For example, the field lines of an ordinary toy magnet emanate from one end of the magnet, loop out into the space around it, and return to the other end. The Earth's magnetic field has a similar shape, as represented in Figure 35.12. A free-floating compass needle will align itself with the depicted field lines.

An important question about any force is, What generates it? Gravitational fields are generated by masses. Magnetic fields are generated by electric currents, either large-scale currents or currents on the scale of atoms—which is why we speak of magnetism as being part of the electromagnetic force. You can demonstrate that an electric current produces magnetism by wrapping a few coils of insulated wire around an iron nail and attaching the wire ends to a battery. The nail will now act like a magnet: It will be able to pick up small pieces of iron and will deflect a compass needle. Microscopic currents on the atomic scale create the magnetism of permanent magnets.

The magnetic field of the Earth is generated by electric currents flowing in its molten iron core—a process called a **magnetic dynamo.** Scientists are still unsure how the electrical currents originate but hypothesize that complex churning and swirling motions are generated by a combination of the Earth's rotational motion and convection in the molten iron core. Studies of the magnetic fields of other Solar System bodies support this view. For example, bodies with weak or no magnetic fields, such as Mars or Venus, are either too small to have a convecting core or rotate very slowly. On the other hand, bodies with large magnetic fields, such as Jupiter and Saturn, rotate very rapidly and like Earth have liquid metallic regions in their interiors.

The gas giant planets have different internal structures and compositions, but under the extreme conditions there, substances like hydrogen can behave like a metal (Unit 43).

The electrical currents that generate a planet's magnetic field need not be aligned with its rotation axis. The orientation of the Earth's magnetic pole has been observed to have shifted by more than 10° in just the last century. The orientation of the magnetic field currently places the north magnetic pole about 9° away from the Earth's rotation axis in northern Canada. The south magnetic pole is about 25° from the axis and lies between Antarctica and Australia. Therefore, a compass needle does not in general point to true north but instead points several degrees away to *magnetic north.*

Geologists have discovered that the orientation of the Earth's magnetic pole has changed by much greater amounts in the ancient past. They can deduce this from measurements of the magnetic field preserved in congealed lavas. When lava cools, the crystals forming in the rock align with the orientation of the Earth's magnetic field at that particular time. By looking at rocks of various ages, we find

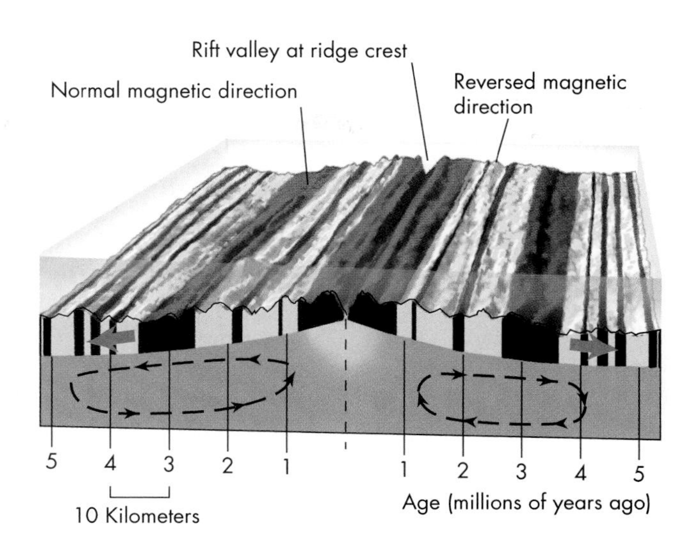

FIGURE 35.13
Magnetic reversals recorded in the ocean floor. As plates spread apart, the upwelling lava cools and records the direction of the magnetic field at that time. Portions of the ocean floor having the same magnetic orientation as we have currently are shown in black, while those with reversed orientation are shown in white.

that the orientation of the Earth's magnetic field shifts gradually over time, and on occasion it has even completely reversed direction.

One of the best-preserved records of magnetic reversals is found on the ocean floor. As lava wells up in the rift between plates that are moving apart, the fresh lava records the orientation of the magnetic field. This is illustrated in Figure 35.13, which shows that there are regions of normal and reversed magnetic field that are symmetrical around the mid-ocean rift.

From the pattern of the magnetic field recorded in the ocean floor, geologists find that reversals happen, on average, every few hundred thousand years. Reversals follow no regular pattern, however, and there has not been a reversal in nearly 800,000 years. Measurements indicate that the Earth's magnetic field has weakened by about 10% since the 1800s, and this has caused some geologists to speculate whether this might signal the beginning of a field reversal. This could potentially have a serious impact on our environment because the Earth's magnetic field helps prevent energetic particles from the Sun and other astronomical sources from reaching Earth's surface. During a reversal the field may be too weak to protect us from this cosmic radiation.

Q What would be the impacts if the magnetic field of the Earth dropped to zero? This probably happens during the magnetic reversals known to occur every few hundred thousand years.

KEY TERMS

convection, 270
crust, 268
inner core, 269
magnetic dynamo, 274
magnetic field, 273
mantle, 268
mountain range, 271
outer core, 269
plate tectonics, 270
P wave, 267
radioactive decay, 269
radioactive element, 269
rifting, 271
seismic wave, 267
silicate, 266
subduction, 271
S wave, 267

QUESTIONS FOR REVIEW

1. How do the ways in which P waves and S waves propagate indicate the internal structure of the Earth?
2. Why is the interior of the Earth and the other inner planets hot?
3. How do we know the Earth has an iron core?
4. What is meant by *plate tectonics?*
5. How does plate tectonics cause earthquakes and volcanoes?
6. What generates the magnetic field of the Earth?
7. What is the evidence that the Earth's magnetic field has undergone dramatic changes over time?

PROBLEMS

1. P waves travel at about 6 km/sec.
   a. At this speed, how long would it take P waves to travel to the opposite side of the Earth from the center of an earthquake?
   b. In the iron core, the speed of P waves is approximately two times faster. Sketch a picture of the interior of the Earth and estimate how much sooner the wave would reach the opposite side of the Earth if the wave traveled this much faster while in the core.
2. Draw a scale model of the Earth (radius 6357 km) and the liquid core (radius 3500 km). How large a region is "shadowed" from S waves? Use a protractor to estimate how many degrees away from the site of the earthquake S waves are still detected.
3. S waves travel slower than P waves—about 4 km/sec versus 6 km/sec. Suppose a seismometer detects P waves at 7:00 P.M. and S waves at 7:05 P.M. How far away was the earthquake? At what time did the earthquake occur?
4. Suppose the Earth were made of nothing but iron (at a density of 7.8 kg/L) and silicates (at a density of 3.0 kg/L).
   a. Calculate the percentage (by volume) of iron and silicates that would be required to give a density of 5.5 kg/L. (Hint: you know that the two volume percentages must add up to 100%, so the percentage of silicates equals 100% minus the percentage of iron.)
   b. The outer radius of the core is approximately 3500 km, while the overall radius of the Earth is 6357 km. What percentage of the Earth's volume is occupied by the core?
   c. The percentage of Earth's *mass* that is composed of iron is estimated to be 32% (see Unit 33). What might explain why this percentage is smaller than the value in (a) and larger than the percentage in (b)?
5. We can get a rough estimate of how fast the rift zone in the Atlantic Ocean is moving the continents of South America and Africa apart simply by assuming that at 240 million years ago Natal, Brazil, and Douala, Cameroon, occupied the same neighborhood. Theses cities are presently 5000 km apart. How fast would they have to be moving, on average, to reach their present separation? Express your result in cm/year.
6. In 2005 the Earth's magnetic field was about 10% weaker than it was in 1845 when German mathematician Carl Friedrich Gauss first measured it. If the magnetic field continues to weaken by 10% of its current value every 160 years, approximately when will it be at half of its 1845 value? When will it be only 10% of its 1845 value? (Note that we are supposing, in this question, that after 320 years the strength of the field is 90% of 90%—or 81% of its original value.)

TEST YOURSELF

1. Our knowledge of the composition of the Earth's core comes from
   a. analysis of earthquake waves.
   b. X-ray pictures taken with a powerful device in Russia.
   c. samples obtained by drilling deep holes.
   d. analysis of the material erupted from volcanoes.
   e. both (c) and (d).
2. Scientists think the Earth's core is composed mainly of
   a. silicate rocks.
   b. uranium.
   c. lead.
   d. sulfur.
   e. iron.
3. What evidence indicates that part of the Earth's interior is liquid?
   a. With sensitive microphones, sloshing sounds can be heard.
   b. We know the core is lead, and we know the core's temperature is far above lead's melting point.
   c. Deep bore holes have brought up liquid from a depth of about 4000 kilometers.
   d. No S-type seismic waves are detectable at some locations after an earthquake.
   e. S-type waves are especially pronounced at all locations around the Earth after an earthquake.
4. The slow shifts of our planet's crust are thought to arise from
   a. the gravitational force of the Moon pulling on the crust.
   b. the gravitational force of the Sun pulling on our planet's crust.
   c. the Earth's magnetic field drawing iron in crustal rocks toward the poles.
   d. heat from the interior causing convective motion, which pushes on the crust.
   e. the great weight of mountain ranges forcing the crust down and outward from their bases.

unit

36 Earth's Atmosphere and Hydrosphere

Background Pathways

Unit 35: The Earth as a Terrestrial Planet 266

Surrounding the solid body of the Earth is a gaseous envelope—the Earth's **atmosphere**; and most of the Earth's surface is covered with liquid water—the **hydrosphere** (Figure 36.1). The mass of all the Earth's water is estimated to be about 1.4×10^{21} kilograms—about 1/4000th of the Earth's total mass. If the rocky surface of Earth were to be ground smooth, the oceans would have an average depth of about 2700 meters (about 1.6 miles) over the entire surface. The mass of the atmosphere is about 5.1×10^{18} kilograms—about one one-millionth of the Earth's total mass. Although these are small fractions of the Earth's overall mass, they are major components of the Earth's surface environment, and they play critical roles in shaping the surface and setting its temperature.

Most planets in the Solar System have an atmosphere, but the Earth's has many unique features. One of the most striking differences between the atmosphere of the Earth and that of other planets is its composition. For example, the atmospheres of Mars and Venus are nearly completely carbon dioxide, while the atmospheres of the Jovian planets are mostly hydrogen and hydrogen compounds. By contrast, Earth's atmosphere is primarily a mixture of nitrogen and oxygen. Nitrogen molecules make up about 78% of the atmosphere, and oxygen about 21%. The remaining 1% is mostly the gas argon but includes trace amounts of carbon dioxide and ozone—gases crucial for protecting us and making life possible.

36.1 STRUCTURE OF THE ATMOSPHERE

FIGURE 36.1
The Earth is covered by a thin layer of liquid and gas that create our environment.

Our atmosphere extends from the ground and remains detectable to an elevation of hundreds of kilometers; but at the highest altitudes the air is extremely tenuous. In fact, the density of the atmospheric gases decreases steadily with height. Gases near the ground are compressed by the weight of gases above them. Thus our atmosphere is a little like a tremendous pile of pillows. The pillow at the bottom is squashed by the weight of all those above (Figure 36.2). Similarly, air near sea level is more compressed and therefore has a greater density than air near the top of the atmosphere. This is why it is so difficult to breathe at 6000 meters (or 20,000 feet), where the air is only about half as dense as at sea level. The atmosphere's density declines to about half its previous value for every 6 kilometers in height from the Earth's surface. Traces of atmosphere can be detected by satellites up to about 800 kilometers above the surface, although the air there is extremely thin and gradually merges with the near vacuum of interplanetary space.

The weight of the atmosphere presses down on the ground below. This **atmospheric pressure** at sea level is often described as 1 **atmosphere,** which in metric units is

FIGURE 36.2
Like the layers of the atmosphere, a stack of pillows compresses the bottom pillow most and the top one least because of the increasing pressure lower down in the stack.

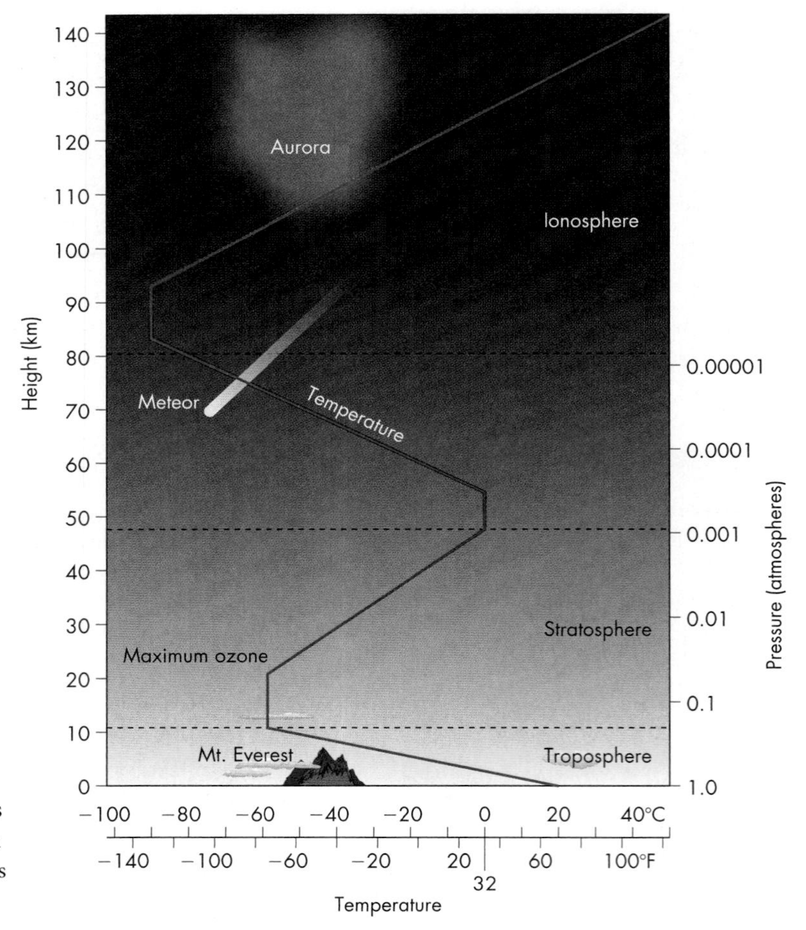

FIGURE 36.3
The Earth's atmosphere contains warm layers and cool layers. The warm layers are those in which solar radiation is absorbed. Cool layers do not absorb much solar radiation.

Q Why is it not possible to suck water up a hose more than about 10 meters (about 33 feet)?

Gaseous H_2O—water vapor—is an important component of the atmosphere, ranging up to 3% of the atmospheric gases. It is not usually counted with the other components because it is so highly variable—depending on location, season, and conditions—and because it is constantly changing its physical state to become part of the far more massive liquid and solid forms of water found on the Earth's surface.

about 1.03 kilograms per square centimeter (about 14.7 pounds per square inch). What this means is that if we could measure a column of air 1 cm × 1 cm all the way up through the atmosphere, its mass would be slightly more than 1 kilogram. Over the roughly 16,000-cm^2 surface area of an adult human's skin, this amounts to about 16 tons of weight! However, because the pressure pushes equally on all sides, the forces balance, and we do not feel any net force. We are so accustomed to a pressure of 1 atmosphere that we often do not notice its effect except in its absence.

When you suck on a straw in a glass of iced tea, you are not "pulling" the liquid up the straw. Actually, you are removing the air from the straw so there is no pressure to counterbalance the pressure pushing down on the surface of the tea in the rest of the glass, so the tea is pushed up the straw by the weight of the atmosphere. The total weight of the atmosphere is equivalent to a layer of water about 10 meters deep. This means that the pressure in the ocean increases by about 1 atmosphere with every 10 meters of additional depth—an important consideration for underwater diving.

The dense lowest layer of the atmosphere, up to about 12 kilometers (7 miles), is called the **troposphere** (Figure 36.3). In this region of the atmosphere water vapor circulates, clouds form, and the air is very turbulent. Above this is a much more stable layer called the **stratosphere,** which extends up to about 50 kilometers (30 miles) above the surface. Airplanes fly in the stratosphere to avoid the turbulence of the troposphere. More importantly, gases in the stratosphere, such as ozone, shield the Earth's surface from most of the short-wavelength radiation from space. Above about 80 kilometers, the nature of the atmosphere is quite different. Most molecules and atoms there have been broken apart into ions in a region called the **ionosphere.** These electrically charged particles interact with long-wavelength

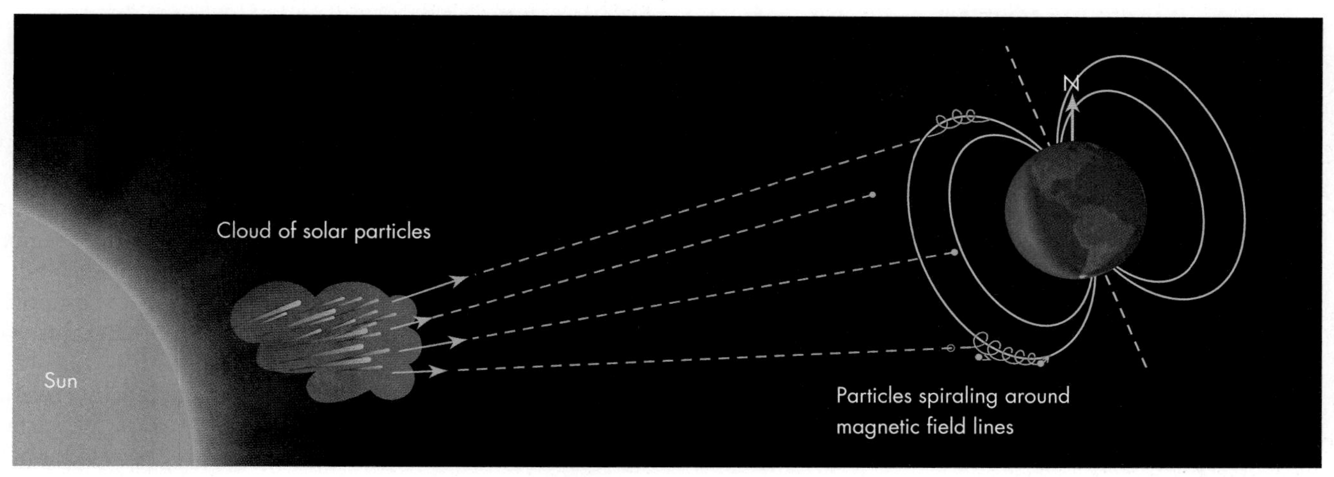

FIGURE 36.4
Electrically charged particles from the Sun spiral in the Earth's magnetic field.

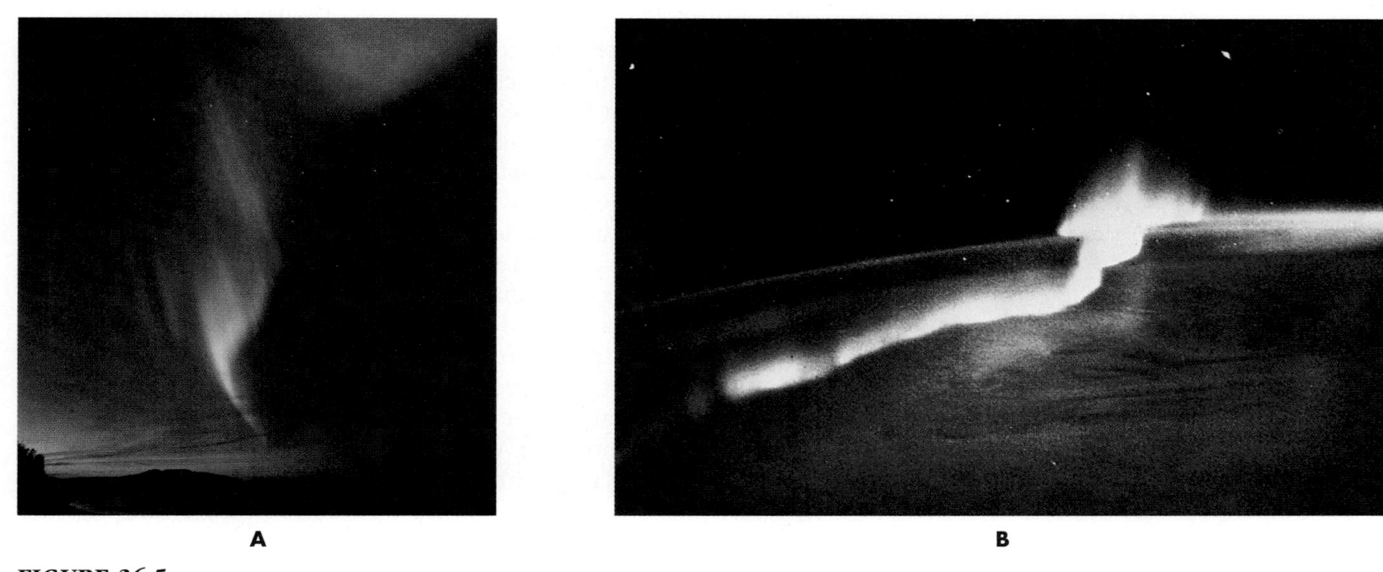

FIGURE 36.5
Photographs of an aurora (A) from Anchorage, Alaska, and (B) from space. The green colors are generally produced by ionized oxygen atoms, while red can be produced by both oxygen and nitrogen.

Energetic charged particles are trapped by Earth's magnetic field in the **Van Allen radiation belts,** doughnut-shaped regions surrounding the Earth, thousands of kilometers above its surface.

radio waves, reflecting AM signals back to the surface and blocking our view from the ground of the longest-wavelength electromagnetic radiation from space.

The Earth's magnetic field (Unit 35) also partially screens us from electrically charged particles emitted by the Sun (Unit 49). Such particles are often energetic enough to damage living cells, so they are potentially harmful to us. Because moving charged particles interact with magnetic fields, they are deflected into spiraling motion around the Earth's magnetic field lines (Figure 36.4). Thus we are spared the full impact of the Sun's charged particles.

As the solar particles stream past the Earth, they apply forces on the magnetic field, and this in turn generates electric currents in the ionosphere. These currents excite the ions to emit the lovely light we see as the **aurora** (Figure 36.5). The exact process by which the aurora forms is still not completely understood, but there is no doubt that its beautiful streamers are shaped by the Earth's magnetic field. The aurora tends to be most visible close to the magnetic poles, where the solar particles can penetrate closest to the Earth's surface as they move along the magnetic field lines.

36.2 THE SHAPING EFFECTS OF WATER

One of the main features that makes the Earth unique among the planets in our Solar System is the great quantity of liquid water on its surface. Oceans cover 71% of the Earth's surface, while lakes, rivers, and streams mark the faces of the continents. Water is vital to the existence of life on our planet, and we see enormous differences in the amount and kinds of plant and animal life in dry and wet regions of the Earth's surface. Water also affects the geology of the Earth's surface.

The Sun's heat powers the cycling of water through the atmosphere by evaporating the water, which rises through the atmosphere, condenses, and then rains over the land before flowing back downhill to the oceans. The constant flow of water over the Earth's surface wears away rock and other surface materials by **erosion,** which over millions of years can level even high mountains. Much of the surface of the continents exhibits the results of such erosion, which carves river channels and grinds rock into sand (Figure 36.6). Moreover, water in the form of glacial ice has ripped enormous scars into the landscape. If the Earth had no geological activity to rebuild mountains, the action of rain, flowing water, and waves would have gradually ground down the continents over the Earth's long history, and the entire surface would be under water.

Water has important chemical effects too. As rain falls through the atmosphere, it absorbs carbon dioxide. In fact, the oceans have about 500 times more carbon dioxide dissolved in them than the total contained in the atmosphere. This carbon may then become incorporated in plants and microscopic creatures that settle into sediments after they die. As the layers of sediment are compressed by the kilometers of material deposited on top of them, they can be converted to rock such as limestone. There the carbon remains locked away until the rock is melted by geological activity, such as a volcanic eruption.

Water itself can react at a molecular level with the elements that make up a rock. As a result, different kinds of rock form in the presence of water when the material is modified by heat and pressure deep in the Earth's crust. Rocks that contain trapped water melt at a lower temperature and are "runnier" when molten than rocks with little or no water. We can see this clearly in Figure 36.7 where we compare Mount Fuji in Japan with Mauna Loa, the currently active volcano

FIGURE 36.6
The Grand Canyon has been carved through rock by the flow of water over millions of years.

FIGURE 36.7
Comparison of Mt. Fuji and Mauna Loa.

in Hawaii. Both Fuji and Mauna Loa have been built by erupted lava, but their lavas differ. Mauna Loa's lava is rich in water, so it remains molten at a lower temperature and disperses more evenly and farther from its point of eruption. As a result, Mauna Loa (like other Hawaiian peaks built similarly) has shallow slopes. Mount Fuji's lava, on the other hand, is relatively dry, so it is "stiff," and its slopes are steep.

36.3 THE ATMOSPHERE, LIGHT, AND GLOBAL WARMING

The Sun is the primary source of heat for the Earth's surface. The presence of an atmosphere over the surface of a terrestrial planet changes how the surface interacts with sunlight and releases thermal energy to space. Some molecules in the atmosphere have little effect, whereas others can dramatically change the energy balance. Two separate phenomena are discussed in this section, one in the stratosphere and one in the troposphere. Both have become the subject of recent concern because of the ways in which humans are altering the chemistry of our atmosphere.

Our atmosphere can block some wavelengths of light from ever reaching the Earth's surface. This is an important effect of oxygen. Indeed, oxygen is vital not only because we breathe it but also because it shields us from harsh solar ultraviolet radiation. Part of that shielding is provided by O_2 (the normal form of oxygen), and part comes from another molecular form of oxygen, O_3 or **ozone.**

Most of the ozone in our atmosphere is located in the ozone layer at an altitude of about 25 kilometers (80,000 feet) in the stratosphere. Ozone is formed because at these upper levels solar ultraviolet radiation is intense enough to split O_2 into individual oxygen atoms. The splitting occurs because the ultraviolet photons are absorbed by the molecules, making them vibrate so energetically that they fly apart into individual oxygen atoms. These atoms then combine with other O_2 molecules to form O_3.

Ozone is important because it is such a strong absorber of ultraviolet radiation; without the ozone layer, much more solar ultraviolet radiation would pour into the lower atmosphere. The short-wavelength, and therefore high-energy, ultraviolet photons can damage the complex, delicate organic molecules of living organisms. Without ozone's protective effects, life on Earth's surface would suffer severe radiation injury, which is why scientists are concerned about damage done to the ozone layer by pollutants. Pollutants that release chlorine atoms in the stratosphere, notably chlorofluorocarbons (or CFCs), appear to be a major source of ozone destruction. The South Polar region experiences conditions that allow chemical pollutants to accumulate in the stratosphere during its winter, followed by a large ozone "hole" when the Sun returns (Figure 36.8). The problems we observe there may be a warning sign of future problems over the rest of the Earth. CFCs were widely used in spray cans and refrigeration systems, but have now been banned, and it appears that the size of the ozone hole has stopped increasing, and perhaps started to reverse.

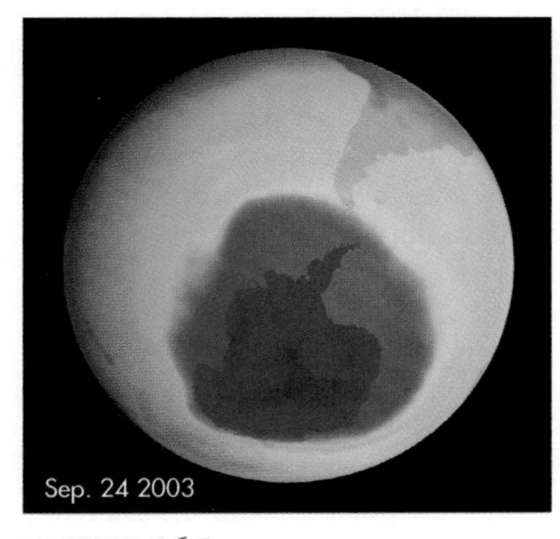

FIGURE 36.8
The Antarctic ozone hole in September 2003. The colors in this image indicate the relative amounts of ozone above different regions of the Earth's surface. The purple region above Antarctica shows only about one-third of its normal amount of ozone.

Ozone is not the only gas that absorbs radiation in our atmosphere. Carbon dioxide and water are also strong absorbers, but they absorb infrared rather than ultraviolet radiation. Relatively little of the Sun's energy arrives at infrared wavelengths, so these molecules do not significantly block the Sun's energy from reaching the surface; but they do play a crucial role in regulating our planet's temperature.

The transparency of the Earth's atmosphere to visible radiation allows sunlight to pass through the atmosphere and reach the surface, where its energy is

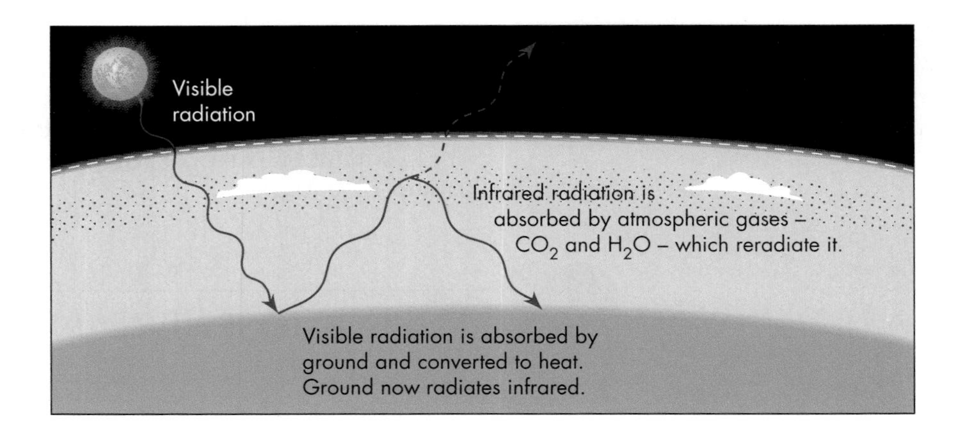

FIGURE 36.9
The greenhouse effect. Radiation at visible wavelengths passes freely through the atmosphere and is absorbed at the ground. The ground heats up and emits infrared radiation. Atmospheric gases absorb the infrared radiation and warm the atmosphere, which in turn warms the ground.

absorbed, heating the ground and ocean. The warmed surface in turn radiates infrared energy, but heat loss is reduced by the opacity (blocking power) at infrared wavelengths of the carbon dioxide and water in the troposphere. This makes the Earth's surface warmer than it would be if its infrared energy could escape freely, a phenomenon shown in Figure 36.9 and known as the **greenhouse effect.**

Note that the greenhouse effect does *not* cause the ozone hole. This is a common point of confusion, but global warming and the destruction of the ozone layer have separate, unrelated causes.

You can get some indication of how effectively water vapor traps heat by noticing how the temperature drops dramatically at night in desert regions or on clear low-humidity nights. All gardeners know that in cold seasons, clear nights (with no clouds and little water vapor) are most likely to produce frost. On cloudy nights, heat is retained.

The greenhouse effect does not generate heat; rather, it slows heat loss to space. The greenhouse effect therefore warms the Earth as a blanket warms you—by slowing the loss of heat. Similarly, water and carbon dioxide simply slow the loss of heat from the ground by absorbing infrared energy. Eventually they reemit it, but some is reemitted back toward the ground, keeping the ground warmer than it would be otherwise. That extra infrared energy reradiated to the ground helps keep it warm at night.

Levels of carbon dioxide have risen by about 25% over the last century—a rise that most scientists attribute to the increased burning of fossil fuels by humans. Human activities such as deforestation have also reduced the places where carbon can be stored. Over the same time period, average ocean and air temperatures have increased by about 0.6°C, a phenomenon referred to as **global warming.** Most climate scientists have concluded that the primary source of global warming is the trapping of heat by the additional carbon dioxide in Earth's atmosphere.

Over the past 160,000 years, higher temperatures correlate with higher levels of CO_2 (Figure 36.10). The Earth passed through much warmer periods, hundreds of millions of years in the past. Some of these warm episodes may have been triggered by massive geological activity releasing greenhouse gases, or variations in the Sun's energy output (Unit 51).

Compared to the last several hundred million years, Earth is currently in one of its coolest periods. During some past warm periods, the polar caps have completely melted and sea level has been more than 200 meters higher. None of the models suggest that a return to such dramatically higher sea levels is imminent. However, some predictions suggest a rise of sea level by as much as 1 meter over the next century or two, and this would have drastic effects on coastal cities and environments. Whatever the ultimate cause of current global warming, it seems likely that reducing greenhouse gases in the atmosphere can help slow the change and reduce the severity of the consequences.

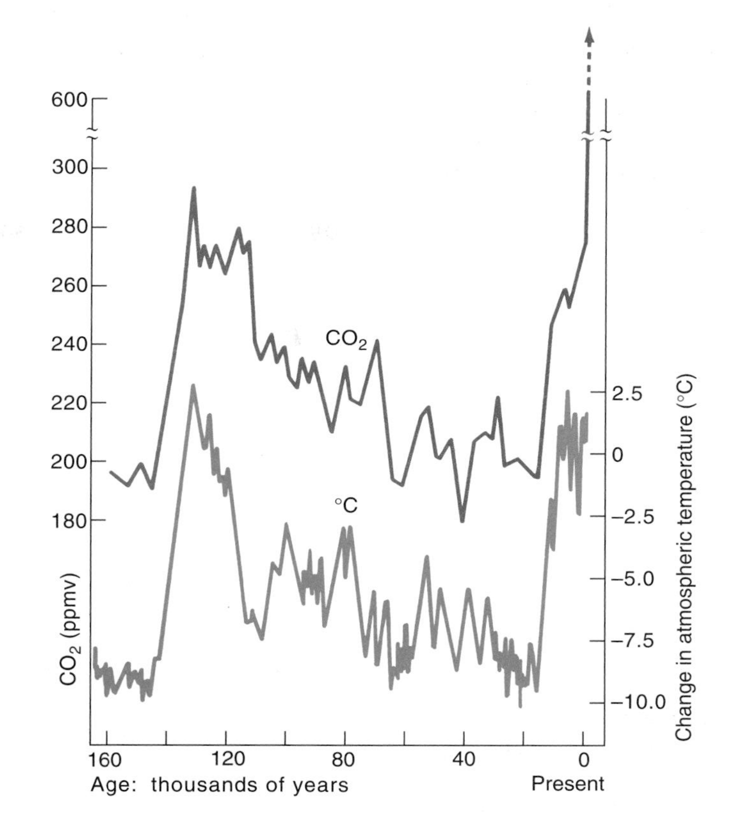

FIGURE 36.10
Carbon dioxide levels (red line) and atmospheric temperatures (blue line) are shown for the past 160,000 years. These values are estimated from gas bubbles trapped in Antarctic ice layers. Temperatures can be estimated from the relative numbers of different isotopes found in the gases.

36.4 AIR AND OCEAN CIRCULATION: THE CORIOLIS EFFECT

If you sit with a friend on a rotating merry-go-round and throw a ball back and forth, you will discover that the ball does not travel in the direction in which you aim but instead curves off to the side. This is largely a matter of perception. Someone standing to the side of the merry-go-round and observing the ball's path would see it as having traveled in a straight line once you released it. The curving path you see on the merry-go-round is actually a reflection of your own curving path. A similar effect faces ocean and air currents sweeping across our rotating planet's surface. They are deflected from their original direction of motion in a phenomenon called the **Coriolis effect.**

Coriolis effect

The Coriolis effect, named for the French engineer who first studied it, alters the paths of objects moving over a rotating body, such as the Earth, other planets, or stars. To understand why the Coriolis effect occurs, imagine standing at the North Pole and throwing a rock toward the equator (Figure 36.11). As the rock arcs through the air, the Earth rotates under it. Thus if you were aiming at a particular point on the equator, you will miss because the surface has turned beneath the rock's path, making the rock turn to the right from the standpoint of someone in the Northern Hemisphere and to the left in the Southern Hemisphere. From the equator, a rock thrown toward either pole will also be deflected because the conservation of angular momentum (Unit 20) causes objects to move faster around the axis of rotation when they move closer to it. The rock therefore must rotate faster than the Earth's surface beneath it as it moves closer to the Earth's pole. This again turns the path

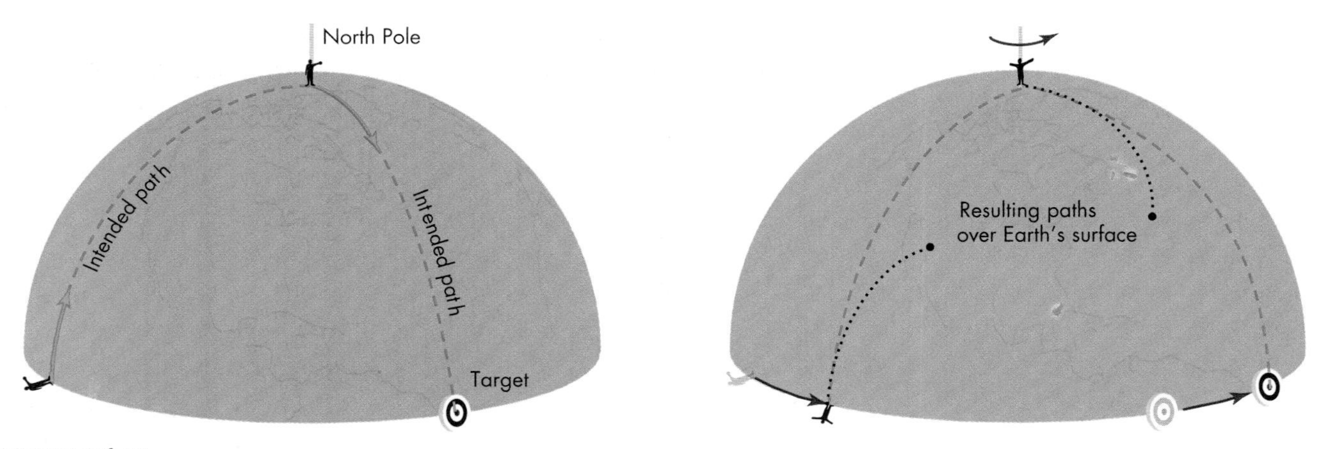

FIGURE 36.11
The Coriolis effect causes the path of material moving over the Earth's surface to curve to the right in the Northern Hemisphere. The figure illustrates the actual paths on the surface for rocks thrown from the pole toward the equator and from the equator toward the pole.

FIGURE 36.12
Weather satellite pictures show clearly the spiral pattern of spinning air around a storm that results from the Coriolis effect. Air drawn in toward the low-pressure region is deflected to the right in this Northern Hemisphere storm, giving it a counterclockwise spin.

to the right from the standpoint of someone in the Northern Hemisphere and to the left in the Southern Hemisphere. Air, water, rockets, or anything else moving across the rotating Earth in any direction will be always be pushed to the right in the Northern Hemisphere and to the left in the Southern Hemisphere.

Tornados, hurricanes, and storm systems generally spin counterclockwise in the north and clockwise in the south because of the Coriolis effect's deflection of the air being drawn inward toward these low-pressure regions. This can be seen in the spiral pattern of clouds of a Northern Hemisphere storm shown in Figure 36.12. Likewise, the Coriolis effect deflects ocean currents, creating flows such as the Gulf Stream. It also creates atmospheric currents such as the *trade winds* that consistently blow in a particular direction, as well as the high-speed *jet streams* in the stratosphere—not only on Earth but on other planets too.

The Coriolis effect is often said to make water spiral down a drain counterclockwise in the Northern Hemisphere and clockwise in the Southern Hemisphere.

Although these are the directions that the effect produces in principle, the way the water is stirred or fills the basin plays a far more important role. Only when water or air flows inward from an area at least many kilometers across is the Coriolis effect likely to play any significant role.

36.5 THE HISTORY OF THE OCEANS AND ATMOSPHERE

The Earth's oceans and atmosphere probably developed fairly late in the planet's formation after its surface began to cool. The Earth was probably too hot to have liquid water on its surface for over 100 million years after it formed. At the high early temperatures, many of the gases that were present would have been hot enough to escape from the Earth entirely. The volatile compounds that make up Earth's atmosphere and hydrosphere today may have come from volcanic eruptions after the Earth's surface began to cool, or they may have been delivered by comets (Unit 33).

Scientists have tried to test these very different hypotheses. It appears that volcanoes could have expelled enough gas and water vapor to have supplied our atmosphere and hydrosphere. It also appears that the proportion of trace elements and isotopes in volcanic gases is similar to those in the air around us. However, volcanic activity is tied to patterns of atmospheric recycling, and present-day volcanoes may not reflect the composition of volcanic gases produced in early times. Likewise, testing the comet delivery hypothesis is difficult because we lack precise knowledge about what frozen gases and ices comets contain and the rate at which they struck the Earth in its early history. It seems likely that comets could have delivered enough material, but spectral-line studies of comets, from both ground-based telescopes and space probes, indicate some compositional differences from Earth's atmosphere.

It is clear in any case that the Earth's atmosphere has undergone a series of changes. Some gases were lost to space. To hold on to atmospheric gases, a planet must have a strong enough gravitational field to prevent the gases from escaping to space. This depends on the strength of the planet's surface gravity, the temperature of the gas, and the mass of the gas molecules. If a gas is hot, the gas molecules move faster and therefore escape more easily. Within a gas of a given temperature, the lightest molecules move fastest. This is why hydrogen gas (the lightest of all gas molecules) is retained only by very large planets in the outer parts of the Solar System, where the cooler conditions cause the hydrogen atoms to move more slowly. At the Earth's current temperature, Earth's gravity cannot retain hydrogen and helium, but the Earth can hold on to most other gases.

None of the models of the origin of the Earth's atmosphere can account for the current abundance of oxygen in our atmosphere. It is possible to generate oxygen from water molecules by splitting them with solar ultraviolet radiation into hydrogen and oxygen. When this happens the lighter hydrogen drifts to the top of the atmosphere and escapes, leaving oxygen behind. However, this mechanism can generate only a small fraction of the amount of oxygen Earth has. Where then did such quantities of this vital ingredient originate?

Chemical analysis of ancient rocks, particularly those rich in iron compounds that react with oxygen, shows that our atmosphere once contained much less oxygen than it does today. It was probably similar to the current atmospheres of Venus and Mars in containing a large amount of carbon dioxide (CO_2). The chemical compositions of ancient rocks indicate that the amount of oxygen in our atmosphere has steadily increased over the past 3 billion years. At the same time, carbon dioxide has decreased life across our planet. Life itself has thus changed the Earth's atmosphere. The interaction of life with a planet's atmosphere is discussed further in Unit 83.

KEY TERMS

atmosphere, 277
atmospheric pressure, 277
aurora, 279
Coriolis effect, 283
erosion, 280
global warming, 282
greenhouse effect, 282
hydrosphere, 277
ionosphere, 278
ozone, 281
stratosphere, 278
troposphere, 278
Van Allen radiation belt, 279

QUESTIONS FOR REVIEW

1. What produces atmospheric pressure?
2. What are the major layers of the Earth's atmosphere, and what are the major features of each layer?
3. What is the greenhouse effect? How does it relate to global warming?
4. How does the ozone layer protect life on the Earth's surface?
5. What causes the aurora?
6. What is the Coriolis effect? What motions does it affect? What motions does it not affect?

PROBLEMS

1. How does the mass of hydrogen in the ocean's H_2O compare with the total mass of the Earth's atmosphere? The ocean's mass is about 1.4×10^{21} kg, which you can assume is entirely water for this problem. The atmosphere's mass is about 5.1×10^{18} kg. (Note: Hydrogen has an atomic mass of 1, whereas oxygen has an atomic mass of 16.)
2. It is estimated that there are about 3.6×10^{16} kg of carbon (C) dissolved in the oceans. If all that combined with oxygen (O) to form carbon dioxide (CO_2), how much mass would it be? What percentage of the Earth's atmospheric oxygen would that be? (Note: Carbon has an atomic mass of 12, whereas oxygen has an atomic mass of 16.)
3. Geologists estimate that the Grand Canyon was created primarily during a period of about 3 million years. Given its current depth of about 1.6 km (1 mile), on average how many centimeters deeper did the canyon become each year during this time period?
4. The colors in the Aurora Borealis are due to the interactions between electrons and atoms or molecules in the atmosphere. At low densities, collisions with oxygen produce light at 630 nm. At higher densities, collisions with oxygen atoms produce light at 558 nm, and collisions with nitrogen molecules produce light at 428 nm. What colors do we associate with these wavelengths? Examine Fig. 36.5A and describe what is producing the aurora seen there.
5. Antarctica holds about 3×10^{19} kg of ice. If all of that ice melted and entered the oceans, approximately how much would sea level rise? Use the fact that the total mass of oceans = 1.4×10^{21} kg, while the average depth of the oceans would be 2.7 km if they covered the entire Earth.
6. As ocean water warms, the water expands in volume by about 0.014% per 1°C rise in temperature. This thermal expansion is small, but it contributes a significant part of the recent rise in sea level. The average depth of the oceans today is 3800 meters. How much has the water risen over the last century due to the 0.6°C rise in ocean temperature? If the temperature rose by 20°C, how much would sea level rise?
7. Air moving along with the surface of the Earth takes 24 hours to circle the Earth's axis. At a latitude of 60°, the air is 3183 km from the axis and its path around the axis is 20,000 km long. Calculate its speed in kph. Air at a latitude of 61° is 3086 km from the axis following a path around the axis that is 19,392 km long. Suppose now that a mass of air at 60° latitude moves to 61° latitude. Use the principle of conservation of angular momentum (Unit 20.3) to estimate how fast air must move around the Earth's axis at the smaller radius there. How fast is this *relative to* stationary air at this location?

TEST YOURSELF

1. Which of the following is the most abundant gas in the Earth's atmosphere?
   a. Oxygen
   b. Nitrogen
   c. Carbon dioxide
   d. Water vapor
   e. Hydrogen
2. Auroras are produced when the Earth's atmosphere receives a blast of
   a. energetic electrically charged particles from the Sun.
   b. debris from comets that pass through the inner Solar System.
   c. gamma rays generated by unknown cosmic sources.
   d. ozone-destroying chemicals, mostly generated by refrigerants.
3. The layer of the Earth's atmosphere in which weather occurs is the
   a. stratosphere.
   b. troposphere.
   c. ionosphere.
   d. hydrosphere.

unit

37

Our Moon

The Moon (Figure 37.1) is our nearest neighbor in space, a natural satellite orbiting the Earth. It is the frontier of direct human exploration, an outpost that we reached more than a quarter century ago but from which we have since drawn back. Despite our retreat from its surface, the Moon remains of great interest to astronomers. Although the Moon is a satellite of the Earth, it is almost six times more massive than Pluto. The Moon is so large that it would be regarded as a planet in its own right if it orbited the Sun separately from the Earth. In this sense the Moon can be regarded as the smallest of the terrestrial planets, at the opposite extreme from the Earth.

Because its mass is only about one-eightieth of the Earth's and its radius is only about one-quarter of the Earth's, the Moon rapidly lost its internal heat and did not have sufficient gravity to retain an atmosphere. It is therefore a dead world, with neither plate tectonic nor volcanic activity. The Moon is a barren ball of rock, possessing no air, water, or life. In the words of lunar astronaut Buzz Aldrin, the Moon is a place of "magnificent desolation."

Background Pathways

Unit 35: The Earth as a Terrestrial Planet 266

37.1 SURFACE FEATURES

The Moon is a world of grays without the vivid colors of Earth's landscape. With just your eyes, you can already see that the Moon's surface has light and dark areas. A small telescope or even a pair of binoculars reveals that the dark areas are smooth while the bright areas are covered with numerous large circular pits called **craters,** as illustrated in Figure 37.1. Craters usually have a raised rim, and they range in size from tiny holes less than a centimeter across to gaping scars in

FIGURE 37.1
Photograph showing the contrasting appearance of the lunar highlands and maria. The highlands are heavily cratered and rough. The maria are smooth and have few craters.

Why do you suppose that the Moon has so many more craters on its surface than the Earth does?

the Moon's crust such as Clavius, about 240 kilometers (150 miles) across. Most lunar craters are named for famous scientists. For example, Cristoph Clavius (1537–1612) was a German astronomer and mathematician.

The craters were made by the **impact** of solid bodies striking its surface. When such an object hits a solid surface at typical orbital speeds of several tens of kilometers per second, it disintegrates in a cloud of vaporized rock and rock fragments. The kinetic energy released in such an impact by an asteroid 1 kilometer in diameter is roughly a million times the energy released by the atomic bomb dropped on Hiroshima. The size of an impact crater reflects the kinetic energy released, and is estimated to be 10 to 100 times the diameter of the impacting object. As the vaporized rock expands from the point of impact, it forces surrounding rock outward, pushing it into a raised rim and making circular craters.

The large, smooth, dark areas that you can see in Figure 37.1 are called **maria** (pronounced *MAHR-ee-ah*), from the Latin word for "seas." However, these regions, like the rest of the Moon, are totally devoid of water. The use of the term *maria* (singular **mare**) comes from early observers who believed the maria looked like oceans and who gave them poetic names such as Mare Tranquillitatis (*MAHR-ay tran-KWIL-ih-TAH-tis*) or "Sea of Tranquility," the site where astronauts first landed on the Moon.

The bright areas that surround the maria are called **highlands.** The highlands and maria differ in brightness because they are composed of different rock types. The maria are **basalt,** a dark, congealed lava rich in iron, magnesium, and titanium silicates. The highlands, on the other hand, are rich in calcium and aluminum silicates. This difference has been found from rock samples obtained by astronauts. Moreover, the samples also show that the highland material is generally less dense than mare rock and considerably older. Highland rocks have been dated radioactively (Unit 33) and have ages of about 4.0 to 4.4 billion years, whereas the rocks from the maria range from about 3.1 to 3.9 billion years old.

Highway surfaces develop "potholes" over time, so you can use the number of potholes as an indication of the "age" of the paving. However, not all roads form potholes at the same rate. What other factors do you think might affect the number of potholes you see?

Not only are the highlands brighter than the maria, they are also more rugged, being pitted with large numbers of craters. In fact, highland craters are so abundant that they often overlap, as shown in Figure 37.2A. The shorter age of the maria is evident from the smaller number of impact craters seen in Figure 37.2B. Geological activity in the maria has erased the oldest craters.

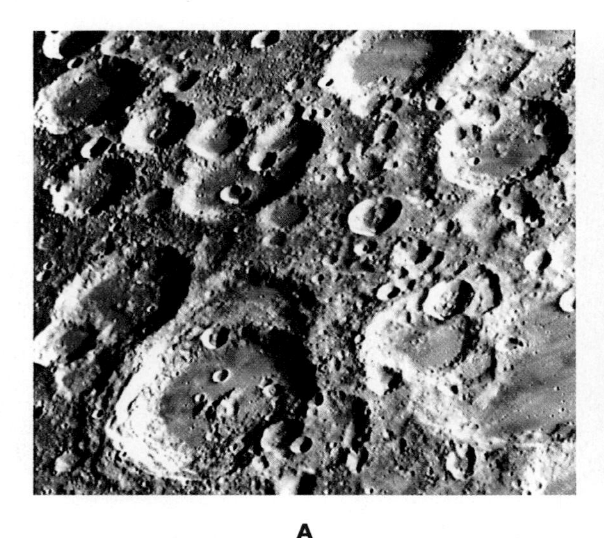

A

B

FIGURE 37.2
(A) Overlapping craters in the Moon's highlands. (B) Isolated craters in the smooth mare.

Craters are not the only features we can see on the Moon. Long, light streaks called **rays** radiate outward from many craters. A particularly bright set of rays is visible in and around a crater named Tycho close to the southern edge of the Moon (see Figure 37.1). The material blasted out of a crater by an impacting body splatters pulverized rock— sometimes melted into small glasslike spheres—outward in various directions. The pattern of the rays was probably determined by the preexisting topography.

Closer examination of the Moon's surface reveals some less common features. One example is the lunar canyons known as **rilles,** perhaps carved by ancient lava flows, which wind away from some craters, as shown in Figure 37.3. Elsewhere, straight rilles gouge the surface, probably the result of crustal cracking. You can observe everyday examples of similar cracks developing by monitoring either drying mud or chocolate pudding left uncovered in the refrigerator.

Sometimes a large impact compresses rock at the point of contact so strongly that afterward the ground rebounds upward, creating a central peak, as you can see in Figure 37.4A. Large impacts may also leave behind multiple concentric rings in the impact crater as the crust solidifies after the impact. These phenomena are analogous to how a pond might look if flash frozen a moment after someone dropped a rock into it (Figure 37.4B).

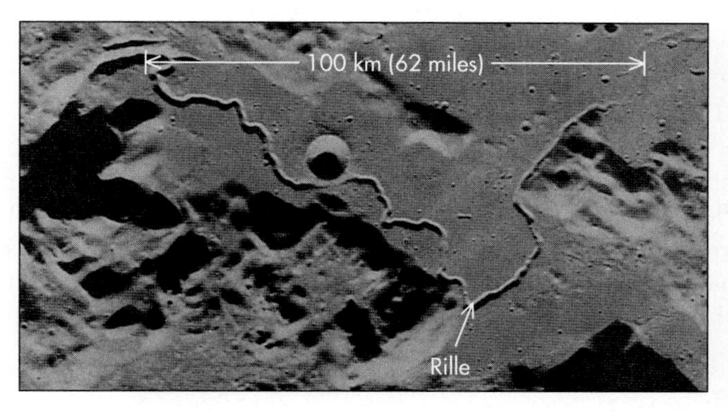

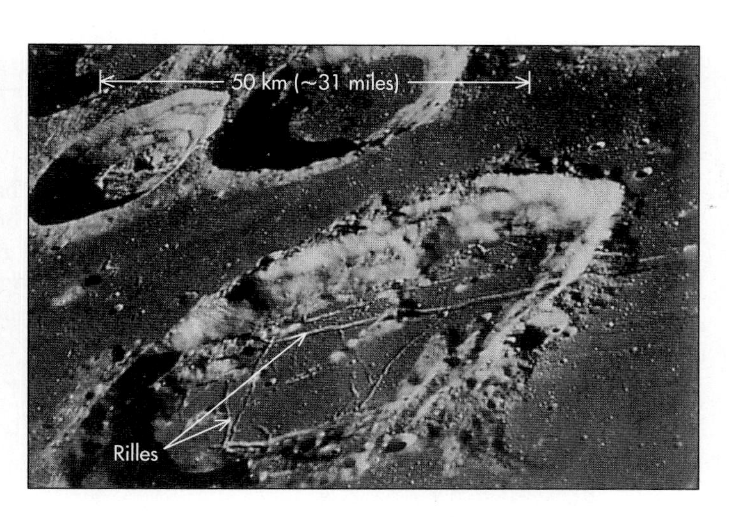

FIGURE 37.3
Photographs of lunar rilles. These channels may have been formed by lava flows.

FIGURE 37.4
(A) Central peak in a crater and slumped inner walls. Apollo astronauts took this photograph of the crater Eratosthenes on the last human flight to the Moon. This crater is 58 kilometers (approximately 36 miles) in diameter. (B) A raindrop falling into water exhibits surrounding rings and central rebound.

FIGURE 37.5
An image made from *Apollo 16* as it orbited the Moon. The left of the image shows part of the side visible from Earth, while the right shows the far side of the Moon. The far side is more heavily cratered and lacks the large maria we see on the near side.

Figure 37.5 shows a view never visible from Earth, revealing part of the far side of the Moon seen by *Apollo 16* astronauts as they orbited the Moon. The Moon always keeps the same side facing the Earth, as is discussed further in Section 37.4. Spacecraft orbiting the Moon show that the far side looks quite different from the side of the Moon that faces the Earth. The far side consists almost entirely of heavily cratered terrain.

The differences between the near and far sides of the Moon are illustrated in Figure 37.6. This map is colored according to elevation and shows the surprising result that the lowest region of the Moon's surface is on the far side, near the south pole. This region, called the Aitken Basin, is as much as 13 kilometers below the surrounding terrain and spans about 2500 kilometers (1550 miles), making it the largest known impact structure in the Solar System. The basin is completely covered with impact craters, so it resembles the highlands regions more than the maria on the near side of the Moon.

Presumably Earth also was battered by impacts in its youth. Although the vast multitudes of these craters have been wiped out by erosion and plate tectonics, a few remain, either in ancient rocks or from more recent collisions (Unit 48). From the small number of such craters on the maria versus the huge number in the lunar highlands, astronomers can deduce that the main period of bombardment must have ended within a few hundred million years of the Solar System's formation.

37.2 CRUST AND INTERIOR

Astronomers think that the maria are also impact features; but to understand their formation, we must briefly describe the early history of the Moon. From the great age of the highland rocks (in some cases nearly 4.5 billion years), astronomers deduce that these rugged uplands formed shortly after the Moon's birth. At that time, much of the Moon was probably molten, allowing denser material to sink to

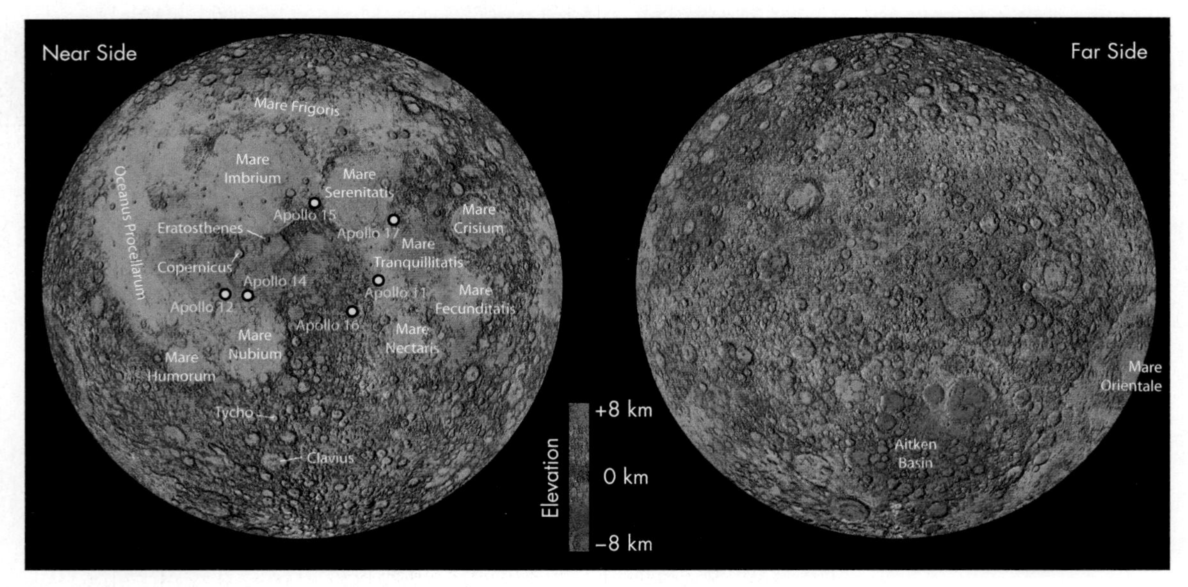

FIGURE 37.6
Topographic map showing the near and far sides of the Moon. The elevations were mapped by the *Clementine* satellite and shown in different colors. The maria are at low elevations, but the largest known impact feature in the Solar System, the Aitken Basin found on the far side of the Moon, is at an even lower elevation. Several major features and the locations of the six Apollo landing sites are labeled.

FIGURE 37.7
Photograph of mountains along the edge of a mare. These mountains were probably thrown up by the impact that created the mare.

its interior while less-dense material floated to the lunar surface. On reaching the surface, the less-dense rock cooled and solidified, forming the Moon's crust. At about the same time, a similar process was probably occurring on Earth, forming its early continents. But during that early formative time, the crusts on both the Moon and the Earth were being heavily bombarded by solid bodies from space, forming many of the numerous craters we still see today on the Moon.

As the Moon continued to cool, its crust thickened. But before the crust grew very thick, a small number of very large bodies—objects over 100 kilometers (60 miles) in diameter—struck the surface, blasting huge craters and pushing up mountain chains along their edges (Figure 37.7). Molten material from within the Moon flooded the vast craters and congealed to form the smooth, dark lava plains that we see now, as illustrated in Figure 37.8. Because the maria formed after most of the impacting bodies were gone—having built the Earth, Moon, and other planets by earlier collisions—few bodies remained to crater the maria. The maria therefore remain relatively smooth to this day.

Craters and maria so dominate the lunar landscape that we might not notice the absence of folded mountain ranges and the great rarity of volcanic peaks—landforms common on Earth. Why have such features not formed on the Moon? The answer lies in the Moon's small size relative to the Earth. Heat escapes more easily from small objects like the Moon than from larger objects like the Earth. Thus, the Moon has cooled more than the Earth. Without a strong heat source, the Moon lacks the convection currents that drive plate tectonic activity such as that on Earth. Over time, tectonic action has erased Earth's earliest blemishes, and thereby the history of most major impacts to its surface. The Moon, with its lack of tectonic activity, preserves its impact history for us to observe on any clear night.

The Moon's interior can be studied via seismic waves, just as the Earth's can. One of the first instruments set up on the Moon by the Apollo astronauts was a seismic detector. Measurements from that and other seismic detectors placed by later Apollo

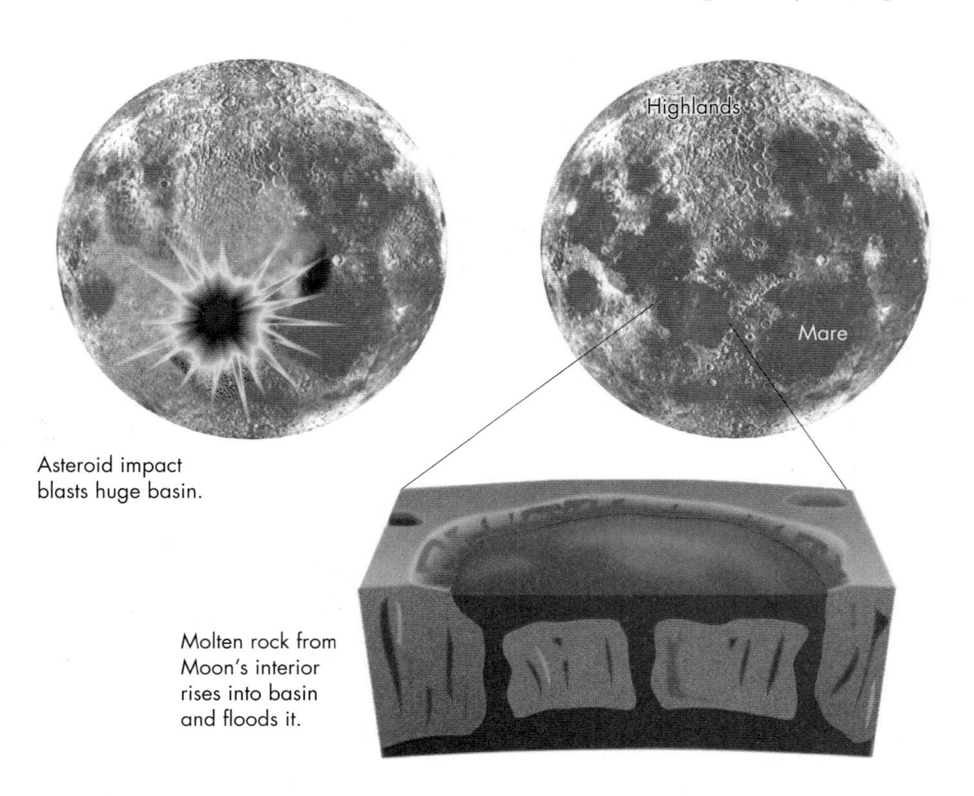

FIGURE 37.8
Large impacts late in the process of the Moon's formation formed huge basins. Lava flooded the basins to make the maria.

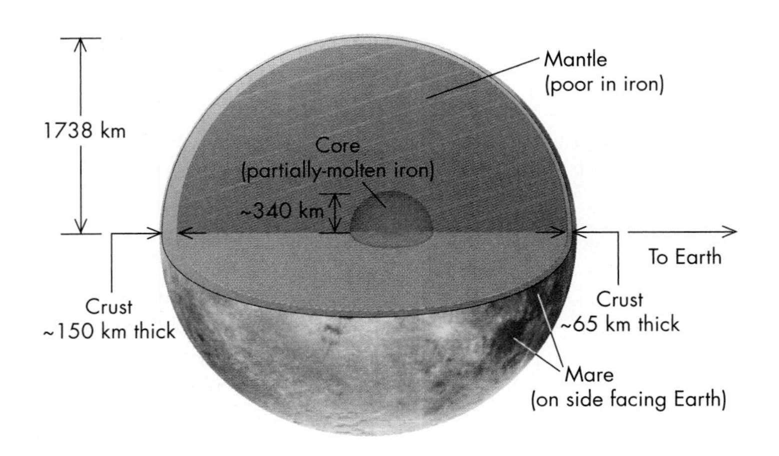

FIGURE 37.9
An artist's impression of the Moon's interior. Notice the thinner near-side crust and the displacement (exaggerated for clarity) of the core toward the Earth. The Moon's iron core is small and probably partially molten, and the surrounding mantle rock may be partially molten too.

Astronauts also generated seismic waves by sending their lunar landers back to crash into the Moon's surface after starting home to Earth.

FIGURE 37.10
Footprint of an astronaut on the Moon.

Q The Earth's distribution of continents is also very asymmetric, with most of the land mass concentrated in the Northern Hemisphere. In what ways is this similar to, and in what ways is it different from, the asymmetry of the Moon's crust?

crews show that the Moon's interior consists of a core, mantle, and crust. These detectors show that the Moon is much less active, and has a much simpler structure, than the Earth, as illustrated in Figure 37.9. "Moonquakes" are relatively infrequent because the Moon is so inactive. The Earth's gravitational pull introduces internal stresses in the Moon, and landslides of crater rims occur occasionally. The quakes allow astronomers to learn more about the Moon's interior, just as on Earth.

The Moon's surface layer is shattered rock that forms a **regolith**—meaning "blanket of rock"—tens of meters deep. The regolith is made of rocky chunks and fine powder, the result of successive impacts breaking rock into smaller and smaller pieces. This powdery nature can be seen in a photograph of an astronaut's footprint on the Moon (Figure 37.10). Below this surface layer of rock and dust rubble is the Moon's crust, averaging about 100 kilometers (60 miles) in thickness. Like the Earth's crust, it is composed of silicate rocks relatively rich in aluminum and poor in iron. Beneath the crust is a thick mantle of solid rock, extending down about 1000 kilometers (600 miles), or about 700 kilometers (400 miles) from the Moon's center. Unlike the Earth's mantle, however, the Moon's mantle appears too cold and rigid to be stirred by the feeble heat found coming from the Moon's interior.

The Moon's low density (3.3 kg/L) tells us that its interior contains relatively little iron. Recall our earlier observation that the Earth's high density (about 5.5 kg/L) indicates that the Earth has a large iron core. Some molten material may lie below the Moon's mantle, but the Moon's core is smaller and contains far less iron than the Earth's. These factors, plus the Moon's slow rotation, cause the Moon to lack a magnetic field. Thus a compass would be of no use to an astronaut lost on the Moon.

One puzzle is that the Moon's crust is quite asymmetric. It is about 65 kilometers (40 miles) thick on the side facing the Earth, and about 150 kilometers (90 miles) thick on the far side. The thicker crust of the Moon's far side probably explains why there are so few maria there—the crust was too thick for magma in the Moon's interior to flood even large impact basins. It is possible that the many large impacts that created the maria played a role in redistributing the matter within the Moon. Then later as the Moon slowed to synchronous rotation, the asymmetric mass distribution aligned with Earth's tidal force.

37.3 THE ABSENCE OF A LUNAR ATMOSPHERE

The Moon's surface is never hidden by lunar clouds or haze, nor does the spectrum of sunlight reflected from it show obvious signs of gases. The Moon has no atmosphere because the Moon's small mass creates too weak a gravitational force for it to

Q If the Moon's mass is only about 1/80th of the Earth's mass, why is its escape velocity one-quarter of the Earth's escape velocity?

retain the gas. To be more precise, its escape velocity (Unit 18) is so low that gases are able to leak away from the Moon over times that are short compared to its age.

If the Moon were currently producing large amounts of gas, it would retain an atmosphere for many millions of years. However, after its formation the interior cooled quite rapidly. It released few gases by volcanic activity, which was probably an important source of much of the Earth's early atmosphere. The Moon's escape velocity is only about a fourth of the Earth's. Therefore atoms in any atmosphere the Moon might have once possessed would have escaped relatively easily.

The lack of atmosphere has several important consequences for the Moon's physical characteristics. Without an atmosphere's greenhouse effect, the average temperature of the Moon's surface is only about 250 K (about −10°F) versus the Earth's mean temperature of about 290 K (about 60°F). Without an atmosphere to retain heat, there are dramatic temperature changes between day and night—reaching about 390 K (about 240°F), hot enough to boil water, during the two-week-long lunar day, and dropping to about 120 K (about −240°F) during the equally long lunar night.

Near the north and south poles of the Moon, there are craters that are never lit by the Sun, and their temperature probably remains close to about 40 K (about −390°F). At such constant low temperatures, it is possible that water ice survives there, perhaps buried below the surface. NASA space probes have found probable evidence for ice in these regions, based on their reflectivity to radio waves and neutrons emitted when high-energy particles interact with hydrogen atoms within about a meter of the surface. The amount of neutrons detected suggests that there is a significant amount of water near both poles. If a large source of water could be found, it could improve the possibilities for someday building a permanent base for humans on the Moon.

37.4 THE MOON'S ROTATION

As it orbits Earth, the Moon keeps the same side facing us, as you can see by watching it through a cycle of its phases. You might think from this that the Moon does not rotate. Figure 37.11 shows, however, that the mountain on the side facing the Earth points to the right when the Moon is at A and to the left when the Moon is at B. Thus the Moon *does* turn on its axis relative to the stars, but with a rotation period exactly equal to its orbital period, a condition known as **synchronous rotation.**

The Moon probably rotated more rapidly in the distant past than it does now. But the Earth and Moon exert gravitational forces on each other that slow their rotation. The Earth's spin is currently slowing due to tidal forces (Unit 18), and the same was presumably true in the Moon's early history. The Earth's tidal force slowed the Moon's spin until the Moon now keeps the same side toward the Earth—similar to how a ball with an off-center mass inside it might settle with its heavier side facing down.

The Moon appears to *librate,* or "wobble," slightly from our perspective on Earth, so that we can see a small percentage of the far side at different times of the month. This effect is mainly due to the Moon's orbital speed varying (because its orbit is elliptical) while its speed of rotation remains almost precisely constant.

The Moon never had oceans that would produce the kinds of tides familiar to us in the Earth's oceans. Instead of shifting the position of highly mobile water as on Earth, the Earth's gravitational pull distorts the shape of the whole Moon from a sphere into an elongated shape. Assuming the Moon was rotating faster early in its history, the tidal effect of the Earth would have pulled on the Moon's bulge, braking the Moon's rotation. Once the Moon's spin slowed to the point where it was in synchronous rotation, no further tidal braking was possible. This left the Moon in the position of appearing not to rotate relative to the Earth, but it still rotates relative to the Sun and stars.

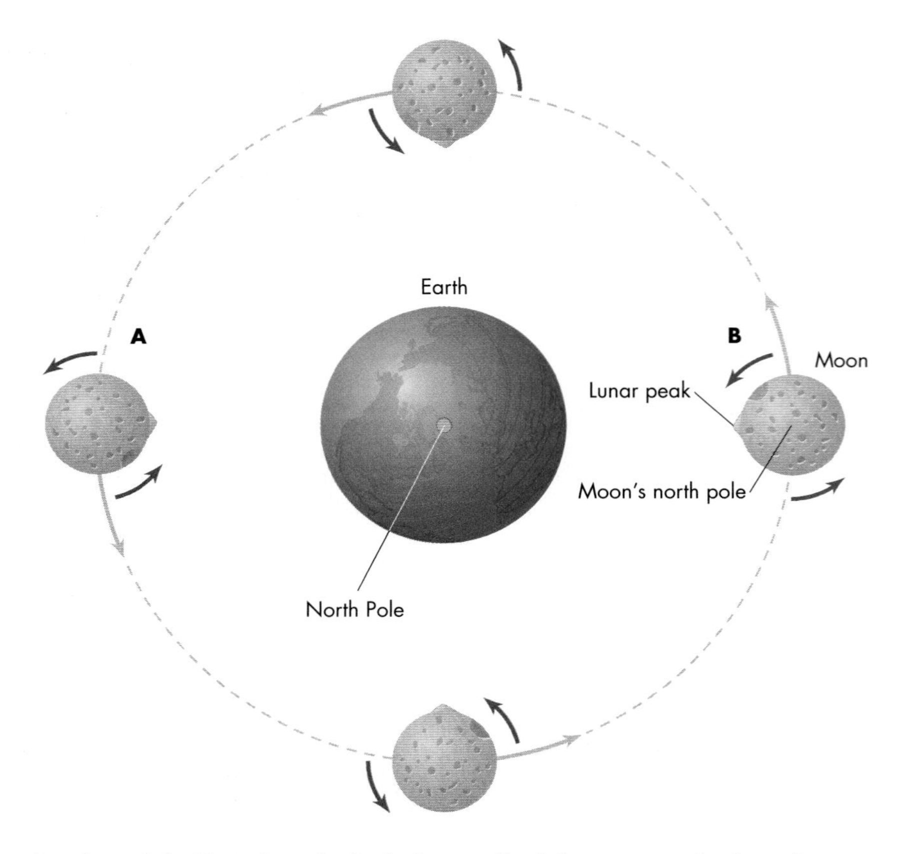

FIGURE 37.11
The Moon rotates once each time it orbits the Earth, as can be seen from the changing position of the exaggerated lunar mountain. Notice that at (A) the lunar peak is to the right, while at (B) it is to the left. Thus, from the Earth we always see the same side of the Moon even though it turns on its axis.

To help you see that the Moon rotates even though it keeps the same face toward the Earth, put a coin on the figure of the Moon and move it around the Earth so that the same edge of the coin always faces the Earth.

Rotation of the Moon

Similar tidal effects have locked almost all of the moons of other planets into synchronous rotation. Tidal braking by the Sun probably is a major reason for the slow rotation of Mercury and Venus (Units 38 and 39).

37.5 THE ORIGIN OF THE MOON

Lunar rocks brought back to Earth by the Apollo astronauts have led astronomers to radically revise their ideas of how the Moon formed. Before the Apollo program, lunar scientists had three hypotheses of the Moon's origin. In one, the Moon was originally a small planet orbiting the Sun that approached the Earth and was captured by its gravity. In another, the Moon and Earth were "twins," forming side by side from a common cloud of dust and gas. In the third, the Earth initially spun enormously faster than now and formed a bulge that ripped away from the Earth to become the Moon.

Each of these hypotheses led to different predictions about the composition of the Moon. For example, had the Moon been a captured planet formed elsewhere in the Solar System, its composition might be very unlike the Earth's. If the Earth and Moon had formed as twins, their overall composition should be the same. Finally, if the Moon was once part of the Earth's surface layers, its composition should be nearly identical to the Earth's crust. When the rock samples returned by the Apollo missions were analyzed, astronomers were surprised that for some elements the composition was the same as the Earth, but for others it was very different. For example, Moon rocks have an abundance of high-melting-point materials such as uranium and a strong deficit of low-melting-point materials such as lead. The Moon also has much less iron than the Earth, as we pointed out when discussing its interior and low density.

Thus the Moon's composition is quite puzzling. It is rich in high-melting-point materials, as if it formed at a higher temperature than the Earth. However, it has a deficit of iron—the most common of all the high-melting-point elements—as if it formed in a region farther from the Sun than the Earth.

Birth of the Moon

Q Would tidal braking in the distant past between the Earth and the Moon have been greater than, equal to, or less than what it is today? Why?

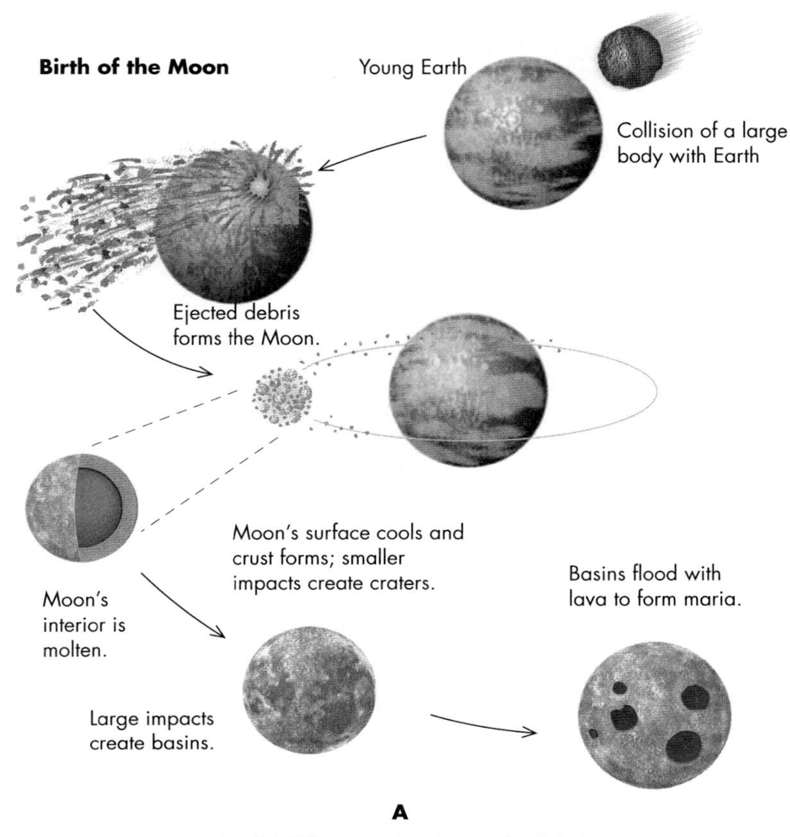

The failure of the lunar surface samples to confirm any of the three hypotheses led astronomers to consider alternatives; and now a different hypothesis for the Moon's origin has emerged after a variety of computer simulations. According to the new hypothesis, the Moon formed from debris blasted out of the Earth by the impact of an object comparable to the size of the planet Mars, as illustrated in Figure 37.12. The great age of lunar rocks and the absence of an impact feature on the Earth indicate that this event would have had to occur during the Earth's own formation, about 4.5 billion years ago.

This hypothesis can explain the compositional abnormalities of the Moon if the collision occurred after the proto-Earth had become differentiated. The material blasted out of the Earth's mantle would have relatively little iron because the iron, being dense, had sunk deep into the interior. The colliding body splattered millions of cubic kilometers of the Earth's lower-density mantle and crust, hurling it into space in an incandescent plume. The strong heating of the blasted material would have baked out the more volatile substances, so that what became incorporated in the Moon would have a deficit of them as well. As the debris cooled, its gravity would have drawn it together into what we now see as the Moon.

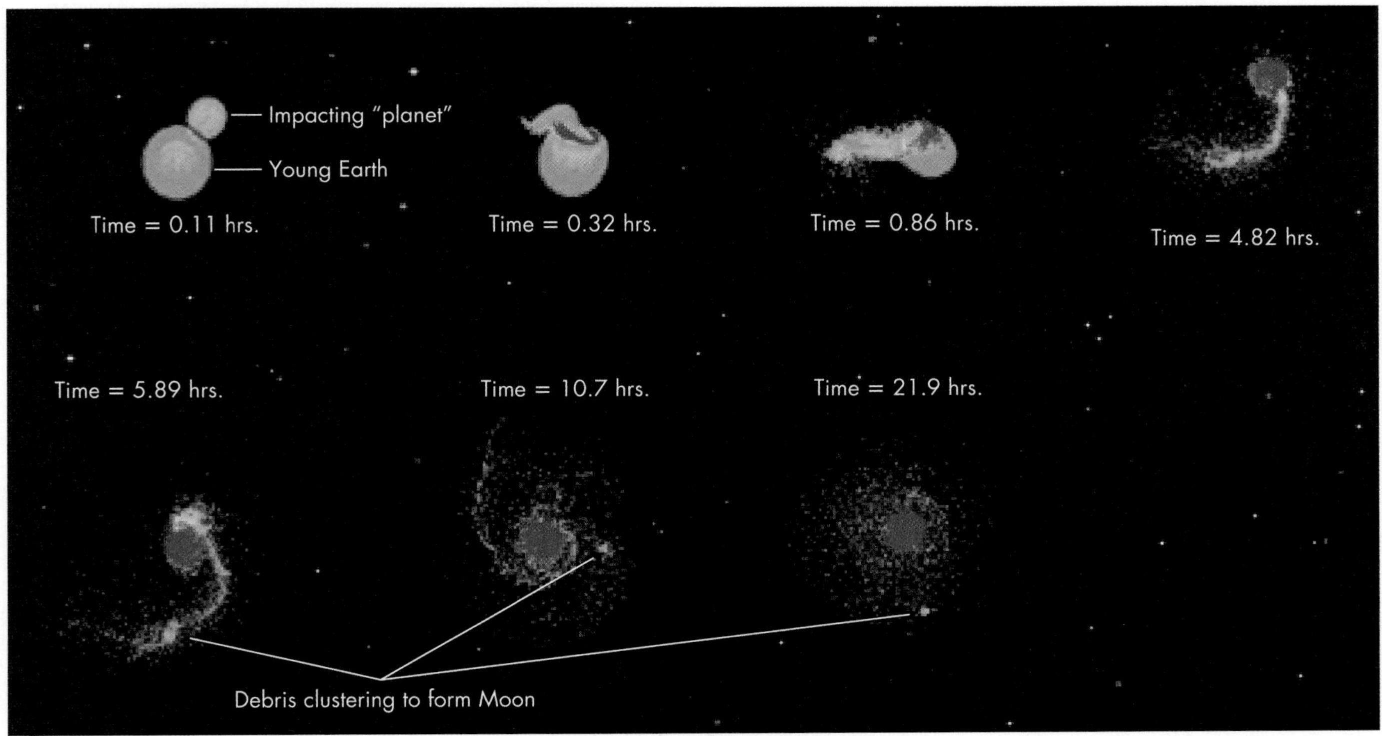

FIGURE 37.12
Origin of the Moon. (A) Artist's sketch illustrating the main stages in the birth of the Moon. (B) This computer simulation shows how the Moon might have formed when a Mars-size object hit the young Earth and splashed out debris that later assembled into the Moon.

KEY TERMS

basalt, 288
crater, 287
highlands, 288
impact, 288
mare (pl. *maria*), 288
rays, 289
regolith, 292
rille, 289
synchronous rotation, 293

QUESTIONS FOR REVIEW

1. Why is the crater created by an impacting object so much larger than the object itself?
2. Why are the maria and highlands so different in appearance?
3. Why is the Moon so much less active than the Earth?
4. What is odd about the Moon's rotation? Why does it spin in this fashion?
5. What are the effects of tides on the Earth and on the Moon?
6. How do astronomers now think the Moon formed? What evidence supports this hypothesis?

PROBLEMS

1. Our ideas about how the Moon formed and how those ideas were gradually discarded are excellent examples of how science works: If a hypothesis is disproved it must be either modified or thrown out. Explain in terms of the origin of the Moon. What were the problems for the early hypotheses?
2. Galileo was able to estimate the height of mountains on the Moon by measuring the length of their shadows. Make a drawing of a tall mountain sitting on a flat plane viewed from the side. With a protractor, draw a line with a 10° angle to the plane that touches the mountain peak. This shows how long the shadow is when the Sun is 10° above the horizon.
   a. If a shadow is 5 kilometers long when the Sun is 10° above the horizon, estimate how tall the mountain is.
   b. Estimate at what angle the Sun must be above the horizon for the shadow to be twice as long as the height of the mountain.
3. The Moon's density is 3.3 kg/L. Suppose that the Moon is made of just two substances: silicates with a density of 3.0 kg/L and iron with a density of 7.9 kg/L.
   a. What fraction of its mass must be iron?
   b. What fraction of its volume must be iron?
4. The Moon is currently moving away from the Earth at about 4 cm per year.
   a. At this rate, how many years would it have taken the Moon to move half the current distance between the two bodies?
   b. Would the effects leading to this recession have been larger or smaller in the past? Why?
5. As the Moon moves away from the Earth, the radius of its orbit grows larger, although its orbital speed grows smaller. The angular momentum of the Moon is equal to the product of its mass and its orbital velocity and radius ($M \times V \times R$). Is the Moon's orbital angular momentum growing larger or smaller? Use Newton's version of Kepler's third law (Unit 17) to determine the Moon's speed as a function of its orbital radius, and show whether the angular momentum is increasing or decreasing.
6. Assume that the density of an average asteroid is 3.0 kg/L, and that the craters on the Moon are roughly 20 times the size of the impacting body.
   a. What was the diameter of the asteroid that created the 240-km-diameter Clavius crater? Using the density of an average asteroid, calculate the mass of this asteroid.
   b. Calculate the kinetic energy of this impact if the speed of the asteroid was 20 km/sec relative to the Moon.
7. The largest impact basin on the Moon is Aitken, near the lunar south pole. It is approximately 2500 kilometers in diameter. Repeat the calculations you made in problem 6. Compare the kinetic energy of this asteroid to the kinetic energy of the Moon in its orbit around the Earth of 1.0 km/sec. (The Moon's mass is 7.35×10^{22} kg.)

TEST YOURSELF

1. Compared to the lunar highlands, the lunar maria are
   a. smoother and older.
   b. smoother and younger.
   c. more cratered and older.
   d. more cratered and younger.
2. If the Moon did not rotate on its own axis, we would observe
   a. both sides of the Moon.
   b. the Moon remaining stationary against the stars.
   c. a lack of tides on the Earth.
   d. the Moon from only one hemisphere of Earth.
   e. Trick question—it doesn't rotate.
3. The lunar maria were formed by
   a. numerous impacts by small meteoroids.
   b. tectonic uplifts of the surrounding highlands.
   c. sedimentation as dust was deposited in shallow seas.
   d. lava flows into large impact basins.

unit

38 Mercury

Mercury is named for the Roman deity who was the speedy messenger of the gods. The name was inspired by the fact that Mercury changes its position in the sky faster than any other planet. Mercury is difficult to observe from Earth because it orbits the Sun at just 0.39 AU and usually lies in the Sun's glare. But every 1½ to 2½ months, it reaches its largest angular separation ("greatest elongation," Unit 11) from the Sun, alternately in our evening and morning skies. At these times you may see it at not much more than a hand's span above the horizon.

Mercury (Figure 38.1) is the smallest of the eight planets, with a radius about one-third of Earth's and a mass 18 times smaller—not very much bigger than the Moon. It has been visited by only two space probes, *Mariner 10* in 1974–1975 and *Messenger,* which began a long-term rendezvous in 2008. Before *Messenger* arrived, less than half of its surface had been seen. Images from these spacecraft show that Mercury's surface resembles our Moon in many ways. However, it also differs from the Moon in several important respects.

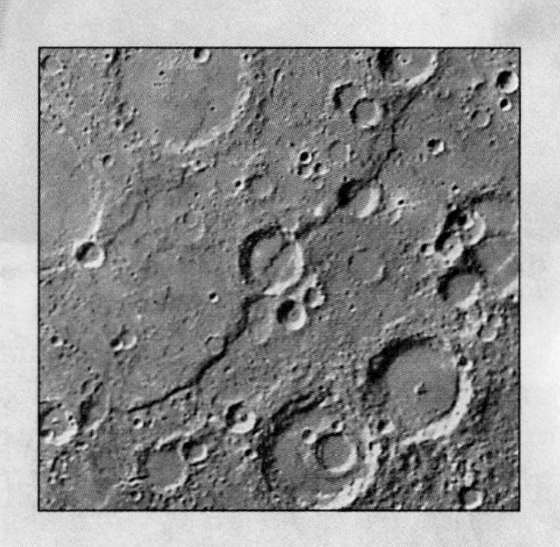

Background Pathways

38.1 MERCURY'S SURFACE FEATURES

At first glance Mercury's landscape (Figure 38.2) is difficult to distinguish from the Moon's; but a careful examination of the craters reveals differences. First, Mercury's impact craters generally overlap less than the Moon's. In addition, the

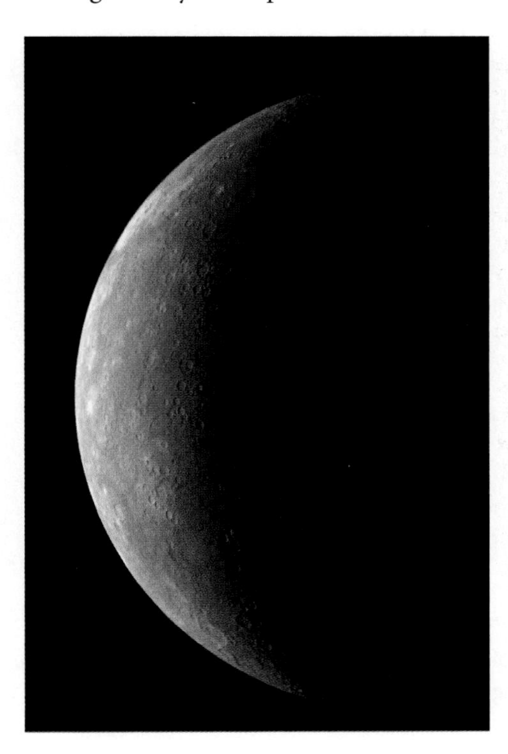

FIGURE 38.1
The planet Mercury imaged by the *Messenger* spacecraft in 2008.

Most features on Mercury are named after artists, writers, and other cultural figures.

The *Messenger* spacecraft must make a series of flybys, designed to minimize the need for rocket fuel while it matches Mercury's orbital velocity. *Messenger* will be placed into orbit around Mercury in 2011.

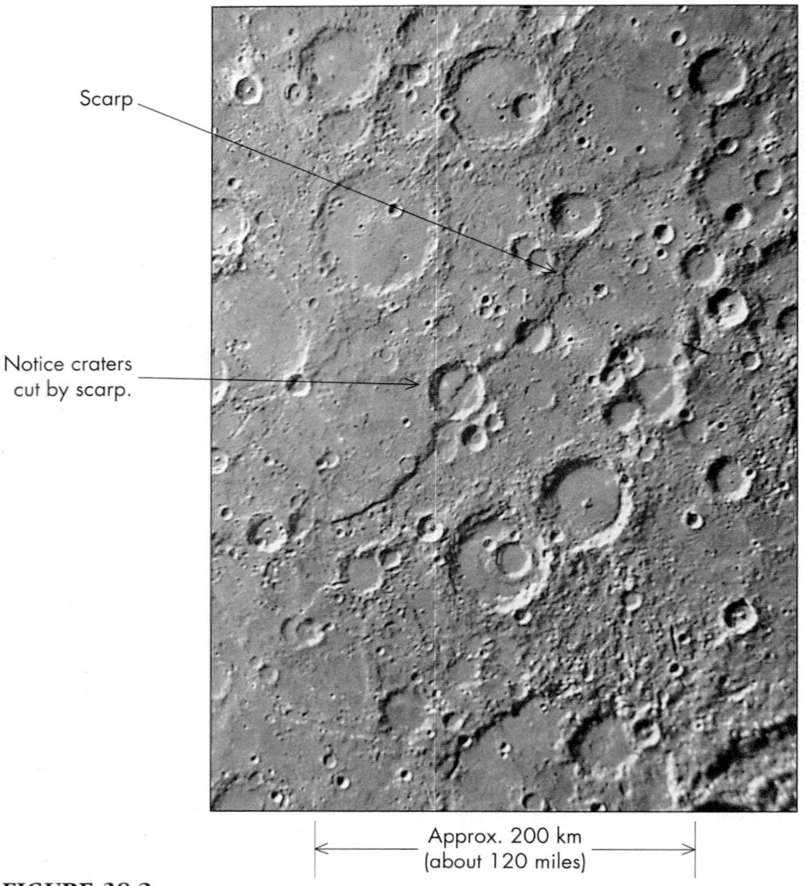

FIGURE 38.2
Picture of Mercury's cratered surface (taken by the *Mariner 10* spacecraft in 1974). Note the scarp crossing several craters, indicating that it formed after the craters.

crater walls tend to be less steep, most likely because Mercury's surface gravity is more than twice as strong as the Moon's, making steep hills less stable on Mercury.

Mercury's surface consists of more than just craters. Congealed lava flows flood not only many of its old craters but much of its surface. On our Moon, these flows are found almost exclusively within the maria. The most unique features on Mercury's surface are enormous **scarps**—cliffs formed where the crust has shifted—as seen in Figure 38.2. Hundreds of scarps cover Mercury's surface. Some run for hundreds of kilometers and range up to 3 kilometers high. Because the scarps cut across many impact craters, they must have formed later in Mercury's history, perhaps as the planet cooled and shrank, wrinkling like a dried apple.

By far, the largest of the craters imaged in 1974 is the vast **Caloris Basin,** only the edge of which was seen by *Mariner 10* (Figure 38.3A). With a diameter of about 1500 kilometers (about 900 miles), this mountain-ringed depression is reminiscent of some of the lunar maria. Moreover, its circular shape and surrounding hills indicate that, like the maria, it was formed by a major impact after the young Mercury's surface had begun to cool.

Astronomers waited more than three decades to see the rest of the Caloris Basin, which was imaged by *Messenger* in 2008. Near its center is an

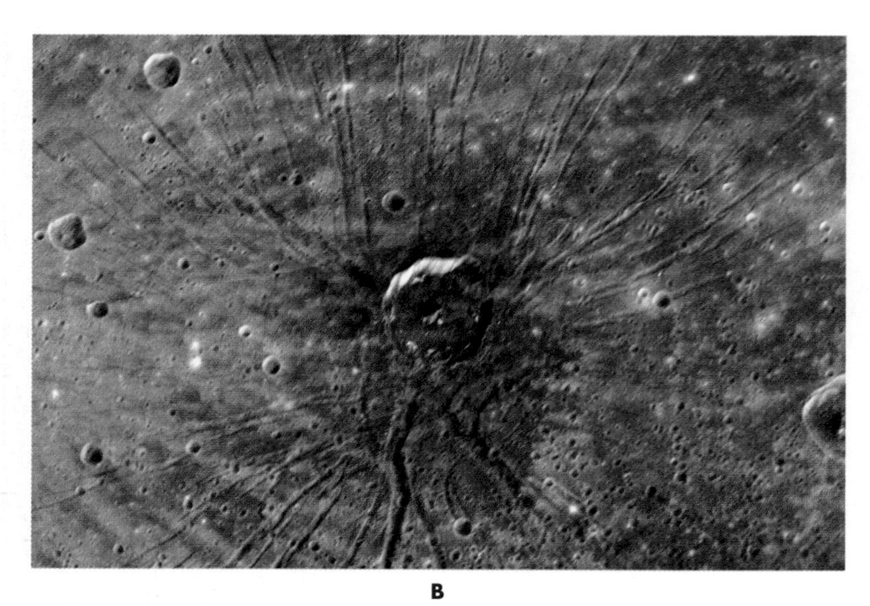

FIGURE 38.3
(A) The edge of the Caloris Basin is seen in the semicircular set of rings surrounding Mercury's largest impact feature. Only this edge of the basin was seen by the passing *Mariner 10* spacecraft in 1974. (B) The rest of the Caloris Basin was imaged by the *Messenger* spacecraft in 2008. Those images revealed a strange spidery pattern of troughs surrounding a crater 40 kilometers in diameter near the center of the basin.

FIGURE 38.4
Mariner 10 image of odd jumbled terrain that lies on the side of Mercury opposite the Caloris Basin.

FIGURE 38.5
False-color image of Mercury made by *Messenger* in 2008. The imager can measure light over a range of wavelengths from the infrared to violet. Subtle differences in color are exaggerated in this image to reveal compositional differences in the surface rock. Lighter blue regions are probably fresh material from relatively recent impacts. The large orange region at upper right is the Caloris Basin.

unusual and poorly understood feature, a spider-shaped pattern of troughs radiating away from a small crater (Figure 38.3B). The crater is too small to have been the result of the impact that formed the basin, so it is probably the result of a later collision. The spidery pattern of troughs suggests that this portion of Mercury's surface has been stretched, as if internal activity in Mercury caused the surface here to bulge outward and split.

The Caloris impact had global effects on Mercury. On the side of the planet opposite the Caloris Basin, the terrain has a hilly, jumbled appearance (Figure 38.4). Astronomers have hypothesized that this was produced by the seismic waves that would have traveled through the planet and along the surface after the Caloris impact event. These waves would have converged from all directions on the far side of the planet almost simultaneously, and might have shaken apart the landscape into its present appearance.

Messenger's imaging system can detect subtle color differences between different portions of Mercury's surface, which are exaggerated in Figure 38.5 to make them clearer. Some fresh impact features, including rays of ejected material, are visible as light-blue features in this image. Different colors of other regions may indicate compositional variations and possibly ancient lava flows in some regions. The Caloris Basin stands out in this image as a large orange-colored region, but the cause of this color difference is not yet known.

Q What could explain why Mercury shows none of the large dark maria that are so prominent on our Moon?

38.2 MERCURY'S TEMPERATURE AND ATMOSPHERE

Mercury's surface is one of the hottest places in the Solar System, and it undergoes some of the most extreme changes. At its equator, noon temperatures can reach approximately 710 K (about 820°F). On the other hand, nighttime temperatures are among the coldest in the Solar System, dropping to approximately 80 K (about −320°F). These extremes result from Mercury's closeness to the Sun and from its lack of atmosphere. As on the Moon, no atmospheric gas is present that could retain heat during the long Mercurian night.

Mercury has the most elliptical orbit of any planet. Its closest approach to the Sun, when Mercury is at its **perihelion,** is 46 million kilometers (29 million miles). Its orbit carries it about 50% farther away at its most distant, about 70 million kilometers (43 million miles), where the heating it receives from the Sun is substantially less. At perihelion, from Mercury's perspective the Sun looks about 50% bigger, and the planet receives about twice as much solar radiation as at its greatest separation. Thus noon temperatures on Mercury's equator reach 710 K at perihelion but "only" about 580 K when the separation is largest. The name *Caloris* (from the Greek for "hot") was given to the large impact feature because it lies near one of two points on Mercury's equator that face the Sun directly when Mercury is at perihelion and therefore reach the highest temperatures.

Mercury lacks an atmosphere for the same reason the Moon does. Although traces of gas, perhaps captured from interplanetary space, have been detected in Mercury's spectrum, Mercury's small mass makes its gravitational attraction too weak to retain any significant amount of gas. Even though Mercury's escape velocity is almost twice as large as the Moon's, Mercury's closeness to the Sun makes its temperature higher. That in turn causes any gas molecules there to move so rapidly that they readily escape into space.

The conditions on Mercury are so severely hot that astronomers were surprised to find evidence of ice at Mercury's poles. This evidence is based on the reflection of radar waves, which are affected in a characteristic way by ice. Astronomers think that Mercury's ice, like that on our Moon, is located in the perpetual shadows within craters near the planet's poles. Where would ice have come from? Astronomers are not certain, but perhaps it was delivered by comet impacts. One of the goals of the *Messenger* spacecraft mission is to understand these observations.

Q If Mercury were able to retain an atmosphere of carbon dioxide, how might its surface characteristics differ?

38.3 MERCURY'S INTERIOR

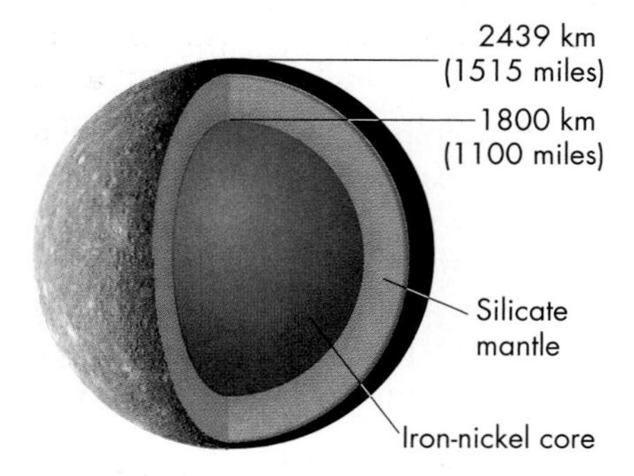

FIGURE 38.6
Artist's depiction of Mercury's interior.

Mercury appears to have a large iron core beneath its silicate crust. Because no spacecraft has landed there to deploy seismic detectors, astronomers base this conclusion primarily on Mercury's density. Although a massive planet's gravity can compress its interior to high density, Mercury is too small for this effect to be significant. Therefore its high density (5.4 kilograms per liter) indicates an iron-rich interior with only a thin rock (silicate) mantle. A model of the probable internal structure is shown in Figure 38.6.

There is only a slight difference in the temperatures at which rock and iron condense. Mercury's richness in iron relative to its rock content is thus larger than expected. One possible explanation for this difference is that Mercury once had a thicker rocky crust that was blasted off by the impact of a large planetesimal early in the Solar System's history (Figure 38.7). A collision like this could also explain Mercury's relatively elliptical orbit. Another hypothesis is that after Mercury formed and differentiated, the young Sun became hot enough at some stage to vaporize the planet's surface rock.

Astronomers used radar measurements of Mercury in 2007 to measure precisely the way that Mercury "wobbles" as it rotates. They found that Mercury behaves as though its core is partially molten. The effect they observed is similar to how a raw egg behaves if you spin it, then touch it momentarily to stop it. After letting go, it will continue to spin a little more. Mercury likewise experiences tugs

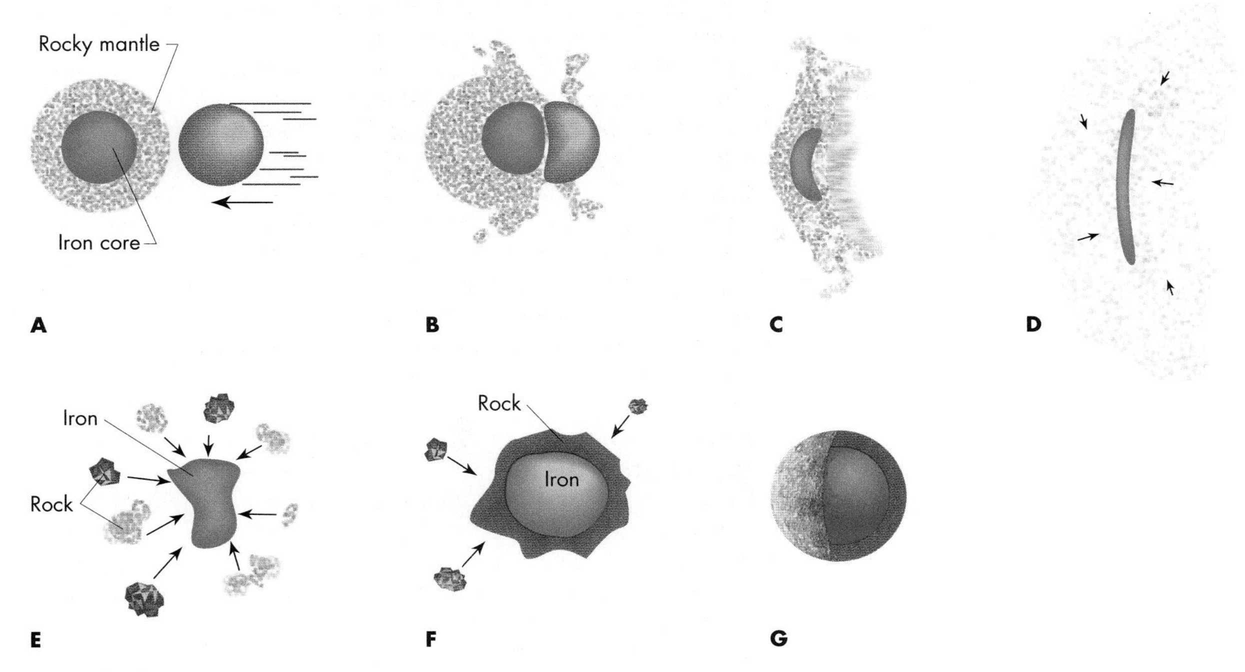

FIGURE 38.7
Computer simulation of a collision between Mercury and a large planetesimal. The impact compresses and flattens the iron core and blasts away most of the outer rocky layers. As the remaining material settles back into the planet, its iron core remains highly distorted, surrounded by a thinner crust of rocky debris. Gravity eventually reshapes the planet into a sphere.

from the Sun that would alter its spin slightly, but it wobbles about twice as much as it would if it were solid all the way through.

Mercury was not expected to have a molten core because of its small mass and radius. Even though it is close to the Sun, this does not provide nearly enough heat to make the core molten. Instead, astronomers now suspect that the outer iron core may be mixed with sulfur, which can lower the melting temperature. This suggests some greater amount of mixing in the solar nebula because sulfur should not have condensed so close to the Sun. This partially molten core also helps explain why Mercury has a magnetic field, albeit one that is only about 1% that of the Earth's. A magnetic field is generated by circulation of electrically conducting material in a planet's interior.

Q If a planet of Mercury's mass had formed at the Earth's distance from the Sun, how might it have differed from Mercury?

38.4 MERCURY'S ROTATION

Mercury spins very slowly. Its rotation period is 58.646 Earth days relative to the stars, exactly two-thirds its orbital period around the Sun of 87.969 Earth days. This means that it spins exactly three times for each two trips it makes around the Sun. This is not just coincidence. Mercury probably spun much faster when it formed, but the Sun's tidal forces slowed Mercury's rotation. This is similar to how the Earth tidally braked the Moon's rotation until the same side always faced us (Unit 19). However, Mercury did not slow down to the point where the same side always faces the Sun.

The difference between Mercury and the Moon in the final result for their rotation is probably a consequence of Mercury's highly elliptical orbit around the Sun.

FIGURE 38.8
Mercury's odd rotation. The planet spins three times for each two orbits it makes around the Sun. Only one orbit is shown here. Note that a feature on the surface has completed one and a half rotations after one orbit.

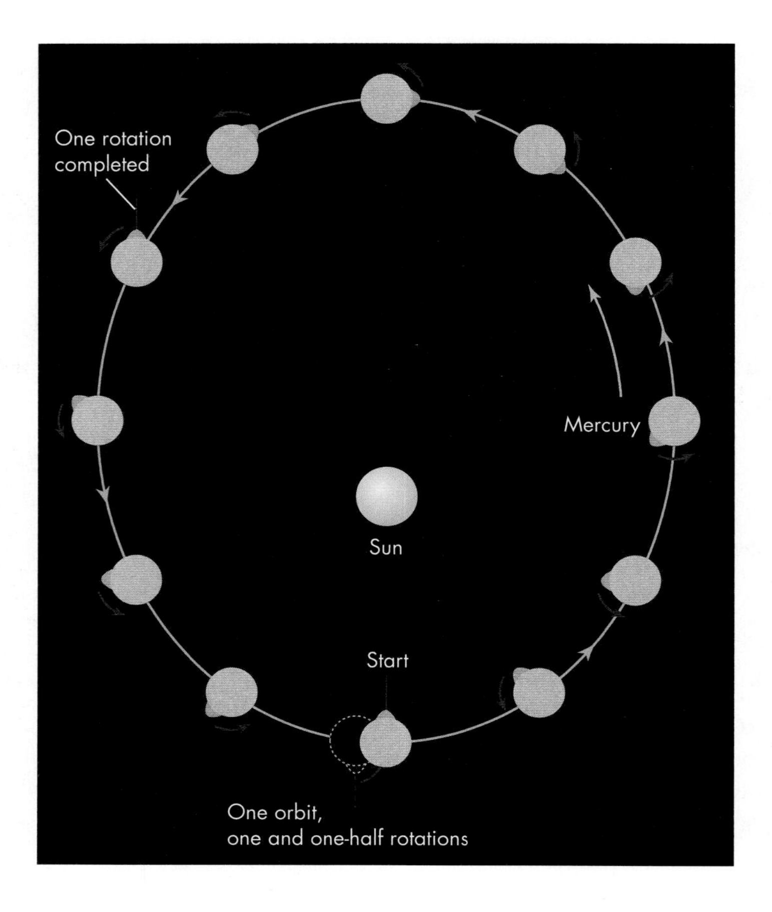

Rotation of Mercury

Because of the ellipticity, Mercury's orbital speed changes as it moves around the Sun, in accordance with Kepler's second law of planetary motion. That changing speed prevents the Sun from locking Mercury into a purely synchronous spin with a spin rate equal to the orbital rate, leaving one side always facing the Sun. Instead, it has slowed Mercury's spin until one side faces toward the Sun on one closest approach and then away from the Sun on the next closest approach (Figure 38.8) when the tidal forces are strongest. This achieves a very slow spin and a stable orientation for the planet.

A secondary effect of this unusual **tidal lock** is that Mercury's day is longer than its year! While Mercury orbits the Sun in 88 Earth days, the time between successive "noons" is two orbital periods, or 176 days. Having a day longer than a year sounds strange, but consider that if one side of Mercury always faced the Sun, the "day" would be perpetual on that side.

If Mercury did not rotate at all with respect to the stars, the Sun would appear to rise in the west and set in the east.

As on Earth, Mercury's spin causes the Sun to rise in the east and set in the west. However, the slow rate of Mercury's rotation combined with its large orbital eccentricity gives rise to a peculiar effect. Near perihelion, Mercury's orbital speed is so large that for about eight days (of Earth time) the speed of its motion around the Sun dominates the effect of the planet's slow rotation. As a result, the Sun travels backward across the Mercurian sky, moving from west to east. If you were watching from the Caloris Basin, you would see this backward motion of the Sun around Mercurian "noon." Forty days after sunrise the Sun would stop traveling to the west, reverse direction for eight days, then begin moving again toward the western horizon, setting 40 days later. From some locations on Mercury, the Sun rises slightly above the horizon for several Earth days of time, then dips back below the horizon before rising again to complete the long Mercurian day.

KEY TERMS

Caloris Basin, 298
perihelion, 300
scarp, 298
tidal lock, 302

QUESTIONS FOR REVIEW

1. How do the craters on Mercury differ from those on the Moon?
2. What are scarps? How might they have formed?
3. What is the evidence for a large iron core in Mercury? What might explain why it is so large?
4. Why does Mercury have no atmosphere? How does this affect Mercury's temperature?
5. What is the evidence for ice on Mercury? How can there be ice there?
6. How can Mercury's day be longer than its year?

PROBLEMS

1. Astronomers think that Mercury's scarps could have formed if the circumference of the planet shrunk by about 3 kilometers. By what percentage would Mercury's 2439-km radius have to have changed in order to produce this reduction in circumference? What percentage decrease in Mercury's volume is this?
2. Find Mercury's escape velocity (Unit 18), given that its radius is 2439 kilometers and its mass is 3.3×10^{23} kg. How does this compare to the Moon's escape velocity of 2.4 km/sec?
3. The highest temperature on Mercury is about 710 K, while the highest temperature on the Moon is about 400 K. The average speed of a molecule in a gas is proportional to the square root of the temperature. The rate of molecules escaping to space depends on the fraction of their average velocity relative to the escape velocity, and Mercury's escape velocity is about twice the Moon's. Given these facts, which would have retained an atmosphere longer, Mercury or the Moon?
4. The average radius of Mercury's orbit is about 58 million kilometers, and it completes an orbit in 88 days.
   a. Find its average speed in km/sec.
   b. Use the conservation of angular momentum to estimate Mercury's speed when it is at perihelion at about 46 million kilometers.
   c. What is its speed at its farthest distance from the Sun of 70 million kilometers?
5. Suppose Mercury's rotation period were three-quarters of its orbital period. How many Mercury days would there be in a Mercury year?
6. Mercury rotates approximately once every 58.6 Earth days relative to the stars, and takes 88.0 Earth days to revolve about the Sun. (It may also help to consult Figure 38.8 for this problem.)
   a. Estimate how fast the Sun would move across Mercury's sky if Mercury were spinning at its same rate, but not revolving about the Sun. In other words, calculate the angular speed (degrees per Earth day) at which the Sun moves across the sky due to Mercury's rotation but not Mercury's orbital motion.
   b. Calculate the average angular speed (degrees per Earth day) at which the Sun moves across the sky due just to Mercury's orbit of 88 days. (This is slower than the motion of the Sun due to its rotation, which is why the Sun rises in the east and sets in the west on Mercury.)
   c. When Mercury is at perihelion, its orbital speed is about 25% faster than its average orbital speed. In addition, because Mercury is closer to the Sun, the angle the Sun shifts by is about 25% greater too. What is the resulting angular speed when Mercury is at perihelion?

TEST YOURSELF

1. Why does Mercury have so many craters and the Earth so few?
   a. Mercury is far more volcanically active than the Earth.
   b. Mercury is much more dense than the Earth, and therefore its gravity attracts more impacting bodies.
   c. The Sun has heated Mercury's surface to the boiling point of rock, and the resulting bubbles left craters.
   d. Erosion and plate tectonic activity have destroyed most of the craters on the Earth.
   e. Mercury's iron core produces a strong magnetic field that attracts other bodies made of iron.
2. Mercury's average density is about 1.5 times greater than the Moon's, even though the two bodies have similar radii. What does this suggest about Mercury's composition?
   a. Mercury's interior is much richer in iron than the Moon's.
   b. Mercury contains proportionately far more rock than the Moon.
   c. Mercury's greater mass has prevented its gravitational attraction from compressing it as much as the Moon is compressed.
   d. Mercury must have a uranium core.
   e. Mercury must have a liquid water core.
3. The scarps that cut across the surface of Mercury probably were
   a. cut by flowing lava.
   b. produced by impacts pushing portions of the crust outward.
   c. formed when the crust buckled as Mercury cooled.
   d. formed when crustal plates ran together during plate tectonics.

unit

39

Venus

Background Pathways

Unit 36: Earth's Atmosphere and Hydrosphere 277

Venus is named for the Roman goddess of love. From our perspective on Earth, Venus's smaller orbit (at about 0.72 AU) never carries it farther than 48° from the Sun. For a period of about 9 to 10 months, we see it as a very bright "evening star" that sets after the Sun by as much as three hours. Then, after Venus moves along its orbit and passes between the Earth and the Sun, it reappears as a "morning star" for another 9 to 10 months. After it passes behind the Sun, Venus again appears as an evening star (Unit 13).

Of all the planets, Venus (Figure 39.1) is most like the Earth in diameter and mass. Its density is thus also similar to the Earth's density, implying that the deep interior of Venus is like the Earth's—an iron core and a rock mantle. Because Venus is so like the Earth in its overall characteristics, we might expect it to be like the Earth in other ways. However, Venus and the Earth have radically different atmospheres and surfaces.

39.1 THE VENUSIAN ATMOSPHERE

The atmosphere of Venus is about 100 times more massive than the Earth's atmosphere. It also differs greatly in composition, consisting mostly (96.5%) of carbon dioxide, whereas Earth's atmosphere consists primarily of nitrogen. Astronomers know the atmospheric composition of Venus from its spectrum and from measurements with space probes. Gases in its atmosphere absorb some of the sunlight falling on the planet and create absorption lines that reveal the composition and density of the gas (see Unit 24). From such observations, we have learned that, in addition to carbon dioxide, Venus's atmosphere contains about 3.5% nitrogen and trace amounts of water vapor and other gases.

Instruments aboard the European Space Agency's *Venus Express* satellite reveal that Venus has lightning storms but no rain. Spectra reveal that the Venusian clouds are composed of sulfuric acid droplets that form when sulfur compounds—perhaps ejected from volcanoes—combine with the traces of water in the atmosphere. These clouds permanently cover the planet and are very high and thick; in fact, they are so thick that no surface features can be seen through them with ordinary telescopes. However, below the clouds, the Venusian atmosphere is relatively clear, and some sunlight penetrates to the surface. The light is tinged orange, because the blue wavelengths are absorbed in the thick cloud layer.

On Venus's surface, the atmosphere exerts a pressure roughly 90 times that of the Earth's, or "90 atmospheres"—equivalent to the pressure you would feel under 900 meters (about 3000 feet) of water. Its atmosphere is also extremely hot—hot enough that lead would melt there. Observations made with radio telescopes from Earth show that the surface temperature is more than 750 K (about 900°F), a value confirmed by spacecraft landers. The dense atmosphere allows the surface temperature to cool by only a few degrees at night.

What makes the atmosphere so hot on a planet that is so similar to the Earth in size and only slightly nearer the Sun? Venus's carbon dioxide atmosphere creates an

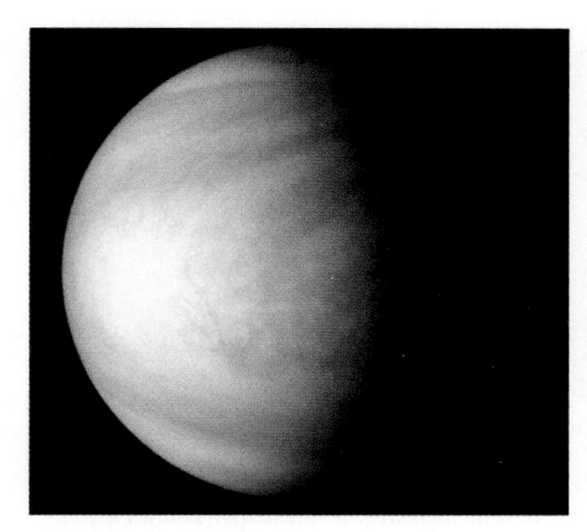

FIGURE 39.1
Photograph through ultraviolet filter of the clouds of Venus. The picture is artificially colored and enhanced to show the clouds clearly.

extremely strong greenhouse effect. We discussed in Unit 36 how carbon dioxide and water vapor in Earth's atmosphere allow sunlight to enter and warm the surface but prevent much of the infrared radiation emitted by the heated surface from escaping to space. On Earth the greenhouse effect is mild. Venus, however, has about 300,000 times more carbon dioxide than Earth, and so its greenhouse effect is dramatically stronger. In fact, it is so effective at trapping heat that Venus's surface is hotter than Mercury's, even though Venus is farther from the Sun. Venus has suffered what scientists term the **runaway greenhouse effect,** in which heating has driven the temperature up, and the heating has led to an even stronger greenhouse effect. We can imagine how this might have happened by considering how a similar situation might occur on Earth.

Suppose the Earth started heating up strongly, as we might hypothesize occurred to Venus early in its history. Although Venus today has a much more substantial atmosphere than Earth, this does not consider the total amount of volatiles on Earth. We should compare Venus's atmosphere to the combined content of the Earth's oceans and atmosphere. As the oceans grow hotter, they generate more water vapor. Water vapor is a very effective **greenhouse gas**—that is, it also absorbs infrared radiation, trapping heat. You may have experienced this on a humid summer night when you noticed little cooling between sunset and sunrise. In addition, a far larger amount of carbon dioxide resides dissolved in Earth's oceans than in its atmosphere, and as the oceans warm they release more of this greenhouse gas as well. With a higher level of greenhouse gases, more of Earth's heat would be trapped, and the ocean temperature would rise. This would put more water vapor and carbon dioxide into the atmosphere, further increasing the heat trapping. These are the ingredients that might set off a runaway greenhouse effect.

There are many complicating factors in this process. For example, water vapor can produce clouds that reflect away some sunlight; so on Earth, for the time being, the process seems to be self-limiting. However, in its distant past, Venus probably also had much more water than its current trace atmospheric amounts, delivered by the same mechanisms that gave Earth its atmosphere, such as comet impacts or volcanic outgassing. When the water (H_2O) vapor drifted into the upper atmosphere, it would have been broken down into its constituent hydrogen and oxygen atoms by ultraviolet photons from the Sun, with the hydrogen atoms probably escaping into space. The oxygen atoms would then have been free to combine with carbon to make carbon dioxide (CO_2).

Q If astronauts were to someday explore the surface of Venus, what kind of protection would they need?

On Earth, the carbon that was originally in the atmosphere has been locked away—dissolved in the oceans and incorporated in rock and soil. If that carbon were instead combined with the oxygen in the oceans and atmosphere, and the hydrogen was lost, Earth would have a carbon dioxide atmosphere, probably even more massive than Venus's.

Could a runaway greenhouse effect ever happen to Earth? Perhaps by better understanding how it happened to Venus, we will someday be able to say.

39.2 THE SURFACE OF VENUS

The surface of Venus is hidden beneath its thick clouds, but several Soviet *Venera* spacecraft have transmitted pictures back to Earth from the Venusian surface (Figure 39.2). The landers generally survived little more than an hour on the planet's surface because of the high temperatures and pressures there. The pictures show a barren surface covered with flat, broken rocks and lit by the pale orange glow of sunlight diffusing through the deep clouds. These robotic spacecraft also sampled the rocks, showing them to have volcanic origins.

Eight Soviet *Venera* landers descended to the Venusian surface between 1970 and 1981, sampling its atmosphere as they parachuted to the surface.

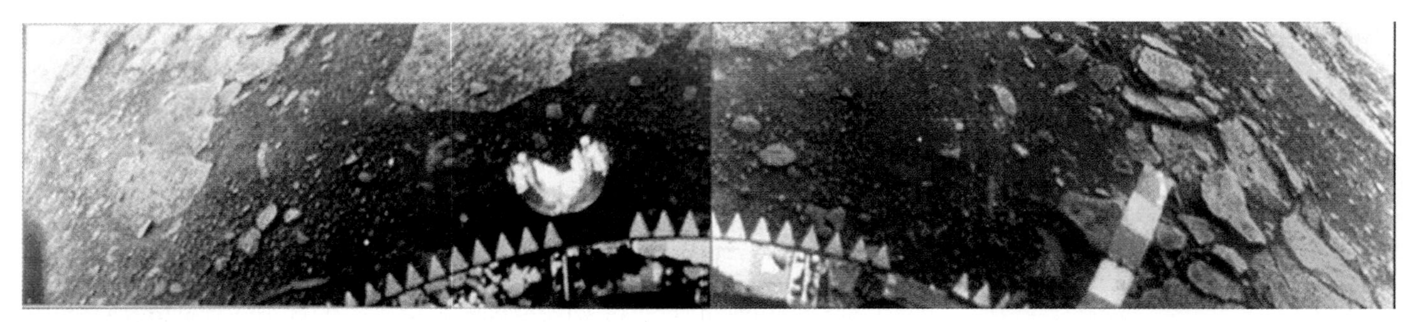

FIGURE 39.2
Picture made on the surface of Venus by one of the Russian *Venera* spacecraft that landed there. Sunlight filtering through the thick clouds gives the landscape its orange color. Through the unusual "fisheye" lens of this camera, the view is distorted, looking down at the lander's base along the middle-bottom portion of the image and off to the horizon in the upper left and right corners.

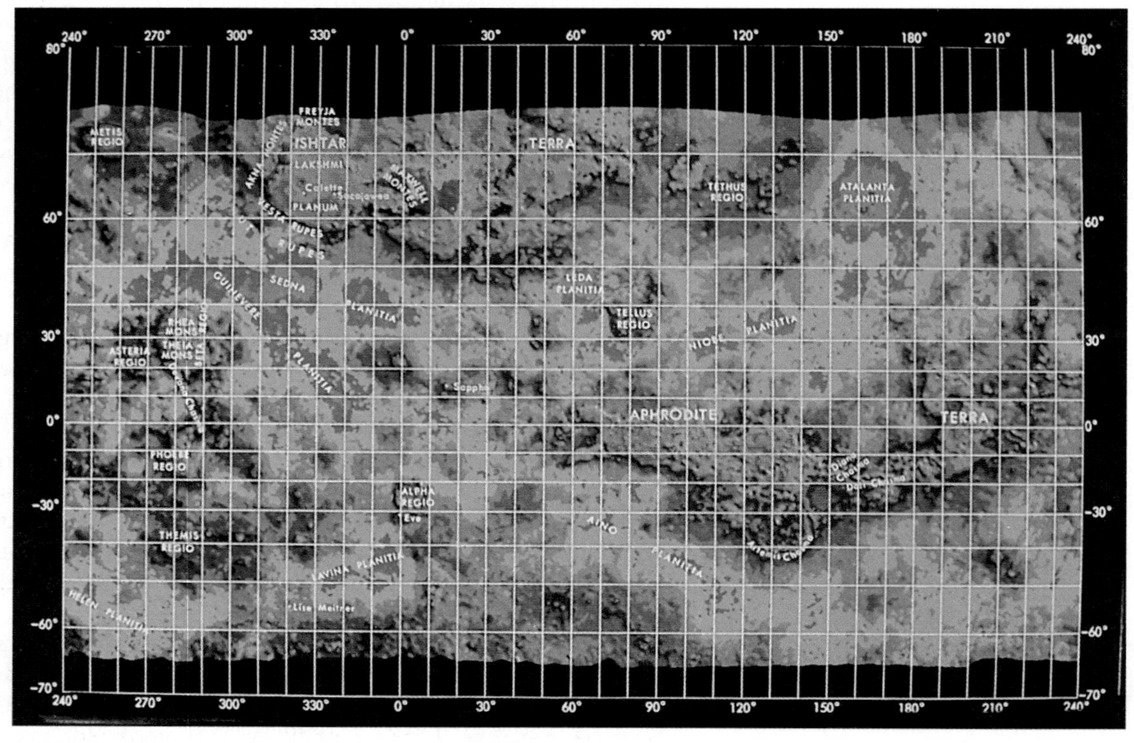

FIGURE 39.3
Global radar map of Venus made by the Venus-orbiting satellite *Pioneer.* Colors indicate the relative height of surface features. Lowlands are blue; mountain peaks are red.

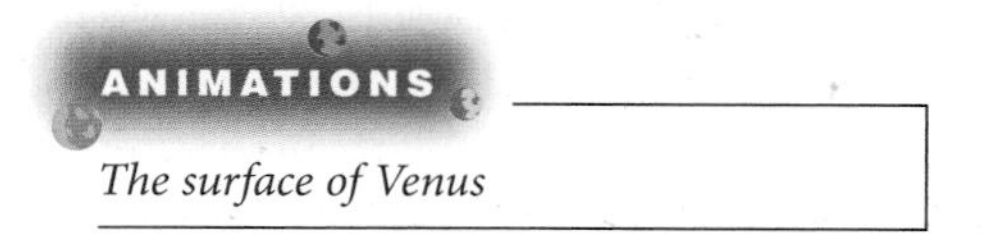

Much of what astronomers know about the geology of the Venusian surface comes from radar maps. Just as radar penetrates terrestrial clouds to show an aircraft pilot a runway through fog, so too radar penetrates the Venusian clouds, revealing the planet's surface. Figure 39.3 shows a radar elevation map of Venus's surface. Such maps reveal that Venus is less mountainous and rugged than Earth, with most of its surface being low, gently rolling plains. The most detailed images of most of Venus's surface were made with *Magellan,* a U.S. spacecraft that orbited the planet in 1993–1994 (Figure 39.4).

Only two major highland regions, **Ishtar** and **Aphrodite,** rise above the lowlands to form landmasses similar to terrestrial continents. Ishtar, named for the

FIGURE 39.4
Global radar map of Venus made by the spacecraft *Magellan* in orbit around Venus. The light areas are regions of high radar reflectivity that are rougher than the dark areas. The picture is artificially colored to simulate lighting conditions observed by the Russian *Venera* landers.

By convention, surface features on Venus are named after real or mythological females. The only feature named in honor of a man is Maxwell Montes—named after James Clerk Maxwell, the English physicist who first worked out the mathematical principles of electromagnetism. This mountain was one of the few features identified on Venus from Earth-based radar before more-detailed radar maps were made by orbiting spacecraft.

Small craters do not provide a clear comparison between planets because smaller impacting bodies may have been "burned up" or substantially reduced in size by friction with Venus's dense atmosphere before they struck the surface.

Babylonian goddess of love, is about the size of Australia and is studded with volcanic peaks, the highest of which, **Maxwell Montes,** rises about 11 kilometers (7 miles) above the surrounding plateau. Aphrodite, named for the Greek goddess of love, is about the size of South America and is located near Venus's equator. (Notice that because no oceans exist on Venus, "sea level" has no meaning as a reference height.) Together Ishtar and Aphrodite compose only about 8% of Venus's surface—a far smaller fraction than for Earth, where continents and their submerged margins cover about 45% of the planet. On the other hand, measurements of the gravitational field of Venus by orbiting spacecraft indicate that Venus's crust is thicker than Earth's.

Several high-resolution radar images of Venus's surface are shown in Figure 39.5. You can see numerous volcanic peaks and channels, and the planet's wrinkled crust. Close-up pictures show many odd and unique structures. Venus is so similar in diameter and mass to the Earth that geologists expected to see landforms there similar to those on the Earth. Instead, Venus has a surface very different from the Earth's. For example, although Venus has some crumpled mountains, volcanic landforms dominate. These include peaks with immense lava flows, "blisters" of uplifted rock called **pancake domes,** and grids of long narrow cracks or **faults.**

Some of the characteristics of rock on Venus and Earth may not be the same because of the differing amounts of water in the rocks of these two worlds. We saw that the behavior of rock on Earth is different when water is present (Unit 36). In the absence of water, rock melts at a higher temperature and can form steeper mountains. This would explain steep mountains like Maxwell Montes on Venus, and it may help explain the differing behavior of the rock in response to internal heating.

Many of Venus's surface features indicate a young and active surface, a deduction borne out by the scarcity of impact craters. We can estimate the age of Venus's surface by comparing the number of the largest craters found there—those with diameters greater than a few kilometers—with the number of those found on

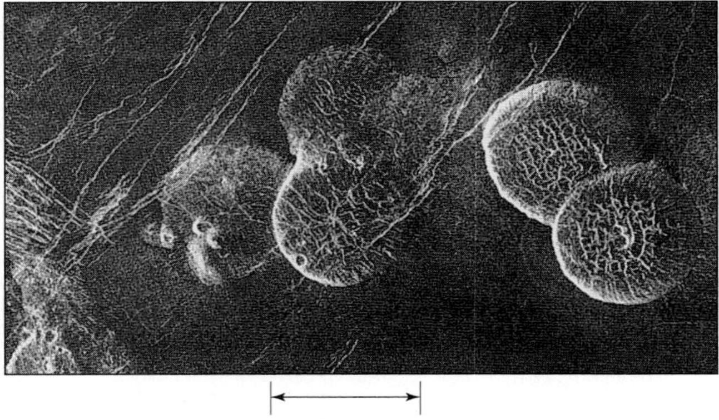

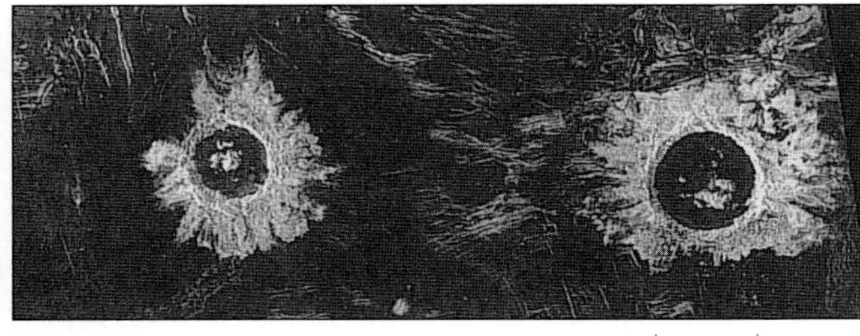

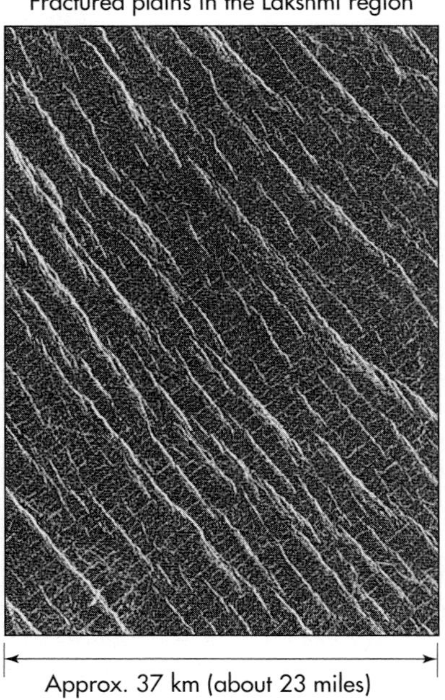

FIGURE 39.5
Gallery of *Magellan* radar pictures. The image at lower left uses the radar data to illustrate height variations, exaggerated vertically to emphasize height differences. The orange color is artificially added to match the color of the landscape observed by the Russian *Venera* landers.

the Earth. From the small number of craters, the average age of Venus's surface appears to be roughly 500 million years, with some regions less than 10 million years old—fairly similar to Earth. This presents us with a puzzle, though. Earth's young surface is a result of plate tectonics (Unit 35); but on Venus we do not see evidence of the same kind of geological activity. Ultimately, the heat in a planet's interior must escape, but geologists disagree about how this happens for Venus and how it might cause the renewal of Venus's surface.

One hypothesis is that rock in Venus's mantle does not circulate in the same kind of large convection currents found in the Earth's mantle. If the mantle is not undergoing convection, there would be no plate tectonics. For example, a few locations on Earth, such as Yellowstone Park and the Hawaiian Islands, appear to be heated intensely but locally by "plumes" of rising heat in the mantle. It may be that Venus's internal geology is dominated by such localized plumes. As heat wells upward, it bulges the crust, stretching and cracking it. This might explain the numerous volcanic peaks, domes, and uplifted and fractured surface regions.

Q: Would it be possible to make Venus a habitable place for humans if we could somehow block enough of the Sun's light so that Venus received an amount similar to what Earth receives?

No direct evidence has been found of current volcanic activity, although some lava flows appear very fresh based on the lack of erosion of features. Given the lack of current volcanic activity, some geologists propose that the activity may be periodic. Venus's thick crust may effectively trap the internal heat. As the interior heats up, it might then gradually melt the bottom of the crust, thinning it and allowing it to break up. This may then produce widespread volcanic activity, flooding much of the planetary surface with lava in a brief time. With much of the internal heat released, the interior cools, and the crust again thickens. The thicker crust causes heat to be trapped once more, and the process repeats at intervals of hundreds of millions of years.

According to another hypothesis, rock is circulating inside Venus's mantle much as in Earth's; but because the crust is thicker and the rock more rigid, it behaves differently. Venus's crust may not so readily break into the crustal plates we find on Earth. At points above where the hot material rises within the mantle, the crust bulges upward, stretches, and weakens; volcanoes form, but without producing the "conveyor belt" motion of plates on Earth. Where the mantle material circulates back down, portions of the crust are pushed together, thickening and crumpling the crust into regions like Aphrodite and Ishtar. This might create a planet with few and small continents and whose surface is active but only in isolated spots. The push and pull on the bottom of the crust by convection flows in the mantle may also explain many of the faults seen on the surface of Venus, and may gradually erase surface features such as impact craters.

39.3 ROTATION OF VENUS

Q: If the Earth spun as slowly as Venus, what effects might that have on our environment?

Venus spins on its axis more slowly than any other planet in the Solar System, taking 243 days to complete one rotation. Moreover, Venus is unique among the terrestrial planets in that it has a **retrograde spin.** That is, compared with the direction of rotation of the other terrestrial planets, it spins backward. Thus the Sun rises in the west and sets in the east. Because Venus's rotational period is longer than its orbital period of 225 days, you might at first expect that Venus's day would be longer than its year. That would be true if Venus's spin were in the same direction as its orbit; but by turning in a retrograde direction, a point on the surface turns back to face the Sun more quickly than if it weren't rotating at all—in about half a Venusian year (117 days), as illustrated in Figure 39.6.

A retrograde spin cannot be produced by tidal braking (Unit 37.4). It is likely, though, that Venus had a faster retrograde spin in the distant past and that the Sun's tidal effects have slowed it down to its current rate. Some astronomers hypothesize that Venus was struck late in the formation process by a large planetesimal that collided in a direction that set the planet spinning backward. Venus's slow rotation, however it was caused, also means that despite Venus's iron core, the dynamo process for generating a magnetic field does not operate strongly there. In fact, Venus's magnetic field is tens of thousands of times weaker than the Earth's.

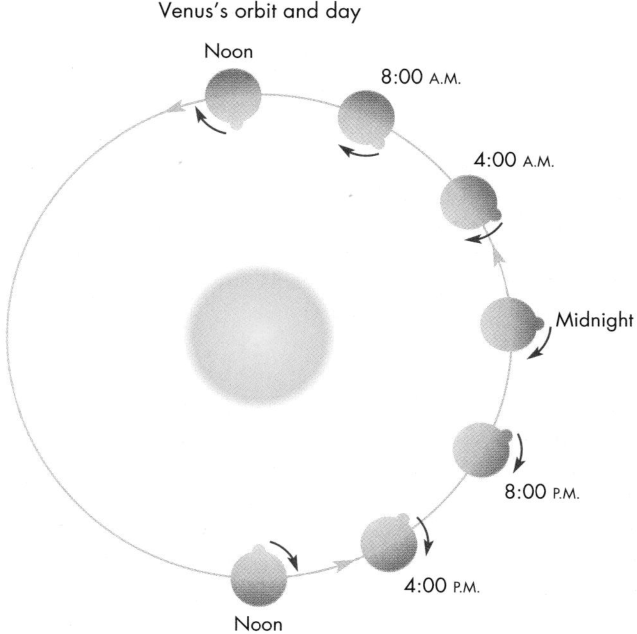

FIGURE 39.6
Venus's rotation is illustrated for a spot on its surface. Venus's slow retrograde spin results in a day that is about 117 Earth days long—more than half of Venus's year.

KEY TERMS

Aphrodite, 306
fault, 307
greenhouse gas, 305
Ishtar, 306
Maxwell Montes, 307
pancake dome, 307
retrograde spin, 309
runaway greenhouse effect, 305

QUESTIONS FOR REVIEW

1. How does Venus compare with the Earth in mass and diameter?
2. How does Venus's surface differ from Earth's? How have astronomers determined what the surface of Venus is like?
3. What is the dominant gas in Venus's atmosphere? How do astronomers know this?
4. What is the runaway greenhouse effect? Could it happen on Earth?
5. What do crater counts on the Earth and Venus imply about these planets' past geological activity? What are the challenges in making comparisons using crater counts?
6. What is unusual about Venus's rotation? What might have caused this?

PROBLEMS

1. Venus's atmosphere contains about 4.6×10^{20} kg of carbon dioxide (CO_2). If all of the oxygen atoms in the CO_2 were combined with hydrogen to make water (H_2O) instead of carbon dioxide, what would the total mass of water be? (Note: Hydrogen has an atomic mass of 1; carbon, 12; oxygen, 16.) How does this compare to the mass of Earth's oceans, about 1.4×10^{21} kg?
2. From your answer to problem 1, how deep an ocean would this water produce if it were the same depth everywhere on the surface? (Note that 1000 kg of water fills a volume of 1 cubic meter.)
3. How does the total mass of nitrogen in Venus's atmosphere compare to the total mass of nitrogen in Earth's atmosphere? (Use data presented in this unit and in Unit 36 to solve this problem.)
4. Venus and the Earth make their closest approach to each other (inferior conjunction, Unit 11) every 583.9 Earth days. Use a diagram like Figure 39.6 or math to show that nearly the same side of Venus is facing Earth at each closest approach.
5. The Earth covers about 1° per day in its orbit about the Sun, and the solar day is slightly longer than the sidereal day (Unit 7). If Earth spun in a retrograde direction like Venus but it still had the same sidereal period (23 hr 56 min), how long would the solar day be? Explain your answer, using a diagram if necessary.
6. Contrast the magnetic fields of Earth, Mercury, and Venus and explain why these magnetic fields are so different.

TEST YOURSELF

1. Why is Venus's surface hotter than Mercury's?
   a. Venus rotates more slowly, so it "bakes" more in the Sun's heat.
   b. Clouds in Mercury's atmosphere reflect sunlight back into space and keep its surface cool.
   c. Carbon dioxide in Venus's atmosphere traps heat radiating from its surface, thereby making it warmer.
   d. Venus is closer to the Sun.
   e. Venus's rapid rotation generates strong winds that heat the ground by friction as they blow.
2. Why are there few spiral patterns seen in ultraviolet images of the clouds of Venus?
   a. The rotation rate of Venus is very small.
   b. The clouds are constantly rising and falling.
   c. The clouds are not made of water droplets.
   d. Venus has no Moon to produce tides that make spiral patterns.
3. What evidence is there that the surface of Venus was covered by giant flows of lava a few hundred million years ago?
   a. The surface is still hot.
   b. There are relatively few impact craters on Venus.
   c. Radioactive dating shows that the surface is a few hundred million years old.
   d. The cracks where lava reached the surface can be seen in radar maps.

unit

40

Mars

Background Pathways

Unit 36: Earth's Atmosphere and Hydrosphere 277

Mars is named for the Roman god of war, presumably because of its "blood red" color. Compared with the harsh environments of Mercury and Venus, Mars seems positively Earth-like. Although its diameter is only about half Earth's and its mass about one-tenth Earth's, its surface and atmosphere are less alien. Mars is colder than the Earth because it orbits 1.52 AU from the Sun; but on a warm day, the temperature at the Martian equator may rise above the freezing temperature of water to about 283 K (about 10°C or 50°F). And although winds sweep dust and patchy clouds of ice crystals through its sky, the Martian atmosphere is generally clear enough for astronomers on Earth to view its surface. Such views show a world with sparkling white polar caps that contrast with the reddish color of most of the planet, as depicted in Figure 40.1.

Mars rotates with a period similar to Earth's, with a day just 40 minutes longer than our own. Mars even has a similar tilt to its polar axis, 25° versus the Earth's tilt of 23.5°. Thus Mars experiences seasonal changes like Earth's during its own year of 1.88 Earth years—which takes 669 Martian days. Mars resembles the Earth in so many ways that many searches have been made for life on the planet, but there is no strong evidence that life ever existed there, as is discussed further in Unit 84. Even though life has not been found on Mars, there is growing evidence that a self-sustaining base on the planet is possible, potentially paving the way for future human exploration.

40.1 THE MAJOR FEATURES OF MARS

Earth-based telescopic views hinted at the intriguing geology of Mars; but space-based pictures, the legacy of many spacecraft missions, reveal the planet's true marvels. The first successful mission to Mars was a flyby in 1965 by NASA's *Mariner 4* (NASA is the U.S. National Aeronautics and Space Administration). Over the subsequent decades, the United States and Soviet Union have succeeded in placing several spacecraft in orbit around Mars and landers on its surface. Such maneuvers and landings are challenging technically, and several of the spacecraft have been lost. The most recent missions have revealed much more detail about Mars. In 2003 the European Space Agency (ESA) placed in orbit the *Mars Express* spacecraft, which, along with NASA's *Global Surveyor, Odyssey,* and *Mars Reconnaissance Orbiter,* has let us take high-resolution pictures and make chemical analyses of the surface from space. In 2004 NASA also landed surface rovers, which have traveled several kilometers about the planet's surface, studying Mars's geology.

FIGURE 40.1
Picture of Mars made by the Hubble Space Telescope orbiting Earth.

Astronomers have a variety of indirect evidence that Mars's interior is differentiated, like the Earth's, into a crust, mantle, and iron core. For example, using spacecraft in orbit around the planet, astronomers have measured Mars's magnetic field and internal structure from detailed measurements of its gravitational field. This evidence implies that Mars has a metallic core with about half the overall radius of the planet. This is very similar to the fractional size of the Earth's iron core, but in Mars the iron core appears to have a significant fraction of sulfur—which helps explain why its overall density is lower. Recent measurements of the way

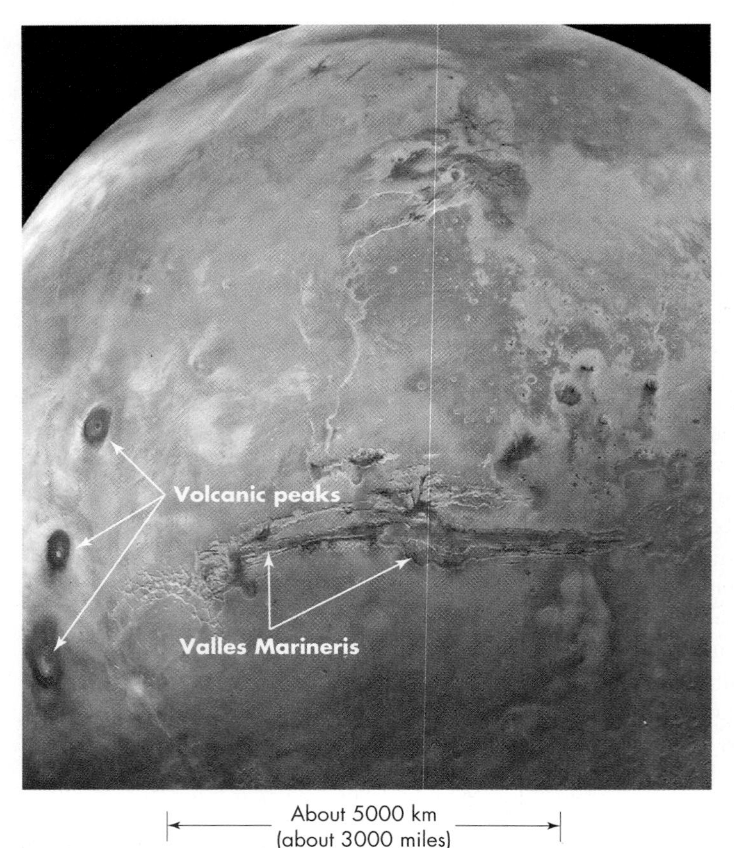

FIGURE 40.2
Photograph of Valles Marineris, the Grand Canyon of Mars. This enormous gash in Mars's crust may be a rift that began to split apart the Martian crust but failed to open further. The canyon is about 5000 kilometers (approximately 3000 miles) long. Were it on Earth, it would stretch from California to Florida.

Mars is distorted by tidal effects of the Sun also indicate that its core is partially molten, but not to the degree of the Earth's core. Spacecraft have also detected the magnetization of ancient volcanic flows, indicating that Mars's core generated a much stronger magnetic field in the past when the geology was more active.

Mars's surface has a number of remarkable geological features. Along the equator runs an enormous rift called **Valles Marineris** (Figure 40.2). Named for the *Mariner* spacecraft whose pictures first showed it, it is 5000 kilometers (3000 miles) long, 100 kilometers (60 miles) wide, and 10 kilometers (6 miles) deep. This chasm would span the continental United States, dwarfing the Grand Canyon, which is less than a tenth as long or a fifth as deep.

An elevation map made by NASA's *Mars Global Surveyor* shows that the entire southern hemisphere has a higher elevation than the northern hemisphere (Figure 40.3). Spanning the equator is an uplands region, about the size of North America, called the **Tharsis bulge.** It is dotted with volcanic peaks, three of which can be seen in Figure 40.2. One end of Valles Marineris lies near the center of Tharsis, but the great chasm continues about one-fifth of the way around the planet, near the equator, into lower elevation regions.

Another volcano on the edge of Tharsis, **Olympus Mons,** is shown in Figure 40.4. It rises to nearly three times the height of Earth's highest peaks, some 26 kilometers (15 miles) above its surroundings, making it the biggest volcano in the Solar System. If interplanetary parks are ever established, Olympus Mons should lead the list!

Mars's immense volcanoes and Tharsis bulge give testimony to an active geological past. However, Mars does not show evidence of large-scale crustal motion, like folded mountain ranges, that plate tectonics on Earth have generated. Astronomers therefore think that Mars cooled relatively rapidly and its crust thickened to perhaps twice the thickness of the Earth's crust. As a result, its now weak interior heat sources can no longer break through to the surface or drive tectonic

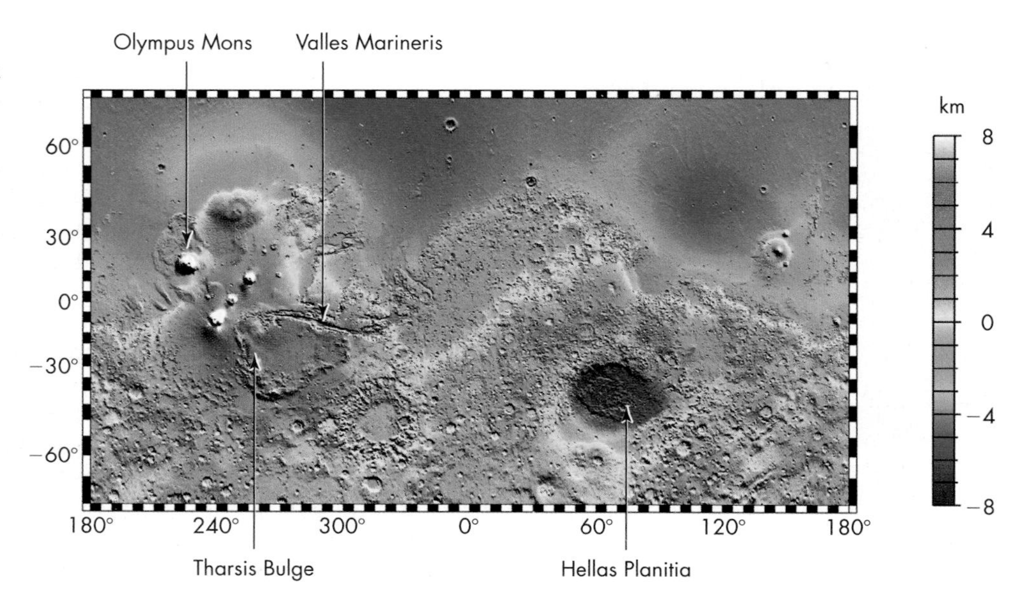

FIGURE 40.3
Topographic map of Mars made with a laser altimeter aboard the *Mars Global Surveyor.* The colors show elevations in kilometers according to the scale on the right.

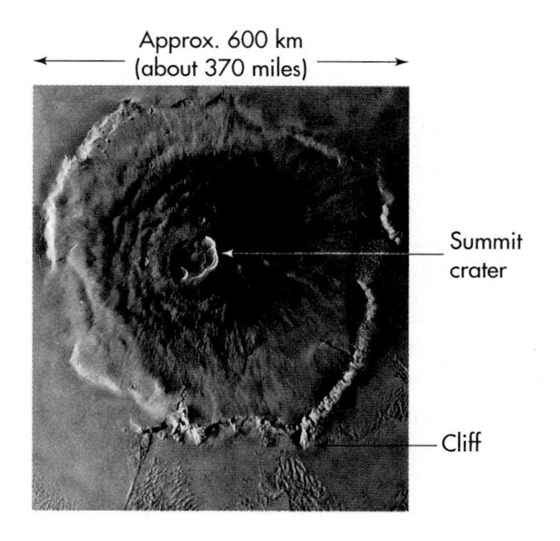

A

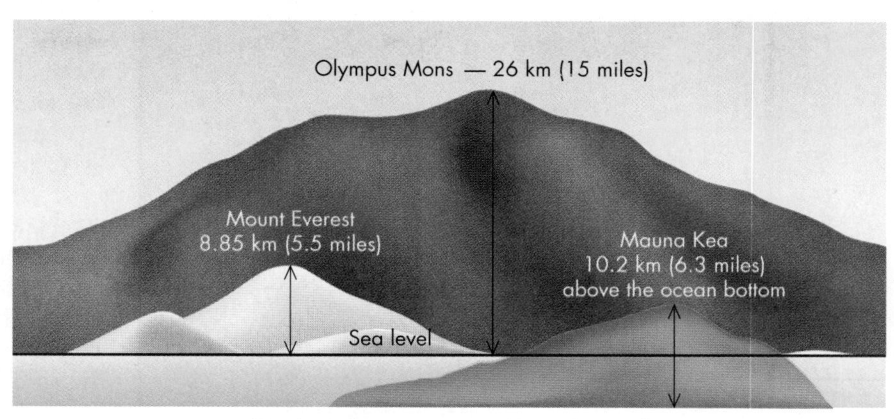

B

FIGURE 40.4
(A) Picture of Olympus Mons, the largest known volcano in the Solar System. (B) Profile of Olympus Mons; Mount Everest, the highest mountain above sea level on Earth; and the Hawaiian volcano Mauna Kea, rising from the sea floor.

Q Examining the photographs in the figures, in what ways can you distinguish between an impact crater and the crater (caldera) of a volcano?

motion. The thick Martian crust may also explain why Mars has a small number of very large volcanoes. The heat from the interior may have found a few weaker regions to break through, in contrast to the Earth, where a thin shifting crust allows less dramatic volcanic activity to occur at many more sites.

Geologists think that the Tharsis region formed as hot material rose from the deep interior of the planet. As this molten material neared the crust, it forced the surface upward. The hot matter then erupted through the crust to form the volcanoes, some of which appear to be relatively young. For example, the small number of impact craters on its slopes implies that Olympus Mons is probably no older than 250 million years and that it may in fact have been active much more recently. Some planetary geologists think the Tharsis bulge may be linked to the formation of Valles Marineris. According to this hypothesis, Valles Marineris formed as the Tharsis region swelled, stretching and cracking the crust. Perhaps this vast chasm is something like the early stages of plate tectonic activity on Earth, circulation in the Martian mantle having begun to split the planet's crust. Unlike on Earth, however, the tectonic activity ceased as the planet aged and cooled.

Mars's current low level of tectonic activity is also demonstrated by the many craters that cover most of its terrain—far more than are seen on either Earth or Venus. The number of those craters implies that most of Mars's surface has been geologically quiet for billions of years. Thus Mars appears to be entering a phase of planetary old age similar to Mercury or the Moon. Unlike them, though, Mars retains an atmosphere, and winds sweep its surface. This produces immense deserts, at midlatitudes, with dunes blown into parallel ridges, as shown in Figure 40.5.

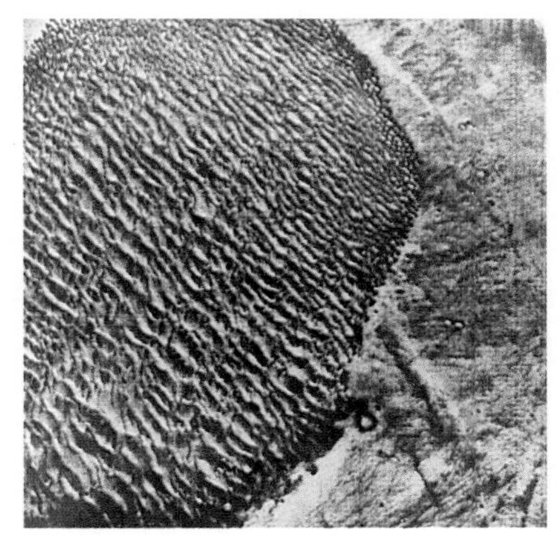

FIGURE 40.5
Picture of dune fields in the Martian desert.

The hemisphere that experiences summer when Mars is closest to the Sun alternates as Mars precesses. This produces especially hot summers in that hemisphere that lead to severe dust storms.

At its poles, Mars has frozen polar caps (Figure 40.6). These frozen regions change in size during the cycle of the Martian seasons. The tilt of Mars's rotation axis produces a pattern of seasons similar to those on Earth. The Martian seasons are more extreme than Earth's because the Martian atmosphere is much less dense than Earth's and therefore does not retain heat as well. Because Mars's seasonal changes are so extreme, its polar caps vary greatly in size, shrinking during the Martian summer and growing again during the winter.

Mars has a much more elliptical orbit than the Earth, so that its distance from the Sun varies by about 45 million kilometers during its orbit. This represents a 20% change in Mars's distance from the Sun and means that the solar heating Mars receives varies substantially during its orbit compared to the Earth, whose distance

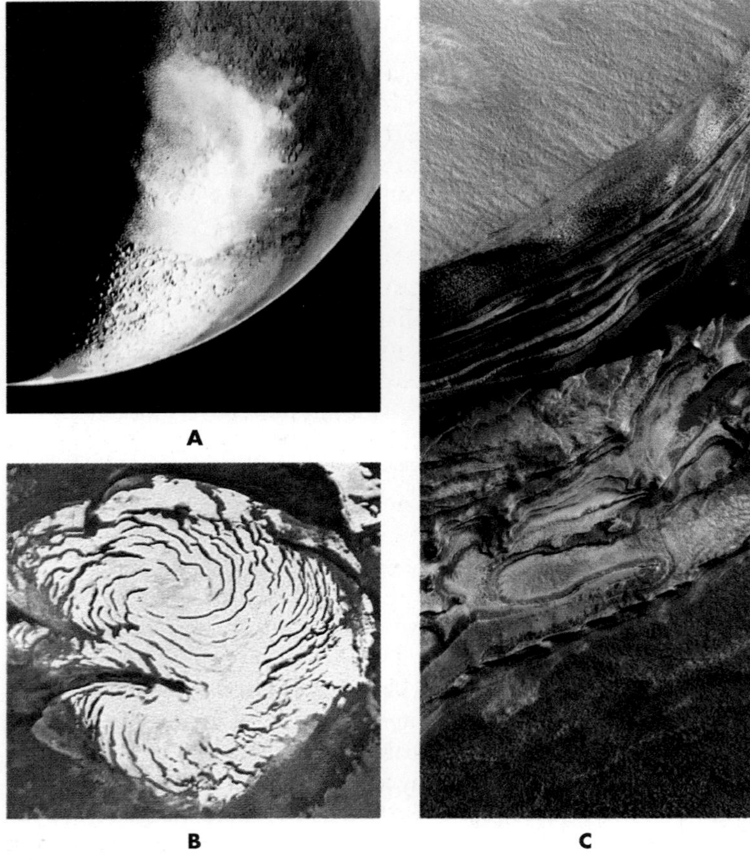

FIGURE 40.6
Pictures of (A) the south Martian polar cap and (B) the north Martian polar cap. (C) An image from the *Mars Reconnaissance Orbiter* shows the wall of a chasm approximately 1 kilometer (0.6 mile) deep in the north polar cap. The image covers a section of the ice cliff about 1.3 kilometer wide. The alternating layers of ice are mixed with more or less dust, laid down during cycles of changing climate conditions.

from the Sun varies by only about 3%. Mars is farthest from the Sun during its southern winter and closest during southern summer. As a result, the seasons in the south are more extreme, and the southern polar cap shows dramatic changes in size. The visible changes in the polar caps appear to be due primarily to frozen carbon dioxide—dry ice. In winter, its frost extends in a thin layer across a region some 5900 kilometers (about 3660 miles) in diameter, from the south pole to latitude 40°, much as snow cover extends to middle latitudes such as New York during our Earth winters. But because the frost is very thin over most of this vast Martian ice cap, it shrinks in the summer to a diameter of about 350 kilometers (approximately 220 miles).

The northern hemisphere seasons are less extreme because the planet is farther from the Sun in summer and closer in winter. The northern cap shrinks less, to a diameter of about 1000 kilometers (about 600 miles). The northern cap consists of numerous separate layers called **laminated terrain,** as can be seen in Figure 40.6C, with each layer representing a separate long-term episode of ice deposition. Astronomers think that these strata are evidence for cyclical changes in the Martian climate. Such change might occur because Mars's polar axis precesses over a period of about 50,000 years. This means that 25,000 years ago Mars's northern hemisphere would have had more extreme seasons and the southern hemisphere less extreme. Patterns of dust storms would vary with these global climatic changes, leading to darker and lighter layers of ice laid down over the course of tens of thousands of years.

Although the polar regions have a surface layer of CO_2, the bulk of the frozen material in the caps is ordinary water ice, which remains almost permanently frozen. An instrument aboard the *Mars Express* orbiter has been able to probe through the surface layers with radar. It showed that the polar caps contain thick layers of ice ranging up to 1.8 kilometers (1.1 miles) in the northern cap and up to 3.7 kilometers (2.3 miles) in the southern cap. Astronomers estimate that the southern cap contains enough water to cover the entire surface of Mars to an average depth of 11 m (36 feet).

40.2 A BLUE MARS?

Perhaps the most surprising features revealed by the first spacecraft to orbit Mars were huge channels and dry riverbeds, such as those seen in Figure 40.7A. Other features that meander across the Martian surface contain "islands" (Figure 40.7B). These images clearly indicate that liquid water once flowed on Mars, even though no surface liquid is present now. Did liquid water ever make Mars a blue planet like Earth? How much water actually remains on the planet today?

A detailed examination of the geological features and minerals on Mars has begun to answer these questions. NASA's *Mars Reconnaissance Orbiter* can resolve

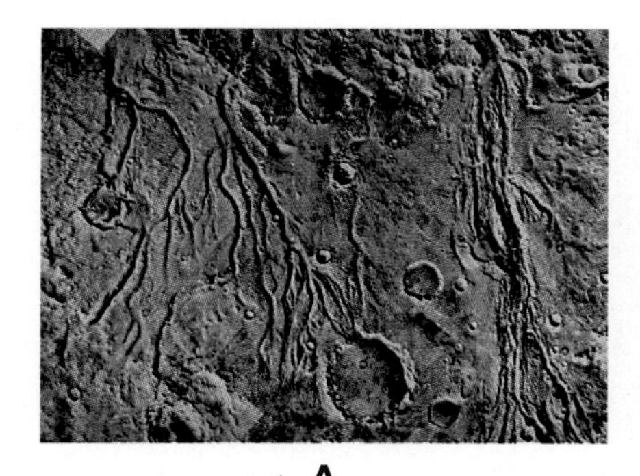

A

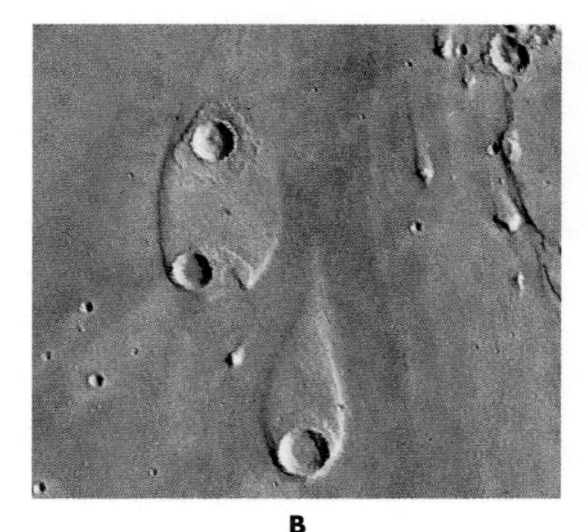

B

FIGURE 40.7
(A) Picture of channels probably carved by running water on Mars. (B) "Islands" formed as water flowed (from bottom toward top of figure) around the rims of craters.

details smaller than 2 meters (about 7 feet). These spacecraft have sent back thousands of detailed images of Mars.

Many astronomers now suspect that huge lakes and small oceans once existed on Mars. Some evidence for these ancient bodies is found in the shape of features seen on Mars's surface. For example, Figure 40.8A shows what may have once been a lake inside an impact crater. The interior of the crater contains terraces that resemble old beaches (lower left of Figure 40.8A), and the crater's rim is cut by narrow canyons that appear to be the inflow channel (upper right) and outflow channel draining into lowland areas (left). In Figure 40.8B, a *Mars Global Surveyor* image shows what appear to be the frozen remains of a river delta and a fan of sediment where it once emptied into a sea. In 2005 the *Mars Express* spacecraft captured an image (Figure 40.8C) of a region that closely resembles the ice "rafts" that are seen breaking off the coast of Antarctica. This may be a picture of an ancient sea that froze solid and is now covered with Martian dust.

Some features demonstrate clearly that ice lies buried below the Martian surface. Figure 40.9A shows a crater from which partially melted matter "squished" out on impact. High-resolution images from the *Mars Global Surveyor* show evidence of seepage in the walls of craters and canyons in many regions of Mars. These probably arise from deep subsurface ice layers that have been cut into by the canyons and craters. The flows into the crater leave gullies and dark stains on the walls that continued to flow so recently that the dark areas have not been coated by dust. In some regions they have been observed to change in appearance in images taken many months apart. However, it is not clear if this is subsurface ice, frozen carbon dioxide, or small landslides.

The spacecraft currently orbiting Mars have studied the surface with spectrometers and infrared imagers. From the spectra, astronomers can deduce what minerals are present at a given point on the Martian surface. Matching the composition of those minerals with data on whether water is needed to produce a given mineral gives additional evidence that Mars was once much wetter. For example, these devices have detected the presence of the mineral hematite, a compound of iron that forms by minerals precipitating in water. This indicates that a region now dry was once covered by a salty lake or sea.

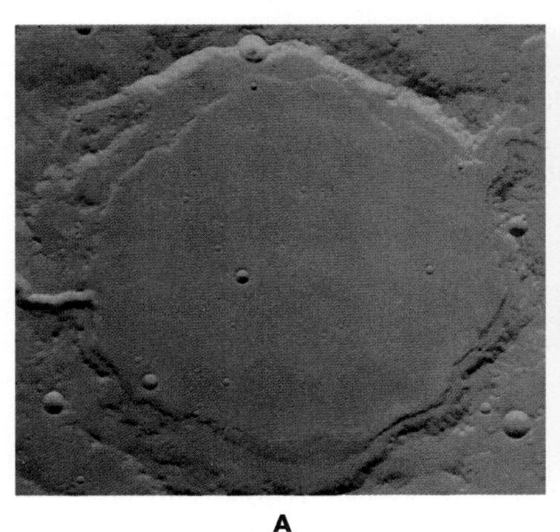

A

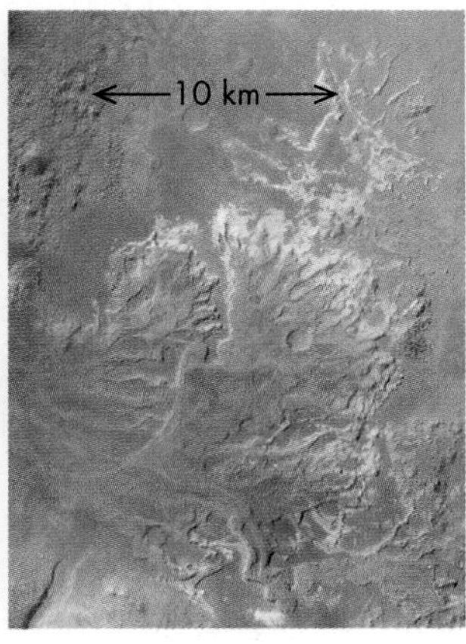

B

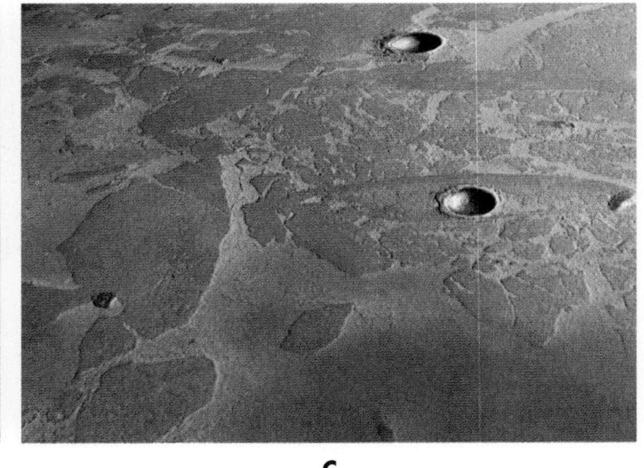

C

FIGURE 40.8
(A) A Martian crater thought to have once been a "crater lake." The crater is roughly 50 kilometers (30 miles) across. Note the inflow channel on the upper right and the outflow channel on the left. The smooth floor (apart from a few small craters) suggests that the crater bottom is covered with sediment left behind as the lake dried out. (B) A view of what appears to be a dried-up river delta. The fan of sediment is about 13 kilometers (8 miles) wide. (C) An image made near the Martian equator by the *Mars Express* spacecraft, possibly showing a frozen sea covered by dust. The irregular flat shapes appear to be ice "rafts" like those seen off Antarctica. The region shown is about 30 kilometers (20 miles) across.

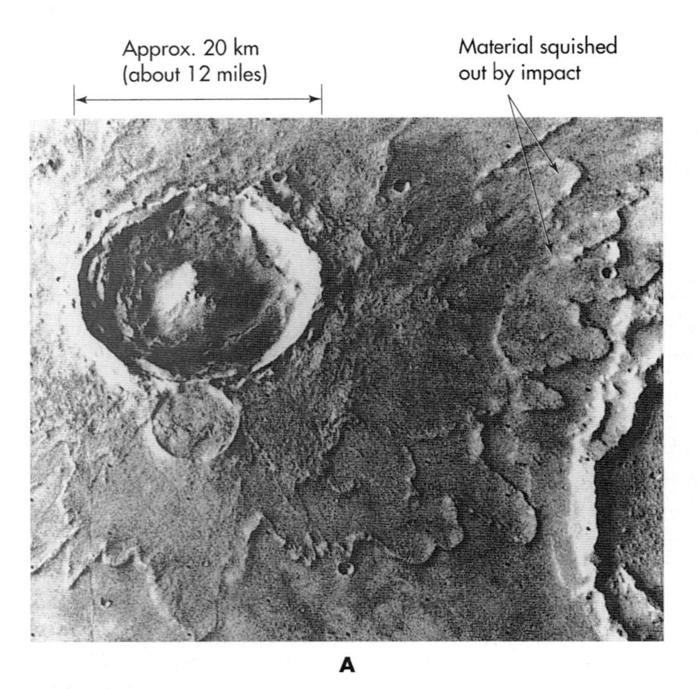

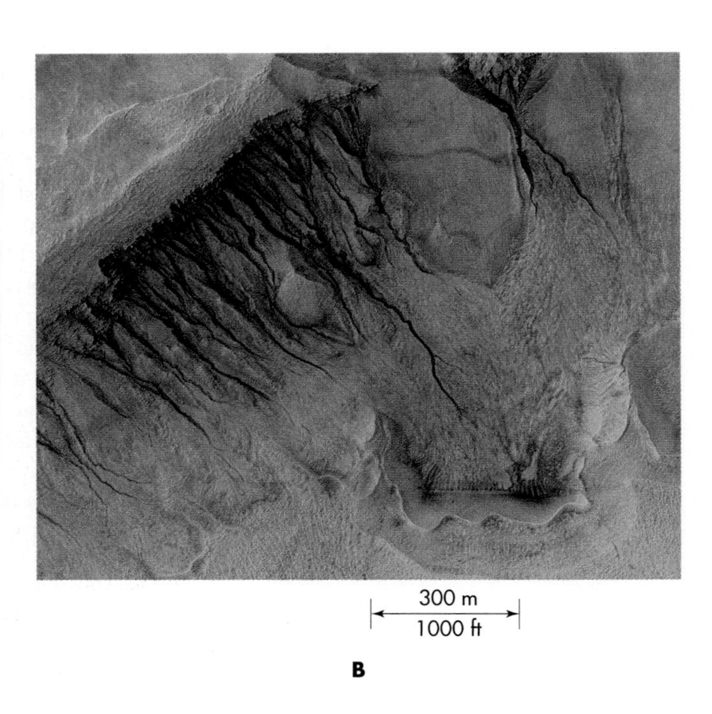

FIGURE 40.9
(A) A crater with surrounding flow patterns. Heat generated by the impact that formed the crater melted subsurface ice, and the muddy material oozed into the surrounding plains. (B) Gullies in the wall of an impact crater appear to have been formed from subsurface water seepage.

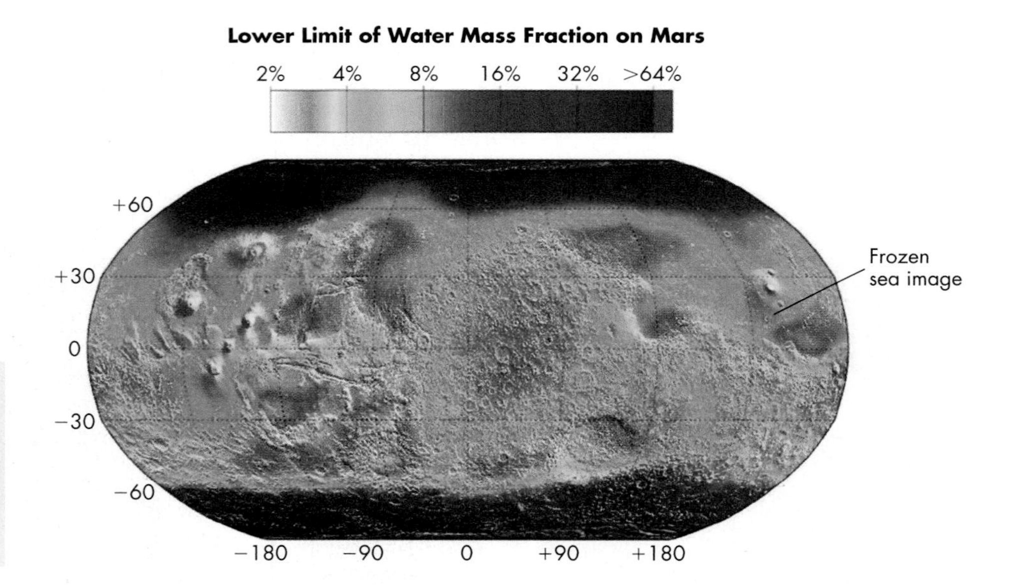

FIGURE 40.10
A map indicating the water content of the top meter of Martian soil. The map was made by the *Mars Odyssey* spacecraft with its neutron detector.

Neutrons can be emitted by various atoms when they are struck by high-energy cosmic rays. The cosmic rays come from many energetic processes throughout space as well as from the Sun.

Q What materials besides water ice might explain the hydrogen detected in Mars's surface by the *Odyssey* spacecraft? Would these be likely?

The most direct evidence of water near the surface comes from the *Mars Odyssey* spacecraft, which arrived at Mars in 2002. It contains a neutron detector that can detect hydrogen in the top meter of the Martian soil. The hydrogen is most likely bound up in water ice—and this suggests that the surface contains a large fraction of ice. Water is present not just in the polar ice caps but at lower concentrations all over the planet, as shown in a global map made by the spacecraft (Figure 40.10). It is not clear whether much of this water is locked within mineral compounds and therefore not easily extracted.

40.3 THE MARTIAN SURFACE

Although a number of the geological features seen in the early images of Mars strongly suggest that water once flowed on the Martian surface, some astronomers argued that Mars did not necessarily ever have standing liquid water on its surface. The features could have been formed when Mars was struck by one or more asteroids whose impact melted large volumes of buried ice, liberating huge amounts of water that flowed in a great flood across the planet, carving the features we see before rapidly refreezing.

Direct analysis of surface rocks can reveal even more. The first successful landers on Mars were NASA's *Viking 1* and *Viking 2* spacecraft, which landed on the surface in 1976 and sent back many pictures and performed chemistry experiments on the Martian soil. NASA's *Pathfinder* was the next successful lander, reaching Mars in 1997. It landed in a flat region that astronomers suspect may have been created by massive flooding billions of years ago. The craft descended by parachute and then deployed air bags to cushion its landing. The following Martian day, *Sojourner,* a small motorized cart, rolled down a ramp from *Pathfinder* and began exploring the Martian surface. *Sojourner* roamed over a path about 100 meters long (about the length of a football field) and closely examined several Martian rocks. At least one rock had rounded lumps similar to those found on Earth in **conglomerates**—rocks that were eroded smooth and then fused together by heat and pressure. Such findings imply that the landing spot was once scoured by water and its constituent rocks were later fused together by geological activity.

In 2004 two much more ambitious NASA rovers, *Spirit* and *Opportunity,* landed on Mars. These small wheeled vehicles were parachuted to sites chosen because pictures from orbit suggested that water had once been present there. *Spirit* landed in Gusev Crater, a smooth-floored feature at the end of a narrow Martian valley. *Opportunity* landed on the flat plains along the Martian equator called Meridiani Planum, on the opposite side of Mars from *Spirit.* In Figure 40.11 you can see the view at the *Spirit* lander site in Bonneville Crater. Both rovers continue to move about the surface more than two Martian years later, and after traveling more than 10 kilometers on the surface.

The rovers are performing a variety of experiments to sample and test rocks and soil, even drilling into the rocks to get beneath their weathered surface. Based on the pictures they send back, the rovers are instructed each Martian day about the path they should follow; but the rovers also have artificial intelligence that allows them to make on-the-spot decisions to find the best paths around rocks or other obstructions. This is essential because the images sent from Mars can take up to 20 minutes to reach Earth; so the time between ground control receiving an image and being able to send back a response can be up to 40 minutes round-trip.

FIGURE 40.11
A panoramic view of the Bonneville Crater, lying in the floor of the much larger Gusev Crater on Mars. This spot is near where the *Spirit* spacecraft landed.

FIGURE 40.12
Rock outcropping at the *Opportunity* landing site. These rocks show thin layers and contain minerals that suggest they were formed on the bottom of a salty lake or ocean.

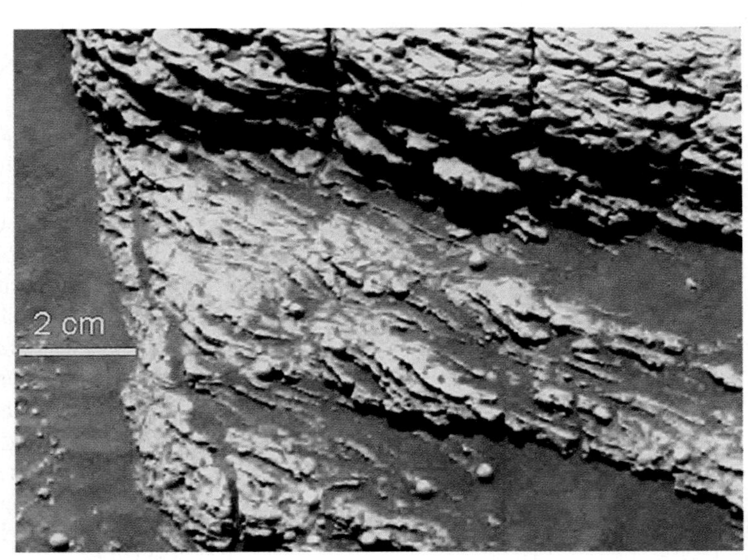

FIGURE 40.13
A close-up image of a rock at the *Opportunity* landing site. The lines in the rock suggest that the rock formed from sediment in flowing water.

Both rovers have been highly successful in their searches. For example, *Opportunity* took the picture, shown in Figure 40.12, of a rock outcropping thought to be material deposited in an ancient ocean that long since dried up. Close inspection of the rocks indicates that they contain layers typical of sediment that has sunk to the bottom of a body of water and later transformed into rock. A close-up photo, again made by the *Opportunity* rover, shows details of the layers (Figure 40.13). *Opportunity* has, however, found more than just rock layers. It has also found in such layers features that are similar to the ripple marks you see at the beach as water washes back and forth over the sand. Moreover, minerals in the rocks at that site have a chemical makeup consistent with their having been deposited in a salty lake or small ocean.

Q One possible plan for sending a human mission to Mars is a one-way trip. Materials to build a base would be sent, but there would be no launch craft for a return trip. From what you know of Mars, would you consider going on such a mission?

Although analysis of the images and the readings from the many instruments carried on these landers is still under way, the evidence so far recovered has already convinced many scientists that the locations on Mars that *Spirit* and *Opportunity* have explored were indeed once under liquid water.

40.4 THE MARTIAN ATMOSPHERE

Clouds and windblown dust are visible evidence that Mars has an atmosphere. Spectra confirm this and show that the atmosphere is 95% carbon dioxide and 3% nitrogen, with traces of oxygen and water. This is similar to the composition of Venus's atmosphere, but Mars's atmosphere is far less dense. Astronomers can measure the density of Mars's gases from the strength of their spectral lines; the density turns out to be very low—only about 1% the density of Earth's atmosphere. This composition and low density have been verified by the spacecraft that have landed on Mars.

Although the Martian atmosphere is mostly carbon dioxide, a greenhouse gas, its density is so low that the carbon dioxide creates only a very weak greenhouse

FIGURE 40.14
(A) Fog in Martian valleys and (B) frost on surface rocks near the *Viking* lander.

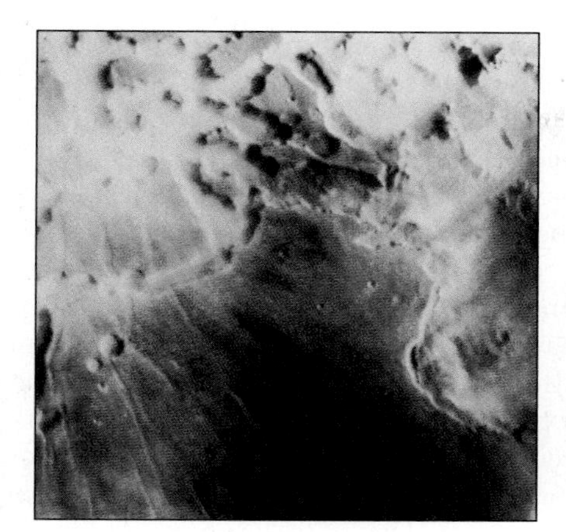
A

B

effect. The consequent lack of heat trapping and Mars's greater distance from the Sun make the planet much colder than Earth. Temperatures at noon at the equator may reach a little above the freezing point of water, but at night they plummet to about 180 K (about −140°F). The resulting average temperature is a frigid 210 K (about −80°F).

Clouds of dry ice (frozen CO_2) and water ice crystals (H_2O) drift through the Martian atmosphere, carried by the Martian winds. These winds, like the large-scale winds on Earth, arise because air that is warmed near the equator rises and spreads toward the poles. However, the Coriolis effect, arising from the planet's rotation, deflects the winds so that they blow around the planet approximately parallel to its equator. These winds are generally gentle, but seasonally in the southern summer and near the poles they can become gales, which sometimes pick up large amounts of dust from the surface. The resulting vast dust storms sometimes completely cover the planet and turn its sky pink.

No rain falls from the Martian sky, despite its clouds, because the atmosphere is too thin and cold. Despite such dryness, however, fog sometimes forms in Martian valleys at night when the air temperature drops so low that water vapor in the atmosphere condenses into low clouds. Frost also condenses on the ground on cold nights, as seen in Figure 40.14. In addition, during the Martian winter, CO_2 "snow" falls on the Martian poles. The combined evidence suggests that Mars has not always been so dry, but if Mars had liquid water present on its surface in the past, it needed to have an atmosphere that was both denser and warmer.

Water is not stable in liquid form at pressures below about 0.6% of Earth's normal atmospheric pressure at sea level. If the pressure on a liquid is low, molecules can break free from its surface, evaporating easily because no external force restrains them. If you have cooked at high altitudes, you have discovered that water boils there at a lower temperature than 100°C (373 K or 212°F)—so it takes longer to hard boil an egg, for example. If you could go to a high enough altitude in Earth's atmosphere, you would find that the boiling temperature dropped all the way to the freezing temperature, and water would go directly from a solid to a gas. This is how carbon dioxide ("dry ice") behaves at Earth's surface pressure—if the pressure was about five times higher, carbon dioxide could liquefy.

Q The dream of many science fiction stories is that it might be possible to "terraform" Mars (to make it more like Earth) so that its environment would be suitable for humans. How might this be done?

The Martian surface pressure is so low that it is very close to the limit where liquid water is possible. So for water to exist on Mars, it must remain frozen below the surface or in the polar caps as solid water ice. If astronauts someday reach

Mars, they will have to bring any ice they find into a higher-pressure environment before they can melt it and drink it.

From the many craters on Mars, astronomers deduce that its surface has not been significantly eroded by rain or flowing water for about 3 billion years. Why has Mars dried out, and where have its water and atmosphere gone? And why did the planet cool so much?

If Mars had a denser atmosphere in the past, as deduced from the higher pressure needed to allow liquid water to exist, then the greenhouse effect might have made the planet significantly warmer than it is now. The loss of such an atmosphere would then weaken the greenhouse effect and plunge the planet into a permanent ice age. Mars's moderately weak gravity probably allowed gas molecules to escape over the first 1 to 2 billion years of the planet's history. Gradually the surface pressure would have been reduced, allowing ever-increasing numbers of water molecules to evaporate and float into space.

Some astronomers hypothesize that Mars lost its atmosphere when it was struck by a large asteroid, or a series of midsize asteroids. A large enough impact could blast the planet's atmosphere off into space.

As its atmosphere dwindled, Mars would have cooled. The remaining water would have frozen solid, and wind might then have buried this ice under protective layers of dust, as happens in polar and high mountain regions of Earth. The remaining water was locked up in subsurface layers, while the low atmospheric pressure no longer allowed liquid water to exist on its surface.

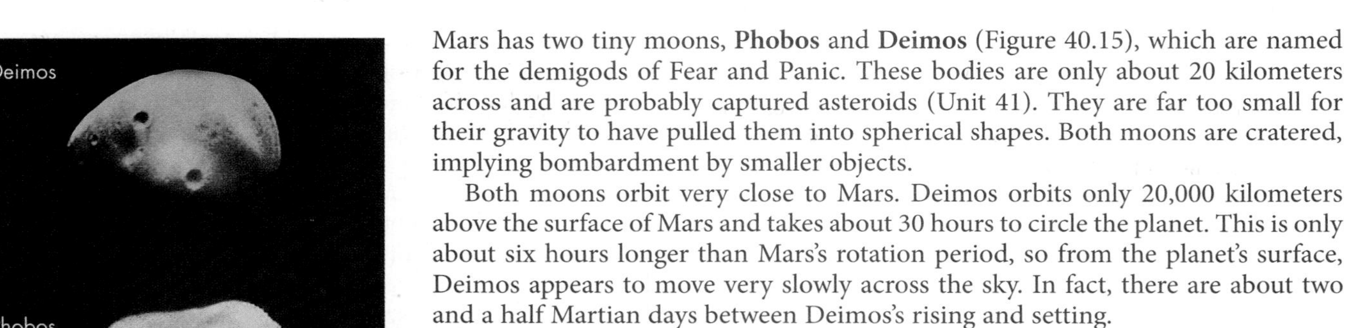

40.5 THE MARTIAN MOONS

FIGURE 40.15
Picture of Phobos and Deimos, the moons of Mars. These tiny bodies are probably captured asteroids.

Mars has two tiny moons, **Phobos** and **Deimos** (Figure 40.15), which are named for the demigods of Fear and Panic. These bodies are only about 20 kilometers across and are probably captured asteroids (Unit 41). They are far too small for their gravity to have pulled them into spherical shapes. Both moons are cratered, implying bombardment by smaller objects.

Both moons orbit very close to Mars. Deimos orbits only 20,000 kilometers above the surface of Mars and takes about 30 hours to circle the planet. This is only about six hours longer than Mars's rotation period, so from the planet's surface, Deimos appears to move very slowly across the sky. In fact, there are about two and a half Martian days between Deimos's rising and setting.

Phobos is in an even lower orbit, only about 6000 kilometers above the surface. It orbits in little more than a quarter of a Martian day. Even though it orbits in the same direction as Mars rotates, because it revolves around the planet faster than Mars rotates, it rises in the west and sets in the east. Phobos is so close to Mars that tidal drag on Phobos will eventually cause it to spiral in and crash into the planet. These unusual orbits are part of the reason that astronomers suspect Phobos and Deimos are captured asteroids (Unit 41).

Phobos and Deimos were discovered in 1877. But centuries earlier Kepler had guessed that Mars had two moons because the Earth has one moon and Jupiter, at least in Kepler's time, was known to have four. He thought that Mars, lying between these two bodies, should therefore have a number of moons lying between one and four, and he chose two as the more likely case. This was a lucky guess—more lucky than he could have known, because Jupiter has far more moons than were known to Kepler. The idea that Mars has two moons even appeared in Jonathan Swift's book *Gulliver's Travels,* nearly two centuries before the moons were discovered. Gulliver stops at the imaginary country Laputa, whose inhabitants include numerous astronomers. Among the accomplishments of these people is the discovery of two tiny moons of Mars.

KEY TERMS

conglomerate, 317
Deimos, 320
laminated terrain, 314
Olympus Mons, 312
Phobos, 320
Tharsis bulge, 312
Valles Marineris, 312

QUESTIONS FOR REVIEW

1. How does Mars compare with the Earth in mass and diameter?
2. What are the major geological features on Mars?
3. What evidence indicates that Mars was once geologically very active but has been geologically inactive for hundreds of millions or billions of years?
4. Valles Marineris on Mars is often compared to the Grand Canyon on Earth. In what ways are these two features similar? In what ways do they differ?
5. What are the Martian polar caps composed of? Why do they show layering?
6. What is the Martian atmosphere like?
7. What is the evidence that Mars once had running water on its surface? Is there evidence for water in any form today?
8. Why is it impossible for Mars to have liquid water on its surface today?

PROBLEMS

1. Mars's orbit is quite elliptical, so at different times when it is in opposition, its distance from Earth can vary substantially. From the data given in this chapter, estimate the largest and smallest distance to Mars at opposition and when it is on the far side of the Sun. Assume that the Earth's orbit is circular at 1 AU.
2. How long would it take an electromagnetic signal sent from Earth to reach a spacecraft on Mars, for each of the distances you calculated in problem 1?
3. From the surface of Earth under very good conditions, we can resolve an angular size of about 1 arc second. When Mars is 0.5 AU away, calculate the linear size (in kilometers) of a feature on the surface that would appear to be 1 arc second across. The Hubble Space Telescope can resolve about 0.04 arc seconds; what size on Mars does this correspond to?
4. What angular size would Phobos and Deimos have as viewed from the surface of Mars? Could either produce a solar eclipse?
5. Earth's atmospheric pressure drops by about a factor of 2 every 5 km higher up you go, so it is one-fourth the surface pressure at 10 km, for example. How high up in Earth's atmosphere would you have to go to reach a pressure similar to the 0.6% atmospheric pressure on Mars? (You can do this mathematically with logarithms or plot the changing pressure with elevation and estimate the height where this pressure is reached.)
6. The average surface pressure on Mars is almost exactly equal to that below which water cannot be liquid. Review the topographic map of Mars in Figure 40.3. If you wanted to search an area that had the greatest probability of any location on Mars for liquid water today, where would you go and why?

TEST YOURSELF

1. Why are daily temperature variations on Mars much larger than we experience on Earth?
   a. Mars spins slowly, so its nights are very long.
   b. Mars is much darker than the Earth, so it absorbs more sunlight.
   c. Mars's atmosphere is too thin to insulate the surface.
   d. Mars has smaller internal heat sources than the Earth.
2. Which of the following has provided evidence that the volcanoes of Mars are relatively young?
   a. The lava fields are warmer than surrounding regions.
   b. The volcanoes have relatively few impact craters on them.
   c. The volcanoes are not carved by river channels.
   d. The volcanoes' tops have only a thin layer of snow and ice.
   e. All of the above.
3. What evidence suggests that liquid water was once present on the Martian surface?
   a. Branching channels in the shape of riverbeds
   b. Flat ice-covered basins that appear to have been lakes
   c. High levels of humidity measured spectroscopically in the atmosphere
   d. Fossils found by the Martian rovers
   e. All of the above

unit

41 Asteroids

Asteroids are small, generally rocky bodies that orbit the Sun. More than 90% lie in the asteroid belt, a region between the orbits of Mars and Jupiter stretching from about 2 AU to 4 AU from the Sun, as shown in Figure 41.1. Several have been visited by spacecraft, but we still have only blurry images of the largest one. A NASA mission to the two most massive astroids will arrive at the first of them in 2011.

Some asteroids appear to be nearly solid metal, and perhaps someday we will mine their ores. Others appear to be carbon-rich, and these are probably very primitive bodies—planetesimals that were never incorporated into the planets. Actually, the process of planet formation continues to this day, and from time to time an asteroid collides with and becomes part of one of the planets. These collisions release an enormous amount of energy that can create dramatic changes in a planet's environment (Unit 48). Most of the nearly 100,000 numbered asteroids are in stable orbits, and perhaps represent a planet that never formed because of the gravitational influence of Jupiter.

41.1 THE DISCOVERY OF ASTEROIDS

The discovery of the asteroids was based on a prediction that astronomers argue about to this day. A pattern in the distances of the planets from the Sun, pointed out in 1766, is known today as **Bode's rule.** It reproduced the distances of the planets known at the time and suggested that there was a gap between Mars and

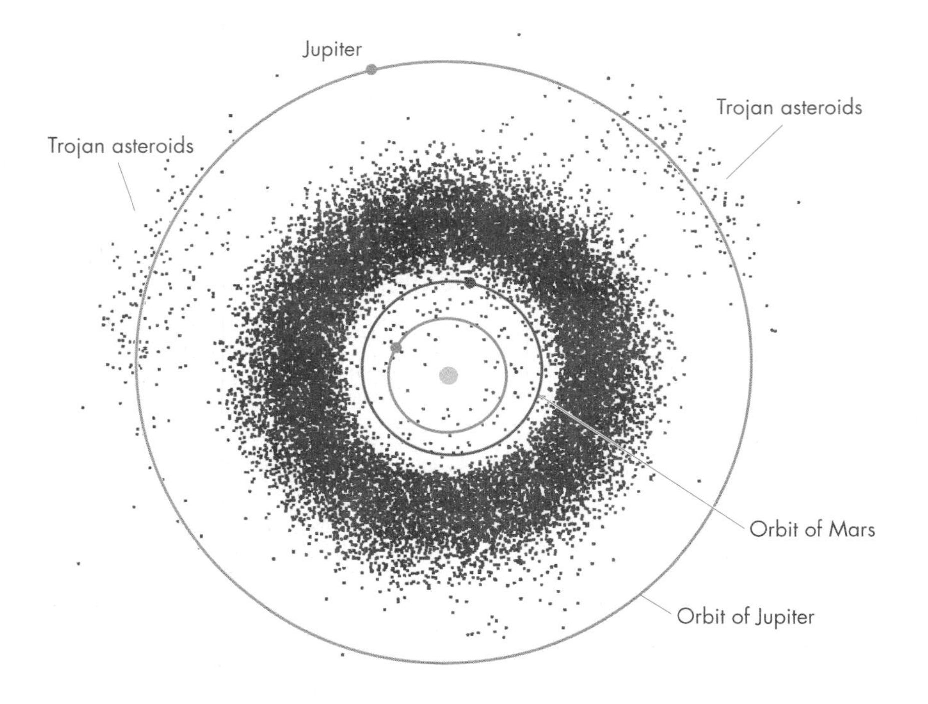

FIGURE 41.1
Diagram showing the distribution of more than 20,000 asteroids—about 1 in every 20 of the total number known. Notice that most lie between Mars and Jupiter, but a small number form two loose clumps—the Trojan asteroids—located along Jupiter's orbit. Making the plotted points large enough to see causes the asteroids to appear far more closely packed than they really are. A spacecraft traveling through the asteroid belt is highly unlikely to encounter an asteroid.

TABLE 41.1 Bode's Rule

Bode's Rule	Number	Planet	True Distance
(0 + 4)/10 =	0.4	Mercury	0.39
(3 + 4)/10 =	0.7	Venus	0.72
(6 + 4)/10 =	1.0	Earth	1.00
(12 + 4)/10 =	1.6	Mars	1.52
(24 + 4)/10 =	2.8	Ceres (dwarf)	2.78
(48 + 4)/10 =	5.2	Jupiter	5.20
(96 + 4)/10 =	10.0	Saturn	9.58
(192 + 4)/10 =	19.6	Uranus	19.2
(384 + 4)/10 =	38.8	Neptune	30.1
(768 + 4)/10 =	77.2	Pluto (dwarf)	39.5
(1536 + 4)/10 =	154.0	Eris (dwarf)	67.7

Jupiter where a planet ought to exist. With what we now know about other planetary systems, most astronomers think Bode's rule does not express a broader universal law; nevertheless, it led to the discovery of the asteroids.

Bode's rule, also known as Bode's law or the Titius-Bode law, was first pointed out by a Prussian astronomer, Johann Titius. Johann Bode confirmed the calculation and published it six years later.

The pattern found among the planets can be expressed by a simple mathematical progression. Bode's rule works as follows: Write down 0, 3, and then successive numbers by doubling the preceding number. That is, 0, 3, 6, 12, 24, and so on. Next add 4 to each and divide the result by 10, as shown in Table 41.1. The resulting numbers are close to the distances of the planets from the Sun in astronomical units, except there was no planet at 2.8 AU and none known beyond Saturn at the time the rule was devised.

The planet Uranus was discovered in 1781, and when astronomers realized that it fit the rule, they began to search systematically for a planet in the "gap" at 2.8 AU. In 1801 Giuseppe Piazzi, a Sicilian astronomer, discovered the largest of the asteroids, which he named **Ceres** in honor of the patron goddess of Sicily. Ceres fit the rule splendidly (Table 41.1), and it was widely accepted as the "missing" planet.

However, within six years, three more large objects were found orbiting in the same general region of the Solar System. Some astronomers called these planets too, but others pointed out that they appeared to be a different class of object. The astronomer William Herschel (who discovered Uranus) noted that unlike the other planets, these objects were so small that they looked like stars—pinpoints of light—through even the best telescopes of the day. That is how they got the name *asteroid*—because they look like stars (*aster* in Latin). They are also sometimes called *minor planets,* reflecting their original status.

Even today, asteroids appear merely as points of light through research telescopes, so their status as asteroids cannot be determined from a single image. Instead, astronomers must track their motion for weeks or months, using their shifting position to determine the size and shape of their orbits. Ceres's and other asteroids' orbits are much less circular and outside the plane of the other planets' orbits, also indicating their difference from the major planets.

Initially asteroids were named after goddesses; but as the names ran out, the discoverers were given wide latitude. There are now more than 10,000 named asteroids. Some have been given the names of discoverers' family members, well-known people, or cities. After an asteroid was named for a pet cat, the International Astronomical Union stepped in and set guidelines for their naming.

By 1850 more than a dozen asteroids had been found, and the number has climbed steadily since. Once an orbit has been confirmed, an asteroid is given a number that indicates the order in which its discovery was recorded, and the discoverer is allowed to name it. Thus Ceres is designated "1 Ceres." The asteroids discovered fourth and second, "4 Vesta " and "2 Pallas," are estimated to be the next two most massive.

The introduction of automated search programs in recent years has dramatically increased the count of asteroids, reaching over 400,000 in 2008. For more than half of these an orbit has not yet been accurately determined; this sometimes requires several years of observations. Astronomers continue to search for these objects, both because they provide valuable information about the formation of the Solar System and because there is a significant possibility that a few may have orbits that could someday cause them to collide with Earth (Unit 48). Many of the asteroids currently being discovered by automated searches are estimated to be under a kilometer in size. These are rarely named, but just given a catalog number and a numeric designation indicating when and how they were found.

The asteroids might have been able to collect together to form a planet at about 2.8 AU, but, probably because of the effect of Jupiter's gravity, these planetesimal-size objects were never able to coalesce into a single body. However, even if all the asteroids present today in the asteroid belt were put together, they would form a body with a mass less than 1% of Earth's mass.

Bode's rule, then, did seem to match a pattern in the planets' spacing. Ironically, though the rule inspired searches that ultimately led to the discovery of Neptune, that planet turned out *not* to continue the pattern (Table 41.1). Today astronomers note in their simulations of planetary system formation that similar patterns of planet spacing often arise, with each planet roughly 1.5 to 2 times farther out than the preceding planet. This appears to be due to each planet's influence on the formation of neighboring planets. However, depending on many factors, such as the particular masses of the forming planets, the final pattern may look quite different from what Bode's rule predicts. These differences in character are also exhibited by the planetary systems found around other stars (Unit 34).

41.2 ASTEROID SIZES AND SHAPES

No spacecraft missions have yet made a close encounter with Ceres, and it is too small to resolve much detail from Earth. However, the Hubble Space Telescope has imaged this tiny body well enough to determine that it is nearly spherical (Figure 41.2) with a diameter of about 930 kilometers. This is a large body, but very small compared to any of the terrestrial planets or even the Moon, which is about 80 times more massive.

NASA's *Dawn* mission will reach Vesta in 2011 and then travel on to orbit Ceres in 2015.

Its shape indicates that Ceres has a gravitational force strong enough to crush the body into a spherical shape. This is one of the criteria for being a planet. However, Ceres fails to meet the other requirement: It is estimated that the rest of the asteroids in the asteroid belt have a combined mass about twice that of Ceres. Ceres has

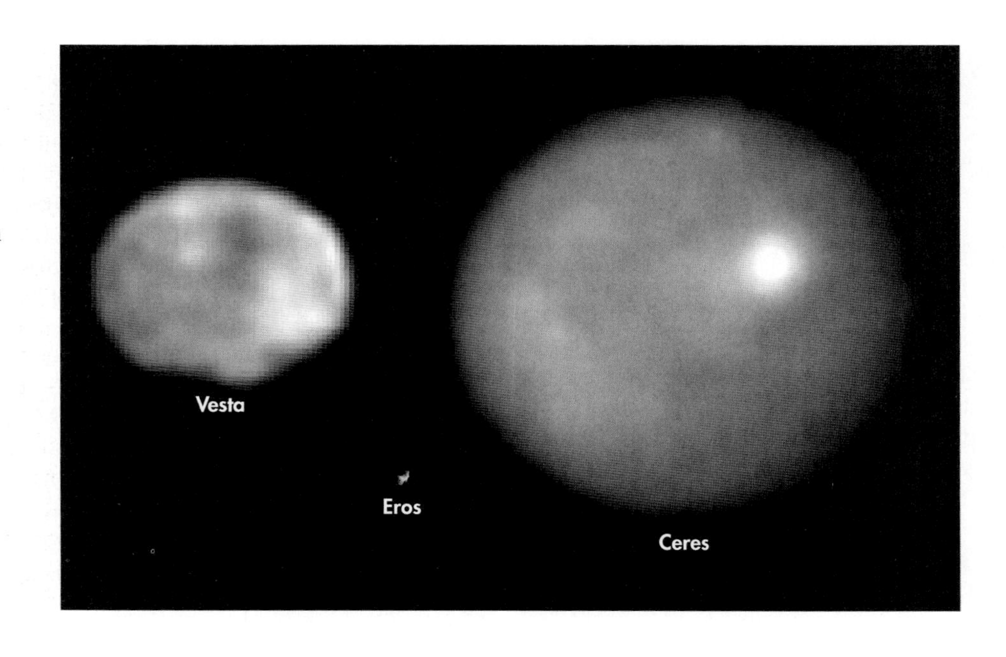

FIGURE 41.2
The two most massive asteroids, Ceres and Vesta, imaged by the Hubble Space Telescope. For comparison, 433 Eros, a near-Earth asteroid that the spacecraft *NEAR* landed on in 2001, is shown to the same scale. Ceres is over 900 kilometers in diameter, and its round shape and density suggest that it may have a thick surface layer of ice.

A

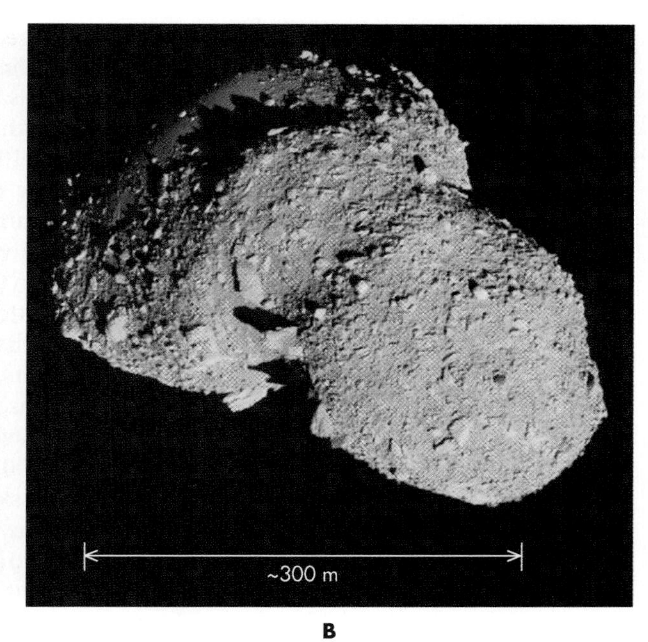

B

FIGURE 41.3
(A) The asteroid Eros, one of the largest near-Earth asteroids, imaged by the spacecraft *NEAR,* which landed on it in 2001. The asteroid is about 33 kilometers (21 miles) long and about 13 kilometers (8 miles) across. (B) The asteroid Itokawa, imaged by the Japanese spacecraft *Hayabusa,* is about 0.5 kilometers (0.3 miles) across. Note the lack of craters and the rough, boulder-strewn surface.

not "cleared its orbit," in the sense that there is a comparable mass in other asteroids that cross Ceres's orbital path. Ceres is therefore designated a "dwarf planet."

Low-resolution imaging is also available for Vesta (Figure 41.2). In addition, observations from many sites on Earth have determined the shape of the third most massive asteroid, Pallas, as it passed in front of a star—effectively casting a shadow on Earth with the size and shape of the asteroid. These smaller asteroids are roughly egg-shaped, each about 500 kilmeters across. Because neither of these appears to have sufficient mass to have pulled it into a round shape, they are not classified as dwarf planets.

To date only a few asteroids have been viewed in detail by spacecraft, and these are all much smaller objects, such as 433 Eros (Figure 41.3A). NASA launched the *NEAR* (Near Earth Asteroid Rendezvous) spacecraft, which landed on the asteroid Eros in 2001. Eros (Figure 41.3A) is shaped like a "fat banana," 33 kilometers (21 miles) long and is one of the largest of the asteroids that come close to Earth's orbit. The mission to Eros showed that it is covered with craters—over 100,000 were cataloged. It is likely that Eros originated in the asteroid belt but was shifted to an orbit between Earth and Mars by some of the collisions that produced its craters. The *NEAR* spacecraft revealed that the surface of Eros is covered with regolith like the Moon. Some of the larger collisions almost certainly had the power to crack the asteroid in two, although it remains held together by its weak gravity. More numerous smaller collisions have left the asteroid pitted and lumpy, and the fragments blasted out by these impacts either have fallen back onto the surface to become part of the regolith or have become other small asteroids.

An even more ambitious mission was carried out by the Japanese spacecraft *Hayabusa,* sending a probe to land on the asteroid 25,143 Itokawa (Figure 41.3B) and attempting to return samples of the asteroid back to Earth. This asteroid is about 0.5 kilometers long and has an orbit that actually crosses the Earth's. Close-up pictures of the asteroid show a surface covered with boulders but no obvious impact craters. Scientists hypothesize that it may have coalesced from two asteroids, erasing evidence of past impacts and creating a "pile of rubble" that is only relatively loosely held together by gravity.

In addition to the images from spacecraft, a few asteroids have passed close enough to Earth to allow radar techniques to map their structure. However, the only information we have about most asteroids is the amount of light they reflect.

This allows us to learn some things about the asteroid. For example, most asteroids vary in brightness in a cyclical fashion. These variations in brightness probably occur because the asteroid is spinning and either it is irregularly shaped or some portions are darker than others. The variations indicate that asteroids typically rotate with a period of 3 to 10 hours.

We can estimate the size of asteroids from the amount of sunlight reflected, but this can be quite inaccurate because a poorly reflective large object will look as bright as a highly reflective small one. For this reason, the thermal radiation (Unit 23) of asteroids is often used to estimate their size. Asteroids are heated by the Sun, and the amount of emitted infrared radiation is a better indicator of diameter; bigger bodies emit more radiation than smaller ones of the same temperature.

From such measurements, astronomers have found that asteroids range tremendously in diameter. Based on the reflection measurements, it appears that about 200 of the asteroids have diameters larger than 100 kilometers. Astronomers estimate there are about 700,000 asteroids larger than 1 kilometer in diameter, based on a recent sensitive sky survey, which provided a statistical sample of the population of asteroids. The smallest asteroids are far too tiny to detect unless they come extremely close to Earth. For example, the diameter of the tiny asteroid 1991 BA, which passed about 170,000 kilometers (less than half the distance to the Moon) from Earth in January 1991, is probably less than 9 meters (30 feet).

Even very small asteroids can have devastating effects if they strike Earth (Unit 48), but a near miss with a planet can also have a catastrophic effect on an asteroid. Because gravity pulls more strongly on the side of the asteroid closer to the planet, material on that side tries to follow a tighter trajectory around the planet than material on the asteroid's far side. This stretching (tidal force) can rip an asteroid into many pieces.

41.3 ASTEROID COMPOSITIONS AND ORIGIN

Q What might be some advantages of mining material from asteroids instead of from the Earth or the Moon?

When sunlight falls on an asteroid, the minerals in its surface create absorption features in the spectrum of the reflected light, from which we can determine the asteroid's composition. Such spectra indicate that asteroids belong to three main compositional groups: carbonaceous bodies, silicate bodies, and metallic iron-nickel bodies. The carbonaceous material is a carbon-rich, coal-like substance. The groups are not mixed randomly throughout the asteroid belt: inner-belt asteroids tend to be silicate-rich, and outer-belt asteroids tend to be carbon-rich. Spectra of some carbonaceous asteroids indicate that minerals on the surface are hydrated, implying the presence of water ice.

Masses have also been estimated for about a dozen asteroids based on the gravitational effect they have on each other and on passing spacecraft. Several asteroids have now been discovered that have small moons orbiting them as well, such as the asteroid 243 Ida and its tiny moon Dactyl (Figure 41.4). A moon's orbit also allows us to estimate the asteroid's mass. The average densities found are typically between about 1.5 and 3.5 kilograms per liter, suggesting that asteroids are mostly rocky and probably include some lower-density solids—such as the carbonaceous compounds and water ice.

The properties of asteroids give us clues to their origin and support the solar nebula theory for the origin of the Solar System. The asteroids are probably fragments of planetesimals, the bodies from which the planets were built (Unit 33). According to the solar nebula theory, because the inner asteroid belt is warmer than the outer belt, bodies that condensed in the inner belt are made primarily of the easy-to-condense silicate and iron materials, whereas the outer belt contains more-volatile substances: carbon-rich materials and water.

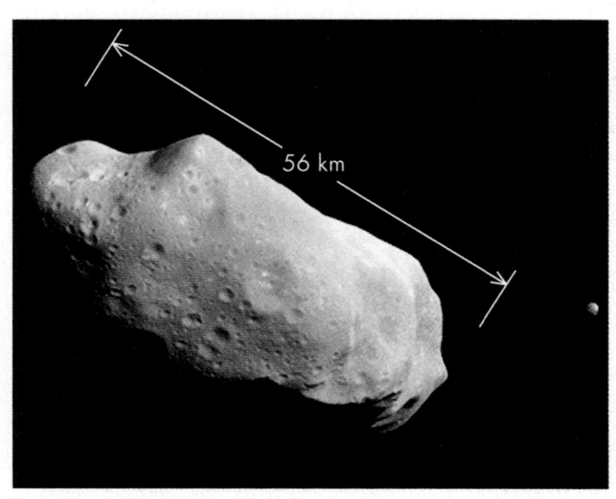

FIGURE 41.4
The asteroid Ida and its moon Dactyl.

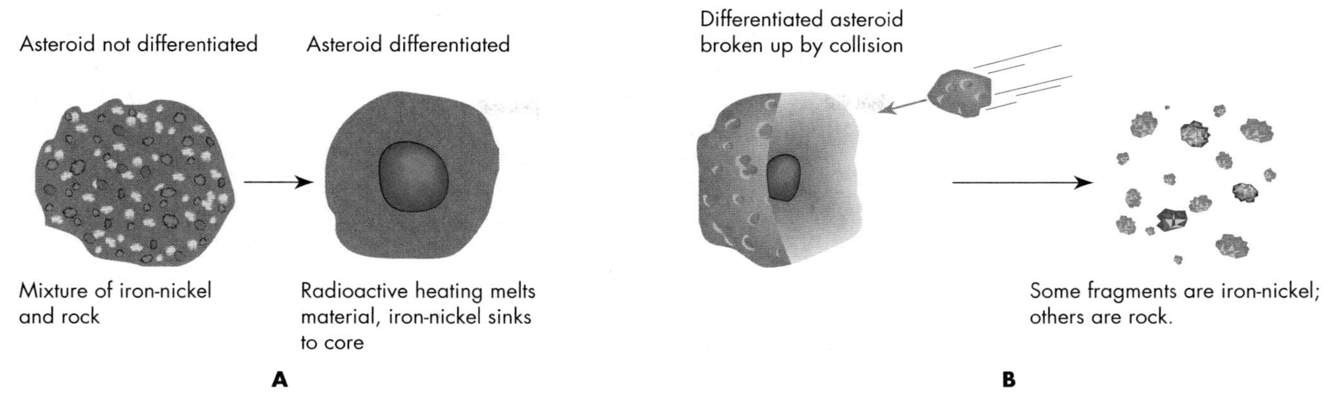

FIGURE 41.5
Sketch depicting (A) differentiation in asteroids and (B) their subsequent breakup by collision to form iron and stony bodies.

Differentiation and breakup of asteroids

The existence of iron asteroids might at first seem to be evidence against the solar nebula theory. However, chemical elements can be separated by gravity if they are in a molten state—in the process called differentiation (Unit 33). When an object that is a mix of rocky and iron-rich material melts (perhaps as a result of naturally occurring radioactive substances heating it), the denser iron will sink to the object's core. We know this has happened to the Earth, and it also appears to have occurred in large asteroids (Figure 41.5A). In fact, some asteroids seem to have been sufficiently heated to have had volcanic eruptions. This activity, now long extinct, is deduced from their spectra, which show the presence of basalt, a volcanic rock.

After differentiation, collisions with neighboring asteroids and close encounters with planets broke up some of these large bodies (Figure 41.5B). The fragments are what we see today: pieces of crust became stony asteroids, whereas pieces of core became iron asteroids.

Q One early hypothesis was that the asteroids were once part of a planet that later broke up. What observational findings argue against this scenario?

41.4 ASTEROID ORBITS

The solar nebula theory also offers an explanation for the asteroid belt being located near Jupiter's orbit. Any small planet beginning to form there would have to compete for material with Jupiter. Moreover, Jupiter's gravitational tidal force would disturb the accretion process, tugging apart clumps of matter in the solar nebula before they could grow very sizable.

Even today Jupiter affects the asteroid belt. Figure 41.6 shows a partial census of the number of asteroids found at specific distances from the Sun within the belt. A clear gap can be seen at 2.5 AU. The seemingly empty region in the asteroid belt is called a **Kirkwood gap** and was discovered by the American astronomer Daniel Kirkwood in 1866. The location of the gap corresponds to the distance at which an orbiting asteroid would have an orbital period of 3.95 years—exactly one-third of Jupiter's orbital period. Orbiting with this period, an asteroid would be at the exact same point in its orbit once every three orbits when it makes its closest approach to Jupiter. Jupiter's proximity would tend to pull the asteroid toward it, and by always tugging outward in the same part of the asteroid's orbit, it would cause the orbit to gradually become increasingly elongated toward Jupiter.

Eventually the orbit would stretch close enough to Jupiter for the asteroid to be flung off in an entirely new direction. Compare this to an orbit that does not have a simple numerical relationship with Jupiter's orbital period. The asteroid is tugged in one direction during one of its closest approaches to Jupiter; but then, in successive orbits, it is pulled in different directions, so that the effects tend to cancel each other out.

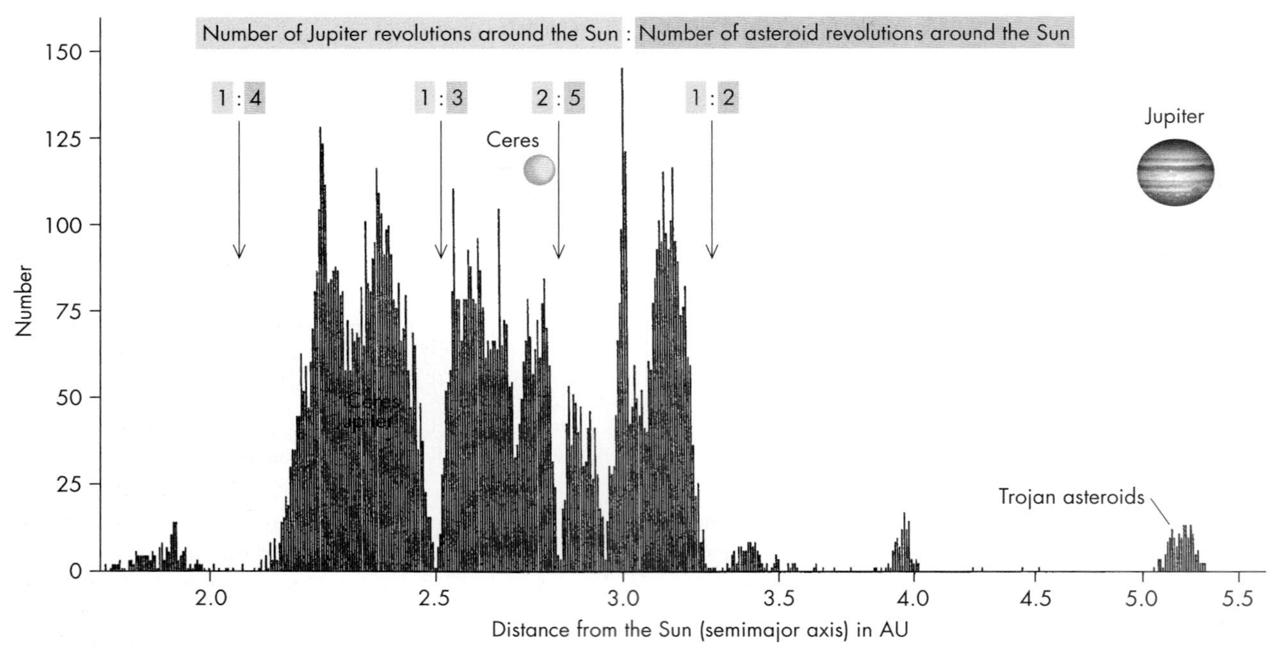

FIGURE 41.6
The number of asteroids at each distance from the Sun within the asteroid belt. Notice the conspicuous gaps where there are few, if any, objects. These empty zones are called *Kirkwood gaps.*

Resonance of orbits

Kirkwood gaps are also found at distances corresponding to ratios of orbital periods of 2:1 and 4:1, as well as at some more complicated combinations such as 5:2—in other words, the asteroid orbits five times while Jupiter orbits twice. These orbits with simple whole-number ratios of their periods have what are called **resonances** with Jupiter's orbit. But resonances with Jupiter's motion not only create gaps; they can also create concentrations of asteroids. For example, a larger-than-average number of asteroids orbit at 4.0 AU. Asteroids orbiting at this distance have a period two-thirds of Jupiter's, and are therefore in a 3:2 resonance.

Not all asteroids are found in the main belt. The *Trojan asteroids* travel along Jupiter's orbit in two loose swarms 60° ahead of and 60° behind that planet, as illustrated in Figure 41.1. These turn out to be spots where gravitational forces from the Sun and Jupiter balance and the asteroids' orbits are stable.

Sometimes asteroids suffer collisions with each other that send fragments into new orbits. By comparing their orbital parameters, astronomers have identified dozens of **asteroid families** that each appear to have originated from a single object that underwent a collision. Supporting this idea, spectroscopic studies of the members of these families show that they have compositions similar to each other.

Many asteroids travel well outside the asteroid belt, possibly because they were disturbed by Jupiter's gravity or a collision with another asteroid, which sent them into highly elliptical orbits. These asteroids can pose a serious threat to Earth. For example, astronomers have identified the Baptistina asteroid family (named after the largest member of the family) as the likely source of the asteroid that killed the dinosaurs (Unit 48). The orbits of the members of this family point to an origin of when a large carbonaceous asteroid broke up in a collision about 160 million years ago.

Q Itokawa is a near-Earth object. If it were on a collision path with Earth, what strategies could we use to avoid the collision?

An assortment of over 4000 other asteroids have orbits that cross or come very close to the Earth's orbit. These are known as **near-Earth objects.** The chance of a collision with Earth is slim, but on average, a body 100 meters in diameter hits the Earth about every 10,000 years. Collisions of even such a "small" asteroid with a planet can have devastating effects, as we will discuss in Unit 48.

KEY TERMS

asteroid, 322
asteroid family, 328
Bode's rule, 322
Ceres, 323
Kirkwood gap, 327
near-Earth object, 328
resonance, 328

QUESTIONS FOR REVIEW

1. What is Bode's rule, and how did it lead to the discovery of asteroids?
2. Where are asteroids located, and why did they not form a planet?
3. What is different about the shape of Ceres compared to other asteroids? Why is there this difference?
4. How do we know what asteroids are made of ?
5. How have the processes of differentiation and impacts led to the variety of asteroid types we observe today?
6. What are the Kirkwood gaps? How were they formed?
7. Where are asteroids found, other than in the main belt? What are their orbits like?

PROBLEMS

1. Asteroids are often discovered on long-exposure images because they are moving relative to the stars during the exposure. They therefore leave a short trail on the image while the stars remain stationary. Suppose that an exposure 25 minutes long is made, and that this image has a "plate scale" of 1.1 arc minutes per mm.
   a. If the asteroid leaves a trail 3 mm long on the image, how many arc minutes did it move during the exposure?
   b. If the asteroid is about 0.5 AU away, what is its velocity through space?
   c. How would Earth's motion affect this estimate? Draw a diagram to support your explanation.
2. Most asteroids are found between 2.1 and 3.3 AU. Use Kepler's third law to calculate the range of orbital periods associated with these orbital sizes.
3. If 100 asteroids, each of diameter 100 km, were combined into a single object, approximately what would its diameter be?
4. The asteroid Icarus has an elliptical orbit that carries it between 0.19 and 1.97 AU from the Sun. What is its semimajor axis? How often does it cross the Earth's orbital radius?
5. The asteroid 2004 FH was discovered on March 15, 2004, by the NASA-funded LINEAR asteroid survey. It is approximately 30 meters in diameter. Through careful observation, astronomers determined its orbital period to be 270.1 days.
   a. What is its semimajor axis?
   b. If its orbit varies from 1.05 AU to 0.58 AU, could this be a potentially hazardous asteroid? Explain your answer.
   c. If it has a density of 2 kg/L, what is its mass?
   d. If this asteroid were to strike the surface of the Earth at 30 km/sec (Earth's orbital velocity), how much kinetic energy would it have?
6. The *NEAR* spacecraft landed on Eros with a velocity estimated at 1.6 m/sec. Convert this speed to kph or mph (whichever is more familiar to you). If the spacecraft had a mass of about 800 kg, what was the kinetic energy of its impact? Discuss how much this would have affected the asteroid's orbit.
7. Ceres has a mass of approximately 8.7×10^{20} kg and a diameter of approximately 930 km.
   a. What would the acceleration of gravity be on its surface (Unit 16)?
   b. What is the escape velocity from its surface (Unit 18)?

TEST YOURSELF

1. The asteroid belt lies between the orbits of
   a. Earth and Mars.
   b. Saturn and Jupiter.
   c. Venus and Earth.
   d. Mars and Jupiter.
   e. Pluto and the Oort cloud.
2. How do astronomers estimate the sizes of most asteroids?
   a. By observing angular diameters from space telescopes
   b. By timing how long they take to pass in front of stars
   c. By determining their gravitational influence on other asteroids
   d. By measuring the amount of light they reflect
3. Which of the following are you *least* likely to find in an asteroid?
   a. Water ice
   b. Rocky materials
   c. Metallic materials
   d. Carbon-based materials

unit

42

Comparative Planetology

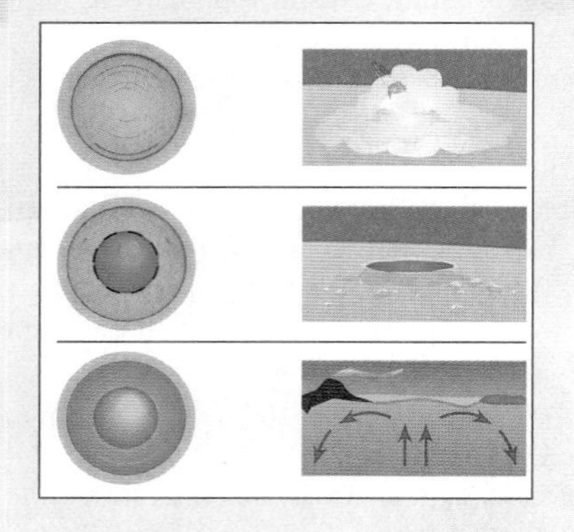

Background Pathways

Unit 32: The Structure of the Solar System 238

The study of the similarities and differences between planets—what astronomers call **comparative planetology**—allows us to better understand all the planets, including our own. Many factors affect a planet, such as its size, distance from the Sun, and internal and atmospheric composition. Seeing how physical processes operate under different conditions leta us test our hypotheses and perhaps better predict the consequences of global changes for our own planet's environment. In this comparison, some planets are clearly more similar to the Earth than others, although all have stories to tell that can help us better understand the Earth.

Because of their similarity to Earth, the planets Mercury, Venus, and Mars are called *terrestrial planets*. Their properties, along with those of the Earth and Moon, are discussed in Units 35–40. Figure 42.1 shows these four worlds along with the Moon. The figure also shows some of their prominent surface features. Note that although the terrestrial planets are roughly similar in diameter, many differences are apparent.

These differences are particularly striking if we compare their atmospheres and surfaces. For example, Mercury and our Moon are essentially airless. Venus, on the other hand, has an atmosphere 100 times more massive than our own. The surfaces of Mercury and the Moon are covered with craters, but they lack most other landscape features. In contrast, Earth and Mars have huge volcanic peaks and canyons, but Mars has many more impact craters.

As great as the differences between the terrestrial planets appear to be, even a brief look at the worlds of the outer Solar System makes it clear just how similar the terrestrial worlds are. Nevertheless, in a few aspects these outer worlds are similar to the terrestrial worlds. We explore these briefly at the end of this unit. These other worlds are discussed further in Units 43–46.

42.1 THE ROLE OF MASS AND RADIUS

Many basic properties of the terrestrial planets are summarized in Table 42.1. These properties play a major role in determining the observed characteristics of each planet.

A planet's mass and radius affect its interior temperature and thus its level of tectonic activity. Low-mass, small-radius planets cool inside more rapidly than larger bodies (Figure 42.2). We see, therefore, a progression of activity from small, relatively inert Moon and Mercury, to slightly larger and formerly active Mars, to the larger and far more active surfaces of Venus and Earth. Mercury's surface still bears the craters made as it was assembled from planetesimals. Mars has some craters, but it also has younger surface features such as volcanoes and riverbeds carved by running water. In contrast, Earth and Venus retain essentially none of their original crust, their surfaces having been modified enormously by interior activity over their lifetimes.

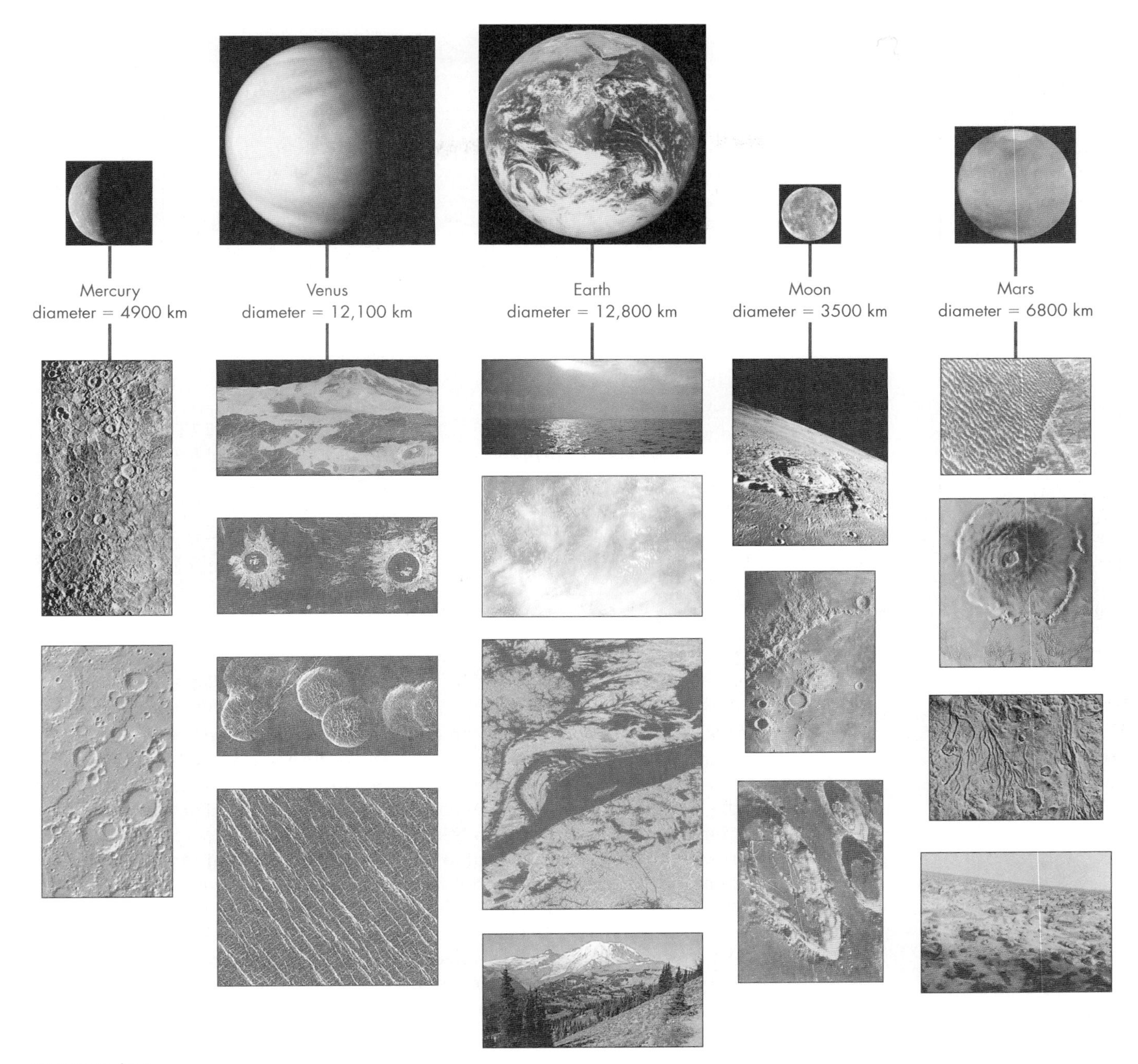

FIGURE 42.1
Pictures (left to right) of Mercury, Venus, Earth, Moon, and Mars, showing representative surface features.

Mars, intermediate in mass and radius between Mercury and Earth, has a surface that bears the traces of an intermediate level of activity. Much of it is cratered, implying that a long time has passed since geological activity ended. Some regions, however, show volcanic uplift like that seen on Venus, implying that Mars's interior was once much hotter than it is now. Studies of the magnetization of surface rocks on both Mars and the Moon show that both of these worlds must have had strong magnetic fields in the distant past, although neither has a global magnetic field like the Earth's anymore. Moreover, Mars's surface shows an amazing variety of landforms: polar

TABLE 42.1 Comparison of the Terrestrial Planets and Our Moon

Property	Mercury	Venus	Earth	Moon	Mars
Distance from Sun	0.387 AU	0.723 AU	1 AU	1 AU	1.524 AU
Orbital period	87.969 Earth days	224.70 Earth days	365.26 Earth days	—	686.98 Earth days
Radius (Earth units)	0.383	0.950	1.000	0.273	0.533
(km)	2439	6052	6378	1738	3398
Mass (Earth units)	0.055	0.815	1.0	0.012	0.107
(kg)	3.30×10^{23}	4.87×10^{24}	5.98×10^{24}	7.35×10^{22}	6.42×10^{23}
Sidereal day	58.65 Earth days	243.02 Earth days	23.9345 hours	27.322 Earth days	24.62 hours
Solar day	176 Earth days	116.8 Earth days	24 hours	29.53 Earth days	24.66 hours
Axial tilt	0°	177.4°	23.45°	5.15°	25.19°
Density (kg/L)	5.43	5.25	5.52	3.34	3.94
Uncompressed density (kg/L)	5.4	4.2	4.2	3.3	3.3
Escape velocity (km/sec)	4.3	10.4	11.2	2.4	5.0
Atmospheric composition	None	CO_2: 96.5% N_2: 3.5%	N_2: 78.1% O_2: 20.9% Ar: 0.9%	None	CO_2: 95.3% N_2: 2.7% Ar: 1.6%
Surface temperature (K)	440	735	287	250	210
Temperature range	80–710	720–750	185–331	120–390	133–293
Pressure at surface (Earth atmospheres)	0	About 90	1.0	0	About 0.007
Magnetic field (relative to Earth)	0.0006	None	1	Remnant	Remnant
Satellites	None	None	1 (Moon)	—	2 (captured asteroids)

Q Why is it surprising that Mercury has a weak magnetic field whereas Venus has none?

caps of ordinary ice and frozen carbon dioxide, immense deserts with dune fields, and canyons created by both crustal cracking and erosion from running water.

The small number of impact craters on Venus and Earth implies that they have young surfaces, extensively altered by volcanic activity. On Earth, the flow of heat from its core has created plate motion. On Venus, perhaps because of its thicker, more rigid crust, the flow of heat to its surface is less uniform. As a result its surface has isolated regions of intense volcanic uplift.

The extent of the atmosphere of a planet is also affected by the planet's size and mass. All of the terrestrial planets and the Moon probably received some gases in the late stages of formation, delivered to them by comets; but volcanic activity was probably an even larger source of atmospheric gases. Volcanic activity is greater in larger planets, and continues to this day for the largest terrestrial planets, Venus and Earth.

A small planet has an additional problem: Any atmosphere it forms or captures is likely to leak away faster than it can be replenished by volcanoes or comets. Mercury and the Moon have so little mass that their escape velocity is low and gases rapidly escape to space. (The escape velocity, atmospheric pressure, and atmospheric content for each planet are listed in Table 42.1.) Mars is an interesting transitional case. It was active once but is now quiescent. To have had liquid water on its surface as the evidence suggests, its atmosphere must once have been quite dense, but today its atmosphere is very thin.

Venus and Mars both have atmospheres rich in carbon dioxide, although Venus's atmosphere is about 100 times more massive than Earth's, while Mars's is about 200 times less massive than Earth's. The great mass of the Venusian atmosphere creates a strong greenhouse effect, and high in the Venusian atmosphere thick clouds of sulfuric acid droplets block our view of the Venusian surface.

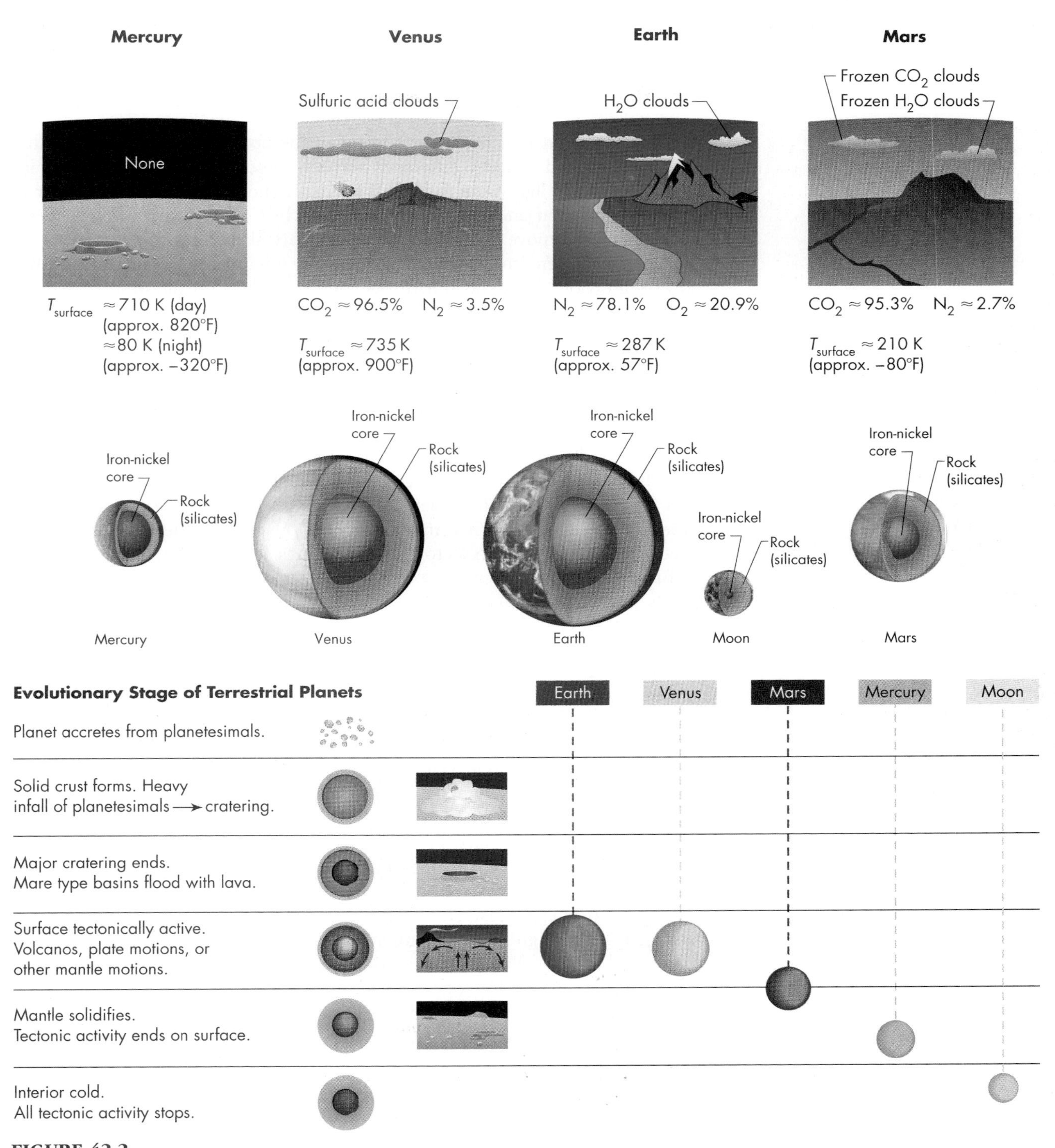

FIGURE 42.2
Gallery comparing interiors, atmospheres, and surfaces of the terrestrial planets and the Moon.

Astronomers think that the atmospheres on Venus, Earth, and Mars have been appreciably changed in composition over time by chemical processes. The atmospheres of all three bodies were probably originally nearly the same: primarily CO_2 with small amounts of nitrogen and water. But these original atmospheres have been modified by sunlight, tectonic activity, water, and, in the case of the Earth, life.

42.2 THE ROLE OF WATER AND BIOLOGICAL PROCESSES

The great difference in the water content of the upper atmospheres of Earth and Venus has created a drastic difference between these planets' atmospheres at lower levels. Ultraviolet light from the Sun is energetic enough to break apart water molecules into oxygen and hydrogen atoms. The hydrogen atoms that are liberated, being very light, move faster than average and are likely to escape into space, while the heavier oxygen atoms remain. Fortunately, other gases in the atmosphere can absorb ultraviolet light and shield water near the surface from destruction. In the warm Venusian atmosphere, water can rise to great heights, so over billions of years it has steadily been dissociated there and almost completely lost. In Earth's cooler atmosphere, however, water has survived, and its transitions between gaseous, liquid, and solid forms are a large part of the complex weather phenomena we experience.

The presence of liquid water permits chemical reactions that can profoundly alter the composition of a planet's atmosphere. For example, CO_2 dissolves in liquid water and creates carbonic acid (which in dilute form we drink as "soda"). In fact, the bubbles in soda are just carbon dioxide that is coming out of solution. As rain falls through our atmosphere, it absorbs CO_2, making the rain slightly acidic even in unpolluted air. When the rain reaches the ground, the carbonic acid reacts chemically with silicate rocks to form carbonates, locking some CO_2 into the rock.

Biological processes can also remove CO_2 from the atmosphere. For example, plants use CO_2 to make the large organic molecules such as cellulose of which they are composed. This CO_2 is usually stored for years or centuries before decay or burning releases it back into the atmosphere. More permanent removal occurs when sea creatures use dissolved carbon dioxide to make shells of calcium carbonate. As these creatures die, they sink to the bottom, where their shells form sediment that may eventually be compressed into rock. Thus carbon dioxide is swept from our atmosphere and locked away, both chemically and biologically, in the crust of our planet. With most of the carbon dioxide removed from our atmosphere, the relatively inactive gas nitrogen is left as the primary component. In fact, Earth's atmosphere contains a total amount of nitrogen comparable to that in the atmosphere of Venus. This suggests that these two planets may have begun with similar overall atmospheres—but the carbon dioxide has not been removed from Venus's atmosphere, so it remains the dominant constituent.

Our atmosphere is also rich in oxygen, a gas found nowhere else in the Solar System in such high relative abundance. Our planet's oxygen is the product of plants and single-celled creatures breaking down CO_2 and H_2O molecules during photosynthesis. Thus much of the cause of the great difference between the atmospheres of Earth and Venus may be that life was able to start on Earth. On our planet, liquid water removed much of the carbon dioxide from our atmosphere and life absorbed most of the rest of it, leaving only the tiny amount (0.03%) we have today. In removing the carbon dioxide, plants also produced the oxygen on which all animal life depends.

What evidence supports this idea that liquid water and living things removed most of the early Earth's carbon dioxide? It is the carbonate rock we mentioned earlier. If the carbon dioxide locked chemically in that rock were released back into our atmosphere and the oxygen were removed to pre-plant-life levels, our planet's atmosphere would closely resemble Venus's.

If water is so effective at removing carbon dioxide from our atmosphere, why does any CO_2 remain? We add small amounts of CO_2 by burning wood and fossil fuels, but the major contribution is from natural processes. Atmospheric chemists hypothesize that tectonic activity gradually releases CO_2 from rock into our atmosphere. At plate boundaries, sedimentary rock is carried downward into the mantle, where it melts. Heating breaks down carbonate rock, releasing its carbon dioxide, which then rises with the heated rock to the surface and reenters the atmosphere.

Carbon dioxide may have been removed from Mars's atmosphere and locked in rocks there during the planet's early history, when it apparently contained much more liquid water. Mars's lower level of tectonic activity, however, prevents its CO_2 from being recycled. Thus, with so little of its original carbon dioxide left, Mars has grown progressively colder. Mars's atmosphere is too thin now for a strong greenhouse effect; but the old river channels, presumably cut by running water, imply that Mars once had a denser and warmer atmosphere, perhaps even resembling Earth's. The great differences between the atmospheres of Venus, Earth, and Mars can probably be explained by the differences in their water content and the fact that life evolved on Earth, leading to plants and their release of oxygen.

Our Earth has retained enough CO_2 in its atmosphere to maintain a moderate greenhouse effect, making our planet habitable. Some geologists hypothesize that living organisms may respond to global changes, such as long-term changes in the Sun's brightness and changes in the Earth's activity, through feedback mechanisms to keep the Earth's climate temperate (Unit 83). Thus, at its location between one planet that is too hot and another that is too cold, Earth has been favored with a relatively stable temperature, one factor in the complex web of our environment to which we owe our existence.

Q Why would the discovery of oxygen in an exoplanet's atmosphere be significant? What opposing hypotheses might explain its presence?

42.3 THE ROLE OF SUNLIGHT

Although the Sun's energy is negligible for a planet's geology, it is the major heating source for the planet's surface and atmosphere. A planet's distance from the Sun determines the amount of sunlight it receives. Thus Venus, which is about twice as far from the Sun as Mercury, receives about a quarter as much energy over each square meter of its surface. The shape of a planet's orbit and the tilt of its rotation axis also affect how sunlight is distributed over the surface and how the amount of sunlight varies during an orbit. A tilted rotation axis, such as Earth's and Mars's, causes seasonal shifts in opposite hemispheres. An elliptical orbit, such as Mercury's and Mars's, adds a significant variation to a planet's global surface heating during each year.

Sunlight affects a planet's atmosphere in several ways. First, of course, it warms a planet by an amount that depends on the planet's distance from the Sun, and so we expect that Venus will be warmer than Earth and Earth will be warmer than Mars. These expected temperature differences are increased, however, by the atmospheres of these bodies. Moreover, even relatively small differences in temperature can cause large variations in physical behavior and chemical reactions within an atmosphere. For example, Earth is cool enough that water vapor condenses to ice at an elevation of about 10 kilometers (about 30,000 feet), making our upper atmosphere almost totally devoid of water. Venus is just enough nearer the Sun that even without a strong greenhouse effect, water would rise much higher in the atmosphere before condensing. Water molecules reaching the upper atmosphere can be broken apart by the Sun's ultraviolet light, and the hydrogen lost from the planet. Without water, carbon dioxide remained dominant in the atmosphere.

Distance from the Sun does not alone determine a planet's temperature. The role of the atmosphere in retaining the energy from the Sun is important too. We see this in the fact that Mercury has lower temperatures than Venus, despite Mercury's being half as far from the Sun. We also see this in the lower mean temperature of the Moon versus the Earth, even though both are at the same average distance from the Sun.

The distance from the Sun in the early history of the Solar System also affected the basic composition of the planets. This is relatively straightforward to see in the overall density of the planets between the inner and outer Solar System (Unit 32). A more refined examination of the terrestrial planets shows that they, too, exhibit a trend of density. This is not obvious at first, because the third planet, Earth, has the highest density—slightly higher than Mercury's. However, the density of a material is affected by the planet's gravitational force. For example, the pressures at the center of a massive planet like Jupiter may crush rock, which has a normal density of 3 kilograms per liter, to a density of 7 or 8 kilograms per liter. In fact, the pressures at the center of Jupiter would be sufficient to squeeze the Earth to half its current volume! Thus, if we compare the density of two planets made of the same materials, the more massive one will be more dense.

The Earth is the most massive of the terrestrial planets, so its gravitational force squeezes its interior most. It is estimated that if Earth's gravity were not squeezing the material within it, Earth's **uncompressed density** would be 4.2 kilograms per liter. This also affects Venus because of its similarly large mass, but affects the smaller planets less. Thus in Table 42.1 we see the highest uncompressed density for Mercury, which contains a large proportion of iron (a substance that condenses at high temperatures), and the lowest uncompressed density for Mars.

Q What features of each planet might be explained by a large collision late in its process of formation?

Although the Moon is at the same distance from the Sun as the Earth, its low density is a partial clue to its unusual history. The Moon is thought to have been torn out from the crust and mantle of an already differentiated Earth by an enormous impact toward the end of the planet-building phase of the young Solar System.

The average density of the Earth and Moon together lies between the densities of Venus and Mars. Geological evidence suggests that Mars has a substantial amount of water frozen at and beneath its surface. Its density likewise suggests it contains a large proportion of low-density substances. In Mars, then, we find hints of some of the characteristics of planets in the outer Solar System.

42.4 THE OUTER VERSUS THE INNER SOLAR SYSTEM

The *frost line* is also sometimes called the *water line* or the *snow line.* "Frost" is perhaps the most descriptive because it is at this distance that gaseous water freezes out into solid ice particles.

Beyond Mars and the asteroid belt, the Solar System is a realm of ice and frozen gas. In this frigid zone far from the Sun, where solar heat is only a vestige of what we receive on Earth, the giant planets formed. Beyond a distance called the **frost line,** the low temperatures—below 180 Kelvin (−140°F)—allowed water (H_2O) to condense. This is critical because hydrogen is the most abundant element, and oxygen the third most abundant, in the universe (Unit 33). Inside the frost line, water remained gaseous, so the terrestrial planets could incorporate only the scarcer rock and iron that condensed there. Giant planets could become giants because they could add to the rock and iron much larger quantities of ice.

The low temperatures beyond the frost line not only provided greater amounts of raw materials to form a planet; they also slowed the movement of gas molecules, making them more likely to be captured. As a result, these cold planets were able to draw from the vast reservoir of uncondensed hydrogen and helium gas in the solar nebula and became far larger than planets that formed near the Sun. Thus, beyond the asteroid belt we see a sudden jump to planet masses ranging from 15 to 320 times the mass of Earth (see Table 42.2).

Planets forming in this way became very different from the terrestrial planets in structure and composition. They consist mostly of hydrogen and its compounds. These gases form a deep atmosphere, often richly colored, that grows denser with depth and eventually becomes liquid. The interiors of Jupiter and Saturn are composed mainly of hydrogen—liquid just below the atmosphere and metallic deeper down. Each has a rock and iron core with a diameter estimated to be similar to the Earth's.

Q Jupiter-size exoplanets (Unit 34) have been found close to a number of stars. Can you think of any scenario in which terrestrial planets might be found far from a star?

Uranus and Neptune may have some liquid hydrogen below their hydrogen- and methane-rich atmospheres, but they are probably mostly water, ammonia, and methane surrounding a smaller rock and iron core. Because Uranus and Neptune are

TABLE 42.2 Comparison of the Four Giant Planets

Property	Jupiter	Saturn	Uranus	Neptune
Distance from Sun (AU)	5.203	9.539	19.19	30.06
Orbital period (Earth years)	11.8622	29.4577	84.014	164.793
Radius (Earth units)	11.19	9.46	3.98	3.81
(km)	71,492	60,268	25,559	24,764
Mass (Earth units)	317.9	95.18	14.54	17.13
(kg)	1.9×10^{27}	5.68×10^{26}	8.68×10^{25}	1.02×10^{26}
Sidereal and solar day (hours)	9.925	10.656	17.24	16.11
Axial tilt	3.12°	26.73°	97.86°	29.56°
Density (kg/L)	1.33	0.69	1.32	1.64
Escape velocity (km/sec)	59.5	35.5	21.3	23.5
Atmospheric composition	H_2: 89.5% He: 10.2% CH_4: 0.3%	H_2: 96.3% He: 3.3% CH_4: 0.4%	H_2: 82.5% He: 15.2% CH_4: 2.3%	H_2: 80.0% He: 18.5% CH_4: 1.5%
Temperature (K) at one atmosphere of pressure	160	130	80	75
Magnetic field (relative to Earth)	19,500	578	47	25
Satellites	63	60	27	13

smaller, their weaker gravity has not allowed them to capture or retain as much hydrogen and helium as Jupiter and Saturn have; but they have retained the heavier gases.

Measurements of the heat rising from the interiors of the gas giants show that they generate more heat than they receive from the Sun. Their cores are even hotter than the Earth's core, and their silicates and iron are probably molten. The hydrogen and hydrogen compounds surrounding the core's molten mass are liquid because they are so strongly compressed, not because they are cold. This white-hot liquid is more dense than rock on Earth's surface. Thus these giant planets have no surfaces and few features we can compare to the terrestrial planets.

The giant planets are probably heated by continued gravitational contraction and settling of heavier matter toward their cores. As this heat flows outward, it generates convective motions that stir the atmosphere. The planets' rotation creates a Coriolis effect on the rising gas, drawing it into the cloud belts (Figure 42.3).

FIGURE 42.3
The gas giants are dominated by their enormous atmospheres. Cloud bands are generated by the Coriolis effect, like Earth's jet streams, but there are few points of comparison with the terrestrial planets.

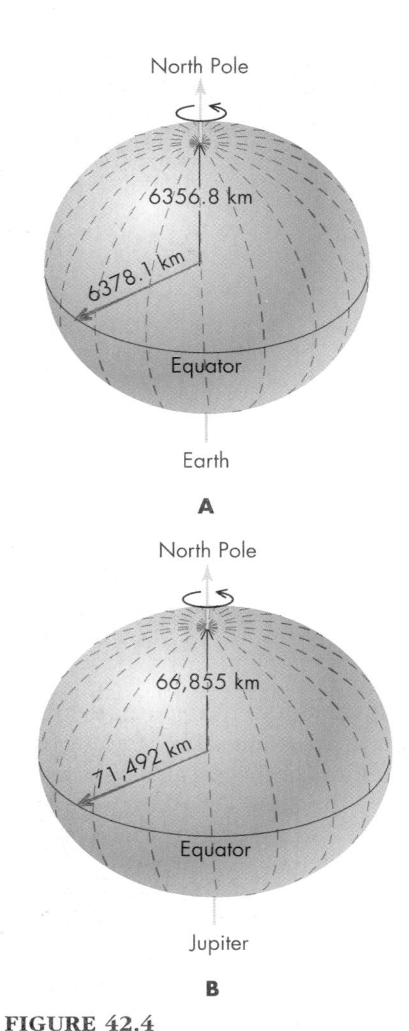

FIGURE 42.4
(A) Rotation makes a planet's equator bulge outward. The Earth's equator bulge is only about 20 kilometers. (B) Jupiter's rapid rotation creates a much greater equatorial bulge that is apparent in most photographs (such as Figure 42.3).

This is similar to the effect that produces high-speed jet streams in the Earth's stratosphere and the spinning motions of our storms, but some of the storm systems on these gas giants are larger than the entire Earth!

All of the gas giants spin more rapidly than the Earth, giving each planet an equatorial radius that is distinctly larger than its polar radius. Rotation also affects the shape of Earth and Mars, although to a lesser extent (Figure 42.4). The outer parts of the planet swing around so rapidly that they move outward, as a child's feet swing outward when you hold the child by the hands and spin around. Rotation provides inertia to material that carries it outward until it reaches a balance with other forces, such as the inward force of gravity.

Each of the four giant planets has a ring system composed of small orbiting particles. The rings are probably debris from small objects that have broken up as the result of either collisions or tidal forces exerted by the planet. Collisions continue to break up small particles, providing fresh ring material. At the same time, some ring materials collide and may fall into the planet's atmosphere. It is possible that the rings come and go, so that if we were to see the Solar System a few billion years from now, Saturn might not have an obvious ring system but maybe Neptune would. Each giant planet also has many orbiting moons. A few of these moons are large, but of the more than 100 cataloged, some are not much more than a kilometer in size. These orbiting systems are similar to mini solar systems (Unit 45).

Although the giant planets have many similarities, closer examination reveals them to be as individual as the terrestrial planets. Some of these differences may have causes parallel to the ones that shape the terrestrial planets. For example, the strong color differences between the giants are related to their distances from the Sun. Ammonia and methane condense at lower temperatures than water, so the chemistry of the outer giants differs from that of the inner giants. The least massive of the giants (Uranus) also seems to generate the least internal heat, again like the terrestrial planets.

One of the interesting aspects of all of the planets is their dissimilarities. Almost every planet has at least one major feature that makes it appear to be an "oddball" of sorts. Venus and Uranus have peculiar backward rotation directions; Earth and Neptune have unusual moons; Mercury and Earth have thin crusts; Saturn, Mars, and Earth have quite tilted rotation axes; Mars has a smaller atmosphere than might be expected. Every one of these oddities has been explained at one time or another as being caused by a major impact late in the formation of the planet. This may be correct. It is hard to imagine the violence of the final collisions between the last few dozen giant planetesimals shortly before the Solar System settled down to its final configuration. The last big collisions probably involved bodies of similar size, such as the collision that is thought to have resulted in the Earth's Moon. Jupiter may be the only "normal" planet because it grew so massive that nothing else was big enough to set its spin in an odd direction, stir up its forming satellite system, or strip out much of its material.

Besides the giant planets, the outer Solar System is teeming with smaller bodies. The largest of these worlds orbit the giant planets, but many more orbit beyond Neptune. Pluto and Eris are the largest trans-Neptunian objects discovered so far, but six moons of the gas giants are larger and more massive than either. In fact, the two largest, Ganymede and Titan, are larger than Mercury, although they have less than half as much mass. The reason they can be so big while still having low masses is that they are "ice worlds," containing large quantities of frozen water, ammonia, and methane surrounding small cores of rock and iron.

One common surface feature of these ice worlds is impact craters that resemble the craters on the terrestrial planets (Figure 42.5). The number of craters is often quite high, suggesting that relatively little geological activity has occurred there to erase them. Given their small masses compared to Mercury, we might not expect to find *any* geological activity, yet some of the larger ice worlds are devoid of impact craters and others have actively erupting volcanoes!

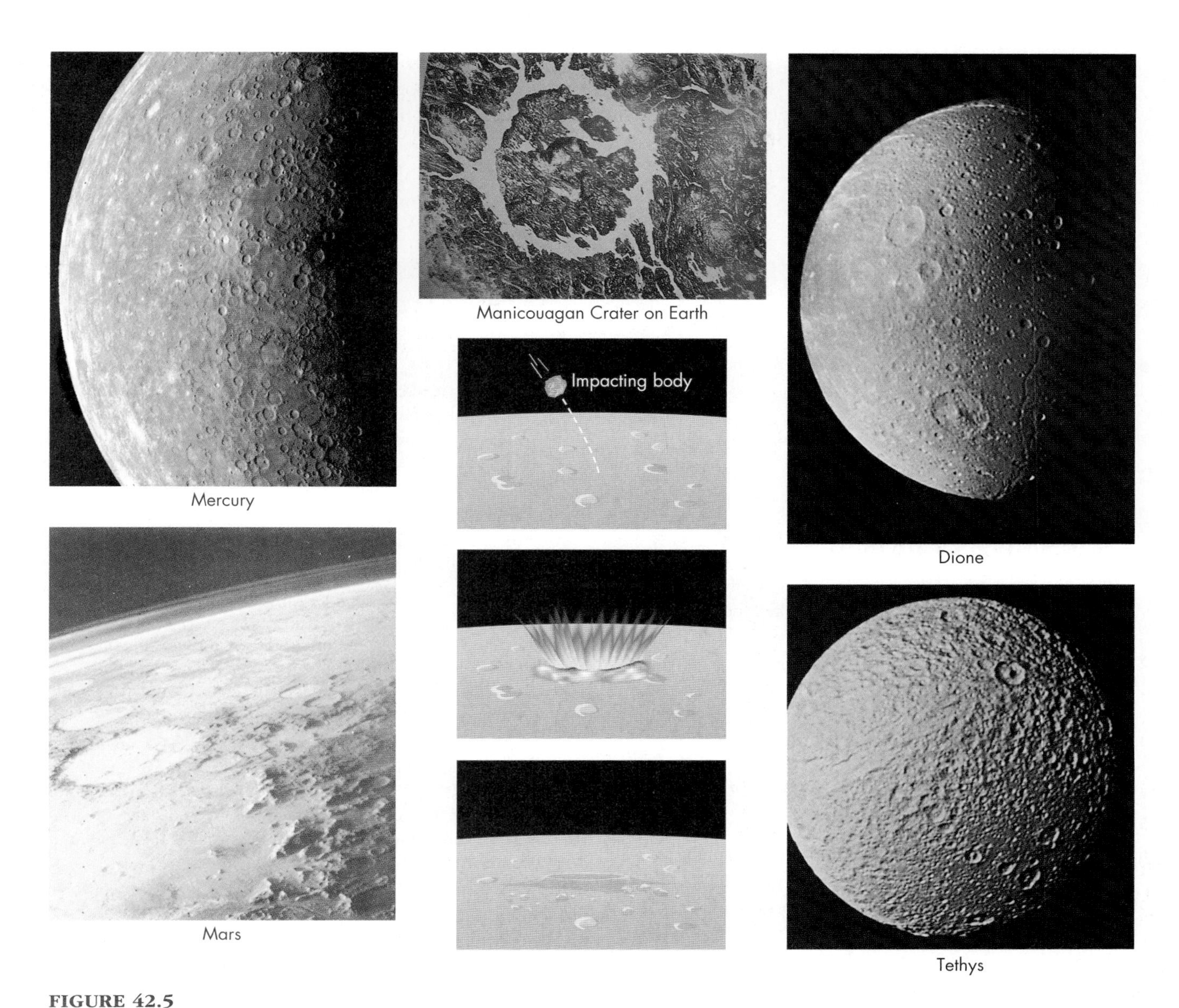

FIGURE 42.5
Pictures taken by spacecraft showing craters on Mercury, Earth, Mars, and two moons of Saturn, Dione and Tethys. Note that objects are shown to different scales.

When we study these ice worlds, we need to keep in mind that at their low temperatures, water ice behaves more like the rock here on Earth. We might even think of ice as the solidified lava of these worlds, and with such "lava" we have the possibility of an icy kind of geological activity. With that readjustment in our thinking, we can see similarities between the geological processes that occur on these ice worlds and more familiar processes here on Earth.

Several of these frozen worlds show signs of an icy kind of tectonic activity. For example, some of the larger moons have faults, smooth uncratered regions, and parallel ridges of mountains analogous to the features of plate tectonics on Earth (Unit 35). This suggests that below the crust, circulation is occurring in a mantle of "molten" ice. The heating source for the tectonic activity and volcanic eruptions on many of these worlds appears be quite different from the terrestrial planets' radioactive heating, caused instead by gravitational tidal forces.

Q What differences between the planets do not appear to agree with the solar nebula theory?

The only ice world that has a substantial atmosphere is Titan, and it is remarkably dense, producing a surface pressure similar to that on Earth. Titan even has lakes and rain—of liquid methane. Other ice worlds have thin atmospheres that freeze back onto the surface at night or seasonally. A few of these worlds have surface features unlike anything seen elsewhere in the Solar System. Indeed, astronomers consider these diverse bodies, virtually unknown before the space age, to be some of the most intriguing members of the Sun's family.

KEY TERMS

comparative planetology, 330
uncompressed density, 336
frost line, 336

QUESTIONS FOR REVIEW

1. Why do we expect to see more volcanic and tectonic activity on Earth and Venus but little or none on the Moon, Mercury, or Mars?
2. Qualitatively compare the crater densities for Earth and Venus with those of the Moon and Mercury. What does this information tell us about geological activity on these Solar System bodies?
3. What factors cause the differences in the terrestrial planet atmospheres?
4. How does distance from the Sun affect a planet?
5. What are the different roles that water can play on a planet?
6. What is meant by the "frost line" in the formation of the planets, and what role did it play in planetary sizes and composition?
7. In what ways are the gas giants and ice worlds similar to the terrestrial planets?
8. How do a planet's distance from the Sun and the presence or lack of an atmosphere combine to determine the temperature of the planet?

PROBLEMS

1. What differences between the planets support the solar nebula theory for the formation of the Solar System?
2. Compare the lengths of the sidereal and solar days for each of the terrestrial planets as given in Table 42.1. Are the reasons for the differences in the lengths of these days the same for each planet? Explain.
3. Table 42.2 lists the sidereal and solar days for the four giant planets. Why are these values the same for each of these planets?
4. The mass of Uranus is about 14.5 times that of the Earth, yet the escape velocity for Uranus is a little less than twice that of Earth's. Explain the seeming discrepancy.
5. Scientists classify objects based upon various characteristics. If we rank the terrestrial planets according to their densities (mass divided by volume), we will get one relationship. If we then rank the terrestrial planets according to their uncompressed densities, we will get a slightly different relationship. Compare the rankings. What do the differences in the rankings tell us about the planets themselves?
6. Examine the data in Tables 42.1 and 42.2, and consider that the average speed of a molecule in a gas with temperature T is approximately

$$V_{mol} = 0.15 \text{ km/sec} \times \sqrt{T/m}$$

where T is the temperature in Kelvin and m is the mass of the molecule in atomic mass units; so $m(H_2) = 2$, and $m(CO_2) = 44$. About how many times smaller than the escape velocity does V_{mol} have to be for a planet to retain a gas?
7. Heating by the Sun decreases with distance. For an object absorbing all sunlight, we can predict that the temperature will be 227 K $\times$ $(1/\sqrt{a})$, where a is the distance from the Sun in AU. Find the expected temperatures at the distances of Venus, Mars, and Jupiter. What factors might cause the temperatures to differ from what this formula yields?

TEST YOURSELF

1. Which of the following planets has the densest atmosphere?
   a. Mercury
   b. Venus
   c. Earth
   d. Mars
   e. Pluto
2. Which of the following planets, at the same distance from the Sun, would be the coolest?
   a. Poorly reflecting in the visible spectrum, radiates well in the infrared spectrum
   b. Poorly reflecting in the visible spectrum, radiates poorly in the infrared spectrum
   c. Highly reflecting in the visible spectrum, radiates poorly in the infrared spectrum
   d. Highly reflecting in the visible spectrum, radiates well in the infrared spectrum
3. Suppose a number of planets all have the same mass but different sizes and temperatures. Which of the following planets is most likely to retain a thick atmosphere?
   a. Small, hot
   b. Small, cool
   c. Large, hot
   d. Large, cool

unit

43 Jupiter and Saturn

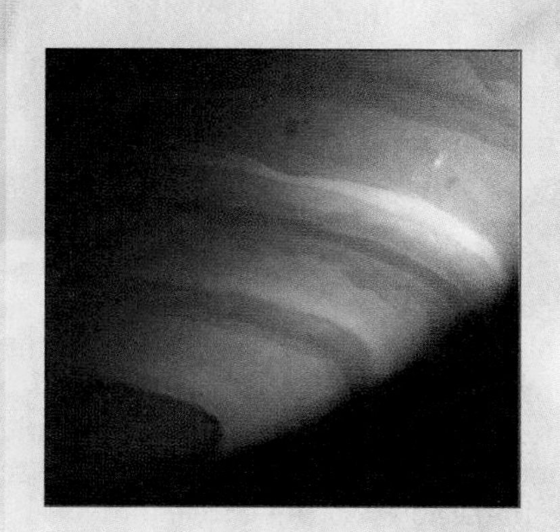

Although we speak of four giant planets in the Solar System, Jupiter and Saturn are by far the largest. Jupiter has more than twice as much mass as all the smaller bodies in the Solar System combined, and the same is true of the smaller Saturn. Our models of both planets indicate that these two giants have similar internal structures, and these two largest planets share many other characteristics despite Saturn's lower mass.

Because of Saturn's spectacular ring system (Unit 45), we sometimes overlook the body of the planet. Both of these giants have no solid surface, but massive atmospheres that dwarf the Earth (Figure 43.1). These behemoths grow progressively hotter and denser deep in their interiors, with cores of molten rock and iron. Jupiter and Saturn also both formed extensive satellite systems with remarkable moons, which have become the focus of recent space missions to these distant systems (Unit 46).

43.1 THE APPEARANCE OF JUPITER AND SATURN

Jupiter, at a distance of about 5 AU from the Sun, has a diameter about 11 times the Earth's and a mass more than 300 times the Earth's. Saturn is almost twice as far from the Sun (9.5 AU) and just slightly smaller than Jupiter, with 9.5 times Earth's diameter and about 100 times its mass.

The ancient Romans did not know that Jupiter was the largest planet in the Solar System, but they nevertheless named it appropriately after the king of the gods. Possibly they chose its name because of its relatively steady brightness and stately motion through the celestial sphere, spending about one year in each constellation of the zodiac before moving to the next. From our Earth-based perspective, only Venus gets brighter than Jupiter at its brightest; but Venus sometimes becomes dimmer than Jupiter's dimmest appearance, and it swings back and forth among the constellations as it orbits the Sun.

FIGURE 43.1
Jupiter as imaged by the Hubble Space Telescope, and Saturn as it appears without its rings. (The rendering of Saturn with the rings removed was made by Björn Jónsson, based on NASA images.)

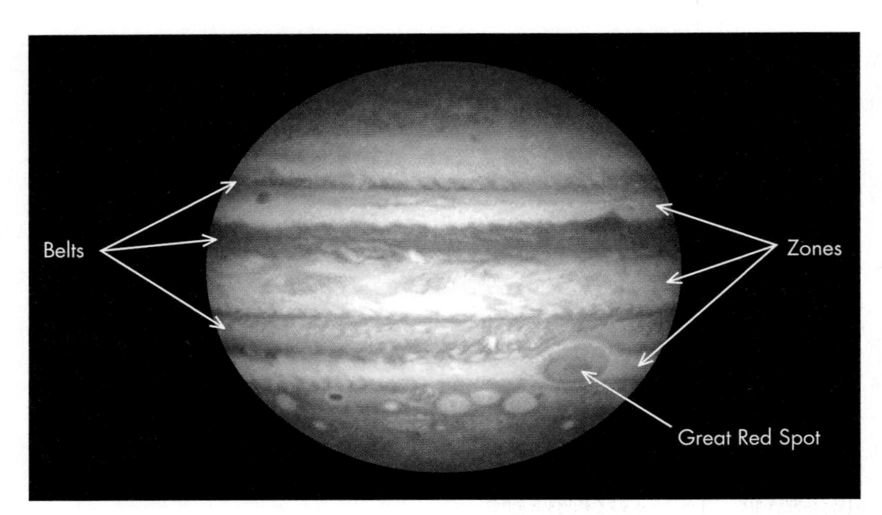

FIGURE 43.2
Jupiter's cloud system has a banded structure of darker belts and light-colored zones, and a "Great Red Spot" which is larger than the Earth.

Saturn bears the name of an ancient Roman god of agriculture who is identified with the Greek god Cronus (also spelled Kronos). Jupiter banished Saturn and the other Titans to the darkest pit, farthest from heaven; perhaps this explains the naming of Saturn, the dimmest, slowest-moving of the planets known before the invention of the telescope.

Through even a small telescope, Jupiter displays parallel bands of clouds ringing the planet, as shown in Figure 43.2. Dark **belts** alternate with light-colored **zones.** Spectra of the sunlight reflected back through Jupiter's atmosphere show that almost 90% of the molecules are hydrogen (H_2) and 10% are helium (He), and there are only traces of other hydrogen-rich gases such as methane (CH_4), ammonia (NH_3), and water (H_2O). These gases were also detected directly in December 1995 when a probe from the *Galileo* spacecraft parachuted into Jupiter's atmosphere. The solid particles and droplets that make up the clouds themselves are harder to analyze, but computer calculations of the chemistry of Jupiter's atmosphere suggest they are made of water, ice, and ammonia compounds. The cause of Jupiter's bright colors remains unclear despite many analyses, but the colors are probably produced by complex organic molecules.

Saturn also shows bands of color, but the variations are much more subtle (Figure 43.1). Spectral analysis of Saturn's atmosphere shows that it is even more dominated by hydrogen, which makes up about 96% of the molecules in its atmosphere. Helium again makes up most of the remainder, with small traces of hydrogen-rich compounds.

Temperatures vary depending on height. These temperatures are estimated for one atmospheric pressure—the pressure at the Earth's surface.

Part of the difference in appearance between these two gas giants may be caused by their different temperatures. At its distance from the Sun, Jupiter's outer atmosphere is heated to about 160 K (about −170°F). Saturn's greater distance from the Sun results in a lower temperature of about 130 K (about −230°F). Saturn is cold enough for ammonia gas to condense into cloud particles that veil its atmosphere's deeper layers. This makes markings in lower layers indistinct, like a colorful object seen through fog.

Rotation of Jupiter

Both planets rotate very rapidly. Jupiter rotates once every 9.9 hours and Saturn once every 10.7 hours. Such rapid rotation results in both planets being visibly flattened. Jupiter's equatorial diameter is about 7% larger than its diameter from pole to pole, while this difference is about 10% for Saturn (see Unit 42.4). Time-lapse photographs show that Jupiter's clouds move swiftly, sweeping around the planet in jet streams that are far faster than those of Earth. The dark belts move westward at speeds hundreds of kilometers per hour slower than the rotation of the planet, while the white zones move hundreds of kilometers per hour faster.

43.2 THE GIANTS' INTERIORS

It is clear from the densities of Jupiter and Saturn that their atmospheres must be extremely deep. A planet's average density is determined by dividing its mass by its volume. The mass can be found by calculating the planet's gravitational attraction on one of its moons (Unit 17), and its radius—and hence its volume—from its angular size and distance (Unit 10). When such calculations are made for Jupiter, we find that its average density is only slightly greater than that of water—1.3 kilograms per liter. For Saturn the density is lower than the density of water—0.7 kilograms per liter. This indicates that the bulk of these planets, not just their surface layers, must be composed mainly of very light elements such as hydrogen and helium.

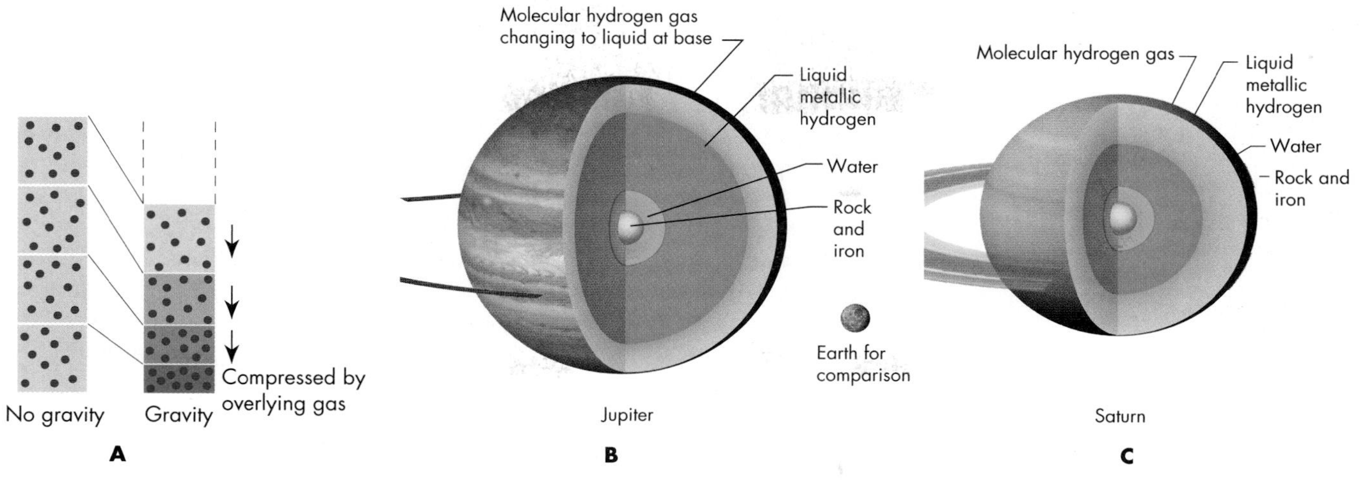

FIGURE 43.3
(A) The density of gas increases with depth because the overlying gas compresses the underlying matter. (B) A sketch of what astronomers think Jupiter's interior is like. (C) The internal structure of Saturn.

Q In the remote future, the Sun will enter a phase in which it may become thousands of times more luminous than it is now. What will happen to the giant planets when this happens?

Astronomers cannot probe the interiors of these massive planets with seismic detectors. However, theoretical models based on observations of the planets' shape and gravitational pull on space probes can provide information on the internal distribution of mass. The models indicate that both Jupiter and Saturn have similar internal structures.

The immense mass of these planets exerts a tremendous gravitational force on their interiors, compressing the gas to very high densities. Near the cloud tops this compression is only slight, so the gas density there is low. Deep in the interior, however, the weight of thousands of kilometers of gas compresses the matter to a high density. Thus the density of the gas increases with depth, as illustrated in Figure 43.3A. Everywhere within the planet, the gas is squeezed until it pushes outward with a force that matches the gravitational force—the weight—of the gas above it. If the forces didn't match, the atmosphere would either expand or contract. This same relationship is true for any other atmosphere, including Earth's.

Deep within Jupiter, the compression created by its gravity presses molecules so close together that the gas changes to liquid. Thus, about 10,000 kilometers below the cloud tops—about one-sixth of the way into the planet—Jupiter's interior is a vast sea of liquid hydrogen. The pressures at this depth are about a million times the pressure we experience on the surface of the Earth. Slightly deeper still, the added weight of the overlying layers compresses the hydrogen below them into a state known as **liquid metallic hydrogen,** a form of hydrogen that scientists on Earth have created in tiny high-pressure chambers for fractions of a second. The hydrogen is "metallic" in the sense that the electrons can move about easily, allowing electric currents to travel through it as in a metal.

Close to the center of Jupiter, the pressure has risen to nearly 100 million times Earth's atmospheric pressure, and liquid is compressed to densities greater than that of rock. Assuming that Jupiter formed from gas with the same overall composition as the Sun, it must contain several times more than Earth's own mass in heavy elements such as silicon and iron. Because of their high density, these elements would have sunk to Jupiter's center, forming a core of iron and rocky material (see Figure 43.3B).

Astronomers infer that Saturn's composition and internal structure are similar to Jupiter's with several minor differences, as depicted in Figure 43.3C. Saturn's equator bulges even more than Jupiter's, despite spinning more slowly. The theoretical models suggest that Saturn's mass is more concentrated toward its center than is Jupiter's. Saturn's core is also probably rock and iron, with many times the mass of the Earth. The pressure in Saturn's core is sufficient to compress this material, though not to the degree that materials are crushed at the center of Jupiter.

Measurements of infrared radiation coming from Jupiter and Saturn show that they both emit more energy than they receive from the Sun. The *Galileo* atmospheric probe also measured this heat as it entered the upper parts of Jupiter's atmosphere. The amount of heat generated implies that Jupiter's core must be hot, perhaps 30,000 K, which is about five times hotter than the Earth's core. Even 10,000 kilometers below the surface, the temperature is already almost as hot as the Earth's core, about 5000 K. The heat rises slowly to the planet's surface and escapes into space as low-energy infrared radiation.

The *Galileo* probe reached about 200 km (about 120 miles) below the cloud tops; it reported a pressure 22 times Earth's atmospheric pressure and a temperature of about 420 K (about 300°F) before the pressure and heat in deeper layers destroyed the craft.

Astronomers are not completely certain what supplies Jupiter's and Saturn's heat, although much of it may be left over from the planet's formation. Planet building is a hot process, heat being generated by the hail of gas and planetesimals pulled into the forming world by its growing gravity. Unlike the Earth, in which a significant amount of energy is released by the decay of radioactive elements, Jupiter and Saturn do not have enough of these heavy elements to explain their heat output.

Additional heat may be generated by slow but steady shrinkage of the planets as they adjust to the strong gravity created by the enormous mass added during their formation. Thus, giant gas planets such as Jupiter may still be shrinking slightly and therefore heating as their matter is squeezed. Yet more heat is released by differentiation, as matter denser than hydrogen, such as helium, sinks toward the planets' cores. Helium gas may condense into droplets, much as water droplets condense in our atmosphere. As the helium droplets fall toward the core, they generate heat by friction as they move through the hydrogen. Unlike much denser rock, which sank rapidly to their cores when the planets were young, helium sinks more slowly. The settling of this helium toward their cores may be continuing to this day.

Q How might Jupiter and Saturn change in the remote future after they stop generating heat by differentiation? Which planet is likely to stop first?

In a relative sense, Saturn is generating more heat than Jupiter, given that its mass is less than one-third of Jupiter's. This extra heating may be related to the observation that Saturn has a lower abundance of helium in its upper atmosphere. Astronomers hypothesize that this low abundance results from the colder temperature on Saturn, which has allowed more of the helium to condense and sink into the interior.

43.3 STORMY ATMOSPHERES

The high temperatures deep inside Jupiter and Saturn drive strong convection currents in the planets' outer layers. The *Galileo* atmospheric probe measured winds faster than 500 kilometers per hour (more than 300 mph). The convection currents carry warm gas upward to the top of the atmosphere. Here the gas radiates heat into space. As the gas cools, it becomes denser and sinks again, as illustrated in Figure 43.4A. Such deep circulation, combined with Jupiter's and Saturn's rapid rotation, makes "weather" on the gas giants very different from that on the Earth.

As material moves on a rotating planet, it is deflected by the Coriolis effect (Unit 36). In particular, winds that move farther away from the planet's axis of rotation (either closer to the equator or upward from the interior) are deflected to the west, and winds that move closer to the axis are deflected to the east, as shown in Figure 43.4B. The rapid rotation of Jupiter and Saturn, along with the deep circulation of gases upward and then back down, creates an extremely powerful Coriolis effect. This deflects rising and sinking atmospheric gases into powerful winds called *jet streams.* The winds in adjacent regions may blow in opposite directions, as shown in Figure 43.4B. We see these winds as the cloud belts, illustrated in Figure 43.4C. Such reversals of wind direction ("wind shear") from place to place also occur on Earth, where equatorial winds generally blow from east to west, while midlatitude winds generally blow from west to east.

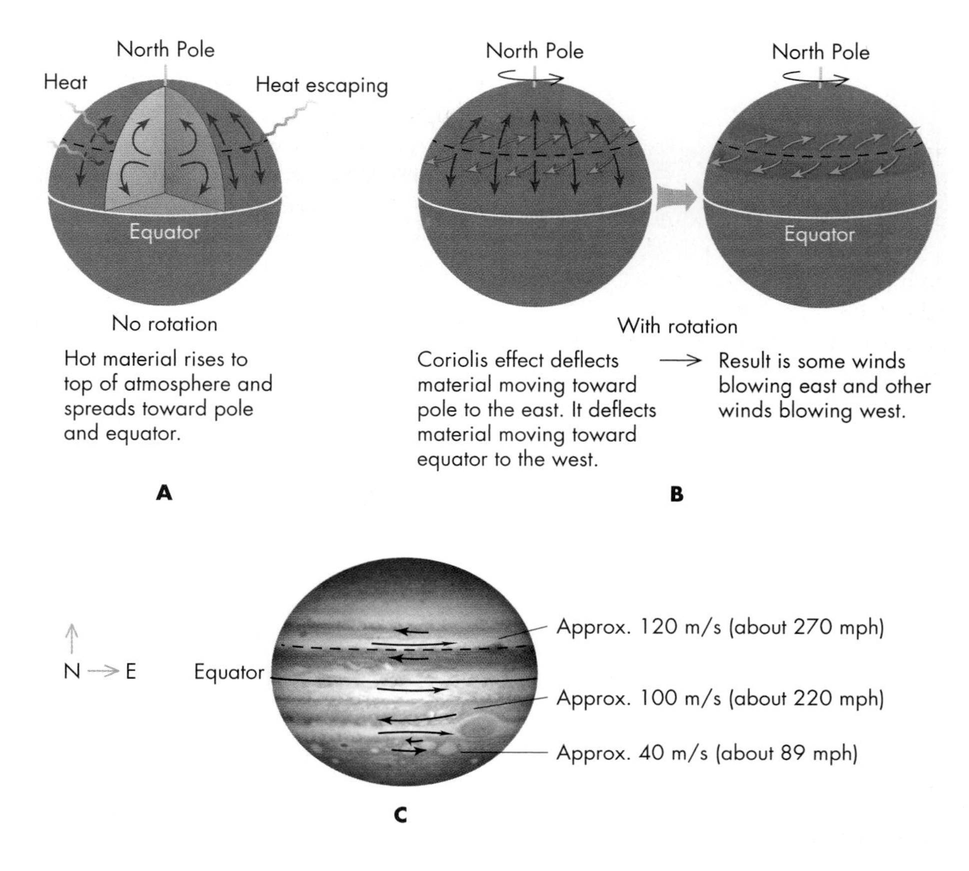

FIGURE 43.4
(A) Rising gas from Jupiter's hot interior cools near the top of the atmosphere and sinks. (B) The Coriolis effect, arising from Jupiter's rotation, deflects the gas, creating winds that blow as narrow jet streams. (C) The wind varies widely in speed and direction from region to region.

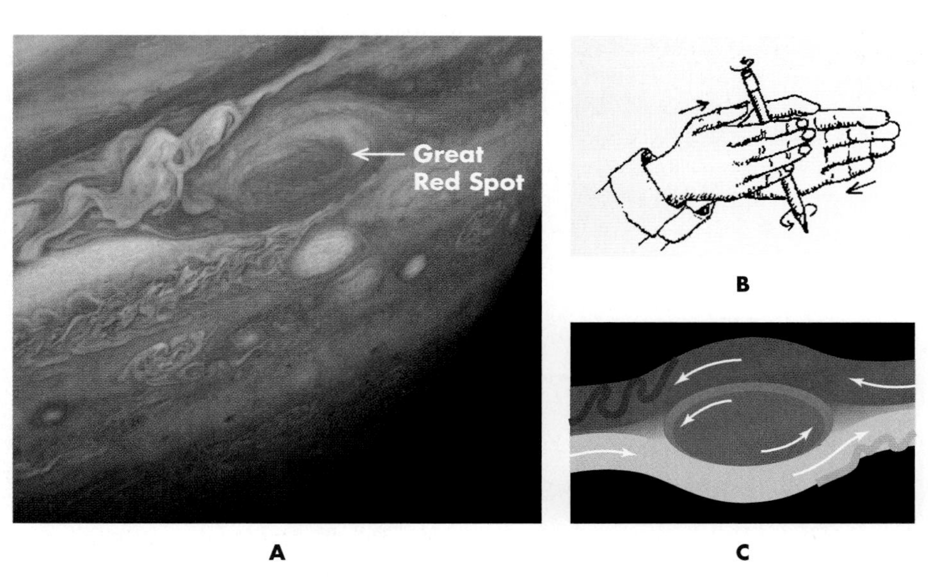

FIGURE 43.5
(A) Vortices form between atmospheric streams of different velocities. Note the Great Red Spot and the white oval below it. (B) The vortex motion is something like the spin of a pencil twirled between two hands. (C) The white zones move around Jupiter faster than the dark belts, creating a relative motion that can produce a vortex.

Q What are some examples of a vortex here on Earth?

As the various jet streams circle Jupiter, gas between them can be spun into a huge, whirling atmospheric **vortex.** The gas in the vortex is given a spin much as a pencil between your palms twirls as you rub your hands together. Some of these spinning regions are brightly colored, as Figure 43.5 shows. Brown and shades of white dominate, but one exceptionally large vortex—bigger across than the Earth—is nearly brick red. Known as the **Great Red Spot,** this vortex was discovered in the seventeenth century and has remained active for over 300 years!

FIGURE 43.6
A high-contrast image of Saturn, made by *Cassini* in 2004, reveals complex cloud belts and storms in Saturn's atmosphere.

The exact chemistry that gives the spots their colors is uncertain, although spectroscopy reveals the presence of methane, ammonia, and more-complex compounds that have not been completely identified. (For complex molecules, spectroscopy can sometimes indicate some of the atomic bonds in the molecules without giving a unique identification.) Infrared imaging indicates that these regions are heated from below.

The Great Red Spot rises to about 50 kilometers above the surrounding regions, and time-lapse images show winds swirling within the spot at about 500 kilometers per hour (300 mph). Over several hundred years of observation, the spot has changed in size and color, but otherwise this giant vortex appears to be very stable. This is unlike hurricanes on Earth, which die out when they are no longer fed by warm water from below. Computer models indicate that as long as Jupiter's interior is able to supply a steady source of heat, the Great Red Spot may persist as a nearly permanent feature of Jupiter's atmospheric circulation.

Saturn's atmosphere looks calmer, but it is much more active than it appears. High haze obscures details of storms deeper in the atmosphere. Figure 43.6 shows a picture of Saturn's atmosphere made by the *Cassini* spacecraft with the color "stretched" to bring out details of the belts as well as a large storm seen in 2004.

43.4 THE MAGNETIC FIELDS

In the deep interiors of Jupiter and Saturn, convection in the metallic liquid hydrogen combines with the planets' rapid rotation to generate magnetic fields. These magnetic fields are driven by a natural dynamo process similar to that which generates the Earth's magnetic field in its own metallic liquid core. Jupiter's dynamo process creates the strongest magnetic field of any planet in the Solar System, almost 20,000 times stronger than Earth's. Saturn's magnetic field is not as strong, but it is still more than 500 times stronger than Earth's.

The Earth's magnetic field steers incoming energetic particles from the Sun into our upper atmosphere, where the particles trigger the lovely pale glow of the northern and southern lights—the **aurora.** Jupiter's and Saturn's magnetic fields do much the same: Particles trapped magnetically in belts far above the planet spiral through the field into the upper atmosphere, where they create an aurora. The Hubble Space Telescope photographed this phenomenon on both Jupiter (Figure 43.7A) and Saturn (Figure 43.7B). Jupiter's aurora is also seen, in Figure 43.7C, as the pale glow along the edge of the planet in a picture taken by the passing *Voyager I* spacecraft.

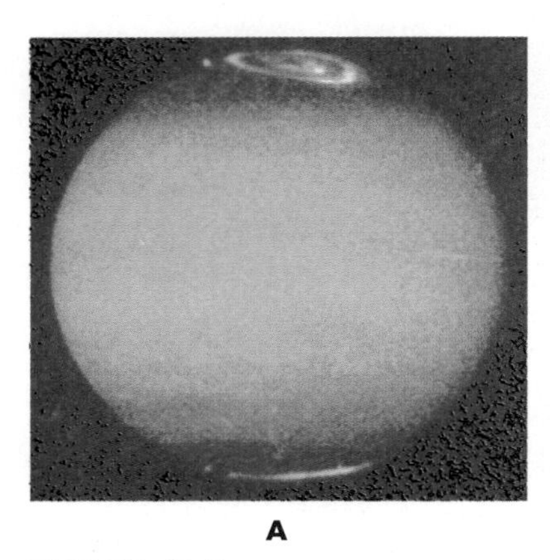

A

B

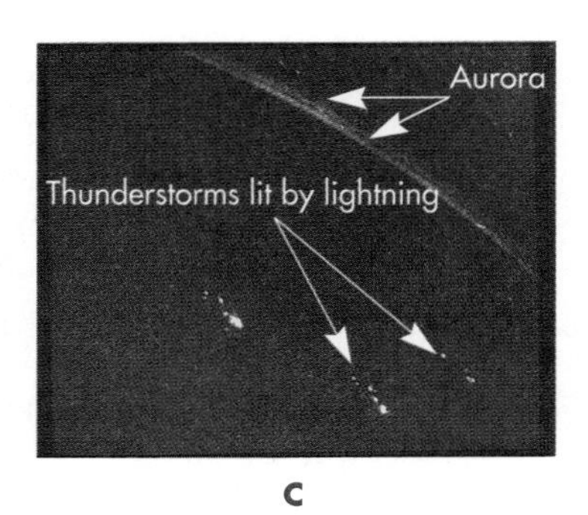

C

FIGURE 43.7
(A) Jupiter's auroral zone as seen by the Hubble Space Telescope. (B) The aurora on Saturn, imaged by the Hubble Space Telescope at ultraviolet wavelengths. (C) Aurora and lightning on Jupiter, as observed by the *Voyager I* spacecraft as it flew over Jupiter's night side.

Another consequence of Jupiter's magnetic field is that the planet is a powerful source of radio emission because the particles that create the aurora generate radio waves as well.

Figure 43.7C depicts another familiar phenomenon occurring in Jupiter's atmosphere: enormous lightning storms that flash on the night side of the planet. These are associated with clusters of Jovian thunderstorms some 1000 km (600 miles) across. Cloud particles carried up and down by the rising and sinking motions in these storm cells collide and generate atmospheric electricity just as such motions do on Earth.

KEY TERMS

aurora, 346
belt, 342
Great Red Spot, 345
liquid metallic hydrogen, 343
vortex, 345
zone, 342

QUESTIONS FOR REVIEW

1. How do Jupiter's and Saturn's masses and radii compare with the Earth's?
2. How are the interior structures of Jupiter and Saturn similar? How do they differ?
3. What is the source of internal heat for Jupiter and Saturn?
4. What sorts of atmospheric motion and activity are observed in Jupiter and Saturn?
5. What is the Great Red Spot? What allows it to persist for centuries?
6. What are the similarities and differences of the Coriolis effect for Jupiter or Saturn compared to the Earth?

PROBLEMS

1. Given its rate of rotation, at what speed does a point on the equator of Jupiter move? Give your result in km/hr.
2. Given its rate of rotation, at what speed does a point on the equator of Saturn move? Compare this speed to that of Jupiter.
3. When discussing the densities of the terrestrial planets, we compared the compressed versus the uncompressed densities (Unit 42). Why don't we do that for the gaseous planets?
4. Compare the volume of Jupiter to the volume of Earth. How many bodies the size of the Earth could fit inside Jupiter?
5. If Saturn were compressed until it had the same density as Jupiter, what would Saturn's new radius be? Compare this to Jupiter's radius. If Saturn were compressed until it had the same density as Earth, what would Saturn's new radius be?
6. In the Sun, hydrogen makes up 90% of the atoms, corresponding to 75% of its mass. Iron and other elements with atomic weights of about 56 make up 0.001% of the atoms, while silicon and other elements with atomic weights of about 28 make up 0.01% of the atoms. These represent the percentage of atoms, not the fraction of the mass. If Jupiter has the same proportions of atoms in its interior, what would the total mass of Jupiter's iron- and silicon-like atoms be? Compare the masses you find to the mass of the Earth.
7. Look up information on the speed, structure, and lifetime of a class 5 hurricane on Earth and compare it to the properties of the Great Red Spot. Describe the differences in their causes.

TEST YOURSELF

1. Astronomers think that the inner core of Jupiter is composed mainly of
   a. hydrogen.
   b. helium.
   c. uranium.
   d. rock and iron.
   e. water.
2. The low average density of Saturn suggests that
   a. Saturn is hollow.
   b. Saturn's gravitational attraction has compressed its core into a rare form of iron.
   c. Saturn contains large quantities of light elements, such as hydrogen and helium.
   d. Saturn is very hot.
   e. volcanic eruptions have ejected all the iron that was originally in Saturn's core.
3. How does Saturn compare with Jupiter?
   a. It is less massive and smaller.
   b. It is more massive but smaller.
   c. It is less massive but larger.
   d. It is more massive and larger.
4. The amount of infrared energy emitted by Jupiter is about twice as great as the amount of sunlight the planet absorbs. What is the significance of this discrepancy?
   a. It implies that the planet is cooler than it should be.
   b. It implies that there are significant energy sources within Jupiter.
   c. It implies that the Sun was once much dimmer than it now is.
   d. It implies that the rotation of Jupiter must be slowing down.

unit

44

Uranus and Neptune

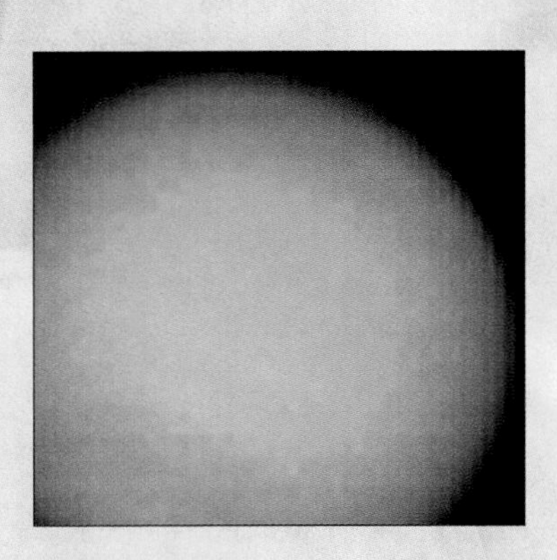

Background Pathways

Unit 43: Jupiter and Saturn 341

Uranus and Neptune are near twins. Both are significantly smaller than Jupiter or Saturn, but they are still giant planets with masses of 15 and 17 Earth masses, respectively, and radii that are both about four times the Earth's. They rotate at about the same rate, with days of length 17 and 16 hours, respectively. Uranus is about 19 AU from the Sun, and Neptune is about 30 AU. Its greater distance from the Sun makes Neptune a little colder than Uranus; but both are close to 80 K in the upper regions of their atmospheres, and they have similar atmospheric structures. Both also have complex systems of rings and moons like Jupiter and Saturn (Unit 43).

To distinguish them from the two larger giants, astronomers sometimes call Uranus and Neptune **ice giants.** This does not describe their internal structure, which is probably as hot as the interior of Earth, but it may describe a difference in how these two giants formed. It appears that the larger giants Jupiter and Saturn gathered a large amount of gas from the solar nebula, but Neptune and Uranus were probably built more from the accumulation of the icy planetesimals in the outer Solar System (Unit 33).

Neither Uranus nor Neptune was known before the invention of the telescope, and both have been visited by a spacecraft only once—the *Voyager II* mission, which flew by Uranus in 1986 and Neptune in 1989, providing our most detailed images of them (Figure 44.1). Much of what we know about these cold blue worlds came from the brief encounters of that mission. Despite the many similarities just described, Uranus and Neptune exhibit some interesting differences, many of which are not yet fully understood.

44.1 THE DISCOVERY OF TWO NEW PLANETS

Uranus was discovered in 1781 by Sir William Herschel, a German émigré to England. A musician by profession, Herschel was at the time an amateur astronomer interested in hunting comets, a task in which he collaborated with his sister Caroline. Uranus is sometimes just barely visible to the unaided eye, but it was with his homemade telescope that Herschel first observed this new planet. He saw a pale blue object whose position in the sky changed from night to night. At first he thought he had discovered a comet, but observations over several months indicated that the body's orbit was nearly circular, and he therefore concluded that he had found a new planet.

For this discovery, King George III named Herschel his personal astronomer, and to honor the king, Uranus was briefly known as *Georgium Sidus* ("George's Star"). The planet was also sometimes called *Herschel* in honor of its discoverer

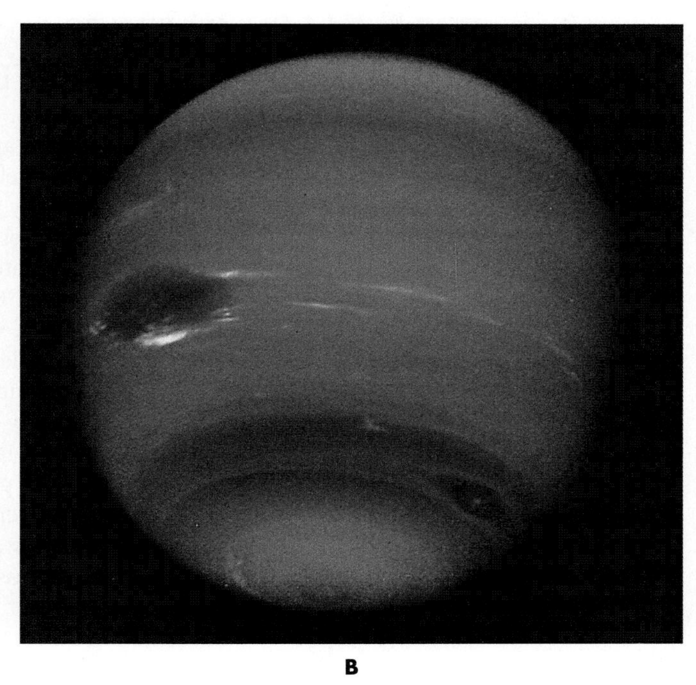

A B

FIGURE 44.1
(A) Uranus as pictured from the *Voyager II* spacecraft. Because of Uranus's odd tilt, we are viewing it nearly pole-on. Note the lack of clearly defined cloud belts. (B) Photograph of Neptune taken by the *Voyager II* spacecraft. Notice the dark blue oval, an atmospheric vortex whose origin is probably similar to that of Jupiter's Great Red Spot.

before astronomers decided that its name should be kept consistent with those of the other planets. The name Uranus seemed appropriate because Uranus was the father of Saturn in Roman mythology, as Saturn was the father of Jupiter. Uranus, the god of the sky, was married to Gaia, the goddess of the Earth, and they ruled the universe until Saturn overthrew his father.

Neptune was discovered in 1846 from predictions made independently by a young English astronomer, John Couch Adams, and a French astronomer, Urbain Leverrier. Adams and Leverrier both noticed that Uranus was not precisely following its expected orbit, and they therefore inferred that its motion was being disturbed by the gravitational force of an unknown planet. From the size of these orbital disturbances, Adams and Leverrier predicted where the unseen body must lie.

Adams completed his calculations in 1845; but when he reported his results, the royal astronomer, Sir George Airy, was unconvinced and gave a low priority to the search for the unseen planet. In 1846, however, Airy was startled to read a paper by Leverrier detailing calculations nearly identical to those made by Adams. This spurred Airy to begin a search in earnest, but by then it was too late: Leverrier had given his predicted positions to Johann Galle, a German astronomer who that same night pointed his telescope to the predicted location and saw Neptune. Assignment of credit for the discovery of the new planet led to a rancorous dispute tinged with national pride that lasted decades. The discovery is now credited equally to Adams and Leverrier. Ironically, Galileo had seen Neptune in 1613 while observing Jupiter's moons. His observation notes record a dim object whose position changed with respect to the stars, as would be expected for a planet. Galileo failed, however, to appreciate the significance of that motion, so Neptune eluded discovery for another two centuries. Having a deep blue color, Neptune was named for the Roman god of the sea after several years of dispute over its name.

Q: In what ways do you suppose the presence of an unknown outer planet might affect the orbit of a planet like Uranus?

44.2 THE ATMOSPHERES OF URANUS AND NEPTUNE

At their large distances from the Sun, more than two and three times Saturn's distance, respectively, Uranus and Neptune are difficult to study from Earth. Both appear as relatively featureless blue disks in most large ground-based telescopes.

Q Methane is a very strong greenhouse gas on Earth. How would you expect the methane in Uranus and Neptune to affect their temperatures?

The atmospheres of both Uranus and Neptune are rich in hydrogen and helium, like Jupiter and Saturn. Their atmospheres both contain a larger amount, about 2%, of methane (CH_4) than the bigger gas giants. We know this from spectra of their atmospheres, which show strong absorption lines of methane. It is methane that gives these planets their deep blue color. When sunlight enters the atmosphere, the methane gas absorbs the incoming red light. The remaining light, now blue, is scattered by cloud particles in the Uranian atmosphere and is reflected into space, as depicted in Figure 44.2. The cloud particles that cause the scattering are thought to consist primarily of crystals of frozen methane. Such crystals can form in Uranus's and Neptune's atmospheres because, being so far from the Sun, their outer atmospheres are extremely cold: mean temperatures of about 80 K (about −320°F) for Uranus and about 75 K (about −330°F) for Neptune.

The interiors of these two planets are probably similar based on the limited information currently available. To study the interior of these planets, astronomers rely on indirect methods, using their density and shape. Uranus has an average density of about 1.3 kilograms per liter. Its density is nearly twice that of Saturn and almost as high as Jupiter's. Neptune's density is even greater, at about 1.6 kilograms per liter. Given their smaller masses, Uranus and Neptune compress the matter in their interiors to a lesser degree, and astronomers therefore deduce that both must contain proportionally fewer light elements, such as hydrogen, than those more massive worlds. This suggests that they did not accumulate as much gas from the solar nebula when the Solar System was forming.

Q Is the increasing density of the planets, from Saturn to Uranus to Neptune, consistent with the solar nebula theory? Why or why not?

The masses of Uranus and Neptune do not produce the extremely large pressures needed to liquefy hydrogen as occurs in Jupiter and Saturn. The best model for their density and atmospheric composition is that they contain a relatively high amount of ordinary water mixed with methane and ammonia surrounding a core of rock and iron-rich material, as illustrated in Figure 44.3. Even though this water

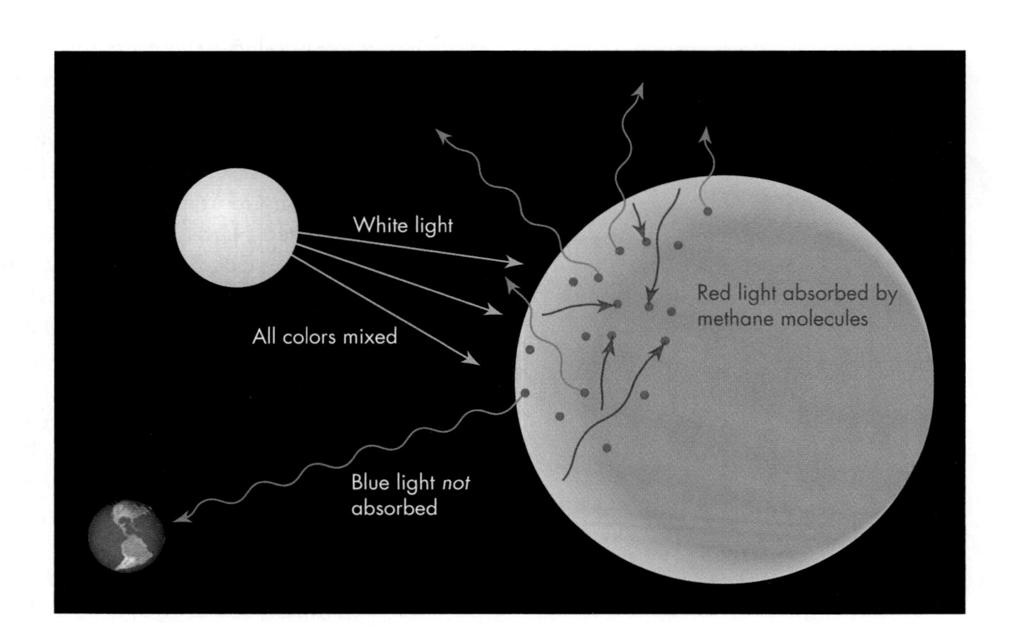

FIGURE 44.2
A sketch illustrating why Uranus is blue. Methane absorbs red light, removing the red wavelengths from the sunlight that falls on the planet. The remaining light—now missing its red colors—is therefore predominantly blue. That blue light scatters off cloud particles in the Uranian atmosphere and returns to space.

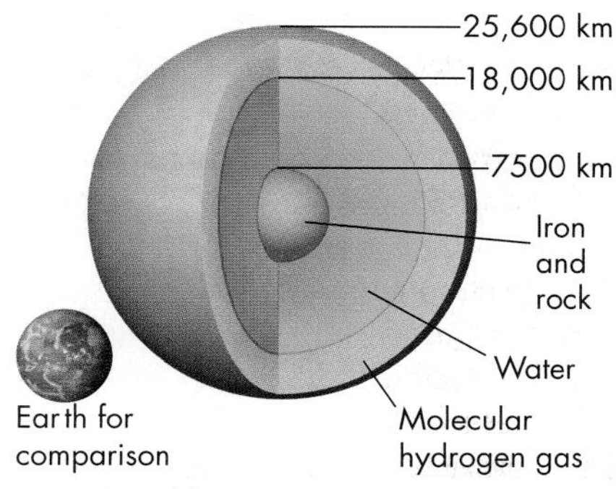

FIGURE 44.3
Artist's view of a recently suggested model for the interior of Uranus. Note how different it is from Jupiter and Saturn, which both contain large regions of liquid and metallic hydrogen.

Rotation of Neptune

is very hot, probably several thousand Kelvin. It is not gaseous because of the large pressure exerted by the atmosphere.

The limited information about Uranus's gravitational field suggests that its material may not be as differentiated as that of the other giant planets. Some astronomers hypothesize that Uranus's core may contain relatively less rock or iron. It is not clear why Uranus should be dissimilar to the other gas giants in composition or degree of differentiation, but such differences might help to explain another Uranian puzzle. Unlike the other gas giants, Uranus emits little heat of its own. It emits only a few percent more than what it receives from the Sun. Because gravitational potential energy is converted to heat when dense materials sink to the core of a planet, it may be that these puzzling differences are interlinked.

Even the high-resolution images made by the *Voyager II* spacecraft as it swung by Uranus reveal few details (Figure 44.1), although computer processing of the images suggests there are faint cloud bands on the planet. Unlike Uranus, however, Neptune has distinctive cloud belts. Pictures taken of it by *Voyager II* in 1989 showed markings reminiscent of Jupiter, as can be seen in Figure 44.4. Cloud bands encircle it, and it even had a "great dark spot," a huge dark blue atmospheric vortex about half the size of Jupiter's Great Red Spot.

Neptune's winds are extremely fast, reaching some of the highest measured speeds in the Solar System: nearly 2200 kilometers per hour (about 1300 mph). As these gales sweep around Neptune, they can create a large, stable vortex. However, Neptune's giant vortex turned out not to be nearly as stable as Jupiter's. By 1994, when the Hubble Space Telescope studied the planet, the spot had disappeared (Figure 44.5); but a new, smaller one had emerged in the opposite hemisphere.

Why do Uranus and Neptune, which in many ways are so similar, have such different cloud formations? Infrared observations offer a partial explanation. Neptune, like Jupiter and Saturn, radiates a substantial amount of internal energy. That energy output may be supplied by denser material still sinking to the planet's core, as it is in Jupiter and Saturn. Whatever the source of the heat, the rate of heat flow in a planet depends on the temperature *difference* between its core and outer atmosphere. Neptune has both a hotter interior and a cooler surface than Uranus, so it generates much stronger convection currents that rise to its outer atmosphere. There the rising gas is deflected into a system of winds by the Coriolis effect (see Unit 43) caused by Neptune's rapid spin—one rotation every 16 hours. The resulting winds create cloud bands similar to those seen on Jupiter and Saturn but tinted deep blue by Neptune's methane-rich atmosphere.

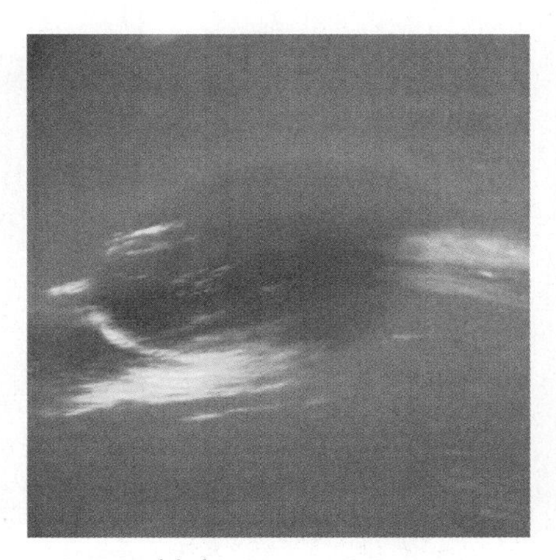

FIGURE 44.4
A dark spot resembling Jupiter's Great Red Spot was seen during *Voyager II*'s flyby.

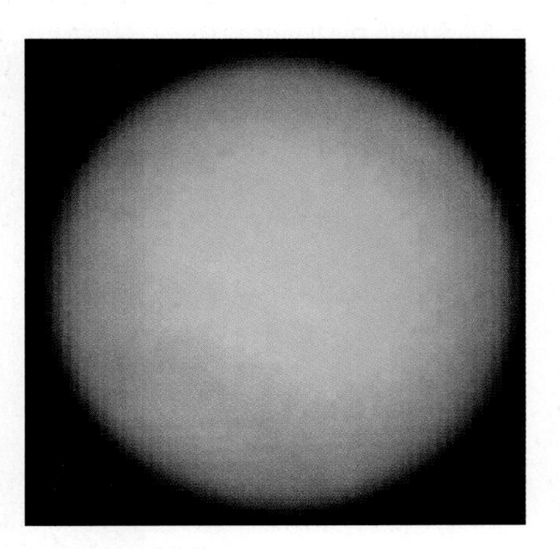

FIGURE 44.5
Neptune as viewed by the Hubble Space Telescope in 1994. When this image was taken, the dark spot had disappeared.

44.3 ODDLY TILTED AXES

Uranus's rotation axis is tipped so that its equator is nearly perpendicular to its orbit. That is, it spins nearly on its side, as illustrated in Figure 44.6. Moreover, the orbits of Uranus's moons are similarly tilted. They orbit Uranus in its equatorial plane; as a result, their orbits are also tilted at the same large angle with respect to the planet's orbit. Some astronomers therefore hypothesize that during its formation Uranus was struck by a large planetesimal and that this impact tilted the planet and splashed out material to create its family of moons.

The strong tilt of its axis gives Uranus an odd pattern of day and night. For part of its 84-year orbit, much of one hemisphere is in "perpetual" day and the other hemisphere is in "perpetual" night. In 1989, at the time of the *Voyager II* encounter, Uranus's south pole was nearly facing the Sun. With this orientation the polar region receives more heating than the equatorial regions, and the whole northern hemisphere received almost no sunlight. This may have contributed to the lack of cloud bands seen during the flyby.

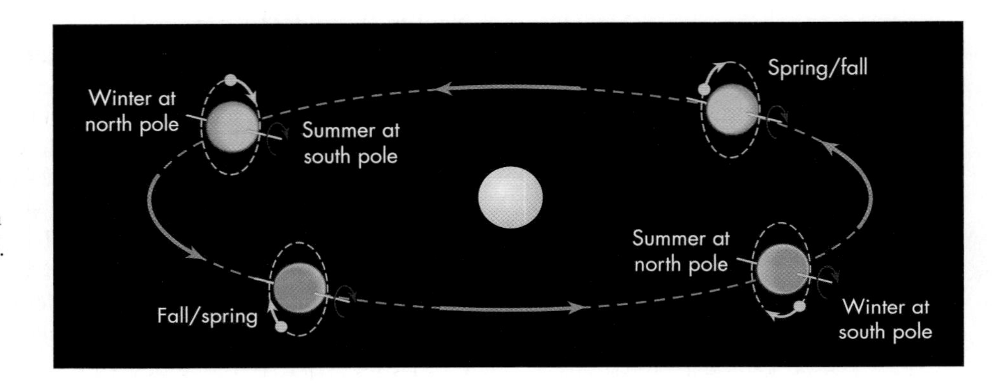

FIGURE 44.6
A sketch of Uranus's odd tilt. Because of this tilt, when Uranus's north pole points toward the Sun (its northern summer), the Sun is above the horizon for many Earth years, while the other pole will be in night. During the Uranian spring and fall, the Sun rises and sets approximately every 17 hours. (Sizes of bodies and orbits are not to scale.)

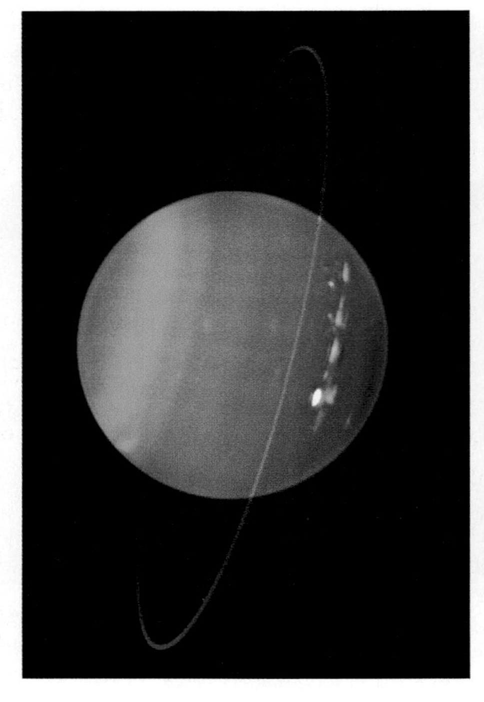

FIGURE 44.7
Uranus observed with the Keck 10-m telescopes in 2004. This is a false-color image made at infrared wavelengths; it shows both the banded structure in the atmosphere and a number of storms in the planet's northern hemisphere.

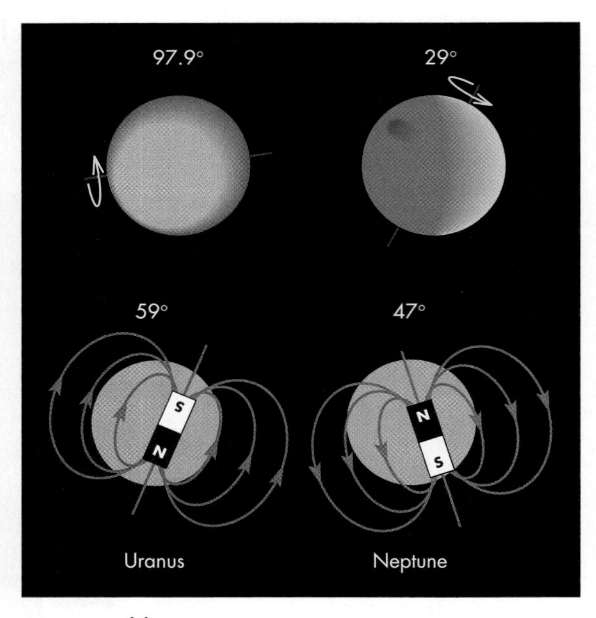

FIGURE 44.8
The magnetic fields of Uranus and Neptune are tilted at a large angle to their rotation axes and are offset from the centers of the planets.

Uranus's polar axis was perpendicular to the Sun in 2007, so the heating it experienced in the decade before and after that is more similar to the solar heat received by the other planets—greatest in the equatorial regions and with typical day/night variations. Markings in its atmosphere were seen around this time with the Hubble Space Telescope and ground-based infrared telescopes (Figure 44.7). These markings appear to be large storm systems, so perhaps Uranus is not always as featureless as it looked during the *Voyager II* flyby in 1986.

Besides Uranus's odd axial tilt, its magnetic field is tilted at a large angle to its rotation axis. The same is true of Neptune's magnetic field, which is also offset by a large distance from the planet's interior, as illustrated in Figure 44.8. Because these planets do not have liquid metallic hydrogen in their interiors, astronomers suspect that the internal electrical currents that create these magnetic fields are generated within the water layer. The magnetic fields generated are not as strong as Jupiter's or Saturn's, but they are still much stronger than Earth's—about 47 times stronger in Uranus and 25 times stronger in Neptune. The mechanisms that produce such highly tilted magnetic fields are not yet understood.

KEY TERMS

ice giant, 348

QUESTIONS FOR REVIEW

1. What are the differences in how Uranus was discovered versus how Neptune was discovered?
2. How do Uranus and Neptune differ from Jupiter internally?
3. Why are Uranus and Neptune blue?
4. What is unusual about Uranus's rotation axis? What might explain this peculiarity?
5. What are seasons like on Uranus?
6. What are the differences between the magnetic fields of Uranus and Neptune and that of Jupiter?

PROBLEMS

1. Find the "surface" gravity on Uranus and Neptune (Unit 16.3). How does this compare with Earth's surface gravity?
2. If Uranus and Neptune had the same density, how many times bigger than Uranus would Neptune be?
3. Why is it not surprising that Uranus and Neptune have proportionally less hydrogen than their more massive companions, Jupiter and Saturn?
4. In order to match the densities of Uranus and Neptune, scientists hypothesize that their interiors are composed of water mixed with methane and ammonia surrounding a core of rock and iron-rich material. A similar mix of a small iron core surrounded by hydrogen, helium, methane, and ammonia could also be constructed to give the same density. Why don't scientists assume that their interiors have that composition?
5. The Uranian day is 17.24 hours. What is the speed of a point on the equator of Uranus due to its rotation? Express your answer in km/hr. If the speed at the Earth's equator were the same, how long would our day be?
6. Assume that Earth's axis is tilted 90° with respect to its orbital plane, like that of Uranus. How would this affect the length of Earth days? What would happen to the seasons? Describe a complete year for where you live.

TEST YOURSELF

1. What makes some astronomers think that Uranus was hit by a large body early in its history?
   a. It goes around the Sun in the opposite direction from the other planets.
   b. Its rotation axis has such a large tilt.
   c. Its composition is so different from that of Neptune, Jupiter, and Saturn.
   d. It has no moons.
   e. All of the above.
2. What gives Neptune and Uranus their blue color?
   a. Objects with such low temperatures take on a blue color.
   b. Large oceans cover both of their surfaces.
   c. Methane gas absorbs the longer wavelengths of visible light.
   d. An unknown blue organic compound covers both planets' surfaces.
   e. The icy blue polar caps of these planets face toward the Earth.
3. Neptune was discovered
   a. by accident while an astronomer was looking for comets.
   b. as it passed in front of a bright star, blocking its light.
   c. by looking at a position predicted by calculations.
   d. in antiquity, but it moves so slowly that people thought it was a star.

unit

45 Satellite Systems and Rings

All of the giant planets have large systems of satellites as well as **rings** of particles too small to be called satellites. Two of the satellites are even larger than Mercury, and are easily seen through a small telescope, but the vast majority of the satellites are much smaller. Technological improvements in telescopes and spacecraft sent to Jupiter and Saturn in recent years have advanced our knowledge of these systems, and currently more than 150 satellites—ranging in size down to as little as 1 kilometer—have been cataloged orbiting the gas giants.

These systems of orbiting bodies bear many resemblances to the Solar System itself. Each gas giant's system has some unique properties, although there are many shared characteristics. The six largest moons are larger than Pluto and Eris and are fascinating worlds in their own right. Their general properties are discussed here, but they are explored in greater detail in Unit 46. The general properties of the satellites and ring systems are the focus of this unit.

45.1 SATELLITE SYSTEMS

When Galileo Galilei first viewed Jupiter with his telescope in 1610, he saw four starlike objects all in a line with the planet (Figure 45.1). These would be bright enough to be seen without a telescope, but they are lost in the glare of the much brighter planet. As Galileo observed the planet night after night, he discovered that these small objects orbited Jupiter, taking between 2 and 17 days to complete their orbits. He realized that they must be satellites of Jupiter, just as our own Moon is a satellite of Earth. They are called the **Galilean satellites** in his honor.

FIGURE 45.1
Jupiter and the Galilean satellites as seen through a small telescope. (This image was made by the Louisiana State Amateur Astronomy Club on March 19, 2004.) The starlike dots are the four large moons of Jupiter. From left to right: Europa, Io, Callisto, and Ganymede.

Over the past four centuries, astronomers have found many smaller satellites. More than 60 satellites of Jupiter are now known, and the total keeps growing as we obtain more-detailed views from spacecraft missions such as *Voyager* and *Galileo*, as well as from large telescopes with steadily improving technologies. Twenty newly discovered moons were added to Jupiter's total in 2003 alone, although most of them are less than a few kilometers across. Currently there is no official minimum size for an orbiting body to be recognized as a satellite, and it is likely that we will catalog many more small objects of this size in coming years.

The distances and sizes of the satellites orbiting each of the gas giants are illustrated in Figure 45.2. The sizes of the satellites are all shown to the same scale along with our Moon for comparison. Note that the spacing between planet and moons in Figure 45.2 is not to a fixed scale but varies to show all the satellites both near and far from each planet.

The Galilean satellites are all large bodies, comparable in size to our Moon. Johannes Kepler suggested naming them after mythological figures who were associated with Jupiter: Io, Europa, Ganymede, and Callisto. This idea eventually

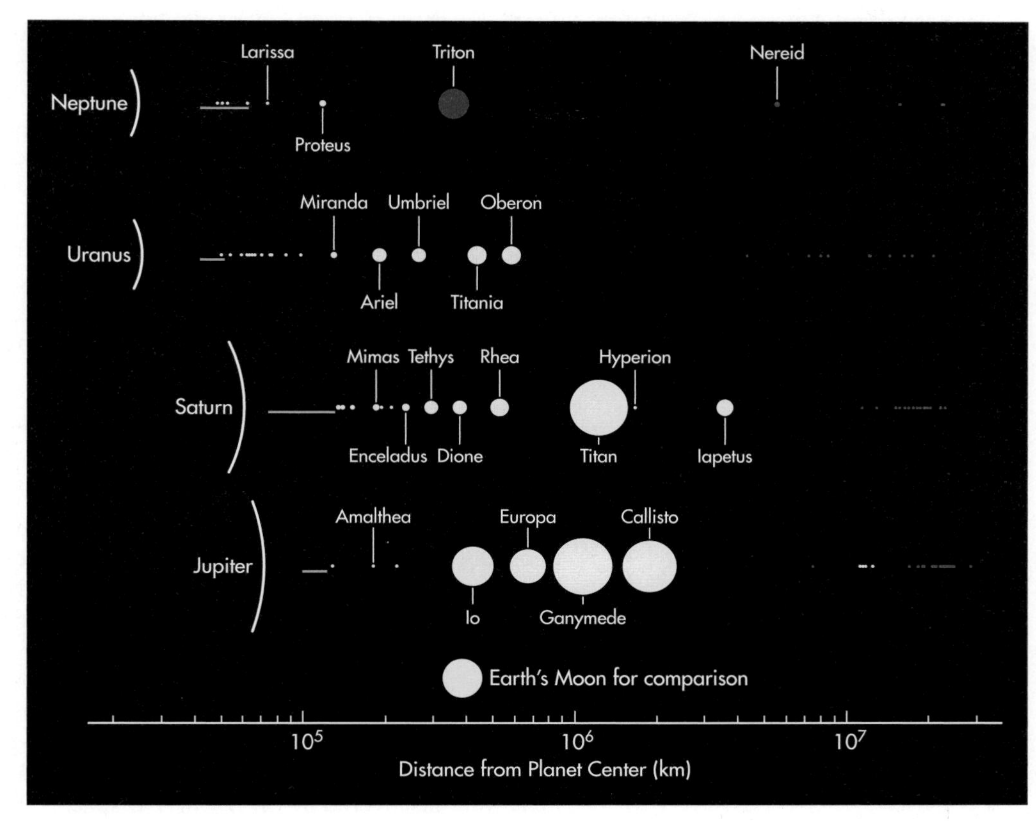

FIGURE 45.2
The satellite systems of the gas giants. The satellites are drawn to the same scale, but their spacing is compressed. The extent of each planet's ring system is shown with a blue line. Satellites with irregular orbits (at a large angle or highly noncircular) are shown in red. Moons with dimensions larger than 200 kilometers (120 miles) are named.

Uranus's satellites are named after characters from the works of Shakespeare and Alexander Pope.

Q One explanation of Uranus's highly tilted axis is that the planet suffered a collision with a large protoplanet late in its formation. What does the regular orbit of the satellites indicate about when they formed relative to any such collision?

took hold for naming the satellites of each planet after related mythological figures. Today we know that two other satellites, Saturn's moon Titan and Neptune's moon Triton, are also quite large. Ganymede and Titan are even larger than Mercury.

Figure 45.2 shows how different each satellite system is. Jupiter has four large moons, but the rest of the satellites have radii smaller than about 100 kilometers. The other planets all have a variety of satellites of intermediate sizes, as well as many small satellites.

Almost all of the satellites orbiting near each planet are in **regular orbits.** Regular orbits are nearly circular, are in the same direction as the planet spins, and are close to the plane of the planet's equator. The satellites with these properties are shown in yellow in Figure 45.2 and include almost all the satellites with radii larger than 100 kilometers. These systems of satellites form a flattened disk like a miniature Solar System. This is true even of Uranus with its oddly tilted rotation axis. This suggests that these satellites formed by a scaled-down version of the process that created the Solar System. That is, they probably aggregated from planetesimals and gas that collected around the gas giants during their formation like a miniature solar nebula (Unit 33).

Many of the smaller satellites discovered in recent years do not have regular orbits. They orbit backward from the planet's spin, or at large angles to the equator, or with highly elliptical orbits. Satellites with these properties are probably captured objects. They may have been asteroids (or icier versions of the same in the outer Solar System) that passed close enough to the planet to undergo a gravitational encounter.

Capturing a satellite is not simple: The same orbit that brings a potential satellite close to a planet will normally carry it away again, following an elliptical trajectory as predicted by Kepler's second law (Unit 12). For a passing body to end up in orbit around a planet, something must slow it down when it is close to the planet. This may occur if it encounters another satellite. It might also occur when the planet is forming and still surrounded by a cloud of material that could slow the object down.

Within these satellite systems, some groups of satellites have similar orbits. Astronomers suspect that these may have a common origin from when a larger body entered the system and was broken up in a collision. This would have created a family of satellites all with similar orbits; over time their orbits have spread out, but they still orbit at similar distances, angles, and directions.

45.2 SATELLITE PROPERTIES

The satellites orbiting the gas giants have many properties in common. Most have densities that imply they are a combination of ices and some rocky material, and their surfaces are peppered with impact craters. Nearly all are "tidally locked," like our own Moon, with the same side always facing the planet they orbit (Unit 37).

The six large satellites are shown to scale in Figure 45.3. Each has a variety of unusual features that are discussed in Unit 46. Medium-size satellites, which still have diameters larger than 200 kilometers (120 miles), are illustrated in Figure 45.4. The images have been obtained by the *Voyager, Galileo*, and *Cassini* spacecrafts. Two of these medium-size satellites, along with many smaller satellites, were discovered by the *Voyager II* spacecraft as it passed Neptune in 1989.

Q Many medium-size satellites are spherical, whereas asteroids of the same diameter are usually irregularly shaped. What might explain this difference?

The satellites with diameters larger than about 400 kilometers are spherical; many smaller ones are irregularly shaped (Figure 45.5). This is illustrated by Jupiter's moon Amalthea (*am-uhl-THEE-uh*), which has a largest extent of about 250 kilometers (150 miles) but is only about half as wide. Similarly, Saturn's moon Hyperion (*hi-PEER-ee-on*) is quite irregular despite an even larger extent of about 360 kilometers (220 miles). The gravity of larger satellites is generally sufficient to pull them into a nearly spherical shape.

Amalthea also illustrates another feature of many satellites. Even though the satellites are made mostly of ices, many are dark. The dark colors appear to be thin layers over the surface of the satellite. For example, Amalthea displays a dark red coating (Figure 45.5A). The color looks like rock, but a small break in its surface at the point of an impact crater reveals a snowy white interior. Measurements of the satellite's

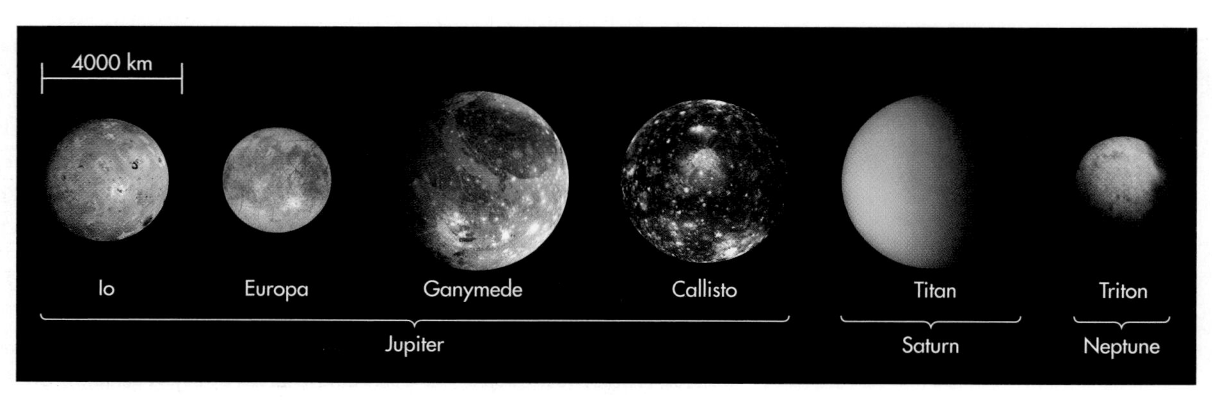

FIGURE 45.3
The large satellites of the gas giants. The Galilean satellites' images are from the *Galileo* spacecraft, Titan's image is from the *Cassini* spacecraft, and Triton's image is from the *Voyager* flyby.

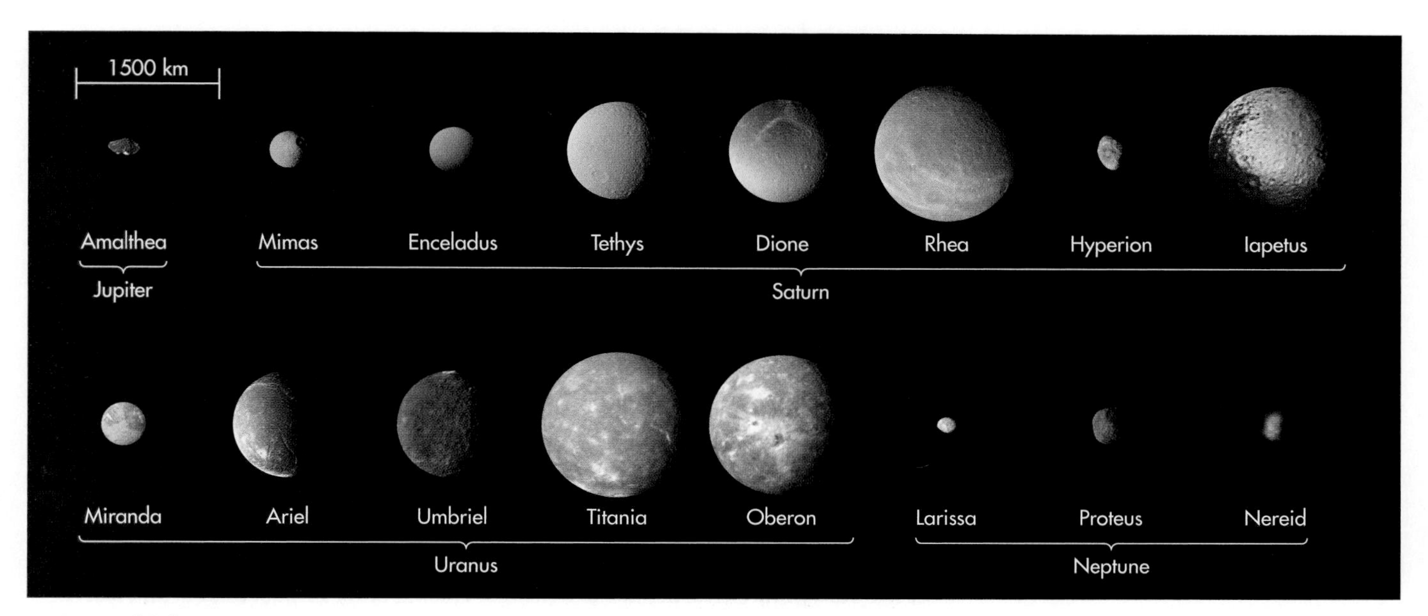

FIGURE 45.4
The medium-size satellites of the gas giants. These are the satellites with sizes between about 200 and 1500 kilometers (120 and 950 miles).

mass by the *Galileo* spacecraft indicate that its density is less than 1 kilogram per liter, suggesting that it is made of ices that are held together only loosely. The dark coating probably comes from Jupiter's moon Io, which has active volcanoes that loft sulfur compounds into a cloud that surrounds Jupiter (Unit 46).

Saturn's moon **Iapetus** (*eye-YAP-ih-tuhss*) is especially peculiar in that one hemisphere is black while the other is white (Figure 45.6). Because Iapetus is tidally locked with one side always facing Saturn, there is also one side that always faces in the leading direction of the orbit—this is the black side of the satellite. It is as if the moon encountered a cloud of dark sooty material that has covered the leading half of the moon as it orbited Saturn. Some astronomers have hypothesized that this dark material may be hydrocarbons from volcanic material that

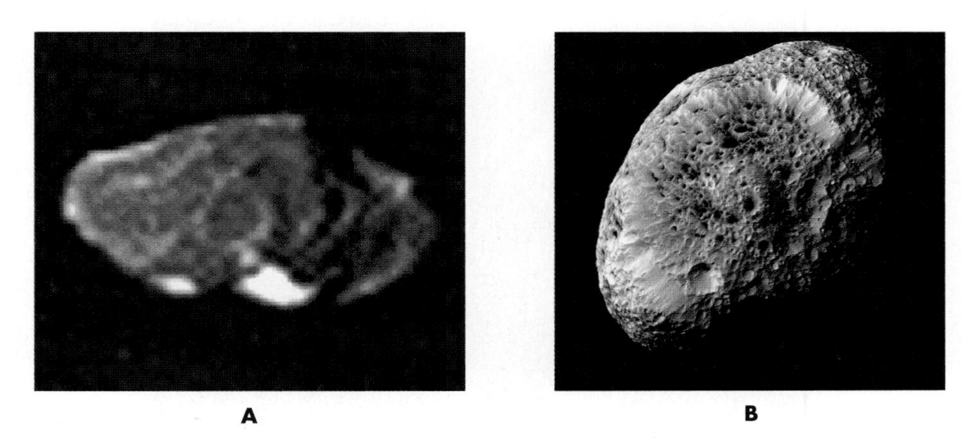

FIGURE 45.5
(A) The satellite Amalthea is Jupiter's next largest after the Galilean satellites—about 250 kilometers (150 miles) in its longest dimension. It is an icy object coated with a layer of sulfur compounds from the moon Io's volcanoes. (B) Saturn's satellite Hyperion, which is 360 kilometers (220 miles) across in its longest dimension. The surface of Hyperion is heavily cratered and it has an extremely low density—just 0.6 kg/liter—suggesting that it may be loosely packed material.

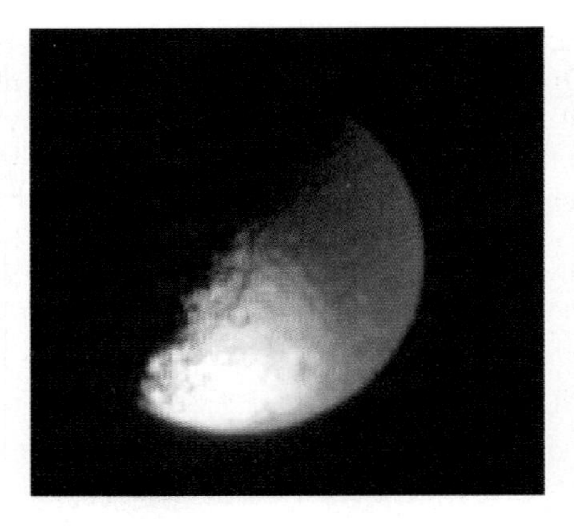

FIGURE 45.6
Two views of Saturn's moon Iapetus photographed by the *Cassini* spacecraft. Half of its surface is snow white, while the other half is coated with a layer of black hydrocarbons whose source is not known. The dark side is so black that it is almost invisible in the first image. The second image is a close view of the dark side, showing a strange ridge along the moon's equator.

was ejected from Iapetus when it was young, or perhaps it is dust from another moon that underwent a large impact.

Smooth areas and long cracks or **faults** appearing in some of the medium-size moons suggest that some internal heat and tectonic activity may have occurred in these bodies. The surface may have melted, erasing older impact craters and leaving a smooth surface. Melting would also allow dusty material to sink and make the surface brighter. There is not as clear a link between size and geological activity as there is for the terrestrial planets. Some of the larger satellites are dark and heavily cratered, suggesting that their surfaces have not been renewed. On the other hand, some smaller satellites, like Saturn's moon Enceladus (*en-SELL-ah-duhss*), have smooth surfaces with long faults, suggesting that extensive geological activity occurred.

Uranus's larger moons are very dark and heavily cratered (Figure 45.4), but the fifth biggest, **Miranda** (*mih-RAN-duh*), has a complex faulted surface (Figure 45.7). Miranda's diameter is only about 470 kilometers (290 miles), but its

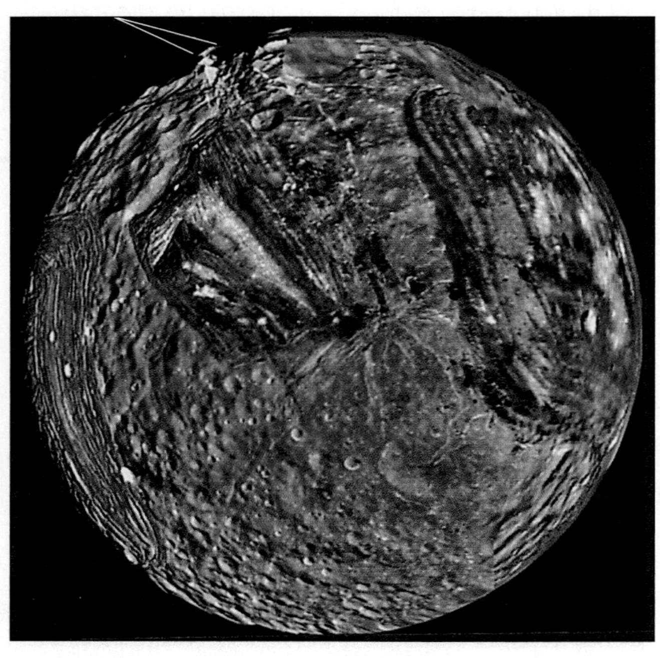

FIGURE 45.7
Miranda, an extremely puzzling moon of Uranus, as observed by the *Voyager II* spacecraft. Note the enormous cliffs glinting in the sunlight at the top of the picture.

surface is broken into distinct areas that seem to bear no relation to one another. One region is wrinkled, while an adjacent one has small hills and craters, rather like our Moon. Miranda has some extremely curious and unexplained surface features such as a set of cliffs, visible in the top corner of Figure 45.7, that are twice the height of Mount Everest. Miranda's patchwork appearance led some astronomers to hypothesize that it had been shattered by impact with another large body, and that the pieces were subsequently drawn back together by their mutual gravity, giving this peculiar moon its jumbled appearance. The leading hypothesis, however, is that the complex pattern arose as the result of strong tectonic activity that broke the surface into "plates," much as occurred on Earth.

45.3 RING SYSTEMS

In mythology, Saturn feared that his children would overthrow him, so he ate them when they were born. However, his son Jupiter was hidden away by his mother and eventually overthrew Saturn and forced him to disgorge the other gods. Galileo mused that the planet Saturn had perhaps swallowed and then disgorged its moons like its mythological namesake!

Saturn's spectacular rings, illustrated in Figure 45.8, were first seen by Galileo; but when he trained his telescope on this planet in 1610, Galileo was baffled by what he saw. Through his imperfect lenses, the rings looked like "handles" on each side of the planet, and he thought there must be two satellites, one on either side of the planet. Two years later Saturn moved into a point of its orbit where the rings were edge-on to Earth and nearly invisible through Galileo's telescope; but then they reappeared several years later, further baffling astronomers of the time.

It was not until 1659 that Christian Huygens, a Dutch scientist, observed that the rings were detached from Saturn and encircled it. Early on, astronomers assumed that the rings were solid; but in 1857 the British physicist James Clerk Maxwell demonstrated mathematically that no material could be strong enough to hold together in a solid sheet of such vast size. He thus deduced that the rings must be swarms of particles. Spectra of the rings support Maxwell's hypothesis: Doppler shift measurements show that each part of the **ring system** orbits at the velocity appropriate to its distance from the planet—fast near the planet and slower farther away. Thus the rings must be composed of myriad individual bodies.

We now know that all the giant planets have rings surrounding them, but none of the others' rings are nearly as substantial as Saturn's. The extent of the main ring systems is illustrated by the blue line in Figure 45.2. These ring systems are very wide but very thin. Saturn's ring system extends out to a little more than twice the planet's radius (136,000 kilometers, or about 84,000 miles). Some faint inner rings can be seen even closer to Saturn, and faint outer rings extend out to about

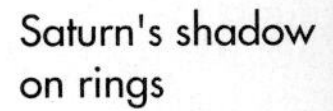

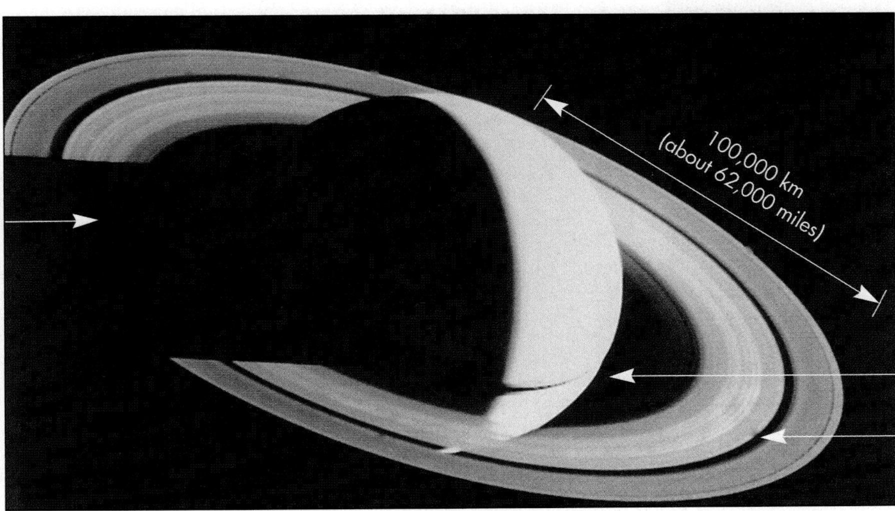

FIGURE 45.8
Rings of Saturn.

FIGURE 45.9
False-color photograph of the ringlets, showing the substructure of Saturn's rings.

eight times its radius. Yet despite the rings' immense breadth, they are probably less than a few hundred meters thick—thin enough that we can see stars through them.

Astronomers have determined that the ring particles are relatively small, only a few centimeters to a few meters across. Although these particles are far too small to be seen individually with telescopes, they reflect radar signals. From the strength of the radar "echo," astronomers can estimate the particle sizes. Spectra of sunlight reflected from the rings indicate that Saturn's ring particles are composed primarily of water ice. However, high-contrast spacecraft pictures, such as Figure 45.9, show that some parts of the rings have different colors, implying that the composition of the rings is not the same everywhere. Particles in the darker ring segments may be rich in carbon compounds similar to those found in some asteroidal material. Figure 45.9 also reveals that the rings are not uniformly filled: They consist of numerous separate **ringlets.** Large gaps in the rings had been seen from Earth, but the many narrow gaps in the rings came as a surprise. What causes these gaps that create the multitudes of ringlets?

Kirkwood also discovered gaps in the distribution of asteroids related to the period of the orbit of Jupiter (Unit 41).

In 1872 the astronomer Daniel Kirkwood noticed that the largest gap in the rings, known as **Cassini's division** (see Figures 45.8 and 45.9), occurs where ring particles orbit Saturn in half the time it takes the moon Mimas to orbit Saturn. This **resonance** (a simple relationship between orbital periods) means that any ring particle orbiting in the gap would undergo a strong and repeated gravitational force from Mimas every other orbit. The cumulative effect of these tugs over long periods would pull particles from the gap and thereby create Cassini's division.

Astronomers think that the many narrow gaps apparent in Figure 45.9 have a different cause. Narrow gaps in the rings probably arise from a complex interaction between the ring particles and the tiny moons that orbit within the rings. As these small moons—some less than 10 kilometers in size—orbit Saturn, their gravitational attraction on the ring particles generates waves. These waves spread through the

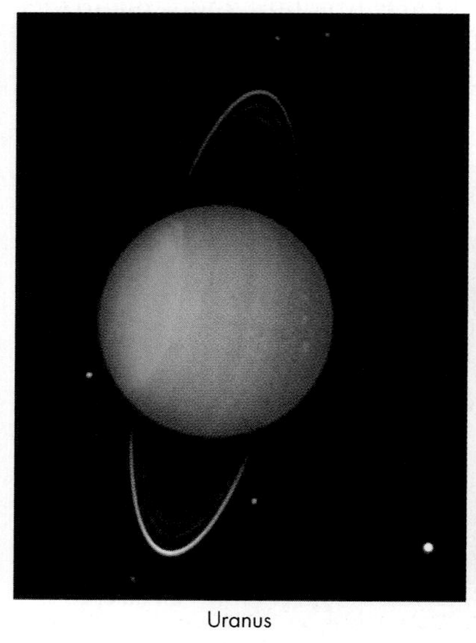

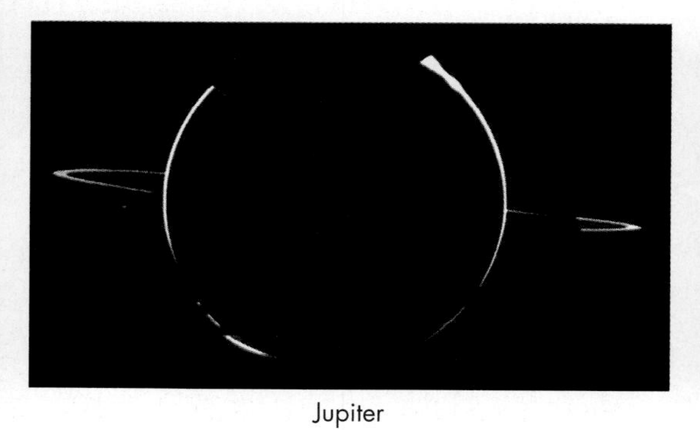

FIGURE 45.10
Ring systems of Jupiter, Uranus, and Neptune. The images of Jupiter's and Neptune's rings were made by the *Voyager II* spacecraft from the night side of the planets, portions of which can be seen in each image. The Uranus image was made by the Hubble Space Telescope.

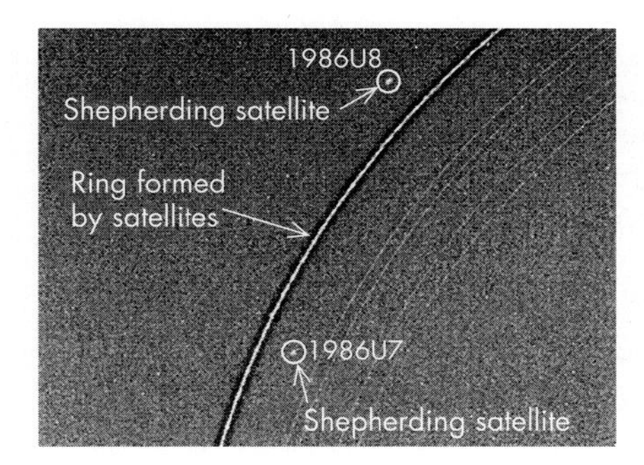

FIGURE 45.11
Two shepherding satellites (the dots inside the small circles) and a portion of the narrow ring around Uranus that they created. The picture's contrast is strongly enhanced to show these faint moons.

rings much like ripples in a cup of coffee that is lightly tapped. Such ripples are circular in a cup or on a still pond; but because the inner part of the ring orbits faster than the outer part, the spreading waves are stretched into a spiral shape. The tightly wound pattern, similar to the spiraling groove in a phonograph record, is called a **spiral density wave.** The crests of these density waves form the narrow rings.

For centuries, astronomers thought that the only Solar System planet with rings was Saturn. But in 1979 the *Voyager I* spacecraft flew by Jupiter. The photographs it took as it passed the planet clearly showed that Jupiter has a ring, although only a very thin one, as illustrated in Figure 45.10. Despite being the "king" of the planets, Jupiter has the least substantial ring of the gas giants. Uranus and Neptune are also encircled by a set of narrow ringlets, as shown in Figure 45.10. The Uranian rings are dark, suggesting that they are not made of, or coated with, ice like the bright rings of Saturn. Instead, they may be rich in carbon particles or carbon-rich molecules.

Thin isolated ringlets can form when two small moons orbit at slightly different radii. The combined gravitational force of these **shepherding satellites** deflects ring particles into a narrow stream between the two moons' orbits (Figure 45.11). Such rings sometimes have complex twisted shapes and are not necessarily circular—and the distribution of particles may be clumpy, as seen in Neptune's rings (Figure 45.10).

45.4 ORIGIN OF PLANETARY RINGS

Until the mid-1900s most astronomers thought that planetary rings were material left over from a planet's formation, perhaps matter that had failed to condense into a satellite. But we now realize that rings are short-lived because they are subject to forces in addition to gravity. For example, gas trapped in a planet's magnetic field may exert a frictional force on the ring particles, gradually causing them to slow and spiral into the planet's atmosphere. Thus, new material must be added to the rings from time to time. Without such replenishment, they would disappear in a few hundred million years.

One source of new material is the satellites orbiting the planet. A moon in a satellite system as complex as those around the gas giants is subject not only to the gravitational force of the planet but also to that of the other moons. The cumulative effect of such forces alters the satellites' orbits and may cause them to collide. Impacts by small bodies may also inject a cloud of debris into the system that becomes part of a ring.

A satellite does not have to undergo a physical collision to be broken apart. If a moon gets too close to its planet, the planet's gravity can rip the satellite apart. This was shown mathematically in 1849 by the French scientist Edouard Roche (pronounced *Rohsh*). Because gravity weakens with distance, a planet pulls harder on the near side of a satellite than on the far side. This is the effect that gives rise to tides, but it can have much more extreme effects. If the difference in this pull exceeds the moon's own internal gravitational force, the moon will be pulled apart, as shown in Figure 45.12. Thus, if a moon—or any body held together by gravity—approaches a planet too closely, the planet raises a tide so large it pulls the encroaching object to pieces.

Roche calculated the distance at which the tide becomes fatally large and showed that for a moon and planet of the same density, breakup occurs if the moon comes closer to its planet than 2.44 planetary radii, a distance now called the **Roche limit.** All planetary rings lie near their planet's Roche limit, suggesting that rings might be created by satellite disintegration. The existence, side by side, of ringlets with different compositions (some rich in ice, others rich in carbon)

The Roche limit

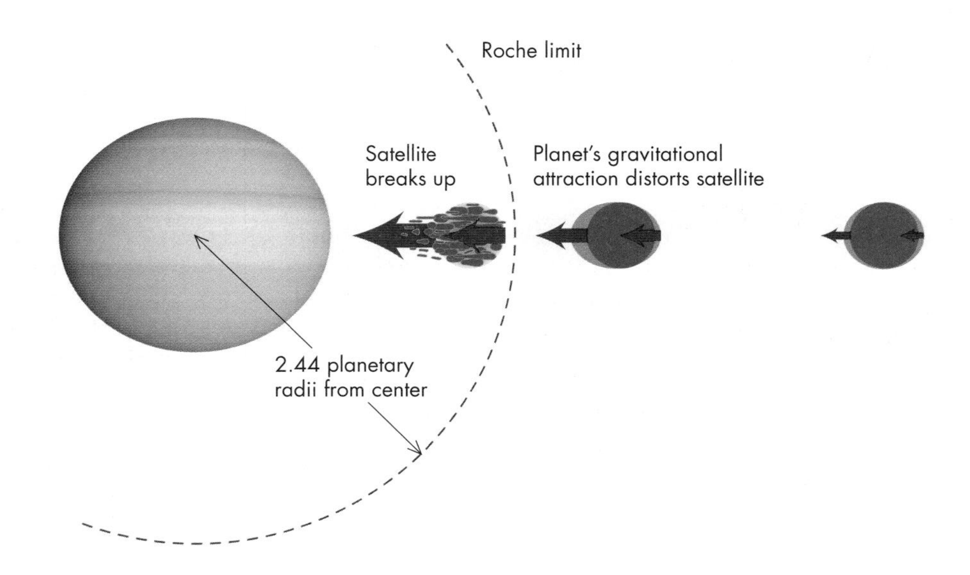

FIGURE 45.12
The Roche limit. A planet's gravity pulls more strongly on the near side of a satellite than on its far side, stretching it. At radii smaller than the Roche limit, the tidal force can overcome the satellite's own gravity, pulling loosely connected pieces of the satellite apart.

is additional evidence that the rings formed from the breakup of many different small objects.

It is important to keep in mind that the Roche limit applies only to bodies held together by gravity. A body will *not* break where forces other than gravity hold it together. For example, the chemical bonds between neighboring atoms are stronger than the tidal force differences across a body. Therefore, an unfractured piece of ice or rock will not be split up by tidal forces, although any loose rubble on the surface will be pulled off. Similarly, most artificial satellites orbiting the Earth are well inside the Roche limit, but they are in no danger of being torn apart. However, a loose part resting on the satellite will be pulled away.

A number of the inner satellites of Jupiter, Uranus, and Neptune orbit so close to their planets that they take less time to orbit than their planets take to rotate. These moons all experience tidal forces that cause their orbits to decay. In addition, friction from the extended atmospheres and material caught in the magnetic fields of these planets puts a drag on these moons. These forces are slowly causing the satellites to spiral closer to the planets, and they will be tugged apart after they cross the Roche limit. Perhaps millions of years from now these planets will have much more visible ring systems around them.

Objects that are not orbiting a planet but happen to pass within the Roche limit can also suffer severe consequences. For example, asteroids or comets occasionally pass too close to a planet. This occurred in 1992 when Comet Shoemaker-Levy 9 (Figure 45.13) passed within about 30,000 kilometers of Jupiter. It was pulled apart into a string of smaller comets, which later collided with the planet.

Q: Why do several of Neptune's satellites overlap the range of distances covered by Neptune's ring system?

FIGURE 45.13
Comet Shoemaker-Levy 9. This image shows 20 or so of the fragments that the comet was broken into by Jupiter's tidal force.

KEY TERMS

Cassini's division, 360
fault, 358
Galilean satellites, 354
Iapetus, 357
Miranda, 358
regular orbit, 355
resonance, 360
ring, 354
ringlet, 360
ring system, 359
Roche limit, 361
shepherding satellite, 361
spiral density wave, 361

QUESTIONS FOR REVIEW

1. What is the range of sizes and distances of satellites orbiting the gas giants?
2. What basic characteristics are common to most satellites of the gas giants?
3. What is a regular orbit for a satellite? Why are satellites that form at the same time as the planet they orbit expected to have regular orbits?
4. Why are large satellites round and small ones irregular in shape?
5. What are the different ways in which satellites affect ring systems? How permanent are ring systems?
6. What is the Roche limit? How does it explain the presence of ring systems around the giant planets?

PROBLEMS

1. During a close flyby, the *Cassini* spacecraft yielded a new estimate of the mass of Enceladus of 1.08×10^{20} kg. Given that Enceladus's diameter is 499 km, calculate its density. The larger satellite Tethys has a density of about 1.0 kg/L. What does this density suggest about Enceladus's composition relative to Tethys's? How might this account for Enceladus's unusual surface features?
2. The Earth's Moon experiences tidal forces that are gradually expanding its orbit by a few centimeters per year (Unit 37). Explain how a moon orbiting faster than its planet rotates will experience a "tidal drag" that causes its orbit to shrink.
3. Scientists have offered two different hypotheses to explain Miranda's tortured-looking surface. It may have resulted from tectonic activity, or it may have sustained a giant impact. If you could send a probe to Miranda, what kind of tests would you make to try to gather evidence that would help choose between these two hypotheses?
4. Figure 45.11 shows a pair of shepherd moons and the ring between them. Assume that the moons and ring particles are orbiting counterclockwise in the image.
   a. Because Kepler's laws apply, compare the relative speeds of the moons and ring particles.
   b. If a passing moon boosts the speed of a ring particle, what will happen to the size of the ring particle's orbit?
   c. If a ring particle experiences a slowing of its speed when it passes a slower moon, what happens to the size of the ring particle's orbit?
   d. Explain how two shepherding moons might drive ring particles into a narrow stream.
5. Examine Appendix Tables 5 and 7 to determine which moons orbit inside the Roche limit, for each of the giant planets. Make a list of these moons and discuss how it is possible for these moons to orbit there. Would you expect these moons to be spherical or irregular? Why?
6. Suppose a moon that was 100 kilometers in diameter broke up into particles just 1 millimeter in diameter.
   a. What was the surface area of the moon before it broke up?
   b. How many particles 1millimeter in diameter could be made from the moon?
   c. What is the total surface area of all of the 1-mm particles you found in part b?
   d. What is the ratio of the surface area in part c to the surface area in part a? Explain why this ratio also represents how much sunlight would be reflected by the particles versus the original moon.

TEST YOURSELF

1. What is remarkable about Uranus's satellite Miranda?
   a. It has several distinct types of surface terrain.
   b. It is more massive than any other satellite of Uranus.
   c. It is so dark that it is not visible in any photographs.
   d. It has a satellite of its own.
2. The particles in Saturn's rings are made of (or covered with)
   a. water ice.
   b. dry ice (frozen carbon dioxide).
   c. metallic hydrogen.
   d. a dark tarlike substance.
3. The Roche limit is
   a. the mass a planet must exceed to have satellites.
   b. the smallest mass a planet can have and still be composed mainly of hydrogen.
   c. the greatest distance from a planet at which its satellites can orbit without falling into the Sun.
   d. the distance at which a moon held together by gravity will be broken apart by the planet's gravitational attraction.
   e. the distance astronomers can see into a planet's clouds.

unit

46 Ice Worlds, Pluto, and Beyond

Six of the satellites that orbit the four gas giants are larger than Pluto and Eris, and two of these satellites even have diameters larger than Mercury's. If they had independent orbits around the Sun, we would call them planets. However, they are quite unlike the terrestrial or Jovian planets. They are made mostly of water ice and rock, because water and other volatile compounds were able to condense from the solar nebula (Unit 33) in the region beyond the asteroid belt.

The surfaces of these icy worlds show evidence of geological processes that resemble what we see on the terrestrial planets. Some of the larger ice worlds have smooth surfaces that have erased the record of impact cratering, indicating geological activity. Some even have active volcanoes and atmospheres. These are worlds that seem familiar in many ways, but the frigid temperatures make ice and water major components of their crustal geology, with behavior resembling that of rock and lava in the terrestrial planets.

Pluto, despite having been called the ninth planet for many decades, is much more like these moons in size and mass. In particular, it appears to be very similar to Neptune's moon Triton. These worlds are mostly ice with a core of rock and iron. In the past few years astronomers have detected many other objects of comparable size, such as Eris, in the outskirts of the Solar System. Quite possibly there are even larger worlds yet to be found in this darkest, coldest part of the Solar System.

In this unit we examine these ice worlds, starting with those that are closest to the Sun. Their sizes relative to each other and to the Earth and the Moon are illustrated in Figure 46.1.

46.1 THE GALILEAN SATELLITES

The four large moons Io, Europa, Ganymede, and Callisto, discovered by Galileo in 1610, are collectively known as the **Galilean satellites.** Each would be considered a planet if it circled the Sun in an independent orbit. Their circular orbits in the plane of Jupiter's equator suggest that they formed along with Jupiter out of the material that made up a disk of gas and planetesimals around the massive planet. At their distance from the Sun, these satellites have temperatures of about 110 to 130 K. The density, brightness of the surface, and level of geological activity are all generally higher for the Galilean satellites that orbit closer to Jupiter. This probably reflects Jupiter's influence during the formation of these moons and over the history of the Solar System.

Of the Galilean satellites, **Io** (pronounced *EYE-oh* or *EE-oh*) lies nearest Jupiter. Because Io is so close to Jupiter, it experiences a strong gravitational tidal force from the planet. That tidal force locks Io's spin, so that one side always faces Jupiter, as our Moon's spin is locked with the same side facing Earth. But Io also

FIGURE 46.1
Ice worlds of the outer Solar System. The large moons of the gas giants are shown to the same scale along with the Earth and Moon for comparison. The images of Pluto and Charon are based on computer models and observational data. Artists' depictions, roughly indicating the size and colors, are shown for several other large trans-Neptunian objects.

undergoes a strong gravitational attraction from Europa, the Galilean satellite next closest to Jupiter, which twists Io from side to side. The result is that Io is subject to strong and changing gravitational forces that distort its shape first in one way then another. This constant flexing heats Io by internal friction, much as bending a paperclip back and forth heats the wire (touch it and it will feel hot).

Over billions of years, the heating has melted not only the ice but also the rocky matter in Io's interior. As molten matter oozes close to the surface, it erupts, creating volcanic plumes that are seen reaching hundreds of kilometers above the surface (Figure 46.2A). Such activity has driven most of Io's water into space, where it is lost because of Io's weak gravitational attraction. As a result, Io's density is 3.5 kilograms per liter, quite similar to the densities of the terrestrial planets.

Io is named for a mythological maiden with whom Jupiter fell in love and whom he changed into a heifer (a young cow) so that his wife, Hera, would not suspect his infidelity.

Sulfur, also common in terrestrial volcanoes, is now the major component of Io's volcanic outpourings, although some eruptions are also rich in ordinary silicate rock. The erupted sulfur and its compounds give Io its rich red, yellow, and orange colors. A recent lava flow can be seen in Figure 46.2B, where some of the molten material is still glowing. The dark regions in the craters of older volcanoes result when molten sulfur cools.

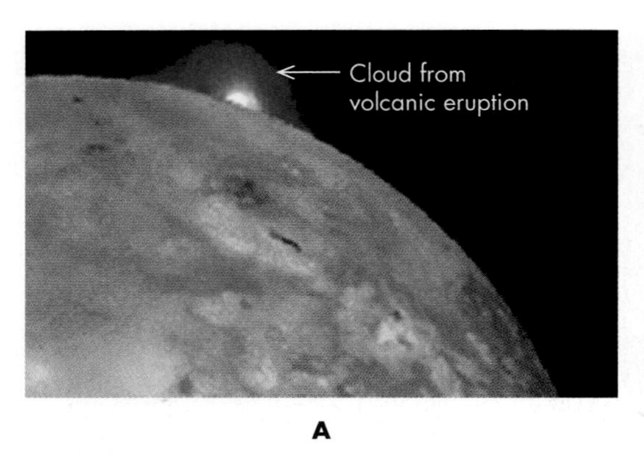

A

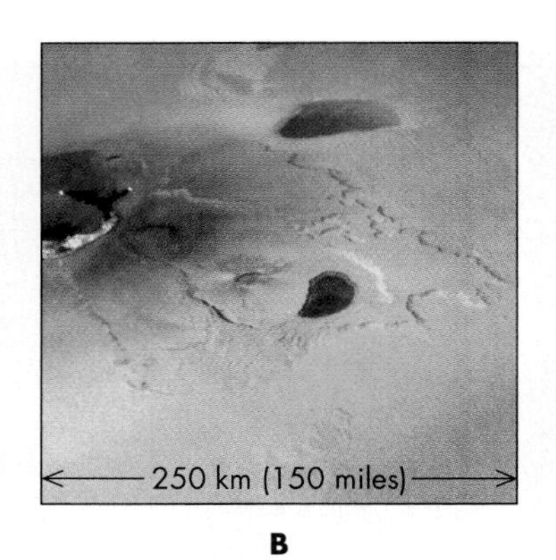

B

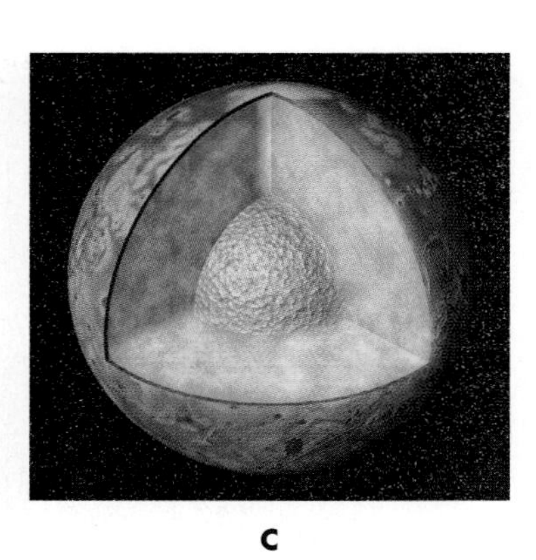
C

FIGURE 46.2
Pictures of Io. (A) The *Voyager I* spacecraft discovered volcanoes erupting on Io. (B) A close-up view of a currently erupting volcano (left side of image) with hot lava seen glowing in the infrared in this false-color image made by the *Galileo* spacecraft. (C) A model of Io's interior.

Europa is named for another maiden whom Jupiter pursued. According to the legend, Jupiter disguised himself as a bull and carried her on his back across the Hellespont from Asia to Europe, thereby giving the continent its name.

Europa (*you-ROPE-ah*), the smallest of the Galilean satellites, looks rather like a cracked egg. Long, thin lines score its surface, as shown in Figure 46.3A. The white material is probably a crust of ice, whereas the red material is probably mineral-rich water that has oozed to the surface through the cracks and frozen. From the small number of impact craters, astronomers estimate that Europa's surface is no more than about 30 million years old, suggesting that its surface has been heated and melted recently, obliterating older impact craters.

Close-up pictures taken by the *Galileo* spacecraft show a complex set of interwoven cracks that are reminiscent of ice floes (Figure 46.3B). In addition, small regions known as "freckles" are spots about 10 kilometers (6 miles) across that appear to have melted. In some freckles, material has been pushed up, whereas in others the region has subsided. These may be the result of a volcanic process in which warmer regions of ice rise and melt—somewhat like the silicate rock in lava fields on Earth.

The cracks in the surface ice and the "freckles" suggest that heat is rising from the interior, shifting and melting Europa's ice crust. Such heating could come from Jupiter's gravitational force deforming Europa, as occurs with Io. Europa, also like Io, has a relatively high density of 3 kilograms per liter, indicating that it has a relatively large rock and iron core; so it may generate heat from the decay of radioactive elements in its interior, like the terrestrial planets. These sources of heat may be sufficient to keep a layer of water melted beneath Europa's crust, forming an ocean (Figure 46.3C). The *Galileo* spacecraft also detected a weak magnetic field in Europa, which might be generated within an electrically conductive salty ocean under the surface. Some astronomers even speculate that Europa's ocean might harbor life, although there is no evidence to support that view.

Ganymede (*GAN-ih-meed*) and **Callisto** (*kal-IH-stoh*) look somewhat like our own Moon. They are grayish brown and covered with craters made during the late stages of their formation, much as our own Moon was pockmarked by debris (Figure 46.4). However, spectroscopy indicates that the surfaces are composed mainly of ice. The darkest areas are apparently the oldest, based on the high density of impact craters, whereas recent impact craters that break through the "dirty" surface layers reveal bright white ice beneath. In addition, large sections of Ganymede are covered with a younger **grooved terrain** (Figure 46.4) that astronomers suspect results from tectonic activity of the icy surface. Long parallel ridges, typically 10 to 20 kilometers apart and up to a kilometer tall, stretch for hundreds of kilometers over the surface. This is somewhat like the linear features seen on Europa. These regions are probably formed by faulting of the icy surface as it is stretched, perhaps driven by convection of "molten slush" layers below.

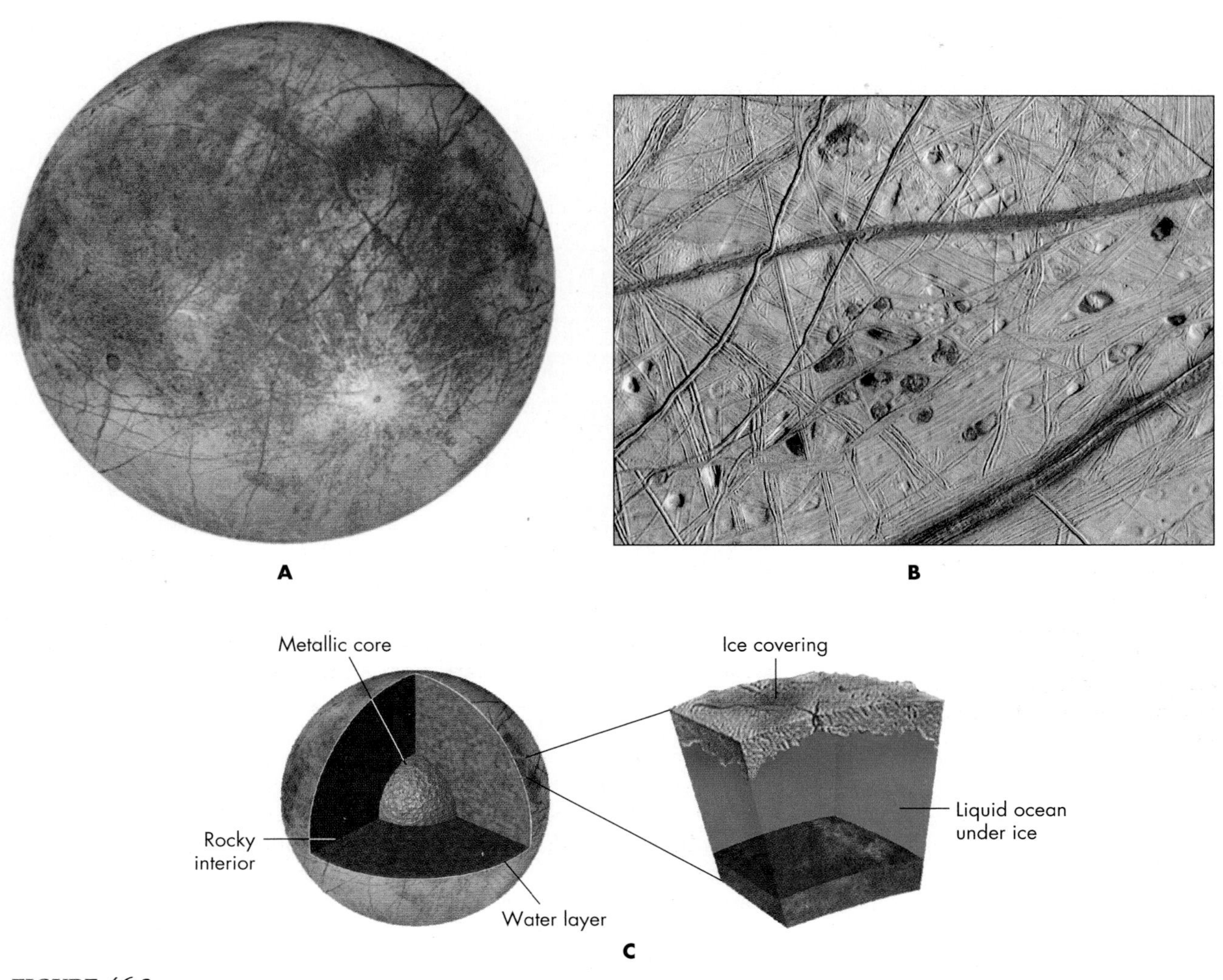

FIGURE 46.3
Pictures of Europa. (A) Europa as imaged by the *Galileo* spacecraft. (B) Close-up picture of Europa's surface showing fractures and "freckles." The dark spots and pits are about 10 kilometers (6 miles) across. (C) A model of Europa's interior. Note the possible liquid water ocean beneath ice.

Ganymede and Callisto have densities of 1.9 and 1.8 kilograms per liter, respectively, intermediate between ice and rock. Although these moons are similar in density, astronomers think their internal structures are quite different. The distribution of the masses of these satellites was studied by carefully measuring the gravitational effects on the *Galileo* spacecraft during close flybys of each moon. The craft's trajectory was affected by the fine details of the mass distribution; and by analyzing dozens of these flybys, astronomers have determined that Ganymede is composed mainly of ice from its surface to a depth of about 800 kilometers (about 500 miles). This ice appears to surround a rocky mantle and probably an iron core. This core may be molten, but astronomers suspect that Ganymede also has a thick layer of water, as illustrated in Figure 46.4C. A molten core or a thick layer of saltwater can conduct electricity, and this would help explain why this moon has a magnetic field.

Heating of these moons during their formation would probably have allowed differentiation to occur, similar to what occurred in our own planet. Thus, iron sank to their cores, surrounded by less-dense silicates and then thick layers of water and ice nearer the surface. The measurements of Callisto suggest that it may

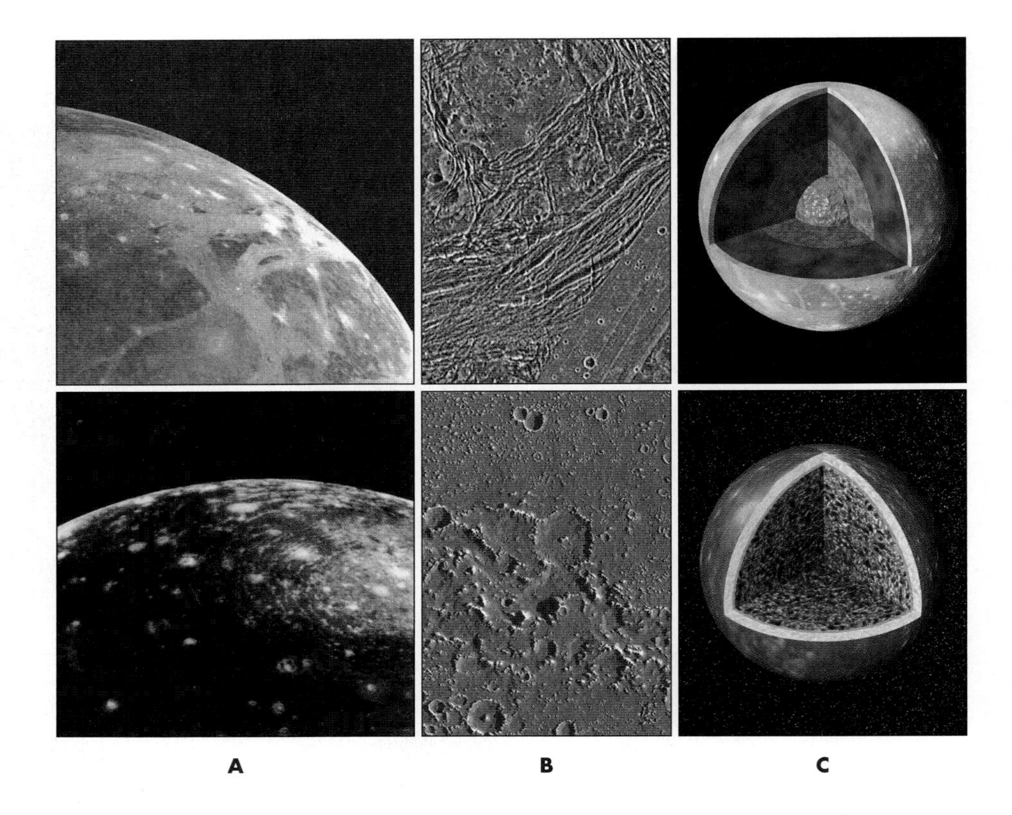

FIGURE 46.4
Images of Ganymede (top) and Callisto (bottom). (A) The surfaces of both satellites are dark, but Ganymede has long linear features. (B) Close-up of regions about 70 kilometers (45 miles) across. (C) Models of the interior of each satellite.

be less differentiated, with rock and ice still intermixed throughout much of its interior. Analysis of the flybys of Io and Europa indicates that they both have large iron and rock cores. Indeed, if their thick icy surface layers were stripped away, Ganymede and Callisto would be similar in density to the inner two Galilean satellites. This may actually be what happened to Io and Europa as a result of the tidal heating they experience—ice and other volatile substances, once vaporized, would escape from the relatively weak gravity of these objects.

Q In what ways are the trends observed among the Galilean satellites similar to those of the planets? In what ways are they different?

46.2 SATURN'S MOON TITAN

Titan is one of the most intriguing moons in the Solar System. It is by far the largest Saturnian moon, containing more than 20 times the mass of all of Saturn's other satellites combined. Titan has a diameter of about 5000 kilometers (3000 miles), making it slightly bigger in diameter than the planet Mercury and comparable in mass and radius to Jupiter's largest moon, Ganymede. However, it is shrouded by a thick atmosphere and dense clouds that hide its surface (Figure 46.5A). Infrared images made by the *Cassini* spacecraft (Figure 46.5B) reveal complex surface features through the outer cloud layers.

Because Titan is farther from the Sun than the Galilean satellites, it is colder—about 95 K (−290°F)—so gas molecules move more slowly and are less likely to escape Titan's gravitational attraction. Titan's atmosphere is so dense that it has a pressure comparable to Earth's. Astronauts visiting there would not need pressure suits, although they would rapidly freeze solid unless they had space suits to keep them warm! Spectra of Titan's atmosphere show that it is mostly nitrogen, like the Earth's, but an astronaut would also need to bring a supply of oxygen.

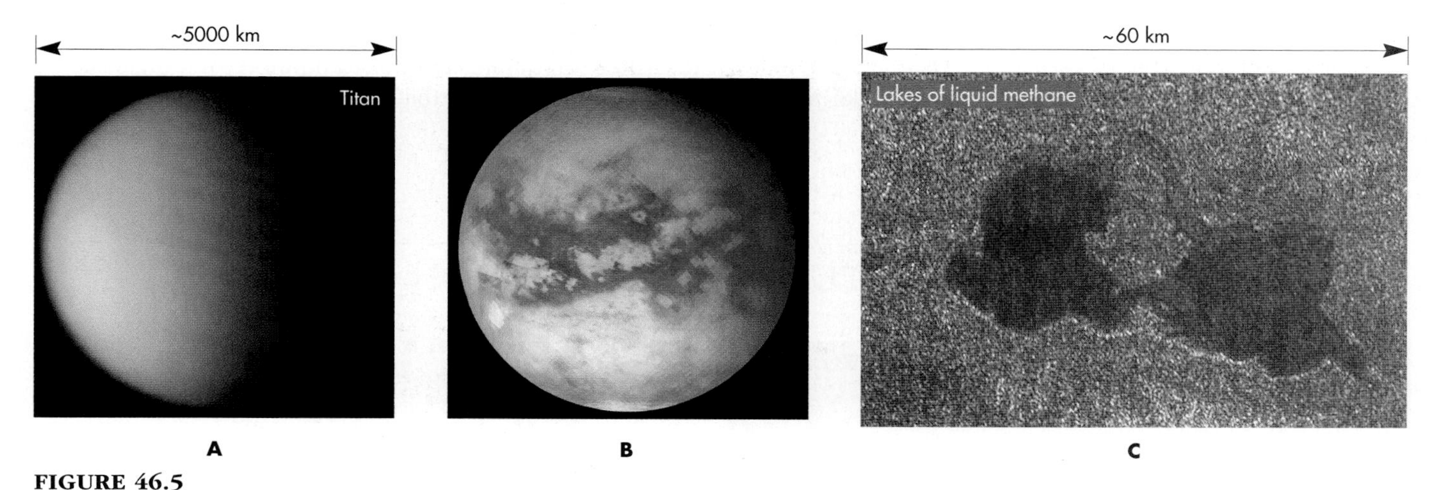

FIGURE 46.5
(A) Haze covers Titan and obscures its surface. (B) A false-color image made by the *Cassini* spacecraft at infrared wavelengths shows some surface features, glimpsed through the clouds. (C) Radar images show the presence of lakes of liquid methane in Titan's polar regions.

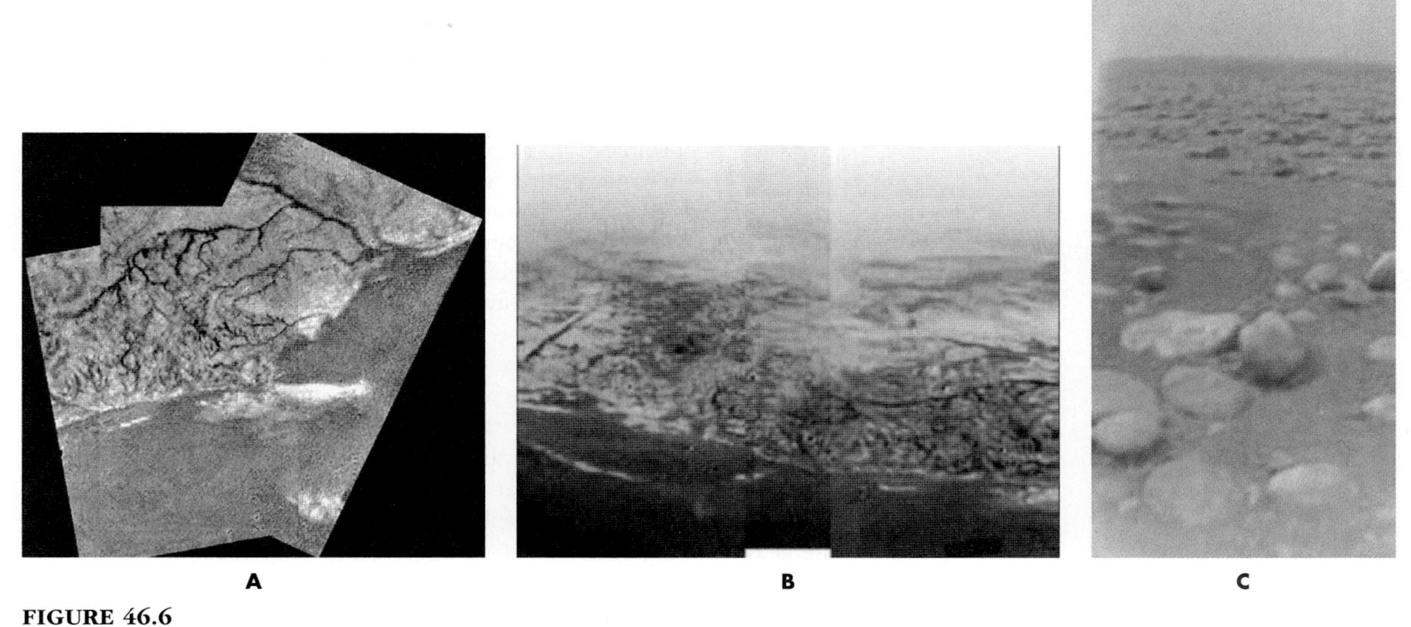

FIGURE 46.6
Images of Titan's surface made by the *Huygens* descent probe in 2005. (A) From about 16 kilometers (10 miles) above the surface, this image shows a region about 10 kilometers (6 miles) across. Riverlike features may have been formed by liquid methane. (B) A panorama of the horizon made when the probe was about 8 kilometers (5 miles) above the surface shows lighter-colored hills and a flat dark region in front of them. (C) The probe took this picture from the surface. The rocklike objects in the foreground are about 10 to 15 centimeters (4 to 6 inches) across.

In 2005 a joint mission of the European Space Agency (ESA) and NASA sent the *Huygens* probe parachuting to the surface of Titan. As it descended, the lander sent back a series of remarkable images (Figure 46.6) showing features that resemble rivers and lakes. Astronomers think that these rivers may have been carved by liquid methane cutting through rock-hard water ice. On the basis of chemical models, some astronomers suspect that the hydrocarbons methane (CH_4) or ethane (C_2H_6) may circulate on Titan, evaporating, raining, and flowing in rivers, much as water does on Earth. A picture from the surface (Figure 46.6C) shows a rubble-strewn field. The "rocks" are probably made of ice, and their dark color may indicate that hydrocarbon compounds are mixed in. Moreover, the bases of the "rocks" show signs of erosion, suggesting that a liquid has flowed through the region.

46.3 NEPTUNE'S MOON TRITON

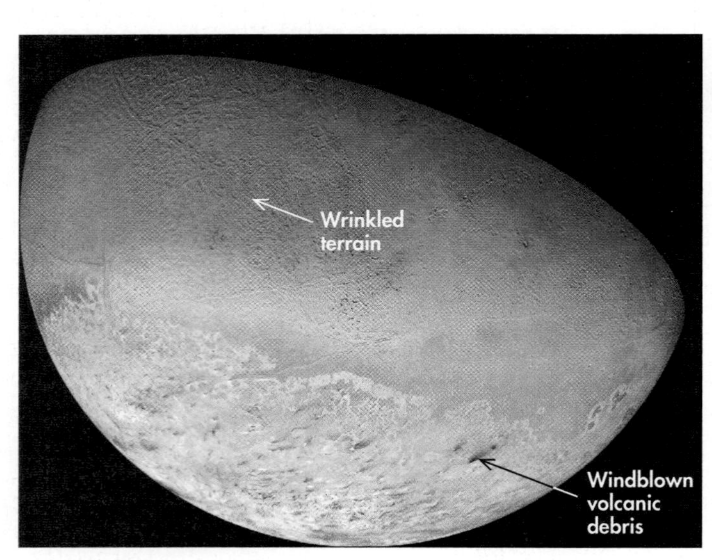

FIGURE 46.7
The surface of Neptune's moon Triton.

Neptune's moon **Triton** is larger than Pluto and nearly as large as Europa. Moreover, its orbit is highly unusual for a large moon. Triton orbits backward relative to Neptune's rotation, in an orbit that is highly tilted with respect to Neptune's equator. The small moons orbiting near the planet are in regular, roughly circular orbits, but the orbit of the medium-size moon **Nereid** is also peculiar, traveling on a highly elliptical orbit.

These orbital peculiarities lead many astronomers to think that Triton was an icy planetesimal that Neptune captured after the planet and its satellite system formed. When Neptune captured Triton, the encounter would have destroyed or expelled most of the outer moons that Neptune originally possessed—and those interactions also would have slowed Triton enough to leave it in orbit around Neptune. Nereid's elliptical orbit carries it far beyond Neptune's other moons, most likely because it was deflected by an amount just short of driving it to escape from Neptune altogether.

Triton intrigues astronomers for more than its orbital oddity. It is massive enough that its gravity, in combination with its low temperature, allows it to retain gases. Triton is thus one of the few moons in the Solar System with an atmosphere. However, its atmosphere is much less dense than Titan's, and it apparently freezes on the night side of the moon, somewhat as water vapor can condense as frost on a cold night on Earth.

Triton's surface has many unusual features, as you can see in Figure 46.7. Wrinkles give much of the surface a texture that looks like a cantaloupe. Craters pock the surface elsewhere, and dark streaks extend from some of them. At least one of these streaks originates from a geyser seen erupting by the passing *Voyager II* spacecraft in 1989. Astronomers think that the matter ejected by the geysers is a mixture of nitrogen, ice, and carbon compounds. Sunlight, weak though it is at Triton's immense distance from the Sun, may warm such matter trapped below the surface and make it expand and burst through surface cracks. The erupted material cools and condenses in Triton's cold, thin atmosphere, where winds carry it and deposit it as a black "soot."

46.4 PLUTO

In the 1920s astronomers suspected there was another large planet in the outer Solar System because the orbits of Uranus and Neptune appeared to be disturbed. The predicted body was dubbed "Planet X," and a young American astronomer, Clyde Tombaugh, carefully searched for the missing object near the ecliptic. In

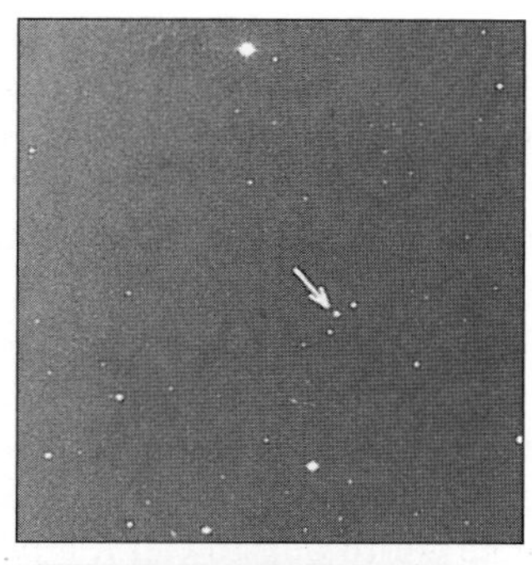

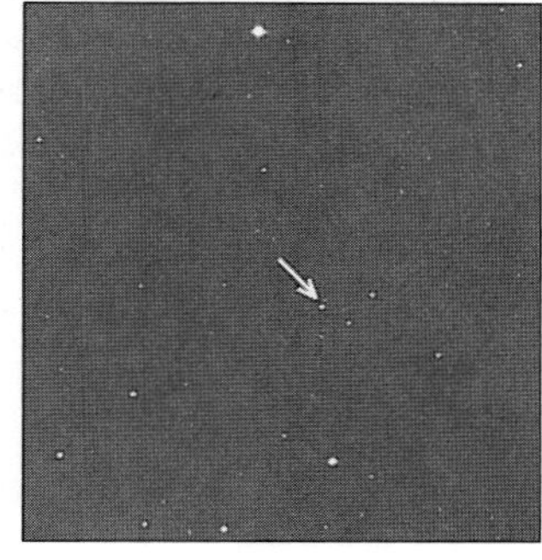

A

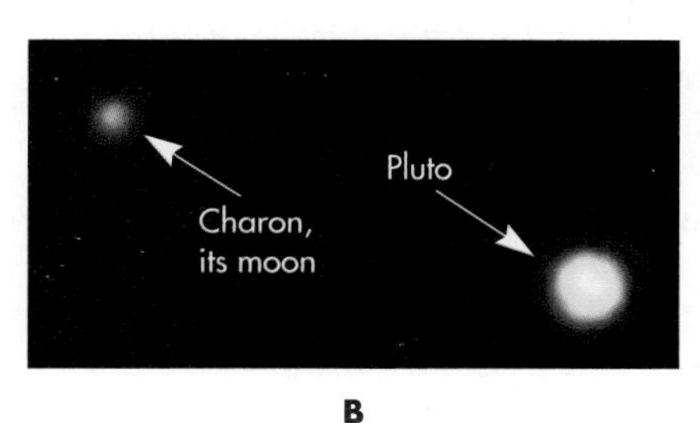

B

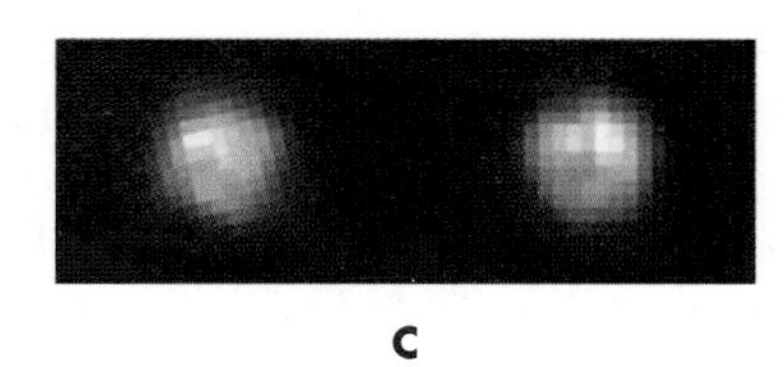

C

FIGURE 46.8
(A) Pluto looks like merely another dim star in this photograph taken at Lick Observatory in California. Only its change in position from night to night shows it to be orbiting the Sun. (B) A picture of Pluto and its moon Charon obtained with the Hubble Space Telescope—free of the blurring effects of our atmosphere. (C) Two pictures of Pluto made with the Hubble Space Telescope.

Two more tiny moons were discovered in 2005. They orbit Pluto two to three times farther from Pluto than Charon.

1930, after scanning millions of pairs of star images and searching for objects whose position changed between the exposures (Figure 46.8A), he discovered an object that seemed to be a good candidate for "Planet X.". The motion between images distinguishes an orbiting body in the Solar System from a star, and the rate of motion implied it was orbiting beyond Neptune. The object Tombaugh discovered was named **Pluto,** after the Greek and Roman god of the underworld.

Pluto's great distance from the Sun made it too dim and too small to measure its angular size. Even in the largest telescopes on the ground, Pluto looked like a dim star. For years after its discovery, astronomers had only a rough idea of Pluto's size. They had thought it must be massive, to explain the disturbance of Uranus's and Neptune's orbits, but careful reanalysis showed that actually there were no disturbances. The search for "Planet X" had been based on a mistake! In the decades following Tombaugh's discovery, the size estimates for Pluto were based on the amount of light it reflected from the Sun. This gave very uncertain results, because Pluto could be a small object that reflected most of the sunlight striking it or a large object reflecting only a small percentage of the light.

The mass of Pluto was not determined until 1978 when James Christy of the U.S. Naval Observatory discovered a moon orbiting Pluto. The moon, Charon (Figure 46.8B), is named for the boatman who, in mythology, ferries dead souls across the river Styx to the underworld. Charon takes just 6.4 days to orbit Pluto, and its orbit is tiny. Its average distance from Pluto is just 19,600 kilometers—only a few times bigger than the Earth's radius, so the entire orbit could fit inside the planet Neptune. From Charon's orbital data we can calculate Pluto's mass using Newton's modified form of Kepler's third law of planetary motion (Unit 17). Such a calculation indicates that Pluto's mass is about 0.002 times the Earth's, showing that Pluto is far less massive than any of the eight planets, and even less massive than Earth's Moon or the six largest satellites of the giant planets. The presence of a large satellite also helped explain why some observers had thought that Pluto was larger than it ultimately proved to be.

Pluto's diameter is a little less than one-fifth the Earth's diameter. Putting together the data on Pluto's mass and radius, astronomers find that its density is approximately 2.0 kilograms per liter—a value suggesting that it must be mainly a mix of rock and ice, quite similar to the large satellites of the giant planets. Images from the Hubble Space Telescope show very little detail on Pluto (Figure 46.8C). Pluto's south pole is brighter than its equator, implying the presence of a polar cap there. Spectra suggest that the cap is frozen methane and the rest of the surface is covered with a mixture of water ice and ices of methane, nitrogen, and carbon monoxide.

Astronomers have also detected a very tenuous atmosphere on Pluto. From measurements of Pluto's temperature and surface composition, they deduce that its atmosphere is mostly nitrogen and carbon monoxide with traces of methane. In the bitter cold (40 K, or about −390°F) of this remote part of the Solar System, molecules move so slowly that, despite its tiny mass, Pluto's gravity is strong enough to retain a thin atmosphere.

Pluto's elliptical orbit, which crosses Neptune's orbit, once led some astronomers to hypothesize that it was originally a satellite of Neptune that escaped and now orbits the Sun independently. Today, however, astronomers think almost the reverse—that Neptune has "captured" Pluto. Pluto's orbital period is 247.7 years, almost exactly 1.5 times Neptune's. Thus, at its orbital radius of 40 AU from the Sun, Pluto makes two orbits around the Sun for every three made by Neptune—a 2-to-3 resonance (see, for comparison, Unit 41). Neptune's strong gravitational attraction on Pluto may have tugged it into its current orbit. The arrangement is stable, and Pluto is in no danger of ever coming close to Neptune.

Pluto was called a planet for many decades after its discovery, but astronomers became increasingly troubled by this designation. It may seem surprising

Q Charon's orbit around Pluto is almost perpendicular to the plane of the Solar System (the ecliptic). Both Charon and Pluto keep the same face toward each other. How would Charon appear to an observer on Pluto in terms of phases and location in the sky?

that there was never a definition for "planet," but it is a term that comes from ancient times, long before formal definitions were developed. When Pluto was first discovered, it was thought to be a large, massive body, but it "shrank" over the decades—it took almost 50 years for a firm measurement of its tiny mass to be made. It became clear finally that Pluto was different from the eight major planets in almost every respect. Its mass was less than many moons' masses, its orbit was more elliptical and more tilted than any planet's orbit, and its orbit even crossed the orbit of another planet.

Despite the obvious differences between Pluto and the eight major planets, for more than 60 years it remained the only known inhabitant of the deep Solar System other than comets (Unit 47), so astronomers continued to call it a planet despite their misgivings. Pluto's discovery is really a tribute to Clyde Tombaugh's skills and perseverance. It was not until 1992, using new telescopes and techniques, that astronomers began to discover more objects beyond Neptune, even displacing Pluto as the most massive object. This ultimately led to the recognition that Pluto was just one of many icy worlds orbiting in the **Kuiper belt,** an orbiting swarm of objects similar to the asteroid belt. And as happened to the asteroids 150 years earlier, Pluto was demoted from the status of planet.

46.5 THE TRANS-NEPTUNIAN WORLDS

Since 1992, astronomers have detected more than a thousand objects orbiting beyond Neptune, now commonly known as **trans-Neptunian objects** or **TNOs**. The largest of these, named **Eris** (*AIR-iss*) after the Greek goddess of strife and discord, is about 27% more massive than Pluto. The name Eris is particularly appropriate because this object's discovery drove astronomers to "demote" Pluto from the status of planet, causing much discord among astronomers and the public.

With the discovery of so many objects orbiting beyond Neptune, Pluto's position began to resemble that of one of the large asteroids in the asteroid belt. History provided a precedent for this situation. The first four asteroids were discovered between 1801 and 1807 and were called planets in most textbooks for about four decades. Then beginning in 1845, new observations began revealing that there were many more asteroids, so in the 1850s astronomers decided on the label *asteroid* for this population of objects.

In 2006 the International Astronomical Union (IAU) met to decide on a definition of "planet" to settle the debate about Pluto's status. They decided that an object orbiting the Sun (and not itself a satellite) needs to meet two criteria to be classified as a planet. First, it has to be massive enough that its gravitational force can overcome its internal solid-body forces and pull it into a round shape. This criterion is met by even many of the moderate-size moons—those larger than about 400 kilometers (250 miles) in diameter are almost all spherical. Second, to be a planet, an object must have "cleared the neighborhood around its orbit." What this second requirement means is that there is substantially less mass of material sharing a similar orbit. The eight planets meet this criterion very readily, while none of the TNOs, including Pluto, comes even close.

Q Many people argued that Pluto should continue to be labeled a planet because it had become established as a planet for so many decades. Do you agree with this? When should popular or historical usage take precedence over creating a scientifically consistent nomenclature?

The IAU decided to call objects that do not meet the neighborhood-clearing condition, but that are otherwise orbiting the Sun on their own and have pulled themselves into round shapes **dwarf planets.** Only the largest asteroid, Ceres (Unit 41), also appears to meet this criterion, but it is possible that in addition to Pluto and Eris, dozens of the TNOs will achieve this designation once observations confirm their round shape. The IAU decided in 2008 that TNOs that are dwarf planets will be designated **plutoids**. Table 46.1 shows the properties of several of the largest TNOs, and Figure 46.9 shows the orbits of the twelve largest TNOs besides Eris and Pluto.

TABLE 46.1 Large Trans-Neptunian Objects

Name	Diameter	Mass	Semimajor Axis	Orbital Eccentricity
Eris	2600 km	1.7×10^{22} kg	67.8 AU	0.44
Pluto	2310 km	1.3×10^{22} kg	39.4 AU	0.25
2005 FY9	~1600 km	?	45.8 AU	0.16
Sedna	~1500 km	?	526. AU	0.85
2003 EL61	~1900 × 1000 km	4.2×10^{21} kg	43.3 AU	0.19
Charon	1210 km	1.5×10^{21} kg	39.4 AU	Satellite of Pluto
Quaoar	~1000 km	?	43.6 AU	0.038
Orcus	~950 km	7.5×10^{20} kg	39.4 AU	0.22

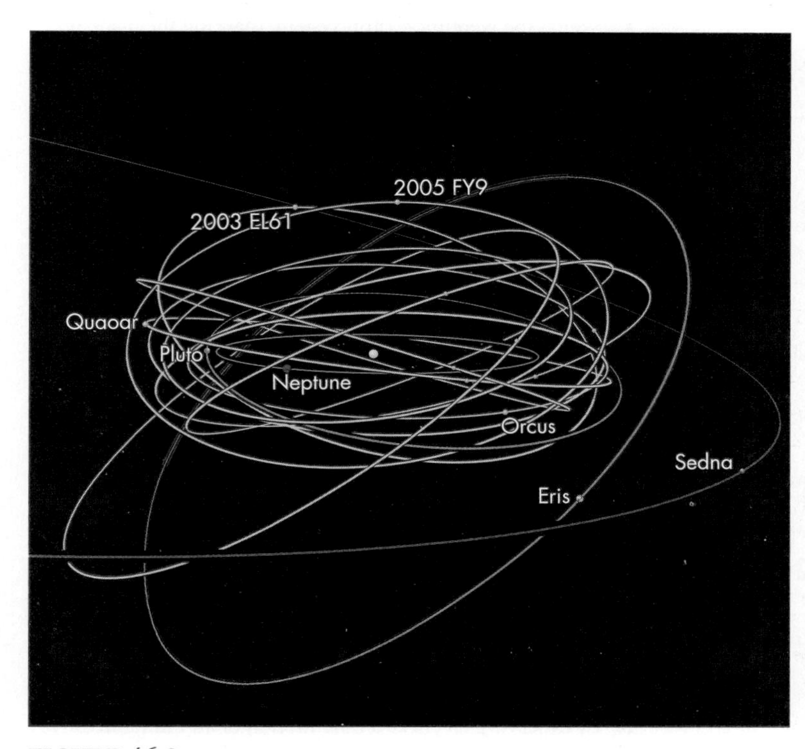

FIGURE 46.9
Orbits of the largest trans-Neptunian objects. The orbits of twelve of the largest known TNOs are shown in addition to Eris and Pluto.

Demonstrating that these objects have round shapes will be challenging because of their great distance, so proving that they are plutoids will be challenging.

The TNOs include objects orbiting at a wide range of distances from the Sun. It is becoming clear that there are classes among them. For example, the members of one category, termed **plutinos,** have orbits similar to Pluto's, with a 2-to-3 resonance with Neptune. There are several hundred plutinos. The largest after Pluto is *Orcus* (*OR-kuss*), with an estimated diameter of about 1500 kilometers (950 miles), about two-thirds of Pluto's diameter.

Other icy worlds lie farther beyond Neptune. Detecting these more distant objects is difficult because they are so dimly lit. Eris, for example, has an elliptical orbit that stretches from about 38 AU to about 98 AU from the Sun. Another object, named **Sedna,** resides extremely far from the Sun. Its highly elliptical orbit carries it from 76 AU (more than twice Neptune's distance from the Sun) to 928 AU from the Sun. Many of these objects have unusual characteristics that are not yet fully understood. Sedna, for example has a color nearly as red as Mars. Another object, 2003 EL61 (not yet officially named), may be spinning so rapidly that it bulges out nearly twice as far at its equator as at its poles.

In recent years astronomers have found that several of these distant objects have companions orbiting around them. Just as for Pluto, this orbital motion permits astronomers to determine the masses of these objects. From the mass, and with an estimate of the radius of the objects, astronomers can calculate their density. Typically the density is about 1 or 2 kilograms per liter, supporting the idea that these objects are similar to Pluto in composition.

It is difficult to estimate how many more large TNOs will be discovered in the outer Solar System as technologies improve. Sedna is currently near its closest approach to the Sun. If it were in the outer part of its orbit, it would not have been detected. We have to wonder how many more objects like Sedna and Eris, or even larger ones, might reside in the outer Solar System. The icy worlds found so far may indeed be just the "tip of the iceberg."

KEY TERMS

QUESTIONS FOR REVIEW

1. Which ice planets have impact craters on them? Which do not? What are the causes of the differences?
2. What are the satellites of the giant planets thought to be composed of?
3. What sort of activity has been seen on Io? What is Io's heat source thought to be?
4. Describe the evidence of a possible liquid ocean underneath the icy surface of Europa.
5. Jupiter's moon Ganymede and Saturn's moon Titan are similar in size. Why does Titan have a thick atmosphere while Ganymede has none?
6. Why do astronomers think that Triton may be a TNO that was captured by Neptune?
7. Why did the discovery of a moon orbiting Pluto cause astronomers to question its status as a planet?
8. What is a TNO? What is a dwarf planet?

PROBLEMS

1. Io, Europa, and Ganymede orbit at distances from Jupiter's center of 421,800 km, 671,100 km, and 1,070,400 km, respectively. Calculate how many times longer it takes for Europa and Ganymede to orbit Jupiter than for Io to orbit. (Hint: You do not need to know Jupiter's mass to calculate the ratio of times.)
2. Use the modified form of Kepler's third law (discussed in detail in Unit 17) to calculate Pluto's mass from the orbital data for Charon given in the text. Be sure to convert the orbital period to seconds and the orbital radius to meters before putting those numbers into the formula.
3. Calculate the density of Charon, given that its radius is approximately 593 km and its mass is about 1.1×10^{21} kg. (Be sure to convert kilometers to meters, and remember that there are 1000 liters per cubic meter.) Is it likely that Charon has a large iron core? Why?
4. Calculate the angular size of Pluto in arc seconds as seen from Earth. Pluto's diameter is 2320 km and its distance is about 39 AU.
5. If Orcus has three-fourths of Pluto's radius and the same density, how many times smaller is its mass?
6. Triton's orbit is tilted about 157° with respect to Neptune's equator, while Neptune's equator is tilted about 28° with respect to the plane of its orbit around the Sun.
   a. Sketch the Neptune-Triton system, marking the plane of Neptune's orbit and setting up the tilts and inclinations correctly.
   b. Tidal forces cause Earth's Moon to gradually spiral outward. Draw a diagram similar to the one shown in Unit 19, Figure 19.5, for Earth's Moon to explain why Triton will instead spiral inward.
   c. Is it possible that any of the giant planets might have previously "swallowed" a moon? Explain your reasoning.
7. Figure 46.6a shows a picture of about 10 kilometers of Titan's surface, taken by the *Huygens* probe as it descended to Titan's surface. Find a picture of river valleys on Earth that looks similar to the Titan image. What differences do you find between the feature seen on Titan and the most similar-looking features on Earth? Is this comparison strong enough evidence to state that Titan has rivers? Support your answer.

TEST YOURSELF

1. How was the mass of Pluto determined?
   a. By measuring the effect of its gravity on the terrestrial planets
   b. By observing its effect on the motion of an unmanned flyby of Pluto
   c. By determining the orbit of its satellite, Charon
   d. By measuring how it bends light rays passing near it
2. The volcanic activity of Io is a consequence of
   a. tidal heating.
   b. radioactive materials in Io's interior.
   c. frequent meteoroid impacts.
   d. frictional heating due to atmospheric winds.
3. In what respect are the atmospheres of Titan and Earth most different?
   a. Chemical composition
   b. Presence of clouds
   c. Density
   d. Temperature

unit

47 Comets

A bright comet is a stunning sight, as you can see in Figure 47.1. Comets have long been feared and revered, and their sudden appearance and rapid disappearance after a few days or weeks of gracing the sky have added to their mystery. Many comets have made only a single appearance in recorded history, whereas others return at regular intervals; **Halley's comet**, for instance, reappears about every 76 years. Throughout history, comets have been blamed for plagues, or believed to foretell disasters, or taken as a sign to go to war.

Comets are generally named for their discoverer or for the year they were first seen. Halley's gets its name because Sir Edmund Halley was the first to propose that some comets move around the Sun like planets. Using methods developed by Newton, Halley found that comets seen in 1531, 1607, and 1682 had similar elliptical orbits. He predicted these were a single object that would reappear in 1759. It did, but Halley did not live to see his prediction verified.

Today we recognize that comets give us a glimpse of the most remote parts of the Solar System. Many comets appear to originate from distances greater than 10,000 AU from the Sun. Comets are probably some of the most primitive bodies in the Solar System, perhaps surviving unchanged from when the Solar System formed.

FIGURE 47.1
Photograph of Comet McNaught as seen from Australia in January 2007.

47.1 STRUCTURE OF COMETS

There is uncertainty about how to pronounce *Halley.* Most astronomers say it to rhyme with *Sally,* although there is some evidence that Halley himself pronounced his name as *haw-lee.*

Comets consist of two main parts, as illustrated in Figure 47.2. The largest part is the long tail, a narrow column of dust and gas that may stretch across the inner Solar System for as much as 100 million kilometers (nearly an AU).

The tail emerges from a cloud of gas called the **coma,** which may be some 100,000 kilometers (60,000 miles) in diameter (roughly 10 times the size of the Earth). However, despite the great volume of the coma and the tail, these parts of the comet contain very little mass. The gas and dust are extremely tenuous, so a liter of the gas contains only a few million atoms and molecules. (In contrast, the air we breathe contains about 10^{22} molecules per liter.) By terrestrial standards, this would be considered a superb vacuum. This extremely rarified gas is matter that the Sun's heat boils off of the heart of the comet, its nucleus.

A typical comet **nucleus** is about 10 kilometers (6 miles) in diameter, made of frozen gases along with a mix of loose rock and dust. The nucleus of a comet has been described as a giant **"dirty snowball,"** and it contains most of the comet's mass. Only a few comet nuclei have been observed up close, the first being the nucleus of Comet Halley, which was observed by the *Giotto* spacecraft during its passage close to the Sun in 1986. The *Giotto* spacecraft was launched by the European Space Agency (ESA) as part of an international study of Halley. It traveled through Halley's coma, approaching to within 600 kilometers of Halley's nucleus. From images sent back to Earth we could measure the size of the nucleus.

The *Giotto* spacecraft was named in honor of the Italian artist who painted a portrait of a comet, possibly depicting the 1301 appearance of Halley's comet, as part of a Christmas scene for a church altar.

Despite its icy composition, the nucleus is extremely dark, as you can see in Figure 47.3A, which is one of the last pictures made by the *Giotto* spacecraft before particles from the comet damaged the camera. Astronomers think that the dark color comes from dust and carbon-rich material coating the surface of the nucleus. Other visible features of the nucleus are its irregular shape and the jets of gas erupting from the frozen surface. The jets form when sunlight heats and vaporizes the icy material.

Figure 47.3B shows a close-up image made by the *Stardust* spacecraft passing by Comet Wild (pronounced *vilt*) 2. The pitted surfaces and irregular shapes of these comet nuclei are caused in part by uneven melting of the nuclei during passage by the Sun on previous orbits.

In mid-2005, NASA's *Deep Impact* mission sent a 370-kilogram (820-pound) probe into a direct collision with Comet Tempel 1. The speed of the impactor

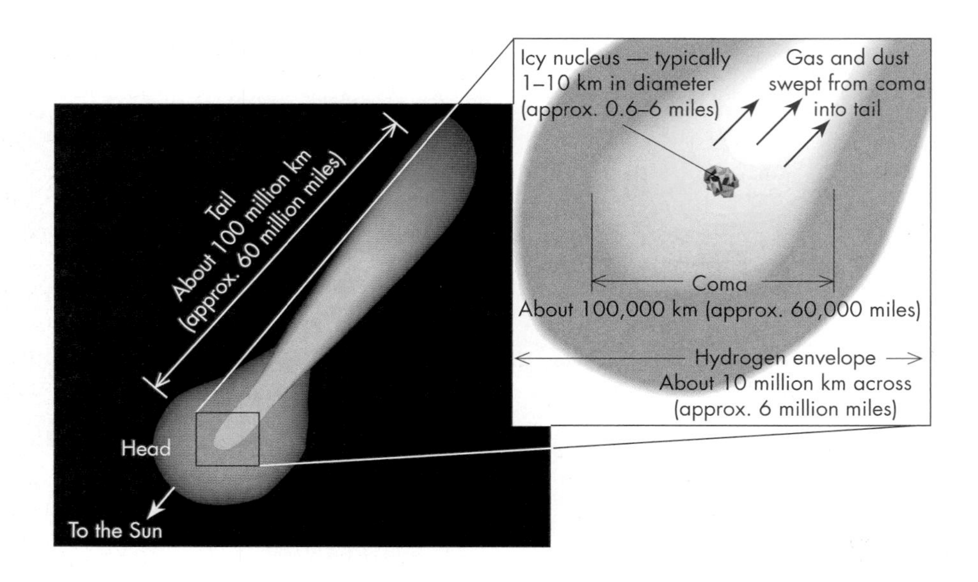

FIGURE 47.2
Artist's depiction of the structure of a comet, showing the tiny nucleus, surrounding coma, and long tail.

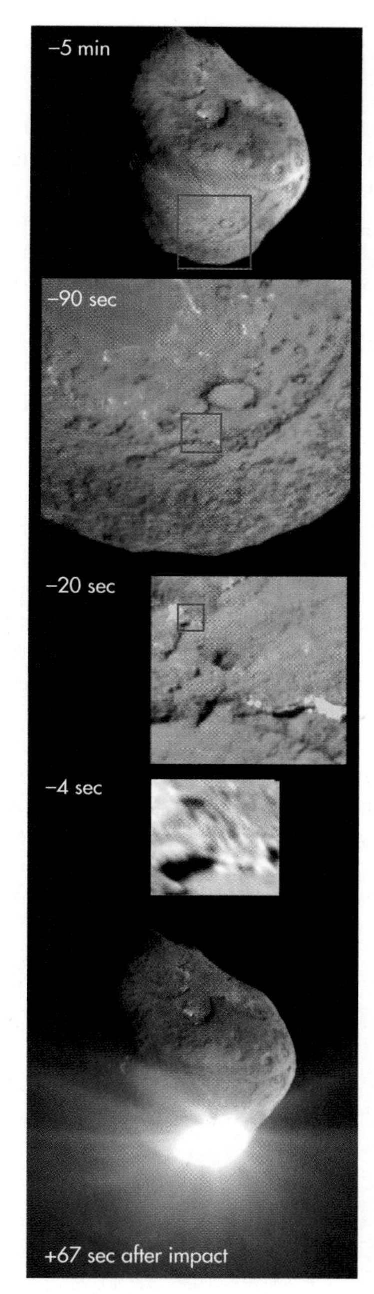

FIGURE 47.4
Images made by the *Deep Impact* mission. The impactor probe took the top four pictures during the final minutes and seconds before it collided with Comet Tempel 1 at a speed of about 10 kilometers per second (6 miles per second). The comet is about 6 kilometers (about 4 miles) across. The impact produced a bright flash, and a plume of dusty material expanded outward from the comet, catching the sunlight as seen in the bottom picture, made about a minute after the collision by the main *Deep Impact* spacecraft.

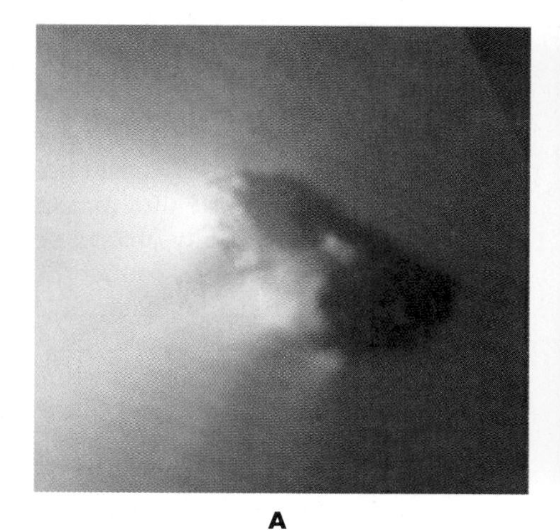

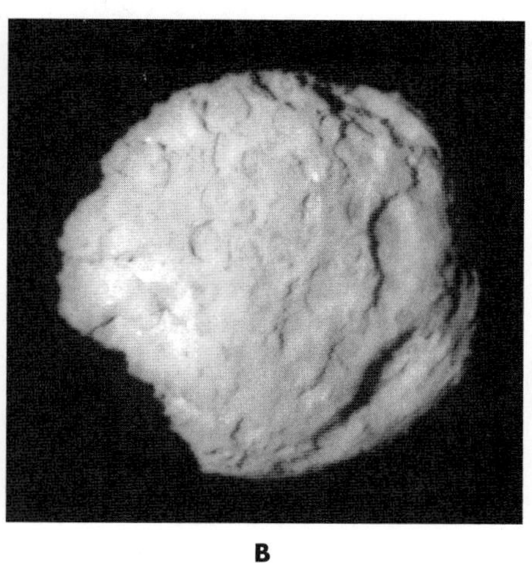

FIGURE 47.3
(A) Picture made by the *Giotto* spacecraft of the nucleus of Halley's comet, which is about 15 kilometers (9 miles) long. The spacecraft was approximately 1200 kilometers (750 miles) from the nucleus—deep inside Halley's coma of gas—when this picture was made. (B) The nucleus of Comet Wild 2, which is about 5 kilometers (3 miles) in diameter. Note the many craters. The image was made by NASA's *Stardust* spacecraft, which collected samples of dust in the comet's coma.

probe relative to the comet was about 10 kilometers per second, creating a collision with the explosive force of about five tons of TNT. Although this sounds like it might have had devastating effects on the comet, the impact was only about the equivalent of a pebble striking a speeding truck.

The impactor probe sent images of the comet until a few seconds before the collision (Figure 47.4). These images show features that look like impact craters on the comet, as well as irregular craters that may be sites where volatile ices have boiled away. By having the probe collide with the comet, scientists were able to examine material blasted out from the comet. The main spacecraft continued past the comet and photographed the impact (bottom image of Figure 47.4) and carried out spectroscopic analyses of the material ejected by the comet. One surprise was that the collision produced an unexpectedly large quantity of fine dust, about the consistency of talcum powder. It is estimated that the impact blasted out a crater about 30 meters deep—deeper than most astronomers expected. This suggests that the comet may be quite loosely constructed or even "fluffy," with a thick layer of dust on its surface.

From the sizes and estimates of the masses of the comet nuclei, their densities can be estimated. This turns out to be in the range of approximately 0.3 to 0.8 kilograms per liter—less than the density of water ice, a value implying that the icy material of the nucleus is loosely packed like snow, not compacted like pure ice. Alternatively, the comet's nucleus may be a "rubble pile" of loose smaller chunks of ice and rock rather than a solid chunk of icy matter. Unfortunately, the mass estimates are difficult to make and uncertain, so the density we infer from them is also uncertain.

The escaped gas from comets offers astronomers a way to probe the comets' composition. Spectra of gas in the comas and tails show that comets are rich in water, CO_2, CO, and small amounts of other gases that condensed from the primordial solar nebula. Evaporating water is broken up by solar ultraviolet radiation to create oxygen and hydrogen gas, and most comets are surrounded by a vast cloud of hydrogen created in this way. If a comet passes by the Sun too often, the escape of gas eventually erodes the comet away, leaving behind only the less volatile materials, perhaps resembling an asteroid.

47.2 ORIGIN OF COMETS

Based on the portion of their orbits that can be observed, astronomers think that comets may come from the **Oort cloud,** a swarm of trillions of icy bodies thought to lie beyond the orbit of Pluto. Astronomers think the Oort cloud formed from planetesimals that originally orbited near the giant planets and were tossed into the outer parts of the Solar System by the gravitational force of those planets. There they form a roughly spherical cloud that completely surrounds the Solar System and extends to about 100,000 AU from the Sun, as shown in Figure 47.5. Astronomers deduce this shape for the Oort cloud from the many comet orbits that are highly tilted with respect to the main plane of the Solar System.

The Oort cloud and Kuiper belt

Some comets, however, come from a flatter, less remote region—the **Kuiper** (*Ky-per*) **belt,** also shown schematically in Figure 47.5. The Kuiper belt begins at about the orbit of Neptune, 30 AU from the Sun, and seems to thin out near 55 AU from the Sun, but its outer boundary is uncertain. Many much larger bodies, such as Pluto, Eris and other dwarf planets (Unit 46), also orbit in the Kuiper belt. Their gravitational interactions may be what sends smaller bodies in the Kuiper belt into the inner Solar System to become comets.

Comet nuclei from the Oort cloud may take millions of years to complete an orbit. With orbits so far from the Sun, these icy bodies receive virtually no solar heating, and calculations indicate that their temperature when so far from the Sun is a mere 3 K (−454°F). Thus, their gases and ices remain deeply frozen.

Such cold and distant small objects are invisible to us on Earth; so generally we can see them only if they travel into the inner Solar System. Astronomers think that the gravitational influence of stars neighboring the Solar System may affect the orbits of bodies in the Oort cloud, altering their paths and causing some to drop in toward the inner Solar System. A single disturbance may shift enough orbits to supply comets to the inner Solar System for tens of thousands of years.

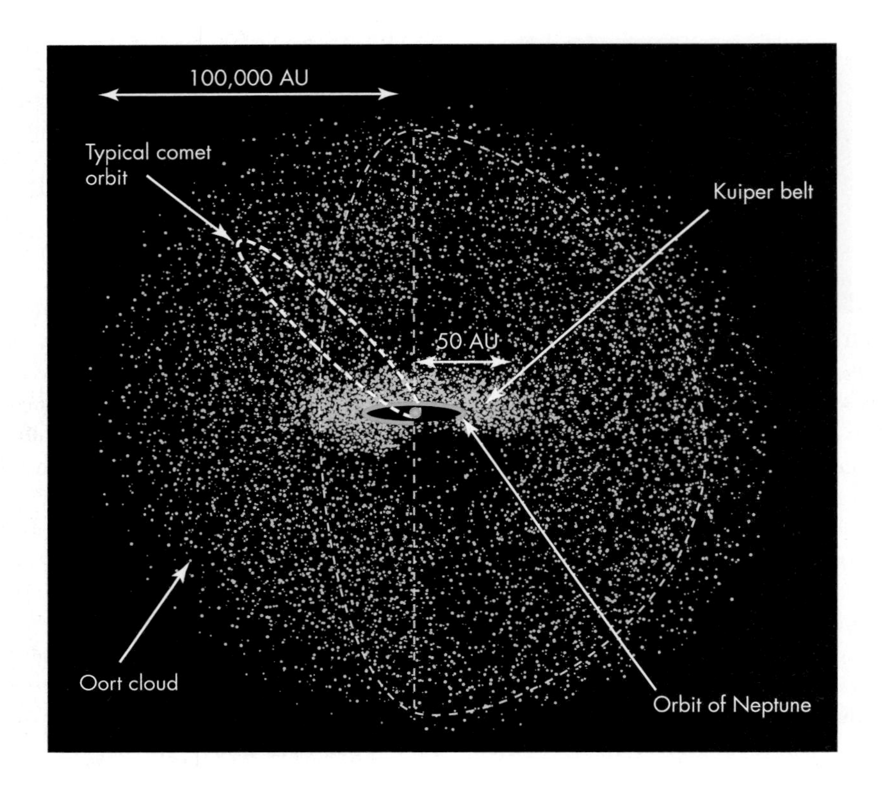

FIGURE 47.5
Schematic drawing of the Oort cloud, a swarm of icy comet nuclei orbiting the Sun out to about 100,000 AU. Also shown is the Kuiper belt, another source of comet nuclei. (The scale is exaggerated for clarity.)

As an icy body falls inward toward the inner Solar System, the Sun's radiation heats it and the ices near the surface begin to **sublimate,** passing directly from a solid to a gaseous state in the vacuum of space. At a distance of about 5 AU from the Sun (Jupiter's orbit), the heat is enough to sublimate the ices of water, methane, ammonia, and other volatile compounds, turning them into gases that escape to form the coma. The escaping gas carries tiny dust grains that were frozen into the nucleus with it. The "dirty snowball" then appears through a telescope as a dim, fuzzy ball—the initial stage of a comet. As the comet falls ever nearer the Sun, its gases sublimate even faster.

Discovering comets is one of the areas in which amateur astronomers have made major contributions to astronomy. Finding a comet before it becomes bright requires careful scanning of large amounts of the sky. One reward is that a comet is named for the person who discovers it. Most comets never brighten enough to be seen without a telescope. However, every few years a comet is discovered that grows to spectacular size, when it travels so close to the Sun that gas sublimates vigorously and the Sun exerts additional forces dragging long streams of gas and dust out of the comet.

47.3 THE COMET'S TAILS

Orientation of comet tails

Sunlight striking dust grains imparts a tiny force to them, a process known as **radiation pressure.** We do not notice radiation pressure when sunlight falls on us, because the force is tiny and the human body is far too massive to be shoved around by solar photons. However, the microscopic dust grains in a comet's coma do respond to radiation pressure and are pushed away from the Sun, as illustrated in Figure 47.6. Because all the grains move in the same direction, away from the Sun, a tail begins to form.

The tail pushed out by radiation pressure is made of dust particles, but Figure 47.7 reveals that comets often have a second tail. That tail is made of gas and ions from the comet pushed by an outflow of gas that streams from the Sun into space, a flow called the **solar wind** (Unit 49). The solar wind blows away from the Sun at about 400 kilometers per second. It is very tenuous, containing only a few atoms per cubic centimeter. But the material in the comet's coma is also extremely tenuous, so the solar wind is able to push it outward into a long plume.

Magnetic fields carried along by the solar wind enhance its effect on electrically charged particles—ions—in the gas from the comet, helping to drag matter out of

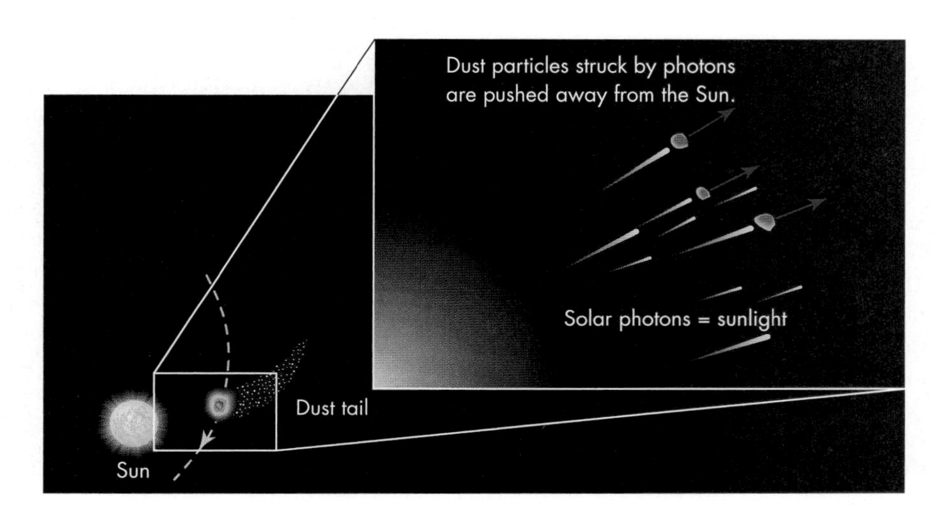

FIGURE 47.6
Sketch of how radiation pressure pushes on dust particles. Photons hit the dust, and their impact drives the dust away from the Sun, forming a dust tail. (Sizes and distances are not to scale.)

FIGURE 47.7
Photograph of Comet Hale-Bopp in the evening sky, showing two tails, one of ions, one of dust. These two tails point in slightly different directions because the ion tail is driven away from the Sun by the extremely fast-moving solar wind, so it points nearly directly away from the Sun. The dust particles are more massive and remain in orbit around the Sun, but radiation pressure produces an outward force that partially counterbalances gravity. As if orbiting in a weaker gravitational field, the dust particles follow orbits that swing wider around the Sun, sailing farther outward from the Sun than the comet nucleus. Therefore the dust tail also extends in a direction away from the Sun, but often along a curved path.

the coma and channel its flow, just as magnetic fields in the Earth's atmosphere channel particles to form the aurora. Thus radiation pressure and the solar wind act respectively on dust and gas in the comet's coma to drive out two tails. The forces act slightly differently, so the **dust tail** and **ion tail** are usually slightly separated from each other. Because those forces are directed away from the Sun, the tails always point away from the Sun—and the tails may even point out ahead of the comet as it moves away from the Sun (Figure 47.8).

It might help you understand this seemingly odd phenomenon if you think of a runner carrying a torch. If the air is still, the smoke and flame from the torch will, of course, trail behind the runner. However, if a strong wind is blowing from behind the runner, the smoke will be carried along in the direction of the wind regardless of which way the runner moves. Likewise, the high velocity of the solar wind (400 kilometers per second versus about 40 kilometers per second for the comet) carries the tail of a comet outward from the Sun regardless of the comet's motion.

The gases and dust swept out into the tail are lit up by the Sun. The dust particles reflect sunlight, and the gases emit light of their own by a process called **fluorescence.** Fluorescence is produced when light at one wavelength is converted to light at another wavelength. A familiar example is the so-called black light that you may have seen for illuminating posters. Black light is really ultraviolet radiation that we have difficulty seeing because of its short wavelength and therefore high energy. When such ultraviolet radiation falls on certain paints or dyes, the chemicals in the pigment absorb the ultraviolet radiation and convert it into visible light.

Some laundry detergents have "whiteners" in them that fluoresce. These chemicals convert ultraviolet radiation into visible light and thereby make a white shirt or blouse look brighter. You can see this effect strongly if you shine a black light (ultraviolet light) on a freshly washed shirt in a darkened room.

A major part of a comet's light is created by fluorescence. A photon of ultraviolet radiation from the Sun lifts electrons in the atoms of the comet's gas molecules to an upper, excited level in a single leap. The electrons then return to their original levels in two or more steps, emitting a photon each time they drop to a lower energy level (Unit 24). Thus, the energy of each emitted photon must be less than that of the original ultraviolet one. That smaller energy then gives the emitted photons longer wavelengths, which we can see with our eyes as the glow of the comet's ion tail. In addition, by observing the precise wavelengths emitted in the spectrum of the fluorescing gas, we can identify what atoms and molecules are present and thereby deduce the comet's composition.

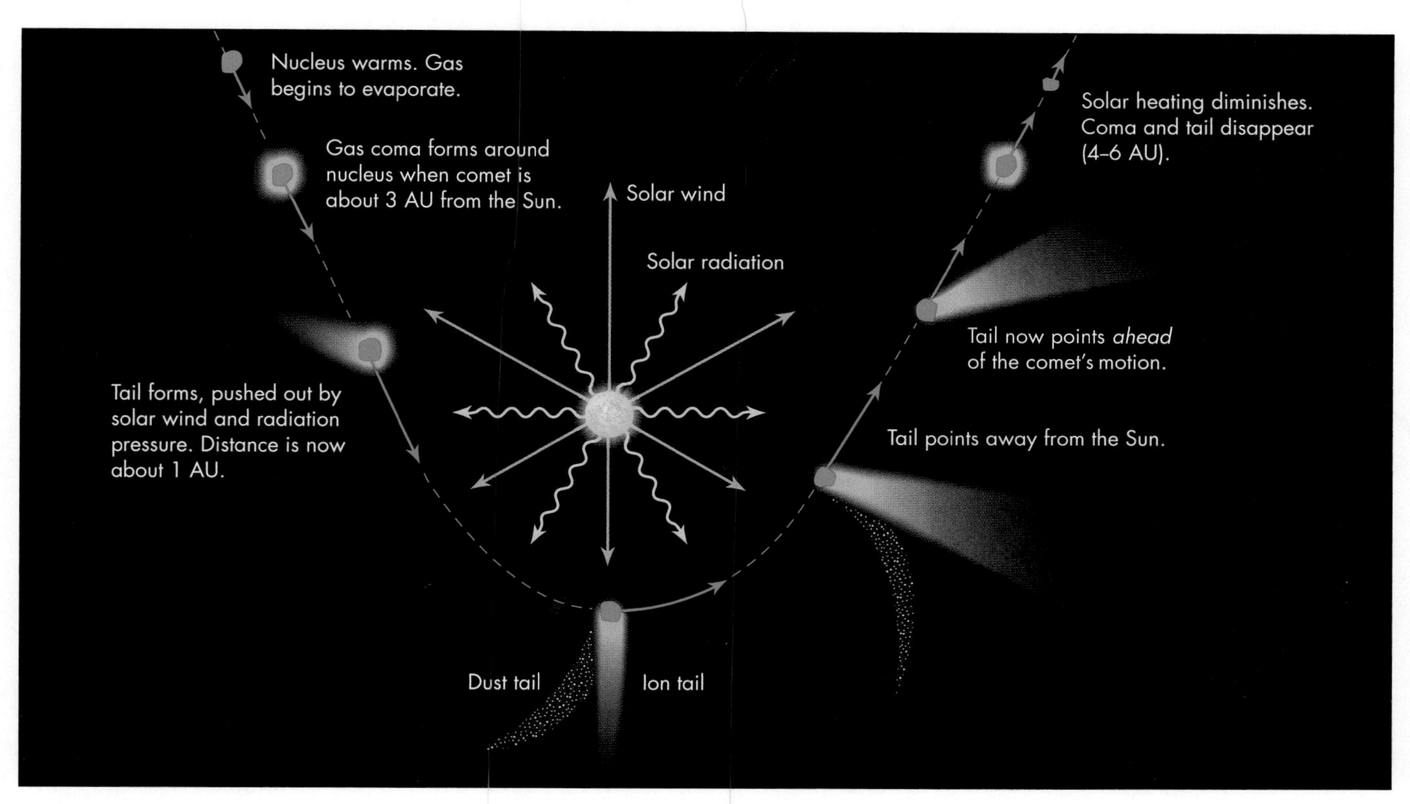

FIGURE 47.8
Sketch illustrating how radiation pressure and the solar wind make a comet's tail always point approximately away from the Sun. (Sizes and distances are not to scale.) Comets may orbit in any direction around the Sun.

47.4 SHORT-PERIOD COMETS AND THE KUIPER BELT

Although most comets that we see from Earth swing by the Sun on orbits that will bring them back to the inner Solar System only after millions of years, a small number of comets reappear at intervals of less than 200 years. These **short-period comets** include Halley's, which has a period of 76 years.

The origin of short-period comets is still under study. At one time it was thought that they come from the Oort cloud. While moving through the region of the Solar System containing the giant planets, their orbits might be shifted by a close encounter with one of the planets into smaller orbits with periods of centuries rather than millennia. Many astronomers now suspect that most short-period comets come from the Kuiper belt, because most of their orbits lie close to the plane of the Solar System and orbit in the same direction as the planets.

Comets that reach the inner Solar System are generally smaller than 10 kilometers. Even before astronomers began to detect large numbers of trans-Neptunian objects, they had hints that other, more massive objects were present in the outer Solar System. For example, in 1977 astronomers discovered an object, *Chiron,* with an orbit that stretches from just inside Saturn's orbit to almost the orbit of Uranus. This was a surprising discovery because orbits in this region are generally unstable, due to the gravitational perturbations of the giant planets. Astronomers thought that perhaps Chiron should be classified as an asteroid because of its unexpected location and very large size compared with known comets: 180 kilometers (about 110 miles) in diameter versus the more typical 10 kilometers. But Chiron has other properties that make it more similar to comets than to asteroids. In particular, it changes in brightness, sometimes flaring up and ejecting gas like a comet. It does not display the familiar comet's tail only because it remains too far from the Sun to be strongly affected by its heat and the solar wind.

Astronomers now estimate that the Kuiper belt may contain tens of thousands of objects 100 kilometers or larger in diameter, and millions as large as a typical comet. Chiron may have originated in this region, later to be deflected into its current orbit by a close passage by Neptune. Objects at this distance are still difficult to study because they are so dimly lit by the Sun, but we can extrapolate from the number of large objects to predict the number of smaller ones. The Kuiper belt probably contains a total mass hundreds of times larger than the asteroid belt between Mars and Jupiter, although it is spread out over a much larger volume of space. These frozen objects, like the rocky asteroids, are survivors of the Solar System's birth—icy planetesimals still orbiting in the disk—but are too widely scattered for gravitational attraction to bring them together to form a planet.

Q How would you expect the spent core of a comet to differ from an asteroid?

47.5 METEOR SHOWERS

The Sun steadily whittles away a short-period comet. With each close passage by the Sun, ices sublimate and carry off solid matter (dust and grit), leaving behind a smaller nucleus. This fate is like that of a snowball made from snow scooped up alongside the road, where small amounts of dirt have been packed into it. If such a snowball is brought inside, it melts and evaporates, leaving behind only the grit accidentally incorporated in it. So too, as a comet sublimates, it leaves behind in its orbit grit that can continue to circle the Sun long after the comet has lost so much of its ices that it no longer produces a visible tail.

If you go outside on a clear night after midnight, far from city lights with an unobscured view of the sky, you will see, on average, one "shooting star" or **meteor** every 15 or so minutes. Most of these meteors are tiny stray fragments of asteroids and comets that arrive at the Earth randomly (Unit 48) and then burn up in its atmosphere.

However, at certain times of the year, instead of one meteor per quarter hour or so, you may observe one every few minutes. Furthermore, if you watch such meteors carefully, you will see that they appear to come from the same general direction in the sky in what is called a **meteor shower.**

The best-known meteor shower is probably the Perseid meteor shower. It reaches its peak each year between August 10 and 14, when meteors rain into our atmosphere from a direction that lies toward the constellation Perseus. The meteors themselves have no association with Perseus. Rather, they are the debris of a comet (named Swift-Tuttle) and are following an orbit around the Sun that crosses the part of our orbital path that the Earth travels through in mid-August (Figure 47.9A).

Q Given how dust particles are affected by solar radiation pressure, where would you expect to find most of the debris from a comet many orbits after the comet expelled the material?

This encounter creates an effect similar to what you observe when you drive at night through falling snow: The flakes seem to radiate from a point in front of you, the location of which depends on a combination of the direction and speed of both the falling flakes and your car. Thus, while the Earth crosses the path followed by the cometary debris, the debris seems to diverge from a common point (Figure 47.9B), called the **radiant.** Meteor showers are generally named for the constellation from which they appear to diverge. Table 47.1 lists several of the brighter and more impressive showers and the date when they peak. Each shower therefore marks when the Earth crosses the path of the comet listed in the

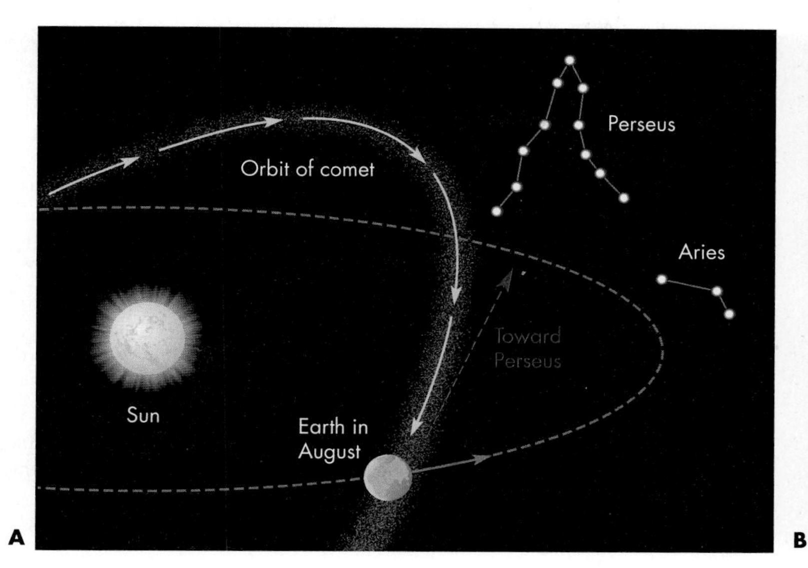

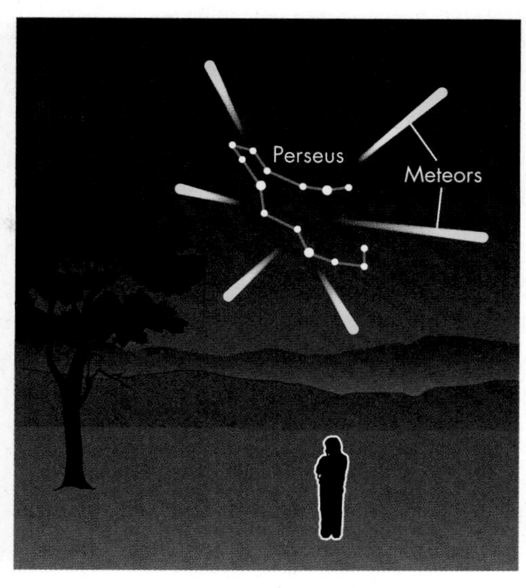

FIGURE 47.9
Sketch showing how (A) in mid-August, at the time of the Perseid meteor shower, the Earth is moving along its orbit. When the Earth crosses the debris strewn along a comet's orbit, the scattered material plunges into our atmosphere, producing (B) the diverging pattern of meteors characteristic of a meteor shower. (Bodies and orbits are not to scale.)

TABLE 47.1 Major Meteor Showers

Peak Date	Shower Name	Maximum Rate	Constellation	Comet
January 3	Quadrantids	30–90 per hour	Bootes	2003 EH1
April 22	Lyrids	10–15 per hour	Lyra	Thatcher
May 6	Et Aquarids	10–20 per hour	Aquarius	Halley
July 28	South Delta Aquarids	15–20 per hour	Aquarius	?
August 12	Perseids	40–100 per hour	Perseus	Swift-Tuttle
October 21	Orionids	15–20 per hour	Orion	Halley
November 17	Leonids	10–? per hour	Leo	Tempel-Tuttle
December 14	Geminids	40–90 per hour	Gemini	Phaethon
December 22	Ursids	10–15 per hour	Ursa Minor	Tuttle

FIGURE 47.10
This time-lapse picture shows meteors from the Leonid meteor shower on November 19, 2002.

table—sometimes even a "dead" one like Phaethon or 2003 EH1, which were only recently discovered.

On rare occasions the Earth passes through a dense clump of material left by a comet. When that happens, thousands of meteors per hour may spangle the sky. Such a display happened in November 2002 during the Leonid meteor shower (so named because it appeared to originate from the constellation Leo). At various spots around the world, observers reported periods with one meteor every second or two (Figure 47.10). This shower comes from the debris of the short-period comet Tempel-Tuttle, which orbits the Sun approximately every 33 years. Comet Tempel-Tuttle is now difficult to see because it has sublimated most of its volatile material in repeated passages around the Sun. The densest clumps of debris remain close to the comet nucleus; so every 33 years or so the meteor display is spectacular. In 1966 in western North America there were observations of a meteor shower, or storm, containing dozens of meteors *per second!* Beginning around 2030, the chances of meteor storms during the Leonid shower will increase once again.

KEY TERMS

coma, 376
dirty snowball, 376
dust tail, 380
fluorescence, 380
Halley's comet, 375
ion tail, 380
Kuiper belt, 378
meteor, 382
meteor shower, 382
nucleus, 376
Oort cloud, 378
radiant, 382
radiation pressure, 379
short-period comet, 381
solar wind, 379
sublimate, 379

QUESTIONS FOR REVIEW

1. Where do comets originate? What causes them to travel into the inner Solar System?
2. What is sublimation, and how does it differ from evaporation or boiling?
3. What creates a comet's tail? In what direction does the tail point?
4. Some comets have two tails—what is the difference between the tails?
5. What is a meteor shower?
6. Why are meteor showers named for a particular constellation?

PROBLEMS

1. Use Kepler's third law to find the semimajor axis of Halley's comet, given that its orbital period is 76 years.
2. With the result of problem 1 and the fact that for a very elliptical orbit the distance farthest from the Sun is roughly twice the semimajor axis, estimate how far from the Sun Halley's comet travels. What planet is about that same distance from the Sun? How does this distance compare with the distance to the Kuiper belt?
3. Use Kepler's third law to determine the period of a comet with an orbit extending from 1AU to 50,000 AU.
4. Use data from the text to estimate the total kinetic energy, in joules, of the impact of NASA's *Deep Impact* probe onto comet Tempel 1. If the comet has a mean diameter of about 6 km and a density of about 0.8 kg/L, and is orbiting at about 30 km/sec, what is its kinetic energy?
5. The temperature, T, of a body decreases as the square root of its distance, d, from the Sun increases: $T = \text{constant}/\sqrt{d}$. If the temperature at 1 AU is 250 K, estimate the temperature of a comet nucleus in the Oort cloud.
6. As a comet orbits the Sun, its temperature changes approximately as described in problem 5.
   a. Calculate the distance from the Sun at which it would reach the sublimation temperature of water (sublimates at $T =$ 180 K). Repeat for ammonia (130 K) and carbon dioxide (80 K).
   b. Based on your results in (a), describe how a comet would behave as it orbits the Sun.
   c. What would be the effect on the comet's temperature if it had dark regions on its surface? Qualitatively, how would this affect its behavior as it orbited the Sun? Explain your reasoning.

TEST YOURSELF

1. The tail of a comet
   a. is gas and dust pulled off the comet by the Sun's gravity.
   b. always points away from the Sun.
   c. trails behind the comet, pointing away from the Sun as the comet approaches it and toward the Sun as the comet moves out of the inner Solar System.
   d. is gas and dust expelled from the comet's nucleus by the Sun's heat and radiation pressure.
   e. Both *b* and *d*.
2. Meteor showers such as the Perseids in August are caused by
   a. the breakup of asteroids that hit our atmosphere at predictable times.
   b. the Earth passing through the debris left behind by a comet as it moved through the inner Solar System.
   c. passing asteroids triggering auroral displays.
   d. nuclear reactions in the upper atmosphere triggered by an abnormally large meteoritic particle entering the upper atmosphere.
   e. none of the above.
3. Astronomers think that most comets come from
   a. interstellar space.
   b. material ejected by volcanic eruptions on the moons of the outer planets.
   c. condensation of gas in the Sun's hot outer atmosphere.
   d. small icy bodies in the extreme outer parts of the Solar System that are disturbed into orbits that bring them closer to the Sun.
   e. luminous clouds in the Earth's upper atmosphere created when a small asteroid is captured by the Earth's gravitational force.

unit

48 Impacts on Earth

If you have spent even an hour looking at the night sky, you have probably seen a "shooting star," a streak of light that suddenly appears and just as quickly fades away (Figure 48.1). Astronomers call this lovely but brief phenomenon a **meteor.** What we observe as a meteor is actually the glowing trail of hot gas and vaporized debris left by a solid object heated by friction as it moves through the Earth's atmosphere at very high speed. Most meteors are produced by the bits and pieces of material that are released by a comet when it passes the Sun (Unit 47)—small debris that totally burns up about 50 kilometers (30 miles) up in the Earth's atmosphere. However, not all meteors are so small, and some may reach the Earth's surface. The solid body, while in space and before it reaches the atmosphere, is called a **meteoroid.** If it reaches the ground, it is called a **meteorite.**

The idea of stones falling from space seemed so strange that for centuries most scientists found the concept unthinkable, suggesting instead that they must be debris ejected from distant volcanoes. By the early 1800s the evidence that these were objects from space became convincing. Many witnesses reported seeing the dramatic fall of meteors and then collecting the associated meteorites, and many of these meteorites had properties unlike any geological rock specimen—some were giant chunks of pure iron, for example. But this still left the mystery of how and where these bodies formed, and what brought them to Earth. Astronomers think that most meteorites are fragments of asteroids, although a few meteorites are now convincingly identified to be fragments blasted out of the crust of Mars and the Moon.

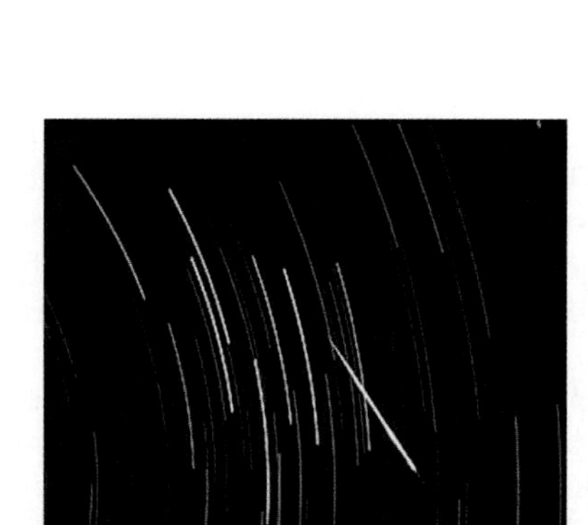

FIGURE 48.1
A time-exposure photograph captures a "shooting star" (meteor) flashing overhead. The curved streaks are star trails.

48.1 HEATING OF METEORS

Meteors heat up on entering the atmosphere for the same reason a reentering spacecraft does. When an object plunges from space into the upper layers of our atmosphere, it collides with atmospheric molecules and atoms. These collisions convert some of the body's energy of motion (kinetic energy) into heat, as illustrated in Figure 48.2. In a matter of seconds, the outer layer of the meteor reaches thousands of degrees Kelvin. The entry speeds are typically from 10 to 60 kilometers per second, depending on the relative directions of the orbits of the Earth and asteroid, which are each moving at about 30 kilometers per second around the Sun. The collisions with air molecules are extremely violent and tear electrons from the air molecules and vaporize material from the surface of the meteoroid. The hot, ionized gas surrounding the meteoroid produces the glow that we see.

If the meteoroid is larger than a few centimeters, it creates a large ball of incandescent gas around it and may leave a luminous or smoky trail. Such exceptional meteors, sometimes visible in daylight, are called *fireballs.*

Meteoroids bombard the Earth continually—a hail of solid particles that astronomers estimate at hundreds of tons of material each day. More strike between midnight and dawn (actually until noon, but they are not usually visible

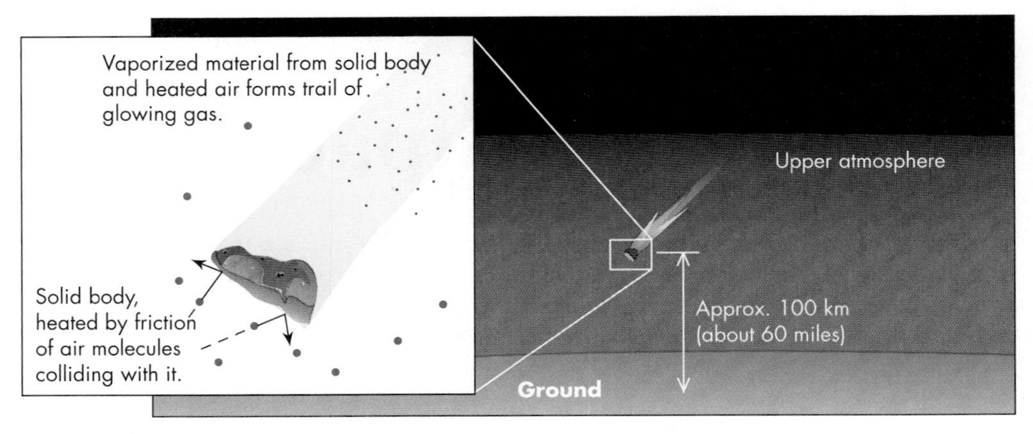

FIGURE 48.2
Sketch depicting how a meteor is heated by air friction in the upper atmosphere.

It may help you recall the meanings of the meteor-related terms if you remember that *meteoroid* rhymes with *void,* whereas the last two letters of *meteorite* are the first two letters of *terrestrial.*

in the daylight) than in the evening hours, so it is best to watch for meteors before dawn. This difference arises for the same reason that if you run through rain, your front side will get wetter than your back. That is, the dawn side of our planet advances into the meteoritic debris near us in space, while the sunset side moves away from it.

Most meteors that we see last no more than a few seconds and are made by meteoroids the size of a raisin or smaller. These tiny objects are heated so much that they completely vaporize. Larger pieces may survive the ordeal and reach the ground, with their surfaces scorched and partially vaporized, to be found as meteorites.

48.2 METEORITES

Most meteorites are small pieces of asteroids, broken off during collisions. Asteroids generally exhibit many impact craters (Unit 41), and because the gravitational attraction of the asteroids is weak, fragments blasted out by the impact can escape the asteroid and travel along new orbital paths around the Sun. Astronomers classify meteorites into three broad categories based on their composition: *stony* (that is, composed mainly of silicate compounds), *iron,* and *stony-iron.*

About 86% of all meteorites are **chondrites** (*KON-drites*). These are stony meteorites that contain small rounded bits of rocky material stuck together as shown in Figure 48.3A. These small spheres of rock are called **chondrules** (*KON-drools*), meaning "grain" in Greek. Experiments show that chondrules form when particles of silicate rocks are rapidly heated to form millimeter-size molten droplets that then quickly cool before they grow large. The source of the heating is not known, although there are many hypotheses. Chondrules may have been melted by outbursts from the young forming Sun, collisions between planetesimals, or the explosion of a nearby star—perhaps an explosion that triggered the collapse of the solar nebula. Radioactive ages (Unit 33) of chondrules indicate that they are some of the oldest material in the Solar System—slightly more than 4.5 billion years old.

In ordinary chondrites, the chondrules, along with flakes of metal, are embedded in a matrix of other rocky material that condensed from the solar nebula. This gives us a glimpse of what the inner parts of the solar nebula must have

FIGURE 48.3
Slices of various types of meteorites. (A) Chondrites are the main kind of meteorite. They contain small rounded chondrules that formed in the solar nebula. Carbonaceous chondrites have a matrix of carbon-rich material and probably condensed in the outer part of the asteroid belt. (B) Meteorites from asteroids that are large enough to have differentiated come in three main types: irons from the core of the asteroid, stony-irons from the core mantle boundary, and achondrites from the rocky surrounding material. The iron slice has been etched with acid to reveal the pattern of crystallization, which only develops when molten iron is allowed to cool gradually over millions of years.

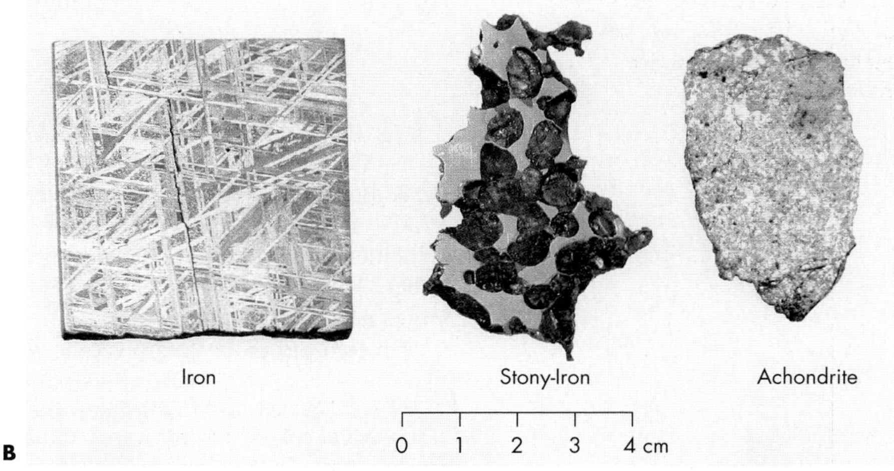

Q Depending on the different hypotheses about the source of melting of chondrules, would you expect chondrules to be more or less abundant in comets?

In astronomical usage, the term *organic* does not indicate that living organisms are involved.

looked like: a disk of hot gas full of these small rocky pebbles and metal flakes. In a small percentage of chondrites, the chondrules are embedded in a black, carbon-rich, coal-like substance (Figure 48.3A), and are therefore called **carbonaceous chondrites.** Spectroscopic comparisons indicate that these came from the carbon-rich asteroids in the outer part of the asteroid belt (Unit 41.3). This carbonaceous matter contains **organic compounds**—carbon-based molecules—as well as water and other volatile chemicals. This demonstrates that natural processes in space can produce some of the raw materials essential for life and that those materials would have been raining down on the Earth (in fact, continue to do so) after the young Earth cooled enough to form a solid crust.

Carbonaceous chondrites contain some fairly complex organic compounds, such as amino acids, which are used by living things to construct proteins (Unit 83). We can be fairly certain that the amino acids found in meteorites were not themselves produced through biological processes, however. When living things on Earth build amino acids, the molecules are always put together in what is called a "left-handed" form. Nonbiological processes produce both the left-handed forms and mirror-image "right-handed" forms in about equal numbers, and this 50-50 mix of left and right is what we find in meteorites.

Although the great majority of meteorites are stony, this is based on the statistics of meteorites that were collected after being observed falling ("falls"). If they were not observed falling, stony meteorites are not easily distinguished from Earth rocks, so most of the meteorites that are found ("finds") tend to be the more unusual iron meteorites (about 5% of falls) and the very rare stony-iron meteorites (about 1% of falls). Slices of these types of meteorites are shown in Figure 48.3B. Iron meteorites must come from asteroids so large that they melted and differentiated (Unit 33), allowing the dense iron to sink to the core. Stony-iron meteorites come from the boundary between the iron core and the rocky mantle. An asteroid probably has to be at least 20 to 50 kilometers in diameter to form an iron core. It then has to be smashed apart in a collision with another large asteroid in order to scatter the metal fragments in space.

The final 8% of meteorites are stony, but are **achondrites,** meaning that they have no chondrules (Figure 48.3B). The achondrites are rocky material from the upper layers of differentiated bodies. The metal flakes are gone, having melted and sunk to the center of the asteroid, and the rock shows evidence of melting and undergoing geological processes. Spectroscopic and chemical studies have even been able to link some of the achondrites to particular large asteroids or even Mars and the Moon. Evidently, some impacts are powerful enough to blast out pieces of one planet's crust and send it colliding into another planet.

48.3 THE ENERGY OF IMPACTS

Based on the number of impact craters found on Earth, every few thousand years a large meteoroid, tens of meters or more in size, strikes our planet. The impact velocities of tens of kilometers per second are more than 10 times faster than a rifle bullet. At such high velocities, meteoroids have a large kinetic energy—that is, a large energy of motion.

A meteoroid more than several centimeters in size may not burn up completely during its passage through the atmosphere, so its remaining kinetic energy is released when it hits the ground. The energy released from large meteoroids can be huge. To calculate the energy released by an object striking the Earth, we use the formula for an object's kinetic energy, E_K, which is given by the expression

$$E_K = \frac{1}{2} m \times V^2$$

where m is its mass and V is its velocity (Unit 20).

As an example, consider the meteorite impact depicted in Figure 48.4. About 100,000 kilograms (100 tons) of iron struck the Earth in Russia in 1947. This was not an especially large meteoroid, since 100 tons of iron would fit in a sphere less than 3 meters (10 feet) in diameter. Supposing it struck the Earth at 30 kilometers per second (= 3×10^4 meters per second = 108,000 kilometers per hour, or almost 65,000 miles per hour), the kinetic energy of impact was

$$\frac{1}{2} \times 10^5 \text{ kg} \times (3 \times 10^4 \text{ m/sec})^2 = 4.5 \times 10^{13} \text{ joules.}$$

This is about as much energy as is released when 10,000 tons of dynamite are exploded, or approximately

FIGURE 48.4
An iron meteorite from the Ural Mountains in Russia. Approximately 100 tons of iron struck the Earth with an impact energy comparable to the atomic bomb dropped on Hiroshima. A commemorative stamp depicts the fireball and smoke trail from a painting by an eyewitness.

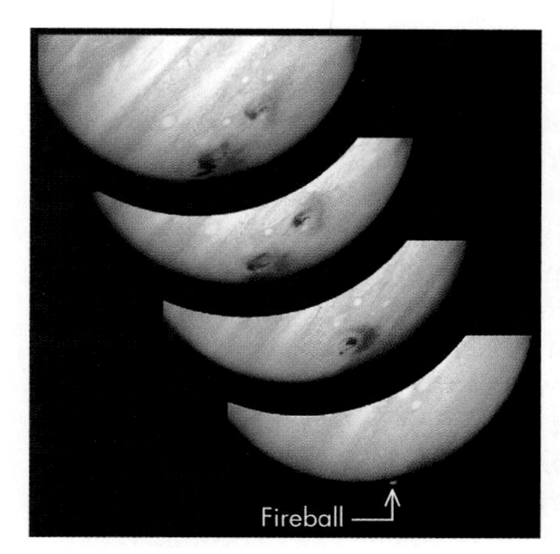

FIGURE 48.5
The impact of Comet Shoemaker-Levy 9 with Jupiter in 1994 was observed with the Hubble Space Telescope. This series of images shows the evolution of the impact of one of the fragments of the comet. The bottom image was taken a few moments after the impact, which occurred on the night side of Jupiter. A fireball rose over 3000 kilometers (1900 miles) above the cloud tops. When the impact site rotated into view about 1.5 hours later (second image from bottom), dark material blasted up from the impact site became visible. The top two pictures show the dispersal of the material over the next several days, along with additional impact sites from other fragments of the comet.

the same as the explosive power of the nuclear bomb that destroyed Hiroshima. Events of similar strength are found every few years by Earth monitoring satellites, which detect the atmospheric fireball from these impacts as a brilliant flash of light. Most of these events occur over the ocean or in unpopulated areas and may not be witnessed from the ground. Were such a body to hit a heavily populated area, it could be lethal.

Astronomers estimate that the meteoroid that produced the 1947 event may have originally had a mass up to ten times greater than the 100 tons of material that reached the ground. About 90% of the meteoroid was vaporized and most of the meteoroid's kinetic energy was released as it was torn apart and vaporized by its high-speed passage through the air. Meteoroids smaller than about 10 meters in diameter release most of their kinetic energy in the fireball produced during their passage through the atmosphere, but larger bodies retain most of their kinetic energy until they strike the Earth's surface

An impact of a comet with Jupiter was observed in 1994. Comet Shoemaker-Levy 9 had previously passed close to Jupiter, and tidal forces pulled it apart into a string of smaller fragments (Unit 45), each estimated to be a few kilometers in diameter. The impact of one large fragment, photographed by the Hubble Space Telescope, is shown in Figure 48.5. Dark "sooty" material generated in the blast made large dark areas as big as the entire Earth, traces of which lasted for months after the impacts. The energy released is estimated to have been the equivalent of about 6 million megatons of TNT. This is more than 400 times the total nuclear arsenal amassed by all countries during the Cold War!

Because a similar impact on Earth would have devastating consequences, astronomers have developed an international monitoring system to hunt for asteroids that might impact the Earth in the future. Such bodies might be deflected away from Earth by altering their orbit, but this requires early detection. Relatively little force could deflect the object if this is done early enough, but the force required rapidly grows larger as the collision time grows nearer. Fortunately, we have been spared such disasters recently; but there have been some close calls and some truly horrific impacts in the distant past.

48.4 GIANT METEOR CRATERS

The **Barringer crater** is one of the first documented meteor impact features on Earth. It is in northern Arizona, about 40 miles east of Flagstaff, and is easily visible from the sky. Geological analysis of the fragments of the impacting body recovered there suggests the impact occurred about 50,000 years ago. The meteoroid is estimated to have been some 50 meters in diameter. Its impact vaporized tons of rock, which expanded and peeled back the ground, creating a crater about 1.2 kilometers across and 200 meters deep (Figure 48.6).

More recently, in 1908 an asteroid broke up in our atmosphere over a mostly uninhabited part of north-central Siberia. This **Tunguska event,** named for the region where it hit, leveled trees radially outward from the blast point to a distance of some 30 kilometers and broke windows up to 600 kilometers away. Eyewitnesses reported seeing a fireball brighter than the Sun traveling through the sky. The blast was followed by clouds of dust that rose to the upper atmosphere. Sunlight reflecting off this dust gave an eerie glow to the night sky for several days. According to some accounts, the blast killed two people. Casualties were few because the area was so remote.

There are false but persistent rumors that radioactivity has been detected at the Tunguska site.

Unfortunately, scientists did not visit the site until about two decades later (or no records of earlier visits survive), because of the political turmoil in Russia at that time. When they did reach the site, they found no crater, just the felled trees. Interestingly, some trees at the center of the damaged area were left standing

A

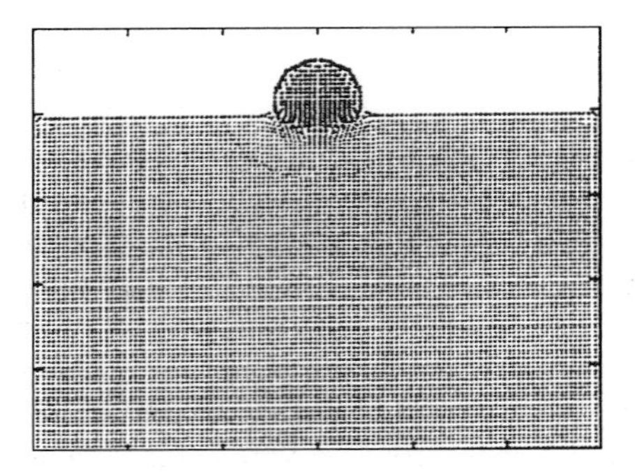

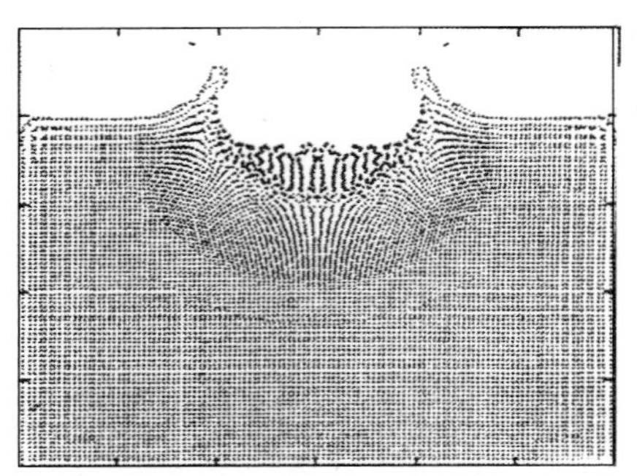

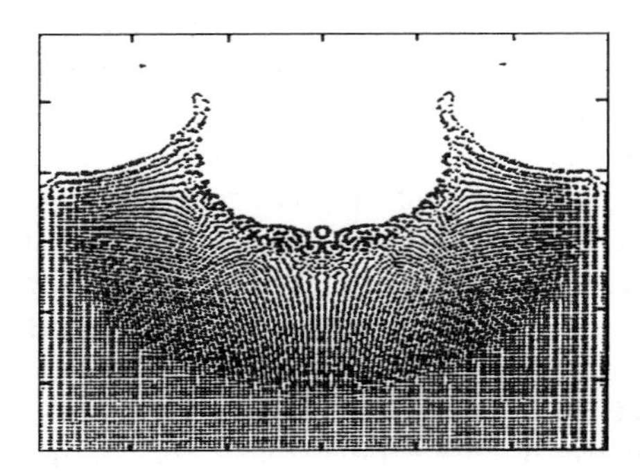

B

FIGURE 48.6
(A) Photograph of the Barringer meteor crater in Arizona. (B) A computer simulation of a crater's formation.

Q Why might some trees directly below the blast be left standing?

vertically but with their branches stripped off (Figure 48.7), suggesting that the explosion occurred in the air.

Hypotheses to explain the Tunguska event offer a good example of how science tests and retests ideas. The lack of a big crater and the absence of meteorites at the site at first seemed like evidence against the idea that it was caused by a small asteroid. This led some astronomers to propose that a comet might have caused the blast. Being icy, a comet would break up more easily—perhaps even while still in the atmosphere—so there would be no crater. Moreover, ice fragments would rapidly disappear, explaining the absence of large rock or iron meteorites.

However, computer simulations of how an asteroid travels at high speed through our atmosphere indicated what at first seemed a surprising result: even stony asteroids can create devastation without leaving a crater or telltale fragments. Depending on an asteroid's orbit relative to the Earth's, it will typically approach Earth at speeds between 10 and 40 kilometers per second. When it hits our atmosphere, it compresses and heats the air ahead of it. The hot compressed air obeys Newton's law of action–reaction and exerts a tremendous force back on the asteroid that may shatter it. The resulting fragments plunge deeper into our atmosphere, where air resistance heats them further until they vaporize, creating a fireball at a height of 10 kilometers (6 miles) or so above the ground. No traces of the asteroid necessarily survive to form a crater or fragments at the ground. All that remains is the 6000-K incandescent air of the fireball that blasts downward and out, crushing and igniting whatever lies below it, exactly as a nuclear bomb burst would.

Tunguska-like events may not be as rare as once thought. Evidence for comparable events in the 1930s has been found from accounts of fireballs and widespread damage in two remote regions of South America. Some geologists now suspect that some tsunamis (tidal waves) have been caused by impacts in the ocean. Evidence has been found of an impact crater off the coast of New Zealand from about 500 years ago, which left deposits of beach sand more than 200 meters above sea level in New Zealand and 100 meters above sea level in eastern Australia. This catastrophic event may explain why the Maori people abandoned many coastal settlements at that time, although this is debatable.

More impact scars have been found in many places on our planet. Wolf Creek crater in northwestern Australia is about 0.9 kilometers in diameter and formed in an impact estimated to have occurred about 300,000 years ago. Geological processes and erosion on Earth hide many older craters, although some remain evident. The ring-shaped Manicouagan Lake in Quebec is about 70 kilometers (approximately 43 miles) in diameter (Figure 48.8). It is a meteor crater estimated to be about 210 million years old, produced by a meteoroid approximately 5 kilometers in diameter. Other impact craters that are completely buried have been

FIGURE 48.7
The site of the Tunguska event photographed by an expedition in 1927.

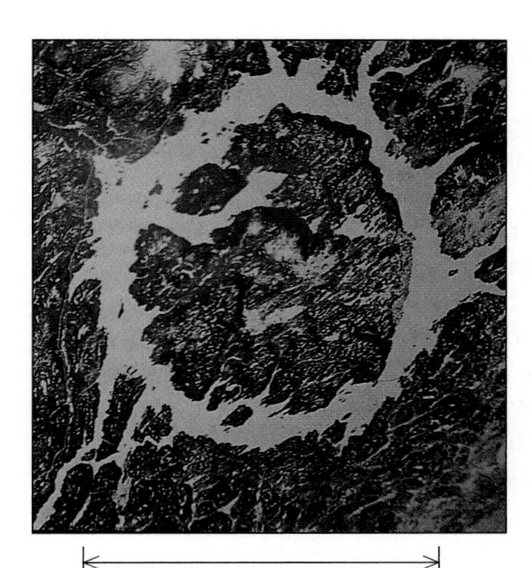

Approx. 70 km
(about 43 miles)

FIGURE 48.8
Picture (from Earth orbit) of the Manicouagan crater in Quebec. This winter view shows the lake that fills the crater frozen and covered with snow.

detected through drilling and seismic studies of underground features. For example, an impact crater from about 35 million years ago has been identified below the mouth of the Chesapeake Bay on the East Coast of the United States. The crater has since been buried by hundreds of meters of sand and silt, but geologists can trace shattered rock and the profile of the buried crater over a region about 90 kilometers (55 miles) in diameter. An even larger impact gave rise to the Sudbury iron mines in southern Canada. They lie in a 1.85-billion-year-old impact crater, now highly eroded, that is about 250 kilometers (150 miles) in diameter.

Many other large arc-shaped features on the surface of the Earth may also have originated as impact features. Two such craters are the vast arc (nearly 500 kilometers across) on the east edge of Hudson Bay and a basin (about 300 kilometers across) in central Europe. Geological activity and erosion so change the Earth's surface that most features become unrecognizable through surface wear, distortion, or burial. The Moon's unchanging surface, however, offers a visual reminder of the crater-forming impacts the Earth also must have suffered during its earliest history.

Astronomers have come to realize that meteor impacts are more common (and more deadly) than once believed. Although there are no recent, clear cases of people being killed by meteorites, there are many documented near misses. Searching old records from Europe and China, astronomers have found past instances of fatalities. One grim report is from a fifteenth-century Chinese chronicle that describes "stones that fell like rain" that killed 10,000 people. Scientists studying the tusks of woolly mammoths from about 35,000 years ago discovered that several were peppered with pockmarks with small bits of iron meteorite embedded in them. It appears that the iron meteoroid burst in the air, sending a spray of possibly lethal shrapnel into a herd of the beasts. If a similar even occurred over a city today, it would have the potential to kill enormous numbers of people.

48.5 MASS EXTINCTION EVENTS

Throughout Earth's history, the geological record shows evidence of mass extinctions of life on Earth. These events are so large that they wipe out entire species, not just a few individuals. Sometimes a relatively small number of species disappear from the geological record, but at other times thousands of species of animals and plants, large and small, are suddenly killed off. For many years the favored explanation for these sudden changes was a geological event, such as massive volcanic eruptions or drastic changes in climate and sea level. Such events may indeed be responsible for some of these episodes of dramatic change on Earth; but in 1980 two scientists from the United States, the father-and-son team of Luis and Walter Alvarez, found compelling evidence that one of the most devastating extinctions was caused by the impact of an asteroid.

FIGURE 48.9
A layer of dark, iridium-rich clay marks the end of the Cretaceous period on Earth. The layer was probably formed when an asteroid struck the Earth. Not only the dinosaurs were wiped out; a wide variety of species were destroyed in the worldwide devastation that followed the impact.

About 65 million years ago, some disaster exterminated the dinosaurs and many less-conspicuous but widespread creatures and plants. The sudden disappearance of large numbers of life forms occurred at the end of the **Cretaceous period.** The evidence that an extraterrestrial body caused this cataclysm comes from the relatively high abundance of the otherwise rare element iridium that the Alvarezes discovered in sediments from that time (Figure 48.9). Iridium, a heavy element similar to platinum, is very rare in terrestrial surface rocks. Early in the Earth's history, when our planet differentiated, the high density of iridium caused that element to sink to the Earth's deep interior. However, iridium is relatively abundant in much meteoritic material, which either did not differentiate or includes the dense material from the core. Thus the presence of so much iridium in a layer of clay laid down 65 million years ago suggests a link to meteoritic material. On the basis of samples gathered worldwide from that layer, scientists

Q Many other major extinction events have been found in much more ancient geological strata. Given Earth's geological activity, how likely is it that the craters from the impacts that might have caused these events can be found?

calculate that it would have taken the explosion of a meteor 10 kilometers (6 miles) in diameter to disperse that much iridium. Therefore, most astronomers and paleontologists believe the Earth was hit by an asteroid of that size.

A 10-kilometer asteroid hitting the Earth would produce an explosion on impact equivalent to that of several billion nuclear weapons. Not only would the impact make an immense crater, but it would also blast huge amounts of dust and molten rock into the air. The molten rock raining down would raise the surface temperature as high as that under an electric broiler and ignite global wildfires. The hot fragments and blast would also create nitrogen oxides, which would combine with water to form a rain of highly concentrated nitric acid. This toxic combination of heat, acid rain, and blast would then be followed by months of darkness and intense cold caused by the dust shroud blotting out the Sun. It seems likely that the biosphere would be devastated, leading to mass extinctions, just as the fossil record shows.

This frightening picture is supported by the finding that this geological layer contains soot that could have been produced by the blast and by burning, as well as a layer of tiny quartz pellets that could have been created by the melting of surface rocks during a violent impact. Material dating to this time was found at high elevations on islands in the Caribbean, most likely deposited by tsunamis more than a kilometer high, and this focused scientists' search for a large impact in this region. Scientists have now identified a buried crater 180 kilometers (110 miles) wide, dated to 65 million years ago, in the Yucatán region of Mexico near **Chicxulub** (pronounced *cheek-shoo-loob* and meaning "flea of the devil" in Mayan). The Chicxulub crater rim can be clearly seen in a *gravity map* made by an oil company to look for subsurface density variations (Figure 48.10).

Astronomers have recently identified a family of asteroids that appears to be the probable source of the dinosaur killer (Unit 41.4).

The Cretaceous mass extinction may have played an especially important role in our own evolution. Before that event, reptiles were the largest animals on Earth. Subsequently, mammals have assumed that niche. Small mammals may have escaped the fury of the heat and acid rain by remaining in burrows, and they may have survived the subsequent cold by virtue of their fur. You may be running your fingers through your hair, rather than your claws across your scales, because of that impact.

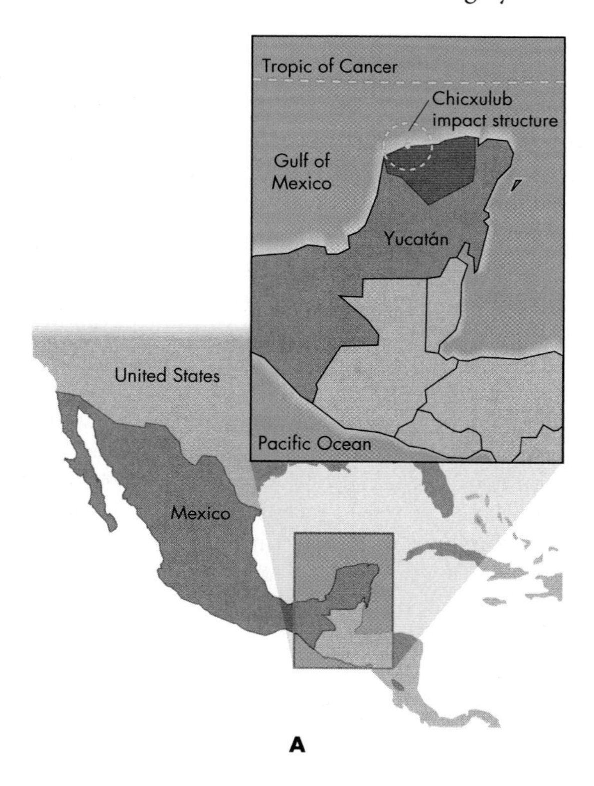

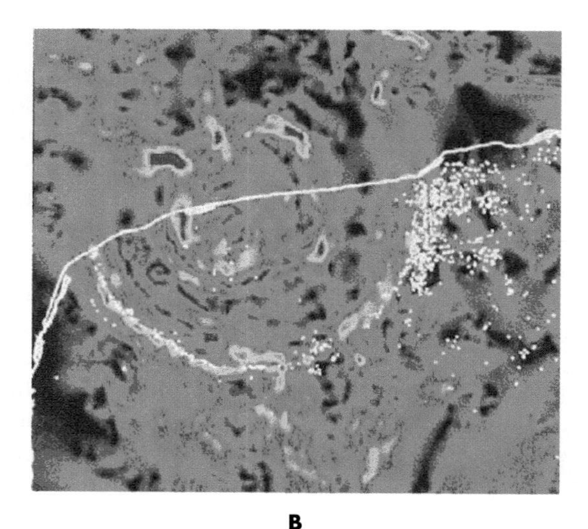

FIGURE 48.10
(A) A large impact took place 65 million years ago at Chicxulub and produced a crater 180 kilometers (10 miles) in diameter. This impact probably caused the widespread extinction at the end of the Cretaceous period. (B) The crater has been buried under a kilometer of sedimentary rock, so it is difficult to trace on the surface. However, maps of local gravity anomalies reveal the crater rim beneath the surface.

KEY TERMS

achondrite, 388
Barringer crater, 389
carbonaceous chondrite, 387
Chicxulub, 392
chondrite, 386
chondrule, 386
Cretaceous period, 391
meteor, 385
meteorite, 385
meteoroid, 385
organic compound, 387
Tunguska event, 389

QUESTIONS FOR REVIEW

1. What are the differences between a meteor, a meteoroid, and a meteorite?
2. What are chondrules, and what do they indicate about conditions in the solar nebula?
3. What kinds of meteorites come from differentiated asteroids?
4. How do we know that asteroids are similar to some meteorites in composition?
5. What evidence is there that the Earth has been hit by asteroids or comets?
6. Why do some scientists think that asteroids and comets played a role in mass extinctions?
7. How might meteorites be involved in the formation of life on Earth?

PROBLEMS

1. What would be the radius of a meteoroid with a mass of 10^6 kg (1000 metric tons) if it were made of silicate rock with a density of 3kg per liter? If it were made of water ice with a density of 1 kg per liter? Which would do more damage? Explain your reasoning.
2. Use the formula for the kinetic energy of a moving body to estimate the energy of impact of a 10^6-kg (1000 metric ton) object hitting the Earth at 30 km/sec. Express your answer in kilotons of TNT, using the conversion that 1 kiloton is about 4×10^{12} joules. *Note:* Be sure to convert kilometers/second to meters/second.
3. Suppose a major impact blasted out rock from a 100-km diameter crater that was, on average, 1 km deep. If all that material became dust that settled out evenly over the entire surface of the Earth, how thick a layer would it form?
4. There is about a 1 in 45,000 probability that the potentially hazardous asteroid named Apophis will impact Earth in 2036. Based on observations, the asteroid is estimated to have a mass of 2.1×10^{10} kg, a diameter of 0.25 km, and a density of 2.6 kg/L. It has been estimated that its impact velocity would be 12.6 km/sec. If the mass may be uncertain by a factor of 3 and the impact velocity uncertain by a factor of 2, what is the range of predicted impact energy in joules?
5. The "rule of thumb" ratio for crater diameter versus impactor diameter is 10:1—the crater's diameter will be 10 times that of the impactor. If there is an uncertainty in the mass of Apophis of a factor of 3, what is the uncertainty in the asteroid's diameter? What range in crater size does this imply?
6. It is most likely that an asteroid will impact in an ocean rather than on land. Here is a simplified formula that gives the height of the wave (h), in meters, at a distance of 1000 km from the impact site:

$$h = 8 \times \sqrt{\left(\frac{D}{406}\right)^3 \times \left(\frac{V}{20}\right)^2 \times \left(\frac{\rho}{3}\right)},$$

where D is the asteroid diameter in meters, V is the velocity in km/sec, and ρ is the asteroid density in kg/L. Calculate the expected tsunami height in meters at 1000 km from the impact if Apophis happens to land in the middle of an ocean. How would you expect the tsunami height to vary with distance from the point of impact?

TEST YOURSELF

1. The bright streak of light we see as a meteoroid enters our atmosphere is caused by
   a. sunlight reflected from the solid body of the meteoroid.
   b. radioactive decay of material in the meteoroid.
   c. a process similar to the aurora that is triggered by the meteoroid's disturbing the Earth's magnetic field.
   d. frictional heating as the meteoroid speeds through the gases of our atmosphere.
   e. the meteoroid's disturbing the atmosphere so that sunlight is refracted in unusual directions.
2. If the Earth were struck by an asteroid 10 km in diameter, which of the following would have the most serious consequences for life on Earth?
   a. The huge crater produced by the collision
   b. The atmospheric shock wave
   c. The dust cloud raised by the impact
   d. Lava released from the Earth's interior through the crater
   e. Iridium and other radioactive substances brought by the asteroid
3. What is the source of most meteorites?
   a. Comets
   b. The Moon
   c. Asteroids
   d. Mars
   e. Condensed matter from the solar wind

Part Four

The content of some Units will be enhanced if you have previously studied some earlier Units in this textbook. These Background Pathways are listed below each Unit image.

Background Pathway: Unit 4

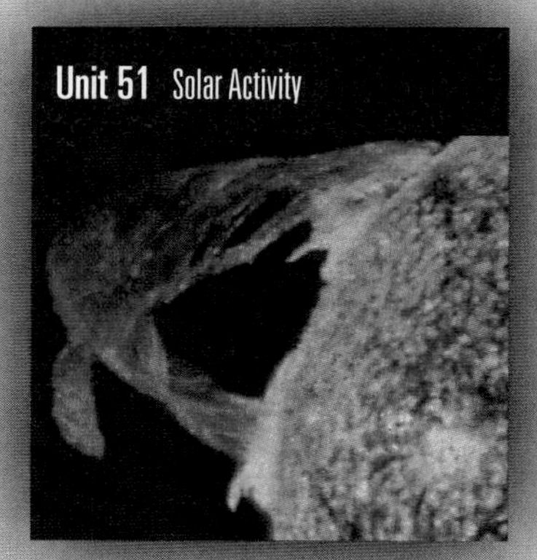

Background Pathway: Unit 49

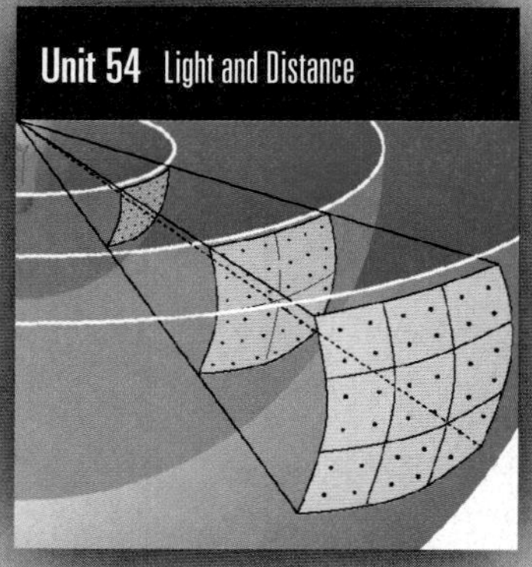

Background Pathway: Unit 21

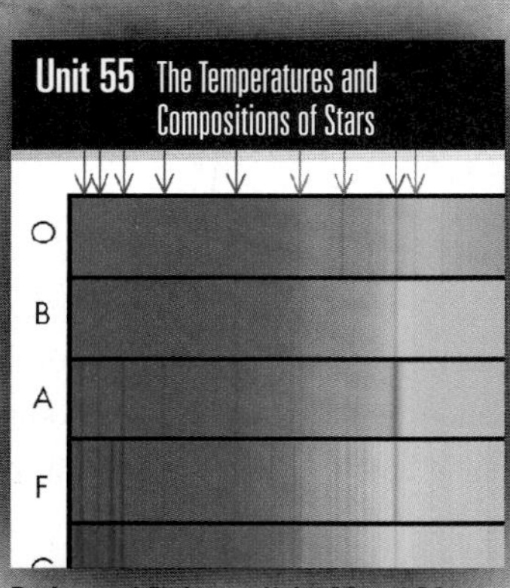

Background Pathways:Units 23 and 24

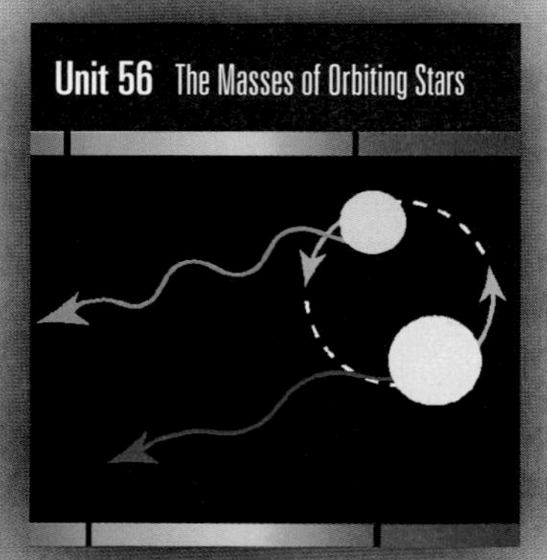

Background Pathway: Unit 17

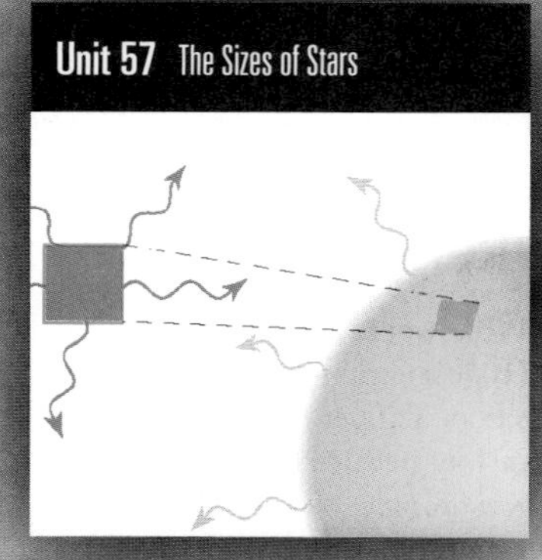

Background Pathways: Units 54 and 55

Background Pathways: Units 54–57

Stars and Stellar Evolution

Unit 59 Overview of Stellar Evolution

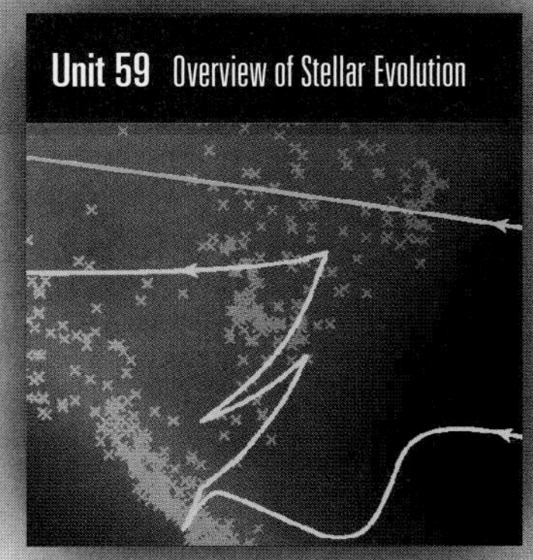

Background Pathways: Unit 58

Unit 60 Star Formation

Background Pathways: Unit 59

Unit 61 Main-Sequence Stars

Background Pathways: Unit 59

Unit 62 Giant Stars

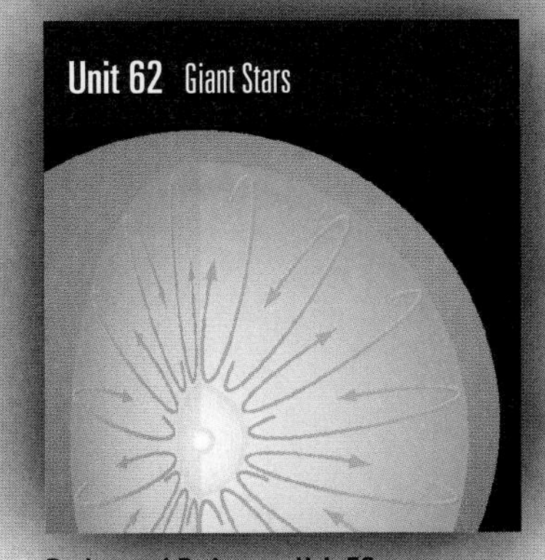

Background Pathways: Unit 59

Unit 63 Variable Stars

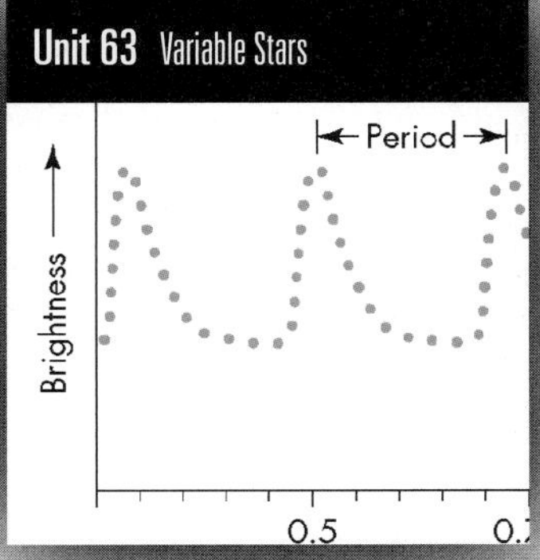

Background Pathways: Unit 57

Unit 64 Mass Loss and Death of Low-Mass Stars

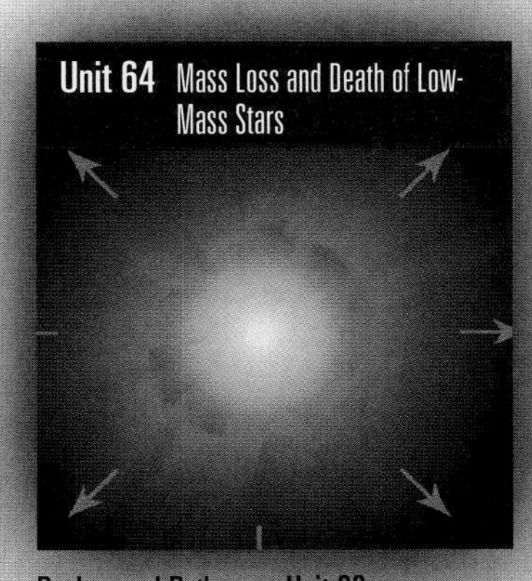

Background Pathways: Unit 62

Unit 65 Exploding White Dwarfs

Background Pathways: Unit 64

Unit 66 Old Age and Death of Massive Stars

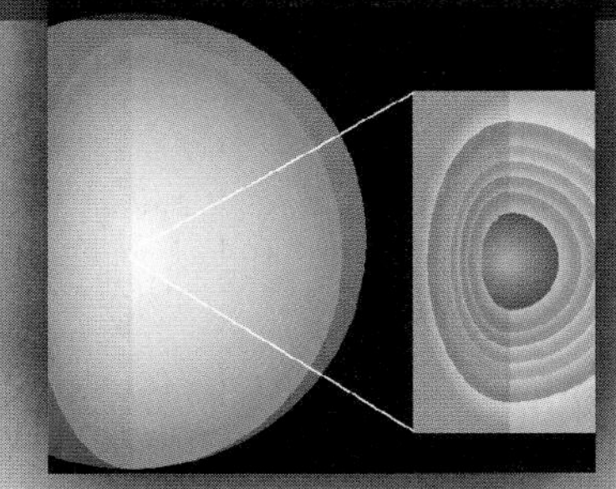

Background Pathways: Unit 59

Unit 67 Neutron Stars

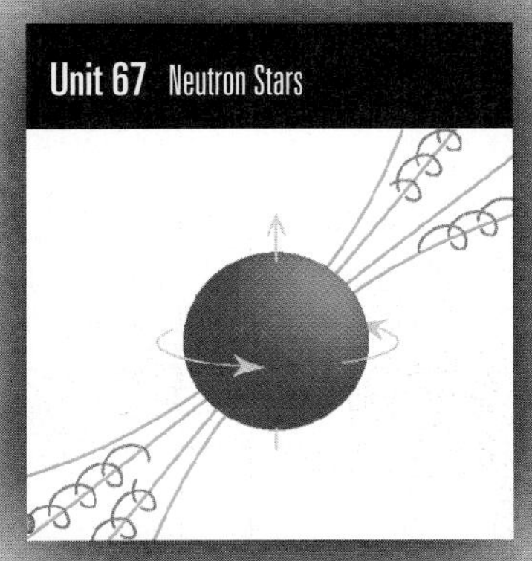

Background Pathways: Unit 66

Unit 68 Black Holes

Background Pathways: Unit 67

Unit 69 Star Clusters

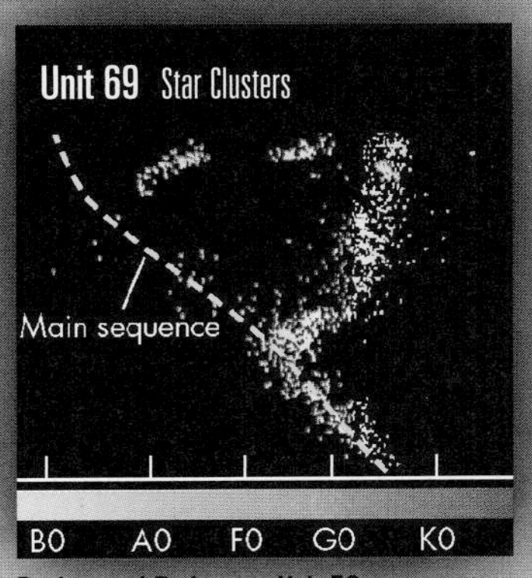

Background Pathways: Unit 59

unit

49

The Sun, Our Star

The Sun is a star—a dazzling, luminous ball of ionized gas. It is so large that a million Earths would fit inside its volume, and it is about 300,000 times more massive than the Earth. It is so bright that it will damage our eyes if we look at it directly. All stars generate power in enormous quantities, lighting up the universe despite their immense distances from one another. For centuries, the source of the Sun's and other stars' power was one of astronomy's greatest mysteries. Solving this mystery requires us to understand almost incomprehensible temperatures and pressures deep in a star's interior.

In its core the Sun releases the energy of 100 billion atomic bombs every second; yet the Sun does not blow itself apart because this enormous power is counterbalanced by enormous gravity. Stars pit titanic forces against one another. If these forces somehow became unbalanced, stars can blast themselves apart or collapse in on themselves. Yet the Sun has remained stable for billions of years. Understanding this balance is one of the main themes of Part IV of this book.

In this unit we explore the structure of our star and the methods astronomers use to study parts of the Sun that cannot be seen directly. In Units 50 and 51 we will examine the source of energy in the Sun and the activity that occurs on its surface. In later units we will discover that stars cannot maintain forever their precarious balance between gravity and pressure; and when the balance fails, cataclysmic events result. Along the way, we will learn of the likely fate of our own star.

49.1 THE SURFACE OF THE SUN

Although the Sun exhibits some features in visible light (Figure 49.1A), observations at other wavelengths reveal that the Sun's surface is violently agitated, with enormous storms blasting out fountains of incandescent gas (Figure 49.1B). A single solar storm may release as much energy in a few minutes as the combined

A

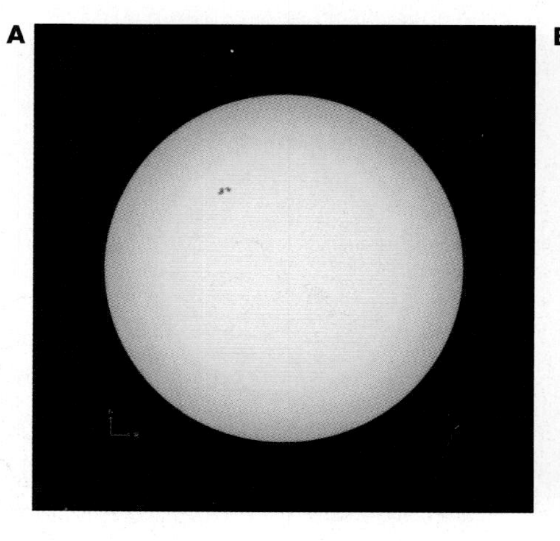

B

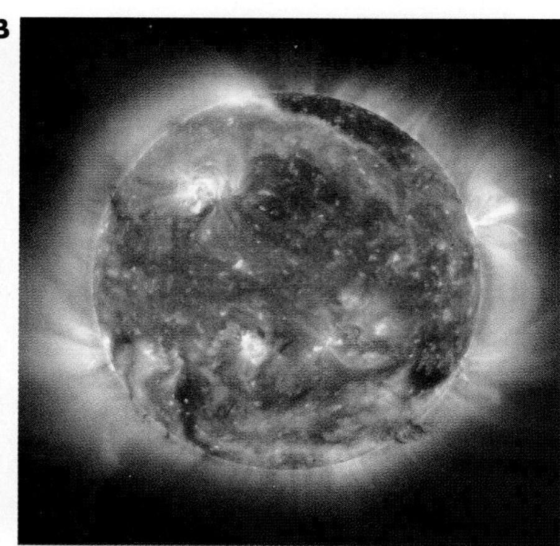

FIGURE 49.1
Images of the Sun made at the same time by the *SOHO* satellite in (A) visible and (B) ultraviolet light. The complex stormy surface of the Sun is revealed in the false-color ultraviolet image. Blue colors in this image indicate the highest-energy photons, whereas red colors indicate the lower-energy photons.

energy of all the earthquakes on Earth over the last 10 million years, and such storms are not rare. Yet despite their vast energy, these storms are dwarfed by the Sun's total flow of energy. Every second the Sun radiates a hundred times more energy than a single one of its solar storms.

When we look at the Sun, we see through the low-density, tenuous gases of its outer atmosphere. Our vision is ultimately blocked, however, as we peer deeper into the Sun, because the material there is compressed to higher densities by the weight of the gas above it. In this dense material, the atoms are sufficiently close together that they strongly absorb the light from deeper layers, blocking our view of them much as a fog obscures our view of what lies beyond a certain distance.

The Sun's "surface" is about 700,000 kilometers (430,000 miles) from its center. This is not a surface in the usual sense of the word. The glowing gas there is thin—only about one thousandth the density of the air we breathe. However, this region is where light can escape into space as the sunlight we see. This layer, where the Sun's gas changes from opaque to transparent—where the photons we see come from—is called the **photosphere.** The photosphere looks like a surface, but it is actually a region about 500 kilometers (300 miles) thick, with gas that grows denser below and thinner above. The location of the photosphere is illustrated in Figure 49.2—this diagram also depicts the other parts of the Sun discussed throughout the rest of this Unit.

From the upper part of the photosphere to its lowest part, the gas density increases by about a factor of 10. If we could see farther into the Sun, we would find that its density rises steadily toward its center, compressing the gas, much as in a pile of laundry the clothes on the bottom are flattened most, while those on top remain puffed up. A similar compression occurs in the atmosphere of the Earth and other planets, but the greater mass of the Sun leads to a vastly greater compression of its gas. Not only does the density of the Sun rise as we plunge deeper into its interior, the temperature rises too. The average temperature of the photosphere is about 5800 K; but at its top, the photosphere's temperature is about 4500 K, while near the bottom, the temperature is about 7500 K.

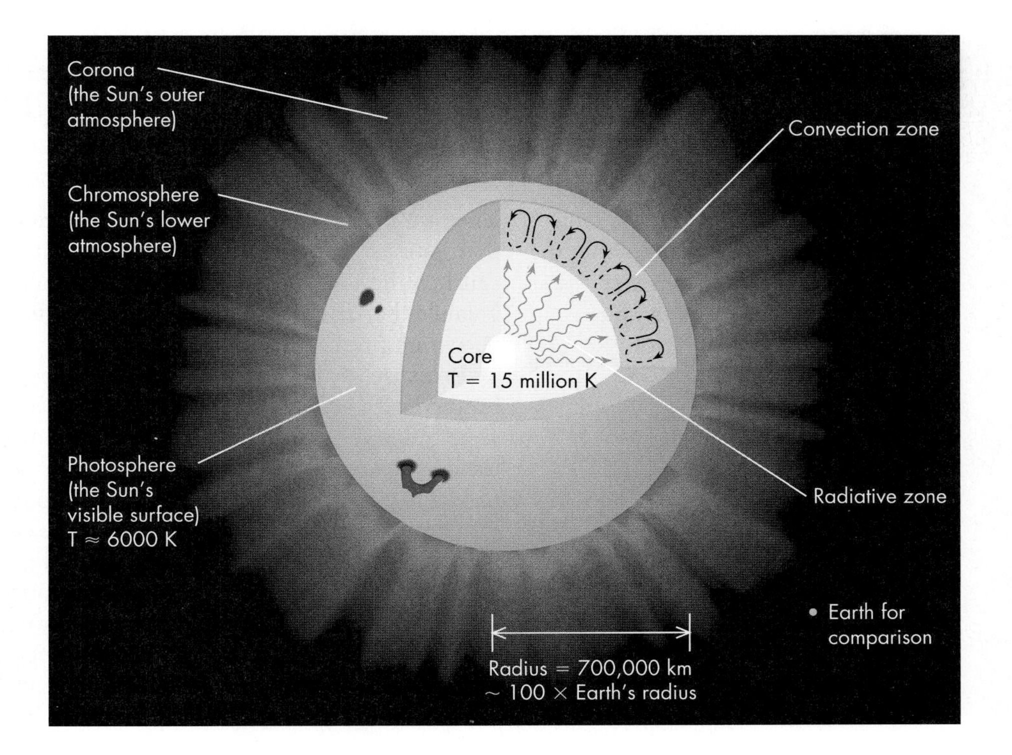

FIGURE 49.2
A cutaway sketch of the Sun.

TABLE 49.1 Properties of the Sun

Property	Value	Method of Determination
Distance	1 AU = 150 million kilometers or 93 million miles	Triangulation or radar (Unit 52)
Radius	7×10^5 km	From angular size and distance (Unit 10)
Mass	2×10^{30} kg	Modified from Kepler's third law (Unit 17)
Average density	1.4 kg/L	From radius and mass
Central density	160 kg/L	Indirectly from need to balance internal pressure and gravity (Unit 49)
Surface temperature	About 5800 K	Color–temperature relation (Wien's law, Unit 23)
Central temperature	15 million K	Indirectly from need to balance internal pressure and gravity (Unit 49)
Composition	71% hydrogen, 27% helium, and 2% vaporized heavier elements such as oxygen, carbon, and iron	Spectra of gases in surface layers (Unit 55)
Luminosity	4×10^{26} watts	From amount of energy reaching Earth and inverse-square law (Unit 54)

Deep in the Sun, the matter is compressed to densities far greater than rock or iron in the Earth. Despite this great density, *the Sun is gaseous throughout* because its high temperature gives the atoms so much energy of motion that they are unable to bond with one another to form a liquid or solid substance. In fact the temperatures are so high that the atoms are ionized, and therefore the matter is properly termed a **plasma.**

Before we discuss how the Sun can remain so hot and dense, we need to understand better some of its overall properties. What forces are at work in its interior? How rapidly does it lose energy? What resources does it have available to supply its energy needs?

Table 49.1 lists a number of the Sun's properties and shows how astronomers measure them. Some of these properties can be calculated in a straightforward way, but others, such as the Sun's internal temperature and density, cannot be observed directly and therefore must be deduced. Astronomers use the known physical properties of gases and computer models to determine these, as we discuss next.

49.2 PRESSURE BALANCE AND THE SUN'S INTERIOR

The Sun dwarfs the Earth and even Jupiter, and its immense mass is what drives it to shine. The gravitational force (Unit 16) of this much concentrated matter crushes the gas in its interior to extremely high temperatures and densities. To balance that crushing force and prevent its own collapse, the Sun must maintain a very high pressure within the gas.

Pressure is measured in terms of the force exerted by a gas over an area. This can be expressed, for example, as newtons per square meter or pounds per square inch (see Unit 3).

Pressure is the force produced by the particles in a gas as they collide and bounce off each other (or the walls of a container). For example, a balloon is kept inflated by trillions upon trillions of molecules bouncing off the balloon's inside surface, each giving a tiny "punch" outward. Pressure and gravity must be in balance in the Sun, or else it would contract or expand. This balance between pressure and gravity is called **hydrostatic equilibrium.**

In the Sun, as in virtually all stars and planets, the balance of hydrostatic equilibrium requires that at every point, the outward force created by pressure exactly balances the inward force due to the body's gravity (Figure 49.3). If the forces were unbalanced, the gases would begin flowing, and within minutes the Sun's shape or size would be altered. To understand the pressure balance in the Sun, we need to discuss in more detail how its pressure arises.

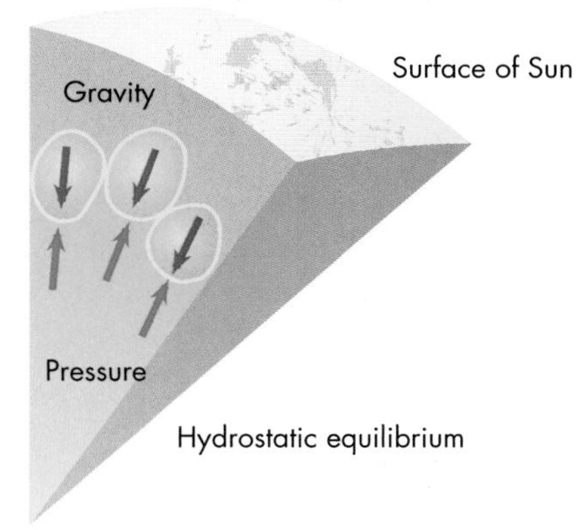

FIGURE 49.3
A sketch illustrating the condition of hydrostatic equilibrium, the balance of pressure (blue arrows) and gravitational force (purple arrows) in the Sun.

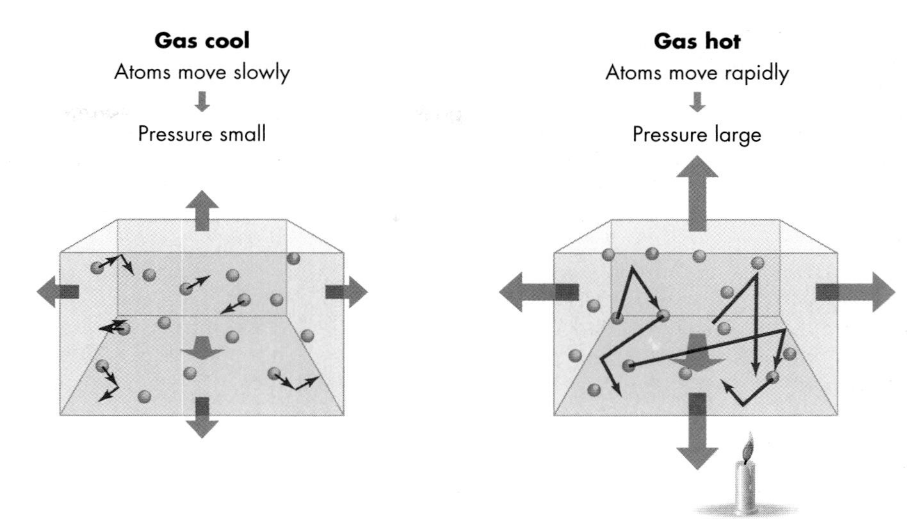

FIGURE 49.4
Sketch illustrating the ideal gas law. Gas atoms move faster at the higher temperature, so they collide both more forcefully and more often than atoms in a cooler gas. Therefore, other things being equal, a hotter gas exerts a greater pressure.

Pressure in a gas comes from collisions among its atoms and molecules. For example, if you squeeze a balloon, the atoms are pushed closer to each other. They collide and rebound more frequently, raising the pressure. When you squeeze a balloon, you experience this as an increasing resistance the harder you squeeze; or perhaps as you squeeze it, a section of the balloon you are not pressing may stretch outward in response to the increased pressure. Temperature also plays an important role. You can raise the pressure in a balloon by heating the air inside it, as in a hot-air balloon. Raising the temperature of the gas speeds up the atoms, making them collide harder and more often (Figure 49.4). Thus, the strength of the pressure is proportional to the density times the temperature of the gas:

$$\text{Pressure} = \text{Constant} \times \text{Temperature} \times \text{Density}.$$

The relationship is known as the **ideal gas law.**

In chemistry, the ideal gas law is often written $PV = NkT$, where P stands for pressure, V stands for volume, N stands for the number of atoms, k is a constant, and T is the temperature. Because N/V is a measure of the density, the equation can be rearranged as we have shown it here.

The ideal gas law tells us how a gas responds to a change in any of the three quantities in the formula—the other quantities must change so that the equation remains true. Some simple examples illustrate the power of this law. For example, if the pressure in a gas is doubled, the product of the density and temperature must also double in response. In some situations, external conditions might hold the pressure steady. In that case if the temperature doubles, the gas must expand until it reaches half its former density. This is how a hot-air balloon works—as it is heated, the gas in the balloon expands until it is in pressure balance with the outside air. The hot air inside the balloon is less dense than the outside air, making the balloon buoyant.

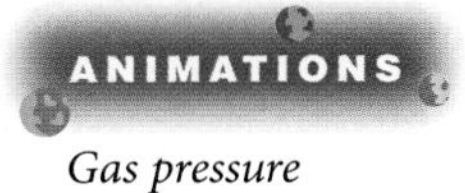

Gas pressure

A huge pressure is needed at the center of the Sun to support the crushing force of gravity inward. The ideal gas law alone cannot tell us whether it is temperature or density that generates the necessary pressure, but astronomers can also examine other properties to determine this. Astronomers use computers to model not only how gases respond in terms of the ideal gas law, but also to consider how the Sun's internal gravity changes based on the density structure in the interior of the star, and how the temperature structure responds to the way the energy travels out from the center. The computer models divide up the Sun's interior into a series of layers like the layers of an onion. The weight of each layer on the layers below it can be calculated to find the pressure needed to balance that weight. The model is constrained by a variety of requirements. For example, the sum of all the layers

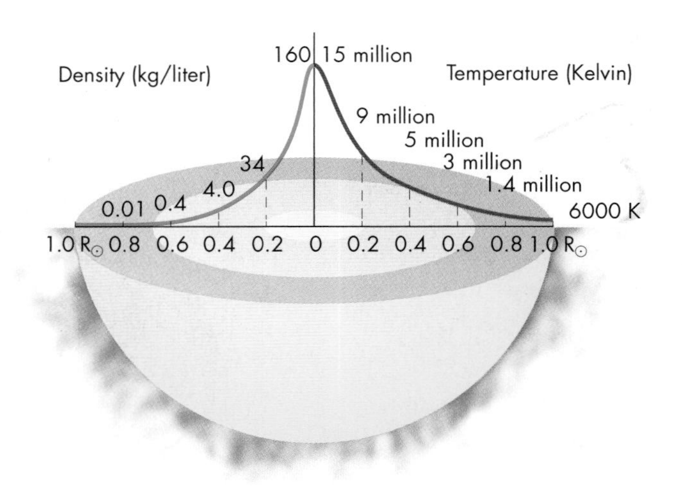

FIGURE 49.5
Plots of how density and temperature change throughout the Sun.

must add up to the total mass of the Sun, and the temperatures must be consistent with the flow of energy seen coming out of the Sun.

Below the photosphere, heat is retained more effectively by the denser gas, and the models show that the temperature climbs until at its core the temperature soars to about 15 million K. And while the Sun has an average density of about 1.4 kilograms per liter (close to that of Jupiter), near its core the Sun's density reaches 160 kilograms per liter—about 20 times the density of steel! Figure 49.5, based on computer models, illustrates how the temperature and density change throughout the Sun.

> Q How do you suppose the interior conditions in a very massive star would differ from the conditions in the Sun?

No space probe is ever likely to make direct measurements of the Sun's interior, but we are confident that the computer models are correct. It is true that the same pressure could be produced at the center of the Sun if the temperature were 10 times lower and the density 10 times higher. However, a model that assumes that the Sun is 10 times more dense would also predict a higher mass, and an assumption that the temperature is 10 times lower would predict a lower energy output. Even the composition of the Sun's core used in the computer models will result in slightly different predictions for the radius and surface temperature of the Sun, so astronomers have high confidence in the accuracy of the models.

49.3 ENERGY TRANSPORT

The high temperature in the Sun's core requires an powerful energy source because energy is constantly being lost from the Sun as it radiates energy out into space. We experience the Sun's lost energy as sunshine. Sunshine is essential for life on Earth, but it is a death warrant for the Sun. The Sun is gradually consuming itself to replace the lost energy, and thereby maintain the high temperature and pressure in its core. When its fuel is used up, the Sun must collapse—a fate facing all stars.

Your own experience and experiments in the laboratory show that heat always flows from hot to cold. Applying this principle to the Sun, we can infer that because its core is hotter than its surface, heat will flow outward from the Sun's center. Near the core, the energy is carried by photons through what is called the **radiative zone,** illustrated in Figure 49.2. Because the gas there is so dense, a photon travels less than a centimeter before it is absorbed by an atom and stopped. The photon is quickly reemitted, but it leaves the atom in a random direction—perhaps even back toward the Sun's center—and then is almost immediately reabsorbed by

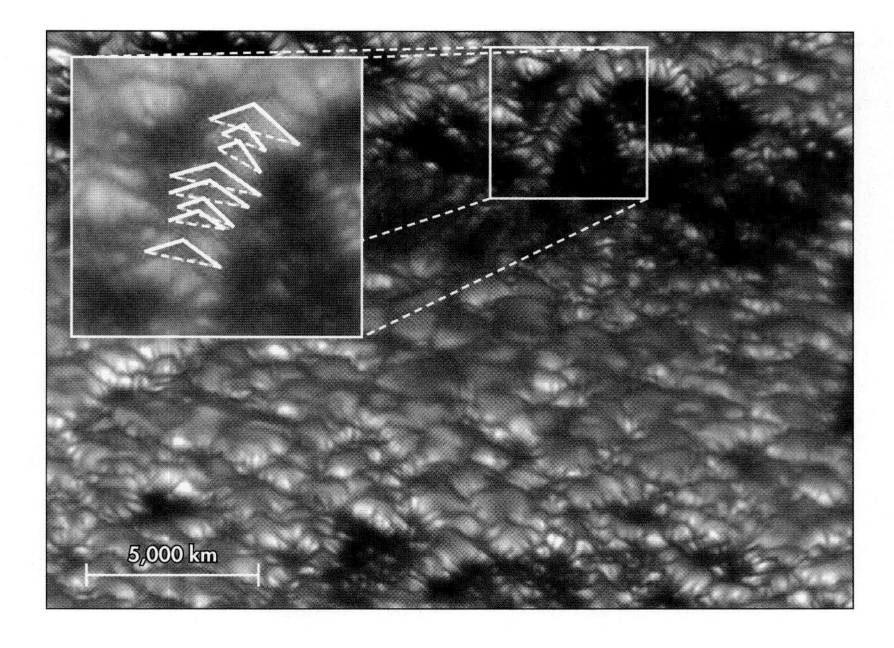

FIGURE 49.6
Image of a portion of the Sun's surface showing the granulation on the surface. This image was made using a telescope with adaptive optics to capture high-resolution details. The inset black-and-white image shows a region where analysis has found the height of the granulation features. The triangles represent the heights of these features, which are from 200 to 450 kilometers tall.

another atom. This constant absorption and reemission slows the rate at which photons escape from the Sun, like people lost in a dense forest walking randomly between trees until eventually they wander close enough to the edge of the forest to see out. Even though photons move at the speed of light between absorptions, it takes them over 10 million years, on average, to travel from the core to the surface. Today's sunshine was born in the Sun's core before the birth of human civilization!

The outward movement of photons is slowed even more in a region that begins about two-thirds of the distance out from the Sun's center. In this region the gas is cooler and the atoms are less completely ionized than in the radiative zone; as a result the atoms are more effective at blocking photons. The energy from the Sun's interior is therefore trapped, and it heats the gas there. Hotter regions in the gas expand, become buoyant, and rise like a hot-air balloon. The gas rises and cools, then sinks back down in a region called the **convection zone** (Figure 49.2).

We can infer the gas's motion in the convection zone by observing the top layer of it—the photosphere. There we see numerous bright regions surrounded by narrow darker zones called **granulation,** as shown in Figure 49.6. The bright areas are bubbles of hot gas, up to about 1000 km (600 miles) across, which rise up through the convection zone. Upon reaching the surface, these hot bubbles radiate their heat to space, causing them to cool. That cooler matter then sinks back toward the hotter interior, where it is reheated and rises again to radiate away more heat. These bubbles of hot gas generally last only 5 to 10 minutes before they sink back into the roiling depths below the visible photosphere.

Q In your experience, where else does radiative energy transfer occur? Where else does convection occur?

49.4 THE SOLAR ATMOSPHERE

Astronomers refer to the lower-density gases that lie above the photosphere as the Sun's *atmosphere.* This region marks a gradual change from the relatively dense gas of the photosphere to the extremely low-density gas of interplanetary space. A similar transition occurs in our own atmosphere, where gas density decreases steadily with altitude and eventually merges with the near-vacuum of space.

FIGURE 49.7
Photograph of a portion of the solar chromosphere during a total solar eclipse.

Q What are the speeds of winds in the most severe storms on Earth? How do these compare with the speeds of the motion in spicules?

The Sun's atmosphere consists of two main regions. Immediately above the photosphere lies the **chromosphere,** the Sun's lower atmosphere. It is usually invisible against the glare of the photosphere, but can be seen during a total eclipse of the Sun as a thin red zone around the Sun (Figure 49.7) that is about 2000 kilometers (1200 miles) thick.

The chromosphere's red color is the source of its name, which literally means "colored sphere." The color comes from the strong red emission line of hydrogen, H-alpha (Unit 24). Telescopic views reveal that the chromosphere contains hundreds of thousands of thin columns or spikes called **spicules** (Figure 49.8). Each spicule is a jet of hot gas that grows during several minutes to be thousands of kilometers tall before it cools and sinks back to the surface.

Astronomers can determine a great deal about a gas by studying the emission lines it produces (Unit 24), and in the chromosphere the emission lines reveal a surprising reversal in the gas temperature. By the top of the photosphere, the temperature has declined to about 4500 K, but in the chromosphere the temperature begins climbing again. At the top of the chromosphere, only about 2000 kilometers above the photosphere, the temperature reaches 50,000 K. Above this, in a thin transition region at the top of the chromosphere, the density drops rapidly and the temperature shoots up to about 1 million Kelvin as we enter the **corona,** the Sun's outer atmosphere.

The corona's extremely hot gas has such low density that under most conditions we look right through it. But like with the chromosphere, we can see it during a total solar eclipse when the Moon covers the Sun's brilliant disk. Then the pale glow of the corona can extend far beyond the Sun's edge to distances several times larger than the Sun's radius (Figure 49.9). Because the corona is so tenuous, it contains little energy despite its high temperature. It is like the sparks from a Fourth of July sparkler: Despite their high temperature, you hardly feel them if they land on your hand because they are so tiny and carry little total heat.

The corona is strongly influenced by the Sun's magnetic field. Scientists do not fully understand how the Sun generates its magnetic field, but magnetic fields

FIGURE 49.8
Photograph of spicules in the chromosphere made at the wavelength of hydrogen's H-alpha spectral line. The spicules are the thin, stringy features that look like tufts of grass.

FIGURE 49.9
Photograph of the corona during a total eclipse of the Sun in 1988.

FIGURE 49.10
Magnetic loops in the Sun's lower atmosphere. Gas trapped along the loops is heated by magnetic activity and in turn heats the Sun's corona. This image was made by the *SOHO* satellite at ultraviolet wavelengths.

arise around almost all spinning bodies that have electrically conductive fluids circulating in their interiors. For example, Earth's magnetic field arises in its liquid iron core through the *magnetic dynamo* process (Unit 35.4). The Sun is made of plasma that conducts electricity, and the circulation of this gas inside the Sun generates a magnetic field.

As we will discuss in greater depth in Unit 51, the Sun's magnetic field varies in strength and intensity, and this changes the appearance of the corona. This is because magnetic fields exert a force on charged particles, such as the ions and electrons in a plasma, that tends to make them follow the direction of the field when they move. Pictures of the Sun made at X-ray wavelengths show streamers in the corona that follow the direction of the Sun's magnetic field (Figure 49.10).

Studies of the ionized gas in the chromosphere and corona show that it is stirred by the magnetic fields, and this may cause the heating of these regions. An analogy may help you understand how magnetic waves can heat a gas. When you crack a whip, a motion of its handle travels as a wave along the whip. As the whip tapers, the wave's energy of motion is transferred to an ever smaller piece of material. Having the same amount of energy but with less mass to move, the tip accelerates and eventually breaks the sound barrier. The whip's "crack" is a tiny sonic boom as the tip moves faster than the speed of sound—about 1200 kilometers per hour (about 700 miles per hour).

A similar speedup occurs in the Sun's atmosphere when magnetic waves, formed in the turbulent photosphere, move into the corona along the Sun's magnetic field lines. As the atmospheric gas thins, the wave energy is imparted to an ever smaller number of atoms, making them move faster, as illustrated in Figure 49.11. Solar physicists find that turbulent motions in the photosphere can

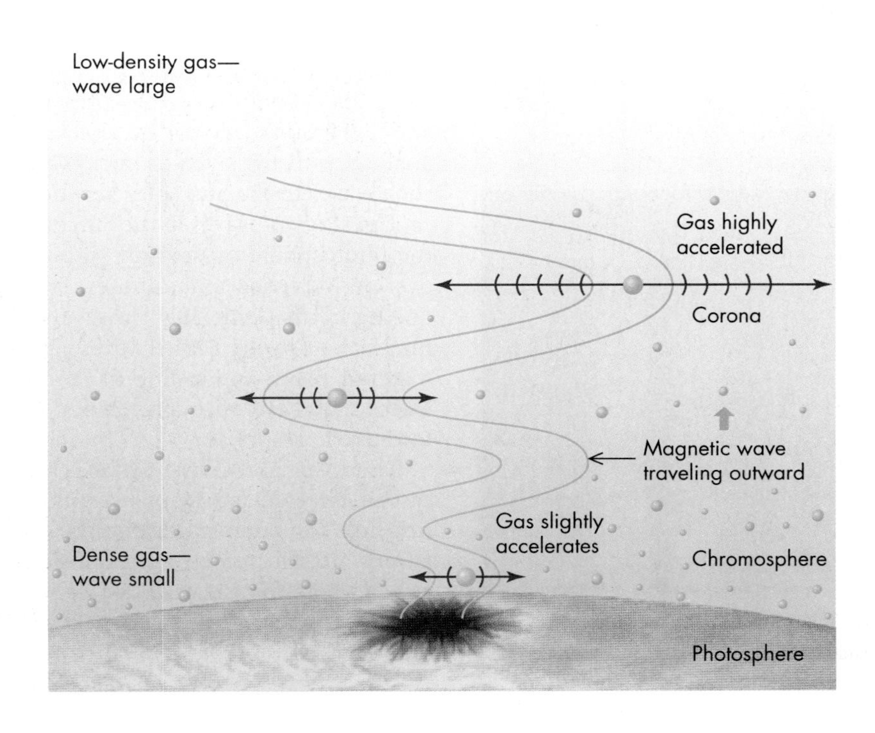

FIGURE 49.11
Diagram illustrating how magnetic waves (blue) heat the Sun's upper atmosphere. As the waves move outward through the Sun's atmosphere, they grow larger, imparting ever more energy to the gas ions (green dots) through which they move. This wave motion accelerates the ions, which collide with other gas atoms, generating heat.

wind up the magnetic field and send rapid twisting motions up into the thin coronal gas. The whipping around of the ions in the Sun's atmosphere produces collisions in the gas that heat it.

The corona contains large cooler regions, called **coronal holes,** where the magnetic field is weak. The corona's high temperature generates pressures high enough to drive gases away from the Sun in the coronal holes because the magnetic field does not trap the gas. The flow of gas away from the Sun is known as the **solar wind,** which results in a gradual loss of mass from the Sun. The expanding gas has a very low density (only a few hundred thousand atoms in a liter, compared to about 10^{22} molecules in a liter of the air we breathe). The amount of material lost from the Sun is "small": about 1.5 million tons each second. This number sounds large, but even after 10 billion years this would amount to only about 0.02% of the Sun's mass.

Near the Sun, the pressure within the corona may be only slightly larger than the Sun's gravitational attraction. The solar wind therefore begins its outward motion slowly. The gravitational pull of the Sun weakens with distance, but the pressure in the corona remains relatively high, so the solar wind gradually accelerates. At the distance of the Earth's orbit, the solar wind speed reaches about 500 kilometers per second (300 miles per second). The flow of the solar wind pushes comets' tails away from the Sun (Unit 47) and interacts with the Earth's magnetic field and ionosphere to create the aurora (Unit 36).

Beyond the Earth, the solar wind coasts at a relatively steady speed that carries it outward until it impinges on the interstellar gas surrounding the Solar System, where the gases collide and heat. The *Voyager I* spacecraft crossed this boundary in 2004 and *Voyager II,* on a slightly slower trajectory, crossed it in 2007, at distances of about 94 and 84 AU from the Sun, respectively—roughly three times Neptune's distance. The spacecraft are now sending back data about this outermost region of the solar wind. Scientists estimate that it will take the spacecraft 10 more years to reach interstellar space.

49.5 SOLAR SEISMOLOGY

Just as scientists can study the Earth's interior by analyzing earthquake waves, or waves in a stream reveal the presence of submerged rocks, astronomers can learn about the Sun's interior by analyzing the waves that travel through the Sun. By analogy with the study of such waves in the Earth, astronomers call the study of such waves in the Sun **solar seismology.**

The study of waves in the Sun began in the 1960s, when astronomers noted that the photosphere of the Sun vibrates, oscillating up and down by several meters over periods of several minutes. These oscillations are similar to earthquake waves on the Earth (Unit 35). However, instead of arising from the shifting of rocky material, as in the Earth, waves in the Sun arise from huge convecting masses of material rising and falling in its outer regions. These motions in turn generate waves that travel through the body of the Sun, causing other portions of the Sun to shake in response.

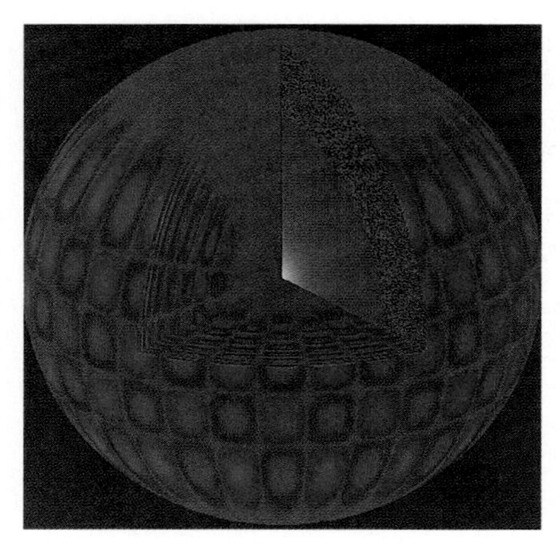

FIGURE 49.12
Computer diagram of solar surface waves.

The rising and falling surface gas makes a regular pattern (Figure 49.12), which can be detected as a Doppler shift (Unit 25) of the moving material. Astronomers next use computer models of the Sun to predict how the observed surface oscillations are affected by conditions in the Sun's deep interior. With this technique, astronomers can measure the density and temperature deep within the Sun. The results provide an independent confirmation of our computer models of the Sun's interior.

KEY TERMS

chromosphere, 402	photosphere, 397
convection zone, 401	plasma, 398
corona, 402	pressure, 398
coronal hole, 404	radiative zone, 400
granulation, 401	solar seismology, 404
hydrostatic equilibrium, 398	solar wind, 404
ideal gas law, 399	spicule, 402

QUESTIONS FOR REVIEW

1. How big is the Sun compared to the Earth?
2. What are the photosphere, chromosphere, and corona?
3. How do densities and temperatures vary throughout the Sun? What is hydrostatic equilibrium?
4. What visible evidence do we have that the Sun has a convection zone?
5. Why is the corona hotter than the Sun's surface? Why is it said that the corona contains relatively little heat?
6. Why does light take so long to travel through the Sun's interior?

PROBLEMS

1. The Sun's angular size is 0.5°, and its distance is 1.5×10^8 km. Use this information to calculate the Sun's diameter.
2. Calculate the Sun's average density in kilograms per liter. The Sun's mass is approximately 2×10^{30} kg. Its radius is approximately 7×10^8 m. Show that its average density is approximately 1.4 kg/L.
3. How much mass would the Sun have to lose each second in order to lose 10% of its mass in 10 billion years?
4. At the rate at which the Sun is losing mass in the solar wind, how many years does it take the Sun to lose as much mass as the mass of the entire Earth?
5. Wolf-Rayet stars have very strong stellar winds, losing as much as $10^{-5}\ M_{\odot}$ per year. If a typical Wolf-Rayet star is $20\ M_{\odot}$, what percentage of its mass might it lose in a million years?
6. Suppose you were an astronomy student on Jupiter. Use the orbital data for Jupiter (distance from Sun 5.2 AU; period 11.8 years) to measure the Sun's mass using the modified form of Kepler's third law.
7. If a packet of gas rising through the interior of the Sun reduces its temperature from 1.5 million K to 10,000 K and has a density decrease from 120 kg/L to 5.5 kg/L, by what factor has the pressure on it changed?

TEST YOURSELF

1. The diameter of the Sun is about how large compared with the Earth's?
   a. Twice as big
   b. Half as big
   c. 10 times as big
   d. 100 times as big
   e. 10,000 times as big
2. According to the ideal gas law, if the temperature of a gas is made 4 times higher, which of the following is a possible result?
   a. Its pressure increases by 4 times and its density remains the same.
   b. Its density increases by 4 times and its pressure remains the same.
   c. Its pressure and density both double.
   d. Its pressure increases by 4 times while its density decreases by 4 times.
   e. Its pressure and density both decrease by 2 times.
3. The solar wind
   a. is caused by the Sun's gravity.
   b. is caused by the corona's pressure.
   c. causes comets' tails to swing toward the Sun.
   d. will deplete the Sun in 10 billion years.
   e. (a), (c), and (d)
   f. None of the above.

unit

50 The Sun's Source of Power

Background Pathways

Unit 4: Foundations of Astronomy 30

Energy that leaves the Sun's hot core eventually escapes into space as sunshine. That energy loss in the core must be replaced, or the Sun's internal pressure would drop and the Sun would begin to shrink under the force of its own gravity. The Sun is therefore like an inflatable chair with tiny leaks through which the air escapes. If you sit in the chair, it will gradually collapse under your weight unless you pump air in to replace that which is escaping. What acts as the energy pump for the Sun?

Although the immense bulk of the Sun hides its core from view, we have a number of clues to the possible sources of energy. From a combination of theoretical models and direct observations, we can deduce what the core's temperature and density must be (Unit 49). From the spectra of its atmospheric gases, we know the Sun's composition. From the amount of sunlight we receive, we know how much energy the core must be generating. Moreover, in recent years a new area of astronomy has been developed based on measuring the numbers of a subatomic particle called a **neutrino** (Unit 4) that the Sun emits. Observations of such neutrinos allow us to "view" reactions occurring in the Sun's interior more directly than ever before.

These astronomical discoveries, combined with the discovery of nuclear energy in the 1900s, lead us to a picture of almost unbelievable violence in the core of the Sun. Every second, more than 4 million tons of matter are annihilated—not just turned into vapor or ionized or ejected into space, but turned into light—so that the Sun's mass gradually declines. Over 40 million years the Sun obliterates a mass equivalent to the entire Earth! This loss may seem large, but the Sun has such a tremendously large mass that it can afford this extravagant destruction of its mass in its fight to keep from collapsing under its own weight.

50.1 THE MYSTERY BEHIND SUNSHINE

Early astronomers believed that the Sun might burn ordinary fuel such as coal. But even if the Sun were pure coal, it could burn only a few thousand years given its prodigious energy output. In the late 1800s another proposal was that the Sun is not in hydrostatic equilibrium, but that gravity slowly compresses it, making it shrink. In this hypothesis, compression heats the gas and makes the Sun shine, much as the giant planets generate heat in their interiors (Unit 43). However, gravity could power the Sun by this mechanism for only about 10 million years, and the Sun would have to be shrinking by about 10 kilometers each year, which modern observations rule out. Therefore, something else must supply the Sun's energy.

We know the Sun has been shining with approximately the same luminosity for billions of years because we have geological and fossil evidence of water on Earth dating back that far. For Earth to have maintained a climate with liquid water

E = energy

m = mass

c = speed of light = 3×10^8 m/sec

Q How do we know what the Sun is made of?

An explosion of 1 kiloton (1000 metric tons) of TNT releases about 4.2×10^{12} joules of energy.

on its surface, the Sun's power output must have remained fairly steady over that whole time. In 1899 T. C. Chamberlin suggested that subatomic energy—energy from the reactions of atomic nuclei—might power stars, but he could offer no explanation of how the energy was liberated. In 1905 Einstein proposed that energy might come from a body's mass. His famous formula,

$$E = m \times c^2,$$

states that a mass, m, can become an amount of energy, E, equal to the mass multiplied by the square of the speed of light, c. It is important to understand that this formula simply states a basic "exchange rate" between energy and mass, similar to the way we might convert between dollars and euros. Einstein's formula indicates how many joules (kg · m²/sec²) of energy (Unit 3) correspond to 1 kilogram of mass. The term c^2 in the equation is the factor we use in calculating the exchange.

For example, if you could convert 1 gram (10^{-3} kg) of mass—about the amount of matter in a paperclip—into energy, it would release an energy of

$$\begin{aligned} E &= 0.001 \text{ kg} \times (3 \times 10^8 \text{ m/sec})^2 \\ &= 9 \times 10^{13} \text{ kg} \cdot \text{m}^2/\text{sec}^2 \end{aligned}$$

or 9×10^{13} joules. This is the equivalent of approximately 20 kilotons of TNT, about as powerful as the atomic bomb that destroyed Hiroshima. You could grind the same paperclip into fine metal powder and burn it—a chemical process. This would release less than one ten-billionth as much energy—about as much as burning a match. With the understanding that mass may be converted into energy, scientists realized that if the Sun can convert even a tiny fraction of its mass into energy, it would have an enormous source of power. Einstein's equation does not say what process might release so much energy, but it provides a clue where to look: reactions that result in a significant change in mass.

In 1919 the English astrophysicist A. S. Eddington, a pioneer in the study of stars, developed a hypothesis for a reaction that might provide a significant source of energy: the conversion of hydrogen into helium. Hydrogen contains one nuclear particle, while helium contains four nuclear particles; however, the mass of helium is significantly less than four times the mass of hydrogen. Eddington recognized that if hydrogen nuclei could be combined to make helium nuclei, it would involve the loss of enough mass to provide the energy necessary to power the Sun. It was not until the late 1930s that the German physicists Hans A. Bethe and Carl F. von Weizsäcker worked out the details of how such a conversion might take place. They showed that the Sun generates its energy by converting hydrogen into helium through a process called **nuclear fusion,** a process that bonds two or more atomic nuclei into a single heavier one.

Under normal conditions hydrogen nuclei repel one another, pushed apart by the electrical charges of their **protons.** Protons are positively charged nuclear particles, and like charges repel each other. This force of repulsion grows greater the closer together the protons get, so they must be moving toward each other at extremely high speeds for the nuclei to come into close contact, about 10^{-15} m apart. When this happens, the nuclear force of attraction—the **strong force** (Unit 4)—overcomes the electrical repulsion between the protons. The strong force can also bind protons together with another kind of nuclear particle, a **neutron,** which has no electrical charge but is otherwise similar to a proton. The strong force is about 100 times stronger than the electrical repulsion between neighboring protons, but it remains strong over only a very short distance, then dies away rapidly at larger distances.

Fusion is possible in the Sun because its interior is so hot. At very high temperatures—above about 5 million Kelvin—atomic nuclei move so fast that they collide at speeds that bring them close enough together where the strong force overcomes the electrical repulsion and pulls them together (Figure 50.1). The two

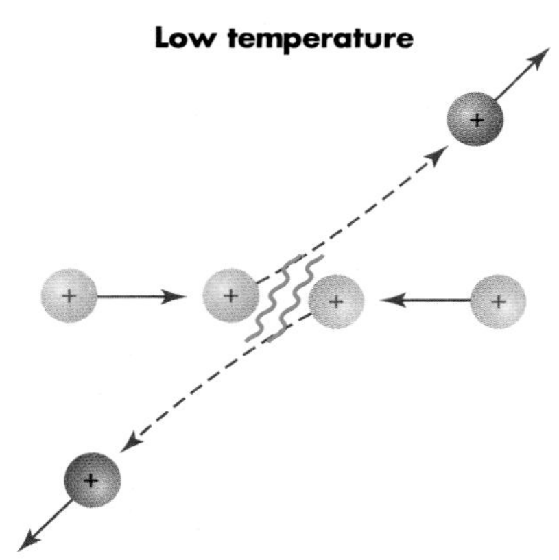

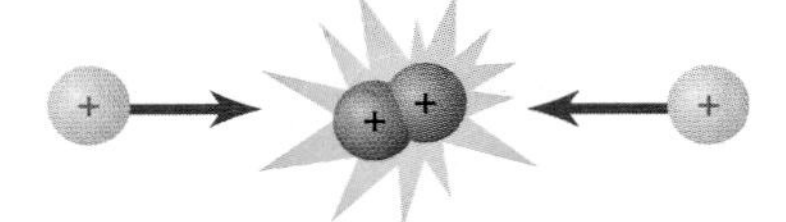

FIGURE 50.1
How a high temperature acts to overcome the electrical repulsion of the nuclei and brings them close enough for the nuclear force to fuse them.

separate nuclei then "fall" together, somewhat like two large bodies pulled together by their mutual gravitational attraction. They crash into each other and release energy when they collide. Similarly the atomic nuclei merge, or fuse, into a single new nucleus, and the potential energy of the strong force is released in the form of a gamma-ray photon. Because the strong force is 10^{40} (ten thousand trillion, trillion, trillion) times stronger than the gravitational force, the energy released is immensely larger. But because this nuclear fusion process requires such a high temperature, the only place in the Sun hot enough for fusion to occur is its core.

50.2 THE CONVERSION OF HYDROGEN INTO HELIUM

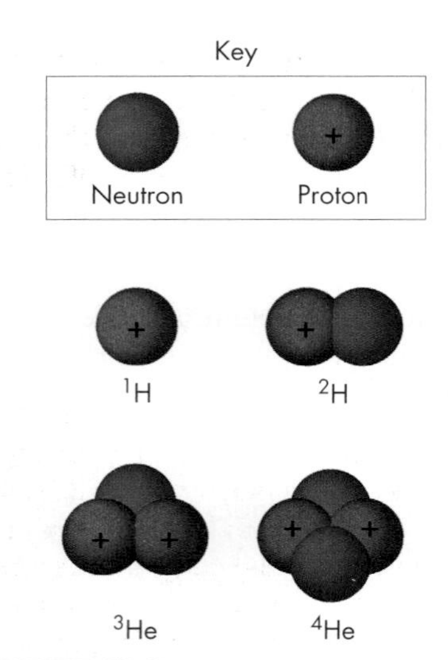

FIGURE 50.2
Schematic diagrams of the nuclei of hydrogen, its isotope deuterium, and two isotopes of helium.

Before we can fully understand how fusion creates energy, we need to look at the structure of hydrogen and helium as they are found in the Sun's core. (See also Unit 4.2.) The common form of hydrogen consists of one proton and an orbiting electron, and that of helium consists of two protons, two neutrons, and two orbiting electrons (Figure 50.2). In the Sun's hot interior, however, atoms collide so violently that the electrons are generally stripped off; that is, the atoms are completely ionized.

Hydrogen and helium always have one and two protons, respectively, but they have other forms, **isotopes,** with different numbers of neutrons. To identify the isotopes, we write their chemical symbol preceded by a superscript that shows the total number of protons + neutrons. The usual form of hydrogen with one proton and no neutrons is written ^{1}H, whereas the form of hydrogen containing one proton and one neutron is ^{2}H. Most isotopes are not given separate names, but ^{2}H is so common that it has been given its own name: **deuterium.** This isotope of hydrogen is more common than all but about a half dozen elements in the universe.

The most common form of helium, with two protons and two neutrons, is written ^{4}He, whereas helium with two protons but only one neutron is ^{3}He. When speaking, we refer to these as "helium four" and "helium three," respectively. These isotopes of hydrogen and helium play a critical role in the energy supply of the Sun.

Hydrogen fusion in the Sun occurs in three steps called the **proton–proton chain.** In the first step, two ^{1}H nuclei (protons) collide and fuse to form the isotope of hydrogen, ^{2}H or deuterium. During the collision, one proton turns into a neutron through the weak force (Unit 4) and two other particles are ejected: a **positron** (the "antiparticle" of the electron, denoted as e^+) and a neutrino (denoted by ν, the Greek letter "nu"). This step is depicted in Figure 50.3A and can be written symbolically like this:

$$^1\mathrm{H} + {}^1\mathrm{H} \rightarrow {}^2\mathrm{H} + e^+ + \nu + \text{Energy}$$

The terms to the left of the arrow are the normal hydrogen nuclei that start the process. The terms to the right are the deuterium, positron, neutrino, and the potential energy of the nuclear bond that is released.

Each positron produced in the first step of the proton–proton chain almost immediately annihilates one of the many electrons contained in the dense gas in the Sun's core. This creates two energetic photons (gamma rays). This side reaction contributes even more energy than the fusion of protons into deuterium.

An indication that energy is released by this reaction is that the mass of ^{2}H (and the by-products of the reaction) is less than the mass of the two ^{1}H that started the process. The amount of energy released can be found from Einstein's formula $E = m \times c^2$. That energy gives the particles large kinetic energies that sustain the high temperature in the core and ultimately end up as thousands of visible-wavelength photons in the sunlight we see. The neutrinos produced in this reaction also carry away a small part of the nuclear energy, but they play no further

Proton–proton chain

role in the Sun's energy generation. We will encounter them again, however: They can be detected when they leave the Sun, helping astronomers to learn more about conditions in the Sun's interior.

Once deuterium is created, the second step of the proton–proton chain proceeds very rapidly. The ^2H nucleus collides with one of the numerous hydrogen nuclei, ^1H, to make the isotope of helium containing a single neutron, ^3He. This process releases a high-energy gamma ray, denoted by the Greek letter gamma, γ. Figure 50.3B shows this step, which can be written as follows:

$$^1H + {}^2H \longrightarrow {}^3He + \gamma + \text{Energy}$$

Here again the resulting particle, ^3He, has a smaller mass than the particles from which it was made, and several times more energy is released than from the first reaction.

The third and final step in the proton–proton chain is the collision and fusion of two ^3He nuclei. Here the fusion results not in a single particle, but rather in one ^4He and two ^1H nuclei. You can think about this reaction as the attempt to form a nucleus with four protons and two neutrons, except that two protons are ejected by their electric repulsion, as shown in Figure 50.3C. This reaction, which releases about twice as much energy as the second reaction, is written as

$$^3He + {}^3He \longrightarrow {}^4He + {}^1H + {}^1H + \text{Energy}$$

where again the final mass is less than the initial mass.

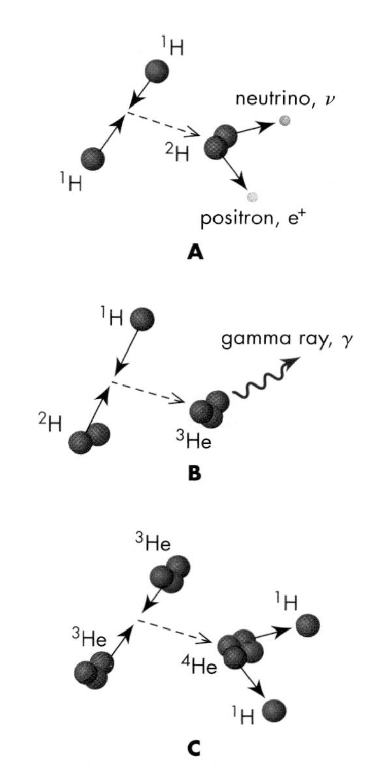

FIGURE 50.3
Diagram of the proton–proton chain. (A) Hydrogen (H) nuclei first combine to make deuterium (^2H). (B) Deuterium and hydrogen combine to make ^3He. (C) ^3He nuclei combine to form ^4He.

Scientists who want to understand the details of fusion in the Sun study each of the steps in the proton–proton chain individually; but for understanding the larger picture of the Sun's energy generation, we are concerned primarily with the overall energy production. We can find the quantity of energy released by comparing the initial and final masses of the reactions and using the mass–energy relationship $E = m \times c^2$. Steps 1 and 2 use three ^1H, but the first two steps must occur twice to make the two ^3He nuclei needed for the last step. Therefore, six ^1H nuclei are used, but two are returned in step 3, and so a total of four ^1H nuclei are needed to make each ^4He nucleus.

The mass of a hydrogen nucleus is 1.673×10^{-27} kilograms, whereas the mass of a helium nucleus is 6.645×10^{-27} kilograms. Comparing their masses, we find the following:

$$\begin{aligned} \text{4 hydrogen} &= 6.693 \times 10^{-27}\text{ kg} \\ \underline{-1\text{ helium}} &= \underline{-6.645 \times 10^{-27}\text{ kg}} \\ \text{Mass lost} &= 0.048 \times 10^{-27}\text{ kg} \end{aligned}$$

Multiplying the mass lost by c^2 gives the energy yield per helium atom made. That is,

$$\begin{aligned} E &= 0.048 \times 10^{-27}\text{ kg} \times (3.0 \times 10^8\text{ m/sec})^2 \\ &= 0.048 \times 10^{-27} \times 9.0 \times 10^{16}\text{ kg} \cdot \text{m}^2/\text{sec}^2 \\ &= 4.3 \times 10^{-12}\text{ joules per helium nucleus created.} \end{aligned}$$

To continue to power itself at its current rate, the Sun needs to generate 4×10^{26} joules of energy per second. This means that each second, about 4×10^{38} hydrogen nuclei are converted into 10^{38} helium nuclei, and about 4×10^9 kg—*4 million metric tons*—of mass are converted into energy each second. The total energy released is equivalent to exploding 100 billion-megaton H-bombs per second! Our sunshine has a violent birth.

Q The term *solar energy* is also used to refer to the generation of power on Earth from sunlight. This is done using collectors that convert the energy of photons from the Sun into electrical current. What other kinds of power generation ultimately rely on energy that came from the Sun?

50.3 SOLAR NEUTRINOS

We saw in the previous section that the Sun makes neutrinos as it converts hydrogen into helium. Neutrinos are unusual subatomic particles. They have no electric charge and only a tiny mass, and they travel at nearly the speed of light. This gives them phenomenal penetrating power. They escape from the Sun's core through its outer 700,000 kilometers, and into space like bullets through a wet tissue. They pass straight through the Earth and anything on the Earth, such as *you*, and keep going. In fact, several trillion neutrinos from the Sun passed through your body in the time it took you to read this sentence.

The number of neutrinos coming from the Sun is a direct indication of how rapidly hydrogen is being converted into helium. This is a more immediate measure of the rate of current nuclear reactions than the light we see, because photons take so long to work their way to space from the Sun's core as they interact with any intervening matter. Thus, if we can measure how many neutrinos come from the Sun, we can directly deduce the conditions in the Sun's core.

The penetrating power of neutrinos is the theme of a poem called "Cosmic Gall" by novelist and poet John Updike:

> Neutrinos, they are very small.
> They have no charge and have no mass
> And do not interact at all.
> The earth is just a silly ball
> To them, through which they simply pass,
> Like dustmaids down a drafty hall
> Or photons through a pane of glass.
> They snub the most exquisite gas,
> Ignore the most substantial wall,
> Cold-shoulder steel and sounding brass,
> Insult the stallion in his stall,
> And scorning barriers of class,
> Infiltrate you and me! Like tall
> And painless guillotines, they fall
> Down through our heads into the grass.
> At night they enter at Nepal
> pierce the lover and his lass
> From underneath the bed—you call
> It wonderful: I call it crass.

Counting neutrinos is extremely difficult. The elusiveness that allows neutrinos to slip so easily through the Sun makes them slip with equal ease through detectors on Earth. However, although they interact weakly, the rate of interaction is not zero. And because so many neutrinos are produced, we need detect only a tiny fraction of them to get useful information. Still, this requires very large detectors.

The detection of solar neutrinos was pioneered by the American physicist Raymond Davis Jr., who advanced the idea of placing tanks of material deep underground that could react with neutrinos from the Sun. Neutrino detectors are buried to shield them from the many other kinds of particles besides neutrinos that constantly bombard the Earth. For example, protons, electrons, and a variety of other subatomic particles shower our planet. These particles, traveling at nearly the speed of light, are called **cosmic rays** and are thought to be particles blasted across space by cataclysmic events, such as when a massive star explodes.

Cosmic rays can penetrate only a short distance into the Earth; so if a detector is located deep underground, nearly all the cosmic rays are filtered out. Neutrinos, on the other hand, are unfazed by a mere mile of solid ground. They have a small but predictable chance of interacting with the matter through which they pass, but that chance is so small that even if they encountered a wall of lead a light-year thick, most of the neutrinos would pass through!

Davis's experiment showed that the count of neutrinos was substantially lower than what physicists had predicted. However, his experiment relied on neutrinos produced in a side process of the proton–proton chain. Newer detectors are capable of detecting the neutrinos produced by the primary proton–proton interaction. Currently the largest neutrino "telescopes" are the Super-Kamiokande detector, located deep in a zinc mine west of Tokyo, and the Sudbury Neutrino Observatory (Figure 50.4) more than a mile underground in a nickel mine in northern Ontario, Canada. The Sudbury detector contains about 1000 tons of *heavy water,* water consisting of molecules in which one of the hydrogen atoms in the molecule is the isotope ^{2}H (deuterium). This heavy water gets its name because the extra neutron in the deuterium nucleus gives the water a slightly higher density.

Raymond Davis and Japanese scientist Masatoshi Koshiba (who headed the first Kamiokande experiment) shared the 2002 Nobel Prize in Physics for their work on solar neutrinos.

The principle behind a neutrino telescope is that occasionally a neutrino collides with a neutron in the water, breaking the neutron into a proton and an electron. As the electron shoots off, it emits a tiny flash of light, which is recorded by photodetectors. These experiments detect solar neutrinos, and it is possible to reconstruct the approximate direction the neutrinos came from. This is shown in Figure 50.5—the first neutrino image of the Sun. Although the image is blurry, it confirms that the neutrinos are coming from the Sun. Like Davis's earlier experiment, these telescopes see only one-third the expected number of neutrinos. What

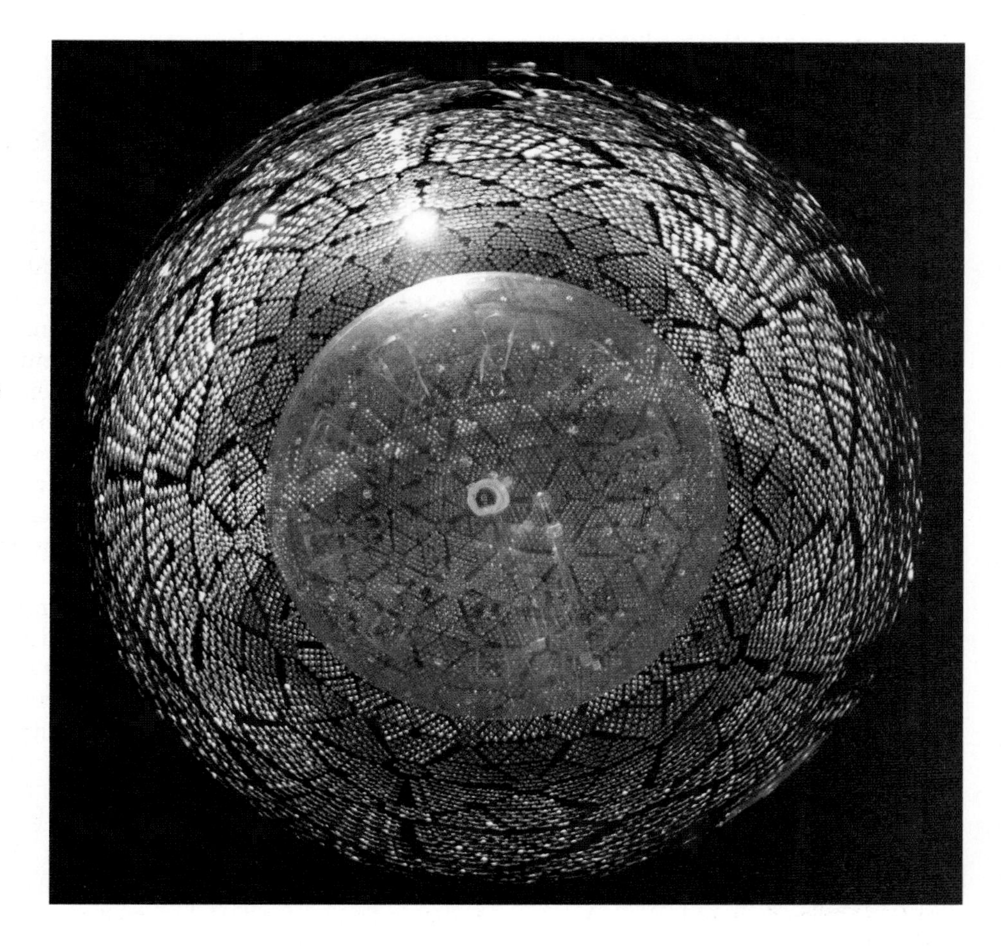

FIGURE 50.4
An inside view of the neutrino detector located in Sudbury, Canada. The sphere is 12 meters in diameter (about 40 feet). It is filled with heavy water, a form of water in which hydrogen atoms contain a neutron as well as the usual proton. When a neutrino strikes one of the neutrons, the neutron may break down into a proton and electron. As the electron streaks off, it produces a tiny flash of light. Ten thousand detectors (which form the grid visible around the sphere) record the emitted light, thereby allowing scientists to detect the neutrino's passage.

FIGURE 50.5
The first neutrino image of the Sun. This image was constructed from 1.3 years' worth of neutrino data collected by the Super-Kamiokande detector. The image covers about one quarter of the sky. Neutrinos come from a small spot at the center of the image, but uncertainties in the measurement of the paths of the neutrinos give rise to the extended shape.

does this low number mean? Is the Sun not fusing hydrogen into helium as predicted? Do neutrinos somehow escape detection? The solution to the puzzle offers us another chance to see the scientific process at work.

As a first step toward resolving the solar neutrino discrepancy, astronomers checked that their calculations for the predicted number of neutrinos were correct. All such checks led to roughly the same result, implying that there was no obvious flaw in their understanding of how the Sun works. Was it possible, then, that neutrinos had undiscovered properties that affected their detectability on Earth or their production in the Sun?

As scientists looked more closely at the properties of neutrinos, they realized that two other kinds of neutrinos were associated with different kinds of subatomic reactions (Unit 4). The nuclear fusion reactions in the Sun produce only one type of neutrino (called an *electron neutrino*), the type of neutrino that the early experiments were designed to detect. The standard theory held that the three kinds of neutrinos could not be converted into one another, but a new hypothesis suggested that they could. This hypothesis additionally predicted that the neutrinos have a mass, whereas the established model held that they had no mass. The new hypothesis suggested that the first experiments detected only one-third the expected number because the rest had been converted into the other types of neutrinos.

To test this idea, scientists built a second generation of neutrino detectors—the Sudbury detector is one of these—that could also detect the two other varieties of neutrino. By 2003, results from Sudbury clearly showed that the total number of neutrinos (of all three types) coming from the Sun agreed with the predictions of nuclear fusion rates. This gives astronomers a greater confidence in their understanding of the Sun. But in confirming the nuclear fusion model for energy generation in the Sun, they have helped to confirm a new hypothesis involving neutrinos, which modifies the previous standard model of subatomic particles. This new work gives us evidence that neutrinos must have a small mass, and that the different varieties can be converted into each other.

50.4 THE FUSION BOTTLENECK

The fusion process in stars depends sensitively on all four of the fundamental forces. We have seen that gravity forces matter together and drives up its temperature to the point where protons can overcome the electromagnetic forces holding them apart. They then come so close to each other that the strong force can take hold. However, there is a problem with this process. Two protons cannot directly combine with each other because there is no stable isotope of helium with just two protons—no "^{2}He."

The first step of the proton–proton chain is actually a highly improbable interaction because in the extremely short moment of time while two protons are colliding and very close to each other, one of them must be transformed into a neutron. This is a process that requires the **weak force** (Unit 4), and such interactions are rare. On average, a particular proton in the Sun is likely to undergo this transformation while colliding with another proton only once in several billion years!

The conversion of a proton into a neutron *requires* energy:

$$^1\text{H} + \text{Energy} \longrightarrow \text{n} + \text{e}^+ + \nu$$

where the "n" denotes a neutron. The amount of energy required for this process is large—more than half as much energy as is released when a proton and neutron combine to make deuterium. This presents a barrier that dramatically slows down the rate of fusion, although it does not prevent the process from occurring, because the overall process still has a net production of energy.

By contrast, the second and third steps of the proton–proton chain do not involve the weak force, and they typically occur just seconds after the first reaction. The weak force introduces a "bottleneck" in the rate of reaction. If the weak force were a bit stronger or weaker, stars would be very different than they are. They would burn at very different rates, or perhaps would not be able to undergo fusion at all. Thus, the weak force—as well as each of the other fundamental forces—is critical for making the Sun release its nuclear energy gradually over billions of years. This long time has allowed life to form and evolve on our planet.

KEY TERMS

cosmic ray, 410
deuterium, 408
isotope, 408
neutrino, 406
neutron, 407
nuclear fusion, 407
positron, 408
proton, 407
proton–proton chain, 408
strong force, 407
weak force, 412

QUESTIONS FOR REVIEW

1. Why is it impossible to explain the Sun's power output as coming from a chemical process such as the combustion of hydrogen and oxygen to form water?
2. What is the meaning of Einstein's formula $E = m \times c^2$?
3. How does hydrogen fuse to form helium? Why is a high temperature necessary?
4. What is the solar neutrino discrepancy and how was it resolved?
5. Why are neutrino detectors located deep underground?
6. How does the weak interaction control the rate of fusion in the Sun?

PROBLEMS

1. In chemical reactions, when energy is released, mass is lost according to Einstein's formula just as it is for nuclear energy (though the amounts lost are so small that for centuries scientists thought that mass was conserved). When a kilogram of coal is burned, it releases about 40 million joules of energy.
   a. What amount of mass is lost?
   b. How many carbon atoms (mass = 2.0×10^{-26} kg) would it take to match the amount of mass that is lost?
2. Worldwide, humans use about 4×10^{20} joules of energy each year. A typical power station generates about 10^9 watts of power, while the Sun's luminosity is about 4×10^{26} watts. (Recall that 1 watt = 1 J/sec.)
   a. How many power stations would be needed to produce power equivalent to the Sun's?
   b. How many typical power stations are needed to supply current human energy usage?
   c. If we could store all of the energy that the Sun produces in 1 second, how long would it last humans at current energy use rates?
3. Using $E = m \times c^2$, calculate how much mass is turned into energy during a year by a typical power station. How much mass is turned into energy each year by worldwide energy usage? (Use the data given in problem 2.)
4. Assume that the Sun generates 4×10^{26} watts throughout its lifetime. (Recall that 1 W = 1 J/sec.)
   a. How much mass will our Sun convert to energy over its entire main-sequence lifetime (10^{10} years)?
   b. What fraction is this of the mass of the Sun?
   c. How does this compare to the mass of the Earth?
5. One "kiloton" of explosive energy is 4.2×10^{12} joules. Using Einstein's formula, calculate the mass that is converted into energy in a 1-megaton H-bomb explosion.
6. How long does it take a neutrino produced in the center of the Sun to reach the Sun's surface?

TEST YOURSELF

1. The Sun produces its energy from
   a. fusion of neutrinos into helium.
   b. fusion of positrons into hydrogen.
   c. disintegration of helium into hydrogen.
   d. fusion of hydrogen into helium.
   e. electric currents generated in its core.
2. The Sun is supported against the crushing force of its own gravity by
   a. magnetic forces.
   b. its rapid rotation.
   c. the force exerted by escaping neutrinos.
   d. gas pressure.
   e. the antigravity of its positrons.
3. The number of neutrinos measured coming from the Sun permits us to measure
   a. how much deuterium is left in the Sun.
   b. the strength of the strong force within atomic nuclei.
   c. the current age of the Solar System.
   d. the risk of radiation exposure on the Earth.
   e. the rate of hydrogen fusion in the Sun's core.

unit

51 Solar Activity

The Sun is a stormy place. The vast amount of energy pouring out of its interior heaves upward enormous masses of gas, which then plunge back down with forces far greater than any storm on Earth. Small hot spots send jets of white-hot ionized gas streaming up into the Sun's atmosphere, and occasional eruptions blast billions of tons of matter into space. Astronomers call these various disturbances **solar activity.** Astronomers infer that other stars with convection zones near their surface also have activity in their chromospheres and coronas, because they too exhibit phenomena such as flarelike brightenings.

The Sun's storminess exhibits dramatic kinds of activity that can take place in as little as a few minutes, or sometimes persist for months. As we look over even longer time scales, we find that the Sun goes through a regular pattern of changes every 11 years, and shows other variations over longer periods still. From our safe distance on Earth, these phenomena are dramatic and lovely. Solar storms often take on forms quite unlike anything we might imagine from our experience on Earth, because the Sun's hot ionized gas interacts with its strong, complex magnetic field. Solar activity can also affect us more directly: It can damage spacecraft, interfere with radio communications, trigger auroral displays, and alter Earth's climate.

Background Pathways

Unit 49: The Sun, Our Star 396

51.1 SUNSPOTS

Sunspots are the most common type of solar activity. They are large, dark regions (Figure 51.1) ranging in size from a few hundred to many thousands of kilometers across. Sunspots last from a few days to over a month. They are darker than the surrounding gas because they are cooler (about 4500 K as opposed to the 5800 K of the normal photosphere), but they actually shine very brightly—they appear dark only by contrast. Sprinkling a few drops of water onto a hot electric stove burner will have a similar effect. Each drop momentarily cools the burner, making a dark spot. Four centuries after Galileo's first telescopic observations of sunspots, solar physicists are just beginning to piece together the complex interactions that drive them.

One clue to the puzzle of sunspots was the discovery in the early 1900s that they contain intense magnetic fields. Astronomers can detect magnetic fields in sunspots and other astronomical bodies by the **Zeeman effect,** a physical process in which the magnetic field causes some of the spectral lines produced by atoms to split into two, three, or more components. The splitting occurs because the magnetic field alters the electron orbitals within an atom, which in turn alter the wavelengths of light the electrons can absorb or emit.

Figure 51.2A shows the Zeeman effect splitting spectral lines in a sunspot. The spectrum was taken from a region that cuts across a sunspot. The spectral line is single outside the spot but triple within. By measuring the strength of such splitting across the Sun's face, astronomers can map the Sun's magnetic field, creating a **magnetogram,** as seen in Figure 51.2B. The colors in a magnetogram show the strength and polarity of the magnetic field. The *polarity* indicates whether the magnetic field is "north" or "south"; in other words, it indicates whether a compass

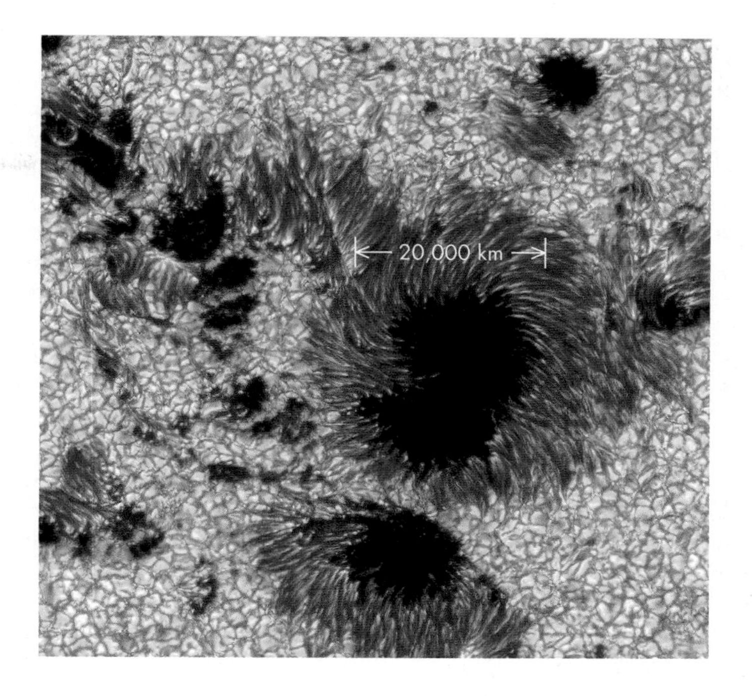

FIGURE 51.1
Image of a large group of sunspots. The darker areas are cooler gas, but they are still bright—they are dark in the image just by contrast to the surrounding hotter regions.

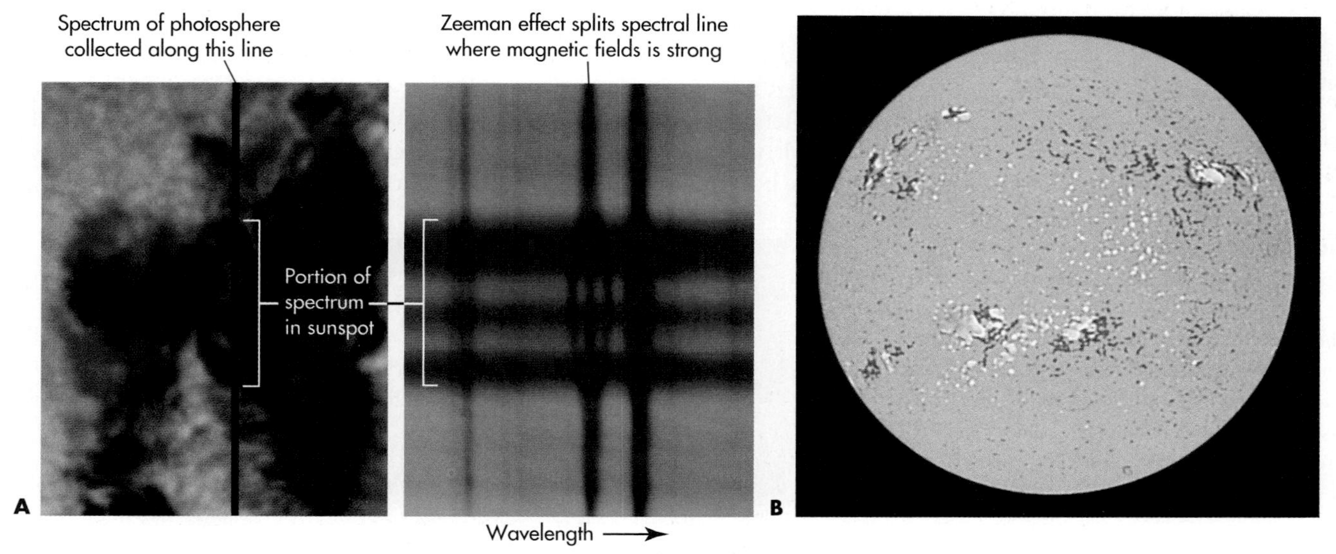

FIGURE 51.2
(A) Spectrum made across a sunspot. The sunspot is shown on the left, and a small part of the spectrum around 525 nm is shown at right. An absorption line produced by iron atoms splits into several slightly different wavelengths in the portion of the spectrum crossing the sunspot. This is because a strong magnetic field slightly alters the atoms' energy levels due to the Zeeman effect.
(B) Magnetogram of the Sun. Yellow indicates regions with north polarity, and dark blue indicates regions with south polarity. Notice that the polarity pattern of spot pairs is reversed between the top and bottom hemispheres of the Sun. That is, in the upper hemisphere, blue tends to be on the left and yellow on the right. In the lower hemisphere, blue tends to be on the right and yellow on the left.

needle would point toward or away from the region. The field is strong around spots and weaker elsewhere.

Curiously, magnetic fields heat the chromosphere and corona (Unit 49.4), but they cool sunspots. This is related to how magnetic fields affect the way hot gas circulates in their vicinity. The magnetic fields of sunspots are sometimes a hundred times stronger than the normal field of the Sun, which is already a hundred times stronger than the Earth's magnetic field. In such intense fields, electrons and

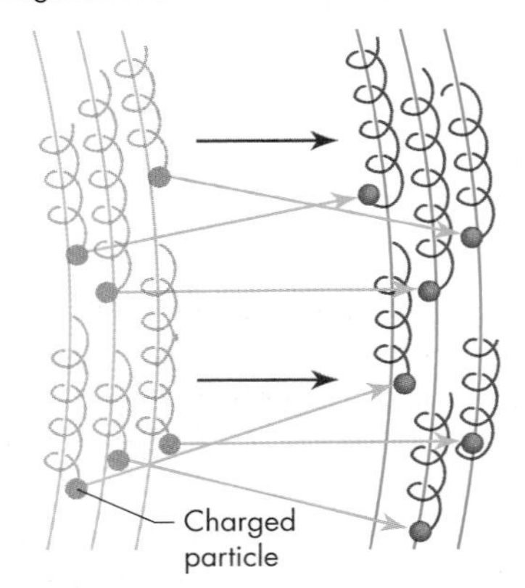

FIGURE 51.3
Charged particles are tied to magnetic field lines, spiraling back and forth along them. If forces move the gas, it can drag the field with it—and vice versa.

other charged particles spiral around the field, tightly tied to it, as illustrated in Figure 51.3. They are thereby forced to follow the direction of the magnetic field as they spiral along it. For example, the Earth's magnetic field deflects particles from striking the Earth in most locations. Instead the particles move along the magnetic fields until they reach low altitudes near the north and south magnetic poles, where they interact with the atmosphere and create the **aurora** (Unit 36.1). Likewise, in the Sun, if the magnetic field does not align with the direction in which the ionized gas is moving, it deflects the gas. Regions of strong magnetic fields may prevent hot gas in the interior of the Sun from rising to the surface, so that the surface cools and becomes comparatively darker, making a sunspot.

A strong magnetic field cannot be the whole explanation for sunspots, though. A region where the magnetic field is strong will normally push itself apart—just as magnets oriented in the same direction repel each other. Why, then, do sunspots persist for days or weeks, instead of rapidly weakening? This was one of the puzzles explored by a joint NASA/European Space Agency spacecraft called the *Solar and Heliospheric Observatory (SOHO).* Using detailed Doppler measurements of the flow and wave patterns of the photospheric gas, scientists were able to construct the three-dimensional image of the motions of the gas below the surface shown in Figure 51.4.

The cooling that occurs in the middle of the sunspot causes the gas there to sink, drawing in gas from the surrounding regions, much as a hurricane or tornado on Earth draws in air from the surrounding regions. *SOHO* measured flow speeds in and down toward the sunspot of about 5000 kilometers per hour (3000 mph). The rapidly flowing gas drags along the magnetic field because, just as the magnetic field exerts a force on the charged particles, the charged particles exert a force on the magnetic field, as required by Newton's law of action and reaction (Unit 15.3). As a result, the flow of gas inward helps trap and intensify the magnetic field, which keeps the hot gas from rising into the region of the sunspot from the layers below.

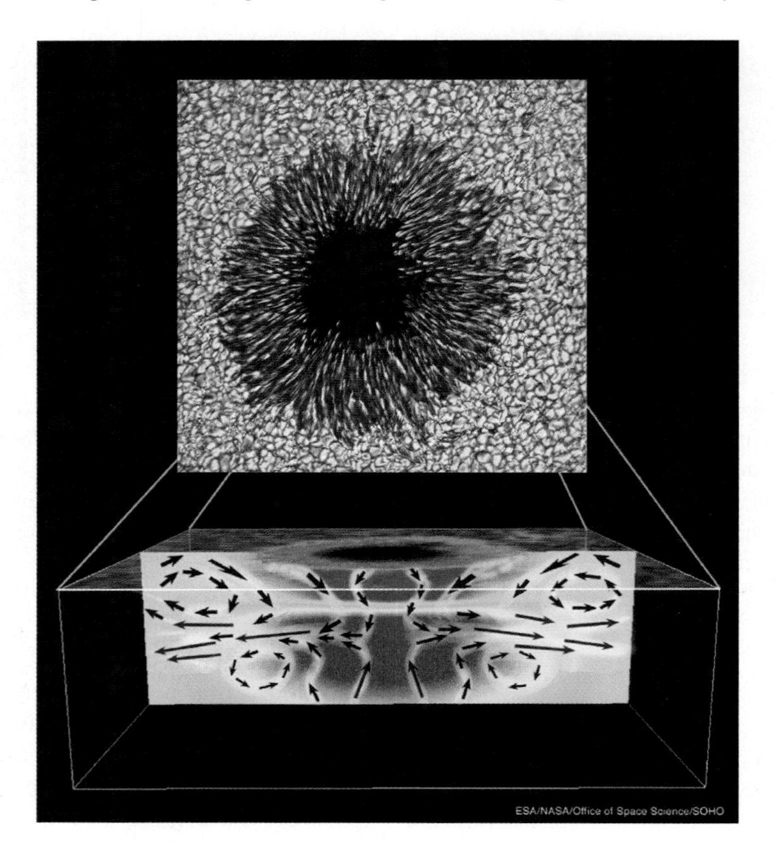

FIGURE 51.4
Flows of gas underneath a sunspot measured by the spacecraft *SOHO* in 2001. The colored diagram at the bottom of the figure shows a slice through the region below the sunspot, indicating where the gas is hottest (red) and coolest (blue). The arrows show the direction of gas flow.

This feedback between the magnetic field and the flow of gas is what makes sunspots persist on the Sun's surface.

51.2 PROMINENCES AND FLARES

ANIMATIONS
Solar prominences

Q Prominences have been found to be less hot than the surrounding coronal gas. How then can they appear brighter than the corona in visible-wavelength images?

The strong magnetic fields associated with sunspots also shape the motions of ionized gas as it flows above the photosphere. **Prominences** are huge plumes of glowing gas that jut from the lower chromosphere into the corona (Figure 51.5A). You can get some sense of their immensity from the white dot in the figure, which shows the size of the Earth for comparison. Time-lapse movies show that gas streams through prominences, sometimes rising into the corona, sometimes raining down onto the photosphere. The flow is channeled by, and supported by, strong magnetic fields that arc between sunspots of opposite polarity. The pressure of the surrounding coronal gas helps confine and support the gas in the prominence (Figure 51.5B–D). Under favorable conditions, the gas in a prominence may remain trapped in its magnetic prison and glow for weeks.

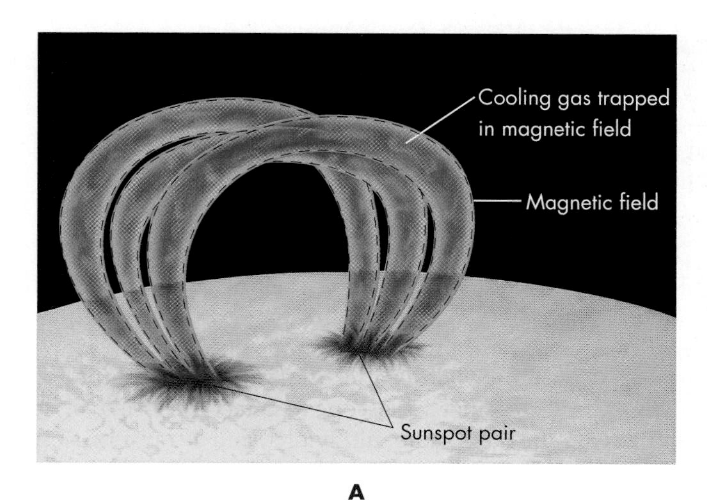

A

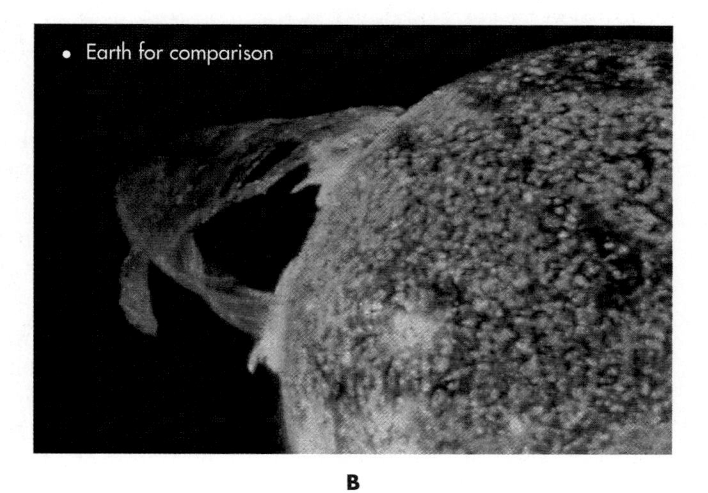

B

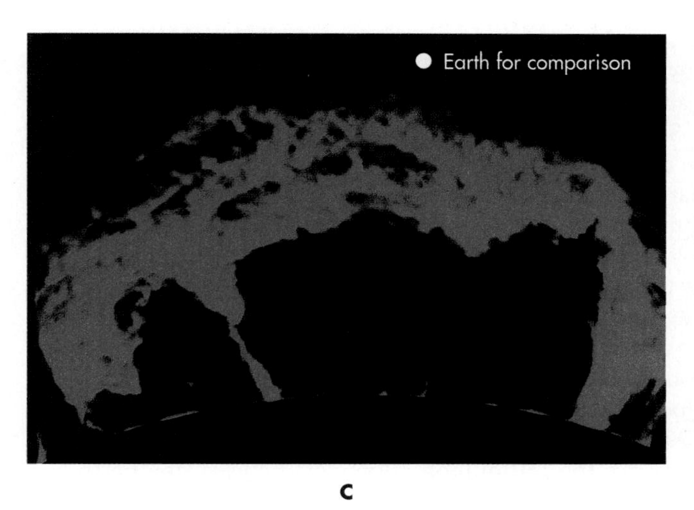

C

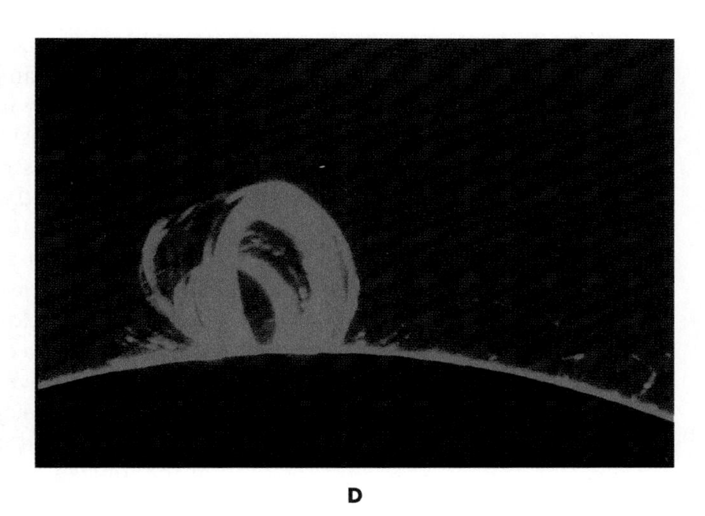

D

FIGURE 51.5
(A) Sketch illustrating how magnetic fields support a prominence. (B–D) Photographs of prominences. Image (B) was made at ultraviolet wavelengths, whereas (C) and (D) were made at the 656-nm-wavelength spectral line of hydrogen (see Unit 24).

FIGURE 51.6
(A) A solar flare (lower right side of image) erupts from the Sun in this ultraviolet image. (B) A coronal mass ejection is recorded at visible wavelengths. A disk blocks light from the Sun, and an ultraviolet image of the Sun made at the same time is superimposed.

Sunspots also give birth to solar **flares,** which are brief but bright eruptions of hot gas in the chromosphere. Over a few minutes or hours, gas near a sunspot may brighten, emitting the energy equivalent of millions of atomic bombs. Such eruptions, though violent, are so localized they hardly affect the visible light output of the Sun at all. Generally you need a specialized telescope to detect the visible light of flares, though they can increase the Sun's radio, ultraviolet, and X-ray emission by factors of a thousand in a few seconds (Figure 51.6A).

One hypothesis suggests that flares arise when the field near a sunspot gets twisted by gas motions. Somewhat like winding up a toy powered by a rubber band, the twisting can go only so far before the rubber band breaks. So, too, the magnetic field can be twisted only so far before it suddenly readjusts, whipping the gas in its vicinity into a new configuration. The sudden motion heats the gas, and it expands explosively. Some gas escapes from the Sun and shoots across the inner Solar System to stream down on the Earth.

Closely related to flares, but on a much larger scale, are **coronal mass ejections.** These are enormous bubbles of hot gas, sometimes containing billions of tons of ionized gas, that are blasted from the corona out into space (Figure 51.6B). The mechanism for ejecting this gas is so powerful that the ionized gas drags the embedded magnetic field along with it. The cause of these ejections is still not well understood, but they seem to be triggered by flares. The most spectacular auroral events on Earth occur when one of these ejections strikes our planet (Figure 51.7).

The aurora usually begins a few days after the ejected outburst. Even though it moves at a speed of 400 kilometers per second, it takes several days for the bulk of the ionized gas to traverse the 150 million kilometers between the Earth and the Sun. However, the most energetic particles from a flare can sometimes reach the Earth in less than an hour. These high-energy "proton storms" are particularly dangerous for astronauts in spacecraft outside of the Earth's magnetic field, and could expose astronauts traveling to the Moon or Mars to dangerous levels of

FIGURE 51.7
Photograph of the great aurora of March 1989. The magnetic storm associated with this aurora shut down the power grid in Quebec, Canada.

radiation. Strong flares and coronal mass ejections can also disrupt communications, disable satellites, and even induce electric currents that force power grids to shut down.

51.3 THE SOLAR CYCLE

Sunspot and flare activity changes from year to year in what is called the **solar cycle.** This variability can be seen in Figure 51.8, which shows the number of sunspots detected over the last 140 years. The numbers clearly rise and fall approximately every 11 years. For example, the cycle had peaks in 1969, 1980, 1990, and 2001. The interval between peaks is not always 11 years: It may be as short as 7 or as long as 16 years. The numbers of flares and prominences also follow the solar cycle.

The solar cycle may be caused by the way the Sun rotates. Gas near the Sun's equator circles the Sun faster than gas near its poles; that is, the Sun spins differentially. This is a property common in many spinning gaseous bodies. The Sun's differential rotation is such that its equator rotates in about

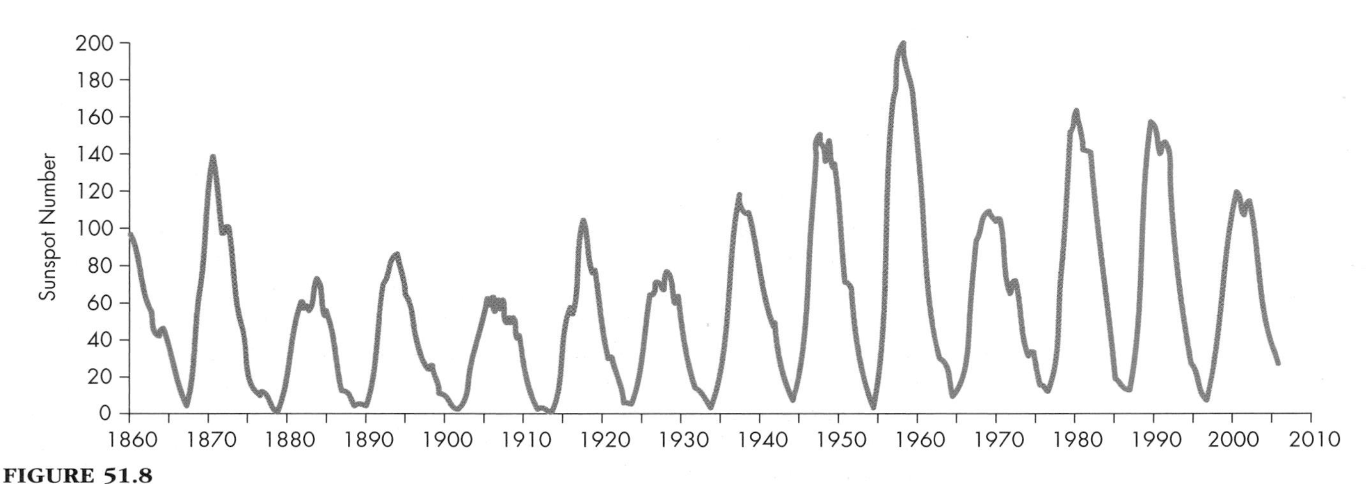

FIGURE 51.8
Plot of sunspot numbers showing the solar cycle. The number of sunspots grows to a peak about every 11 years.

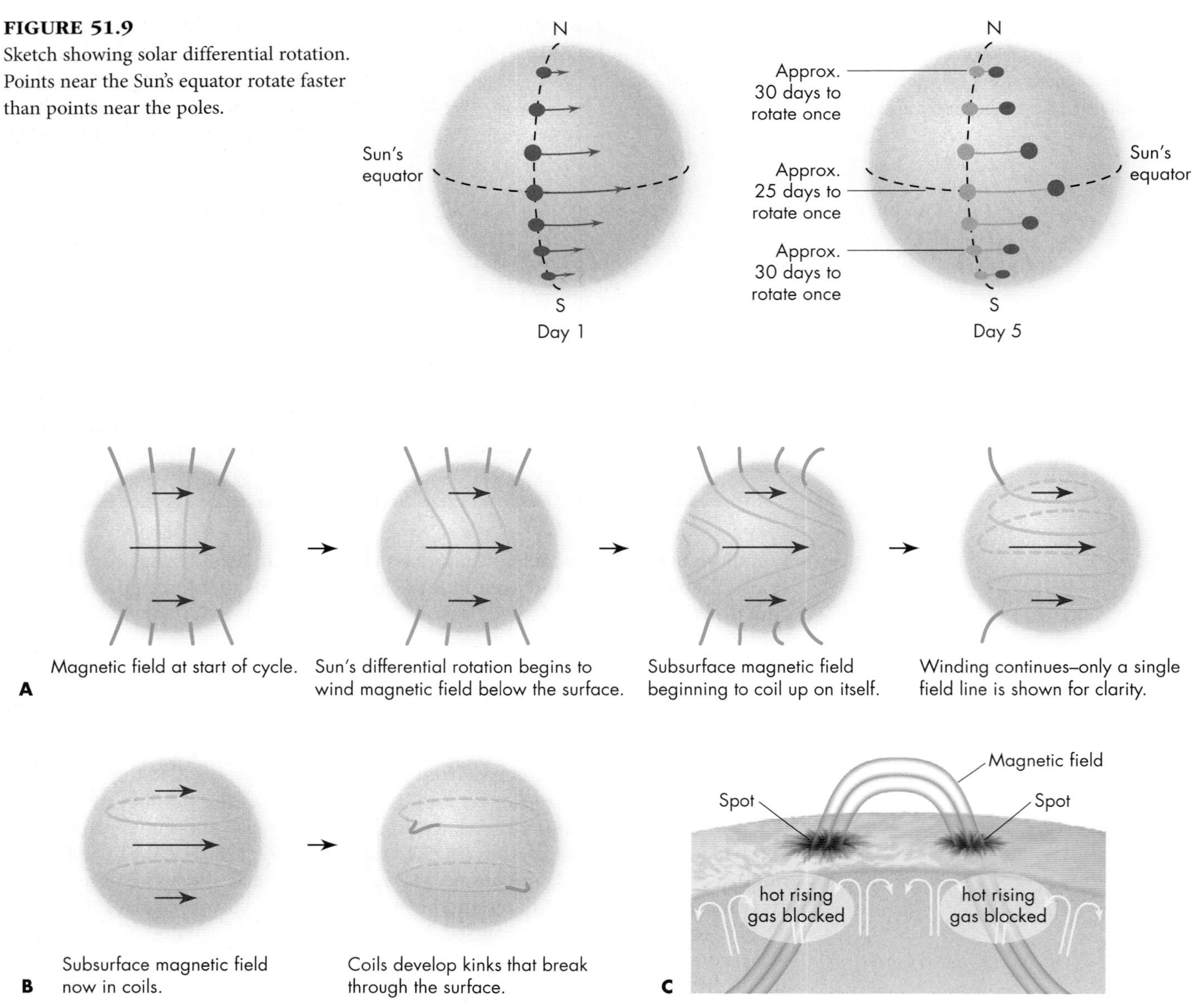

FIGURE 51.9
Sketch showing solar differential rotation. Points near the Sun's equator rotate faster than points near the poles.

FIGURE 51.10
Sketch showing (A) the possible winding up of the subsurface magnetic field; (B) fields penetrating the Sun's surface; and (C) formation of a spot pair.

Twisting of Sun's magnetic field

25 days and its poles in 30. So a set of points arranged from pole to pole in a straight line would move, over time, into a curve, as shown in Figure 51.9.

Differential rotation should similarly distort the Sun's magnetic field, "winding up" the field below the Sun's surface. Although the exact mechanism is still not well understood, astronomers think this winding of the Sun's magnetic field causes the solar cycle. According to one hypothesis, the Sun's rotation wraps the solar magnetic field into coils below the surface (Figure 51.10), making the field stronger and increasing solar activity: spots, prominences, and flares. The wrapping occurs because the Sun's magnetic field is twisted around by the differential rotation of the ionized gas. Sunspots form where the twisted magnetic field rises to the Sun's surface and breaks through the photosphere (last panel of Figure 51.10).

From time to time claims are made that something like, for example, the stock market correlates with the sunspot cycle. Do you think there *could* be a way for sunspots to cause changes in such things? How could you test your ideas?

The 11-year cycle appears to represent the time it takes differential rotation to twist the Sun's magnetic field to the breaking point. When the field becomes too strained, the twisted field lines break and reconnect into a simpler pattern. These events, when the magnetic field suddenly restructures itself, may produce flares and coronal mass ejections. After several years of these violent eruptions the magnetic field has relaxed into an untwisted pattern again. Interestingly, the overall polarity of the Sun's magnetic field reverses after each of these peaks of activity, so that the "north magnetic pole" is in the opposite direction. Somehow the twisting of the magnetic fields reverses the internal circulation of the electric currents that produces the Sun's magnetic field.

The mechanism causing these reversals of the magnetic field is not fully understood, but it is probably similar to what makes the Earth's magnetic field also flip direction. On Earth this occurs every 250,000 years, on average, as indicated by geological evidence (Unit 35.5). Earth's magnetic field reversals are far more sporadic than the Sun's—the last one occurred almost 800,000 years ago, so we are long "overdue" for one. Perhaps through understanding the Sun's magnetic field, we will come to understand the Earth's magnetic field better.

51.4 THE SOLAR CYCLE AND TERRESTRIAL CLIMATE

Climatologists have long wondered how solar activity might affect Earth's climate. Outflows of ionized gas interact with the Earth's magnetic field, but a more important effect may be that as the solar activity varies, the Sun's total power output varies as well. It is counterintuitive, but as the number of sunspots increases, the Sun becomes slightly more luminous by a fraction of a percent. Although sunspots are relatively dim, the surrounding photosphere actually grows slightly brighter, increasing the net output.

The slight changes in the solar luminosity measured during a single 11-year cycle are minor compared to changes that occur over longer time periods. The evidence for such long-term changes is based in part on the work of E. W. Maunder, a British astronomer who studied sunspots. Maunder noted in 1893 that, according to old solar records, almost no sunspots were seen between 1645 and 1715 (Figure 51.11). He concluded that the solar cycle turned off during that period. The period is now called the **Maunder minimum** in honor of his discovery.

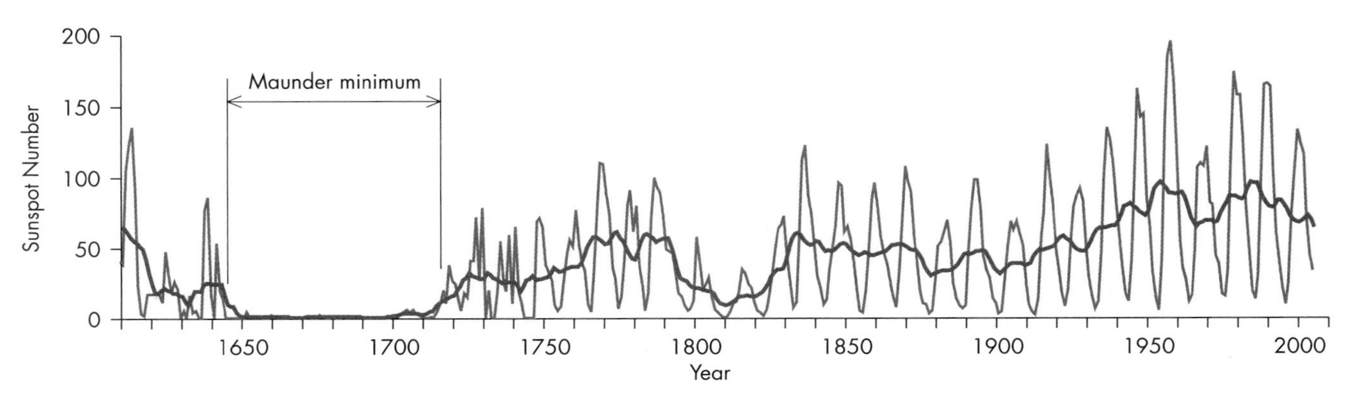

FIGURE 51.11
Plot of the solar cycle extended back to 1610 pieced together from early records. The dark line shows the number of sunspots averaged over 11 years, showing how the Sun's overall level of activity changes with time. During the period of the Maunder minimum, the Sun became almost completely inactive, and Earth experienced a period of cooling.

The Maunder minimum coincides with an approximately 70-year spell of abnormally cold winters in Europe and North America. Glaciers in the Alps advanced; rivers froze early in the fall and remained frozen late into the spring. The Maunder minimum is in the middle of a period of several centuries when the Earth appears to have been cooler—meteorologists call this time the "little ice age." If only one such episode were known, we might dismiss the sunspot–climate connection as a coincidence; but three other cold periods have also occurred during times of low solar activity. This strengthens the case for solar activity affecting our climate. It is suspected that the Sun's luminosity decreases during these periods of solar activity, but there may be other effects as well. A recent hypothesis suggests that energetic particles from the Sun during periods of solar activity stimulate cloud formation on Earth, thereby slowing the escape of heat from our planet.

Although scientists are still unsure about what creates the link between solar activity and our climate, few now doubt that such a link exists. However, there is substantial debate about whether it has as large an impact on climate as other sources. Figure 51.12 shows how the ocean surface temperature (expressed as deviations from the normal average) changed from 1860 to 2005. The figure also shows how the number of sunspots, averaged over the 11-year cycles, changed over the same time span. Until the mid-1980s, solar activity was rising, and its rise showed some degree of correlation with the rise in sea surface temperature. This led some scientists to propose that solar activity was the main cause of global warming. However, the Sun showed fewer sunspots in the last two solar cycles, while sea temperature has continued to rise. For the great majority of scientists, this suggests that other factors, such as human production of greenhouse gases (Unit 36.3), is probably the main cause of global warming.

Q Graphs like those in Figure 51.12 can be quite controversial. What kinds of changes in measurement techniques have occurred over the last 150 years? How might these changes have affected the values plotted?

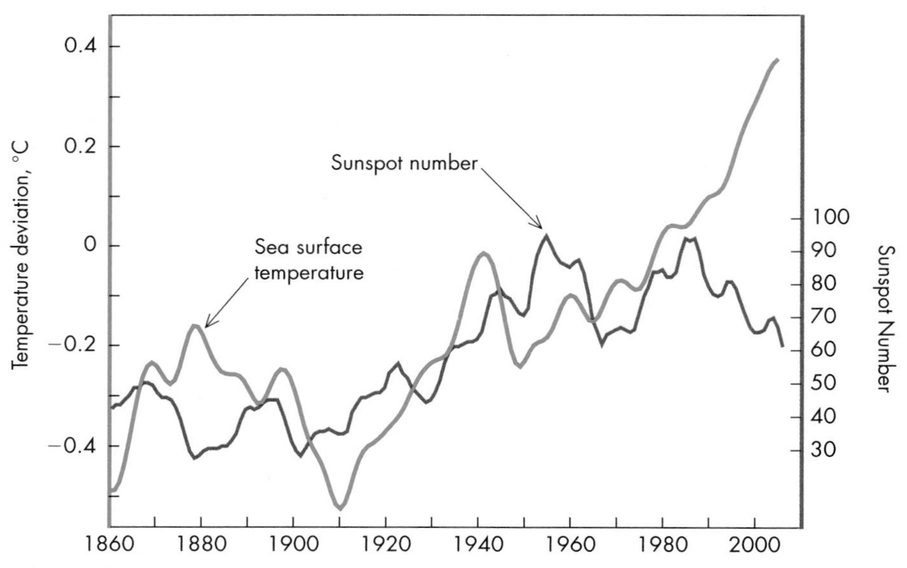

FIGURE 51.12

Curves showing the change in ocean surface temperature on Earth and the change in the average sunspot number since 1860. Notice that the curves change approximately in step until the 1980s. Over the last two decades the oceans have continued to warm while sunspot counts have declined. Astronomers deduce from these curves that solar activity affects our climate, but other factors must explain the recent temperature rise.

KEY TERMS

aurora, 416
coronal mass ejection, 418
flare, 418
magnetogram, 414
Maunder minimum, 421
prominence, 417
solar activity, 414
solar cycle, 419
sunspot, 414
Zeeman effect, 414

QUESTIONS FOR REVIEW

1. Why do sunspots appear dark?
2. What roles does magnetic activity play in solar activity?
3. How do we know there are magnetic fields in the Sun?
4. What are solar flares? What effects do they have on Earth?
5. What is the solar cycle? What happens to the Sun's magnetic field during the solar cycle?
6. What is the Maunder minimum? Why is it of interest?
7. How might solar activity affect the climate?

PROBLEMS

1. If the surface of the Sun near the equator completes one rotation every 25 days, compared to a 30-day rotation near the poles, how long does it take for the surface near the equator to complete one extra rotation around the Sun compared to the poles? After 11 years, how many extra turns would this result in?
2. What is the speed of material ejected by a solar flare if it takes four days to reach the Earth?
3. How long does it take the material ejected by a solar flare to reach the orbit of Neptune?
4. The amount of light emitted by a hot region is proportional to the temperature to the fourth power ($L \propto T^4$). If a sunspot is at a temperature of 4500 K, how much less luminosity does it generate than if it were at 6000 K?
5. Using the reasoning in problem 4, how much more luminous is a solar flare at a temperature of 1 million K than a similar-size region of the Sun at 6000 K? Why don't we see this with our own eyes as an obvious brightening of the Sun?
6. Describe three different impacts of solar activity on Earth. Discuss which of these is the most important in the short term and in the long term.

TEST YOURSELF

1. About how many years elapse between times of maximum solar activity?
   a. 3
   b. 5
   c. 11
   d. 33
   e. 105
2. Sunspots are dark because
   a. they are cool relative to the gas around them.
   b. they contain 10 times as much iron as surrounding regions.
   c. nuclear reactions occur in them more slowly than in the surrounding gas.
   d. clouds in the cool corona block our view of the hot photosphere.
   e. the gas within them is too hot to emit any light.
3. Differential rotation results in
   a. the solar wind.
   b. a wound-up magnetic field.
   c. the Maunder minimum.
   d. the Sun's generation of energy.
   e. All of the above.

unit

52 Surveying the Stars

Determining the distance of a remote object is one of the fundamental problems that astronomers must tackle. Knowing stars' distances is necessary for determining many of their most basic properties such as their diameters, masses, and the amount of light they emit.

In the last few decades, space exploration has allowed us to measure many distances within the Solar System directly. Spacecraft have traveled to most of the planets and their moons. As a result, we can time how long signals take to travel (at the speed of light) back and forth between the object being visited and the Earth. Ground-based technology has also advanced, so we do not have to rely on space probes to measure distances accurately. Instead, we can bounce radar signals off an asteroid, for example, and time how long the signals take to travel there and back. With this method we can determine a distance accurate to within a few meters (Figure 52.1).

Beyond the Solar System, however, distances are so vast that such direct techniques are impossible and will remain so for centuries to come. The situation is not hopeless, though. We can take a cue from the ancient Greeks, who were able to estimate the distances to the Moon and Sun by applying geometry (Unit 10) with only the crudest of astronomical instruments. The same geometric methods that a surveyor might use to measure distances on the Earth allow us to measure the distances of nearby stars, and such methods provide a fundamental stepping-stone to learning about the objects that lie far beyond the Solar System.

52.1 TRIANGULATION

Long before radar provided us with high-accuracy distance measurements, and long before space probes ventured out into the Solar System, some planetary distances were very accurately determined by a geometric method, used by surveyors, called **triangulation.** We construct a triangle in which one side is the distance we seek, but cannot measure directly, and another side is a distance we can measure, as shown in Figure 52.2A.

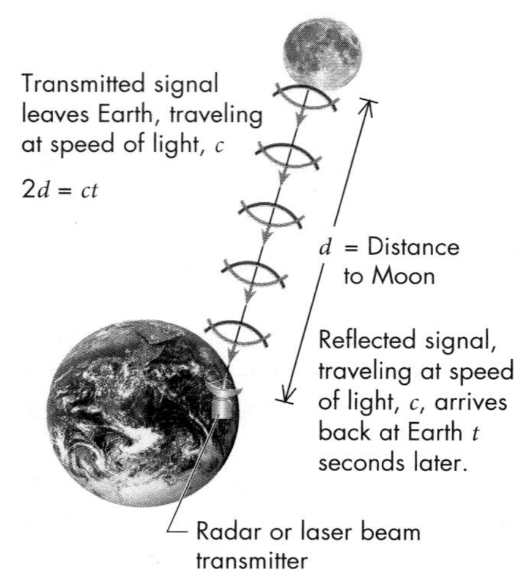

FIGURE 52.1
The distance to the Moon can be measured very precisely by bouncing radar or laser signals off the Moon and timing how long they take to return.

Suppose we want to measure the distance across a deep gorge. We can construct an imaginary triangle to a point on the far side of the gorge. Two sides of this triangle span the gorge, while we can choose the length and direction of the third side—the **baseline**—along the edge we are on, as shown in Figure 52.2B. It simplifies our calculations if we choose the baseline so it is perpendicular to the line straight across the gorge. Then, by measuring the length of the baseline and the angle α between the baseline and the remaining side (the *hypotenuse*), we can determine the distance across the gorge from a scale drawing of the triangle.

Solving this type of problem is the motivation behind the area of mathematics called *trigonometry* (literally meaning "the measuring of triangles"). Trigonometric formulas allow us to calculate the length of the sides without requiring us to make precision scale drawings of the triangle. In the figure, the baseline is 5 meters, and the angle α is 63.4°. Constructing a right triangle with one of the angles at this

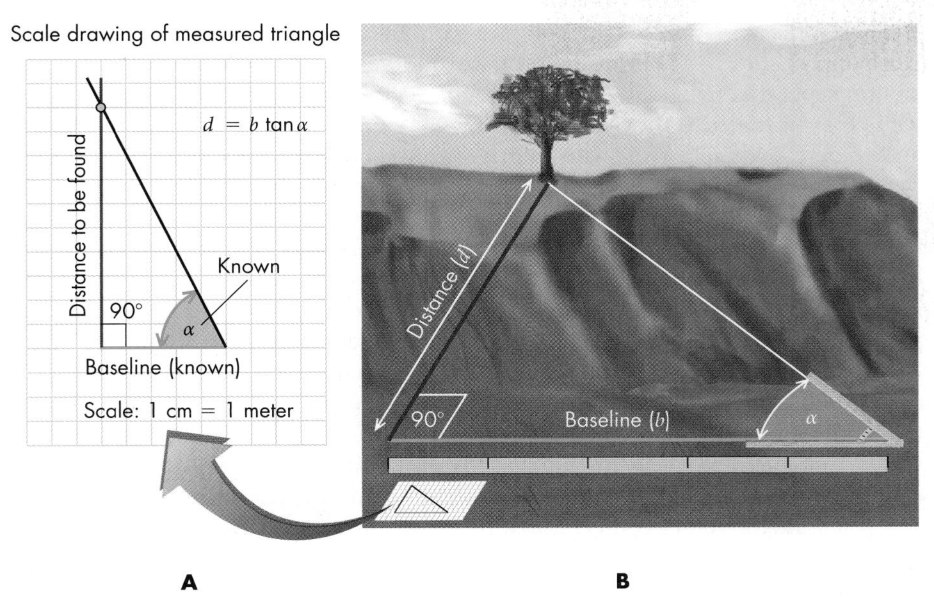

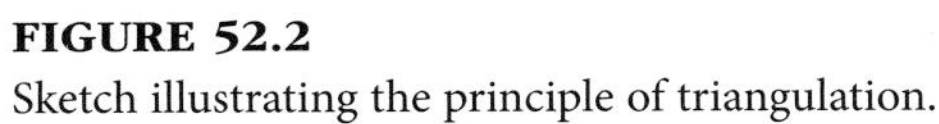

FIGURE 52.2
Sketch illustrating the principle of triangulation.

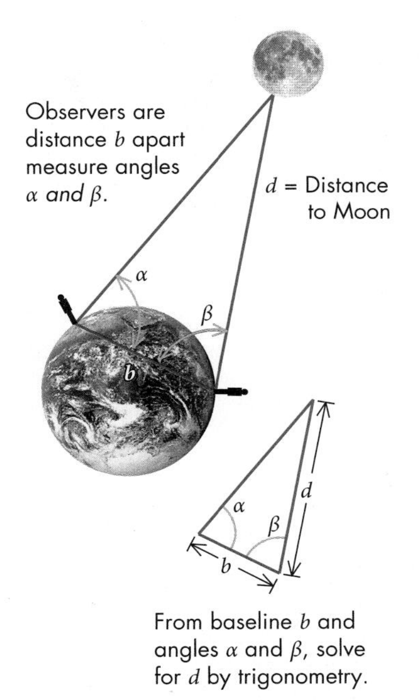

FIGURE 52.3
Finding the distance from the Earth to the Moon by triangulation.

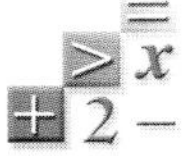

Using trigonometry, the distance, *d*, across the gorge equals the baseline length, *b*, times the "tangent" of the angle *A*: $d = b \times \tan \alpha$.

Q To make triangulation measurements more accurate, surveyors usually make the baseline as big as possible. Why would this improve the results?

The exact values of the distances to Venus and Mars at closest approach vary because the orbits are elliptical, but Kepler also determined these shapes.

value, we find that the unknown distance across the gorge must be two times the baseline, or 10 meters.

As early as about 140 B.C.E. the Greek astronomer **Hipparchus** used triangulation to estimate the distance to the Moon. He compared the angle of the Moon with respect to the Sun, as seen from two locations during a solar eclipse. A similar experiment can be conducted today by two observers at different locations, simultaneously measuring the position of the Moon in the sky, who then compare their observations. From distant points on the globe, this difference can be as large as about 2°, an angle that can be measured with fairly simple instruments (Figure 52.3). The principle is no different from the idea of measuring the distance across a gorge. However, determining the length and angle of the actual baseline when the Earth's surface is curved makes this a more difficult calculation.

Triangulating distances to objects more distant than the Moon requires greater accuracy. Because of the much larger distances involved, the angles may be only a small fraction of an **arc minute.** An arc minute is 1/60th of a degree, or about 1/30th of the angular size of the Moon. Nonetheless, the ancient Greeks attempted to find the Sun's distance from Earth. Their result was far too small because instruments at the time were simply too crude to make a sufficiently accurate measurement of the angles. Because the Earth–Sun distance—known as the **astronomical unit** or AU—is so fundamental to our knowledge of the scale of the universe, we will discuss it in some detail here.

It turns out that we do not have to measure the distance to the Sun to determine the size of the AU. Johannes Kepler's discoveries in the early 1600s established the *relative* distances of the Sun and planets (Unit 12). For example, Kepler's law relating orbital size and the period of the orbit established that Mars's orbit is, on average, 1.52 times larger than the Earth's—1.52 AU—while Venus's is 0.72 AU. Therefore Mars is about 0.52 AU away from Earth at its closest approach and Venus is about 0.28 AU. So if we could triangulate the distance to Mars during one of its close approaches, we could determine the size of the AU.

Such a measurement of the distance to Mars was made during one of Mars's close approaches to Earth in 1672 by astronomers in Paris and South America, at sites separated by about 7000 km (about 4000 miles). We know today that the value they obtained was correct to about 10%, but it implied a distance to Mars—and therefore a size for the astronomical unit—that was almost 10 times greater than had been believed previously. This led many to question the accuracy of the value obtained. If the new distance was correct, the distance to the Sun that had been used for centuries was far too small. Recall that sizes of objects are measured by the angular size versus distance formula (Unit 10.4). If the distance was actually 10 times larger, then the size of the Sun would have to be 10 times larger as well. Not only was the Sun even bigger than previously thought, the bigger value of the AU also implied that Jupiter and Saturn were much larger than the Earth—a challenging idea to accept in an age when even the concept of the Earth orbiting the Sun was new. Because of the controversial nature of these results, astronomers sought independent measurements to confirm these findings. It took almost a century, however, before the distance of Venus could be determined.

Q Why do you suppose scientists are usually reluctant to accept new results that are very different from previously accepted results?

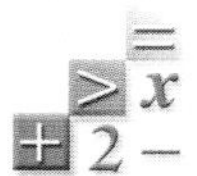

Halley's technique is based on the time it takes Venus to cross the face of the Sun seen from different locations north and south on Earth. By comparing the duration, we can determine how much closer Venus was to crossing near the Sun's equator or pole from each location, giving the precise angular shift north or south relative to the Sun.

Even though Venus comes closer to Earth than Mars, triangulating its distance is difficult because when it is closest to Earth, it is lost in the glare of the Sun. However, the British astronomer Edmund Halley (of comet Halley fame) developed a method for determining both Venus's and the Sun's distance. He did this by timing how long Venus takes to move across the face of the Sun, from different locations on Earth, during one of Venus's rare **transits.** Because its orbit is tilted relative to our own, Venus usually lies slightly "north" or "south" of the Sun's disk as it passes between the Earth and Sun. But twice every 120 years or so it crosses directly in front of the Sun. This last happened in 2004 (Figure 52.4) and will next happen June 6, 2012. After that, it will not occur again until 2117.

For the transit of 1761, British astronomers Charles Mason and Jeremiah Dixon sailed to the southern tip of Africa, even braving cannon fire (Britain was at war with France) to get to their destination. The observations of Mason and Dixon provided the long baseline needed for this astronomical surveying project. (Other astronomers were not so lucky, spending years traveling to remote parts of the globe, only to have clouds prevent them from making the critical measurements on the day of the transit.) The results of the Venus transit measurements confirmed the Mars measurements, yielding a size for the astronomical unit to within about 1% of the modern value.

Mason and Dixon were summoned to the American colonies in 1763 to settle a dispute about the boundary between Pennsylvania and Maryland. This became known as the "Mason–Dixon Line," which figured prominently in the history of the United States in disputes between slave and free states.

In the end, however, the triangulation method of surveying is limited by the size of the baseline that we can obtain on the Earth. Even the most accurate measurements are nowhere near precise enough to tell us about the vast distances to stars if our baseline is at most the 12,800-km diameter of the Earth. A much bigger baseline is needed.

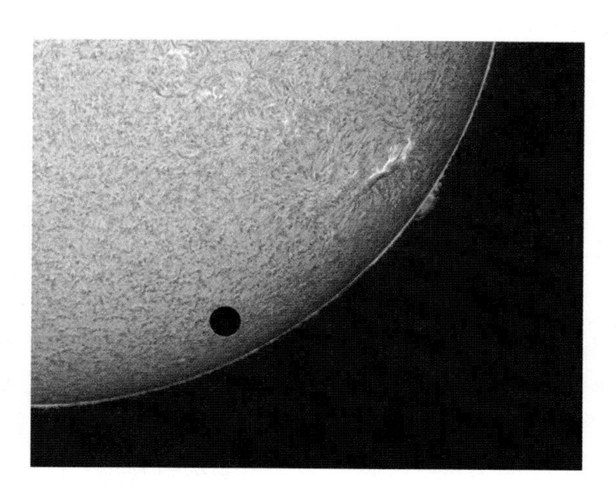

FIGURE 52.4
This image shows Venus (the round, dark circle) passing between us and the Sun on June 8, 2004. By precisely timing how long Venus takes to transit the Sun's disk from different locations on Earth, we can find Venus's distance.

52.2 PARALLAX

INTERACTIVE

Stellar parallax

ANIMATIONS

Parallax

There is a baseline available to us that is more than 20,000 times larger than the size of the Earth—the Earth's orbit. As we move around the Sun, we shift our location in space by about 300 million kilometers (185 million miles) from one side of our orbit to the other. Even the ancient Greeks recognized that if the Earth moved around the Sun, stars should show shifts in their positions called **parallax** (see Unit 10).

An easy way to demonstrate the principle of parallax is to hold your thumb pointed up, motionless, at arm's length, and shift your head from side to side. Your thumb seems to move against the background even though in reality it is your head that has changed position. This simple demonstration also illustrates how parallax gives a clue to an object's distance. If you hold your thumb at different distances from your face, you will notice that the apparent shift in your thumb's position—its parallax—is also different. Its parallax is larger if your thumb is close to your face, and smaller if it is at arm's length (Figure 52.5). That is, for a given motion of the observer:

Distance is inversely proportional to parallax.

This law is just as true for stars as it is for your thumb.

Using parallax to find distance may seem unfamiliar, but parallax creates our stereovision—the ability to see things three-dimensionally. When we look around with two eyes, each eye sends a slightly different image to the brain. Your brain processes the two images and determines the distances to various objects in your field of view. The effect of parallax can be seen if you extend your arm in front of you, with your thumb pointed up. Look first through one eye and then the other, while keeping your head stationary. Notice that your thumb seems to shift to the left when looking with the right eye and then to the right when looking with the other eye.

You can demonstrate the importance to your brain of comparing the two images simultaneously by trying some experiments that involve covering one

A

B

FIGURE 52.5
When you hold your thumb at arm's length and shift your head from side to side or look through one eye and then the other, you see your thumb shift by a small amount relative to the background. This shift is *parallax*. If you hold your thumb close to your face and do exactly the same thing, your thumb shifts more against the background. This demonstrates that parallax is larger for nearer objects.

eye. Cover either eye, then drop a coin on a table in front of you. Hold your arm straight out in front of you, and with your first finger pointing down, try to lower it so that the tip of your finger comes down directly on the center of the coin. If you try this again with both eyes open, you'll see this is much easier using two eyes than one. With one eye closed, you may find yourself moving your head side to side (another way to estimate parallax) or looking for other visual cues to compare and adjust your hand's distance from you as you move it downward. Another experiment you can do begins by standing in front of a hanging cord (from a lightbulb or window shade, for example). Hold one arm out to the side of the cord with a finger pointed toward it. Now bring your pointed finger in from the side, and try to touch the cord with the tip of your finger. Without stereovision, this is surprisingly difficult.

Even after the Copernican revolution established that the Earth moves around the Sun, stellar parallax was not detected until centuries later. The failure to detect stellar parallax was one of the reasons some astronomers continued until the 1700s to believe that the Earth did not move through space.

The stellar parallax technique is a little different from the triangulation method. Instead of measuring the angle between two views made along different lines of sight at the same time, we need to compare measurements made six months apart. For even the nearest star, the change in position, or **angular shift,** relative to distant stars is tiny.

To achieve such precision for observations made at different times, astronomers observe a star and carefully measure its position against background stars, as shown in Figure 52.6A. They then wait six months, until the Earth has moved to the other side of its orbit, and make a second measurement. As Figure 52.6B shows, the star will have a different position when compared with the background of stars, as seen from the two different points six months apart. The amount by which the star's apparent position changes depends on its distance from the Earth, just as the angular shift of your thumb depends on how far you hold it away from your eyes.

In the late 1500s, from his observatory at Uraniborg, Tycho Brahe carried out measurements of star positions that could have detected an angular shift of about 1 arc minute. When he failed to detect a shift, he concluded that the Earth must be stationary. Astronomers recognized that this failure to detect parallax might be because the stars are very far away, but this implied distances to the stars that seemed absurdly large to many of the time. In fact, it required the invention of the telescope and more than two centuries of improvements in optics and measurement techniques before the parallax shift was finally detected.

The distances of stars are so large that the parallax effect is extremely small—so small that it is measured in the fractions of a degree called **arc seconds.** One arc second is 1/3600th of a degree, or 1/60th of an arc minute. It may help you visualize how tiny an arc second is if you keep in mind that 1 arc second is equivalent to the angular size of a U.S. penny from a distance of 4 kilometers (2.5 miles).

52.3 CALCULATING PARALLAXES

Q If you lived near one of these stars and could look back and observe the Earth orbiting the Sun, how would the motion of the Earth compare to the parallactic motion of your star as seen from Earth?

As the Earth orbits the Sun, stars appear to move in the sky in a direction opposite to the Earth's orbit. Each of them moves along a small circle, ellipse, or line, depending on their position relative to the Earth's orbital plane. For all but a few of them, the distances are so large that their positional shifts are virtually undetectable. For some nearer stars, however, we can detect a shift relative to the background stars.

Mathematically, astronomers define a star's parallax, p, as the angle by which a star shifts to each side of its average position (see Figure 52.6C). Because

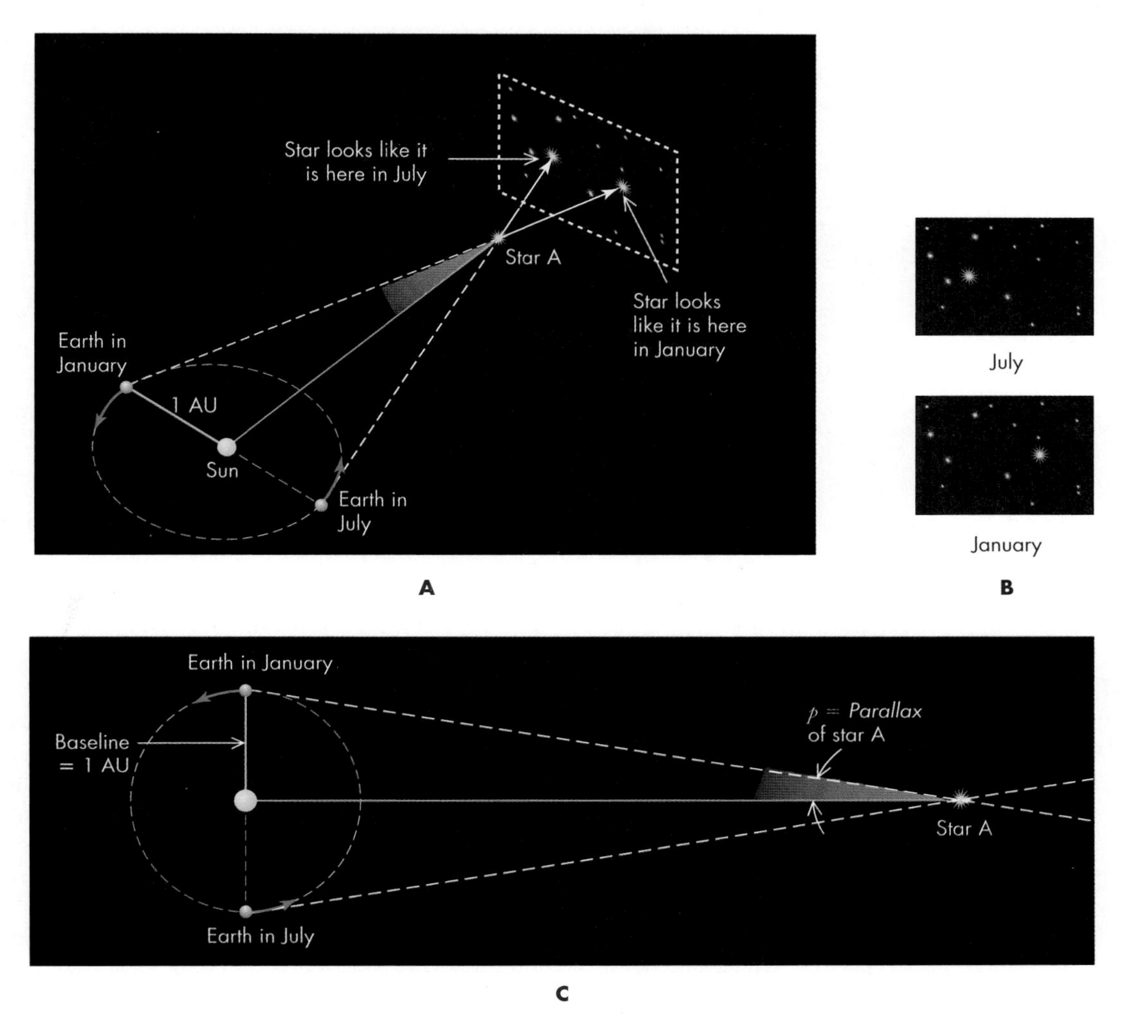

FIGURE 52.6
(A) Triangulation to measure a star's distance. The radius of the Earth's orbit is the baseline. (B) As the Earth moves around the Sun, the star's position changes as seen against background stars. (C) Parallax is defined as one-half the angle by which the star's position shifts. Sizes of bodies and their separation are exaggerated for clarity.

the distance d is inversely proportional to parallax, as discussed earlier, we can write

$$d = \frac{\text{constant}}{p}$$

where the constant depends on the size of the baseline (2 AU) and the units in which we measure angles and distances.

To make this equation easier to use, astronomers invented a new distance unit, called the **parsec,** by setting the constant in this equation equal to 1. If we measure p in arc seconds and set the constant to 1, the value of d from this formula comes out as parsecs. That is,

$$d_{\text{pc}} = \frac{1}{p_{\text{arcsec}}}$$

We have added the subscripts "pc" (for parsecs) and "arcsec" (for arc seconds) as reminders that for this formula to work, the parallax must be measured in arc seconds, which will result in a distance in units of parsecs.

Figure 52.7 shows how we can determine the size of a parsec from the geometry of a right triangle with a baseline of 1 AU and a parallax of 1 arc second. From the

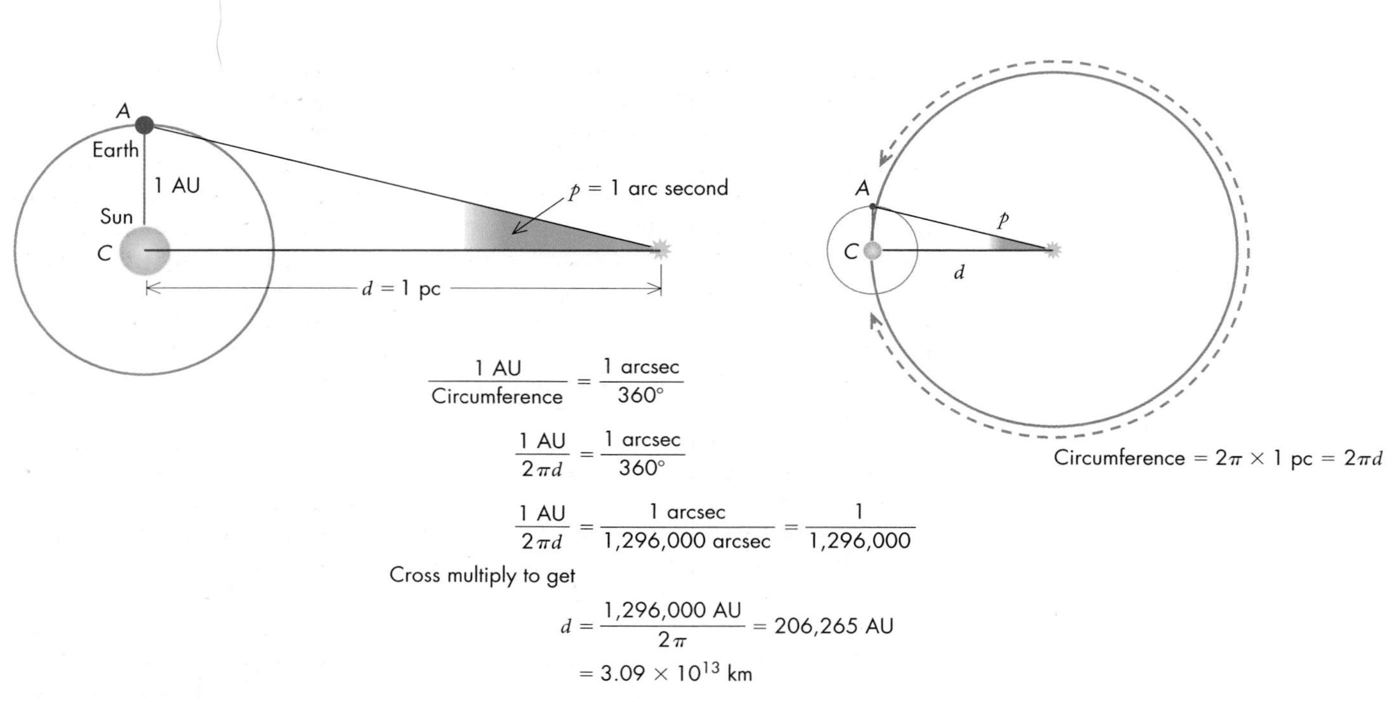

FIGURE 52.7
How to determine the distance of a star with a parallax of 1 arc second.

way we have defined a parsec, someone one parsec from the Sun would see the Earth move 1 arc second to either side of the Sun as it orbited. Therefore, 1 astronomical unit is to the circumference of a 1-parsec-radius circle as 1 arc second is to 360°, as shown in the figure. Using the known size of an AU then allows us to determine that the size of a parsec is

$$1 \text{ parsec} = 3.09 \times 10^{13} \text{ kilometers.}$$

It turns out that the parsec is just a few times larger than the size of the light-year:

$$1 \text{ parsec} = 3.26 \text{ light-years.}$$

The word *parsec* comes from a combination of *parallax* and *arc second.* Most astronomers use parsecs when discussing distances—they are easy to calculate from parallax measurements. However, light-years are also frequently used, particularly when we discuss the time light takes to travel from an object to Earth—the conversion between travel time and distance is easy to calculate. It is a little like the difference in describing distances in meters versus feet.

To see how we can use this relation, suppose, for example, that we find from the shift in position of a nearby star that its parallax is 0.25 arc seconds. Its distance is then $d = 1/0.25 = 4$ parsecs. Similarly, a star whose parallax is 0.1 arc second is at a distance $d = 1/0.1 = 10$ parsecs from the Sun. Parallax measurements of nearby stars reveal that stars are typically separated by a few parsecs.

The star **Proxima Centauri** has the largest known parallax: 0.772 arc seconds. This indicates that it is at a distance $d = 1/0.772 = 1.30$ parsecs (or 4.22 light-years) from the Sun. Proxima Centauri is a dim companion of two slightly more distant stars, in orbit about each other, that have a parallax of 0.742 arc seconds, indicating a distance of 1.35 parsecs (or 4.40 light-years) from us. The larger of these two companions, Alpha Centauri A, is the third brightest star in the sky and is very similar to the Sun. This system of stars rises above the horizon only from latitudes south of 30° N and is too far south to be seen from most of the Northern Hemisphere.

Recognizing that these three stars are, in fact, close to one another in space is already telling us something interesting. Not all stars are as isolated from other stars as the Sun is. Surveying other nearby stars, we see that some are isolated but most are members of small systems of stars.

Although the parallax–distance relation is mathematically a simple formula, obtaining a star's parallax to use in the formula is difficult because the angle by which the star shifts is extremely small. It was not until the 1830s that the first parallax was measured by the German astronomer Friedrich Bessel. Even now, the method fails for most stars farther away than about 100 parsecs. This is because the Earth's atmosphere blurs the tiny angle of their shift, making it very difficult to measure. Astronomers can avoid such blurring effects by observing from above the atmosphere. The European Space Agency launched the satellite *Hipparcos*—a clever acronym standing for HIgh-Precision PARallax COllecting Satellite in honor of the ancient Greek astronomer Hipparchus—which measured the parallax of almost 120,000 stars from space. With its data, astronomers can accurately measure distances to stars as far away as about 500 parsecs.

The tiny parallaxes of stars tell us that they are so far away that it is hard to even imagine the distances involved. Astronomers throughout history have had as difficult a time understanding the immensity of these distances as have *any* of us, when first hearing of such numbers. Modern measurements tell us that the Sun is about 150 million kilometers (about 93 million miles) away, and that the next nearest star is about 42 trillion kilometers (about 25 trillion miles) away. One way of thinking of these huge distances is in terms of the time light takes to travel that far. Light moves so fast that it could circle the Earth almost eight times in 1 second. Light takes about 1.3 seconds to reach us at this speed from the Moon, and about 8½ minutes to reach us from the Sun. But even from the very nearest star, Proxima Centauri, light takes 4.22 years to arrive here at Earth, and for most of the stars we can see at night, their light takes hundreds or thousands of years to reach us.

52.4 MOVING STARS

Although it took more than two centuries after Tycho Brahe before telescopes and techniques were accurate enough to detect parallax, astronomers discovered other motions of the stars over a century earlier. Observed over many human lifetimes, the positions of stars appear to remain fixed on the celestial sphere. This was one of the reasons it was easy for early astronomers to suggest that what we see at night are glowing dots on a celestial sphere that surrounds us. However, if we could watch the stars for millions of years, we would see them moving about like bees in a swarm.

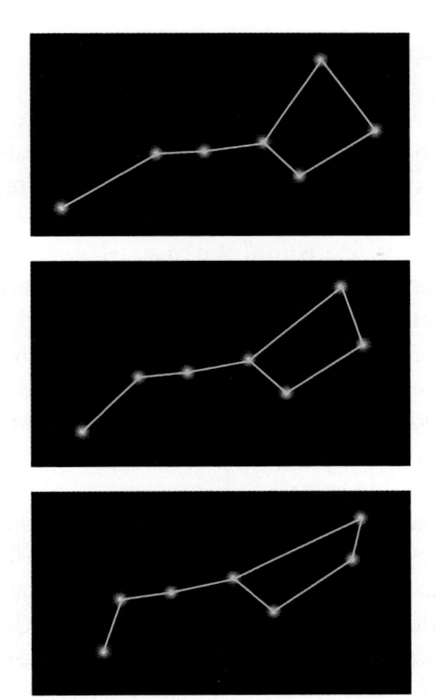

FIGURE 52.8
Proper motion of stars gradually alters their positions on the sky. The three figures show the pattern of these stars 50,000 years ago (top), today (middle), and 50,000 years in the future (bottom).

In the short history of human records of the sky, the motion of stars is almost imperceptible. Nonetheless, in 1718, by comparing Tycho Brahe's star positions to those listed in ancient catalogs, Edmund Halley discovered that stars were shifting positions. This effect is called **proper motion**—the term *proper* here indicates a property of the star, as opposed to an *apparent* effect caused by the Earth's own motion.

As a result of proper motion, the pattern of stars we see gradually changes. The configuration of constellations 50,000 years from now will be significantly different from their appearance today. The changing appearance of the Big Dipper, for example, is illustrated in Figure 52.8.

Because they are close to us, some stars move relatively rapidly over time on the celestial sphere. For example, Barnard's Star moves more than 10 arc seconds each year, and other nearby stars, such as Proxima and Alpha Centauri, also move by several arc seconds each year. Most stars, however, shift position much more slowly—not necessarily because their speed through space is slow, but because they are so far away that they must move a large distance for us see a significant angular shift. This is the same effect we see when a nearby car moves past us rapidly, while an airplane overhead seems to move much more gradually, even though it is traveling at a much greater speed than the car.

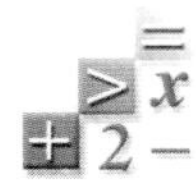

Because the radial and transverse velocities are perpendicular to each other, the magnitude of a star's space velocity can be determined by the Pythagorean theorem:

$$V = \sqrt{V_R^2 + V_T^2}$$

The actual speed of the star across, or *transverse to,* our line of sight is called its **transverse velocity**. The transverse velocity, V_T, can be determined from the proper motion and distance of the star. Astronomers usually use the Greek letter μ (mu) to represent the proper motion in arc seconds per year. In that case the transverse velocity turns out to be

$$V_T = 4.74 \text{ km/sec} \times \mu_{\text{arcsec/yr}} \times d_{\text{pc}}$$

where μ is measured in arc seconds per year and d is measured in parsecs as indicated by the subscripts after each.

By the early 1800s, careful examination of stars' positions also showed that some stars orbited each other (see Unit 56); and by the late 1800s astronomers were also able to detect stars' motions from their spectra, using the Doppler shift (Unit 25). This is the change in wavelength of light from a source as the source moves toward or away from us. If a source moves toward us, its wavelengths are shortened, whereas if it moves away, they are lengthened. The amount of wavelength shift depends on the source's speed along our line of sight—its **radial velocity.** Note that a star will have a radial velocity only if its motion has some component toward or away from us, and it will have a transverse velocity only if it has a motion perpendicular to its direction from us. The overall **space velocity** of the star can be determined by vector addition of the star's radial and transverse velocities.

Q If a star has proper motion, is orbiting another star, and has a measurable parallax, what will its motion through the sky look like to us? How can we distinguish the different motions?

What we learn of stars' motions by measuring Doppler shift and radial velocity can give us clues about an individual star's past. For example, we have found some stars that are moving at many hundreds of kilometers per second relative to all other stars. There is evidence that some of these "runaway" stars were flung out of orbit after a second star they were orbiting exploded!

52.5 THE ABERRATION OF STARLIGHT

The Earth's own space velocity has a surprising effect on the positions of stars. This was discovered in 1729, over a century before parallax was detected. James Bradley, an English astronomer, discovered a shift in the position of stars during the course of the year. At first he believed that this was the long-sought parallax, but it proved to be a different effect of Earth's motion around the Sun.

Bradley observed that all stars in the sky exhibit a shift in different directions depending on the time of year. The shift reaches a maximum of about 20 arc seconds for all stars at certain times during the year depending on their position on the sky. This is about 25 times larger than the largest parallax shift, and it does not depend on a star's distance. Instead it is equally large for all stars in the same part of the sky. This makes it a difficult effect to observe directly because we do not see it as a shift of one star relative to other stars seen near it.

The shift of 20 arc seconds represents the angle of a right triangle with one side represented by c = 300,000 km/sec and another by the speed of the Earth $V_\oplus$ = 30 km/sec. This angle is $360° \times (V_\oplus/c)/(2\pi) = 0.0057° =$ 20.6 arc seconds.

What Bradley saw is easiest to visualize for a star that passes straight overhead. Imagine a telescope that is cemented into the ground and points straight up to the zenith. We would expect a star that passes straight overhead to pass through the middle of the field of view of the telescope once every rotation of the Earth. An example of the effect we might see instead is that the star was shifted 20 arc seconds to the north when it passed through the field of view in September. Then in December the star crosses through the field of view a little late; in March the star passes 20 arc seconds to the south of the zenith position, and in June it crosses through a little early.

If we marked all stars' positions throughout the year, we would see that they move along little ellipses or circles on the sky. This is similar to the kind of motion that is caused by parallax, but the stars always shift in a direction *ahead* of the Earth's motion through space as the Earth orbits the Sun. To center a star in a telescope, you always have to point the telescope a little ahead of the actual direction toward the star.

A star will pass through the center of the field of view of a stationary telescope once every sidereal day (Unit 7) of 23 hours 56 minutes.

Today astronomers call this shift in the positions of stars the **aberration of starlight,** and they recognize that it occurs because of the finite speed of light. We must tilt a telescope so it points ahead of the direction to a star because between the time light enters the front end of the telescope and the time it reaches the back end of the telescope, the Earth has shifted position in space (Figure 52.9). This is similar to the reason you should hold an umbrella tilted slightly ahead of you if you are running through vertically falling rain. To keep the lower part of your body dry, you need to block the raindrops that are falling where your body *will* be.

If you ran through the rain around a circular racetrack, you would tilt your umbrella ahead of you in whatever direction you ran, in the same way astronomers must tilt their telescopes to compensate for the Earth's orbital motion. Note that you would have to tilt the umbrella only slightly if you were running slowly, but by a larger amount if you were running faster. Likewise, we tilt the telescope forward by an amount determined by the ratio of the Earth's speed through space and the speed of light. Even though it took another century to detect parallax, Bradley's discovery settled the question of whether the Earth is moving around the Sun.

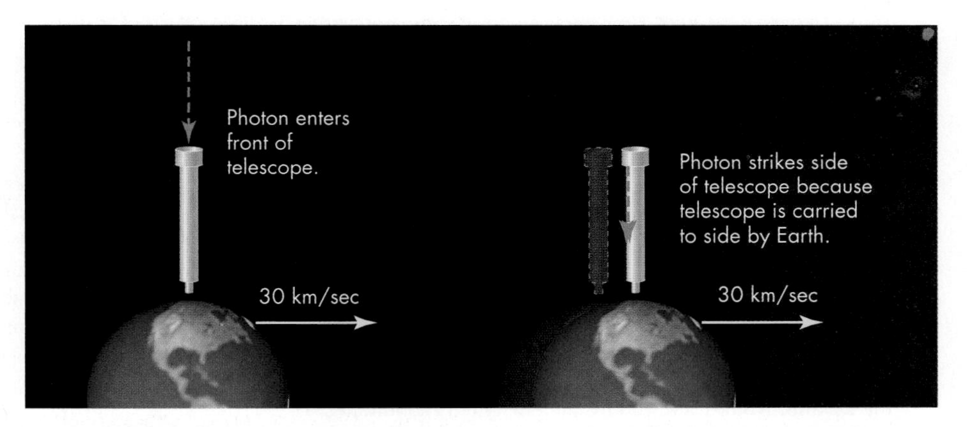

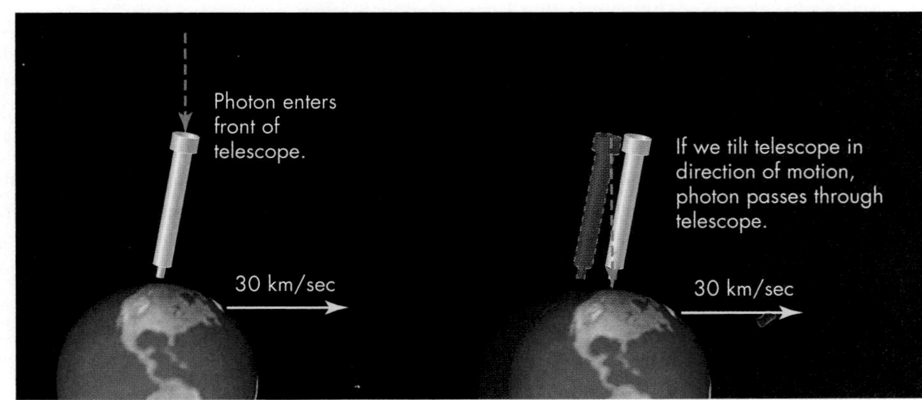

FIGURE 52.9
The "aberration of starlight" is an effect of the Earth's motion through space. If a telescope is pointed toward the actual direction of a star (upper panel) the light won't reach the back of the telescope. We must tilt telescopes into the direction of the Earth's motion (lower panel) so that photons entering the front of the telescope will reach the back of the telescope. Six months later, when the Earth is traveling in the opposite direction, we must tilt the telescope in the opposite direction.

Q Would the aberration of starlight be larger or smaller on Mars than on Earth? Why?

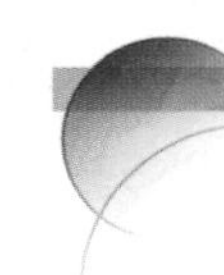

KEY TERMS

aberration of starlight, 433
angular shift, 428
arc minute, 425
arc second, 428
astronomical unit, 425
baseline, 424
Hipparchus, 425
parallax, 427
parsec, 429
proper motion, 431
Proxima Centauri, 430
radial velocity, 432
space velocity, 432
transit, 426
transverse velocity, 432
triangulation, 424

QUESTIONS FOR REVIEW

1. How can distances be measured by triangulation?
2. How are stars' distances measured? How do we know that Proxima Centauri is the closest star to Earth?
3. What is the largest distance we are able to measure for stars? Why can't the distances of farther stars be measured?
4. How is a parsec defined? How big is a parsec compared with a light-year?
5. What is proper motion? How is it related to "space velocity"?
6. Why would a star with a large proper motion be likely to be close to Earth?
7. Why does the speed of the Earth's motion through space make stars' positions appear to be shifted?

PROBLEMS

1. Using triangulation, make an approximate measurement of the distance from where you live to a neighboring building. Make a scale drawing, perhaps letting 1 cm represent 1 m (or you might let 1 inch represent 10 feet), and choose a baseline. Construct a scale triangle on your drawing and find how far apart the buildings are from one another.
2. Sirius has a parallax of 0.377 arcsec. How far away is it?
3. A dim star is believed to be 5,000 pc away. What should its parallax be?
4. If a star's parallax is 0.1 arcsec on Earth, what would it be if measured from an observatory on Mars?
5. Suppose a satellite is developed that can measure the parallax of stars with an uncertainty of 2×10^{-5} arcsec. Suppose a star is measured with a parallax of 3×10^{-5} arcsec with this satellite. What range of distances could the star be at and still fall within the range of uncertainty of the measurement?
6. Suppose a star 25 pc distant is observed to have a proper motion of 0.2 arcsec per year. What is the tangential velocity of the star? If measurements of its Doppler shift show it is moving toward us at 32 km/sec, what is the star's overall speed through space?
7. How much greater is the angle of aberration of starlight measured from Mercury compared to measured from the Earth? What would the angle of aberration of starlight be as measured from Saturn?

TEST YOURSELF

1. A star has a parallax of 0.04 arcsec. What is its distance?
   a. 4 ly
   b. 4 pc
   c. 40 pc
   d. 25 pc
   e. 250 pc
2. Star A is 10 times closer than star B. Star A's parallax is therefore ________ star B's.
   a. 10 times larger than
   b. 100 times larger than
   c. 10 times smaller than
   d. 100 times smaller than
   e. the same as
3. The most favorable conditions for observing the angular shift of a star occur when
   a. a large star moves across a field of smaller stars.
   b. a small star moves against a field of larger stars.
   c. Venus transits the Sun.
   d. a nearby star moves against a field of farther stars.
   e. nearby stars move relative to one another.

unit

53 Special Relativity

The history of astronomical study of the Solar System progressed from remote observation to direct exploration of individual bodies with space probes. The same will never be true of our study of the stars. Today's fastest space probes would take hundreds of generations to reach other stars. Even supposing a space probe were already orbiting Proxima Centauri, the distance is so vast that if we sent out a message saying "Are you there?" today, we would have to wait nine years to hear back "Yes." The Solar System is isolated not just by distance, but by the time it takes signals to travel over these immense distances. As we examine more remote objects in the rest of this book, it will become increasingly important to recognize that we are also looking farther into the past because of the finite speed of light.

Understanding how light moves led Albert Einstein to discover the surprising result that *nothing* can travel through space faster than 299,792 kilometers per second—the speed of light. His discovery, called **special relativity,** has changed some of our most basic ideas about the nature of space and time. One of the central discoveries of special relativity is that motion fundamentally alters the way space and time behave. It also has implications for the possibility of ever traveling to distant stars.

The real universe is very different from the popular vision of science fiction movies. There can be no dialogues at interstellar distances, and high-speed travelers will find themselves always traveling rapidly into the future of the stars around them. As strange as special relativity may seem, it is now as solidly confirmed as that the Earth is not flat. The ideas of relativity begin from simply trying to explain how light from moving stars behaves.

53.1 LIGHT FROM MOVING BODIES

The universe is full of motion at every scale. As the age-old notion of the Earth being fixed at the center of things crumbled, it soon became clear that the Sun cannot be assumed to be stationary either. In fact, we now know that all stars move at tens or even hundreds of kilometers per second relative to one another, and that the large systems of stars known as galaxies move at even greater speeds.

Astronomers began detecting the motions of stars in the early 1800s (Unit 52). When they measured the motion of a star, they measured it relative to the Sun. They used our vantage point in the Solar System to define a local **rest frame**—a coordinate system tied to our position and motion. With respect to this frame we can measure the positions of other objects in space and the speeds at which they are moving relative to us. In our rest frame we observe, for example, that some stars are moving through space at high speed. However, it is important to recognize that an alien living in a "high-speed" star system could define its own rest frame. To this alien, it would appear that the Sun is moving by *it* at high speed. Is there some way to decide what is stationary and what is not?

Perhaps we might determine what is stationary by seeing how light moves through space. Consider the light reaching us from a star that is moving toward

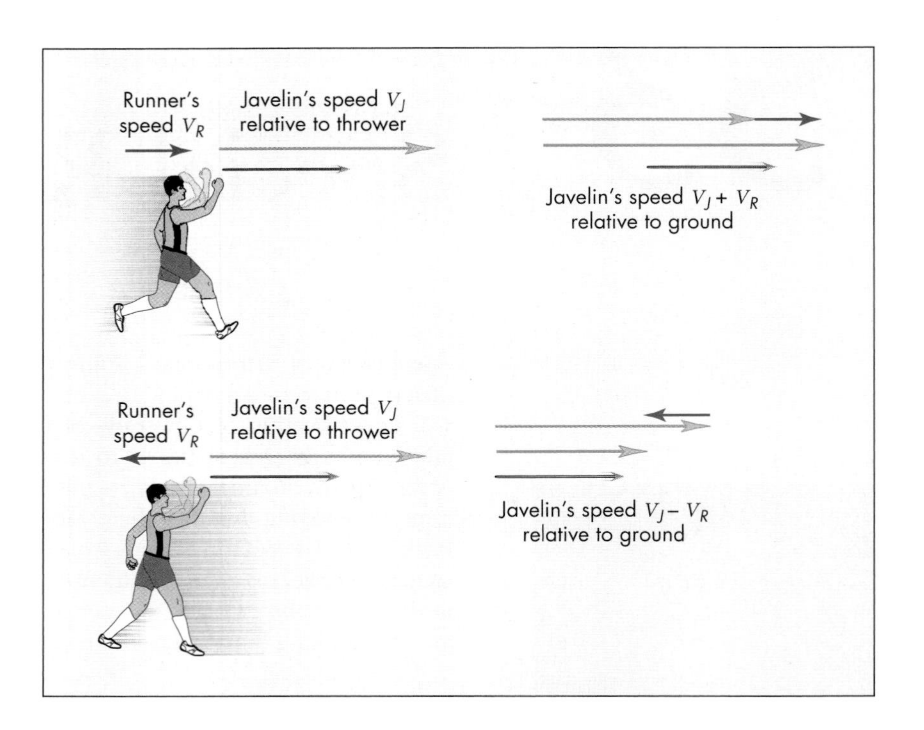

FIGURE 53.1
The speed of a javelin throw may be the same relative to the thrower no matter what direction he or she runs, but the speed of the javelin relative to the ground depends on the direction of motion.

us. We know that motion toward or away from us can be detected by the Doppler shift of light (Unit 25). We might expect that if space itself had its own rest frame, light's motion in space would be affected like sounds carried by a wind.

To understand how space might affect light, first consider the situation if there is no fundamental rest frame. We might then predict a boost or reduction in the speed of photons emitted by a source depending on whether it is moving toward or away from us. For example, you can throw a baseball to first base faster if you are running toward the base rather than away from it. Similarly, you will find that you can throw a javelin faster and therefore farther if you are running forward when you release it (Figure 53.1). If you are running *backward,* your throw will be weak and short even if you throw the javelin just as hard. This is because by running toward the target, you add your speed of running to the speed of the javelin. However, when you are running away from the target, your speed subtracts from the speed of the javelin. Thus, with respect to the ground, the javelin is traveling more slowly.

Q What are other examples of situations when different speeds add and subtract from each other?

Using the term we introduced at the beginning of this section, the javelin is traveling faster or slower in the target's rest frame depending on whether you move toward the target or away from it. In your own rest frame, the javelin moves just as fast (you throw it just as hard), but the difference in your rest frames combines with the speed of the javelin to produce the overall speed.

The addition of relative speeds (you plus javelin) in this fashion is an example of what is sometimes called **Galilean relativity.** Galilean relativity gets its name because Galileo was one of the first scientists to recognize how motions add and subtract from one another, and how the measurements of these motions depend on the rest frame of the observer.

If light's speed depended on the radial velocity of a star, we would observe an increase in the amount of aberration of starlight (Unit 52.5) for stars moving away from us. The resulting speed of light would be slow, so the Earth's motion would be more significant in shifting its position as the light passed down through the telescope. This would be easily detectable with modern instruments, but no such effect is seen.

We might expect similar behavior for photons emitted by a star that is moving toward or away from us. If Galilean relativity applied to photons from a star moving toward us, we would expect the photons to have an increased speed toward us and to reach us in a shorter time. Likewise, if the star were moving away from us, we would expect the photons to have a decreased speed.

A difference in light speed would be obvious for a star in orbit around another star. Astronomers first detected stars that orbit each other in the early 1800s (Unit 56). An orbiting star will move toward us during one part of its orbit, and

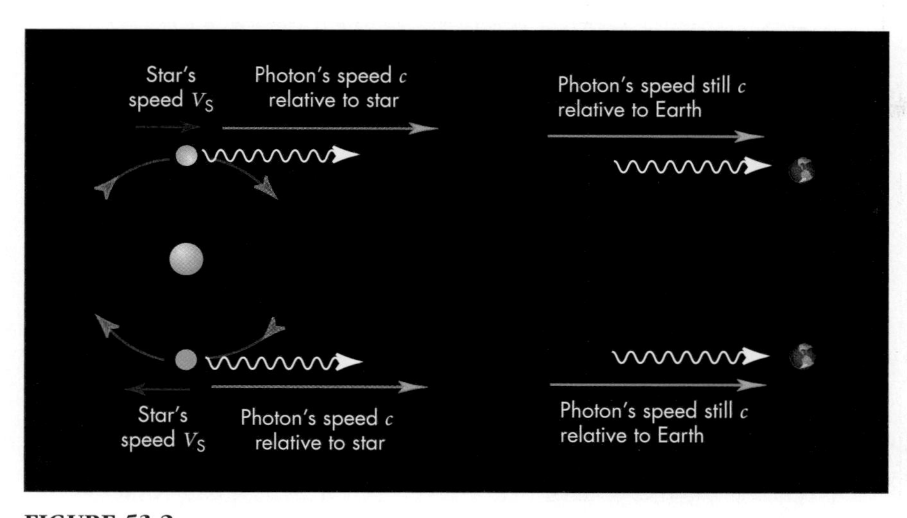

FIGURE 53.2
Light from a star in orbit reaches Earth at the same speed whether the star is moving toward or away from us.

away from us half an orbit later. Such motion is easily detected from the stars' changing Doppler shift. If the photons traveled faster when the star was moving toward us, they could overtake photons that left the star earlier when it was moving away from us. The result would be that we would see the star in different parts of its orbit in a scrambled order. We might even see the star in several different parts of its orbit simultaneously!

The light we see does nothing of the sort; we see no changes in the apparent position of the star that would be expected if light's speed could change. Light therefore must move through space at the same speed no matter what the speed of the star (Figure 53.2). This is not as odd as it might first seem. For example, when a boat makes waves on a lake, the boat may move toward or away from us, but the waves travel through the water at the same speed. In the 1800s scientists concluded that space must behave something like the water in this example and have its own rest frame. They believed space must contain some substance that light waves travel through, always at the same speed. They called this hypothesized substance the **aether.** However, as we will see in the next section, the behavior of light in other situations cannot be explained by the aether. It requires instead a completely new understanding of space and time.

53.2 THE MICHELSON-MORLEY EXPERIMENT

By the late 1800s physicists were able to measure the speed of light with good precision, and the intriguing possibility of discovering the absolute motion of the Earth through the hypothesized aether became a possibility. As the Earth moved through the aether, from the perspective of the moving Earth the aether should be "flowing" by us. This flowing aether should affect the motions of photons moving through it much as flowing water affects the motion of boats traveling through it. In other words, the photons should travel at different speeds depending on the direction of their motion relative to the flow.

FIGURE 53.3
Photograph of the experimental apparatus devised by Albert Michelson and Edward Morley. An optical assembly on a granite slab floated on a pool of mercury, allowing extremely sensitive measurements of the speed of light in different directions. No speed differences were found, no matter what direction the light traveled.

American scientists Albert Michelson and Edward Morley designed an apparatus in 1887 that could compare the speed of light moving in two perpendicular directions simultaneously (Figure 53.3). This was an extremely sensitive apparatus built on a massive granite slab that floated in a pool of mercury to isolate it from any vibrations, and allowing them to rotate the slab. They reasoned that if the slab was oriented in the right direction, then along one path through the apparatus, the light would travel parallel to the flow of the aether. Along the other path the light would travel perpendicular to the aether. Therefore, the speed of the light ought to be different along the different paths.

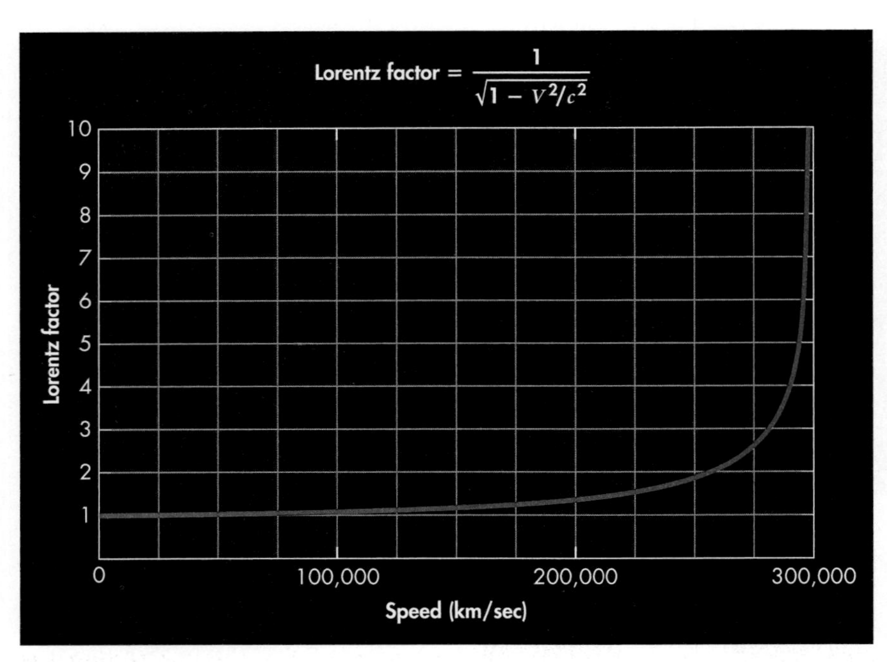

FIGURE 53.4
The Lorentz factor expresses the amount by which objects appear to compress in the direction of their motion. It is also the factor by which a moving clock appears to slow down, and the factor by which a moving object's mass appears to increase.

However, when Michelson and Morley conducted measurements using their apparatus, they detected no difference in the speed of light along the perpendicular paths no matter how they oriented the apparatus. They realized that this could just be bad luck during the first experiment—the combination of the Earth's motion around the Sun and the Sun's motion through space might just happen to make the Earth stationary relative to the aether at that point in its orbit. So they repeated their experiment many times over a year. At some point they should have easily detected the motion of Earth relative to the aether, but they found no sign of any relative motion at all.

The Michelson-Morley experiment has been called the most famous "failed" experiment in history, because it led to a revolution in physics. The results implied that there was no aether regulating the speed of light. However, if light was not moving relative to the aether, then physicists could not explain the constancy of the speed of light reaching us from sources moving at different speeds toward or away from us.

All experiments made then and now show that no matter how fast the source generating the light is moving, and no matter how fast the observer measuring the light is moving, the speed of light is always measured to be c = 299,792,458 meters per second. How can this be?

The idea of a contraction caused by motion was first proposed by Irish physicist George Fitzgerald. The Dutch physicist Hendrik Lorentz developed a model for explaining how the contraction might arise.

One explanation offered in the late 1800s was that motion through space somehow caused matter to contract in the direction of motion. If matter contracts when it is moving, this would change our perception of length so that we might be tricked into thinking the speed of light had not changed. For example, Michelson and Morley were searching for a difference in the speed of light in two perpendicular directions. If the apparatus were compressed in the direction it was moving through space but not in the perpendicular direction, the path the light traveled would be shorter. Such a contraction could potentially cancel out the effect that Michelson and Morley were searching for. The factor by which the apparatus would need to contract is known today as the **Lorentz factor.** The Lorentz factor, usually denoted by the Greek letter gamma (γ), hypothesizes that an object's length shortens in the direction in which it is moving by a factor equal to

γ = Lorentz factor
V = speed relative to observer
c = speed of light (3.00×10^8 m/sec)

$$\gamma = \frac{1}{\sqrt{1 - V^2/c^2}}$$

where V is the speed of the object and c is the speed of light. The value of γ for different speeds V is plotted in Figure 53.4.

The Lorentz factor is close to 1 at small speeds. For example, at the 30 km/sec speed at which the Earth is orbiting the Sun, the Earth would contract by only a few centimeters in its direction of motion. However, at high speeds the contraction factor becomes very large, growing to infinity if the speed V were to reach the speed of light. The Lorentz contraction factor explained the Michelson-Morley experiment, but it could not explain the results of a variety of other experiments. It contains, however, an important idea that grew into a whole new concept about the nature of motion.

53.3 EINSTEIN'S THEORY OF SPECIAL RELATIVITY

FIGURE 53.5
Albert Einstein (1879–1955).

In 1905 a 26-year-old graduate student named Albert **Einstein** (Figure 53.5) took on the problem of the seemingly inexplicable measurements of the speed of light. He was completing his physics degree while working in the Swiss patent office and supporting a new family. Yet in that one year alone, he completed his doctorate degree and wrote four papers in several areas of physics. Physicists widely agree that three of these papers were each worthy of a Nobel Prize! Einstein was little known at the time and had few colleagues with whom to discuss his ideas; but nonetheless in one of these papers he came up with a brilliant new approach to the question of the motion of light.

Einstein began by concentrating on the findings that light travels at the same speed no matter what the speed of its source or of the observer measuring the light. Even though many experiments had come to this conclusion, most physicists had assumed it was an impossibility and so were seeking other explanations—such as errors in the experiments or the Lorentz contraction. Einstein, instead of thinking of a constant speed of light for all observers as an impossibility, accepted this finding as correct, and proceeded to work out its consequences. He found that this led inevitably not only to a Lorentz-like contraction of space, *but also to a stretching of time by the same factor.* Even more important, he found that this contraction was not relative to some imagined aether filling space, but that these effects depended on just the relative motions of any two objects.

These alterations of space and time affect everything we see that moves relative to us. If a rocket moves by us at high speed, we will see it squashed in its direction of motion by the Lorentz factor (Figure 53.6). If we could watch a clock tick or

FIGURE 53.6
When a spacecraft travels by us, its length is contracted by the Lorentz factor and clocks on board run more slowly by the same factor. If the spacecraft is moving at 87% of the speed of light relative to Earth, it has a Lorentz factor of $\gamma = 2.0$. As a result, the spacecraft looks 2.0 times shorter, and each tick of its clock appears to take 2.0 seconds. From on board the spacecraft it appears that the Earth is compressed and time on Earth is running slowly.

measure the rate of an astronaut's heartbeat in the spaceship, we would discover that all these processes occur more slowly by the Lorentz factor.

What is even more remarkable is that the mathematics Einstein worked out showed that the situation is exactly symmetrical. The astronaut moving by us at high speed will sense herself being the one who is stationary and see us on the Earth as moving by her at high speed. She will see us and the Earth contracted in the direction of "our" motion, and she will see our clocks and our hearts running slowly (Figure 53.6). So what two observers in relative motion see is parallel—each would find that the other was the one undergoing the distortions of space and time.

An especially important feature of Einstein's work is the behavior of light. Suppose we or the passing astronaut shine a light beam at each other. We will each measure that the light is moving past us at the same speed, *c*. However, as we watch each other making these measurements, we will each think that the other is measuring a shorter distance and using a clock that runs too slow.

This theory of special relativity is far-reaching in its implications. The theory is the basis for Einstein's discovery of the relationship between energy and mass: $E = m \times c^2$ (Unit 50). Other fundamental quantities also change for moving objects. For instance, a moving object grows more massive, also by the same Lorentz factor that describes how lengths grow shorter and time slows.

These effects mean that nothing can reach the speed of light in our rest frame, let alone exceed it. The rocket ship traveling by us can fire its rockets and accelerate forever—but every time it goes a little faster, its time slows down more, and the rocket exhaust comes out more slowly and does not travel as far because of the contraction in length of the speeding ship. Meanwhile the ship's mass grows heavier and heavier, so it picks up less and less speed. The ship may approach the speed of light, but because the Lorentz factor goes to infinity, it would require infinite energy to reach the speed of light.

Q How do science fiction stories you have seen or read treat motion at speeds approaching *c*? Do any of these stories correctly depict the effects of special relativity?

53.4 SPECIAL RELATIVITY AND SPACE TRAVEL

The theory of special relativity may seem strange, but it has been tested by a century of high-precision experiments. There is not a single verified contradiction to it, and its predictions about such things as the slowing of time have been verified directly.

Special relativity is also about more than just perceived differences in space, time, and mass. For example, when atomic clocks (the most accurate clocks, used for establishing time worldwide) are flown between locations, it is found that their travel in airplanes leaves them a little bit slow relative to the network of fixed clocks maintained around the world. The moving clocks must be readjusted after each trip. The Lorentz factor for traveling at airplane speeds of about 1000 kilometers per hour (600 mph) is just $\gamma = 1.0000000000004$, so every tick of the clock on an airplane takes about 4 ten-trillionths of a second longer than a tick of the clock in the ground-based network; but after several hours of flying the effect is measurable.

More intriguing is what happens at such high speed that the Lorentz factor is large. Experiments have demonstrated, for example, that a subatomic particle called a *muon* (Unit 4) normally has a lifetime of only about 2 millionths of a second before it decays. However, when muons are traveling at 99% of the speed of light (and therefore have a Lorentz factor of about 7), they live about 14 millionths of a second. This much longer lifetime allows them to travel distances that would be impossible for them within their normal lifetimes.

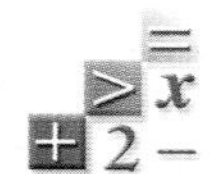

For $V = 0.99\,c$ the Lorentz factor is

$$\gamma = \frac{1}{\sqrt{1-(0.99c)^2/c^2}} = \frac{1}{\sqrt{1-0.99}} = \frac{1}{\sqrt{0.0199}} = 7.089$$

When speaking of microseconds, this change seems minor; but the same factor would apply for human space travel at 99% of the speed of light. If a spaceship could be built that traveled that fast, time would effectively run seven times more slowly on board compared with time here on Earth. If astronauts had food and air supplies for a 10-year trip, the Lorentz factor of $\gamma \approx 7$ would mean they could travel for $\gamma \times 10$ years ≈ 70 years in Earth time. At their speed relative to Earth of 0.99 *c* they would be able to reach a distance of almost 70 light-years (Figure 53.7A) before they ran out of supplies. At speeds even closer to *c* the Lorentz factor becomes even larger (see Table 53.1) and the potential distances greater.

From the perspective of astronauts on a craft traveling at 0.99 *c*, it would not seem that time was passing any more slowly than normal. Nor would they feel that they or their ship was foreshortened or more massive. From their perspective the Earth and the star they are visiting *and the distance between the two* are contracted by the Lorentz factor (Figure 53.7B). The distance would look like 70 light-years when they were in the rest frame of Earth and the distant star; but at their high speed, the distance would look only one-seventh as large!

Theoretically, at a high enough speed, time slows down so much that you could travel a million light-years away within your lifetime. This is far beyond current technologies, but it is also important to realize that such travel would have major challenges beyond simply reaching such high speeds. From the perspective of your spaceship traveling among the stars at near the speed of light, every atom and every dust particle in space along the ship's path has a mass increased by the Lorentz factor, and it is heading toward the ship at near the speed of light! This can impart to a pebble the impacting force of a ship-destroying asteroid.

TABLE 53.1 The Lorentz Factor at High Speeds

Speed	Lorentz Factor
0.87 *c*	2.0
0.97 *c*	4.1
0.99 *c*	7.1
0.999 *c*	22.4
0.9999 *c*	70.7
0.99999 *c*	223.6

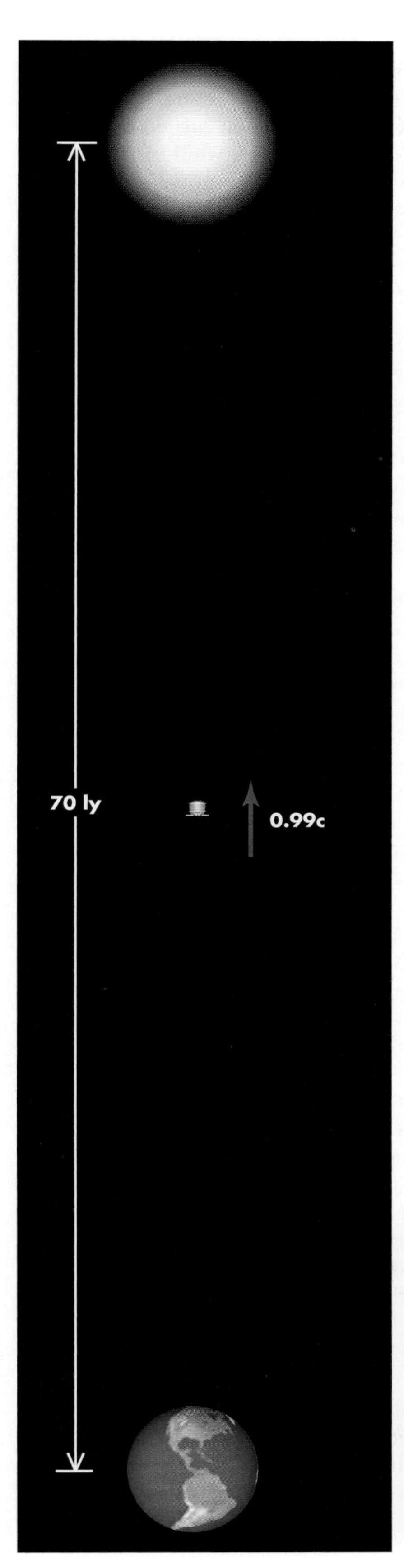

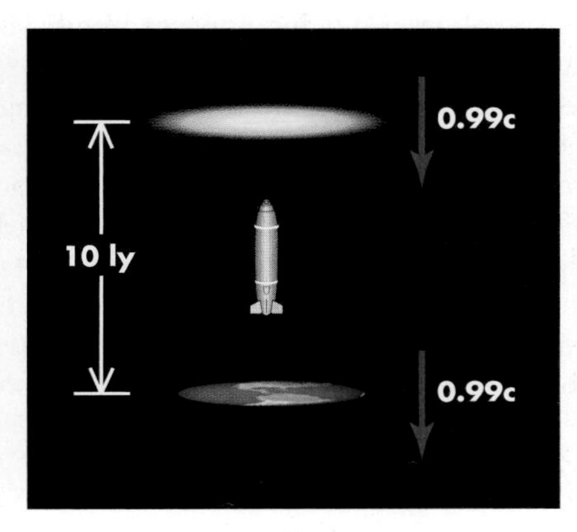

FIGURE 53.7
(A) From the rest frame of Earth and a star 70 light-years away, it appears that a spaceship traveling at 99% of *c* is shortened to about one-seventh of its original length. (B) On the spaceship (once it is up to speed) it appears that the Earth and star are both moving at 99% of *c*. They and the distance between them are shortened to about one-seventh of their original length.

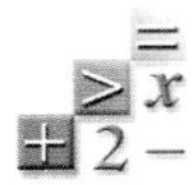

At a speed of $V = 0.99\ c$, in 70 years a spaceship would travel $0.99 \times 70 = 69.3$ light-years—nearly 70 light-years.

Also, as intriguing as these possibilities are, they offer no time savings for the rest of us back on Earth. Consider again the astronauts traveling at 99% of the speed of light for 10 years to visit a distant star. If they then turned around and came home in another 10 years (by their reckoning), they would find that 140 years had passed on Earth. Everyone they knew would have grown old and died. They might even be younger than their great-great-grandchildren!

53.5 THE TWIN PARADOX

Something may seem wrong about the description of space travel in the previous section, because when we first introduced relativity we noted that the time stretching appears symmetrical. While from the rest frame of the Earth it appears as though the astronauts' time is running slowly, from the rest frame of the astronauts it appears that time on Earth is running slowly. This is sometimes presented as the **twin paradox:** If one of the astronauts left a twin back on Earth, how can we say that one ages more than the other?

The explanation of this seeming paradox is that the situations are not actually symmetrical. Astronauts aboard a spacecraft experience accelerations as they first speed up their ship and then slow down at the star they visit. They turn their ship around and experience more accelerations as they again speed up and later slow down when they reach Earth again. The people remaining back on Earth experience none of these accelerations because they always remain in the same rest frame, and so the progression of time remains constant on Earth. By contrast, the astronauts' rest frame keeps changing, and in the end they return to the rest frame of the Earth.

Imagine, for example, that one of the astronauts sends messages once each day to her twin back on Earth, and meanwhile the twin on Earth sends messages once each day to his astronaut sister on the ship. As they part, they each receive the other's messages much more slowly—both because of the Lorentz factor *and* because the separation between the ship and Earth is growing larger and the messages take longer to reach their recipients. When the astronaut twin reaches the distant star, she sends her 10th-year message. Until this time, the situations are symmetrical. Both have received only a small fraction of each other's messages because of the Lorentz factor and growing separation.

Q With everyone's time running at different speeds, can you imagine a story line that would make a good movie if it portrayed space travel accurately?

The astronaut twin turns around and starts back, receiving the messages from her brother that were sent years earlier and have been on their way to her across the 70 light-year gap. They come more quickly now because she is approaching Earth, cutting the distance each message has to travel. She reads about her brother getting older and older, now seemingly very rapidly. In the meantime, her brother back on Earth is still reading her messages from the outgoing trip. He will die before the arrival of the 10th-year message announcing that the ship had reached the other star. At the speed of light in Earth's rest frame, that 10th-year message would take 70 years to reach Earth. In fact, the spaceship, traveling at 99% of the speed of its messages beamed back toward Earth arrives just shortly after the message announcing they had reached the star.

In the movies, space travel is fast and everyone ages at the same rate. In reality, traveling at high speeds means that the travelers must leave behind not only their homes but their own times and the people in them.

KEY TERMS

aether, 437
Einstein, Albert, 439
Galilean relativity, 436
Lorentz factor, 438
rest frame, 435
special relativity, 435
twin paradox, 442

QUESTIONS FOR REVIEW

1. What is Galilean relativity? When is it reasonable to use this concept?
2. What did the Michelson-Morley experiment fail to detect?
3. What quantity is always constant in special relativity?
4. What is the Lorentz factor?
5. How are length, time, and mass altered according to special relativity?
6. Why does special relativity rule out the possibility that a spacecraft could travel at the speed of light?
7. If it is impossible to travel faster than light, why is it possible (in principle) for you to travel millions of light-years away from Earth within your own lifetime?

PROBLEMS

1. Looking at the javelin thrower in Figure 53.1, would the effect of special relativity as observed by a bystander result in a speed for the javelin slower than Galilean relativity would predict, or faster? Explain.
2. What is the Lorentz factor for a spaceship traveling at 0.5 c?
3. Assuming that the Lorentz factor becomes significant if it affects measured lengths and times at the 10% level (a Lorentz factor value of 1.1), how fast must the observed speed of an object be to reach this threshold?
4. What speed would give a Lorentz factor of about 1 million (10^6)?
5. The Earth is traveling at about 30 km/sec as it orbits the Sun.
   a. How much more slowly does time run on the Earth than in the Sun's rest frame?
   b. Why will the time be measured as slower on the Earth than in the Sun's rest frame? (Why is it *not* "just relative"?)
6. Suppose a 1-gram paperclip struck a spaceship at 99% of the speed of light. Special relativity indicates that the paperclip's energy goes from $\gamma \times mc^2$ and drops down to its "rest mass energy" of mc^2 once it stops moving.
   a. How much energy is released in the collision?
   b. How much energy is this in kilotons of TNT?
7. How fast must a spacecraft travel to reach the center of our Galaxy (26,000 ly away) in 100 years "ship time"?

TEST YOURSELF

1. When a spaceship is traveling at 99% of the speed of light (Lorentz factor 7), an astronaut on board the ship will find that
   a. everything in the ship weighs seven times more.
   b. the ship has become very small—only one-seventh its original length.
   c. everyone on the ship is talking seven times more slowly than normal.
   d. All of the above.
   e. None of the above. Everything seems normal to the astronaut on board.
2. Suppose Tom and Molly are flying in spaceships toward each other at half the speed of light (0.5 c). If Tom shines a light toward Molly, what speed will Molly measure for the light coming toward her?
   a. 0.25 c
   b. 0.5 c
   c. 1.0 c
   d. 1.5 c
   e. 2.0 c
3. If Bob travels at close to the speed of light to another star and then returns, he will find that his twin sister Alice who remained on Earth is
   a. younger than him.
   b. older than him.
   c. the same age as him.
   d. He cannot return to Earth because it would violate special relativity.

unit

54 Light and Distance

Background Pathways

Unit 21: Light, Matter, and Energy 157

It is hard to believe that the stars we see in the night sky as tiny glints of light are in reality huge, dazzling balls of gas similar to our Sun. In fact, many of those gleaming specks are vastly larger and brighter than the Sun. Stars look dim to us only because they are so far away—several **light-years** (tens of trillions of kilometers or miles) to even the nearest. Such remoteness creates tremendous difficulties for astronomers trying to understand the nature of stars (Figure 54.1). Direct probing of such remote objects is impossible; but by studying the light they emit, we can learn about many properties of stars despite their vast distance.

Astronomers have discovered that most stars are like the Sun in many ways. They are composed mostly of hydrogen and helium, and they have similar masses. Most fuse hydrogen in their cores and emit visible-wavelength photons into space. A small percentage of stars have more than 30 times the Sun's mass (30 $M_\odot$). These stars are substantially hotter than the Sun and are bluish. Other stars are much less massive than the Sun—as little as one-tenth its mass—and are cool, red, and dim. Moreover, even stars similar to the Sun in composition and mass may differ enormously from it in their diameter. For example, some giant stars have diameters hundreds of times larger than the Sun's—so big that were the Sun their size, it would extend beyond the Earth's orbit and swallow the inner planets. On the other hand, some dwarf stars are only about the size of the Earth.

Astronomers have learned about these properties of stars by using physical laws and theories to interpret measurements made from the Earth. This area of astronomy, known as **astrophysics**, has the remarkable ability to give us detailed "pictures" of stars even though nearly all of them look like points of light under even the highest magnifications of the largest telescopes.

FIGURE 54.1
The brightest star in the image, slightly below center, is Alpha Centauri, one of the nearest stars to us, at a distance of about 4.3 ly. The star nearly as bright as Alpha Centauri, just up and to the right from it, is over 500 ly away. Comet Halley, on the left side of this photograph taken in April 1986, looks almost as bright as the stars but was at a distance of under 1 AU.

54.1 LUMINOSITY

One critical property for understanding a star is the total amount of energy it radiates out into space each second: its **luminosity**, usually abbreviated as L. A star's luminosity tells us how much energy is being generated within it, which is one of the most important differences between star types.

An everyday example of luminosity is the wattage of various lightbulbs. A typical table lamp bulb has a luminosity of 100 watts, whereas a bulb for an outdoor parking lot light may have a luminosity of 1500 watts. By comparison, stars are almost unimaginably more luminous. The Sun, for example, has a luminosity of about 4×10^{26} watts. This "wattage" is fairly typical of stars, although they may range up to millions of times larger or smaller.

These numbers are so enormous that it is more convenient to use the Sun's luminosity as a standard unit, where $1\ L_\odot = 4 \times 10^{26}$ watts. When we speak, then, of a star with a luminosity of $2\ L_\odot$ or $100\ L_\odot$, we mean it has a luminosity of 8×10^{26} or 4×10^{28} watts, respectively.

Stars generate their luminosity by "burning," or more accurately fusing, the nuclei of atoms together in their cores. A star's luminosity indicates not only the rate at which it is emitting energy into space, but also the rate at which it is fusing atoms in its core. Over thousands of years of observation, nearly all of the stars we see in the night sky have remained steady in their brightness. From this constancy, we conclude that the energy generation process deep inside stars must be very stable, steadily replacing the energy radiated to space.

The Sun must convert 4 million tons of matter into energy every second (Unit 50). One of the most luminous stars known, discovered by the Hubble Space Telescope in 1997, has 10 million times the Sun's luminosity. At that phenomenal rate of energy release, every few months the star is annihilating as much mass as is contained in the entire Earth.

54.2 THE INVERSE-SQUARE LAW

When we look at stars at night, some look brighter than others, but this is not necessarily because they are more luminous. We all know that a light looks brighter when we are close to it than when we are far from it. As light travels outward, its energy spreads uniformly in all directions, as shown in Figure 54.2A. If you are standing near a light source, the light will have spread out only a little, and so more light enters your eye; if you are farther away, less light enters your eye, which you will perceive as the light source being dimmer.

This dimming with increasing distance may be easier to understand if you think in terms of photons. Photons leaving a star or other light source spread out along straight lines in all directions. If you imagine a series of progressively larger spheres drawn around the light source, the same number of photons pass through each sphere in one second. However, because more-distant spheres are larger, the number of photons passing *through a fixed area* on that sphere grows smaller, as shown in Figure 54.2B. As you can see, fewer photons pass through 1 square meter on a sphere at each successive distance from the source. As a result, observers would collect fewer photons if they were far from a star than they would with the same size telescope if they were near the star.

Astronomers also use the terms *flux* and *apparent luminosity* to describe the brightness.

The amount of light reaching us from a star is called its **brightness** and is a function of distance. This is different from a star's luminosity, which is the total amount

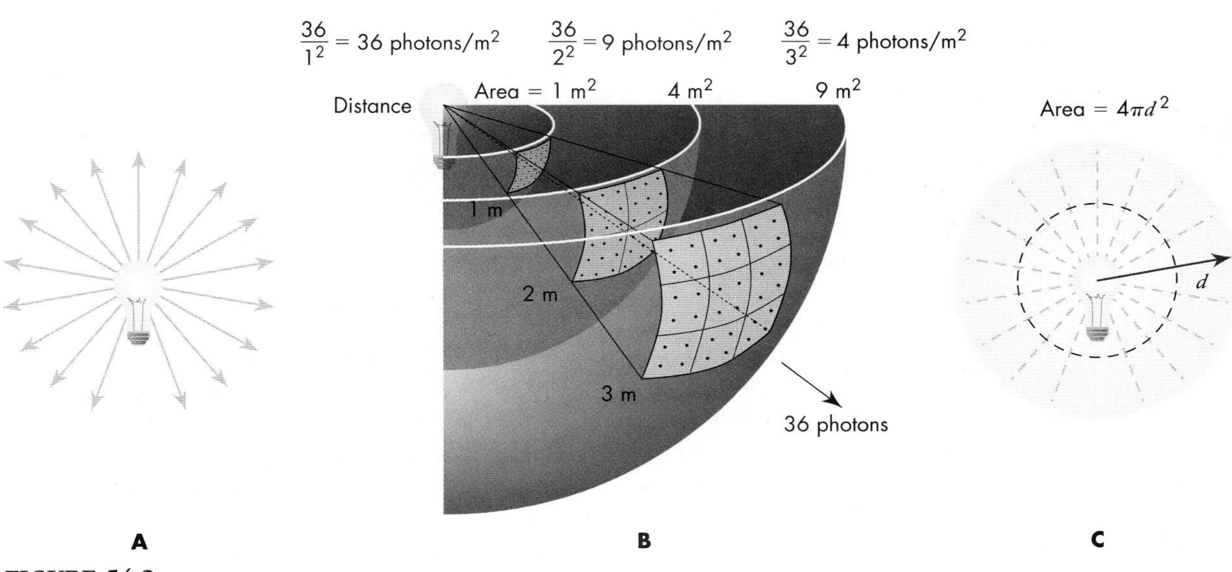

FIGURE 54.2

The inverse-square law. (A) Light spreads out from a point source in all directions. (B) As photons move out from a source, they are spread over a progressively larger area as the distance from the source increases. Thus a given area, here one square meter, intercepts fewer photons the farther it is from the source. (C) At a distance d from a source of light, the area over which the light is spread equals $4\pi d^2$.

Inverse-square law

of light emitted. The luminosity is expressed by the number of watts of light energy generated by the star, but we describe brightness as the wattage received *per square meter*. That is, if we had a telescope with a collecting area of one square meter, and we collected 10 watts of light, we would conclude that the brightness was 10 watts per square meter. If we measured the same light source with a smaller telescope, with a collecting area of just 0.1 square meters, we would collect only one watt of light.

To picture how the brightness of a star depends on its distance from us, imagine all the photons emitted by a star in a single moment. At some later time, those photons will have traveled a distance d, forming a sphere surrounding the star. The star's luminosity L will be spread over the surface of a sphere of radius d, as shown in Figure 54.2C. We use d to stand for radius here to emphasize that we are speaking of a distance. The surface area of the sphere is given by the geometric formula $4\pi d^2$, so the brightness we observe, B, is calculated as follows:

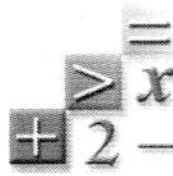

B = Brightness of received light (watts/meter^2)

L = Luminosity of star (watts)

d = Distance to star (meters)

$$B = \frac{L}{4\pi d^2}.$$

This relationship is called the **inverse-square law** because distance appears in the denominator as a square. The inverse-square law relates an object's luminosity (the total energy it emits) to its distance and its brightness (how bright it looks to us). Thus, the inverse-square law puts into a mathematical form the everyday experience that distant objects look dimmer.

A standard 40-watt lightbulb produces about 1 watt of luminous power at visible wavelengths. The difference between the electrical power consumption (40 watts) and the light delivered (1 watt) occurs because most of the power is radiated as heat at infrared wavelengths we cannot see. Stars also generate some of their light at infrared and ultraviolet wavelengths, but a star like the Sun emits most of its power at visible wavelengths.

To get an idea of the brightness levels of stars, you could perform the following experiment. Take a light source that generates 1 watt of luminosity and place it at various distances until it looks similar in brightness to a bright star. It turns out that you will have to place it about 2 kilometers away to look about as bright as the brightest stars in the sky. By applying the brightness formula to this known source, you would find that its brightness is

$$B = \frac{1 \text{ watt}}{4\pi(2000 \text{ m})^2} = 2 \times 10^{-8} \text{ watt/m}^2.$$

This tells us that the brightest stars in the sky deliver only 2 hundred-millionths of a watt per square meter. Because the pupil of your eye is at most about 8 millimeters in diameter, your eye is sensing only about a *trillionth* of a watt when you look at a bright star.

The inverse-square law is one of the most powerful tools of astrophysics. We can measure B for a star by using a **photometer,** a device similar to the electronic exposure meter in a camera. A photometer measures the amount of electromagnetic energy that strikes it each second. When calibrated for the size of the telescope, this can be converted into watts per square meter. If we can also determine the star's distance, d, using a technique such as parallax (Unit 52), for example, then we can calculate the star's luminosity, L, by rearranging the inverse-square law:

$$L = B \times 4\pi d^2.$$

Such measurements indicate that although many stars have luminosities comparable to the Sun's, some stars are millions of times more or less luminous than the Sun. Determining what causes this wide range of luminosities is one of our challenges in understanding stars.

54.3 DISTANCE BY THE STANDARD-CANDLES METHOD

We use brightness to estimate distance in many situations. For example, suppose you look at two street lights that have bulbs of the same wattage, but one street light is close and one is far away. From how bright they appear, you can estimate how much farther away the dim one is than the bright one (Figure 54.3). In fact, if you drive at night, your life depends on making such distance estimates when you see traffic lights or oncoming cars. Astronomers use a more refined version of this idea to find the distances to stars and galaxies.

The inverse-square law provides a way to determine a star's distance if we know the star's luminosity and can measure how bright it appears. Finding the distances of stars based on their luminosities works only if we know the luminosities with some assurance. For example, if we determine that the properties of some star are almost identical to the Sun's, we might reasonably infer that its luminosity would be the same as the Sun's too. Then, knowing the star's luminosity, *L*, we can measure its brightness, *B*, and solve for its distance using the inverse-square law. Rearranging that equation, we get

$$B \times \left(\frac{d^2}{B}\right) = \frac{L}{4\pi d^2} \times \left(\frac{d^2}{B}\right)$$

$$\sqrt{d^2} = \sqrt{\frac{L}{4\pi B}}$$

$$d = \sqrt{\frac{L}{4\pi B}}$$

$$d = \sqrt{\frac{L}{4\pi B}}.$$

For example, Alpha Centauri A is a star whose color and spectrum have been found to be very similar to the Sun's. Using a photometer, we find that its brightness is about 3×10^{-8} watts per square meter. If we assume it has the same luminosity as the Sun, we can plug the numbers into the above equation:

$$d = \sqrt{\frac{4 \times 10^{26} \text{ watts}}{4\pi \times 3 \times 10^{-8} \text{ watts/m}^2}} = \sqrt{1 \times 10^{33} \text{ m}^2} = 3 \times 10^{16} \text{ m}.$$

Converting this to parsecs, we find a distance of about 1 parsec, which is slightly less than Alpha Centauri's measured distance of 1.35 parsec. The match is not perfect because Alpha Centauri is slightly more luminous than the Sun; however, the distance is fairly close to the correct value.

Before parallaxes were measured, astronomers made very early light measurements by eye (discussed in detail in the next section) and reached similar conclusions about the distances of stars. Of course, most stars are *not* as similar to the Sun as is Alpha Centauri A, so the particular distances derived for them were often much less accurate than the distance we just found. Nevertheless, astronomers began to appreciate just how far away the stars must be for us to be receiving so little light from them compared with the Sun.

FIGURE 54.3
Street lights in this photograph illustrate that the brightness of light we see from sources with the same luminosity diminishes with distance.

In this early history of light measurement, or *photometry*, scientists used as their standard of light generation a particular kind of candle that burned wax at a specified rate. These were well-calibrated light sources and were known as "standard candles." Today, if we can find any particular type of star whose luminosity is well determined and can therefore serve as a reference for other stars, we call them **standard candles** too. The Sun is *not* considered a good standard candle because many stars look quite similar to it, yet have substantially different luminosities than the Sun. On the other hand, some stars have peculiar characteristics—such as unusual aspects of their spectra, or curious ways in which their brightness pulsates—that allow them to be identified as unique. Tests have shown that some of these unique classes of stars have well-determined luminosities, making them good standard candles.

We can use standard candles to find distances just as we did for Alpha Centauri. This distance-finding scheme is referred to as the *method of standard candles.* This is a powerful method for the following reason: If we identify a set of stars that all have the same luminosity, *we can determine the distances to all of them if we can find the distance to any one of them.*

Care must be taken, though! The accuracy of the standard-candle method relies on the degree to which the stars actually have the same luminosity. If we mistakenly group stars of dissimilar luminosities, we will get incorrect distances. This kind of mistake has happened several times in the history of astronomy, forcing us to revise distance estimates, and even forcing us to revise our notions of the size of the entire universe!

Q How could you determine the light output of a candle compared to that of a lightbulb?

54.4 THE MAGNITUDE SYSTEM

The magnitude system is convenient in some ways—it avoids the need to express brightness and luminosity with scientific notation—but it is complicated to use in many other ways. It is not necessary to use it, although it is handy to know a little bit about the values used because star brightness levels are often reported in magnitudes.

About 140 B.C.E., the Greek astronomer Hipparchus measured the apparent brightness of stars using units he called *magnitudes.* He designated the stars that looked brightest "magnitude 1" and the dimmest ones he could just barely see "magnitude 6." For example, Betelgeuse, a bright red star in the constellation Orion, is magnitude 1, and there are only about two dozen first-magnitude stars over the whole celestial sphere. The somewhat dimmer stars in the Big Dipper's handle are approximately magnitude 2, whereas the stars in the bowl of the Little Dipper range from magnitude 2 to magnitude 5 (Figure 54.4). From a dark clear sky, you may be able to see about 4500 stars, about two-thirds of which are magnitude 6.

Astronomers still use Hipparchus's scheme to measure the brightness of astronomical objects, but they now use the term **apparent magnitude** to emphasize that they are measuring only how bright a star *appears* to an observer. A star's apparent magnitude is a way of indicating its brightness. Astronomers use the magnitude system for many purposes (for example, to indicate the brightness of stars on star charts, like the foldout chart in the back of this book), but it has several confusing properties. First, the scale is "backward" in the sense that bright stars have "smaller" magnitudes while dim stars have "larger" magnitudes. Moreover, modern measurements show that Hipparchus underestimated the magnitudes of the brightest stars, and so the magnitudes now assigned to the brightest few stars are negative numbers!

Q Where else do we use systems like the magnitude system, where a step is used to indicate a change by some factor?

Magnitudes are not easy to work with because *differences* in magnitudes correspond to *ratios* in brightness. In particular, a difference of five magnitudes corresponds to a ratio of 100 in brightness. So when we say one star is five magnitudes brighter than another, we mean it is a factor of 100 brighter. That is, if we measure the brightness of a first-magnitude star and a sixth-magnitude star, the first-magnitude star is 100 times brighter than the sixth.

Each magnitude difference corresponds to a factor of about 2.512 (the fifth root of 100) in brightness. A first-magnitude star is 2.512 times brighter than a second-magnitude star and is 2.512×2.512, or 6.310, times brighter than a third-

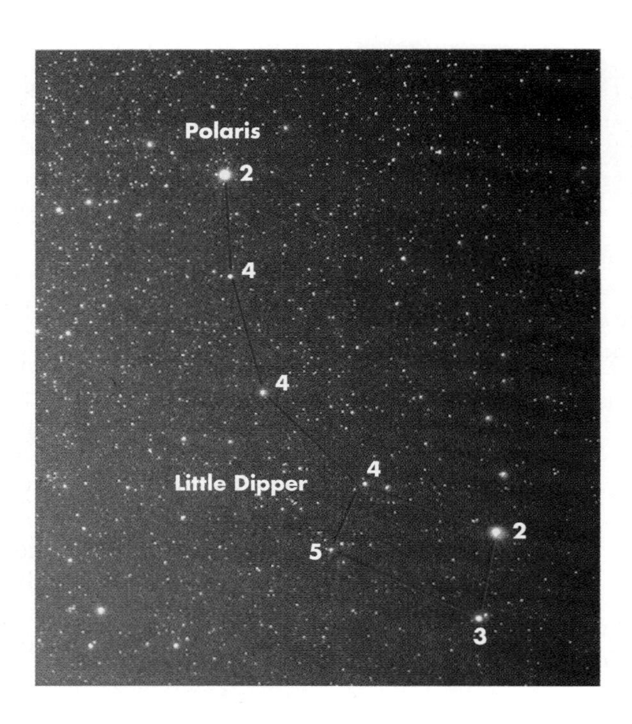

FIGURE 54.4
In Ursa Minor, the bowl of the "Little Dipper" is made of stars of magnitude 2, 3, 4, and 5 at each of its corners. Polaris is also a second-magnitude star.

TABLE 54.1 Relating Magnitudes to Brightness Ratios	
Magnitude Difference	Ratio of Brightness
0	$2.512^0 = 1{:}1$
0.1	$2.512^{0.1} = 1.10{:}1$
0.5	$2.512^{0.5} = 1.58{:}1$
1	$2.512^1 = 2.512{:}1$
2	$2.512^2 = 6.310{:}1$
3	$2.512^3 = 15.85{:}1$
4	$2.512^4 = 39.81{:}1$
5	$2.512^5 = 100{:}1$
10	$2.512^{10} = 10^4{:}1$
20	$2.512^{20} = 10^8{:}1$

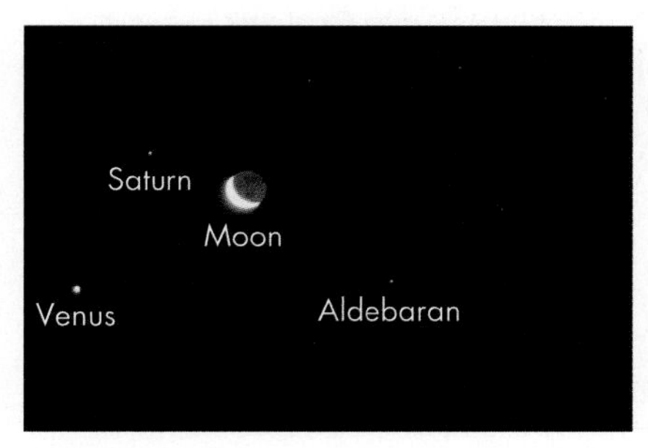

FIGURE 54.5
Venus, Saturn, and the Moon in close alignment with the star Aldebaran. Aldebaran has an apparent magnitude of about 1. When this photograph was taken, Venus had apparent magnitude −4, Saturn about 0, and the Moon about −11.

magnitude star. It is possible to make finer gradations by measuring fractions of a magnitude. For example, 0.1 magnitudes is a change of about 10% in brightness. Table 54.1 lists the ratios that correspond to various differences in magnitude.

Figure 54.5 shows a picture of the star Aldebaran when Venus, Saturn, and the crescent Moon were all close to it in the sky. When this photograph was taken, Venus was at about its brightest, with an apparent magnitude of about −4; Aldebaran's apparent magnitude is 1. The difference in their magnitudes is $1 - (-4) = 5$. We see from Table 54.1 that Venus at its brightest appears 100 times brighter to our eye than Aldebaran.

When using magnitudes, it is important to understand the distinction between what astronomers call **absolute magnitude** and the apparent magnitude we observe. Absolute magnitude is defined as the apparent magnitude a star would have if it were situated at a distance of 10 parsecs from us. Although we cannot, of course, move stars to whatever distance we choose, absolute magnitude serves as a way of indicating a star's luminosity in units that are mathematically compatible with the apparent magnitude system. Table 54.2 illustrates how absolute magnitude is related to a star's luminosity.

TABLE 54.2 Relating Absolute Magnitude to Luminosity	
Absolute Magnitude	Approximate Luminosity in Solar Units
−5	10,000
0	100
5	1
10	0.01

KEY TERMS

absolute magnitude, 449
apparent magnitude, 448
astrophysics, 444
brightness, 445
inverse-square law, 446
light-years, 444
luminosity, 444
photometer, 446
standard candle, 447

QUESTIONS FOR REVIEW

1. What is the difference between luminosity and brightness?
2. What units are appropriate for measuring the brightness of a star?
3. What is the inverse-square law? What does it say about how brightness varies with distance?
4. What makes a star a good "standard candle"?
5. Explain why magnitudes decrease for a brighter object.
6. What is the difference between apparent magnitude and absolute magnitude?

PROBLEMS

1. The faintest visible stars deliver only about 10^{-14} watts to your eye. A typical visible-wavelength photon has an energy of about 3×10^{-19} joules. How many photons per second are you seeing?
2. In the text we estimated the distance to Alpha Centauri by assuming that it has the same luminosity as the Sun. Use the measured brightness (3×10^{-8} watts/m^2) and its distance (4.4 ly) to calculate Alpha Centauri's actual luminosity. How many times bigger than the Sun's luminosity is this?
3. Suppose star A is 36 times brighter than star B, but we have reason to believe that they have the same luminosity.
   a. Which is more distant?
   b. How many times more distant is it than the other star?
4. Star C and star D are at the same distance from us, but star D is 10,000 times more luminous than star C. How do their brightness levels compare?
5. How do the magnitudes of stars C and D in problem 4 compare?
6. The star Deneb has an apparent magnitude of 1.25 and an absolute magnitude of -8.5. What two statements can you make about it, based on this data?
7. Star A has a magnitude of 3.5. Star B has a magnitude of -1.5. Which star is brightest, and by what brightness ratio?
8. If you observed the Sun from Antares, what apparent magnitude would it have?
9. Proxima Centauri is approximately the same distance from Earth (slightly less than 4.3 ly) as Alpha Centauri A, but it has an apparent magnitude of $+11.05$, compared to Alpha Centauri A's magnitude of -0.01. How much more luminous is Alpha Centauri A than Proxima Centauri?

TEST YOURSELF

1. Brightness is used by astronomers as a measure of
   a. watts received per square meter.
   b. a star's total luminosity in watts.
   c. a star's distance.
   d. absolute magnitude.
   e. a star's surface temperature.
2. The luminosity of a star depends on
   a. the star's distance from us.
   b. the inverse-square law.
   c. the energy the star generates each second.
   d. the star's rate of hydrogen to helium fusion.
   e. (c) and (d).
3. If a star is 5 *magnitudes* brighter than another, it is a *factor* of _____ brighter.
   a. 2.512
   b. 5
   c. 10
   d. 100
   e. None of the above

Starry Night

See the Pathways *2nd edition ARIS site for online Starry Night™ planetarium exercises.*

unit

55

The Temperatures and Compositions of Stars

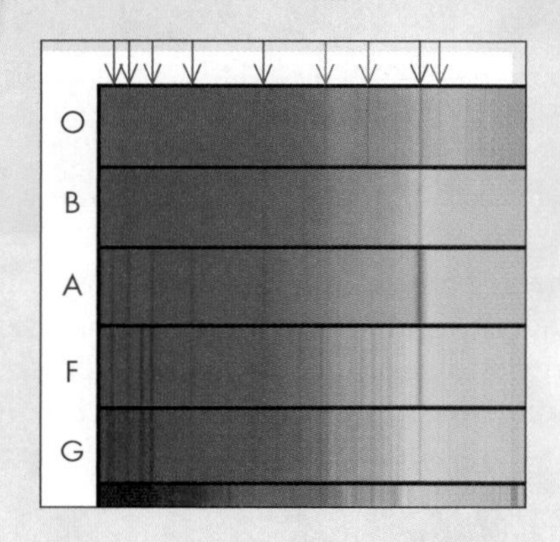

Background Pathways

Unit 23: Thermal Radiation 173

Unit 24: Atomic Spectra: Identifying Atoms by Their Light 179

If we were studying flowers or butterflies, we would probably first make a list of their properties, such as their appearance, size, shape, colors, and structure. So, too, astronomers want to know the sizes, colors, and structures of stars. Such knowledge not only helps us better understand the nature of stars, but also is vital in unraveling their stories. Obtaining such knowledge is not straightforward, though, because stars are so far away that we cannot examine them directly.

A star's **spectrum** is perhaps the single most revealing thing we can examine to learn about the star. The spectrum depicts the amount of light emitted at each wavelength (Units 23 and 24) and tells us directly about a star's temperature and composition, as we shall explore in this unit. Beyond this, detailed analyses of the spectrum can tell us a star's luminosity, its velocity in space, and information about its mass and radius. This wealth of information is available to us because the physical conditions at the surface of a star determine the properties of the light emitted and absorbed by the atoms there.

55.1 INTERACTIONS OF PHOTONS AND MATTER IN STARS

Photons are generated deep in a star's extremely hot interior and then move outward through successive layers of ionized gas that have lower temperatures and densities. As the photons make their way outward, they typically travel less than a centimeter before interacting with one of the numerous tightly packed electrons and ions. At each layer, the photons exchange energy with the matter there, and all reach a shared temperature.

The last stop photons make in their journey outward from the star's interior is in the **photosphere** (Unit 49). The photosphere is the layer where photons can escape to space—essentially because the gas above this region is thin enough for photons to pass through with little likelihood of additional interactions (Figure 55.1). The photons that we see come from the photosphere, so that is what we see as the "surface" of the star, even though there is gas of progressively lower densities continuing out beyond this radius. When we examine starlight, its properties reflect the temperature and composition in the photosphere.

Not all photons have an equal chance of escaping to space from the photosphere. The photosphere is relatively cool (compared to the interior of the star), so some electrons recombine with atoms there. Some stars' photospheres are so cool

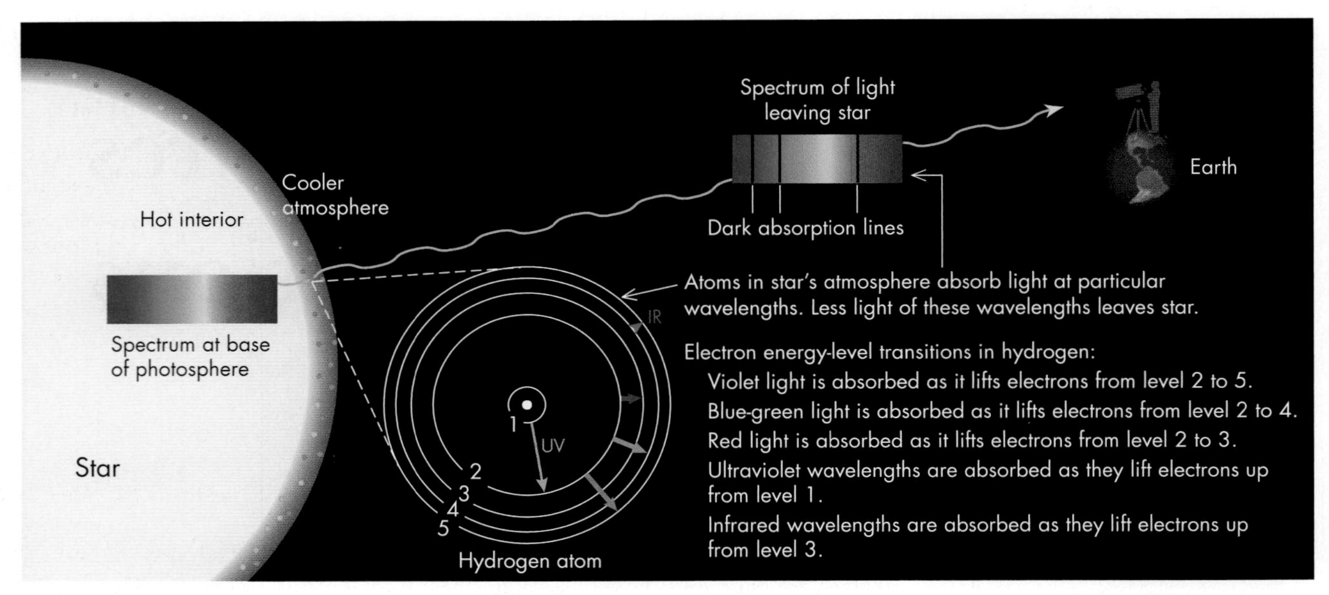

FIGURE 55.1
Formation of stellar absorption lines. Atoms in the cooler atmospheric gas absorb radiation at wavelengths corresponding to jumps between electron orbitals. Absorption lines of hydrogen are illustrated.

Q: Some of the photons we detect from a star have wavelengths matching the wavelengths of absorption lines in the star's atmosphere. What elevation would you expect these photons to come from, compared to photons that are far from any absorption line?

that atoms even combine to form molecules there. These atoms and molecules will absorb photons whose energies match the energies needed to lift electrons into higher energy levels. Because a photon's energy is determined by its wavelength, or color, when we observe the spectrum of light from a star, we find that certain colors are more strongly blocked. This creates dark **absorption lines** in the star's spectrum, as shown in Figure 55.1.

Each type of atom—hydrogen, helium, calcium, and so on—absorbs photons with a unique set of wavelengths. For example, hydrogen absorbs photons with wavelengths of 656.3, 486.1, and 434.1 nanometers, which are, respectively, in the red, blue-green, and violet parts of the spectrum. Gaseous calcium, on the other hand, absorbs strongly at 393.3 and 396.8 nanometers, producing a dark double line in the violet portion of the spectrum. Because each type of atom absorbs a unique combination of wavelengths of light—its "fingerprint"—the dark lines in the spectrum depend on the composition of the gas in the photosphere.

It is possible, therefore, to measure a star's composition by comparing the absorption lines in its spectrum to the wavelengths of lines made by each kind of atom. When we find matching absorption lines, we infer that the element exists in the star. To find the quantity of each atom in the star—each element's abundance—we use the darkness of the absorption line. A darker line generally implies a greater amount of that particular element, although some spectral lines are naturally stronger than others, so laboratory and theoretical work is needed to interpret the line strengths. The temperature of the atmosphere is also important because the hotter the gas, the more frequent and higher the speed of collisions between the atoms. Collisions knock the electrons into different energy levels, changing which spectral lines we observe. We therefore need to know the temperature in a star's photosphere to interpret its spectrum.

55.2 STELLAR SURFACE TEMPERATURE

To determine how hot a star is, we use the same method used in judging the temperature of an electric stove burner or a piece of glowing charcoal. A bright orange glow indicates a higher temperature than a dull red glow (Unit 23.2). Even the coolest stars have surface temperatures hotter than a stove burner can become. They are about as hot as the yellow-white tungsten filament in a bright lightbulb. Many stars are even hotter and glow with a blue or even violet color.

Thus a star's surface temperature can often be deduced from the color of its emitted light, hotter stars emitting shorter wavelengths more strongly:

Hotter stars emit more blue light, and cooler stars emit more red light.

You can see such color differences if you look carefully at stars in the night sky. For example, Rigel and Betelgeuse, the two brightest stars in the constellation Orion, have distinctly different colors (Figure 55.2). Rigel has a blue tint whereas Betelgeuse is reddish, so even our eyes can tell us that stars differ in temperature.

We can use color in a more precise way to measure a star's temperature with Wien's law (Unit 23.4). Wien's law relates an object's temperature to the wavelength, λ_{max}, at which it radiates most strongly, as shown in Figure 55.3. Wien's law lets us calculate the temperature T from λ_{max} using this formula:

T = Temperature of continuum source (Kelvin)

λ_{max} = Wavelength of maximum brightness (nanometers)

$$T = \frac{2.9 \times 10^6 \text{ K} \cdot \text{nm}}{\lambda_{max}}$$

This law applies specifically to sources of thermal radiation, and the photospheres of stars are fairly good examples of this.

As a demonstration of Wien's law, consider Rigel again. Detailed measurements of its spectrum show that it radiates most strongly (has a wavelength of maximum brightness λ_{max}) at about 240 nanometers, in the ultraviolet range. Its temperature is therefore about

$$T(\text{Rigel}) = \frac{2.9 \times 10^6 \text{ K} \cdot \text{nm}}{240 \text{ nm}} = \frac{2{,}900{,}000}{240}\text{K} = 12{,}000 \text{ K}.$$

The spectrum of light for 12,000 K thermal radiation is illustrated in Figure 55.3. Notice that even though the spectrum reaches a maximum outside the range of visible light, visible light is produced. The visible light is stronger toward the blue end of the spectrum, so Rigel looks blue to us.

The star Betelgeuse in Orion looks red to our eyes, but its wavelength of maximum brightness is actually in the infrared, at about 970 nanometers (Figure 55.3). So if we calculate its surface temperature, we find:

$$T(\text{Betelguese}) = \frac{2{,}900{,}000}{970}\text{K} = 3000 \text{ K}.$$

FIGURE 55.2
The constellation Orion contains both a bright red star, Betelgeuse, and a bright blue star, Rigel. The colors show that Betelgeuse has a surface temperature of about 3000 K, while Rigel's is about 12,000 K.

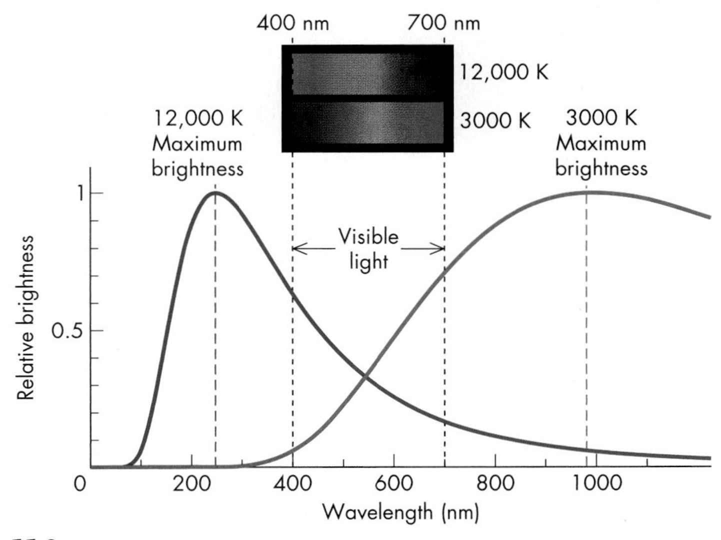

FIGURE 55.3
Spectra of a 12,000 K source and a 3,000 K source. The wavelength at which a spectrum peaks allows us to estimate the source's temperature. The curves shown extend from ultraviolet to infrared wavelengths. Over the range of visible wavelengths, the brightness of the emission climbs from one end to the other, giving stars with these temperatures a blue-white or red-white color overall.

This is about the temperature of the filament inside a standard 100-watt lightbulb. Sometime when you observe stars, compare the color of a distant incandescent bulb with Betelgeuse or another of the reddish stars you see in the sky. That reddish color also helps explain why photographs taken using indoor lighting often look yellow or orange compared with those taken in sunlight. Like Betelgeuse, lightbulbs emit radiation mostly in the infrared. We see their visible light because they also emit electromagnetic radiation at shorter wavelengths than λ_{max}.

Wien's law is important for interpreting stellar colors, but for many stars the maximum brightness lies outside the visible-wavelength range. In addition, light absorption by atoms and molecules in a star's atmosphere may make it difficult to determine the peak wavelength of the thermal radiation. Fortunately, studies of the spectra of stars have revealed another way to determine stars' temperatures.

55.3 THE DEVELOPMENT OF SPECTRAL CLASSIFICATION

The classification of stellar spectra is important to astronomers today because it offers a quick way to determine a star's temperature. But this discovery took decades of study. **Stellar spectroscopy**—the study and classification of spectra—originated early in the nineteenth century when the German scientist Joseph Fraunhofer discovered absorption lines in the spectrum of the Sun and later in other stars.

In the late 1800s, Henry Draper, a physician and amateur astronomer, began recording spectra with the then-new technology of photography. By observing a field of stars through a thin prism, he spread each star's light into a spectrum (Figure 55.4). On his death, Draper's widow endowed a project at Harvard to create a compilation of stellar spectra. In this work, stars were assigned to alphabetic types running from A to Q, collecting together similar-looking spectra in each type. These classifications were based mostly on the strength of the spectral lines—how dark they were—particularly the lines of hydrogen that are seen in the visible part of the spectrum. A-type stars have the strongest lines of hydrogen, while O-type stars, for example, have very weak lines.

FIGURE 55.4
This is a photograph of the Hyades cluster made through a thin prism. The prism spreads the light of each star into a spectrum.

About 1901, Annie Jump Cannon (Figure 55.5), the astronomer doing most of the classification for the Draper Catalog, determined that the types fell in a more reasonable sequence if she rearranged them by temperature. She eliminated a number of types that had similar temperatures and reordered the rest into the sequence, from hottest to coolest, of O-B-A-F-G-K-M. Her work is the basis for the stellar **spectral types** we use today.

Application of Wien's law and theoretical calculations show that temperatures range from more than 30,000 K for O stars to less than 3500 K for M stars. A-type stars have temperatures ranging from about 7500 to 10,000 K, and G stars, such as our Sun, range between 5000 and 6000 K. Because a star's spectral type is set by its temperature, its type also indicates its color, ranging from violet-blue colors for O and B stars to orange and red colors for K and M stars.

To distinguish finer gradations in temperature, astronomers subdivide each type by adding a numerical suffix—for example, B0, B1, B2, . . . , B9—with the smaller numbers indicating higher temperatures. With this system, the temperature of a B0 star is almost 30,000 K, whereas that of a B5 star is about 15,000 K. Our Sun is a G2 star with a surface temperature of about 5800 K.

Stellar spectroscopy

Q If the Sun had a temperature like that of Rigel or Betelgeuse, do you think our eyes would have evolved differently? Why or why not?

An example of the differences between spectra is given in Figure 55.6, which shows the spectra of stars with three different temperatures. The B-type spectrum is for a hot star like Rigel that produces mostly violet and blue light. The G-type spectrum is similar to the Sun's, with a temperature of about 6000 K and radiation peaks near the middle of the visible-wavelength range. The spectra are shown in Figure 55.6 both as they appear through a prism and as line tracings showing the relative intensity at each wavelength. The M-type spectrum is for a cool star like Betelgeuse. The B-type spectrum has only a few lines, but the lines are strong, and their pattern indicates they come from hydrogen. The M-type spectrum, however, shows a welter of lines with no apparent regularity.

The complicated history of the development of the spectral types resulted in the odd nonalphabetical progression for the spectral types—O-B-A-F-G-K-M. So much effort, however, had been invested in classifying stars using this system that it was easier to keep the types as assigned, with their odd order, than to reclassify them. Cannon, over her lifetime, classified nearly a quarter million stars!

In the late 1990s, astronomers started using sensitive infrared detectors that allow detection of stars even cooler than the M-type stars. Some of these stars are so cool that not only do molecules form in their atmospheres, but solid dust particles condense there as well. To continue the progression of spectral types, astronomers have designated two new types—L and T—to describe these cool objects. The full spectral type sequence is therefore now **O-B-A-F-G-K-M-L-T,** from hottest to coolest.

As a help to remember the peculiar order of spectral types, astronomers and students often use a mnemonic. Before the discovery of the L and T types, the most widely known mnemonic was "Oh, Be A Fine Girl/Guy. Kiss Me," which was printed in textbooks throughout the twentieth century. Perhaps you can develop a better mnemonic for the twenty-first century.

FIGURE 55.5
Annie Jump Cannon (1862–1941) founded the modern system of classifying stars.

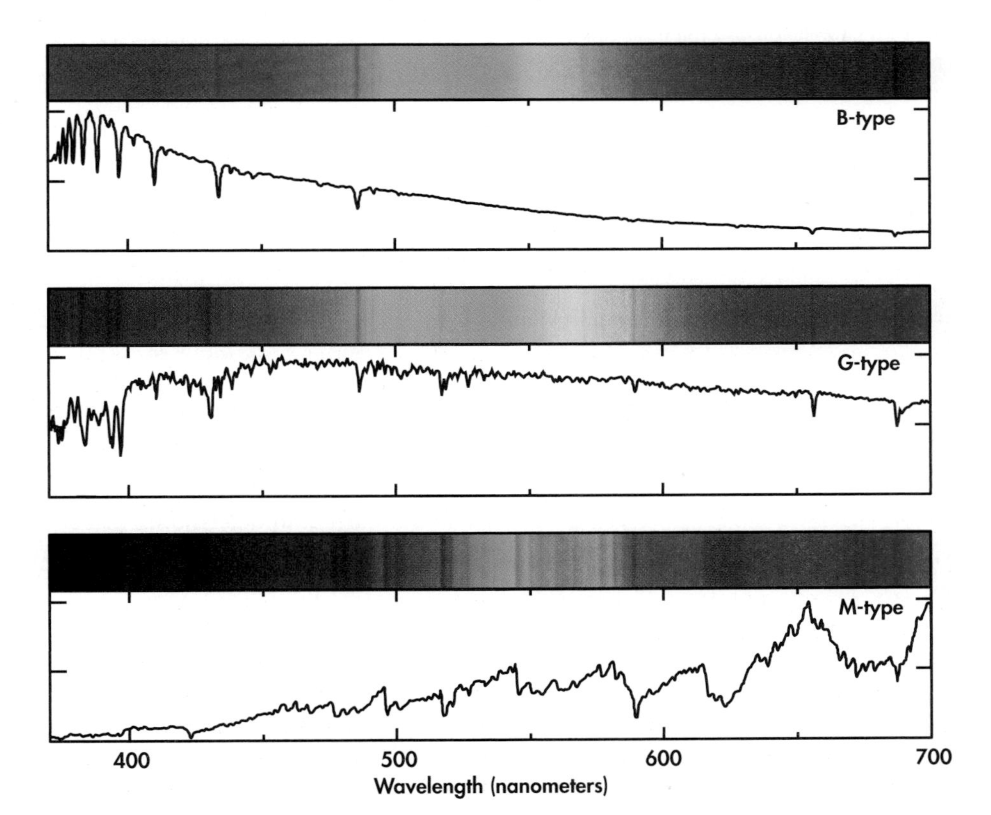

FIGURE 55.6
Spectra of stars of type B, G, and M. The B star is the hottest of the three, while the M star is the coolest. Cooler stars tend to have more absorption lines, but some lines are darker in hotter stars.

55.4 HOW TEMPERATURE AFFECTS A STAR'S SPECTRUM

FIGURE 55.7
Cecilia Payne (1900–1979) discovered that hydrogen is the most abundant element in stars. She went on to become the first woman to be made a full professor at Harvard.

Balmer lines are named for Johann Balmer, a German scientist who first studied their pattern.

To see why temperature affects the lines in a star's spectrum, recall that a photon can be absorbed by an atom only if the photon's energy exactly matches the energy difference between two electron orbitals. For an atom to absorb the photon, it must already have an electron in the orbital with the correct lower energy level. An atom may be abundant in a star's atmosphere but create only weak lines at a particular wavelength simply because the gas is so hot or so cold that its electrons are in the "wrong" level to absorb light at that wavelength.

This was first worked out in the 1920s by the U.S. astronomer Cecilia Payne (later Payne-Gaposhkin; Figure 55.7). She pointed out that the strength of the hydrogen lines in a star's spectrum depend strongly on the star's temperature. She recognized that it would be a mistake to assume that a star had little hydrogen if these lines were weak, unless the appropriate corrections for temperature had been made.

The absorption lines of hydrogen that are observable at visible wavelengths are made by electrons orbiting in the hydrogen atom's second level (see Figure 55.1). These lines are sometimes called the **Balmer lines** to distinguish them from other hydrogen lines with ultraviolet and infrared wavelengths. The Balmer lines occur at wavelengths where light has exactly the amount of energy needed to lift an electron from a hydrogen atom's second energy level up to the third or higher levels. These lines are no more important than other transitions in hydrogen that occur at infrared or ultraviolet wavelengths, but because their wavelengths are in the visible spectrum they are much more easily observable.

If the hydrogen atoms in a star have very few electrons orbiting in level 2, the Balmer absorption lines will be weak, even if hydrogen is the most abundant element in the star. This may happen in a cool star, because most of the electrons in hydrogen atoms are in level 1, the lowest energy level. In a hot star the atoms move faster, and when they collide, electrons get excited ("knocked") into higher energy orbitals: the hotter the gas, the higher the orbital. In fact, in very hot stars, electrons may be knocked out of the atom entirely, in which case the atom is said to be *ionized.* As a result of this excitation, proportionally more of the electrons in a very hot star will be in level 3 or higher. Hydrogen has the maximum number of its electrons in level 2 when its temperature is about 10,000 K—corresponding to an A0 star. Hence it is in this temperature range that the Balmer lines are strongest. If we are therefore to deduce correctly the abundance of elements in a star, we must correct for such temperature effects.

After accounting for these effects, Payne discovered that virtually all stars are composed mainly of hydrogen. This idea was controversial at the time because most astronomers thought that stars were made primarily of heavy elements, and she was discouraged from publishing her conclusions by her thesis adviser. However, several years after her thesis was published, her ideas gained wide acceptance. Payne's idea led to our current understanding that about 71% of a star's mass is hydrogen, 27% is helium, and the remaining 2% is a mixture of all the other elements known to exist.

55.5 SPECTRAL CLASSIFICATION CRITERIA

Astronomers have put the stellar spectral types in order of temperature by using the pattern of lines in a star's spectrum. The similarity in most stars' compositions means that the main cause of differences in the spectra is the temperature of the photosphere. Figure 55.8 shows spectra of the seven main types—O through M—and illustrates that the differences in the line patterns are easy to see.

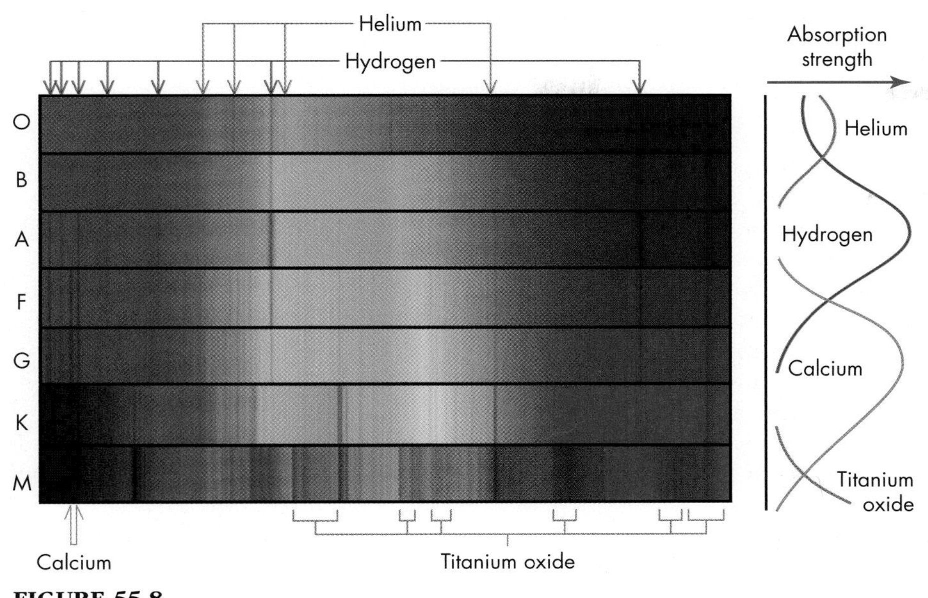

FIGURE 55.8
The stellar spectral types. The types are shown in order from hottest (O) to coolest (M). Several absorption lines are indicated, and their changing strengths are illustrated at right.

For example, A-type stars show strong hydrogen Balmer lines because they are near the ideal temperature for hydrogen electrons to be in energy level 2. O stars have weak absorption lines of hydrogen but detectable absorption lines of helium, the second most abundant element, because they are so hot. At their high temperature, O stars' hydrogen atoms collide so violently and are excited so much by the star's intense radiation that the electrons are stripped from most of the hydrogen, ionizing it. With its electron missing, a hydrogen atom cannot absorb light. Because most of the O stars' hydrogen is ionized, such stars have extremely weak hydrogen absorption lines. Helium atoms, on the other hand, are more tightly bound and typically retain at least one electron, allowing them to produce absorption lines.

Lines of some elements, like ionized calcium, become more prominent at lower temperatures, but in cooler stars the hydrogen Balmer lines grow much weaker. In this case the lines are weak because the hydrogen's electrons are mostly in level 1 and therefore require much more energetic ultraviolet photons to raise them to a higher energy level. The K and M stars have such cool atmospheres that some molecules, like titanium oxide, are able to form. These molecules often have rich and complex patterns of absorption lines.

Q What might a cool star's spectrum look like if it contained no elements other than hydrogen and helium? What might a hot star's spectrum look like if it contained a large fraction of heavy elements?

The even cooler L stars show strong molecular lines of iron hydride and chromium hydride, while T stars show strong absorption lines of methane. These features, and those found in the other spectral types, are summarized in Table 55.1. Comparing the strength of spectral lines to determine a spectral type is now often done automatically by computers that scan a star's spectrum and match it against standard spectra stored in memory.

Finally, we note that a small number of stars have spectra and temperatures that do not fit into the normal classification sequence. They have compositions or temperatures that are quite different from those of stars in the standard spectral sequence, and they are sometimes assigned their own classification letters. Some of these are now understood to be stars that have undergone dramatic changes late in their lifetimes; others remain mysteries that are still being studied.

TABLE 55.1 Summary of Spectral Types

Spectral Type	Temperature Range (K)	Features
O	Hotter than 30,000	Ionized helium, weak hydrogen
B	10,000–30,000	Neutral helium, hydrogen stronger
A	7500–10,000	Hydrogen very strong
F	6000–7500	Hydrogen weaker, metals—especially ionized Ca—moderate
G	5000–6000	Ionized Ca strong, hydrogen weak
K	3500–5000	Metals strong, CH and CN molecules appearing
M	2000–3500	Molecules strong, especially TiO and water
L	1300–2000	TiO disappears. Strong lines of metal hydrides, water, and reactive metals such as potassium and cesium
T	900?–1300?	Strong lines of water and methane

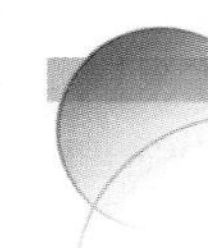

KEY TERMS

absorption line, 452
Balmer lines, 456
O-B-A-F-G-K-M-L-T, 455
photosphere, 451
spectral type, 454
spectrum, 451
stellar spectroscopy, 454

QUESTIONS FOR REVIEW

1. What different ways are there to measure a star's temperature?
2. Why do stars have dark lines in their spectra?
3. Where in the star's atmosphere are the dark lines produced?
4. What are the stellar spectral types? Which are hot and which are cool?
5. What distinguishes the spectral types of stars?
6. If a star has a peak emission in the infrared part of the spectrum, is it hotter or cooler than the Sun? What color would the star be?
7. Why do both O and M stars have weak absorption lines of hydrogen in their spectra?

PROBLEMS

1. Suppose a star radiates most strongly at about 200 nm. How hot is it?
2. The bright southern star Alpha Centauri radiates most strongly at about 500 nm. What is its temperature? How does this compare to the Sun's?
3. A star radiates most strongly at 850 nm. What spectral type is it?
4. If a T star has a surface temperature of 1000 K, at what wavelength should it be brightest?
5. If a star has very faint Balmer lines, what are the next steps involved in determining its spectral type?
6. When our star reaches the end of its main-sequence lifetime, it will become an M-class red giant (Unit 64) with a surface temperature of approximately 3300 K. How will its wavelength of peak radiation differ from the current value?
7. Suppose a star radiates most strongly at 330 nm. Would you expect to see strong absorption lines in its spectrum? Why or why not?

TEST YOURSELF

1. A star radiates most strongly at 400 nm. What is its surface temperature?
   a. 400 K
   b. 4000 K
   c. 40,000 K
   d. 75,000 K
   e. 7500 K
2. Which of the following stars is hottest?
   a. An M star
   b. An F star
   c. A G star
   d. A B star
   e. An O star
3. A star has lines of ionized helium. Ionizing helium requires a very high temperature. The star is therefore most likely to belong to spectral type
   a. O.
   b. A.
   c. G.
   d. M.
   e. T.

unit 56

The Masses of Orbiting Stars

Background Pathways

Unit 17: Measuring a Body's Mass Using Orbital Motion 135

Many stars are not alone in space; rather, they have one or more stellar companions held in orbit about each other by their mutual gravitational attraction (Figure 56.1). Astronomers call such stellar pairs **binary stars.** Binary stars are extremely useful, because from their orbital motion astronomers can determine the stars' masses. To understand how, recall that the gravitational force between two bodies depends on their masses. The gravitational force, in turn, determines the stars' orbital motion. Therefore, if we can measure that motion, we can work backward to find the mass.

A star's mass is not just another piece of data in our catalog of stellar properties; it is the most critical property for determining the structure and fate of a star. A star's mass tells us how much weight squeezes the interior of the star, thereby driving the nuclear reactions that generate the heat and pressure that hold the star up (Unit 49). The mass also tells us how much hydrogen "fuel" is potentially available for the nuclear reactions needed to keep the star from collapsing under its own weight (Units 50 and 61).

Binary star systems are quite common—at least 40% of all stars known have orbiting companions, and the percentage may be much higher. Typically the stars in a binary system are a few AU apart. In many cases more than two stars are involved. Some stars are found in triplets, others in quadruplets, and at least one six-member system is known. The masses we determine for these stars give us mass estimates for stars of similar type that do not have companions.

56.1 TYPES OF BINARY STARS

A few binary stars are easy to see with a small telescope. Mizar, the middle star in the handle of the Big Dipper, is a good example and is illustrated in Looking Up #2: Ursa Major. When you look at this star with just your eyes, you will also see a dim star, Alcor, near it, but this is not the binary companion. Alcor and Mizar form an **apparent double star:** two stars lying in the same direction but not in orbit around each other. With a telescope, however, we can resolve Mizar to reveal that it consists of two much closer stars in orbit around each other, Mizar A and Mizar B: a **true binary.**

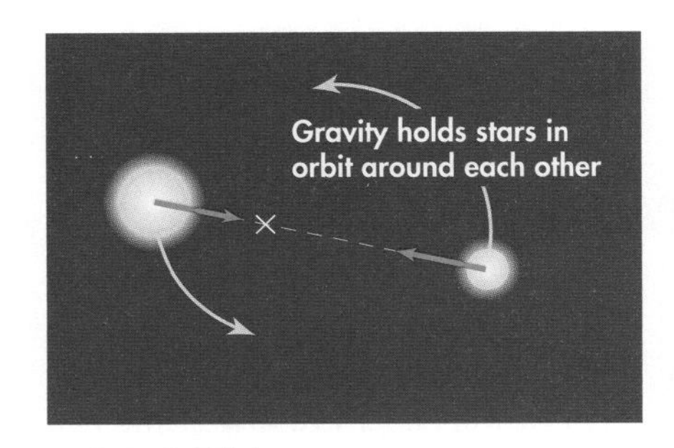

FIGURE 56.1
Two stars orbiting in a binary system, held together as a pair by their mutual gravitational attraction.

We can directly observe the orbital motion of Mizar A and Mizar B around each other by comparing images made years apart. Such binary stars are called **visual binaries** because we can see two separate stars and their individual motions. In visual binary systems we can usually observe the two stars completing an orbit around each other over several years. Some widely separated binary stars take a very long time to complete an orbit. For example, binary stars Mizar A and B have completed only a fraction of their orbit since their discovery in 1617; astronomers

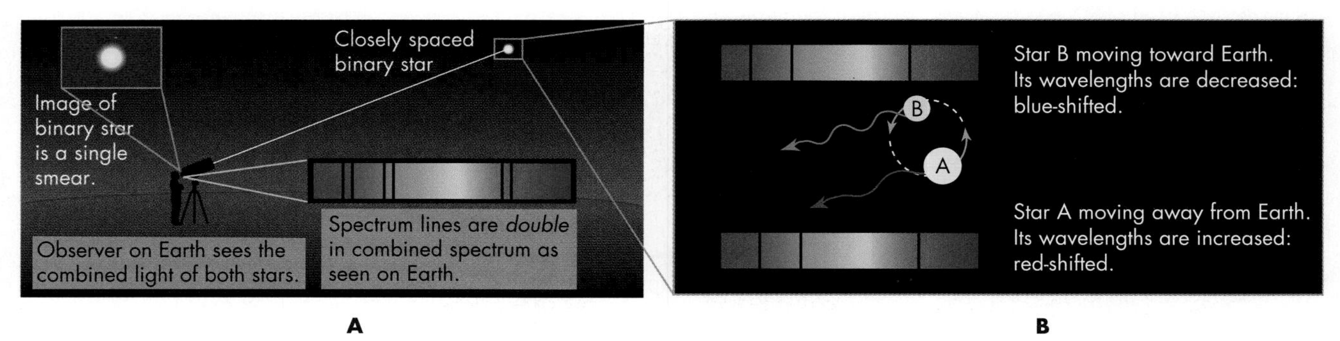

FIGURE 56.2
Spectroscopic binary star. (A) The two stars are generally too close to be seen separately by even the most powerful telescopes. (B) Their orbital motion creates a different Doppler shift for the light from each star. Thus, the spectrum of the stellar pair contains two sets of lines, one from each star, shifted relative to each other by different amounts depending on the stars' movements.

INTERACTIVE
Binary stars

ANIMATIONS
Spectroscopic binary

can trace a partial arc of the orbit, and they estimate that the total length of the orbital period is a few thousand years.

Some binary stars are so close together that their light blends into a single blob that defies separation with even the most powerful telescopes. In such cases their orbital motion cannot be seen directly, but it may nevertheless be inferred from their combined spectra. As each star moves along its orbit, it alternately moves toward and then away from the Earth. This motion creates a Doppler shift, and so the spectrum of the star pair shows two sets of spectral lines that shift relative to each other. While one star's spectral lines shift to longer wavelengths (because the star is moving away from us), the other's spectral lines shift to shorter wavelengths (because it is approaching us). Then, in a cyclic fashion, half an orbit later, the pattern reverses (Figure 56.2). Astronomers call such star pairs **spectroscopic binaries,** and by observing a full cycle of their spectral shifts, we can determine the orbital motion of the stars. These stars are typically close together, and the orbital periods are short. For example, Mizar A is itself a spectroscopic binary with a period of just 20 days.

Eclipsing binaries are pairs of stars whose orbits are oriented exactly edge-on to our line of sight; so the stars eclipse each other sequentially. From the duration of the eclipses, it is possible to determine the sizes of the stars (Unit 57). When an eclipsing binary is also detected as a spectroscopic binary, we can determine the parameters of the system better than for other spectroscopic binaries. This is because we need to know the orientation of the orbits to know whether the Doppler shift reflects the full speed of orbital motion or just the part of the motion that happens to be along our line of sight toward the star. In an eclipsing system we know that the orbit is edge-on to us.

Q: When we observe spectroscopic binaries, we often do not know how the plane of the orbit is oriented to our line of sight. How will the Doppler shift change if the same orbit is oriented edge-on, face-on, or somewhere in between?

56.2 MEASURING STELLAR MASSES WITH BINARY STARS

From their knowledge of orbital motion as just described, astronomers can find the mass of a stellar pair using a modified form of Kepler's third law (Unit 17). Kepler demonstrated that the time required for a planet to orbit the Sun is related to its distance from the Sun. If P is the orbital period and a is the semimajor axis (half the long dimension) of the planet's orbit, then $P^2 = a^3$, a relation called Kepler's third law.

Q: When Kepler devised his third law, he did not include a term describing mass. Why was he able to omit mass for the Solar System?

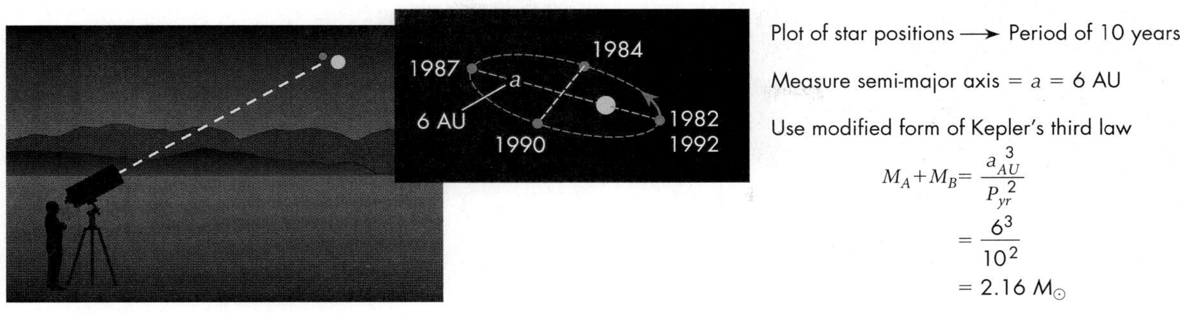

FIGURE 56.3
Measuring the combined mass of two stars in a binary system using the modified form of Kepler's third law.

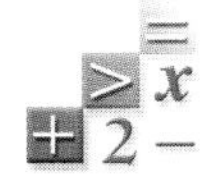

M_A = Mass of star A in solar masses
M_B = Mass of star B in solar masses
P_{yr} = Period of orbit in years
a_{AU} = Semimajor axis of orbit in AU

Newton discovered that Kepler's third law could be generalized to apply to *any* two bodies in orbit around each other. As shown in Unit 17, the sum of their masses, $M_A + M_B$, can be found from the period of their orbit and their separation:

$$M_A + M_B = \frac{a_{AU}^3}{P_{yr}^2}$$

where a_{AU} is expressed in astronomical units, P_{yr} in years, and the masses are in solar masses. This relationship is our basic tool for measuring stellar masses.

To find the mass of the stars in a visual binary, astronomers first plot their orbital motion, as depicted in Figure 56.3. It may take many years to observe the entire orbit, but eventually we can determine P, the time required for the stars to complete an orbit. From the plot of the orbit, and with knowledge of the stars' distance from the Sun, astronomers next measure the semimajor axis, a, of the orbit of one star about the other.

Consider, for example, the orbit of the two stars that compose the visual binary star Alpha Centauri (shown in Looking Up #8: Centaurus and Crux, the Southern Cross). The two components have an orbital period of 68 years and a semimajor axis of 20.6 AU. We can then find their combined mass, $M_A + M_B$, by inserting the measured values for P and a, giving

$$M_A + M_B = \frac{20.6^3}{68^2} = 1.9\ M_\odot$$

That is, the combined mass of the stars is 1.9 times the Sun's mass.

56.3 THE CENTER OF MASS

Kepler's law allows us to find the combined mass of two stars that orbit each other. Additional analysis of stars' orbits allows us to find their individual masses. We can tell their masses relative to each other by the amount each star moves. If one star is much more massive than the other, then the massive one will hardly move while the less massive one exhibits nearly all of the motion—like a planet orbiting the Sun. If the two stars are equal in mass, they will move the same amount.

Even in the Solar System, where we usually speak of the planets "orbiting the Sun," we should properly say that the Sun and planets orbit their common center of mass. The Sun is so much more massive than the planets that the center of mass turns out to be inside the Sun. The Sun "wobbles" because of the planets, primarily Jupiter. This kind of wobble seen for other stars is what allows us to detect planets orbiting them (Unit 34).

FIGURE 56.4
The location of the center of mass of two bodies depends on their relative mass. (A) If the masses are equal, the center of mass is halfway between. (B) If one body is more massive than the other, the center of mass is closer to the more massive body.

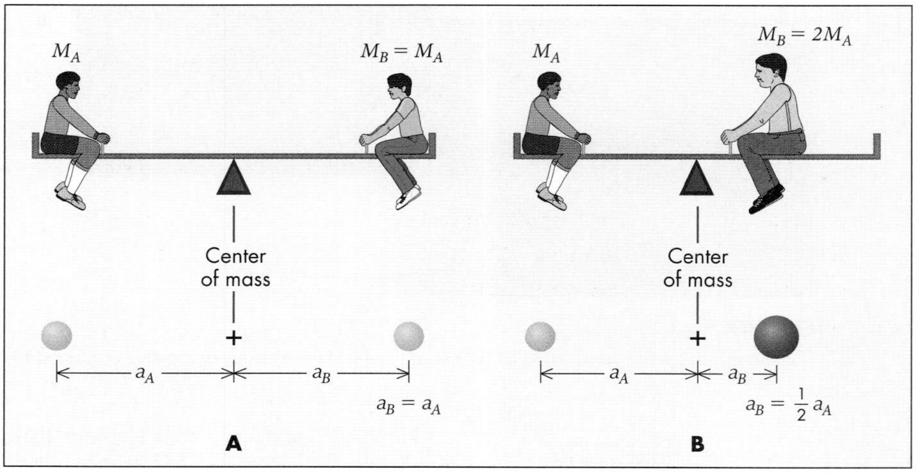

Q The Greek scientist Archimedes is supposed to have said, "Give me a lever and a place to stand and I will move the Earth." Might this be possible? How?

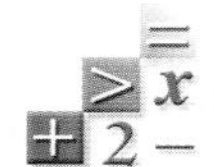

Note that $a_A + a_B$ equals the semimajor axis a used in Kepler's formula.

Pairs of stars do not have such disparities in mass, though, so they orbit a point more nearly equidistant between them called the **center of mass,** as depicted in Figure 56.4. This is located along a line joining the two stars at a position that depends on the stars' relative masses. It is like the point of balance on a children's playground "seesaw" or "teeter-totter." If one star is three times more massive than the other, the point of balance will be three times closer to the more massive star.

We can express this idea mathematically as follows. If one star has mass M_A and is orbiting at a distance a_A from the center of mass, and the other has mass M_B and is orbiting at a distance a_B (Figure 56.4), then

$$M_A \times a_A = M_B \times a_B.$$

The *larger* mass has the *smaller* distance from the center of mass. If two stars are equal in mass ($M_A = M_B$), then $a_A = a_B$, and they will orbit a point exactly halfway between them. On the other hand, if star *B* is two times less massive than star *A* ($M_B = ½ \times M_A$), then star *B* will orbit two times farther from the center of mass than its orbital companion ($a_B = 2 \times a_A$). Similarly, an adult who weighs twice as much as a child will overbalance the child on a seesaw unless he sits half as far from the pivot point as the child.

For spectroscopic binaries, we can compare the relative speed of each star. Each orbits the center of mass in the same period of time, but the more massive one has less distance to move in its orbit. So, for example, we might detect star *A* moving toward us at 2 kilometers per second while star *B* moves away from us at 8 kilometers per second (Figure 56.5). Half an orbit later we see star *A* moving away from us at 2 kilometers per second while star *B* moves toward us at 8 kilometers per second. This shows us that star *B* is moving 4 (= 8/2) times faster and farther than star *A*. Therefore, star *B* is four times less massive than star *A*.

For Alpha Centauri, measurements show that star *A* orbits about 0.7 times as far from the center of mass as star *B*. So $a_A = 0.7 \times a_B$. We therefore know that $M_B = 0.7 \times M_A$. From the previous section we also know that their combined mass equals 1.9 solar masses: $M_A + M_B = 1.9\ M_\odot$. Therefore,

$$M_A + 0.7 \times M_A = 1.7 \times M_A = 1.9\ M_\odot.$$

So Alpha Centauri A must be about 1.1 solar masses, while Alpha Centauri B is about 0.8 solar masses.

From analyzing many star pairs, astronomers have determined that most stars have masses from about 0.1 to 30 $M_\odot$. A few rare stars are even more massive, ranging up to more than 100 $M_\odot$. Our star, the Sun, is fairly average in mass, as it is in many other ways.

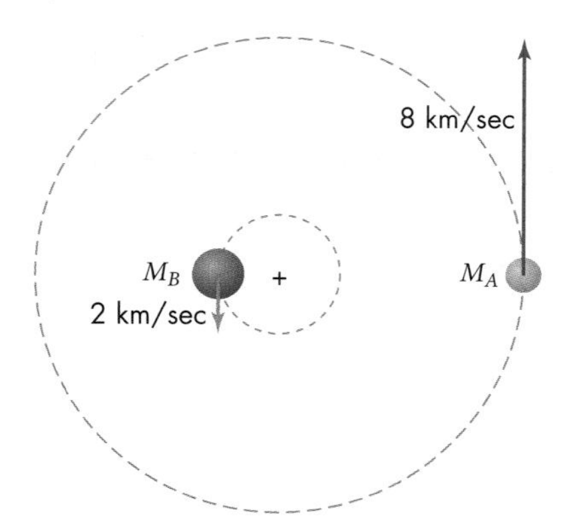

FIGURE 56.5
As two stars orbit their center of mass, the more massive star completes a smaller orbit about the center of mass and therefore travels at a slower speed than the less massive star. The Doppler shift of the more massive star is also smaller in proportion to the stars' relative masses.

KEY TERMS

QUESTIONS FOR REVIEW

1. What is a binary star?
2. How do visual and spectroscopic binaries differ?
3. What are eclipsing binaries? Under what circumstances do binary stars appear to us to be eclipsing binaries?
4. What properties of stars can astronomers determine from binary stars?
5. What is the center of mass?
6. Applying Kepler's third law to binary stars gives us only the sum of the stars' masses. How do we find the individual masses?

PROBLEMS

1. Two stars are in a binary system. Find their combined mass if they have
   a. an orbital period, P, of 5 years and an orbital separation, a, of 10 AU.
   b. an orbital period, P, of 2 years and an orbital separation, a, of 4 AU.
2. Mizar A and B are about 400 AU apart, currently. Their orbital period is estimated to be between 2000 and 4000 years. What range of masses does this suggest for the pair?
3. Two stars are orbiting each other, and the sum of their masses is 6 $M_{\odot}$. It is found that star A of the pair is orbiting 3 arcsec from the center of mass, while star B is orbiting 6 arcsec from the center of mass. What are the two stars' masses?
4. Find the distance from the center of the Sun of the center of mass between the Sun ($M_{\odot} = 1.99 \times 10^{30}$ kg) and the following planets:
   a. Earth ($M = 5.97 \times 10^{24}$ kg; $a = 1.50 \times 10^{8}$ km)
   b. Jupiter ($M = 1.90 \times 10^{27}$ kg; $a = 7.78 \times 10^{8}$ km)
   c. Saturn ($M = 5.68 \times 10^{26}$ kg; $a = 1.43 \times 10^{9}$ km)
5. When Jupiter and Saturn are on the same side of the Sun, is the center of mass inside or outside the Sun's radius ($R_{\odot} = 6.96 \times 10^{5}$ km)?
6. Two stars are orbiting each other, both 4.2 arcsec from their center of mass. Their orbital period is 420.3 years and their distance from the Earth is 104.1 ly. Find their masses.
7. Using Kepler's third law, we find that the combined mass of a binary system is 55.5 $M_{\odot}$. The stars have separations of 6.3 arcsec and 2.5 arcsec from the center of mass, and the system is at a distance of 64 ly from the Earth. Find their distances from the center of mass of the system, and their individual masses.

TEST YOURSELF

1. Calculating a star's mass from its binary orbital motion is important because
   a. such systems are rare, so we don't often get to collect this information.
   b. then we can predict the star's ultimate fate.
   c. then we can tell which star is orbiting the other.
   d. it's the only way we can tell if there's a second star in the system.
   e. it's the only method available for detecting apparent double stars.
2. Eclipsing binaries
   a. can be spectroscopic binaries.
   b. can be visual binaries.
   c. are edge-on to our line of sight.
   d. in some cases allow us to determine a pair's exact orbital parameters.
   e. All of the above.
3. Which of the following is not a class of orbiting star systems?
   a. A sequential binary
   b. An eclipsing binary
   c. A spectroscopic binary
   d. A true binary
   e. A visual binary

unit

57

The Sizes of Stars

A property of stars as basic as diameter seems like it should be easy to measure. However, stars are at such great distances from Earth that this is one of the most challenging stellar properties to determine through direct observation. Astronomers have devised a variety of techniques for determining stars' sizes from observations, but measuring stellar sizes remains near the limits of our present technologies.

Astrophysics offers another solution. A great success of astrophysics in the 1800s and 1900s was the discovery of a physical law that describes the amount of light generated by hot objects. This law, discovered in the late 1800s, is called the *Stefan-Boltzmann law* (Unit 23.5). It can tell us the radii of stars if we know their temperatures and luminosities.

We begin this unit by reviewing techniques for making direct measurements of stars' sizes. We conclude by showing how we can use the physical laws determined in Earth's laboratories to give us this important information about the sizes of distant stars.

Background Pathways

57.1 THE ANGULAR SIZES OF STARS

In principle, we can measure a star's radius from its angular size and distance (Unit 10). Unfortunately, the angular sizes of all stars except the Sun are extremely tiny because they are so far from the Earth. Even under the highest magnification in the most powerful telescopes, stars generally look like smeary spots of light. The size of the smeared spot has nothing to do with the star's size, but is caused by the blurring effects of our atmosphere and by a physical limitation of telescopes called *diffraction* (Unit 29).

Astronomers have developed techniques for reducing the blurring effect of the Earth's atmosphere, or they can avoid it altogether by putting telescopes into space (Unit 30). However, diffraction presents a more fundamental limit. Bending of light at the edge of a telescope's aperture smears the light by an amount that depends inversely on the telescope's diameter. That is, diffraction effects are less severe in bigger telescopes. To measure the angular size of stars, however, we need telescopes with truly immense diameters. For example, measuring the angular size of a star like the Sun, if it were 15 parsecs away, would require a telescope 300 meters in diameter, about three times the size of a football field—nearly impossible with current technologies, and unimaginably costly to build!

To avoid the need for enormous telescopes, astronomers have devised an alternative way to measure the angular size of stars: by using not one huge telescope, but two (or more) smaller ones separated by a large distance (for example, several hundred meters). This method, called **interferometry** (Unit 29), allows astronomers to measure angular sizes with a precision almost equal to that of a single telescope whose diameter is equal to the distance that separates the two smaller ones. Two smaller telescopes separated by 300 meters can measure details almost equivalent to those from a single 300-meter diameter telescope. A computer can combine the information from the two telescopes to determine the angular size of the star,

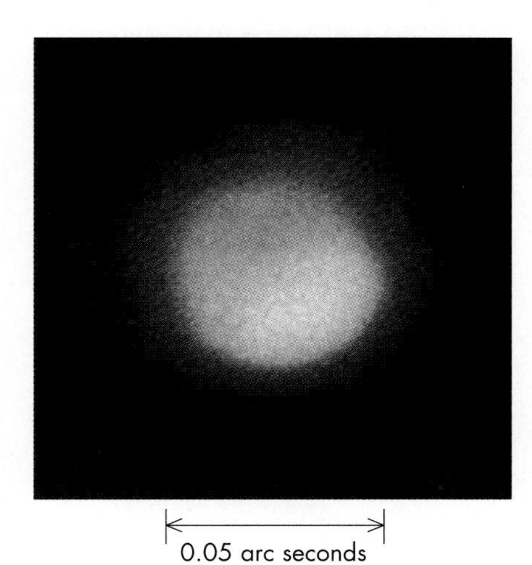

FIGURE 57.1
Image of the star Betelgeuse made by speckle interferometry. Betelgeuse is about 130 pc distant. The lack of detail in this image reflects the current difficulty in observing the disk of any star but our Sun.

although it does not provide a detailed image. New interferometers, using some of the largest optical telescopes in the world, only in the last several years have succeeded in determining the sizes of our nearest neighbors, the stars of the Alpha Centauri star system. These are stars similar to the Sun, but at their distance of about 1.3 parsec, they are less than one one-hundredth of an arc second in diameter.

Another technique for determining stellar diameters is called **speckle interferometry.** In speckle interferometry, a single telescope is used to capture many short-exposure images of the object of interest. Each separate image is called a "speckle"—it is actually an image of the star formed by momentary focusing of light in the turbulent atmosphere of the Earth. By combining all of these images with a computer, astronomers can compensate for much of the blurring effects of the atmosphere. This can be contrasted to adaptive optics (Unit 30.2), which attempts to adjust the position of the image on the detector as the atmosphere changes. You can see a typical example of speckle interferometry in the "deblurred" image of Betelgeuse, shown in Figure 57.1. Betelgeuse has the largest detected angular size of any star, but this is still only about 0.05 arc second, so the image has very little detail. Observations of other stars, which have smaller angular sizes, reveal even less detail.

With interferometry, astronomers have measured the radii of hundreds of stars, but something is quite unusual about the stars that have been detected. The majority are *much* larger than the Sun. For example, Betelgeuse's diameter can be determined using the angular size formula (Unit 10) because its distance is known from parallax measurements. The result is a diameter of more than 6 AU—bigger than the orbit of Mars! If the Sun were that big, all of the inner planets would be swallowed up and destroyed. The great majority of stars are far smaller, and they require less-direct techniques to determine their sizes.

57.2 USING ECLIPSING BINARIES TO MEASURE STELLAR DIAMETERS

Astronomers have found that many stars are in orbit around a companion star, much as the Earth orbits the Sun. Such gravitationally bound star pairs are called **binary stars** (Unit 56), and astronomers can measure the stars' masses by observing their orbits. Some binary star pairs orbit almost exactly edge-on to our perspective from the Earth. As the stars orbit, one will eclipse the other as it passes between its companion and the Earth. Such systems are called **eclipsing binary stars,** and if we watch such a system, its light will periodically dim. The star Algol (Arabic for "Demon Star") in the constellation Perseus (see Looking Up #3: M31 and Perseus) is the brightest example of this phenomenon. Normally Algol is fairly bright—it is the second brightest star in Perseus—but once every 68 hours it dims to one-third its usual brightness for 10 hours, becoming the seventh brightest star in that constellation.

Eclipsing binary

During most of the orbit of an eclipsing binary system, we see the combined light of both stars; but at the times of eclipse, the brightness of the system decreases as one star covers part or all of the other. This produces a cycle of variation in light intensity called a **light curve.** Figure 57.2A is a graph of such a change in brightness over time.

The duration of such an eclipse depends on the diameter of the stars. Figure 57.2B shows why. The eclipse begins as the edge of one star first lines up with the edge of the other, as shown in the figure. Dimming increases until the smaller star is completely in front of the larger star; the light level remains low until the smaller star reaches the other edge of the large star. Then the process reverses until

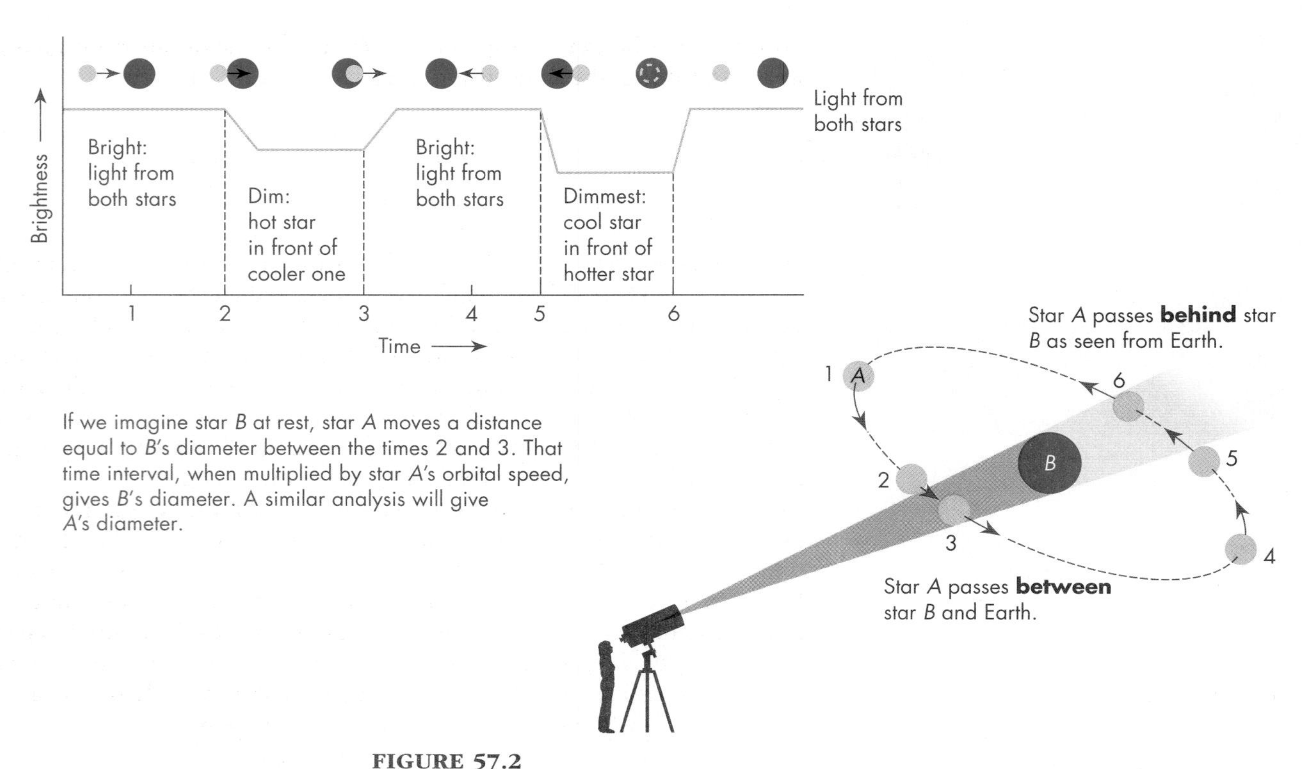

FIGURE 57.2
An eclipsing binary and its light curve. From the durations of the eclipses, the diameter of each star may be found. In this illustration, the yellow star is hotter than the red star.

we see all the light from both stars again. A dip in brightness is also seen when the smaller star passes behind the larger star. The relative size of these dips depends on which star has the hotter surface temperature—not necessarily on which star is bigger. For example, when a large cool star is eclipsed by a smaller, hotter one, the reduction of light may be quite small. The pattern of the light curve also depends on whether the stars pass completely behind each other, as we have assumed here, or if they only partially eclipse each other.

For a circular orbit, when the stars are aligned for an ecliipse, we will not see any speed difference. We must measure the speed of the stars a quarter orbit later when the stars appear farthest apart.

If Doppler measurements (Unit 25) tell us how fast the stars are orbiting each other, we can estimate the size of each star. Suppose the stars are moving around each other at a speed of 100 kilometers per second and the light takes 10,000 seconds (about 3 hours) to dim from its full brightness at the start of the eclipse to its dim state at full eclipse. This is how long it takes for the smaller star's disk to move sideways by its own diameter across the edge of the larger star—either in front of or behind it. The star's diameter is therefore equal to the product of its speed and the time for the dimming. In this example the distance is 100 km/sec × 10,000 sec = 1,000,000 km. The total duration of the eclipse is the time it takes for the disk of the smaller star to move from being entirely uncovered on one side of the larger star, across the larger star, to being completely uncovered on the other side—which is the sum of the diameters of both stars. Suppose this takes 100,000 seconds (about 28 hours); then the sum of the diameters is 100 km/sec × 100,000 sec = 10,000,000 km. Combining the two results, we find that the larger star has a diameter of 9,000,000 km.

Q Would you expect the kinds of stars that occur in eclipsing binary systems to be typical stars?

This method of using eclipses lets us estimate the sizes of stars that are too small to detect with an interferometer. It has drawbacks, though. For example, it applies only to the small subset of stars that happen to be in binary systems with orbits oriented in just the right way. Moreover, the method has uncertainties because the orbit may not be exactly edge-on, in which case the duration of the eclipse does not reflect the total diameter of the star.

57.3 THE STEFAN-BOLTZMANN LAW

In the late 1800s scientists discovered how the luminosity of a hot object depends on its temperature. They found that bodies radiate an amount of energy that depends on their temperature. Each square meter of a body radiates a luminosity equal to σT^4, where σ is a constant and T is the temperature, as illustrated in Figure 57.3A. This relationship, named for its discoverers, is called the **Stefan-Boltzmann law** (Unit 23.5). It expresses the fact that most matter—such as the relatively dense gas of a star's photosphere—emits an amount of electromagnetic radiation that increases rapidly as the temperature climbs.

We can use the Stefan-Boltzmann law to determine a star's radius by comparing the star's total luminous output to the amount expected from each square meter. According to the law, if a star has a temperature T, each square meter of its surface radiates an amount of energy per second given by σT^4. We can find the total energy the star radiates per second—its luminosity, L—by multiplying the energy radiated from one square meter (σT^4) by the number of square meters of its surface area (Figure 57.3B). If we assume that the star is a sphere, its surface area is $4\pi R^2$, where R is its radius. Its luminosity, L, is therefore

$$L = 4\pi R^2 \times \sigma T^4.$$

That is, a star's luminosity equals its surface area times σT^4. The relationship between L, R, and T may at first appear complex, but the equation's meaning is

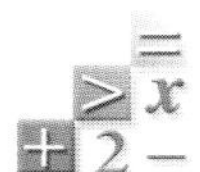

L = Luminosity of star (watts)

R = Radius of star (meters)

σ = Constant = 5.67×10^{-8} watts per meter^2 per Kelvin^4

T = Surface temperature of star (Kelvin)

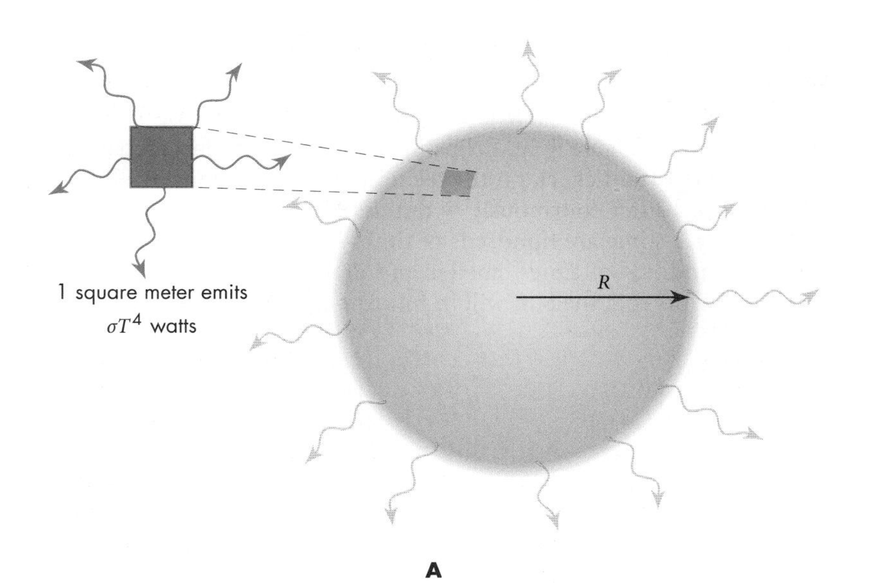

FIGURE 57.3
The Stefan-Boltzmann law can be used to find a star's radius. (A) Each part of the star's surface radiates σT^4. (B) Multiplying σT^4 by the star's surface area ($4\pi R^2$) gives its total power output—its luminosity, $L = 4\pi R^2 \sigma T^4$. To find the star's radius, we solve this equation to get $R = \sqrt{L/(4\pi\sigma T^4)}$. Finally, we can use this equation and measured values of the star's L and T to determine R.

Q What other equations have you used that have three variables? Can you always solve for one variable if you know the other two?

$L \div (4\pi\sigma T^4) = 4\pi R^2 \times \sigma T^4 \div (4\pi\sigma T^4)$

$$R^2 = \frac{L}{4\pi\sigma T^4}$$

Taking square root of both sides gives

$$R = \frac{\sqrt{L}}{\sqrt{4\pi\sigma} \times T^2}$$

A convenient form of the Stefan-Boltzmann equation is to solve for the radius in terms of the Sun's measured properties:

$$\frac{R}{R_\odot} = \sqrt{\frac{L}{L_\odot}} \times \left(\frac{T_\odot}{T}\right)^2$$

as follows: Increasing either the temperature or the radius of a star makes it more luminous. Making T larger makes each square meter of the star brighter. Making R larger increases the number of square meters.

The Stefan-Boltzmann equation has three "variables"—the quantities luminosity, radius, and temperature—that may vary from star to star or even from one time to another in a particular star. Whenever we have an equation with three variables, if we can measure two of them, we can find the third one from the equation. This means that we can find a star's radius from the Stefan-Boltzmann equation because we have ways of determining a star's luminosity (from its brightness and distance) and temperature (from its color or spectrum). Mathematically, this permits us to solve the equation for R if we know T and L.

We can rearrange the Stefan-Boltzmann equation algebraically and solve for a star's radius:

$$R = \frac{\sqrt{L}}{\sqrt{4\pi\sigma} \times T^2}.$$

We can test this formula on the Sun. The Sun's luminosity is measured to be $L_\odot = 3.83 \times 10^{26}$ watts, while its surface temperature is measured to be $T_\odot = 5800$ K (Unit 49). We calculate then that

$$R_\odot = \frac{\sqrt{L_\odot}}{\sqrt{4\pi\sigma} \times T_\odot^2} = \frac{\sqrt{3.83 \times 10^{26}}}{\sqrt{7.13 \times 10^{-7}} \times 5800^2} = \frac{2 \times 10^{13}}{2.8 \times 10^4} = 6.9 \times 10^8 \text{ m}$$

or 690,000 kilometers. This is quite close to the Sun's measured radius of 696,000 kilometers, demonstrating the power of the Stefan-Boltzmann law.

To gain a feel for this equation, think about two different cases: (1) when the temperature remains the same or (2) when the luminosity remains the same. If two stars have the same temperature, the more luminous one has the larger radius. On the other hand, for two stars that have the same luminosity, the hotter one must be smaller (because T is in the denominator). We can use this form of the Stefan-Boltzmann equation to solve for the radius of any star whose distance, brightness, and temperature are known—which is many hundreds of thousands of stars.

The Stefan-Boltzmann law is very useful, but it relies on some assumptions about how stars generate light. Eclipsing binary and interferometry measurements provide an important check on the technique. Observations by all three methods show that stars differ enormously in radius. Although many stars have radii similar to the Sun's, some are hundreds of times larger; astronomers call them *giants.* Some are hundreds of times smaller and are called *dwarfs.* We will discover in subsequent units that the Sun will in future stages turn into both!

KEY TERMS

binary star, 465
eclipsing binary star, 465
interferometry, 464
light curve, 465
speckle interferometry, 465
Stefan-Boltzmann law, 467

QUESTIONS FOR REVIEW

1. How does interferometry work to measure stars' diameters? What sort of stars does this method work for?
2. What is speckle interferometry?
3. What is a light curve?
4. What are eclipsing binaries? How can they be used to measure stellar sizes?
5. The Stefan-Boltzmann law is based on what assumptions?
6. How does the Stefan-Boltzmann law allow us to measure stars' sizes?
7. For two stars of the same surface temperature, how does their relative luminosity depend on their radii?

PROBLEMS

1. Alpha Centauri is actually two stars that orbit each other. Recent measurements show that Alpha Centauri A has an angular diameter of 0.0085 arcsec, while Alpha Centauri B has an angular diameter of 0.0060 arcsec. The pair are at a distance of 4.4 ly. What are their diameters, compared to the Sun's?
2. A star with the same color as the Sun is found to produce a luminosity 81 times larger. What is its radius, compared to the Sun's?
3. A star with the same radius as the Sun is found to produce a luminosity 81 times larger. What is its surface temperature, compared to the Sun's?
4. The surface temperature of Arcturus is about half as hot as the Sun's, but Arcturus is about 100 times more luminous than the Sun. What is its radius, compared to the Sun's?
5. If a star's surface temperature is 30,000 K, how much power does a square meter of its surface radiate?
6. Betelgeuse is roughly the same temperature as Proxima Centauri, the closest star to the Earth aside from the Sun. However, it is 10^9 times brighter. How much larger in radius is Betelgeuse than Proxima Centauri?
7. A star is five times as luminous as the Sun and has a surface temperature of 9000 K. What is its radius, compared to that of the Sun?
8. Using the information that Betelgeuse has an angular size 0.05 arcsec and is 130 pc from the Earth, show how you can calculate its diameter in kilometers.

TEST YOURSELF

1. If the surface temperature of a star is doubled, but its radius remains the same, its new luminosity is _____ its old luminosity.
   a. 16 times smaller than
   b. 4 times smaller than
   c. the same as
   d. 4 times larger than
   e. 16 times larger than
2. The Stefan-Boltzmann law applies to
   a. individual atomic spectral lines.
   b. a star's chromosphere.
   c. a star's photosphere.
   d. clouds of interstellar gas.
   e. All of the above.
3. If a star's luminosity increases, we can conclude that
   a. its radius has increased.
   b. its temperature has increased.
   c. its temperature or its radius has increased.
   d. its temperature and its radius both must have increased.
   e. None of the above.

unit

58 The H-R Diagram

Background Pathways

By the early 1900s astronomers had learned how to measure stellar temperatures, masses, radii, and compositions, but they understood little of how stars worked. We have now reached a similar point in our study of stars. We know that stars exhibit a wide range of properties. But why does such variety occur?

An important method in science for clarifying a wide array of data is to look for patterns and relationships among the different variables. A basic tool for doing this is to plot one variable against another in an *X–Y* diagram. Astronomers early in the twentieth century discovered that a plot of temperature versus luminosity gave new insights into the nature of stars. That plot is now known as the **H-R diagram**—the letters *H* and *R* standing for the initials of the Danish astronomer Ejnar Hertzsprung and the U.S. astronomer Henry Norris Russell, who independently discovered its usefulness.

To construct an H-R diagram of a group of stars, astronomers plot each star according to its spectral type (or temperature) on the *X* axis and its luminosity on the *Y* axis, as shown in Figure 58.1. The H-R diagram shows that stars do not have just any combination of temperature and luminosity. Instead, there are patterns in this diagram, suggesting relationships between these quantities for certain classes of stars. The H-R diagram is like a crime detective's bulletin board. We pin up the various pieces of information we have about stars and see what they suggest about the stars. For example, it takes many billions of years for a star like the Sun to be born and to die. So we cannot follow a single star in its lifetime. But if we look at all the stars with masses similar to the Sun's, we can begin to piece together clues about how a star with the Sun's mass might appear at different stages of its life.

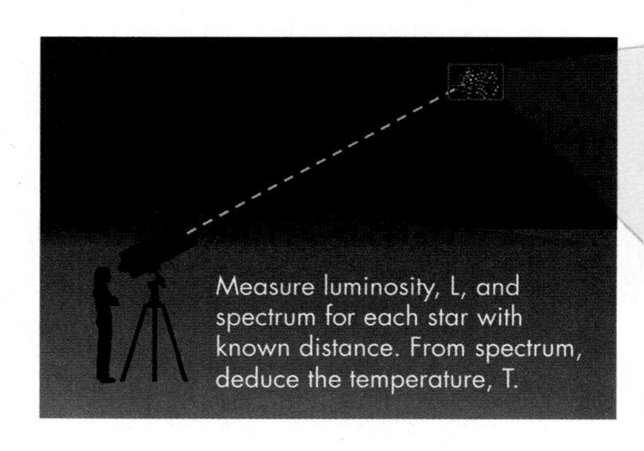

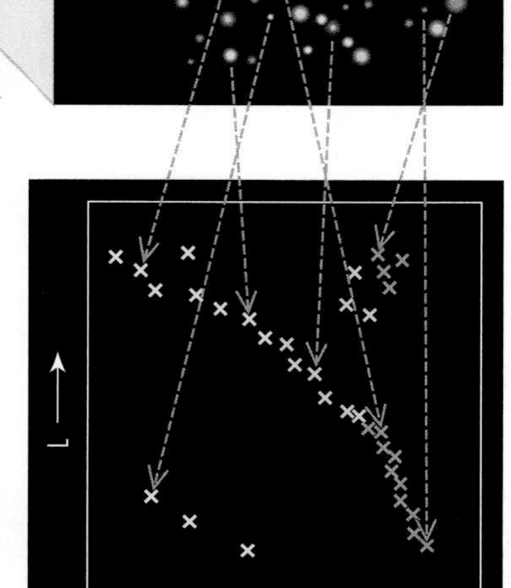

To make H-R diagram of star group, plot *L* and *T* of each star.

FIGURE 58.1
Constructing an H-R diagram from the spectra and luminosities of a set of stars.

58.1 ANALYZING THE H-R DIAGRAM

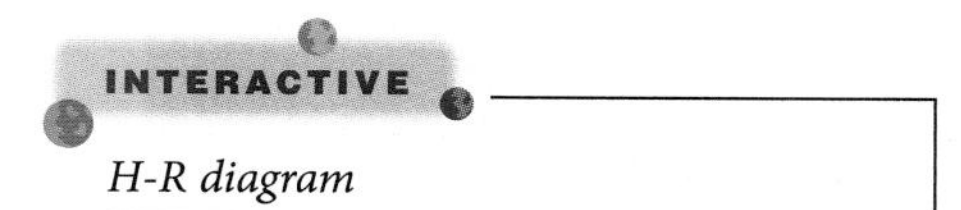

Figure 58.2 shows an H-R diagram based on modern measurements of parallax, brightness, and color from the *Hipparcos* satellite. These data have been converted into luminosities and temperatures, and one point is plotted for each star. It is important to keep in mind that the H-R diagram does *not* depict the position of stars at some location in space. It shows how stellar properties correlate, much as a height-weight table does for people, where we can see the trend that in general taller people are heavier than shorter people.

By tradition, the H-R diagram is plotted with bright stars at the top and dim stars at the bottom—O stars (hot, blue) on the left and M stars (cool, red) on the right. Notice that temperature therefore increases to the left, which is the opposite of the usual convention in graphs. The majority of the stars lie along a diagonal band in the diagram that runs from upper left (hot and luminous) to lower right (cool and dim). This line is called the **main sequence,** and stars within this region of the diagram follow a trend that you might expect: The hotter stars are brighter than the cooler stars.

Although about 90% of the stars lie along the main sequence, some others are in the upper right part of the diagram, where stars are cool but very luminous, and some are below and to the left of the main sequence, where stars are hot but dim. To understand what makes stars in these two regions different, we turn to the Stefan-Boltzmann law (Unit 57). This equation shows that if two stars have the same surface temperature but differ in their size, the larger one emits more energy—that is, it is more luminous—than the smaller one. For stars of the same temperature, more luminous implies a larger radius.

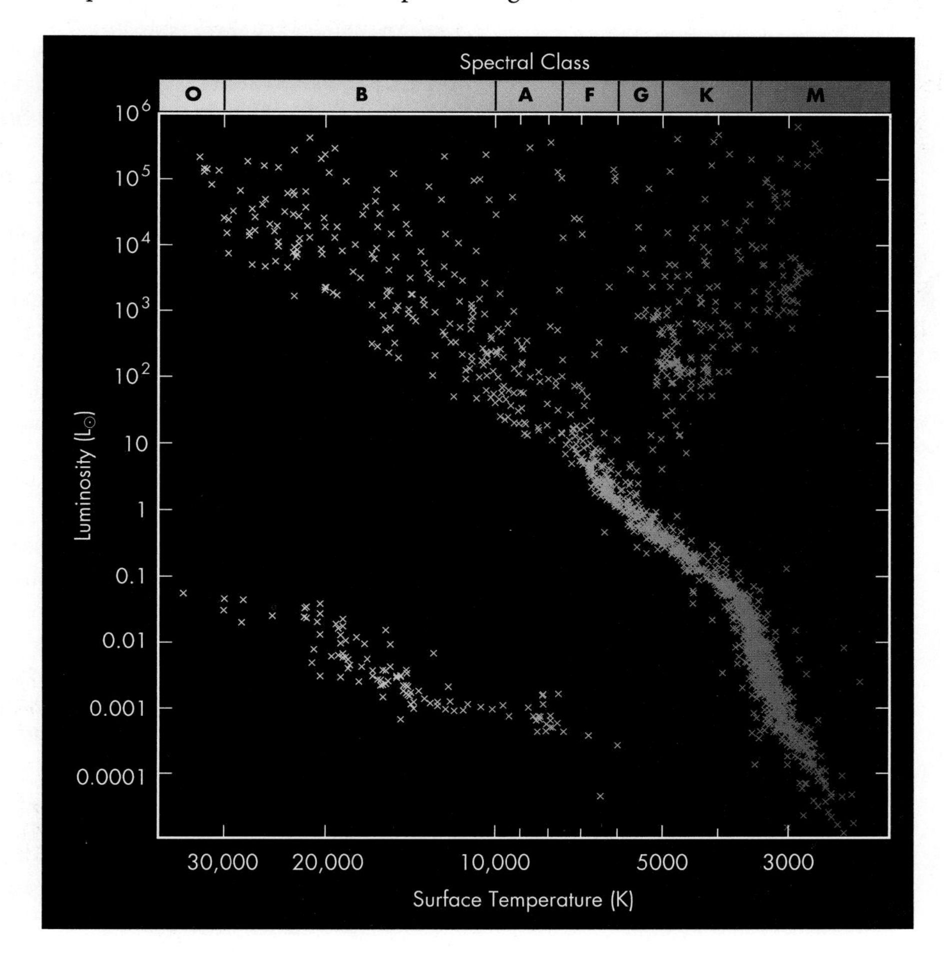

FIGURE 58.2
Hertzsprung-Russell diagram of bright and nearby stars with accurately measured distances. Each "×" is plotted according to one star's estimated temperature and luminosity. The symbols are colored according to the approximate color of the star. Notice that bright stars are at the top of the diagram and dims stars are at the bottom. Also notice that hot (blue) stars are on the left and cool (red) stars are on the right.

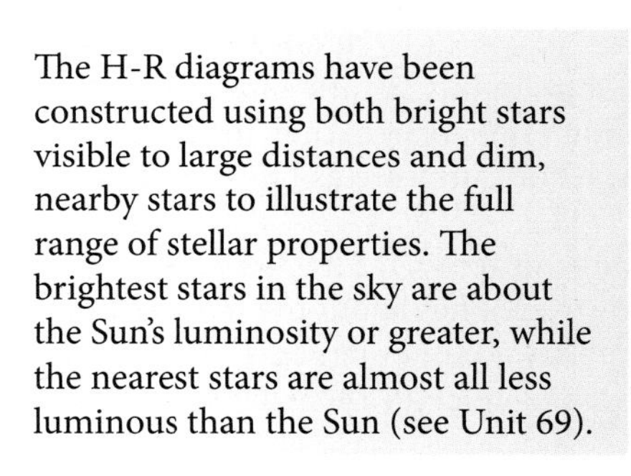
The H-R diagrams have been constructed using both bright stars visible to large distances and dim, nearby stars to illustrate the full range of stellar properties. The brightest stars in the sky are about the Sun's luminosity or greater, while the nearest stars are almost all less luminous than the Sun (see Unit 69).

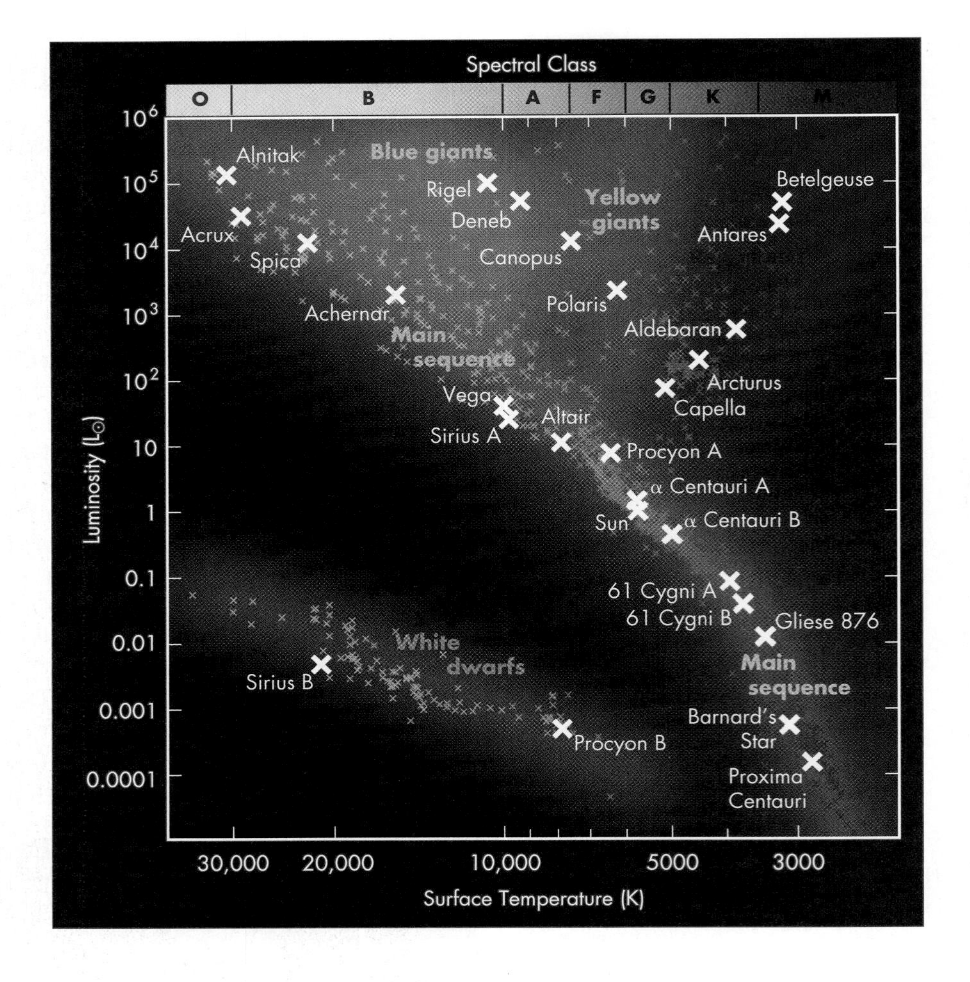

FIGURE 58.3
Regions of the H-R diagram. Most stars lie along a diagonal band from upper left to lower right, known as the main sequence. Giant stars lie above the main sequence and are sometimes labeled according to their color. White dwarf stars lie along a narrow band near the bottom of the diagram. Some well-known stars are marked on the diagram.

Q Think of two properties that describe humans. If you made a plot of these two properties, what would the diagram look like? What could you learn about people from it?

We can apply this reasoning to the H-R diagram. In Figure 58.3 we have labeled a number of the better-known stars. Consider first the stars on the right (cool) side of the diagram. For example, Proxima Centauri and Antares lie along an approximately vertical line because they have the same surface temperature. However, Antares, being nearer the top of the diagram, is much more luminous than Proxima. Because these two stars have the same surface temperature, each square meter of their surfaces emits the same amount of light. Therefore, to emit more light, Antares must have a larger size. Because Antares is large and red, it is called a **red giant.**

We can go one step further, however. By comparing how much more luminous one of these stars is, we can calculate how much larger its radius is. For example, we see that Antares has a luminosity about 100 million times larger than Proxima. This implies that Antares has surface area that is 100 million times larger. If the stars are spheres, then their surface area depends on the square of their radii according to the formula $4\pi R^2$, where R is the star's radius. A little math shows that if these two stars differ by a factor of 100 million (10^8) in area, they must differ by a factor of $\sqrt{100{,}000{,}000} = 10{,}000$ in radius. If the Sun were as big as Antares, it would swallow the Earth.

Not all giant stars are red. They span a range of temperatures and colors. For example, the North Star, Polaris, is a giant. However, it has about the same temperature and color as the Sun—yellow—but it is about 50 times larger. To distinguish between different classes of giants, we sometimes refer to them by their actual color—a blue or yellow giant, for example.

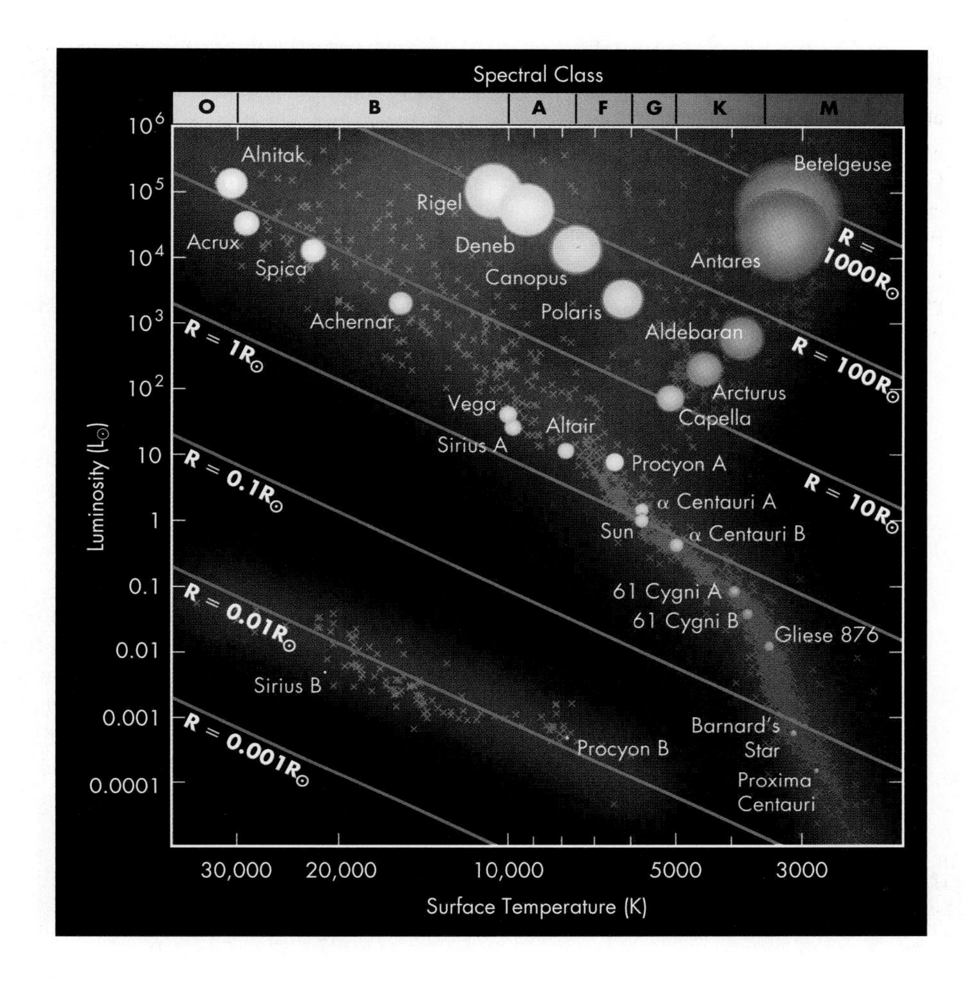

FIGURE 58.4
Sizes of stars in the H-R diagram. Green lines show where stars of a given radius will lie. The sizes and colors of some well-known stars are suggested in the figure. However, the size range has been greatly compressed because of the enormous difference between the largest and smallest stars (about a factor of 100,000).

Q Would you expect most of the stars visible to you in the night sky to be giants or main-sequence stars? What factors determine which is more common?

A similar analysis shows that the stars lying below the main sequence must have small radii if they are both hot and dim. Because these stars are so hot compared to main-sequence stars of the same luminosity, they are much bluer. Of the first of these stars to be discovered, many were very hot, glowing with a white heat, and these stars were therefore called **white dwarfs.** With modern observations we can detect some that have surface temperatures cooler than the Sun; but nonetheless, stars falling in this region below the main sequence are all called white dwarfs (Figure 58.3). For example, Sirius B, a dim companion of the bright star Sirius, is a white dwarf. Its radius is about 0.008 the Sun's—a little smaller than the Earth's radius—and its surface temperature is more than 20,000 Kelvin.

We can illustrate the range of star diameters by drawing lines in the H-R diagram that represent a star's radius. That is, we use the Stefan-Boltzmann equation ($L = 4\pi R^2 \times \sigma T^4$) and set R constant. As we make T larger, L must get larger. The result is a line for each value of R that slopes from upper left to lower right as shown in Figure 58.4. Any stars lying along such a line have the same radius. If we move up or to the right in the H-R diagram, we find stars of progressively larger radii. Thus the largest stars lie in the upper right of the diagram, while the smallest stars lie in the lower left. We can see that the giants span a wide range of radii. By contrast, the white dwarfs are all about the same size, about 1/100th of the Sun—about the size of the Earth. The main-sequence stars increase gradually from about under 1/10th the Sun's radius at the cool end to nearly 10 times larger at the hot end.

58.2 THE MASS–LUMINOSITY RELATION

Some of the stars we have plotted in the H-R diagram are in binary systems, allowing us to determine their masses (Unit 56). When we label the stars in the H-R diagram by their masses, we notice that the masses are not randomly scattered around the diagram. As on a detective's bulletin board, we see certain patterns emerging. Stars near the same position on the main sequence all have approximately the same mass. Thus, main-sequence stars of the same spectral type as the Sun all have approximately the same mass.

Farther up and left along the main sequence we find stars of steadily larger masses, while farther down and right on the main sequence we find that the stars have smaller masses. From top to bottom along the main sequence, the star masses decrease from about 20 $M_{\odot}$ to 0.2 $M_{\odot}$ as shown in Figure 58.5A. If we expand our sample of stars, we find that we can extrapolate this trend to even more luminous stars that lie along the line of the main sequence to masses of more than 50 $M_{\odot}$, while there are extremely dim stars extending down from the main sequence that have masses under 0.1 $M_{\odot}$.

This trend does not hold for stars in the rest of the H-R diagram. The giants have a wide range of masses, similar to that of the stars on the main sequence but in no particular order. White dwarfs, on the other hand, tend to all be around one solar mass or less, but again in no organized sequence. We will note these facts about the giants and white dwarfs, but set these types aside for the moment and concentrate on the main-sequence stars.

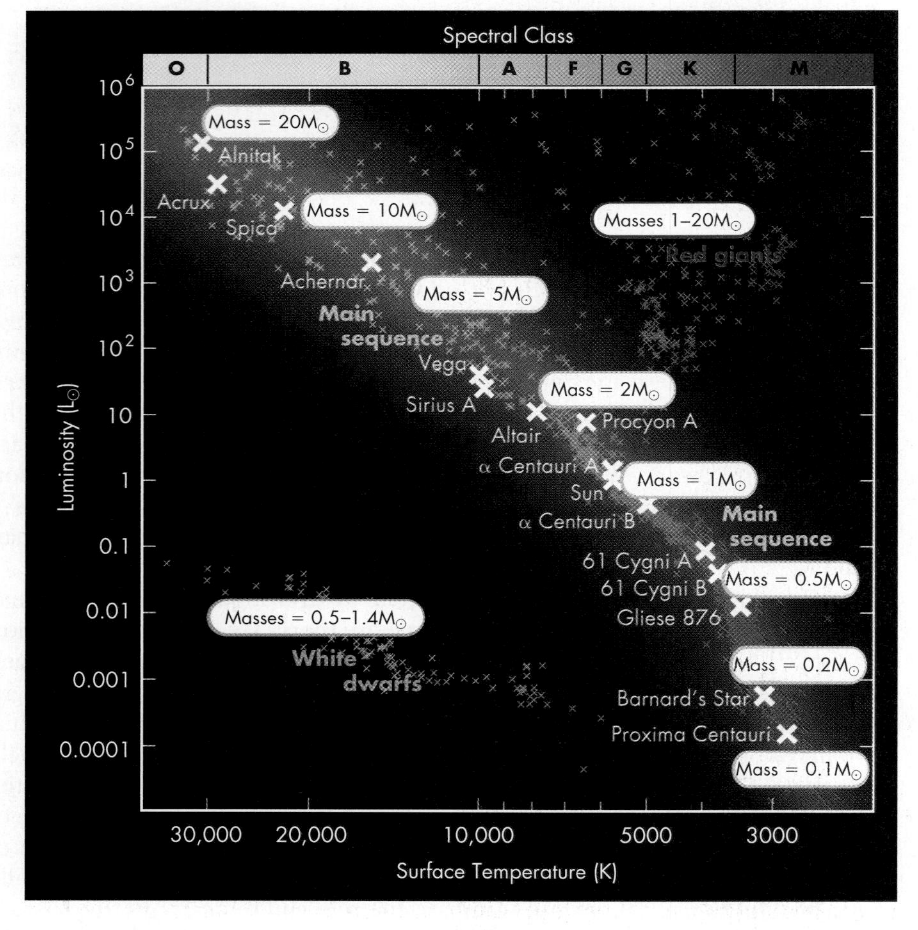

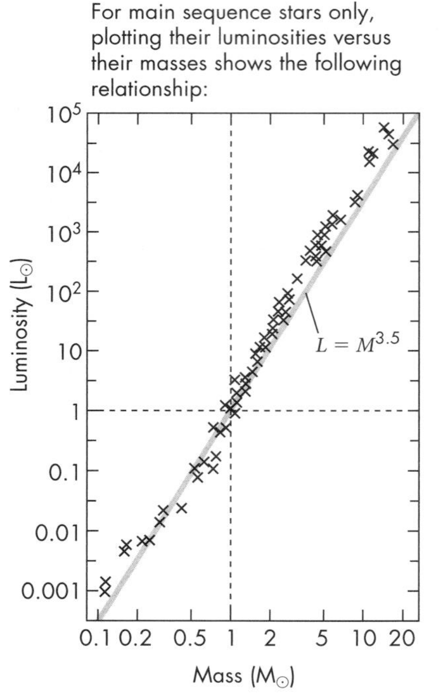

FIGURE 58.5
(A) On the H-R diagram, high-mass stars lie higher on the main sequence than low-mass stars. The giants and white dwarfs exhibit a range of masses in no particularly obvious order. (B) The mass–luminosity relation shows that along the main sequence more-massive stars are more luminous. The relationship does not hold for red giants or white dwarfs.

The trend of masses among main-sequence stars is illustrated in Figure 58.5B, where the stars' luminosities are plotted against their masses. This pattern was first recognized in 1924 by the English astrophysicist A. S. Eddington. The relationship between mass and luminosity can be expressed by a formula. If M and L are given in solar units, then for main-sequence stars

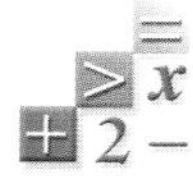

L = Luminosity in solar luminosities
M = Mass in solar masses

$$L \approx M^{3.5}$$

This **mass–luminosity relation** is approximate and varies between $L \approx M^3$ and $L \approx M^4$ over different portions of the main sequence, but it gives a fairly accurate description of how the luminosity of a main-sequence star is related to its mass. If we apply this relation to the Sun, whose mass and luminosity are each 1 in these units, the relation is obeyed because 1 raised to any power still equals 1.

Note that taking M to the power of 3.5 is the same as multiplying $M \times M \times M \times \sqrt{M}$. On many calculators you can carry out this calculation using a key labeled x^y. To find $10^{3.5}$ you would type this into the calculator: 1 0 x^y 3 . 5 =

If we consider a more massive star—for example, one 10 times as massive as the Sun—then the mass–luminosity relationship tells us that $L \approx 10^{3.5} \approx 3 \times 10^3$. That is, its luminosity is about 3000 $L_\odot$. Alternatively, a star with a mass of just 1/10th the Sun's mass would have a luminosity $L \approx 0.1^{3.5} \approx 3 \times 10^{-4}$. That is, its luminosity is about 0.0003 $L_\odot$.

We can understand in general terms how such a mass–luminosity relationship arises. For increasingly larger stellar masses, the gravitational pull among all the parts of a star squeezes the interior more strongly so that the gas is driven to a higher temperature. A higher temperature generates more of the energetic collisions between hydrogen nuclei that can result in fusion, so more massive stars are more luminous.

58.3 LUMINOSITY CLASSES

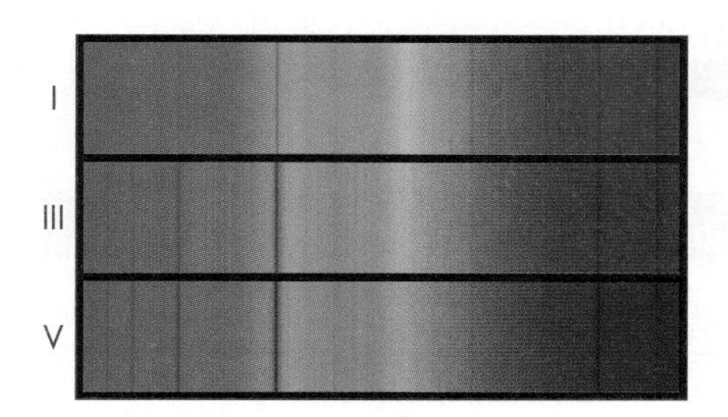

FIGURE 58.6
Spectral lines are narrow in the most luminous stars (I) and wide in main-sequence (dim) stars (V).

Because luminosity is so important in characterizing a star, astronomers have sought other ways to estimate it. We cannot always measure stars' luminosities, because their distances may be unavailable. Fortunately, in the late 1800s Antonia Maury, an astronomer at Harvard, discovered that the more luminous the star, the narrower the absorption lines in its spectrum, as illustrated in Figure 58.6.

The width of spectral lines is affected by the density of the gas producing them. This occurs because in a dense gas the atoms collide more frequently, "jostling" the electrons' orbitals. This slightly alters each atom's energy levels, making the atoms able to absorb slightly larger photon energies. The net effect in the entire star's atmosphere is to smear the range of photon energies that are absorbed in the photosphere and to broaden the absorption lines. By contrast, when the atoms are spread farther apart, they collide less frequently and tend to absorb only the exact energies for that type of atom, leading to narrow absorption lines.

Maury's discovery implies that more-luminous stars are less dense. We can show why this must be true for giant stars. Recall that the average density of a body is its mass divided by its volume. Giant stars have masses similar to those of main-sequence stars. So with similar masses but much larger volumes, the giant stars are much less dense. For example, the average density of main-sequence stars is roughly 1 kilogram per liter, whereas the density of a typical giant star is about 10^{-6} kilogram per liter—about a million times less dense. The atmospheres of giant stars are also much less dense than the atmospheres of main-sequence stars.

Using the relationship between spectral line width and luminosity just described, astronomers divide stars into five luminosity classes, denoted by the Roman numerals I through V. Class V stars are the densest and dimmest (for a given surface temperature), and class I stars are the largest and brightest. Class I stars—**supergiants**—are split into two classes, Ia and Ib. Figure 58.7 and Table 58.1 show the correspondence between class and luminosity. Although the scheme is not precise, it allows astronomers to get an indication of a star's luminosity from

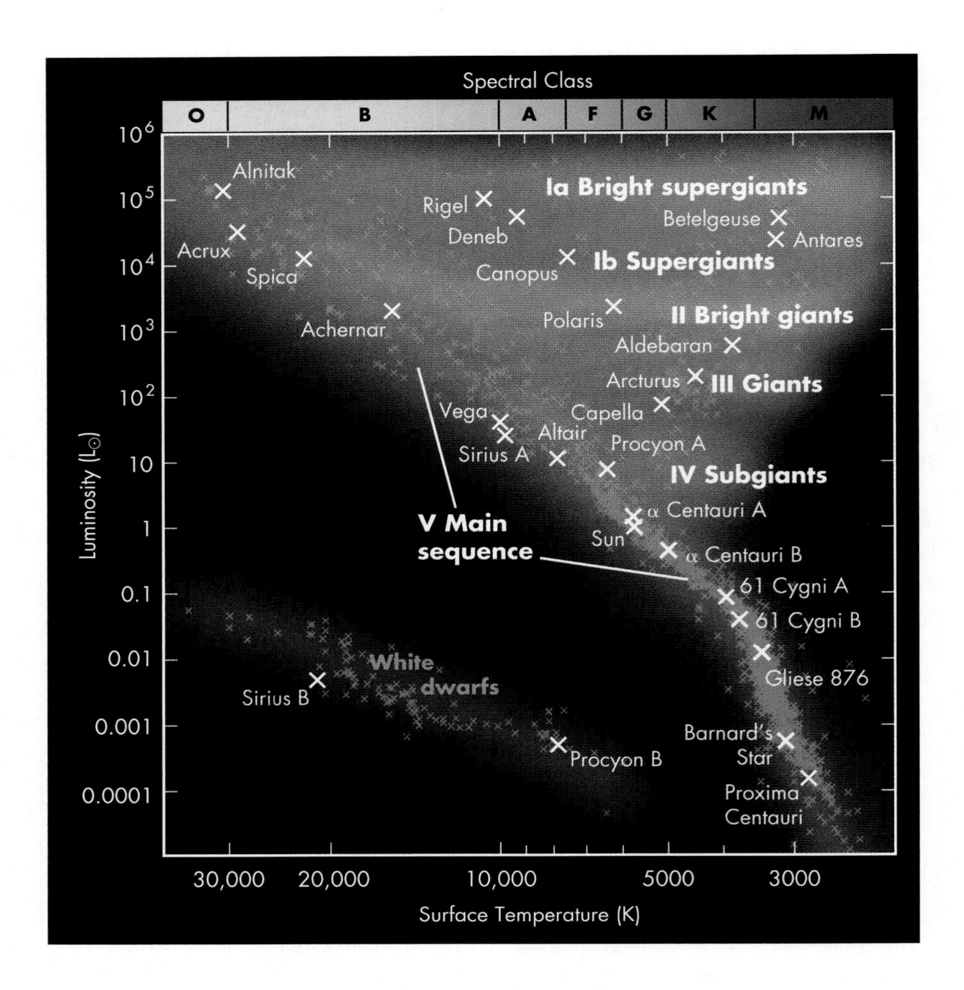

FIGURE 58.7
Stellar luminosity classes in the H-R diagram.

TABLE 58.1 Stellar Luminosity Classes

Class	Description	Example
Ia	Bright supergiants	Betelgeuse, Rigel (brightest stars in Orion)
Ib	Supergiants	Antares (brightest star in Scorpius)
II	Bright giants	Polaris (the North Star)
III	Ordinary giants	Arcturus (brightest star in northern constellation Boötes)
IV	Subgiants	Procyon A (brightest star in Canis Minor)
V	Main sequence	The Sun, Sirius A (brightest star in sky, in Canis Majoris)

its spectrum. A star's luminosity class is often added to its spectral type to give a more complete description of its light. For example, our Sun is a G2V star, whereas the blue supergiant Rigel is a B8Ia star.

The high luminosity of the giants is a surprise, given their low densities. After all, in discussing the mass–luminosity relation we argued that the more-luminous main-sequence stars became that way because of the greater compression of their cores. The giants seem at first to defy this relationship—growing more and more luminous at lower and lower densities. The key to this puzzle is that the density of a star's outer atmosphere does not necessarily tell us about the density deep in a star's core. Giant stars differ in structure significantly from the main-sequence stars, the giants having far denser cores. As we will explore in subsequent units, this difference is a result of how stars age—one of many critical clues that will help us unravel the stories of stars.

Q: If two stars have atmospheres of the same density, but one of the stars has a higher temperature, how would this affect the width of the absorption lines of the hotter star? Why?

KEY TERMS

H-R diagram, 470
main sequence, 471
mass–luminosity relation, 475
red giant, 472
supergiant, 475
white dwarf, 473

QUESTIONS FOR REVIEW

1. What is the H-R diagram? What are its axes?
2. Where would a very hot and very luminous star be located on the H-R diagram?
3. What is the main sequence?
4. How do we know that giant stars are big and dwarfs are small?
5. How does mass vary along the main sequence?
6. What is the mass–luminosity relation?
7. How do spectral lines vary between main-sequence stars and giant stars of the same temperature? Why?

PROBLEMS

1. Estimate the luminosity and temperature of the star Acrux by examining its vertical and horizontal position in the H-R diagram (Figure 58.3). (You should make an estimate of these values by interpolating between the values marked on each axis.) Using the values you estimated, determine the size of Acrux as compared to the Sun, using the Stefan-Boltzmann law (Unit 57).
2. Alnitak is the brightest O-class star in the night sky, located at the eastern end of Orion's belt. Proxima Centauri is the nearest M-class main-sequence star. Estimate the luminosities and temperatures of these two stars from their positions on the H-R diagram (Figure 58.3) as in the previous problem. Calculate the ratio of their sizes using the Stefan-Boltzmann law.
3. From your data in problem 2, use the mass–luminosity relationship to estimate the ratio of masses of these two stars.
4. If a main-sequence star has a luminosity of 3000 $L_{\odot}$, what is its mass in relation to the Sun's?
5. Estimate the luminosity of a main-sequence star that has a mass twice the Sun's. By what factor is such a star brighter than the Sun?
6. How massive would a main-sequence star need to be, to be 100 times brighter than the Sun?
7. Use the mass–luminosity relationship and the Stefan-Boltzmann equation from Unit 57 to work out the surface temperature of a star with half the radius of the Sun and one-eighth the mass.

TEST YOURSELF

1. A star that is cool and very luminous must have
   a. a very large radius.
   b. a very small radius.
   c. a very small mass.
   d. a very great distance.
   e. a very low velocity.
2. In what part of the H-R diagram might you find a white dwarf?
   a. Upper left
   b. Lower left
   c. Upper right
   d. Lower right
   e. Just above the Sun on the main sequence
3. In a sample of nearby stars, about what percentage will lie on the main sequence?
   a. 3.5%
   b. 9.0%
   c. 35%
   d. 50%
   e. 90%

unit 59

Overview of Stellar Evolution

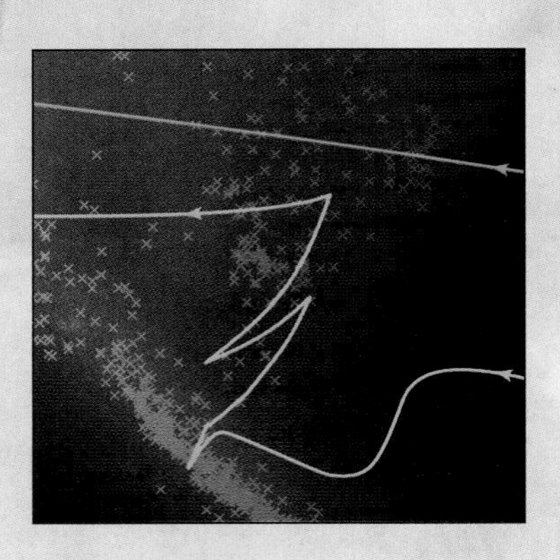

Background Pathways

Unit 58: The H-R Diagram 470

Note that astronomers and biologists use of the term *evolution* differently. In astronomy, *stellar evolution* refers to the gradual changes that occur during a single star's lifetime, whereas in biology *evolution* refers to the gradual changes that occur between generations as a result of natural selection.

To us, the stars appear permanent and unchanging; but they, like us, are born, grow old, and die. Over their lifetimes they slowly change, transforming themselves in many basic ways but taking millions or billions of years to do so. Over the time span of one human life, it is rare to see change in any one star. Even during recorded human history, few stars have been seen to change significantly. We would need to go back to times before humans evolved to notice major changes in the properties of the stars we see today. Astronomers refer to the changes that occur while stars age as **stellar evolution.**

The English astronomer Sir William Herschel offered an analogy to the effort of humans seeking to understand the evolution of stars, and in particular our own Sun. He suggested it is like being allowed to spend a brief time in a forest, and then trying to deduce the life story of trees from the information you've collected. One minute in the forest compared to the lifetime of a tree would be about the equivalent of the entire human record of astronomical observations compared with the lifetime of the Sun.

Figuring out a tree's life story from a minute's observation might sound like an impossible task, but think of all we could see in a forest. We could match the leaves and bark to identify and track the life stages of a particular type of tree. We might see some trees of that type as saplings, others that were fully grown, and yet others as rotting stumps. We might find seeds on the tree and on the ground, and perhaps even some seedlings. We might see a leaf fall. If we piece together our snippets of information correctly, we could develop a description of a tree's life without ever having witnessed an individual tree change at all.

This, then, is the challenge of studying stellar evolution—to piece together the story of stars from the vast array of data we have gathered. In this unit we summarize the story of stellar evolution; the individual pieces of the story are explored in more detail in the units that follow.

59.1 STELLAR EVOLUTION: MODELS AND OBSERVATIONS

From their observations, astronomers can deduce some features of how a star evolves. For example, the existence of main-sequence stars, red giants, and white dwarfs suggests to astronomers a picture of how stars age, much as a snapshot of a baby with its parents and grandparents depicts human aging. However, we need not wait vast spans of time to watch a star age. We can "see" a star's aging with computer calculations that solve the equations that govern the star's physics. **Stellar models,** as such calculations are called, allow us to trace a star's life from birth to death.

The stellar models suggest that the entire life story of a star is largely determined at its birth. How a star evolves and then dies depends primarily on how much material the star contains—its mass. Mass is the most critical factor in a star's evolution because it determines the force with which the matter in a star is squeezed, and it also indicates the total amount of fuel that is available for the star to consume in its lifetime.

A star evolves in response to the gravity that is constantly squeezing it. At each moment, the star must have a balance between the gravitational forces that compress it and the internal pressure forces that resist compression. The gravitational forces and pressures inside an object as massive as a star are intense. For the Sun, the stellar models indicate the matter at its core is squeezed to a density 20 times greater than that of steel (Unit 49), and there are stars that reach much higher densities.

In the competition between gravity and pressure, the force of gravity never runs down or gets used up. It crushes the matter together inexorably. Meanwhile, a star's heat energy steadily escapes into space (mostly as starlight), so the gas pressure will drop unless the star can replenish its lost energy.

Technically, when we refer to an object as a "star," we are speaking of an object like the Sun that uses nuclear fusion to generate the heat that maintains the pressure resisting gravity (Unit 61). Nuclear energy is a finite resource, and over its lifetime a star will exhaust all possible sources of nuclear fuel (Units 62 and 66). As a star is forming, when it is a *protostar,* the heat is generated from the process of contraction itself (Unit 60). At the end of its lifetime, a star becomes a **stellar remnant** that is supported by forces between the atoms and particles that it is composed of (Units 64 and 67)—more like the forces that keep the Earth from collapsing under its own weight.

The stellar models show that any change in the source of energy or in the way a star produces its internal pressure will result in a different internal structure of the star. Observations show that sometimes these changes are gradual, but sometimes they occur violently.

59.2 THE EVOLUTION OF A STAR

Figure 59.1 depicts what we might see if we could watch the changes that occur in a star like the Sun over many billions of years. This figure is based on the findings of observations and stellar models over the last century. We summarize these findings here.

Evolution of a low-mass star

The formation of a star begins when a region of interstellar gas—an **interstellar cloud**—starts to contract. Initially this contraction is very slow, and other events may disrupt it. Eventually, though, if the knot of material grows dense enough, the gravitational force increases to a point where the material falls together, collapsing in on itself as the gravitational force grows steadily stronger. The gas heats up as it falls together and collides. The hot gas generates enough pressure to slow down the collapse to a more gradual contraction.

The cloud at this point has become a **protostar.** A protostar gradually contracts under its own weight. The contraction converts gravitational potential energy (Unit 20) into thermal energy or heat. This is the same energy that is released when you drop a stone to the ground, but on a vastly larger scale. During the protostar stage, inflowing matter becomes redistributed, sometimes in surprising ways. For example, observations show that some of the gas is heated and driven back outward in **bipolar flows,** driven off in opposite directions from the protostar. That heat provides the pressure that counters gravity's inward pull, but it is a losing battle. Because the star must keep shrinking to generate this heat, gravity's hold grows steadily stronger. The formation process is described further in Unit 60.

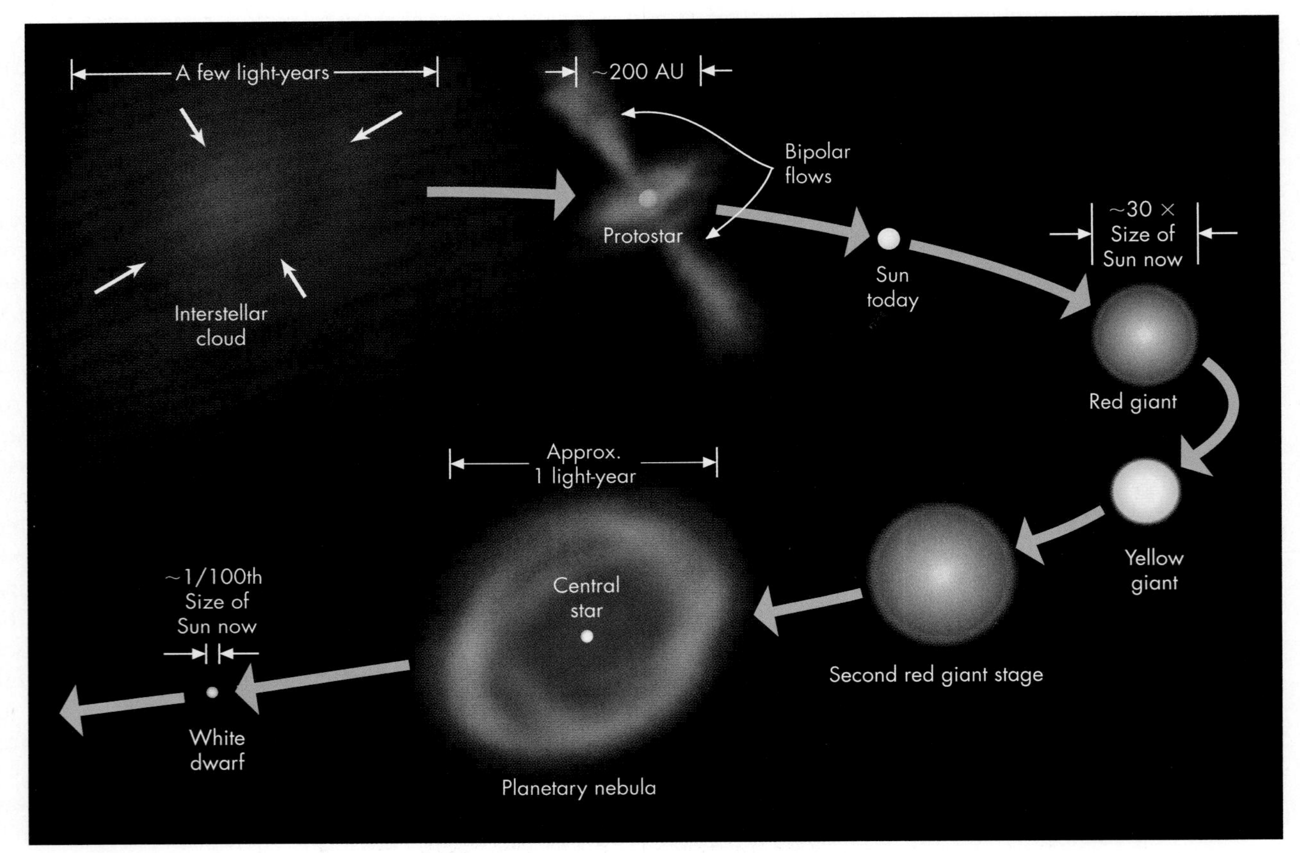

FIGURE 59.1
Evolution of a 1-solar-mass star such as the Sun, based on observations and computer calculations.

A protostar becomes a **main-sequence star** when its core grows hot enough to start fusing hydrogen into helium (Unit 61). The amount of energy available from fusion reactions is enormous, so the star becomes stable for a long time. However, eventually it will consume all the hydrogen in its core and so must resume the process of compressing its core. Compressing a gas causes it to heat up, and this heating can raise the temperature in regions surrounding the core so that they now become hot enough to fuse hydrogen. The core may grow hot enough for helium fusion to begin there as well. This yields even more energy than before, which makes the star brighter yet. As the energy floods through the star's outer layers, they swell, and the outermost layers are pushed so far from the core that they become cooler. The star becomes a **red giant** (Unit 62).

Stars differ slightly in many details of their evolution, depending on their exact mass, but the differences grow greater late in the stars' lifetimes. In particular, the ultimate fate of stars is much different for stars whose mass is more than eight times greater than the Sun's. We will therefore divide stars into two groups—**low-mass stars,** such as the Sun, and **high-mass stars,** whose mass is greater than about eight times the Sun's mass.

High-mass stars are blue during their main-sequence stage because their surfaces are hotter than those of low-mass stars, but the more important differences lie deep inside. Whereas a star like the Sun consumes its core's hydrogen in about 10 billion years, high-mass stars may use up their hydrogen in only a few million years (Unit 61). High-mass stars fuse their fuel quickly to supply the energy they need to support their great weight. The sequence of stages that high-mass stars experience as they evolve beyond the main-sequence stage also differs from that of low-mass stars, as illustrated in Figure 59.2. High-mass stars evolve through these stages very rapidly.

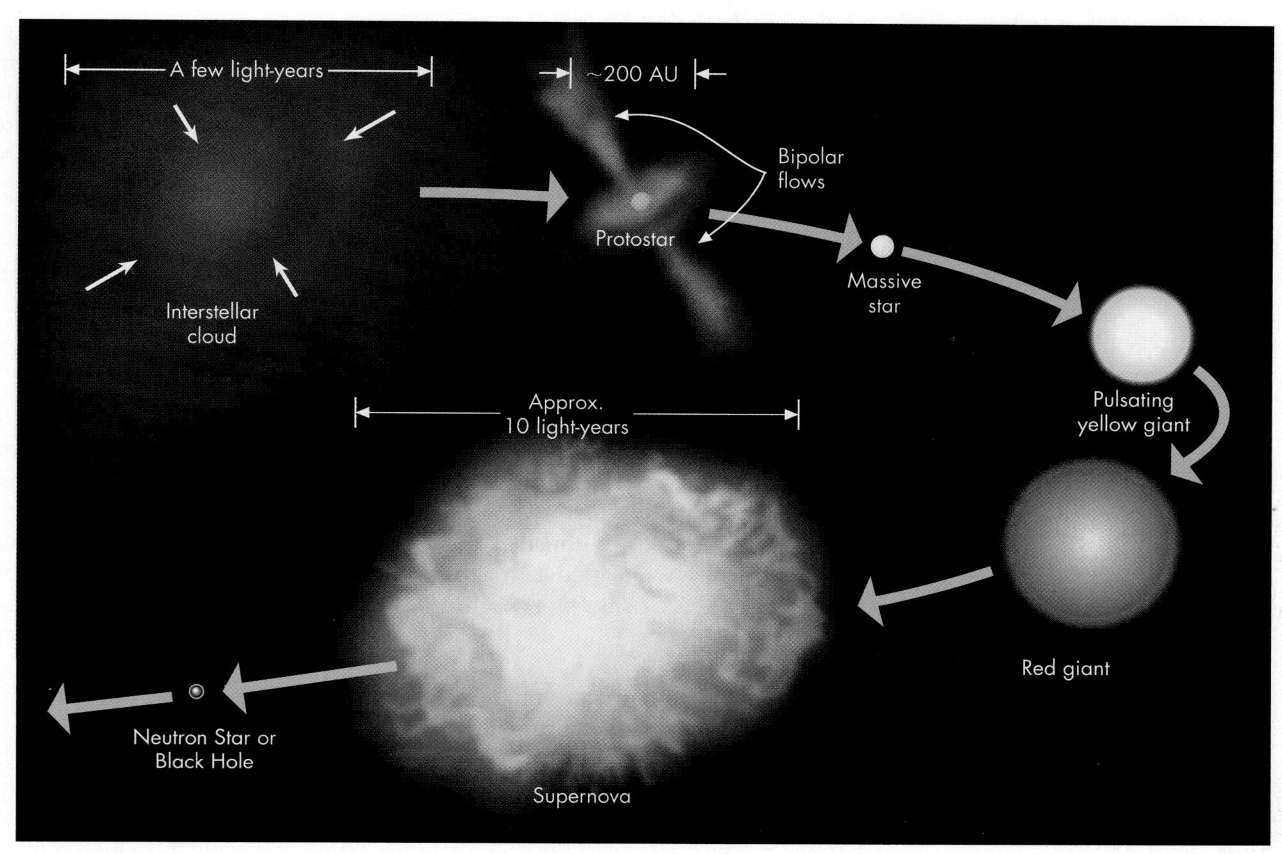

FIGURE 59.2
Evolution of a high-mass star, based on observations and computer calculations.

In both high- and low-mass stars, the core eventually compresses and heats to the point where the helium can begin fusing, although this happens in somewhat different sequences in the two kinds of stars. In low-mass stars, the core takes a long time to reach helium fusion temperatures. During this time the star grows into a red giant. But when helium fusion begins, the energy released expands the core, and the outer layers shrink, making the star's surface hotter and hence more yellow than red. The cores of massive stars are able to reach helium fusion temperatures fairly rapidly, allowing them to expand directly into *yellow giants.* The surfaces of these yellow giants are often seen to pulsate in and out, and their total light output varies, making them **variable stars** (Unit 63).

When a star exhausts its helium, its core again shrinks, and the star grows into a red giant—for the second time, in low-mass stars. These red giants are so luminous now that their radiation drives off the outer layers of gas in a **stellar wind.** In low-mass stars, the stellar wind carries away so much material that the intensely hot core of the star is laid bare. The core's energetic photons may ionize the inner parts of this wind of gas flowing away from the star. This faintly luminous cloud of gas surrounding the dying star is called a **planetary nebula.** The core remains as a small, dense stellar remnant: a **white dwarf** (Unit 64), which gradually cools and dims. However, a stellar remnant may be brought back to life in explosive fashion if a companion star deposits material on its surface. If enough material is deposited, the white dwarf may collapse and blast itself apart in a **supernova** explosion (Unit 65).

The term *planetary nebula* can be easily misinterpreted. The name arose because to astronomers with early telescopes these objects looked somewhat like the planets, but they have no relation otherwise to planets.

High-mass stars fuse and generate in their cores a sequence of heavier and heavier elements—such as carbon, oxygen, and iron. They fuse these fuels rapidly to supply the energy they need to support their great weight. However, they exhaust their fuel and die quickly (Unit 66). Thus, the stars that begin their lives on the blue end of the main sequence burn out quickly and therefore die "young" by the standards of other stars. They also die more violently because of the tremendous forces at work inside them. At the end of their lives, high-mass stars undergo dramatic implosions in their cores that lead to supernova explosions. The stellar remnants left behind, **neutron stars** and **black holes,** are some of the strangest objects in the universe (Units 67 and 68).

59.3 TRACKING CHANGES WITH THE H-R DIAGRAM

A good way of tracking the changes a star undergoes as it evolves is to plot its "path" in the H-R diagram. A star's position in the diagram changes whenever its surface temperature or luminosity changes, which it usually does in response to any changes in its interior.

It may be helpful to consider a similar sort of diagram for a population of people. Suppose we plotted a "height–strength diagram" for a random sample of people. On the *X* axis we plot height from tall to short (in the backward tradition of the H-R diagram), while on the *Y* axis we plot the amount of weight the person is able to lift.

The analogy to our traditional H-R diagram would be to find a large sample of people, measure their heights and the amount of weight they can lift, and plot one point for each person. A hypothetical diagram like this is shown in Figure 59.3.

If we took a random sample of people, everyone from infants to the elderly, we would generally find a trend that taller people can lift more weight, something like the main sequence in the H-R diagram. Some especially muscular people can lift a much larger weight than we would predict from their heights, while others

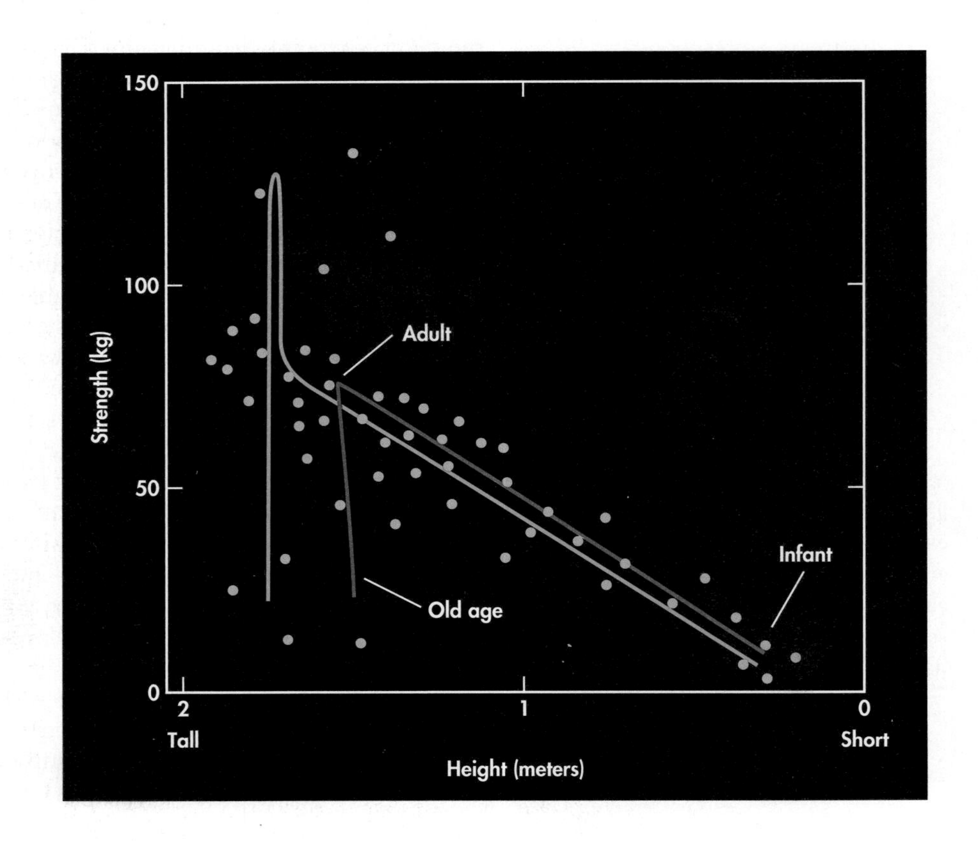

FIGURE 59.3
A hypothetical diagram of the heights of people versus their strength, as measured by the amount of mass they can lift. Each person in a sample is marked by a green dot. Paths of two individuals throughout their lifetimes are also shown.

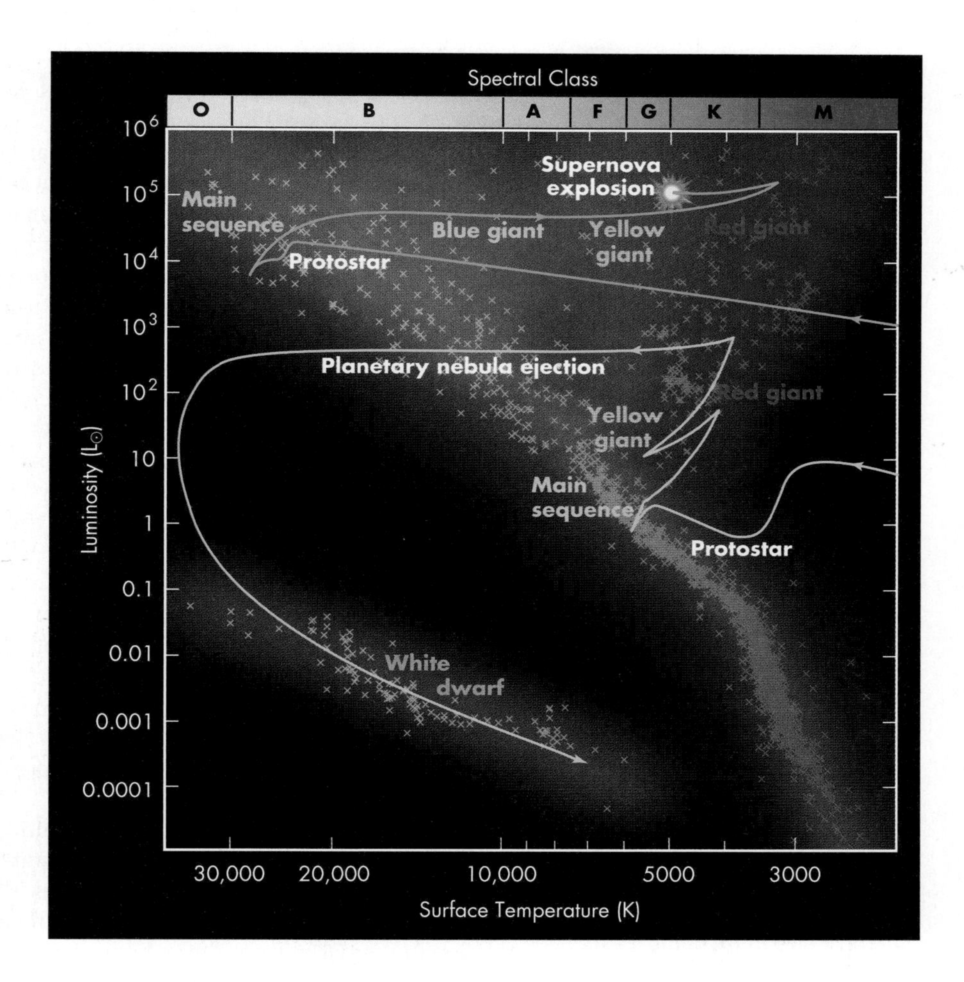

FIGURE 59.4
Schematic diagram of the evolutionary tracks of a low-mass star (yellow) and high-mass star (blue) through the H-R diagram.

Q If stars move through almost all parts of the H-R diagram during their lifetimes, why do we see most of them concentrated in isolated regions of the diagram when we plot the temperatures and luminosities of a sample of stars?

Q Try drawing what you would expect a height–strength diagram to look like for different populations (such as preschoolers, college students, retirees). What characteristics would allow you to identify the age of the people plotted in such a diagram if you were not told their ages in advance?

are particularly weak for their height—analogous to the giant and dwarf regions of the H-R diagram.

If you traced various individuals' heights and strengths throughout their lifetimes, you would see that they "move" through this diagram. Perhaps they follow a relatively normal pattern like that illustrated by the red line in Figure 59.3, growing up and gaining strength as they grow taller, but losing strength (and maybe even a little height) in old age. Some individuals might have jobs that require a lot of lifting, and therefore they grow stronger after reaching adulthood, moving above the "main sequence" into the upper part of the diagram. Perhaps they suffer debilitating injuries, which reduce their strength drastically so that they drop below the main sequence. Every person's track through this diagram might be a little different, although the common features of their "evolution" will create certain patterns in the diagram.

Similarly, an individual star's path through the H-R diagram can be quite complex. The paths of a low-mass and a high-mass star through the H-R diagram are illustrated in Figure 59.4. Compare these paths to Figures 59.1 and 59.2. Each is a way of illustrating a star's lifetime.

Finally, we can also find clues to the changes that result with aging if we examine groups of stars or people who are all of a similar age. If we draw samples of humans from a nursery school, a college, and a retirement community, we find very different distributions of height and strength in each group. Similarly, we can test our ideas about stellar evolution if we can identify groups of stars that are all of similar ages, where physical traits will also differ within the group. We will return to the idea of testing sets of stars of the same age in Unit 69.

59.4 THE STELLAR EVOLUTION CYCLE

A star forms from gas in an interstellar cloud that is drawn together by gravity. This process heats the gas to the point where it can fuse hydrogen into helium. Low-mass stars fuse the helium into carbon and sometimes oxygen before dying, but high-mass stars can fuse a series of heavier elements.

Stellar winds from low-mass stars during their red giant phase (Unit 64) return some of these newly created elements to become part of the interstellar gas. The internal structure of some stars does not allow much mixing between layers of stars, so most of the material returned to interstellar space by the winds is the outer atmospheric gases—which are changed little from when the star formed. However, the more massive of the low-mass stars eject more than 80% of their mass before they shut down, and this may include a significant amount of carbon and oxygen. Others blast all the heavy elements they formed into space when a companion star dumps material onto their surface, inducing them to undergo a supernova explosion.

High-mass stars build in their cores a series of layers of heavy elements: carbon, neon, silicon, and eventually iron. Once an **iron core** forms in a high-mass star, the star must soon die, because iron cannot release energy through fusion to continue supporting the star. When the iron core collapses under its own weight, it triggers a supernova explosion. The heavy elements produced during the high-mass star's life are blasted into space. Some heavy elements, such as gold, are formed during the explosion itself. Supernovae therefore enrich the gas in interstellar clouds with heavy elements.

In the end, some of the mass that initially formed stars is locked away in stellar remnants—the white dwarfs, neutron stars, and black holes—that remain after the stars die. However, most of the material is returned to interstellar space, enriching the gas there with heavy elements formed through nuclear fusion. The material flowing outward from these dying stars may also run into interstellar clouds, compressing them and triggering new episodes of star formation. In this way, dying stars may initiate the birth of new generations of stars as well as provide elements vital to human life, such as carbon, oxygen, and iron. Each new generation of stars incorporates a greater amount of these heavy elements, so that stars in later generations, such as the Sun, are able to have planets made of heavy elements, such as the Earth, and carbon-based life forms, such as us. We therefore owe our existence to stellar evolution.

KEY TERMS

QUESTIONS FOR REVIEW

1. What is a stellar model?
2. What makes a star's mass the most important parameter for predicting its future evolution?
3. What two forces need to be balanced inside a star to keep it at a constant size?
4. What determines when a star becomes a main-sequence star?
5. What is a protostar? What is a planetary nebula?
6. How can we track stellar evolution in an H-R diagram?
7. What kinds of stars end up as neutron stars or black holes?

PROBLEMS

1. Visit a spot where trees are growing. List the different things you find there and the evidence you can put together about their relative place in the life cycle of trees.
2. The solid line that illustrates the evolutionary path of a star like the Sun in Figure 59.4 is a bit deceptive. It gives the impression that we are as likely to see stars like the Sun in their protostar phase as in their main-sequence phase. However a star moves through some phases rapidly and others much more slowly. If instead of a line you put down one point for every 10 million years, you would get a better idea of where you would be likely to see stars like the Sun in the H-R diagram. Lightly sketch or trace the Sun's evolutionary path in an H-R diagram and then use the following times to estimate how many dots to put down in each portion of the Sun's evolutionary path: protostar phase ~20 million years (Myr); main-sequence phase ~10,000 Myr; first red giant phase ~1000 Myr; yellow giant phase ~100 Myr; second red giant phase ~10 Myr; planetary nebula phase 0.01 Myr; white dwarf phase ~3000 Myr. (The star remains a white dwarf beyond this time, but fades from visibility.)
3. Examine the evolutionary path of the Sun in Figure 59.4. Compare the luminosity of a low-mass star on the main sequence to its luminosity as a red giant and as a white dwarf. How many times brighter and fainter will the Sun become than it is currently? What effects will this have on the Earth?
4. Examine the evolutionary path of the Sun in Figure 59.4. Estimate the size of the Sun when it becomes a red giant and white dwarf by referring to Unit 57. If the Sun is currently 0.5° across, how big will it appear to be when it is in its later stages of evolution?
5. Suppose you visited three different locations and measured the heights and strengths of everyone at each location. Imagine what kinds of heights and strengths you would likely find, then produce a height–strength figure for the population of people you predict that you would find at each of the following locations:
   a. a kindergarten class
   b. a college classroom
   c. a shopping mall

   What differences are there between your diagrams? Why?
6. Imagine that a large collection of stars with a wide assortment of masses forms in a group. Draw sketches of the Hertzsprung-Russell diagram at three times: (a) soon after they formed; (b) at the end of the main-sequence life of a star similar in mass to the Sun; (c) at the end of the main-sequence life of a star much less massive than the Sun. Describe in words the differences between the three sketches.

TEST YOURSELF

1. Suppose we could measure one property of a protostar. Which property would tell us most about its future evolution?
   a. Its temperature
   b. Its radius
   c. Its color
   d. Its luminosity
   e. Its mass
2. Blue main-sequence stars
   a. may end their lives as supernovae.
   b. have longer lives than most other types of stars.
   c. have cores that are hotter than a red giant's.
   d. are blue because they are fusing helium.
   e. All of the above.
3. A star with half the mass of the Sun has
   a. too little mass to make it onto the main sequence.
   b. enough mass to end as a supernova.
   c. too little mass to fuse carbon in its core.
   d. enough mass to become a black hole.
   e. All of the above.

unit

60

Star Formation

We described in Unit 33 the idea that the Solar System formed from a large cloud of gas, contracting because of its gravity. We saw how it spun faster as it contracted, much like water going down a drain, but with the bits of material that did not fall into the star becoming the planetary system.

This hypothesis that the Solar System formed from a gas cloud was proposed in the 1700s, long before astronomers had any evidence of gas between the stars (Unit 33). But the idea made sense because it successfully explained so many features of the Solar System. Today we can study interstellar gas clouds directly and hunt for clouds that may be forming stars. The idea that a collapsing cloud of gas can produce a star is convincingly borne out by our current observations, and with these data we find a rich variety of unexpected details in the ways stars form.

Background Pathways

Unit 59: Overview of Stellar Evolution 478

60.1 THE ORIGIN OF STARS IN INTERSTELLAR CLOUDS

Stars—whether they are of high or low mass—form from **molecular clouds.** These clouds are the densest and coldest kinds of interstellar clouds, in which atoms combine to form molecules and heavier elements combine to form small solid particles called **interstellar dust.** Molecular clouds can be hundreds of light-years across and can contain anywhere up to 10,000 solar masses of cold gas and dust. Some of the most active star formation occurs in **giant molecular clouds,** which can have masses up to about 1,000,000 $M_{\odot}$.

Molecular clouds orbit along with the stars inside our Galaxy (Unit 72). The gas is mostly hydrogen (71%) and helium (27%); the dust is composed of solid, microscopic particles of silicates, carbon, and iron compounds with a coating of ices. The presence of dust in these clouds blocks most light from entering the clouds, while the molecules can radiate energy even at low temperatures. The combination of these effects allows the gas to cool to low temperatures, which is essential for star formation.

Molecular clouds cool down to about 10 Kelvin—not far above absolute zero. At such low temperatures, atoms and molecules in the gas cloud move too slowly to generate much pressure. As a result, the pressure may be barely enough to support the gas cloud against its own gravity, so if the gas cloud grows even slightly more dense, it can collapse. Such a collapse might be triggered by a collision with a neighboring cloud, strong stellar winds from a nearby star, or even the explosion of such a star. Whatever disturbance triggers the collapse, the cloud does not collapse as a whole. Instead, denser regions in the cloud will begin to contract on their own, and these may fragment even further because the rate of collapse is faster wherever the density is higher.

At any given time, many clumps exist within a molecular cloud. Astronomers map the structure of these clumps at radio wavelengths by observing spectral lines from molecules in the cold gas (Figure 60.1). The radiation coming from these molecules allows us to trace where the gas is densest. The emission of this

FIGURE 60.1
False-color radio map of a star-forming molecular cloud in the constellation Taurus. The Taurus molecular cloud is about 450 light-years distant, and the region shown in the figure covers an area of about 60 ly × 100 ly. The brightest regions in the image indicate the densest concentrations of molecules, where stars are beginning to form.

FIGURE 60.2
Bok globules seen in silhouette against the background glow of a hot gas cloud by the Hubble Space Telescope. The two largest globules, which overlap each other at the top of the image, are each over 1 light-year across. They contain about 15 solar masses of gas. The many smaller dark clouds in the image may eventually dissipate or perhaps form objects too small to become hydrogen-fusing stars.

radiation is also significant because it is one of the main ways the clouds lose energy, so they can continue to collapse.

Gravitational energy is released whenever something falls. For example, if a cinder block falls onto a box of tennis balls, the impact scatters the balls in all directions, giving them kinetic energy—energy of motion. As an interstellar cloud contracts, the atoms and molecules collide and scatter like the tennis balls. These randomly directed motions are what we mean by thermal energy or heat. This type of compression heating is something you can observe for yourself with a bicycle pump—the pump cylinder grows warmer as you compress the gas to high pressure to fill a tire. Gravity squeezes the star, much as you use your weight and muscles to squeeze the gas inside the pump cylinder, and the gas responds by heating up.

If a cloud had no way to radiate the heat produced by this gravitational collapse, the cloud would grow hotter until its pressure was high enough to stop it from contracting any further. The emission of radiation allows the cloud to continue contracting. Because it radiates away energy, this is sometimes called "cooling" radiation. However, in allowing the cloud to contract, the gravitational force grows stronger so the collisions grow more violent, and the cloud grows hotter.

As the cloud contracts, the dust becomes so concentrated that virtually no visible light can pass through. The resulting dense, dark blobs are called **Bok globules** in honor of Bart Bok, the Dutch-born U.S. astronomer who first studied them in detail. Because these small dark clouds emit virtually no visible light, they are most easily seen in silhouette against the background glow of hot gas, as in Figure 60.2. (Also see Looking Up #9: Southern Circumpolar Constellations.)

A large, cold molecular cloud may break up into thousands of separate clumps, each clump collapsing on its own. The time that it takes a clump to collapse to a star ranges from hundreds of thousands of years to hundreds of millions of years—short by astronomical standards. The result is that protostars generally form in groups, not in isolation, and all the stars within a group form at approximately the same time. These groups of stars often persist for millions of years, or longer, after the stars form, and they can tell us a great deal about star formation (Unit 69).

60.2 PROTOSTARS

In the constellation Orion, one star visible to your unaided eye is no star at all. The middle star of Orion's sword (see Looking Up #6: Orion) is actually a cloud of gas in the process of forming several thousand stars. Because stars do not form in isolation and are not all born simultaneously, star-forming regions are complex in appearance. The glowing core of the Orion Nebula, heated by new massive stars there, is still surrounded by the original dark cloud, as can be seen in Figure 60.3. The firstborn stars have begun heating up and ionizing gas in the cloud even while other clumps of gas are still in the early stages of contraction.

The transformation from molecular cloud to star proceeds in several distinct stages, illustrated in Figure 60.4. Initially a dense clump within a much larger molecular cloud begins to collapse as described in the previous section. When the clump of gas becomes hot and dense enough to begin holding in the radiation produced by the heat generated during its collapse, it is called a **protostar.** In its initial *accretion phase* the protostar is surrounded and hidden by the cloud of infalling material. When the cloud of infalling material is mostly used up, the protostar becomes visible for the first time, in what astronomers often refer to as the protostar's *pre-main-sequence phase.* Although they do not yet fuse hydrogen into helium, the surfaces of these protostars are as hot as many main-sequence stars and can generate even higher luminosities. Protostars are often surrounded by a disk of gas and dust, as illustrated in Figure 60.4 and seen in the inset image for a

FIGURE 60.3
The Orion Nebula (M42) as imaged with the Hubble Space Telescope. Dark, dusty gas still surrounds the central glowing region where young massive stars have "hollowed out" the center of the cloud. The image shows a region about 11 light-years across. The nebula is about 1300 light-years distant, but visible to the unaided eye as the middle star of the sword in the constellation Orion.

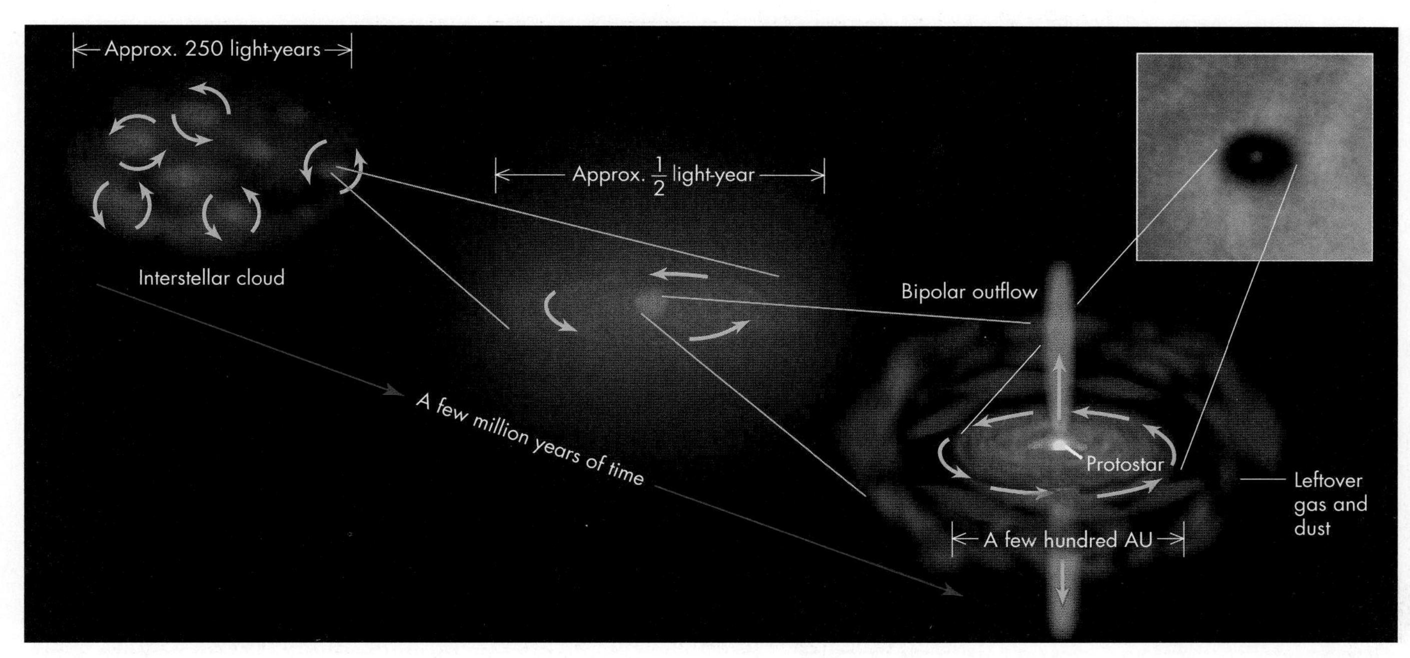

FIGURE 60.4
An artist's depiction of the birth of a star. A Hubble Space Telescope image of a protostar with a dusty disk of gas surrounding it is inset at right. The disk is seen in silhouette against glowing gas in the Orion Nebula.

Q If parallaxes are not available, how might it be possible to decide whether stars seen at infrared wavelengths are inside a molecular cloud instead of behind it?

protostar in the Orion Nebula. Protostars can be highly variable in brightness, and high-speed jets of gas are sometimes seen flowing from them.

The turbulent life of a protostar is powered by the energy released as a result of its gravitational contraction. In the accretion phase, gravitational compression causes the protostar to grow from a gas clump, just slightly warmer than the interstellar cloud, to a dense ball with a temperature of about 3000 K. The thermal radiation of the growing protostar is substantial. An accreting protostar that will evolve into a star like the Sun has a luminosity about 10 times greater than the Sun's. Throughout most of the accretion phase, the radiation from the protostar is absorbed by the surrounding cloud of infalling material and re-radiated at infrared and radio wavelengths. Computer models suggest that most of the surrounding material accretes during several hundred thousand years. The protostar at the end of the accretion stage is several times larger than its eventual size on the main sequence, but otherwise it does not look much different from an ordinary star.

When the surface temperature of the protostar has reached about 4000 K, ionization causes it to become much more opaque to the radiation generated in its hot interior than it was when it was cooler. The star's radiation loss declines, so the star contracts more slowly while keeping a nearly constant surface temperature. The contraction of a pre-main-sequence protostar to a star like the Sun is estimated to take tens of millions of years.

The Hubble Space Telescope photograph of the Eagle Nebula in Figure 60.5A shows a star-forming region that is estimated to be about 2 million years old. A few young stars are visible, but many more remain hidden in the dark pillars of gas and dust in the image. Protostars inside these dark clouds are not completely hidden, however. Astronomers can detect infrared emission from them, as seen in Figure 60.5B. This image was made with one of the 8-meter-diameter telescopes of the VLT. Many of the stars in this false-color image lie behind the Eagle Nebula. But comparison of the visual and infrared images reveals a number of protostars

A

B

FIGURE 60.5
The Eagle Nebula (M16), a region where stars are forming from molecular clouds. (A) Dust hides many protostars in dense portions of the cloud in this visible-wavelength image made with the Hubble Space Telescope. (B) A false-color infrared image of the Eagle nebula made with the VLT (Very Large Telescope) 8-m telescope in Chile. At infrared wavelengths, some stars hidden within the dust become visible.

(appearing bright orange in this false-color image) that are hidden by the dark veil of dust in the visible-wavelength image.

Throughout the accretion process of the protostar, as the cloud shrinks, any small amount of rotation in the initial gas cloud becomes magnified because of the conservation of angular momentum (Unit 20.3). This makes the rotating cloud flatten into a disk as it shrinks, with a rapidly spinning protostar at its center. The disk material is in orbit, perhaps eventually to turn into planets (Unit 33); this remaining material can fall onto the star only if friction or collisions slow it down enough. In fact, the rotation of all the material that comes together to make the protostar poses a problem for star formation. The protostar cannot continue to contract, because the outer layers are held up by the rapid spin. The protostar must lose some of its angular momentum to continue to contract. This is difficult for astrophysicists to model precisely, but it appears that magnetic fields play a critical role. The rapidly rotating star drags its magnetic field through the slower-rotating disk, and this produces a reaction force on the star that slows its spin down.

When the temperature in a protostar's core reaches about 1 million Kelvin, some nuclear fusion begins. This temperature is not sufficient to begin the proton–proton chain (Unit 50), but fusion of naturally occurring deuterium (the isotope of hydrogen with one proton and one neutron) begins. Because naturally occurring deuterium is less than one ten-thousandth ($<10^{-4}$) as abundant as hydrogen, it is not a long-lived energy source. Nevertheless, this fusion sustains the temperature and pressure in the core, slowing further contraction for a while.

The protostar phase is not quiet. Protostars sometimes eject gas in opposite directions called a **bipolar outflow,** as shown in Figure 60.6. In these images, protostars still hidden in dark clouds are ejecting gas that is traveling outward at high speed in opposite directions, carving out a path through the surrounding interstellar gas. Doppler measurements of bipolar outflows shows that the gas is shooting outward at speeds of several hundred kilometers per second. Where the jets of material run into surrounding interstellar clouds, they heat and ionize the gas, producing bright glowing regions. The cause of these outflows and jets is not yet fully understood. They are probably powered by energy that is released as matter falls from the inner parts of the disk surrounding the star, crashing into the protostar's surface layers and becoming very hot. The jets' narrow shape may be caused by the star's magnetic field, which confines ionized gas to flow out along the directions of the magnetic poles, much as such fields confine prominences on the Sun (Unit 51). The disk and infalling matter surrounding a protostar may also funnel the hot gas into the direction of least resistance, perpendicular to the disk.

Another phenomenon observed among some protostars is a very strong wind. This outward flow of gas is similar to the solar wind, but far more intense, and it seems to occur only episodically during a protostar's life. This outflow may be powered by the spinning star's magnetic field. In the same process that slows the star down, some of the hot ionized gas surrounding the star is accelerated to such a high speed that it is flung outward—carrying away some of the protostar's angular momentum.

Finally, pre-main-sequence protostars are also often seen to vary erratically in brightness. Such variable stars are called **T Tauri stars** after a star in the constellation

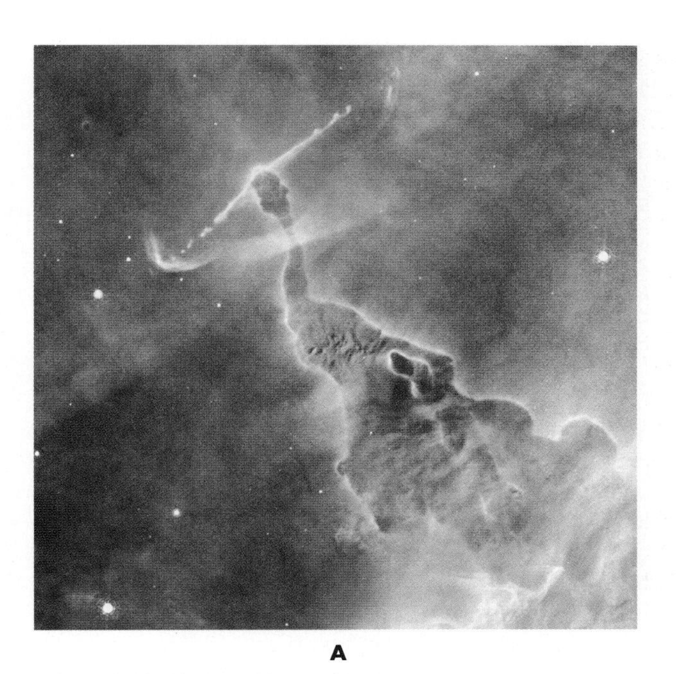

A

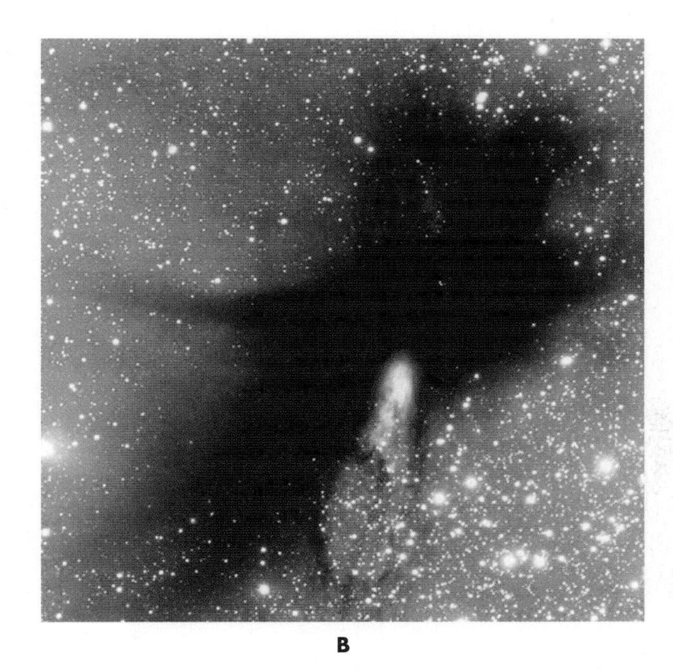

B

FIGURE 60.6
Bipolar outflows. (A) A protostar lies hidden at the tip of a "pillar" of gas and dust in the Carina Nebula in this Hubble Space Telescope image. The protostar is generating jets of gas that extend more than one-fourth of a light-year in each direction. At the end of the leftward jet, the outflow strikes interstellar gas, which is heated and glows. (B) A protostar hidden inside this Bok globule is generating a bipolar outflow that is "blowing bubbles" in the dark cloud in this VLT image. The globule is about 1 light-year across, and infrared observations reveal several protostars forming in the cloud in addition to the one producing the visible outflow.

Taurus, which is one of a group of young stars that exhibit this behavior. Some of this variability is probably caused by magnetic activity, much like that observed on our Sun. In addition, it is suspected that the inflow of material from the disk onto the star may occur irregularly—the gas in the disk may be "lumpy"—and this could contribute to the variability as well.

60.3 PROTOSTARS IN THE H-R DIAGRAM

When pre-main-sequence stars are plotted in an H-R diagram, they lie a little above the main sequence because they are still shrinking and are therefore larger and brighter than when they become main-sequence stars. Figure 60.7 shows the H-R diagram for a young group of stars and a photograph of the star-forming region. The region contains a number of T Tauri stars and Bok globules, and is still surrounded by gas and dust left over from the period of star formation. The H-R diagram for this and other young star-forming regions shows that the massive stars reach the main sequence before the lower-mass stars.

Computer models indicate that the mass of the forming star is the most important variable in determining how long the protostar stage lasts. An example of the predicted evolutionary tracks for a range of star masses is illustrated in Figure 60.8. The protostars begin their contraction at low temperatures far off to the right of the diagram. The dashed lines show the late stages of accretion as the protostars approach their final masses. The accreting protostars grow hotter at a fairly steady luminosity. During their pre-main-sequence phase, lower-mass protostars like the Sun finish accretion then drop in luminosity at a relatively constant temperature. The star settles down to a configuration where it is hot enough in its core to stably

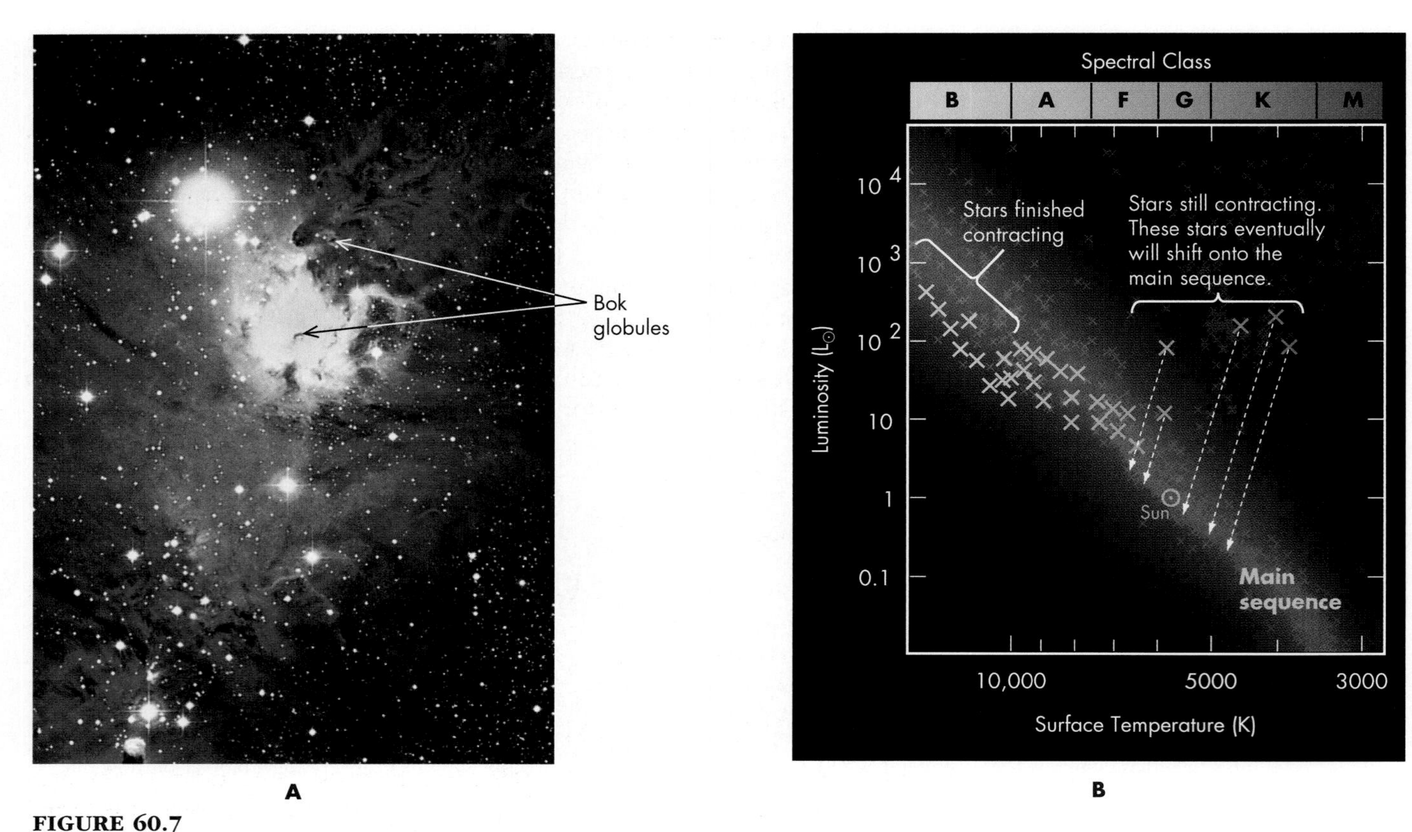

FIGURE 60.7
(A) This glowing region is named NGC 2264. The gas is heated mostly by recently formed massive stars. The low-mass stars are still in their protostar phase, some still in small dark clumps called Bok globules. (B) The H-R diagram for young stars and protostars in NGC 2264.

FIGURE 60.8
Evolutionary tracks for protostars in the H-R diagram. The solid red lines show the changing state of the surface temperature and luminosity for stars of various masses. The white dashed lines show where a star can be found along these tracks 1 million and 10 million years after the cloud begins to collapse. Note that the massive stars contract to become stars faster than the lower-mass stars.

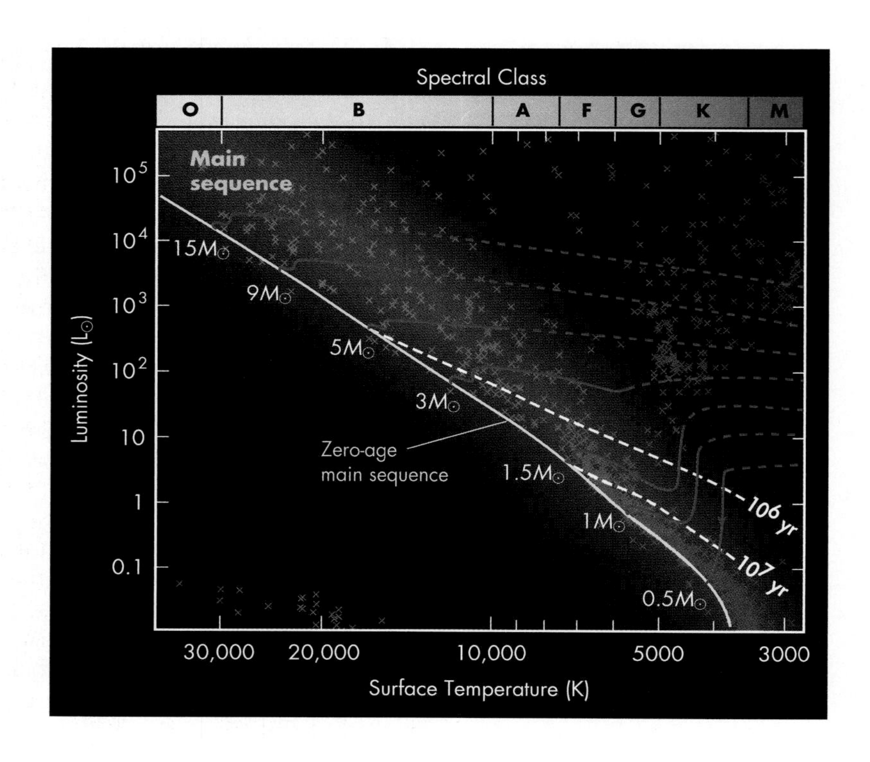

The nearly vertical paths in the late stages of protostars like the Sun are known as *Hayashi tracks,* named after the Japanese astrophysicist who predicted them.

Q If protostars lie above the main sequence, why are they not seen filling this region in most H-R diagrams (such as those shown in Unit 58)?

fuse hydrogen to helium. The curve of stellar temperatures and luminosities where fusion begins is known as the **zero-age main sequence.**

For a star like the Sun, the protostar phase lasts approximately 20 million years. For high-mass stars it may be less than a tenth as long, whereas for very low-mass stars it may be more than 10 times longer. The difference in the duration of the protostar stage explains why the higher-mass stars have already reached the main sequence in Figure 60.7B, while lower-mass stars in that star-forming region are still in their protostar stage.

The varying timescales for formation imply that the more massive stars may also influence the formation of lower-mass stars. In most star-forming regions, the high-mass, high-luminosity stars form rapidly and begin to ionize and heat up the surrounding gas. The radiation and stellar winds from these massive stars may limit the growth of the slower-forming low-mass stars.

60.4 STELLAR MASS LIMITS

Observationally, we find few stars outside the range of about 0.2 $M_\odot$ to 20 $M_\odot$. Lower-mass stars are dim and hard to detect, while higher-mass stars are rare. However, it is not just observational limitations that determine this range. Theoretical studies of the physical conditions necessary to allow a star to form indicate that stars must have masses between 0.08 $M_\odot$ and 150 $M_\odot$.

Stars below 0.08 $M_\odot$ do not occur because such mass is too small to compress the gas in the core enough to reach hydrogen fusion temperatures. Although lower-mass objects form, they never establish a stable balance between gravity and fusion's internal heat production. Some unstable fusion of hydrogen may occur in slightly less massive objects, and fusion of deuterium may occur in objects down to about 0.016 $M_\odot$ (about 17 times Jupiter's mass). However, naturally occurring deuterium is too scarce for this to be a stable state. Objects with masses between 0.016 $M_\odot$ and 0.08 $M_\odot$ fall between our definitions of planet and star and have been dubbed **brown dwarfs.** These are dim, cool objects that infrared technologies began detecting only in the mid-1990s. Most of the objects in the L and T spectral types (Unit 55) appear to fall in this mass range. These objects are probably very numerous, but astronomers are still in the early stages of characterizing this population of "failed stars."

Q How would young brown dwarfs differ in appearance from protostars with masses greater than 0.08 $M_\odot$?

By contrast, the upper limit to star masses might be described as too much of a good thing! A star more massive than 150 $M_\odot$ never stabilizes for the following reasons: If a giant cloud of gas begins to contract, the intense gravitational compression drives up its temperature and luminosity very rapidly. The center of a contracting clump of gas more massive than 150 $M_\odot$ emits such intense radiation that it heats the surrounding gas to an extremely high temperature. This raises its pressure so much that no additional material can fall in, setting a limit on the amount of material the largest stars can accumulate. Furthermore, once formed, high-mass stars become so luminous (recall the mass–luminosity relation) that their radiation rapidly drives gas from their outer layers into space in a strong stellar wind. The most massive stars—or would-be stars—essentially "shine" themselves apart.

In reality, conditions governing the protostar stage are rarely conducive to the formation of stars much larger than 30 solar masses. Moreover, such massive stars burn themselves out extremely rapidly (Unit 61). Nevertheless, some isolated examples have been found, like the star Eta Carina (see Looking Up #8: Centaurus and Crux, the Southern Cross), which is estimated to have a mass 100 to 150 times the Sun's mass.

KEY TERMS

bipolar outflow, 490
Bok globule, 487
brown dwarf, 493
giant molecular cloud, 486
interstellar dust, 486
molecular cloud, 486
protostar, 488
T Tauri star, 490
zero-age main sequence, 493

QUESTIONS FOR REVIEW

1. What are molecular clouds, and how are they related to star formation?
2. What are Bok globules, and how are they relevant to stars?
3. At what wavelengths is it best to observe a protostar when it is still accreting gas? Why?
4. What generates the heat in a protostar?
5. What is a T Tauri star? What are bipolar outflows?
6. What determines the upper and lower mass limits of stars?
7. What is a brown dwarf?

PROBLEMS

1. Use Wien's law to find the wavelength of maximum radiation from a 2000-K protostar. To what part of the electromagnetic spectrum does this wavelength belong?
2. Consider a 2000-K protostar that is the same diameter as the Sun. What is its luminosity, compared to the Sun? How many times bigger than the Sun would it have to be to emit 10 times more luminosity than the Sun?
3. The jets from a T Tauri star are measured to have speeds of about 300 km/sec. At this speed, how long would it take the ejected material to travel 1 ly away from the protostar?
4. Suppose a molecular cloud begins with a mass of 1 $M_\odot$ and has a radius of 1 ly. Imagine a dust grain at the outer edge of the cloud pulled straight inward, accelerating the entire way. When it reaches a radius of 1 million km from the center of the forming star, how fast will it be moving? (*Hint:* This problem is much easier to answer through the conservation of energy.)
5. Compare the average density of a newly formed star of mass 20 $M_\odot$ (radius $\approx$ 10 $R_\odot$) and 0.1 $M_\odot$ (radius $\approx$ 0.1 $R_\odot$) to the Sun's density. You should find that the higher-mass star has a lower density. Relate these relative densities to the formation processes that determine the upper and lower mass limits for stars.
6. Assuming that a 0.08-$M_\odot$ brown dwarf has the same density as Jupiter (see Appendix Table 5 for data), what is its radius? Compare this radius to the Sun's radius.

TEST YOURSELF

1. Which of the following is *not* true about molecular clouds?
   a. They contain helium.
   b. They contain iron.
   c. Their temperature is near absolute zero.
   d. Some of their particles are coated in ice.
   e. The entire cloud contracts as a whole.
2. T-Tauri stars
   a. lie on the main sequence.
   b. vary in brightness.
   c. are dimmer than main-sequence stars.
   d. are typically found in older clusters.
   e. All of the above except (a).
3. Why are there no stars less massive than 0.08 $M_\odot$?
   a. With such a low mass they take longer than the age of the universe to contract.
   b. Their mass is so small that deuterium fusion blasts them apart.
   c. They exist, but they are so dim that we cannot detect any.
   d. They cannot compress their cores to hydrogen fusion temperatures.
   e. All of the above

Starry Night

See the Pathways *2nd edition ARIS site for online Starry Night™ planetarium exercises.*

unit 61

Main-Sequence Stars

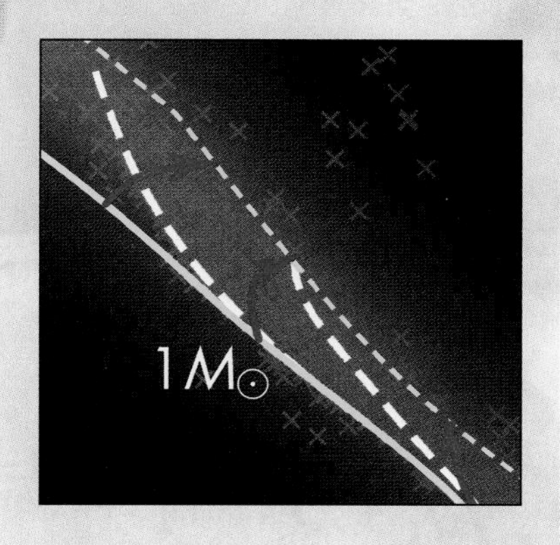

Background Pathways

Unit 59: Overview of Stellar Evolution 478

A star stops being a protostar and becomes a **main-sequence star** when it begins fusing hydrogen into helium in its core. Fusion generates the heat needed to maintain pressure in the core, which stops the protostar's contraction. At this stage of its life, the star's interior structure is much like the Sun's: the star contains a core, where hydrogen is fusing into helium, and an envelope of gas around the core, where energy is transported to the star's surface. The properties of the core and envelope, however, depend greatly on the star's mass.

61.1 MASS AND CORE TEMPERATURE

In examining the physical properties of stars along the main sequence, we find that the higher the mass of a star, the hotter and more luminous it is. This relation between mass, temperature, and luminosity results from the fact that the gravity and pressure inside a star must be in balance, a condition called **hydrostatic equilibrium.** The principles governing the structure of a star are illustrated in Figure 61.1. With computer modeling of stars, we can determine how these processes balance with one another and can learn many details about a star's internal structure—its density, temperature, pressure, and energy generation.

The lowest-mass stars take the longest time to complete the protostar phase and enter the main-sequence phase. Their weaker gravity causes them to contract so

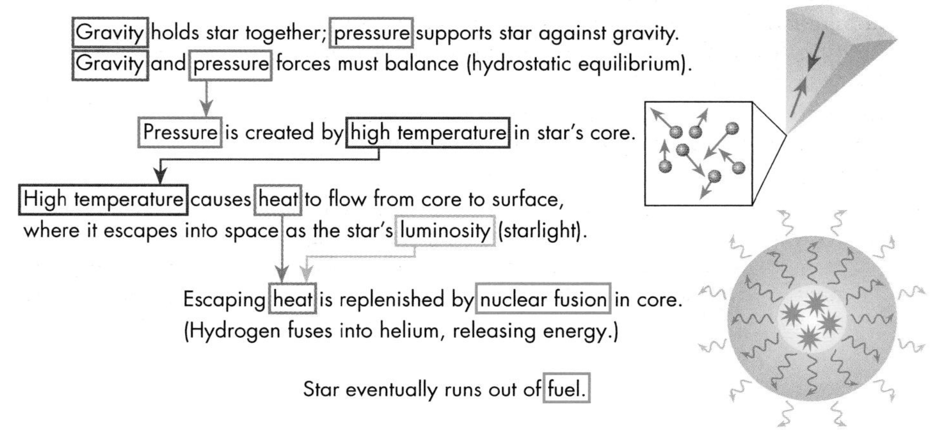

FIGURE 61.1
Outline of the processes that govern the structure of stars.

slowly that their core temperature rises very gradually. As a consequence, a 0.1-$M_\odot$ star must compress the gas in the core to a very high density—about 500 kilograms per liter—just to reach a temperature where hydrogen fusion can begin. By contrast, massive protostars contract rapidly and reach high temperatures in their cores without compressing the gas to nearly as high a density. A 10-$M_\odot$ star's core reaches a density of only a few kilograms per liter when fusion begins.

The higher density of low-mass stars may seem a little surprising at first: It would seem that high-mass stars would compress matter more. Later in their lifetimes that does eventually prove true; but during the main-sequence phase, the highest-mass stars are lowest in density. They achieve hydrostatic equilibrium despite their lower density because they reach a higher temperature in their cores, and that produces the balancing pressure. At the bottom of the main sequence, the cores may be only 5 million Kelvin. At the upper end, the core temperature is about 10 times hotter. Thus high-mass stars have hotter, lower-density cores than do low-mass stars.

The rate of nuclear fusion rises rapidly with temperature, and additional fusion mechanisms become possible at high temperatures as well. The high temperatures in high-mass stars thus allow nuclear reactions to generate the *very* large luminosities observed in massive stars.

61.2 STRUCTURE OF HIGH-MASS AND LOW-MASS STARS

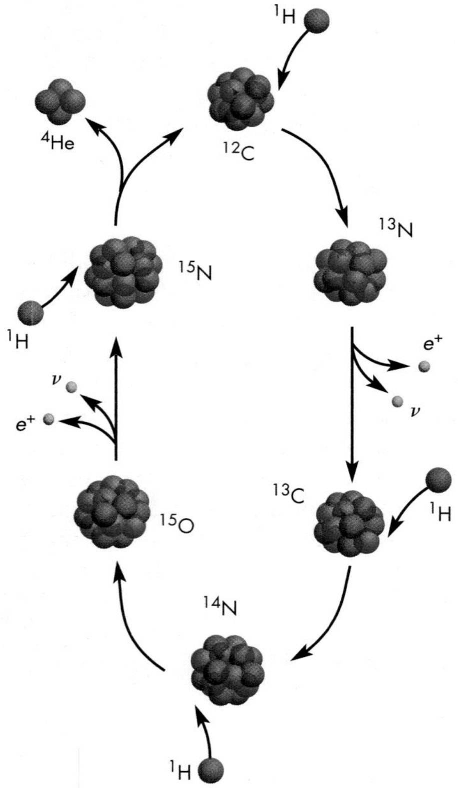

In the Sun and other low-mass stars, hydrogen is converted to helium through the **proton–proton chain** (Unit 50). In the proton–proton chain, two hydrogen nuclei, or protons, fuse to form hydrogen isotope ^{2}H (deuterium). A third proton combines with the deuterium to make ^{3}He, and the ^{3}He fuses to form ^{4}He.

At the higher temperatures in the cores of stars more massive than about 2 $M_\odot$, a different set of fusion reactions converts hydrogen to helium. In these larger stars, where core temperatures rise above about 20 million Kelvin, hydrogen fusion takes place primarily by means of the **CNO cycle.** In this cycle, illustrated in Figure 61.2, carbon atoms already present in the star's core act as catalysts to aid the reaction through a series of steps that build nitrogen and oxygen nuclei. No carbon, nitrogen, or oxygen is created or destroyed in the end by the CNO cycle, and the net energy release is essentially the same per created helium atom as in the proton–proton chain. However, at high temperatures the CNO cycle is much faster than the proton–proton chain reaction.

The high rate of energy production at these high temperatures makes the internal structure of high-mass stars different from that of low-mass stars. The photons (gamma rays) generated by the fusion reactions can travel only in short hops, only a centimeter or so at a time, before interacting with another atom. Because the photons cannot carry the energy away quickly, clumps of gas in the core region become superheated. These clumps rise toward the star's surface, traveling a sizable fraction of the star's radius, and then sink back down after releasing their heat and cooling—in the process of **convection.**

Convection occurs wherever the rate of energy flow by photons becomes too slow. Similar convection motions occur just below the surface in stars like the Sun, where the cooler surface gas strongly absorbs the photons and impedes the flow of energy by radiation (Unit 49). In stars less massive than the Sun, the surfaces are even cooler, and the convection zone reaches proportionately deeper into the star. These structural differences are illustrated in Figure 61.3.

A star's core is the region where temperatures are high enough for hydrogen to fuse into helium. The core typically contains only about one-tenth of the star's

FIGURE 61.2
The nuclear reactions of the CNO cycle. Four hydrogen nuclei (^{1}H) combine with carbon, nitrogen, and oxygen (C, N, and O) in a cycle whose net result is to produce one helium (^{4}He) nucleus.

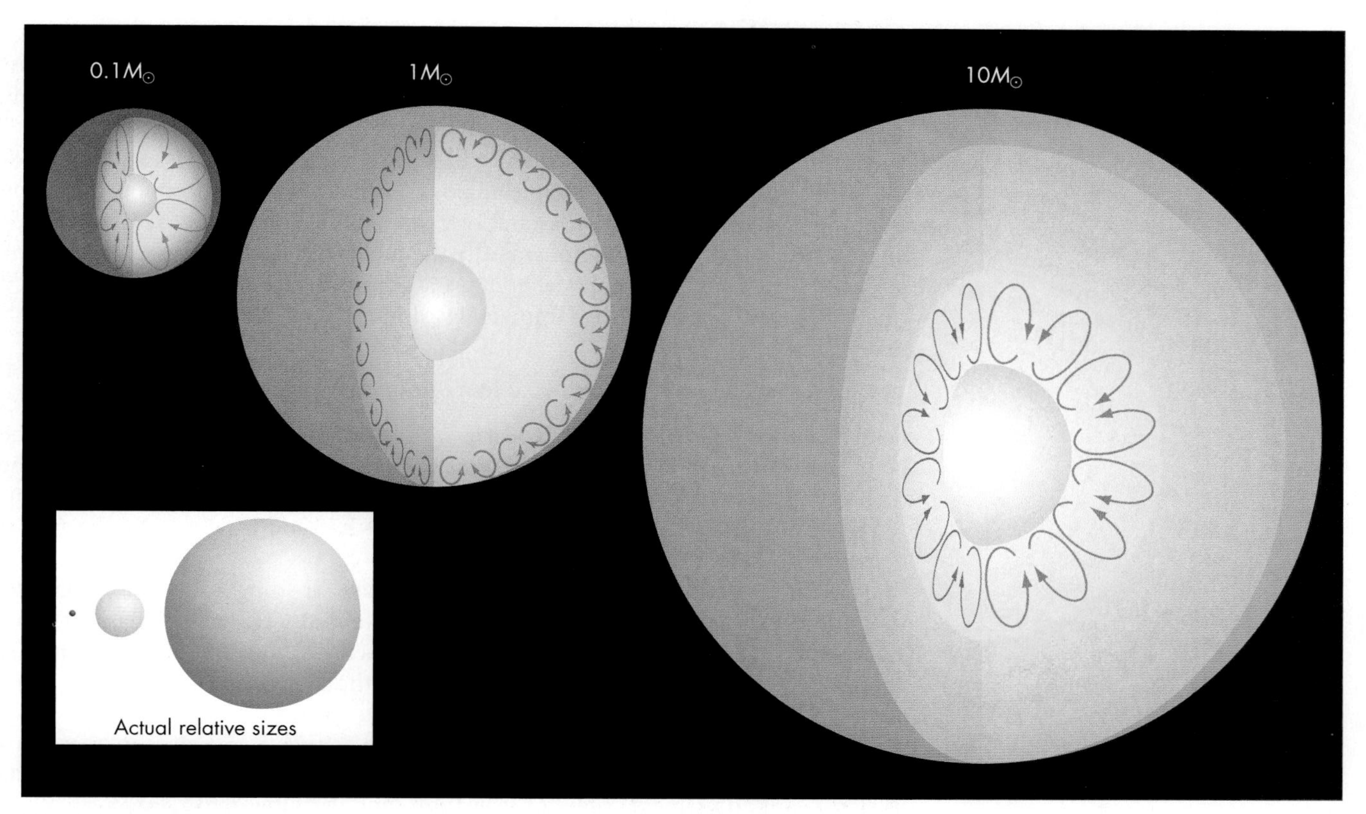

FIGURE 61.3
The comparative structure of main-sequence stars of different masses. A massive star undergoes convection in the neighborhood of its core, while low-mass stars undergo convection in their surface layers. For very low-mass stars (less than about 0.4 $M_{\odot}$) the convection reaches all the way from the surface to the core of the star. The inset box shows the stars' true relative sizes.

total mass. For a star like the Sun, the convection zone does not carry gas all the way from the surface into the core. So as the hydrogen fuel is gradually consumed in the core, the Sun's convection is unable to reach deep enough to mix unfused hydrogen from the outer parts into the core. Therefore the core can run out of hydrogen fuel, even though there is a large amount in surrounding regions.

Computer models indicate that the highest- and lowest-mass stars can replenish their cores with hydrogen to varying degrees. Convection within the inner regions of massive stars can mix some of the unburned hydrogen from the surrounding layers into the core and thereby increase, by up to several tenths of the star's mass, the amount of hydrogen that is available for fusion. In stars less massive than about 0.4 $M_{\odot}$, the gas in the outer layers mixes down into the core completely. Such stars are said to be **fully convective.** These stars can fuse most of their hydrogen. By contrast, more massive stars, whose surface gas does not mix down into the core, can be starved for hydrogen in their cores even though their outer layers remain hydrogen-rich.

Q If all stars were fully convective, how would we be able to determine this from studying stars' spectra?

61.3 MAIN-SEQUENCE LIFETIME OF A STAR

While a star is fusing hydrogen to helium in its core, its structure remains nearly constant. Therefore, it remains in almost the same position in the main sequence on the H-R diagram. The length of time a star spends fusing hydrogen to helium in its core is called its **main-sequence lifetime.** This also explains why the main sequence stands out so clearly when we plot the luminosities and temperatures of a random set of stars. Because all stars spend most of their lives in the main-sequence phase, and because the external properties of these stars remain nearly constant, we find most stars in the main-sequence region of the H-R diagram.

Any individual star's main-sequence lifetime depends on its mass and luminosity. To see why, we can use a simple analogy. Suppose we want to know how long a camping lantern will run on a tank of propane fuel. That time clearly depends on how much fuel it has and how rapidly the fuel is consumed. For example, if the tank contains 2 liters of propane, and at maximum brightness the fuel burns at a rate of 0.2 liters per hour, the lamp will last for

$$t = \frac{\text{Amount of fuel}}{\text{Rate of burning}} = \frac{2 \text{ liters}}{0.2 \text{ liters/hour}} = \frac{2}{0.2} \text{ hours} = 10 \text{ hours}$$

before running out of propane. Clearly we can increase the amount of time the lamp will run if we increase the amount of fuel or dim the light so it burns the fuel more slowly. We can apply the same formula to a star to determine its main-sequence lifetime if we know how much fuel it has and how rapidly the fuel is being consumed.

The amount of fuel a star has is set by its mass, or to be a little more precise, the amount of a star's mass that is hydrogen—about 71% of the mass of most stars. This represents the total amount of potential fuel. But only within the core—approximately the central one-tenth of the star's overall mass, for most stars—are temperatures high enough for fusion to occur. Thus only one-tenth of 71%, or about 7%, of a star's mass is available as fuel. Some stars increase this amount by mixing gas from the outer parts of the star into the core, but this results in significant adjustments only for very high- and very low-mass stars.

Q Suppose in the remote future we wanted to extend the Sun's lifetime. A government official proposes collecting interstellar hydrogen and dumping it on the Sun. If this were possible, would it be a good idea? Why or why not?

The rate at which a star consumes its fuel is directly indicated by its luminosity. More-luminous stars fuse their fuel faster. This makes sense because the energy that a star puts out is supplied by its fuel. Again, it is like a lantern. If it is brighter, it is burning the fuel faster; if it is dimmer, it is burning the fuel more slowly.

Thus a star's lifetime, t, is proportional to its mass divided by its luminosity, M/L. To work out the actual times involved, we have to calculate what mass of hydrogen is needed to produce a given quantity of light. This is determined by the amount of nuclear energy released in the fusion process. For the Sun, astronomers have determined that 360 million tons of hydrogen are being fused each second to generate its observed luminosity (Unit 50). This creates 356 million tons of helium and "4 million tons of light"—that is, according to Einstein's law, $E = m \times c^2$, 4 million tons of mass are turned into energy each second.

Taking the amount of hydrogen in the Sun's core (7% of its total mass of 2×10^{30} kg) and dividing it by the amount of fuel consumed each second (360 million tons per second), we find that the total number of seconds in the Sun's main-sequence lifetime is

$$t_{\odot} \approx \frac{0.07 \times 2 \times 10^{30} \text{ kg}}{360 \times 10^{9} \text{ kg/sec}} = 3.9 \times 10^{17} \text{ sec} = 1.2 \times 10^{10} \text{ yr.}$$

For the values given, we obtain a lifetime for the Sun of about 12 billion years on the main sequence. Other, more detailed stellar evolution models indicate that slightly less hydrogen will be fused in the core before the Sun leaves the main-sequence phase, so the Sun's main-sequence lifetime is rounded off to about 10 billion (10^{10}) years. Given its present age, the Sun is roughly halfway through its main-sequence lifetime.

t = Main-sequence lifetime
M = Star's mass (in solar masses)
L = Star's luminosity (in solar luminosities)

We can estimate the main-sequence lifetimes of other stars by comparing them with the Sun. If we measure M and L in solar units, a star's main-sequence lifetime is

$$t = \frac{M}{L} \times 10^{10} \text{ years.}$$

The mass–luminosity relation (Unit 58) states that $L \approx M^{3.5}$ for a star whose mass and luminosity are measured in solar units. Using this relationship to substitute for the luminosity, we find that a star's lifetime is approximately

$$t \approx \frac{M}{M^{3.5}} \times 10^{10} \text{ years}$$

$$\approx \frac{1}{M^{2.5}} \times 10^{10} \text{ years.}$$

This assumes that all stars fuse the same fraction of their mass as the Sun does, so it will underestimate somewhat the lifetimes of the highest- and lowest-mass stars; but it gives a good lifetime estimate for most stars.

We might expect lifetimes to be longer for massive stars because they have more fuel to fuse. However, the luminosities of massive stars go up by even more than their masses. For example, the star Sirius has an estimated mass of about 2 $M_{\odot}$, but its luminosity is about 20 $L_{\odot}$. Therefore it has 2 times more fuel than the Sun, but it is fusing it 20 times faster, so its lifetime will be 2/20ths as long as the Sun's. In other words, Sirius will have a main-sequence lifetime, *t*, of

$$t(\text{Sirius}) = \frac{2}{20} \times 10^{10}\text{years} = 10^{9}\text{years.}$$

So Sirius is expected to live on the main sequence only about a tenth as long as the Sun. For a star whose mass is 10 $M_{\odot}$ and whose luminosity is 10^4 $L_{\odot}$, the main-sequence lifetime is a mere 10^7 years.

These results demonstrate that massive stars have much shorter lifetimes than the Sun. Despite having more fuel, they fuse it much faster to supply their greater luminosity. In other words, massive stars are like gas-guzzling cars that, despite having large fuel tanks, run out of fuel sooner than fuel-efficient cars with smaller tanks. This brevity of massive stars' lives implies that those we see must be relatively "young" by the standard of the Sun. Because massive stars are blue, we can conclude that, in general,

Blue stars have formed recently.

Because of their recent formation, we can understand why this type of star is often seen associated with clouds of interstellar gas. They die so quickly that there has been little time for the interstellar gas to disperse in response to stellar winds, or for the star to drift away from the cloud if it has any relative motion. Therefore blue stars are frequently embedded in matter left over from their formation, as you can see in Figure 61.4. In these recently formed groups of stars, the hot blue stars produce ultraviolet light that ionizes the interstellar hydrogen, producing the pink glow of interstellar hydrogen seen in the image.

Q If a star is red, what can you say about its age?

FIGURE 61.4
Photograph of the young star cluster NGC 2264. The massive, short-lived blue stars are still surrounded by the interstellar gas from which they formed. NGC 2264 is also called the "cone nebula" for the shape of the dark column of gas and dust, the top of which is ionized by the bright blue star at the center of the image.

61.4 CHANGES DURING THE MAIN-SEQUENCE PHASE

Stars change relatively little as they age during their main-sequence phase, but they do change. Stellar structure models tell us that as a star depletes the hydrogen in its core, the hydrogen nuclei become more spread out and less likely to collide. The decline in energy production causes the star's core to contract slightly, compressing and heating the core. This drives up the temperature until the overall rate of reactions once again generates enough heat to support the star. Over the long lifetime of a star in its main-sequence phase, the star will gradually increase in luminosity by a factor of about 3. The stellar models show that the star's atmosphere expands in response to the more luminous core. Tripling the luminosity is not a minor change, but on the H-R diagram, which displays a factor of over a million in luminosity, the change leaves the stars in a narrow band (Figure 61.5).

It is possible to estimate how much a star has aged by its shift from the expected initial characteristics when it began fusing hydrogen—also known as its **zero-age main sequence** (or **ZAMS**) position in the H-R diagram. This requires such precise measurements of a star's mass, luminosity, temperature, and chemical composition that we have accurate age estimates for relatively few stars.

Astronomers estimate that the Sun is about 1.4 times more luminous today than when it began on the ZAMS. Its luminosity will continue to increase, approximately doubling from its current value over the remaining 5 or 6 billion years of its main-sequence lifetime. As the Sun's core temperature climbs, the CNO cycle will become the dominant source of its energy production.

Q As the Sun's power output continues to increase, even the complete removal of greenhouse gases will not prevent the Earth from heating up. What else could humans do to compensate for the Sun's increasing luminosity?

Although these changes are small compared to the range of other stars' luminosities, they are very significant for the environment of the Earth and other planets. Remarkably, through most of the Earth's history the temperature has remained stable, apparently due to the steady removal of "greenhouse gases" by microscopic life forms (Units 36 and 83). The consequences for the Earth of the future increase

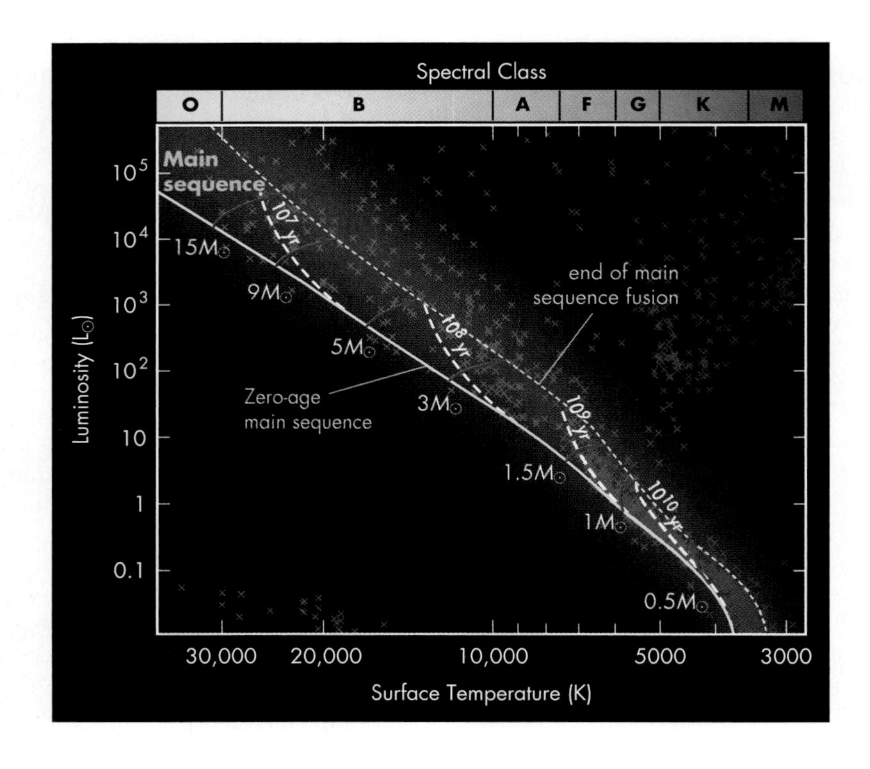

FIGURE 61.5
Stellar evolution while stars are on the main sequence. Luminosities typically increase by about a factor of 3 during this phase. Tracks of stars of different masses are shown in red. The white dashed lines indicate where the stars will be along the tracks at different times after they begin main-sequence fusion.

in the Sun's luminosity are not clear. Just a few percentage points of increase in the Sun's luminosity would drive the Earth's temperature up by about 2°C (about 4°F), enough to cause polar ice cap melting, which would dramatically raise ocean levels. Unless compensating effects occur, the Earth's climate and habitability will change in the distant future.

KEY TERMS

CNO cycle, 496
convection, 496
fully convective, 497
hydrostatic equilibrium, 495
main-sequence lifetime, 497
main-sequence star, 495
proton–proton chain, 496
zero-age main sequence (ZAMS), 500

QUESTIONS FOR REVIEW

1. What determines when a star becomes a main-sequence star? What is the ZAMS?
2. Why do low-mass stars tend to be denser than high-mass stars?
3. What factors determine how long a star stays on the main sequence?
4. Why do high-mass stars last a shorter time on the main sequence than low-mass stars?
5. How does convection differ in very low-mass stars, low-mass stars, and high-mass stars?
6. How does a star's luminosity change during the star's main-sequence lifetime? What consequences does this have for Earth?

PROBLEMS

1. The star Alpha Centauri A is of the same spectral type as the Sun, but it is about 20% more luminous. Is it more or less massive than the Sun? Is it older or younger than the Sun? Explain your reasoning.
2. Calculate the luminosity of the 2.0-$M_\odot$ star Sirius predicted by the mass–luminosity equation (Unit 58). Its observed luminosity is about 20 $L_\odot$. How might you explain the difference?
3. Calculate the main-sequence lifetime of a 25-$M_\odot$ star, using the mass–luminosity relationship (Unit 58) to estimate its luminosity.
4. The main-sequence star Proxima Centauri has a mass of 0.1 $M_\odot$. Calculate its luminosity and main-sequence lifetime.
5. Using the main-sequence lifetime equation and the mass–luminosity relationship to estimate the mass of a star that would have a main-sequence lifetime of 500 million years.
6. According to a recent computer model of the Sun's evolution, when the Sun was young its surface temperature was 130 K cooler (than its current value of 5800 K) and its luminosity was 0.71 $L_\odot$. When the Sun is approaching the end of its main-sequence lifetime 5.5 billion years from now, its surface temperature will be 650 K cooler than it is currently, and its luminosity will be 2.02 $L_\odot$. Use the Stefan-Boltzmann law (Unit 57) to estimate the Sun's diameter at these past and future times.
7. Suppose the Sun had a luminosity of 0.71 $L_\odot$ when it was young. Compare the brightness of sunlight (Unit 54) on Venus (at 0.72 AU) when the Sun was young to the brightness of sunlight on Earth today.
8. How much more luminous would the Sun have to become for Mars to receive the same brightness of sunlight as Earth currently receives?

TEST YOURSELF

1. A star whose mass is 2 times larger than the Sun's has a main-sequence lifetime
   a. many times longer than the Sun's.
   b. a few percent longer than the Sun's.
   c. a few percent shorter than the Sun's.
   d. many times shorter than the Sun's.
2. Blue main-sequence stars have
   a. a large mass and a short lifetime.
   b. a large mass and a long lifetime.
   c. a low mass and a short lifetime.
   d. a low mass and a long lifetime.
   e. You can't say, because colors of main-sequence stars are unrelated to their masses or lifetimes.
3. *Hydrostatic equilibrium* means that
   a. a star is not convecting in its interior.
   b. a star has achieved a state where water is a fusion by-product.
   c. electrons are in their highest possible energy levels before ionization.
   d. gravity and pressure are in balance.
   e. fusion is occurring at a steady rate.

unit

62

Giant Stars

Although the main-sequence stage of a star's life is a time of relative stability, the next stage of its life is one of constant change. When hydrogen runs out in the core of a main-sequence star, the force of gravity remains, and the star's core resumes the gravitational contraction that was taking place during the protostar phase. But an unexpected thing happens. Even though the core of the star begins contracting and heating again, the surface of the star expands and cools, and the star grows into a **giant**. This peculiar behavior needs to be explained so we can understand many of the unusual features of the giant phase of a star's life.

The giant phase never establishes the stability of the main-sequence phase, because its new sources of fuel run out rapidly and the star is constantly readjusting its structure. Stellar models indicate that a star's time as a giant is about 10% to 20% as long as its time as a main-sequence star.

Background Pathways

Unit 59: Overview of Stellar Evolution 478

62.1 RESTRUCTURING FOLLOWING THE MAIN SEQUENCE

During a star's main-sequence phase, its structure is stable. This stability occurs because the nuclear fusion in the star's core acts as a thermostat. For example, if a star contracts slightly, the compression of the gas in its core heats it, driving up the rate of nuclear reactions. Consequently, the core produces more heat and pressure, reversing the contraction. A slight expansion of a star reduces nuclear reactions, leading to less pressure, which reverses the expansion. A main-sequence star is therefore quite stable. However, once nuclear reactions cease in the core, the thermostat is broken, and this leads to unusual effects, which are the hallmarks of the next phase of a star's life.

When a main-sequence star has consumed the hydrogen in its core, the energy production there halts. With no energy generation in the core, the weight of the star's outer layers pressing downward overwhelms the pressure supporting its core. The core is squeezed smaller. This compresses the gas and therefore heats it (as discussed also in Units 49 and 60). The result for the star is that its core temperature rises. However, because the core contains little hydrogen, the star has no way to generate energy to halt this compression process. As a result, the core of the star resumes the contraction that stopped when it began to fuse hydrogen as a new main-sequence star.

Although the core itself has no hydrogen to fuse, regions outside the core are still rich in hydrogen because they never reached temperatures high enough for hydrogen to fuse. Now, however, as the interior of the star shrinks and is compressed, the regions immediately surrounding the core are also compressed and heated to fusion temperatures. Hydrogen fusion begins in a layer outside the core in a process that is commonly called hydrogen **shell fusion** (Figure 62.1). Although this shell fusion generates heat and pressure, it is outside the core, and therefore it cannot provide the internal pressure needed to stop the core's contraction.

This situation is unlike the thermostat effect that makes a main-sequence star so stable. The core continues contracting and growing hotter, causing more and more hydrogen to fuse in regions surrounding the core. The star grows steadily

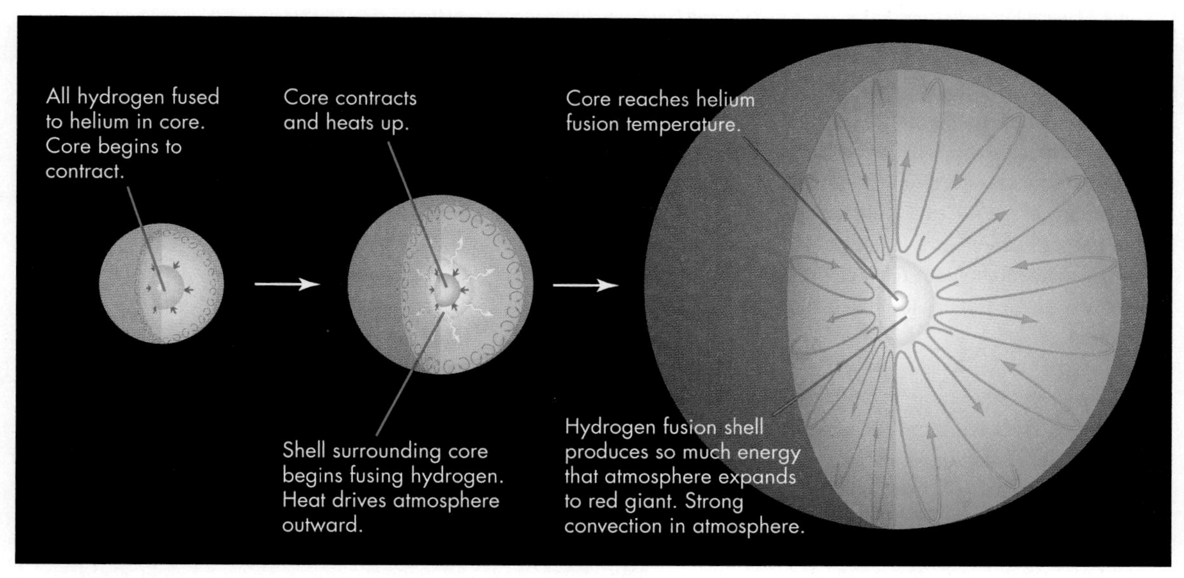

FIGURE 62.1
(left) A star's core begins to shrink as the star uses up the hydrogen in its core. This compresses and heats the core. (middle) The hotter core ignites the surrounding gas and the outer layers of the star expand, turning it into a red giant. (right) When the core temperature rises to about 100,000,000 K, helium fusion begins.

Evolution to red giant phase

Some energy is also released from gravitational potential energy as the star's core contracts.

more luminous, but the rise in energy production now does nothing to reverse the contractions. For low-mass stars like the Sun, the luminosity may increase by a factor of a thousand. In other words, it will be fusing 1000 times more hydrogen each second in its shell region during its giant phase than it fused each second in its core during its main-sequence phase.

By contrast, when high-mass stars complete hydrogen fusion in their core, they change relatively little in luminosity. They have already consumed more of their hydrogen outside the core because of greater convection activity during their main-sequence lifetimes, so the transition to shell fusion is more gradual. The shell fusion during this phase does not represent a substantial increase over their vigorous main-sequence rate of fusion. Indeed, even while on the main sequence, these massive stars already have the luminosities of supergiants (Unit 58)

Shell fusion pushes the surface layers outward for both low- and high-mass stars, and this is reflected by a significant drop in surface temperature. This drop in surface temperature is perhaps the most counterintuitive aspect of star evolution after the main sequence. The core shrinks and heats up, but the outer layers of the star expand and cool.

We can understand this unusual behavior as follows. The shell fusion region raises the temperature and pressure of the gas in layers of the star that are above it. That stronger pressure pushes the surrounding gas outward, causing the star's surface to expand. The star may grow in radius by a factor of anywhere from five to several hundred, depending on the star's mass. This expansion leaves the outer layers much farther from the source of the star's luminosity, the fusion in and around its core.

At this larger distance from the core, even if the star has grown more luminous, the intensity of radiation is reduced. This is a consequence of the inverse-square law (Unit 54.2): $B = L/4\pi d^2$, in which the brightness of light is inversely proportional to the square of the distance from the source of light. The luminosity coming from the interior of the star is thus spread out over a much larger area because the distance d from the core has grown much larger. The decreased intensity of radiation results in surface layers that are cooler, and because cooler bodies radiate more strongly at longer wavelengths, the star becomes redder. Thus the main-sequence star has evolved into a red giant or supergiant.

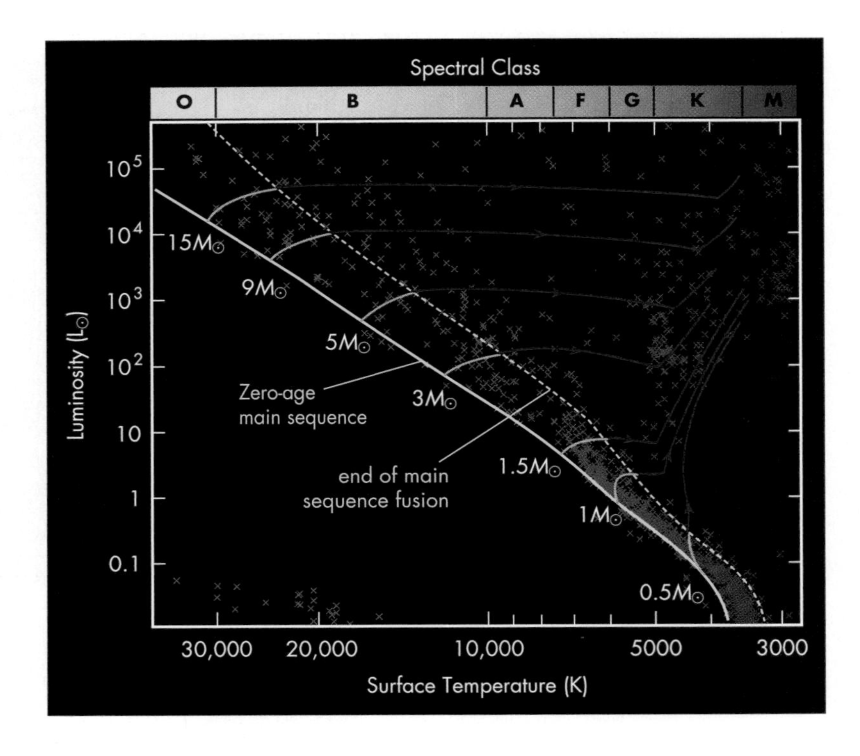

FIGURE 62.2
Post-main-sequence evolutionary tracks followed by stars with masses from 0.5 to 15 times the Sun's mass in the H-R diagram. The tracks show how stars' outward characteristics change from the end of hydrogen fusion in the core to the start of helium fusion. Very low-mass stars, 0.5 $M_\odot$ and lower, never achieve helium fusion before their cores become too dense and stop contracting.

In the H-R diagram (Figure 62.2), low-mass stars move up and to the right of the main-sequence region along a part of its evolutionary track called the **red giant branch.** A star like the Sun will grow steadily larger and more luminous over nearly a billion years as it searches for a new equilibrium, its core gradually shrinking and heating throughout this time. High-mass stars rapidly drop in surface temperature at a relatively steady luminosity, so their evolutionary tracks traverse the H-R diagram along a nearly horizontal path.

62.2 HELIUM FUSION

A red giant star has a very different structure from that of a main-sequence star. As its name implies, its surface is cool and its radius is large: in some cases more than a thousand times larger than the Sun's. But this enormous size is deceptive. Most of the star's volume is taken up by its very low-density atmosphere, with a tiny, hot, compressed core within the central 1% of its radius. Hydrogen shell fusion supplies most of the energy during the beginning of the giant phase, but if the star's core becomes hot enough, it will begin to generate energy by the fusion of helium.

Helium nuclei can fuse to form heavier elements, but the process requires a much higher temperature than hydrogen fusion does. Two nuclei fuse when they are brought close enough together for the strong force (Unit 4) to bind them. That force, which is what holds the protons and neutrons together in a nucleus, operates only over very short distances. If nuclei are more than a few diameters apart, the nuclear strong force is too weak to bond them. In fact, they will be repelled by the electrical force between the similarly charged protons, as illustrated in Figure 62.3. The electric repulsion grows as the number of protons increases. The repulsion is smaller for hydrogen with its single proton, but larger for heavier elements with more protons. It is more difficult, therefore, for helium nuclei, with their two protons, to fuse than for hydrogen atoms to do so.

Helium fusion occurs when three ^{4}He nuclei combine to make a carbon nucleus, ^{12}C. This is called the **triple alpha process.** Helium nuclei are known as

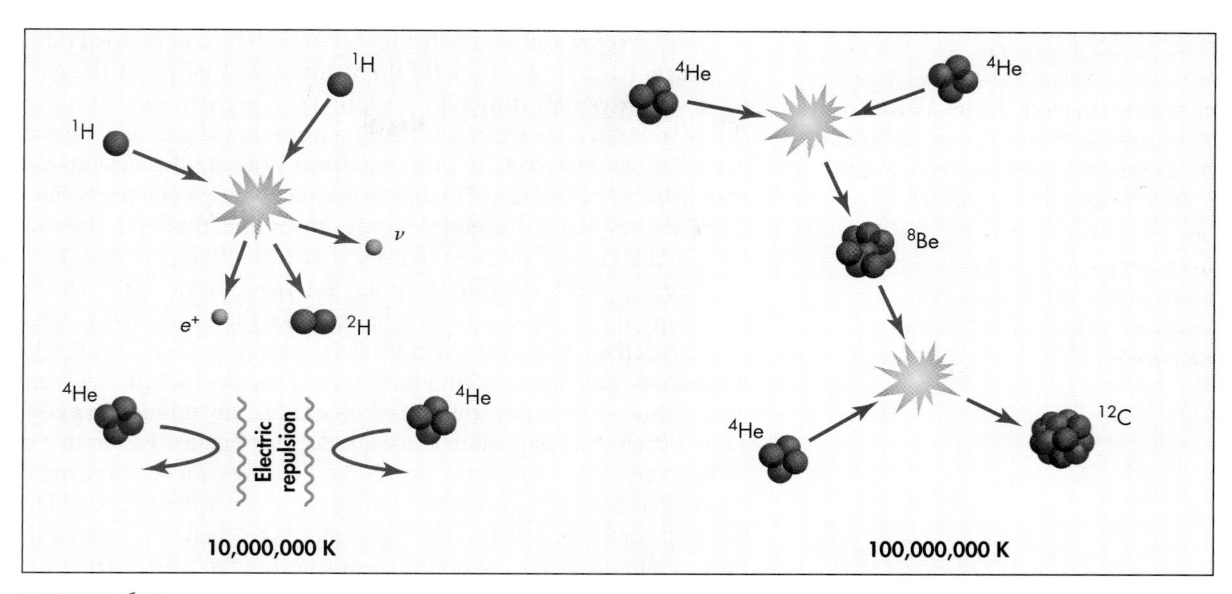

FIGURE 62.3
At 10 million K, hydrogen nuclei can fuse, but the larger electric repulsion between helium nuclei keeps them from fusing. At 100 million K, helium nuclei can fuse, making first beryllium and then carbon.

Q The triple alpha process requires two steps in rapid succession. First, two helium nuclei fuse to make a beryllium nucleus, ^{8}Be. However, this nucleus is unstable and breaks back apart into the helium nuclei in only about 2 millionths of a second, so another helium nucleus must collide with it in that short interval. If ^{8}Be were stable, how might this change stars' giant phase?

alpha particles because they were detected in some of the first nuclear radioactivity experiments, but their identity was unknown for several years, so they were given this alternative name. For the triple alpha process to become a steady source of energy, a temperature in excess of 100 million K is required. However, despite operating at such a high temperature, the triple alpha process releases much less energy than hydrogen fusion—only about one-tenth the amount of energy per kilogram of fuel as hydrogen fusion yields. Thus, although the initiation of helium fusion stabilizes the core against further contraction, the triple alpha process does not produce the majority of a giant star's luminosity.

Calculations indicate that the compression that occurs in stars more massive than about 0.5 $M_\odot$ is sufficient to reach temperatures that allow helium fusion, but the process begins very differently in low- and high-mass stars. For example, a 10-$M_\odot$ star needs to compress its core to only about 100 times its main-sequence density before helium begins to fuse, because its core is already extremely hot. A low-mass star like the Sun, however, must compress its core by about a factor of *10,000* before it becomes hot enough for helium fusion. This requires the Sun's core to reach a density of about 1 million kilograms per liter. A teaspoonful of such matter would weigh as much as a truck! As you might suppose, such densities strongly affect the properties of the gas. To see how, we need to look more closely at the nature of extremely dense gases.

62.3 ELECTRON DEGENERACY AND HELIUM FUSION IN LOW-MASS STARS

To see why gases behave differently when they are densely packed, imagine a few dozen tennis balls inside a large box. If you shake the box, the tennis balls bounce about inside the box. This is similar to the behavior of the particles in a normal gas. You can make the box smaller, and the tennis balls will still bounce—until you make the box so small that the balls are packed tightly against one another. At that point the balls no longer bounce, and they prevent you from making the box any smaller.

Gases display a similar behavior—they can be compressed until the particles are so closely packed that they fill the volume. At that point the gas becomes almost rigid and extremely difficult to compress. The particles that resist being packed so closely together are the electrons in a hot ionized gas. The behavior of subatomic particles like electrons is different from how solid balls interact. The electrons obey what is called an **exclusion principle:** No two electrons can occupy the same space if they have the same energy. Note that this rule *does* allow electrons to occupy the same volume of space if they have *different* energies.

If a gas is compressed to the point where electrons of the same energy are trying to occupy the same space, they behave like the tennis balls, resisting coming any closer together. A gas in which its electrons are packed together like this is called a **degenerate gas.** This packing limit makes the gas as "stiff" as a solid. As a result, a degenerate gas does not contract when it cools, any more than a cooled brick shrinks. This stiffness has important consequences for nuclear fusion in a degenerate gas.

For very low-mass stars of 0.5 $M_{\odot}$ or less, the compression makes the core degenerate before it grows hot enough to start helium fusion. Hydrogen fusion in the shell continues briefly after degeneracy is reached, but without the rising temperature coming from the continued contraction of the core, there is nothing to keep the surrounding gas hot enough for fusion to continue in the shell. As the core cools, hydrogen fusion slows in the shell and then halts. The star then eventually cools and dims.

The cores of stars with masses from about 0.5 $M_{\odot}$ to 2 $M_{\odot}$ get compressed to the degeneracy limit when they are close to the helium ignition temperature, and their fate is quite different. Normally the thermostatic feedback behavior of nuclear fusion in a gas prevents the nuclear reactions from becoming too violent. If the gas becomes hot, the increase in nuclear fusion heats the gas, raising its pressure so the gas expands. The expansion cools the gas and reduces the rate of nuclear fusion. This is not true when electron degeneracy pressure supports the core: In that case the pressure resisting contraction remains almost independent of the temperature.

When the core becomes degenerate, it no longer compresses easily, but compression is not completely eliminated. Normally the electrons in a gas are all in the lowest energy level possible, so a degenerate gas cannot be compressed more unless some of its electrons are raised to higher energies. Whenever the electrons in the core absorb energy, they have a different energy from their neighbors, so according to the exclusion principle they can be packed more tightly. As the core is heated by gravitational contraction and the radiation from the surrounding fusion shell, the electrons absorb some of this energy. The core shrinks, the gravity grows more intense, and the greater compression on the hydrogen fusion shell makes it hotter so it generates more energy, which continues the process—although the compression proceeds more slowly than before the electrons became degenerate.

Under these circumstances, the initiation of helium fusion can occur explosively. When the helium fusion temperature is reached, the energy released by the fusion reactions is soaked up by the electrons. This relieves some of the degeneracy, so the core shrinks. But this drives the temperature up, causing more-rapid helium fusion, which gives the electrons more energy, allowing more compression and heating. This runaway process accelerates the star's energy production explosively in what is called a **helium flash.**

During the helium flash, a star's energy production increases by many thousand times in just a few minutes. The outburst is totally hidden from our view by the star's outer layers, just as a firecracker set off under a mattress creates little visible disturbance. When the released energy heats the core enough that the gas is no longer degenerate, it stabilizes. With continued energy generation from helium fusion, the gas can expand and adjust to maintain a more stable pressure balance with the gravitational weight of the layers pressing down on it.

The star's outer layers adjust in turn—shrinking and rising in temperature even as the core expands and cools. The star's surface grows hotter and therefore

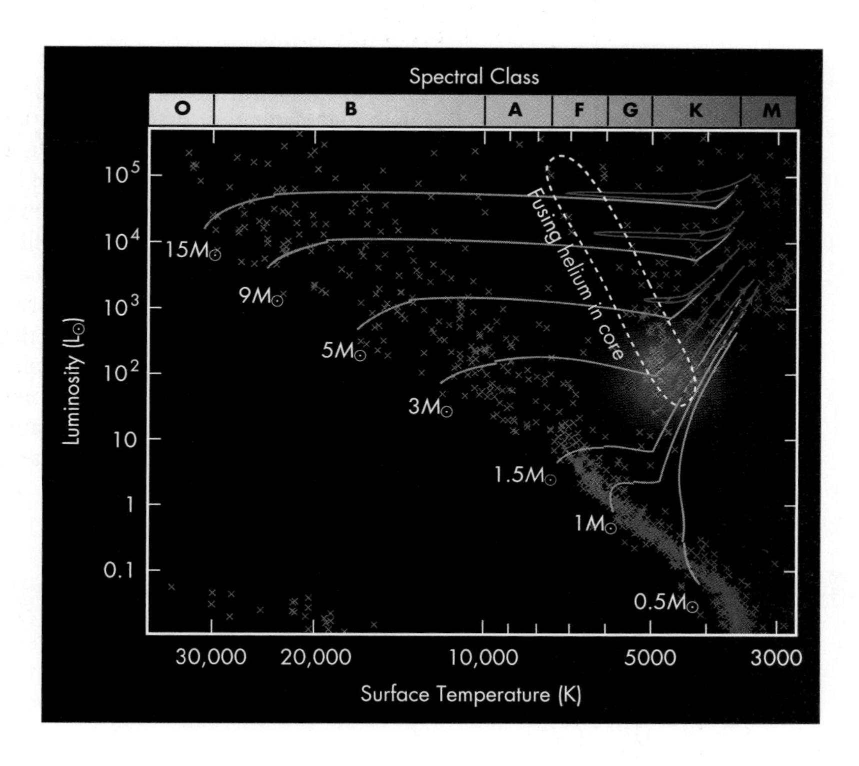

FIGURE 62.4
Evolutionary tracks (in red) of stars in the H-R diagram after the start of helium fusion. With the beginning of helium fusion in the core, stars rapidly readjust and stabilize in the region outlined by the dashed white line. After helium is used up in the core, helium begins fusing in the region surrounding the core, and stars begin climbing back up to the upper right part of the H-R diagram. Many red giants are found in the region highlighted by a red cloud because this is where the common class of stars with masses similar to the Sun are positioned while fusing helium.

changes color from red to yellow, and the star drops slightly in luminosity. In the H-R diagram (Figure 62.4) these stars shift abruptly down and to the left (dashed lines) as they adjust to helium fusion in the core.

For about 100 million years these stars become relatively stable again as they fuse helium in their cores. These stars all remain at fairly steady luminosities of about 100 $L_{\odot}$ and are similar in temperature to each other, so they show up as a "clump" or "branch" in the H-R diagram.

When the helium in the core is exhausted, the core begins contracting again, just as after the main sequence. The luminosities climb and temperatures decline, and the stars travel back up in the H-R diagram, growing to be even more luminous red giants than before. However, the cores quickly become degenerate and halt the contraction. This final bright stage marks the end of fusion within stars like the Sun.

Q When the Sun becomes a red giant, it will be about a thousand times more luminous than it is now. What effect will this have on the outer planets?

62.4 HELIUM FUSION IN HIGH-MASS STARS

When stars more massive than 2 $M_{\odot}$ exhaust the hydrogen in their cores, they leave the main sequence and shell fusion begins. Meanwhile, their surfaces expand and cool, just like in the low-mass stars. However, the cores of high-mass stars are already so hot that they ignite helium with relatively little compression, and so they do not become degenerate at this stage of their lives. After helium fusion begins, these stars also readjust, typically shrinking in size and luminosity, but they do this in a controlled way rather than an explosive helium flash.

The time it takes massive stars to adjust to begin fusing helium is only about 1% of their main-sequence lifetime. As a result, their positions in the H-R diagram shift rapidly from the main-sequence region over to the low-temperature side of the plot. After their cores are compressed sufficiently to initiate helium fusion, their luminosity generally stabilizes at a slightly higher value than when the star was on the main sequence. Luminosity is generated by both core helium fusion and hydrogen shell fusion. The surface layers of these stars all stabilize around a temperature of about 4000 K. The radii, however, may grow to a few hundred times the original radius—the larger the mass, the larger the radius. This relatively stable helium fusion phase lasts for a time 10% to 20% as long as the star's main-sequence lifetime lasted, before the helium in the core is consumed.

The large compressive forces in these stars may even raise the temperature in the core high enough, to about 200 million K, for more *alpha capture* fusion reactions to take place, providing an additional source of energy. For example, an additional ^{4}He nucleus can fuse with ^{12}C to form oxygen, ^{16}O. This tends to occur where both carbon and helium are present—at the inner boundary of the hydrogen fusion shell where newly formed helium is deposited onto the core, most of which has already been converted into carbon. Convection inside these stars can drag some of this material to the surface, where it may be expelled in winds (Unit 64). This is one of the major sources of oxygen and carbon—elements essential to life.

KEY TERMS

degenerate gas, 506
exclusion principle, 506
giant, 502
helium flash, 506
red giant branch, 504
shell fusion, 502
triple alpha process, 504

QUESTIONS FOR REVIEW

1. What makes a star move off the main sequence?
2. Where do main-sequence stars end up as they evolve?
3. Why do high- and low-mass stars evolve differently as they become giants?
4. What are the ingredients and the products of the triple alpha process?
5. What is the exclusion principle for electrons, and why is it important for the evolution of giant stars?
6. What is the helium flash triggered by?

PROBLEMS

1. When the Sun expands to be a red giant, its radius may be 100 $R_{\odot}$. What will the average density of the Sun be at that time?
2. When the Sun becomes a red giant, its core region (containing about 20% of its mass) may contract to 0.01 $R_{\odot}$. What will the density of the core be at that time?
3. By how many times its current radius would the Sun have to expand before its outer atmosphere reached the Earth?
4. If the Sun becomes a red giant that has a radius 100 times larger than its current radius and a luminosity 1000 times its current luminosity, what will its surface temperature be? (Use the Stefan-Boltzmann equation from Unit 57 and solve for T.)
5. We observe a red giant with a luminosity 127 times the Sun's and a wavelength of maximum radiation of 675 nm. What is its radius? (Use Wien's law, Unit 55, to find its temperature.)
6. Estimate the radius of the 15-$M_{\odot}$ star shown in the H-R diagrams in this unit from when it is initially a blue supergiant to when it becomes a red supergiant. (You will need to estimate luminosities and temperatures from the H-R diagram and then use the Stefan-Boltzmann equation from Unit 57 to solve for radius.)

TEST YOURSELF

1. As a star like the Sun evolves into a red giant, its core
   a. expands and cools.
   b. contracts and heats.
   c. expands and heats.
   d. turns into iron.
   e. turns into uranium.
2. High-mass stars entering the giant phase
   a. intensify greatly in luminosity.
   b. contract significantly.
   c. have most of the hydrogen left in their cores.
   d. experience a drop in their surface temperatures.
   e. All of the above
3. Helium fusion
   a. begins as soon as hydrogen is depleted in a star's core.
   b. can never happen once a core is degenerate.
   c. happens only with high-mass stars.
   d. eventually happens with all stars.
   e. None of the above.

unit

63 Variable Stars

Many stars pass through a stage sometime during their lives in which their luminosity varies. For many stars this stage occurs after they enter the red giant phase. Astronomers call stars whose luminosities change **variable stars.** All stars vary slightly in brightness—even the Sun varies by a few percentage points because of its magnetic activity cycle (Unit 51)—but this kind of variation is detectable only with careful photometric measurements. All stars also experience gradual changes in their luminosity and temperature while they are on the main sequence, and faster changes while they are giants.

Although virtually all stars vary to some degree, traditionally the stars cataloged as variables change in brightness over a decade or less. Furthermore, these changes can be seen just by looking through a telescope and comparing the variable star with neighboring stars. An example of a variable star is shown in Figure 63.1—the star Mira (*MY-rah*) in the constellation Cetus. For hundreds of years Mira has been known to regularly brighten and fade in a little less than a year.

Background Pathways
Unit 57: The Sizes of Stars 464

FIGURE 63.1
These two photographs show the star Mira in the constellation Cetus. Mira varies over a period of 332 days from being visible to the unaided eye to being about 100 times fainter.

63.1 CLASSES OF VARIABLE STARS

The first variable stars identified were stars that brightened dramatically, often from previous invisibility, before fading again. A "new star" suddenly appearing this way in 1572 was called a *nova stella* by Tycho Brahe (Unit 12). Today this is shortened to **nova.** These are associated with explosive events sometimes marking the final stages of a star's lifetime, and they may remain visible for weeks or months before fading away. Actually the nova seen in 1572 was an especially luminous type that today is called a **supernova.** Tycho Brahe's measurements of the brightening star's position relative to other stars showed that it had no parallax—demonstrating that it must belong to the celestial sphere. Tycho's discovery was revolutionary at the time, because it proved that even the highest celestial realm was changeable, running counter to beliefs of the day.

To characterize a star's variability, astronomers measure its brightness at frequent intervals and plot these against time. This results in a graph called a **light curve.** Over a hundred types of variable-star patterns have been identified, but these can be divided into a few classes. Stars like novae belong to the class of **irregular variables,** which undergo unpredictable outbursts of brightness that do not follow a repeating pattern. Another example of an irregular variable is a T Tauri star (Unit 60), which is a stage some protostars pass through. These objects probably brighten as material falls irregularly onto the forming star. Dozens of patterns have been identified in the irregular class, and these are often associated with very young or very old stars.

Some stars also vary because of periodic eclipses by other stars in orbit about them (Unit 56), but the majority of cataloged variable stars fall in the class of **pulsating variable.** These stars change in luminosity in a rhythmic pattern. Based on the shapes of their light curves, astronomers have identified more than a dozen types of regularly pulsating variable stars. For example, some stars pulsate with a

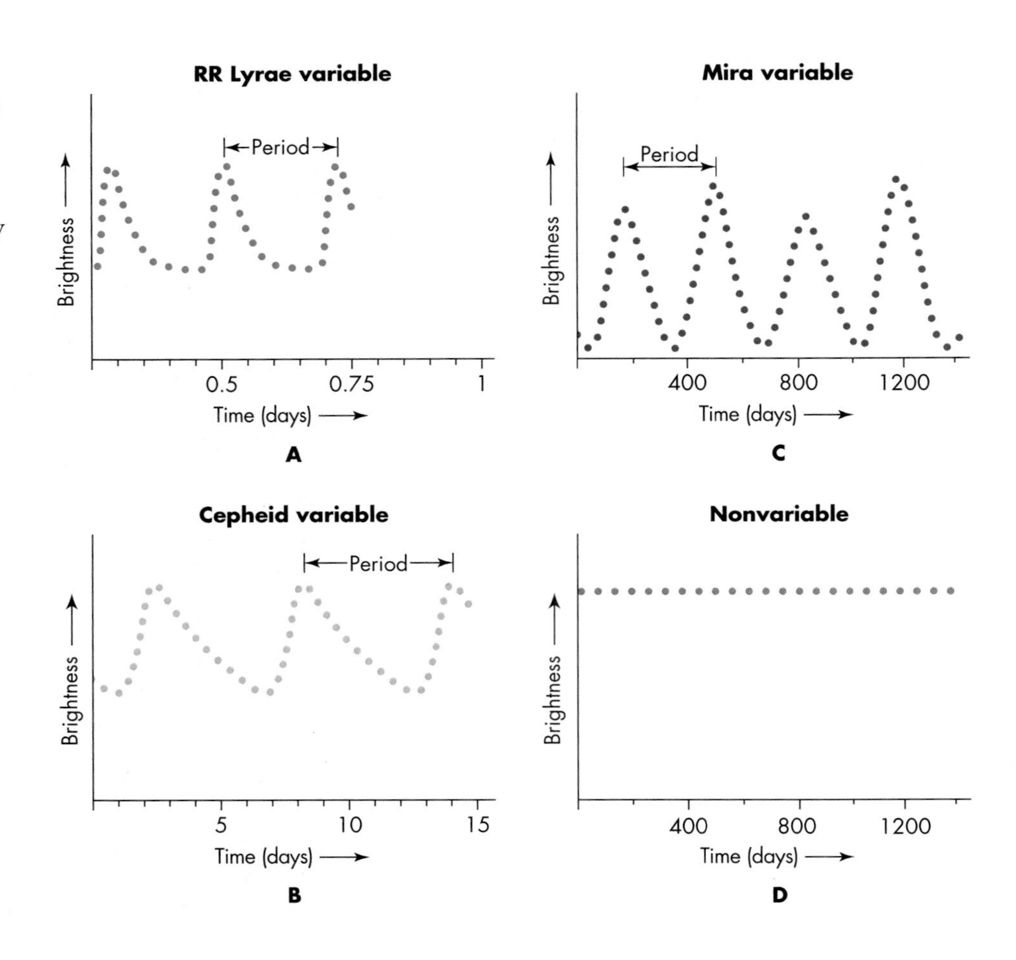

FIGURE 63.2
Schematic light curves of (A) an RR Lyrae variable; (B) a Cepheid variable; (C) a Mira variable; and (D) a nonvariable star. The timing and unique pattern of variability allow astronomers to identify these stars.

The Latin heritage of *nova* is remembered by most astronomers in the plural form *novae,* although *novas* is also acceptable as a plural. The same applies to other Latin-derived words, such as *supernovae* versus *supernovas.*

period of about half a day, while others take more than a year to complete a cycle of pulsation. The interval of time it takes the pattern of brightness to repeat is called the **period,** as shown in Figure 63.2.

When we analyze the temperatures and luminosities of pulsating variable stars, we find that these stars also vary in size, rhythmically swelling and shrinking. These changes in diameter are caused by the pressure rising below a star's surface, which makes the surface expand outward. Some classes of these stars may more than double in size over their period, but the size variations are more commonly less than 20%. Such variations in size also cause a star's luminosity to change. If the star expands, its surface area increases so that it has more area to radiate. This would make the star more luminous. However, as the gas expands, its temperature usually drops, sometimes to half its former value. This would make the star dimmer. The result of both changes is that, as the star pulsates, its luminosity changes according to the Stefan-Boltzmann law (Unit 57) in a more complex way. Stars like Mira can vary by more than a factor of 100 in luminosity, although the majority of variables exhibit a factor of two or less.

63.2 YELLOW GIANTS AND PULSATING STARS

Pulsating variable stars are important to astronomers because they can serve as "standard candles" (Unit 54) and thereby offer a way to measure distances within the Milky Way and even to other galaxies (Unit 74). When such stars are plotted on the H-R diagram according to their average luminosities and temperatures,

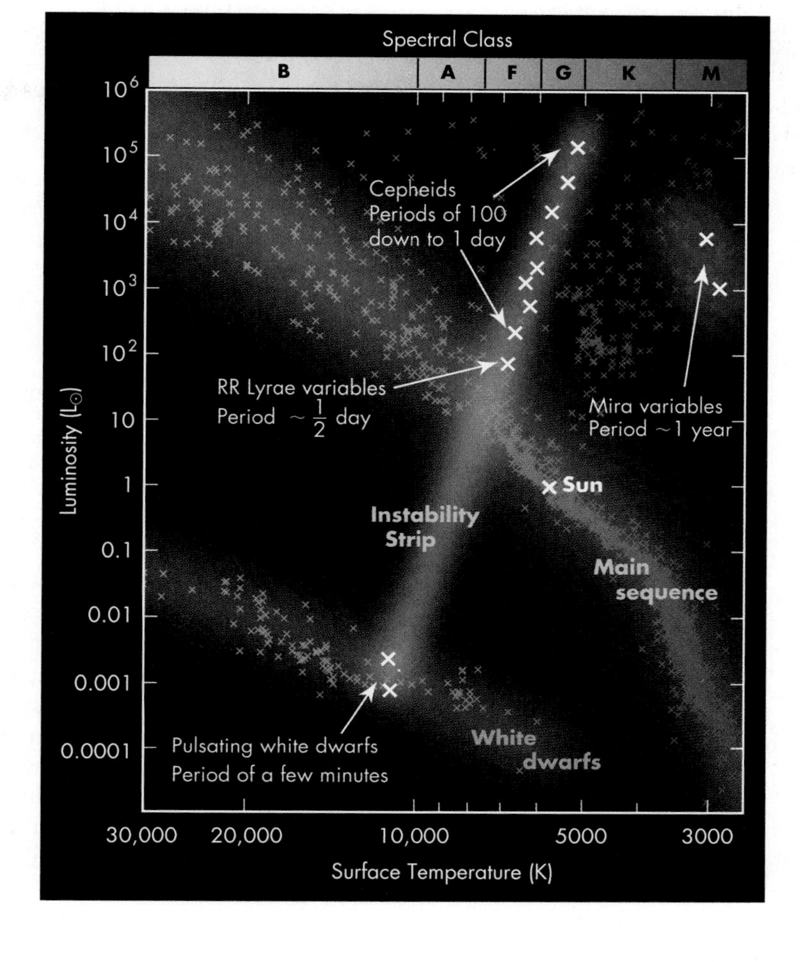

FIGURE 63.3
The locations of several types of pulsating variables in the H-R diagram. Many stars pulsate along the narrow region known as the "instability strip" in the diagram.

many of them lie within a narrow region called the **instability strip** (Figure 63.3). Because many of these stars have temperatures of about 5000 to 6000 K, they are yellow. Most pulsating variables are giants (Unit 62) and are sometimes called **yellow giants** or **yellow supergiants,** according to their luminosity.

One type of pulsating star useful for finding distances is the **RR Lyrae** (pronounced *LIE-ree*) **variable.** RR Lyrae variables have a mass comparable to the Sun's and are yellow giants with about 40 times the Sun's luminosity. Their pulsation cycle lasts about half a day. They are named for RR Lyrae, a star in the constellation Lyra, the harp, which was the first star of this type to be identified. From their known luminosity, we can apply the standard-candles method (Unit 54) to RR Lyrae variables. Knowing that their luminosity is $40 \times L_\odot = 1.6 \times 10^{28}$ watts, we can plug this value into the equation for distance ($d = \sqrt{L/4\pi B}$), and if we measure the brightness B we can find the distance.

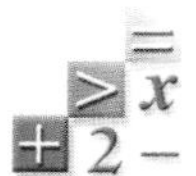

If an RR Lyrae variable is measured to have a brightness of 10^{-11} watts/m^2, we would calculate a distance of

$$d = \sqrt{L/4\pi B} = \sqrt{1.6 \times 10^{28}/10^{-11}}$$
$$= \sqrt{16 \times 10^{38}} = 4 \times 10^{19}\text{m},$$

or about 1300 parsecs.

Other pulsating stars important for finding distances are the **Cepheid** (pronounced *SEF-ee-id*) **variables.** Cepheid variables are yellow supergiants that are more massive than the Sun, typically with about 10,000 times its luminosity. They are named for the star Delta Cepheus (see Looking Up #1: Northern Circumpolar Constellations), and their periods range from about 1 to 70 days. Their luminosities vary widely, but they obey a relationship (discussed in the next section) that allows us to determine individual luminosities.

Giant stars pulsate because their atmospheres trap some of their radiated energy. This heats their outer layers, raising the pressure and making the layers expand. The expanded gas cools and the pressure drops, so gravity pulls the layers downward and recompresses them. The recompressed gas begins once more to absorb energy,

FIGURE 63.4
Schematic view of a pulsating star.

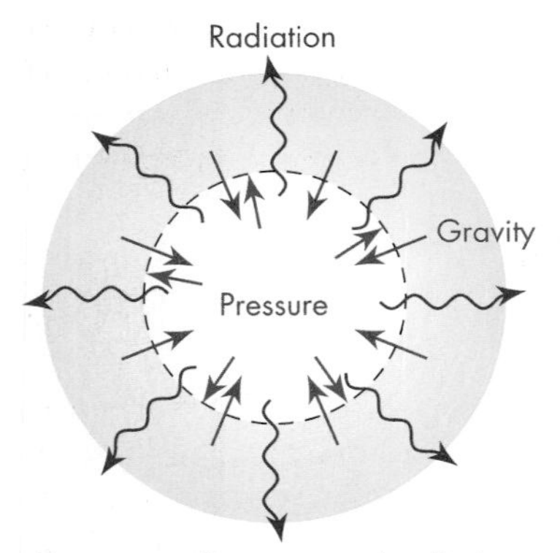

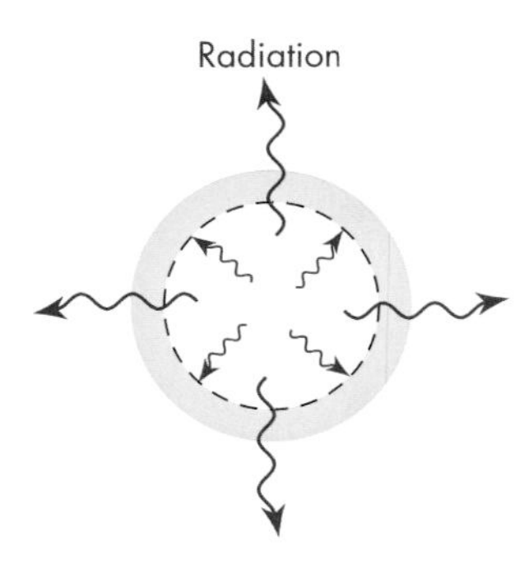

Radiation (purple wiggles) partially trapped. Pressure (blue arrows) increases, overcomes star's gravity (red arrows), and makes it "inflate" and expand.

Expansion allows trapped radiation to escape. Star cools and pressure decreases. Gravity now compresses star back to original size.

Cycle begins again.

leading to a new expansion. These stars continue alternately to trap and release the energy, and so they continue to swell and shrink, as shown in Figure 63.4.

A covered pan of boiling water behaves similarly. The lid will trap the steam so that pressure inside rises. Eventually the pressure becomes strong enough to tip the lid, and steam escapes. The pressure decreases, and the lid falls back. It again traps the steam, the pressure builds up, and the cycle is repeated. In pulsating stars, the role of steam in this process is played by the star's radiation, and the role of the lid is played by partially ionized helium gas in the star's atmosphere. This is sometimes called the **valve mechanism,** because it is like a valve that opens to relieve excessive pressure.

Q If you put a heavy lid on a pan of boiling water instead of a light lid, how would this change the frequency of the steam escaping? What does this suggest about the variability of a star with a greater depth of atmosphere that must be adjusted?

Stars in the instability strip have surface temperatures of about 5000 K. The reason these stars are susceptible to pulsation is that stars of this temperature have an unstable region, not far below their surface, where partially ionized helium is able to absorb and then release the energy flowing out from the center of the star. This unstable region can alternately push the outer layers of the star upward and release the pressure so they fall back. The instability strip shifts to lower temperatures for higher-luminosity, larger-radius giants because their surface gravity is weaker, so the partially ionized helium layer can be deeper in the star and still provide enough pressure to push the surface layers upward.

After stars complete their main-sequence phase, they move along evolutionary paths through the giant region of the H-R diagram (Unit 62). When a star crosses through the instability strip, it begins to pulsate, and it continues to pulsate until its surface temperature changes enough to remove it from the instability strip. When a low-mass star crosses the instability strip, it becomes an RR Lyrae variable. When a high-mass star crosses, being more luminous, it lies above the RR Lyrae variables in the instability strip and instead becomes a Cepheid variable.

The amount of time a given star spends in the instability strip depends on its mass. Cepheids evolve across the strip in less than 1 million years. RR Lyrae variables spend more time in the strip, perhaps a few million years. In either case, stars pulsate for only a brief portion of their lives.

Astronomers have identified several other types of pulsating variables besides Cepheids and RR Lyrae stars. For example, the ZZ Ceti stars are pulsating white dwarfs with periods as short as a few minutes. These lie along an extrapolation of the instability strip to the other side of the main sequence in the lower part of the H-R diagram (Figure 63.3).

Another kind of variable star, Mira variables, have pulsation periods of about a year. These are bright red giants that lie in the upper right portion of the H-R diagram. It is thought that the Sun will become a Mira variable close to the end of its lifetime. Mira variables are stars that have already completed both hydrogen and helium fusion in their cores, and now have shell fusion of both. It appears that the pulsation of these stars has a different mechanism, with light being blocked by molecules that form in the cool atmosphere.

63.3 THE PERIOD–LUMINOSITY RELATION

Cepheids and some other classes of pulsating variable stars obey a law that relates their luminosity to their period—the time it takes them to complete a pulsation. Observations indicate that the more luminous a Cepheid is, the slower it pulsates. This **period–luminosity relation** is plotted in Figure 63.5.

Why does a more luminous star pulsate more slowly? We can trace the reason for this back to the gravitational force for stars of different sizes. Because Cepheids all lie in the instability strip and have similar surface temperatures, a higher luminosity implies a larger radius according to the Stefan-Boltzmann law (Unit 57). Because an object's gravity weakens as the distance from the center increases, a large-radius star has weaker surface gravity than a small-radius star. So if the layers of a star with a large radius are pushed outward, the feeble gravity pulls them inward more slowly than in a small-radius star. Therefore, the pulsations in a big star take longer than in small, less luminous stars.

The surface gravity also depends on the masses of the Cepheids, but the range of masses is not large enough to be as important as the differences in radii.

The period–luminosity law is one of the astronomer's most powerful tools for determining distances. A Cepheid with a period of a few months has a luminosity of tens of thousands of solar luminosities. Such a luminous star can be seen even in galaxies millions of light-years distant, and it can be identified by its characteristic light curve. Once its period is determined, its distance can be recovered by using the standard-candle formula (Unit 54.3).

Some other classes of variables, such as Miras, also obey a period–luminosity relationship, although a different law than the one Cepheids obey. On the other hand, RR Lyrae stars have a nearly fixed luminosity, independent of their pulsation period. In short, if we can identify the class of a variable star from its light curve, we can determine the star's luminosity from the period–luminosity law appropriate to it. If the variable star is associated with another object or other stars, we can learn the distance to them as well. This makes variable stars one of the astronomer's best tools for finding distances to objects too far away to measure by parallax.

Q At optical wavelengths, Miras sometimes vary in brightness by a factor of 10,000. At infrared wavelengths the variation is much smaller. What might explain this difference?

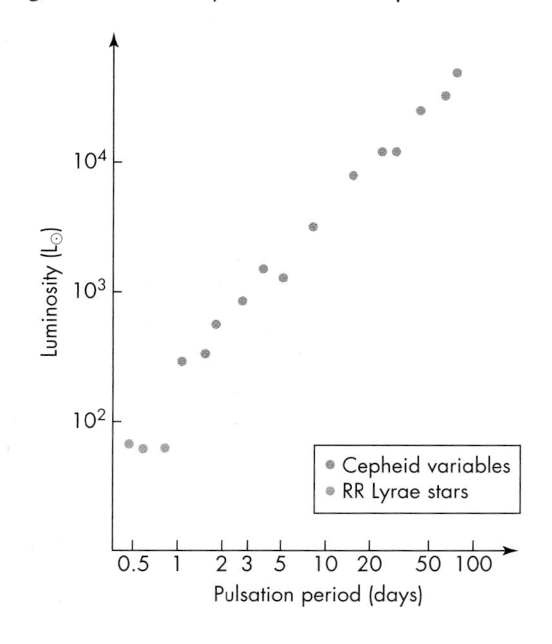

FIGURE 63.5
The period–luminosity relation. More-luminous Cepheid variables tend to pulsate more slowly. RR Lyrae variables lie along an extrapolation of this relationship.

Q If you extrapolated the Cepheid period–luminosity relationship to a luminosity like the Sun's, how fast would you expect it to vary?

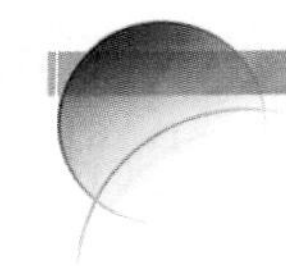

KEY TERMS

Cepheid variable, 511
instability strip, 511
irregular variable, 509
light curve, 509
nova, 509
period, 510
period–luminosity relation, 513
pulsating variable, 509
RR Lyrae variable, 511
supernova, 509
valve mechanism, 512
variable star, 509
yellow giant, 511
yellow supergiant, 511

QUESTIONS FOR REVIEW

1. What is a variable star?
2. How is a pulsating variable different from other types of variables?
3. What is meant by the "period" of a variable star?
4. What is the "instability strip"?
5. What is the difference between an RR Lyrae star and a Cepheid variable?
6. What makes the period–luminosity relationship so valuable to astronomers?

PROBLEMS

1. An interferometer measured the change in radius of one Cepheid as about 10% during its 10-day period. The star was measured to have an average radius of 40 million km. How fast was the surface of the star moving outward, on average, if it took 5 days to expand by 10%?
2. The same interferometer as in problem 1 measured the change in radius of an RR Lyrae star to be 10% during its 10-hour period. If the star was measured to have a radius of 10 million km and took half a period to expand by this amount, what would be the speed of the surface of the star as it expands?
3. A variable star varies in luminosity and temperature throughout its cycle according to the values listed in Table 63.1. The *phase* indicates the fraction of time between one maximum and the next. Graph the luminosity and temperature change, then plot the positions of the star in the H-R diagram. Connect the sequential points with a line to show the pattern of variation.
4. Using the information from Table 63.1, calculate the average luminosity of the variable star at the four distinct phases listed (the phase listed as 1.00 is back around to the same phase as 0.00). Using your average luminosity, estimate the period this variable should have if it is a Cepheid from Figure 63.5.
5. Use the information from Table 63.1 along with the Stefan-Boltzmann law to calculate the radius of the variable star at each phase. When is the star largest? When is it smallest?
6. Suppose a Cepheid has a period of 5 days. Estimate its luminosity from Figure 63.5, then determine its distance if its measured brightness is 4×10^{-12} watts/m^2.

TEST YOURSELF

1. Cepheid variables are important to astronomers because
   a. they are one of the only types of stars whose radii are known.
   b. they demonstrate the presence of spots on stars other than the Sun.
   c. the precise rate of their pulsation allows them to be used as time standards.
   d. we can use them to estimate accurate distances.
   e. All of the above.
2. The period–luminosity relation indicates that
   a. a dimmer star pulsates more slowly.
   b. a brighter star rotates more quickly
   c. a larger star rotates more quickly.
   d. a dimmer star rotates more slowly.
   e. a brighter star pulsates more slowly.
3. Variable stars change in
   a. luminosity.
   b. temperature.
   c. size.
   d. density.
   e. All of the above.

Starry Night

See the Pathways *2nd edition ARIS site for online Starry Night*^{TM} *planetarium exercises.*

TABLE 63.1 Temperature and Luminosity of a Variable Star

Phase	Luminosity	Temperature
0.00	10,000 $L_\odot$	6700 K
0.25	7,800 $L_\odot$	5900 K
0.50	6,100 $L_\odot$	5400 K
0.75	5,000 $L_\odot$	5500 K
1.00	10,000 $L_\odot$	6700 K

unit 64

Mass Loss and Death of Low-Mass Stars

A low-mass star (less massive than about 8 $M_{\odot}$) reaches the end of its life when its core can no longer contract. This prevents its core temperature from rising further, and as a result the star's nuclear fusion ends. The end does not come quietly, however.

In its terminal stages such a star ejects great quantities of matter, forming dramatically shaped and brightly colored objects (see Figure 64.1) known as **planetary nebulae.** How a star ejects its outer layers is still not fully understood, but in this unit we discuss some of the ideas proposed by astronomers to explain what happens when a low-mass star approaches the end of its life. The planetary nebula signals the death of a star. The remnant it leaves behind, a **white dwarf**, is initially hot but does not generate power, so it gradually cools and fades from view.

Background Pathways

Unit 62: Giant Stars 502

64.1 THE FATE OF STARS LIKE THE SUN

When a star like the Sun begins to fuse helium in its core, it is nearing the end of its life. As death approaches for the Sun, its evolution will speed up. This is shown in Figure 64.2, which illustrates the Sun's evolutionary track through the H-R diagram. A similar sequence is followed by other low-mass stars, although, like for all other stages of stellar evolution, the time taken to pass through the phases depends on mass—taking several times longer for lower-mass stars than for higher-mass stars.

We can summarize the Sun's life as follows. The Sun will spend approximately 10 billion years in its main-sequence phase, consuming the hydrogen in its core. It will then spend about one-tenth as long—about a billion years—as a red giant. Because its core will become degenerate (Unit 62.3) before helium fusion begins, the Sun will spend most of its red giant phase just fusing hydrogen into helium in a shell surrounding the core. However, its core will finally grow hot enough to fuse helium. The onset of helium fusion is marked by the *helium flash* (Unit 62). After that event, the Sun's structure will rapidly change from that of a red giant to that of a smaller yellow giant. The evolution during that phase will be even faster because helium yields less energy than hydrogen when it is fused, so the helium must be fused faster to produce enough energy to support the star. Helium fusion will occur in the core for about the last 100 million years of the Sun's giant phase.

When the helium in the Sun's core is consumed, the core will again be unsupported, as it was at the end of main sequence. The core, now primarily made of carbon, will contract and heat. Similar to the onset of the red giant phase, a surrounding shell of helium will begin fusing, and a hydrogen shell around that will continue fusing. The Sun's atmosphere will re-expand to an even greater diameter than it had when it first entered the red giant stage—160 times its current diameter, according to one calculation.

FIGURE 64.1
Matter ejected from a dying star can produce complexly shaped planetary nebulae.

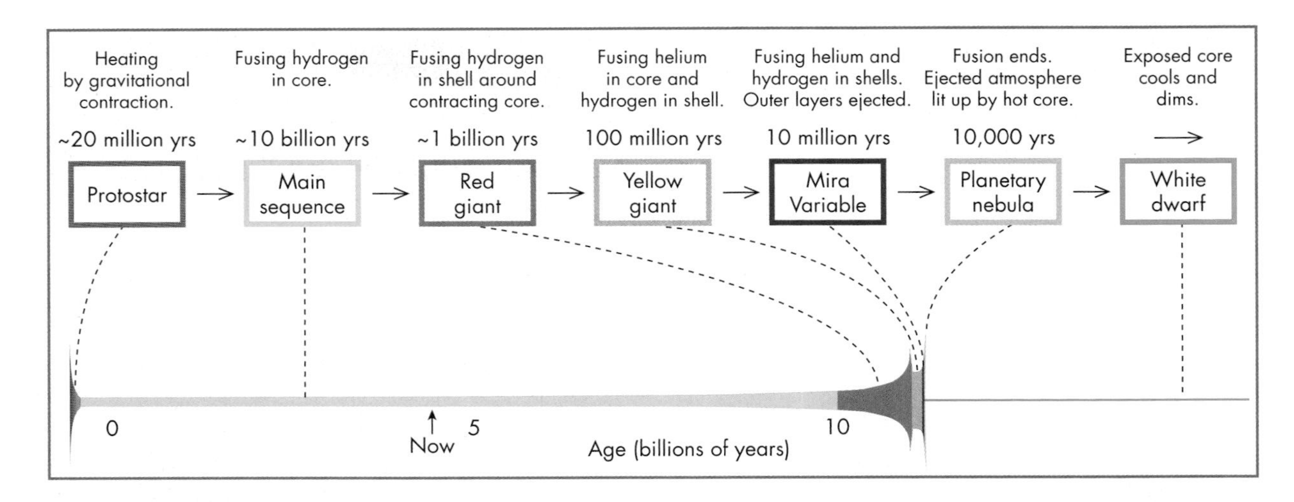

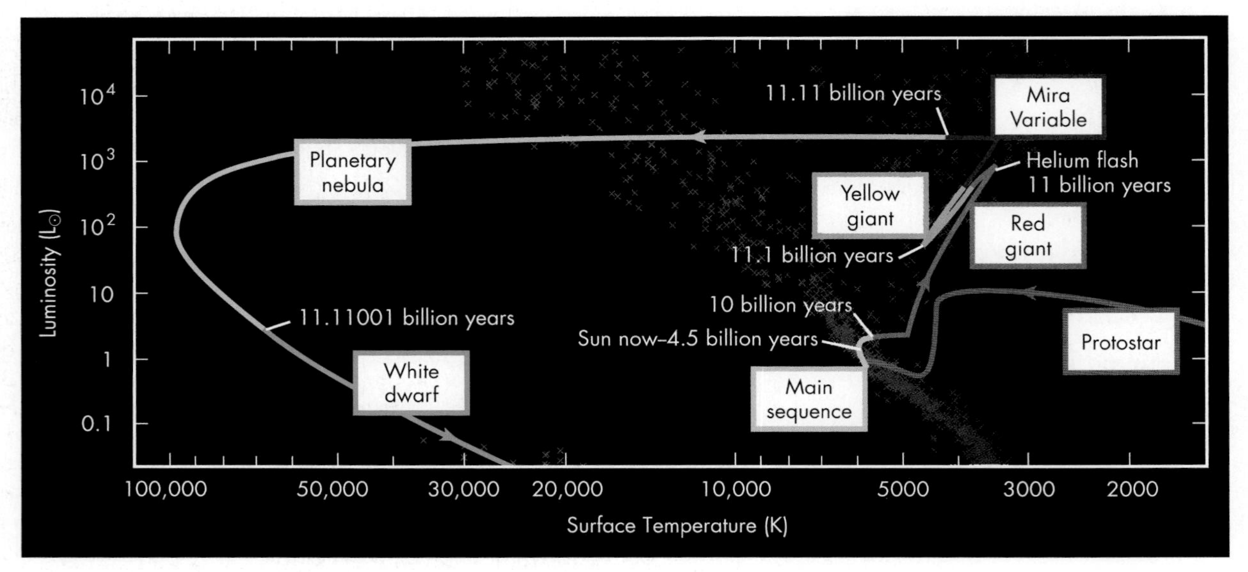

FIGURE 64.2
The evolution of a low-mass star such as the Sun plotted in an H-R diagram. The time line along the top illustrates the relative times spent in each stage and provides a sense of the changing size of the star (not to scale). The fuel powering the star in each stage is described above the time line.

The Sun will expand so much by the end of this phase that it will be classified as a bright giant with a luminosity about 2000 times greater than it has now (Unit 58). However, the core's contraction will not raise temperatures enough to begin fusing carbon. Contraction will cause the core's density to climb to more than 2 million kilograms per liter—approximately 30 tons per cubic inch! This phase has an even shorter duration of about 10 million years. By the end of this phase, electron degeneracy pressure will halt the compression, and the Sun's core will begin to cool.

During this final phase of intense fusion of hydrogen and helium in shells, the Sun will become a Mira variable (Unit 63). These luminous red giants generate very strong stellar winds, far stronger than the solar wind. Over a few million years, these winds may carry away most of the star's mass (discussed in the next section). In fact, so much material is removed from the star over time that the gravitational force squeezing the core drops substantially, greatly altering the star's evolution. This too will happen to our Sun as it nears the end of its life.

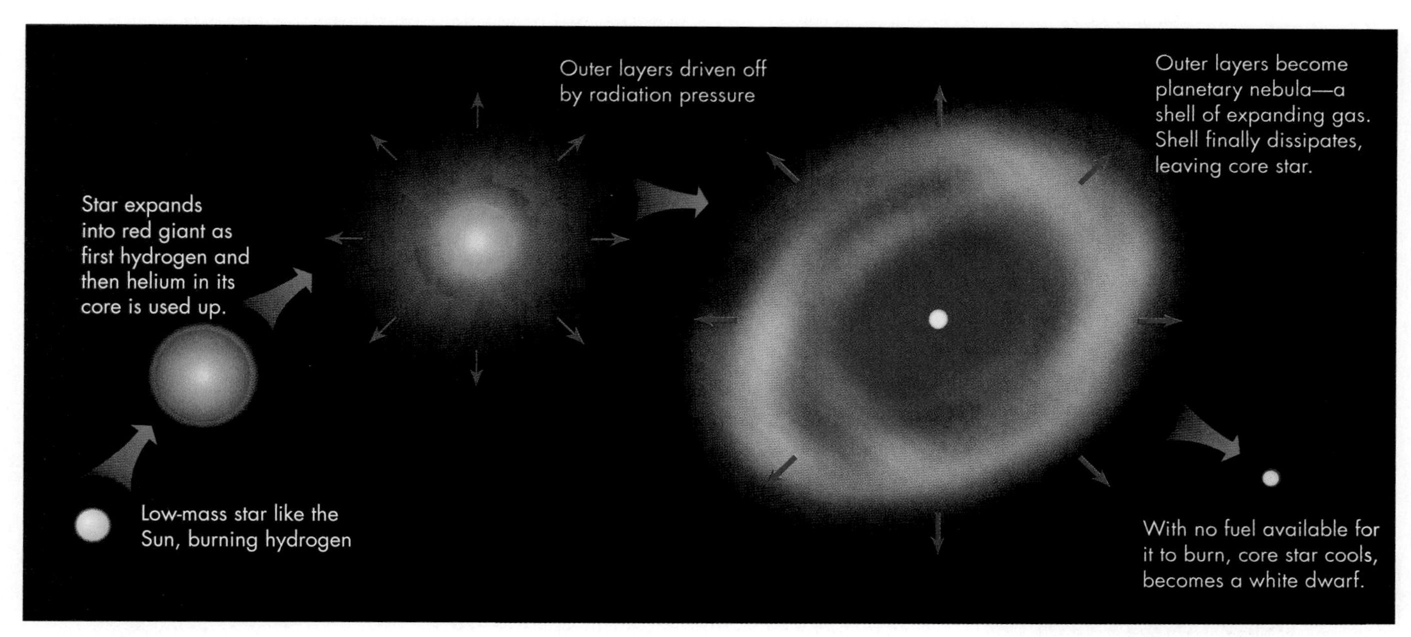

FIGURE 64.3
Final stages of the evolution of a low-mass star. After helium is used up in its core, the star expands to a very luminous red giant, gradually driving off its outer atmosphere. The ejected gas forms a shell around the star, lit up by the still-hot core, creating a planetary nebula. The remaining core of the former star becomes a white dwarf.

Although the Sun may not expand quite enough to swallow the Earth, the intense luminosity will heat the Earth's surface to be nearly molten. More and more of the Sun's outer layers will flow into space, thinning out the remainining atmosphere and exposing deeper levels of what remains of the Sun. Ultraviolet photons from the very hot interior of the Sun will now be able to travel out large distances, ionizing some of the Sun's former atmosphere—now expanded far beyond Pluto's orbit. This gas will glow briefly as a planetary nebula (Section 64.3). Finally the remainder of the core will be exposed and will gradually cool to become a white dwarf (Section 64.4). These final stages are illustrated in Figure 64.3.

When the Sun is driving off its atmosphere at the end of its Mira variable stage, it will shift rapidly all the way across the H-R diagram horizontally from right to left before dropping into the white dwarf region (Figure 64.2). Stars maintain their luminosity during the brief transition to the planetary nebula phase, but the observed color and surface temperature of the central star will rapidly climb as deeper layers are exposed. Once exposed, though, the core—now a white dwarf—steadily cools, dropping in luminosity and surface temperature.

64.2 EJECTION OF A STAR'S OUTER LAYERS

After helium is exhausted at the core of a yellow giant, the star's core contracts again. It heats to temperatures well over 100 million K, allowing helium and hydrogen shell fusion to occur in layers surrounding the core. As a result, the star's atmosphere may grow to be more than a hundred times larger than its main-sequence diameter.

At this huge size, the surface gravity of the star is quite weak, so heated gas may reach escape velocity, helping to drive a **stellar wind** of gas flowing away from the star like the solar wind (Unit 49). However, another mechanism for producing a stellar wind may be even more important for red giants. As the star swells, its outer layers cool to about 2500 K—so cool that carbon and silicon atoms condense and form **grains,** much as water in our atmosphere forms snowflakes when it cools. These are not like compact grains of sand, however, but rather are expected to be like flakes of carbon and "rock," very loosely assembled as their atoms and molecules stick together.

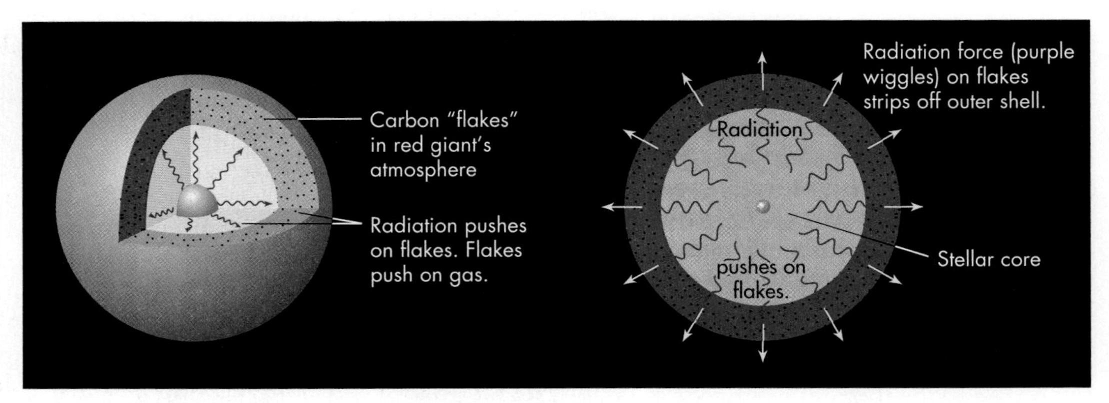

FIGURE 64.4
Diagram of how strong stellar winds are produced around luminous red giants. Solid carbon flakes form in the cool gas of the outer atmosphere. Radiation pressure pushes on the particles and drives off the gas as well.

FIGURE 64.5
Ultraviolet image of the variable red giant Mira made by the *Galex* space telescope. The star is located in the clump of gas at the right of the image. The former atmosphere of Mira has been ejected into space, and as Mira moves through interstellar gas clouds, the ejected gas trails behind it. The portion of the tail seen in this image is about 4 light-years long.

The grains do not fall into the star. As illustrated in Figure 64.4, they are pushed outward by the flood of photons pouring from the star's luminous core, just as the dust in a comet is blown by the Sun's radiation into a tail that streams away from the Sun (Unit 47). The rising grains in a red giant drag gas with them, driving it into space. This forms a strong stellar wind, in which as much as about 10^{-4} solar masses of material flow each year into space, typically at tens of kilometers per second. If the flow remained constant at this rapid rate, a star like the Sun would lose its entire mass in just 10,000 years, although the flow is not always so strong. The stellar winds eventually leave what remains of the star's dense inner regions in the middle of a growing expanse of gas—gas that used to be the star's atmosphere.

The results of this process can be seen around the red giant Mira (Figure 64.5). Using an ultraviolet space telescope called *Galex,* in 2007 astronomers discovered a long tail of gas trailing behind Mira. The star is moving through interstellar gas, and the atmosphere it has expelled is left behind like the smoke from a train. The "lumpiness" of the tail probably indicates how the outflow of atmosphere from Mira has varied over many thousands of years.

The gas expelled from stars like Mira is rich in carbon and oxygen. Although these two elements are produced by fusion reactions in and around the core, strong convection during the red giant phases carries them out into the atmosphere (Unit 62). Stellar winds from Mira-type variables are thought to be one of the main sources of carbon and oxygen in the universe.

64.3 PLANETARY NEBULAE

As the atmosphere of a star flows away into space, the pressure compressing the core of the star decreases. The core of the star is still furiously hot, and fusion of hydrogen and helium is taking place in shells around the core. However, with the declining compression, these inner regions begin to expand and cool enough that the rate of fusion declines. At the same time, the expanding atmosphere thins and becomes progressively more transparent, which allows high-energy photons from the hot core to travel farther within the remaining atmosphere. These energetic photons drive what is called a *fast wind,* pushing much of the remaining atmosphere of the star outward at high speed. The fast wind runs into the slower-moving gas of the star's earlier stellar wind, clearing out the region surrounding the star and exposing the star's core.

Because the core is so hot, its radiation is rich in ultraviolet light. Photons at these energetic wavelengths heat and ionize the inner portion of the expanding cloud of gas around it, causing it to glow. Astronomers have observed many such glowing shells around dying stars (Figure 64.6) and call them *planetary nebulae.* This term is unfortunate because planetary nebulae have nothing to do with planets. The usage survives from times when astronomers had only poor telescopes, through which planetary nebulae looked like small disks, similar to planets.

The unusual colors of planetary nebulae come from a mix of emission lines (Unit 24) of a number of elements, but particularly oxygen (green) and nitrogen (red), which have been enriched by fusion reactions during the star's late stages. These spectral lines occur in addition to the more common emission lines of hydrogen that give rise to the pink-colored glow from many ionized gas clouds (the Balmer lines, Unit 24). Initially astronomers had difficulty identifying many of the spectral lines in planetary nebulae. The oxygen and nitrogen lines are sometimes called *forbidden lines* because they do not normally occur under laboratory conditions. Even low-density gases studied in a lab are much denser than the gases in a planetary nebula. When the gas is very diffuse, uncommon electron energy transitions can occur in the long time between atoms colliding with each other.

For many years, astronomers thought that the typical planetary nebula was ejected uniformly in all directions, so that it formed a huge "bubble" around the star's core, as you might deduce from Figure 64.6A. Astronomers knew of oddly shaped planetary nebulae but thought them to be the exception. This view has changed markedly over the last decade, especially now that more detailed images (such as Figure 64.6C and D, made with the Hubble Space Telescope) are available. Most of the planetary nebulae revealed by these photos show that the shell is not

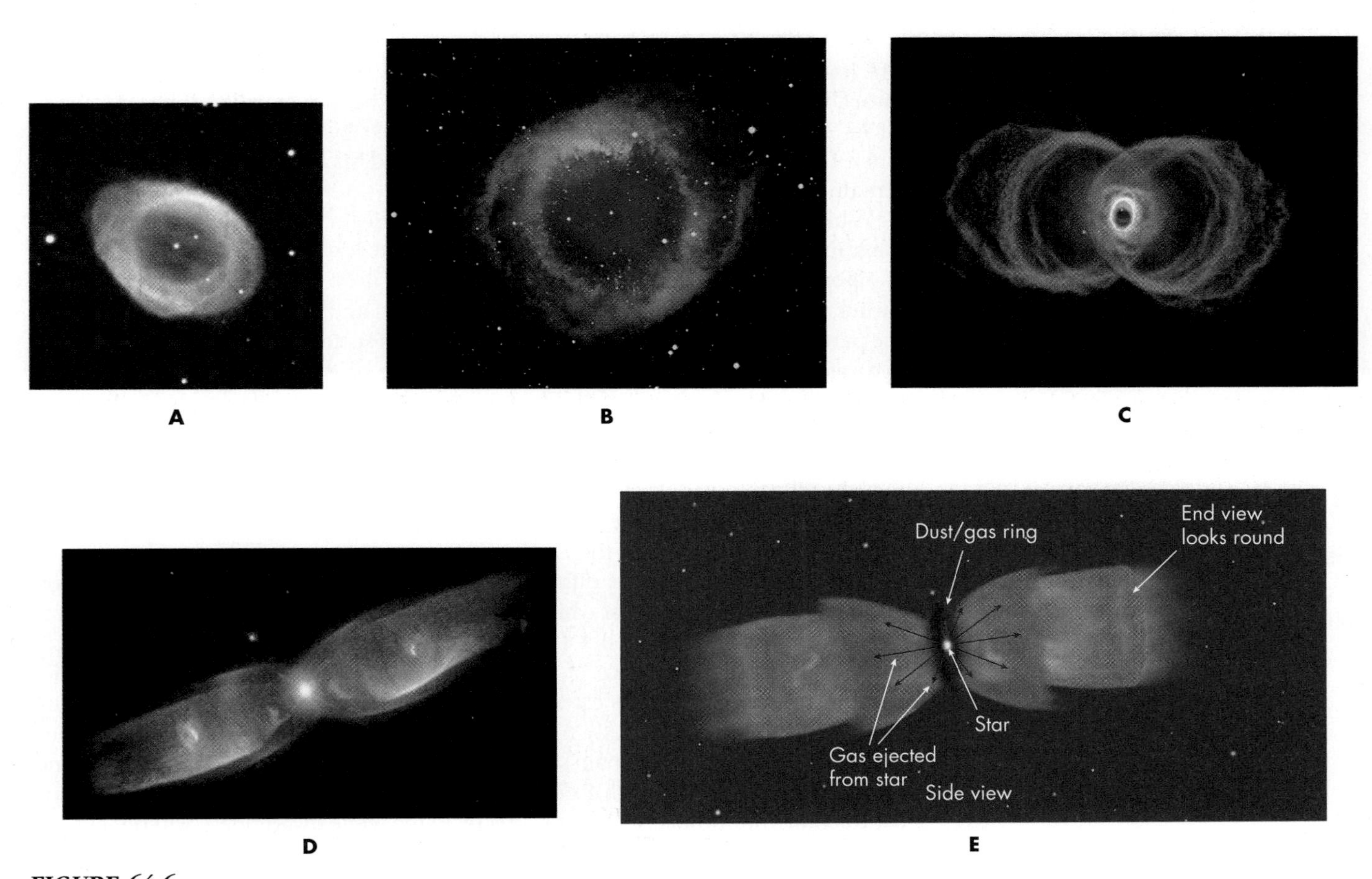

FIGURE 64.6
Pictures of several planetary nebulae. (A) The Ring Nebula. (B) The Helix Nebula. (C) The Hourglass Nebula. (D) The Butterfly Nebula. Notice the central star in each. Other stars that look as if they are inside the shell are foreground or background stars. (E) Sketch of a planetary nebula.

Q Take a clear, cylindrical glass and shine a light through it so it casts a shadow. What kinds of shapes can you make with the shadow?

spherical, but that instead the gas has been ejected primarily in two directions. Such ejection creates not a bubble but two oppositely directed cones, as sketched in Figure 64.6E. Why this happens is not entirely clear, although a ring of dust and gas in orbit around the star—perhaps part of a planetary system—may physically block the ejection in those directions. Magnetic fields in the stars probably also play a role, much as they are thought to affect the bipolar flow seen ejected from some protostars (Unit 60). Some of the more complex geometries may be caused as the stellar wind interacts with a companion star. Spherically shaped planetary nebula, perhaps the result of isolation from other bodies, may be the exception, because we suspect that few stars are truly isolated.

Planetary nebulae eventually grow so big and diffuse that light from the central star no longer ionizes the gas much, and the nebula fades to invisibility. This occurs on a very short time scale by the standards of stellar evolution. In a mere 15,000 years, gas traveling at 10 kilometers per second in the expanding envelope around the star will have expanded to a light-year in diameter—about the largest size seen among planetary nebulae. We can directly detect this expansion in less than a human lifetime by careful comparison of pictures made several decades apart. The expanding gas mingles with the general interstellar gas, and the cycle of a star's life is complete: gas to star and back to gas. Only the hot core of the star remains: a tiny, glowing ball that gradually cools.

64.4 WHITE DWARFS

The hot core of the former star at the center of a planetary nebula becomes a white dwarf. A white dwarf's dim light does not come from fusion, for it has exhausted all its fuel supplies. Rather, the light is produced by the residual heat left from when the star was fusing hydrogen and helium. This light slowly drains away the remaining heat from the star's interior.

White dwarfs are tiny but hot objects that lie in the lower part of the H-R diagram, as illustrated in Figure 64.2. A white dwarf typically has the same composition as the core of the star from which it evolved. As we have discussed earlier, such cores are composed mainly of carbon and oxygen, the end products of the parent star's nuclear fusion. The white dwarf's surface is a very hot, thin layer of hydrogen and helium, but its carbon-oxygen core has too little mass, and therefore too little gravity, to contract and heat itself to the ignition temperature of carbon. So, unable to fuse any further, the star's "corpse" gradually cools and grows dim.

Q It is estimated that the Sun will have lost about half of its mass by the time it becomes a white dwarf. How will this mass loss affect the orbits of the planets that survive its red giant phases?

The stars that reach this final stage may begin with masses up to 8 $M_{\odot}$, but with the mass lost from stellar winds during the red giant phase, the remaining mass is found to be less than 1.4 $M_{\odot}$. The diameter of a white dwarf is about a hundredth of the Sun's—roughly the same as the Earth's. Their tiny size gives white dwarfs so small a surface area that they are very dim despite having hot surface temperatures up to about 100,000 K. Even the brightest white dwarfs are about 100 times fainter than the Sun.

Very low-mass stars (about 0.5 $M_{\odot}$ or less) become white dwarfs following a somewhat different sequence of events than other low-mass stars. The convection currents in these small stars extend throughout the stars' interior, mixing their core gas and outer layers (Unit 61.2). As a result, there is little or no hydrogen left anywhere in the star at the end of the main sequence phase. Moreover, their weak gravity cannot compress their cores enough to reach helium's fusion temperature. After some core contraction, they probably progress directly to the white dwarf phase without ejecting a planetary nebula. Because very low-mass stars evolve so slowly, however, they require longer than the current age of the universe to reach this point in their evolution.

What happens to a white dwarf as time passes? Although it is initially very hot (it was, after all, the core of a star), it has no fuel to fuse to stay hot. So, like a dying ember, it simply cools. Astronomers calculate that it takes about 10 million years for the temperature of a white dwarf to drop to 20,000 K. Subsequent cooling becomes progressively slower, requiring billions of years to drop to the current temperature of the Sun. Some astronomers are hunting for the dimmest of these stars as a way to test theories about the age of the universe. Eventually white dwarfs evolve toward what astronomers call **black dwarfs**—the dead, dark corpses of former stars. In the remote future, our Galaxy may contain tens of billions of these former stars, darkly orbiting their common center of mass.

KEY TERMS

QUESTIONS FOR REVIEW

1. What phases will the Sun pass through in the future?
2. Why do red giants have strong winds?
3. What is a planetary nebula?
4. What is left when a planetary nebula dissipates?
5. Why do white dwarfs take so long to cool?
6. What is the energy source for a white dwarf?
7. What is a black dwarf?

PROBLEMS

1. As a dying star moves across the H-R diagram during its planetary nebula ejection phase, it maintains an almost constant luminosity as successively deeper and hotter layers inside the star are exposed. If it begins in the upper right part of the H-R diagram as a 3000-K, 100-$R_{\odot}$ red giant, use the Stefan-Boltzmann law (Unit 57.3) to calculate the central star's radius when its surface temperature is
   a. 6000 K. b. 30,000 K. c. 100,000 K.
2. How long will it take a planetary nebula shell moving at 20 km/sec to expand to a radius of one-fourth of a light-year?
3. The tail of gas behind the star Mira spans about 13 ly in total. The star's proper motion indicates that it is moving at 130 km/sec. How long ago was Mira at the position of the start of the tail? Give your answer in years.
4. How does the luminosity of an old, cool white dwarf (20,000 K) compare to that of a hot, young white dwarf (100,000 K)? What would be the wavelength of peak radiation of the two white dwarfs?
5. Estimate the average density (in kg/L) of a planetary nebula, assuming that a star like the Sun loses half its mass to a spherical nebula that expands to a light-year in diameter.
6. Using the result of problem 5 and the fact that the mass of a hydrogen atom is 1.7×10^{-27} kg, calculate the average number of hydrogen atoms in a planetary nebula per liter and compare it to the average density of the interstellar medium given at the beginning of Unit 72. Comment on your comparison—does it make sense?

TEST YOURSELF

1. Which of the following sequences correctly describes the evolution of the Sun from birth to death?
   a. White dwarf, red giant, main-sequence, protostar
   b. Red giant, main-sequence, white dwarf, protostar
   c. Protostar, red giant, main-sequence, white dwarf
   d. Protostar, main-sequence, white dwarf, red giant
   e. Protostar, main-sequence, red giant, white dwarf
2. A planetary nebula is
   a. another term for the disk of gas around a young star that forms planets.
   b. the cloud from which protostars form.
   c. a shell of gas ejected from a star late in its life.
   d. what is left when a white dwarf star explodes as a supernova.
   e. the remnants of the explosion created by the collapse of the iron core in a massive star.
3. How does a 4 solar mass star end up as a 1 solar mass white dwarf?
   a. A stellar wind carries away most of the star's mass when it is a red giant.
   b. Helium fusion turns most of the star's mass into energy.
   c. When an object contracts, its mass turns into density instead.
   d. Neighboring stars and planets pull the star apart when it is a giant.
   e. It spins so much faster as it contracts that much of the gas is thrown off.

See the Pathways *2nd edition ARIS site for online Starry Night™ planetarium exercises.*

unit

65 Exploding White Dwarfs

Background Pathways

Unit 64: Mass Loss and Death of Low-Mass Stars 515

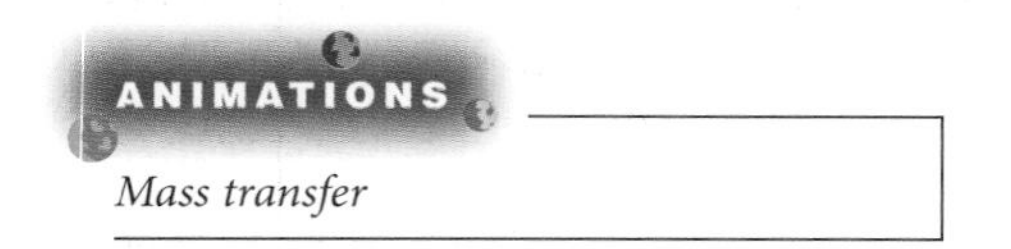

Mass transfer

After a low-mass star has run out of nuclear fuel, the stellar remnant left behind is a **white dwarf.** This fate awaits the Sun about 6 billion years from now (Unit 64). Gravity will crush its remains into an object about 100 times smaller than the present Sun, or roughly the size of the Earth. Initially hot, the white dwarf will gradually cool and fade.

Because most stars eventually reach the white dwarf stage, the Galaxy is littered with billions of these tiny dead stars. They shine only with heat left over from their previous phase. Their matter, too, is unusual. In crushing a white dwarf to its small dimensions, gravity squeezes the matter to densities far higher than anything we can create in Earth laboratories. For example, a piece of white dwarf material the size of an ice cube would weigh about 20 tons.

Most white dwarfs are extremely dim. None are visible to the naked eye, and most are challenging to detect even telescopically. Over time, isolated white dwarfs cool off and disappear from sight to become *black dwarfs*. However, some white dwarfs have an opportunity for a new life—and a spectacular death—if they accrete fresh matter.

65.1 NOVAE

A white dwarf can accrete matter when it has a nearby companion star. This is a possibility for a fairly large number of white dwarfs because many stars begin life in multiple-star systems (Unit 56). For example, the nearest white dwarf, Sirius B, orbits the brightest star in our sky, Sirius A. Stellar winds from Sirius A may deposit material on the white dwarf in a process astronomers call **mass transfer.**

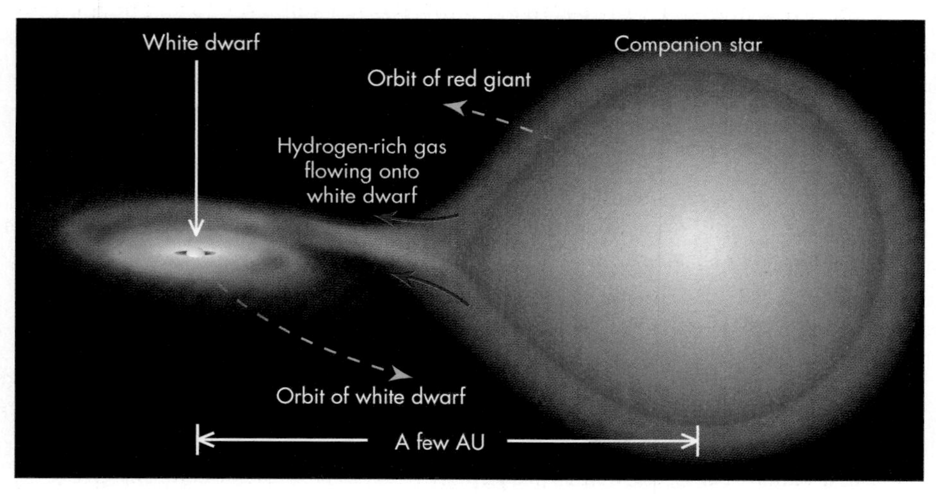

FIGURE 65.1
A white dwarf accreting (pulling in) mass from a binary star companion, which in this case is shown as a red giant. The typical separation of the stars is a few AU.

Mass transfer can grow strong if the white dwarf's companion enters the red giant phase, as illustrated in Figure 65.1. As the atmosphere expands, it can eventually expand out to a distance where the gravitational pull from the white dwarf exceeds the pull of the red giant itself. The red giant's atmosphere can grow to fill a volume called the **Roche** (pronounced *rohsh*) **lobe** before it starts "spilling" onto the companion star. This teardrop-shaped volume is named after the French astronomer Edouard Roche, who calculated the gravitational effects in a rotating binary system. The process of sweeping up additional matter tends to add a drag

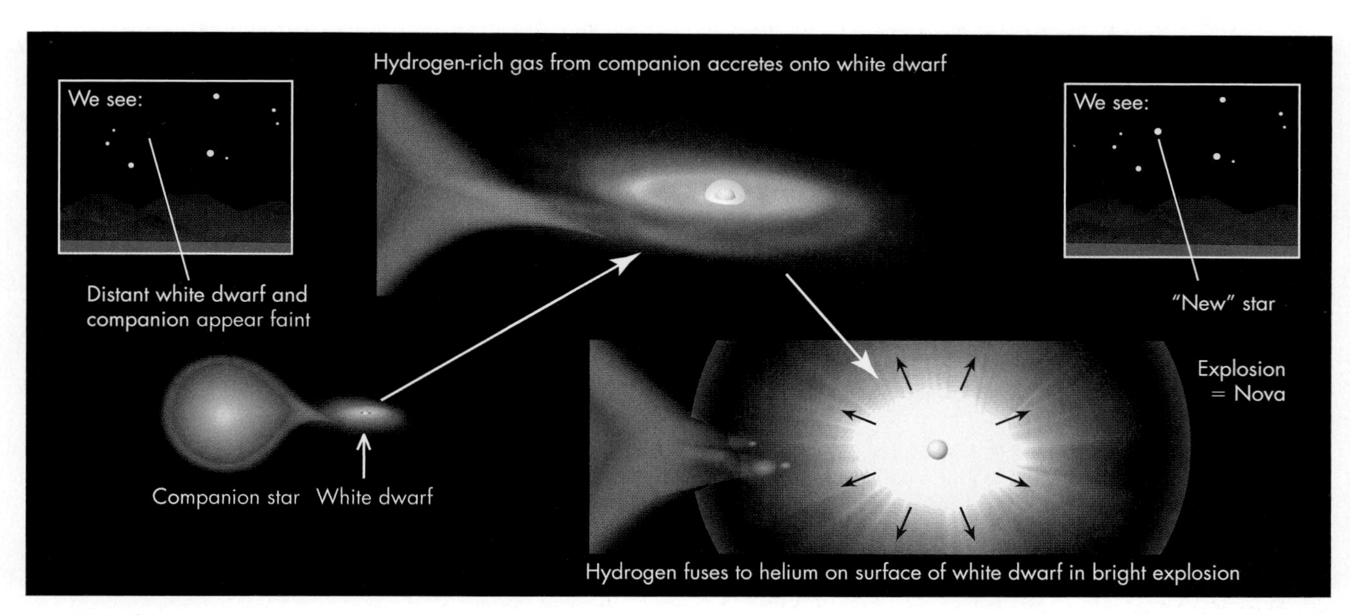

FIGURE 65.2
A nova explosion. The hydrogen that falls onto the surface of a white dwarf from its companion may suddenly fuse into helium. For several weeks the explosion typically generates a luminosity hundreds of thousands times greater than that of the Sun.

Q In what ways are the disks around forming stars similar to and different from white dwarf accretion disks?

on the white dwarf's orbit, gradually drawing the two stars closer together and further increasing the rate of transfer.

Coming from the companion's outer layers, the gas deposited on the white dwarf is generally rich in hydrogen. This gas forms a layer on the white dwarf's surface, where gravity compresses and thereby heats it. Additional heating occurs as more matter falls onto the star's surface. Such matter gains a great deal of kinetic energy as it falls and releases that energy as heat when it strikes the star's surface (Unit 20).

Most of the gas falling toward a white dwarf from a companion star spirals around the white dwarf before it reaches the white dwarf's surface. Astronomers call this spiraling gas an **accretion disk.** Gas in the accretion disk orbits the white dwarf rapidly, and collisions occur violently, strongly heating the gas as it spirals in toward the star's surface.

Gas accumulates on the white dwarf's surface, and it may eventually grow hot and dense enough to begin fusing hydrogen into helium. Even if the fusion begins only in a small region initially, the heat released will trigger fusion in surrounding areas. As a result, more and more of the hydrogen ignites in a **thermonuclear runaway,** a detonation that blasts some of the gas into space. This forms an expanding shell of hot gas, as Figure 65.2 illustrates. This shell of gas radiates far more energy than the white dwarf itself. In fact, the luminosity may grow to 100,000 times its previous level, so that the explosions may be visible to the naked eye.

When early astronomers saw such an event, they called it a *nova stella,* from the Latin for "new star," because the explosion would make a bright point of light appear in the sky where no star was previously visible (see also Unit 63 on variable stars). Today we shorten *nova stella* to **nova.** Modern observations allow us to observe the expanding shell of material blasted outward as a nova by the white dwarf (Figure 65.3).

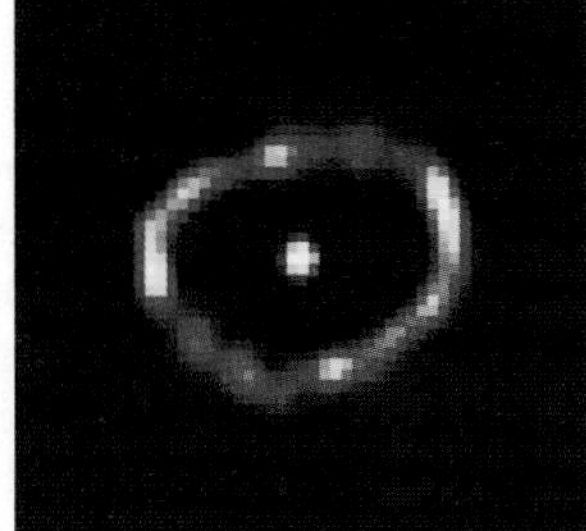

FIGURE 65.3
Two images of a nova explosion taken 7½ months apart using the Hubble Space Telescope. They show a shell of gas expanding away from a white dwarf.

Novae (plural form, pronounced *NO-vee*) may occur repeatedly for the same white dwarf, depending on the rate at which material is transferred from the companion to the white dwarf. **Recurrent novae** generally occur at intervals of a decade to a century, and it is possible that many novae that have been recorded only once will repeat in the future. The timing depends on how rapidly material accumulates on the surface, whether it has time to cool, and how much the rate of mass transfer fluctuates—in other words, many factors that are difficult to predict.

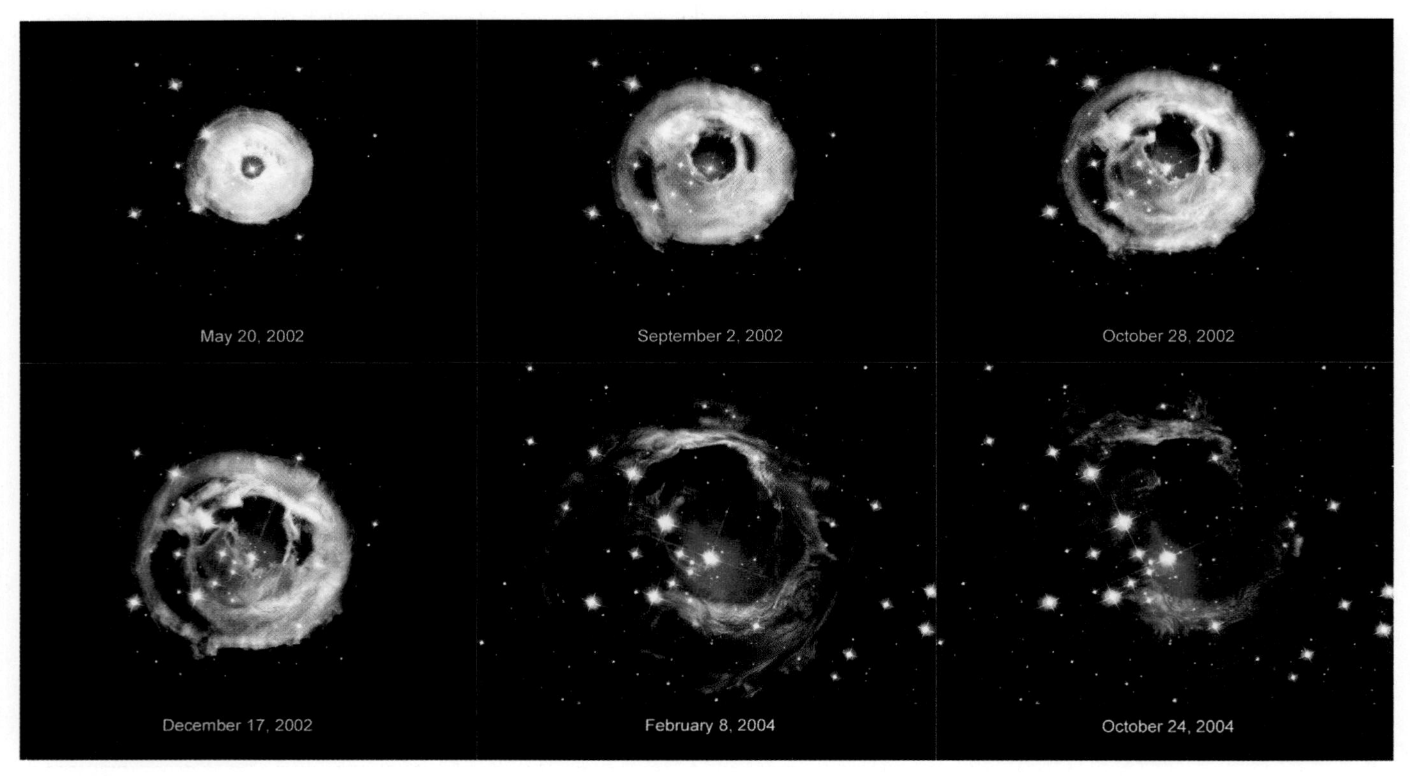

FIGURE 65.4
Hubble Space Telescope images of the nova V838 Mon. This sequence of images does *not* show an expanding gas cloud. Instead, we are seeing the expanding "light echo" off of material surrounding a star that underwent an unusual type of nova. As the flash of light from the nova travels outward, it illuminates more-distant parts of the cloud of gas and dust that surrounds the star.

Not all novae necessarily explode by this same mechanism. The exact cause of a nova in the constellation Monoceros (the unicorn) first seen in 2002 is still debated by astronomers. It initially appeared to be a fairly ordinary nova, but exceptionally luminous—becoming about a million times more luminous than the Sun. The initial burst of light was then followed by two more brightenings that did not resemble a standard nova. Astronomers have offered a variety of hypotheses: a series of planets swallowed by an expanding red giant; a collision and merger of two stars; or perhaps a more standard nova that triggered some further explosions in the star or its companion. One of the most remarkable aspects of this nova is that the intense flash of light is illuminating the surroundings of the star through a "light echo." As the flash of light travels outward, it lights up more and more distant regions surrounding the star (Figure 65.4), providing a detailed glimpse of the gas and dust that may have been ejected during the star's red giant phase.

65.2 THE CHANDRASEKHAR LIMIT

Despite the ability of a nova to blast away some of the material accumulated from a companion star, a white dwarf gradually increases its mass through mass transfer. This gradual accumulation of mass may have dire consequences for the white dwarf.

Because white dwarfs are very compact and have no fuel supply, their structure differs significantly from that of ordinary stars. Although they are in hydrostatic equilibrium, with pressure balancing gravity, their pressure arises from **electron degeneracy**, the peculiar interaction that prevents electrons of the same energy from occupying the same volume (Unit 62.3). This gives white dwarfs an odd property: Adding material to them makes them shrink. Even more crucial, however, white dwarfs must maintain a mass below a critical limit, or they collapse. To see why, we must consider conditions in a white dwarf's interior.

Q Some astronomers have suggested that cooled white dwarfs are made of diamond. If we could travel to such a white dwarf, why might it be impractical to mine it?

White dwarfs are very dense, having formed from the dense cores of their parent stars. If we divide the mass of a white dwarf by its volume, we find that its density is about 10^6 kilograms per liter—one ton per cubic centimeter (about 16 tons per cubic inch). This high density packs the star's particles so closely that they are separated by less than the normal radius of an electron orbit. Even as the white dwarf cools, then, the electrons cannot drop into lower-energy orbits around their atomic nuclei, as we would normally expect when matter cools.

If mass is added to a white dwarf, the extra mass increases the star's gravity. This increases the white dwarf's gravitational potential energy and drives some electrons into higher energy levels. The electrons that shift into higher energy levels can now overlap with neighboring electrons, allowing the white dwarf to compress into a slightly smaller volume. When the white dwarf's mass grows large enough, however, any further additional mass will cause the white dwarf's gravity to increase so much that it will contract more and more in a runaway process. The mass at which this occurs was determined by theoretical calculations made in 1931 by the young Indian astrophysicist Subrahmanyan Chandrasekhar (pronounced *Chahn-drah-say-car*). Chandrasekhar won the Nobel Prize for physics in 1983 for these calculations and related work. The limiting mass of a white dwarf is now called the **Chandrasekhar limit** in his honor.

For stars made of a helium-carbon mix—the expected composition of many white dwarfs—the Chandrasekhar limiting mass is about 1.4 $M_{\odot}$. If a white dwarf's mass exceeds this limit, it begins to contract uncontrollably—until other forces come into play. Chandrasekhar's work has been confirmed observationally by the fact that all white dwarfs for which we can measure the mass (those in binary systems) always have masses below the limit he determined theoretically.

65.3 SUPERNOVAE OF TYPE IA

As a white dwarf accumulates mass from a companion star, it may eventually approach the Chandrasekhar limit. When the white dwarf exceeds the limit, the white dwarf collapses in on itself. The resulting compression of the matter drives the temperature up, rapidly climbing to several billion Kelvin.

A star becomes a white dwarf because it did not originally have enough mass to compress and heat its core to temperatures that would allow carbon or heavier elements to fuse. It is therefore full of fuel. Carbon and oxygen in the collapsing white dwarf rapidly reach their ignition temperature and begin nuclear fusion. Carbon and oxygen can fuse to form silicon either directly ($^{12}C + {}^{16}O \rightarrow {}^{28}Si$) or through a chain of reactions (as in the core of a high-mass star; see Unit 66.1). Much of the silicon in turn fuses into nickel ($^{28}Si + {}^{28}Si \rightarrow {}^{56}Ni$). The energy released in these fusion reactions is enough to blow the entire white dwarf apart, resulting in a **Type Ia supernova** (Figure 65.5). The isotope of nickel, sprayed into space by the

Q: Are there any circumstances under which the Sun could become a nova or supernova?

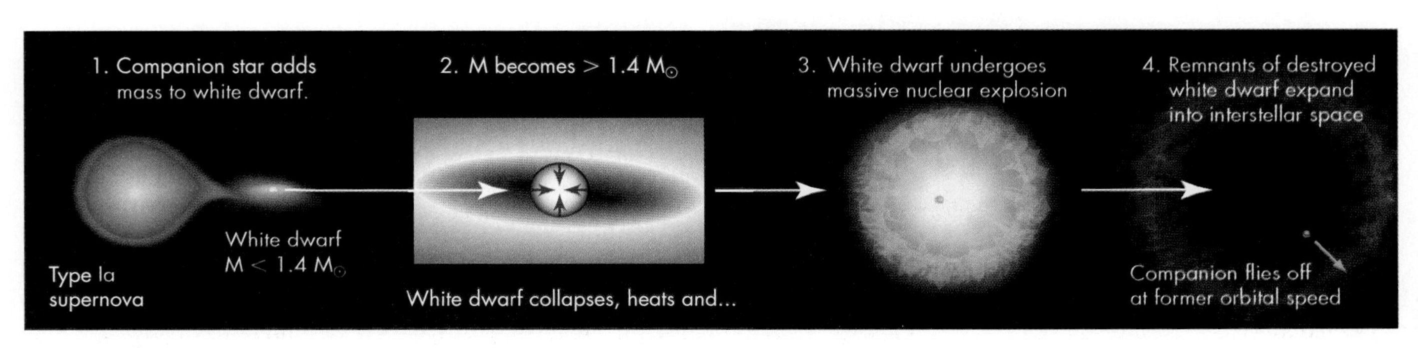

FIGURE 65.5
A white dwarf exploding as a Type Ia supernova. If too much hydrogen from a companion accumulates on the white dwarf, it may raise its mass to a value above the Chandrasekhar limit (about 1.4 $M_{\odot}$). A white dwarf more massive than this limit collapses, heats from the resulting compression, and explodes.

FIGURE 65.6
A Type Ia supernova in a distant galaxy. The brilliant white supernova seen in 2002 shines as brightly as its galaxy of approximately a hundred billion stars.

At their brightest, Type Ia supernovae are about two times brighter than Type II supernovae in visible light; however, the total power output is greater in Type II, in which much more of the power goes into the energy of expansion and neutrinos.

explosion, is highly radioactive and rapidly decays into cobalt (^{56}Co), which then decays to iron (^{56}Fe), each decay adding additional energy to the outburst.

A Type Ia supernova explosion can produce nearly 10^{10} times the Sun's luminosity at its brightest. This makes the supernova nearly as bright as an entire galaxy of stars. Figure 65.6 is a photograph of a distant galaxy with a Type Ia supernova: The exploding star gleams like a beacon. Supernovae can also occur when the core of a massive star is no longer generating heat and pressure from fusion in its core. When the core becomes massive enough, it too will collapse (Unit 66.3). This also produces supernovae that are called by a variety of names: Type Ib, Type Ic, and Type II supernovae (Unit 66).

Type Ia supernovae have several features that indicate their distinctive origin from a collapsing white dwarf. First, the spectra of Type Ia supernovae show no signs of hydrogen because they are the remnants of stars that have lost their outer atmospheres of hydrogen in the stellar wind and planetary nebula phases. Second, the spectra contain strong lines of silicon, which formed from the rapid fusion of carbon and oxygen during the fatal last moments. Finally, the characteristics of the Type Ia **light curve**—the way the luminosity changes over time (Unit 63)—closely match theoretical predictions for the thermonuclear runaway of a white dwarf of this mass. The light curve of a Type Ia supernova is illustrated in Figure 65.7.

These characteristics not only confirm the white dwarf origin of the Type Ia supernovae, but allow astronomers to distinguish them when they occur. Studies of other galaxies indicate that Type Ia supernovae are the most common. No supernova has been directly observed in our own Galaxy for over 400 years, although the last two were Type Ia—one in 1572 known as Tycho's supernova and another in 1604 known as Kepler's supernova. Both occurred before the invention of the telescope.

Type Ia supernovae appear to be excellent "standard candles" (Unit 54.3), all having almost identical maximum luminosities. This is because the pattern of their collapse and explosion occurs the same way each time. The mass gradually increases to the Chandrasekhar limit, and then the collapse occurs. In other words, the Chandrasekhar limit is a leveling factor, which causes each Type Ia supernova to ignite in the same way, with roughly the same raw materials at its disposal.

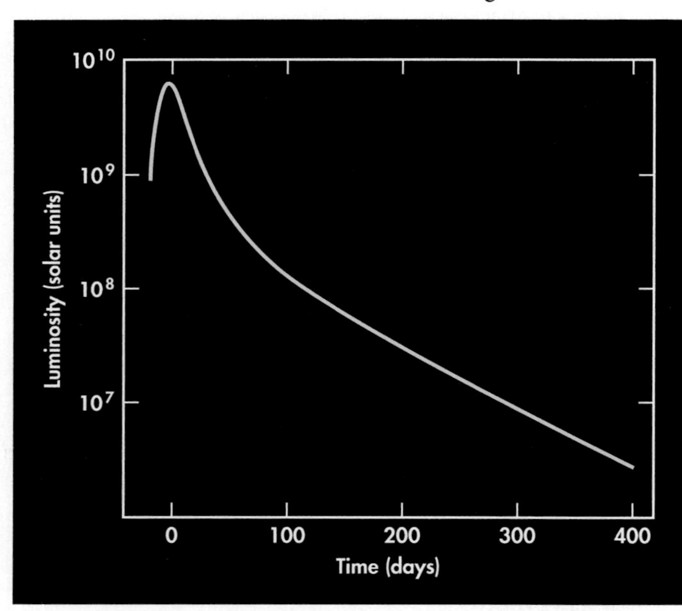

FIGURE 65.7
The light curve of a Type Ia supernova. The exploding white dwarf initially shines many billion times brighter than the Sun, then dims steadily following a curve characteristic of the decay of the radioactive elements produced in the explosion.

Type Ia supernovae leave no star behind except the former companion of the white dwarf. The matter from the white dwarf becomes a rapidly expanding cloud of extremely hot gas, known as a **supernova remnant.** An X-ray image of Tycho's supernova remnant is shown in Figure 65.8 along with a remnant from a supernova seen exploding in the year 1006. Astronomers have also detected a star moving away from the position of Tycho's supernova at over 100 kilometers per second. Apparently this is the former companion of the white dwarf, flung out of orbit when its partner exploded.

The supernova remnant is rich in leftover carbon and oxygen, plus the silicon, iron, and other elements made during the final moments of nuclear fusion during the explosion. Magnetic fields in the rapidly expanding remnant can accelerate some atomic nuclei to close to the speed of light. These high-energy particles, or **cosmic rays,** travel vast distances throughout space. Currently cosmic rays from millions of past supernovae are detected reaching the Earth's surface, and they pose a serious radiation risk to astronauts in space, who are not shielded by the Earth's own magnetic field. Although supernova explosions pose some radiation risks, they are also essential for our planet and life. The silicon in a chunk of rock, and the iron in your blood, were probably made when a white dwarf met its doom.

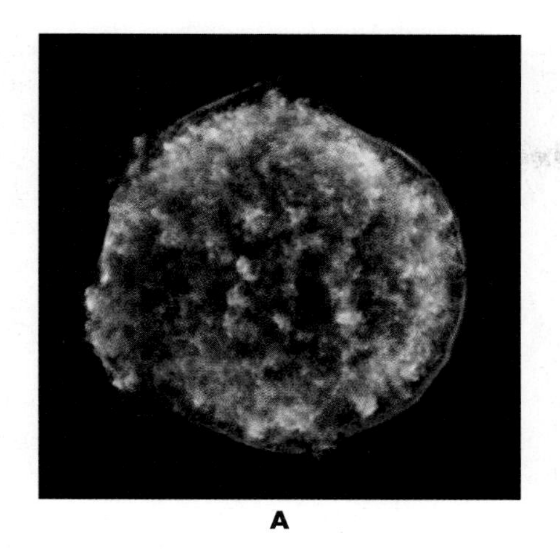

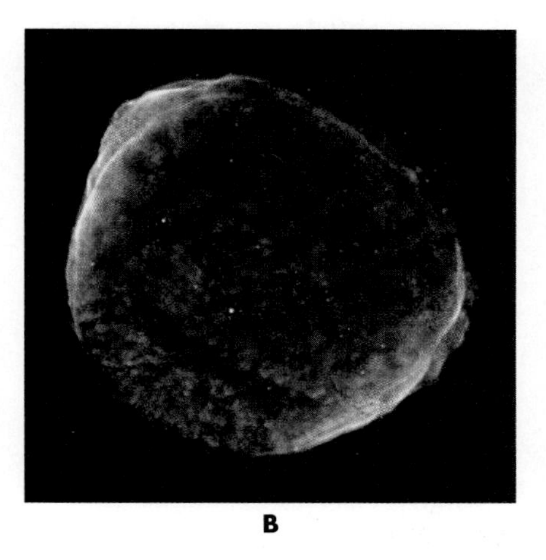

FIGURE 65.8
X-ray false-color images of two Type Ia supernova remnants. (A) Tycho's supernova was seen exploding in 1572 and is now about 20 light-years across. (B) A supernova seen exploding in the year 1006 is now about 60 light-years across. The images are colored blue for the highest energy X-rays, yellow for intermediate energies, and red for the lowest energies.

KEY TERMS

accretion disk, 523
Chandrasekhar limit, 525
cosmic rays, 526
electron degeneracy, 524
light curve, 526
mass transfer, 522
nova, 523
recurrent nova, 523
Roche lobe, 522
supernova remnant, 526
thermonuclear runaway, 523
Type Ia supernova, 525
white dwarf, 522

QUESTIONS FOR REVIEW

1. Where did the term *nova* originate?
2. What causes a nova explosion? What is a Roche lobe?
3. What is meant by *electron degeneracy?* What is the Chandrasekhar limit?
4. What is a supernova? Why are Type Ia supernovae "standard candles"?
5. What is distinctive about the appearance of Type Ia supernova—an exploding white dwarf—that we can use to distinguish it from other supernovae?
6. How bright is a Type Ia supernova?

PROBLEMS

1. Calculate the density of a 1-$M_\odot$ white dwarf that has a radius of 10^4 km.
2. How does the acceleration of gravity (Unit 16) on the surface of a 1-$M_\odot$ white dwarf, which has a radius of 10^4 km, compare to *g*, the acceleration due to gravity on the surface of the Earth?
3. What is the escape velocity (Unit 18) from the surface of a 1-$M_\odot$ white dwarf that has a radius of 10^4 km?
4. If the maximum luminosity of a Type Ia supernova is about $10^{10}\ L_\odot$, how far away would one have to be for the supernova to appear as bright as the full Moon (about 3×10^{-3} watts/m^2)? Express your answer in light-years.
5. If the maximum luminosity of a Type Ia supernova is $10^{10}\ L_\odot$, and the supernova remains at this brightness for 15 days, estimate how long our Sun would take to emit the same amount of energy.
6. IK Pegasi, a white dwarf/A-type main-sequence binary system 150 ly away, is a candidate to become a Type Ia supernova in the future. Assuming that it will reach a luminosity of $10^{10}\ L_\odot$, how bright will it appear from Earth? How many times brighter than an average star in our sky would it appear?

TEST YOURSELF

1. If mass is added to a white dwarf, which of the following does *not* occur?
   a. Its radius increases.
   b. Its mass increases.
   c. Its density increases.
   d. It may exceed the Chandrasekhar limit and collapse.
   e. Some of its electrons gain higher energies.
2. Why do Type Ia supernovae have no hydrogen in their spectra?
   a. They have converted all of their hydrogen into iron.
   b. They have lost their hydrogen atmospheres during their planetary nebula phase.
   c. They do not have enough power to excite the hydrogen lines.
   d. They are so powerful that they fuse all of the hydrogen in the explosion.
   e. They become black holes, which swallow up the hydrogen.
3. Which of the following is *not* part of the reason for a white dwarf exploding?
   a. Mass transfer
   b. Fusion of oxygen and carbon
   c. Decreasing gravitational pressure
   d. Exceeding the Chandrasekhar limit
   e. Exceeding the pressure of electron degeneracy

unit

66

Old Age and Death of Massive Stars

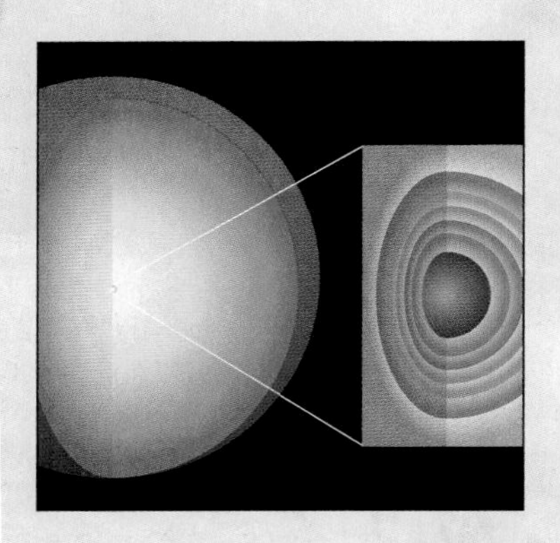

Background Pathways

Unit 59: Overview of Stellar Evolution 478

High-mass stars (masses greater than about 8 $M_{\odot}$) have violent ends. Like low-mass stars, they shed mass as they age. And like low-mass stars, as a high-mass star approaches the end of its life, the temperature in its core grows hotter. It grows so hot, in fact, that it can fuse not only helium but also carbon and heavier elements to provide the energy it needs—for a while. Less energy is available from these fuels, and the demands of the star grow steadily larger.

Once a massive star forms iron in its core, the end is near, because elements heavier than iron do not release energy when they fuse. The inability of an iron core to provide energy triggers the collapse and explosion of these stars. When massive stars explode, they spew into space the products of their nuclear fusion, adding heavy elements to the interstellar medium. They also leave behind some of the densest objects in the universe—neutron stars and black holes—which are explored in Units 67 and 68.

66.1 THE FATE OF MASSIVE STARS

The life of a high-mass star is plotted in the H-R diagram in Figure 66.1 from its birth on the main sequence up to its spectacular death in a supernova. Similar tracks, with minor variations, are followed by all stars more massive than about 8 $M_{\odot}$.

The massive star begins its life on the upper main sequence as a luminous blue star. When its core hydrogen is consumed, it leaves the main sequence. The core contracts, hydrogen shell fusion begins, and the surface swells. This is similar to the pattern followed by lower-mass stars, but everything happens much faster because fuels are consumed so quickly. This is shown in Table 66.1, which lists the various nuclear fusion reactions that take place in a 25-$M_{\odot}$ star. Main-sequence fusion of hydrogen to helium lasts only about 7 million years in such a star. By comparison, the Sun's main-sequence lifetime is more than a thousand times longer.

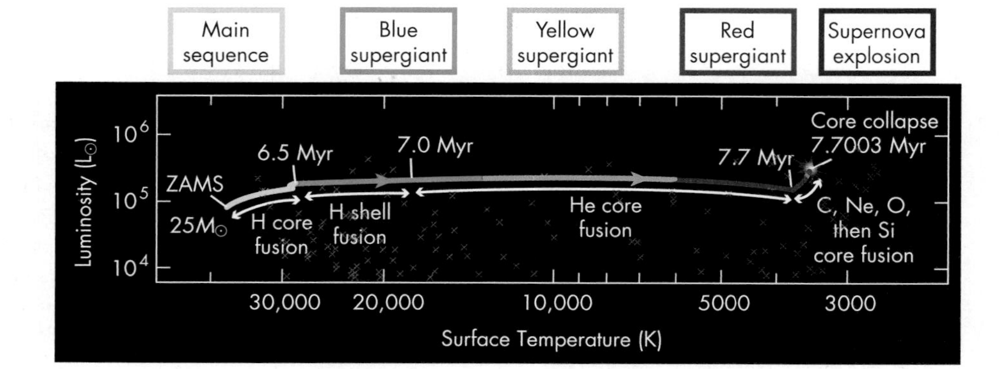

FIGURE 66.1
A massive star's evolution through the H-R diagram. Massive stars usually begin fusing helium in their cores while they are still moving across the H-R diagram, but depending on their mass and heavy-element composition, this may occur when the star is in its blue, yellow, or red supergiant phase.

TABLE 66.1 Major Fusion Reactions in Stars

Reaction	Primary Fusion Products	Minimum Temperature	Energy Released (% of Mass)	Duration of Fusion in 25 $M_\odot$ Star
Hydrogen fusion	$4\,^{1}H \rightarrow {}^{4}He$	5,000,000 K	0.71%	7,000,000 yr
Helium fusion	$3\,^{4}He \rightarrow {}^{12}C$	100,000,000 K	0.065%	700,000 yr
Alpha capture	$^{12}C + {}^{4}He \rightarrow {}^{16}O$	200,000,000 K	0.016%	
Carbon fusion	$^{12}C + {}^{12}C \rightarrow {}^{20}Ne + {}^{4}He$	600,000,000 K	0.021%	300 yr
Neon fusion	$^{20}Ne + {}^{20}Ne \rightarrow {}^{24}Mg + {}^{16}O$	1,500,000,000 K	0.012%	8 months
Oxygen fusion	$^{16}O + {}^{16}O \rightarrow {}^{28}Si + {}^{4}He$	2,000,000,000 K	0.032%	3 months
Silicon fusion	$^{28}Si + {}^{28}Si \rightarrow {}^{56}Fe$	2,500,000,000 K	0.034%	1 day

Key: H = hydrogen; He = helium; C = carbon; O = oxygen; Ne = neon; Mg = magnesium; Si = silicon; Fe = iron.

After the massive star leaves the main sequence, its surface is still so hot that it has a blue color, and it is so luminous that it is labeled a blue supergiant (Unit 58). The contracting core reaches helium fusion temperature relatively quickly because its core reached such a high temperature during its main-sequence phase. This may begin even while the atmosphere is just beginning to swell and cool. The star continues to expand, and on the H-R diagram it crosses from the blue through the yellow to the red supergiant regions. By the time it is a red supergiant, the star may have a diameter more than a thousand times larger than the Sun—so much larger that in the Sun's place it would swallow Jupiter. When the helium in the core is consumed, the core contracts and heats again. When it reaches 600 million K, it can begin to fuse carbon and, after the carbon is consumed, a series of heavier elements (Table 66.1). Now the changes in the core happen so fast that the bloated surface of the star has too little time to react to the series of new fuels being ignited and exhausted.

Throughout all of these changes, the star's luminosity remains approximately constant, but the star changes in both size and surface temperature. Computer models yield a variety of results for these stages, for several reasons: The amount of mass lost during the supergiant stage is uncertain, and it appears that factors like the star's rotation rate and initial composition may have major effects. For some mass ranges and chemical compositions, the models indicate that a massive star may loop back to the blue side of the H-R diagram; observational evidence that this occurs is discussed in Section 66.3. As a result, a massive star may follow a path back and forth in the H-R diagram, similar to the path of a low-mass star in its red giant phase but at a higher luminosity.

As the star consumes one nuclear fuel after another, the "ashes" of one set of nuclear reactions become the fuel for the next set. Oxygen, neon, magnesium, and silicon are formed; but because progressively higher temperatures are needed for each new fusion process, each new fuel is confined to a smaller and hotter region in the star's core. The star develops a layered structure, as illustrated in Figure 66.2. Hydrogen remains plentiful in the star's atmosphere, because the temperature has remained too low there for nuclear reactions to have occurred. At the base of the hydrogen atmosphere, the temperature is high enough for hydrogen fusion, and the hydrogen fusion shell "eats" its way out into the surrounding

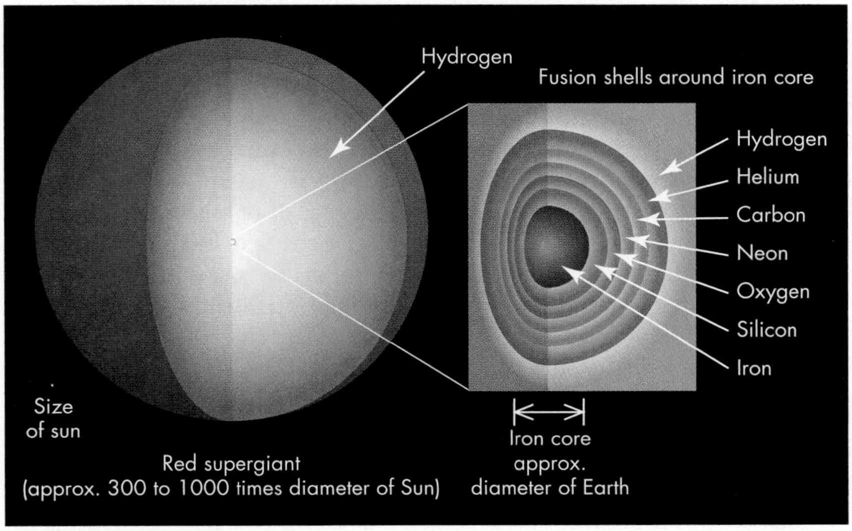

FIGURE 66.2
The structure of a massive star at the end of its life. Shells of lighter elements forming heavier elements surround an inert iron core.

atmosphere, depositing the helium it produces. At the base of the helium layer, helium fusion produces carbon and some oxygen (through the alpha capture process listed in Table 66.1). At the base of that layer, carbon fusion produces a neon layer, and so on—a series of nested shells, each made of an element heavier than the one surrounding it.

By the time the star is fusing silicon into iron, its core has been compressed to a size smaller than the Earth, to drive the temperature up to about 2.5 billion K. At such temperatures, the silicon is being consumed at a fantastic pace. To appreciate how rapidly the star consumes its silicon, consider the following: A massive star of 25 $M_{\odot}$ takes almost 8 million years from its birth to form an iron core. The main-sequence phase lasts about 7 million years; the helium fusion phase lasts only about 700,000 years; the carbon fusion phase lasts just a few hundred years; and the silicon fusion phase lasts only about one day!

The time decreases for each phase for several reasons. First, the star must produce ever larger amounts of heat and energy to support the core against the greater intensity of gravity as the core compresses. Second, in most of the heavy-element fusion reactions, less energy is released per kilogram of material fused than in hydrogen or helium fusion. Third, much of the energy released in these heavy-element reactions is carried away by neutrinos, which travel out of the star, reacting very little with its gases—and therefore offers little help in either heating the gas or providing the pressure needed to support the star.

When silicon fuses to form an iron core, the death of a star is very near. It is the end of the line because iron cannot fuse to provide the energy needed to keep the star going. To see why, we need to examine how atomic nuclei interact.

66.2 THE FORMATION OF HEAVY ELEMENTS

Q The even-numbered elements are generally more common throughout the universe than the odd-numbered elements. How might this be understood from the fusion reactions we have examined?

To understand massive stars, we need to understand atomic nuclei. The fusion of lighter nuclei into heavier nuclei is what powers stars. Astronomers call the formation of heavy elements by such nuclear fusion processes **nucleosynthesis.** The details of these reactions were first worked out in the 1950s by a team of astronomers—E. Margaret Burbidge, Geoffrey R. Burbidge, William A. Fowler, and Fred Hoyle. They hypothesized that all of the chemical elements in the universe heavier than helium were made by nuclear fusion in stars. Detailed calculations and measurements since then confirm that the observed abundances of the elements in the universe agree extremely well with this hypothesis. The fact that oxygen and carbon are the third and fourth most common elements in the universe, and that the Earth is made largely of elements such as iron and silicon, are direct consequences of nucleosynthesis in stars.

The process of nucleosynthesis depends on a balance of repulsive and attractive forces between nuclei. Because protons are positively charged, atomic nuclei repel each other. However, if nuclei can come close enough to each other, the short-range *strong force* (Unit 4) takes over and the nuclei can fuse. The fusion of larger nuclei requires successively higher speeds. To see why, consider the electrical repulsion between two carbon nuclei: With 6 protons each, the repulsion is $6 \times 6 = 36$ times stronger than the repulsion between two hydrogen nuclei, which have only 1 proton each. The collision speed needed to overcome this repulsion can be achieved if the gas is hot enough, because the hotter a gas, the faster its particles move. To fuse progressively heavier nuclei, the star's core must grow progressively hotter. For example, fusing carbon requires a temperature of about 600 million K, whereas fusing heavier elements may require temperatures greater than a billion K.

The superscript number listed before an element symbol equals the number of protons + neutrons. The element name indicates the number of protons. So ^{13}C, or "carbon-13," is element #6 and must therefore have 6 protons and $13 - 6 = 7$ neutrons (also see Unit 4).

When nuclei fuse, the sum of the number of protons and neutrons always remains the same. However, the *weak force* (Unit 4) allows neutrons and protons to

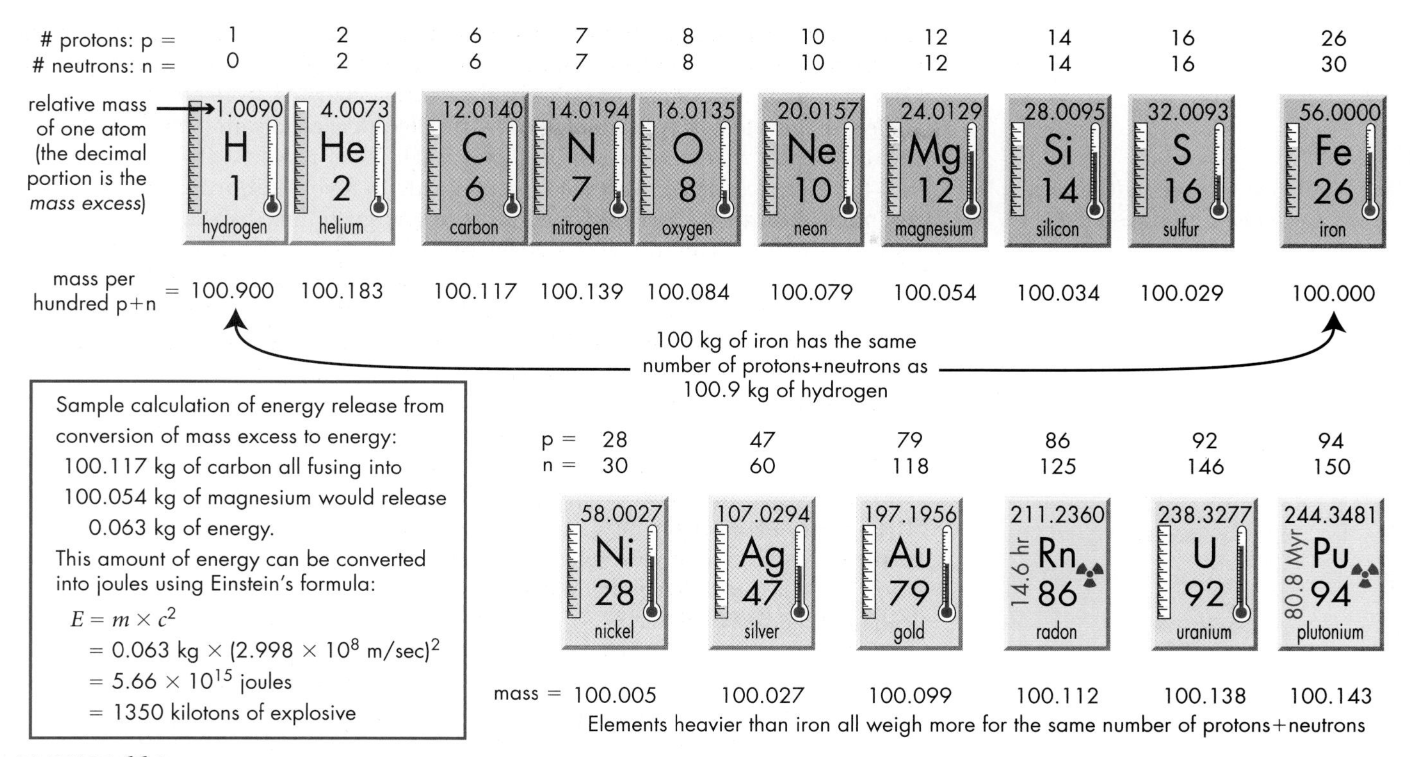

FIGURE 66.3
Some of the most common elements in the universe (top row) and an assortment of elements that are heavier than iron. The boxes are taken from the periodic table, with color coding and other symbols that indicate a number of element properties—see the periodic table on inside back cover of this book. The figure shows how the element masses reflect the mass excess of each element and how this can be used to calculate the energy released by a nuclear reaction.

be converted into each other. For example, suppose two ^{28}Si nuclei fuse. Silicon has 14 protons and 14 neutrons, so the fusion reaction begins with 28 of each nuclear particle. Their fusion ultimately results in the production of ^{56}Fe, which contains 2 fewer protons, but 2 more neutrons. The number of protons and neutrons for the main isotopes involved in these reactions are shown in Figure 66.3, which also shows the properties of an assortment of other elements in the periodic table.

The most common nuclear fusion reactions occurring in stars are listed in Table 66.1, along with the approximate temperatures required for these fusion processes to take place. Besides the reactions listed in the table, a wide variety of less common interactions create other combinations of nuclei. For example, as you can see in the table, oxygen fusion produces the element silicon, but it may also produce sulfur, although normally in smaller quantities. Note that helium nuclei (alpha particles) are created in a number of these fusion reactions, and these are usually quickly "captured" in further fusion interactions with the heavier elements. By tracing through all of the nuclear interactions, it is possible to predict the amount of each element that is produced, and this agrees remarkably well with the mix of elements we observe in the universe.

Each nuclear reaction converts some of the nuclei's mass into energy, in accordance with Einstein's law $E = m \times c^2$. The percentage of mass turned into energy is listed in the second to last column of Table 66.1. Nearly 0.7% of the fusing nuclei's mass is converted into energy during hydrogen fusion. However, the sum of all of the other reactions converts only about 0.2% of the matter's mass into energy.

The declining amount of energy available from heavier elements is reflected in the masses of the elements. This is indicated in Figure 66.3, by the **mass excess** for each of the elements, which is the amount of mass—or equivalently, energy, according to Einstein's formula—potentially available from an element through nuclear reactions. The mass of each element is given relative to ^{56}Fe, which has the lowest-energy configuration of protons and neutrons of any nucleus. In nuclei larger than the iron-56 nucleus, the attraction of the strong force between nuclear particles begins to be counterbalanced by the electrical repulsion between protons. This is because the strong force weakens much more rapidly with distance than does the electrical force. As a result, attempting to fuse additional nuclei onto an iron nucleus weakens the bonds and absorbs energy, rather than releasing it.

Q Why do you suppose that heavier atoms' nuclei tend to have a larger proportion of neutrons than is found in lighter atoms?

The formation of an iron core signals the end of a massive star's life. Iron cannot fuse and release energy. Thus, nuclear fusion stops with iron, and a star with an iron core is out of fuel. Nucleosynthesis does not quite end here, though. In one final burst, these stars create the rest of the elements beyond iron in their final collapse.

66.3 CORE COLLAPSE OF MASSIVE STARS

When an iron core forms in a star, the reaction of the star at first looks the same as when previous fuels were exhausted. The star's core shrinks and heats, and just as for low-mass stars, the shrinkage presses the matter tighter and tighter until electron degeneracy (Unit 62.3) finally halts the contraction. However, the pressure is so enormous, and the most energetic electrons have so much energy, that a new reaction can occur. Protons and electrons may themselves combine to become neutrons by this reaction:

$$p + e^- + \text{Energy} \rightarrow n + \nu$$

Besides converting a proton (p) and electron (e^-) into a neutron (n) and neutrino (ν), this reaction *absorbs* energy, and that absorption triggers a catastrophe for the star.

To understand why a reaction that absorbs energy can have such dire consequences, consider the following: Each layer of the star is supported by the one below it, with all of them finally supported by electron degeneracy at the very center, where the density reaches over 1 million kilograms per liter. When the electrons are swallowed up by protons to become neutrons, the pressure provided by electron degeneracy disappears. The center of the star begins to implode.

Neutrons also have a degeneracy limit, but the volume each degenerate neutron occupies is about a billion times smaller than the volume the electrons filled. This effectively creates a hole in the middle of the star. In a fraction of a second, the core is thus transformed from a sphere of iron, with a radius of 10,000 kilometers, into a sphere of neutrons with a radius of just 10 kilometers held up by neutron degeneracy pressure. This is something like pulling out the bottom brick from an enormous stack of loose bricks. Nothing remains to support the star, and so it begins falling inward.

Q Hold a small rubber ball on top of a basketball, and drop them together toward the floor. What happens to the small ball? How is this similar to a type II supernova?

The plunging outer layers of the star strike the neutron core, reaching speeds of over 100,000 kilometers per second—about one-third of the speed of light. Some of the material that smashes into the extremely dense core "bounces" back outward at nearly the same speed, colliding with other infalling matter. The impact heats the infalling gas to billions of degrees. Several solar masses of matter are almost instantaneously raised to fusion temperature and detonate all at once. Meanwhile an enormous quantity of neutrinos is pouring out from the core—one neutrino for each of the protons that turned into a neutron. Normally neutrinos interact so weakly that they escape freely from a star; but at the enormous densities of the collapsing material, many of the neutrinos flooding out from the core collide with the infalling matter and give up their energy to it. This causes a range of nuclear reactions, further heats the infalling matter, and drives it outward.

All of these processes generate more energy in a few seconds than the Sun generates in its entire lifetime. A powerful blast wave travels outward from the core into the atmosphere of the star. The blast wave tears through the surface of the

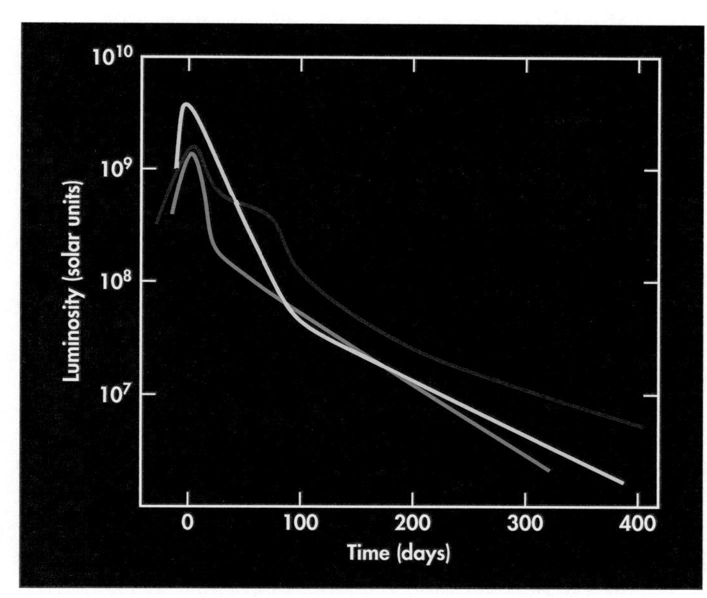

FIGURE 66.4
Some examples of the light curves of supernovae of stars undergoing core collapse.

supergiant in an eruption that sends most of the star's atmosphere flying outward into space. This catastrophe marks the death of a high-mass star in a titanic supernova explosion.

We saw in Unit 65 that the white dwarfs left behind by lower-mass stars can also undergo supernova explosions if a companion star deposits enough matter on them. This process is similar in the sense that the mass grows too large for the core to support the weight. The collapse of a white dwarf is called a Type Ia supernova, whereas a massive star undergoes a **Type II supernova.** The types of supernovae were defined before the mechanisms were known, and astronomers classified these explosions according to the spectrum of the light they generated. Type II supernovae show spectral lines of hydrogen, produced by the superheated gas of the star's former atmosphere, whereas Type I do not exhibit hydrogen lines.

The lack of hydrogen in a Type Ia supernova is a result of that element's having been completely consumed in the core of the star that ended up as a white dwarf. However, some massive stars explode without hydrogen lines visible in their spectra. These supernovae, which are classified as Type Ib and Type Ic, appear to be massive stars undergoing core collapse that have lost their hydrogen atmospheres to a companion star or stellar winds. Astronomers can distinguish these from Type Ia supernovae because they do not exhibit some of the secondary characteristics of Type Ia, such as silicon or helium lines. Supernovae of types Ib, Ic, and II are sometimes collectively referred to as **core-collapse supernovae.**

Core-collapse supernovae come from stars that have a wide range of masses and histories. A star of 20 $M_\odot$ will build up the critical mass for the iron core to collapse much more quickly than a star of 10 $M_\odot$. When the iron core in each of these stars is about to collapse, their atmospheres are structured in significantly different ways in response to their many layers of shell fusion and the amount of atmosphere lost in stellar winds. Core-collapse supernova blasts may therefore look quite different for stars of different masses. These differences are illustrated by a wide variety of light curves—the amount of light emitted as a function of time—as illustrated in Figure 66.4.

In death, a massive star releases more energy in a few minutes than it generated by nuclear fusion during its entire existence. Core-collapse supernovae are generally slightly less luminous at visible wavelengths than the Type Ia supernovae, but they are more energetic. Most of their energy is carried by the burst of neutrinos. By some estimates, almost 99% of the energy of a Type II supernova is carried away by the neutrinos produced when the protons and electrons combine. Several neutrino detectors around the world detected just such a burst in February 1987 when a supernova—SN 1987A—blew up. Several hours later, light from the exploding star became visible as the blast wave emerged through the surface of the supergiant star.

Images of this star had been obtained before it exploded, making it the first supernova for which we had "before" and "after" photographs (Figure 66.5). A surprise for astronomers was that the star that exploded was a blue supergiant. Before that time it had been widely expected that supernovae would occur in red supergiants. It is now understood that some massive stars may "loop back" to bluer colors in the H-R diagram after their red supergiant phase (Figure 66.1), and can have almost any color when the iron core develops.

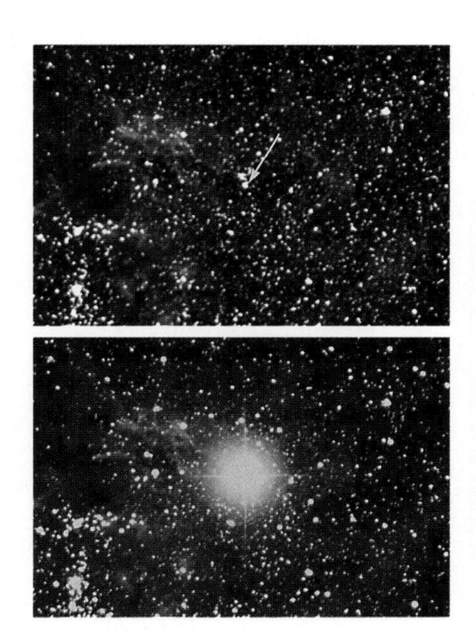

FIGURE 66.5
Photograph of a portion of the Large Magellanic Cloud (a small, nearby galaxy) before Supernova 1987A exploded, and a second photograph taken during the event.

Supernova 1987A did not even occur in our own Galaxy, but in the Large Magellanic Cloud, a small galaxy neighboring our own (Unit 76). Nonetheless, even though the blast took place about 160,000 light-years away—hence 160,000 years ago—trillions of the neutrinos from that explosion passed through every person on Earth!

66.4 SUPERNOVA REMNANTS

Gas ejected by a supernova blast plows into the surrounding interstellar space, sweeping up and compressing other gas that it encounters. Because massive stars have such short lives, they are often surrounded by large amounts of the gas and dust in the interstellar cloud from which they formed (Unit 60). The **supernova remnant**—a huge, glowing cloud of stellar debris and interstellar gas—may expand to a diameter of several light-years within a century, but its expansion slows as it runs into more surrounding gas. Figure 66.6 shows photographs of three Type II supernova remnants of different ages. Notice how ragged the remnants in Figure 66.6 look compared with the smoothly ejected bubbles of planetary nebulae (Unit 64), which reflects the more violent death of massive stars.

The explosion of a massive star mixes the elements synthesized by nuclear fusion during the star's evolution with the star's outer layers, and blasts them into space. This incandescent spray expands away from the star's collapsed core at more than 10,000 kilometers per second. Depending on the star's mass, many solar masses of matter may be flung outward, ultimately mixing with surrounding material and, in time, forming new generations of stars. Interstellar gas is thereby enriched in heavy elements, the atoms needed to build the rock of planets and the bones of living creatures here on Earth.

Some of the light seen coming from supernovae in the weeks and months following the explosion has been shown to originate from heavy radioactive elements. These must have been produced in the explosion because they decay rapidly.

The supernova outburst itself generates even heavier elements than were originally created in the star's core. Elements such as silver, gold, and uranium are created in a supernova explosion from the abundant energy of its blast. This occurs because the explosion creates free neutrons, which rapidly combine with atoms from the star to build up elements heavier than iron.

One such violent outburst was seen almost a thousand years ago in 1054 by astronomers in China and elsewhere in the Far East (Unit 26.4). This is perhaps the most famous of all supernova remnants, now known as the Crab Nebula (Figure 66.6B). With even a small telescope, you can still see the glowing gases ejected from that dying star (see Looking Up #5: Taurus). Another intriguing supernova shown in Figure 66.6A, Cassiopeia A, was first seen as a radio source—the brightest radio source in the sky outside the Solar System. Astronomers have determined from its expansion rate that it exploded about 300 years ago, making it one of the most recent known supernovae in our Galaxy. However, there are no

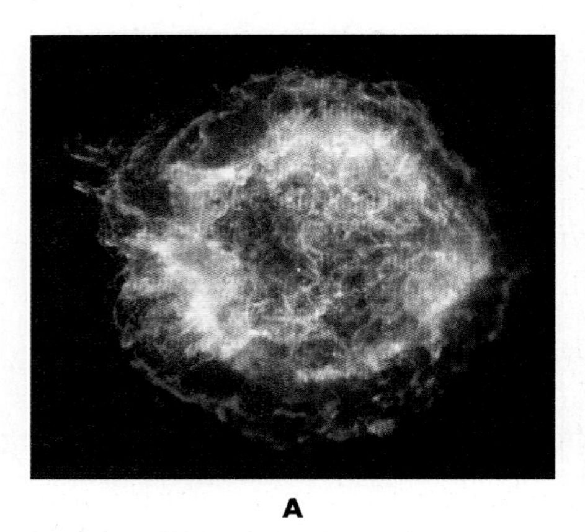

A

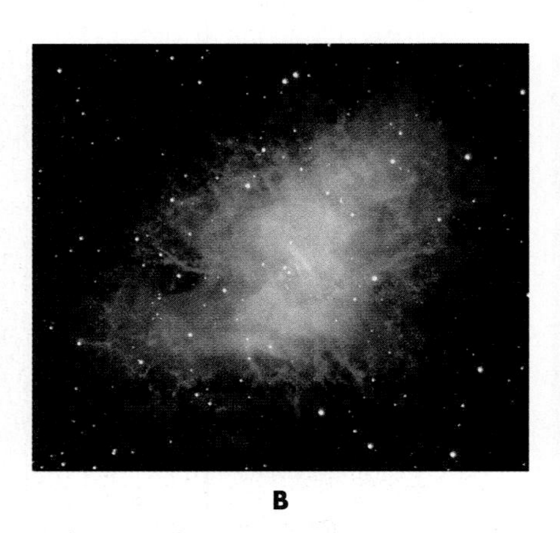

B

C

FIGURE 66.6
A sequence of Type II supernova remnants of increasing age. (A) Cassiopeia A is about 300 years old; (B) the Crab Nebula (also known as Messier 1 or M 1) is about 1000 years old; (C) the remnant of a supernova that is several thousand years old. The image of Cassiopeia A is a false-color X-ray image—red corresponds to X-rays of lower energies, while blue corresponds to the highest energies. Scientists have shown that energetic cosmic rays are generated in the blue regions.

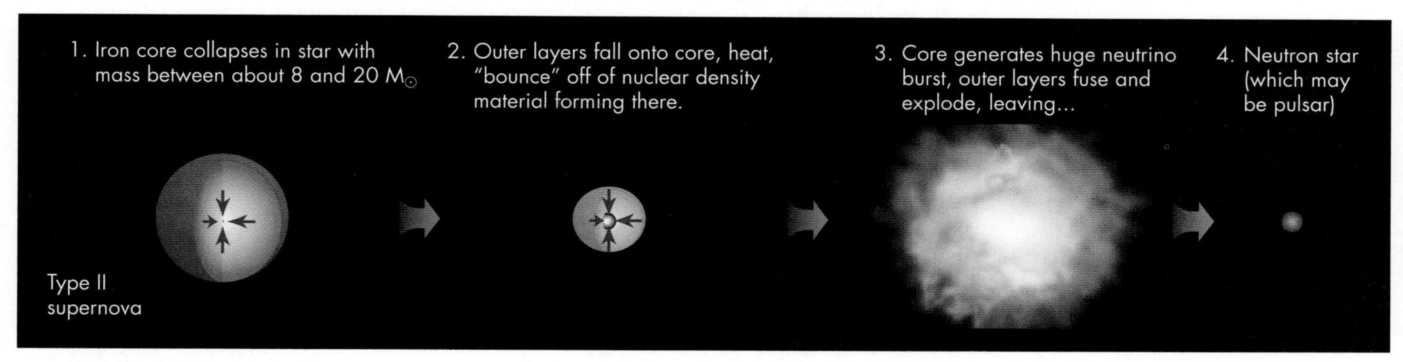

FIGURE 66.7
Formation of a supernova explosion and a neutron star by the collapse of the iron core of a star with a mass between about 8 $M_{\odot}$ and 20 $M_{\odot}$.

clear records that its explosion was seen. The explosion appears to have been hidden from view by dense dust surrounding the young massive star.

A supernova explosion marks the death of a massive star, but a part of the star—the collapsed core of neutrons that initiated the explosion (Figure 66.7)—may survive. Such *neutron stars* (Unit 67) have been detected, for example, in the middle of the expanding debris of the three supernova remnants in Figure 66.6. In some other supernova remnants, no remaining core is detected. It is possible that these stars destroyed themselves entirely, like the Type Ia supernovae; or perhaps they remain but are hidden from sight in the form of an even more compressed body—a strange object known as a black hole (Unit 68).

66.5 GAMMA-RAY BURSTS AND HYPERNOVAE

Astronomers have made many of their most exciting discoveries when new telescopes have allowed them to observe the sky at wavelengths not previously detectable. Gamma-ray astronomy began in 1965 with a small and (by modern standards) primitive satellite designed to detect cosmic gamma rays. Gamma rays are the electromagnetic waves with the highest of energies, so naturally they are associated with the most energetic events in the cosmos. By the 1970s astronomers had discovered that many energetic sources, such as the center of our Galaxy and other galaxies, the Crab Nebula, and the remnants of other exploded stars, emitted gamma rays. Ironically, and unknown to astronomers, the most intriguing gamma-ray sources had already been discovered accidentally in the 1960s.

In 1967 the United States placed several military surveillance satellites in orbit to watch for the gamma rays produced when a nuclear bomb explodes. The satellites were designed to monitor the United States–Soviet Union ban on nuclear bomb tests in the atmosphere. Curiously, on a number of occasions, the satellites detected gamma-ray bursts coming not from the Earth but from space. This caused great concern among political and military leaders, but fortunately some scientists involved had enough background to recognize these as astronomical sources and avert a crisis. Unfortunately for astronomers, the discovery of the bursts was top secret at the time and was not made public until 1973.

Strangely, gamma-ray bursts did not appear to be associated with any identifiable object. They would appear suddenly in otherwise blank areas of the sky, flare in intensity for a few seconds, and then fade to invisibility. Despite intense study, it was not until 30 years after the first gamma-ray burst was observed that astronomers could even say whether the bursts came from nearby or far away. Without a distance, it was impossible to know how powerful these bursts were.

The breakthrough came in 1997, when astronomers detected a gamma-ray burst that coincided with a distant galaxy. They achieved this by rapidly pointing an X-ray telescope to the direction of the sky where the burst was detected. At the time, gamma-ray telescopes had very poor resolution, but the X-ray telescope saw the fading afterglow of the burst and was able to pinpoint its location. Over the following decade, astronomers succeeded in making hundreds more identifications with improving detail, even observing the afterglow of bursts at visible and radio wavelengths. A number of these events are associated with supernovae explosions.

Based on their brightnesses and distances, gamma-ray bursts herald the most luminous explosions in the universe, even brighter than normal supernovae. A hypothesis is emerging that explains most of their features in terms of an explosion called a **hypernova.** A hypernova is an intense burst generated during the core collapse of a star so massive that it collapses not to a neutron star but to a black hole—an object with such intense gravity that nothing, not even light, can escape (Unit 68).

Instead of forming a dense core of neutrons, the core and surrounding material collapse inward and disappear into a growing black hole. Models suggest that a black hole is likely to form upon the death of stars that began their life with more mass than about 20 $M_{\odot}$. Without the core bounce and huge outpouring of neutrinos, there cannot be a core-collapse supernova like that which we described earlier.

If all of the matter fell into the black hole, we would see absolutely nothing at all, so it may seem odd that these hypernovae are so bright. However, not all of the matter can fall into the black hole. The black hole is only about 20 kilometers across, and material falling in from thousands or even millions of kilometers away is very unlikely to fall straight in. This is illustrated in Figure 66.8. Because of the conservation of angular momentum (Unit 20), as the outer parts of the star collapse inward toward the black hole, they will spin faster and flatten out. A large percentage of the material will end up in a flattened disk of material, similar in appearance to the disk around a protostar (Unit 60). The material collapsing toward the new black hole will crash together at near the speed of light and rise to extremely high temperatures. A huge thermonuclear explosion occurs, with jets of gamma rays escaping as the explosion bursts through the infalling matter along the direction of the poles of the former star.

Not all gamma-ray bursts are produced by hypernovae. Other proposed events that might produce such bursts are collisions between neutron stars or the collapse of a neutron star to a black hole when too much material falls on it from a companion star (analogous to a white dwarf supernova).

Because stars more massive than 20 $M_{\odot}$ are rare, hypernovae are also rare. Furthermore, the burst of gamma rays can be seen only by observers who happen to lie close to the direction of the star's polar axis. If we do not lie along the axis, we would not see the gamma-ray burst, but we might see something more similar to other types of supernova explosions. Astronomers have identified some peculiar supernovae that they suspect may be such events. It is a good thing that we are unlikely to see the gamma rays from these bursts. Estimates suggest that one of these gamma-ray jets could cause mass extinction of life on the Earth if it occurred within 1000 light-years of us.

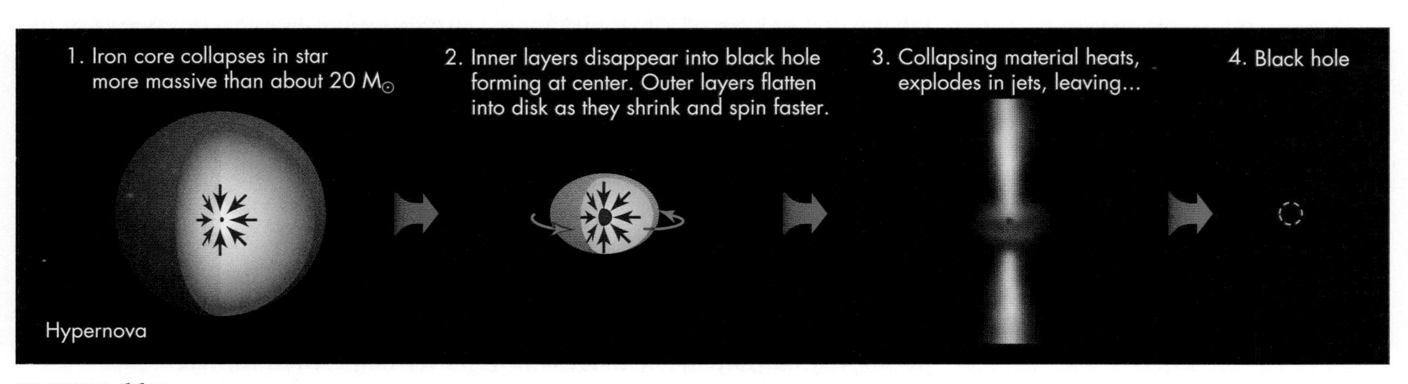

FIGURE 66.8
Formation of a hypernova blast and a black hole by star more massive than about 20 $M_{\odot}$.

KEY TERMS

core-collapse supernova, 533
hypernova, 536
mass excess, 531
nucleosynthesis, 530
supernova remnant, 534
Type II supernova, 533

QUESTIONS FOR REVIEW

1. What is nucleosynthesis?
2. Why can't a star fuse iron in its core?
3. What happens that causes a high-mass star's core to collapse?
4. What is a core-collapse supernova?
5. What is the observational difference between Type I and Type II supernovae?
6. How are Type Ib and Type Ic supernovae thought to be related to Type II supernovae?
7. What role do neutrinos play in a core-collapse supernova?
8. What is a hypernova? How does it relate to gamma-ray bursts?

PROBLEMS

1. The Crab Nebula is 6,300 ly away. The supernova that formed it was observed from Earth in 1054 A.D. When, in fact, did the supernova event occur?
2. If every proton in the core of a massive star turns into a neutron and releases one neutrino, how many neutrinos are produced? Assume the core contains 2 $M_\odot$, and half the mass is made of protons ($M_{proton} = 1.67 \times 10^{-27}$ kg).
3. Assume that 10^{12} neutrinos passed through every square meter of surface perpendicular to the radiation from Supernova 1987A (which was 160,000 ly away). How many neutrinos were produced by the explosion?
4. How long does it take the blast wave from the center of a star undergoing a supernova to reach the star's surface if the blast wave is traveling at 10,000 km/sec and the star is about 100 times the size of the Sun?
5. Assuming an average collapse speed of 50,000 km/sec, how long does a star take to collapse during the supernova stage if it starts off with a surface radius of 7 AU and ends up at the surface of the neutron star core, 10 km from the center of the star?
6. The Crab Nebula is today measured to be about 10 ly across. If it exploded in 1054, what has its average speed been in kilometers per second?
7. Using the information in Table 66.1, what is the luminosity of a giant star in the silicon fusion stage in solar luminosities? Assume that the core of this star is 1 M¤
8. It is estimated that in the final stages of fusion, a giant star produces a core of iron more massive than the Sun. From the energy release given in Table 66.1,
   a. how much energy is this (in joules)?
   b. how long would it take the Sun to emit this much energy at its rate of 4×10^{26} joules/sec?

TEST YOURSELF

1. Stars like the Sun probably do not form iron cores during their evolution because
   a. all the iron is ejected when they become planetary nebulae.
   b. their cores never get hot enough for them to make iron by nucleosynthesis.
   c. the iron they make by nucleosynthesis is all fused into uranium.
   d. their strong magnetic fields keep their iron in their atmospheres.
   e. None of the above.
2. What is the heaviest element that fusion can produce in the core of a massive star?
   a. Carbon
   b. Gold
   c. Iron
   d. Nickel
   e. Silicon
3. Protons combine with electrons to form neutrons
   a. just before the formation of an iron core.
   b. when neutrinos are replaced by neutrons.
   c. just before a supernova explosion.
   d. just after a supernova explosion.
   e. (b) and (d).

Starry Night

See the Pathways *2nd edition ARIS site for online Starry Night™ planetarium exercises.*

unit

67

Neutron Stars

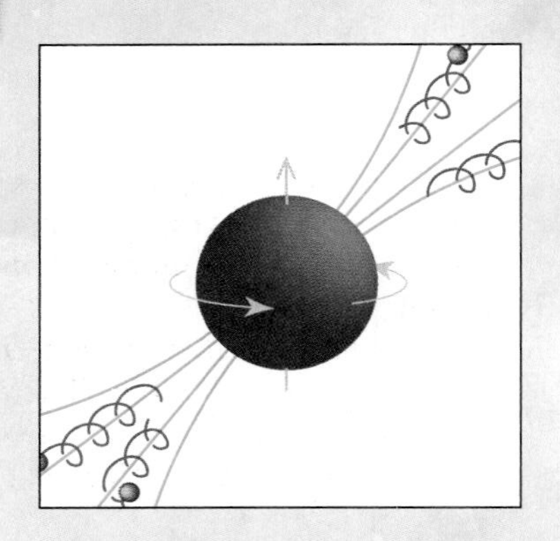

Background Pathways

Unit 66: Old Age and Death of Massive Stars 528

When a massive star forms an iron core late in its lifetime, the pressures at the center of the star can force electrons to merge with protons to make neutrons, and the whole star collapses to an extraordinarily small size. The resulting star resembles a giant atomic nucleus, with a density a billion times greater than a white dwarf—about a trillion tons per liter!

In 1934 Walter Baade (pronounced *BAH-deh*) and Fritz Zwicky, astrophysicists at Mount Wilson Observatory and the neighboring California Institute of Technology, respectively, suggested that when a massive star reaches the end of its life, its gravity will crush its core and make the star collapse. They predicted that the collapse of the core would trigger a titanic explosion, which they named a *supernova* (Unit 66). Almost as an afterthought, they speculated that the collapsed core might be so dense that the star's protons and electrons would be driven together and merge into neutrons, forming a **neutron star**.

Astronomers readily accepted Baade and Zwicky's idea that a massive star dies as a supernova, because such violent explosions had been seen (Unit 66); but they paid little attention to their suggestion that a neutron star might be born in the blast. Such stars would have radii of only about 10 kilometers (about 6 miles), incredibly tiny compared with even white dwarfs, and probably unobservable if they existed at all. The idea of these tiny, collapsed stars lay dormant for more than 30 years.

67.1 PULSARS AND THE DISCOVERY OF NEUTRON STARS

In 1967 a group of English astronomers led by Anthony Hewish was observing fluctuating radio signals from distant, peculiar galaxies. Jocelyn Bell, a graduate student working with the group, noticed an odd radio signal with a rapid and astonishingly precise pulse rate of one burst every 1.33 seconds. The precision of the pulses led some astronomers to wonder whimsically if they had perhaps stumbled onto signals from another civilization, and informally the signal became known as LGM-1, for "little green men #1." Over the next few months Hewish's group found several more pulsating radio sources, which came to be called **pulsars** for their rapid and precisely spaced bursts of radiation (Figure 67.1).

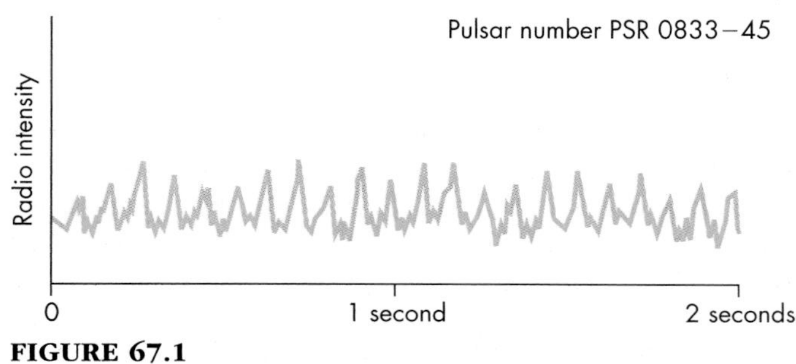

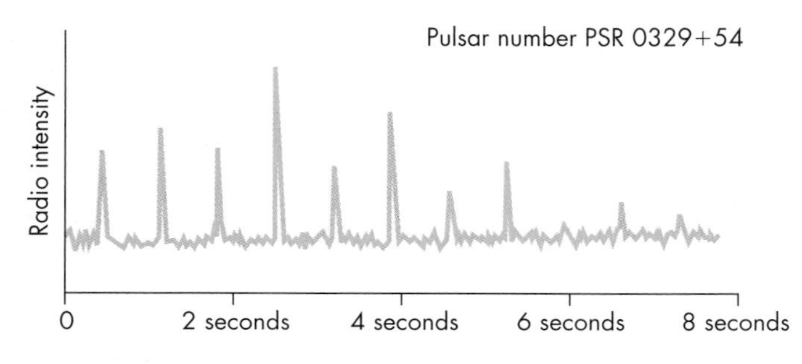

FIGURE 67.1
Pulsar signals recorded from a radio telescope.

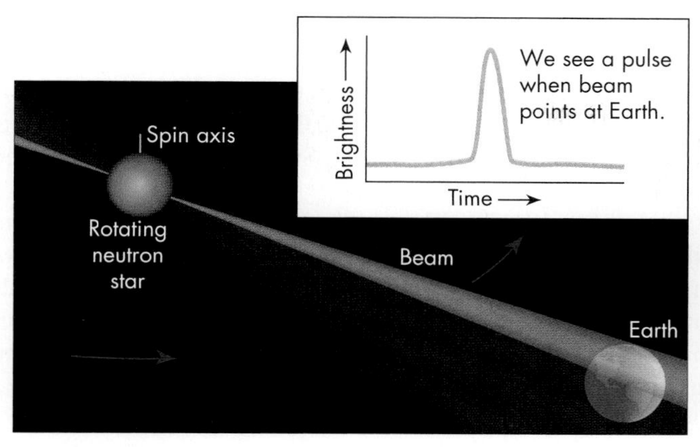

FIGURE 67.2
A pulsar's pulses are like the flashes of a lighthouse as its lamp rotates.

The ensuing discovery of many more pulsars, and the unchanging repetitiveness of their signals, convinced astronomers that they were observing a natural phenomenon. But it was difficult to come up with a plausible hypothesis of an astronomical phenomenon that could be so regular and rapid.

The solution to this puzzle came from the work of the Italian astronomer Franco Pacini and the Austrian-born British astronomer Thomas Gold. Their work led to the idea that a pulsar is a rapidly spinning neutron star. This idea soon gained support from the discovery of a pulsar at the center of the Crab Nebula, a supernova remnant.

The pulses from a spinning neutron star are produced by its magnetic field. Collapsed to such a small radius, the star's magnetic field is squeezed into a far smaller volume, amplifying the field's strength to about 1 trillion times that of the Earth's magnetic field. A neutron star is thus an extremely powerful magnet. Its intense magnetic field causes it to emit radiation in two narrow beams along its north and south magnetic poles (Section 67.2). Like the Earth's and most other astronomical bodies' magnetic fields, a pulsar's magnetic field is not aligned with the star's rotation axis; so as the star spins, its beams sweep across space. We see a burst of radiation at radio wavelengths when a beam points at the Earth. A pulsar shines like a cosmic lighthouse whose beam swings around as its lamp rotates, as illustrated in Figure 67.2.

Pulsar model

But what could make a neutron star spin so fast? For example, the pulsar at the center of the Crab Nebula rotates 30 times per second. By comparison, the Sun and most other stars take weeks to make a single rotation. However, when a massive star collapses to a tiny radius, the **conservation of angular momentum** (Unit 20) requires it to spin faster. This is the same effect that allows ice skaters to spin rapidly by pulling their arms and legs close to their axis of rotation.

An object's angular momentum is approximately given by the body's mass, M, times its equatorial rotation velocity, V, times its radius, R:

$$\text{Angular momentum} \approx M \times V \times R$$

The principle of conservation of angular momentum states that the product of these three quantities must remain constant unless some force brakes or accelerates the spin. Therefore, in the absence of external forces, if a fixed mass contracts, it must increase its rotational speed by the same factor by which its radius decreases—if R shrinks, V must increase to keep the angular momentum constant. Therefore the contracting core of a star must spin faster and faster.

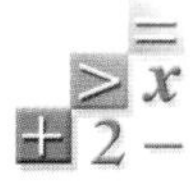

To convert days to seconds, multiply by the number of hours per day, the number of minutes per hour, and the number of seconds per minute: 25 days $= 25 \times 24 \times 60 \times 60$ seconds $=$ 2,160,000 seconds.

Imagine the Sun contracted from its current radius of about 700,000 kilometers to just 7 kilometers, similar to the radius of a neutron star. Because its radius would be 100,000 times smaller, conservation of angular momentum requires that the speed of material rotating at its equator would be 100,000 times greater. The period—the time it takes a point on the equator to come back around again—would therefore be $(100{,}000)^2 = 10{,}000{,}000{,}000 = 10^{10}$ times faster. So instead of rotating once every 25 days, the collapsed Sun would rotate once every 0.0002 seconds—about 5000 times per second!

Such a rapid spin is what is expected when the iron core of a massive star collapses. Pulsars probably begin spinning at such a rate, and then are slowed down gradually by friction. The Crab Nebula pulsar period, for example, is currently observed to be slowing down by about 10^{-5} seconds each year, so when the nebula formed nearly a thousand years ago, it may have been spinning as fast as our calculation here suggests.

67.2 EMISSION FROM NEUTRON STARS

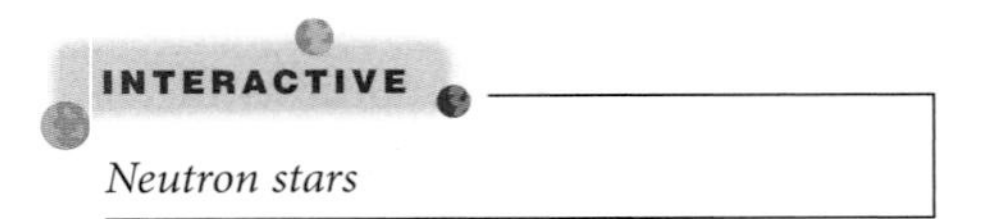

When a magnetic field moves, it creates an electric field—a principle we use here on Earth by spinning magnets in dynamos to generate electricity. Similarly, the rapid spin of a neutron star and its intense magnetic field generates powerful electric fields. These fields rip positively and negatively charged particles off the star's surface and accelerate them to nearly the speed of light. The electrically charged particles are channeled by the pulsar's intense magnetic field to travel along the magnetic field lines. This is similar to how the Earth's magnetic field directs solar wind particles, except in the Earth's case they are directed *inward,* creating the aurora near the magnetic poles. The magnetic field of a spinning pulsar, in a like manner, generates two narrow beams of charged particles flowing outward at the magnetic poles.

As the charged particles move along the pulsar's magnetic field, they generate radio waves along their direction of motion. This is somewhat like a radio broadcast antenna on Earth, where electric currents are pulsed through the antenna, accelerating electrons in it; the accelerated electrons in turn produce the radio waves we detect. In a pulsar, too, the charges radiate as they accelerate, in this case along the magnetic field, spiraling as they go (Figure 67.3A). The pulses of radiation we see are the collective emission from myriad charges pouring off the neutron star's surface. This radiation is beamed because the charges are traveling along the field lines that emerge from the star's surface at each magnetic pole. Because of this beaming, we see only neutron stars that have a magnetic pole that points toward Earth at some time as they rotate. Pulsars radiating in other directions are invisible to us, so many more neutron stars must exist than the approximately 1500 we have detected to date.

The emission created by the accelerating charges is called **synchrotron radiation,** named for a kind of particle accelerator used in physics experiments that accelerates charged particles using a magnetic field. This kind of radiation produces electromagnetic waves across a broad and continuous range of wavelengths. Synchrotron radiation is different from thermal radiation, which also produces radiation over a broad range of wavelengths (Unit 23), because its properties depend on the charged particles' acceleration and on the strength of the magnetic field, rather than on the temperature of a heated gas.

Most pulsars generate synchrotron radiation that is detectable only at radio wavelengths. However, some young pulsars, like the Crab Nebula, generate synchrotron radiation across the entire electromagnetic spectrum, including visible light and gamma rays. A high-speed camera has captured the flashes of visible light from the Crab Nebula's pulsar as it spins (Figure 67.3B).

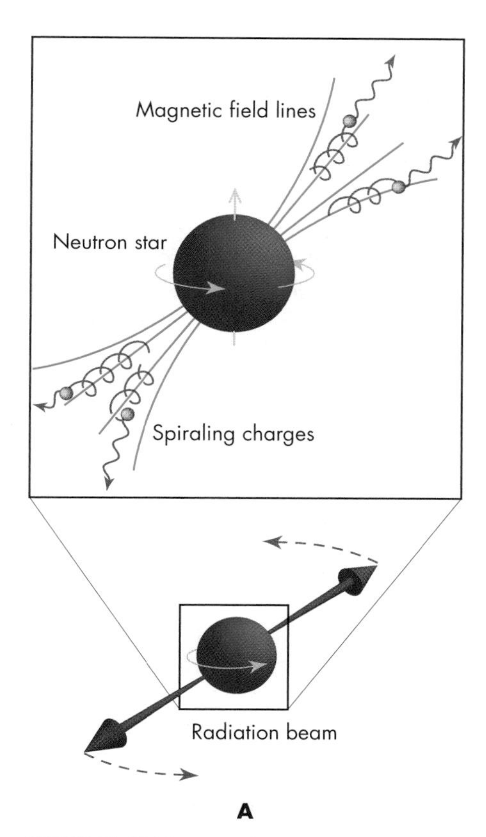

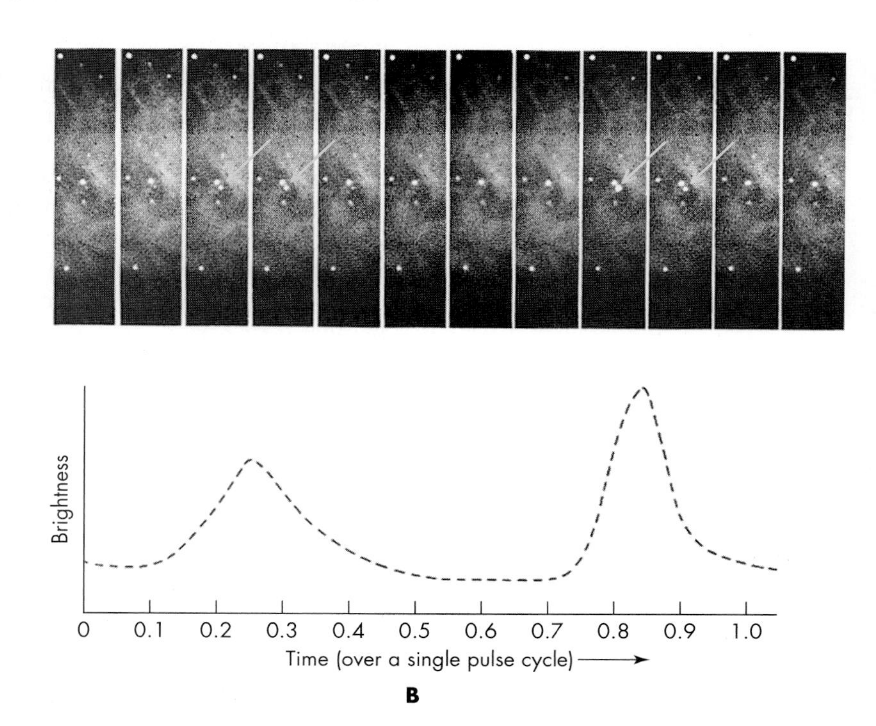

FIGURE 67.3
The strong magnetic field of a spinning pulsar generates a powerful electric field, which strips charges from its surface. (A) The charges spiral in the star's magnetic field and emit radiation along their direction of motion. Because the charges stream in a narrow beam confined by the the magnetic field, their radiation is also in a narrow beam at each magnetic pole of the star. (B) The rapid rotation of the pulsar in the Crab Nebula makes the star appear to turn on and off 30 times per second as its beams of radiation sweep across the Earth.

Q Astronomers suspect that the slowdown rate of pulsars is probably greater when they are young. Why might this be?

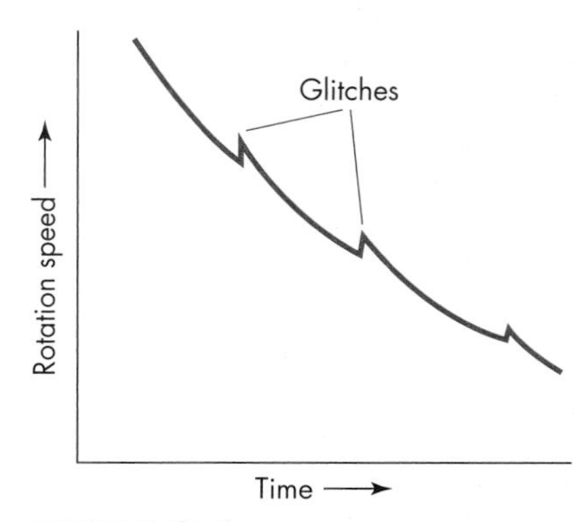

FIGURE 67.4
A pulsar's rotation speed gradually slows, but occasionally it suddenly speeds up during a glitch. This probably occurs as the crust readjusts its shape.

In everyday life, spinning objects slow down. Pulsars are no exception. As a pulsar spins, it drags its magnetic field through the particles that boil off its surface into the surrounding space. That drag slows down the pulsar. Astronomers measure this **spindown** of a pulsar by precisely timing the interval between pulses. Such measurements indicate that the time interval—the period of a spinning pulsar—gradually lengthens. The slowing rotation also reduces the energy of the radiation that the pulsar emits. Thus young, rapidly spinning pulsars emit electromagnetic waves from visible light to radio waves, whereas old, slowly spinning ones generate only radio waves.

Although pulsars gradually slow down overall, occasionally they suddenly speed up by a small amount, as illustrated by the graph in Figure 67.4. Such jumps in rotation speed are called **glitches,** and they can tell us about the interior structure of the neutron star. The sudden speedup occurs because the neutron star's shape changes during the star's overall slowdown. When it is spinning rapidly, the pulsar is larger at its equator because of the centripetal force outward. The Earth, Sun, and other spinning planets all bulge outward at their equators because of this same effect (Unit 42), but at the high spin rate of a pulsar, the distortion is much larger. As the spin slows, the neutron star settles toward a more spherical shape, and therefore its equatorial radius shrinks slightly. Because the matter moves closer to the axis of rotation, conservation of angular momentum causes the star to speed up a small amount before continuing its gradual slowdown.

Pulsar glitches reveal that the crust of the neutron star is rigid, because it does not deform gradually. Computer models suggest it is probably made of iron, a few hundred meters thick, as illustrated in Figure 67.5. The material inside appears to be a "fluid" of neutrons. The models suggest the neutron star also has a gaseous atmosphere—about 1 millimeter thick.

As the pulsar's rotation slows, its radiation weakens, and ultimately the star becomes undetectable. Thus, a pulsar "dies" by becoming invisible. But just as a white dwarf can be resurrected with infalling gases from a nearby companion star, so too can a pulsar in a binary system be reawakened. As material from the companion spirals into a pulsar, it adds angular momentum to the pulsar, causing it to spin faster and faster. The fastest pulsars known are called **millisecond pulsars** because their periods are just a few milliseconds long. They all appear to have been "revived" by this process.

Millisecond pulsars tend to be the exception, however, because pulsars are not frequently found in binary star systems. The supernova explosion that creates a pulsar usually expels so much mass that the gravitational attraction between the stars is no longer great enough to hold them together at their existing orbital speed. The sudden loss of mass can be compared to a ball spun around at the end of a string: If the string breaks, the ball flies off with the speed at which it was revolving. After the mass loss of a supernova explosion, the pulsar may fly off at its orbital speed—perhaps several hundred kilometers per second. It eventually escapes from the debris of the explosion that spawned it. Such speeding pulsars have been detected moving through our Galaxy at hundreds of kilometers per second.

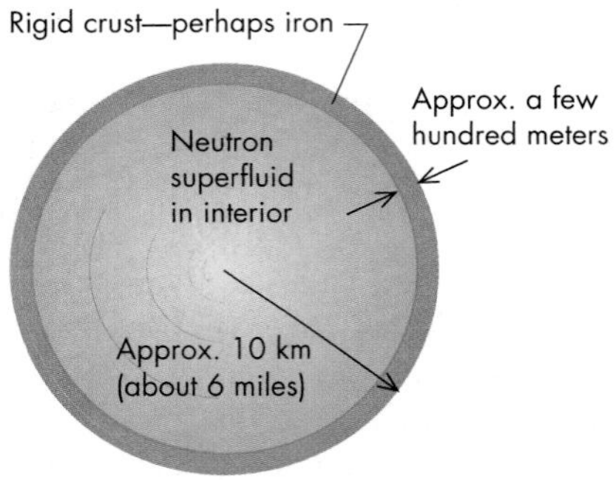

The nearest known neutron star to the Solar System probably escaped from a binary star system after it became a supernova. This object has the catalog name RX J1856.5–3754, and its fast motion is reflected by its shifting position seen in Hubble Space Telescope measurements made several years apart. Its distance is measured to be about 60 parsecs (about 200 light-years). RX J1856.5–3754 is not a pulsar, but the extremely faint thermal emission from its hot surface has been detected at visible and X-ray wavelengths. Its surface temperature is estimated to be about 700,000 K. Even though it is so hot, it is thought to be 1 million years old, based on its trajectory and distance from the nearest star-forming region.

Q If the Earth's polar caps melt, will the Earth's rotation speed change? How does this relate to the changes of rotation speed of a pulsar?

Rigid crust—perhaps iron

Approx. a few hundred meters

Neutron superfluid in interior

Approx. 10 km (about 6 miles)

FIGURE 67.5
Schematic structure of the interior of a neutron star.

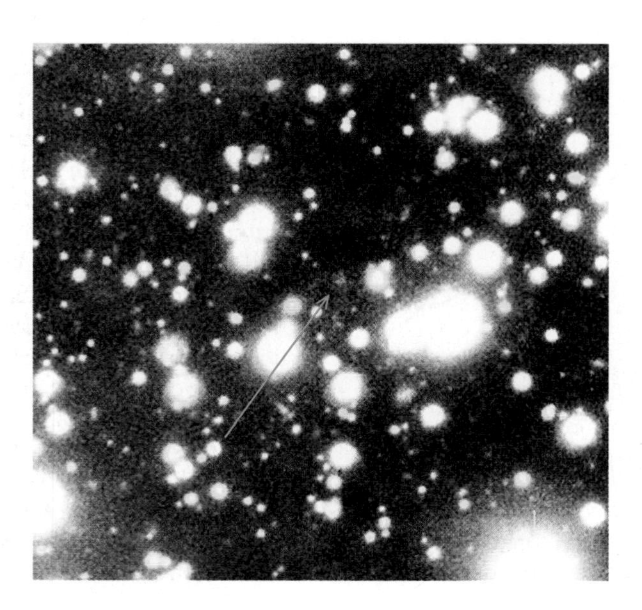

FIGURE 67.6
The nearest known neutron star to the Solar System, RX J1856.5–3754, is an extremely faint object shown in this image made with the VLT. An arrow points to the neutron star and shows its direction of motion through space. As the star plows through interstellar gas at high speed, it creates a "bow wave," seen here in pink from hydrogen emission.

Observations with the VLT 8-meter telescope in Chile show that it is creating a "bow wave" like a boat as it plows through interstellar gas (Figure 67.6). One intriguing aspect of this neutron star is that its diameter appears to be smaller and its temperature higher than most models have predicted. According to some computer models, these observations suggest that the neutrons in the star have split up into their constituent quarks (Unit 4). If true, this would be unlike any form of matter previously studied.

67.3 HIGH-ENERGY PULSARS

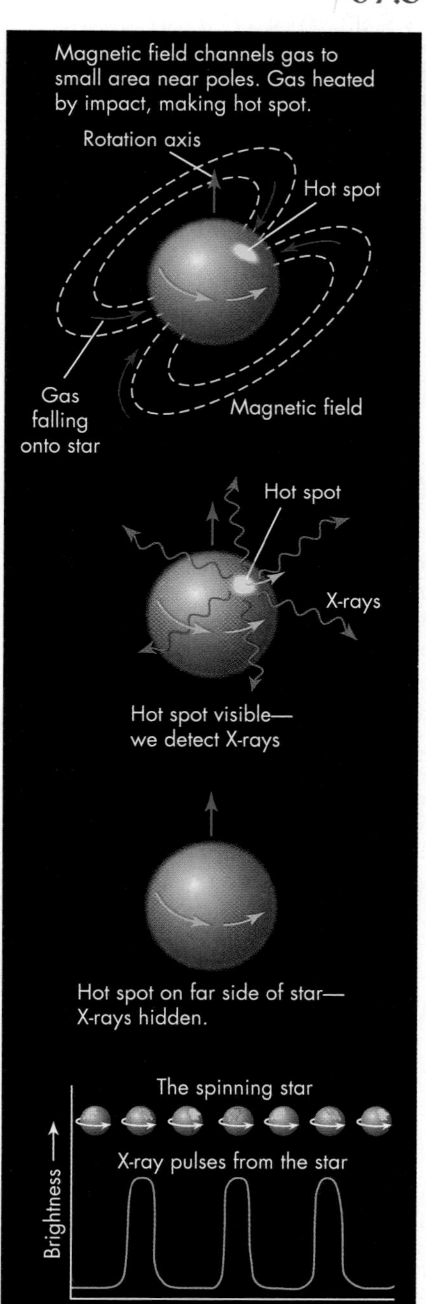

Astronomers have identified some additional classes of neutron stars that produce very high-energy pulsed radiation at X-ray and gamma-ray wavelengths. These appear to be powered by different sources than normal pulsars are—by either accreting matter or extremely strong magnetic fields.

X-ray pulsars generate pulses of X-ray radiation, as their name implies. However, besides the higher energy of the radiation, their pulsations are more erratic. The strength of the pulses is sometimes fairly steady and sometimes arrives in strong bursts, and the rate of pulsation is seen to speed up sometimes and to slow down at others. These appear to be neutron stars that remain in a binary star system after the supernova explosion.

An X-ray pulsar generates its emission from small "hot spots" near its magnetic poles. Gas from a companion star flows in along magnetic field lines. As the gas falls toward the neutron star, the star's intense gravity accelerates the gas. As it crashes onto the neutron star at the magnetic poles (Figure 67.7), the gas is compressed and heated intensely, causing it to emit X-rays. As the neutron star spins, the hot regions near each pole move into and then out of our field of view, creating the X-ray pulses that we observe. These X-ray pulsars are often measured to be speeding up—and perhaps someday will be millisecond pulsars.

Another type of pulsar is known as a **magnetar.** These are neutron stars with extremely intense magnetic fields. The magnetic field of a magnetar may be a thousand times stronger than the already intense magnetic fields of a normal pulsar, and more than 10 billion times stronger than the Sun's magnetic field. Magnetars have been identified from extremely intense bursts of X-ray and gamma-ray radiation. These bursts last for minutes, pulsing during the burst at the rate at which the neutron star is known to be spinning. During a burst, the magnetar can generate as much energy as the Sun produces in hundreds of thousands of years. A magnetar outburst in December 2004 was so powerful that it briefly shut down a number of satellites and ionized gas in the Earth's upper atmosphere. These effects occurred despite the huge distance to the magnetar—about 50,000 light-years.

Only a few magnetars have been identified to date, perhaps because their strong magnetic fields cause them to spin down much faster than normal pulsars. The leading hypothesis for explaining the enormous outbursts is that these neutron stars undergo massive glitches as they slow down, and these create huge shifts and rearrangements of the magnetic field. These might be analogous in some ways to the flare activity seen on the Sun, but vastly stronger because of the enormous strength of the magnetic field.

FIGURE 67.7
Gas falling onto a neutron star follows the magnetic field lines and makes a hot spot on the star's surface, creating X-rays. As the star rotates, we observe X-ray pulses.

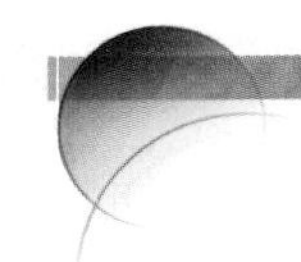

KEY TERMS

conservation of angular momentum, 539
glitch, 541
magnetar, 543
millisecond pulsar, 542
neutron star, 538
pulsar, 538
spindown, 541
synchrotron radiation, 540
X-ray pulsar, 543

QUESTIONS FOR REVIEW

1. What are the mass and radius of a typical neutron star, compared with those of the Sun?
2. What is a pulsar? In what sense does it pulsate?
3. How does a pulsar generate the pulses we see?
4. When in its lifetime does a neutron star produce pulses? What determines how rapid the pulses are?
5. What is a pulsar "glitch"? How does it affect the pulsar?
6. What is a magnetar? How can we tell magnetars from normal pulsars?

PROBLEMS

1. Calculate the escape velocity (Unit 18) from a white dwarf and a neutron star. Assume that each has 1.4 $M_{\odot}$. Let the white dwarf's radius be 10^4 km and the neutron star's radius be 10 km.
2. Suppose a neutron star with a radius of 10^4 km is spinning with a period of 0.001 sec. How fast is its equator moving? Compare this speed to the neutron star's escape velocity calculated in problem 1.
3. The mass of a neutron is 1.67×10^{-27} kg. How many neutrons are in a neutron star that has 1.4 $M_{\odot}$?
4. The volume of a neutron is about 10^{-45} m^3. Suppose you packed the number of neutrons you found in problem 2 into a sphere of radius R. The volume of a sphere is $4/3\ \pi R^3$. What is R in kilometers?
5. Suppose a neutron star is spinning with a period of 1.000 sec. If its radius shrinks by 1%, what will its new period be?
6. Assuming that a 1.4-$M_{\odot}$ pulsar has a radius of 10 km, calculate the acceleration due to gravity at the surface (see Unit 16). Compare this value to that of the Earth. How much would you weigh if you stood on this pulsar?

TEST YOURSELF

1. Which of the following has a radius (linear size) closest to that of a neutron star?
   a. The Sun
   b. The Earth
   c. A basketball
   d. A small city
   e. A gymnasium
2. What causes the radio pulses of a pulsar?
   a. The star vibrates like the quartz crystal in a watch, which sends electromagnetic waves through space.
   b. As the star spins, beams of radio radiation from it sweep through space. If one of these beams points toward the Earth, we observe a pulse.
   c. The star undergoes nuclear explosions that generate radio emission.
   d. The star's dark orbiting companion periodically eclipses the radio waves emitted by the main star.
   e. Convection inside the pulsar produces twisted magnetic fields much like what occurs during the Sun's sunspot cycle, but at a much faster rate.
3. If a spinning star suddenly collapses to a radius one-tenth of its former value, then
   a. its gravitational pull also decreases by a factor of 10.
   b. its mass also decreases by a factor of 10.
   c. the speed of rotation at the equator increases by a factor of 10.
   d. the period of rotation increases by a factor of 10.
   e. All of the above.

unit

68

Black Holes

Background Pathways

Unit 67: Neutron Stars 538

The iron core of a massive star is supported by pressure between the electrons in its interior. If enough mass is added to the core, the gravitational force grows so large that the star implodes into a neutron star. What happens if we gradually add mass to a neutron star? Will it collapse to a body with a yet smaller size? The answer appears to be that something fundamentally stranger happens. The nature of space and time around the object changes in such a way that the collapsing object disappears from view. It becomes an object that astronomers call a **black hole,** from which nothing can escape, not even light.

This strange fate is not limited to massive stars, but occurs for any concentration of enough mass in one place. When gravity grows large enough, it changes the nature of space in a way that prevents any other possible force from supporting the star. The effect that mass has on space was another of the extraordinary discoveries made by Albert Einstein in the early 1900s. He developed the **general theory of relativity** to describe these effects. General relativity is a theory of gravity. It explains not only what happens to turn an object into a black hole, but how any amount of mass or energy alters the nature of space and time in its vicinity.

Black holes are the most extreme example of how space can be altered by matter. This is the fate that awaits the most massive stars. Some astronomers think that stars that begin life more massive than about 20 $M_{\odot}$ collapse directly into black holes, with no accompanying supernova explosion. Likewise, a neutron star that grows too massive may simply disappear into the black hole that forms. Collisions of matter *outside* the black hole may be extremely violent, so the region immediately outside of black holes is full of radiation. However, general relativity makes it clear that no radiation, neutrinos, or any other particles can come out once they get closer than a certain distance from the center of the black hole. Anything that gets within that distance is lost from the universe as we know it.

Just as for Einstein's theory of special relativity (Unit 53), the term *theory* is used to describe general relativity to indicate that it is a thoroughly tested and verified hypothesis. The theory of general relativity is not just speculation, but has been verified by a wide range of experiments. Some modern technologies like Global Positioning System (GPS) units would not work if general relativity were not taken into account.

68.1 THE ESCAPE VELOCITY LIMIT

To understand black holes, we need to review the concept of escape velocity (Unit 18). **Escape velocity** is the speed a mass must acquire to avoid being drawn back by another object's gravity. For a body of mass M and radius R, the escape velocity, V_{esc}, for an object moving away from that body is

$$V_{esc} = \sqrt{\frac{2GM}{R}}.$$

In this equation G is Newton's gravitational constant, which has a value of 6.67×10^{-11} if V_{esc} is in meters per second, R is in meters, and M is in kilograms. For the Earth, the escape velocity is about 11,000 meters per second; for the Sun, it is about 600,000 meters per second.

You can see from the formula that because R is in the denominator, the escape velocity for an object of a particular mass will be larger as its radius shrinks. This is because the same mass packed into a smaller radius creates a larger force of gravity at its surface, making it more difficult to escape from the object.

A white dwarf whose mass is the same as the Sun's has an escape velocity much larger than the Sun's 600 kilometers per second, because the white dwarf's radius is so small. In fact, because its radius is about 100 times less than the Sun's, and because the escape velocity depends on the square root of the radius, its escape velocity is larger by a factor of $\sqrt{100} = 10$, making it about 6000 kilometers per second. An object leaving the surface of a white dwarf at a velocity of 6000 kilometers per second would therefore just overcome the white dwarf's gravity and be able to escape into space.

Let's now make the same calculation for a neutron star. If the neutron star has a radius 10^5 times smaller than the Sun's (or about 10 km) and it has a mass about twice the Sun's, its escape velocity is about $\sqrt{2 \times 10^5}$ larger than the Sun's escape velocity, or about 270,000 kilometers per second. This is approximately 90% of the speed of light! Hence if a neutron star is compressed just slightly, its escape velocity could exceed the speed of light. Such an object becomes a black hole.

As long ago as 1783, the English cleric John Michell discussed the possibility that objects with escape velocities exceeding the speed of light might exist. A little more than a decade later, the French mathematician and physicist Pierre Simon Laplace entertained the same idea. Following their logic, we can calculate the radius, R_S, at which a body of a given mass would become a black hole, by equating its escape velocity to the speed of light. We find then that a mass M becomes a black hole if it is compressed to a radius

R_S = Schwarzschild radius of black hole (m)

M = mass of object (kg)

G = Gravitational constant $= 6.67 \times 10^{-11}\ \mathrm{m^3/(kg \cdot sec^2)}$

c = speed of light $= 3.00 \times 10^8$ m/sec

$$R_S = \frac{2 \times G \times M}{c^2}.$$

We label this radius R_S and call it the **Schwarzschild radius** in honor of the German astrophysicist Karl Schwarzschild, who studied how to modify Newton's theory of gravity to account for effects near the speed of light and when gravity is extremely strong. His results, published in 1915, showed that the simple approach we have adopted here provides the correct result.

If we set the escape velocity to the speed of light, $c = \sqrt{2GM/R_S}$, then we can solve for the Schwarzschild radius, first by squaring both sides of the equation:

$$c^2 = \frac{2GM}{R_S}$$

$$c^2 \times \frac{R_S}{c^2} = \frac{2GM}{R_S} \times \frac{R_S}{c^2}$$

$$R_S = \frac{2GM}{c^2}$$

In principle, a body of any mass can be turned into a black hole if it is compressed to a small enough radius. For example, suppose we wanted to find out how small we would need to squeeze the Sun to make it a black hole. The Sun has a mass of 1.99×10^{30} kilograms, so if we plug in this mass and the speed of light (3.00×10^8 m/sec), we find that the Schwarzschild radius for the Sun is

$$\begin{aligned} R_S(\text{Sun}) &= \frac{2 \times G \times M_\oplus}{c^2} \\ &= \frac{2 \times 6.67 \times 10^{-11} \times 1.99 \times 10^{30}}{(3.00 \times 10^8)^2} \\ &= 2950\,\text{m} = 2.95\,\text{km}. \end{aligned}$$

So if the Sun could be compressed to a radius slightly smaller than 3 kilometers (1.9 miles), it would become a black hole. Similarly, a 2-$M_\odot$ would become a black hole if it were compressed to a radius of about 6 kilometers; a 3-$M_\odot$ object would become a black hole if it were compressed to a radius of 9 kilometers; and so on. So the radius of a black hole approximately equals 3 kilometers times its mass in solar masses.

Stars can be turned into black holes if they are compressed to sizes on the order of kilometers. However, black holes can be created, in principle, for any size mass. We could even turn the Earth ($M_\oplus = 5.97 \times 10^{24}$ kg) into a black hole if it could be compressed to a radius

$$\begin{aligned} R_S(\text{Earth}) &= \frac{2 \times 6.67 \times 10^{-11} \times 5.97 \times 10^{24}}{(3.00 \times 10^8)^2} \\ &= 0.0088\ \text{m} = 8.8\ \text{mm}. \end{aligned}$$

Q Thinking about the tidal effects of gravity (the effects of gravitational force at different distances), why might passing close to a small black hole be more destructive to a spaceship than passing close to a large one?

Thus an Earth-mass black hole would have a radius of just under 1 centimeter—about the size of a marble! However, it would require forces and circumstances far beyond anything we think exists today to turn the Earth into a black hole. Nonetheless, some astronomers hypothesize that early in the history of the universe such forces might have existed and might have formed Earth-mass black holes that survive to the present. At the other extreme, there is strong evidence inside galaxies for black holes with masses millions of times greater than the Sun's, although the mechanism for forming these massive objects is not yet clear.

The pathway for a high-mass star to become a black hole is much clearer. Suppose we consider a 2-$M_\odot$ neutron star that formed after a supernova explosion. If it had a companion star that remained in orbit with it after the explosion, mass transfer could begin adding mass to the neutron star, just as happens to some white dwarfs. Material falling onto the neutron star would be crushed to neutron star densities by the intense gravity, and gradually the neutron star's mass would increase. The escape velocity from the surface of the neutron star would grow larger and larger until the mass reached about 3 $M_\odot$, and then some last bit of matter would fall on the surface, pushing the escape velocity over the speed of light.

At that point, an odd thing happens. When we think of particles like neutrons pushing against each other, the "push" is a force between the neutrons. These forces must themselves travel from one point to another, and they can travel no faster than the speed of light. Anywhere inside the Schwarzschild radius, where the escape velocity exceeds the speed of light, none of the neutrons can communicate its "push" to the particle above it. No force can be communicated outward, so the star collapses inward, disappearing from sight like a boat sinking below the surface of water. What happens next is unobservable because it occurs inside the black hole, but it is thought that the collapse continues until the matter collapses to a single point—what astronomers call a **singularity.** Black holes are remarkable objects, and they reveal that our understanding of gravity based on Newton's law (Unit 16) is incomplete. To better understand how matter behaves in gravitational fields, we need to look in a little more detail at the relation between mass, gravity, and space.

68.2 CURVED SPACE

In 1916 Albert Einstein proposed his theory of general relativity, which provided a new mathematical and physical understanding of gravity. General relativity describes how mass alters the nature of space, creating what astronomers call **curvature of space.** In particular, according to general relativity, a black hole is a place where the curvature of space has become so extreme that a hole forms.

An analogy may help illustrate how gravity and curvature of space are related. Imagine a water bed on which you have placed a baseball (Figure 68.1). The baseball makes a small depression in the otherwise flat surface of the bed. If a marble is now placed near the baseball, it will roll along the curved surface into the depression. The bending of its environment made by the baseball therefore creates an "attraction" between the baseball and the marble. Now suppose we replace the baseball with a bowling ball. It will make a bigger depression, and the marble will roll in farther and be moving faster as it hits the bottom. We therefore infer from the analogy that the strength of the attraction between the bodies depends on the amount by which the surface is curved. Gravity also behaves this way, according to the general theory of relativity. According to that theory, mass creates a curvature of space, and gravitational motion occurs as bodies move along the curvature.

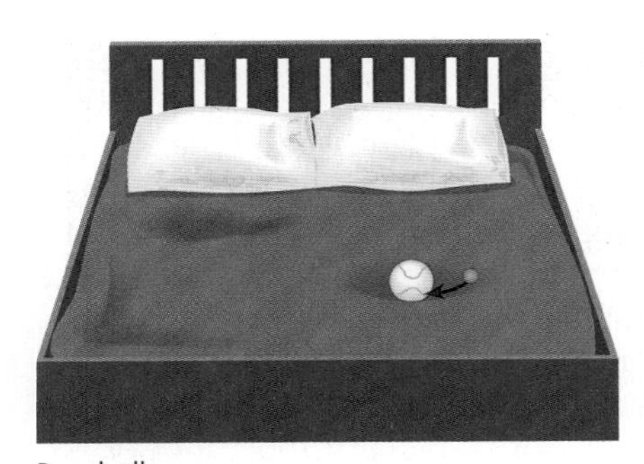

Baseball:
Marble rolls into depression

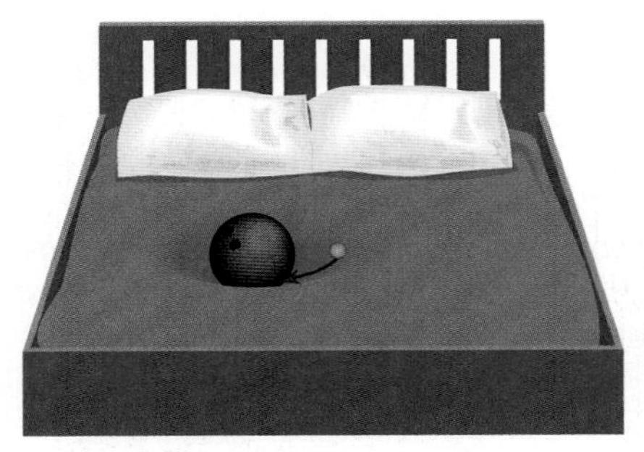

Bowling ball:
Marble rolls in faster

Big rock:
Marble disappears into hole

FIGURE 68.1
Objects on a water bed make depressions analogous to the curvature of space created by a mass. According to the general theory of relativity, the curvature of space produces the effect of gravity. We can see the similarity by placing a marble at the edge of the depression and watching it roll inward, as if attracted to the body that makes the depression. Bigger bodies make bigger depressions, so a marble rolls in faster. However, a very big body may tear the water bed, creating an analog of a black hole.

We can extend our analogy to black holes by supposing that we remove the bowling ball and put on the water bed a large boulder. The boulder presses down so hard on the vinyl that it tears a hole in it. A marble placed on the water bed will roll into the depression made by the boulder and disappear into the hole it made. So too, a black hole creates a "rip" in space where the curvature has become so strong that the structure of space is disrupted.

General relativity is a major departure from Newton's ideas about gravity. For example, in the old idea of a black hole, like that envisioned by Michell and Laplace, photons trying to leave the black hole should behave like rocks thrown upward at less than the escape velocity. That is, they would reach some maximum height and then fall back down again. If this were the correct description of how a black hole works, a black hole would be surrounded by photons that had traveled out and were momentarily stopped before they fell back in again. This contradicts observations that photons must always travel at the speed of light (Unit 53).

Problems such as this are resolved by general relativity. Another way of describing the curvature of space is that space can have a motion of its own. To see how, imagine a huge lake in which the water steadily drains out of a hole deep in its middle. Suppose you were in a motorboat on the lake: You would find the boat being pulled in toward the middle of the lake. At the edges of the lake this pull would be gentle, but as you drifted closer to the position of the drain, the current would grow faster. If the flow were powerful enough, you might even find that at a certain distance from the drain your boat was being pulled in by a flow of water so rapid that, even when operating at its top speed, the boat could make no progress away from the drain.

At the Schwarzschild radius of a black hole, space is flowing inward at the speed of light, just as the water in the imaginary lake is flowing toward the drain. A photon inside the Schwarzschild radius finds itself, like the boat, fighting a current moving faster than it can travel. Even though the photon is moving at the speed of light, the space through which it is traveling is "falling" into the hole at the speed of light, and thus the photon makes no progress outward and cannot escape (Figure 68.2A).

Black holes are not the only astronomical objects that curve space. The inward motion of space occurs around all masses. If you were in a motorboat traveling across the imaginary lake in our example, you would find that even a small flow of water toward the drain would bend the path of your boat into a curve. Photons behave similarly as they pass by any massive object. As a photon travels through space, its path is bent by the inward flow of space toward the mass, deflecting the direction of the light as illustrated in Figure 68.2B.

A total solar eclipse in 1919 gave astronomers their first opportunity to test general relativity's prediction that light would be deflected. Astronomers had accurate positions of the stars that would be seen near the Sun during the eclipse. They discovered that the stars were all deflected by exactly the amount predicted by Einstein's theory. Today astronomers do not need to wait for an eclipse to see objects whose light passes close to the Sun where the deflections can be measured. Distant objects producing radio waves, for example, show these same deflections. Astronomers have also detected the deflection of light around many other bodies. The deflections can sometimes even focus and brighten the light from a background object, as a lens does. This allows astronomers to detect objects that might not otherwise be visible (Unit 78).

The extreme curvature of space in a black hole, which prevents light from escaping from it, creates a kind of boundary around it that astronomers call the **event horizon.** Just as the curvature of the Earth's surface blocks our view of what lies beyond the horizon, so too the curvature of space at the Schwarzschild radius prevents our seeing beyond the event horizon into the interior of the

FIGURE 68.2
(A) Curved space around a black hole produces such a strong inward motion of space at the Schwarzschild radius that a photon moving outward at this radius cannot make any progress away from the black hole, and all photons and matter inside this radius must flow inward. (B) Curved space around the Sun bends a ray of light passing near the Sun.

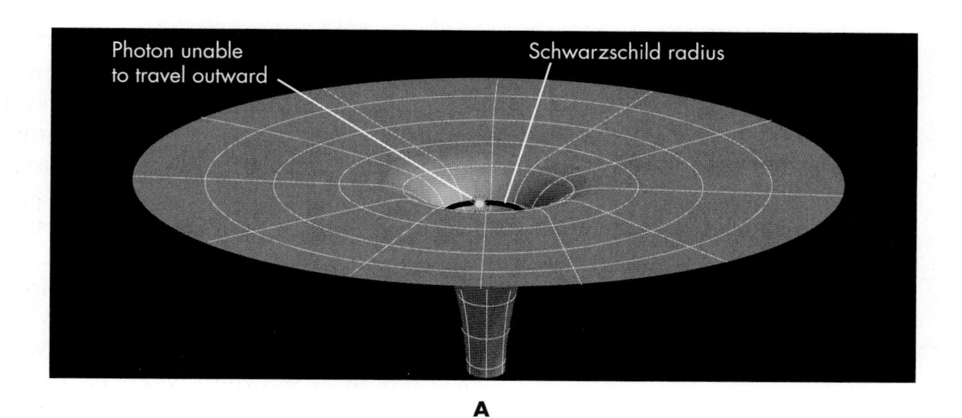

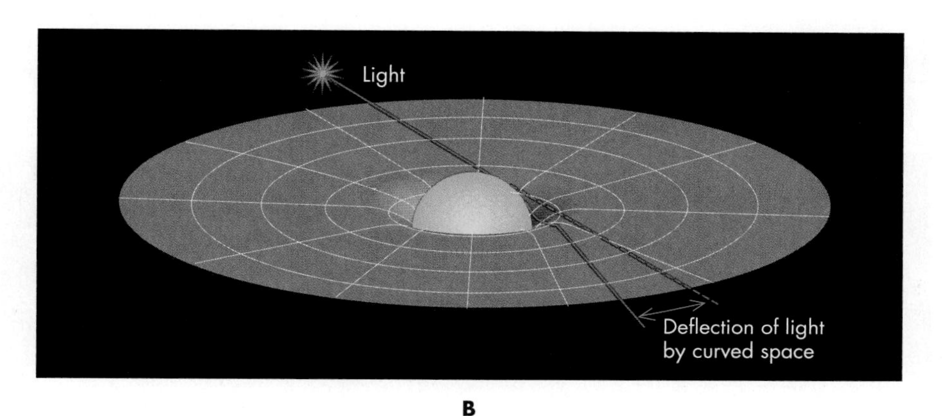

The motion of space into a black hole does not "remove space" from the universe like water flowing down a drain. Another analogy for this motion is an escalator that carries matter downward unless it is traveling upward at a greater speed than the escalator is moving down.

Besides having a gravitational field detectable beyond the Schwarzschild radius, black holes may also have an electric charge if, for example, an excess of positive charges falls into them. They may also have a spin—which will change the shape of the event horizon to be fatter at the equator than at the poles, like in other spinning masses.

black hole. All that happens within the black hole is hidden forever from our view. No radiation of any sort, nor any material body—rocket, spacecraft, or other object—can break free of its gravity if it travels past the event horizon. Because we cannot observe the interior of a black hole even in principle, there are only a limited number of physical properties we can ascribe to it. For example, it is meaningless to ask what a black hole is made of. It could be made of neutrons or cornflakes—only the amount of mass is important, not what it is composed of.

We can measure the mass of a black hole because its mass generates a gravitational field. In fact, at large distances the gravitational field generated by a black hole is no different from that generated by any other body of the same mass. For example, if the Sun were suddenly to become a black hole with the same mass it has now, the Earth would continue to orbit it without any change: It would *not* be pulled in. What makes a black hole so special is that it is so small for its mass. The small size means that it is possible for other objects to get to within an extremely small distance of the center. And when the separation between objects becomes small, gravity's force grows extremely strong because the gravitational force depends inversely on the separation.

68.3 OBSERVING BLACK HOLES

An object that emits no light or other electromagnetic radiation is not easy to observe. But just as you can "see" the wind by its effect on leaves and dust, so too astronomers can see black holes by their effects on their surroundings.

Suppose a massive star in a binary system undergoes core collapse, directly forming a 10-$M_\odot$ black hole. Gas from the companion star may be drawn toward the hole by its gravity, just as we know happens for neutron stars and white dwarfs. The infalling matter swirls around the black hole and forms an **accretion disk** whose inner edge lies just outside the Schwarzschild radius, as depicted in Figure 68.3. Here, where the disk orbits at nearly the speed of light, turbulence and friction heat the swirling gas to a furious 10 million K, causing it to emit X-rays and gamma rays.

As the black hole's companion star orbits it, the X-ray-emitting gas may disappear from our view as it is eclipsed by the companion star. An X-ray telescope trained on such a star system will show a steady X-ray signal that disappears at each eclipse, as shown in Figure 68.4. Such a signal might be the sign of a black hole, and at least three cosmic X-ray sources bear that signature.

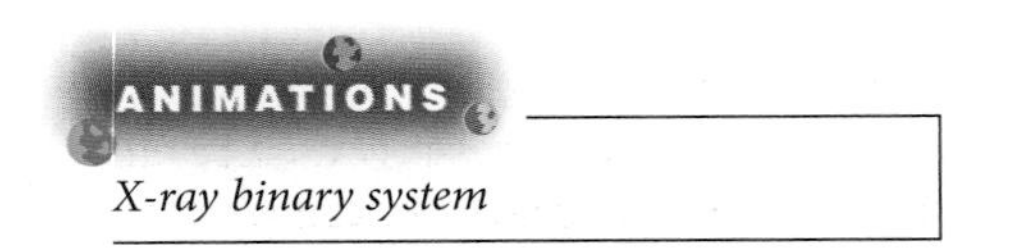

But how do we know an X-ray source is not just a neutron star? The answer involves the escape velocities we calculated earlier. If X-ray-emitting gas surrounds a body we cannot see, but whose mass exceeds 3 $M_\odot$, we can be reasonably

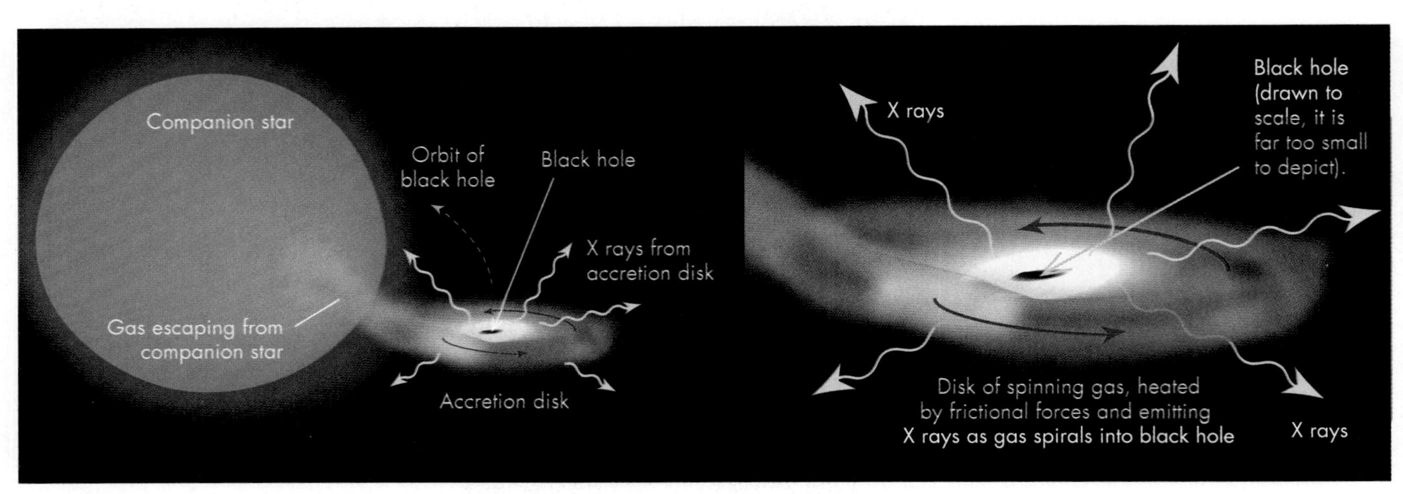

FIGURE 68.3
Black holes may reveal themselves by the X-rays emitted by gas orbiting them in an accretion disk. (Sizes and separations are not to scale.)

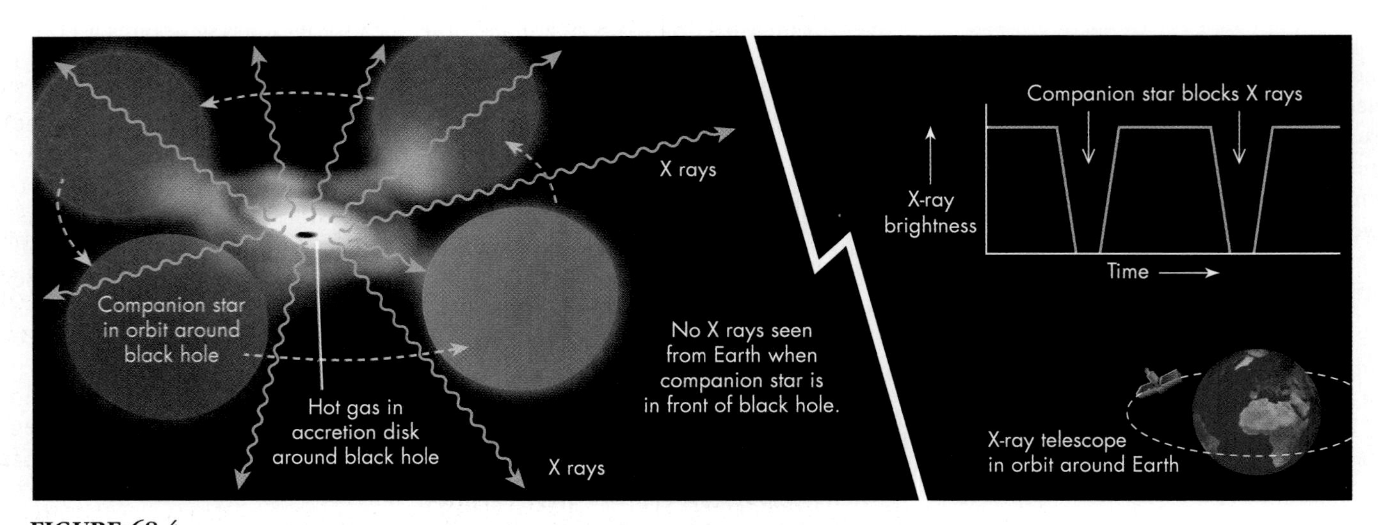

FIGURE 68.4
X-ray emission from gas around a black hole. The signal drops as the hole is eclipsed by its companion star. (Sizes and separations are not to scale.)

confident that the invisible object is a black hole. This is because if more than 3 $M_{\odot}$ of neutron-density matter is packed together, its radius will be smaller than the Schwarzschild radius and it therefore must become a black hole. No special transition is necessary for the neutron matter to become a black hole, simply enough of it has to be collected in one place.

Astronomers have good evidence for several such black holes. For example, Cygnus X-1, the first X-ray source detected in the constellation Cygnus (see Looking Up #4: Summer Triangle), consists of a B supergiant star and an invisible companion whose mass—based on an application of the modified form of Kepler's third law—is at least 6 $M_{\odot}$. Nothing we know of but a black hole can be so massive and yet invisible. An even better candidate is A0620-00, an X-ray-emitting object in the constellation Monoceros, the Unicorn. For A0620-00, the invisible star's mass is estimated to be 16 $M_{\odot}$.

Although no radiation can leave the interior of a black hole, the English physicist Stephen Hawking discovered that black holes emit radiation through a different process. Hawking noted that at the subatomic level, empty space undergoes constant fluctuations. For example, a particle and its antiparticle (such as an electron and a positron—see Unit 4) will spontaneously form. Such particle–antiparticle pairs are called **virtual particles** because they exist as a tiny energy fluctuation, although for only a minuscule fraction of a second. Conservation of energy requires them to rapidly recombine, annihilating each other and returning the energy they have momentarily "borrowed."

When such a virtual particle–antiparticle pair forms outside the event horizon of a black hole, one of the pair may fall into the black hole while the other escapes into space. This prevents them from recombining. But because energy must be conserved, some energy must be lost from the black hole, with the consequence that the black hole drops in mass according to $E = m \times c^2$. The particles created by this process do not actually come out of the black hole, but from a distance it would appear to an observer that they did. According to this model, the predicted **Hawking radiation** is extremely weak, and it has never been detected. However, such energy loss from a black hole would mean that a black hole would not last forever.

68.4 STRETCHING OF SPACE AND TIME BY GRAVITATION

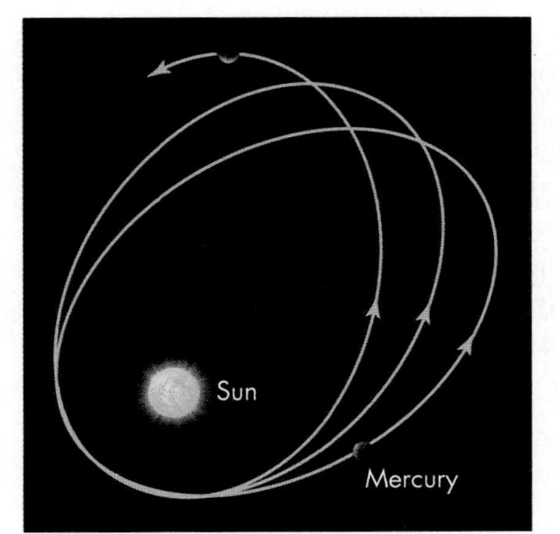

FIGURE 68.5
The orientation of Mercury's orbit shifts slightly each orbit by an amount predicted precisely by general relativity.

Einstein's theory of general relativity has become one of the cornerstones of modern physics. It predicts a number of surprising phenomena that have now been measured and precisely agree with Einstein's predictions, such as the gravitational bending of light discussed in the previous section.

When Einstein introduced the theory, he showed that it solved a long-standing problem in explaining a small deviation in the orbit of Mercury. Mercury's orbit does not repeat itself precisely as predicted by Kepler's laws (Unit 12), but instead shifts direction slightly with each orbit. Astronomers tried to explain this shift using Newton's laws to adjust for the influence of the other planets, but this still left an unexplained deviation.

General relativity predicts that the space near the Sun—deeper in the Sun's gravitational field—is stretched more than the space far away. In addition, general relativity predicts that time runs slower where gravity is stronger. Because Mercury's orbit is elliptical, when it is closest to the Sun at its perihelion (Unit 38), it is traveling through stretched-out space at a rate that is slightly too slow. As a result it does not travel as far as expected, making the orientation of the orbit shift slightly, as illustrated in Figure 68.5.

FIGURE 68.6
Time runs slower where the gravitational field is stronger, so electromagnetic waves emitted at a particular frequency from the surface of a star will have a lower frequency (longer wavelength) at larger distances. The waves lose energy as they travel away from the star, similar to how a ball loses energy rolling up a hill.

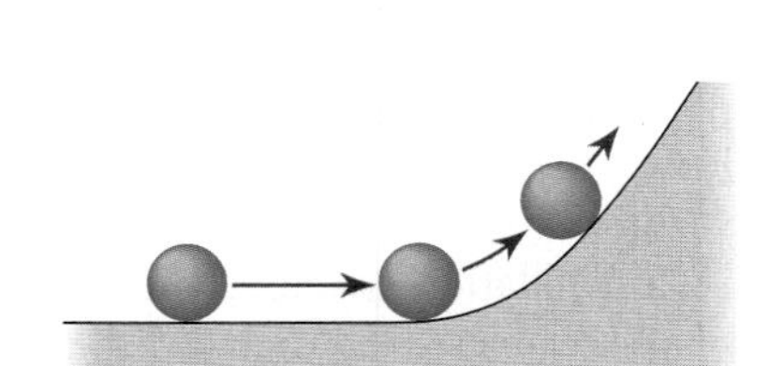

Ball rolling up a hill loses energy and slows down.

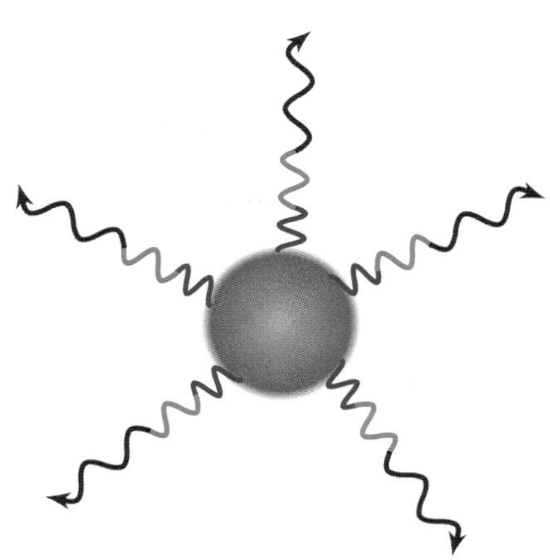

Light moving away from a star loses energy.

Q Would it be possible for an astronaut to visit the surface of a neutron star? If you could stand on a neutron star, what might be the effect on your body if time ran more slowly at your feet than at your head?

The distortion of time by gravity is also visible in the way gravity affects the wavelength of light emitted in a region where gravity is strong. Physical processes, including the frequency of vibrations in atoms, all occur more slowly where gravity is strong. When light is emitted from the surface of a white dwarf, for example, the spectral lines are all found to be at lower frequencies (longer wavelengths) than we would normally expect. This effect is called a **gravitational redshift,** because it shifts visible light toward the red end of the spectrum. For example, hydrogen on the surface of a white dwarf may emit from its surface a spectral line of a known frequency; but when that light reaches a distant observer whose clock is running faster, the waves have a lower frequency, as illustrated in Figure 68.6. Physics experiments have directly measured gravitational redshift, even between the top and bottom of a building. This effect also results in measurable differences between the time measured on a satellite and the time measured on the ground.

This stretching of time means that if one astronaut stayed in a spaceship and watched her twin brother travel down to the surface of a neutron star, the twin would appear to be moving, speaking, and breathing more slowly than normal when he was in the strong gravitational field. On the other hand, to the twin visiting the surface of the neutron star, it would appear that his sister on the spaceship was moving, speaking, and breathing faster than normal. When he returned to the spacecraft, he would discover that less time had passed according to his clocks than had passed according to the clocks on the spaceship. He would be slightly younger than his sister.

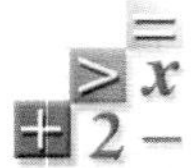

There is a gravitational Lorentz factor γ_G that is mathematically similar to the Lorentz factor of special relativity (Unit 53):

$$\gamma_G = \frac{1}{\sqrt{1 - R_S/R}}$$

where R is the distance from the center of a massive object, and R_S is the object's Schwarzschild radius. This factor describes the amount by which both time and space are stretched by gravity.

The degree by which time slows depends on how close you get to the Schwarzschild radius. To an outside observer, time appears to stop at the Schwarzschild radius; while far from any masses, time runs at its normal rate. In principle then, if you could spend some time just outside the Schwarzschild radius of a black hole, when you came back out (*if* you could get back out) hundreds of years might have passed in the rest of the universe!

68.5 GRAVITATIONAL WAVES

Another prediction of Einstein's theory of general relativity is that when one object orbits another, the motion of the two objects generates **gravitational waves.** Just as ripples spread away from a stone tossed into a pond, so too gravitational waves spread across space, stretching and distorting the space through which the waves

FIGURE 68.7
Gravitational waves generated by a rapidly orbiting pair of neutron stars. Such waves are "ripples" in space—and like ripples in water, they make objects move slightly as they pass.

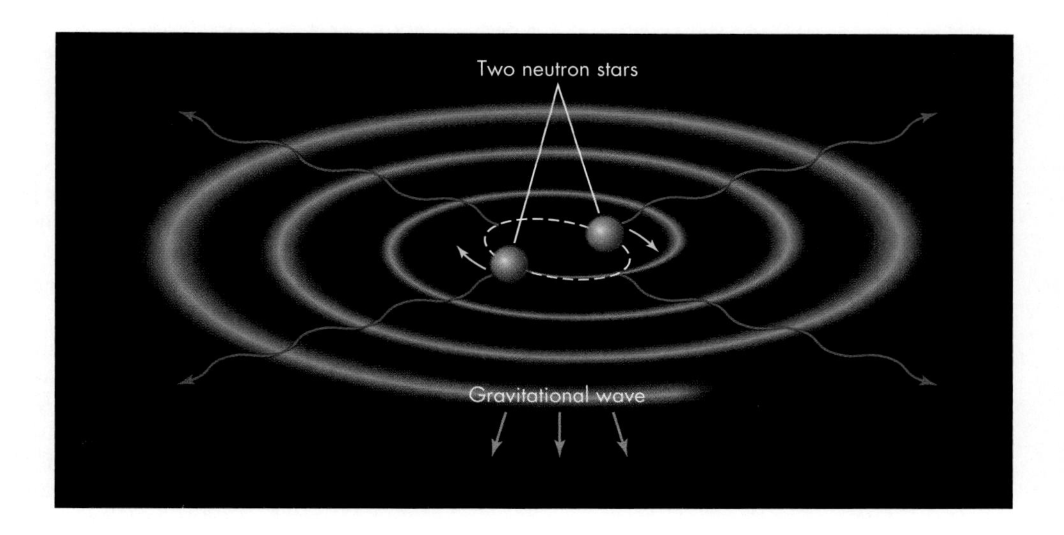

move (Figure 68.7). These waves have not yet been directly detected, but U.S. scientists have recently built a large gravitational wave detector called LIGO, the Laser Interferometer Gravitational Wave Observatory, to search for them. LIGO consists of vacuum tunnels, 4 kilometers long, in which light is reflected back and forth between mirrors. If a gravitational wave of sufficient strength passes through the detector, it will shift the distance between the mirrors in a predictable way, slightly altering the light patterns between the mirrors. The detector has begun operating, and its sensitivity is being steadily increased. It is expected to eventually have the sensitivity to measure these passing distortions of space and time.

Although gravitational waves have not yet been directly detected, indirect evidence of such waves has been observed from rapidly orbiting neutron stars. Such a detection is possible because as the waves carry gravitational energy away from the orbiting stars, the stars spiral inward toward each other. Exactly such behavior was detected in 1974 by the U.S. astronomers Joseph Taylor and Russell Hulse, who discovered a pulsar in a binary system with another neutron star. Their observations showed that as the neutron stars orbit each other, they are gradually losing energy and the two stars are spiraling toward each other. Taylor and Hulse were able to measure this inward motion extremely precisely because the pulsar member of the pair acts as an extraordinarily precise clock.

Taylor and Hulse showed that the rate of orbital decay exactly matches the rate of predicted energy loss in the form of gravitational waves. This has become one of the strong tests supporting the theory of general relativity. For this discovery, Taylor and Hulse were awarded the 1993 Nobel Prize in physics.

Q Suppose you were traveling in a spaceship and calculated that your trajectory might carry you within a million kilometers of the center of a 5-$M_\odot$ black hole, or the same distance from the center of a 5-$M_\odot$ main-sequence star. Why would the main-sequence star present the bigger danger of your crashing?

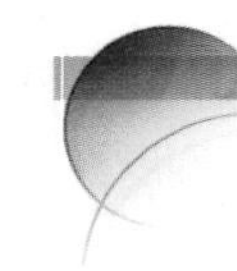

KEY TERMS

accretion disk, 550
black hole, 545
curvature of space, 547
escape velocity, 545
event horizon, 548
general theory of relativity, 545
gravitational redshift, 552
gravitational wave, 552
Hawking radiation, 551
Schwarzschild radius, 546
singularity, 547
virtual particle, 551

QUESTIONS FOR REVIEW

1. What is a black hole?
2. What is the Schwarzschild radius? In what sense is this an "event horizon"?
3. Why is Newton's law of gravity insufficient to explain black holes?
4. What mass stars are expected to become black holes? What mass of neutron stars must turn into black holes?
5. What would you observe if you watched a friend travel into a black hole?
6. What happens to time near the event horizon of a black hole?
7. In addition to black holes, what other phenomena does general relativity explain?

PROBLEMS

1. Calculate the Schwarzschild radius of a 3-$M_\odot$ object.
2. Calculate *your* Schwarzschild radius. How does that compare to the size of an atom? How does it compare to the size of a proton?
3. Some galaxies show evidence of very massive black holes at their centers. Calculate the radius of a billion-$M_\odot$ black hole. Calculate its density in kilograms per liter. (*Reminder:* There are 1000 liters in a cubic meter.)
4. Calculate the density of a black hole with a billion (10^9) times the mass of the Sun. (Use $4/3\ \pi R_S^3$ as the volume of the black hole.) Compare your result to the density of water, and explain whether larger black holes have higher or lower densities.
5. If a neutron star is 10% bigger than its Schwarzschild radius, how slowly would a clock run on its surface? (Use the formula $\gamma_G = 1/\sqrt{1 - R_S/R}$ to calculate the gravitational time-stretching factor.)
6. Suppose you observe two friends. One takes her spaceship to an orbit around a black hole a distance of 1.1 R_S from the singularity (i.e., 10% farther away from the singularity than the event horizon is). The other moves in a straight line at 0.1 c. Which friend has the greatest time-stretching factor (as observed by you)? (See Unit 53 for the special relativity Lorentz factor).
7. What is the radius and density of A0620-00, assuming that it is a 16-$M_\odot$ black hole?
8. Neutron-density matter is close to being incompressible. Assuming that it is, we can calculate how big a mass of neutron-density matter turns into a black hole.
   a. Calculate the density ρ of a 1.4-$M_\odot$ neutron star, which has a radius of 10 km.
   b. The mass of a neutron star is equal to its volume times its density. Using the density you found in (a), multiply it by the volume ($4/3\ \pi R^3$) and use the Schwarzschild radius formula to determine the radius when $R = R_S$.
   c. Use the radius you found in (b) to calculate the mass of this neutron star.

TEST YOURSELF

1. What evidence leads astronomers to believe that they have detected black holes?
   a. They have seen tiny dark spots drift across the faces of some distant stars.
   b. They have detected pulses of ultraviolet radiation coming from dark regions of space.
   c. They have seen X-rays from accretion disks orbiting dark massive objects.
   d. They have seen a star suddenly disappear as it was swallowed by a black hole.
   e. They have looked into a black hole with X-ray radar telescopes.
2. The Schwarzschild radius of a body is
   a. the distance from its center at which nuclear fusion ceases.
   b. the distance from its surface at which an orbiting companion will be broken apart.
   c. the maximum radius a white dwarf can have before it collapses.
   d. the maximum radius a neutron star can have before it collapses.
   e. the radius of a body at which its escape velocity equals the speed of light.
3. If by some unknown process the Sun suddenly collapsed in on itself and became a black hole tomorrow, the planets would
   a. move off into space in the direction they were moving when the Sun collapsed.
   b. be dragged inward and sucked into the black hole.
   c. be blasted into pieces by the neutrinos emitted by the black hole.
   d. continue orbiting the black hole just as they orbit the Sun now.
   e. also collapse into black holes because of the gravitational waves generated.

unit 69

Star Clusters

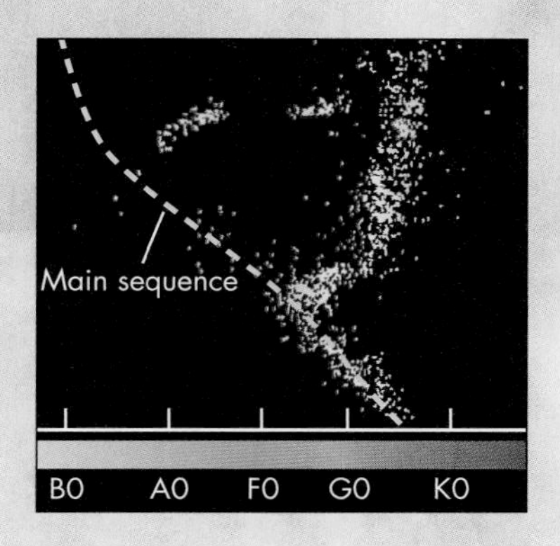

Background Pathways

Unit 59: Overview of Stellar Evolution 478

Groupings of stars called **star clusters** provide a unique environment to test our ideas about stellar evolution. They also provide us with an important transition for understanding the behavior and dynamics of the much-larger collections of stars known as galaxies, which we examine beginning in Unit 70.

Stars usually form in clusters out of huge clouds of interstellar gas (Unit 60). When a cloud containing thousands or even millions of solar masses of gas and dust begins to contract under its own gravity, smaller clumps begin to form. Each clump forms a star or system of stars, and possibly planets as well. Some clumps are disrupted by the firstborn stars before they too can form a star, but typically hundreds or thousands of stars are born almost simultaneously out of nearly identical material.

Each star in a cluster moves along its own orbit about the center of mass of the cluster, held to the cluster by the gravitational attraction of all the other stars. Many clusters are loosely bound, and after the remnants of the original gas cloud are driven away—by stellar winds, radiation pressure, and exploding stars—the stars may drift apart. Because the stars within a cluster form at approximately the same time, they are all nearly the same age. This allows us to determine what kinds of stars die more quickly than others, and we can learn about the relative numbers of stars of each type that normally form.

69.1 TYPES OF STAR CLUSTERS

Star clusters range from loose associations of tens of stars to dense concentrations of millions of stars. The spacing between the stars in a cluster varies enormously. In some clusters the stars are loosely packed, so the cluster is only slightly denser than its surroundings. In other clusters, however, the stars are so closely spaced that their separation may be as little as a tenth of a light-year. Yet even in such dense star clusters, the spacing between stars is still large compared with the sizes of stars. Thus, although the stars in a cluster sometimes look quite crowded in a photograph, the likelihood of a physical collision between them is near zero.

The star cluster nearest to us is about 75 light-years away and is about 30 light-years across. It is so close that it covers a large portion of the sky. Called the **Ursa Major group,** it includes most of the stars of the Big Dipper (see Looking Up #2: Ursa Major) and a number of stars from neighboring constellations. About a hundred stars have been identified as members of this group by their common location and motion through space, including the brightest star in the constellation Corona Borealis and the second brightest in Auriga.

With your unaided eyes you can see several clusters that are richer than the Ursa Major group. One of the most obvious is the **Pleiades** (Figure 69.1A), or the "Seven Sisters," named for the daughters of the giant Atlas, who in Greek mythology carried the world on his shoulders. The Pleiades (pronounced *PLEE-ah-deez*) are visible as a tiny group of stars north of the V in the constellation Taurus (see Looking Up #5: Taurus). The V in Taurus is another cluster of stars called

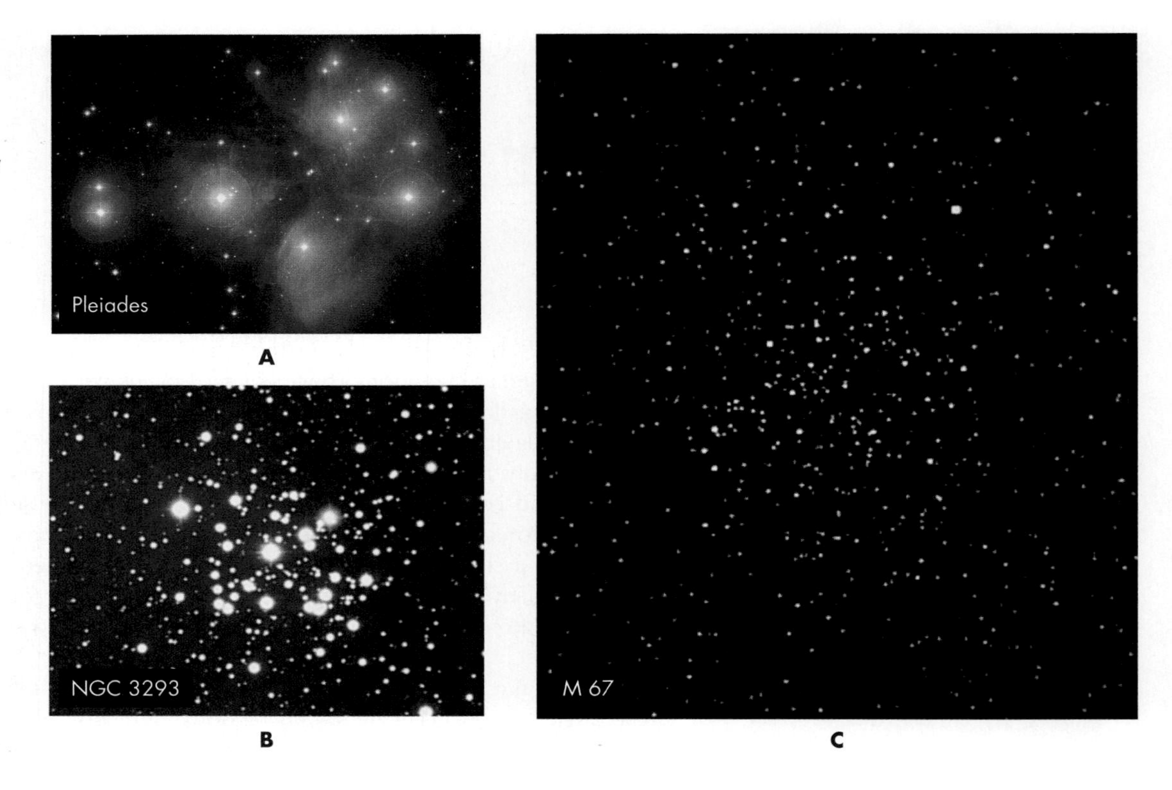

FIGURE 69.1
Open clusters. (A) The Pleiades in the constellation Taurus; (B) NGC 3293 in Carina; (C) M67 in Cancer.

Note that the brightest stars in the Pleiades form almost a miniature "dipper" shape. They are sometimes confused with the "Little Dipper" of the constellation Ursa Minor.

the Hyades (pronounced *HI-ah-deez*), named for the half-sisters of the Pleiades in mythology. The two clusters both happen to lie in the same constellation, but they are not otherwise associated. The Hyades are much closer to us—about 150 light-years away—while the Pleiades are 440 light-years distant. These clusters are highest in the evening sky in December and January. With binoculars you can see that the Pleiades and Hyades each contain hundreds of stars—far more than the half dozen or so stars visible to the unaided eye.

More than a thousand other **open clusters** are cataloged in our Galaxy; they contain up to a few thousand stars in a volume with a radius of typically 7 to 20 light-years. They are called *open* because their stars are scattered loosely, as you can see in the sampling of other open clusters shown in Figure 69.1. Astronomers think that open clusters form when giant, cold interstellar gas clouds are compressed and collapse, breaking up into hundreds of stars whose mutual gravity binds them into the cluster. Once formed, the stars of an open cluster continue to move through space together; but over hundreds of millions of years the stars gradually drift off on their own, so the cluster eventually dissolves. The high number of blue stars in the Pleiades suggests that this cluster is fairly young, perhaps only some tens of millions of years old. The Hyades and the Ursa Major group, by contrast, are both estimated to be many hundreds of millions of years old, and they appear to be less tightly bound than the Pleiades. Our own Sun was probably a member of such a star group, but its companion stars have long since scattered across space.

Other stars sometimes occur in loose groups called **stellar associations** that are a few hundred light-years across. Associations typically spread out from a single large open cluster near their center and may contain other, smaller star groupings. Moreover, the stars in associations are usually still mingled with the massive clouds of dust and gas from which they formed. The stars in associations probably form at about the same time, perhaps from the same triggering event. They have at most a very weak gravitational link to one another and often have different motions, so they disperse rapidly.

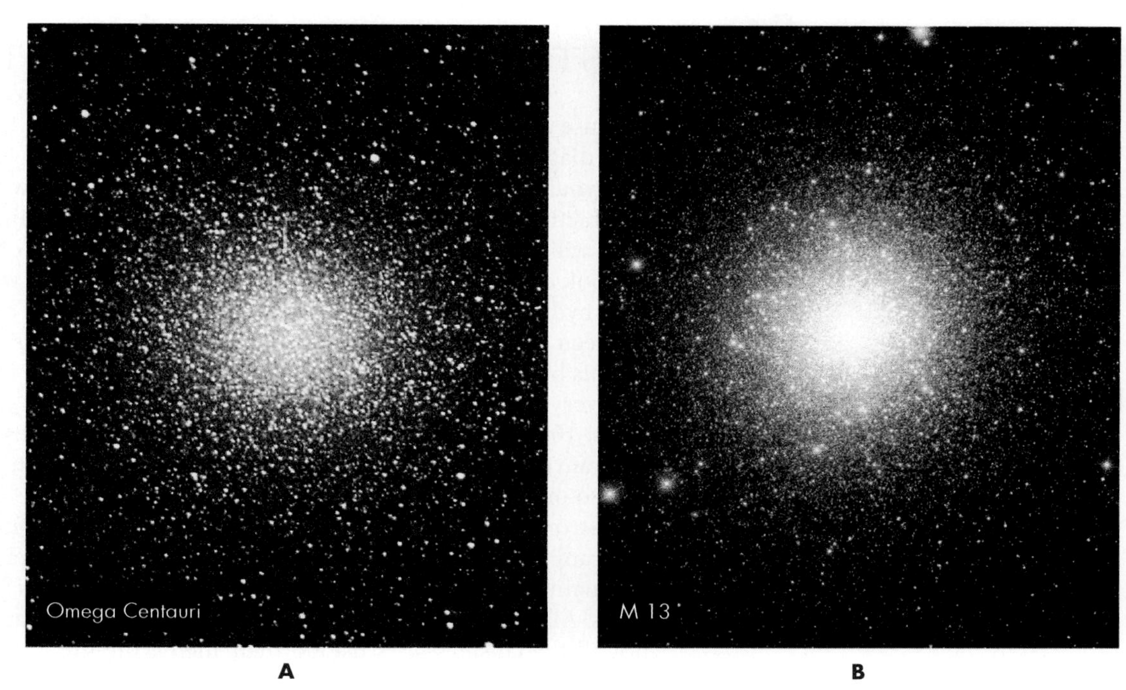

FIGURE 69.2
Globular clusters. (A) Omega Centauri in the constellation Centaurus; (B) M13 in Hercules.

Q What would the sky look like if you lived on a planet orbiting a star within a globular cluster?

A far denser type of cluster, called a **globular cluster,** is much more strongly bound together, and is larger in size and stellar content than an open cluster. These clusters contain from a few hundred thousand to several million stars and have radii of about 40 to 160 light-years. The stronger gravity in globular clusters pulls their stars into a dense ball, as you can see in Figure 69.2. About 150 globular clusters have been cataloged. Although they are part of our Galaxy, most of them are so distant that despite the many stars they contain, they are not easily seen without a telescope. A notable exception to this is the globular cluster Omega Centauri in the southern constellation Centaurus (Figure 69.2A; also see Looking Up #8: Centaurus and Crux, the Southern Cross). Cataloged as the 24th brightest "star" in the constellation, it is clearly fuzzy even to the naked eye. In northern skies in the constellation Hercules, a globular cluster cataloged as M13 can sometimes be seen by eye as a faint fuzzy patch (Figure 69.2B). A small telescope shows both of these clusters to be a large swarm of stars.

Table 69.1 summarizes some properties of open clusters, globular clusters, and stellar associations.

TABLE 69.1 Properties of Clusters and Associations

Type	Number of Stars	Radius*
Open cluster	Tens to a few thousand	7–20 ly (2–6 pc)
Globular cluster	10^5–10^6	40–160 ly (12–50 pc)
Associations†	5–70 O or B stars	65–325 ly (20–100 pc)

*Because star clusters do not have sharp edges but instead gradually thin out from their center, quoted dimensions differ substantially.

†Astronomers identify several other types of associations as well. For example, T associations are regions with above-average numbers of T Tauri stars.

69.2 TESTING STELLAR EVOLUTION THEORY

Because the stars in a given cluster are approximately the same age, a cluster's H-R diagram shows a snapshot of the state of evolution of its stars. For example, in a young cluster, we expect that more-massive stars will have completed their contraction stage from protostars and will lie on the main sequence, and that no stars will have had time to become red giants and shift off the main sequence. If we look at an older cluster, however, some massive stars will have consumed their core hydrogen and evolved off the main sequence. We can use such differences between cluster H-R diagrams to check our theory of stellar evolution. We can do this by calculating evolutionary tracks (such as those in Figures 64.2 and 66.1) for every star on the main sequence to show where each star will be at 10 million years, 100 million years, and so on. The resulting curves show us what the H-R diagram of the entire cluster should look like at 10 million years, 100 million years, and so on after its birth.

Astronomers can deduce the age of a star cluster from the pattern of its H-R diagram. To understand why a cluster's H-R diagram reveals its age, recall that the main sequence is determined by the location of stars fusing hydrogen in their cores. All the stars in a newly formed cluster lie on or near the main sequence. But not all the cluster's stars use up their fuel at the same rate. Massive stars use up their fuel more rapidly (to maintain their high luminosity) than low-mass stars do. With their hydrogen used up, high-mass stars leave the main sequence and turn into red supergiants. The low-mass stars, on the other hand, still have hydrogen to fuse, so they remain on the main sequence longer.

Because the high-mass stars in an older cluster turn into red giants (and therefore lie off the main sequence), a line in the H-R diagram connecting the position of the cluster's stars bends away to the right, as shown in Figure 69.3. The point where that line bends away from the main sequence is called the **turnoff point.** A star just below the turnoff point is not yet old enough to have used up the hydrogen in its

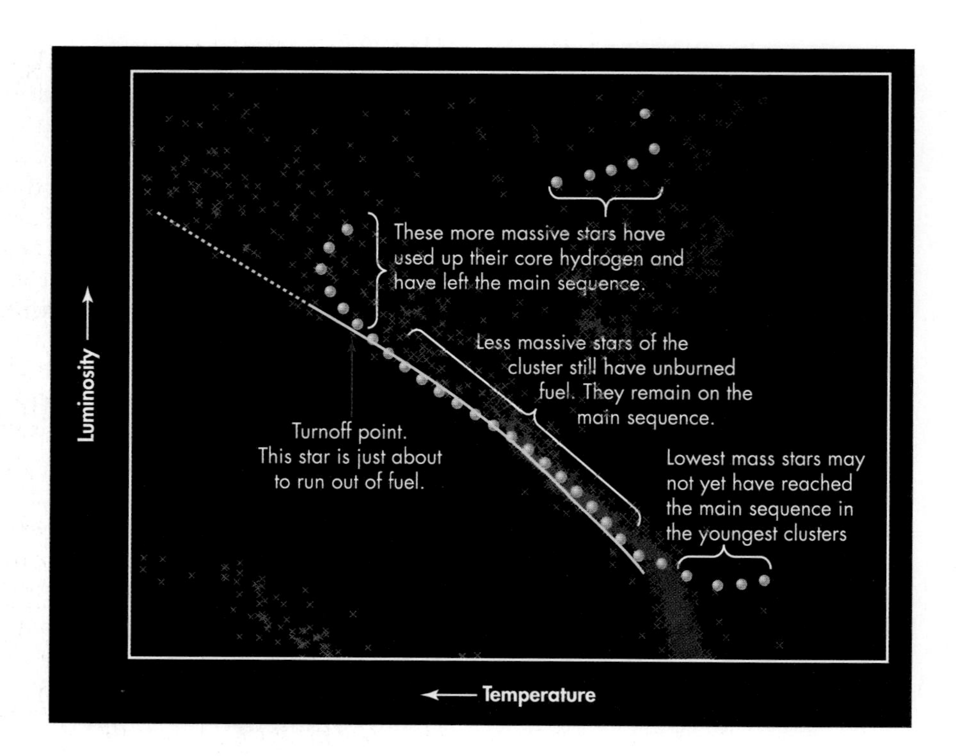

FIGURE 69.3
The pattern of stars in the H-R diagram of a star cluster indicates its age.

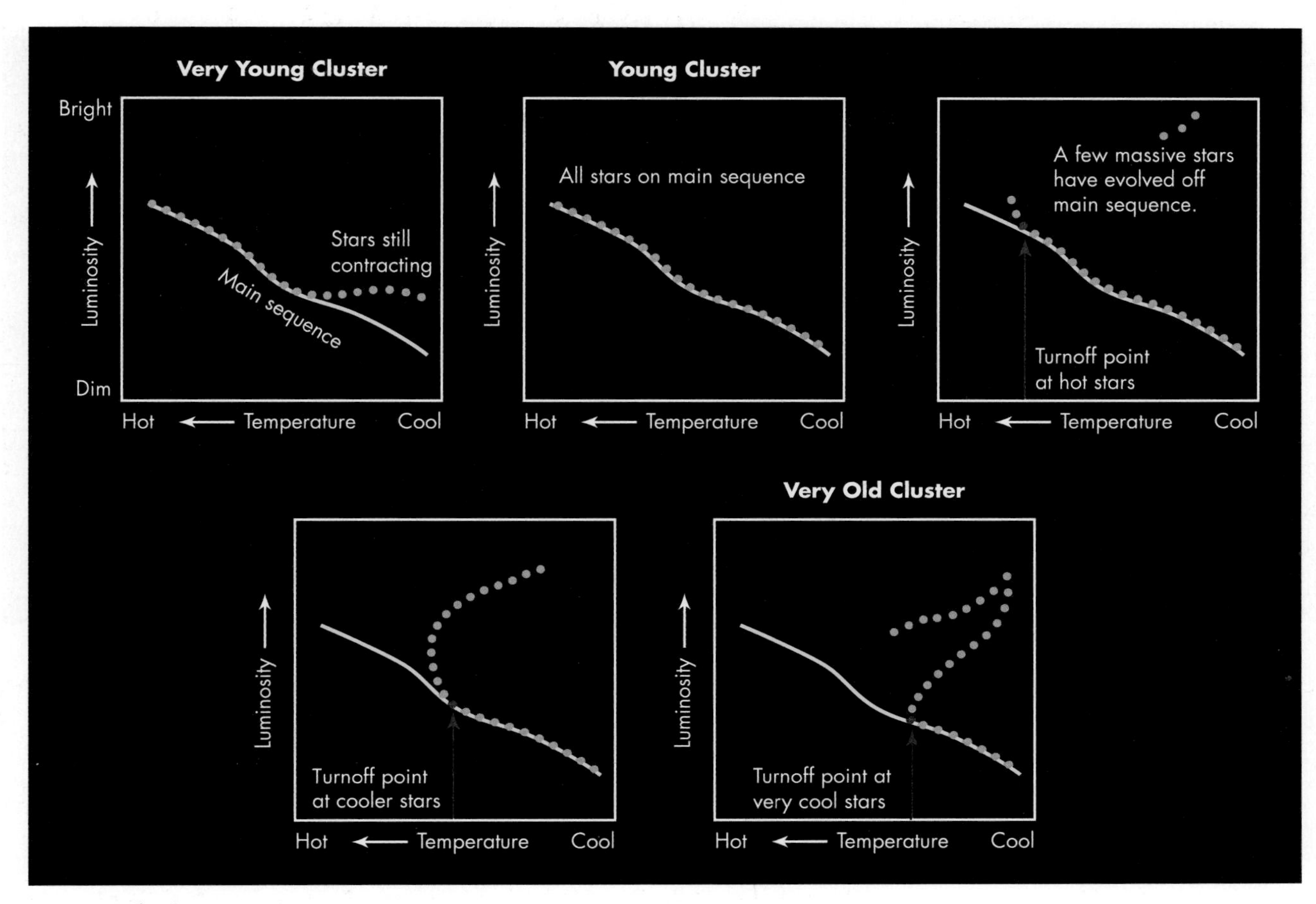

FIGURE 69.4
Schematic H-R diagrams of clusters of different ages illustrating the turnoff point and how it shifts down and to the right for older clusters.

core, but it is about to run out. That is, its age is just a tiny bit less than its total main-sequence lifetime. But all stars in the cluster are the same age—namely, the age of the cluster. To get the cluster's age, we therefore determine how long a star at the turnoff point can live by calculating its main-sequence lifetime from its mass and luminosity, as described in Unit 61. The answer we get is the age of the star cluster.

Figure 69.4 shows a series of schematic H-R diagrams for clusters ranging from very young to very old (a few million years for the youngest to more than 10 billion years for the oldest). Notice that old clusters have few, if any, stars on the upper part of the main sequence. On the other hand, short-lived stars are still present on the upper main sequence of young clusters.

When we compare such curves with H-R diagrams of actual star clusters, as in Figure 69.5, the match is excellent. The examples in the figure are very similar to the first and last panels of Figure 69.4, and their estimated ages are 10 million and 7 billion years for the open cluster and globular cluster, respectively. If our theory of stellar evolution were wrong, the shapes would be unlikely to agree so well. In addition, the models help astronomers understand why so many kinds of stars exist. For example, the models not only offer a natural explanation of features such as main-sequence, red giant, and white dwarf stars; they also help us see how totally different objects such as pulsating variables, planetary nebulae, and supernovae fit into the overall scheme of stellar evolution. This success in interpreting such stellar diversity is evidence that our theory of stellar evolution is essentially correct. Moreover, this success lets astronomers measure the age of a star cluster.

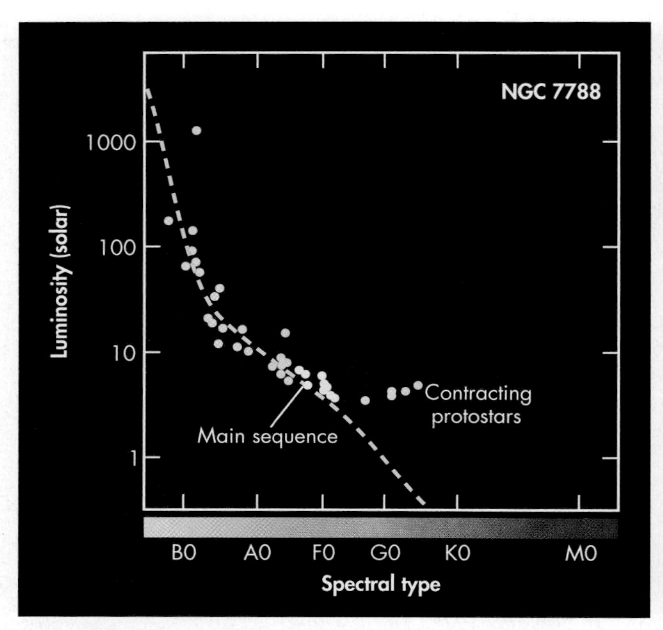

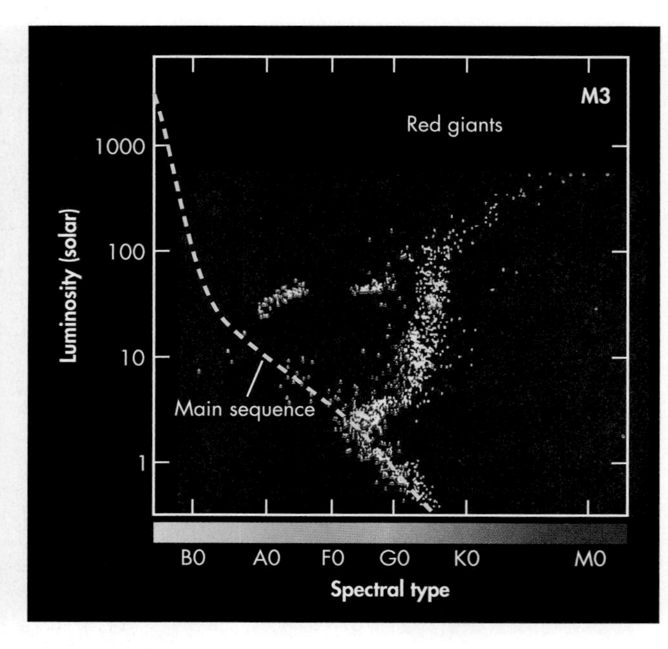

FIGURE 69.5
The H-R diagrams of two clusters. NGC 7788 is an open cluster in Cassiopeia with an estimated age of about 10 million years. M3 is a globular cluster in Canes Venatici with an estimated age of 7 billion years.

69.3 THE INITIAL MASS FUNCTION

The stars we see at night turn out to be very unrepresentative of the general stellar population. When we look at the night sky, we see the stars whose brightness is great enough for us to detect their light. We primarily see the stars that are unusually luminous and that, therefore, stand out relative to the others. As a result, when we look at the night sky, the stars we see tend to be giants. This is illustrated in Figure 69.6, which shows an H-R diagram of all the stars in the sky that are visible to the unaided eye. Notice that the Sun is one of the least luminous among all of these stars.

If star A is a million times more luminous, then $L_A = 10^6 \times L_B$. If star B is a thousand times closer than star A, then $d_A = 10^3 \times d_B$. Comparing these, we find the following:

$$B_A = \frac{L_A}{4\pi\, d_A^2} = \frac{10^6 \times L_B}{4\pi(10^3 \times d_B)^2} = \frac{10^6 \times L_B}{10^6 \times 4\pi d_B^2} = \frac{L_B}{4\pi d_B^2} = B_B$$

Stars of the class we see at night are rare; but because we can see them out to very large distances, we see greater numbers of them than the far more numerous stars that are too dim to be visible. How bright a star looks to us is determined by its luminosity and distance, as we saw in discussing the inverse-square law (Unit 54.2): $B = L/4\pi d^2$. From this law we can see that if star A is a million times more luminous than star B, but star B is a thousand times closer than star A, they will appear to have the same brightness because $1000^2 = 1{,}000{,}000$. So according to the inverse-square law, star A will appear just as bright to us even if it is a thousand times farther away than star B. That is, great luminosity can compensate for distance.

Many giant stars are *more* than a million times more luminous than the low-luminosity stars. As a result, when we see a set of stars that looks bright to us, we are looking at the most luminous stars out to distances more than a thousand times greater than those of the low-luminosity stars. This biases what stars we are able to see heavily in favor of luminous stars. Astronomers call this problem of sample biasing a **selection effect.**

When astronomers observe a star cluster, they avoid this problem because all of the stars are at (nearly) the same distance. If we can count all the stars within the cluster, we have an unbiased sample. The relative numbers of stars of different luminosities in the sample will then indicate the true proportion of stars of each luminosity. If we use the mass–luminosity relation (Unit 58) to deduce the masses

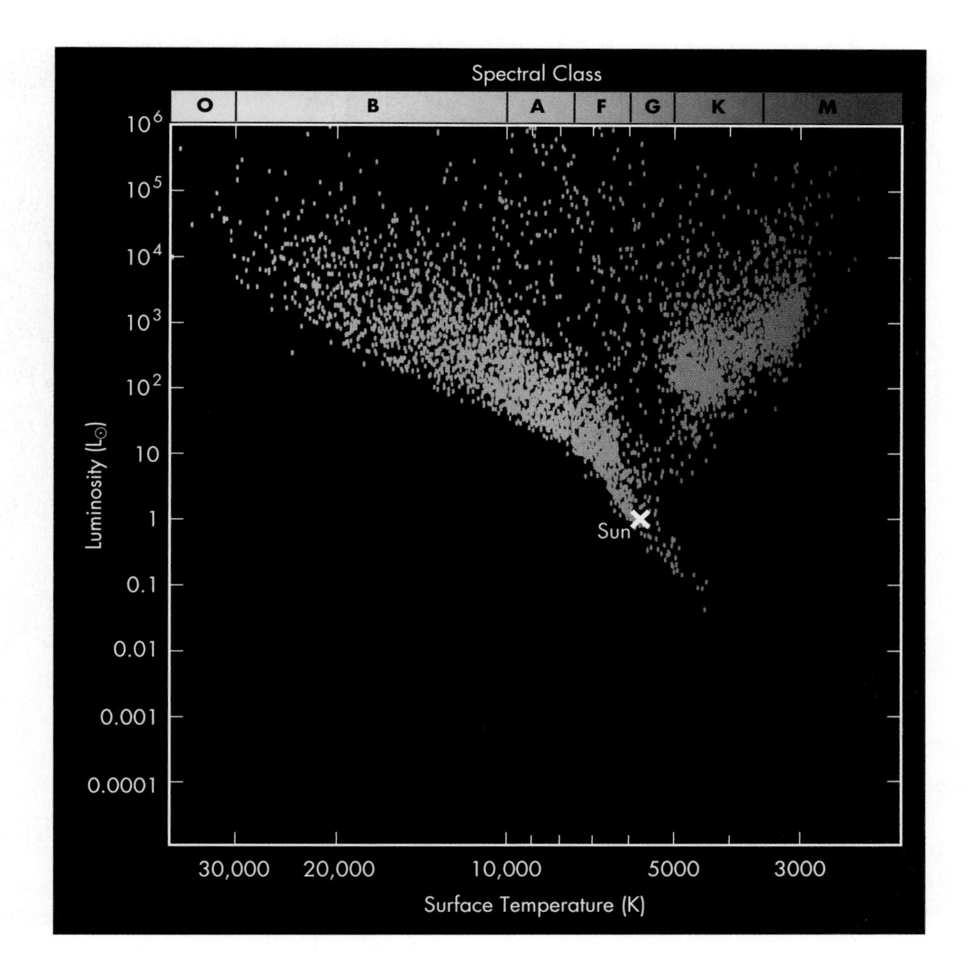

FIGURE 69.6
The H-R diagram of the stars that are visible to the unaided eye in the night sky. Nearly all are more luminous than the Sun.

of the main-sequence stars from their luminosities, we can estimate how many stars of each mass were born when the cluster formed. Astronomers call such a census of a cluster's stars its **initial mass function.**

Studies of star clusters show that stars similar to or smaller than the Sun vastly outnumber more massive stars, as shown by the initial mass function plotted in Figure 69.7. The most numerous stars turn out to be dim, cool, red dwarfs—stars that lie on the main sequence but whose mass is below 0.5 $M_\odot$. Astronomers do not yet understand what determines the distribution of stellar masses, but it appears to be similar in most clusters, so it must be a property of how a collapsing interstellar cloud breaks up into stars.

The initial mass function found from these studies indicates that the number of stars having a certain mass increases sharply as mass decreases. Typically there are about 20 times more stars with masses between 1 $M_\odot$ and 2 $M_\odot$ than between 10 $M_\odot$ and 20 $M_\odot$. The relative numbers of stars of different masses are also suggested by Figure 69.8, which shows an H-R diagram for all of the stars currently known within 10 parsecs of the Sun.

Despite their great numbers, low-mass stars are hard to see because they are so intrinsically dim. Not a single main-sequence M-type star is visible to the unaided eye, even though these are by far the most common type of star. Even the star nearest to the Solar System, the M-type star Proxima Centauri (see Looking Up #8: Centaurus and Crux, the Southern Cross), is more than 10,000 times too dim to be seen with the unaided eye. The initial mass function suggests that M-type stars (masses ranging from 0.08 $M_\odot$ to

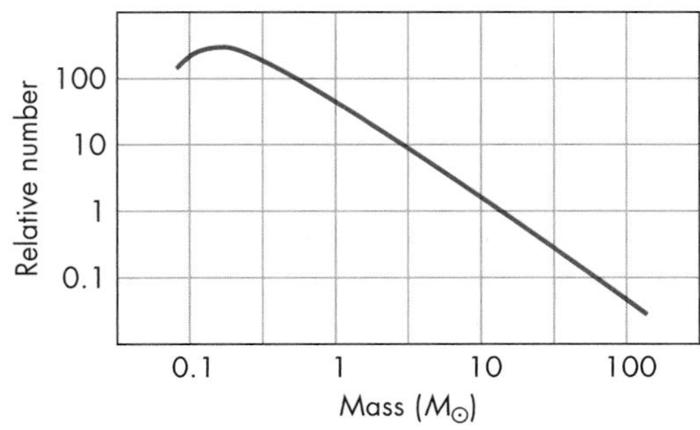

FIGURE 69.7
The initial mass function of stars indicates the relative number of stars of each mass. Studies of star clusters indicate that there are *many* more low-mass stars than high-mass stars, as indicated by the curve.

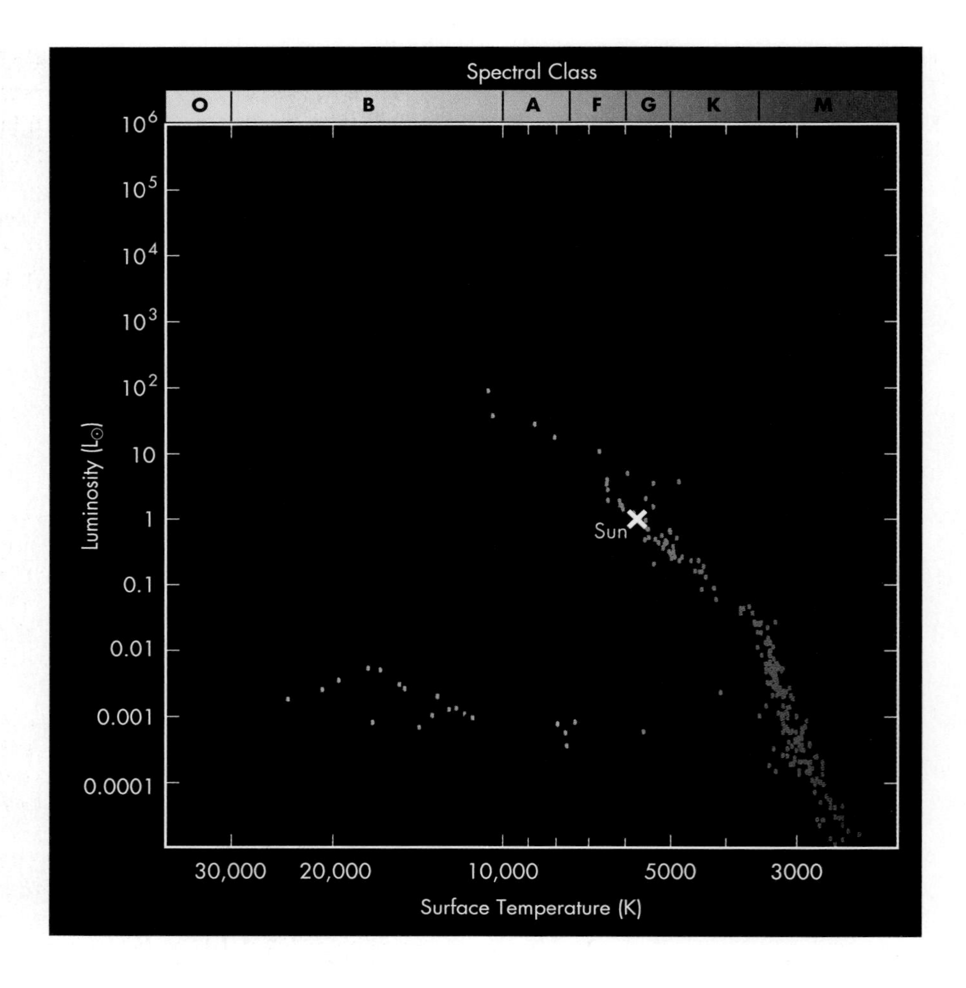

FIGURE 69.8
The H-R diagram of the known stars within 10 pc of the Sun. Few of the nearest stars are more luminous than the Sun.

Q What does the number of white dwarfs in Figure 69.8 versus the number of red giants suggest about the relative duration of these stages of stellar evolution?

Q Can you think of everyday situations in which choosing people based on one set of criteria might give you an erroneous sense of the characteristics of the whole population?

0.48 $M_\odot$) outnumber all other types of stars combined by a factor of 10. In fact, the combined mass of all stars of this dim spectral type is approximately half the combined mass of all types of stars.

Objects less massive than about 0.08 $M_\odot$ were once also believed to be rare, but astronomers now think they may be common and were simply missed in previous stellar surveys. How could astronomers miss these objects if they are so abundant? Recall that low-mass stars tend to be cool. In fact, if a contracting mass of gas is less massive than about 0.08 $M_\odot$, it is so cool that it is unable to fuse hydrogen into helium and is termed a *brown dwarf* (Unit 60). These are in some ways more like huge planets than tiny stars, and their extreme dimness and small size make them difficult to detect. However, searching the sky at the wavelengths at which brown dwarfs are brightest greatly increases the chance of finding them. What wavelengths are best? Because they are so cool, brown dwarfs are brightest in the infrared, so it is with infrared surveys such as 2MASS (2-Micron All Sky Survey) that astronomers are now finding large numbers of these "failed" stars. Studies of the Pleiades and other young open clusters at infrared wavelengths have revealed many probable brown dwarfs. Astronomers now think brown dwarfs may outnumber ordinary stars, making them one of the most common astronomical objects in our Galaxy, although it appears that they do not contain as much mass in total as is contained in ordinary stars.

In almost any unbiased sample of stars, we find that most of the mass is contained in the lowest-mass objects, yet most of the light comes from the highest-mass objects. In this wide range of properties, the Sun is sometimes called a "typical" star. It is true that the Sun sits somewhere in the middle of the range of stellar properties, yet it is typical neither in mass nor in luminosity.

The Sun is more massive than about 97% of all stars, yet more than 99% of the total luminosity in a typical collection of stars is produced by stars more massive than the Sun. However, when we add up the total mass and the total luminosity of stars in the neighborhood of the Sun, it happens that the ratio of mass to luminosity is similar to that of the Sun. In other words, if a thousand solar masses are contained in the collection of stars, then this star collection will generate about a thousand solar luminosities of light. Hence, the mass and luminosity of the stars in our neighborhood balance in a way that makes the Sun seem fairly average.

KEY TERMS

globular cluster, 557
initial mass function, 561
open cluster, 556
Pleiades, 555
selection effect, 560
star cluster, 555
stellar association, 556
turnoff point, 558
Ursa Major group, 555

QUESTIONS FOR REVIEW

1. How do star clusters form?
2. What are the different types of star clusters?
3. How are star clusters useful for testing stellar evolutionary theory?
4. What kind of cluster did our Sun form in? Do we know what other stars were born in this cluster?
5. Why does the H-R diagram of a cluster have a "turnoff point"? Why don't we see a turnoff point when we make an H-R diagram of all the stars we see in the night sky?
6. Is the Sun a typical star in terms of mass? in terms of luminosity? Why is it called a typical star?

PROBLEMS

1. Sketch an H-R diagram for three clusters of stars. Make your sketches for one that is 10^8 years old, one that is 10^9 years old, and one that is 10^{10} years old. In what parts of the diagrams is the distribution of stars similar? in which parts is it different? Explain what causes the differences.
2. We can see associations of O and B stars in the Andromeda Galaxy, 2 million ly away. What would be the brightness of an association of 70 O stars at this distance?
3. The Hyades is 150 ly away from Earth, and is 75 ly in diameter. The globular cluster Omega Centauri is 18,000 ly away from Earth, and is 200 ly in diameter. What are the angular diameters of these two objects as seen from Earth?
4. A globular cluster contains no main-sequence stars with surface temperatures higher than 5000 K and luminosities greater than 0.80 $L_{\odot}$. How old is the cluster?
5. If a globular cluster contains 1 million stars in a volume with a radius of 100 ly, what is the average separation between stars? Calculate this by assuming that each star fills one millionth of the total volume of the cluster. The diameter of the sphere it occupies is then the average separation. (The volume of a sphere is $4/3\ \pi R^3$.)
6. Suppose we were observing the globular cluster in problem 5 through a telescope. At the distance of the globular cluster, its radius appears to be 100 arcsec, so it covers an area on the night sky of $\pi \times (100\ \text{arcsec})^2$.
   a. Supposing each star has the radius of the Sun, what is its radius in arc seconds?
   b. What fraction of the area of the globular cluster is covered by the summed area of the million stars?

TEST YOURSELF

1. The most numerous type of star is
   a. about half the Sun's mass.
   b. about the Sun's mass.
   c. about twice the Sun's mass.
   d. about 20 times the Sun's mass.
   e. None of the above. The stars are fairly evenly distributed as far as mass goes.
2. When looking at the night sky with our unaided eyes, most of the stars we see are
   a. brown dwarfs.
   b. lower in mass than the Sun.
   c. about equal in mass to the Sun.
   d. more luminous than the Sun.
   e. young, hot O stars.
3. Most of the mass in stellar populations comes from the
   a. most luminous stars.
   b. lowest-luminosity stars.
   c. most massive stars.
   d. hottest stars.
   e. red giants.

Starry Night

See the Pathways *2nd edition ARIS site for online Starry Night™ planetarium exercises.*

Part Five

The content of some Units will be enhanced if you have previously studied some earlier Units in this textbook. These Background Pathways are listed below each Unit image.

Unit 70 Discovering the Milky Way

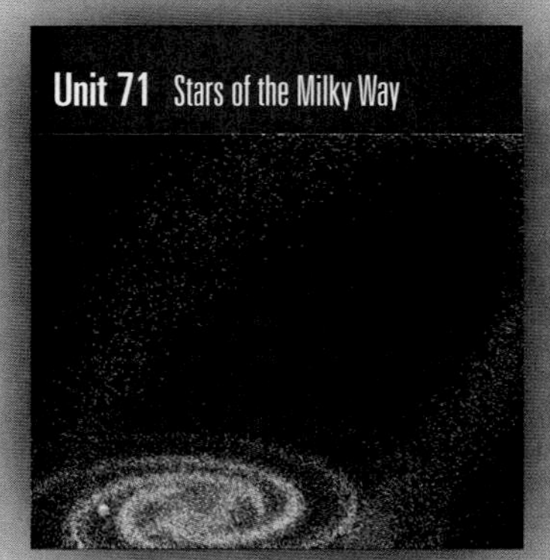

Unit 71 Stars of the Milky Way

Background Pathways: Unit 70

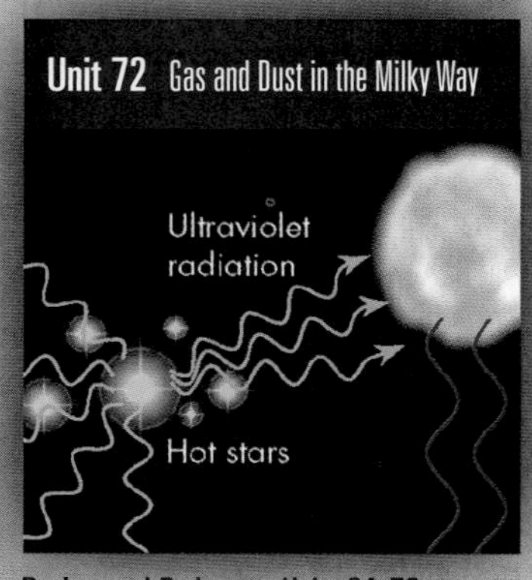

Unit 72 Gas and Dust in the Milky Way

Background Pathways: Units 24, 70

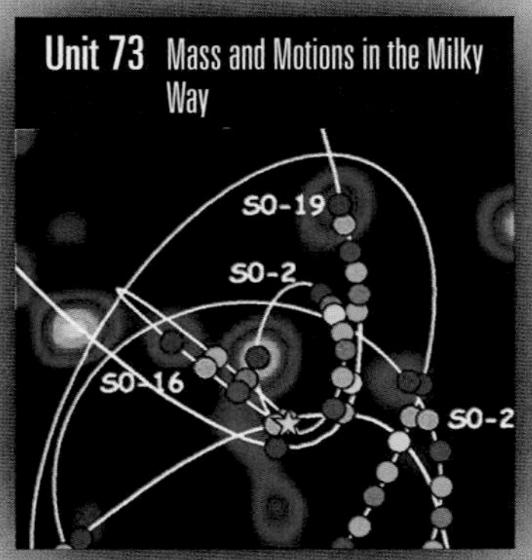

Unit 73 Mass and Motions in the Milky Way

Background Pathways: Units 17, 70

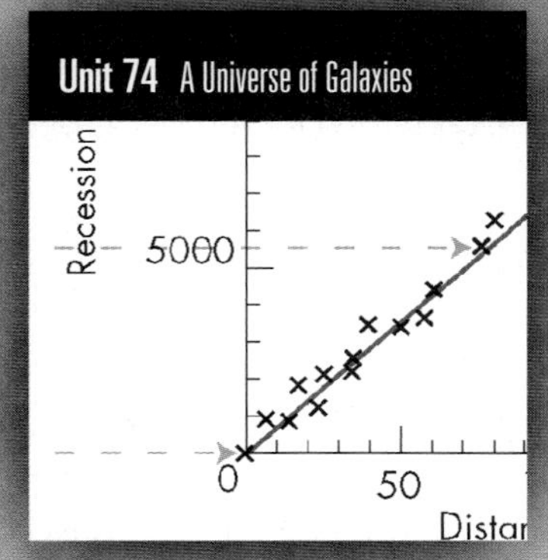

Unit 74 A Universe of Galaxies

Background Pathways: Unit 70

Unit 75 Types of Galaxies

Background Pathways: Units 71, 74

Unit 76 Galaxy Clustering

Background Pathways: Unit 75

Unit 77 Active Galactic Nuclei

Background Pathways: Unit 74

Galaxies and the Universe

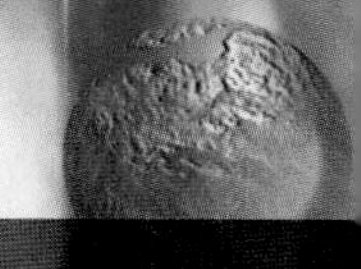

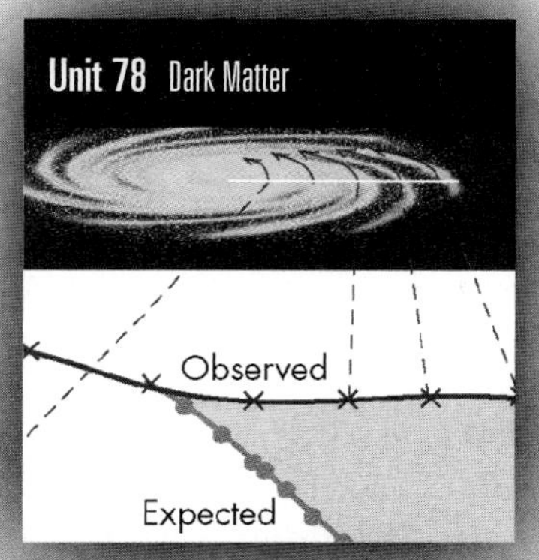

Background Pathways: Units 73, 76

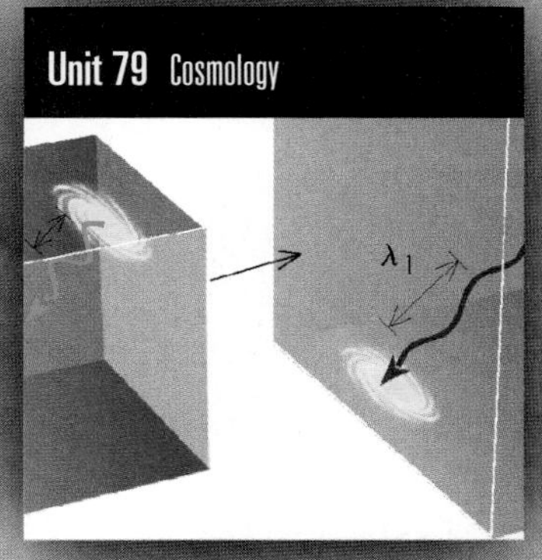

Background Pathways: Unit 74

Background Pathways: Unit 79

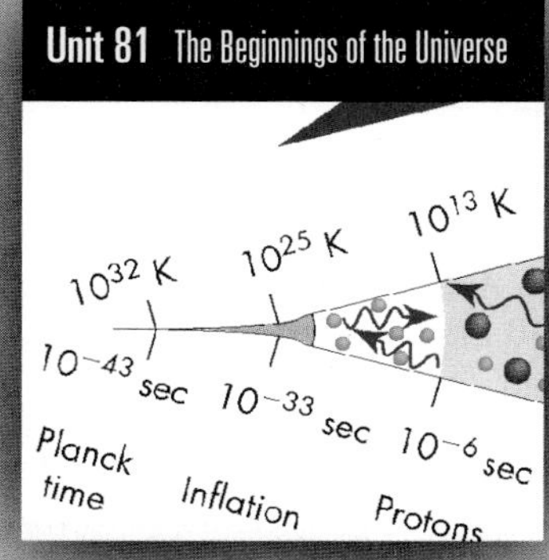

Background Pathways: Unit 79

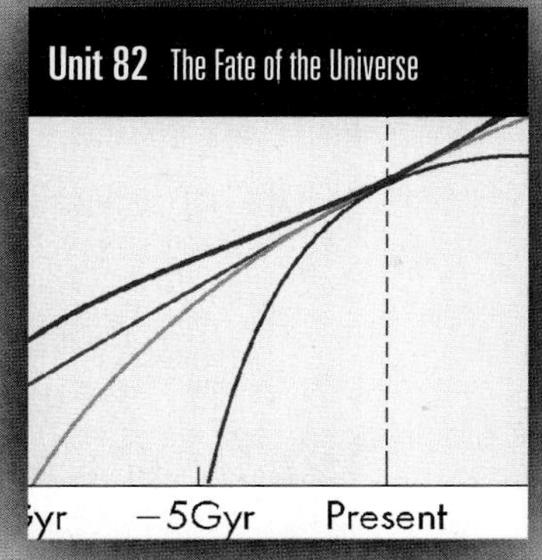

Background Pathways: Unit 78

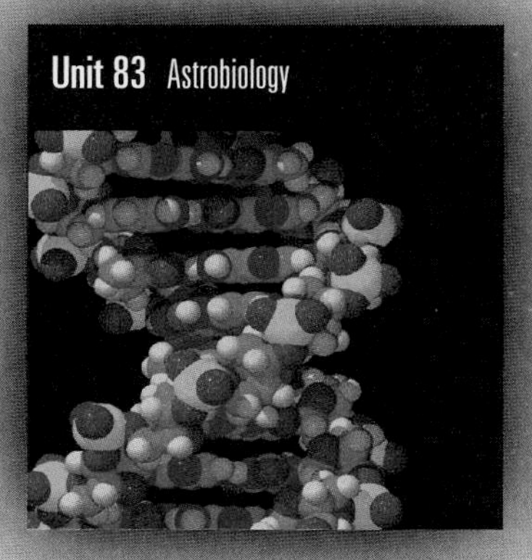

Background Pathways: Unit 83

unit 70

Discovering the Milky Way

On a clear, moonless night, far from city lights, you can see a pale band of light spangled with stars stretching across the sky (Figure 70.1). The ancient Hindus thought this shimmering river of light in the heavens was the source of the sacred river Ganges. To the ancient Greeks, this dim celestial glow looked like milk spilled across the night sky, so they called it the **Milky Way.** Astronomers also call this our **Galaxy,** from the Greek word for "milk" (*galactos*).

A view of the Milky Way on a clear, dark night is one of nature's finest spectacles. The band stretches in a full circle around us on the celestial sphere, but it is at a different angle from either the Earth's equator (the celestial equator) or the Solar System (the ecliptic). The orbits of the planets as well as the Earth's equator are tilted by about 60° with respect to the circle of the Milky Way. As a result, when you observe the Milky Way it crosses the sky at different angles, depending on when or where you see it.

Superimposed on the dim background glow are most of the bright stars and star clusters that we can see, which all belong to our Galaxy. Here and there dark blotches interrupt the glowing backdrop of stars, as you can see in Figure 70.1. The Incas of ancient Peru, who observed the Milky Way from their temple observatories in the Andes, gave these dark areas names, just as peoples of the classical world named the star groups. Today we know the dark regions are clouds of dust and gas that give birth to new stars.

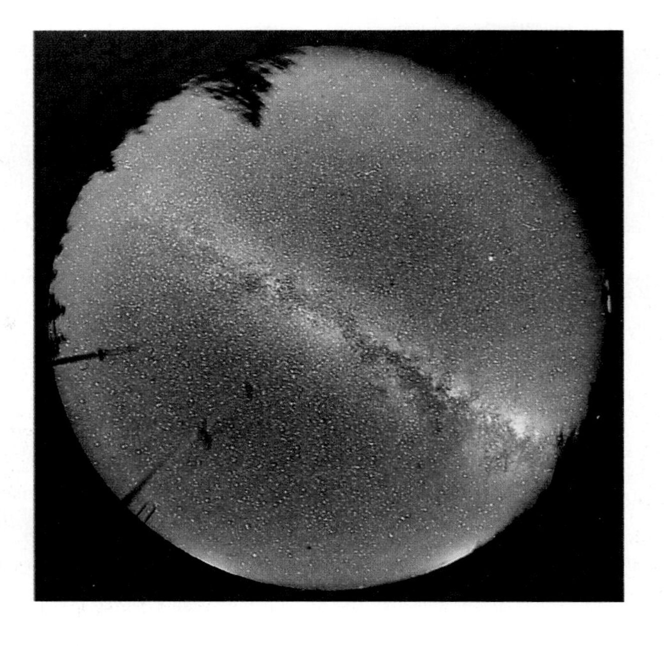

FIGURE 70.1
Wide-angle photo of the Milky Way taken from Mount Graham in Arizona. The dark ring around the edge of the picture is the horizon. The white regions along the horizon at the bottom of the figure are distant city lights.

The pattern of the Milky Way can be seen on the foldout chart at the back of the book. It is shown as an irregular S-shaped blue region crossing north and south of the celestial equator.

Our Galaxy is an enormous system of stars—home to the Sun and hundreds of billions of other stars. It contains huge interstellar clouds, stars that are forming and stars that are dying. Yet the Galaxy is more than just a collection of all these other things. It has its own structure on a scale far larger than stars or solar systems. Understanding that structure provides clues to why and where stars form and the Sun's relationship to other stars.

70.1 THE SHAPE OF THE MILKY WAY

When Galileo Galilei pointed a telescope at the Milky Way in 1609, he discovered that it contains millions of stars too dim to be seen as individual points of light with the unaided eye (Figure 70.2). This had been hypothesized 2000 years earlier by the Greek philosopher Democritus, who also proposed the idea that matter was made of atoms. In the twenty-first century we know that these stars, along with our Sun, form a huge, slowly revolving disk—our Galaxy.

The Milky Way

Our understanding of the Milky Way as a star system developed further in the mid-1700s when Thomas Wright, an English astronomer, and Immanuel Kant, a German philosopher, independently suggested that the Milky Way is a flattened swarm of stars. They argued that if the Solar System were near the center of a spherical cloud of stars, we would see roughly the same number of stars in all directions. However, from inside a disk-shaped system, we would see vastly more stars in directions toward the outer edge of the disk than in directions perpendicular to it, as illustrated in Figure 70.3. This explains why the stars of the Milky Way appear to stretch around us in a great circle.

A simple analogy may help you better understand this argument. Imagine making a lemon gelatin fruit salad in a big, round flat dish. Imagine that you have spread blueberries throughout the dish. If an ant fell into the salad and was clinging to a blueberry in it, the ant would see few berries if it looked up through the top of the salad or down through the bottom. However, it would see lots of blueberries if it looked in directions that lie in the plane of the dish. So too it is with us and stars in the disk of the Milky Way.

FIGURE 70.2
Through a telescope the Milky Way is resolved into millions of stars. Several clusters of stars, some pink from the emission lines of hydrogen, are seen scattered across this image.

FIGURE 70.3
If the Milky Way had a spherical shape, we would see about the same number of stars in every direction. However, because it is a disk, we see stars concentrated into a band around the sky.

70.2 STAR COUNTS AND THE SIZE OF THE GALAXY

Q If the Earth's axis were tilted 90° with respect to the circle of the Milky Way, how would the Milky Way appear in our sky? What if the Earth's axis pointed toward the Milky Way?

The first quantitative attempts to determine the Milky Way's size and shape were made in the 1780s. William and Caroline Herschel seem an unlikely pair to have revolutionized astronomy, but this brother and sister team of musicians made a hobby of studying the skies. William (1738–1822) and Caroline (1750–1848) spent their free nights scanning the stars with high-quality telescopes built by William, discovering comets and the planet Uranus (Unit 44). Their growing fame won them support from the king of England, and they became full-time astronomers.

William decided to attempt to measure the shape of the Milky Way by observing hundreds of areas over the sky and counting the stars. He reasoned that the number of stars that he could see within his telescope's field of view would tell him the extent of the star system in that direction. When Caroline cataloged and indexed the results of William's counts, there were typically several hundred stars in each field throughout the band of the Milky Way, but on average only a few stars when looking at directions 90° away from it. William assumed that he could see the stars all the way to the edge of the Milky Way and that the stars were equally spaced on average. In that case, the number of stars should be larger in proportion to the volume encompassed by his telescope's field of view out to the edge of the Milky Way. Based on this, he produced the cross-sectional diagram shown in Figure 70.4. The overall size of this star system could be estimated by assuming that the other stars were like the Sun, and knowing how light dims with distance (Unit 54.2). This suggested that the Milky Way was a disk about 2500 parsecs in diameter, with the Sun near the center, and about one-fifth as thick as it was wide.

The left side of this diagram illustrates one of the problems with this method. The stars extend out a large distance, except right in the middle, where their distribution looks like a pair of open alligator jaws. This odd shape was caused by a dusty interstellar cloud that blocked light in that direction—however, astronomers did not learn about interstellar clouds until 150 years later. William recognized another problem after he had built larger telescopes. He could see even more stars, and he realized his earlier observations had not seen to the edge of the Milky Way. Nevertheless, for over 100 years this remained the best model of the star system in which we live.

It was not until the early 1900s that much more extensive studies gave us a clearer idea of the size of the Milky Way. By this time it was clear that stars did not all shine with the same luminosity as the Sun, and that stars had to be studied to much fainter limits to detect the "edge" of the Galaxy. The Dutch astronomer Jacobus C. Kapteyn (pronounced *CAP-tine*) carried out an extensive study along the lines of the Herschels', but with modern instruments, photography, and knowledge of the different types of stars. Because most of the stars were too distant to make direct parallax estimates, Kapteyn made his distance estimates for various types of stars by determining how much they appeared to shift their position on

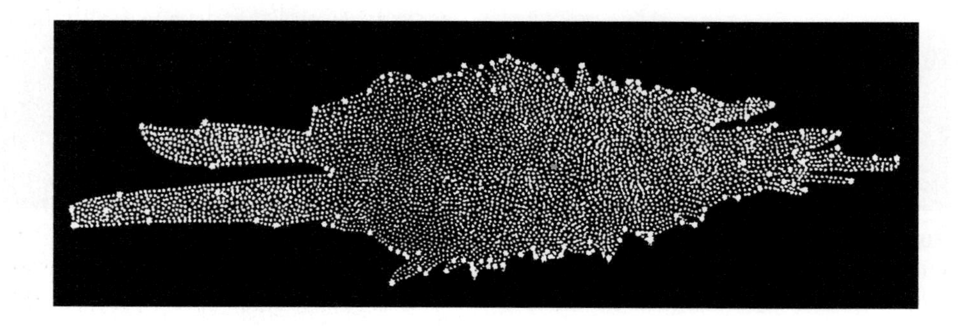

FIGURE 70.4
A copy of William Herschel's cross-sectional diagram of the Milky Way made in 1785. This was based on the number of stars Herschel could see in different directions. From this picture, he correctly deduced that the Milky Way is wider than it is thick. However, Herschel *incorrectly* concluded that the Sun (the orange dot) is near the Milky Way's center. He was led astray by his lack of knowledge that dust clouds blot out distant stars and prevent us from seeing our Galaxy's true extent.

FIGURE 70.5
"Kapteyn's Universe," a model of the Milky Way as a roughly disk-shaped system of stars about 3 kpc thick, with the Sun near its center.

the sky over the course of many years (their *proper motion*—Unit 52.4). If two stars are moving through space at the same speed relative to the Sun, the one that is farther away will appear to have a smaller motion across the sky, just as a ball tossed past your head moves across your field of view more rapidly than one thrown at the far end of a playing field. This method avoids the problem of assuming that stars all have the same luminosity.

Kapteyn's resulting model (Figure 70.5) proposed a greatly revised size of the Milky Way, suggesting that it was 18,000 parsecs in diameter, again with the Sun fairly near the center. This model became known as **Kapteyn's Universe,** because at the time it was not known that there were other galaxies beyond the Milky Way. Galaxy dimensions are so huge that even parsecs are inconvenient for measuring their size, so astronomers often use **kiloparsecs** (kpc) for that purpose. One kiloparsec = 1000 parsecs = 3300 light-years. Thus, the Herschel model of the Milky Way was just 2.5 kpc in diameter, while the Kapteyn Universe was 18 kpc in diameter.

In both the Kapteyn model and the Herschel model, our Galaxy is depicted essentially as a disk of stars. Midway between the two faces of this disk is what astronomers call the **Galactic plane.** Within the disk, different kinds of stars are concentrated more or less tightly around the Galactic plane. Young stars, gas, and dust are found close to the plane, on average, but older stars span a much larger range of distances from the plane, extending "above" or "below" to several kiloparsecs.

70.3 GLOBULAR CLUSTERS AND THE SIZE OF THE GALAXY

At about the same time as Kapteyn was working, the U.S. astronomer Harlow Shapley argued that the Milky Way was even larger—about 100 kiloparsecs across—and that the Sun was not near the center but rather was about two-thirds of the way out in the disk.

To determine the Milky Way's size, Shapley used a method entirely different from the methods used by the Herschels or Kapteyn. He studied the locations of **globular clusters** (Unit 69)—these are dense groupings of up to a million stars (an example of a globular cluster is shown in Figure 70.6). Because these clusters contain so many stars, they are very luminous and can be seen at large distances—across the Galaxy and beyond. Moreover, many of them have orbits that carry them far outside the Galactic plane, so we have a clear view of them above or below the dense disk of the Milky Way. Shapley argued that these massive star clusters must orbit the center of the Milky Way.

Shapley noticed that the globular clusters are *not* scattered uniformly across the whole sky but are concentrated in the direction where the Milky Way looks brightest to us, toward the constellation Sagittarius (see Looking Up #7: Sagittarius). He hypothesized that the middle of the Galaxy lay somewhere in the direction of Sagittarius, and the system of globular clusters was centered on it. By mapping where the clusters lay, he could deduce our distance from the center and therefore the size of the Milky Way.

To map the positions of the clusters, Shapley needed to estimate their distances from us. He did this by observing the variable stars in them (Unit 63). Certain kinds of variable stars have predictable luminosities, and globular clusters often contain stars of a class known as RR Lyrae variables. From the luminosity and apparent brightness of these stars, he could calculate their distance using the inverse-square law. The distance to the stars in a cluster gave the cluster's distance. With good estimates of the clusters' distances, Shapley plotted where they lay in the Milky

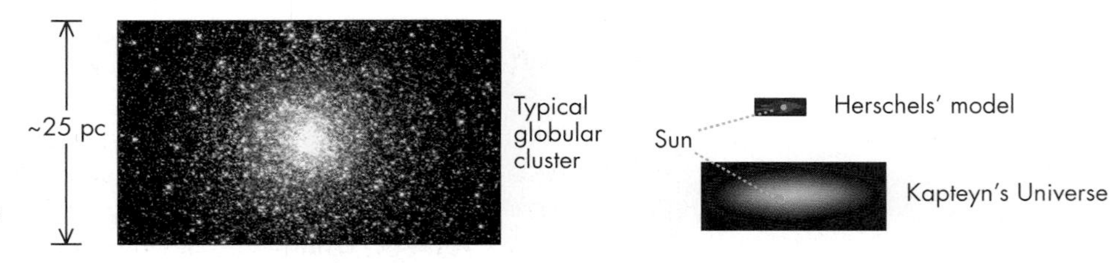

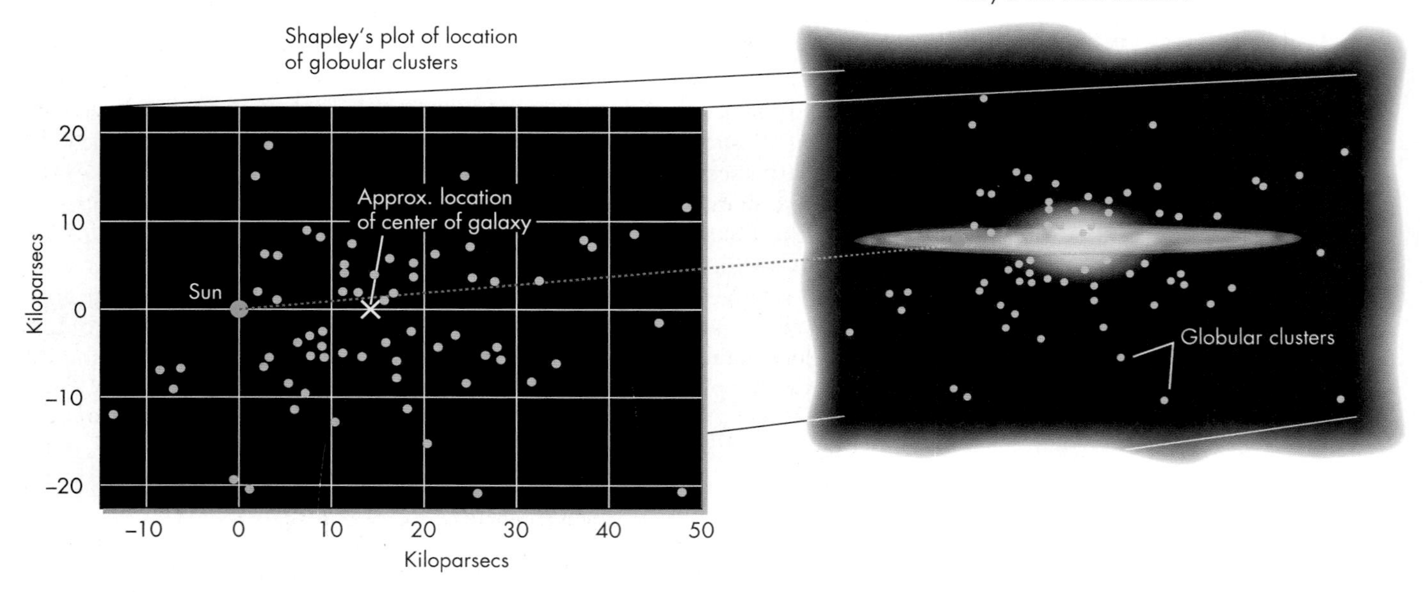

FIGURE 70.6
Schematic version of Shapley's plot of globular clusters, from which he inferred the size of the Milky Way and the Sun's location in it. Notice that the clusters fill a roughly elliptical region and that the Sun is *not* at the center of their distribution.

Way. Figure 70.6 shows his results. The clusters fill a roughly elliptically shaped region, and the Sun lies *not* at the middle of the region, but about two-thirds of the way from its center. Shapley therefore concluded that our Galaxy is roughly 100 kiloparsecs in diameter and that the Sun is nearer to its edge than to its center.

The great difference between Shapley's and Kapteyn's findings about the Milky Way created a major controversy among astronomers. As in so many other controversies, both sides were partially right and partially wrong. However, Shapley's results about the Milky Way were closer to the truth. The main reason both results differed from what we know today is that neither recognized the effects of **interstellar dust**—small particles of solid matter that are found in deep space (Unit 72). Close to the Galactic plane, the dust is so dense that it blocks light from distant stars, which limited Kapteyn's observations to the portion of the Milky Way's disk that surrounds the Sun, making it appear that the Sun is near the center. Most globular clusters, on the other hand, are located far "above" or "below" the Galactic plane where there is little dust, so their light is dimmed only slightly. Still, this dimming makes them look a little more distant than they really are, so Shapley overestimated the distance to the globular clusters and deduced that our Galaxy is bigger than it really is. Several decades later, astronomers also discovered that variable stars of the class Shapley used to estimate the cluster distances were less luminous than he had assumed, which caused him to further overestimate the distances.

Modern estimates place the center of the Galaxy about 8 kiloparsecs from us in the direction of the constellation Sagittarius. Observations also indicate that the outer part of the Galaxy has no distinct edge—the density of stars steadily declines, becoming so sparse that it is difficult to detect stars at about twice the Sun's distance from the center. Overall, though, the Milky Way's disk has a diameter of about 30 to 40 kiloparsecs, or roughly 100,000 light-years.

70.4 GALACTIC STRUCTURE AND CONTENTS

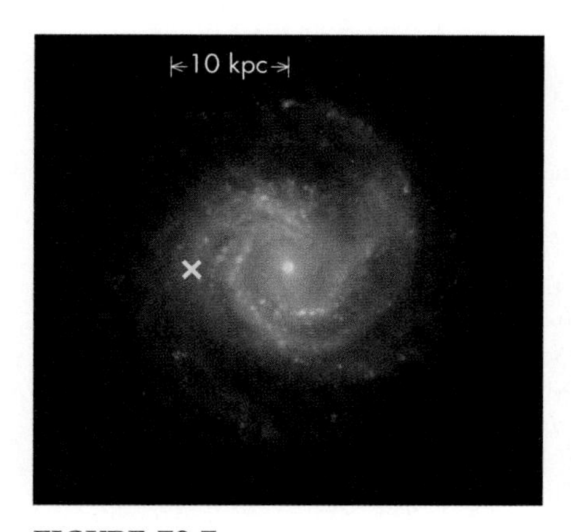

FIGURE 70.7
Image of the spiral galaxy NGC 4303 (also known as M61), illustrating its spiral arms. This galaxy is thought to be fairly similar in size and shape to the Milky Way. The X marks a position about 8 kpc from the galaxy's center.

Q What would the Milky Way look like in the night sky to an observer on a planet at the very edge of the Galaxy? What would it look like to an observer at the Galactic nucleus?

Today our understanding of the structure of the Milky Way is aided greatly by studies of other galaxies. In the early 1900s it was not widely accepted that there *were* other galaxies; this is explored further in Unit 74. Because we live inside the Milky Way, we cannot observe our Galaxy in its entirety—it is difficult "to see the forest for the trees." Having other examples of galaxies gives us a better idea of the kinds of structures we might find in the Milky Way.

Many external galaxies are flat disks with conspicuous **spiral arms** (Unit 75), as shown in Figure 70.7. Such arms are hard to detect in our own Galaxy because from our location in the Milky Way we cannot get an overview of its disk. Nevertheless, given that the Milky Way is a flat disk and that other disk galaxies have spiral arms, it is reasonable to infer that ours does too, and observations of the locations of interstellar gas, dust, and bright stars are consistent with this picture. By combining observations of our own Galaxy with the general features known to occur in other galaxies, astronomers can assemble a more detailed model for the Milky Way, as described next.

The Milky Way consists of three main parts, illustrated in Figure 70.8: a **disk** about 30 to 40 kiloparsecs in diameter, a more spherically distributed component called the **halo,** and a flattened, somewhat elongated **bulge** of stars at its center. Within the disk, numerous bright young stars collect into spiral arms that wind outward from near the center. Our Solar System lies about 8 kiloparsecs from the center in a region between spiral arms. You may find the following analogy helpful in visualizing the scale of our Galaxy: If the Milky Way were the size of the Earth, the Solar System would be the size of a large cookie.

Mingled with the stars of the disk are huge clouds of gas and dust that amount to about 15% of the disk's mass. We can see some of these clouds by the visible light they emit, whereas others reveal themselves because their dust blocks the light of background stars, as can be seen in Figure 70.2. In fact, dust scattered throughout the Galaxy prevents us from seeing farther than about 3 kiloparsecs away from the Sun in almost any direction close to the Galactic plane. For example, we cannot see our Galaxy's **nucleus**—its core—at visible wavelengths. However, radio, infrared, and X-ray telescopes can "see" through the dust and reveal that the core of our Galaxy contains a dense swarm of stars and gas as well as a supermassive black hole (Unit 73).

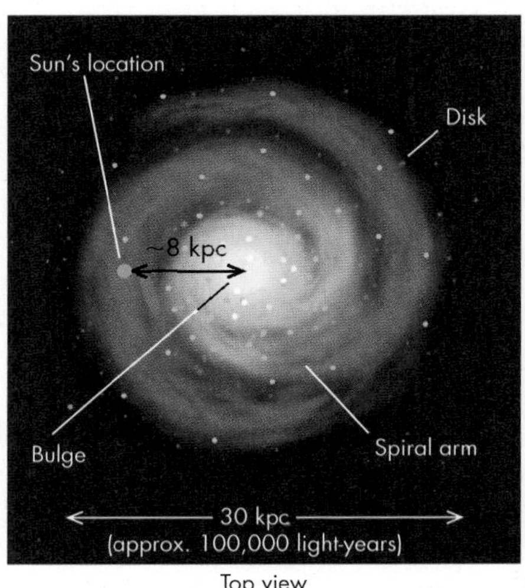

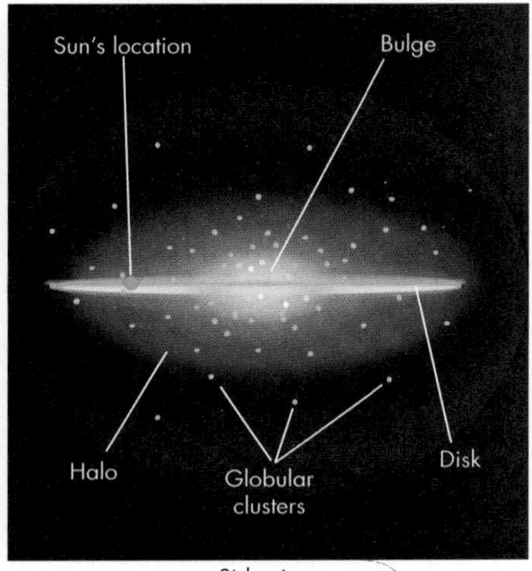

FIGURE 70.8
Artist's sketch of the Milky Way, showing top and side views, illustrating the disk, bulge, and halo. Notice how thin the disk is and how the halo surrounds the disk much as a bun surrounds a hamburger. Because the Milky Way's stars gradually thin out and do not just stop at some distance, its size is labeled only approximately.

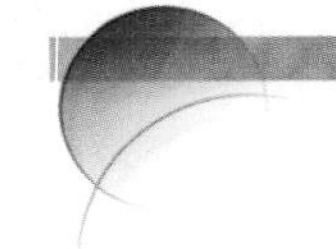

KEY TERMS

bulge, 571
disk, 571
Galactic plane, 569
Galaxy, 566
globular cluster, 569
halo, 571
interstellar dust, 570
Kapteyn's Universe, 569
kiloparsec, 569
Milky Way, 566
nucleus, 571
spiral arm, 571

QUESTIONS FOR REVIEW

1. What does the Milky Way look like in the night sky?
2. How do we know that the Sun is not located at the edge of the Milky Way?
3. How do we know our Galaxy is a flat disk?
4. How did star counts determine the shape of the Milky Way?
5. What are globular clusters, and how do we determine their distances from the Sun?
6. How did Shapley deduce the Milky Way's size and the Sun's position in the Milky Way?
7. What effects did dust have on the determination of the size of the Milky Way?
8. What are the major components of the Milky Way?

PROBLEMS

1. If the total mass of the Milky Way is 10^{12} $M_{\odot}$, and 10% of that is in the disk, how many solar masses of dust and gas are in the Galaxy?
2. How long does it take light from the center of the Milky Way, 8 kpc away, to reach the Solar System?
3. Traveling at 60,000 kph, how long would it take us to reach the center of the Milky Way?
4. Suppose a galaxy like the Milky Way is about 2 million ly away (the distance to the nearest galaxy similar in size to the Milky Way). What would its angular size be, given that its disk is about 100,000 ly across?
5. If we made a model of the Milky Way that had the diameter of the Earth, how big would the Earth itself be in this shrunken model? (Assume that the Milky Way's diameter is 35 kpc for this problem.)
6. Kapteyn used observations of the smaller shift in the position of stars that are farther away to estimate their distances. This is essentially a consequence of the angular size versus distance relationship discussed in Unit 10, as illustrated by the following: (a) If stars move at a typical speed of 20 km/sec, how far will they move in 10 years? (b) If a star is 10 pc distant, what will its angular shift be over these 10 years? (c) If a star is 100 pc distant, what will its angular shift be?
7. If Shapley had found that his variable stars had half the luminosity he'd thought they had, how would this have affected his estimate of the size of the Milky Way?

TEST YOURSELF

1. One way astronomers deduce that the Milky Way is a disk is that they
   a. see stars arranged in a circular region around the north celestial pole.
   b. see far more stars along the band of the Milky Way than in other directions.
   c. see a large dark circle silhouetted against the Milky Way in the Southern Hemisphere.
   d. see the same number of stars in all directions in the sky.
   e. None of the above.
2. Our Solar System is approximately _______ from the center of the Milky Way Galaxy.
   a. 4.3 pc
   b. 8 kpc
   c. 30 to 40 kpc
   d. 1 Mpc
   e. 13.7 Mpc
3. One of the reasons Kapteyn underestimated the size of the Milky Way and Shapley overestimated it is that they did not recognize the
   a. effects of the motion of the Sun from the center.
   b. dimming effect of interstellar dust.
   c. age of the globular clusters.
   d. existence of dark matter.

unit

71

Stars of the Milky Way

The Milky Way contains many types of stars: giants and dwarfs, hot and cold, young and old, stable and exploding. All of these star types combine to define the overall population of stars in our Galaxy. Such studies of stellar populations reveal that despite the wide range of star types and the enormous range of luminosities, the typical star in the Milky Way is rather small, dim, and cool.

Astronomers can also make a "stellar census" in different regions of the Milky Way by counting the relative numbers of each type of star. The types of stars and their relative numbers reveal similarities and dissimilarities in when and how many stars formed in these regions. By putting this information together, astronomers can deduce some of the history of our Galaxy, much as archaeologists can learn about life in an ancient city from the locations and kinds of buildings it contained.

Background Pathways

Unit 70: Discovering the Milky Way 566

71.1 STELLAR POPULATIONS

Hidden in the great diversity of star types is an underlying simplicity that astronomers first noted in the 1940s. At that time, the 100-inch reflector at Mount Wilson Observatory, located outside of Los Angeles, California, was the largest telescope in the world and the best for observing galaxies. The glow from city lights made it hard to see faint galaxies, but blackouts during World War II darkened the night sky, and Walter Baade (*BAH-deh*), an astronomer at Mount Wilson, took advantage of the darkness to make a series of photographs of neighboring galaxies. He noticed that stars in these nearby galaxies were segregated by color. Red stars were concentrated in the bulges and halos of the galaxies, whereas blue stars were concentrated in their disks and especially in their spiral arms. To distinguish these groups, Baade called the blue stars in the disk **Population I**, and the red stars of the bulge and halo **Population II** (Pop I and Pop II, for short).

While pursuing his studies of the stellar populations of other galaxies, Baade realized that Milky Way stars showed the same division. Furthermore, when he examined the properties of Population I and II stars in our Galaxy, he found further differences between the two populations. They differ not only in color and location in the Galaxy, but in their age, motion, and composition. Pop I stars are young, typically less than a few billion years old, and many of them are blue. They lie in the plane of the Galaxy's disk and follow approximately circular orbits, as shown by the blue orbits in Figure 71.1. Their atmospheric composition is like the Sun's: mostly hydrogen and helium, with a few percent of their mass consisting of **metals**, which for an astronomer means any element heavier than helium. Thus, Pop I stars are relatively rich in metals.

Population II stars are generally red and more than about 10 billion years old. They lie in the bulge and halo of the Galaxy, moving along highly elliptical orbits

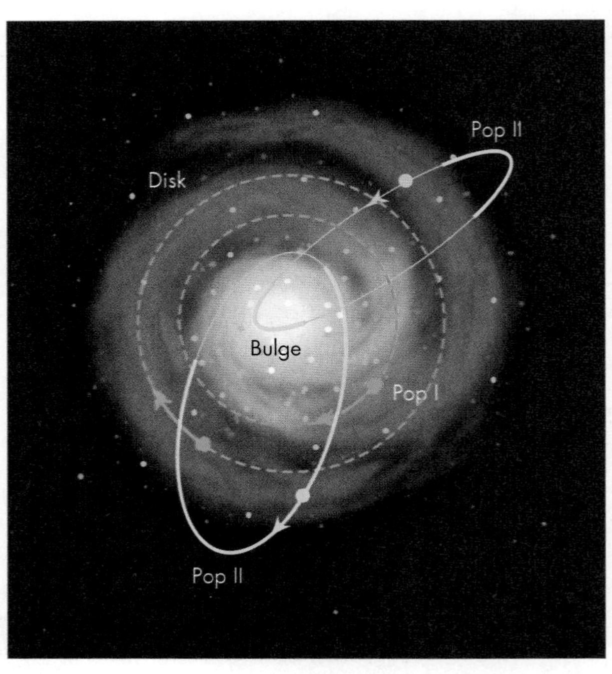

FIGURE 71.1
Stellar orbits in the Milky Way. Population I stars orbit in the disk (blue lines). Population II stars orbit in the halo (yellow lines). Notice the elongation and inclination of their orbits.

ANIMATIONS
Pop I and II

Q Suppose the Sun had an orbit similar to that of a Pop II star in the halo of the Milky Way. In what ways would that make studying the Milky Way easier? Is it likely there are Earth-like planets orbiting halo stars?

that are often tilted strongly with respect to the Galactic disk, as illustrated by the yellow orbits in Figure 71.1. Pop II stars are almost entirely hydrogen and helium, with only a few hundredths percent of their mass composed of heavy elements—roughly a hundred times less than stars like the Sun. Table 71.1 summarizes some of these properties.

Division of all stars into two broad categories is an oversimplification. For example, the Sun's age does not fit precisely into either category, but the Sun is considered a Pop I star because of its heavy element content and approximately circular orbit in the disk. To avoid forcing stars into population categories they do not fit, astronomers often refer instead to disk, bulge, and halo stars, recognizing that the great majority of disk stars are Pop I, while most of the bulge and halo stars are Pop II.

The fairly sharp distinction in properties between the two population categories appears to relate to major changes in the structure of the Milky Way during its history. The distinct populations suggest that star formation has not occurred continuously in the Milky Way. Population II stars probably formed in a major burst

TABLE 71.1 Properties of Population I and II Stars

Property	Pop I	Pop II
Location	Disk and concentrated in arms	Halo and bulge
Age	Young (< few billion years)	Old (> 10 billion years)
Color	Blue (overall)	Red
Orbit	Approximately circular in disk	Plunging through disk on approximately elliptical orbit
Heavy element (metals) content	High (a few percent—similar to Sun's)	Low (10^{-2} to 10^{-3} times Sun's)

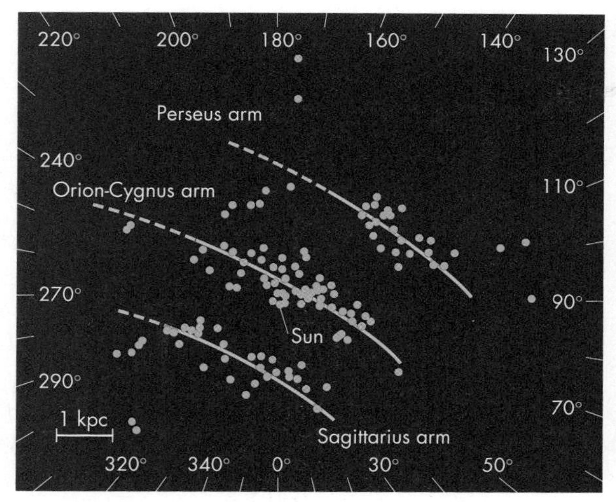

FIGURE 71.2
Young star clusters and HII regions show parts of the spiral arms in the Sun's neighborhood. The arms are named after a constellation in which they appear bright from our vantage point in the Solar System.

at the Galaxy's birth during its initial collapse, whereas Population I stars began forming later and continue forming today. This hypothesis explains the differences between the two populations, as we will discuss in greater detail in the next section.

One of the first uses astronomers made of differences between stellar populations was to map the Milky Way's spiral arms. As Baade noted, the spiral arms of other galaxies gleam with the blue light of their Pop I stars. The blue color of Pop I is produced primarily by the luminous stars with spectral types O and B (Unit 55) that use up their fuel rapidly, "burning out" in just tens of millions of years. Therefore, by measuring the location of O and B stars near the Sun, astronomers can make a picture of the spiral arms in the Milky Way's disk. This method is limited, however, because dust in space prevents us from seeing even the most brilliant O and B stars if they are farther from the Sun than about 3 kiloparsecs. Nevertheless, maps such as that shown in Figure 71.2 provided some of the first direct evidence that we live in a spiral galaxy.

The stars in open clusters and globular clusters (Unit 69) also exhibit the differences between Populations I and II. Open clusters are almost all Pop I, located within the Milky Way's disk, whereas globular clusters are always Pop II and orbit in the Milky Way's halo. This difference in their stellar populations makes clusters especially useful to astronomers for studying the structure of the Milky Way: Globular clusters outline the halo and bulge, while young, open clusters trace the Galaxy's arms.

71.2 FORMATION OF OUR GALAXY

One of the major research topics in astronomy today is how galaxies form and evolve. The process is presumably a large-scale version of star formation (Unit 60)—that is, a gas cloud collapses under the influence of gravity and breaks up into stars. But this does not explain why the Galaxy contains two main categories of stars, Population I and II, that differ so greatly in their properties. Based on an analysis of the orbits of Pop I and Pop II stars, the British astronomer Donald Lynden-Bell and the U.S. astronomers Olin J. Eggen and Allan R. Sandage proposed in 1962 a two-stage collapse model to explain the birth of the Milky Way.

The structure of the Solar System also provides clues to its origin. Its disklike shape, the common age of the planets, and the existence of different families of planets allowed astronomers to develop the solar nebula theory (Unit 33).

According to the two-stage collapse model, our Galaxy began as a vast, slowly rotating gas cloud, a few million light-years (a million parsecs) in diameter, containing several hundred billion solar masses of gas. The cloud was composed of almost pure hydrogen and helium. As gravity began shrinking this immense cloud, clumps of gas within it grew in density, forming stars, as illustrated in Figure 71.3A.

These first stars would have had a composition unlike stars today. The heavy elements are produced by nuclear fusion inside stars, so the first stars would have been composed of essentially pure hydrogen and helium. The most massive of these stars would have evolved quickly and blown up as supernovae (Units 65 and 66), adding the first heavy elements to the gas cloud even before it had finished collapsing. It may seem odd that stars could form and die before the cloud's collapse was complete. But the collapse of such an immense cloud takes hundreds of millions of years, whereas massive stars can evolve and die in less than a few tens of millions of years.

As the cloud's collapse continued, the next generation of stars now contained at least a few heavy elements. Formed from gas falling in from all directions, the resulting inward trajectories of these bodies would have given them highly elliptical orbits. Hence their orbits continue to carry them in and out of the inner part of the

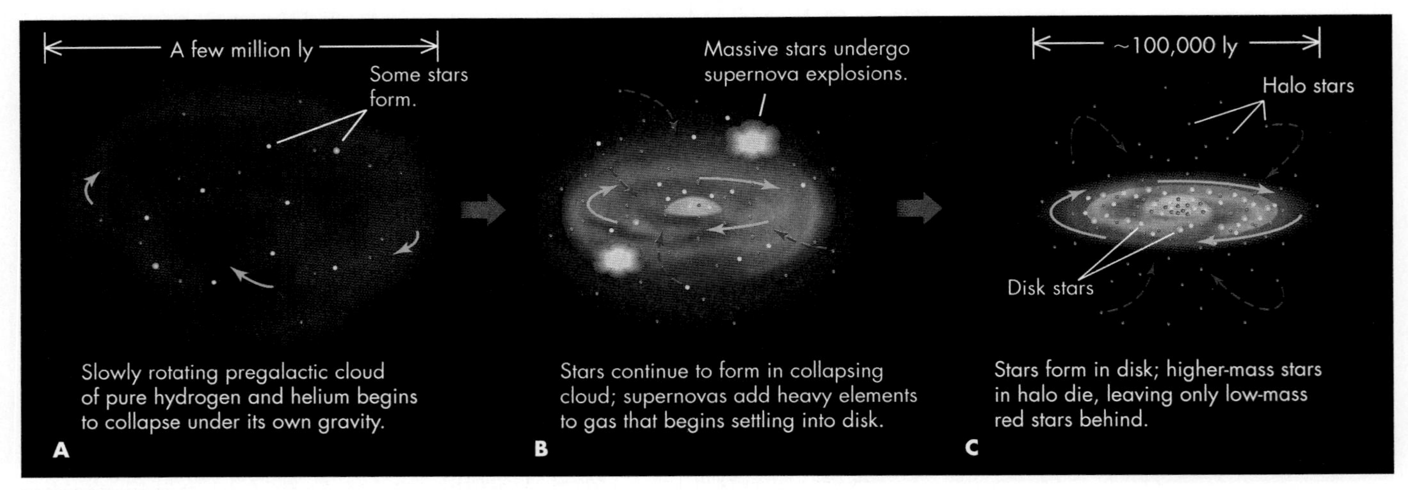

FIGURE 71.3
(A) Birth of the Milky Way as a gas cloud collapses. A first generation of stars (Population II) forms. (B) Collapse continues with more Population II stars forming and some exploding as supernovae. Heavy elements created by the stars exploding as supernovae are mixed into the gas. Gas not used in making stars settles into a rotating disk. (C) Population I stars form in the disk.

Galaxy, distributed all around the Galaxy in the halo. These stars, many of which formed in globular clusters, are the Population II stars that we see today.

We can use the globular clusters to estimate the Milky Way's age. Using the techniques for measuring stellar ages in star clusters (Unit 69), astronomers calculate that our Galaxy's most ancient globular clusters are about 13 billion years old, a number we will take as the approximate age of our Galaxy.

Q If you randomly selected two Population I stars, one red and one blue, which would you expect to have higher metallicity?

Star formation during this initial stage did not turn all of the collapsing gas into stars. The remaining clouds of gas continued to fall inward. Clouds and stars behave very differently under these circumstances. Stars are so small, relative to the vast space between them, that it is extremely improbably that they will collide, and so nearly all continued past each other along their original trajectories. On the other hand, clouds fill a large fraction of the space they occupy, and as they fell toward the center of the growing Galaxy, they became even more crowded. These clouds collided with each other, losing orbital energy in the process, but conserving angular momentum (Unit 20.2), and settling into a smaller rotating disk of gas, as sketched in Figure 71.3B. This is similar to how drops of water might slide down a shower floor toward the drain in the middle, where they merge into a small spinning pool.

The infalling gas that formed the disk was enriched in heavy elements from the death of additional stars formed during the collapse. Therefore, even the first Population I stars born in the disk had a higher level of heavy elements. Continuing the motion of their parent gas, they circled the center of the young Galaxy in the disk (Figure 71.3C).

The two-stage collapse model explains the difference in the properties of Pop I and Pop II stars quite well, but it also suggests that there was a third, even earlier population of stars: the very first stars that formed from pure hydrogen and helium. These first, or **Population III**, stars would presumably also have orbits much like Pop II stars, but they would be even older and would have no metals at all.

Astronomers have been searching for "Pop III" stars for decades, but the closest they have come are two stars discovered in 2002 and 2005 that have less than $1/100{,}000^{\text{th}}$ the heavy element abundance of the Sun. They have merely $1/1000^{\text{th}}$ the typical metal content of Pop II stars, but they do still contain heavy elements.

It is possible that no Population III stars survive today. The only ones that might remain today would have to be less massive than about 0.9 $M_{\odot}$. Only such low-mass stars fuse their hydrogen slowly enough to survive for 13 billion years (Unit 61).

Q How might our Galaxy have been different if stars had formed at a significantly faster rate? at a significantly slower rate?

One hypothesis is that conditions in the initial cloud of gas may have prevented low-mass stars from forming; for example, the gas may have been too hot or turbulent, or a cloud of pure hydrogen and helium may collapse differently from the clouds of gas we observe today. According to this idea, only massive stars formed from the pure hydrogen and helium gas, and all of those stars have already died.

Another possibility is that Population III stars may exist but are masquerading as Population II stars. For example, the two stars with extremely low metal content just noted are estimated to be about 0.7 $M_{\odot}$ to 0.8 $M_{\odot}$ in mass and 13 billion years old. Such low-mass stars take more than 100 million years to contract and form—longer than the entire lifetime of massive stars forming out of the same collapsing cloud (Unit 60). Their outer layers may have been contaminated by gas ejected at the death of their more massive brethren. Stars like these may, then, be the only survivors from the initial collapse of the Milky Way.

71.3 EVOLUTION THROUGH MERGERS

Some features of the Milky Way call for refinements to the two-stage collapse model we have just described. First, according to the model, all Population II stars should be about the same age, having formed during the relatively brief period of the Galaxy's initial collapse. But observations of Pop II stars show that they formed over a significantly longer time span than the model predicts. Second, studies of the ages and distribution of stars in the disk and halo show some complex patterns that do not obviously result from the collapse model. For example, there appear to have been episodes of intense star formation, and there are streams of stars that follow unusual orbits within the Galaxy.

One hypothesis for explaining some of these anomalies is that the Milky Way has collided with other galaxies throughout its history. As we saw regarding the formation of our Galaxy, galaxy collisions are unlike the collision of two solid bodies. Because of the large spaces between stars, when two galaxies collide, their individual stars would almost never strike each other, although the gas and dust between the stars *would* collide. The compression of the gas clouds in both galaxies could lead to huge bursts of star formation. Computer modeling of such collisions suggests that two colliding galaxies often merge into a single larger system, a process that astronomers call **galactic cannibalism**.

FIGURE 71.4
Astronomers have mapped out a stream of stars that were pulled out of a small galaxy in its encounter with the Milky Way. Based on the estimated distances of the stars in the stream, the figure shows what the stream of stars (colored red) might look like from outside the Milky Way (shown as the blue spiral). The small galaxy is in the process of being swallowed up by the Milky Way.

We see evidence that some small galaxies are merging with the Milky Way even now. For example, astronomers recently detected a small "dwarf" galaxy on the far side of the Milky Way. The dwarf galaxy is elongated, probably because the Milky Way's gravity has stretched it out, and a long stream of stars within the Milky Way's halo appears to have been pulled from the smaller galaxy (Figure 71.4). Other streams of stars seen in the Milky Way may be all that remains of other smaller galaxies that have been pulled apart and merged with our Galaxy. Astronomers conclude that large galaxies like the Milky Way may have begun life with the collapse of a single gas cloud, but they continue to grow by swallowing smaller galaxies.

The interpretation of these observations is greatly aided by computer simulations. Computers can track the simultaneous effects that large numbers of stars and gas clouds have on each other. Modeling gravitational interactions with computers is, in principle, straightforward. The computer simply needs to calculate the forces and reactions between every pair of objects, using Newton's laws. The computer then projects where these objects will have moved in response to these forces after a short period of time (for example, one year) and then, for the new configuration, carries out the calculation again. Of course, with hundreds of billions of stars in the Milky Way, there are a lot of calculations to carry out! Computers cannot yet model every star in the Milky Way, but they have become fast enough to carry out such calculations for millions of stars at a time, which appears to be enough to sample many of the effects that occur.

Even the most advanced supercomputers, though, do not have the capacity to simulate all of the details of collisions between interstellar clouds. Such collisions generate rapid pressure changes, shock waves, heating, and turbulence, as well as a complex string of subsequent events. As the gas cools, it might become gravitationally unstable and form new stars. By treating a cloud as millions of small interacting "blobs" of gas, modern simulations are able to reproduce many known results. In fact, modern simulations attempt to add in the effects of stellar evolution, supernova explosions, and metal enrichment in the interstellar clouds. The great power of computer simulations is that they allow us to observe the consequences of large numbers of interactions and to predict the changes that will result over very long periods of time.

Computer simulations support the two-stage collapse model and merger model for the birth and evolution of the Milky Way. For example, Figure 71.5 shows how an

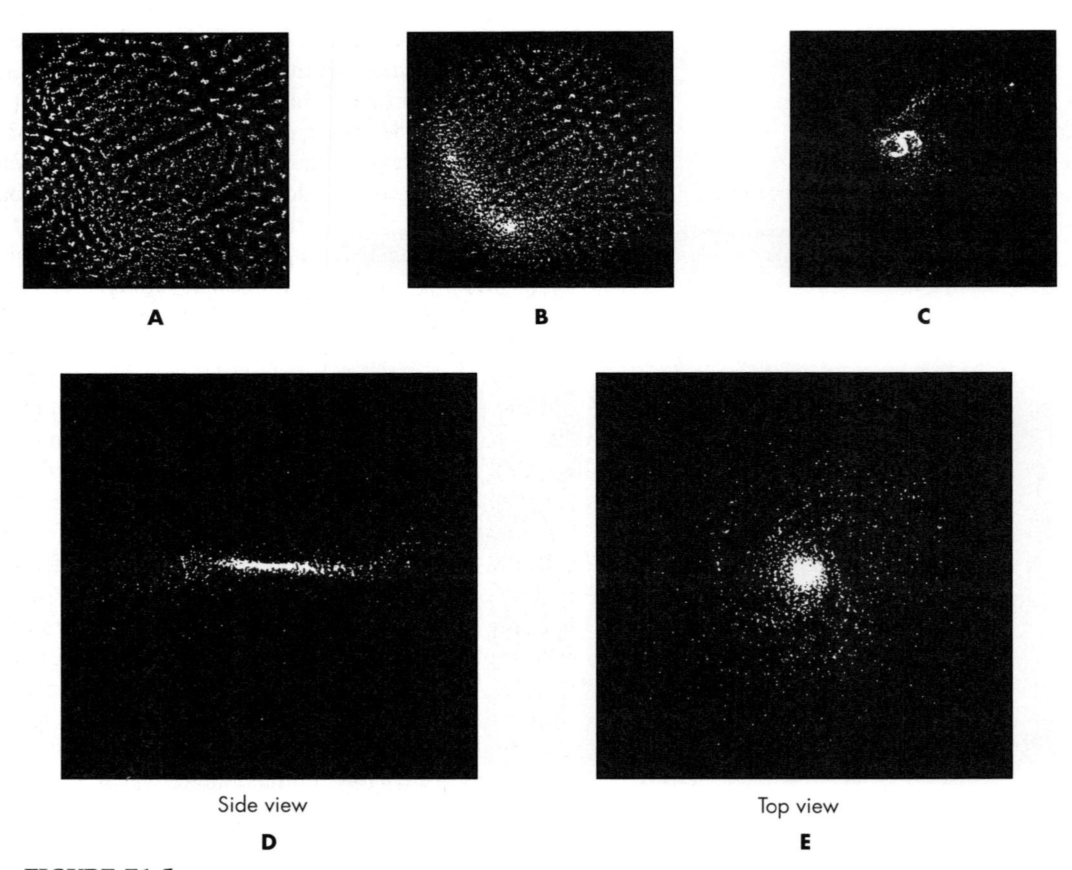

FIGURE 71.5
Computer simulation of the birth of a galaxy similar to the Milky Way. The top three frames (A), (B), and (C) show the initial state of the gas and the development of a clump that becomes the galaxy, illustrated in the bottom two frames (D) and (E). Notice the thin disk and spiral arms.

unstructured region of gas, of the appropriate size and rotation speed, collapses. You can see that it forms a thin disk with spiral arms, surrounded by a halo. Simulations are an important part of astronomy today, but it should always be kept in mind that the simulations only reflect the aspects of the physical interactions that were designed into them. If, for example, magnetic fields are important in these collisions, a new set of computer models would need to be developed to incorporate the effects of magnetic interactions.

71.4 THE FUTURE OF THE MILKY WAY

Throughout the Milky Way's history, as stars form, the amount of gas remaining is depleted. With a smaller reservoir of gas, the rate of star formation must also steadily decline, except for the occasional influx of new material caused by a merger event. Dying stars return a large fraction of their material to space as planetary nebula shells, supernova remnants, and stellar winds; but with each generation of stars, more and more of the Galaxy's mass is locked away from further use, trapped in stellar remnants such as white dwarfs, neutron stars, and black holes (Unit 59). An even larger fraction remains tied up in the very low-mass stars that can live even longer than the current age of the Galaxy.

Throughout this process, the interstellar gas becomes steadily more enriched with heavy elements, so that later generations of stars have a greater percentage of metals (elements heavier than helium). At the time of our Galaxy's birth, the birthrate of stars was probably very high and has since declined steadily. The enrichment of metals in the Sun suggests that it is about a "fifth generation" star, in the sense that its enrichment of metals would require that the gas had cycled through about five stars before forming the Sun and Solar System.

Q When the Milky Way is very old, the interstellar gas may contain little hydrogen. In what ways might this affect the evolution of stars?

By dividing the number of stars in the Milky Way by its age, astronomers estimate that, on average, about 10 stars have been born each year. A current star formation rate of a few stars per year is consistent with studies of star-forming clouds today. This seemingly small number of new star births will be sufficient to keep our Galaxy shining for many billions of years into the future. At current rates the Milky Way may run out of material for making new stars about 10 billion years from now. The lowest-mass stars will continue to shine for several hundred billion years, and the Galaxy will grow dimmer and redder. In about a trillion years even these low-mass stars will have died, and all that will remain will be the dark remnants of former stars. The Milky Way will fade, slowly spinning in space, a dark disk of stellar cinders.

The future will not be entirely quiet. Large merger events can be anticipated. Using computer simulations, astronomers predict that in about a billion years, two fairly large satellite galaxies—the Large and Small Magellanic Clouds—will spiral into our Galaxy and eventually merge with it. When they do, they will bring fresh gas and will probably initiate a period of strong star formation. In about 5 billion years, the Milky Way will probably collide with a nearby galaxy larger than our own, the Andromeda Galaxy (Unit 76), and eventually merge with it. And so this cannibalism will continue, the galaxies consuming one another even as their stars fade and die.

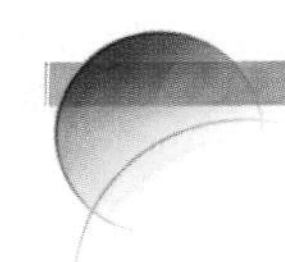

KEY TERMS

galactic cannibalism, 577
metals, 573
Population I (Pop I), 573
Population II (Pop II), 573
Population III (Pop III), 576

QUESTIONS FOR REVIEW

1. What are some differences between Pop I and Pop II stars?
2. How do the orbits of Pop I stars differ from the orbits of Pop II stars? Why?
3. Why must O and B stars be Pop I?
4. Why are spiral arms outlined by Pop I stars?
5. Describe one model for the origin of the Milky Way. How does this model explain the differences between Pop I and Pop II stars?
6. What are Pop III stars?
7. How are elements heavier than helium produced?
8. What is the eventual fate of the Milky Way?

PROBLEMS

1. A Pop II star has a highly eccentric orbit 12 kpc in diameter. What is its orbital period?
2. Conservation of angular momentum (Unit 20) requires that the product of the speed at which an object is orbiting times the distance from the point around which it is orbiting remain constant. If the Sun is currently orbiting at 220 km/sec, 8 kpc from the center of the Milky Way, how fast did the material that formed the Sun rotate if the gas began 400 kpc distant (about halfway to the next large galaxy)?
3. A 10^9-$M_\odot$ galaxy merges with a 10^{10}-$M_\odot$ galaxy. If 30% of their combined mass is lost in the collision, what is the total mass of the newly formed galaxy?
4. Suppose that after x billion years, the Milky Way converted half of its gas into stars. In the next x billion years, it converted half of the remaining gas into stars—and so on, each x billion years. The Milky Way is estimated to be about 13 billion years old, and only about 15% of its disk remains today in the form of gas.
   a. What value of x would give the Milky Way's current gas fraction? (*Hint:* You can calculate this mathematically, or you can draw an approximate graph of the declining gas fraction and estimate x.)
   b. In how many billion years will the gas fraction be just 1%?
5. Suppose we found a nearby Pop II star that is on a highly elliptical orbit, but it is currently near its most distant point from the center of the Milky Way in its orbit. To stars like the Sun in orbit in the disk, what would be the approximate relative speed of this Pop II star? In what direction?

TEST YOURSELF

1. A young blue star moving along a circular orbit in the disk is
   a. a Pop I star.
   b. a Pop II star.
   c. a Pop III star.
   d. not observable from Earth.
   e. none of the above.
2. Which of the following may explain the failure to observe Pop III stars?
   a. We do not expect to observe Pop III stars because pure hydrogen and helium stars never formed.
   b. The surfaces of Pop III stars were contaminated by remnants of massive stars.
   c. The first stars to be formed in the young Milky Way were very massive and therefore short-lived.
   d. All of the above.
   e. Only (b) and (c) are acceptable explanations.
3. The greater number of heavy elements seen in the spectra of Pop I stars relative to Pop II stars is explained by the fact that Pop I stars
   a. are older, and therefore have fused more hydrogen into heavy elements.
   b. formed more recently, and therefore have been made from enriched interstellar gas.
   c. have planetary systems, and debris from these fall onto the star's surface.
   d. are colder and therefore exhibit strong "metal" lines.

unit

72

Gas and Dust in the Milky Way

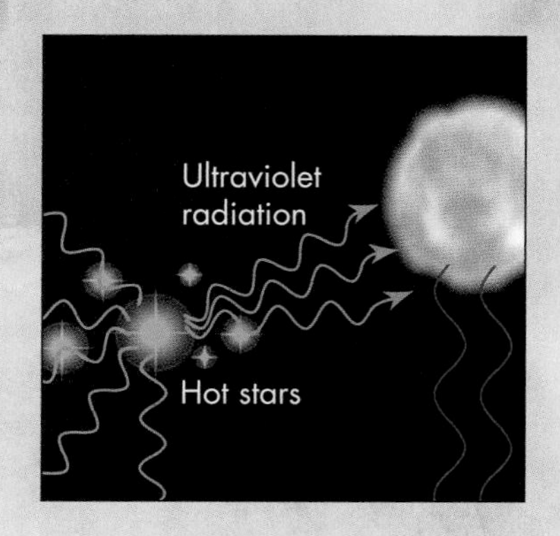

Background Pathways

Unit 24: Atomic Spectra: Identifying Atoms by Their Light 179

Unit 70: Discovering the Milky Way 566

The space between stars is not empty but contains gas and dust particles that compose what astronomers call the **interstellar medium**, or **ISM**. By terrestrial standards, this space is almost a perfect vacuum. On average, each liter contains only a few thousand atoms of gas. For comparison, the air we breathe contains about 10^{22} atoms per liter. The density of gas in interstellar space compared to the Earth's air is like having one marble in a box 8 kilometers (5 miles) on a side compared to the same box being filled completely with marbles.

Despite the interstellar medium's very low density, it plays several important roles in the Milky Way. First, it is the material from which new stars form. With no interstellar gas, star birth would stop. Second, we would not even exist if there had been no interstellar dust, because the atoms that make up the Earth and our bodies came primarily from such material.

One property of interstellar matter is that it limits what we can observe within the Milky Way. Interstellar dust blocks our view of distant parts of our Galaxy and impedes our ability to accurately measure the size of the Milky Way and distances to the stars within it. Before we can understand how astronomers overcome these observational difficulties, it will help to look more closely at the nature of interstellar matter.

72.1 PROPERTIES OF THE INTERSTELLAR MEDIUM

Interstellar matter is not spread smoothly throughout the Milky Way. Gravity pulls most of it into a thin layer in the disk, as can be seen by the dark band of dust in other galaxies (Figure 72.1). Within this layer, gravity and gas pressure cause the gas and dust to clump into clouds, which are embedded in a very low-density background gas.

FIGURE 72.1
Interstellar matter, outlined by dark dust clouds, is concentrated in the plane of a galaxy, forming a dark band across it. This galaxy is sometimes called the "Sombrero Galaxy." Technically, it is called either NGC 4595 or M104.

A

B

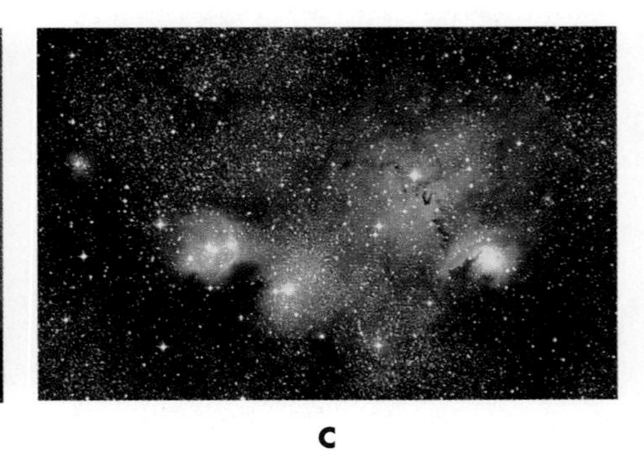
C

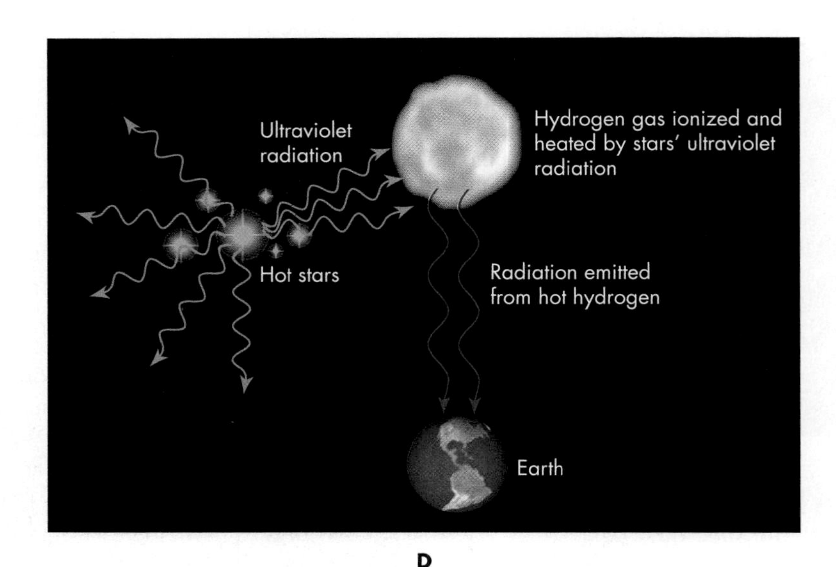

D

FIGURE 72.2
(A), (B), (C) Emission nebulae. Notice the bright (and hot) stars near the glowing gas. (D) These stars emit ultraviolet radiation, which heats and ionizes the nearby hydrogen gas. Electrons dropping from hydrogen's third to second energy level generate the pink light.

Stars may be scattered into new orbits when they make close gravitational encounters with other massive objects in the galaxy, so their orbits tend to spread out rather than settle into the plane.

Pink dots in the photographs of other galaxies are usually emission nebulae as well.

Interstellar matter behaves quite differently from stars as it orbits within the Galaxy, because the gas and dust particles fill space and collide with each other—losing energy and settling down into the common orbital plane. Stars, by contrast, are so dense relative to the gas that they can pass through the gas near the Galactic plane, which offers too little "friction" to slow them down significantly.

Sometimes the gas in an interstellar cloud is hot and emits visible light, as you can see in Figure 72.2. Such a glowing cloud is called an **emission nebula** (*nebula* is the Latin word for "cloud"). The power source causing the gas to light up is usually young, luminous O and B stars. They produce ultraviolet radiation that can ionize the hydrogen and other atoms in these clouds. An example of such a gas cloud is the Orion Nebula, which is easy to see with a small telescope or even a pair of binoculars (see Looking Up #6: Orion). If you look at the middle star in Orion's sword, you will see a pale, fuzzy glow created by luminous interstellar gas surrounding a dense star cluster.

Sometimes the gas in an interstellar cloud is not hot enough to emit visible light of its own, but dust particles in it may reflect light from a nearby star, in which case it is called a **reflection nebula** (Figure 72.3). The reflection effect is something like the beams of light you see coming from the headlights of a car as it drives through fog.

If a cold interstellar cloud is sufficiently dense, the dust in it may block the light of background stars and appear as a **dark nebula** against the starry background (Figure 72.4). Clouds like these are too cold to emit visible light, but they emit lower-energy radiation at infrared or radio wavelengths. Several dark nebulae are

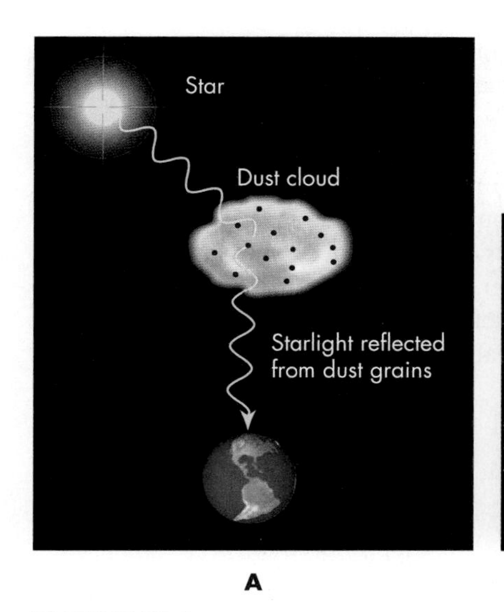

A

B

FIGURE 72.3
Reflection nebulae. (A) Light from a star near a dust cloud is reflected from the dust, making it visible. (B) The Pleiades star cluster. Note the bluish reflection nebula around many of the brighter stars.

FIGURE 72.4
A dark nebula (called Barnard 86) in the Milky Way. Dust blocks our view of the background stars. The stars that appear to be within the outline of the dark nebula lie between us and the nebula.

Q Under what circumstances could an interstellar cloud include all three types—emission, reflection, and dark nebulae?

also visible against the emission nebulae seen in Figure 72.2. Dark nebulae are the starting point for the formation of the next generation of stars.

Radiation is both emitted and absorbed by gas in an interstellar cloud, giving astronomers a way to measure many of the cloud's properties. An atom or molecule will absorb the specific wavelengths of light that correspond to energy-level changes of its electrons (Unit 24). Therefore, when astronomers observe the spectrum of a star behind an interstellar cloud, they see the dark lines of the interstellar gas's absorption superimposed on the star's own spectral signature (Figure 72.5). On the other hand, if the gas has been heated, it produces emission lines. By studying the absorption and emission lines of interstellar gas, astronomers can deduce the composition, temperature, density, and motion of the cloud. Such spectra show that interstellar gas has a composition similar to that of the Sun and other stars—about 71% hydrogen and 27% helium, with the remainder made up of heavier elements.

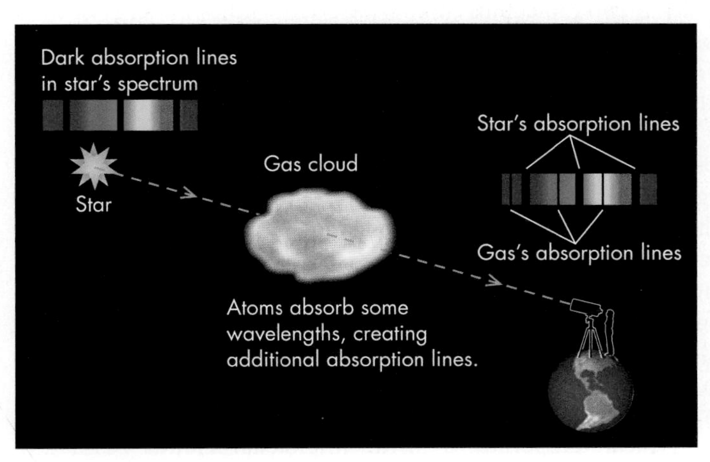

FIGURE 72.5
Gas atoms in an interstellar cloud between us and a distant star absorb some of the star's light, adding their own absorption lines to those in the star's spectrum.

Dust particles also absorb and emit light. But because dust particles are solids, they do not generate clear spectral signatures as a gas does. Still, astronomers can deduce dust particles' compositions from the elements that are "missing" from the gases detected in cool interstellar clouds versus hot interstellar gas. Based on the chemical properties of these elements, it is likely that the dust particles are made of silicates and carbon compounds, coated with a thin layer of ices made of molecules such as water, carbon monoxide, and methyl alcohol. Based on the way they interact with light, these particles must be extremely tiny, ranging from about a micrometer to a nanometer or so in diameter (not much larger than some large molecules).

Although interstellar clouds are similar in overall composition, they differ greatly in size, ranging from a fraction of a light-year to a few hundred light-years in diameter.

They also range widely in mass from about one to millions of solar masses. Finally, they also have a wide range of density and temperature. For example, some regions have temperatures of millions Kelvin while others are only a few degrees above absolute zero.

The complexity of the interstellar medium is a little like "weather." If we could watch a high-speed movie of interstellar clouds, we would see that they constantly form and dissolve, driven by powerful physical phenomena such as stellar winds and supernova explosions. Although matter in interstellar space undergoes huge changes in temperature and density, this occurs on timescales of millions of years. Our observations of interstellar clouds reveal a momentary snapshot of a dynamic and turbulent medium.

72.2 HEATING AND COOLING IN THE ISM

The temperature of any object is set by a balance between the energy it receives (heating) and the energy it loses (cooling), and interstellar clouds are no exception. Gas can be heated in several ways. Two especially important heat sources are radiation and winds from hot stars, and blasts and radiation from supernovae. These processes add energy to the cloud's gases. Atoms and molecules are physically set in motion, and the high-energy photons from hot stars or explosions give atoms so much energy that they are ionized and eject electrons at high speed. If we could watch microscopically, we would see this energy translated into the random motions of thermal energy as electrons, atoms, and molecules collide with one another and rebound in different directions.

Heating of interstellar gas is strongly affected by the distance from the source and by shielding. Greater distances spread out the impact of photons, winds, and blast waves—the energy being spread over a larger and larger surface area with distance. **Shielding** occurs wherever material near a star "casts a shadow" that blocks energy from reaching interstellar clouds that are farther away. Dust is a particularly important source of shielding. Just as dust blocks the light of some stars from us, it can shield material behind it.

Cooling occurs when the gas emits radiation of its own. Radiation often results from collisions between gas atoms or molecules. To see how the thermal energy generated by a collision is turned into radiation energy, we look again at the interactions microscopically. If two molecules collide, part of the kinetic energy of their collision may be absorbed by electrons being knocked into higher energy levels. Energy is conserved because the two molecules rebound from each other with slightly lower speeds than when they collided, reducing the thermal energy. The electrons later drop down to lower energy levels, emitting one or more photons.

The net effect of collisions between molecules is to transfer thermal energy into radiation energy. Indeed, the radiation that allows us to see a cloud is also the means by which the cloud cools. Because collisions are more frequent where the atoms or molecules are closer together—that is, where interstellar matter is denser—high-density regions tend to cool more rapidly. This is especially interesting because cool high-density regions are also where gravitational forces are largest, so these regions tend to cool even more rapidly as they grow denser.

Dust grains also heat up when they absorb radiation or undergo collisions. The grains cool by emitting thermal radiation, which is generally at infrared wavelengths for the typical temperatures inside interstellar clouds (Unit 23). In dense clouds containing lots of dust, the combination of high density and shielding allows the gas and dust deep in the cloud to cool down to only a few degrees Kelvin (about −270°C or −450°F).

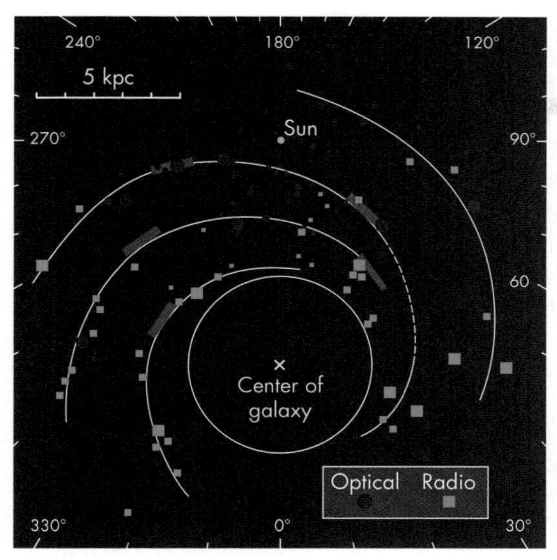

FIGURE 72.6
Map of the Milky Way made by combining radio observations (radio can penetrate dust and thereby allow us to see all the way across the Milky Way's disk) and optical observations of the position of HII regions. Such regions outline the spiral arms of other galaxies. In our Galaxy, too, they form a spiral pattern.

Applying these ideas, we find that near hot, blue stars (spectral type O or B, Unit 55), gas and dust will itself be hot, typically about 10,000 K. Such temperatures may even be hot enough to vaporize the dust grains or boil off some of their surface molecules. O and B stars are especially effective at heating nearby material because they emit large amounts of ultraviolet photons, which have shorter wavelengths and much more energy than visible photons. Ultraviolet light with a wavelength less than 91.2 nanometers is so energetic that when it is absorbed by hydrogen, it tears the electron free of the nucleus, ionizing the gas. The free electrons collide with atoms of oxygen, nitrogen, and other elements and excite them, making them emit light of their own. Eventually the electrons recombine with the ionized hydrogen atoms and emit more radiation as they drop from higher-energy orbitals down to lower-energy orbitals. Visible-wavelength photons are produced when the electrons drop down to the second energy level from higher levels, producing the pink light of hydrogen Balmer lines (Unit 55.4). This gives these nebulae their characteristic pink color, as seen in Figure 72.2. Because hydrogen in these hot gas clouds is ionized, they are sometimes called **HII** (pronounced *H-two*) **regions**—the *H* denoting the hydrogen and the *II* indicating that it is ionized.

The free electrons in HII regions are themselves a powerful source of radio emission. Astronomers can detect that emission and thus see, in radio wavelengths, HII regions that are otherwise hidden from us by dust in the interstellar medium. Because these HII regions are generally located in spiral arms, radio maps of their location help to reveal the spiral structure of our Galaxy. In fact, maps made by combining radio and optical observations of HII regions provide some of the best maps we have of the Milky Way's spiral structure (Figure 72.6).

72.3 INTERSTELLAR DUST: DIMMING AND REDDENING

Although only about 1% of interstellar matter is in the form of dust grains, the effects of dust are strong. Not only does dust dim the light of distant stars, it also alters the light's color, much as haze in our atmosphere dims and reddens the setting Sun. These effects occur because visible light interacts strongly with dust.

When a photon strikes a dust particle, it might be absorbed or reflected in a random direction in a process called **scattering.** Even if there is no absorption, scattering dims a star's light because some of the radiation heading toward us is sent off in random directions and is lost to our sight. A thin cloud of dust will scatter just a fraction of the light headed in our direction, which we observe as a reflection nebula (Figure 72.3). However, if a dust cloud is extremely thick or dense, the photons are scattered repeatedly and their energy is eventually absorbed by the dust grains. This may dim the light of background stars to the point where they become invisible in a dark nebula (Figure 72.4).

A dust particle has the strongest scattering effect on electromagnetic waves whose wavelength is close to or slightly smaller than the size of the particle. If the dust particle is much smaller than the wavelength of radiation, it is much less effective at scattering the radiation. An analogy to this is that a boat will block short-wavelength ocean waves, which "slap" against the side of the boat, while it does little to affect long-wavelength waves, which raise and lower the boat and continue past it.

A sunny day shows the effect that small particles in our atmosphere have on sunlight. Here oxygen and nitrogen molecules scatter short-wavelength blue photons of sunlight much more strongly than longer-wavelength red photons. The sky is not blue because the molecules in the atmosphere are glowing with this color. Instead, the blue photons we see are part of the mix of all wavelengths coming from the Sun. Because of the size of Earth's atmospheric molecules, the Sun's blue wavelengths are scattered more strongly than longer, redder wavelengths that pass through to the surface. This process is similar to the effect of small dust grains in interstellar space, and this is why reflection nebulae usually have a bluish color (Figure 72.3).

Q Suppose we observe a main-sequence star behind a thin dust cloud, so its light is slightly reddened and dimmed. If we used its color to predict its luminosity, would we tend to overestimate or underestimate the luminosity? What if we tried to use its brightness to estimate its distance?

Because shorter-wavelength photons are scattered in random directions, the balance of the remaining light contains a relatively larger number of longer-wavelength photons. This effect is called **reddening** because the color balance shifts toward the red end of the visible spectrum. It might more accurately be called "de-bluing" because no red color is added to the light, but instead the bluer colors are subtracted by scattering. The effect of reddening by dust in the Earth's atmosphere can make the Sun look yellow, orange, or even red as more of the short-wavelength photons are scattered. The effect is strongest at sunset when the sunlight reaching us has traveled a long path through our atmosphere, so that more of the short-wavelength photons are scattered away from their path toward us.

Small interstellar dust grains are also very effective at scattering short-wavelength blue and ultraviolet light. As a result, when we look at a star through a cloud containing dust, the light appears reddened. Light is scattered and absorbed at all wavelengths to some degree, so there is an overall dimming, or **extinction**, of the light as well. If the cloud is thick enough, the extinction may be so strong that little or no light of any wavelength will pass through it, and the cloud will be seen as a dark nebula. Both reddening and extinction can be seen around the edges of the dark nebula shown in Figure 72.7.

The extinction effects of dust on visible wavelengths of light make distant parts of our Galaxy hard to study. However, the longer wavelengths of infrared and radio waves are relatively unaffected because they are much larger than the dust

A

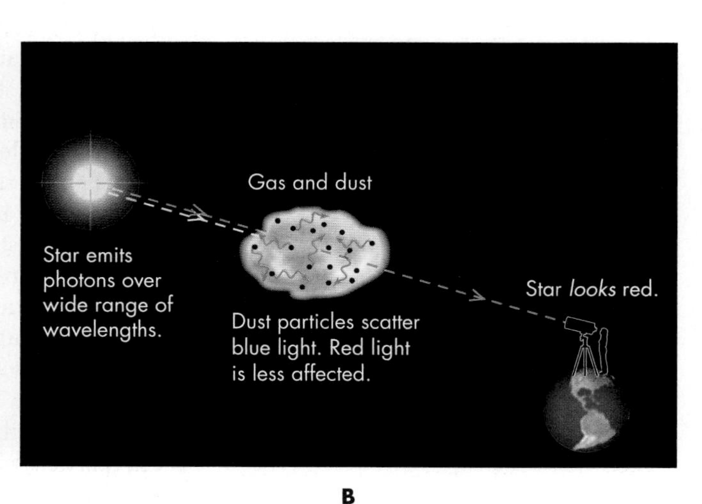

B

FIGURE 72.7
(A) A string of dark clouds called the "Snake Nebula" blots out the light from background stars. Short-wavelength light is blocked more strongly than long wavelengths, so near the edge of the clouds, the background stars are reddened. (B) Dust in an interstellar cloud scatters the blue light from a distant star, removing it from the radiation reaching Earth and making the star look redder than it really is.

FIGURE 72.8
An all-sky view of our Galaxy, as we see it from Earth. The image was made by plotting on a map of the whole sky the millions of stars observed in the infrared by the 2-Micron All Sky Survey (2MASS). It is therefore like a map of the Earth flattened out to show all continents. The map clearly reveals the flatness of our Galaxy and the bulge of its central regions.

particles. As a result, infrared pictures reveal the structure of the Milky Way far better than visible-light pictures can, as shown in Figure 72.8. The disk and central bulge of our Galaxy are easy to see in this wide-angle computer image of stars mapped by infrared cameras.

72.4 RADIO WAVES FROM COLD INTERSTELLAR GAS

Unless an interstellar cloud is within a few parsecs of a hot star, it may cool down to just a few degrees Kelvin. Such cold material emits no visible light because it has too little energy to generate visible photons. However, even at such low temperatures it may still emit the low-energy, long-wavelength radiation that we can detect with radio telescopes.

One of the most important sources of radio emission is cold hydrogen atoms, which radiate at a wavelength of 21 centimeters. This radiation is called **HI** (pronounced *H-one*) **emission** because cold hydrogen atoms with their electrons attached are called HI to distinguish them from ionized hydrogen atoms, HII.

HI emission arises because subatomic particles such as protons and electrons possess a property, known as *spin,* that makes them behave like tiny magnets. The energy of the magnetic interaction between the proton and electron is a tiny bit higher in the hydrogen atom when the proton and electron spin in the same direction. If the electron flips its spin direction, as shown in Figure 72.9, the energy is slightly lower. The atom emits the energy difference as a radio wave with a wavelength of 21 centimeters.

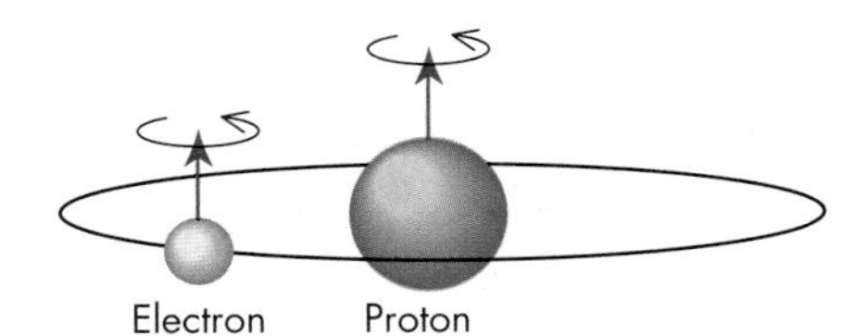

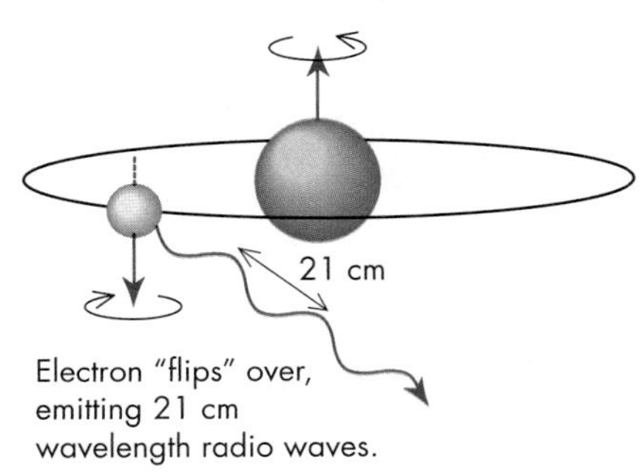

FIGURE 72.9
Radio radiation at 21 centimeters is emitted by cold hydrogen as an electron "flips" and spins in the opposite direction.

This **21-centimeter radiation** has proved extremely valuable for studying the Milky Way and other galaxies. First, this type of radiation is not absorbed by interstellar dust. Second, hydrogen is abundant in space, so the signal of 21-centimeter radiation is strong. From the strength of this signal, astronomers can deduce the amount of hydrogen in a region. If the gas is moving, the wavelength will be Doppler-shifted (Unit 25), and the shift will enable us to determine the gas's motion toward or away from us. Radio observations thereby allow astronomers to map not only where the gas is concentrated in the Milky Way but also how it is moving. Astronomers still face complications in interpreting these measurements, because the 21-centimeter radiation coming from a particular direction in the Milky Way is a blend of the signals from nearby and more distant gas along our line of sight. However, by modeling known motions in the Galaxy, astronomers can estimate the distances and produce maps of the entire Milky Way at the 21-centimeter wavelength. Such maps show that HI gas is confined to a thin disk, and its distribution is suggestive of a spiral pattern, confirming the picture of our Galaxy deduced by other means.

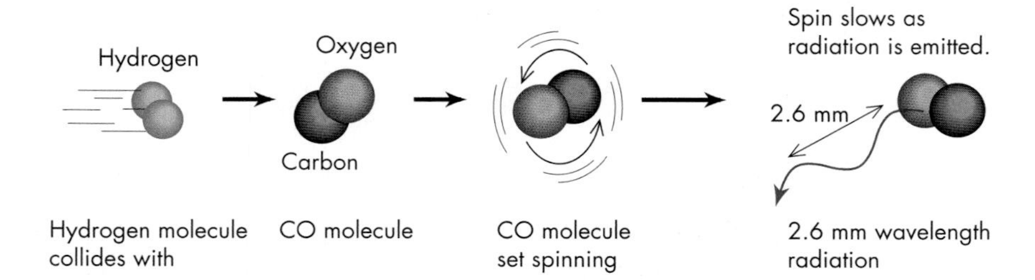

FIGURE 72.10
Emission of radio waves by spinning molecules in cold interstellar clouds. Collisions set the molecules spinning. When their rotation speed changes, they emit radio waves. The wavelength emitted depends on the kind of molecule and by how much its rotation speed changes.

In addition to 21-centimeter radiation, many other wavelengths of radio radiation are emitted by interstellar gas. In the densest and coldest clouds, molecules form. These regions are known as **molecular clouds**, and molecules within them emit primarily at radio wavelengths. More than 100 interstellar molecules have been identified so far, including common compounds such as molecular hydrogen (H_2), carbon monoxide (CO), formaldehyde (HCHO), and ethyl alcohol (CH_3CH_2OH), along with even more complex molecules. Table 72.1 lists a number of the more common molecules detected.

Q Why are H, C, N, and O the most common constituents of interstellar molecules?

Collisions between molecules may make them spin or vibrate. As they slow down, the decrease in energy of spinning or vibration is emitted as radio wavelength radiation. For example, carbon monoxide is a common interstellar molecule, and it can emit at wavelengths of 2.6 millimeters, 1.3 millimeters, and a series of shorter wavelengths as its spin slows (Figure 72.10). Astronomers can analyze these other waves, just as they do 21-centimeter radiation, to learn about gas in cold dense clouds. In fact these clouds do not usually emit much 21-centimeter radiation because their atomic hydrogen has combined to form molecular hydrogen. The observations of molecules therefore provide information about the densest interstellar environments, which complements the 21-centimeter observations.

TABLE 72.1 Some Interstellar Molecules

Molecule	Chemical Formula
Molecular hydrogen	H_2
Hydroxyl radical	OH
Cyanogen radical	CN
Carbon monoxide	CO
Water	H_2O
Hydrogen cyanide	HCN
Ammonia	NH_3
Formaldehyde	HCOH
Acetylene	HCCH
Formic Acid	HCOOH
Methyl alcohol (methanol)	CH_3OH
Ethyl alcohol	CH_3CH_2OH

KEY TERMS

21-centimeter radiation, 587	interstellar medium (ISM), 581
dark nebula, 582	molecular cloud, 588
emission nebula, 582	reddening, 586
extinction, 586	reflection nebula, 582
HI emission, 587	scattering, 585
HII region, 585	shielding, 584

QUESTIONS FOR REVIEW

1. How do we know interstellar matter exists?
2. What are the differences between emission, dark, and reflection nebula?
3. How do we know that some interstellar matter is dust?
4. How does the gas of the ISM cool?
5. How does interstellar dust affect our observations of stars and the Milky Way?
6. What type of radio radiation does the ISM emit? How?
7. What evidence makes astronomers believe the Milky Way has spiral arms?

PROBLEMS

1. What is the density of a gas cloud that is 10 ly in diameter and contains 100 $M_{\odot}$ of material?
2. If a gas cloud shrinks to half its size, how much change is there in its density? in its temperature?
3. Convert the following radio wavelengths to frequencies (see Unit 22):
   a. 21-cm line of atomic hydrogen
   b. 2.6-mm line of carbon monoxide
4. Find the ratio of the energy of a 21-cm photon to the energy of an electron shifting from the second energy level of hydrogen to the ground state, which produces a photon of 122-nm wavelength.
5. On average, a hydrogen atom emits a 21-cm photon once every 15 million years. How many 21-cm photons does a 100-$M_{\odot}$ gas cloud emit per second?
6. The dust in an interstellar cloud blocks blue light in the following way: For every 1 pc light travels through the cloud, only 90% of the light continues. Thus, after 2 pc, 90% of the remaining 90%, or 81% ($0.9 \times 0.9 \times 0.81$), remains.
   a. You might initially expect that after 10 pc, the ten 10% reductions would have removed all of the blue light. How much blue light actually remains?
   b. The same cloud removes about 7% of the red light every parsec, so after 1 pc, the ratio of blue to red has dropped to $90/93 \approx 97\%$ of its unreddened value. What is the ratio of blue to red after 5 pc? after 10 pc?

TEST YOURSELF

1. Astronomers know that interstellar matter exists because
   a. they can see it in dark clouds and clouds that absorb light.
   b. the matter creates narrow absorption lines in the spectra of some stars.
   c. they can detect radio waves coming from atoms and molecules in the cold gas.
   d. spacecraft have sampled clouds near Orion.
   e. All of the above except (d).
2. What is the source of energy that makes emission nebulae glow?
   a. Heat from hot white dwarfs within the nebula
   b. Nearby hot O and B type stars
   c. Nearby black holes
   d. Intense radio waves
   e. Gravitational compression
3. The typical size of interstellar dust particles is ____; and they consist of ______.
   a. 1 cm; silicates and carbon compounds
   b. 1 mm; hydrogen and helium
   c. about a micrometer or less; silicates and carbon compounds
   d. about a nanometer or less; hydrogen and helium

unit

73

Mass and Motions in the Milky Way

Astronomers' measurements have found that our Sun and neighboring stars move around the center of the Milky Way at a speed of about 220 kilometers per second. The Sun's large orbital speed (more than seven times the speed of the Earth around the Sun) may make you think the Milky Way spins rapidly; but because our Galaxy is so huge, the Sun takes approximately 220 million years to complete one trip around it. Since the extinction of the dinosaurs about 65 million years ago, our Galaxy has made less than one-third of a revolution.

Were it not for the collective gravity of all the components of the Milky Way, the motions of its stars and gas would cause the Galaxy to fly apart and disperse within a few billion years. At the same time, were it not for the motion, gravity would draw all the stars and gas inward, causing the Galaxy to collapse. The motion and gravity are in balance, and astronomers can use this balance to determine the mass of our Galaxy by studying the motions of the stars within it.

Background Pathways

Unit 17: Measuring a Body's Mass Using Orbital Motion 135

Unit 70: Discovering the Milky Way 566

73.1 THE MASS OF THE MILKY WAY AND THE NUMBER OF ITS STARS

The rotation of matter in a flattened disk around the Galaxy's center is somewhat similar to the orbital motion of the planets around the Sun. Unlike a spinning solid disk, such as a wheel or a Frisbee, different parts of the Milky Way complete their orbits in different amounts of time. This is similar to how planets near the Sun complete their orbits faster than planets farther out—so too do stars near the center of the Galaxy complete their orbits faster than stars at the edge of the Galaxy. The phenomenon of the inner and outer parts of the Galaxy taking different times to complete an orbit is called **differential rotation.**

We can describe the unique properties of the Milky Way's rotation by observing stars at a variety of distances from the Galactic center to obtain a relation between orbital velocity and radius. Astronomers call this relation a **rotation curve.** Rotation curves for a solid disk and the Solar System are illustrated in Figure 73.1A and B. When we measure the speeds of stars rotating around the center of the Milky Way, though, we see neither a steadily rising rotation speed nor a steadily declining one. The stars orbit at about the same speed at most radii (Figure 73.1C).

The Milky Way's rotation curve reflects the distribution of mass in our Galaxy. It is very different from the distribution of mass in the Solar System, where more than 99% of the mass is concentrated in one central object (the Sun). A planet orbiting the Sun is subject to the gravitational pull essentially of just that one object. In the Milky Way, on the other hand, an orbiting star is held in its orbit

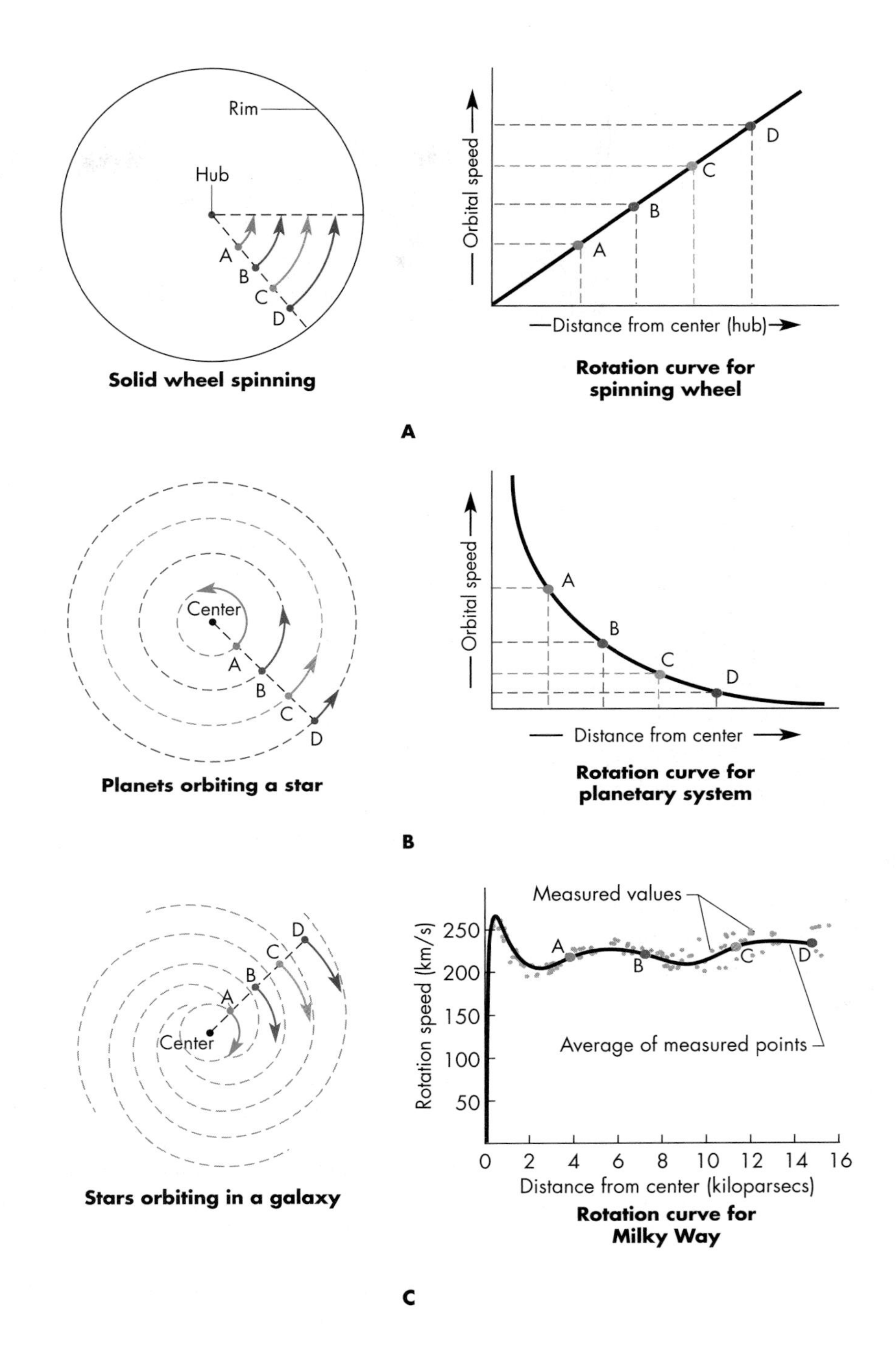

FIGURE 73.1
(A) When a wheel rotates, the outer parts move faster, creating a rising rotation curve. (B) Planets orbiting a star have slower speeds, the farther they are from the star. (C) The Milky Way's rotation speed remains nearly constant at all radii, as illustrated by actual data measured within our Galaxy.

by the collective gravitational pull of all the stars and other matter lying inside its orbit. For example, there might be a total of 1 billion solar masses out to a radius of 1 kiloparsec, but 2 billion solar masses within a radius of 2 kiloparsecs of the center, so a star orbiting at 2 kiloparsecs will be feeling the pull of much more mass. As a reminder that the amount of mass depends on the radius out to which we measure it, we will write $M_{<R}$. The subscript "$<R$" indicates "within a radius smaller than R." This peculiar notation will be helpful when we discuss the Milky Way and other galaxies, because modern observations suggest that galaxies have no clear edge and that their masses increase as we measure them out to larger radii.

The mass outside of a star's orbital radius has a gravitational pull, but it pulls outward in a variety of directions, canceling the net force.

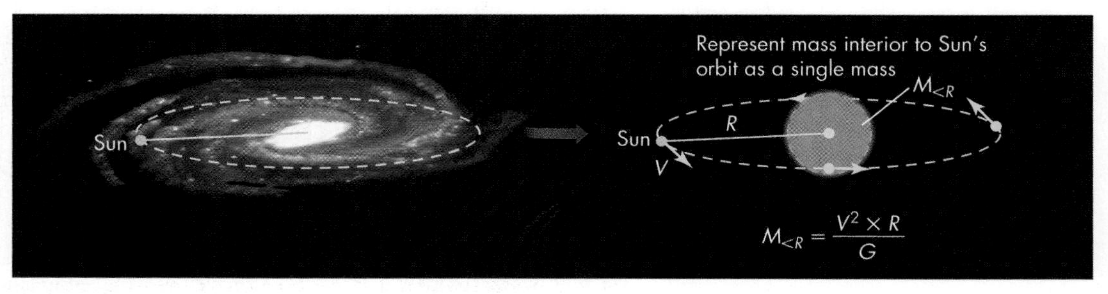

FIGURE 73.2
The Sun orbits all of the mass interior to its orbit $M_{<R}$. We can represent all of this mass by a single mass (the orange circle in the figure) and apply Newton's law to find the mass within the Sun's orbit.

$M_{<R}$ = mass within radius R (kilograms)
R = radius (meters)
V = velocity of rotation (meters per second)
$G = 6.67 \times 10^{-11}$ m^3/kg-sec^2

Astronomers can calculate the Milky Way's mass from the gravitational attraction needed to hold the Sun and other stars in orbit (Figure 73.2). This is yet another application of Newton's law of gravitation (Unit 17). From the speed V of a star orbiting at radius R from the center of the Galaxy, we can calculate the mass interior to its orbit as follows:

$$M_{<R} = \frac{V^2 \times R}{G}$$

where G is Newton's gravitational constant. By looking at the rotation speeds of other stars at different distances from the center, we can build up a picture of the mass structure of the Galaxy.

If we put in the rotation speed of the Sun (220 km/sec = 2.2×10^5 m/sec) and distance from the Galaxy's center (8 kpc = 2.5×10^{20} m), we find that the Milky Way's mass out to 8 kiloparsecs is

$$M_{<8\text{ kpc}} = \frac{(2.2 \times 10^5 \text{ m/sec})^2 \times 2.5 \times 10^{20} \text{ m}}{6.67 \times 10^{-11} \text{ m}^3/\text{kg} \cdot \text{sec}^2}$$
$$= 1.8 \times 10^{41} \text{ kg}.$$

This is 9×10^{10} times the Sun's mass ($M_\odot = 1.99 \times 10^{30}$ kg). Thus, there are 90 billion solar masses of material in the Milky Way out to the Sun's orbit—the total of stars, gas, and any other matter.

Because 16 kpc = 5.0×10^{20} m, we find that:

$$M_{<16\text{kpc}} = \frac{(2.2 \times 10^5 \text{ m/sec})^2 \times 5.0 \times 10^{20} \text{ m}}{6.67 \times 10^{-11} \text{ m}^3/\text{kg} \cdot \text{sec}^2}$$
$$= 3.6 \times 10^{41} \text{ kg}$$
$$= 2 \times M_{<8\text{ kpc}}$$

This enormous total of material within the Sun's orbit is somewhat higher than estimates of the number of stars and other known material in the inner part of the Milky Way. The level of disagreement is not considered significant, given the uncertainties in our observations because of problems such as interstellar extinction. What is peculiar, though, is that outside the Sun's orbit, the rotation speed remains high even though the number of stars declines rapidly toward our Galaxy's outer reaches. At twice the Sun's distance from the center (16 kpc), the measured rotation speed remains at about 220 kilometers per second. So in the calculation of mass interior to 16 kiloparsecs, everything remains the same except that we multiply by a radius that is two times larger. The mass within 16 kiloparsecs is therefore twice the mass within 8 kiloparsecs, even though there appears to be only a fraction as many stars in this outer region.

We can even estimate the Milky Way's mass out to about 100 kiloparsecs from its center—beyond where we detect any stars—by examining the speeds of small satellite galaxies orbiting the Milky Way. We do not have very complete information about their orbits, but the Milky Way's gravitational effect on them implies that the Milky Way contains approximately 2 trillion solar masses ($2 \times 10^{12} M_\odot$). This is about 20 times more mass than we just found within the Sun's orbit, which is difficult to understand because we see almost nothing at these large distances that could account for so much mass. This discrepancy leads astronomers to conclude that the Milky Way's halo contains much more matter than the disk, and that most of this material emits no as-yet detectable light. They therefore describe this unseen mass as **dark matter.**

Q If there were no mass beyond the Sun's orbit around the center of the Milky Way, what would the rotation curve look like?

Measurements of the speed of rotation at different radii suggest that dark matter is present everywhere in the Milky Way. Its density is highest near the Galaxy's center, but in those regions the density of normal matter is much greater than the dark matter density. Based on rotation curve measurements, it appears that the overall fraction of the mass that consists of dark matter is greater in the outer parts of the Galaxy, eventually constituting the majority of mass. The dark matter apparently continues well beyond the area where we detect the outermost stars and gas.

In the vicinity of the Sun, astronomers have another clue about the nature of dark matter, from the way stars move within the Galactic disk. As the Sun and other stars orbit the center of the Milky Way, their orbits do not remain precisely in the midplane of the disk. Instead small "vertical" motions through the disk carry them up and down through the plane, similar to how horses on a merry-go-round go up and down as they all circle the center. This occurs because as a star travels away from the midplane, the concentration of mass there pulls on it until it falls back. It then crosses through to the other side of the plane, and the process repeats, producing an oscillating motion. The Sun, for example, crosses through the midplane about once every 30 million years.

It was recognized by the 1930s that the speed of these oscillations gives us a way to estimate the mass of material in the disk. These estimates suggest that there is about twice as much mass in the disk as is evident based on the known stars, gas, and dust. This factor is not as big as in the outer parts of the Milky Way, but it is nearby where telescopes can detect such things as dim, cold stars, old white dwarfs, or neutron stars. After decades of searches, the failure to identify the source of all this mass even near to the Solar System indicates that the dark matter must be something quite unusual.

Q Would it be difficult to detect planets elsewhere in the Galaxy? Are they likely to make up much of the dark matter? Why or why not?

Some of the most powerful evidence for the nature of dark matter comes from studying other galaxies and the beginnings of the universe. These suggest that dark matter probably is unlike the matter we are familiar with—and that the Milky Way may be sitting at the center of a vast cloud of dark matter particles that contain most of the mass of our Galaxy. We return to the question of dark matter in Unit 78.

Despite the uncertainty about the Milky Way's total mass, in the inner portions of the disk it appears that stars dominate the overall mass, so we can still estimate the number of stars from the rotation speed. From the rapid decline in stars seen at radii larger than the Sun's, we can deduce that nearly all of the Milky Way's stars are contained within about twice the Sun's distance from our Galaxy's center. In total, the disk contains about 100 billion stars, along with about an equal amount of mass consisting of interstellar matter and stellar corpses. Each object moves along its own orbit around the center of the Galaxy, held in its path by the collective gravity of all the other objects.

73.2 THE GALACTIC CENTER AND EDGE

Within this spinning cloud of stars, the Sun is nearly lost, like a speck of sand on a beach. But unlike grains of sand that touch others, stars are widely separated. For example, in the vicinity of the Sun, stars are typically more than a parsec apart. The Sun's nearest neighbor is 1.3 parsec away, a separation akin to sand grains spaced 10 kilometers (6 miles) apart. Near the core of the Milky Way, stars are packed far more densely, with a separation roughly 100 times smaller than near the Sun. Here the typical separation can be represented by sand grains placed at opposite ends of a football field.

The density of stars at the Sun's orbital radius can be measured in terms of the number of stars within a cube 1 parsec on a side (stars/pc^3). This is found to be about 0.1 stars/pc^3 or equivalently one star every 10 cubic parsecs. At twice the Sun's orbital radius (about 16 kpc) from the center, stars are spread even more thinly—less than 0.01 stars/pc^3—much as our atmosphere thins out as it merges into space. Thus, the Milky Way has no sharply defined outer edge.

FIGURE 73.3
The central region of the Milky Way as observed at X-ray wavelengths by the orbiting Chandra X-ray telescope. The supermassive black hole at the Milky Way's core lies inside the bright white patch at the center. The image shows an area 900 light-years across. In this false-color image, the colors represent X-rays of different energies. The small bright spots are white dwarfs, neutron stars, and probably gas surrounding a few black holes. The diffuse glow is from extremely hot gas surrounding our Galaxy's core.

Toward the center of the Galaxy, the density in the disk rises to about 1 star/pc^3 just outside the bulge—about 2 kpc from the center. Within the bulge the density is about 10 times higher still. At the center of the Galaxy the density rises sharply, exceeding 100,000 stars/pc^3 in the central few parsecs, and the stars there orbit around the center quite rapidly.

What lies at the very center of the Milky Way Galaxy? The core is difficult to observe because interstellar dust clouds almost completely block the visible light emitted by objects there (see Looking Up #7: Sagittarius). But radio, infrared, X-ray, and even gamma-ray telescopes allow astronomers to see through the dust, and they reveal a dense swarm of stars and gas swirling around the center of our Galaxy (Figure 73.3). Moreover, these telescopes reveal that at the core is an intense source of radio waves known as **Sagittarius A***, which is abbreviated "Sgr A*" (but usually pronounced *sadj ay star* by astronomers). The position of this mysterious object is shown in Figure 73.4, an infrared image of the stars within the central 2 light-years of the Galaxy.

FIGURE 73.4
This infrared image shows the inner 2 light-years of the core of the Milky Way. The position of Sagittarius A*—the center of the Galaxy—is shown by two small arrows.

Sgr A* appears to mark the very center of the Milky Way. Gas and stars orbit it, reaching speeds of thousands of kilometers per second. Many of those orbits have been tracked with infrared cameras over the last decade (Figure 73.5). We can use these stars' orbital motions to estimate the mass of Sgr A* using Newton's law of gravitation, just as we did to find the mass of the Galaxy. For example, the object SO-2 in Figure 73.5 was measured to be traveling at 6000 kilometers per second (6×10^6 m/sec) when it was about 20 billion kilometers (2×10^{13} m) from Sgr A*. The mass that SO-2 is orbiting, lying within this distance, should thus be

$$M_{\text{Sgr A*}} = \frac{(6 \times 10^6 \text{ m/sec})^2 \times 2 \times 10^{13} \text{ m}}{6.67 \times 10^{-11} \text{ m}^3/\text{kg} \cdot \text{sec}^2}$$
$$= 1 \times 10^{37} \text{ kg}$$

This is about 5 million solar masses. Yet no object is visible there, only a strong source of radio emission. That emission comes from a very small object, less than 10 AU in diameter (about the size of Jupiter's

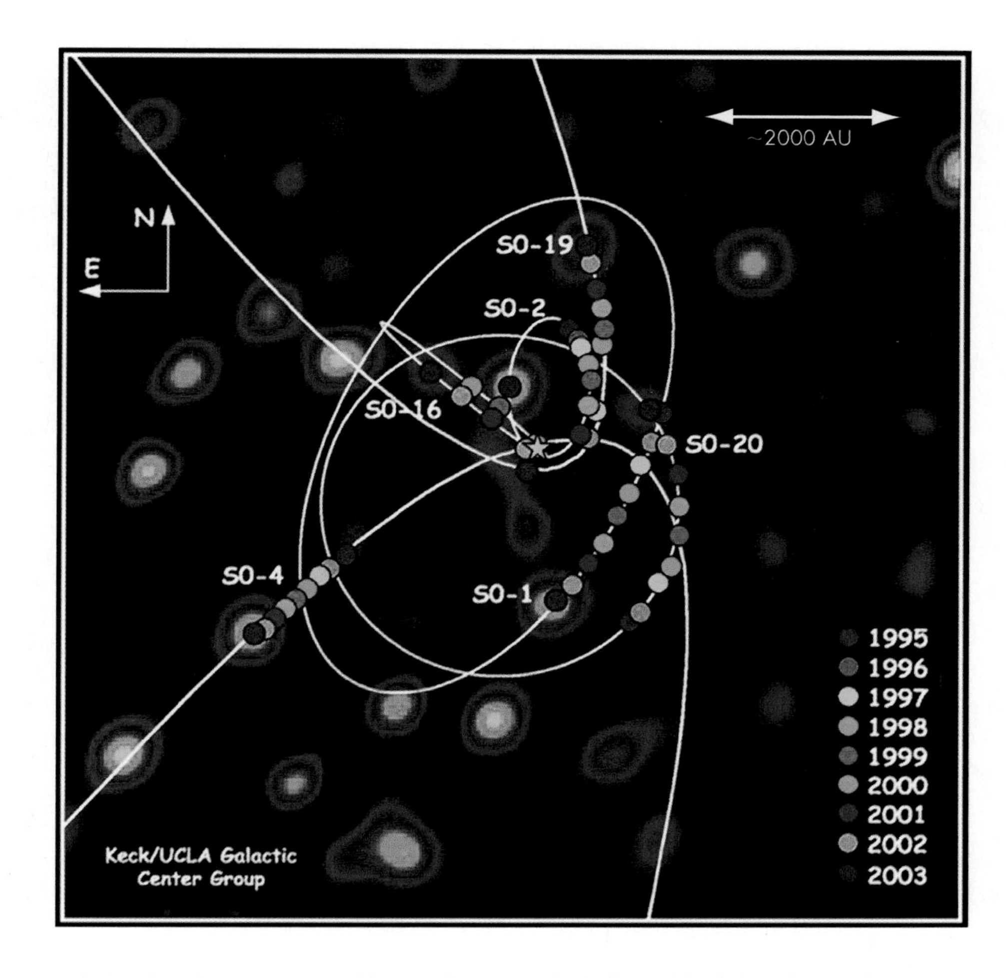

FIGURE 73.5
A plot of the position of several stars as they orbit the massive black hole at the core of the Milky Way. The stars are colored to show their positions in successive years. (The yellow "star" marks the center of the Milky Way.)

orbit). The almost inescapable conclusion is that a huge black hole lies at the center of the Milky Way and holds SO-2 and the other rapidly orbiting stars in its grip. In fact, these observations are perhaps the best evidence yet that black holes exist.

How might such a supermassive black hole form? When a massive star reaches the end of its life, it may collapse, leaving a black hole as a remnant. Such black holes contain only 10 or so solar masses; but in the central regions of a galaxy, where there is a greater density of objects, ordinary black holes may grow vastly more massive by drawing in nearby gas with their gravitational attraction. In normal star clusters, stars virtually never collide. But stars are packed more tightly in the center of the Milky Way than within clusters, and they may have been even more tightly packed at its center when the Milky Way formed. In such a crowded environment, stars may have collided, tearing away one another's gas, which then fell into the growing black hole. Once a black hole starts growing, it will not stop until it runs out of available matter, and some computer simulations suggest that a $10\text{-}M_{\odot}$ black hole could grow to $10^6\ M_{\odot}$ in less than 1 billion years.

73.3 DENSITY WAVES AND SPIRAL ARMS

The motions of stars and gas give rise to the structure of the Milky Way. Although the stars in the Milky Way's disk orbit its center approximately in circles, we also know that the Milky Way displays concentrations of stars in its spiral arms. Spiral arms appear to be a common feature of galaxies, so some process must drive stars to collect together in this pattern.

The winding dilemma

The presence of a feature like spiral arms is puzzling, because as we noted earlier, the stars in the Milky Way exhibit differential rotation—stars take different periods of time to orbit the center of the Milky Way. The Sun takes about 220 million years, but a star at a radius of 4 kiloparsecs takes about 110 million years, and a star at 12 kiloparsecs takes about 330 million years. Yet a spiral arm may stretch from 4 kiloparsecs to 12 kiloparsecs or even farther—so within a few hundred million years it should be torn apart by the different rotation speeds. Astronomers call this the **winding dilemma,** because it seems that differential rotation should cause spiral arms to become tightly "wound up" as the inner parts wrap around the Galaxy several times for every single orbit of the outer part of the arm.

A sound wave passing through the air is another kind of density wave—molecules in the air are alternately pressed closer together and then spread farther apart.

We know that gravity holds a galaxy together, so we might guess that the gravity of the stars in a spiral arm might hold the arm together. Gravity does in fact help hold a spiral arm together, but not as a fixed rotating mass of material. Instead the stars and matter in different regions become alternately more or less tightly packed together due to gravity's effects, creating a phenomenon known as a **density wave.**

In the density-wave model, waves of stars and gas sweep around the Galactic disk. The waves are not an up-and-down motion like an ocean wave; instead, they are places where the density of stars and gas is large compared to the surroundings. The densest region is analogous to the wave crest, with the concentration of stars in the wave crest resulting in what we observe as a spiral arm. Spiral waves are not unusual in rotating matter; for example, liquid in a food blender often forms a spiral wave with the liquid moving through it.

Q Where else do we see spiral patterns? What causes them? Are there any connections to the spiral pattern in the Milky Way?

As stars circle the Galaxy, a group of closely spaced stars at a given location will make the local gravity slightly stronger than elsewhere in the disk. That excess gravity draws stars toward the region so that the clump grows. Now imagine that clump stretched out into a spiral arm. Stars cannot remain in the arm continuously because their orbits move around the center of the Galaxy at speeds different from the **pattern speed** at which the arm goes around the center. However, the gravitational pull of the arm first hurries stars toward the arm as they approach it and then slows them as they depart it. Because all of the stars linger a little longer in that portion of their orbits where they are closest to the arm, the higher density of the arm is sustained.

When interstellar clouds move into an arm, they become compressed and more crowded, raising the density of the clouds and increasing the probability that they will collide with one another. This may in turn trigger the collapse of a cloud so that it forms a cluster of stars as illustrated in Figure 73.6. The brilliant blue O and B stars that form illuminate the arms and make them stand out against the more ordinary stars of the disk. These massive stars remain concentrated in the arms because they die before they have a chance to move past the area of clumping.

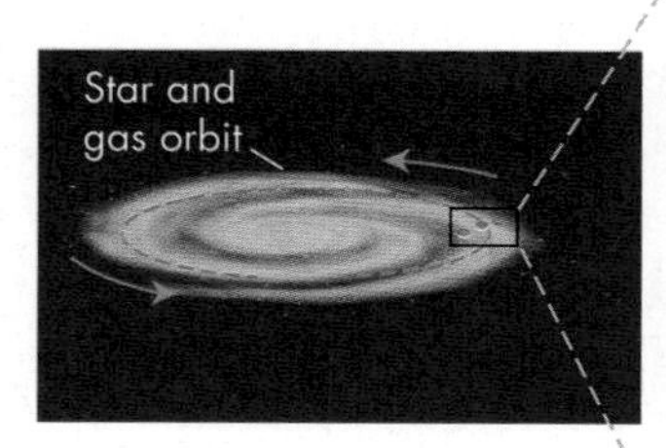

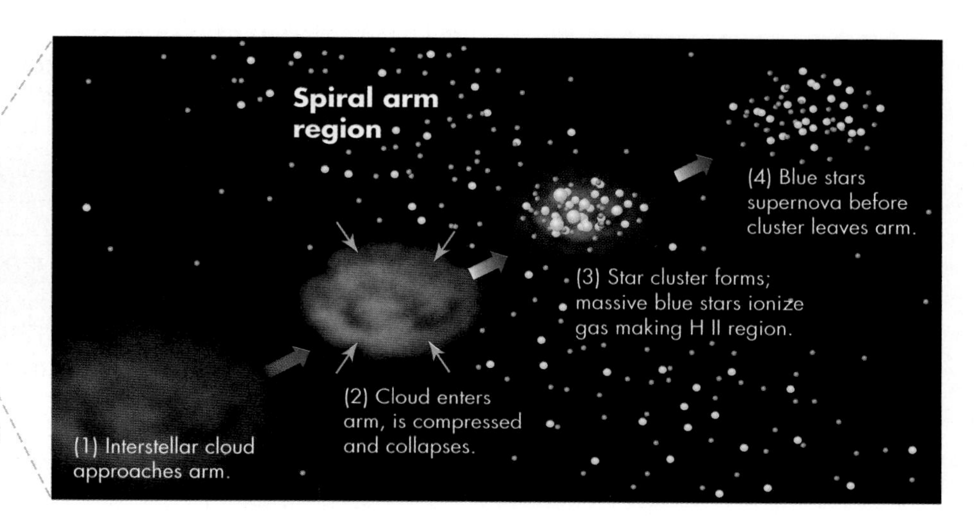

FIGURE 73.6
Sketch of a density wave in a galaxy, showing the progression from dust and gas to stars across the arm.

Longer-lived stars are born there too, but they survive long enough to continue orbiting within the disk, so they are not so obviously concentrated in the arms.

To better understand how arms form in the Milky Way, consider the following analogy. As cars move along a freeway, most move at nearly the same speed and keep approximately the same separation. If one car moves slightly slower than the others, however, traffic will begin to bunch up behind it. Cars can pass the slower vehicle but must change lanes to do so. The result is a clump of cars behind the slow one. But that clump is not composed permanently of the same cars. Cars join the clump from behind, pass the slower car, and then leave the clump at its front. Of course, in the spiral arms no stars move more slowly than all the rest; rather the *concentration* of stars in the arm is moving slower than the stars themselves.

The density-wave model explains several features of spiral arms. For example, in the galaxy shown in Figure 73.7A, the gas and stars are orbiting faster than the pattern speed, so they enter the arms on their inner edge. The raised density of the clouds makes them more opaque to starlight, producing a dark edge to that side of the arm. Star formation makes the region just behind this dark edge bright, and then the arm gradually fades because the bright, short-lived stars die out as they cross the arm. Radio wavelength studies of the Doppler shifts of the gas also clearly show the gas moving into the arm on one side and leaving on the other.

Although many spiral galaxies exhibit a distinct spiral pattern, others have much more ragged arms, such as the galaxy illustrated in Figure 73.7B. An alternate hypothesis has been proposed to explain these ragged arms called the **self-propagating star formation** model.

In this second model, star formation starts at some random point in the disk of a galaxy when a gas cloud collapses and turns into stars. As the stars drive winds into the gas around them and explode as supernovae, the pressure they generate makes the surrounding gas clouds collapse and turn into stars. The original stars die, but the new stars—formed as the old ones evolve and explode—trigger more gas clouds to collapse and form additional stars. In this fashion, the region of star formation spreads across the galaxy's disk much as a forest fire burns outward in a circle through a forest. But unlike a forest, where the trees stand still, stars orbit in a galaxy, and those near the center orbit in less time than those farther out. The zone in which star formation occurs is drawn into a spiral shape by the difference in rotation rates between the inner and outer parts of the disk. The result is a spiral pattern without clearly delineated arms, as illustrated in Figure 73.8.

Based on what we can determine about the Milky Way's structure, astronomers suspect that both density waves and self-propagating star formation are important, but in many spirals just one of these processes may be dominant. As we shall see in Unit 75, there are yet other types of galaxies that have no spiral structure at all, indicating that star formation can occur in a variety of ways. Our understanding of how galaxies form and evolve is an area of active research and current debate.

A

B

FIGURE 73.7
Classic spiral density waves are visible in the Galaxy M81 (A), while the ragged arms in the spiral galaxy M63 (B) are perhaps the result of self-propagating star formation. The spiral arms of the Milky Way probably have an appearance intermediate between these two galaxies, which are both similar in size to our Galaxy.

FIGURE 73.8
A computer model of self-propagating star formation illustrating how this process can create the appearance of spiral structure despite the lack of clearly defined arms.

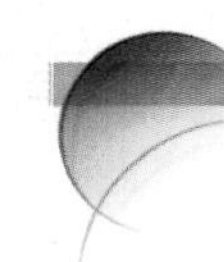

KEY TERMS

dark matter, 592
density wave, 596
differential rotation, 590
pattern speed, 596
rotation curve, 590
Sagittarius A* (Sgr A*), 594
self-propagating star formation, 597
winding dilemma, 596

QUESTIONS FOR REVIEW

1. How can we determine the Milky Way's mass?
2. What does differential rotation tell us about the mass distribution in the Milky Way?
3. What is dark matter? Why do astronomers conclude that the Milky Way may contain such unobserved material?
4. What types of material and densities are found close to the Galactic center?
5. What is the evidence for a black hole at the center of the Milky Way?
6. What are possible causes of the Milky Way's spiral arms?
7. How does a density wave induce star formation?
8. How do supernovae induce star formation?

PROBLEMS

1. Given that the Sun moves in a circular orbit of radius 8 kpc around the center of the Milky Way, and its orbital speed is 220 km/sec, work out how long it takes the Sun to complete one orbit of the Galaxy.
2. If the Milky Way were to double in mass, what would the Sun's velocity be?
3. Near the center of the Milky Way, the rotation speed climbs in proportion to the radius; so at 400 pc from the center, the speed is twice as fast as at 200 pc from the center.
   a. What does this imply about the mass within 400 pc versus the mass within 200 pc?
   b. Is there any differential rotation in this region? Explain.
4. Suppose there were no additional mass in the Milky Way beyond the distance of the Sun's orbit. What would the rotation speed be at a radius of 16 kpc (twice as far from the center as the Sun)?
5. What is the Schwarzschild radius of a black hole with the mass of the black hole thought to be at the center of the Milky Way? (See Unit 68.)
6. If spiral arms are about 2 kpc in width, how long does the Sun, at a distance of 8 kpc from the Galactic center and an orbital speed of 220 km/sec, spend inside a spiral arm on its motion around the Milky Way?
7. Cut three circles of paper or cardboard whose diameters are 8, 10, and 12 cm. Punch a hole in the middle of each and fasten them together through their centers. Fill in a circle near the edge of the 10-cm disk that extends across the visible portion of this disk onto the two other disks. This dot represents a large region of star formation. To model the differential rotation of our Galaxy, assume the disks rotate with a "flat rotation curve." If the 8-cm disk makes half a rotation, estimate how much the other two disks will have rotated in the same time. Draw how the star-forming region has become distorted. How does this relate to ideas for spiral structure in galaxies?

TEST YOURSELF

1. Astronomers think the Milky Way has spiral arms because
   a. they can see them unwinding along the celestial equator.
   b. radio maps show that gas clouds are distributed in the disk with a spiral pattern.
   c. young star clusters, H II regions, and associations outline spiral arms.
   d. globular clusters outline spiral arms.
   e. Both (b) and (c) are correct.
2. The modified form of Kepler's third law allows astronomers to determine the Milky Way's
   a. mass.
   b. age.
   c. composition.
   d. shape.
   e. number of spiral arms.
3. Astronomers think that spiral arms form because
   a. of shock waves from the black hole at the center of the Galaxy.
   b. younger stars travel more slowly than older stars.
   c. dust and gas do not orbit at the same speed as the stars.
   d. density waves create a stellar pileup.
   e. stars must travel up and down instead of in circles along a flat plane.

unit

74

A Universe of Galaxies

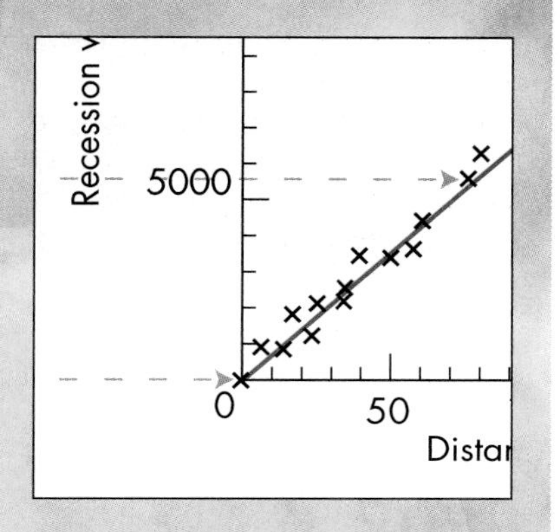

Background Pathways

Unit 70: Discovering the Milky Way 566

Beyond the edge of the Milky Way and filling the depths of space are billions of other star systems similar to our own. These remote, immense star clouds are called "external galaxies" or simply galaxies. When they were first discovered they were called "spiral nebulae" because of their unusual appearance, but it was not yet clear that they were objects like the Milky Way. Indeed, until early in the 1900s the Milky Way was thought to be the entire universe, the only galaxy. When it was realized in the 1920s that the spiral nebulae were in fact huge star systems, one name that was suggested was "island universes," which was an apt description because each is an island of stars in a vast dark space.

Today we have the technology to detect more galaxies than the total number of stars in the Milky Way, and every one of these galaxies itself contains enormous numbers of stars. It is interesting to realize that when Einstein made his great discoveries about space and time in the early 1900s (Units 53 and 68), he thought that the universe consisted of a flattened collection of just a billion stars surrounded by a vast emptiness. It was not until a few years later that the picture of the universe as we understand it today began to take shape. In this picture, each **galaxy** is an immense cloud of hundreds of millions to trillions of stars, with each star moving along its own orbit, held within its galaxy by the combined gravitational force of all the other stars and matter. Thus each galaxy is an independent, isolated star system.

74.1 EARLY OBSERVATIONS OF GALAXIES

Astronomers normally capitalize *Galaxy* when referring to the Milky Way and refer to other galaxies with the lowercased *galaxy*.

All other galaxies are extremely distant from us, generally millions or billions of light-years away. Their great distance makes galaxies look dim to us, and only a few can be seen with the unaided eye. For example, on clear, dark nights in autumn and winter in the Northern Hemisphere, you can see the Andromeda Galaxy (Figure 74.1) with just your eyes (see Looking Up #3: M31 & Perseus), which is over 2 million light-years away. It looks like a pale elliptical smudge on the sky. Early observers such as the tenth-century Persian astronomer Al-Sufi noted this object, and it appeared on early star charts as the "Little Cloud." From the Southern Hemisphere, you can easily see by eye the Large and Small Magellanic Clouds, two small satellite galaxies of the Milky Way that are "only" 150,000 light-years away from us (Figure 74.2).

One thing you can learn when looking at these galaxies by eye or through a telescope is that galaxies, even though magnified, still shine with a pale glow—similar to the pale light from the Milky Way itself. Even though they contain billions of stars, they do not appear bright. The reason for this is that the large distances between stars in any galaxy mean that the light is always spread out. The galaxies also look pale through a large telescope, because although the starlight is magnified, so are the vast spaces between the stars. Astronomers refer to this property as

FIGURE 74.1
Photograph of M31, the Andromeda Galaxy, the nearest spiral galaxy to us.

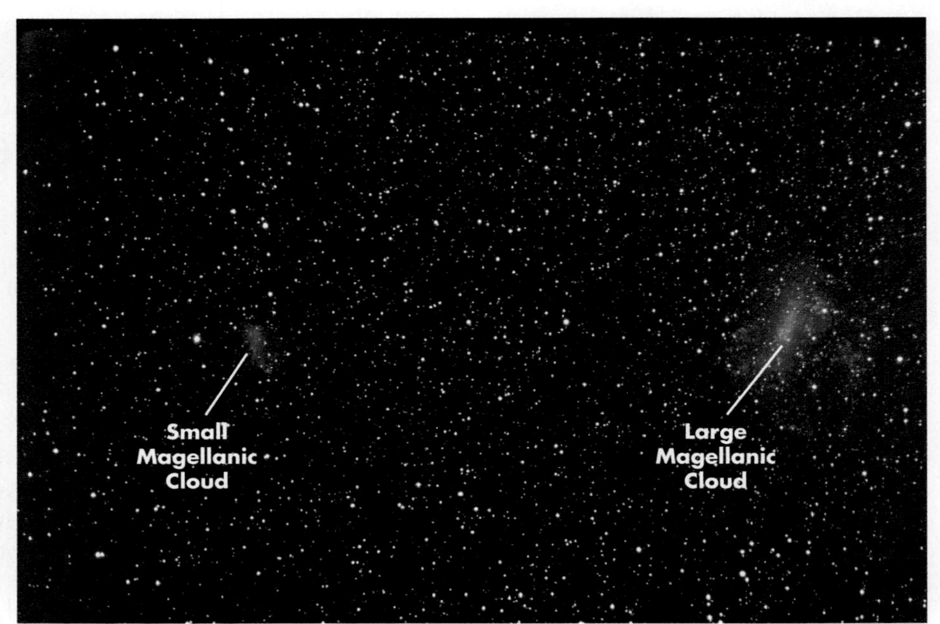

FIGURE 74.2
Photograph of the Large and Small Magellanic Clouds, two small galaxies that orbit the Milky Way.

the **surface brightness** of galaxies (also see Unit 31.6). The surface brightness does not change with distance or magnification because the tiny fraction of a galaxy's total area that is covered by the area of stars remains the same.

The faintness of galaxies' surface brightness makes them challenging to observe, even with a good telescope. The study of external galaxies began in the eighteenth century when the French astronomer Charles Messier (pronounced *MESS-yay*) accidentally discovered many galaxies during his searches for new comets. He realized that some of the faint, diffuse patches of light never moved; so to avoid confusing them with comets, he assigned them numbers and made a catalog of their positions. Although many of Messier's objects have since been identified as star clusters or glowing gas clouds in the Milky Way, several dozen are galaxies. These and the other objects in the Messier catalog are still known by their Messier number, such as M31, the Andromeda Galaxy mentioned earlier.

In the early part of the nineteenth century other astronomers, such as William and Caroline Herschel, began to map faint objects in the heavens systematically, finding and cataloging numerous astronomical objects, including galaxies. The Herschels' work was continued by William's son John, and in the late 1800s was revised and added to by John Dreyer to include many thousands of galaxies and additional nebulae. This compilation is now known as the New General Catalog (or NGC for short), and it gives many galaxies their name, such as NGC 1275. Many galaxies appear in more than one catalog and so bear several names. For example, Messier's M82 is the same galaxy as NGC 3034.

For these early observers, the nature of the galaxies was a mystery. They looked different from other types of cataloged objects—planetary nebulae, globular clusters, emission nebulae, and so on—and because individual stars could not be resolved, many believed they were clouds of glowing gas rather than the merged light of billions of stars. One notable aspect of galaxies was that they were found only in portions of the sky away from the plane of the Milky Way. Even a modern catalog of deep-sky galaxy observations contains very few galaxies near the galactic plane, as shown in Figure 74.3A. Many astronomers once interpreted this **zone of avoidance** around the Milky Way as evidence that

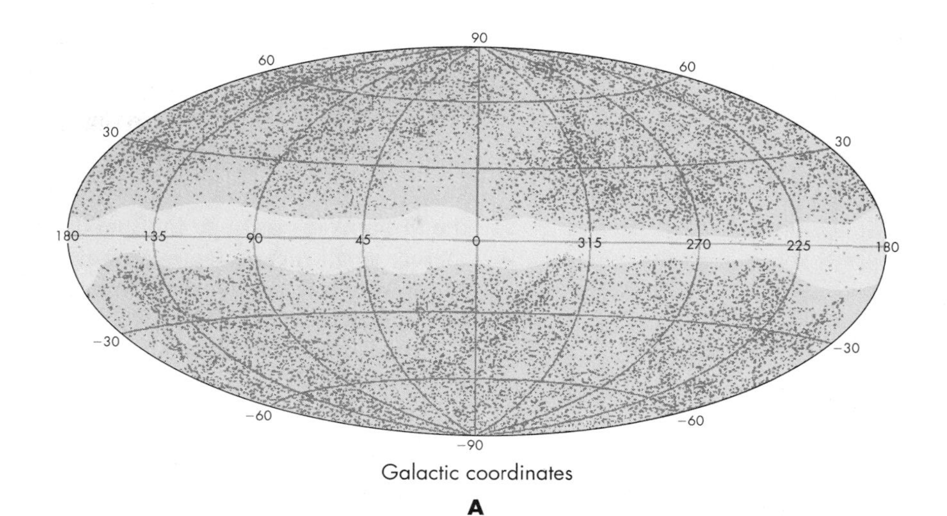

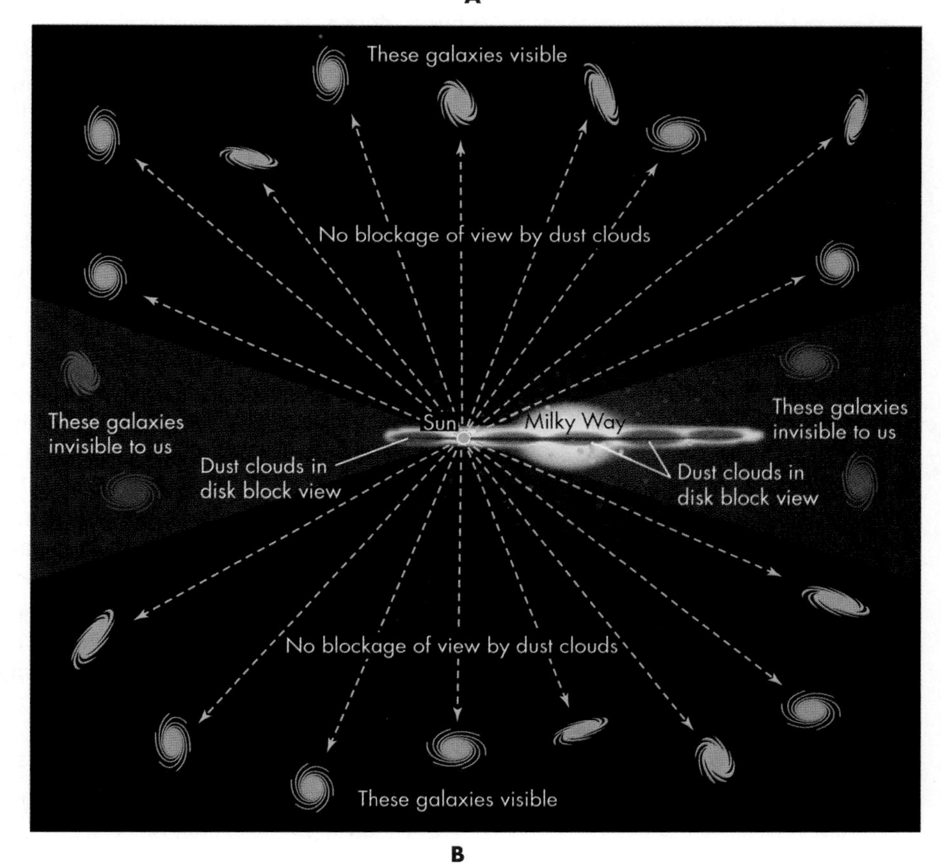

FIGURE 74.3
Dust limits our view in the disk of our Galaxy and creates the zone of avoidance. We see no galaxies in the plane of our own Galaxy because dust in the disk blocks our view. (A) A plot of where galaxies appear to be distributed on the sky. Notice how few lie along the central line (the galactic equator). This blank region is the zone of avoidance. (B) A sketch illustrating why we see so few galaxies along the plane of the Milky Way. The gray blobs represent dust clouds that block our view toward directions that lie close to Milky Way's plane. (Note that the galaxies are in reality about the same size as the Milky Way and are much farther away than can properly be shown in the figure.)

the galaxies must somehow be interacting with (and therefore be near) the Milky Way. They asked, If these objects were far away, how could they "know" to avoid this direction? What astronomers did not realize at the time was that the zone of avoidance is simply an effect of the dust in our own Galaxy blocking the light from distant galaxies (Figure 74.3B). This is the same effect that blocks our view of the center of the Milky Way.

As early as 1755, the German philosopher Immanuel Kant, whose views on the origin of the Solar System were far ahead of their time, suggested that galaxies might be remote star systems—island universes—similar to the Milky Way. However, until the 1920s there was little observational evidence to clearly show whether galaxies were huge distant systems of stars like the Milky Way or were small nearby objects associated with our own Galaxy, and astronomical opinion was strongly divided.

This controversy was aired in detail during a series of debates organized by the National Academy of Sciences in which two U.S. astronomers, Harlow Shapley (Unit 70) and Heber Curtis, took opposing sides. Shapley argued that the Milky Way was huge and that other galaxies (then called spiral nebulae) were merely small, nearby companions. Curtis, on the other hand, argued that the Milky Way was smaller than Shapley claimed, that the nebulae were star systems like our own, and that they were immensely distant from us.

The debate reflects the challenges of an observational science like astronomy. For example, Curtis argued that objects such as M31 must be far away based on a supernova that had been observed in 1885. Assuming this was as luminous as the supernovae that had been observed in our own Galaxy hundreds of years earlier (Unit 65), he found that M31 must be very distant; and if it was distant, then M31 must be huge. Shapley countered that the object seen in 1885 was not a supernova but a nova—a much less luminous explosion—and therefore M31 was nearby, perhaps at a distance more like he had found for the globular clusters orbiting the center of the Milky Way. Both arguments were reasonable based on the evidence available at the time, and both fit into the model each side had developed.

Shapley based part of his argument on a claimed detection of motions inside galaxies—observations that were later found to be incorrect. If such motions had been detectable, they would have ruled out the possibility that the galaxies were far away. Curtis correctly doubted these reported motions, even though the report was made by a well-respected astronomer of the time. As in any mystery story, not all clues are useful or correct.

Historians differ on who "won" the debate. Shapley was more nearly correct about the Milky Way's size and structure (also see Unit 70). Curtis was right about the nebulae being distant systems similar in structure to the Milky Way.

Q If the intergalactic space between the Milky Way and M31 contained dust, how would that have affected Curtis's measurement of M31's distance?

One of the biggest problems for both sides in the debate was the lack of understanding about the dimming caused by dust in the Milky Way. Curtis adopted *Kapteyn's Universe,* a model of a small Milky Way (Unit 70), whereas Shapley interpreted the zone of avoidance as evidence that the galaxies were nearby. The other major problem was the lack of a definitive method for measuring the galaxies' distances. Breakthroughs in the 1920s solved both of these problems. Astronomers discovered how starlight was dimmed by dust in the Milky Way, and they were also able to find *standard candles* (Unit 54.3) in external galaxies. With these discoveries, the true nature of galaxies as enormous systems containing billions of stars was finally confirmed.

74.2 THE DISTANCES OF GALAXIES

Galaxies' distances are so enormous that we cannot measure them using the same methods that we use to find distances to stars. Parallax cannot be used to measure such vast distances because the angle by which the galaxy's position changes as we move around the Sun is too tiny to detect. Individual stars within an external galaxy are extremely difficult to detect because the angular separations are so small that atmospheric blurring (Unit 30.3) blends their light together. To apply the method of standard candles, then, astronomers must find stars of known luminosity that are extremely bright, so that their light can be picked out from the sea of surrounding stars.

In using the standard-candles method to find a galaxy's distance, astronomers use the principle that the farther from us an object is, the dimmer it looks.

A

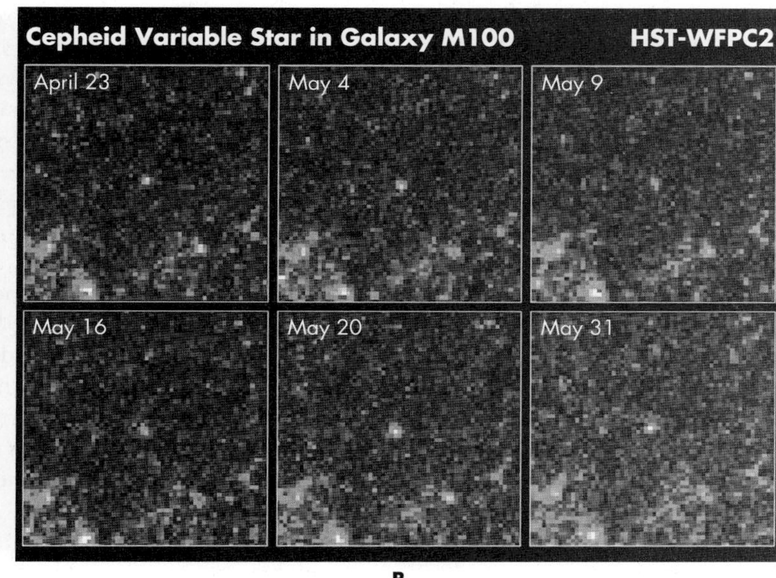

B

FIGURE 74.4
(A) Cepheids in the Milky Way have known distances, d_*, so we can determine their luminosities. Knowing those luminosities, we can find the distance to a galaxy, d_g, if we can measure the brightness of a Cepheid in it. (B) Hubble Space Telescope observations of a Cepheid variable in the galaxy M100.

More precisely, if we know the luminosity of a light source—our standard candle—and how bright it looks to us, we can find its distance by using the inverse-square law.

Astronomers use many different astronomical objects as standard candles, but one of the most reliable is the class of luminous, pulsating variable stars known as Cepheids (Unit 63). Cepheids have luminosities more than 10,000 times the luminosity of the Sun, and because of that brilliance it is possible to detect them in nearby galaxies. Furthermore, because their brightness changes as they pulsate, their light can be distinguished from other stars. The period–luminosity relation (Unit 63.3) then allows us to use a variable star's period (the time it takes to go from bright to dim to bright again) to determine its luminosity.

Once astronomers know the variable star's luminosity, they can find its distance from a measurement of how bright it appears. For example, suppose we observe a Cepheid in a distant galaxy, as illustrated in Figure 74.4A. From observations of Cepheids in our own Galaxy, we know what luminosity a Cepheid has based on its period, which gives us the period–luminosity relationship. Therefore we can determine the luminosity L from the period of variation, and we can measure the brightness B directly. With these values we can determine the distance to the Cepheid using the standard-candle method:

$$d_g = \sqrt{\frac{L_c}{4\pi B_c}}$$

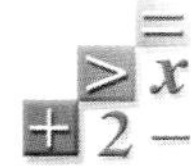

d_g = Distance to Cepheid and galaxy (meters)

L_c = Luminosity of Cepheid from period–luminosity relationship (watts)

B_c = Measured brightness of Cepheid (watts/m^2)

Astronomers generally measure the distances of many Cepheids within the same galaxy to get an accurate average for the galaxy's distance.

Cepheids are very luminous, but few galaxies are close enough that we can make successful ground-based measurements of their light. The first such measurement was achieved by the U.S. astronomer Edwin Hubble in 1923 for M31, finally proving that it was millions of light-years away and even larger than the Milky Way. Over the subsequent half century, Cepheids were observed in only a handful more galaxies. It required the Hubble Space Telescope to carry out

Q: Why do multiple measurements give a more accurate value?

similar observations for dozens of more distant galaxies, such as that shown in Figure 74.4B. The Cepheid in the figure is in the galaxy M100 and was measured to have a period of 51 days. The period–luminosity relationship indicates that this Cepheid has a luminosity of $3 \times 10^4\ L_\odot$ (or 1.2×10^{31} watts). The apparent brightness measured with the Hubble Space Telescope was 3.4×10^{-18} watts per square meter. Therefore, the distance to the Cepheid (and its galaxy) is

$$d_g = \sqrt{\frac{1.2 \times 10^{31} \text{ watts}}{4\pi \times 3.4 \times 10^{-18} \text{ watts/m}^2}} = 5.3 \times 10^{23} \text{ m} = 1.7 \times 10^7 \text{ parsecs.}$$

Q When Hubble originally measured the distance to M31, the values of the luminosity in the period–luminosity relationship he was using were four times lower than they are known to be today. How would this have affected his distance measurement?

Thus, the galaxy M100 is about 17 million parsecs (56 million light-years) distant. Even for the Hubble Space Telescope, though, Cepheids are undetectable in any but the nearest few hundred galaxies.

To measure greater distances, astronomers must use even more-luminous standard candles, such as supernovae. However, we cannot predict when a supernova of known luminosity might occur in a galaxy of interest. This might have been an insurmountable problem but for a very strange discovery.

74.3 THE REDSHIFT AND HUBBLE'S LAW

In the early 1900s the U.S. astronomer Vesto Slipher discovered that the spectral lines of nearly all of the galaxies he observed were strongly shifted toward longer wavelengths—to the red end of the visible spectrum. This so-called *redshift* is seen in all galaxies, except for a few nearby ones whose motion is influenced by the gravitational attraction of the Milky Way and its neighbors.

Redshift is defined as the fraction by which the light's wavelength changes—that is, the change in wavelength divided by the original wavelength. For example, suppose we observed a galaxy's spectrum and saw that the pattern of spectral lines was shifted to longer wavelengths (Figure 74.5A). Among these we might identify an absorption line that normally occurs at 500 nanometers, but is shifted to

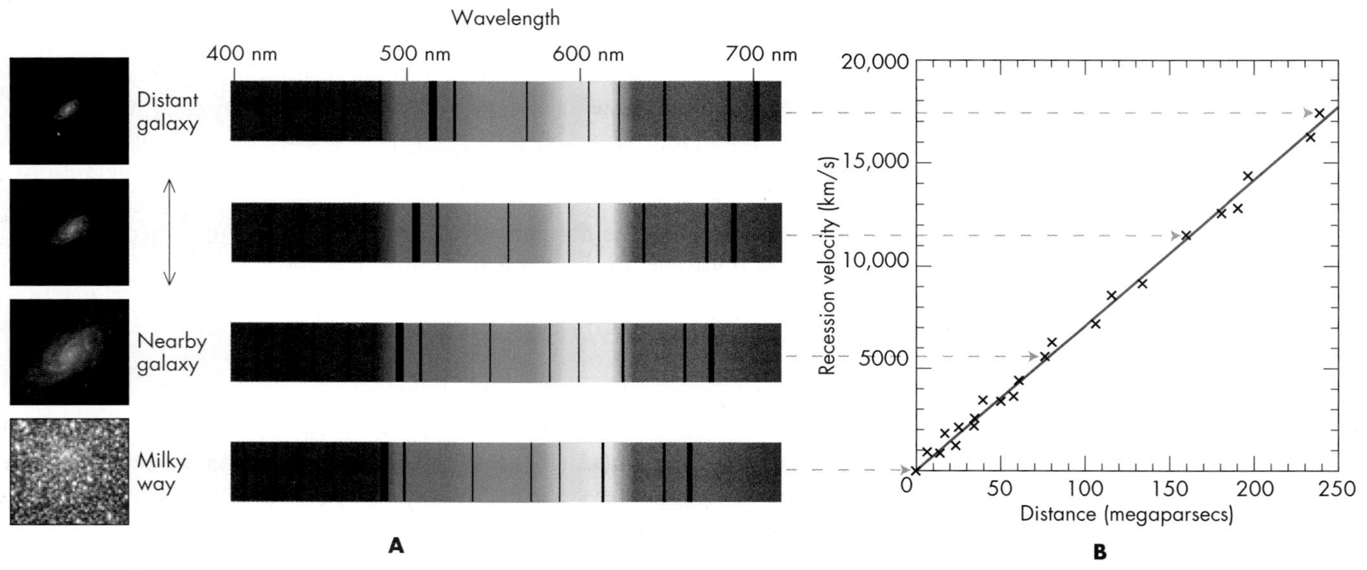

FIGURE 74.5
(A) Galaxies show a larger redshift of their spectral lines the more distant they are. The spectrum of the integrated light from the Milky Way (which has zero redshift) is shown for comparison. (B) Hubble's law (shown by the red line) indicates that the recession velocity increases in proportion to the redshift.

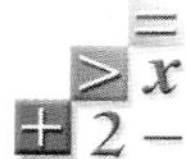

z = Redshift

$\Delta\lambda$ = Change in wavelength from its original value

λ = Original value of the wavelength

505 nanometers in this galaxy. The line has thus been shifted by 5 nanometers, 1% of its normal wavelength of 500 nanometers. This galaxy's redshift is 1%, or 0.01.

Astronomers use the letter z to stand for redshift, and mathematically we can define the redshift as follows:

$$z = \frac{\Delta\lambda}{\lambda}$$

In this formula, $\Delta\lambda$ stands for the change in wavelength, and λ is the original, unshifted wavelength. If we observed a second galaxy for which the wavelength of the 500-nanometer line had shifted to 515 nanometers, we would say $\Delta\lambda = (515 \text{ nm} - 500 \text{ nm}) = 15$ nm, and $\Delta\lambda = 500$ nm, so

$$z = \frac{15 \text{ nm}}{500 \text{ nm}} = 0.03.$$

This galaxy's redshift is 0.03, or 3%. Incidentally, we chose a 500-nanometer line for this example for numerical simplicity. If we looked at any other spectral line, we would find that they have all been shifted by 3% relative to their normal value, whether they were generated at radio wavelengths or X-rays.

The discovery of galaxy redshifts proved to be even stranger when Edwin Hubble and others found that the redshift was larger for more-distant galaxies. They interpreted the redshift as a Doppler shift (Unit 25), and the increasing redshifts then implied that the speed with which a galaxy moves away from us—its **recession velocity**—increases with distance, as Figure 74.5 shows. That is,

The faster a galaxy recedes, the farther away it is.

If $V = c \times z$, then for $z = 3\%$, we find

$V = 3.00 \times 10^8$ m/sec $\times$ 0.03

$= 9 \times 10^6$ m/sec $=$ 9000 km/sec.

Assuming the Doppler effect, Hubble calculated the recession velocity $V = c \times z$. A 1% redshift implies a recession velocity of 1% of the speed of light, or about 3000 kilometers per second. A 3% redshift implies a recession velocity of about 9000 kilometers per second.

Galaxy redshifts look like Doppler shifts, but today we understand from Einstein's theory of general relativity (Unit 68) that the redshifts are caused by a different effect: the expansion of the space between the galaxies as the light travels through it. Does this sound like nothing more than a semantic difference? It turns out to be fundamentally important for understanding the universe as a whole. We will use the notation and language of velocities here that were adopted by Hubble and are still used by astronomers today, but we must keep in mind that astronomers today interpret them quite differently. In particular, astronomers now view redshift as arising from the expansion of space. We will return to the implications of the redshift for understanding the universe in Unit 79.

In 1920 Edwin Hubble discovered that a simple formula relates the recession velocity, V, and the distance, d. When he graphed the recession velocities and distances of a number of galaxies (Figure 74.5B), he found that they lay along a straight line. The line can be described by the equation

V = Recession velocity of galaxy (km/sec)

H = Hubble's constant (km/sec per Mpc)

d = Distance to galaxy (Mpc)

$$V = H \times d$$

where H is a constant. Because of his discovery, this relation is called **Hubble's law,** and H is called **Hubble's constant.**

The value of the Hubble constant depends on the units used to measure V and d. Astronomers generally measure V in kilometers per second and d in **megaparsecs** (abbreviated **Mpc**), where 1 megaparsec is a million parsecs (= 3.26 million light-years). In these units, H is about 70 km/sec per Mpc.

As recently as the 1990s, astronomers could not agree on the value of H to within a factor of 2. The reason for this controversy was that to determine H, we must know both the distance and the velocity of at least a few galaxies to calibrate the law. H is simply the average value of each calibration galaxy's velocity divided by its distance. However, as we have just seen, accurately

Hubble's constant is often written H_0 to indicate its value as measured today at redshift $z = 0$.

measuring the distance to even a nearby galaxy is difficult. Based on galaxies for which Cepheid variable distances have been calculated, the current estimate of Hubble's constant is $H = 70$ km/sec per Mpc, with an uncertainty possibly as large as 5 km/sec per Mpc.

Once a value for H is known, we can turn the method around and use the Hubble law to find the distance to a galaxy. The method is as follows: Take a spectrum of the galaxy whose distance you want to know. From the spectrum, measure the shift of the spectral lines. From the Doppler-shift formula, calculate the recession velocity. Finally, use that velocity in the Hubble law to find the distance. Mathematically, we divide both sides of the Hubble law by H, obtaining

$$d = V/H.$$

We insert the measured value of V and our choice of H and solve for the distance. The power of this formula is also why getting H right is so important—if H is incorrect, our whole scale of the universe will be off.

As an example of applying Hubble's law, suppose we find from a galaxy's spectrum that its recession velocity is 49,000 kilometers per second. We find its distance by inserting this value for V in the previous equation and then divide by H:

$$d = \frac{49{,}000 \text{ km/sec}}{H}$$

For $H = 70$ km/sec per Mpc, the value of d is

$$d = \frac{49{,}000 \text{ km/sec}}{(70 \text{ km/sec})/\text{Mpc}} = 700 \text{ Mpc}.$$

The galaxy's recession velocity implies a distance of about 700 Mpc.

In applying this method, we must be careful about units. When we divide V in kilometers per second by H in km/sec per Mpc, the answer we get will be in megaparsecs. The 700 we just found is not in parsecs or light-years but in megaparsecs. To get the distance in light-years, we multiply by 3.26 light-years per parsec to get about 2300 million light-years, or 2.3 *billion* light-years. Detecting a galaxy this far away was time-consuming until the 1990s, but newer technologies can measure thousands of galaxies at these distances in a single night of observation.

The Hubble law does not work for nearby objects. For example, M31, the spiral galaxy nearest to us, is moving *toward* the Milky Way at about 90 kilometers per second, so it has a "negative redshift." We cannot apply Hubble's law to it, or we would get the absurd result that M31 has a negative distance. Similarly, we cannot apply Hubble's law to a nearby star that is moving away from the Sun, even though its spectrum is redshifted. Its redshift is caused by orbital motion inside our Galaxy, and therefore, again, Hubble's law does not apply.

Hubble's law applies only to redshifts caused by the general motion of galaxies away from each other—what astronomers term the *overall expansion of the universe*. When we are observing distant galaxies, the redshift we measure is caused primarily by the expansion of the universe, but in part by the galaxy's orbital motion around another galaxy or system of galaxies. To make the distinction clear, astronomers refer to the first as the **expansion redshift,** and to the second simply as the *Doppler shift*. If we assume the redshift we measure is entirely due to expansion, the distance we calculate from Hubble's law might be too high or too low by an amount depending on the size of the Doppler shift. For an isolated galaxy, the Doppler shift motion is usually under 100 kilometers per second. So if we calculate a velocity of 700 kilometers per second for such a galaxy from its redshift, the true recession velocity might be anywhere in the range of 600 km/sec to 800 km/sec. By using 700 km/sec in the Hubble law, we will find this distance:

$$d = \frac{V}{H} = \frac{700 \pm 100 \text{ km/sec}}{70 \text{ km/sec per Mpc}} = 10 \pm 1.4 \text{ Mpc}$$

The uncertainty in the expansion redshift produces an uncertainty in the distance of the orbital speed divided by H. This uncertainty may cause significant errors in the distances of galaxies orbiting within a large system of galaxies where orbital speeds in excess of 1000 km/sec have been observed. If we assumed these galaxies' redshifts were caused entirely by expansion, we could make an error of up to $(1000/70) \approx 14$ Mpc in calculating their distances. This is one reason it was so difficult to determine Hubble's constant accurately: the nearby galaxies with good distance measurements all have uncertainties in their expansion redshifts.

Thanks to the more accurate value of Hubble's constant now determined, we can use the redshift to find the distance to a galaxy several hundred megaparsecs away with an uncertainty of only a few megaparsecs if the galaxy is isolated. No other method for finding the distances of remote galaxies is as easy to use or as accurate—provided that the rate of expansion has remained constant.

KEY TERMS

expansion redshift, 606
galaxy, 599
Hubble's constant, 605
Hubble's law, 605
megaparsec (Mpc), 605
recession velocity, 605
redshift, 604
surface brightness, 600
zone of avoidance, 600

QUESTIONS FOR REVIEW

1. How do astronomers measure the distances to nearby galaxies?
2. What is Hubble's law?
3. Why are astronomers uncertain about the precise value of Hubble's constant?
4. Why is a galaxy difficult to see with the naked eye even though its apparent brightness may be greater than that of stars we can see?
5. What is the zone of avoidance?
6. How do astronomers measure the distance to nearby galaxies?
7. Why is parallax not used to determine the distances to galaxies?
8. What is an expansion redshift?

PROBLEMS

1. How long has it taken the light from a galaxy 1 Mpc away to reach us?
2. Suppose that a galaxy's 21-cm line of atomic hydrogen is shifted to 22 cm.
   a. What is the redshift of the galaxy?
   b. What is its recession velocity?
3. Two identical galaxies have redshifts of 0.036 and 0.072. Which has the larger angular size? What is the ratio of their apparent sizes?
4. A galaxy has a recession velocity of 28,000 km/sec. What is its distance in megaparsecs?
5. What is the approximate recession velocity of a galaxy that we are seeing as it looked 1 billion years ago?

TEST YOURSELF

1. A galaxy's spectrum has a redshift of 7,000 km/sec. If the Hubble constant is 70 km/sec per Mpc, how far away from Earth is the galaxy?
   a. 10^6 Mpc
   b. 100 Mpc
   c. 0.01 Mpc
   d. 10^8 Mpc
   e. 10^{-6} Mpc
2. The nearest galaxy to ours is approximately how far away?
   a. 1,000,000,000 ly
   b. 1,000,000 ly
   c. 150,000 ly
   d. 100 ly
   e. 4.3 ly
3. Which of the following would an astronomer be likely to refer to as a standard candle?
   a. The Sun
   b. An S0 galaxy
   c. A Cepheid variable
   d. The Andromeda Galaxy
   e. A wax cylinder approximately 1000 ly tall

unit

75 Types of Galaxies

Sa
Sb
S0
SBa
SB

Background Pathways

Many galaxies are flattened systems of stars with spiral arms, similar to the Milky Way in size and shape. Other galaxies are distinctly different from our own. For example, some are not disk-shaped at all. Rather, they are shaped more like a rugby ball or a distorted sphere, with their stars distributed in a vast, smooth elliptical cloud surrounding a dense central core of stars. Others are neither disks nor smooth elliptical shapes but are completely irregular in appearance.

Galaxies differ in more than shape; they also differ in their content. Some contain mostly old stars, but others contain predominantly young stars. Some are rich in interstellar matter, whereas others have hardly any at all. Some galaxies—despite having as much total mass as the Milky Way—have formed far fewer stars, leaving them dim and scarcely visible to us. Others have a small but tremendously powerful energy source at their core that emits as much energy as the entire Milky Way but from a region about 0.001% its size, less than a few light-years in diameter. Many of these *active galactic nuclei* eject gas in narrow jets at nearly the speed of light (Unit 77).

Astronomers can see all this diversity in galaxies, but we are only beginning to understand how it arises. Unlike stars, whose structure and evolution are well understood, many aspects of galaxies remain a mystery. Astronomers do not all agree, for instance, on what causes some galaxies to be disk-shaped while others are more nearly egg-shaped. Although this lack of certainty can be confusing, it offers us a chance to see how scientists work. In particular, we will learn how astronomers begin with basic observations and construct hypotheses to explain what they observe. For example, many galaxies appear to be undergoing collisions with their neighbors. Could past interactions of this kind account for the different galaxy shapes?

75.1 GALAXY CLASSIFICATION

The first objects recognized as potentially being other star systems similar to the Milky Way were the "spiral nebulae," but by the early twentieth century it became clear that galaxies had a variety of other shapes. Edwin Hubble, working first as a graduate student at the University of Chicago and later at Mount Wilson in California in the 1920s, developed one of the first schemes for classifying galaxies on the basis of their shape. He sorted them into three main types: spirals, ellipticals, and irregulars.

The first type has two or more arms winding out from the center. Astronomers call these **spiral galaxies,** a term often abbreviated to simply **S.** The Milky Way is of this type, and Figure 75.1 shows photographs of several others.

The second galaxy type shows no signs of spiral structure. These galaxies have a smooth, featureless appearance and a generally elliptical shape, as can be seen in Figure 75.2. Accordingly, astronomers call them **elliptical galaxies,** abbreviated as **E.**

Galaxies of the third major type show neither arms nor a smooth uniform appearance. In fact, they generally have stars and gas clouds scattered in random patches. For this reason, they are called **irregular galaxies** (Figure 75.3), abbreviated as **Irr.**

In addition to these three main types, Hubble recognized two additional subtypes of the spiral systems. The first of these is called a **barred spiral galaxy,**

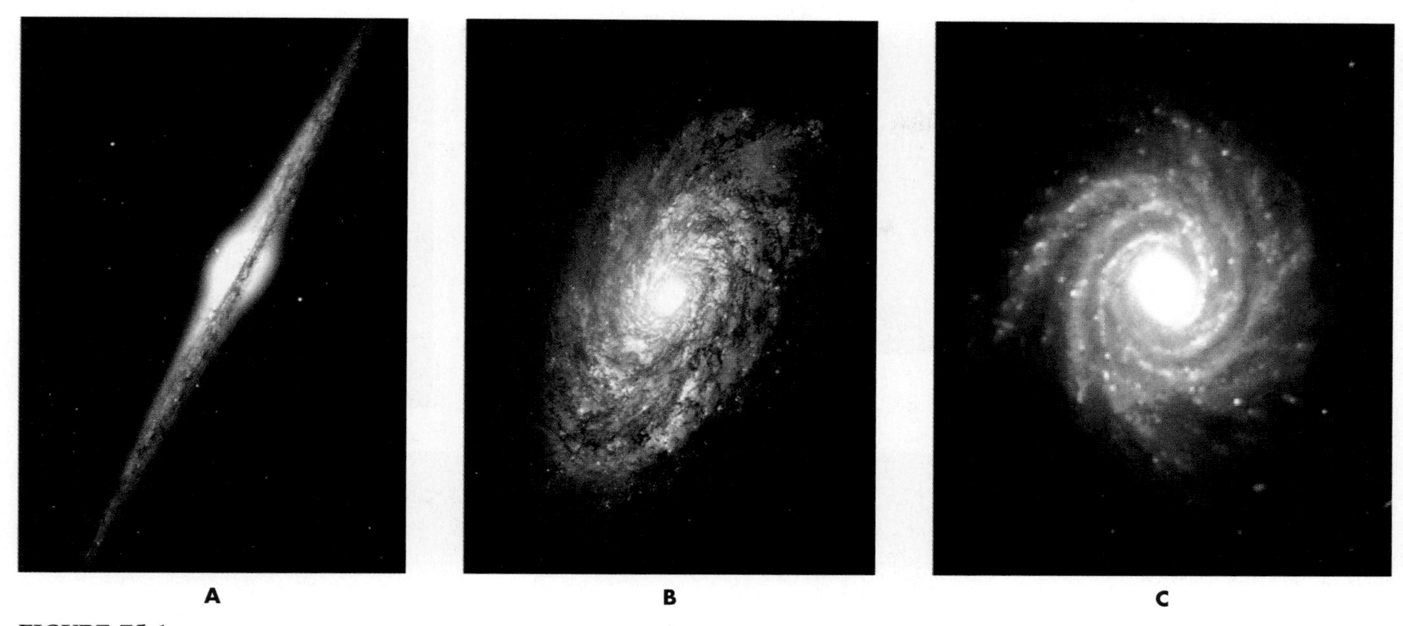

FIGURE 75.1
Photographs of typical spiral galaxies. (A) NGC 4565; (B) NGC 4414; and (C) NGC 1288. These examples show spiral galaxies with their disks oriented from nearly edge-on (A) to nearly face-on (C).

FIGURE 75.2
Photographs of a few elliptical galaxies: NGC 205, M49, and M87.

FIGURE 75.3
Photographs of two irregular galaxies (NGC 6822 and the Small Magellanic Cloud).

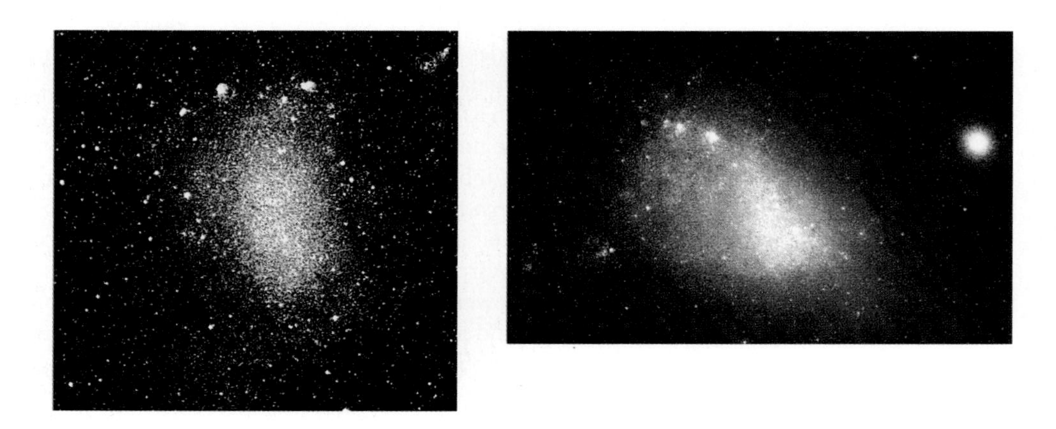

FIGURE 75.4
Barred spirals. NGC 1365 has a prominent bar, while NGC 5236 (M83) has a weaker bar—more like the one thought to be at the center of the Milky Way.

FIGURE 75.5
Two S0 galaxies. Although they have no spiral arms, S0 galaxies are different from ellipticals because they show evidence of a disk of stars.

abbreviated as **SB.** These have arms that emerge from the ends of an elongated central region, or bar, rather than from the core of the galaxy (Figure 75.4).

Hubble thought that bars might be permanent features of these galaxies, but astronomers now think that bars are just temporary features that can arise in any spiral galaxy. This is what is found in computer models of galaxies—the stars can form a bar-shaped pattern that may last for hundreds of millions of years. This happens particularly when a galaxy undergoes a gravitational disturbance as the result of a close encounter with a neighboring galaxy. Infrared observations of the stars in the center of the Milky Way show that it currently has a weak bar, and this might be due to its gravitational interactions with small companion galaxies.

Hubble also identified another subtype of galaxy similar to spirals, but which oddly show no spiral structure. These disk systems with no evidence of arms (Figure 75.5) are called **S0** ("*S-zero*") **galaxies.** Astronomers think S0 galaxies are spiral galaxies whose gas has been removed. For example, if a spiral galaxy collides with an intergalactic gas cloud, the impact can "sweep" gas and dust out of the moving spiral, much as a leaf-blower cleans debris from a sidewalk. Without gas, the disk galaxy can make no new stars, and hence shows no obvious spiral structure (see Unit 76.2)

Hubble developed a sequence of gradations among his major types, seeking a pattern that might lead to an understanding of why galaxies exhibit such diversity in their appearance. He subdivided the ellipticals into classes E0 to E7 according to how circular or elongated they appeared. Differences among spirals are indicated by lowercase letters, ranging from Sa for spirals with a large bulge and tightly wound arms, to Sd for a spiral with a small bulge and loosely wrapped arms.

Hubble hypothesized that a sequence of galaxy types that showed a smooth transition of appearances might represent an evolutionary sequence of galaxies.

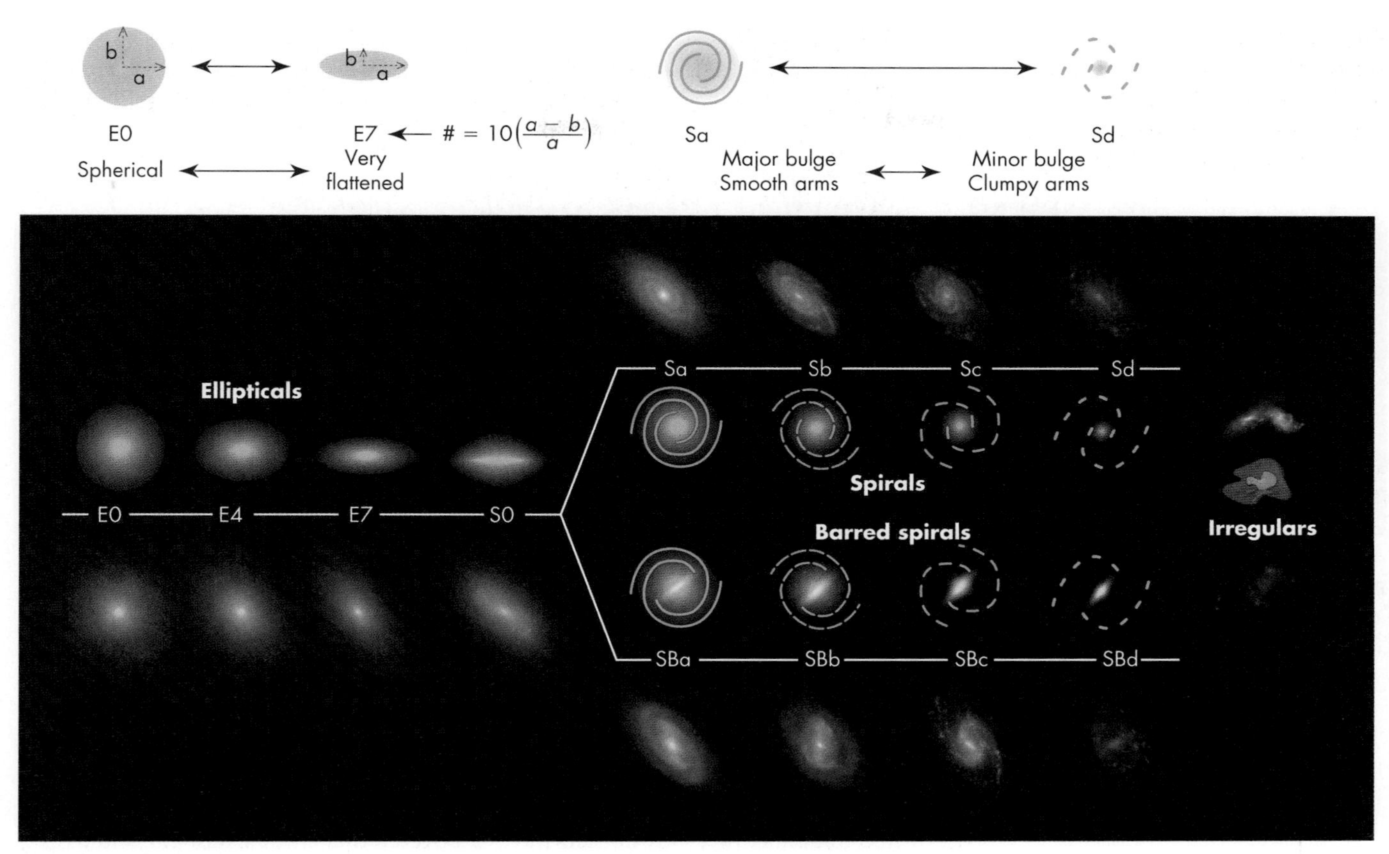

FIGURE 75.6
The Hubble tuning-fork diagram. Elliptical galaxies are subdivided according to how "flattened" they look. Regular and barred spirals are subdivided into subclasses according to how large their bulges are and how tightly wound their arms are. The images are drawn from the Sloan Digital Sky Survey with foreground stars removed to make the galaxies' structures clearer. The spiral galaxies all have flat circular disks, but galaxies were selected that are tilted at similar angles to make comparisons easier.

He developed such a sequence, starting with spherical E type galaxies, then moving to flatter E galaxies, to S0 galaxies, then to the sequence of spirals, and finishing with irregulars. Hubble proposed that as a galaxy ages, its type might evolve through this sequence. This hypothesis seems plausible if you look at the "tuning fork" diagram that Hubble proposed (Figure 75.6), but astronomers now think it is incorrect. Nevertheless, the diagram is still widely used today because it offers a convenient way to organize the galaxy types and subtypes.

Although Hubble's classification system categorizes the outward appearance of galaxies quite well, it does not describe their size or brightness. The galaxies that Hubble studied were generally large and had formed large numbers of stars. However, modern observing instruments reveal that there are many **low surface brightness galaxies** (dim, often large-diameter systems with far less star formation than ordinary galaxies) and **dwarf galaxies** (perhaps the building blocks of ordinary galaxies). Examples are shown in Figure 75.7. These types of galaxies are very numerous, but their properties make them harder to detect, so few were known before modern observing technologies developed. Many of these galaxies can still be categorized according to Hubble's scheme according to the pattern of stars within them, but a dwarf elliptical may contain a million times fewer stars than a typical giant elliptical that Hubble studied.

Q: Surface brightness is discussed in Unit 31. Unlike the apparent brightness of a galaxy, why does the surface brightness not change with distance?

FIGURE 75.7
(A) Dwarf galaxy, Leo I, a nearby galaxy found in 1977. (B) Comparison between two distant galaxies of matching size. The one on the left is a low surface brightness spiral of the type often missed in visible-wavelength surveys.

75.2 DIFFERENCES IN STAR AND GAS CONTENT

Since Hubble's time, astronomers have discovered that spiral, elliptical, and irregular galaxies differ not only in their shape but also in the types of stars they contain. For example, spiral galaxies contain a mix of young and old stars (Population I and II—see Unit 71.1); but elliptical galaxies contain mostly old (Pop II) stars, and irregulars contain mostly young (Pop I) stars.

This difference in the kind of stars in the different galaxy types is understandable in terms of their different gas content. To make young (Pop I) stars, a galaxy must contain dense clouds of gas and dust. Spiral systems typically have at least 10% of the mass of their disk in the form of such interstellar clouds, and so they can make young stars that light up their spiral arms. On the other hand, ellipticals contain much less interstellar matter. In fact, astronomers used to think that the E galaxies contained virtually no interstellar gas or dust. However, more recent observations made with X-ray telescopes show that many E galaxies contain very low-density but very hot (10^7 K) gas. Such gas contains few or no high-density cold clumps that might collapse gravitationally to form stars, so it is no surprise that elliptical galaxies rarely contain young, blue stars like the ones we see in the disks of spiral systems.

Q Are there circumstances in which you might expect to find young stars in elliptical galaxies? Are there likely to be many planetary systems in elliptical galaxies?

Although young, blue stars are rare in elliptical systems, such stars are common in irregular galaxies. These Irr systems also often contain large amounts of interstellar matter, amounting sometimes to more than 50% of their mass. Accordingly, astronomers believe that these galaxies are "young" in the sense that they have not yet used up much of their gas in making stars.

Apart from their different star and interstellar matter content, S, E, and Irr galaxies have few other differences to generalize about. For example, if we look at the mass and radius of galaxies, we do not find that all ellipticals are huge and all spirals are small. Instead, we discover that elliptical galaxies range enormously in size. Some are not much larger than globular clusters and contain only 10^7 solar masses of material; others are monster galaxies, 10 to 100 times the mass of the Milky Way. This spread in sizes can be seen in Figure 75.8, which shows an assortment of galaxies all scaled to their correct relative size.

FIGURE 75.8
An assortment of galaxies showing their actual relative sizes and providing examples of the range of appearances that a galaxy may have and still be classified as the same type. All types of galaxies are found in a wide range of sizes, containing anywhere from fewer than 10^8 stars to more than 10^{12}. There are many times more of the smallest galaxies than the larger galaxies—just a few are shown for size comparison. The images are drawn from the Sloan Digital Sky Survey. Foreground stars have been digitally removed to make the galaxies' structures clearer.

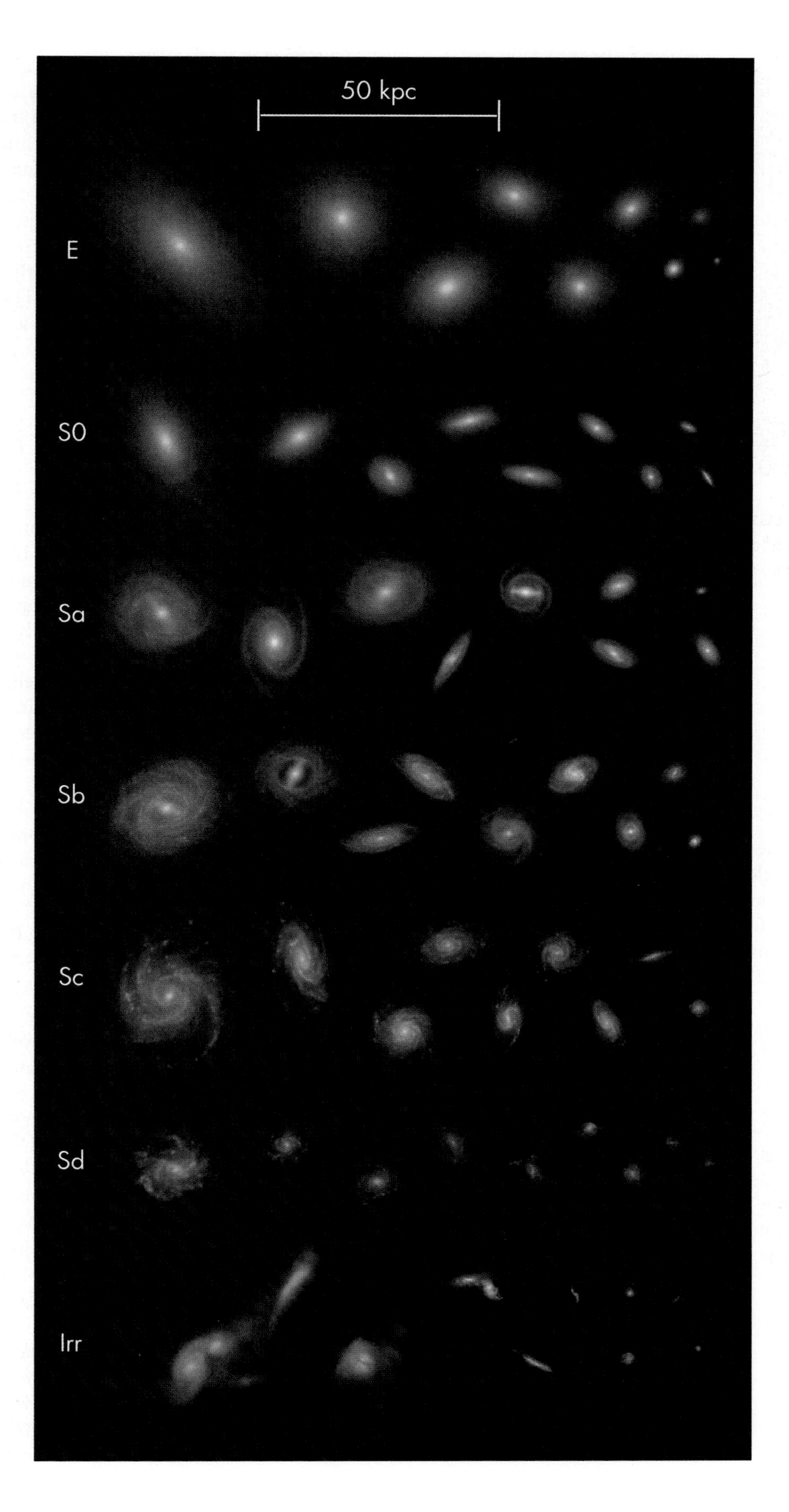

What fraction of galaxies falls into each type classification? Even such a basic question has no clear answer. If we take a census of the galaxies near the Milky Way, we find that most galaxies are dim dwarf E and Irr systems, sparsely populated with stars. These dim galaxies are undetectable at greater distances, where we can see only the more luminous galaxies. If we take a census of those luminous galaxies, we find that the majority of galaxies are spirals, but the percentage depends on whether we look at galaxies that occur in large groups or in more sparsely populated regions. In the largest groups of galaxies, fewer than 50% of the members are spirals or S0 galaxies, whereas in the sparsely populated regions, the proportion of spirals is about 80%. The diversity of galaxy form and content requires an explanation, and a major question in extragalactic research today is why galaxies come in such distinct and different forms.

75.3 WHAT CAUSES THE DIFFERENT GALAXY TYPES?

For many decades astronomers thought that rotation played the main role in determining how galaxies evolved to their present-day types. When they measure the rotation speed of galaxies, astronomers find that the disks of spirals rotate relatively rapidly compared with ellipticals of the same size. On the other hand, the halos and bulges of the spirals do *not* rotate especially fast. In the model where a spiral galaxy's halo forms first during the initial stages of collapse, it is not obvious that there is much difference between a spiral and an elliptical in their early stages. Some factors other than rotation must also play a role in determining a galaxy's type.

The Hubble Space Telescope provides a way of studying galaxies when they were very young, allowing us to see how galaxies have changed throughout the history of the universe. Because light takes a finite time to reach us, when we observe a very distant galaxy, we are seeing it as it was when the light left it. This was the idea behind the Hubble Deep Field project—a very long observation of a small "blank" patch of sky. By collecting light for over 100 hours, it was possible to detect extremely distant and therefore very young galaxies. A section of the Hubble Deep Field is shown in Figure 75.9.

The position of the Hubble Deep Field is shown in Looking Up #2: Ursa Major. If you hold a grain of sand at arm's length, you could completely cover the area observed.

The time it takes the light to reach us is the distance divided by the speed of light: $t = d/c$. Solving for the time is particularly easy if we write the distance in light-years, because the distance in light-years also tells us how long ago the light left the galaxy. For example, suppose we observe a galaxy that is about 3000 megaparsecs (3 billion parsecs) distant. Multiplying by 3.26 light-years per parsec, we find that the galaxy is about 10 billion light-years away from us. Thus, the light we see today left the galaxy 10 billion years ago, and we are seeing the galaxy as it looked at that time. At such a distance a galaxy has a very small angular size; but with the Hubble Space Telescope, which is unencumbered by the blurring effects of our atmosphere, astronomers can see details in such a very distant—and thus very young—galaxy.

Although present observations do not yet reveal to us the very first stages of the collapse of primordial gas clouds into galaxies, the Hubble Space Telescope has allowed astronomers to observe galaxies just past the collapse stage, less than a billion years after the galaxies first formed. Striking pictures of these distant galaxies made with the Hubble Space Telescope (Figure 75.10) suggest that irregular galaxies constituted a higher proportion of the galaxy population in the past than today. Interestingly, these very young galaxies are much smaller than systems such as the Milky Way, but there were far more galaxies than we find today. These earliest galaxies must have somehow disappeared and changed form.

How can galaxies disappear? Astronomers suspect that they may have collided and undergone **mergers** into bigger systems. The large number of galaxies within

FIGURE 75.9
The Hubble Deep Field is a small region observed for more than 100 hours to detect extremely distant galaxies. The galaxies in this image are not close together in space, but are spread out over more than 10 billion light-years along our line of sight.

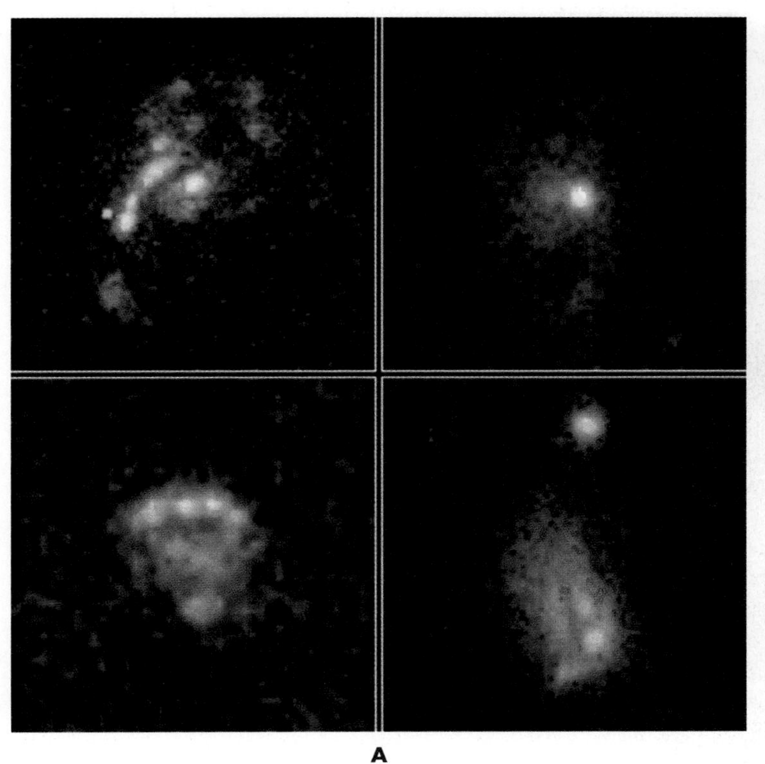

A

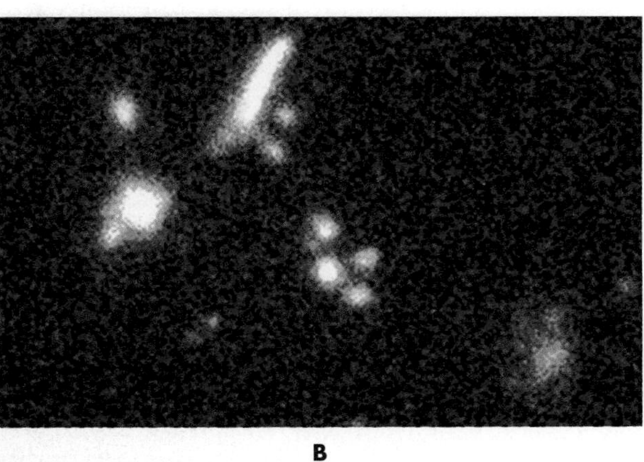

B

FIGURE 75.10
(A) Picture of young, forming galaxies taken with the Hubble Space Telescope. (B) Young galaxies possibly merging to form a larger system.

a given volume at these early times would have made interactions more likely than now. Signs of such collisions in these distant galaxies are seen as bent arms, twisted disks, and other irregular shapes in Figure 75.10. A series of mergers could account both for the formation of large systems such as the Milky Way and for the smaller number of galaxies we see today. But a major question still remains: What causes one of these merging systems to eventually become a spiral galaxy, whereas another becomes an elliptical galaxy?

75.4 GALAXY COLLISIONS AND MERGERS

FIGURE 75.11
A merger in which two spirals are colliding and forming an elliptical galaxy. The contrast in this image has been greatly enhanced to bring out very faint features of the galaxy NGC 1316 (also known as Fornax A) that indicate the galaxy underwent a merger.

Our own Galaxy shows signs of past mergers. For example, astronomers can see streams of stars in our Galaxy's halo, which are suspected to be debris from small galaxies that have fallen into the Milky Way and have been torn apart by its gravity (Unit 71). Could mergers also give rise to elliptical galaxies? The answer seems to be yes.

For example, Figure 75.11 shows two spiral galaxies whose gravitational force has drawn them together and merged them into a single new galaxy that looks much like an elliptical system. Observational evidence such as this and computer simulations have led astronomers to propose that elliptical galaxies form from mergers of smaller galaxies. The basic idea is that most galaxies are born as small spiral or irregular systems with essentially no bulge. Subsequently, collisions and mergers between these systems cause the stars to take on orbits in all different directions, creating the egg-shaped distribution of stars that we see in elliptical galaxies.

The merger hypothesis can also explain the origin of large spiral galaxies such as our own Milky Way. In less direct collisions, small galaxies can merge without completely disrupting the orbits of stars within a disk system, allowing a spiral galaxy to grow in size. The gradual buildup of a smaller number of large galaxies from an initially large number of small galaxies is shown schematically in Figure 75.12.

This process of galaxy growth through mergers helps explain the colors and luminosities of galaxies shown in the graph in Figure 75.12. A *blue sequence* of gas-rich galaxies is explained by mergers of gas-rich galaxies. However, galaxies eventually lose their gas for several possible reasons. The gas may be used up in the formation of stars, or be swept out when galaxies collide, or supernova explosions may drive the gas out of the galaxy. Minus its gas, a galaxy rapidly grows to have a

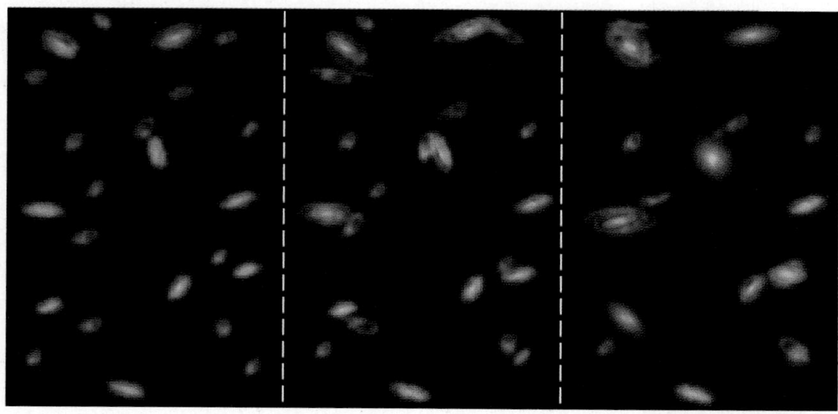

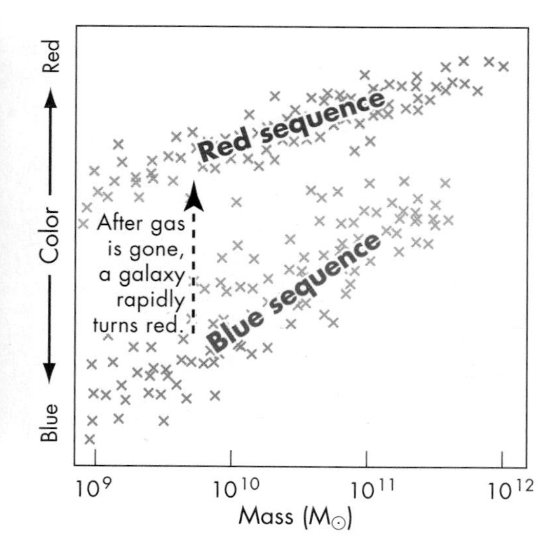

FIGURE 75.12
A possible scenario of how galaxies evolve. The first galaxies were mostly small and gas-rich (left panel). Because they were actively forming stars, they contained short-lived massive stars and were blue in color. Larger galaxies formed through mergers, as illustrated by the next two panels. Some of these galaxies consumed their gas or lost it during collisions, and ceased forming stars, so they grew redder in color. Others merged to form large spiral galaxies that continue to form stars and remain blue in color. When we plot the colors and magnitudes of galaxies today (right panel), galaxies are distributed along a "blue sequence," which are probably galaxies that grew through mergers without losing their gas. After losing their gas, they shift to the "red sequence."

Q Hubble proposed that his tuning-fork diagram might represent evolutionary changes, with E galaxies gradually changing to S galaxies. What evidence, based on the amount of interstellar matter and the kinds of stars in E and S galaxies, indicates that this is unlikely?

red color like an elliptical, because the massive blue stars die off rapidly. These gas-poor galaxies also merge, producing a *red sequence* in the color–mass graph.

This hypothesis has the added attraction of explaining a puzzling feature of our Galaxy. Astronomers at one time thought that the stars in our Galaxy's bulge formed in one single, massive event at the Galaxy's birth. However, stars in the bulge have a range of ages, suggesting that the bulge grew gradually over time. Such gradual growth is easy to explain by mergers. Each time a large galaxy (such as the Milky Way) swallowed a smaller companion, some of the gas that was in the companion sank to the center of the larger system, where it formed new stars. Successive mergers led to successive generations of stars, so that the bulge stars have the wide age range that is observed.

A final feature of the merger model is that it even allows some ellipticals to turn back into spirals. This can happen if an elliptical captures a gas-rich galaxy. The gas and young stars of the captured spiral then become the disk of a "new" spiral while the original elliptical becomes the bulge and halo.

Computer simulations indicate that even when a merger does not occur when two galaxies pass near each other, the galaxies' gravitational force alters the orbits of their stars and thereby changes the shape of each galaxy. Figure 75.13 shows a simulation of such a close passage, and you can see stars torn from each galaxy by their mutual gravitational force and flung outward in long arcs. The photograph (the fifth picture in the sequence) shows two real galaxies with just such plumes of stars.

Galaxies with tails

Although interactions of this type may radically alter the appearance of galaxies, they do not destroy the individual stars. Stars are so far apart within a galaxy that almost all of them move harmlessly past one another. Galaxy interactions are much like tossing two handfuls of sand together: Most of the sand particles simply pass by one another without colliding. Dust and gas clouds in the galaxies are not so lucky. Filling each galaxy's space much more completely than stars, clouds do collide, and their impact may compress them enough to trigger a burst of star formation. Such **starburst galaxies,** as these collisionally-stimulated galaxies are called, are among the most luminous galaxies known (Figure 75.14A).

Galaxy interactions can create other bizarre forms. Figure 75.14B shows a picture of what can happen when a small galaxy plows directly into a large spiral galaxy. The small galaxy has punched a hole through the larger one, creating what astronomers call a **ring galaxy.** The hole results not because the smaller galaxy destroyed the stars in the other galaxy's disk. Rather, it made the hole by shifting the orbits of stars that were near the center of the larger galaxy into wider orbits farther from the

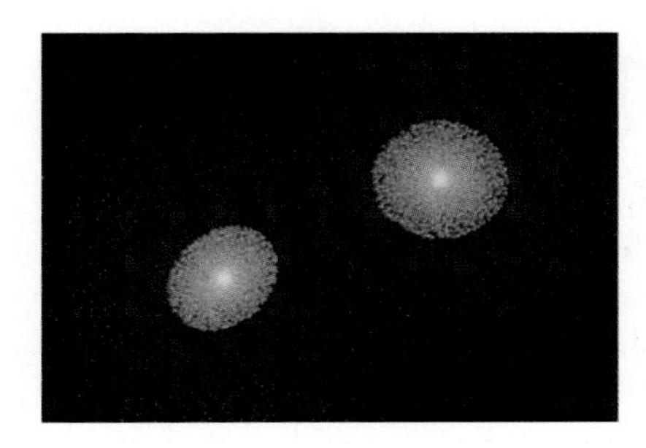

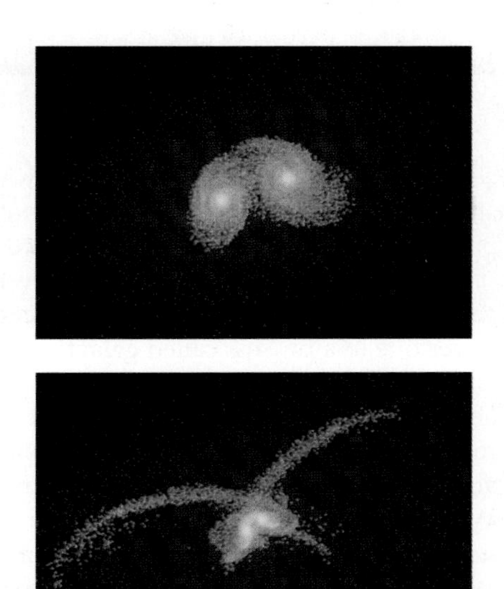

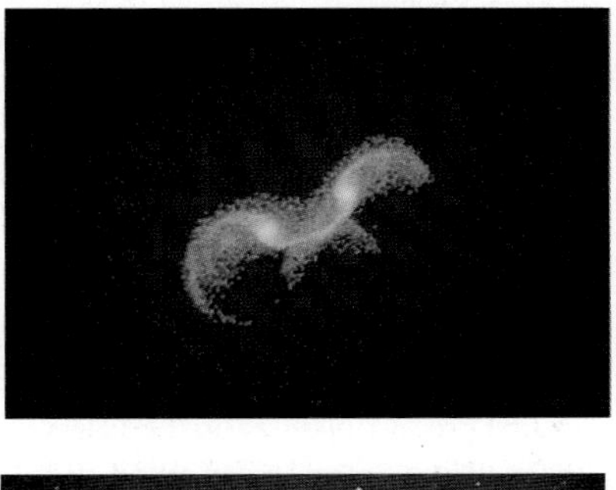

FIGURE 75.13
Computer simulation of two galaxies colliding, and a photograph of a system undergoing just such a collision.

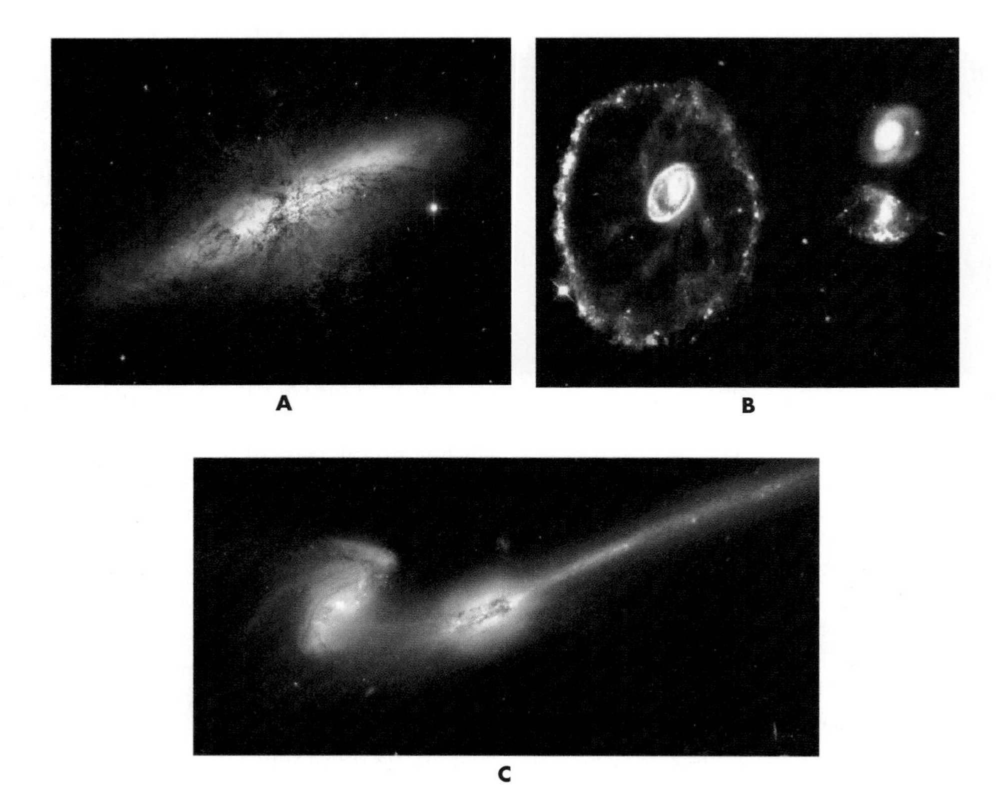

FIGURE 75.14
(A) A starburst in the central portion of the galaxy M82 may be the result of a galaxy collision. (B) The "cartwheel" galaxy shows a ring shape that was probably caused when one of the small galaxy's at right crossed through its center. (C) "The Mice" is a pair of galaxies that have passed very near each other and pulled out a long tidal stream of stars and gas.

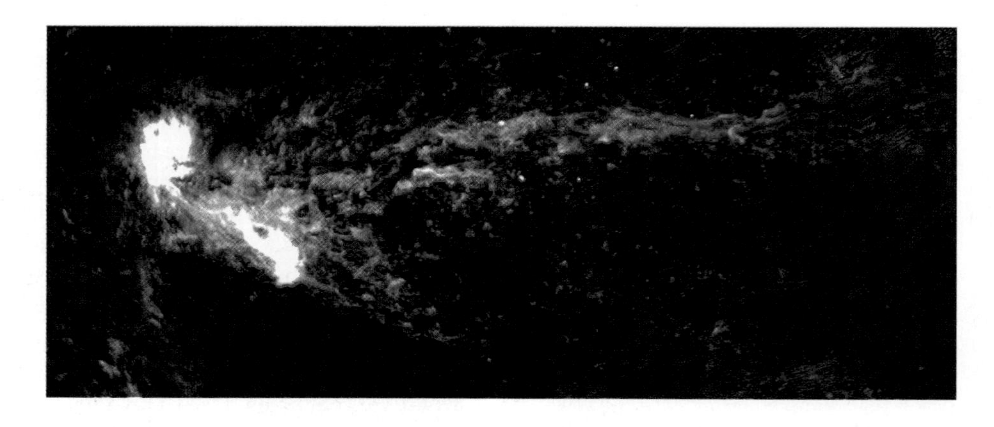

FIGURE 75.15
A false-color radio map of hydrogen gas being stripped out of the Large and Small Magellanic Clouds (which are located in the two brightest regions) by the gravitational tidal pull of the Milky Way. The tail of gas extending to the right is about 100 kpc long.

Q A number of galaxies can be found in Looking Up #1: Northern Circumpolar Constellations; #2: Ursa Major; and #8: Centaurus and Crux, the Southern Cross. What type of galaxy does each appear to be? Do you see any evidence of merging?

center—more like the motion of waves that travel out from the spot where a rock strikes a pond. Figure 75.14C shows another bizarre result of galactic interaction: a pair of galaxies—"The Mice"—distorted by each other's gravity.

Mergers probably affect just about every galaxy sometime during its life. If a large galaxy collides with a small galaxy, the small one is captured and absorbed by the larger one in a process called **galactic cannibalism.** Repeated cannibalism by a galaxy may turn it into a giant, which probably explains why some galaxy groups have an abnormally large elliptical galaxy at their center. In addition to evidence of past mergers, astronomers have detected evidence that our own Galaxy is currently devouring its smaller neighbors. For example, radio observations show that the Milky Way is in the process of tearing apart the Magellanic Clouds (Figure 75.15). Our neighboring galaxy, M31, also appears to have recently cannibalized a small galaxy. With the Hubble Space Telescope, astronomers have detected a small bright clump of stars near the center of M31 that may be the remains of a galaxy it has absorbed.

KEY TERMS

barred spiral (SB) galaxy, 608
dwarf galaxy, 611
elliptical (E) galaxy, 608
galactic cannibalism, 618
irregular (Irr) galaxy, 608
low surface brightness galaxy, 611
merger, 614
ring galaxy, 617
S0 galaxy, 610
spiral (S) galaxy, 608
starburst galaxy, 617

QUESTIONS FOR REVIEW

1. What are the three main types of galaxies?
2. How do the basic galaxy types differ in shape, stellar content, and interstellar matter?
3. What does it mean for a galaxy to have a bar?
4. What criteria distinguish between types of spiral galaxies?
5. What happens to their stars when two galaxies collide?
6. Why do bursts of star formation occur during galaxy interactions and mergers?

PROBLEMS

1. For each of the following distances, if a galaxy the size of the Milky Way (about 30 kpc in diameter) were at that distance from the Milky Way, what would its angular size (Unit 10.4) appear to be? Express your answers in arc minutes.
   a. 10 Mpc
   b. 1000 Mpc
2. If the typical distance between galaxies is 1 million pc (1 Mpc), how many galaxies are found in a sphere of radius 1 million Mpc?
3. For the majority of galaxies astronomers observe, the light travel time is less than a billion years. This is only a fraction of the Sun's age, and therefore we do not expect to find many differences in the galaxies' properties due to their age. What would the redshift be for a galaxy that we are seeing as it was 1 billion years ago?
4. The Hubble Deep Field covers an area of about 5 arcmin^2 and contains about 3000 galaxies. If we observed the entire sky to this depth, about how many galaxies would we find?
5. The radial velocity of M31 toward the Milky Way is about 100 km/sec. If it maintains that speed, how long will it take to travel the 700-kpc distance between the two galaxies?
6. Assume that the disk of the Milky Way has a diameter of 40 kpc and has a surface area R^2. Likewise, each star covers an area determined by R^2.
   a. If the Milky Way contains 100 billion stars each the size of the Sun, what fraction of the galaxy's disk is actually "covered" by a star?
   b. The fraction found in (a) expresses the approximate chance that when a single star passes through the Milky Way's disk, it will strike a star. If another galaxy with the same number of stars passed through the Milky Way, how many stars would you estimate would undergo direct collisions?

TEST YOURSELF

1. A large galaxy contains mostly old (Pop II) stars spread smoothly throughout its volume, but it has little dust or gas. What type of galaxy is it most likely to be?
   a. Irr
   b. S
   c. SB
   d. E
   e. All of the above are possible.
2. In starburst galaxies we do not observe
   a. high luminosity.
   b. stars colliding and exploding.
   c. two galaxies colliding.
   d. a greater proportion than usual of stars forming.
3. Which of the following would be a probable result of a collision between two galaxies?
   a. Complete destruction of all the stars
   b. Formation of a giant spiral from two elliptical ones
   c. Formation of a single open star cluster
   d. Formation of a ring galaxy
   e. All of the above

Starry Night

See the Pathways *2nd edition ARIS site for online Starry Night™ planetarium exercises.*

unit

76 Galaxy Clustering

Galaxies are not spread uniformly across the sky. Just as stars often lie in clusters, so too galaxies often lie in **galaxy clusters**, and galaxy clusters themselves lie in clusters of galaxy clusters called **superclusters.** Galaxy clusters and superclusters are held together by the mutual gravity of the galaxies and other matter within them. Each galaxy within a cluster moves along its own orbit, just as within a galaxy each star moves on its own orbit.

The clustering of galaxies can be seen when we plot the brightest galaxies on the sky. Each dot in Figure 76.1 represents one galaxy. The clusters seen in this figure are typically 10 million light-years across and may contain anywhere from a handful to a few thousand member galaxies. These immense collections of galaxies are not themselves stationary. They are separating from one another, moving apart in the overall expansion of the universe discovered by Edwin Hubble in 1920 (Unit 74).

In this unit we take the next steps in the revolution that began when Copernicus recognized that the Earth is just one of the planets. In the early 1900s we did not know about the universe beyond the Milky Way. We find now that our Galaxy is not even the largest galaxy in its own rather small galaxy cluster.

Background Pathways

Unit 75: Types of Galaxies 608

76.1 THE LOCAL GROUP

On the small end of galaxy clustering, astronomers often refer to collections of galaxies as **galaxy groups.** Our own Galaxy, the Milky Way, belongs to the (unimaginatively named) **Local Group.** The Local Group contains over 40 known members, as sketched in Figure 76.2. Its three largest members are spiral galaxies: M31, the Milky Way, and M33. The smaller members include the satellite galaxies orbiting M31 and the Milky Way, and dozens of small, low-luminosity **dwarf galaxies** scattered about the group.

The most famous of the dwarf galaxies are the **Magellanic Clouds,** which are close to the south celestial pole (see Looking Up #9: Southern Circumpolar

FIGURE 76.1
An all-sky plot of the positions of over a million galaxies detected by 2MASS, an infrared survey of the sky. Each white dot represents a galaxy; in clusters and superclusters of galaxies, they merge to form brighter areas. The blue shaded portion of the diagram shows where the plane of the Milky Way crosses through this map.

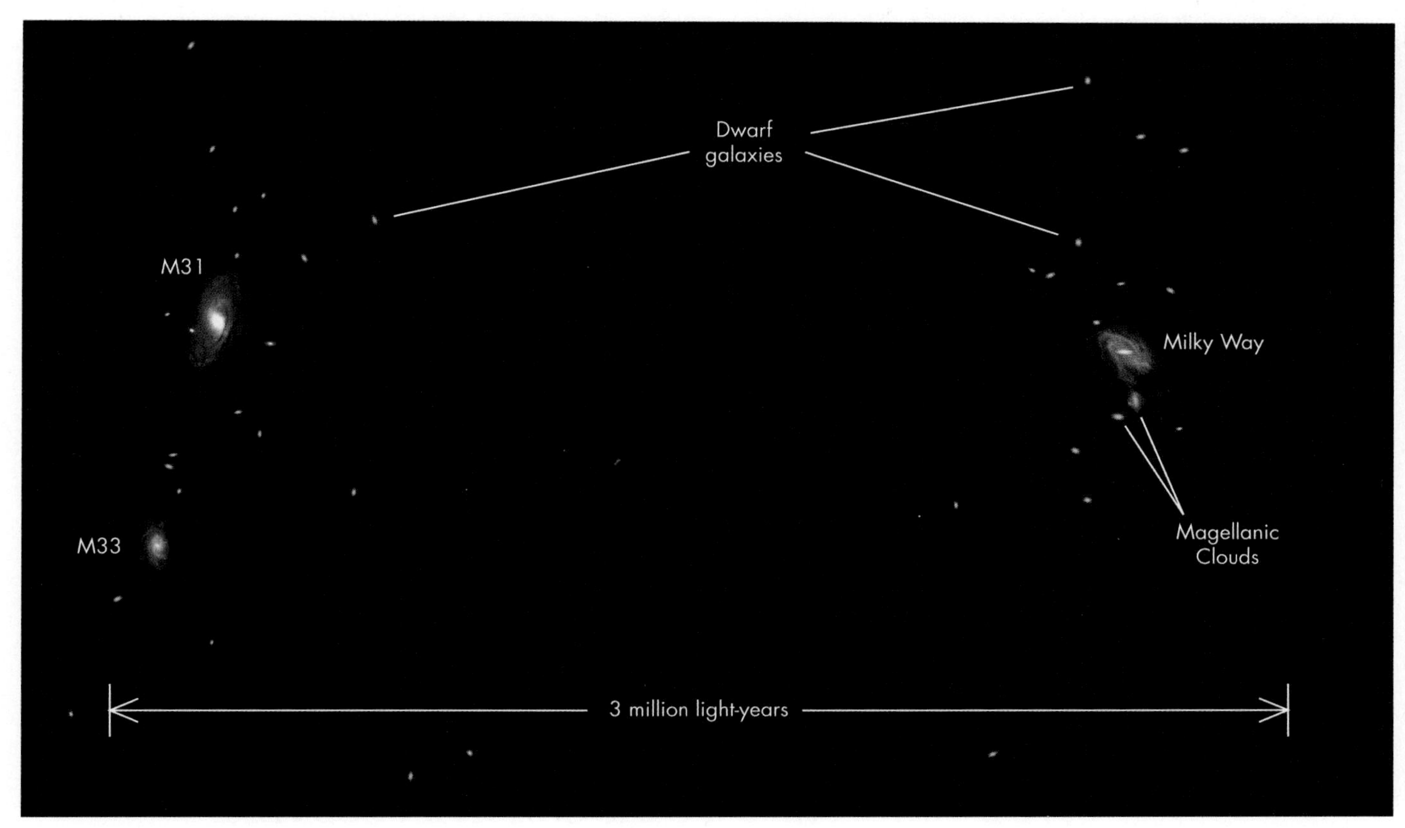

FIGURE 76.2
Images of the galaxies in the local group show what the Local Group might look like from outside. The Local Group's largest member is M31. The Milky Way is intermediate in size between it and M33. An image of galaxy M67 is used to represent the Milky Way because it is thought to be similar in appearance. Two small satellites of M31 are visible in this picture, as are as the Magellanic Clouds. The sizes of several dozen other dwarf galaxies in the group are also illustrated.

Constellations). They are about one-tenth the size of the Milky Way. They appear to be irregular galaxies, but infrared photographs of the large cloud show that it has a faint barlike structure. Both galaxies are approaching the Milky Way in highly elliptical orbits, and the shapes of these two galaxies are being distorted by the Milky Way's gravity (Unit 75).

The nearest galaxy to the Milky Way was discovered in 2003 (Figure 76.3). Known as the Canis Major dwarf galaxy, it orbits only about 13 kiloparsecs (42,000 light-years) from the center of the Milky Way. It was not detected earlier because it is merging with the disk of the Milky Way and therefore is largely hidden by dust in our Galaxy. At least 8 more small galaxies have been detected in recent sensitive imaging surveys, and several of these appear to be in the process of merging with the Milky Way.

Q How could we determine whether most of the mass of the Local Group is in the three spiral galaxies or in the several dozen smaller galaxies?

The Local Group is particularly interesting to astronomers, not only because it is our home galaxy cluster, but also because it allows us to study the demographics of galaxies. That is, we can see the true population of galaxies in this region of space, close to home. For example, most of the galaxies in the Local Group are faint dwarf galaxies. Even with modern telescopes, such galaxies are so dim that they are difficult to detect beyond the Local Group. Without the Local Group to let us observe the abundance of these small systems, we might grossly underestimate their true numbers in other galaxy clusters. When we examine larger clusters, we may find that they contain hundreds or in some cases thousands of galaxies that

FIGURE 76.3
The nearest galaxy to the Milky Way is called the Canis Major dwarf galaxy. It is hidden from view at visual wavelengths because it is so close to the galactic plane that it is obscured by dust. The image here is based on infrared data from the 2MASS survey. This galaxy appears to be merging with the Milky Way.

are as big as the three largest in the Local Group. Based on what we see in the Local Group, it is likely that there are at least 10 times more uncounted dwarf galaxies.

76.2 RICH AND POOR GALAXY CLUSTERS

The largest groups of galaxies are called **rich clusters** because they contain hundreds to thousands of member galaxies (Figure 76.4). The great mass of such a galaxy cluster draws its members into an approximately spherical cloud, with the most massive galaxies near the center.

In contrast, galaxy clusters that contain relatively few members, such as the Local Group, are called **poor clusters.** Poor clusters have so little mass that their gravitational attraction is relatively weak, and their member galaxies are not held in so tight a grouping. These clusters generally have a ragged, irregular appearance, with no central concentration.

FIGURE 76.4
The Hercules Cluster, a rich cluster of galaxies.

FIGURE 76.5
The central 1-Mpc region of the Virgo Cluster, centered around the giant elliptical galaxy M87. Giant ellipticals at the centers of galaxy clusters grow through mergers of galaxies within the cluster. The other galaxies visible in this picture will eventually merge with M87.

Not only do rich and poor clusters differ in their numbers of galaxies, but they also tend to contain different types of galaxies. Rich clusters contain mainly elliptical and S0 galaxies. Moreover, the few spiral galaxies that they contain tend to be found in the outer parts of the cluster. On the other hand, poor clusters tend to have a high proportion of spiral and irregular galaxies. Spirals are rare in the inner regions of rich clusters because of galaxy collisions. In the core of a rich cluster, galaxies are close together, so collisions between them are frequent. As we discussed in Unit 75, collisions may change spiral galaxies into ellipticals. Thus, although rich clusters may have once contained many spirals, today they contain few, and those are the "lucky" ones that escaped running into a neighbor. The many collisions in the crowded inner parts of these clusters also create **giant elliptical galaxies** by cannibalism as small galaxies merge with the larger galaxies they are orbiting. A giant elliptical galaxy is shown in Figure 76.5. It is at the center of the rich cluster nearest to us, the Virgo Cluster at a distance of about 15 megaparsecs (50 million light-years).

Rich and poor clusters differ in yet another way. Observations with X-ray telescopes indicate that rich clusters often contain large amounts (10^{12} to 10^{14} $M_{\odot}$) of extremely hot intergalactic gas that emits X-rays (Figure 76.6A). In fact, some clusters contain 10 times more matter in hot gas than they do in the stars of all the galaxies within the cluster. Astronomers are not certain where that gas originates. Some of it probably came from within the cluster's galaxies. For example, when a star explodes as a supernova, some of the gas it ejects may leave the star's galaxy and end up in the intergalactic space within the cluster. On the other hand, some of the gas may be material left over from when the galaxies formed, or it may be interstellar gas stripped from galaxies when they collide.

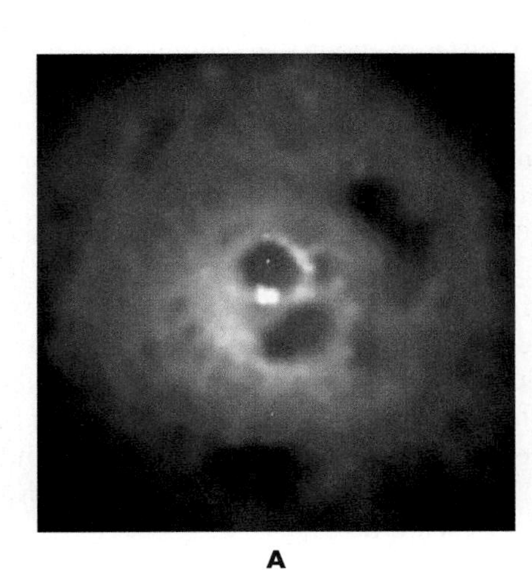

A

B

FIGURE 76.6
(A) A false-color X-ray image of the Perseus galaxy cluster. The violet glow is hot gas in the cluster. The picture shows a region about 300,000 light-years across. (B) A combined false-color image of a distant cluster in the constellation Hydra. X-ray emission is shown in violet, while the optical emission from galaxies is shown in other colors. The picture spans a region about 3 million light-years across. This cluster is more than 8 billion light-years away, .so it shows the appearance of the cluster when it was less than half its current age.

Galaxy clusters form in regions that are dense relative to the overall average in the universe. Hubble's law (Unit 74) indicates that galaxies' motion away from each other is small if they originated close to each other. At the same time, gravity grows weaker with distance. Therefore, if two galaxies form near each other, they will be moving away from each other slowly but will feel each other's gravitational pull strongly. This combination of effects tends to make the universe very clumpy, forming groups and small clusters early in its history.

Over time, these groups pull in more galaxies and merge with other groups to make larger clusters, eventually growing into the rich clusters we see today. Much of the evidence for this process comes from our ability to "see back in time" by looking out in space. In particular, galaxies in remote (younger) clusters, generally have a clumpier distribution than in nearby (older) clusters, as if they have recently been pulled in and have not had time to spread themselves smoothly. Moreover, X-ray observations of these young clusters (Figure 76.6B) show clumps of hot gas that appear to be falling toward one another to form the cluster.

76.3 SUPERCLUSTERS

The great mass of galaxy clusters creates gravitational interactions between them and holds them together into clusters of clusters, or *superclusters.* These large structures may contain many dozen galaxy clusters spread throughout a region of space, ranging from tens to hundreds of millions of light-years across. For example, the Local Group is part of the **Local Supercluster,** which contains the Virgo galaxy cluster and more than a dozen other small clusters spanning nearly 30 megaparsecs. It is sketched in Figure 76.7.

The presence of large concentrations of galaxies can also be deduced from the motions of galaxies. In general, galaxies are moving away from one another as part of the overall expansion of the universe, and their distances follow Hubble's law (Unit 74.3). However, in regions where there is a large concentration of mass, we see deviations from the velocities predicted by Hubble's law. For example, from our location in the Local Group, we measure a recession velocity relative to the Virgo Cluster of about 1000 kilometers per second, yet based on its distance we would expect the recession velocity to be about 1200 kilometers per second. The motion of the Local Group away from the Virgo Cluster has been slowed by the mass concentration there. The Local Group might eventually stop and then actually fall into the cluster—but not for hundreds of billions of years, and possibly never.

The effect on member galaxies is different in clusters and superclusters. In clusters, the gravitational pull and small separations have combined to pull the galaxies into collisions with each other. By contrast, the gravitational pull within superclusters has had only enough time to partially slow down the motion of clusters away from each other. This makes the region of a supercluster more dense than its surroundings, but it will probably be a very long time before the outermost members of the supercluster reverse direction and start moving toward each other.

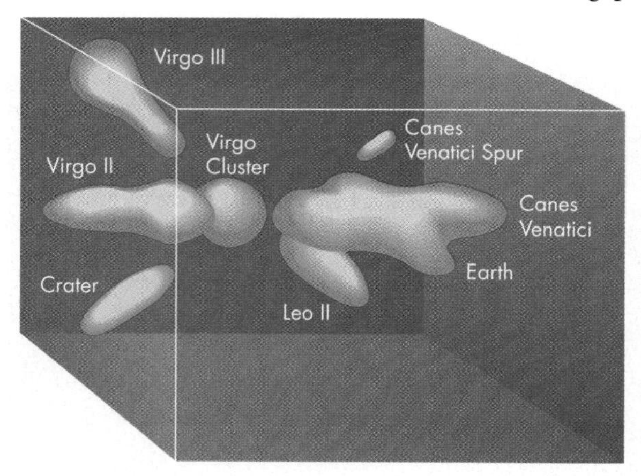

FIGURE 76.7
A depiction of the structure of the Local Supercluster. A number of groups and moderately rich clusters surround the rich Virgo Cluster, which contains most of the supercluster's mass.

If the universe is filled with dark energy, as is suggested by recent observations, the Local Group will probably be kept from falling into the Virgo Cluster (Unit 82).

The technique of studying deviations from Hubble's law has revealed that there are probably even larger structures in our vicinity. In the 1990s astronomers discovered a deviation in the motion of the Virgo Cluster, the Local Group, and other galaxies and clusters out to a distance of about 50 megaparsecs from the Milky Way. These peculiar motions suggest the presence of a huge concentration of mass in the direction of the constellation Centaurus. The whole Local Supercluster has a deviation of over 500 kilometers per second in that direction. To cause an entire supercluster to move at such a speed would require an enormous gravitational pull. The source of this enormous pull has been named the **Great Attractor** by astronomers. The direction of our motion is toward a position near the center of the all-sky map shown in Figure 76.1. There is a fairly large concentration of galaxies in this direction, although many galaxies are hidden from our view in this direction because of obscuration by the plane of the Milky Way.

The Great Attractor is probably not a single large mass, but an assemblage of many superclusters covering a large region beyond 50 megaparsecs (160 million light-years) away from us. It is estimated to contain about $3 \times 10^{16}\ M_{\odot}$—the equivalent of 10,000 or more galaxies like the Milky Way—in excess of the mass in comparably sized regions in other directions. The Great Attractor may hint at an even larger kind of structure than a supercluster—a cluster of superclusters! It is not expected that such structures will ever pull their farthest-flung members back toward the center of the attractor region, but they alter the motion of galaxies from that predicted by Hubble's law.

76.4 LARGE-SCALE STRUCTURE

Outside of our local region, astronomers have traced a variety of immense structures out to the limits of where we have surveyed galaxies. Figure 76.8 shows a map of the distribution of galaxies out to a distance of a few billion light-years. On this map each galaxy is represented by a single dot. Each galaxy's distance from us is based on its recession velocity (Unit 74). This map represents a thin slice of the sky in which the redshifts of about 100,000 galaxies have been observed by a large team of astronomers. It gives us a cross-sectional look at the distribution of galaxies out to a redshift (z) greater than 0.2—a distance of about 900 megaparsecs.

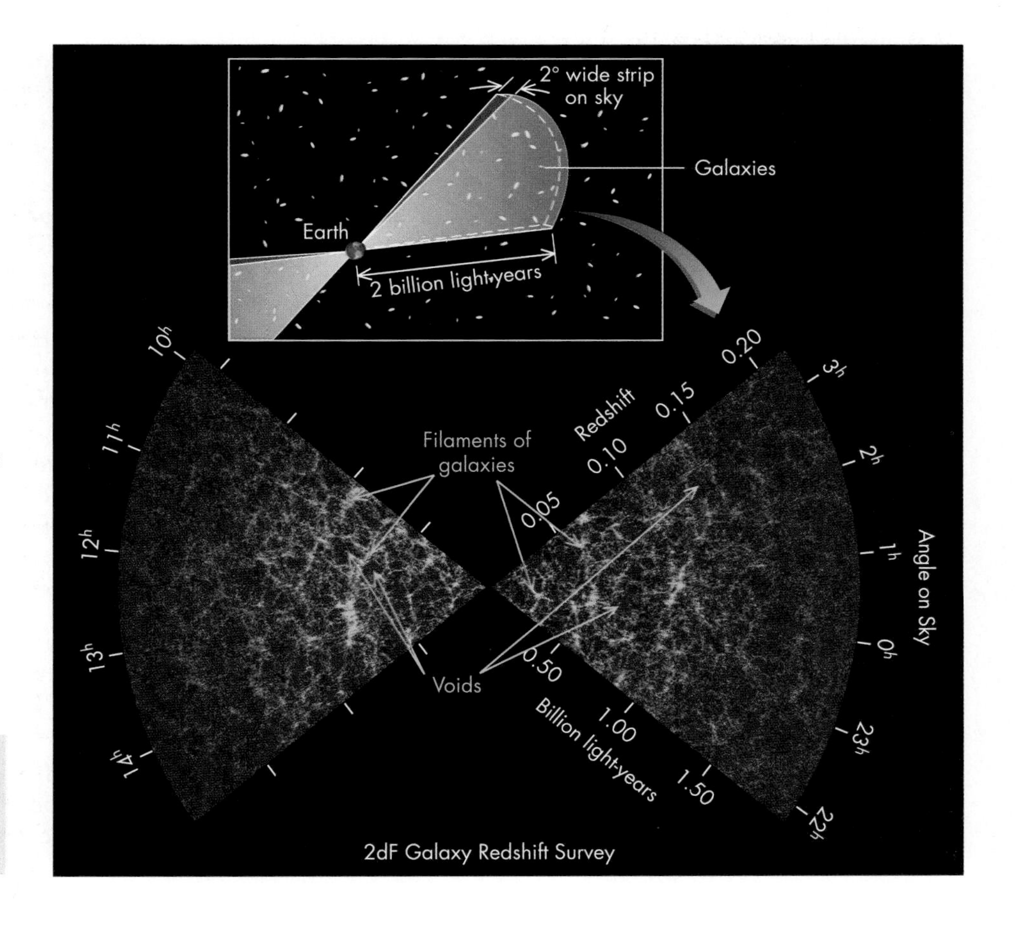

FIGURE 76.8
A map of how galaxies are distributed in space. The shape of the region shown in the figure is like two slices of pizza, tip-to-tip, as illustrated in the upper sketch. Each dot in the lower figure represents one of some 100,000 galaxies that lie in the wedge-shaped regions that extend out to about 2 billion light-years from Earth. Note the nearly empty regions (voids), the stringy structures, and the overall appearance sometimes described as the "cosmic web." The distances to the galaxies were measured from their redshifts. (The angles refer to Right Ascension coordinates on the sky.)

Q What is the difference between the information shown in a map like that in Figure 76.8 and that in Figure 76.1?

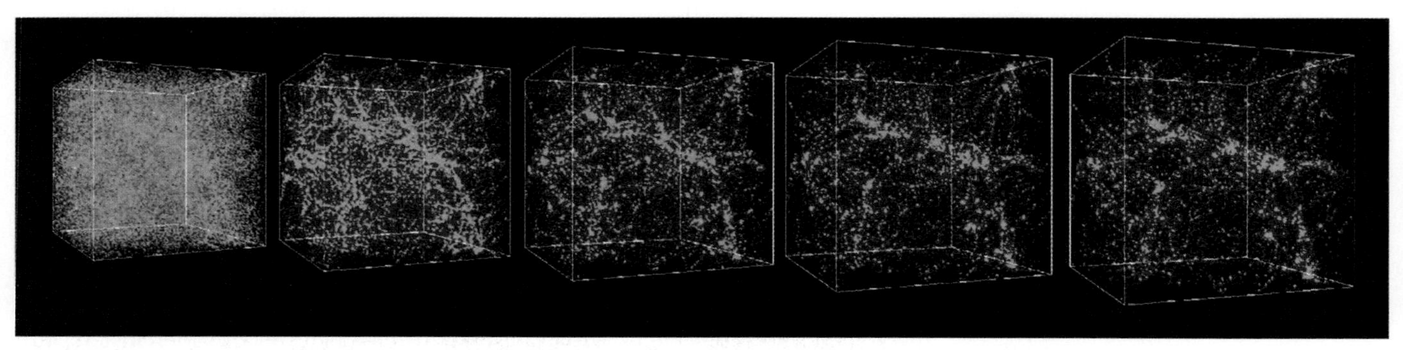

FIGURE 76.9
A computer simulation of the growth of structure as the universe expands. You can see how an initially smooth distribution of matter becomes stringy, much like the features in Figure 76.8. Each box spans about 140 million light-years. The first box on the left represents a time when the universe was roughly one-tenth its present age. The last box on the right shows that same piece of the model of the universe as it would be today.

On this scale, individual galaxies and clusters are no longer obvious, but we *can* see that many appear to be arranged into long filaments or shells surrounding regions nearly empty of galaxies. These latter spaces are called **voids,** and although they are not totally empty, the few galaxies they contain are generally small and dim. Some of these voids are immense. One found in 2007 is estimated to be 300 megaparsecs (about a billion light-years) across—a vast region of space nearly empty of galaxies.

Large-scale structures such as superclusters, voids, and filaments are at the frontier of our knowledge of the distribution of galaxies in space. The patterns of large-scale structure we see at this scale are probably the consequence of the initial conditions in the universe, and astronomers have proposed a variety of hypotheses to explain the features we see. For example, when the huge voids were first found, some astronomers suggested that they might be evidence that even before galaxies began forming, huge explosions pushed the gas out of these regions. Other astronomers have hypothesized that these features were probably produced by gravitational interactions of dark matter (Unit 78).

To attempt to understand large-scale structure, astronomers have turned to computer simulations. In such simulations, astronomers write out the equations that govern the motion of matter (such as Newton's laws) and then solve the equations, assuming that the material filling space is expanding so as to mimic the general expansion of the universe. They then watch what happens as the gravitational force of the material acts within the simulated volume. Figure 76.9 shows a series of images from such a simulation. You can see that matter that initially had a relatively smooth distribution is drawn by gravity into a network of strings. Astronomers identify the places where strings meet as locations where galaxies form. They adjust the equations by varying the amount of ordinary and dark matter and initial lumpiness, thus experimenting to discover what mix best creates structures similar to what is actually observed. The outcome of such simulations by many teams of astronomers confirms that no explosions are needed; but without dark matter, the distribution of galaxies fails to match what we observe.

76.5 PROBING INTERGALACTIC SPACE

The large-scale structure of galaxies indicates where the most luminous galaxies are concentrated, but astronomers suspect that a large amount of matter must lie between the galaxies in intergalactic space. Observations using a variety of

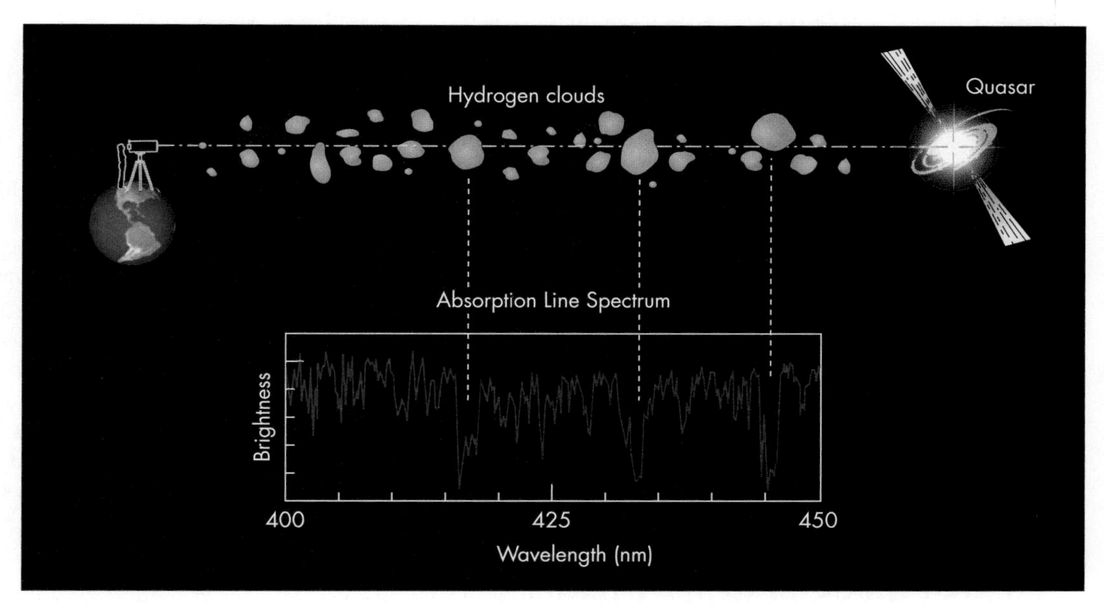

FIGURE 76.10
The light from a distant quasar produces a uniform light source against which we can see hundreds of absorption lines produced by hydrogen clouds along the line of sight to the quasar.

techniques are beginning to uncover large reservoirs of matter that never formed into the collections of stars we call galaxies.

We can see the hot intergalactic gas inside rich clusters of galaxies from the X-rays it emits. Outside these regions, intergalactic gas is cooler and more difficult to detect. On the other hand, gas that is cool enough can be detected spectroscopically by its absorption against a bright background source. **Quasars** are bright sources of light that are extremely distant and therefore are useful as probes of intergalactic space along our line of sight toward them. Quasars are intriguing objects in their own right (Unit 77), but for absorption studies we are interested only in the fact that they are extremely distant light sources.

Because quasars are so distant, their light passes through many other galaxies and intergalactic gas clouds as it travels to Earth. Even if an intervening galaxy or cloud is too dim to see, its matter will imprint a signature on the quasar's spectrum, allowing us to detect these otherwise invisible objects. Because these absorbing objects all have different redshifts, each produces a set of absorption lines shifted to different wavelengths. This results in a whole "forest" of absorption lines, each one corresponding to a cloud of gas along the line of sight, as illustrated in Figure 76.10.

Q How might the absorption lines produced by a spiral galaxy in front of a quasar look different from those produced by a small cloud of gas?

Most of these absorption lines come from clouds that have not yet formed into galaxies themselves but are condensing and cooling. Studies of these clouds suggest that they follow the large-scale structure seen in the pattern of galaxies. The intergalactic gas is located in huge sheets, filaments, and voids out to the very largest distances we can see—and therefore the earliest times we can see in the universe's history. The amount of this gas appears to decline with time—in other words, there is less of it in the regions nearer to us. Simulations suggest that over time intergalactic gas accretes onto galaxies, providing a source of additional gas for galaxies and perhaps affecting their evolution. These simulations also suggest that such accretion may continue even to the present, and astronomers are searching for evidence of this very tenuous gas and its possible effects on the Milky Way and other nearby galaxies.

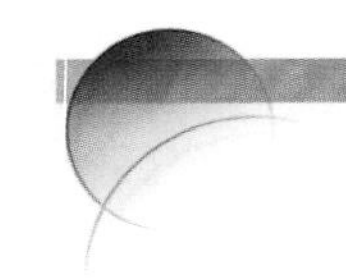

KEY TERMS

dwarf galaxy, 620
galaxy cluster, 620
galaxy group, 620
giant elliptical galaxy, 623
Great Attractor, 625
large-scale structure, 626
Local Group, 620
Local Supercluster, 624
Magellanic Clouds, 620
poor cluster, 622
quasar, 627
rich cluster, 622
supercluster, 620
void, 626

QUESTIONS FOR REVIEW

1. What is the Local Group?
2. What kind of galaxies are the neighbors to the Milky Way?
3. What are the differences between rich and poor clusters?
4. Over what distance scales do galaxies cluster together?
5. What is the source of the X-ray emission from clusters of galaxies?
6. What is the Local Supercluster?
7. What is large-scale structure?
8. What produces quasar absorption lines?

PROBLEMS

1. Some galaxies in the Virgo Cluster are found to be moving at 1000 km/sec through the center of the cluster. How long would it take to move 1 Mpc at that speed?
2. The Great Attractor, located some 50 Mpc away, is suspected to have caused the Local Group to accelerate to a speed of 500 km/sec over the last 13 billion years.
   a. What is the average acceleration over that time, expressed in m/sec^2?
   b. Using Newton's law of gravitation, how big a mass would be needed at that distance to produce the observed acceleration?
3. The spectral line of hydrogen producing the quasar absorption lines in most studies has a laboratory wavelength of 122 nm. If a hydrogen cloud is at a distance of 1 billion ly,
   a. use Hubble's law to find the recession velocity of this cloud.
   b. what is the redshift of the cloud?
   c. at what wavelength will the absorption occur?
4. Suppose that the typical galaxy has a mass of $10^{11}\ M_{\odot}$. If the total mass of a cluster of 100 galaxies is $10^{15}\ M_{\odot}$, how much dark matter is contained in the cluster by the percentage of the total mass (ignoring the hot X-ray gas)?
5. Same as problem 4, but now the hot X-ray gas has a mass 10 times greater than the galaxies. How much dark matter is contained in the cluster by the percentage of the total mass?
6. A spectral line of hydrogen has an absorption laboratory wavelength of 122 nm. It is observed in the spectrum of a quasar at 130 nm, 180 nm, and 330 nm. What are the redshifts (z) of these clouds? What would be their distances, assuming Hubble's law (and using $V = c \times z$)?

TEST YOURSELF

1. The Local Group
   a. contains about 30 member galaxies.
   b. is a poor cluster.
   c. is the galaxy cluster to which the Milky Way belongs.
   d. mostly contains galaxies much smaller than the Milky Way.
   e. All of the above.
2. Why are quasars used to observe the absorption lines produced by intergalactic gas clouds?
   a. Because this is the gas left over from when quasars exploded.
   b. Because quasars and the gas are of similar age and hence are near each other.
   c. Because the hotter gas in galaxies bends the quasar's light away from us.
   d. Because quasars are attracting the clouds.
   e. Because quasars are very distant and bright.
3. The majority of galaxies in rich clusters are ____ , whereas poor clusters contain high proportions of ____ galaxies.
   a. elliptical and S0 type; spiral and irregular
   b. elliptical and spiral; irregular
   c. spiral; irregular
   d. elliptical and irregular; spiral and S0 type
   e. spiral and irregular; elliptical and S0 type

unit

77

Active Galactic Nuclei

Background Pathways

Unit 74: A Universe of Galaxies 599

If our eyes could see radio wavelengths instead of visible light, the night sky would look completely different to us. Very few of the brightest sources of visible light produce much radio emission, and many of the brightest sources at radio wavelengths vary in brightness, so that if we worked out a pattern of constellations, we would find in a few years' time that some "stars" had become so much brighter, and some so much fainter, that the constellations would no longer look the same. And when we tried to measure the distances to these "stars," we would find that many of the brightest are among the most distant objects in the universe.

The strangeness of the radio sky became apparent during the 1950s and 1960s as the early technology for making radio observations steadily improved. Astronomers found that some radio sources were associated with objects already cataloged at optical wavelengths (such as galaxies, nebulae, and the galactic center). But many others—including some of the brightest radio sources—appeared to have no optical counterpart. Some of the strongest radio signals were later identified with dim "stars" that proved to be extremely distant, and such large distances implied that these objects were the most luminous objects in the universe.

Today we recognize these luminous sources as **active galactic nuclei**, or **AGNs**. They are tiny regions in the centers (nuclei) of galaxies that emit abnormally large amounts of energy. Not only is the emitted radiation intense, but it usually fluctuates in intensity as well. At least 10% of all known galaxies have active nuclei, and many of these galaxies exhibit intense radio emission and other activity outside their central region in addition to their AGNs.

77.1 ACTIVE GALAXIES

FIGURE 77.1
The Seyfert galaxy NGC 7742, imaged with the Hubble Space Telescope, is a spiral galaxy with an extremely luminous nuclear region.

Over the past half century different classes of galaxies with unusually vigorous energy output have been discovered. Their discovery occurred in a variety of ways at different wavelengths; at first they were thought to be independent classes of objects. Although these galaxies vary in the amount of activity they exhibit, the past two decades' evidence suggests that most of this activity probably shares a common explanation. Here we describe two major classes of activity found in nearby galaxies.

A **Seyfert galaxy** is a spiral galaxy whose nucleus is abnormally luminous (Figure 77.1). These unusual galaxies are named for the U.S. astronomer Carl Seyfert, who first drew attention to their peculiarities in the 1940s. The core luminosity of a typical Seyfert galaxy is immense, amounting to the entire radiation output of the Milky Way but coming from a region less than a parsec across. Moreover, the radiation is at many wavelengths: optical, infrared, ultraviolet, and X-ray. Despite its immense luminosity, the radiation from a Seyfert galaxy's core fluctuates rapidly in intensity, sometimes changing appreciably in a few minutes.

In addition to the intense source of radiation in their nuclei, Seyfert galaxies also contain gas clouds moving at speeds of about 10,000 kilometers per second near their cores. A second type of Seyfert galaxies was defined, which was similarly luminous in the core, but did not show the same evidence of high-speed gas. Subsequent observations suggest that these are actually the same type of active nucleus, but observed at an angle where dust in the galaxy's disk has blocked our view of the very central region.

Another type of activity is seen in **radio galaxies.** As their name indicates, they emit large amounts of energy in the radio part of the spectrum. Some emit millions of times more energy at radio wavelengths than do ordinary galaxies. This energy is generated primarily in two types of regions: the galaxy's nucleus and enormous regions outside the galaxy and on opposite sides of it, as you can see in Figure 77.2. This figure shows false-color images in which the intensity of radio emission is shown in red while the visible emission (showing light from stars in the galaxy) is shown in blue. The **radio lobes,** as the regions of strong emission outside the galaxy are called, may span hundreds of thousands of parsecs (about a million light-years). On the other hand, the core source is typically less than one-tenth of a parsec across.

What causes the intense emission of radio energy from these galaxies? Astronomers can tell from its spectrum that the emission is **synchrotron radiation,** generated by electrons traveling at nearly the speed of light and spiraling around magnetic field lines. The electrons are part of a hot ionized gas that is shot out of the core in narrow **jets** (such as the one shown in Figure 77.2C) and eventually collides with intergalactic gas in the vicinity. There the jets of ionized gas spread out to form the lobes that emit radio waves.

Clearly there are some similarities between radio galaxies and Seyfert galaxies. They both involve rapidly moving gas and a small, bright nuclear region. Furthermore, some Seyfert galaxies have been found that have jets emanating from the nucleus, though smaller than those seen in radio galaxies. Some further clues about such strange behaviors of galactic nuclei come from some of the most distant objects we know of in the universe.

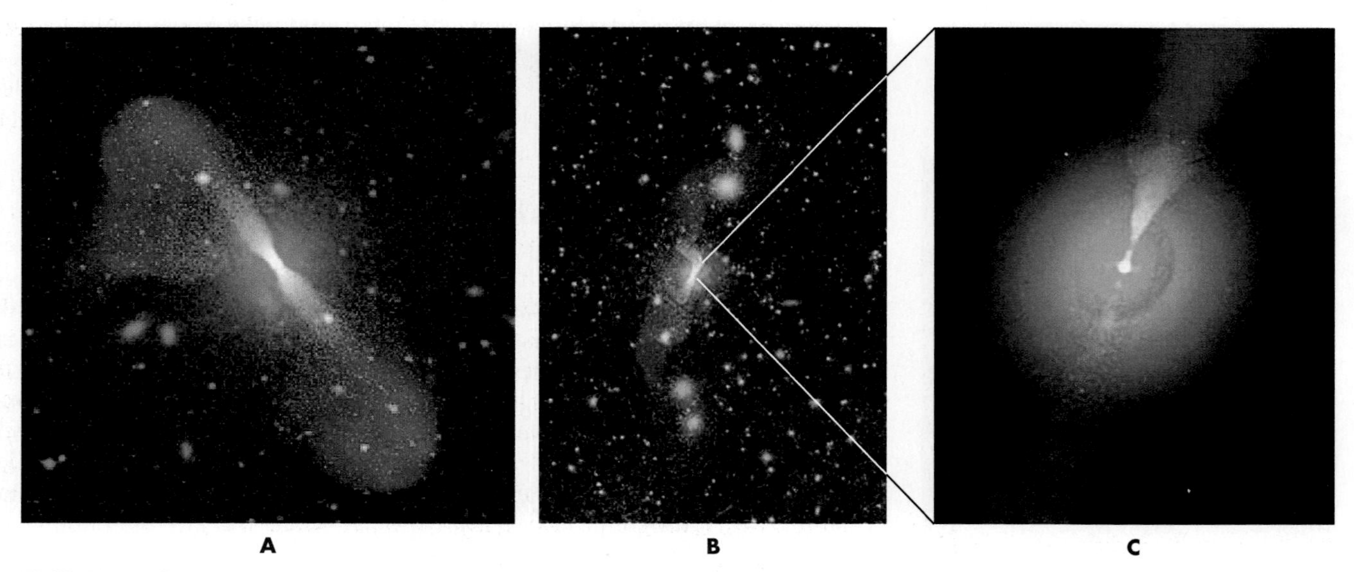

FIGURE 77.2
These images show two radio galaxies, NGC 5532 (A) and NGC 383 (B & C). The figures show the radio emission in red—the brightness indicating its intensity—superimposed on an optical image, shown in blue. The extended red regions in A and B are called radio lobes. Image (C) is a close-up of the central region of the radio emission in NGC 383, showing the bright nuclear source and jets of hot plasma being shot out of the center of the galaxy.

77.2 QUASARS

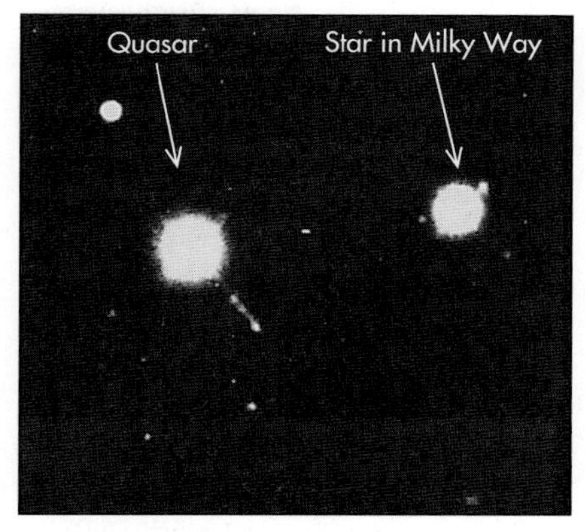

FIGURE 77.3
The quasar 3C273 from a ground-based telescope image. It is the brightest visually of any quasar, but is a dim "star." Even though it is one of the brightest radio sources in the sky, it is a thousand times fainter than the dimmest star visible to the unaided eye. "3C" stands for the Third Cambridge Catalog of Radio Sources.

Q If a quasar is seen as it looked 13 billion years ago, was it 13 billion light-years distant when the light was emitted? Is it this far away now?

FIGURE 77.4
The quasar PG 0052 251 imaged with the Hubble Space Telescope. The image reveals that the quasar lies in the nucleus of a spiral galaxy. ("PG" stands for Palomar Green survey.)

Quasars are extremely luminous, extremely distant, active galactic nuclei. They were originally identified because many of them are among the brightest radio sources found in early surveys of the "radio sky." Quasars get their name by contraction of the term *quasi-stellar radio source,* where *quasi-stellar* (meaning "almost starlike") refers to their appearance in the first photographs taken of them (Figure 77.3), in which they look like stars—that is, like points of light. Early photographs also revealed that some quasars have small wispy clouds near them or tiny jets of hot gas coming from their cores, like the one seen in Figure 77.3.

Pictures of quasars lack detail because these objects are so far away. We deduce their immense distance from the huge redshift of their spectra. (Recall from Unit 74.3 that a galaxy's distance can be found from its redshift using Hubble's law.) In fact, quasars as a class have the largest redshifts known for any kind of astronomical object. The redshifts of quasars are so large that their visible spectra are often shifted completely to infrared wavelengths, and what we observe at visible wavelengths was originally emitted by the quasar in the ultraviolet. The visible-light spectra astronomers observed were therefore unfamiliar to astronomers. In 1963 the Dutch-American astronomer Maarten Schmidt realized that they were highly redshifted and therefore extraordinarily distant objects.

On the basis of its redshift, the most distant quasar yet found emitted the light we see today nearly 13 billion years ago, less than 1 billion years after the universe began (Unit 79). To be visible at such an immense distance, an object must be immensely luminous, and quasars turn out to be about 1000 to 100,000 times more luminous than the Milky Way. In fact, the brightest quasars have a power output equivalent to a supernova explosion occurring every hour!

During the first decades after quasars were discovered, their extraordinary properties caused some astronomers to question the large distances that had been deduced for them. Alternative hypotheses were offered, suggesting that quasars were nearby stars and that their observed redshifts had a different cause—such as a large gravitational redshift (Unit 68). However, the Hubble Space Telescope was able to settle the question in the 1990s by making high-resolution images of quasars' surroundings (Figure 77.4). These images revealed that quasars lie at the centers of galaxies, like Seyfert and radio galaxies. They resemble radio galaxies in other ways too, producing jets of material and radio lobes, for example, although quasars are generally much more luminous.

Despite their huge power output, quasars appear to be very small. Similar to Seyfert galaxies, quasars fluctuate in brightness (Figure 77.5). Slow changes occur over months, but short-term flickering also occurs in periods as brief as hours. Rapid changes of this sort give clues to the size of the emitting object.

For example, suppose an object is one "light-hour" in diameter (about 7 AU). Light from its near side therefore reaches us one hour earlier than the light from its distant side. Even if every part of the object lit up simultaneously, after light from the near side reached us, light from the distant portions of the object would not reach us for an additional hour (Figure 77.6). And if it turned off all at once, it would again take an hour for the object to drop back down to its original brightness. An object's light cannot fully appear or disappear in less time than it takes light to travel across it. We can write this as a formula:

$$\text{Diameter} < c \times \Delta t$$

where c is the speed of light and Δt (pronounced "delta t") is the time interval over which the quasar's brightness changes substantially. We write that the diameter is less than $c \times \Delta t$ because the actual diameter can be smaller than this limit if the luminosity changes are not instantaneous.

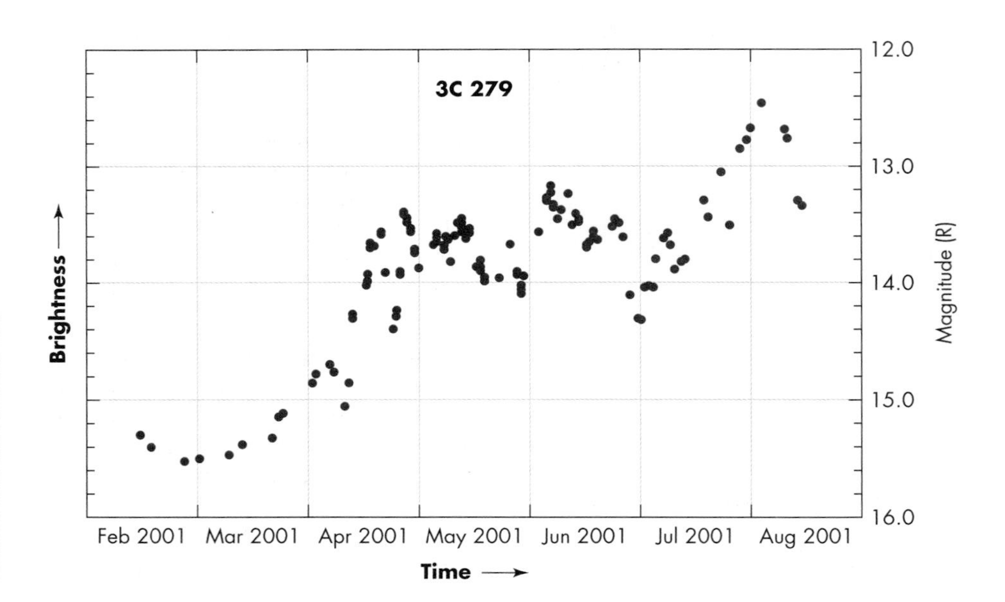

FIGURE 77.5
Plot of light variation of a quasar.

Sound waves are affected by the size of their source much like light waves. Imagine, for example, everyone in a football stadium snapping their fingers simultaneously. The sound you would hear outside the stadium would be spread out by the sound travel time across the stadium.

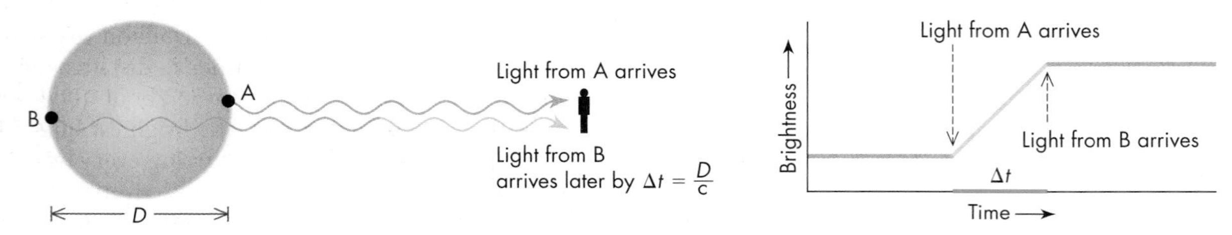

FIGURE 77.6
The effect of size on light arrival times. If a large source lights up instantly, we see the light from its near side before we see the light from its far side. For the objects that we encounter in everyday life, the delay is so short we do not notice it. However, for astronomical bodies, the effect can create delays ranging from days to months, from which we can deduce that the emitting body is light-days or light-months across.

The light variability intervals imply that many quasars are smaller than our Solar System. If a quasar shows substantial variations in $\Delta t = 10$ hours (36,000 seconds), its diameter is smaller than

$$c \times 36{,}000 \text{ sec} = 3.0 \times 10^8 \text{ m/sec} \times 3.6 \times 10^4 \text{ sec} = 1.1 \times 10^{13} \text{ m}.$$

An object of this diameter would fit within the orbit of Pluto. Astronomers have confirmed the remarkably small size of a number of quasars using radio interferometer telescopes. The incredible power output of a quasar comes from a very small object.

Today we recognize a near continuum of properties spanning the range from nearby radio galaxies and Seyfert galaxies to quasars. What, though, could possibly explain such extraordinary luminosity from such a small object? Many ideas have been tried. For example, at one time astronomers thought the luminosity was produced by a single star 10^8 solar masses that had managed to form in the center of the galaxy. Astronomers have also considered more exotic explanations. For example, a few scientists have proposed a new class of objects called "white holes." These would be the opposite of black holes in the sense that they would spew material out into outer space rather than draw it in. This is an intriguing idea, but no other evidence for white holes exists, and we gain little understanding of an extraordinary phenomenon if we resort to inventing a fantastic new kind of object.

77.3 CAUSE OF ACTIVITY IN GALAXIES

Because radio galaxies, Seyfert galaxies, and quasars share many common features, astronomers have sought a unified model for their activity that could explain all three. Any successful model for active galactic nuclei must explain how such a small central region can emit so much energy over such a broad range of wavelengths. For example, at least one quasar is a powerful source of both radio waves and gamma rays. Astronomers therefore hypothesize that the core must contain something very unusual. No ordinary single star could be so luminous. No ordinary group of stars could be packed into so small a region.

One kind of object that is very small but that can also emit intense radiation is an **accretion disk** around a black hole. Most matter falling toward a black hole does not fall straight in, but ends up orbiting just outside the black hole in an accretion disk. There it collides with other matter and emits intense X-ray radiation.

In jets that point nearly toward the Earth, the high-speed motion can sometimes make clumps of gas appear to be moving faster than the speed of light. This is an illusion of timing because the gas is moving toward us, so light emitted from it at later times has less distance to travel before reaching us.

The need for an extreme object in the cores of these objects is demonstrated by the very high speeds of jets measured in a number of quasars and at least one Seyfert galaxy. Clumps of gas shoot out from the AGN at speeds clocked at greater than 90% of the speed of light! Whatever is in the core must be able to accelerate matter to almost the speed of light.

Based on the accumulated evidence—the huge energy output, the small size, the enormous speeds—astronomers have ruled out all "ordinary" objects. The best remaining hypothesis is that active galactic nuclei contain immense black holes. According to this model, the black hole would contain millions of solar masses or more, around which swirls a huge accretion disk of gas that produces the energy we see. Such a black hole would have a radius about the size of the Earth's orbit around the Sun, while the accretion disk would have a radius of tens to hundreds of AU. Gas within the disk spirals toward its center as it loses gravitational energy through collisions, and is ultimately swallowed. Drawn in by the black hole's fatal attraction, the gas orbits faster and faster, growing hotter and hotter as it is pulled deeper into the gravitational field of the black hole—converting the gravitational energy into kinetic and thermal energy. The gas is heated to incandescence by frictional forces in the disk and emits the light and jets that we observe.

Figure 77.7, a picture made with the Hubble Space Telescope, supports this model. The picture shows just such a disk in the central regions of the radio galaxy NGC 4261. Gas falling into the disk releases gravitational energy, which heats the material to temperatures perhaps as hot as several million Kelvin. Although most

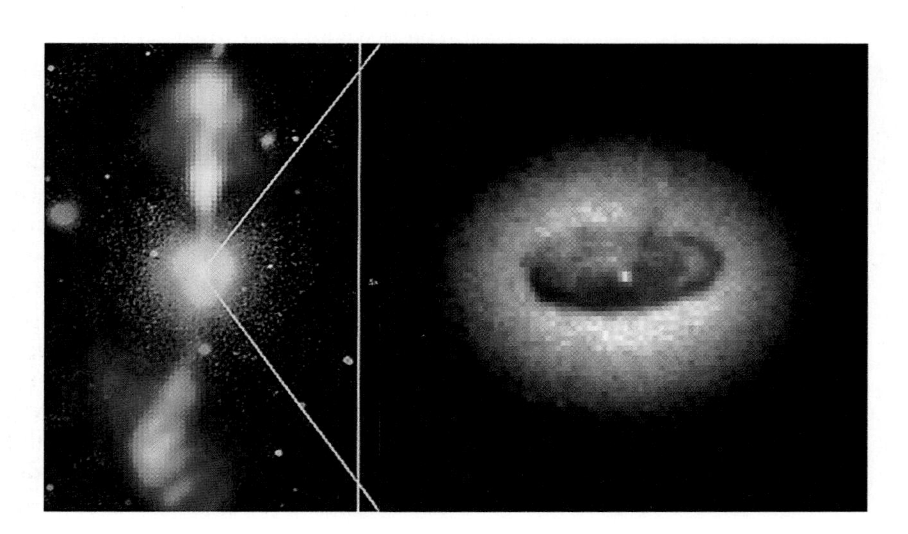

FIGURE 77.7
An image (at far right) made with the Hubble Space Telescope of an accretion disk in the center of the radio galaxy NGC 4261. The inset image on the left shows a wider view of this galaxy's central region and shows the jets of hot gas that were ejected from the nuclear region.

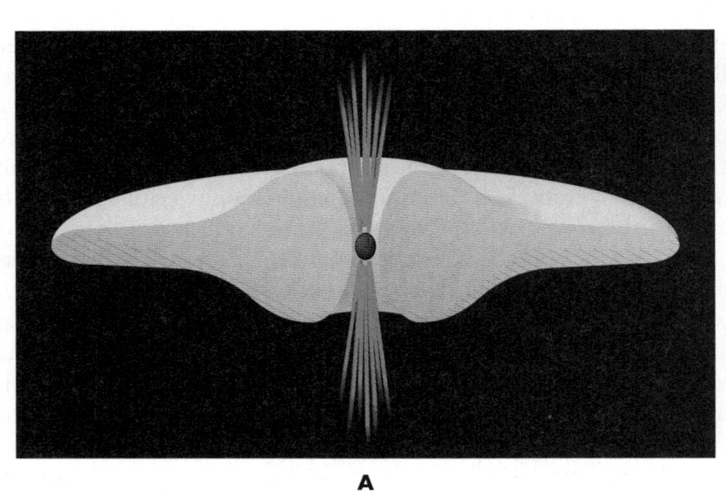

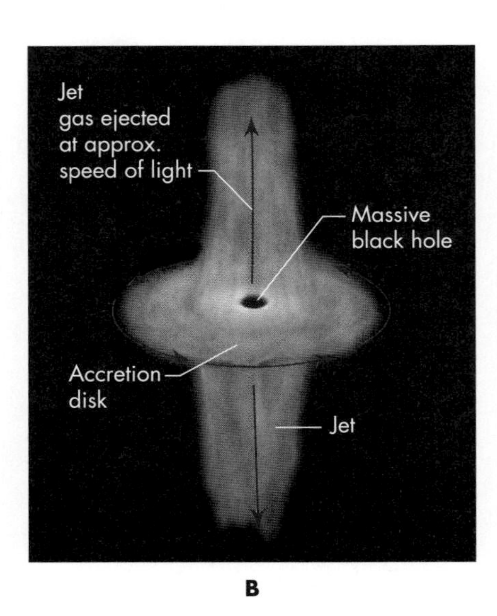

A B

FIGURE 77.8
(A) Sketch of an accretion disk surrounding a black hole, which helps funnel the hot gas into jets emitted perpendicular to the disk. (B) Magnetic fields, twisted up in the hot plasma, may help keep the jet of hot gas tightly confined.

of the mass ultimately falls into the black hole, some material from the accretion disk boils off into space. If a thick accretion disk surrounds the black hole, it may channel such escaping gas into outward flows from the top and bottom surfaces of the disk (Figure 77.8).

Although this model explains the overall features of active galactic nuclei, it cannot fully explain why the jets remain in such narrow beams. Astronomers have proposed a model in which the accretion disk has a strong magnetic field that is twisted as the disk spins. The magnetic field traps charged particles that exist in the hot ionized gas, in much the same way as also occurs to prominences in the magnetic field of the Sun (Unit 51). The spinning and twisting of the magnetic field allows it to capture hot gas from the accretion disk and confine its outward motion along the axis of rotation. This model also provides a powerful supply of energy for the jet—the rotational energy of the spinning black hole—although it is estimated that only a few percent of the infalling matter is ejected through this process.

Over time, if no new sources of matter are added, the accretion disk gradually drains into the black hole. Once this supply is exhausted, the activity ceases. The disk, however, may be replenished with matter from at least two sources. Sometimes stars within the galaxy may approach too closely to the black hole and be torn apart by its tidal forces. Their gaseous debris then forms a new accretion disk, providing a new source of energy. Alternatively, collisions with neighboring galaxies may occasionally add new material as the active galaxy cannibalizes its smaller neighbors. Each such event can give the central black hole a new source of matter for its accretion disk, causing the active galactic nucleus to increase in brightness. Thus AGNs appear to be quasars that have been "reinvigorated" when some fresh food has been thrown in their direction.

Our own Galaxy has a black hole of several million solar masses at its center (Unit 73.2). According to one hypothesis, the Milky Way's central black hole formed in several steps. First, a massive star in the central region of the Galaxy reached the end of its life, exploded as a supernova, and formed a black hole perhaps as small as 5 $M_{\odot}$. Normally a black hole like this has little opportunity to grow—it is only 30 kilometers in diameter, so the odds of anything hitting it are small. However, at the center of the galaxy, matter is packed densely and was falling toward this

Q Suppose you currently lived in one of the galaxies that we are presently observing as a quasar. What would your galactic center look like to you today? What would the Milky Way look like to you?

region during the early stages of galaxy formation. As a result, an initially small black hole at our Galaxy's center had a special opportunity to grow by swallowing infalling matter. Bit by bit, the mass of the black hole (and therefore its radius) increased, making it easier for the hole to attract and swallow yet more material. Eventually the hole probably became large enough to swallow entire stars, at which point it grew rapidly to a million or more solar masses. During its growth the hole was surrounded by an accretion disk. Our Milky Way probably had an active galactic nucleus. As the Milky Way aged, its black hole consumed the gas around it, and the activity subsided to the low level we see today.

We just described how such a huge black hole might have formed at the center of the Milky Way by the steady accumulation of mass into an initially small black hole. However, some galaxies have central black holes that are far larger than the central black hole in the Milky Way, some with masses more than a billion times larger than the mass of the Sun. The evidence for these large masses is similar to that for the black hole at the center of the Milky Way. Astronomers measure the speed at which stars and gas orbit the cores of the galaxies. Material moving at such high speeds could only be kept in orbit by an enormous mass. Yet this huge mass seems to emit no starlight, and black holes are the only known objects that are both massive and dark enough. Astronomers think it is improbable that such large masses could be reached through a process of gradual growth like that described for the Milky Way.

A recent discovery suggests that the growth of supermassive black holes must be closely linked to the overall growth of their host galaxies. Observations made with the Hubble Space Telescope show that where central black holes can be measured, they generally have a mass of a few tenths of a percent of the mass of the host galaxy's central stellar bulge (Figure 77.9). Why does such a correlation exist? It is difficult to see how black holes that form in a region less than 1 parsec across are "aware" of the total mass of stars extending outward many kiloparsecs. Astronomers are not yet certain, but the correlation may be explained if central black holes grow at the same time galaxies grows through mergers. When a galaxy cannibalizes a small neighbor, some of the inflow of extra mass could reach the central black hole in the larger galaxy, allowing it to grow. In addition, the massive black holes of both galaxies may sink toward the center of the merged galaxy and themselves merge. Some X-ray observations support this model, showing a galaxy that appears to contain a pair of massive orbiting black holes in the process of merging.

Even though only about 10% of galaxies show signs of activity, astronomers suspect that most (perhaps even all) large galaxies have a **supermassive black hole** at their cores following the relation found in Figure 77.9. The central black hole is

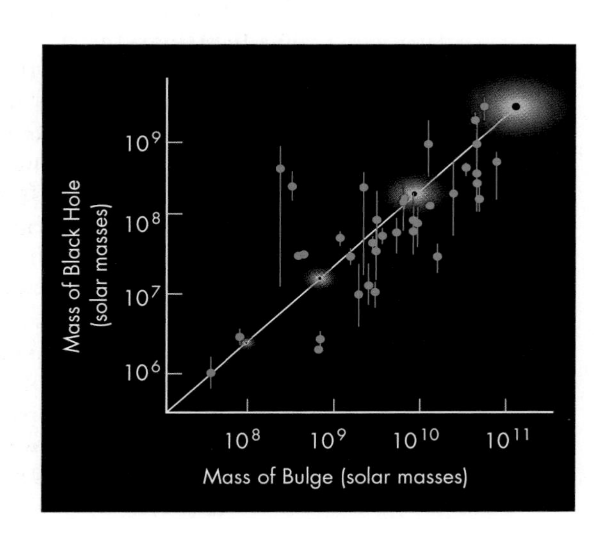

FIGURE 77.9
The correlation between central black hole mass and the masses of the galaxies' central bulge. The green dots indicate the masses of central black holes and bulges in more than 30 galaxies studied with the Hubble Space Telescope. The vertical green bars indicate the uncertainty in the measurements. The mass of central black holes grows with galaxy mass as illustrated by the background diagrams of galaxy bulges.

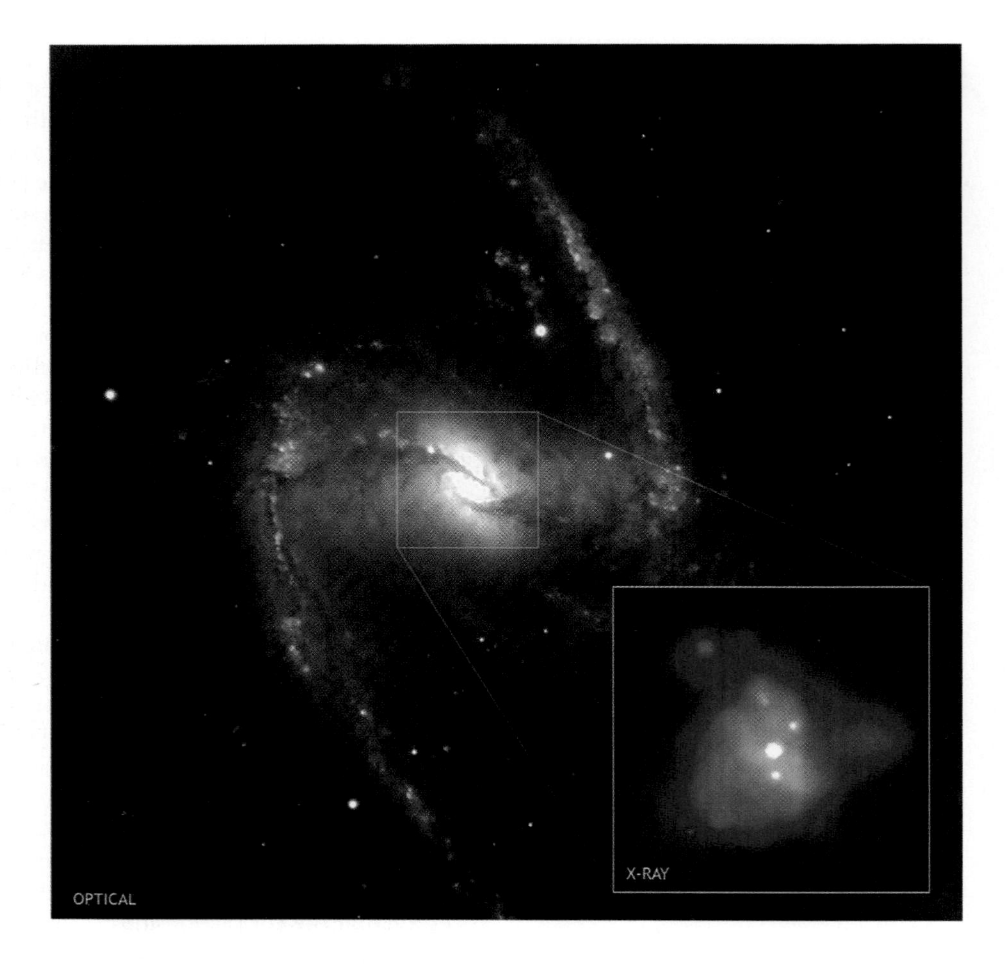

FIGURE 77.10
In 2007 the Chandra X-ray space telescope observed a dark cloud eclipsing the X-rays coming from the accretion disk around the black hole in this galaxy. From the rate of speed of the dark cloud, they were able to determine that the accretion disk is less than 7 AU in diameter.

Quasars often have redshifts greater than 1. This does not mean that they are traveling at speeds greater than the speed of light; instead it indicates that the universe has more than doubled in size since light left the quasar (see Unit 79).

active mainly when the galaxy is young and there has substantial matter accreting onto the black hole. Thus, quasars are mostly found at relatively large redshifts, where we are seeing back to a time when galaxies were young. After a few billion years, any material that has an orbit carrying it near the black hole will have been swallowed by the black hole or ejected by the process that produces the AGN jets. With no more matter feeding the black hole, it grows quiescent. However, each time the galaxy undergoes a collision with another galaxy, additional matter may end up in orbits that lead it close to the central black hole, reawakening the activity as a Seyfert or radio galaxy as the black hole grows in size.

A recent observation made with the Chandra X-ray space telescope has provided even more direct evidence of the supermassive black hole at the center of one galaxy. A dark cloud was observed as it passed in front of and eclipsed the accretion disk around the central black hole (Figure 77.10). From the timing of the eclipse, astronomers determined that the diameter of the accretion disk is only 7 AU. This is about 10 times larger than the event horizon of the black hole estimated to be in the center of this galaxy, which is about the size astronomers predicted for such an accretion disk.

Supermassive black holes at the centers of galaxies are not only a product of the way galaxies form, they also generate *feedback* that affects how galaxies evolve. Recent observations and computer simulations of AGN jets suggest that the furious winds they drive into a galaxy's interstellar medium may impede star formation. It is beginning to appear that there is a coevolution linking the growth of each galaxy's tiny black heart to its grander structure.

KEY TERMS

accretion disk, 633
active galactic nucleus (AGN), 629
jet, 630
quasar, 631
radio galaxy, 630
radio lobe, 630
Seyfert galaxy, 629
supermassive black hole, 635
synchrotron radiation, 630

QUESTIONS FOR REVIEW

1. What are the three main types of active galaxies?
2. In what wavelength ranges do active galaxies emit their energy?
3. What are radio galaxies?
4. What is the source of radio emission from active galaxies?
5. What is a quasar?
6. Why do some astronomers think that quasars are very distant?
7. How is it known that active galaxies have small core regions?
8. What mechanism has been suggested to power active galaxies?

PROBLEMS

1. If a quasar is generating 10,000 times more luminosity than a galaxy, how many times farther away would it be if it appeared to be the same brightness as the galaxy?
2. The radius of a black hole is approximately 3 km times the mass of the black hole in solar masses (Unit 68).
   a. Find the diameter in AU of a black hole with a mass of 1 billion $M_{\odot}$.
   b. For an accretion disk 10 times the radius of the black hole in (a), how many light-hours across would it be?
3. Hydrogen has a spectral line at 122 nm in the ultraviolet portion of the spectrum. What are the largest and smallest redshifts a quasar could have for this line to be seen in the visible portion of the spectrum (400 to 700 nm)?
4. An AGN is observed to double in brightness over seven days and then to grow dim again in the next seven days. What is an upper limit to the probable size of the AGN? Give your result in AU.
5. A quasar has a brightness that corresponds to apparent magnitude 15 (Unit 54.4). If it is 1 billion pc away, how many times brighter than the Sun is it? How does this compare to the brightness of the Milky Way?
6. The hydrogen line of gas orbiting in the nucleus of an active galaxy is Doppler shifted on the right side of the AGN to 123 nm, and on the left side to 121 nm. How fast is the gas rotating?

TEST YOURSELF

1. A spiral galaxy has a small bright central region, and its spectrum shows that it contains hot, rapidly moving gas. It is most likely a ____________ galaxy.
   a. barred spiral
   b. Seyfert
   c. radio
   d. dwarf
   e. quasar
2. What produces the synchrotron radiation seen in radio galaxies?
   a. High-speed electrons spiraling around the magnetic field lines
   b. Supernova explosions
   c. Visible starlight that has been heavily redshifted
   d. Hydrogen molecules
   e. Radiation from a massive black hole at the center of the galaxy
3. Because the large redshifts of quasars arise from the expansion of the universe, we can conclude that
   a. quasars must be very small.
   b. quasars must be within the Local Group.
   c. quasars must be single stars with extremely large masses.
   d. quasars must be moving toward Earth with a large radial velocity.
   e. quasars must be very luminous.

unit

78

Dark Matter

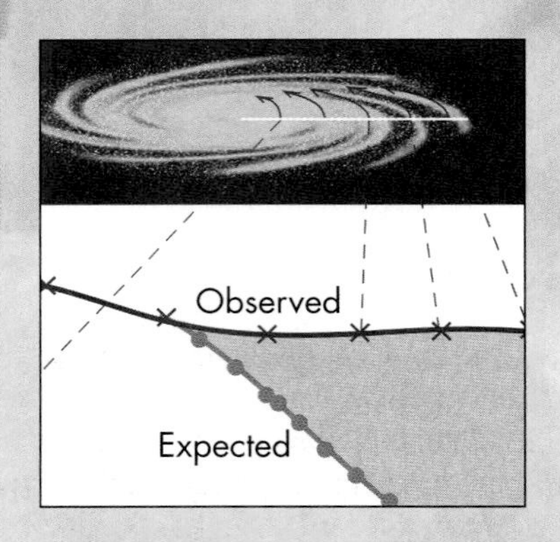

Background Pathways

Unit 73: Mass and Motions in the Milky Way 590

Unit 76: Galaxy Clustering 620

When we map the positions of galaxies throughout the universe, it is something like studying the lights seen from an airplane flying over a city at night. The pattern of lights suggests the general shape of things, but there is a whole landscape that can only be guessed at. Astronomers have begun to realize that galaxies hint at a very strange landscape, in which most of the matter in the universe is of a type unlike anything we know.

We can deduce the existence of this other kind of matter because it exerts a gravitational force on the things we see, such as stars and galaxies, even though it is not seen in visible light. Initially astronomers labeled this as **dark matter** to indicate that they knew mass was present but that it emitted no light. But a growing body of evidence says that it is a kind of matter that in fact *cannot* be seen. This unusual matter has mass that produces gravity, but it cannot coalesce into stars. It may be made of particles that do not interact with photons, and therefore it may be all around us without being detected.

It is interesting to consider how far we have moved from our Earth-centered view of the universe in our exploration of galaxies. We have learned that the Sun lies in the outskirts of a galaxy, which is not a particularly significant galaxy, which is in a minor cluster of galaxies. And now we are realizing that the kind of matter that makes up everything we know is just a minor kind of matter in the universe. This is the Copernican revolution taken to extremes! The idea that dark matter dominates the universe is revolutionary, and revolutionary ideas require exceptionally strong evidence. But the evidence that has accumulated in favor of this idea is so strong that it is now widely accepted among astronomers. In the following sections we will see how astronomers have come to this conclusion. But before we describe the search for dark matter, it will be helpful to review how astronomers measure a galaxy's mass.

78.1 MEASURING THE MASS OF A GALAXY

We saw in Unit 73 that when astronomers use orbital motion to measure the mass of the Milky Way, they find a puzzling discrepancy between the mass they calculate and the mass they can account for among the stars and other material they detect. The discrepancy in the Milky Way is difficult to quantify in some ways, however. Because of our location within the disk, dust blocks our view of many stars and distances can be difficult to determine.

We can measure the masses of other galaxies the same way we measure the masses of stars or planets: by applying the modified form of Kepler's third law (Unit 17). This method uses the principle that the orbital motion of one object around another is set by their mutual gravity. Their gravity is in turn determined by their combined mass. Thus, from the orbital motion of objects we can find their masses.

Studies of other galaxies reveal that the discrepancy found in the Milky Way is a much more universal phenomenon. Almost without exception, the mass calculated from the modified form of Kepler's law is much larger than the mass of stars and interstellar matter detectable by optical, infrared, and radio telescopes.

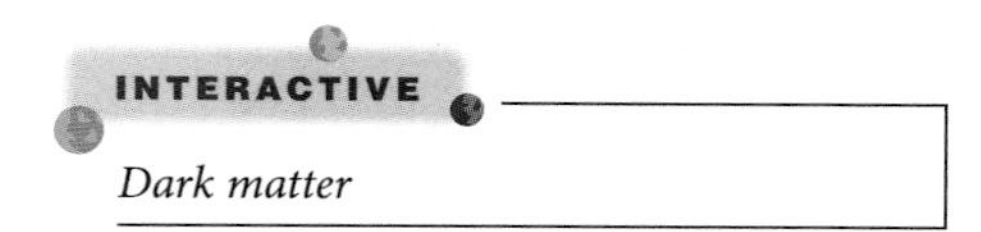

This discrepancy is very large, typically amounting to 10 times a galaxy's visible mass. This means that when we look at the starlight from a galaxy, we are observing only one-tenth of its matter.

The most direct evidence for dark matter comes from **rotation curves** of spiral galaxies. A galaxy's rotation curve is a plot of the orbital velocity of the stars and gas moving around it at each distance from its center (curve A in Figure 78.1). A rotation curve reveals how a galaxy's mass is distributed because the speed at each radius from the galaxy's center reveals how much mass must lie interior to the orbit at that distance (Unit 73). We can understand how that dependence arises by thinking about what keeps a star in orbit.

According to Newton's first law, if an object does not move in a straight line (for example, if a star moves in a circular orbit around a galaxy), a force must be acting on it. We know that for a star orbiting in a galaxy, that force is supplied by gravity. We also know that the strength of the gravitational force depends on the mass of the attracting body (in this case, the galaxy). It turns out, however, that to a good approximation, only the material inside of the star's orbit contributes to the net gravitational force. If the speed of rotation is V at radius R, we can write this as an equation for the mass interior to that radius as in Unit 73:

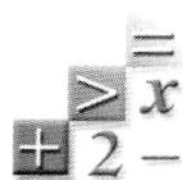

$M_{<R}$ = Mass within radius R (kilograms)

R = Radius (meters)

V = Velocity of rotation (meters per second)

$G = 6.67 \times 10^{11}\ \mathrm{m^3/kg \cdot sec^2}$

$$M_{<R} = \frac{V^2 \times R}{G}$$

where the subscript "$<R$" is used as a reminder that the measurement represents the mass interior to this radius.

Stars near the center of a galaxy will therefore have only a small net force acting on them, because there is little mass between them and the center. Because the force acting on them is small, such stars do not orbit very rapidly. On the other hand, stars orbiting very far from the center should also feel a small gravitational force acting on them from the galaxy. They feel the effect of the galaxy's full mass, but the gravitational force on them is nevertheless weak because the force of gravity grows rapidly weaker at greater distances from a mass. Therefore we would predict that the outermost stars should also be orbiting slowly if they are not to fly out of the galaxy. Detailed mathematical models bear out this analysis and indicate that a galaxy's rotation curve should start at small velocities near the core. The curve should then rise to higher velocities at middle distances and finally drop to lower velocities again far from the center, as indicated in curve B of Figure 78.1.

However, when astronomers measure the rotation curves of galaxies, they do not find this behavior. In work pioneered by the U.S. astronomer Vera Rubin, it was discovered that the rotation curves do rise near the center, but they then flatten out and almost never drop off to low velocities again. These **flat rotation curves** are found in nearly all spiral galaxies studied, as shown by a number of examples in Figure 78.2. Rotation curves have been studied for many galaxies, to distances far beyond those we have been able to study in the Milky Way, and those galaxies maintain flat rotation curves out to radii exceeding 50 kiloparsecs.

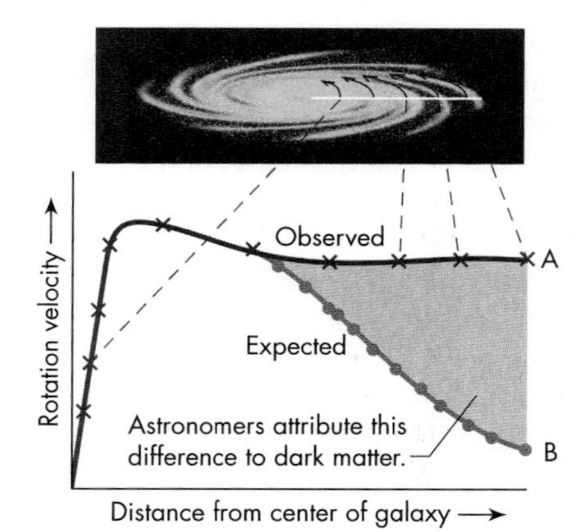

FIGURE 78.1
A schematic galaxy rotation curve. The line with dots represents the curve expected if the galaxy's mass comes only from its luminous stars. The line with crosses represents the observed curve, implying that the galaxy contains dark matter.

A constant rotation speed in the mass formula implies that the mass climbs steadily at larger radii. This is because with a constant value of V, the mass formula indicates that the interior mass rises in proportion to the radius R. For example, the amount of matter measured out to 10 kiloparsecs from a galaxy's center might be found to be $10^{11}\ M_\odot$. Then out to 20 kiloparsecs the constant velocity implies there is $2 \times 10^{11}\ M_\odot$; out to 30 kiloparsecs there is $3 \times 10^{11}\ M_\odot$; and so forth. The amount of mass measured rises steadily, the farther out we look.

This can be contrasted with a galaxy's light curve. In a typical spiral galaxy, about 90% of the light comes from the inner 10 kiloparsecs, while 99% is contained within about 20 kiloparsecs. The number of stars in the outer regions declines very rapidly, so stars cannot contribute enough additional mass at larger

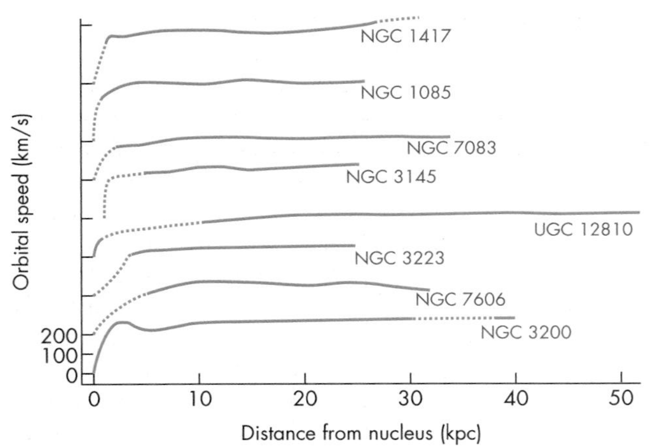

FIGURE 78.2
The rotation curves of an assortment of spiral galaxies. The curves shown are offset vertically for clarity. Each rises to a rotation velocity of about 200 km/sec and remains there to the limits of the observations.

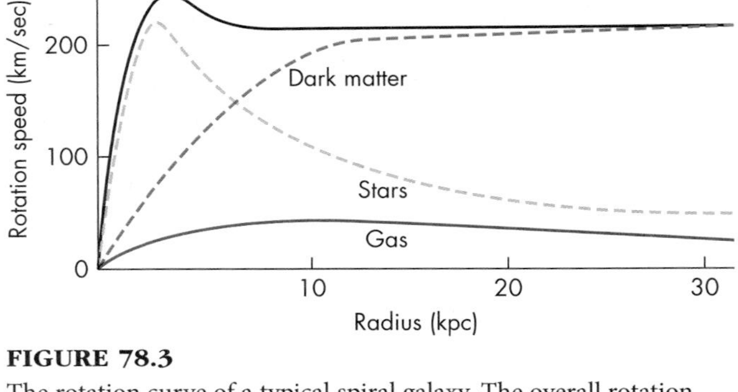

FIGURE 78.3
The rotation curve of a typical spiral galaxy. The overall rotation speed is shown in black. The amounts of rotation produced by the detected stars and gas are shown. Dark matter becomes dominant beyond several kiloparsecs from the center of the galaxy.

radii to explain the flat rotation curve. Astronomers have hunted for other components, such as interstellar gas, which often extends out farther than the stars, but this also provides too little mass. The predicted rotation speeds based on stars and gas alone are illustrated in Figure 78.3.

Analyzing elliptical and irregular galaxies is more complex because the stars and gas do not orbit in circles. However, a similar mass formula with modifications for random orientations of the velocities can be applied. Again, the results indicate that stars alone cannot account, by about a factor of 10, for the orbital speeds observed.

What can explain such a major discrepancy between theory and observation? If our laws of physics are correct, only one thing can explain such behavior: Galaxies must contain large amounts of unseen mass. This unseen mass appears to be present throughout a galaxy, but its presence is particularly evident in the outermost parts of the galaxy where its huge cumulative mass exerts a gravitational force that holds the stars in orbit despite their large velocity. We deduce that what we describe as a galaxy is a small pocket of "normal matter" embedded in a large **dark matter halo** about 10 times more massive than the visible galaxy.

Q If there is the same amount of dark matter between radii of 10 and 20 kpc as between 40 and 50 kpc, what does that imply about the *density* of the dark matter at different radii?

The term *halo* can be a little confusing because it sometimes makes people think of a ring. The dark matter is actually distributed throughout a galaxy. The rotation curve studies imply that the dark matter density is highest at the center of the galaxy. Nonetheless, it represents only a fraction of the mass in the inner parts of a galaxy where the stars and gas there are packed tightly. At larger radii, however, the mass of stars and gas drops off rapidly, while the dark matter content declines only gradually. In the outermost parts of the galaxy, dark matter is left as the primary component.

78.2 DARK MATTER IN CLUSTERS OF GALAXIES

Galaxies in clusters often have speeds of more than 1000 kilometers per second with respect to one another. These great speeds reflect the enormous mass present in the cluster, pulling whole galaxies into high-speed orbits. If we add up the masses of all the galaxies in a cluster, we come up far short of what is needed to hold the cluster together. Just as the stars within a galaxy orbit too rapidly for the gravitational force that can be attributed to its luminous mass, so too the galaxies in a cluster have speeds too large to be kept from flying away by the observed mass.

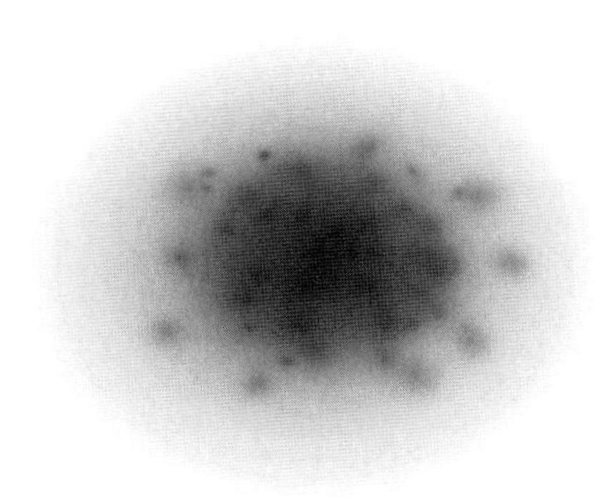

FIGURE 78.4
Artist's depiction of what the dark matter distribution might look like in a galaxy cluster. Individual galaxies reside within each smaller dark matter "halo," but dark matter is also spread throughout the cluster.

This stark discrepancy was discovered in the 1930s by the Swiss astronomer Fritz Zwicky. Once again, applying the mass formula, we can estimate the mass interior to any radius in the cluster by measuring the speeds of the galaxies at that radius. The discrepancy is immense—a typical cluster must have a mass more than 100 times larger than would be predicted from its combined starlight. Galaxy clusters, like galaxies themselves, must contain huge amounts of dark matter.

For decades astronomers studied clusters at other wavelengths to search for matter that might explain the discrepancy. Hot intergalactic gas was one possibility, and X-ray telescopes do reveal the presence of a lot of such gas in clusters (Unit 76). The amount of mass estimated to be within this hot gas is actually larger than that contained within the galaxies' stars, but it is still only about a tenth of the amount of mass predicted from the galaxies' orbital speeds.

Hot gas actually provides additional evidence for the dark matter in galaxy clusters. To hold the hot gas within the cluster and prevent its expansion, the cluster must exert an inward gravitational pull on the gas that matches the pressure within the gas. (This is the same condition of *hydrostatic equilibrium* that pertains to our Sun and other stars—Unit 49.) The gravity needed to confine the gas depends on the cluster's total mass; and the calculated mass is consistent with the mass calculated from the motions of the galaxies.

Combining the evidence for dark matter in galaxies and galaxy clusters, it appears that these systems contain about 10 times more dark matter than any kind of normal matter that is detectable via electromagnetic emission. The dark matter appears to be distributed relatively smoothly within these systems; it is most dense toward the center of each galaxy as well as at the center of a cluster, but it extends out much farther than the galaxy's stars and other matter. The picture that emerges is that if we could see dark matter, it would look something like the artist's depiction in Figure 78.4. The stars and gas that we can detect are concentrated in the regions where the dark matter is most dense.

78.3 GRAVITATIONAL LENSES

Dark matter may not be observable by electromagnetic radiation, but it can be detected in a different way. It not only affects star and galaxy orbits, but it can change the path of light through space. This gravitational bending of light is one of the predictions of *general relativity,* Albert Einstein's theory of gravity (Unit 68).

The bending of light was detected soon after Einstein developed his theory of gravity, and in 1937 Zwicky suggested the possibility of a **gravitational lens.** A gravitational lens would focus the light from an object behind it, making background objects look brighter. A gravitational lens does not require dark matter in principle, but generally the masses of galaxies and galaxy clusters without it were known to be too small to produce detectable lensing effects. As a result, the idea was mostly forgotten for several decades.

Then, in the late 1970s, a quasar was discovered that seemed to have a nearby companion quasar with an essentially identical spectrum but with a slightly different brightness and shape. The existence of two so nearly identical quasars so close together is extremely improbable. Today several dozen such quasar pairs and even "quintuplets" are known (Figure 78.5A). They are not actually companions. Instead, they are images of a single quasar created by a gravitational lens.

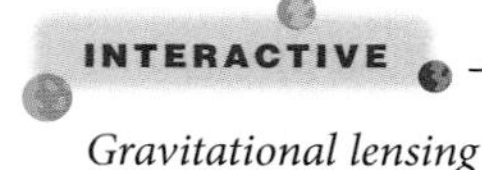

INTERACTIVE

Gravitational lensing

An ordinary lens forms an image because light bends as it passes through the lens's curved glass. A gravitational lens forms an image because light bends as it passes through the curved space around a massive object such as a galaxy. The galaxy's gravitational force bends the space around it so that light rays that would

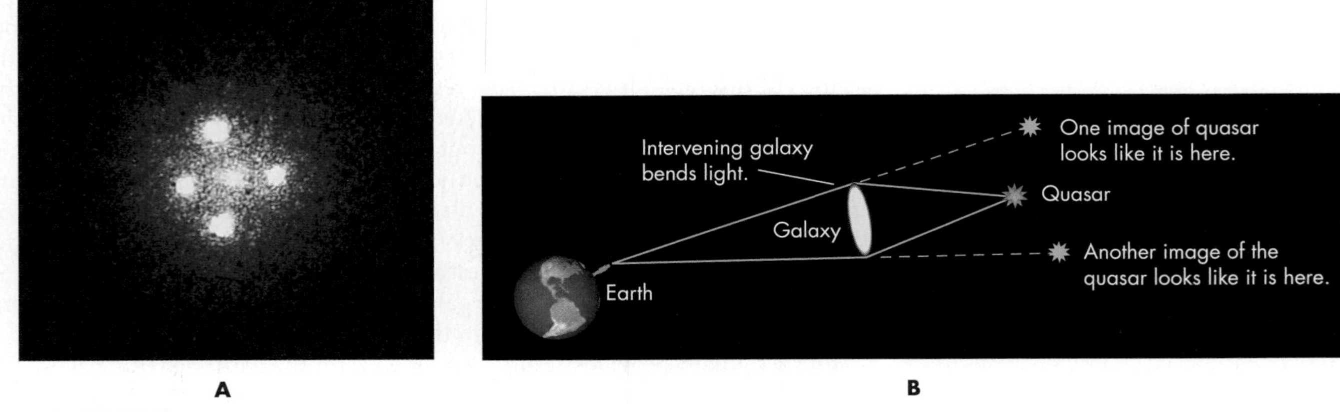

FIGURE 78.5
(A) An Einstein cross, a complex image of a quasar created by a gravitational lens. (B) Sketch of how a gravitational lens forms an image.

FIGURE 78.6
The arcs in this picture of the galaxy cluster Abell 2218 are the images of distant galaxies distorted by the gravitational lens effect created by the galaxy cluster. (One of the arcs is quite red, visible to the left of the center of the image.)

otherwise travel off in other directions and never reach the Earth are bent so that they do reach us, as depicted in Figure 78.5B.

Because the matter is distributed unevenly, a gravitational lens is similar to a magnifying glass made of wavy glass. It produces magnified but distorted images of background objects. For example, Figure 78.6 shows the arcs created when light from a distant galaxy was bent into elongated shapes as it passed through a galaxy cluster. The shape of these arcs depends on the mass of material within the cluster, allowing astronomers to measure the cluster's mass. The strong bending of light that is seen requires large amounts of dark matter, confirming the high mass implied by the high speeds of the galaxies.

Measurements of gravitational lensing provide strong confirmation of the dark matter hypothesis. It also appears to rule out some other possibilities that astronomers had previously considered. For example, one hypothesis argued that the rapid motions in clusters did not indicate the amount of mass present because the clusters were not held together by their gravity, but were instead expanding.

FIGURE 78.7
Dark matter (the blue blobs) revealed in the collision of two galaxy clusters. The image has been made by superposing three images: a picture made at optical wavelengths (showing the galaxies in the two clusters); an image made at X-ray wavelengths (showing in red the hot gas in each cluster); and a map of the dark matter in the two clusters (the blue blobs) as deduced from the gravitational lensing of the background galaxies. In the collision, the dark-matter blobs have passed through each other because the dark matter interacts with itself and other matter only gravitationally. The gas in the two clusters, on the other hand, collides and has clumped up in a bullet-like form. The merged cluster is officially known as 1E 0657–56, but more popularly as the *Bullet Cluster.*

Another hypothesis suggested that the laws of motion developed by Newton behaved differently at the large distances in galaxy clusters, so the formula we used for finding the mass would not apply. Neither of these hypotheses predicted the large bending of light seen. The agreement of the light-bending measurements with other results shows that there is in fact an immense amount of unseen mass within the clusters.

Recent observations of a pair of galaxy clusters that have collided with each other provide further insights into the characteristics of dark matter. When clusters collide, the gas from the two clusters slams together and heats up to millions of Kelvin and emits X-rays. Astronomers observed this hot gas with the Chandra X-ray telescope in the "Bullet Cluster" in 2006, allowing them to trace where the gas was located after the clusters collided. This is shown in red in Figure 78.7.

The dark matter is, of course, invisible, but its location can be mapped by its gravitational lensing effect on the light of background galaxies. The derived distribution is shown in blue. Unlike the gas, the separate masses of dark matter in the two clusters simply passed through each other as the clusters collided. This is similar to the way that stars pass by each other when galaxies collide (Unit 75.4), yet the bulk of the known matter in clusters is in the hot gas. Thus, dark matter must be something that produces no light and that does not interact with other matter in collisions. Astronomers have described the dark matter as behaving like a "collisionless gas," although it challenging to determine whether it is a "gas" made up of particles as massive as stars, or as small as subatomic particles.

78.4 WHAT IS DARK MATTER?

What could this dark "collisionless gas" be? Astronomers have ruled out normal gas because it does collide, and searches with radio telescopes would have revealed cold gas, and optical, radio, or X-ray telescopes would have revealed hot gas.

We know that stars in galaxies are so widely separated that they can travel through a galaxy with little likelihood of colliding. We might speculate then that

Q Suppose dark matter is made of old white dwarfs and neutron stars. What would this predict about the amount and kinds of stars that formed early in the Milky Way's history?

Q Can planets around other stars be detected with the lensing technique?

dark matter is composed of vast numbers of "dark stars" that add up to all of the mass that we do not see. However, astronomers can rule out ordinary dim stars because such objects would emit at least some detectable infrared radiation. Even the very dim brown dwarfs (Unit 69) have been ruled out in the Milky Way's halo by deep observations with infrared-sensitive telescopes.

It is possible to imagine other objects whose radiation would be much more difficult to detect, such as planet-sized bodies, very old and cold dead white dwarfs, old neutron stars that are no longer pulsars, or black holes. This idea is sometimes called the **MACHO (massive compact halo object)** hypothesis. This would require some peculiar sort of star formation scenario in which the formation of all normal long-lived stars was suppressed when galaxies were young, leaving behind an enormous population of dark objects.

Astronomers are currently searching for MACHOs within the Milky Way's halo using the gravitational lensing technique. Although the most dramatic gravitational lenses are created by massive galaxy clusters, astronomers have developed techniques sensitive enough to detect lensing caused by a single star or even a planet (Unit 34.4). If a dark MACHO in the Milky Way's halo passes in front of a background star as seen from Earth, the gravitational force of the MACHO will bend the light of the background star and focus it, brightening the light we see.

The probability of one star aligning closely enough with another to produce a measurable brightening has a low probability, so astronomers have set up telescopes to monitor the light from millions of stars within the Magellanic Clouds, the small companion galaxies of the Milky Way (Unit 76.1). If a MACHO passes in front of one of those stars, it focuses the star's light for several days until it moves out of alignment. The brightening caused by lensing follows a known pattern, and several dozen massive objects have been detected in this way.

Even though the foreground objects are themselves too dim to see, astronomers can estimate their mass from the lensing effect they create. Moreover, from a statistical analysis of the number of such events, astronomers can deduce the number of objects creating the lens effect. From the evidence so far available, the objects creating the lens effect are primarily dim, low-mass stars rather than a new population of dark objects. It does not appear that MACHOs can account for more than about 10% of the halo mass.

In the novels of Sherlock Holmes, the fictional detective often states that "when you have excluded the impossible, whatever remains, however improbable, must be the truth." The nature of dark matter appears to be one of these improbable truths. Having excluded objects made of normal kinds of matter, the prevailing hypothesis today is that dark matter is entirely unlike the matter we are familiar with.

Consider, for example, how difficult it is to detect neutrinos (Unit 4). Neutrinos can pass through the Sun and Earth, almost as if they were totally transparent. Evidence from studies of solar neutrinos (Unit 50) indicates that neutrinos have mass, but this mass is so tiny that neutrinos cannot themselves account for the dark matter. However, some physicists think there may be more massive particles like the neutrino, that also interact very weakly with normal matter. This has been dubbed the **WIMP (weakly interacting massive particles)** hypothesis. Such particles are proposed in various models of particle physics, and physicists around the world have designed experiments to detect them.

If dark matter is made of WIMPs, then it would be more accurate to describe a galaxy as a dense region of these particles. After all, the rotation curve measurements indicate that the dark matter represent about 90% of a galaxy's mass. The normal matter—the stuff that stars, planets, and we are made of—was probably attracted by the gravity of this dark matter and fell into the region, collecting at the center. Normal matter passes through this sea of WIMPs, registering its presence only by the effect its mass has on the matter we can detect. Hence, WIMP particles would be present within the Solar System, but their density is so low as to be virtually undetectable.

The WIMP hypothesis and other similar hypotheses, based on currently undetected particles, are the leading ideas for explaining dark matter. However, the case is not yet settled, and the nature of dark matter remains one of the central mysteries in astronomy today.

KEY TERMS

dark matter, 638
dark matter halo, 640
flat rotation curve, 639
gravitational lens, 641
massive compact halo object (MACHO), 644
rotation curve, 639
weakly interacting massive particle (WIMP), 644

QUESTIONS FOR REVIEW

1. How do we use gravity to measure the mass of a galaxy?
2. What is a galaxy rotation curve?
3. What is meant by dark matter?
4. How do we measure the mass of clusters of galaxies?
5. What is a gravitational lens?
6. How do we know that dark matter is not made of stars or gas?
7. What are MACHOs and WIMPs?

PROBLEMS

1. Suppose we measure two galaxies to the farthest extent that any matter can be detected. One has a rotation velocity of 300 km/sec at 4 kpc, and the other has a rotation velocity of 150 km/sec at 20 kpc. Which has the larger measured mass? What is the ratio of those masses?
2. What is the mass-to-light ratio of a galaxy with $10^{11}\ M_\odot$ in stars and gas and $5 \times 10^{11}\ M_\odot$ of dark matter?
3. A spiral galaxy is observed in which the rotation speed remains 250 km/sec from 1 kpc out to the largest distances at which we can detect anything. How much mass exists out to 10 kpc? 11 kpc? 12 kpc? Can you generalize your results as to how much additional matter there is for each kiloparsec farther out you look?
4. In the Virgo Cluster there are galaxies measured to be traveling at about 1000 km/sec at a distance of 500 kpc from the center of the cluster. If you apply the mass formula to this velocity and radius, what interior mass do you find?
5. Suppose the Milky Way contains 10 billion $M_\odot$ of dark matter at radii between 8 and 9 kpc from its center.
   a. What is the density of matter in the "shell" between the radii of 8 and 9 kpc? (*Hint:* Subtract the volume of a sphere of radius 8 kpc from the volume of a sphere of radius 9 kpc.)
   b. How much mass of dark matter would there be in a spherical volume with a radius equal to the Earth's radius? Compare your result to the mass of the Sun.
6. Most spiral galaxies show a decline in the amount of light they produce that drops by a factor of 2 every 2 kpc farther out. For example, $3 \times 10^{10}\ L_\odot$ are produced within 2 kpc of the center; $1.5 \times 10^{10}\ L_\odot$ are produced between 2 and 4 kpc of the center; half as much again between 4 and 6 kpc; and so on. Compare the amount of luminous matter found this way to the total amount of matter contained within these same regions if the galaxy has a flat rotation curve with a rotation speed of 250 km/sec. Find the ratio of dark matter to luminous matter within 2 kpc of the center; between 2 and 4 kpc from the center; between 10 and 12 kpc from the center.

TEST YOURSELF

1. Astronomers think that dark matter exists because
   a. they can detect it with radio telescopes.
   b. the outer parts of galaxies rotate faster than expected on the basis of the material visible in them.
   c. the galaxies in clusters move faster than expected on the basis of the material visible in them.
   d. it is the only way to explain the black holes in quasars.
   e. Both (b) and (c) are correct.
2. Most of the mass of a galaxy is contained in
   a. the massive O and B stars in the galaxy.
   b. the cold interstellar gas.
   c. the central black hole of the galaxy.
   d. the dark matter of the galaxy.
   e. the disk of the galaxy.
3. Gravitational bending of light does *not*
   a. show that space can be curved.
   b. give the illusion that quasars have companions.
   c. allow astronomers to detect MACHOs.
   d. provide a means to measure the mass of galaxy clusters.
   e. produce the appearance of constant rotation speeds in spiral galaxies.

unit

79

Cosmology

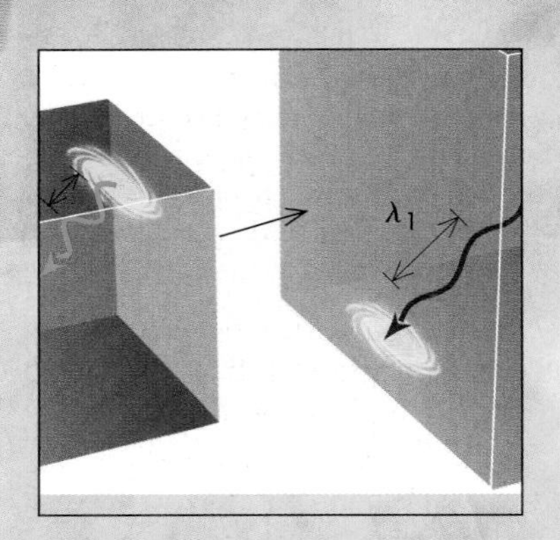

Background Pathways

Unit 74: A Universe of Galaxies 599

Cosmology is the study of the structure and evolution of the universe as a whole. Cosmologists ask: Is the universe infinite? Does it have an edge? Has it existed forever, or does it have a definite age? How did it form? What will happen to it in the future? Given our insignificant size in the cosmos, such questions may seem futile or even arrogant, but most cultures have tried to answer them. Many of the attempted answers have become part of humanity's religious heritage. In cosmology we address some of these same questions based on astronomical evidence and the scientific method. It is important to set aside personal beliefs when examining what the weight of scientific evidence indicates.

We have made many surprising discoveries. Current evidence indicates that the universe was born about 13.7 billion years ago out of a hot, dense, violent state of matter and energy called the **Big Bang.** It has been expanding ever since, and it is filled with radiation from the early stages of the explosion. That radiation carries an imprint of information about the earliest stages of structure formation that we will describe in Unit 80. Within the last few decades, cosmologists have begun to extend our knowledge of the universe to the very beginning of time. They have discovered that the Big Bang may have been born out of even more turbulent events known as the *inflationary stage* (Unit 81), when the entire universe that we see today may have fit in a volume smaller than a proton! Our cosmological models even allow us to predict the ultimate fate of the universe (Unit 82). Understanding how the evidence leads us to such remarkable deductions is the goal of this and the next three units.

79.1 EVOLVING CONCEPTS OF THE UNIVERSE

Over the centuries, our understanding of the universe has steadily changed. For millennia the idea of a central, stationary Earth seemed natural—a moving Earth seemed absurd, both to scientists and to others expressing their common sense. As concepts such as inertia (Unit 14)developed, the idea of a moving Earth no longer seemed physically impossible, and by the early 1600s the idea of planets orbiting the Sun became plausible. Such a model explained the planet's motions and other observed phenomena more simply, although it required a break with common beliefs.

At the time this idea was not yet supported by strong evidence—such as the predicted parallax effect—and therefore it was not accepted by all. Some astronomers continued to work with a geocentric model, which was still taught in many textbooks until the early 1700s. However, additional evidence accumulated, such as the aberration of starlight and parallax of stars (Unit 52), that so strongly contradicted the old Earth-centered model that the model was finally abandoned.

Even as the debate over geocentric and heliocentric models continued, others began proposing even more complex cosmologies that recognized the Sun as one of many stars. One of the most commonly accepted cosmologies was that the universe consisted of the stars of the Milky Way with the Sun near its center, and that this island of stars was surrounded by an empty void. An alternative idea was proposed by the German philosopher Immanuel Kant in the mid-1700s, anticipating some of our current ideas. He suggested that the Milky Way was merely one of a multitude of galaxies. This idea was not widely accepted until the 1920s, when astronomers could demonstrate that galaxies were millions of light-years from the Milky Way and comparable to it in size (Unit 74).

The pace of changes in our models of the universe has grown more rapid in the last century. Even as we were just learning to accept the idea that the universe contains billions of galaxies, the Milky Way being just one of them, observations revealed a strange phenomenon—that the other galaxies are moving away from us at high speeds (Unit 74.3). Edwin Hubble first demonstrated this in the 1920s, and it meshed with ideas about the nature of space and gravity developed by Albert Einstein around the same time. This new picture, of an expanding universe, had implications for how the universe must have begun.

The extraordinary conclusions we have reached about the universe today are rooted in observations that began as we asked simple questions about how the things we see came to be, followed a train of reasoning, and then pursued observations that led to the present model. Given the history of changing ideas about the universe, it needs to be stressed that we are speaking of models of the universe, rather than certain knowledge about a clearly defined "Universe," which we might write with an uppercase "U." This is *not* to say that the scientific evidence supporting modern cosmology is weak, but rather that our understanding of what the entirety of the universe comprises continues to evolve and be refined.

The scientific approach requires that we weigh the evidence and work with the best current model, *and* that we accept that the model may change as new evidence accumulates. In addressing the fundamental questions about the universe, it would be unscientific to work toward "proving" one model of the universe. This is a subtle point: Individual astronomers do attempt to argue the merits of particular models, but as scientists they accept that their models may be inaccurate or incomplete, and that a model should not be selected because of personal beliefs.

79.2 THE RECESSION OF GALAXIES

Nearly all galaxies have a redshift of their spectral lines. Astronomers in the early 1900s interpreted this as meaning that most of them are moving away from us (Unit 74). Hubble showed that the redshift of a galaxy increases with distance, d, in such a way that the recession velocity, V, of a galaxy obeys Hubble's law:

$$V = H \times d$$

where H is Hubble's constant.

The nature of this equation is startling. It seems to say that all the galaxies in the universe are flying away from us, making it appear as if the Milky Way is at the center of the universe—that a vast explosion (the Big Bang) has sent galaxies flying away from us in all directions. But this is incorrect.

A simple example illustrates why the Milky Way's position is not special. Imagine a string of galaxies distributed evenly along a line, as shown in Figure 79.1. Each galaxy is separated from its nearest neighbor by 10 megaparsecs. From the galaxy marked as the Milky Way in the figure, we would observe that each galaxy has a recession velocity given by Hubble's law: $V = (70 \text{ km/sec per Mpc}) \times (\text{distance})$. That is, A recedes from us at 700, B at 1400, and C at 2100 km/sec. Next, suppose we could communicate with an alien in galaxy C and ask what it sees. It would

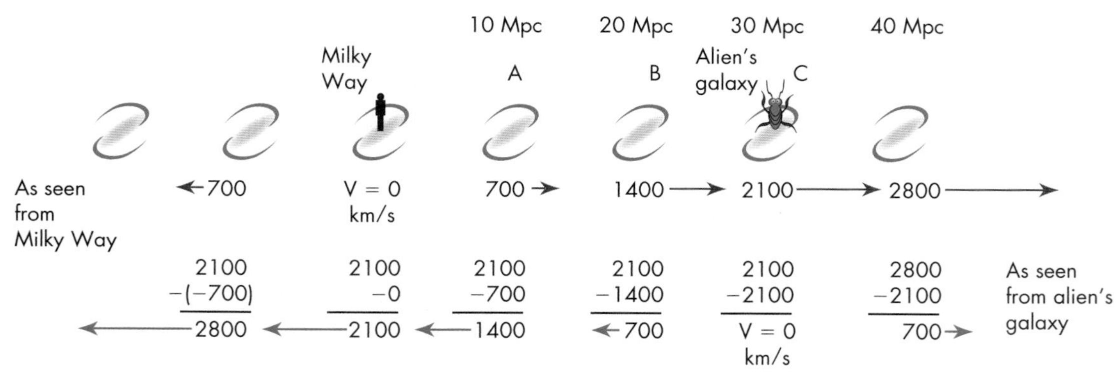

FIGURE 79.1
A line of galaxies illustrating that in a universe that obeys a Hubble law of expansion ($V = H \times d$), an observer in any galaxy sees the same law.

see us receding at 2100 km/sec, galaxy A receding at 2100 − 700 = 1400 km/sec, and galaxy B receding at 2100 − 1400 = 700 km/sec. In fact, the alien would see galaxies receding from it in exactly the same way as we see galaxies receding from us. Furthermore, the galaxies seen by the alien will obey precisely the same form of Hubble's law.

Q If astronomers calculated Hubble's constant when the universe was half its current age, would the value they found for *H* be the same?

A similar argument can be made for an observer in any of the galaxies along this line. Each will see all of the other galaxies flying away from it. This argument can be extended to two or three dimensions, and in each case for whatever galaxy we imagine being "home," we will observe all the other galaxies "flying away" from us with speeds that agree with Hubble's law. A universe obeying Hubble's law has no preferred "center" of expansion. Astronomers describe this lack of a preferred location as the **cosmological principle.** It is a statement of cosmic modesty—an extension of the Copernican revolution (Unit 11.4): There are no special positions in the universe. *We* are not at the center of the universe, nor is anyone else.

If there is no central location from which galaxies are moving away, then it does not make sense to think of the universe's expansion as being like the explosion of a bomb, sending fragments in all directions from a central blast point. But how then *do* we explain this motion? Einstein's **general theory of relativity** offers an answer.

Einstein developed general relativity in 1916, just a few years before the nature and motion of galaxies were discovered. He was trying to solve other puzzles having to do, for example, with gravity and the way it affects light (see Unit 68). The resolution of these problems led him to develop the idea that space can itself have motion.

The idea that space can move is alien to us, and it may even seem a bit silly. How can "nothing" be moving? But this idea is needed to explain many observed phenomena. We can detect changes in the arrival of light signals from galaxies that seem to imply that the light has changed speed as it moved through space. Yet we can measure the speed of light extremely accurately, and every experiment shows that it moves through space at a constant speed. The explanation for the changes in arrival time is that the light traveled through a region where the space itself was moving relative to us. Its arrival time is advanced or delayed much as an airplane's arrival time may be affected if it is traveling through air currents that are moving toward or away from us—even though the airplane maintained a steady speed relative to the air throughout its trip.

As remarkable as it may seem, the more distant galaxies we can see are moving away from us at greater than the speed of light! More properly we should say it is the region of space that those galaxies are in that is moving so fast. For

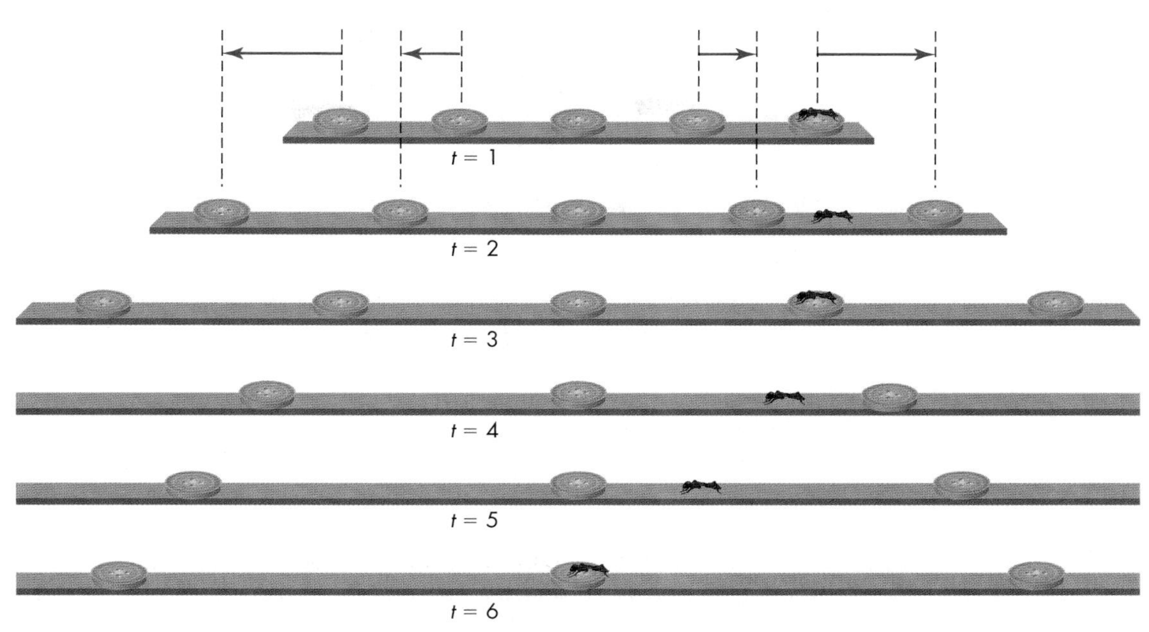

FIGURE 79.2

Galaxies in expanding space are like buttons on a stretching rubber band. An ant walking at a steady speed on the rubber band is like a photon traveling through expanding space: It moves at a speed relative to a distant button that is a combination of its own speed and the speed of the space it is traveling through at that moment.

One effect of the stretching of space is that distant galaxies look much bigger than we might expect. In the example given, the galaxy has an angular size corresponding to what we would expect at 4 billion light-years instead of being several times smaller if it were seen at 12 billion light-years distance and space were not expanding.

example, the most distant galaxies we have detected are currently moving away from us at about twice the speed of light, and they were moving away even faster at the time the light was emitted. At the time they emitted the light we are currently receiving, they were less than 4 billion light-years from the Milky Way, but because of the motion of space, the light has taken over 12 billion years to finally reach us.

How can we understand these bizarre findings? Returning to the example of the galaxies located along a line (Figure 79.1), one way we might picture the idea of space having motion is to think of the galaxies as buttons strung along a rubber band, as shown in Figure 79.2. If the rubber band is steadily stretched, each button remains stationary relative to the rubber band, but the space *between* the buttons expands. Buttons that are farther apart on the rubber band move away from each other faster because there is more of the stretching rubber between them—following a Hubble-like law.

The effect of expanding space on a photon moving through space can be illustrated by an ant walking along the rubber band in our analogy. The ant starts walking from a button on one end of the rubber band with its ultimate destination being the button middle, but what was initially a small distance grows steadily larger. Because the rubber band is stretching faster than the ant can walk, the ant is at first carried farther from its goal by the stretching band. As it keeps walking, though, the ant reaches parts of the rubber band that are not moving away from its target so quickly, and then it makes more rapid progress toward its goal.

A photon traveling through expanding space between galaxies is like the ant. Even though the photon is traveling steadily at a constant speed c through space, it makes slower progress than we would expect if space were motionless. In motionless space, a photon traveling from one galaxy to another would need to traverse only the original distance separating the two galaxies. This would be like the ant in Figure 79.2 hopping off the rubber band and walking on a tabletop to reach

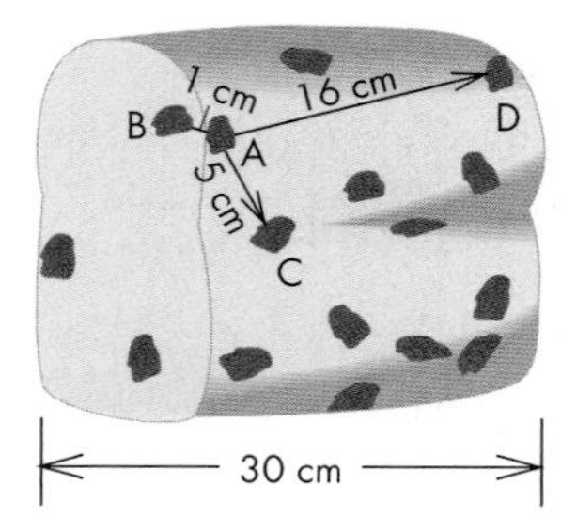

Raisin bread dough before rising

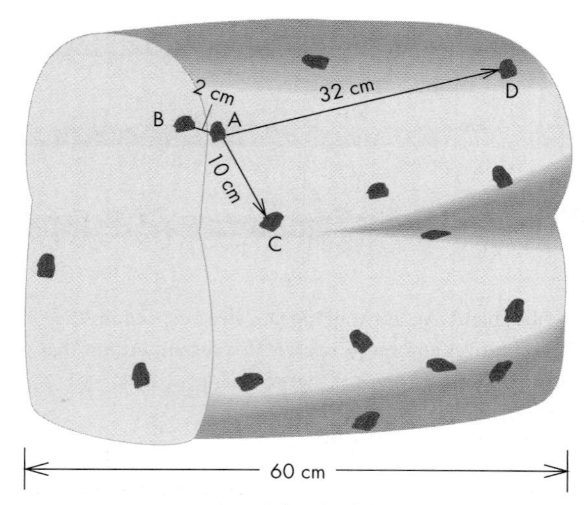

Raisin bread dough after rising

FIGURE 79.3
As a loaf of raisin bread dough rises, the raisins move farther apart from each other at a rate proportional to their separation.

the "stationary" middle button—and it would get there much more quickly.

The rubber-band analogy can be extended to two dimensions by imagining a stretching rubber sheet with the buttons glued to it. A three-dimensional analogy would be the relative motion of raisins in the dough of rising raisin bread (Figure 79.3). The separations between buttons on the rubber sheet or raisins in the dough grow because of the expansion of the medium in which they are set.

However, these analogies present a problem. The rubber band, rubber sheet, and the dough all have edges. Based on these analogies, then, we would expect there to be some galaxies close to the edge of the universe; if we lived in such a galaxy we would see galaxies in one direction but not the opposite. Because we do not see such an imbalance from the Milky Way, does this imply that our location is close to the center of the universe after all? We can avoid a conflict with the cosmological principle that this might imply by hypothesizing that the universe is infinite, or at least so large that none of the edges is anywhere close to being visible to us. (We will examine some alternative possibilities in Unit 80 by using the idea that space can be curved.)

In a similar fashion as these analogies, general relativity predicts that space itself is expanding, and galaxies are carried apart by that expansion, not by their own motion *through* space. However, although space expands, matter is not "glued" in place like the buttons or raisins in our analogies. Galaxies, stars, planets, and other bodies can move within the expanding space. In fact, the force of gravity that attracts them toward other nearby objects is often strong enough to overcome the expansion of space that would otherwise carry them apart. As a result, these objects can move toward each other despite the expansion of space, gathering into clusters, galaxies, and solar systems. Finally, there is no contradiction between general relativity and special relativity because although a region of space may be moving away from us at greater than the speed of light, no object can travel through space at a speed greater than c.

79.3 THE MEANING OF REDSHIFT

The expansion of space affects more than just motion. It also produces an observable effect on the wavelengths of light. As light waves travel through expanding space, they are stretched out by its expansion. The redshift we see from distant galaxies is caused by this stretching. Hubble interpreted this redshift as the recession velocity V used in Hubble's law (Unit 74); but as we discussed in the previous section, this interpretation does not explain the light travel time.

The stretching of wavelengths is the same effect you would find if you drew a wiggly line on the rubber band in the earlier analogy. As the rubber band is stretched, the crests and troughs of the wiggles become more widely spaced. The ant walking on the stretching rubber band would similarly feel its feet being spread apart as it walked (although it would pull them together after each step). The effect on a light wave traveling between galaxies is illustrated in Figure 79.4—the wavelength of the light is stretched by the same amount as the space itself.

Understanding that the redshift we observe is caused by the stretching of wavelengths in expanding space gives us a better way to interpret redshift. The redshift indicates how much the universe has expanded since light left the object we are

Expansion and redshift

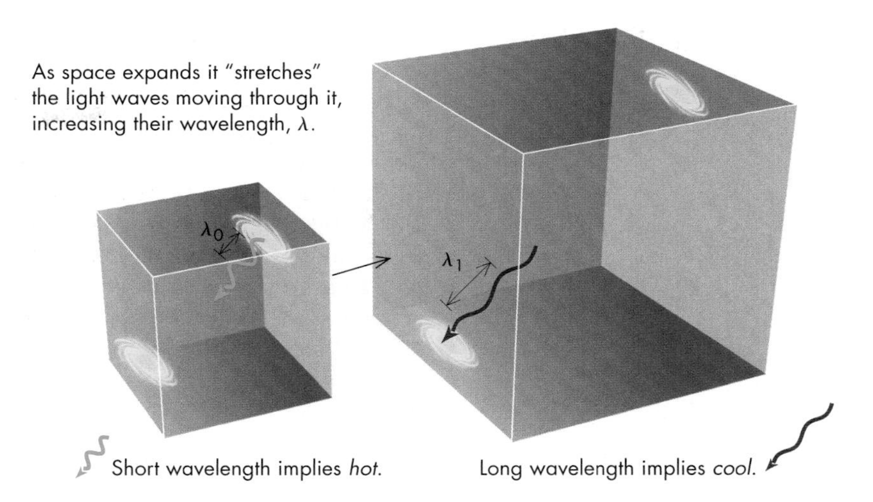

FIGURE 79.4
As space expands, it stretches radiation moving through it, making the wavelengths longer. Because longer wavelengths are associated with cooler objects, the stretching of the radiation has the effect of cooling it.

observing. Astronomers determine the redshift z from the fractional change in the wavelength λ of light emitted by a galaxy. We can calculate the redshift from the following formula:

$$z = \frac{\text{Change in wavelength}}{\text{Normal wavelength}} = \frac{\Delta\lambda}{\lambda}$$

$$= \frac{\text{Change in size of universe}}{\text{Size of universe when light left galaxy}}$$

For example, if we detect a spectral line that was emitted at 500 nanometers but is redshifted to 550 nanometers from a distant galaxy, the change in wavelength $\Delta\lambda$ is 50 nanometers. Therefore the redshift is

$$z = \frac{550\,\text{nm} - 500\,\text{nm}}{500\,\text{nm}} = \frac{50}{500} = 0.1,$$

and we can say that the universe has expanded by 10% since the light left that galaxy.

Hubble interpreted the redshift as a Doppler shift and would have said $z = 0.1$ implies galaxy was traveling away from us at 10% of the speed of light, or about 30,000 km/sec.

Astronomers find it helpful to use z when they describe the expansion of the universe because the redshift is related to the changing size of the universe. In particular, z compares the size of the universe when we receive the light from a galaxy to the size of the universe when the light was emitted by the galaxy. That relation is

$$\frac{\text{Size of universe today}}{\text{Size of universe when light left galaxy}} = z + 1$$

If the size of the universe today is R_0 and the size of the universe when light was emitted from a galaxy was R, then

$$z = \frac{R_0 - R}{R} = \frac{R_0}{R} - 1$$

Adding 1 to both sides of the equation, we find that:

$$\frac{R_0}{R} = z + 1.$$

Because the universe has no measurable edge, for *size* here we might substitute "the average distance between galaxies." In the preceding example, with $z = 0.1$, we can say that the universe is 1.1 times bigger today than when light left the galaxy.

We observe some extremely distant galaxies and quasars for which $z > 1$. This does *not* necessarily mean that they are traveling faster than the speed of light. Rather, the result is telling us that $z + 1 > 2$—that is, space has expanded by more than a factor of 2 since the light we are now seeing left those objects. If a galaxy has a redshift $z = 1$, then the size of the universe has exactly doubled since the light left the galaxy. For $z = 2$, space has expanded by $(z + 1) = 3$ times since the light left the galaxy, or we might equivalently say that the universe was three times smaller when the light left that galaxy.

The current record holder for redshift is a galaxy with $z = 6.96$. For this redshift, the universe today is 7.96 times larger than when the light was emitted.

This means that the matter in the universe was packed much more tightly together when the light we see left that quasar. The average separation between galaxies was just 1/7.96 = 0.126 times the current separation. The smaller separation would have been in every direction, so the *volume* of space (length × width × height) would be just 0.126 × 0.126 × 0.126 = 0.0020 times what it is today.

There are hints of galaxies at even higher redshifts, when galaxies were even closer together, but the signals are very weak, and the identification of spectral lines has not yet been confirmed. Galaxies at high redshifts look very different from nearby galaxies, in part because the matter was just in the beginning stages of gathering together to form a galaxy (Unit 75.4). If we look out farther and farther, to earlier and earlier times in the age of the universe, ultimately we should see a time when the whole universe was packed together at extremely high density.

Q The Solar System formed about 4.5 billion years ago—which is about one-third of the age of the universe. Was the separation between the Sun and Earth smaller then? What about the size of the Earth?

79.4 THE AGE OF THE UNIVERSE

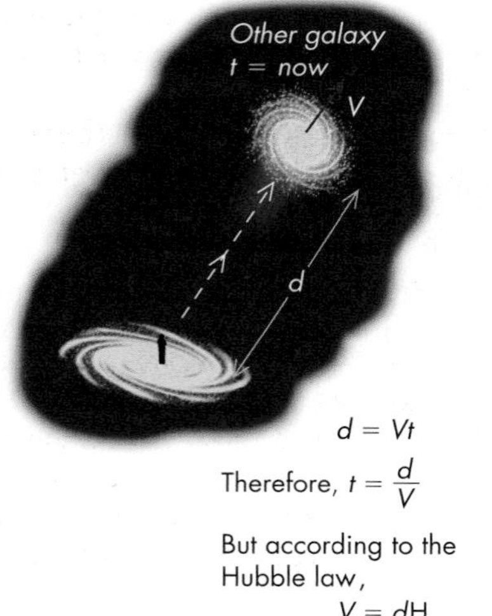

FIGURE 79.5
Estimating the age of the universe by the recession of one galaxy from another galaxy.

The model of the universe we have just discussed, as an expanding space peppered with galaxies, allows cosmologists to estimate its age and to calculate conditions near the time of its birth. Alexander Friedmann, a Russian scientist, made such calculations in the 1920s, but his work attracted little attention. Not until 1927, when Abbé Georges Lemaître, a Belgian cosmologist and priest, independently made similar calculations, did astronomers appreciate that conditions at the birth of the universe could be deduced from what is observed today.

Lemaître pointed out that because the separations of galaxies are growing today, the galaxies must in the past have been closer together, as we saw in the last section. In fact, if you simply follow the expansion implied by Hubble's law back in time, there must have been a period long ago when all of the galaxies in the universe were crowded together. Galaxies and stars as we know them today could not have existed in such a dense environment. With so much mass in such a relatively small volume, the matter that now composes the far-flung galaxies and their stars must have been packed into an extremely dense ball that Lemaître called the *Primeval Atom*. From this dense state, the universe must have expanded at a tremendous speed—the Big Bang.

It is straightforward to estimate the time since the Big Bang as follows. We know the rate at which space is expanding between any pair of galaxies—how fast they are moving away from each other. We solve for how long it would take them to get this far apart moving at their current speed, just as we might solve for how long a car takes to travel 200 kilometers at a speed of 50 kilometers per hour.

Consider two representative galaxies separated by a distance d and separating with a velocity V (Figure 79.5). We will assume that V has remained constant and call the time we calculate that it took the galaxies to reach this separation the **Hubble time,** t_H. We can calculate the Hubble time as follows: The time it takes to travel a particular distance, d, is the distance divided by the speed. (For example, 200 km divided by 50 km per hour gives 4 hours.) Therefore, the Hubble time is

Because Speed = Distance/Time, or $V = d/t$, multiply both sides of the equation by t/V to get $t = d/V$.

$$t_H = d/V.$$

Hubble's law, however, tells us that a galaxy's velocity grows with distance according to the formula $V = H \times d$. Substituting this into our equation, we find

$$t_H = \frac{d}{H \times d} = \frac{1}{H}.$$

TABLE 79.1 Relationships Between Redshift, Distance, and Time

Redshift (z)	Size of the Universe Compared to Now	Distance when Light Was Emitted	Length of Time Ago (Light Travel Time)	Time After Big Bang
0	1	0	Today	13.7 billion years
0.1	0.91	1.2 billion light-years	1.3 billion years	12.4 billion year
0.5	0.67	4.1 billion light-years	5.0 billion years	8.7 billion years
1.0	0.50	5.4 billion light-years	7.7 billion years	5.9 billion years
2.0	0.33	5.7 billion light-years	10.3 billion years	3.4 billion years
3.0	0.25	5.3 billion light-years	11.5 billion years	2.2 billion years
5.0	0.17	4.3 billion light-years	12.5 billion years	1.2 billion years
7.0	0.13	3.6 billion light-years	12.9 billion years	800 million years
10	0.09	2.9 billion light-years	13.2 billion years	500 million years
1000	0.001	45 million light-years	13.7 billion years	400,000 years

The distance d cancels out, so it doesn't matter which two galaxies we choose—they began moving apart a time equal to $1/H$ ago. In other words, the inverse of Hubble's constant indicates a length of time that approximately measures the age of the universe.

To get a value for t_H in years, we have to carry out some unit conversions because H is expressed in units of kilometers/second per megaparsec. One megaparsec is 3.09×10^{19} kilometers, so if Hubble's constant is 70 km/sec/Mpc, this gives

$$t_H = \frac{1 \text{ Mpc}}{70 \text{ km/sec}} = \frac{3.09 \times 10^{19} \text{ km}}{70 \text{ km}}\text{sec} = 4.41 \times 10^{17} \text{ sec}.$$

Because there are 3.16×10^7 seconds per year, $1/H = 1.40 \times 10^{10}$ years. Therefore we find that the age of the universe is about 14 billion years.

In this simple calculation we assumed that the expansion of space has remained constant. Gravity can slow the expansion, but other effects on space can cause acceleration (Unit 82.3). Detailed calculations that take into account these different effects give an answer that is very close to the age we have calculated: 13.7 billion years.

Table 79.1 shows how much the universe has expanded for different values of the redshift according to the best current model. There are some surprising consequences of the higher speed of expansion when the universe was younger. A galaxy at redshift 7 was actually closer to us at the time when the light was emitted than most galaxies we see at lower redshift. We cannot see where that galaxy is *today,* but if we could, we would see that the redshift 7 galaxy is farther away than any galaxy at lower redshift. This can be confirmed in reference to Table 79.1, by multiplying the distance when the light was emitted by $(z + 1)$, the factor by which space has expanded. This means, for example, that the most distant material that telescopes can detect, radiation from a redshift $z = 1000$ (Unit 80.2), is currently at a distance of about 45 billion light-years.

The complications of an expanding universe create an ambiguity for what we mean by the "distance" of a distant galaxy. Its distance when the light was emitted? Its distance when the light was received? Some other distance? To avoid confusion, astronomers often use the **light travel time distance** when speaking of the distance to remote galaxies. The light travel time distance measures how far the light would have traveled in the time since it was emitted if the universe were not expanding. With that definition of distance, we can then say that a galaxy that we observe today and whose light was emitted 10 billion years ago, is 10 billion light-years away from us.

From Table 79.1, an object at $z = 1000$ emitted the light when it was only 45 million light-years from our region of space. Because of the expansion of space by a factor of $z + 1 = 1001$, that region is now 1001 times farther away than it was then.

Q What were your beliefs about the nature of the universe prior to reading about modern cosmology? Are you willing to change any of them if the scientific evidence disagrees with them?

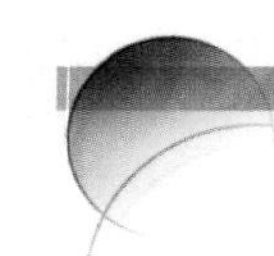

KEY TERMS

Big Bang, 646
cosmological principle, 648
cosmology, 646
general theory of relativity, 648
Hubble time, 652
light travel time distance, 653

QUESTIONS FOR REVIEW

1. What does cosmology study?
2. Why do astronomers think the universe is expanding?
3. What is the cosmological principle?
4. Why is there no place we can call the center of the universe?
5. What causes redshift?
6. How do we measure the size of the universe?
7. How old is the universe? How is its age found?

PROBLEMS

1. When Hubble first estimated the Hubble constant, galaxy distances were still very uncertain, and he got a value for H of about 600 km/sec per Mpc. What would this have implied about the age of the universe? What problems would this have presented for cosmologists?
2. A galaxy at a distance of 25 Mpc has a recession velocity of 1,875 km/sec. What is Hubble's constant based on this one galaxy? Why is this not a good way to determine Hubble's constant?
3. Hubble's constant is still not precisely known. Calculate t_H for two other values of Hubble's constant currently favored by some astronomers:
   a. $H = 71$ km/sec per Mpc.
   b. $H = 65$ km/sec per Mpc.
4. What wavelength would visible light (400 to 700 nm) have if it were seen from a galaxy formed just half a billion years after the Big Bang? What if it came from matter when the universe was just 400,000 years old? (Refer to Table 79.1.)
5. Plot the light travel time (Y axis) versus the redshift (X axis) based on the data from Table 79.1. Estimate the redshift for galaxies that are seen as they were 1, 2, and 3 billion years ago.

TEST YOURSELF

1. From what evidence do astronomers deduce that the universe is expanding?
   a. They can see the disks of galaxies getting smaller over time.
   b. They see a redshift in the spectral lines of distant galaxies.
   c. They see the edge of the universe moving away from us.
   d. They can see distant galaxies dissolve, pulled apart by the expansion of space.
   e. All of the above.
2. If we discovered tomorrow that the distances we measured to galaxies were incorrect, and that Hubble's constant is 35 km/sec per Mpc instead of 70 km/sec per Mpc, we would conclude that the universe's age is______ previously thought.
   a. 4 times older than
   b. 2 times older than
   c. the same as
   d. 2 times younger than
   e. 4 times younger than
3. If a galaxy has a redshift $z = 3$, what fraction of the universe's current size was the universe when the light was emitted from that galaxy?
   a. One-fourth
   b. One-third
   c. One-half
   d. Three-quarters
   e. One-ninth

unit

80

The Edges of the Universe

Background Pathways

Unit 79: Cosmology 646

Some simple questions can have profound answers. One such question puzzled astronomers for centuries: Why is it dark at night? At first glance this question seems to have an obvious answer. However, answering it fully requires a surprising amount of cosmology, involving the expansion and curvature of space, the speed of light, the cosmic density of matter, and the age of the universe.

To understand why the sky is dark at night, we need to know what conditions were like when all of the matter in the universe was packed together just after the expansion of the universe began. Extrapolating to what the physical conditions must have been like at these times, cosmologists have been able to make several predictions about characteristics of the universe. The confirmation of these predictions has provided strong support for the Big Bang theory of the universe's origin.

In this unit we explore the light that has been traveling through the universe since shortly after the Big Bang. As we look out into space, we see galaxies in every direction out to the limits of our technology. As we probe these great distances, we are also looking out into the past, and we can see back to a time just a few hundred thousand years after the Big Bang. Before then all of the matter in the universe was merged into a great ocean of hot gas, and light could not travel freely through space. We can even detect small fluctuations in this hot gas that provide clues about how galaxies first formed.

Light comes to us from all directions from hundreds of billions of galaxies that have been emitting light for more than 13 billion years. We begin then by trying to understand the question we posed: Why is the sky dark at night?

80.1 OLBERS' PARADOX

Our discussion of why the sky is dark at night requires that we look first at the distribution of the objects that generate visible light. We show a map of galaxies on the largest scales in Figure 80.1A. This wide-angle view shows the distribution of nearly 1 million galaxies as seen looking up out of the disk of our own Galaxy, the Milky Way. Around the edge of the picture, there are fewer galaxies because dust in the Milky Way hides the galaxies in that part of the sky. However, the unobscured portion of the picture provides an important clue about the structure of the universe.

No matter what direction you look (ignoring the dust layer of the Milky Way), you see approximately the same number of galaxies. That is, averaged over a very large area, the universe is more or less the same in all directions, and galaxies are spread throughout the universe much as raisins are spread throughout raisin bread. They may clump a bit in one region or another, but overall, they are fairly evenly distributed (Figure 80.1B). If we look over large enough distances, even the superclusters and voids of large-scale structure (Unit 76.4) average out, and

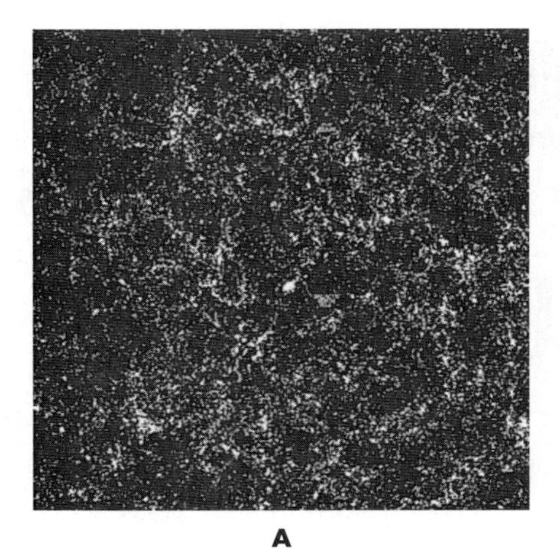

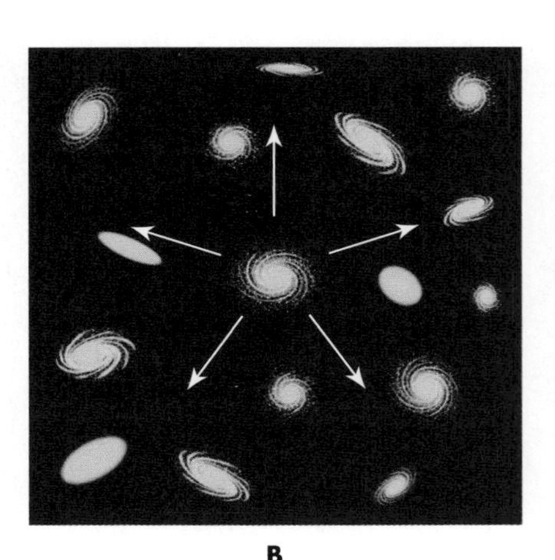

A B

FIGURE 80.1
(A) A computer-generated picture showing the location on the sky of nearly a million galaxies. This is the view looking up and out of the Milky Way. Note the relatively uniform distribution of the galaxies.
(B) A sketch illustrating the uniformity of the universe. Roughly the same number of galaxies occur in all directions.

the matter in the universe appears to be approximately smoothly distributed, or **homogeneous,** out to the largest distances we can measure.

This homogeneity presents a problem, though. If there are galaxies in every direction, then our sky should be completely covered by them. In fact, if the universe has no edge, then in every direction we look, our line of sight should eventually encounter a star. So the sky should be bright everywhere. This is known as **Olbers' paradox** after the German astronomer Heinrich Olbers, who in 1823 popularized the problem. He calculated that if the universe extends forever and has existed forever, the night sky should be blazingly bright. The paradox arises from reasonable premises (or so it seemed at that time) that lead to a conclusion clearly at odds with the simple fact that the sky is dark at night.

Although the paradox bears Olbers' name, it was discussed by various astronomers over the two previous centuries. Kepler pointed out that in an infinite universe, the night sky should be bright, and he therefore concluded the universe was finite.

The argument is easiest to understand from a simple analogy. Suppose you stand in a small grove of trees and look out between the tree trunks to the surrounding landscape. If the grove is small, your line of sight will be blocked by trees in a few directions, but it will pass between the trees in most directions (Figure 80.2). If the grove is larger, more distant trees in it will block your view in what previously were clear gaps. In that case, no matter where you look, your line of sight will be intercepted by a tree trunk.

Now suppose that rather than looking between trees, you look out into space through the galaxies and stars that compose the universe. If space extends sufficiently far and is populated with galaxies homogeneously, then no matter what direction you look, every line of sight will ultimately intercept a star—just as in a sufficiently large forest every line of sight ultimately hits a tree. Even though the stars in distant galaxies are faint to us, there are so many of them that their collective brightness should be large. Therefore, the sky should be covered with starlight, glowing brilliantly with no dark spaces in between. In other words, the night sky should not be dark.

Q Olbers' paradox was originally phrased in terms of individual stars, not galaxies. How would casting it in terms of galaxies have altered the thinking about the paradox?

This argument must contain a false assumption, because the most obvious observation one can make about the sky is that it is dark at night. That is the paradox: The night sky should be bright, but it is not. Where, therefore, is the error in our reasoning?

The first way the paradox can be avoided is if there are no stars beyond some distance. That is, just as we can see out of the woods if there are only a few trees, so too we will see a dark night sky if we run out of stars and galaxies beyond some distance. However, we do not see any evidence of an edge to the universe.

There can be an edge in a second sense—an edge in time. Even if the universe is infinite in extent, if it has a finite *age,* then we can see only the stars whose light has had time to reach us within the time limit set by the age of the universe. That is, if the universe is 13.7 billion years old, we can see light from a star that is 10 billion light-years away, but not one that is 20 billion light-years away. In other words, the universe has to have existed long enough for the light to arrive. The maximum

In a small grove of trees, only a few block your view.
Lots of space to view between trees.
A

In a larger forest, more distant trees block your view.
No open space visible between trees.
B

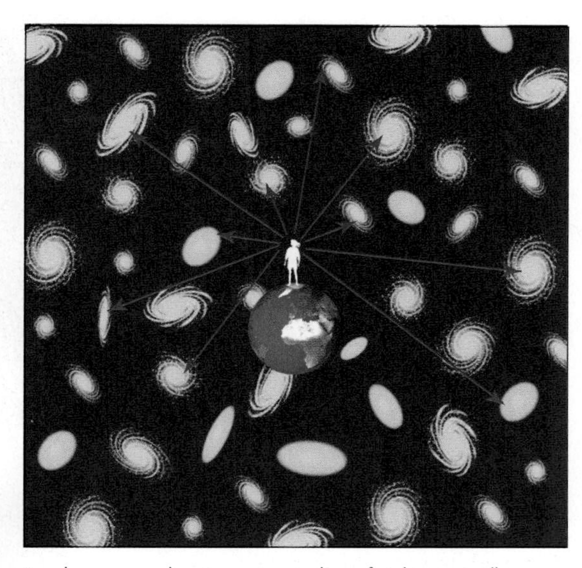

In a large enough universe, every line of sight eventually encounters a star within a galaxy
C

FIGURE 80.2
(A) An observer in a small grove of trees can see out through gaps between the trees to the surrounding countryside. (B) In a larger forest, distant trees block the gaps so that no matter where you look, your line of sight ends on a tree. (C) So, too, in a universe that extends forever and has existed forever, your line of sight will always end on a star. Therefore, the night sky should be bright.

distance that light can travel in the universe's age defines the **cosmic horizon.** Astronomers call the space within the cosmic horizon the **visible universe** (Figure 80.3). As the universe ages, the visible universe grows larger. At present, we don't know how large the entire universe might be, but it is suspected to be far, far larger than the visible universe. This would suggest that if we could wait long enough, in the future the night sky *might* become bright.

The distance to the cosmic horizon can be measured in different ways. For example, the most distant regions whose light we can see today emitted that light when they were only about 50 million light-years away. Those regions are today about 50 *billion* light-years away.

However, realizing that we are looking back in time presents another problem. In every direction we look, we should eventually see back to early times when the universe was filled with hot gas from the Big Bang, and this should be just as bright as the surface of a star. Olbers' paradox predated the discovery that the universe expanded from a fiery hot state, but it appears that we need to understand the effects of that expansion to complete our explanation of why the sky is dark at night.

80.2 THE COSMIC MICROWAVE BACKGROUND

The idea that we should see light from the Big Bang was first suggested by George Gamow (pronounced *GAM-off*), a Russian astrophysicist who fled to the United States before World War II. Gamow built upon Lemaître's idea that the young

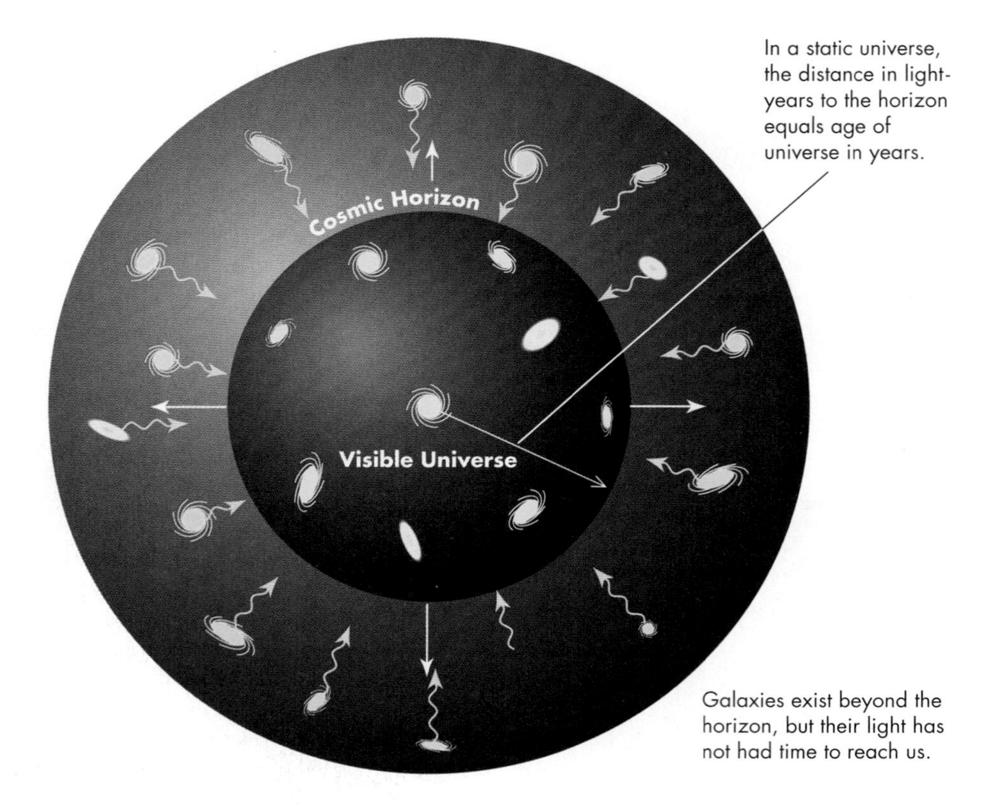

FIGURE 80.3
We cannot, even in theory, see beyond the cosmic horizon. Light from there takes more time than the age of the universe to reach us.

universe was extremely dense (Unit 79.4), and he went on to conclude that it must also have been extremely hot. Gamow's argument was based on the observation that compression heats a gas, and so the enormous compression of the early universe must have heated the matter to very high temperatures at the earliest times.

When we observe out to large enough distances, then, because we are seeing the universe as it was when it was young, we should see evidence of initially high temperatures. However, the radiation from this hot gas has been diluted and redshifted to long wavelengths by the expansion of the universe. When we look out at night, every line of sight *does* encounter a hot glowing surface, and the sky *is* bright, but at radio wavelengths. Thus, the darkness of the night sky results from the limitations of our senses. If we look at radio wavelengths, we should see the afterglow of the Big Bang.

This idea was developed in 1948 by two of Gamow's collaborators, the U.S. physicists Ralph Alpher and Robert Hermann. They showed that this radiation would come not directly from the first moments of the Big Bang but rather from a later time, after the universe had been expanding for several hundred thousand years.

Just minutes after the Big Bang, the temperature of the universe would have been several billion Kelvin. After a few hours it cooled to 100 million K; after about 10 years, about 100,000 K; and after 400,000 years, about 3000 K. This cooling is straightforward to predict from laboratory observations of expanding gases. During this whole time, the high temperature of the gas kept it ionized. In an ionized gas, photons interact strongly with the free electrons and, as a result, can travel only a short distance (Figure 80.4A). Because light could not travel far through this hot gas, the gas was essentially opaque.

The last scattering epoch is also sometimes called the *recombination epoch* in reference to the electrons and ions combining to form atoms. This name can be misleading, though, because this was actually the first time these particles ever combined.

Throughout the first few hundred thousand years of the universe's existence, the temperatures and densities of the original gas were similar to those inside a star. As time progressed, the gas temperature and density dropped, making the conditions more similar to those farther from the core of a star. The conditions at 400,000 years finally reached temperatures like those at the surface of a star—low enough that the

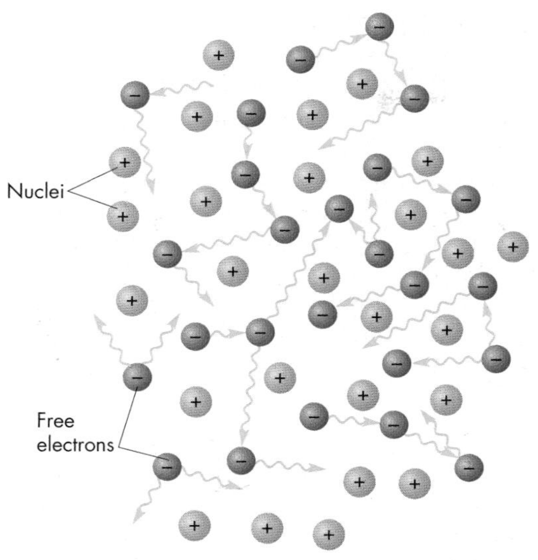

Before recombination: The universe was opaque.

A

After recombination: The universe was transparent.

B

FIGURE 80.4

(A) In the young universe, free electrons in the hot ionized gas constantly interacted with photons, making the universe opaque. (B) When the universe became cool enough that electrons combined with the nuclei to make neutral atoms, the universe became transparent.

ionized hydrogen and helium combined with the free electrons, and the universe became transparent (Figure 80.4B). This period is called the **last scattering epoch** because photons had their final interaction with the expanding gas of the Big Bang.

Even though we cannot see to times before 400,000 years, a variety of clues have survived, and these help take us back to less than one second after the Big Bang occurred. These clues are explored in Unit 81.

Thus, after 400,000 years the photons from the Big Bang were able to move freely through space—unless they hit something, like a telescope on Earth! These photons are the oldest radiation we can see. Just as we cannot see photons directly from the core of the Sun, we cannot see the photons from the Big Bang itself. In both cases the photons travel through a hot ionized gas that constantly interacts with the photons for hundreds of thousands of years before the photons can move freely through space. Through these interactions the photons exchange energy with the gas particles and reach the same temperature as those particles. Eventually the photons reach a place (in the Sun) or a time (in the history of the universe) where the gas they encounter has thinned out, and they are then free to travel through space. The photons travel away, carrying with them energies set by the temperature of the last atoms they interacted with.

So what should this radiation look like? According to Wien's law, the thermal radiation from the photosphere of a star or the last scattering epoch should have a spectrum that is most intense at a wavelength determined by the temperature (Unit 23). In particular, hot matter radiates most strongly at short wavelengths. Just after the last scattering epoch, the radiation would have been at a temperature of about 3000 K and therefore looked much like the surface of an M-type star like Betelgeuse (Unit 55). If we lived then, we would have experienced Olbers' picture of a sky everywhere as bright as the surface of a star. The radiation at that time would have peaked at a wavelength of about 1000 nanometers.

The expansion of space modifies this radiation. The redshift of the wavelengths makes the radiation appear cooler. The universe has expanded by a factor of about 1000 since the last scattering epoch, so the wavelength of the radiation has also stretched by a factor of 1000 (Unit 79). According to Wien's law, the temperature of thermal radiation drops in proportion as the wavelength grows longer, so the

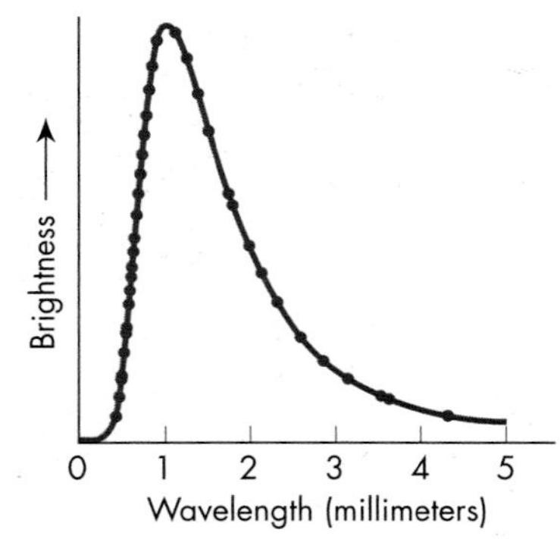

FIGURE 80.5
Spectrum of the cosmic microwave background. The shape of the spectrum matches perfectly a thermal spectrum for a temperature of 2.726 K. Notice the wavelength at which it peaks. Red dots show data points schematically.

apparent temperature of this radiation would drop by a factor of 1000—from 3000 K down to about 3 K today.

If the universe began with a Big Bang, it should today be filled with long-wavelength radiation created when the universe was very young. This prediction was confirmed in 1965, when Arno Penzias and Robert Wilson of the Bell Telephone Laboratories accidentally discovered cosmic microwave radiation having just that property.

Penzias and Wilson were trying to identify sources of background noise on telephone satellite links. They detected a radio signal with the unusual property that its strength was constant no matter what direction in the sky they pointed their detector. They soon demonstrated that the signal came not from isolated objects such as stars or galaxies but rather from all of space. Initially mystified by the signal, they discovered that scientists at Princeton University had repeated Alpher, Hermann, and Gamow's calculations and were searching for the predicted radiation. Once aware of that work, Penzias and Wilson realized that the background interference they had found was radiation created in the young universe. For their discovery of the **cosmic microwave background,** or **CMB** for short, they won the 1978 Nobel Prize in Physics.

Figure 80.5 shows measurements of the intensity of the CMB and demonstrates that it has exactly the thermal spectrum predicted by theory. Moreover, the radiation is most intense at about 1 millimeter (10^6 nanometers), in the microwave part of the radio spectrum. Knowing this wavelength and using Wien's law, we find that the temperature describing the radiation is about 3 K, close to the value predicted nearly 20 years earlier. A more precise measurement shows that the temperature that describes the CMB is 2.726 K, only a little warmer than absolute zero, the lowest temperature anything can have.

Today the cosmic microwave background results provide one of the cornerstones that support the Big Bang theory. In earlier years, though, Gamow's predictions were considered fantastic. The high temperature and rapid expansion from a hot, dense state led the English astronomer Sir Fred Hoyle (a proponent of a competing theory called *steady state cosmology*) to jokingly refer to this theory of the birth of the universe as the *Big Bang,* a name that has stuck despite his sarcasm.

80.3 THE ERA OF GALAXY FORMATION

After the cosmic microwave background scattered off the rapidly thinning and cooling gas, the universe entered what some astronomers have dubbed the "dark ages." The radiation from the microwave background was rapidly growing more and more redshifted, and no matter had yet condensed into stars, so the universe grew darker and darker. After a few hundred million years, the first stars and galaxies formed, and their radiation began to re-ionize much of the gas distributed throughout intergalactic space—lighting up the universe again. How this process occurred is one of the current areas of research in astronomy.

On the basis of the ages of stars within galaxies, astronomers deduce that the Milky Way and other galaxies are at least 10 to 12 billion years old. We can observe galaxies when they were very young in deep images such as the Hubble Deep Field (Unit 75) and Ultra Deep Field (Figure 80.6), where we find some extremely faint galaxies at redshifts that imply they were formed less than 1 billion years after the Big Bang. Galaxies must therefore have formed fairly rapidly once the universe had cooled enough. To understand how the universe evolved from a sea of hot gas so quickly into galaxies, astronomers are carrying out simulations using the most advanced supercomputers available.

Presumably gravity pulled gas clouds into protogalaxies, much as gravity pulls interstellar clouds into protostars; but there appear to be important differences.

FIGURE 80.6
The Hubble Ultra Deep Field. This image shows a portion of the sky in the constellation Fornax that you could easily cover with this dot (•) held at arm's length. The image shows about 10,000 galaxies. Some are so far away that we are seeing them the way they appeared nearly 13 billion years ago, when the universe was only about one-twentieth of its present age.

In particular, the amount of matter observed in the universe exerts too feeble a gravitational attraction for this process to have been able to form galaxies within the age of the universe, let alone within the first billion years. Most astronomers therefore conclude that additional matter must be present that provided a strong enough gravitational pull to speed up galaxy formation. This need for unseen matter to aid galaxy formation strengthens astronomers' conclusion that dark matter (Unit 78) plays a major role in the structure of the universe.

The dark matter's gravity could draw the dark matter into clumps, which then might pull in ordinary matter to form galaxies. Computer simulations show that the dark matter could have begun clumping even before the last scattering epoch, precisely because it does not interact with electromagnetic radiation as ordinary matter does. As a result, the large-scale organization of matter into denser regions was able to begin sooner than 400,000 years after the Big Bang, and continued creating the cosmic web of large-scale structure throughout the dark ages. These higher-density regions began pulling in normal matter as soon as this was no longer being buffeted by high-energy photons. The clumping of the dark matter itself responded to variations in density that arose during even earlier moments immediately after the Big Bang (see Unit 81).

Computer simulations show that the clumping of dark matter would make some regions slightly more dense than others. The greater compression of the gas in these regions would make them slightly hotter and brighter—producing small fluctuations in the temperature of the CMB. Astronomers have found exactly such clumps, as shown in Figure 80.7. This figure displays the CMB over one region of the sky, the bright regions being the dense clumps where clusters of galaxies are going to form.

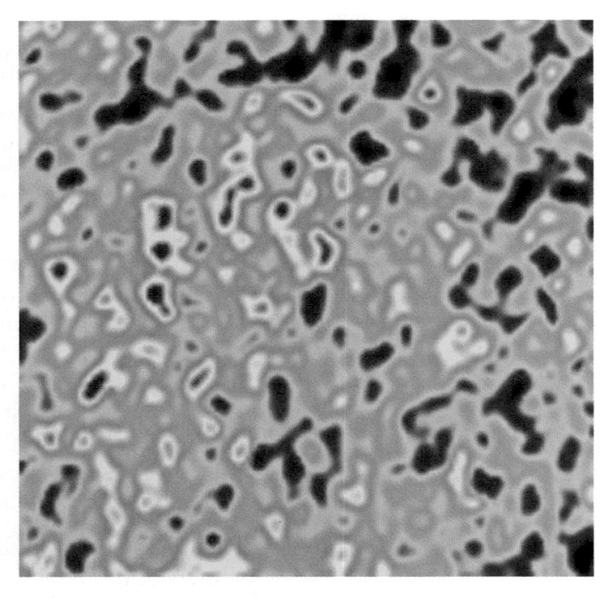

FIGURE 80.7
A portion of the cosmic microwave background radiation. The red bumps are slightly denser and hotter regions, while the dark blue regions on the map are lower-density, cooler regions. At this redshift, 1° is about 800,000 light-years.

The size of these clumps is determined by the size of the cosmic horizon during the last scattering epoch. The cosmic horizon would have had a radius of about 400,000 light-years at this time because that was as long as the universe had lived. This means that regions of gas and dark matter could have interacted with each other—sharing their heat and experiencing each other's gravitational pull—only if they were both within a radius of 400,000 light-years of a common point. The clumps in Figure 80.7 are each regions of about this radius.

In the clumps of dense matter visible in the CMB, we are seeing the distribution of matter in distant regions of space as they were before any galaxies formed. It is possible that those regions today, almost 13.7 billion years later, contain galaxies in which astronomers are looking in our direction and seeing a bright region where the Milky Way and thousands of other galaxies are going to form.

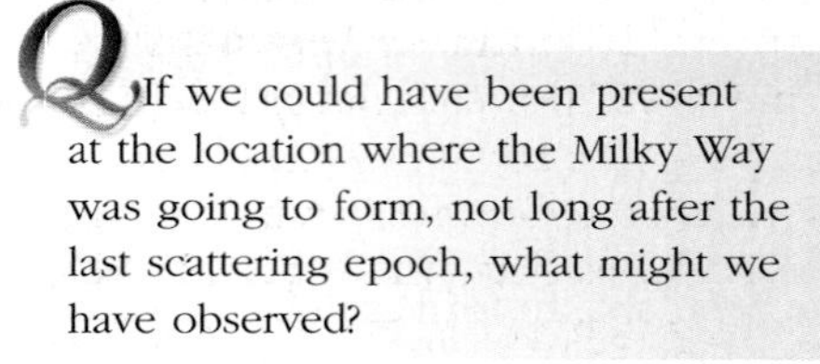
Q If we could have been present at the location where the Milky Way was going to form, not long after the last scattering epoch, what might we have observed?

The angular size of these clumps provides another clue about the properties of the universe, because the path of the light that we see is affected by the overall gravitational field of the universe. The entire universe can be curved by gravity, and this can profoundly alter our notions about what it means for space to have an edge, as we examine next.

80.4 THE CURVATURE OF THE UNIVERSE

Einstein's general theory of relativity has shown that the mass and energy of the universe curve its space. Such bending can be difficult to picture. In some ways it is easier to picture the bending of space by an object like a black hole (Unit 68.2) or a cluster of galaxies (Unit 78.3), because the strong bending is localized to one region. But in fact, the enormous mass and energy of the entire universe can change how space connects to itself, bending and shaping the universe.

We are a little like an ant that crawls into a rubber hose lying on the ground. If the hose is straight, representing uncurved space, the ant walking in a straight line will reach the other end of the hose and emerge. However, suppose the hose is bent and twisted in large loops. If the curves are gradual, an ant crawling through the hose may still feel like it is traveling in a straight line, even though its direction is changing. And if the other end of the hose was brought around so that the ends met, the ant could crawl forever in a seemingly straight line and would never reach the hose's end.

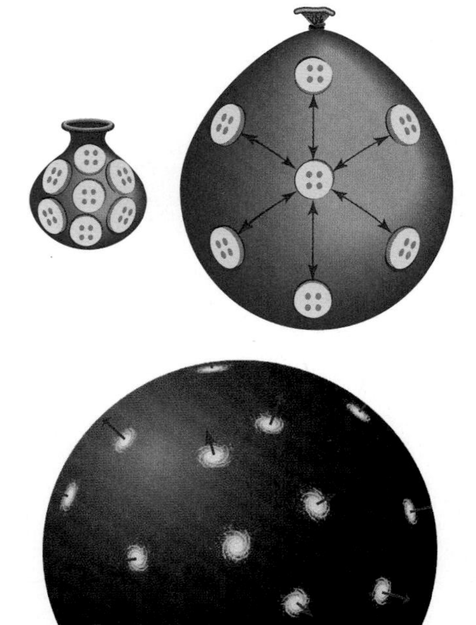

FIGURE 80.8
Representation of a positively-curved, expanding universe. Curved space can expand between galaxies similar to how the space expands between buttons on a balloon as the balloon is inflated. The space between galaxies grows proportional to distance, just as Hubble's law indicates. This analogy represents space by a two-dimensional surface, so nothing could travel off the surface. Space itself would be curved in a dimension that is beyond the normal three dimensions we perceive.

We experience curved space on the surface of the Earth. Athletic fields and parking lots appear flat, and if we ignore hills, the surface of the Earth from horizon to horizon certainly looks flat. We know, however, that if we could walk in a "straight" line along this "flat" surface, we would return to our starting point. Perhaps the expanding universe is similarly curved.

One way of picturing the expansion of space is that it is something like the stretching of a rubber sheet, with buttons on it representing galaxies (Unit 79.2). Imagine, though, that the rubber sheet curved around so that the edges connected to each other, forming a balloon. This gives us an analogy for galaxies in an expanding curved universe: buttons glued to a balloon (Figure 80.8). As the balloon inflates, the space between buttons expands, and the buttons separate at a rate proportional to their distance from each other, as in Hubble's law. Moreover, no button is near an edge even though space is finite.

In the balloon analogy, only two of the dimensions of space can be represented by the balloon's surface. A photon traveling from one button to the next travels along the curved surface of the rubber in the balloon. General relativity predicts that the three dimensions of space can also curve back around on themselves as in the balloon analogy. Such a space looks as if it extends forever in all directions; but if we could travel in a spaceship along an apparently straight line, we would eventually return to our starting position. We could move through this space forever without coming to an end point. A universe with this property is said to have **positive curvature** and is called a **closed universe** because it closes back on itself. The Earth's surface is also positively curved because if we travel in a straight line on it, we return to our starting point. Positively curved space has other properties as well: Parallel lines meet when extended, and the sum of the interior angles of a triangle drawn on a positively curved surface is greater than 180°, as illustrated in Figure 80.9A.

General relativity holds open the possibility that the universe might have other types of curvature. It might be **flat** (that is, have no curvature) or have **negative curvature.** Flat space is what most people picture when they think of space. In it, parallel lines do not meet, and a triangle's interior angles add up to exactly 180°, as illustrated in Figure 80.9B. Negatively curved space is harder to visualize, but you can think of it as being bent into a saddle shape, as shown in Figure 80.9C. Such curvature corresponds with bending space one way along one line and the opposite way along another line. In negatively curved space, the sum of the interior angles of a triangle is less than 180°.

Note that the paths of light through curved space can have surprising properties. If the universe had a very strong positive curvature, we might be able to see all the way around the universe to see ourselves—or perhaps our own Galaxy as it

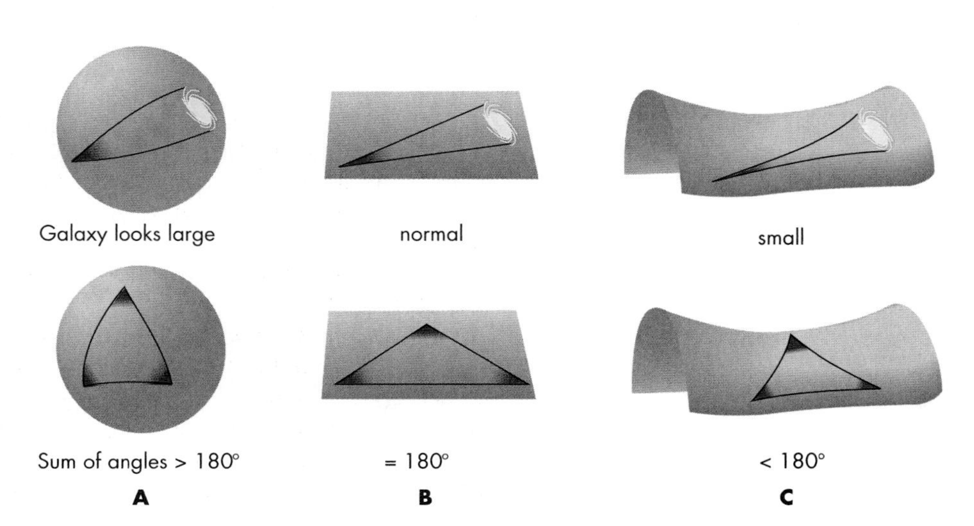

FIGURE 80.9
Figures illustrating how the shape of a surface determines the geometry of a triangle on it. (A) The interior angles of a triangle on the surface of a sphere add up to more than 180°. (B) The interior angles of a triangle on a flat surface add up to 180°. (C) The interior angles of a triangle on a saddle-shaped surface add up to less than 180°. The top row of figures illustrates that a distant object will appear larger if the curvature is positive (A) and smaller if the curvature is negative (C).

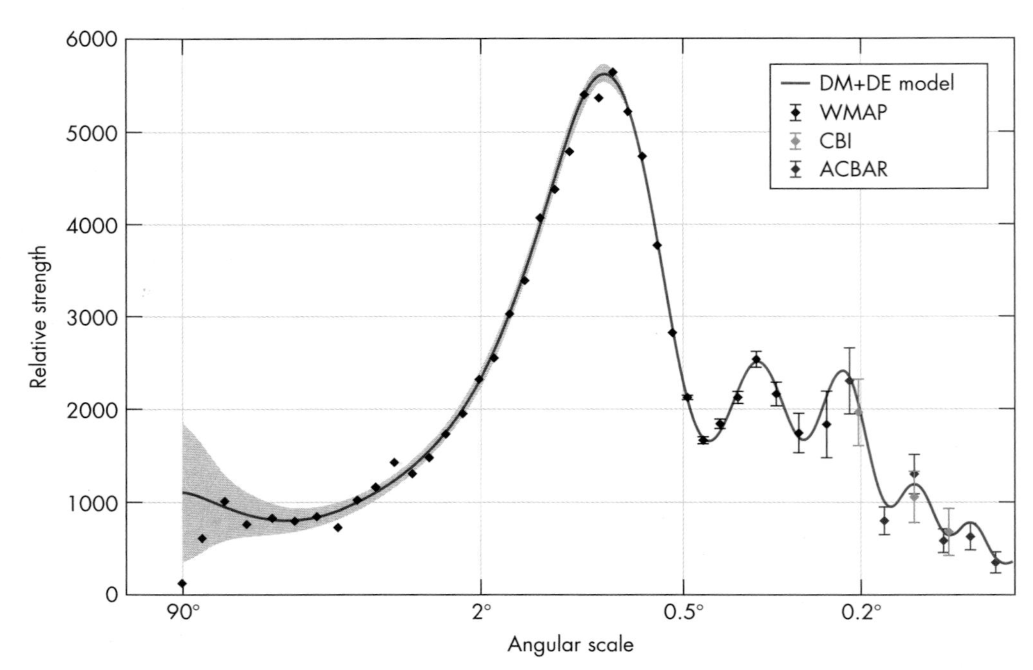

FIGURE 80.10

Plot showing the strength of fluctuations of different angular sizes measured in the cosmic microwave background (CMB). The fluctuations in brightness are caused by the clumping of matter at the last scattering epoch. The position of the first, strongest peak at about an angular size of 1° implies that the geometry of the universe must be very nearly flat. Other peaks in the graph depend sensitively on the proportions of different substances in the universe. The red line shows a plot of the predicted size of the fluctuations for the best-matching model containing dark matter and dark energy.

looked long ago. Positive curvature, like the curvature of light around a black hole or galaxy cluster, also acts as a lens (Unit 78.3), making distant objects look larger, as illustrated in Fig. 80.9. On the other hand, if space is negatively curved, it will make distant objects look smaller than they would look in flat space. Perhaps like a rearview mirror, a negatively curved universe should have a label saying "galaxies in a negatively curved universe are closer than they appear"!

The sizes of clumps on the CMB map give us a way to determine the curvature of the universe. Because the largest clumps should each have a radius of about 400,000 light-years, we can compare this to the size predicted for the different geometries. The measurements of the CMB show that the strongest fluctuations are about 1° across (about twice the apparent size of the full moon). From detailed analyses, this indicates that the universe's geometry must be very nearly flat.

By analyzing the distribution of sizes of these fluctuations, astronomers can estimate H and other cosmological parameters. The relative strengths of different sizes of fluctuations in the CMB (Figure 80.10) depends sensitively on the different ways that normal matter, light, and dark matter interact. Computer simulations can predict the relative strengths of different fluctuations, depending on the relative abundance of each component. Astronomers have been able to obtain beautiful matches to the distribution of sizes of these clumps in their simulations, but only if dark matter is included in the mix, and contributes about 10 times more of the universe's mass than normal matter. This provides additional evidence that dark matter exists. The CMB fluctuations also provides hints of an even more mysterious quantity known as "dark energy," which will affect the ultimate fate of the universe, as we will discuss in Unit 82.

Q Is our cosmic horizon the same as that of alien astronomers living in another galaxy? Why or why not?

KEY TERMS

closed universe, 663
cosmic horizon, 657
cosmic microwave background (CMB), 660
flat universe, 663
homogeneous, 656
last scattering epoch, 659
negative curvature, 663
Olbers' paradox, 656
positive curvature, 663
visible universe, 657

QUESTIONS FOR REVIEW

1. What is meant by a cosmic horizon?
2. What is Olbers' paradox?
3. What is the cosmic microwave background? What is its origin?
4. How was the cosmic microwave background discovered?
5. What is special about the thermal spectrum of the CMB?
6. What is the significance of fluctuations in the cosmic microwave background?
7. What are the three possible geometries for the way space curves in the universe?

PROBLEMS

1. The temperature of the universe needs to be less than 100 K before stars can form. At what redshift does this occur?
2. When our Solar System formed about 4.5 billion years ago, approximately what would the temperature of the CMB radiation have been? Would the structure in the CMB have looked the same as today? Explain your answer.
3. Imagine that there were so many galaxies that 30-K dust inside their interstellar clouds blocked our view in every direction. What would their redshift have to be for the dust to have a temperature like the CMB? What problems are there with this model for explaining the observed properties of the CMB?
4. At the time of the last scattering epoch (at a redshift $z \approx 1000$),
   a. what was the average density of normal matter, if today it is about 3×10^{-31} kg/L?
   b. how large a sphere would have contained $10^{11}\ M_\odot$, about as much normal matter as in the Milky Way Galaxy? Express your answer in parsecs.
5. If the average density of matter in the universe is 3×10^{-31} kg/L, what is the total mass of the universe within our current cosmic horizon? How many galaxies of $10^{11}\ M_\odot$ does this represent?
6. On a flat surface, the angles of a triangle always add up to 180°, but on a spherical surface they may add up to a larger number.
   a. Find a triangle on a spherical surface that contains three 90° angles. Illustrate your example.
   b. What's the largest sum of the angles you can find for a triangle on a spherical surface?
7. Table 79.1, in Unit 79, shows that material we see at a redshift of 1000 was emitted when the material was 45 million ly away from us. This distance determines the angular size of regions at this redshift, as given by the angular size formula in Unit 10.4.
   a. Show that a region that is 1° across at this redshift has a diameter of approximately 800,000 ly.
   b. How big is that region today?

TEST YOURSELF

1. The cosmic background radiation comes from a time in the evolution of the universe
   a. when protons and neutrons were first formed.
   b. when the Big Bang first began to expand.
   c. when the first quasars began to shine.
   d. when X-rays had enough energy to penetrate matter throughout the universe.
   e. when electrons began to combine with nuclei to form atoms.
2. In which part of the sky is the cosmic background radiation brightest?
   a. It is equally bright in all parts of the sky.
   b. Toward the Virgo Cluster
   c. Toward the center of the universe
   d. Toward the spot where the Big Bang took place
   e. Around the locations of quasars
3. Which of the following is important for explaining why the sky is not bright at night?
   a. Dust in our Galaxy blocks the light from distant galaxies.
   b. The universe has a finite age.
   c. There are no stars beyond a distance of about 10 Mpc.
   d. The universe did not expand from a dense hot state, but instead had a temperature close to absolute zero when it exploded.
   e. Galaxies are much farther apart than was believed at the time Olbers proposed the paradox.

unit

81

The Beginnings of the Universe

10^{32} K 10^{25} K 10^{13} K
10^{-43} sec 10^{-33} sec 10^{-6} sec
Planck time Inflation Protons

Background Pathways

Unit 79: Cosmology 646

The early universe was hot and dense. During the first 400,000 years of its existence, the universe was full of radiation that interacted constantly with the matter. This period is hidden from direct view because of this constant interaction—the universe became transparent only after the *last scattering epoch,* when the density and temperature had dropped low enough that electrons combined with protons to form neutral hydrogen atoms. However, it is possible to reconstruct the history at these early times from a variety of evidence.

Figure 81.1 depicts our current understanding of the history of the universe from its beginning, 13.7 billion years ago. If we could journey back to the times before the last scattering epoch, we would find that the radiation from the Big Bang actually "weighed" more than ordinary matter! Energy generates a gravitational influence just as mass does—the relationship between the two is based on Einstein's formula, $E = m \times c^2$.

Today, when we add up all the energy carried by all of the photons in the cosmic microwave background, it amounts to about one one-thousandth of the mass in ordinary matter. At earlier times, though, the balance was shifted. As the universe expanded, the photons and matter spread out equally, but the photons lost energy as their redshift stretched them out to longer wavelengths. At a redshift of 1000, at about the same time as the last scattering, the photons' equivalent mass was 1000 times bigger than today, so the photons matched the mass in ordinary matter. When the universe was younger the importance of radiation was even greater. The first 50,000 years after the Big Bang was a time that astronomers call the **radiation era,** when radiation had even more gravity than dark matter.

During the radiation era, there were none of the large gravitational structures we see today. The universe was a sea of subatomic particles and photons at extremely high temperatures. The behavior of matter at these high temperatures has been explored by physicists with *particle colliders.* These are machines that accelerate subatomic particles to nearly the speed of light, then crash them together, and for an instant create conditions similar to those in the universe just after the Big Bang. This allows us to predict what happened back then and to search for confirming evidence.

In this unit we explore the radiation era, working our way toward the very earliest times that physics allows us to understand—to just a tiny fraction of a second after time began, as depicted in Figure 81.1. Although scientists are confident about the first few of these steps back in time, we reach the limits of our current understanding in the final steps back, when the entire universe visible today was far smaller than an atom.

FIGURE 81.1
A sketch depicting the history of matter and radiation throughout the history of the universe.

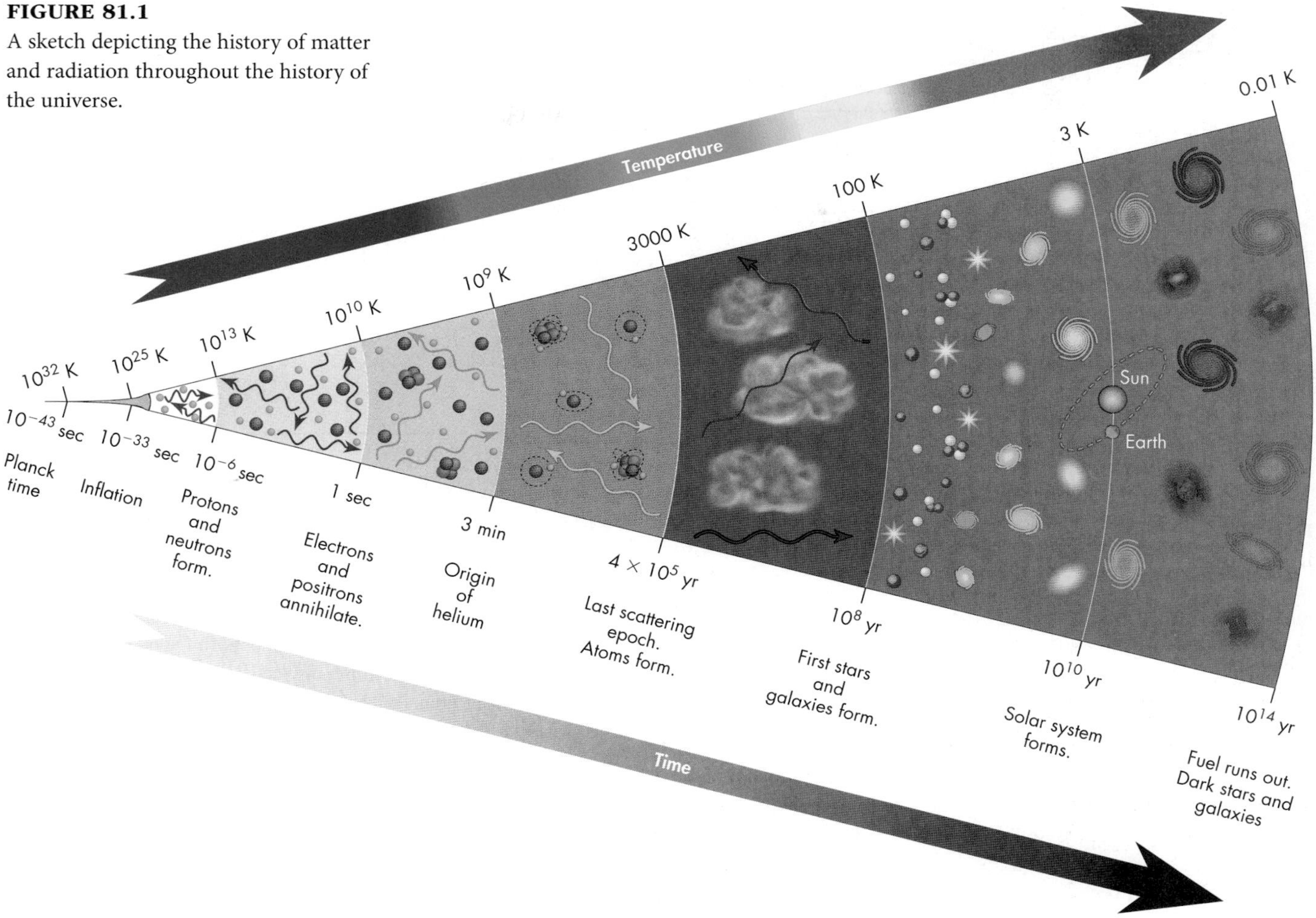

81.1 THE ORIGIN OF HELIUM

When astronomers extrapolate back to times before the last scattering epoch, their first major prediction is that a significant fraction of the hydrogen in the newborn universe would have fused to form helium. We can see why from the following argument: If we take all the matter and radiation in the universe today and calculate how compressed it was a few minutes after the Big Bang, the temperature and density throughout the universe would have been similar to that in the core of a massive star. Therefore, fusion should have occurred, just as it does in the cores of stars.

Such reasoning led George Gamow to hypothesize in 1948 that helium and heavier elements should have been created in the young universe—if the Big Bang model was correct. Gamow's hypothesis is supported by more recent detailed simulations as well as by observations.

We now estimate that about 1 second after the universe began, the temperature of the universe would have been 10 billion (10^{10}) Kelvin, and its matter would have had a density of about 0.1 kilogram per liter. This may not sound very dense, but it is roughly a thousand-trillion-trillion times denser than the universe is at present. At these densities and temperatures, all of the matter in the universe would have broken apart into its constituent particles: electrons, protons, and neutrons. In gas this hot, particles might collide and stick together momentarily, but the intense radiation would have broken them apart as fast as they could form.

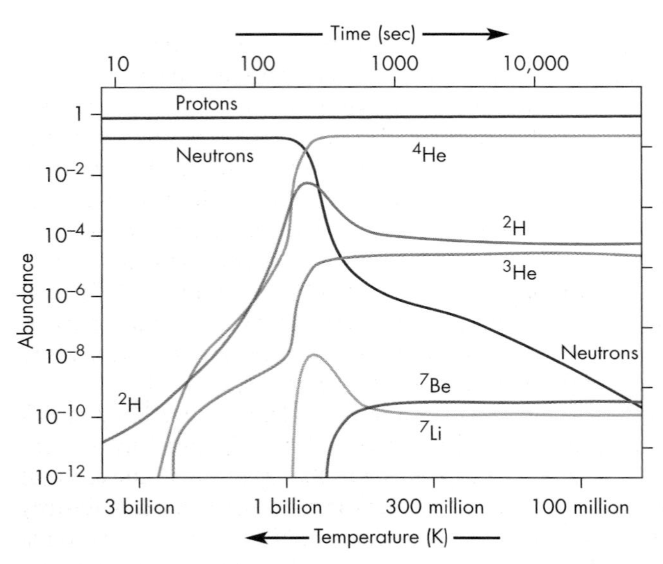

FIGURE 81.2
During the first few minutes after the Big Bang, the universe was hot and dense enough to fuse hydrogen into heavier elements. Only protons and neutrons existed at the start of this period. The production of different elements as they combined took place over the next few minutes. Neutrons and ^{7}Be are both unstable, and they decay after this period of nuclear reactions ends. The predicted amounts of the elements closely match what we observe today.

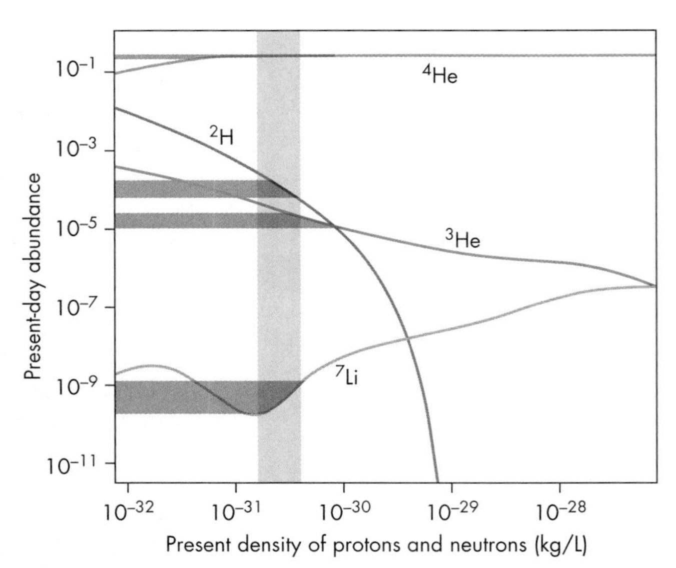

FIGURE 81.3
The amount of different elements expected to be produced by nuclear fusion in the young universe. The curves show how the amount of each element varies depending on the density of matter in the universe. The values along the bottom axis show the density of normal matter in the universe today, while the vertical axis shows the amount of each element compared to the amount of hydrogen that would be produced for that density. The red horizontal bars indicate the best estimates of the amount of each element measured in primitive systems, while the vertical blue bar shows the range of densities consistent with these measurements

In stars there are no free neutrons, so fusion begins with the rarer nuclear reaction during which a proton is converted to a neutron through the weak reaction as the protons collide.

T = temperature of matter and radiation during the early history of the universe.

t_{sec} = time after the Big Bang in seconds

As the universe expanded, it steadily cooled, and the radiation became weaker. Protons and neutrons could fuse, forming hydrogen's isotope deuterium (^{2}H), so deuterium nuclei could begin building up after 1 second, as shown in Figure 81.2. Creating deuterium is also the first step in the process of hydrogen fusion in stars (Unit 50.2).

Throughout the radiation era, the temperature dropped as the inverse square-root of the time, so if we measure the time in seconds, the temperature is:

$$T \approx \frac{10^{10}\ \mathrm{K}}{\sqrt{t_{sec}}}.$$

After 10 seconds the temperature of the universe had therefore cooled to about 3 billion ($10^{10}/\sqrt{10} \approx 3 \times 10^{9}$) Kelvin, which is cool enough for deuterium nuclei to begin fusing to form helium. By the time the universe was 100 seconds old, expansion had cooled the temperature to about 1 billion Kelvin, and particles of deuterium were rapidly fusing into helium-3 (^{3}He) and helium-4 (^{4}He). The universe was not only cooling, it was also rapidly growing less dense. By 100 seconds, the density of matter was only about 10^{-4} kilograms/liter, so the rate of collisions became infrequent, and fusion slowed and then stopped after a few more minutes.

The total amount of helium that formed was determined by how hot and dense the universe was and how long those conditions lasted. These conditions were in turn determined by how rapidly the universe was expanding. Big Bang models predict that about 24% of the matter should have been fused into helium. Stars also fuse hydrogen into helium, but in regions where there has been little star formation the percentage of helium is always found to be very close to this value.

Trace amounts of other light elements would also have been left over from the fusion during these first few minutes after the Big Bang. The amount left of the different elements depends in detail upon the density of matter in the universe. For example, at lower densities, collisions leading to fusion would have been less likely. In this circumstance, slightly less ^{4}He would have had a chance to form, leaving substantially more unfused deuterium and ^{3}He. Higher densities would have allowed more complete fusion, resulting in much lower amounts of ^{3}He and deuterium, and providing an opportunity for more of the heavier elements such as lithium to form. The consequences of these different densities of matter (expressed in terms of the density we would find today) are shown in Figure 81.3.

Measuring the levels of these elements as they were after the Big Bang is difficult, because stars also fuse elements in their cores and then "pollute" interstellar space when they die and expel their atmospheres. On the other hand, the amount of deuterium seen today tends to be lower because processes in stars destroy deuterium. Deuterium survives from the Big Bang only because the universe expanded and cooled too quickly for the normal completion of the fusion process. Astronomers have been able to measure the amounts of deuterium in intergalactic clouds by observing quasar absorption lines (Unit 76.5). This measured amount of deuterium—about 1 deuterium for every 30,000 hydrogen nuclei—is our most sensitive measure of the density of matter in the Big Bang.

Astronomers have also been able to measure the level of ^{3}He in gas clouds where little star formation appears to have occurred. In addition, measurements of the amounts of ^{4}He and the element lithium (^{7}Li) have been made by examining the atmospheres of very old stars, whose composition reflects conditions relatively early in the history of the universe. Figure 81.3 shows that the amount of each of these elements is consistent with the denisty of the universe estimated from the deuterium measurement.

Therefore, from the elements that they observe in the most primitive gas clouds and stars, astronomers can deduce the properties of the expansion of the early universe; and from the properties of the expansion, they can deduce the total mass of normal matter in the universe. When these measurements were first made, the density of matter they implied was puzzling. It was much smaller than most astronomers had expected, given the amount of matter that appeared to be present from the strength of gravitational forces between galaxies. It is now understood that this seeming discrepancy was yet another consequence of the prevalence of dark matter in the universe. The dark matter makes up most of a galaxy's mass, but because it is not composed of protons and neutrons, it could not participate in the fusion that took place shortly after the Big Bang.

The excellent consistency between predictions about Big Bang helium production and the percentages of all the light elements in the oldest matter in the universe provides further strong support for the Big Bang theory. It is interesting that in more than 13 billion years since this early epoch, stars have succeeded in fusing only about 3% more of the matter in the universe into helium. All of the fusion in all of the stars in the whole history of the universe has produced little more than a tenth of what occurred during a few minutes after the Big Bang!

Q If there had not been any fusion in the early universe, how might the evolution of stars have been altered?

81.2 RADIATION, MATTER, AND ANTIMATTER

Next let us take another step closer to the beginning of the universe. During the first second after the Big Bang, the universe was so hot that matter and radiation mingled without the sharp distinction we see between these two entities today.

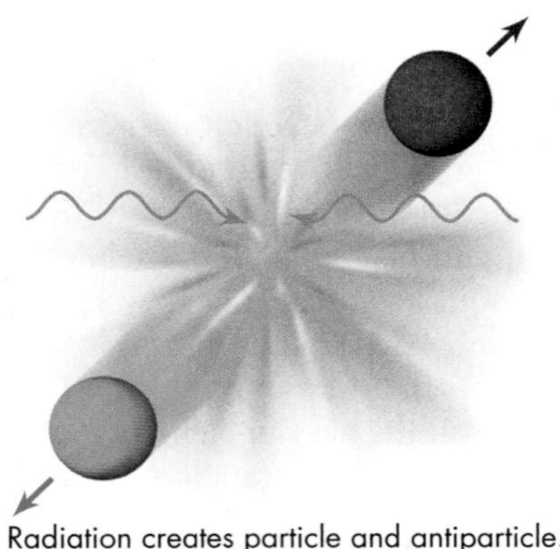

Radiation creates particle and antiparticle.

A

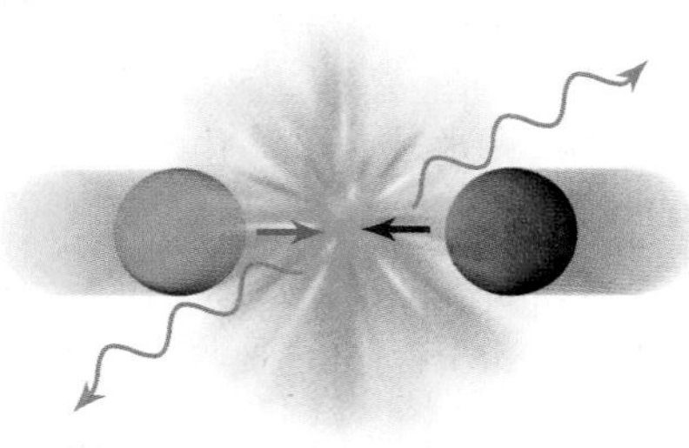

Particle and antiparticle annihilate, creating radiation.

B

FIGURE 81.4
(A) The energy of electromagnetic radiation can be converted into mass with the formation of a particle and its antiparticle. (B) The collision of a particle and its antiparticle leads to their annihilation and the conversion of their mass back into energy.

Q Physicists can generate antimatter in particle colliders. How do you suppose they can keep it from immediately destroying their experimental apparatus?

Even the weakly interacting neutrinos were constantly colliding with other particles during this early time. Astronomers call this period the **early universe.** It was a time when matter and radiation behaved almost as a single entity.

The distinction between matter and radiation can become blurred because of the equivalence between an amount of energy, E, and an amount of mass, m, as given by Einstein's equation $E = m \times c^2$. (The term c^2 is the conversion factor between mass and energy—the speed of light, c, squared.) This equation lets us calculate, for example, the amount of energy that is released when hydrogen fuses into a slightly smaller mass of helium in the core of a star (Unit 50). The reverse reaction is also possible: If an amount of energy E is concentrated in one place, it may be turned into a particle whose mass equals E/c^2. Scientists observe matter created in this way from high-energy gamma rays in laboratory experiments.

In the early universe, extremely high-energy radiation filled all of space. Creation of matter from energy would have occurred everywhere. Such transformations of energy into mass, however, cannot occur in just any way. Laboratory experiments show that the conversion of energy into matter always creates a pair of particles, as illustrated in Figure 81.4A. Moreover, the particle pair has two special properties: The particles must have opposite charge (or no charge), and one of the pair must be made of ordinary matter, whereas the other must be made of what is called **antimatter**. For example, the electron has an antimatter particle, the **positron,** with identical mass but opposite electric charge (also see Unit 4).

Antimatter has the important property that a particle and its antiparticle are annihilated on contact, leaving only high-energy radiation, as depicted in Figure 81.4B. Thus, radiation (electromagnetic energy) can become matter in the form of a particle–antiparticle pair, and matter can become radiation. In the early universe, this shifting between mass and energy happened continually as particles formed and were annihilated.

Energetic radiation cannot turn into just any pair of particles: The pair's combined mass multiplied by c^2 must be less than the radiation's energy. For example, to become an electron–positron pair, the radiation needs a wavelength shorter than about 0.0012 nanometers. Any excess energy of the photon becomes the kinetic energy of the electron and positron, which race away from each other until they encounter and annihilate another of their antiparticle partners. Radiation with such large energies occurs only at extremely high temperatures. To have enough energy to make an electron–positron pair, the temperature must be more than about 6×10^9 Kelvin, which was the universe's temperature for about the first second of its existence.

More massive particle–antiparticles pairs were created at earlier times. Protons, for example, have a corresponding antiparticle called the antiproton. To become a proton–antiproton pair, the radiation needs roughly 1800 times more energy than to create the electron–positron pair. This requires a temperature 1800 times higher, or about 10^{13} Kelvin. Such high temperatures do not occur in ordinary stars, but they were reached about 1 microsecond (10^{-6} sec) after the Big Bang. At this time, the universe we can presently see was packed into a volume about 10 times bigger than the Solar System. At earlier times, the universe was hot enough for its radiation to be converted not only into protons and antiprotons but even into **quarks** and antiquarks, the particles that make up protons and neutrons.

The early universe can be traced back to about 10^{-33} seconds after the Big Bang. At the high temperatures found at these early times, even more massive fundamental particles (and their antiparticles) were present along with extremely intense high-energy radiation. At 10^{-33} seconds, everything in the universe that we can see today was jammed into a volume less than a meter across, expanding at enormous speeds.

We can picture how this progressed as time moved forward. At 10^{-33} seconds the universe was a dazzling sea of particles, antiparticles, and radiation constantly interacting. As time progressed, the universe expanded and cooled. Soon the radiation did not have enough energy to create the most massive particle–antiparticle pairs. Those that existed when the universe had cooled to this point continued to travel until they encountered one of their antiparticle partners and were annihilated. This happened for successively lower mass particles at later times until, at 10^{-6} seconds, the radiation was no longer energetic enough to create the lowest-mass quark–antiquark pairs (the so-called "up" and "down" quarks—see Unit 4). Thereafter quarks annihilated antiquarks, and the quarks that remained combined to make protons and neutrons.

It is a surprise, however, that any quarks remained at all because quarks and antiquarks should have been created from energy in equal numbers. For some reason, though, an imbalance left an excess of quarks over antiquarks. This is fortunate for us, because otherwise there would have been no matter in the universe at all. There would have been nothing in the universe but radiation.

Q If some regions of the universe had been left with a surplus of antimatter instead of matter, atoms would form out of antiprotons, antineutrons and antielectrons. How could we tell whether a distant galaxy was made of antimatter instead of matter?

Physicists have been using high-energy collider experiments to try to gain an understanding of the origin of this imbalance. They have discovered that the weak force (Unit 4) has what they describe as an **asymmetry.** An asymmetry can allow a pathway for an antiquark to decay, for example, but no equivalent pathway for its partner quark to decay as well. The result is that at the end of the first microsecond of the universe's history there was a tiny excess of quarks and electrons—just one in a billion—surviving with no antiquark or positron to annihilate.

The ordinary matter and radiation continued to expand, and for the first second the universe remained hot enough to create electron–positron pairs. At the end of that brief interval (about the time for one heartbeat), the universe cooled below 10 billion Kelvin, and the positrons annihilated the electrons, generating more photons. These photons, along with all of the others generated by these final annihilations, eventually became part of the cosmic microwave background, about a billion photons for every proton or neutron that survived.

The annihilations during the first second of the universe also would have produced a similar number of neutrinos. These neutrinos are today redshifted to energies that are currently too low to detect, only about 2 Kelvin above absolute zero. Someday, new detectors might allow us to observe these neutrinos, giving us a direct glimpse of the first moments of the universe. If eventually we can detect these primordial neutrinos, it will help us to better understand the early universe, much as the detection of solar neutrinos has proved to be helpful for understanding the inner workings of the Sun (Unit 50.3).

81.3 THE EPOCH OF INFLATION

The time before 10^{-33} seconds remains at the frontier of our understanding. Particle colliders cannot reach the energies of such early times, but a variety of evidence suggests what may have happened in the first instants of the universe.

The very earliest time to which we can extrapolate our current understanding of the forces of nature is known as the **Planck time.** The Planck time is a moment about 10^{-43} seconds after the universe began, when the entire universe we can see today would have been packed into a volume far smaller than a proton, and the temperature would have been greater than 10^{33} Kelvin. At the Planck time we encounter an interface between our understanding of the subatomic nature of matter and the way space is curved by gravity. Today physicists studying the subatomic behavior of matter observe **quantum fluctuations,** where the existence of particles and their properties can be described only according to probabilities rather than certainties. With the huge density of energy at the Planck time, extremely dense bits of matter would have fluctuated in and out of existence, causing space to curve in an unpredictable way that seems almost impossible to reconcile with our current understanding of the nature of gravity.

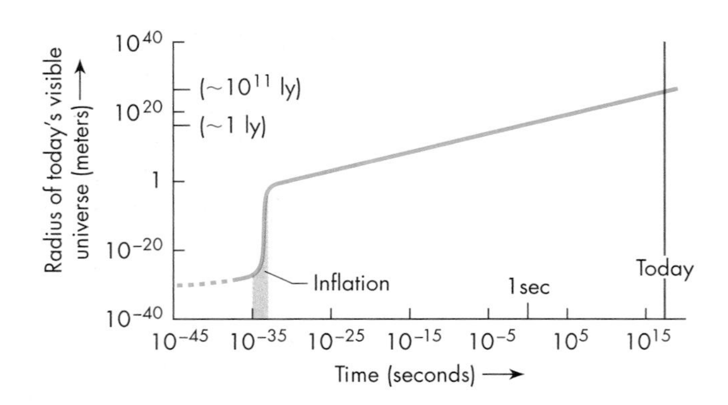

FIGURE 81.5
Changes in the size of the universe through time. The size of the total portion of the universe visible to us today is indicated by the blue curve. During the inflation epoch, from about 10^{-35} to 10^{-33} second, the universe increased in size by about a factor of 10^{30}.

Experiments that physicists have conducted with particle colliders give us a number of clues about conditions after the Planck time. The universe consisted of a boiling "soup" of energy so intense that all types of subatomic particles flashed in and out of existence—their masses almost nothing in comparison to the enormous concentration of energy present. It appears possible that the fundamental forces of nature (Unit 4) interacted at these extremely high energies quite differently from the way they do today. The transition of the strong, weak, and electromagnetic forces into their current form may have provided the mechanism that "launched" the rapid expansion of the universe in a process that cosmologists call **inflation.** Physicists hypothesize that during this brief time the size of the universe visible to us today expanded by a factor of about 10^{30} times its previous size—increasing in size far faster than the expansion before or after, as illustrated in Figure 81.5.

Inflationary models are based on hypotheses that attempt to unify the basic forces of nature. For example, in the 1970s it was discovered that the electromagnetic force and the weak force, which is responsible for radioactive decay, are really two aspects of the same force. The primary difference between these forces is in the mass of the carriers of the force. Photons, which carry the electromagnetic force, have no mass and can therefore be generated easily and travel large distances. W and Z particles, which are corresponding carriers of the weak force, have a very large mass. They therefore require high energies to be created and rarely travel more than a subatomic distance before decaying and giving up their energy. However, in the early universe, when there was so much free energy, W and Z particles could be created virtually as easily as photons. Physicists have found that the carriers of the electromagnetic and weak force form a closely linked family that becomes almost indistinguishable above about 10^{15} Kelvin. Today physicists generally refer to the two forces in combination as the *electroweak force* to emphasize the concept of unification.

Temperatures in the inflation epoch were far higher than can be reached with current experiments, but cosmologists have developed hypotheses for how the strong force unifies with the electroweak force at the temperatures found shortly after the Planck time. For an analogy to how these forces change at different energy levels, imagine a BB hitting a pane of glass. If you flick the BB at the glass with your finger, the level of energy is so low that the BB bounces off the glass as if repelled. However, if the BB is shot from a pellet gun, the energy level is high enough that the BB pierces the glass—the repelling effect vanishes at high energy. So, too, in the high-energy environment of the early universe, forces may have interacted and behaved differently from the way they do today in our low-energy world.

Currently there are several competing hypotheses of unification, called **grand unified theories,** or **GUTs.** According to most GUT models, the strong and electroweak forces would have remained unified until the universe's temperature dropped to about 10^{27} Kelvin. This would have happened about 10^{-35} seconds after the Big Bang. At this temperature the strong force began to "freeze out" of the unified mix. The breakdown of unification had extremely important consequences. GUT models suggest that the separation released a vast amount of energy.

The energy release that occurs when the strong force freezes out can be compared to the heat released when a liquid crystallizes. A crystalline solid must absorb energy to melt, and it releases the same amount of energy when it crystallizes. You may have experienced this phenomenon if you have ever used a "heat pack"—these are plastic pouches of a liquid that solidify and heat up when used. The liquid in the pack would normally be crystalline at room temperature, but it has been gently cooled and kept liquid. It requires just a small trigger to cause the liquid to crystallize and release the stored-up energy.

As the universe began this transition to a separate strong force, it was caught briefly in a peculiar state similar to the heat pack. The strong force remained part of the mix of forces after the universe had cooled to temperatures below which it would normally freeze out. With the transition to a separate strong force, the creation of energy everywhere led to an accelerating expansion of the universe that drove space to expand more and more rapidly. In each 10^{-35} second, the universe approximately doubled in size, growing exponentially. A region far smaller than a proton at first grew slowly, but after doubling and doubling again perhaps a hundred times over, by 10^{-33} second it had grown to the size of a beach ball.

Regions of space at the end of this inflationary expansion were moving away from one another at speeds far exceeding the speed of light. This might sound like an impossibility, but it is not. Motions of different regions of space relative to one another can have any speed, even greater than the speed of light. No light or matter can travel *through* any part of space at greater than light speed, and no observers will ever measure a photon traveling *by* them through space at any speed other than *c*. But different portions of the universe can travel extremely fast *relative* to one another, carrying along the matter within them.

One of the intriguing possibilities suggested by some inflation models is the existence of other universes forever separated from our own. It appears possible that inflation may not have occurred everywhere at the same time. After the Planck time, different regions of space may have inflated independently, and today they may be completely separated from one another with no possible contact between them. Such isolated regions might well be considered separate universes, independent of ours and completely unobservable by us. Furthermore, according to inflationary theories, these other universes may be either expanding at different speeds or even have expanded a while and then contracted back on themselves.

Inflation may create such sections of space much as bubbles form in a pot of boiling water. As water begins to boil, regions of the liquid suddenly change from liquid to gas (steam) and begin expanding. We see that region of steam as a bubble. Some bubbles expand, rise to the surface, and burst, whereas others collapse. So, too, in inflationary theories, entire universes may form and dissolve, entirely unknowable to us (Figure 81.6). It would be amazing indeed if all the wonderful intricacy and beauty of our universe is merely a single bubble in an even vaster sea of space, a space that we can never see.

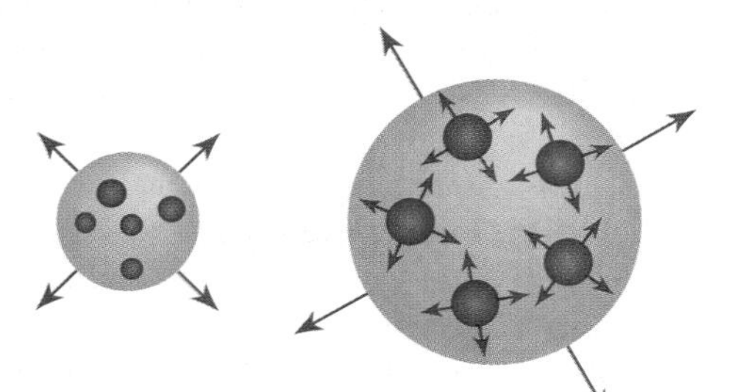

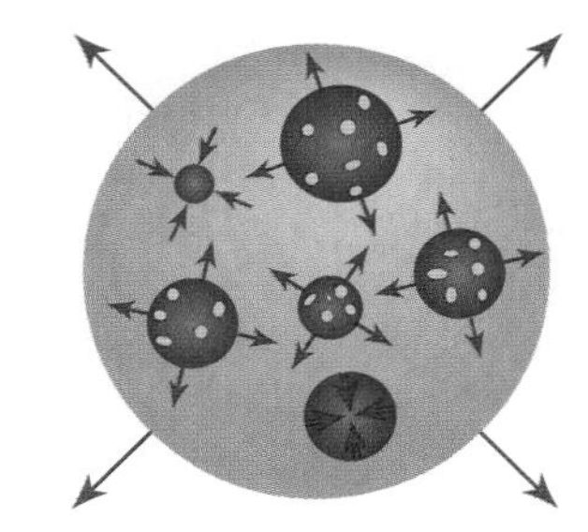

FIGURE 81.6
A sketch depicting bubbles that might become separate universes forming in an inflating universe. Each inflating bubble is completely outside of the region that will ever be visible within other universes.

81.4 COSMOLOGICAL PROBLEMS SOLVED BY INFLATION

The rapid acceleration generated during the inflation epoch offers an explanation of how the Big Bang might have been launched. It also explains several features of our universe that had previously seemed like mysterious coincidences.

The first of these features is the uniformity of the cosmic microwave background, as depicted in Figure 81.7A. In every direction we look, the radiation we observe has the same temperature, 2.726 Kelvin to within about 1 part in 100,000. This is much smoother than the basic Big Bang model predicts because at the time of the last scattering epoch (when the CMB radiation last interacted with matter) regions we see separated by more than 1° never had an opportunity for their radiation and matter to mix and reach a common temperature. Astronomers call this the **horizon problem** because regions outside each other's cosmic horizon (Unit 80.1) are totally independent and there is no obvious reason they should have the same temperature.

The horizon problem is sometimes called the *uniformity problem.*

With the inclusion of an inflation epoch, however, the visible universe we see today was originally one-thousandth-trillionth the size of a proton and was not expanding rapidly at first. Even though only 10^{-35} second had gone by in the history of the universe, light had traveled back and forth millions of times across the region that became our present visible universe. This allowed the temperature to even out. That smoothness was preserved during inflation and is retained to this day in the high degree of uniformity of the cosmic microwave background. The faster-than-light expansion speed of inflation carried these regions so far apart in a billionth-trillionth-trillionth of a second that the regions remained out of sight of each other for billions of years. Cosmic microwave background photons (Unit 80) are only now reaching us from regions that, before inflation began, had been closer to us than the distance between two quarks in a proton.

Although inflation makes the universe uniform on large scales, it also leads to the development of smaller irregularities. The cosmic microwave background

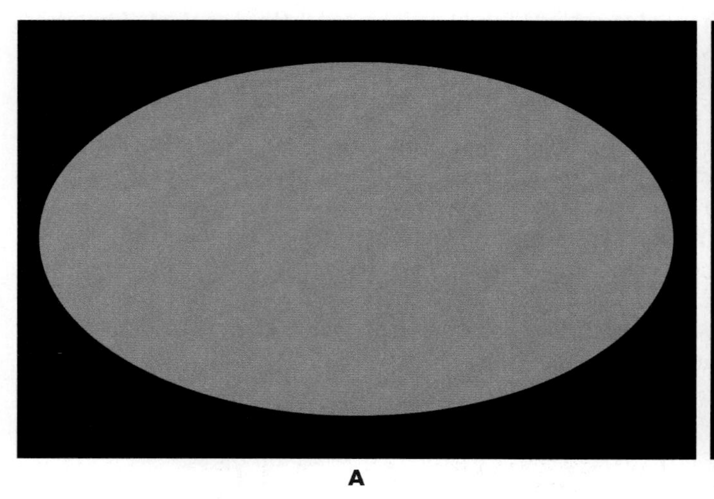

A

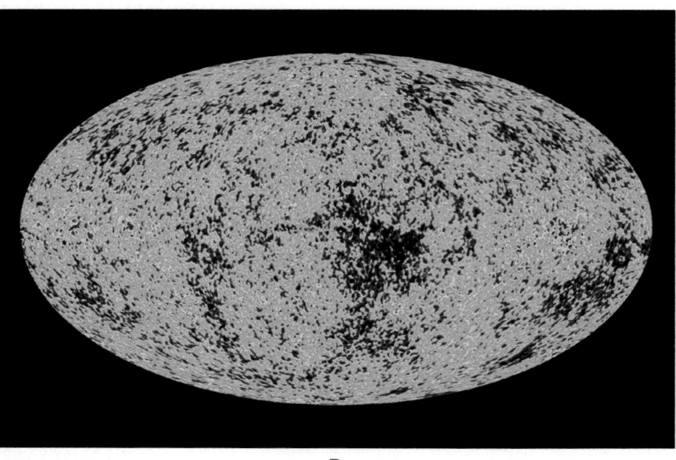

B

FIGURE 81.7
A map of the cosmic microwave background, made by the Wilkinson Microwave Anisotropy Probe satellite (WMAP). The map shows the brightness distribution of the cosmic microwave background across the entire sky, much as a map of the Earth depicts its features on a flat surface. (A) This map depicts the uniformity of the radiation seen—constant everywhere to within 1 part in 100,000. (B) The small deviations from uniformity are brought out in this map. Red regions are brighter (warmer by a few millionths of a degree) and blue regions are fainter (colder). They show irregularities in the temperature of the universe at the time of recombination, when the universe was about 400,000 years old. The satellite that gathered the data for the map was named for David T. Wilkinson, a pioneer in the study of the cosmic microwave background. The term *anisotropy* refers to the lack of uniformity from point to point across the sky.

FIGURE 81.8
An initially small region with obvious curvature looks flatter and flatter if it is inflated by a large factor.

is the same in every direction except for tiny variations that amount to about 1 part in 100,000. Just as water will spread and level out to a uniform height—sea level—it still contains small fluctuations—waves—that deviate from this uniform level. The map shown in Figure 81.7B shows the whole sky, spanning 360° across its center. There are many cool and warm regions that are far bigger than the 1° limit for communication at the time of the CMB. These large-scale fluctuations must have been produced before inflation began.

The inflation model suggests that microscopic quantum fluctuations from the time before 10^{-35} second were inflated tremendously, growing to the size of superclusters of galaxies today (Unit 76.3) or even larger. Based on the inflationary model, cosmologists have predicted a range and distribution in the sizes of denser regions generated by quantum fluctuations and then preserved in the patterns of large-scale structure much later. This closely matches what is seen imprinted on the CMB. The slightly higher-density regions acted as the sites where dark matter began to congregate, and where normal matter collected later on. Thus, the first steps toward the formation of large-scale structure in the universe (Unit 76.4) appear to have been taken when the universe was smaller than an atom.

Inflation also helps to explain a second problem, known as the **flatness problem**. Einstein's general theory of relativity predicts that space may be curved in different ways by gravity (Unit 80.4), depending on the overall density of matter and energy in the universe. A universe arising from a Big Bang might equally well have any positive or negative curvature, yet measurements show that the universe is almost precisely "flat"—having zero curvature. Inflation actually drives the universe toward flatness because the huge expansion removes whatever curvature the universe initially had. Inflation removes this curvature much as inflating a balloon makes its surface less curved. While the balloon is small, the curvature of its surface is obvious. However, if the balloon could be inflated to the size of the Earth, its surface would appear nearly flat, just as the surface of the Earth appears to a local observer as nearly flat (Figure 81.8). The rapid inflation of the early universe does the same to space, taking any initially curvature and expanding it until it appears flat over the size of our visible universe.

Inflationary models of the first moments of the universe mark the frontier of our understanding of the cosmos. They offer tentative explanations of some of its features, and they offer some intriguing possibilities about the nature of the universe beyond what we can see. Some models suggest that the universe may have been created literally from nothing. Some suggest that space has more dimensions than the four we are familiar with: the three of space and one of time. In fact, according to some ideas, called *string theories,* space has as many as 10 or 11 dimensions, but perhaps only our familiar dimensions expanded during inflation.

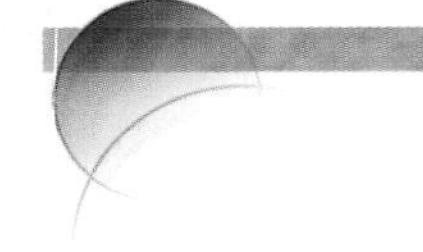

KEY TERMS

antimatter, 670
asymmetry, 671
early universe, 670
flatness problem, 675
grand unified theories (GUTs), 672
horizon problem, 674
inflation, 672
Planck time, 671
positron, 670
quantum fluctuation, 671
quark, 670
radiation era, 666

QUESTIONS FOR REVIEW

1. What leads astronomers to think that the early universe was hot and dense?
2. Why do astronomers call the times before the universe was about 50,000 years old the radiation era?
3. How was helium formed from hydrogen in the early universe?
4. What is antimatter?
5. What is the importance of asymmetry in the production of matter?
6. What are the basic requirements for the creation of matter from energy?
7. What is "unified" in grand unified theories?
8. What problems of the Big Bang theory are resolved by inflationary models?

PROBLEMS

1. One second after the Big Bang, the density of the universe was about 0.1 kg/L.
   a. How big a volume of the universe contained as much mass as the Solar System?
   b. Estimate approximately how much bigger the universe would have been after three minutes. How big would the Solar System mass region be then? Compare this to the present size of the Solar System.
2. How much energy is released when an antiproton and a proton collide? Compare this energy to the energy released when four protons combine to make a helium nucleus in a star.
3. As inflation began, the size of the universe (the currently visible universe) expanded from about 10^{-30} m to twice that size in 10^{-35} sec. Calculate Hubble's constant at that time, converting it to the same units used today. In the next 10^{-35} sec the size of this region doubled again. Show that Hubble's constant remains the same.
4. Examine Figure 81.7B. The map in the figure is 360° wide and 180° tall. An angle of 1° on this map corresponds to about 250 kpc.
   a. Approximately how big are the largest high-density regions you see, in degrees? How large is this in kiloparsecs?
   b. How large a region would that be today?
   c. Relate this size to components of the large-scale structure we see today (Unit 76).
5. Cosmologists have commented that the universe expanded more in the 10^{-33} sec of inflation than in the rest of the history of the universe. What they are referring to is the relative amount of expansion—the ratio of size at the end to the initial size. Compare the amount of expansion that occurred during inflation (doubling in size 100 times over) to the expansion of the present visible universe (13.7 billion ly in radius) from a region with just a 1-cm radius.
6. Compare the amount of energy released when a proton and antiproton annihilate each other to the energy released when four protons fuse to form helium.

TEST YOURSELF

1. What is meant by *inflation* in the early universe?
   a. The force of gravity suddenly grew stronger in the distant past.
   b. Protons expanded to the size of stars, which was how our Sun formed.
   c. The universe increased dramatically in size in an extremely brief period.
   d. The number of galaxies that we see at large distances is much greater than the number we can see near us.
   e. The diameter of distant galaxies is much greater than the diameter of galaxies near us.
2. Which of the following statements about the first few minutes of the Big Bang is true?
   a. The universe was transparent.
   b. The universe was very cool.
   c. Hydrogen was converted to helium.
   d. Galaxies began to form.
   e. The cores of stars began to form.
3. Which of the following observations is explained by a brief period of extremely rapid inflation?
   a. The high uniformity of the cosmic background radiation
   b. The large number of black holes in the universe
   c. The fact that some globular clusters are older than the universe
   d. The presence of helium in the universe
   e. The existence of more matter than antimatter

unit

82 The Fate of the Universe

yr −5Gyr Present

Background Pathways

The universe is currently expanding. But will it expand forever? This question is fundamental to the nature of the universe, because according to the Big Bang theory, all the hydrogen in the universe originated with the Big Bang. If the universe expands forever, one by one its stars will consume their hydrogen and die, eventually leaving the universe a black, cold, empty space.

But what is the alternative? If the universe does not continue expanding, perhaps its gravity will someday force it to stop expanding and fall back together. The overwhelming force of gravity means that all the objects and atoms within it will be compressed to higher and higher densities, crushing everything that has ever existed in what some astronomers have called the **Big Crunch**.

Neither fate is particularly appealing. Are there any alternatives? Cosmologists are exploring the possibilities by tracing the history of the universe's expansion and seeing how well it matches the trajectories predicted for simple expansion or future collapse. The latest results show that there may be a third, more bizarre fate in store for the universe.

82.1 EXPANSION FOREVER OR COLLAPSE?

Will the universe expand forever, or will the gravitational pull between parts of the universe cause it to collapse back on itself? This is similar to the question of whether a stone thrown upward will fall back down. If you throw a stone upward, the Earth's gravitational force slows its rise and eventually stops it and then the stone falls back to Earth. On the other hand, if you could throw the stone faster than the Earth's escape velocity, it would have enough energy to travel away from the Earth forever. The stone's behavior depends on the strength of gravity and the upward impulse given to it. So too with the expansion of the universe.

Moments after the universe's birth, it appears that inflation began the universe's rapid expansion (Unit 81). Since then the gravitational forces between galaxies and all the other contents of the universe have slowed its expansion. If the universe's gravitational force is weak enough, or if the initial impetus of expansion is strong enough, we would expect the universe to expand forever. On the other hand, if the gravitational force is strong or the initial impetus is weak, we expect that the expansion will stop and the universe will collapse.

To measure the relative strength of these effects for the universe, we can compare the gravitational energy holding the universe together to the energy of its expansion. If gravity dominates, the expansion will stop. If the expansion energy dominates, the expansion will continue forever (Figure 82.1). Therefore, we can learn (at least in principle) whether the universe will expand forever or collapse back on itself by seeing if the universe has a gravitational pull strong enough to stop its expansion.

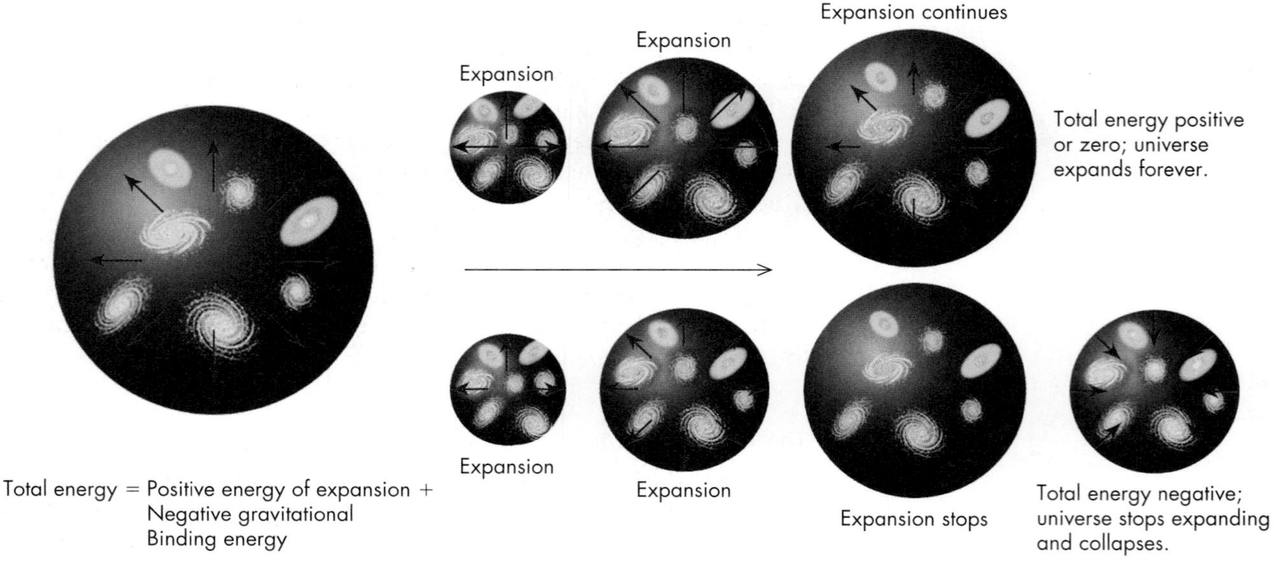

FIGURE 82.1
A sketch of how gravity and the energy of expansion determine the behavior of the universe. If the energy of expansion is small compared to the gravitational pull, the universe eventually collapses back on itself. If the energy is large enough, however, the universe expands forever.

Q Do you have a personal preference about the eventual fate of the universe? What is it, and why?

To see how strong the effect of gravity is, astronomers use Newton's law of gravity, modified to account for relativity theory. Essentially, gravity's effect depends on the universe's mass. It turns out, however, that it is far easier to work with the universe's density (the amount of mass contained within a given volume). The reason is simple: Astronomers can measure the density of the universe but not its mass. To measure its mass, we would need to observe all of space. To measure its density, we need only measure the mass within some large, representative volume of the cosmos.

82.2 THE DENSITY OF THE UNIVERSE

To find the density of the universe, astronomers choose a large volume of space and attempt to find all of the galaxies within it. Next we measure the mass of each galaxy, add the masses, and divide by the volume. For example, to measure the local density, astronomers might choose the Local Group, which contains three large galaxies and about two dozen smaller ones. The total mass of gas and stars in the Local Group is estimated to be about 10^{12} solar masses, which we can convert to kilograms if we multiply by the Sun's mass, 2×10^{30} kilograms. This gives a mass for the group of about 2×10^{42} kilograms.

Next that mass is divided by the Local Group's volume, assumed to be a sphere whose radius is the distance from the center of the Local Group to the next nearest galaxy group, which is about 3 Mpc (about 9×10^{22} meters) away. Using the formula for the volume of a sphere ($\frac{4}{3}\pi R^3$) gives a volume for the Local Group of about 3×10^{69} cubic meters. Dividing this volume into the mass gives a density of about 7×10^{-28} kilograms per cubic meter, or 7×10^{-31} kilograms per liter.

The vicinity of the Local Group is somewhat richer than the average environment, however. To get a more representative sample, we must look at a much larger region that includes both clusters and voids, and we must include intergalactic gas, particularly in clusters. A similar calculation for these much-larger volumes of the universe gives a slightly smaller value of about 3×10^{-31} kilograms per liter, or, on average, roughly 2 hydrogen atoms per 10 cubic meters. This indicates that, on average, the universe today has a fantastically low density.

To determine whether the universe will expand forever or recollapse, astronomers compare this observed density to a theoretically calculated **critical density**, which is written with the Greek letter rho as ρ_c. If the actual density is greater than the critical density, the universe will recollapse; if it is less, the universe will expand forever. We can calculate the critical density by comparing the gravitational potential energy in a volume to the kinetic energy of expansion in that same volume. At the critical density, the two energies are equal. It turns out that mathematically the critical density ρ_c is

ρ_c = Critical density
H = Hubble's constant
G = Gravitational constant

$$\rho_c = \frac{3H^2}{8\pi G}$$

where H is the Hubble constant and G is Newton's gravitational constant. Without carrying out the full derivation, we can see why the equation has this form. The gravitational potential energy depends on the density and Newton's gravitational constant, whereas the kinetic energy depends on the square of the expansion speed, which can be related to Hubble's constant. Thus, we are equating values proportional to $G \times \rho$ and H^2 and then solving for ρ.

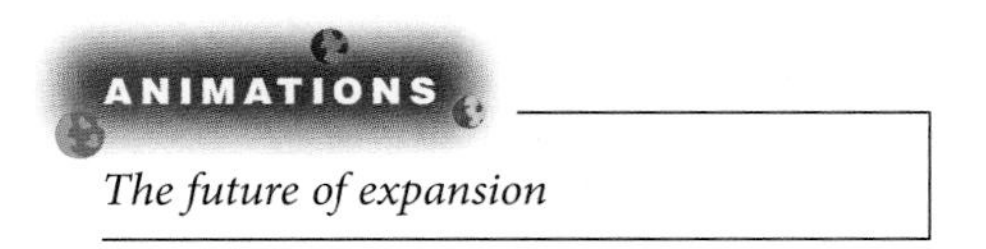

The future of expansion

Cosmology

Astronomers use a quantity called **Omega (Ω)** to indicate how close the observed density is to the critical density. This is because, as the last letter of the Greek alphabet, omega is sometimes used to suggest the "end of things." For the amount of matter in the universe, then, cosmologists define Ω_M as the value of the actual density of matter divided by the critical density:

$$\Omega_M = \rho/\rho_c$$

With this notation:

If $\Omega_M > 1$, the universe will eventually collapse.
If $\Omega_M < 1$, the universe will expand forever.
If $\Omega_M = 1$, the universe is exactly at the critical density.

We specifically note that this is the value of Ω for *matter* because a variety of quantities can contribute to the overall value of Ω, and not all of them have the same effect on the universe's expansion, as we will see later in this unit.

The past and future sizes of the universe are illustrated for various values of Ω_M in Figure 82.2. Each of these models is assumed to match the present size and rate of expansion of the universe. The rate of expansion is known from Hubble's constant, and it is indicated by the slope of the curve today—steeper slopes indicating faster expansion. If Ω_M is larger than 1, the universe expands rapidly after the Big Bang, but the large density of matter causes the universe to slow its expansion quickly. In this case the expansion halts sometime in the future, and the universe then collapses back in on itself, ending in a "Big Crunch."

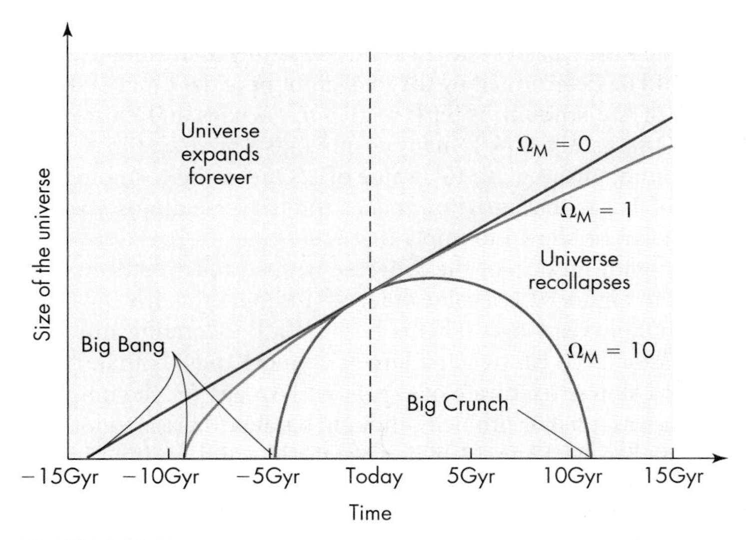

FIGURE 82.2
The size of the universe in the past and future as calculated for different densities. Curves are shown for three cases in which: there is no matter ($\Omega_M = 0$), so the universe expands without deceleration; the pull of gravity and the expansion rate balance ($\Omega_M = 1$), so the universe constantly slows down but never recollapses; and when the pull of gravity greatly exceeds the energy of expansion ($\Omega_M = 10$), so the universe recollapses in about 12 billion more years. All curves have the same slope today, which is determined by the Hubble constant. Curves lying in the blue region of the diagram have enough energy to expand forever, while those in the yellow region will eventually collapse back on themselves.

If Ω_M is 0, the expansion of the universe is not slowed at all, and the rate of expansion remains constant. The time since the Big Bang in this type of universe is equal to the Hubble time $1/H \approx 14$ billion years (Unit 79.4) because the speed of expansion is the same today as it was in the past. In models of the universe with values of Ω_M greater than 0, the time since the Big Bang is smaller than $1/H$.

The $\Omega_M = 1$ case is interesting because it implies that the universe's gravitational potential energy exactly cancels its kinetic energy—a condition that some theories of the beginnings of the universe suggest makes the most sense. In the $\Omega_M = 1$ cosmology, the universe's expansion slows, approaching zero speed in the infinite future but never quite stopping.

When we compare the observed and critical densities, what do we find? If we use the value for H of 70 kilometers/second per megaparsec, the critical density is 9.3×10^{-30} kilograms per liter. This tiny density—less than a billionth-billionth of the density of the best vacuum ever created in a laboratory on Earth—would be sufficient to halt the universe's expansion. As small as this density is, it is about 25 times larger than the estimated density of visible stars and gas throughout the universe. This low value of the density of normal matter is also consistent with the estimates determined from the amount of fusion that occurred during the first few minutes of the universe (Unit 81.1). Therefore the gravitational pull of all the normal matter in the universe is too weak to stop the expansion. This can be written as $\Omega_{VM} = 0.04$, where VM stands for visible matter.

However, this result does not take into account dark matter. Observations of galaxies and clusters imply large quantities of unseen matter, many times more than is directly observed (Unit 78). To explain the rapid rotation of the outer parts of spiral galaxies and motions within galaxy clusters, astronomers estimated that there might be anywhere from 5 to 20 times more dark matter than normal matter, so $\Omega_{DM} = 0.2$ to 0.8. Combined with the estimates for visible matter, and given the uncertainties in the measurements, the majority of astronomers in the 1990s suspected that the total amount of matter, $\Omega_M = \Omega_{VM} + \Omega_{DM}$ was equal to 1 or less.

This was the state of cosmology at the end of the twentieth century. Data on the density of matter in our vicinity, combined with the measured expansion rate of the universe, indicated that the galaxies should continue sailing apart forever. The fate of this universe would be determined by the total amount of dark matter. Even though most estimates of the amount of dark matter in galaxies and clusters of galaxies suggested that Ω was still below 1, many cosmologists predicted that more dark matter would be found, pushing Ω_M to a value of 1. One of the cosmologists' arguments for a value of 1 was that inflation makes the universe almost exactly "flat" (Unit 81.4), which can be shown to imply that $\Omega = 1$.

Q If the universe contained nothing but the visible matter we have observed, what would its expansion look like compared to the models shown in Figure 82.2?

There were problems with models of the universe with so much dark matter, however. As you can see in Figure 82.2, a value of $\Omega_M = 1$ implies that the Big Bang occurred less than 10 billion years ago. This is because a decelerating universe used to be expanding faster, so it can reach a large size more quickly than can a universe that maintains a slower fixed rate of expansion. An age for the universe of under 10 billion years was a major problem, though, because there are globular clusters that are estimated to be 12 to 13 billion years old—and nothing in the universe can be older than the universe itself.

Things were not adding up right, so it seemed that something must have been left out of our calculations.

82.3 THE SUPERNOVA TYPE IA FINDINGS

To improve our predictions about the eventual fate of the universe, cosmologists have developed different means of examining how the universe is expanding. Besides measuring the current rate of expansion and density of the universe, another way is to look at the history of expansion. Figure 82.3 focuses on the history of expansion that was illustrated in Figure 82.2, showing again the same three models of the

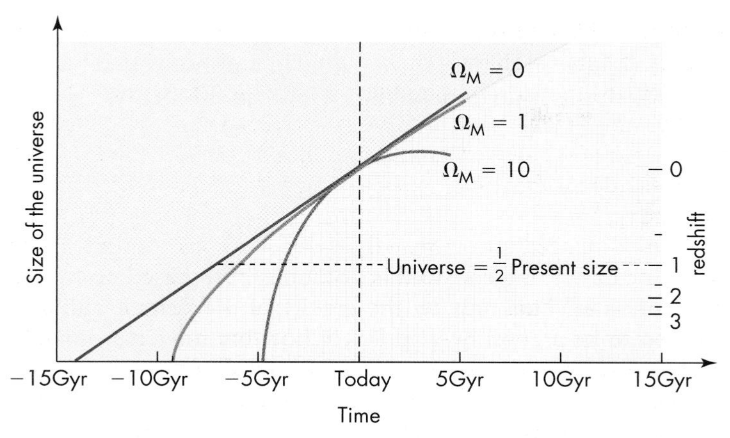

FIGURE 82.3
By studying the rate of expansion at large redshifts, such as at $z = 1$ (when the universe was half its current size), we can determine which cosmological model applies. For the $\Omega_M = 10$ model, for example, light from an object at $z = 1$ will not have had to travel as far as it would have in the $\Omega_M = 0$ model.

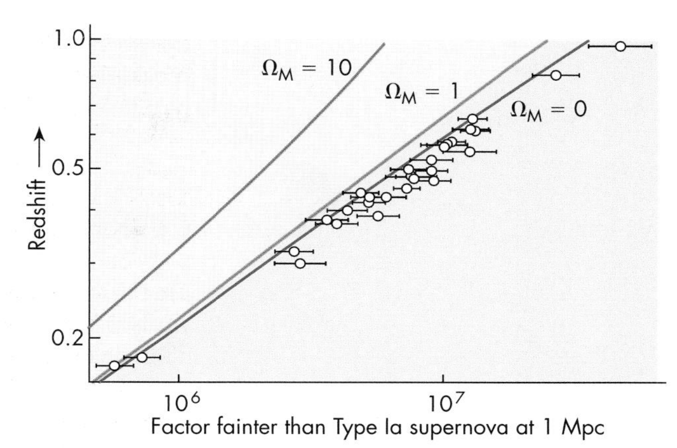

FIGURE 82.4
A graph of the redshift and brightness of Type Ia supernovae in distant galaxies. The brightness is labeled according to how many times fainter the supernovae appeared compared to how bright they would have been if they were in a galaxy 1 Mpc away. The curves show the predicted brightness if the universe has expanded according to models of the universe in which $\Omega_M = 0$, 1, or 10 (as in Figure 82.2). The horizontal lines through each data point show the uncertainty for each supernova distance measurement. The graph is based on observations of S. Perlmutter and collaborators. Note that the distant galaxies tend to lie below any of the models—indicating that the light is dimmer than any model in which the universe has undergone only deceleration.

history of expansion between the Big Bang and the present. Although the three models predict different ages for the universe, this is difficult to measure directly. We can instead look for differences between the models at different redshifts.

When we observe a galaxy at a redshift of $z = 1$, the wavelength of the light coming from it has been stretched by a factor of 2 (Unit 79.2). This means the universe was only half its current size when the light left the galaxy. The horizontal dotted line in Figure 82.3 indicates when the universe was half as big as it is now.

In low-Ω_M models, the rate of expansion slows down only a little; so when the universe was about half as big, it was about half as old as it is now, and the light has taken about 7 billion years to reach us. In the high-Ω_M model, the universe contains much more mass, which makes the universe slow down quickly. In this model the light from an object at $z = 1$ travels for less than 4 billion years. In other words, the distance to a galaxy with a particular redshift will be measured to be smaller if Ω_M is larger. In principle, then, we can determine the value of Ω_M if we can measure the redshifts and distances of distant galaxies. This test is straightforward, but doing it well enough to test the expansion theory requires precise distance measurements to faraway galaxies.

Supernova observations allow such measurements, though they are often difficult to obtain for such remote galaxies. In the late 1990s, using the Hubble Space Telescope, astronomers were able to observe supernova blasts in some very remote galaxies. In particular, they detected supernovae of Type Ia (Unit 65), which result from the explosion of white dwarfs. These have been found to be excellent standard candles, so we can determine accurate distances based on their known luminosity and observed brightness. This allows astronomers to compare the brightnesses of supernovae at different redshifts out to very large distances.

The results are shown in Figure 82.4, and they have delivered quite a shock! The lines drawn in the figure show the predicted supernova brightnesses based on the same three models shown in Figures 82.2 and 82.3. It was expected that the measured distances would all lie close to the green $\Omega_M = 1$ line. Instead, the measurements lie to the right even of the $\Omega_M = 0$ line. In a universe that has experienced only deceleration, this is an impossibility. How can this be?

82.4 DARK ENERGY

Our discussion of the fate of the universe to this point has been based on the assumption that expansion is affected only by the gravity of the matter within it. This originally appeared to be a good description of how the universe expands; but gravity's pull can only slow down the expansion, and this is not what the supernova result shows.

The best current explanation of this strange result comes from Einstein's early efforts to develop the general theory of relativity. Einstein developed equations to describe how matter and energy curve space and create the force of gravity. When he solved the set of equations he had put together to describe gravity, his mathematical solution showed an extra term. This extra term was called the **cosmological constant** because the mathematics of general relativity suggested that it should be the same everywhere and at all times.

The cosmological constant can be described as an energy that fills all of space. This energy is unlike energies we are familiar with. It remains constant everywhere, existing even where there is nothing but space, and it does not grow more dilute as space expands. This is different from how matter and electromagnetic energy behave—they spread out and become more dilute as the universe expands.

No measurements in Einstein's time could determine the value of the cosmological constant. However, if the cosmological constant is not zero, it can counteract the effects of gravity. This constant level of energy everywhere creates a kind of cosmic repulsion, driving space to expand faster. Astronomers have given the cosmological constant the descriptive name **dark energy** because it is a kind of counterpart to dark matter.

Actually, Einstein developed general relativity before Hubble had discovered the expansion of the universe, and at that time the universe was thought to remain static rather than expand or contract. General relativity predicted that the universe should be in motion, so Einstein suggested that the cosmological constant might be quite large and balance gravity's attraction, making the universe static. Then, within a few years of Einstein's suggestion about the cosmological constant, astronomers discovered that the universe was in fact expanding, and Einstein concluded he should have set the cosmological constant to zero all along. He called this his "greatest blunder" because he might have actually predicted the expansion of the universe in addition to his many other remarkable discoveries. Since that time, cosmologists have noted the possibility of a cosmological constant, but for decades most assumed that it was zero.

Q The discovery that the universe can accelerate as well as decelerate opens up the possibility that the universe could have been "stationary" for some period in the past before resuming its expansion. If there had been such a period, would galaxies seen from that time have zero redshift?

The recent findings that the universe's expansion may be speeding up have revived interest in the cosmological constant. If dark energy in fact pervades all of space, it has major implications for our understanding of the universe. For example, because dark energy speeds up the universe's expansion, our estimates of the age of the universe must be modified. The result is an age that agrees better with the ages of the oldest known stars. The current best estimate of the universe's age, accounting for both gravity and the cosmological constant, is 13.7 billion years. This model uses a value for Hubble's constant of 71 kilometers/second per megaparsec, and it assumes that there is almost three times as much dark energy in the universe today as the total of normal and dark matter. The comparison of energy and matter in this way is based on Einstein's formula $E = m \times c^2$, so matter (both normal and dark) represents about 2.5×10^{-30} kilograms per liter, while dark energy is about $6.8 \times 10^{-30}\ c^2$ kilograms per liter $\approx 5 \times 10^{-13}$ joules per liter.

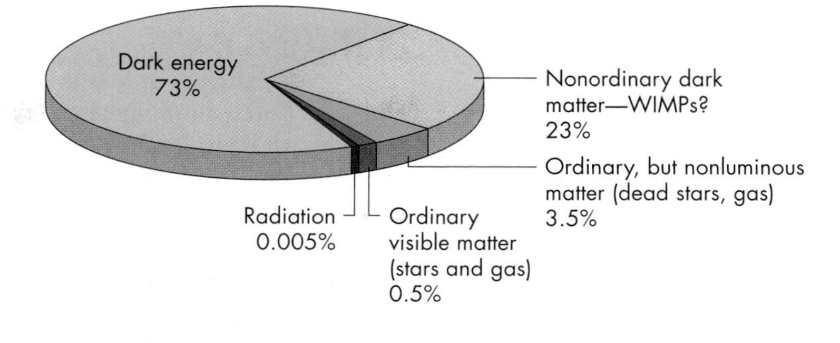

FIGURE 82.5
The makeup of the universe as deduced from observations of the brightness variations in the cosmic microwave background and other data. The percentages have been rounded off, so they do not add up to exactly 100%.

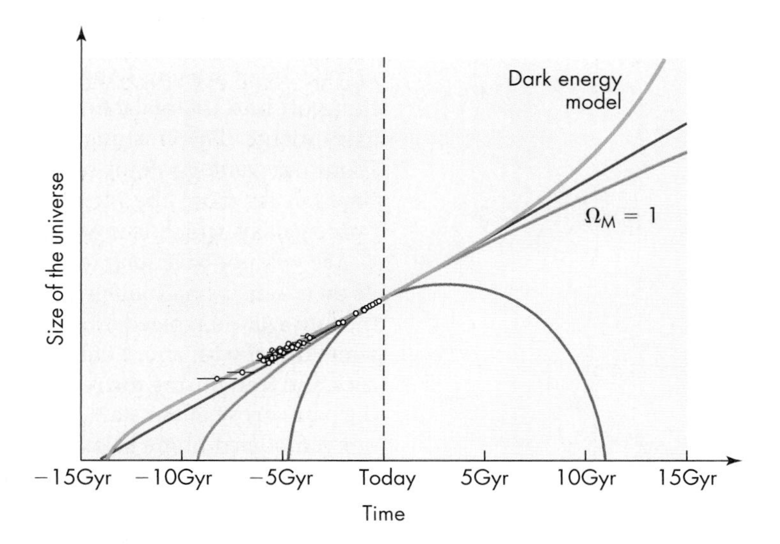

FIGURE 82.6
The data from Type Ia supernova is shown in a graph of the expansion of the universe. At times in the past, the data lie above curves based on expansion with no dark energy. The best-fit curve, in which the universe contains nearly three times as much dark energy as matter, is shown in blue. A universe with dark energy will accelerate more and more rapidly in the future.

In the last few years, additional evidence for the existence of dark energy has accumulated from further supernova measurements as well as from studies of the cosmic microwave background (Unit 80). Detailed observations of the background reveal tiny variations in brightness from place to place. Those tiny variations come from slight "lumps" in the universe's density shortly after its birth. The physical sizes of these lumps are determined largely by the density of both visible and dark matter, and the best fit to these results indicates that $\Omega_M = 0.27$. However, the *apparent* size of these lumps (their angular size) depends on the curvature of the universe (Unit 80.4), which cosmologists have shown implies that $\Omega = 1$. The difference between these results for Ω and Ω_M shows that something else must be contributing to the curvature of the universe.

One of the intriguing aspects of dark energy is that although it works against gravity to speed the expansion of the universe, the amount of dark energy should be *added* to the density of matter in the universe in determining its curvature. If dark energy contributes the missing fraction of Omega—that is, $\Omega_{DE} = 0.73$—then both supernova results and the curvature results can be explained. The contributions of the various components of the universe to its overall composition are shown in Figure 82.5.

The pattern of expansion predicted for a universe with the current estimates of dark energy and dark matter is shown in Figure 82.6. In this model, the universe initially expanded rapidly, but the gravitational pull of its matter began to slow it down. As the universe expanded, the matter was thinning out and its gravitational pull was growing weaker. Meanwhile, though, the dark energy would have remained constant, so its repulsive effect began to take over to accelerate the universe. The universe has been picking up speed since then, and if we extrapolate into the future, the universe should expand faster and faster.

The inflation of the universe in the first tiny fractions of a second was another episode of exponentially increasing expansion (Unit 81.4).

82.5 THE END?

With our understanding of the processes taking place in stars, galaxies, and the universe, we have reached a point where we can attempt to predict the ultimate fate of the universe.

For many billions of years to come, things will look fairly unchanged. The Sun will remain on the main sequence and the Solar System will orbit the center of the Milky Way about 20 more times. If humans survive on Earth through this time, they might notice that the Andromeda Galaxy (M31) appears to be much larger. It will be detectably redder as well, as fewer young stars are being born from the diminishing amount of interstellar gas. We are heading for a collision with M31, and just before we collide the galaxy will fill much of the sky.

The Solar System, lying in the outer parts of the Milky Way, will probably be flung off into intergalactic space during the collision, sailing outward as the galaxies merge. The crashing galaxies will stimulate a great burst of star formation. From our vantage point outside the galaxy, we will watch these fireworks as we first sail far from the interacting system before plunging back toward our new home galaxy, which is now looking more like an elliptical galaxy.

We will not have long to watch our Galaxy transform before the Sun undergoes its own transformation. First becoming a red giant, swallowing Mercury and Venus, the Sun will then bleed away its outer atmosphere, finally leaving behind a cooling white dwarf with about half its current mass. The Earth, turned to a ball of molten rock and iron during the red giant phase, may survive as an ice-cold cinder orbiting the pale corpse of our star. As the remnants of the Solar System plunge back through our remodeled home galaxy, distant encounters with stars will overcome the Sun's own weakened pull. The outer planets and comets will probably be stripped away.

The same story will play out over and over again as other stars first shine brightly in their final stages before dying. Over the next trillion (10^{12}) years the most distant small galaxies and clouds of gas in the Local Group will fall into and merge with our elliptical galaxy. These might lead to a few more bursts of star formation, but the amount of hydrogen fuel remaining has become small and the bursts more feeble. In fact, supernova explosions from the occasional massive star that forms will drive out almost all of the remaining gas into intergalactic space. And then, one by one, the stars will blink out, with no new stars to replace them. Our galaxy will grow steadily dimmer and redder, the last dim red stars fading away in 100 trillion (10^{14}) years, based on projections of stellar lifetimes and the amount of hydrogen fuel that galaxies have available.

What will remain is a galaxy of black holes, neutron stars, and cold white dwarfs orbiting in darkness. Occasional close encounters between stars will fling remaining planets from their systems, and some of the close encounters will send one star flying out of the galaxy, while the other sinks toward the galaxy's center. There the star may pass too close to the supermassive black hole at the galaxy's center, now grown to over a billion times the Sun's mass. The shredded star will shine brightly for a short while as its material careens around the massive black hole before eventually being drawn in.

In this bleak future, it might still be possible for intelligent life to survive. Pockets of hydrogen fuel might remain in diffuse clouds or isolated planets, but they would be extremely far apart and hard to gather. If technology permits it, perhaps even some asteroids or planets could be nudged toward black holes, where their matter will shine brightly for awhile as a hot accretion disk of infalling matter around the black hole. Inevitably, though, the available fuel will be consumed, converted into heat from which no further energy can be extracted. This is described as the **heat death** of the universe—not because it is hot, but because what energy there is has all been converted to the same low-grade disordered motion from which no further power can usefully be extracted.

A curious thing would occur as this future unfolded. The expansion rate of the universe would be growing faster and faster because of dark energy. If there had been no dark energy, we might have expected the Local Group itself to plunge into the Virgo Cluster in a few hundred billion years. Instead, after about 100 billion (10^{11}) years, dark energy will cause the expansion of space to overcome the Milky Way's gravitational attraction toward the Virgo Cluster, and then the separation will grow exponentially faster.

In this future time, our galaxy will be essentially alone in the universe. Strangely enough, it will not be far different from the idea, widely accepted before 1900, that the Milky Way was an island of stars in a vast empty universe (Unit 70). For the next few hundred billion years, if there were future Edwin Hubbles, they would not note much difference in the rate at which nearby galaxies appear to be moving away from our galaxy, but there would be steadily fewer of them to be found as the universe aged. The light from more distant galaxies would become enormously redshifted, and the light from them would take much longer to reach us because the space between us would be expanding so rapidly. An astronomer living trillions of years from now would probably detect no nearby galaxies outside of our galaxy, but she would still be able to see faint signals from distant galaxies when they were young and still forming stars. Their light would be so redshifted that they might be visible only at infrared and radio wavelengths. The sky would be dark at visible wavelengths.

In the long term, much of the matter in our galaxy will end up in black holes. Chance collisions will eventually collect matter in these gravitational traps. Orbits themselves must gradually shrink because they lose energy through gravitational radiation (Unit 68.5). After about 10^{20} years, the orbits of any remaining planets orbiting stars will shrink until the objects collide and coalesce. These events will shine brightly for a few years before fading away again.

Matter that does not end up in a black hole faces a more insidious problem. Physicists suspect that matter itself may not be stable. One of the consequences of grand unified theories (Unit 81.3)—the same theories that predicted inflationary expansion in the very early universe—is that they predict **proton decay.** It used to be thought that protons are stable particles, but if the forces are unified, there should be pathways by which a proton can break down into lighter particles. Physicists have conducted many experiments hunting for this phenomenon without success, but this may be because of the extremely improbable nature of such decays. It is still highly uncertain, but protons may live 10^{40} years before disintegrating.

Even black holes will not last forever. As Hawking showed (Unit 68.3), they can radiate away energy through quantum effects. However, the bigger the black hole, the slower this process. It will take in excess of 10^{60} years for this to occur for stellar-mass black holes and 10^{100} years for the supermassive black hole at the center of our galaxy. In the end, then, the universe might contain nothing but low-energy highly redshifted radiation. As the poet T. S. Eliot wrote,

> This is the way the world ends
> Not with a bang but a whimper.

This dark cold end to the universe currently appears to be the most likely scenario, but it is not a certainty by any means. For example, we still do not know what the dark matter or dark energy are composed of. They may undergo changes of their own that will alter our future. Some cosmologists suggest that the dark energy may grow with time. In that case, it would not just gradually pull galaxies away from each other. In this alternative fate, called the "Big Rip," space on smaller and smaller scales begins to expand rapidly. Dark energy would increase until it tears apart our Galaxy, then even our Solar System, and eventually even smaller structures. In the Big Rip, space expands so fast that every part of the universe is eventually pulled away from every other part at speeds so fast that they all disappear from each other's sight.

Some physicists think that the accelerating expansion of the universe hints at additional dimensions of space beyond the three we know. In these scenarios, our universe may undergo collisions with other universes. All of these different ideas illustrate that astronomers are in the early stages of studying dark energy. They are still developing tests to detect it more directly. The science of cosmology is beginning, however, to sketch the overall structure, history, and future of the universe. It is a universe in which the stars and galaxies—all of the kinds of matter that we are familiar with—are just minor players. We are along for the ride in a universe driven by far larger forces generated by dark matter and dark energy.

KEY TERMS

Big Crunch, 677
cosmological constant, 682
critical density, 679
dark energy, 682
heat death, 684
Omega (Ω), 679
proton decay, 685

QUESTIONS FOR REVIEW

1. What are the possible fates for the universe?
2. Why do astronomers think the universe will expand forever?
3. How is the density of the universe determined?
4. What is the critical density?
5. How was dark energy discovered?
6. How is dark energy related to Einstein's cosmological constant?
7. What is meant by heat death of the universe?

PROBLEMS

1. The critical density depends on the value of Hubble's constant H. If H turns out to be 65 km/sec per Mpc, calculate how much the critical density changes compared to $H = 71$ km/sec per Mpc.
2. The following equations can help us begin to understand why the Local Group is predicted to survive the effects of dark energy, but more distant galaxies are expected to be pulled away:
   a. Using the values of mass and size in the text, what is Ω_M for the Local Group? Note that in the text, this is calculated out to the distance of neighboring groups.
   b. Instead calculate the density inside of the central 1-Mpc radius region of the Local Group (where nearly all of the group's mass resides). How does this density compare to critical density?
3. Dark energy retains a constant density even as the universe expands. Today it has a value of $\Omega_{DE} = 0.7$. Matter, by contrast, grows more dilute as space expands, so its density would have been larger in the past by a factor that depends on the redshift, as $(z + 1)^3$. At what redshift would dark energy and matter (currently $\Omega_M = 0.3$) have had the same density?
4. If dark energy in the universe currently is about 73% of the critical density, then how many joules of energy are there in each liter of space? How many 1-mm-wavelength microwave photons would this correspond to in each liter?
5. If the Solar System were at the critical density, what would be its total mass? (Assume that the Solar System is spherical with a radius of 40 AU.) How many joules of energy does this correspond to? How many Joules of dark energy are there within this volume?

TEST YOURSELF

1. Dark energy
   a. is the same as the cosmological constant.
   b. is repulsive.
   c. is distributed evenly through space, never dissipating.
   d. causes space to expand ever faster.
   e. All of the above.
2. Which of the following is true of the expansion of the universe if there is no dark energy?
   a. The expansion rate will increase.
   b. The expansion rate will decrease.
   c. The expansion rate will remain constant.
   d. The expansion rate will first increase and then decrease.
3. What causes the universe to curve around onto itself?
   a. Its own mass and energy
   b. The cosmic microwave background
   c. The distribution of galactic superclusters
   d. Black holes sucking in all the space around them

unit 83

Astrobiology

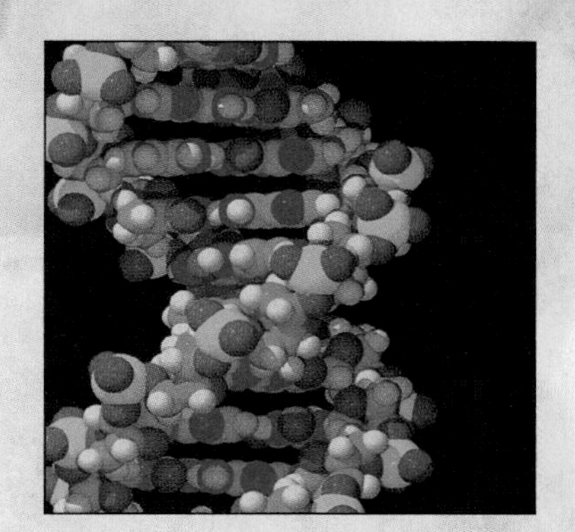

The evolution of the universe from Big Bang to galaxies, stars, and planets has along the way created us. This evolutionary path from quark soup to thinking beings who seek to understand how they fit into this grand scheme—what eighteenth-century scientists called "the great chain of being"—is one of many marvels of the universe.

To understand the possibilities for life elsewhere in the universe, we turn to life on Earth and how it may have formed and developed. We will see that the existence of life on our own planet suggests the possibility of its existence elsewhere.

We begin our discussion by examining the long history of life on Earth, looking at the factors thought to play a role in its development here. Finally, we will speculate about the role of life in the history of our planet—the so-called Gaia hypothesis—and whether our existence may even say something about the nature of the universe itself.

83.1 LIFE ON EARTH

Life has existed on Earth for most of the history of our planet. The first indications of life are found in rocks as old as 3.8 billion years in the composition of particles of carbon. These particles show an unusual mix of carbon isotopes that is found to be particular to living organisms today. Carbon has two stable isotopes, ^{12}C and ^{13}C, which have 6 and 7 neutrons in their nuclei, respectively. In photosynthesis and other biological activities, microscopic organisms prefer the lighter isotope ^{12}C. The remains of living organisms therefore show a deficit of ^{13}C relative to its naturally occurring abundance.

Some microscopic fossils as old as 3.5 billion years are suggestive of organisms that look similar to modern day blue-green bacteria or **cyanobacteria** (Figure 83.1). Few

A

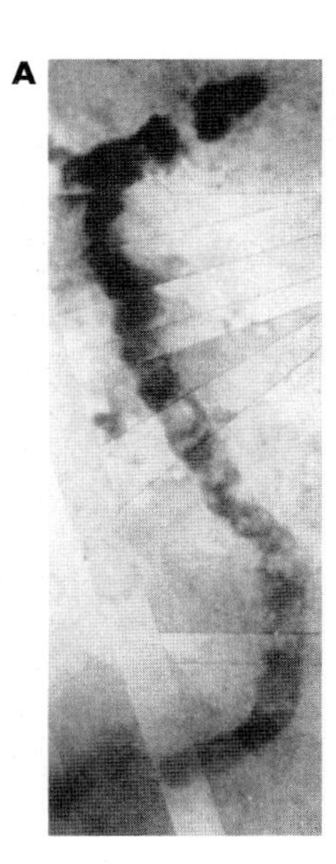

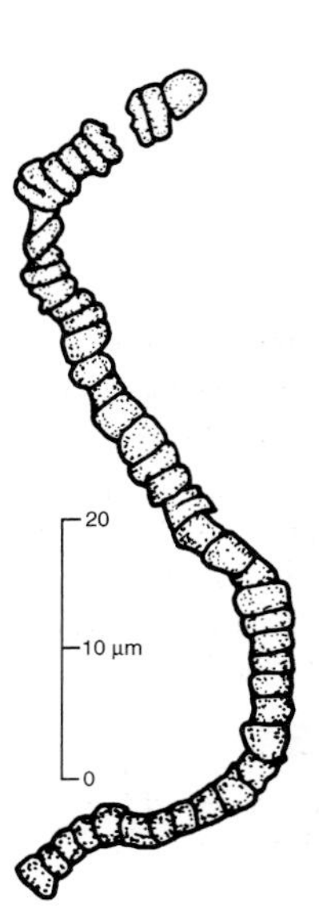

B

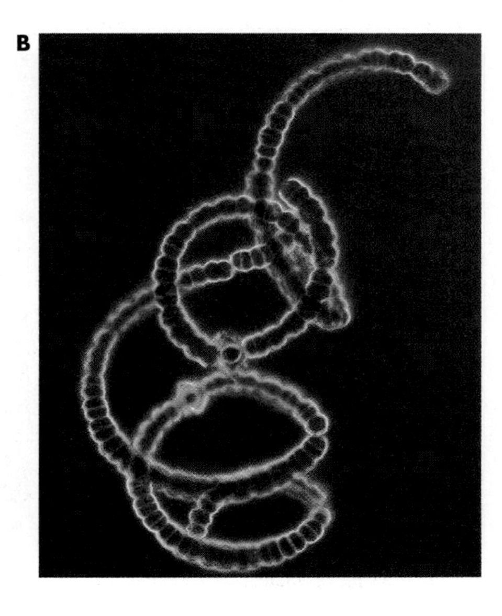

FIGURE 83.1
(A) Photograph and drawing of probable microfossil cyanobacteria in rock 3.5 billion years old. (B) Microscopic picture of modern cyanobacteria. Cyanobacteria used to be known as blue-green algae, although they are actually a kind of photosynthesizing bacteria. ($10 \mu m \approx 0.004$ inch)

Not all scientists agree that the microscopic features in these 3.5-billion-year-old rocks are fossil bacteria. There are more-widespread fossils with ages about 1 billion years younger, as well as possible isolated examples with intermediate ages.

FIGURE 83.2
Stromatolites on the western coast of Australia, seen at low tide, are formed by colonies of a cyanobacteria. They trap sediments, forming a layered structure in this characteristic dome shape. The inset shows a 3.4-billion-year-old stromatolite fossil, sliced to show the layered growth pattern.

ancient rocks survive that have not undergone substantial modification in the billions of years since those early times, so these microfossils are difficult to interpret. Some scientists point out that similar-shaped microscopic features might be produced by nonbiological processes.

Less controversial are the traces left behind when microscopic creatures like cyanobacteria grow into dome-shaped colonies called **stromatolites.** These are formed as successive generations of the bacteria trap sediments and then a new layer forms on the surface. Stromatolites are found today in shallow, warm, salty bays such as those along the coast of western Australia, as shown in Figure 83.2. Fossils with very similar structures have been found with ages up to about 3.4 billion years, as shown in the inset in Figure 83.2. Stromatolites can be traced throughout the geologic record.

The Earth formed about 4.5 billion years ago, but very few rocks older than 4 billion years survive today. This may be because the Earth was undergoing heavy bombardment and large parts of its surface were molten for several hundred million years. The isotopic indications of life in some of the oldest rocks suggest that life began quite quickly. Perhaps we can appreciate this speed by examining Figure 83.3, which shows a time line for the history of life on Earth. This diagram shows the epoch in which various life forms first appear in the fossil record. Their ages have been measured by radioactive dating of their associated rocks (see Unit 33.1). Along with the ages, the figure also illustrates some of these ancient life forms.

Scientists conclude from the available evidence that only extremely simple life forms, such as cyanobacteria and other single-celled creatures, were present for more than three-fourths of the Earth's history. The only nonmicroscopic indications of life would have been stromatolites until about one billion years ago, when we find the first fossils of multicellular algae. Then about 500 million years ago, only a little more than one-tenth of the Earth's lifetime, shells and crustaceans appeared, followed by an explosion of new forms of life. Mammals and dinosaurs both appeared roughly 250 million years ago, but the latter were wiped out about 65 million years ago in the great Cretaceous extinction, caused, we think, by a totally chance event—an asteroid's collision with the Earth (Unit 48). Hominids, our immediate ancestors, appeared roughly 5.5 million years ago; but our own species, *Homo sapiens,* evolved only about 500,000 years ago. Earth has therefore existed as a planet roughly 10,000 times longer than we have as a species. If all of

FIGURE 83.3
A time line illustrating the history of the Earth and life on it.

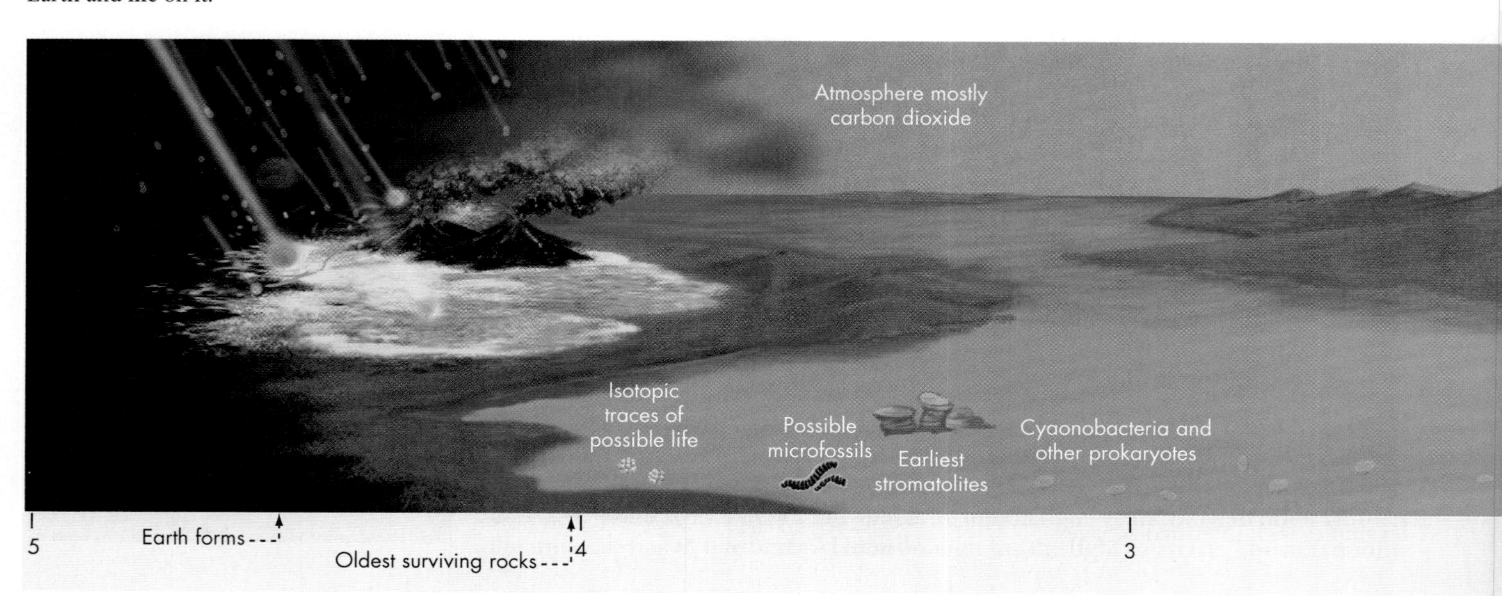

Earth's history were compressed into a year, humans would appear only in the last hour, and what we call civilization only in the last few minutes.

Life on Earth shows a dazzling variety of complex forms, from butterflies to whales, from mushrooms to trees, and so on (Figure 83.4). Despite that variety, all living beings on Earth have an amazing underlying unity of structure, reproduction, and metabolism. For example, all living organisms use primarily the same kinds of atoms for their structure and function: hydrogen, oxygen, carbon, and nitrogen. These atoms are not only those most commonly used by Earth's life forms, but they are also some of the most abundant throughout the universe. Even widely differing organisms use the same chemical substances. Life tends to draw on the materials with which our planet is most richly endowed; we are made primarily of the same substances that make up our ocean and atmosphere, the same elements that are the primary products of nucleosynthesis in stars.

The chemical elements that compose living things are linked into long-chain molecules for their structure and function. Many of these chain molecules are built up from various arrangements of some 20 amino acids. **Amino acids** are organic molecules containing atoms of hydrogen, carbon, nitrogen, oxygen, and sometimes sulfur. Their structure is the same except for a "side chain" that makes each one slightly different (Figure 83.5).

An organic molecule is one that contains carbon atoms. Organic molecules do not require biological organisms for their creation.

Moving up to the next level of complexity, we find that all living things use amino acids as the structural units of more complicated molecules called proteins. Protein molecules give living things their structure. For example, the scales of reptiles and fish are composed of the protein keratin, and the stiff cartilage of our own bodies is composed of the protein collagen. Proteins not only give cells their structure but also supply their energy needs. That energy supply reveals yet another underlying

FIGURE 83.4
A few of Earth's many complex life forms.

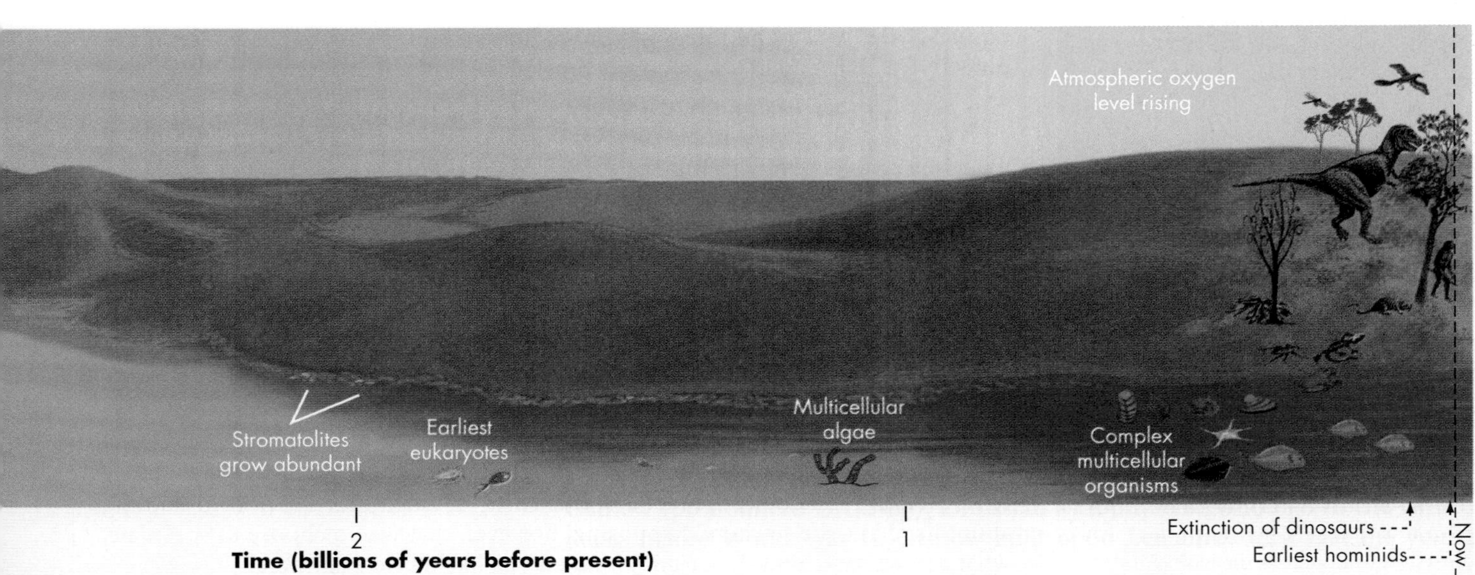

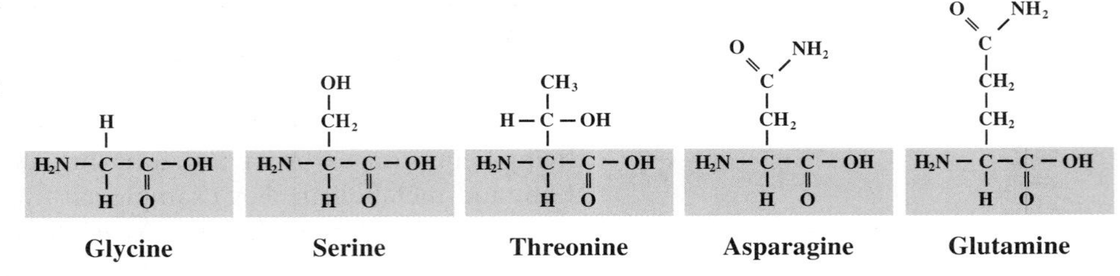

FIGURE 83.5
Diagrams illustrating the structure of several amino acids. H, C, N, and O represent atoms of hydrogen, carbon, nitrogen, and oxygen. The basic building block (shown in yellow) of amino acids is the same.

unity among living things: All cells use the same kind of molecule, adenosine triphosphate (abbreviated as ATP) to supply energy for action and growth.

This unity of structure and energy supply is echoed in the reproduction of life forms. All single-celled organisms divide in two to create new cells; all multicellular organisms produce egg cells, which, once fertilized, divide to create new cells. Moreover, in all cases, parents pass on genetic information to their offspring by the same molecule, **DNA.**

DNA stands for *deoxyribonucleic acid.* It is a massive molecule consisting of two long chains of smaller molecules wound around each other and linked by molecules called base pairs. DNA has the appearance of a twisted ladder with the role of the connecting rungs played by the base pairs (Figure 83.6). When a cell reproduces, the DNA separates into two individual chains. Each chain then adds a new base pair at each point along its length to match the one previously present on the other chain. After division and replication, two new DNA molecules exist that are identical to the original one. Nonetheless, copying errors—*mutations*—sometimes occur. Some can be harmful and kill the organism, but the nonlethal ones make evolution possible. That all terrestrial organisms use DNA for their reproduction gives life a truly remarkable unity.

Unity can also be seen in how more-complex organisms function, such as the chemical processes that move biologically important molecules into and out of cells. Complex organisms also use similar molecules for different purposes. For example, plants derive their food from carbohydrates they manufacture during photosynthesis. The chlorophyll they use in this process is structurally similar to the hemoglobin that transports oxygen in the blood of animals. These two molecules differ primarily in that chlorophyll is built around a magnesium atom, whereas hemoglobin is built around an iron atom. Similarly, cellulose, the structural material of plants, has a molecular structure very similar to chitin, the structural material in insect bodies and crustacean shells.

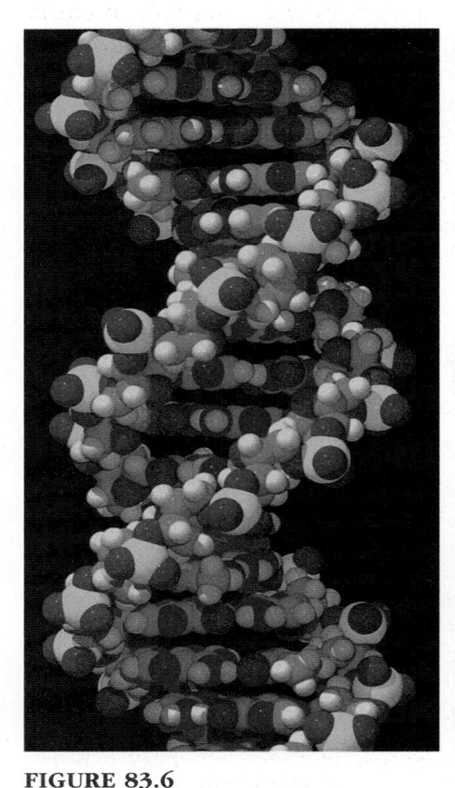

FIGURE 83.6
A diagram illustrating the structure of the DNA molecule. The blue horizontal structures in this model are the base pairs.

At another level, life builds upon life. To survive, complex organisms depend on **symbiosis,** mutually beneficial relationships with other organisms. For example, the human body contains hundreds of kinds of bacteria, many of which are essential to our survival. Bacteria in our intestines break down food to a point where we are able to extract nutrients essential to our own cells. Without those nutrients we would starve. Indeed, there are approximately 10 times more bacteria in our bodies than all of our other cells combined. From the bacterial point of view, a human body is a bag of oceanlike water that travels about in order to feed them!

What conclusions can we draw from the chemical and biological similarities of living beings and their ancient history as deduced from the time line? First, such chemical similarity almost inescapably suggests that all life on Earth had a common origin. Second, the antiquity of life suggests that under suitable conditions, life develops rapidly from complex molecules. Therefore, we might infer that wherever such conditions occur, so too will life.

We should treat these conclusions with great caution, however, because they are based on only a single case: life on Earth. For example, if the first time you played a lottery you won the grand prize, that would not mean that you would be likely to do so again. Likewise, the fact that life exists on Earth may say next to nothing about its chances elsewhere. Supposing, however, that we do choose to speculate, what *can* we say about the origin of life and its likelihood elsewhere?

83.2 THE ORIGIN OF LIFE

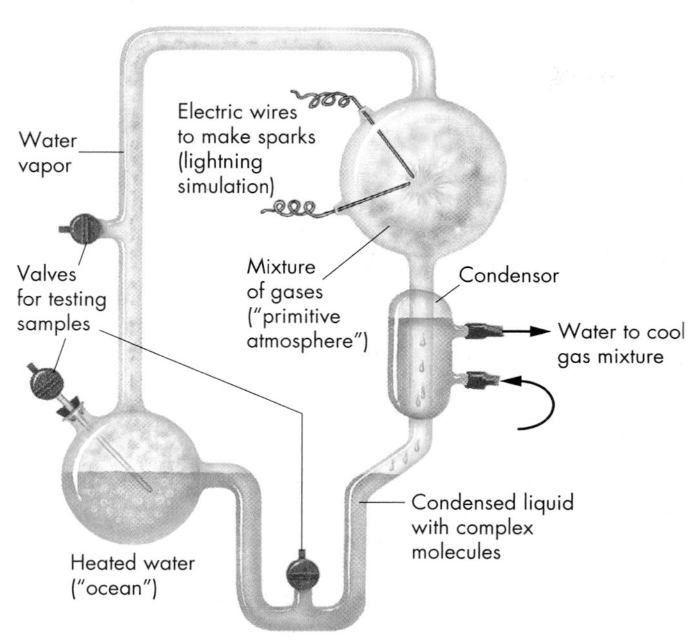

FIGURE 83.7
The Miller-Urey experiment attempted to simulate conditions on the primitive Earth. A mixture of gases received sparks (simulating lightning) and were repeatedly condensed and evaporated as they might have been in the young oceans. A variety of complex organic molecules were generated.

Most scientists today think that terrestrial life originated from chemical reactions among complex molecules present on the young Earth. This idea dates back to at least the time of Charles Darwin, who speculated that life arose "in some warm little pond." The idea was strongly bolstered in 1953 by an experiment performed by Stanley Miller—then a graduate student at the University of Chicago—and his professor, Harold Urey. Miller and Urey filled a sterile glass flask with water, hydrogen, methane, and ammonia—gases thought to be present in the Earth's early atmosphere. They then passed an electric spark through this mixture, as illustrated in Figure 83.7. The spark generated both visible and ultraviolet radiation, which triggered reactions in the gas and water mixture. At the end of a week they analyzed the mixture, finding in it a variety of organic molecules as well as five of the amino acids used to make proteins. They therefore concluded that at least some of the ingredients necessary for life could have been generated spontaneously on the early Earth. The success of the **Miller-Urey experiment** motivated other researchers to perform similar but more complex and realistic experiments, and these have demonstrated that a variety of conditions can generate amino acids.

The U.S. chemist Sydney Fox took complex organic molecules, such as those made by Miller and Urey, and by repeatedly heating and cooling them in water was able to create short strands of proteins called **proteinoids.** Intriguingly, Fox discovered that these proteinoids spontaneously formed small spheres reminiscent of cells (Figure 83.8). These spheres had no power to reproduce, nor did they show any of the complex structure commonly found in living cells; but they demonstrated that organic material can spontaneously give itself structure and wall itself off from its environment, a step toward the first living things.

Even the energy needs of living things can be produced in Miller-Urey types of experiments. For example, the Sri Lankan–American biochemist Cyril Ponnamperuma and the U.S. astronomer Carl Sagan showed that ATP can be synthesized from compounds similar to those generated by Miller and Urey. ATP is used by cells to store and transport energy, demonstrating one additional step in the creation of life from nonliving material.

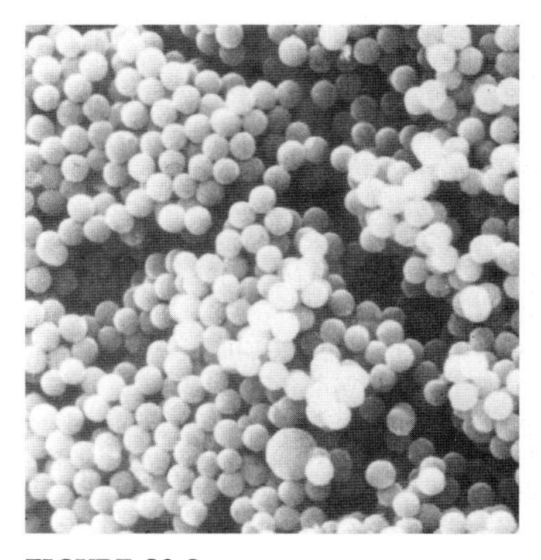

FIGURE 83.8
Spheres about 1 to 2 microns in diameter form when proteinoids are heated and cooled. Similar proteinoids have been discovered around undersea vents of Hawaiian volcanoes.

While the Miller-Urey and subsequent experiments show that organic material could have developed on the young Earth, it is possible that this material was present on our planet right from the start. Interstellar clouds in fact contain a rich mix of organic molecules (Unit 72.4), including several that are precursors to amino acids. Because the Solar System and Earth formed from such clouds, complex organic molecules must have been present on or near the Earth from its birth. Even if these molecules were destroyed during the Earth's formation, fresh molecules may have been carried to its surface by planetesimals at the end of the period of heavy bombardment (Unit 33.5). In fact, carbonaceous chondrite meteorites—probably surviving fragments of these ancient planetesimals—are rich in organic material, and some even contain amino acids (Unit 48.2). Finally, some scientists have demonstrated experimentally that the compression and heating of gases resulting from planetesimal bombardment may be as fruitful as the electric discharges used by Miller and Urey for forming complex organic materials. For a variety of reasons, then, astronomers are confident that organic molecules were common on the ancient Earth.

Q One objection to the Miller-Urey type of proposal for the origin of life is that chemicals in the ocean are far too dilute to undergo long series of interactions. In what kinds of environments on the young Earth do you imagine chemicals might have become concentrated?

Regardless of the origin of organic matter on the early Earth, whether created by lightning, ultraviolet radiation, or planetesimal bombardment, many steps are needed to progress from droplets filled with amino acids to living cells. For example, the structures created must be able to reproduce, a feature not yet seen in any origin-of-life experiments.

Cells in Earth's present life forms replicate using the double-chained molecule DNA. Some researchers have speculated, however, that the earliest forms of life used the slightly simpler molecule called *ribonucleic acid* (RNA). In fact, biochemists have demonstrated that if certain enzymes are present, RNA can replicate in the absence of living cells. Moreover, they have also discovered several other molecules far simpler in structure than RNA that can replicate in a mechanical sense. That is, if such molecules are put in a solution containing the necessary raw materials, the ensuing chemical interactions allow the molecules to spontaneously produce copies of themselves. Furthermore, one particular mix of molecules not only replicates but also can "mutate" into a slightly better replicator.

Life developed when the atmosphere was quite different from today's. Analysis of ancient rocks indicates that Earth's atmosphere lacked oxygen, making it more similar to the other planets. The absence of oxygen may have been crucial for the development of life. Oxygen is highly corrosive to unprotected organic molecules, breaking them down. Thus, the most essential components of Earth's atmosphere from the human perspective today would probably have prevented life from forming if it had always been present.

The first cellular life on Earth had no internal structures, such as a nucleus to contain the DNA, making them similar to modern bacteria. **Prokaryotes,** as these cells without nuclei are known, appear to have arisen relatively rapidly, and then remained the only form of life on Earth until less than 2 billion years ago.

Some of the most primitive prokaryotes known today are found in low-oxygen environments. One of the most intriguing of these locations is around sea floor vents deep in the ocean. These environments are rich in sulfur minerals, and today they support a variety of life forms, as shown in Figure 83.9. The bacteria that feed off these sulfur minerals are simpler than cyanobacteria, because they did not need to develop photosynthesis to supply their energy. It is not difficult to imagine that this kind of bacteria could have thrived on the young Earth. After a few hundred million years the surface was cool enough for liquid water, at which time the oceans formed. The mineral-rich regions around vents would have been conducive to the production of amino acids, proteinoids, and the cell-like spheres found in Fox's experiments.

FIGURE 83.9
(A) Undersea volcanic vents expel hot sulfur-rich minerals, which provide the nutrient source for very primitive types of bacteria. (B) These bacteria in turn are at the bottom of the food chain for colonies of unusual undersea creatures at such depths that sunlight never reaches them.

More-complex prokaryotes, able to tap into the energy in sunlight through photosynthesis, probably developed within the first billion years of Earth's history. These ancestors to cyanobacteria probably account for the fossil record of stromatolites. Stromatolites grew steadily more common until about 3 billion years into the Earth's history, and their success radically altered Earth's environment. One of the waste products of photosynthesis is oxygen. At first this oxygen was absorbed by the Earth, oxidizing minerals such as iron. By about 2.5 billion years (2 billion years ago), oxygen levels had risen to about 1% of present levels. Not until about 500 million years ago was there enough oxygen that we might be able breathe.

The rise of oxygen also permitted the development of an ozone layer, shielding the Earth's surface from the Sun's ultraviolet light. The new atmosphere was caustic to the bacteria that created it, but the presence of oxygen offered opportunities for more complex organisms to develop and eventually spread to the land surfaces of Earth. Cells today take advantage of oxygen to enhance their metabolism, but they needed to develop more-complex structures to handle this caustic chemical. Cells in plants and animals today contain a *nucleus* in which genetic information is stored in long strands of DNA, as well as *mitochondria,* which are tiny bodies that convert food into energy that the cell uses. They also have a complex *membrane*

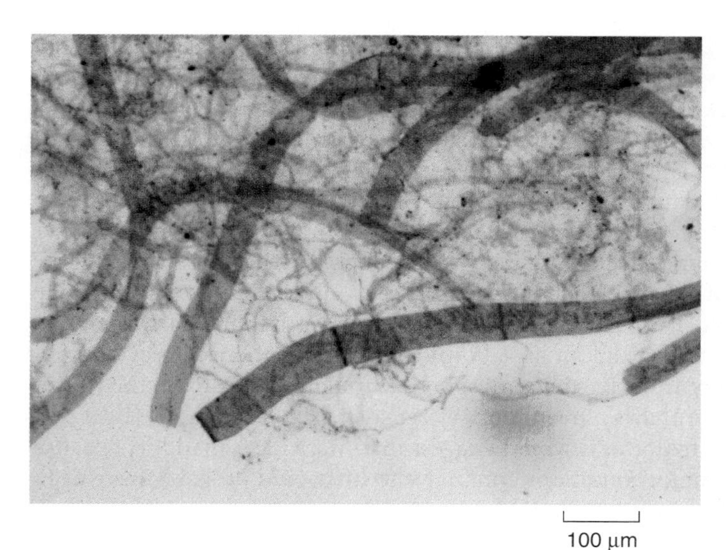

FIGURE 83.10
A 1-billion-year-old fossil of eukaryote algae. The earliest eukaryotes show up in fossils from about 1.8 billion years ago.

that holds them together and allows food to enter and waste products to leave. The development of more complicated cells may have begun when one prokaryote merged with another, producing a symbiosis that benefited both cells by allowing them to grow and reproduce more rapidly. For example, a cell good at storing energy might have combined with a cell good at reproducing. From such simple beginnings, **eukaryotes** (cells with nuclei) probably evolved. Such a scenario would also explain why mitochondria carry their own DNA.

Microscopic fossils of eukaryotes, mostly algae, have been found that are up to 1.8 billion years old (Figure 83.10). Gauged by the amount of time it took for prokaryotes to appear on Earth (probably much less than 1 billion years), compared to the 1.5 to 2 billion years for complex cells to appear, this next step must have been a difficult one. From the timing of their appearance, it seems that the development of eukaryotes was spurred by the development of an oxygen-rich atmosphere. The eukaryotes proved much more adaptable than the prokaryotes, and by about 1 billion years ago, stromatolites nearly disappear from the fossil record.

FIGURE 83.11
A fossil impression of a multicellular animal from about 600 million years ago.

The final significant step in the evolution toward the complex beings of today required individual cells to work cooperatively as a single organism. Microfossils of multicellular algae appear up to 1.2 billion years ago, but larger multicellular plants and animals began to appear only about 600 million years ago, seen today in fossil impressions of soft-bodied creatures that resemble sea worms and jellyfish (Figure 83.11). The development of ways in which cells could cooperate to form multicellular organisms was apparently another difficult step in our evolution, given that it took about a billion years. Once multicellular plants and animals appeared, however, there was an explosion of diverse types of life in an extraordinarily brief time, geologically speaking. The distance between bacteria and sea worms is apparently far larger than the distance between sea worms and humans!

The greater complexity of multicellular organisms would in some cases allow them to exploit resources unavailable to simple cells. For example, an organism that can move from nutrient-poor to nutrient-rich environments may be able to reproduce more prolifically than a stationary creature. Such reproductive advantages, combined with changing environments, have led to the incredible diversity of life that we see around us today. In fact, reproductive advantage (*natural selection*) and random mutations are crucial to evolution and the development of more complex life forms, as Charles Darwin explained in 1859 in *On the Origin of Species by Means of Natural Selection,* a work of monumental importance.

83.3 THE GAIA HYPOTHESIS

In 1974 James Lovelock, a British chemist, and the U.S. microbiologist Lynn Margulis suggested that life creates a single "larger entity" with a planet, a symbiosis of life and planet that they called the **Gaia hypothesis.** They chose this name from the Greek goddess of the Earth, Gaia (pronounced *GUY-uh* in this context but *JEE-uh* elsewhere).

Q Think about your everyday activities. What is the smallest environment you impact? The largest?

According to the Gaia hypothesis, life does not merely respond to its environment but actually alters its planet's atmosphere and temperature to make it more hospitable. Plant life has almost certainly done just that on Earth. For example, by photosynthesis, plants have created an oxygen-rich atmosphere on Earth that shields them from dangerous ultraviolet radiation. Similarly, photosynthesizing organisms have altered our planet's temperature by removing CO_2—a greenhouse gas—and helped lock it away in soil and ultimately rock. This means that plants can modify the Earth's temperature by adjusting the greenhouse effect (Unit 36). If the planet is too cold, plant metabolism is slowed and less CO_2 is converted to oxygen. The greater abundance of CO_2 warms the planet and enhances plant growth. Conversely, if the planet gets warmer, plants grow faster and produce lots of oxygen, reducing CO_2 abundance and thereby also reducing greenhouse warming.

Changing levels of CO_2 on Earth have almost certainly prevented a catastrophe for life on our planet. It is estimated that over the last 4 billion years, the Sun's luminosity has increased by about 40% (Unit 61.4). If greenhouse gases had not been removed steadily from Earth's atmosphere throughout this time, the planet might have suffered a runaway greenhouse effect such as that on Venus (Unit 39). By contrast, the Gaia hypothesis would suggest that Mars never had life, because if life had ever taken hold, conditions conducive to life would probably have been maintained.

Q If the runaway greenhouse effect were to take hold, what processes might help bring it into check?

Lovelock and Margulis have also proposed that many other facets of the environment, such as humidity, salinity of the oceans, sea level, and even plate tectonics may be controlled by living organisms so as to optimize their reproduction. Such an intimate linkage between the living and nonliving is an appealing idea. It almost suggests that the planet is "alive."

But is the Gaia hypothesis correct? Some of the mechanisms described certainly exist, but many scientists are skeptical of the hypothesis's more extreme claims (for example, that life can control plate tectonics). Nevertheless, living beings exert a remarkable control over their local environment, and many unsuspected links between life and the physical world surely remain to be found.

83.4 THE ANTHROPIC PRINCIPLE

In 1961 Robert Dicke, a physicist at Princeton University, wrote a paper concerning some remarkable cosmological coincidences. Dicke noted, for example, that the age of the universe is not much different from the lifetime of stars like the Sun, and he went on to argue that this "coincidence" is in fact not remarkable at all. Rather, it is a *necessity,* if cosmologists are to exist to note it.

Dicke pointed out that life requires elements such as carbon, silicon, and iron that are made in massive stars. So for life to form, enough time must pass for massive stars to evolve and make the heavy elements and then eject them into space with a supernova explosion. Moreover, additional time is needed for the ejected material to be incorporated back into interstellar clouds and to form new stars. Only with this second generation of stars does life become possible, so intelligent life cannot evolve in a universe only a few billion years old. Finally, he argued that life requires stars, and so it can exist only in a universe young enough that stars have not all died. Therefore, the existence of an intelligent observer in the universe requires that the age of the universe fall within certain limits.

In 1974 the English physicist Brandon Carter took Dicke's idea further and proposed what he called the **anthropic principle.** According to the anthropic principle, "what we can expect to observe must be restricted by the conditions necessary for our presence as observers." Since Carter's work, many scientists have shown how "fine-tuned" the universe must be for life to exist. Thus, no matter how unlikely some aspects of the universe may appear to us, if those aspects are necessary for life, then that is what we will observe. For example, we might argue how truly marvelous it is that conditions on Earth are just right for life. According to

Q It has been suggested that if the physical constants were slightly different, life in the universe might have been impossible. Can you think of examples where this might be true?

the anthropic principle, the rude reply is, "Of course conditions are just right for life. If they were not, there would be no life here to do the marveling."

In some respects the anthropic principle suggests that there are many potential Earths, or even potential universes, in which life might have formed. Among all these possibilities, life arises only where all the necessary conditions are present. Perhaps there are billions of other almost-Earths where life did not succeed, or barren universes that spawned no one to marvel at them. If this is true, then our universe and Earth may be the ones that just happened to achieve conditions where life like us would inevitably form.

KEY TERMS

amino acid, 689
anthropic principle, 694
cyanobacteria, 687
DNA, 690
eukaryote, 693
Gaia hypothesis, 693
Miller-Urey experiment, 691
prokaryote, 692
proteinoid, 691
stromatolite, 688
symbiosis, 690

QUESTIONS FOR REVIEW

1. What is the earliest evidence for life on Earth?
2. How old are the earliest life forms on Earth thought to be?
3. For how long has complex life existed on Earth?
4. What features of life suggest a common origin?
5. What is the importance of the Miller-Urey experiment?
6. What is the role of prokaryotes and eukaryotes in the evolution of life?
7. What is the Gaia hypothesis?
8. What is the anthropic principle?

PROBLEMS

1. For what percentage of the Earth's lifetime has *Homo sapiens* existed?
2. A bacterium has a mass of about 10^{-15} kg. It can replicate itself in one hour. If it had an unlimited food supply, how long would it take the bacterium, doubling in number every hour, to match the mass of the Earth? (Hint: Solve for the number of hours N where 2^N gives the number of bacteria you estimate.)
3. A human brain has a mass of about 2 kg. If an individual brain cell has a mass of about 10^{-13} kg, how many cells does a brain contain? How does this number compare to the number of stars in our Galaxy? to the number of bits of information in the largest computers today?
4. From 1950 to 2000, the world population increased from 2.5 billion to 6 billion. If this trend continues, what will be the world population in 2050? in 2100?

TEST YOURSELF

1. Evidence that life on Earth is very ancient comes from
   a. fossil algae in rocks more than 3 billion years old.
   b. fossils buried thousands of feet below the surface.
   c. fossil algae in rocks a few million years old.
   d. the discovery of silicon-based life forms in ancient rocks.
   e. the discovery of silicon-based life forms in a Miller-Urey type of experiment.
2. The Miller-Urey experiment demonstrated that
   a. very simple bacteria can be created from the chemicals present in the atmosphere of the ancient Earth.
   b. Earth's early atmosphere was too harsh for life to have formed until recently.
   c. conditions on the early Earth were suitable for the creation of many of the complex organic molecules found in living things.
   d. if simple life forms landed on the early Earth, they could have survived.
   e. early life forms were silicon-based.
3. According to the anthropic principle,
   a. we are the only civilization in the universe.
   b. without human beings to observe the universe, the universe would not actually exist.
   c. human beings are the highest form of life.
   d. the universe could not be very different, or we would not be here to observe it.
   e. the greenhouse effect is beneficial to us.

unit

84

The Search for Life Elsewhere

Background Pathways

Unit 83: Astrobiology 687

Humans have speculated for thousands of years about life elsewhere in the universe. For example, about 300 B.C.E., the ancient Greek philosopher Epicurus wrote in his "Letter to Herodotus" that "there are infinite worlds both like and unlike this world of ours." Epicurus went on to add that "in all worlds there are living creatures and plants." In a similar vein, the Roman scholar Lucretius wrote about 50 B.C.E. that "it is in the highest degree unlikely that this Earth and sky is the only one to have been created." Such views were not universal, however. For example, Plato and Aristotle argued that the Earth was the only abode for life, a view that prevailed through most of the Middle Ages.

Another possibility that has been debated for centuries is whether terrestrial life descended from organisms created elsewhere in the universe. This is sometimes called **panspermia.** According to this hypothesis, simple life forms (perhaps bacteria) from some other location drifted from their place of origin across space to Earth. However, panspermia does not really simplify the problem of the origin of life: It just shifts it elsewhere. Moreover, it further requires getting the life to Earth, a perilous voyage even for a bacterium.

Today, however, influenced by science fiction movies, television, and books, many people have no difficulty at all in believing that extraterrestrial life exists. In fact, supermarket tabloids make silly claims almost weekly about encounters with aliens. If we look at the facts, however, there is not a shred of evidence that life exists elsewhere. This does not mean there is no extraterrestrial life; it simply means we do not know whether there is. Thus, with no evidence, what can we say meaningfully about life elsewhere in the universe?

Life on Earth arose quite soon after Earth's formation, but billions of years passed before a species arose that could wonder if we are alone in the universe. Do other intelligent civilizations exist? Astronomers are strongly divided about such possibilities. Some argue that such life is highly probable, whereas others argue that it is highly improbable. One goal of this unit is to describe the evidence that goes into forming such conclusions.

84.1 THE SEARCH FOR LIFE ON MARS

Scientists have long wondered whether living organisms developed on Mars. Several hundred years ago, after it was realized that the Earth was a planet, many scientists and philosophers assumed that the other planets must have been made as abodes for life.

Mars became the particular focus of attention after it was discovered in the 1700s to have polar caps and an atmosphere and to rotate at about the same rate as the Earth (Unit 40). In 1820 the German mathematician Karl Friedrich Gauss even

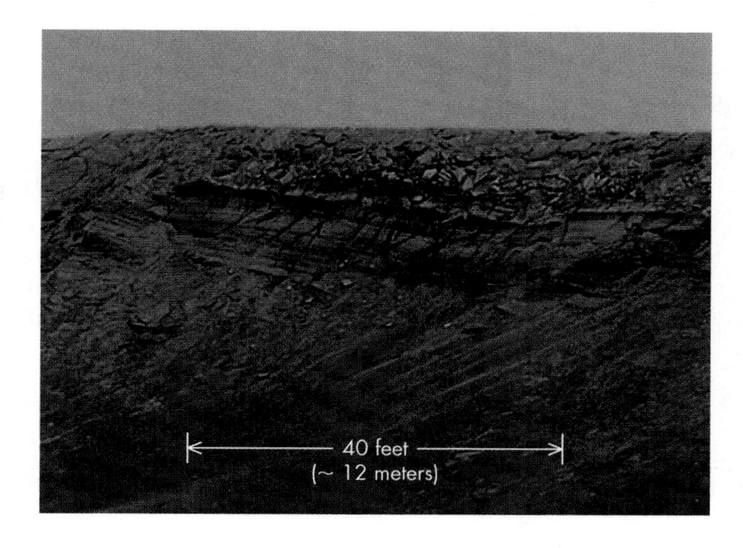

FIGURE 84.1
Cliffs on the inner wall of Endurance Crater, which was visited by *Opportunity,* one of the Mars rovers. The layers (strata) that you can see here suggest that this area of Mars once was covered with water. Because water is vital to life here on Earth, scientists are eager to explore regions on other planets and moons where water is, or was once, present—in the hope of finding traces of present or past life on other worlds.

proposed sending a mathematical message to the Martians by planting wheat fields and pine trees in a huge right triangle. He calculated that the triangle would be visible to the Martians and would show them that intelligent life was present on Earth.

Q Based on their chemical and geological composition, many meteorites have been traced back to several of the asteroids, the Moon, Mars, and possibly Mercury. Why do you suppose none have been identified as coming from Venus?

The belief in Martian life took on a new twist based on a linguistic misinterpretation of observations made in 1877 by the Italian astronomer Giovanni Schiaparelli. Schiaparelli saw what he took to be straight-line features on Mars and called them *canali,* by which he meant "channels." In English-speaking countries the Martian **canali** became called *canals,* with the implication of intelligent beings to build them. Some astronomers thought they could see a complex system of interconnected canals just at the limits of telescope resolution. Most astronomers could see no trace of the alleged canals, but they did note seasonal changes in the shape of dark regions—changes that some interpreted as the spread of plant life in the Martian spring.

Popular belief continued to fuel speculation about life on Mars into the mid-1960s, when the first of a series of *Mariner* spacecraft sent back pictures of a far more desolate world than had been previously imagined. The images of Mars made by the early *Mariner* spacecraft (Unit 40) demonstrated that the straight "canal" features were not real except for the giant Valles Marineras canyon, and that the changes in darkness were caused by widespread dust storms. When the United States landed two *Viking* spacecraft on the planet in 1976, all tests for signs of carbon chemistry in the soil or metabolic activity in soil samples were negative or ambiguous. However, subsequent missions have found geological evidence that there was once fairly abundant water on Mars (Figure 84.1), and Mars may have been able to support life during its early history.

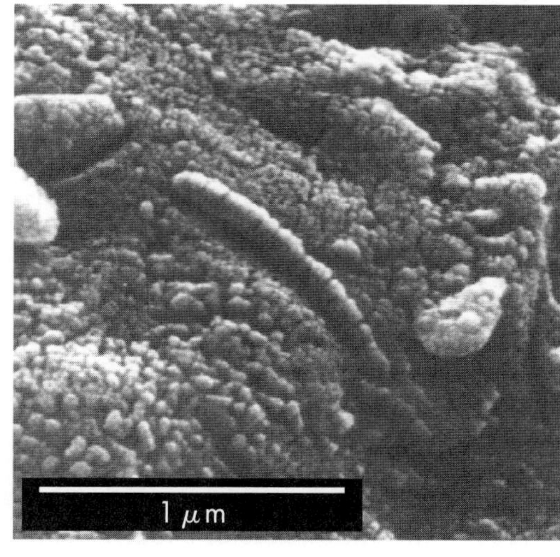

FIGURE 84.2
Fossils of ancient Martian life? The tiny elongated structure in this microscopic image is about 1 μm long. It is much smaller than most bacteria, although similar to some bacterial traces found in rocks on Earth. However, some scientists think these structures formed chemically.

In 1996 a group of American and English scientists reported possible signs of fossil life in a rock from Mars. The rock was a meteorite found in Antarctica. It arrived on Earth after being blasted off the surface of Mars by the impact of a small asteroid. Such impacts are not uncommon, but most fragments are scattered in space or fall back to Mars. Detailed analysis of the composition of this and several other meteorites shows a close match with Martian rock, and they are unlike any other known source.

Microscopic examination of samples from the interior of one such meteorite revealed tiny, rod-shaped structures (Figure 84.2). These look like ancient terrestrial bacteria but much smaller. To some scientists they look like fossilized primitive life, but others have shown that ordinary chemical weathering can form similar structures. As a result, most scientists today doubt that any meteorite yet studied shows evidence of Martian life.

Nothing found by the *Spirit* and *Opportunity* rovers provides any evidence of life either. In 2008, the *Phoenix* lander carried out chemical tests at Mars's north pole. These tests established the presence of water and showed that the soils are nutrient rich, but provided no direct evidence for life. If any life exists or once existed on Mars, it probably lies buried well below the hostile surface of the planet.

The meteorite discovery does raise an interesting issue, however. If evidence of life is ever found on Mars, we will have to consider the possibility that panspermia occurred. A bacteria-laden rock might have been blasted off Earth's surface in the remote past and reached Mars when it was more habitable. Or perhaps if life formed on Mars first, our own ancestors might have been Martians!

84.2 LIFE ON OTHER PLANETS?

Is there any evidence for life elsewhere in the universe? Not really. The absence of hard evidence may seem discouraging; but as the British astronomer Sir Martin Rees has said, "Absence of evidence is not the same as evidence of absence."

We have so far only analyzed rock and soil specimens from a tiny portion of the universe: some half dozen spots on the Moon, six on Mars (by robot space landers), and some asteroid fragments picked up on Earth. The Moon's lack of atmosphere and water makes it so inhospitable that astronomers did not expect to find life there, nor did they. Mars seemed more promising because its environment, though harsh, is more like the Earth's than that of any other planet, and laboratory experiments indicated that some terrestrial organisms (bacteria and lichens) might survive Martian conditions, even though they did not reproduce or grow.

The limited sample of planets available to explore in the Solar System has led scientists to consider other ways to search for signs of life on other planets. One interesting method involves looking for unexpected gases in the atmosphere of a planet—gases that might be produced by living things.

In the late 1960s James Lovelock, a British chemist (see also Unit 83.3), noted that several of the gases making up the Earth's atmosphere would disappear if Earth had no living beings to replenish them. For example, oxygen is highly reactive and readily combines with surface rock. Lovelock argued that the presence of such gases in a planet's atmosphere is an indication of life there. Rather than sending robot spacecraft to explore a planet's surface, astronomers might find evidence of life by analyzing the light from a planet's atmosphere.

Detecting an Earth-mass **exoplanet** (a planet orbiting another star—Unit 34), let alone detecting oxygen from its atmosphere, remains challenging. One exoplanet (HD 209548b), with a mass close to Jupiter's, is in a tight orbit around its star, passing between us and the star during each orbit. Astronomers have detected oxygen and carbon in its atmosphere, seen in absorption against the star's light. This finding is quite unlikely to signify life, however. The planet orbits so close to its star that the planet's atmosphere is boiling away. The oxygen and carbon were probably the by-products of other molecules being broken down when they were heated by the star and driven away from the planet. What we probably detected was a large cloud of gas streaming off the planet, similar to a comet's tail. Nonetheless, this first indication of oxygen in the atmosphere of another planet gives us encouragement that the next generation of space telescopes may eventually lead to the discovery of more hospitable planets.

Finally, we note that life may have taken very different routes in places that would be incompatible with life as we know it. For example, Jupiter's moon Europa, and possibly Ganymede and Callisto, probably have giant oceans of liquid water hidden beneath their frozen crusts. In addition, we detect complex organic molecules in Titan's atmosphere, but temperatures are so cold there that its atmosphere rains liquid methane that cuts rivers into its frozen ice surface—conditions inhospitable to Earth-based life. Could life have formed in these environments? Could intelligent life have evolved? These environments are so alien to us that it is difficult to grasp the possibilities.

84.3 ARE WE ALONE?

Q Suppose astronomers found clear evidence of civilization on another planet. What effect do you think this would have on our society? What effect would it have on you?

Do other intelligent life forms exist in the universe? Hundreds of light-years from Earth, is there perhaps a "student alien" who is right now reading a book discussing the possibility of life on planets such as Gzbhλx? Scientists do not know, and in fact they are strongly divided into two groups. One group of scientists (let us call them **many-worlders**) thinks that millions of planets with life exist in the Milky Way, and that many of these may have advanced civilizations. The other group (let us call them **loners**) argues that we are the only intelligent life in the galaxy. We will discuss below how these two radically different points of view arise.

Many-worlders argue as follows: Earth-like planets are common. Life has formed on Earth. Therefore, it would be surprising if life did not exist elsewhere. In fact, even if only a small fraction of such planets have life, many of them have probably developed advanced civilizations. This may even vastly underestimate the number of planets inhabited by intelligent beings, because life may be able to arise in environments intolerable to terrestrial organisms.

Drake originally wrote his equation in terms of the rate of star formation in our Galaxy and estimated the number of civilizations that were currently sending out communications that we might detect.

To make this argument more quantitative, the U.S. astronomer Frank Drake showed how we can estimate the number of intelligent civilizations by multiplying together the probability of each condition necessary for them to exist. He developed a mathematical estimate of the number of such civilizations known as the **Drake equation.** We will follow a modified form of Drake's method here. We can estimate the number of planets in our Galaxy that contain intelligent life, N_I, by taking the number of suitable stars, N_*, and multiplying that by the fraction of them that have a habitable planet, f_P, and the probability of life forming on such a planet, f_L, and the probability of intelligent life forming given that life had formed, f_I. This can be written in equation form as follows:

$$N_I = N_* \times f_P \times f_L \times f_I.$$

Unfortunately, it is difficult to establish accurate values for each of the probabilities that go into this equation, but this approach helps focus our discussion. To see how such numbers are deduced, let us begin by estimating the number of Earth-like planets in the Milky Way.

We start by asking how many Sun-like stars are in the Milky Way. We limit ourselves to such stars because luminous blue ones fuse their fuel so rapidly that they die and explode as supernovae before life has much chance to evolve on any orbiting planets, whereas dim red stars are so cool that their planets would have to be very close to them to be warm enough for a terrestrial type of life. Given this restriction on star types, we turn to census counts of stars, which reveal that stars like the Sun make up about 10% of our Galaxy's hundred billion stars, for a total of about $N_* = 10^{10}$ such bodies.

Of these 10^{10} stars, we next ask, How many of them have Earth-like planets? Our own system has two planets out of a total of nine that might harbor life as we know it: Earth and Mars. Let us be conservative and say that most other systems are not quite so lucky with respect to having habitable planets. We know that most stars are in multiple-star systems (Unit 56), and of the planetary systems detected so far, most do not resemble the Solar System. We might therefore estimate that 1 in 10 have a habitable planet orbiting them, so $f_P = 0.1$.

How many of these habitable planets might actually have life on them? That depends on how easy it is for life to form, a probability that the Miller-Urey experiments (Unit 83) suggests may be high. However, some astronomers have argued that it is quite low. In fact, the British astrophysicist Fred Hoyle compared the likelihood of life arising spontaneously to the likelihood of a box containing several hundred tons of aluminum being shaken and by chance assembling itself into a 747 jet. Although many scientists find such a comparison silly, it illustrates the divergence of opinion about the likelihood of life forming in the universe. The difficulty in assessing the probability of life forming is that we have only the single case of life here on Earth as our basis. Many-worlders point to the rapidity with which life developed on Earth. They also argue that the success of the Miller-Urey experiment in making molecular precursors to life in just a few days indicates that the chance of life starting is fairly high: 1 in 100, say, so that $f_L = 0.01$.

We now go to the next step and ask, On how many of these life-bearing planets do intelligent beings arise? There is no certain way to know whether the probability of more-complex life evolving is high or low. Moreover, even if life succeeds in starting, perhaps it will be annihilated if its star dies before intelligence develops. One intriguing idea is that natural disasters, such as asteroid impacts, provide opportunities for new species to develop *until* intelligence arises. After all, if an asteroid had not killed the dinosaurs (Unit 48), we might not have gotten a chance to evolve. Now, thanks to human intelligence, we are developing technologies that could protect us from an asteroid collision, and we may potentially be able to colonize other planets in the future. Thus, natural disasters might provide a means of natural selection for intelligence. Suppose we are conservative and say that if life forms, it has a 1 in 1000 chance of developing an intelligent, technological civilization. This sets our final factor, the fraction of life-bearing planets where intelligence arises, to $f_I = 0.001$.

The many-worlder carries out this calculation and estimates that the number of intelligent civilizations in our Galaxy is approximately

$$N_I = 10^{10} \times 0.1 \times 0.01 \times 0.001 = 10{,}000.$$

Therefore, with conservative estimates (from a many-worlder's perspective), there should be many intelligent civilizations within our Galaxy with whom we might communicate.

By contrast, the loners suggest we are the only advanced life in the Milky Way. The loner argument was first proposed by the Italian physicist Enrico Fermi and is sometime called the **Fermi paradox.** Fermi argued that, based on the rapidity with which our species has developed technology, it is just a matter of time before we spread across the galaxy, or at least fill it with the radio waves of our own transmissions. Even if it takes a technological civilization thousands of years to travel to other stars, once they have mastered space flight they will need at most a few million years to colonize the entire galaxy. A few million years is a brief time compared to the billions of years since such civilizations could have arisen, so evidence of their presence should be easy to find if they existed.

Accordingly, loners argue that because no other civilization has been seen, no other technological civilization has already appeared. Therefore, they argue, we are probably alone in the Milky Way, or nearly alone with at most only a handful of other civilizations. However, this argument, like that of the many-worlders, rests on a number of assumptions that are difficult or impossible to substantiate. For example, it assumes that (1) civilizations are driven to colonize, (2) civilizations seek rather than avoid contact, and (3) it is possible to master interstellar travel.

The loners' argument is not based solely on our failure to have already found such a civilization, though. They point out that the various probabilities for life forming may be significantly lower than the many-worlders suggest. For instance, the Earth and Solar System have many unusual attributes. Perhaps a large moon is critical for stabilizing a planet's axis. Or maybe a large outer planet is necessary to perturb the orbits of water-bearing comets, thereby delivering a critical late shower of water to a planet. In addition, among the exoplanetary systems so far detected (Unit 34), it is clear that the nearly circular, regularly spaced orbits of the planets in the Solar System are unusual. What if such orbits are critical for creating stable conditions for the formation of life? It may be that the fraction of planets that are suitable for life f_P is much smaller than the many-worlders suggest.

What if a civilization exhausted its supply of fossil fuels and had to live on limited energy supplies? How important has an abundant energy resource been for human development of technologies?

Perhaps a star's location in the Milky Way is critical for the survival of life. Because the Sun is fairly far from the galactic center, where most of the stars are concentrated, the Earth has not been exposed to as much radiation from nearby nova and supernova explosions. Earth also has a much stronger magnetic field than either Mars or Venus, and maybe that also has been critical for protecting life from high-energy particles. Maybe the great majority of exoplanets are sterile because of the radiation they suffer, and therefore the fraction where life forms f_L is smaller too.

All in all, there could be enough factors such as these to greatly reduce the probability of life forming. Supposing f_P and f_L were each just 1% of what the many-worlders estimate, we would find

$$N_I = 10^{10} \times 0.001 \times 0.0001 \times 0.001 = 1.$$

By the loners' calculation, there is just one planet with intelligent life in the galaxy—the Earth.

84.4 SETI

On the chance that there may be technologically capable life elsewhere in our Galaxy, some astronomers are searching for radio signals from other civilizations. Such signals might be either deliberate broadcasts sent to us or communications directed elsewhere that we could simply overhear. On Earth, radio transmissions have been generated for about a century, so our signals might be picked up by an alien civilization on another planet if it is within 100 light-years of us.

We can estimate how far away the next nearest civilization might be by assuming that the N_I civilizations we predict to be present are spread evenly throughout the disk of the Milky Way, as depicted in Figure 84.3. We can calculate the average distance, d, separating them as follows: Imagine drawing a sphere of radius $d/2$ around each civilized world. Then we ask what the size of the average sphere would have to be for N_I of them to cover the Milky Way's disk. If we make the approximation that the Milky Way's disk is a circle of radius R_{MW}, then its area is given by the formula $\pi R_{MW}{}^2$. So we set the area covered by the N_I spheres (each of radius $d/2$) equal to the area of the galaxy's disk:

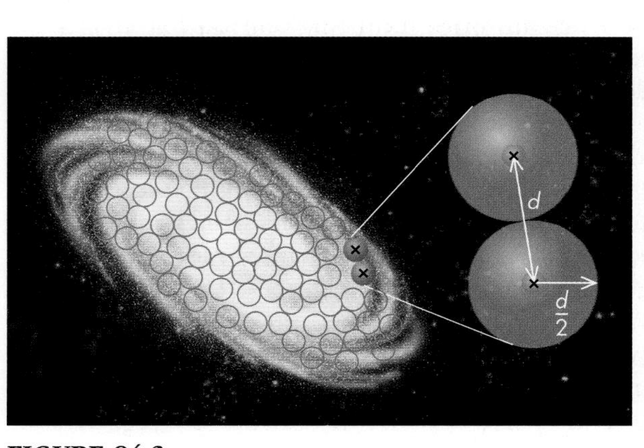

FIGURE 84.3
A sketch illustrating how to estimate the distance between potential civilizations in the Milky Way, if there are N_I civilizations spread evenly within our Galaxy.

$$N_I \times \pi \left(\frac{d}{2}\right)^2 = \pi(R_{MW})^2.$$

We can cancel π on both sides of the equation, multiply both sides by $(2^2/N_I)$, and take the square root of both sides to give

$$d = \frac{2R_{MW}}{\sqrt{N_I}}.$$

The Milky Way's radius R_{MW} is approximately 40,000 light-years, and let us suppose that the number of intelligent civilizations N_I is 10,000, the value deduced from the many-worlder argument. These choices give a value for d of approximately 800 light-years. Therefore, even if 10,000 civilizations exists in our Galaxy, they

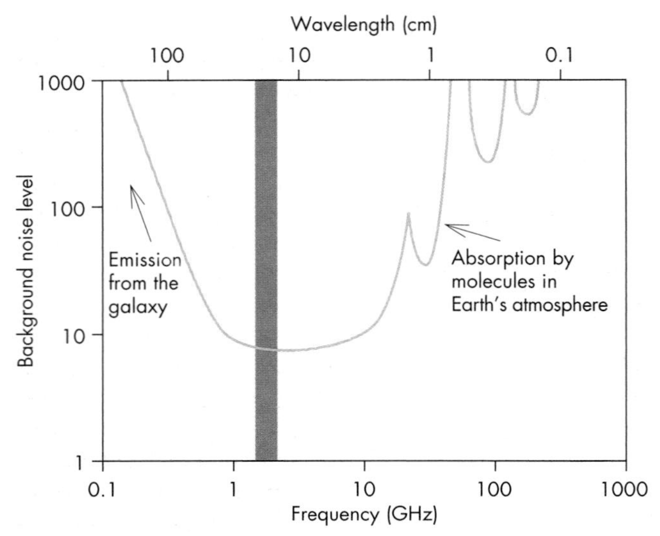

FIGURE 84.4
The *waterhole* is a range of wavelengths between prominent spectral lines of H and OH at 21 and 18 centimeters. It falls in a clear region of the radio spectrum that is free of interference from natural sources of radio emission and is not blocked by absorption in an Earth-like atmosphere.

FIGURE 84.5
When completed, the Allen Telescope Array, being constructed in California, will be able to simultaneously carry out astronomical surveys while it conducts a deep, targeted search for signals from other civilizations. The array will be made up of hundreds of small radio telescopes, using interferometry to simultaneously examine large areas of the sky over a wide part of the radio spectrum.

are all likely to be very far from us, and none has likely yet seen any of the transmissions we have generated.

Other civilizations may have begun transmitting a much longer time ago than we did, so there may be detectable signals passing by Earth right now. The electromagnetic spectrum is very wide, and potential communications signals very narrow, so to have any hope of success, astronomers need to choose the most likely wavelengths to monitor in the *Search for Extra-Terrestrial Intelligence* (**SETI**). For example, at very long wavelengths, interstellar gas in the galaxy emits strongly, overwhelming all but the most powerful signals. On the other hand, at very short wavelengths the molecules in our atmosphere block such signals.

It turns out that the optimum wavelengths for detection are near 10 centimeters (Figure 84.4). This clear window for radio signals spans a range that includes the 21-centimeter line emitted by neutral hydrogen and the 18-centimeter line emitted by the hydroxyl (OH) molecule in the cold gas clouds of our Galaxy. These two wavelengths are distinctive, and any other technologically capable civilization would probably have discovered their significance, so they would have equipment that could receive or broadcast at these wavelengths. Therefore, many searches have focused on signals in the range of 18 to 21 centimeters. This wavelength range is sometimes called the **waterhole** because it is bracketed by the spectral lines of H and OH which combine to make H_2O, an important molecule for life.

Project Ozma was named for the fictional princess in the series of stories of the *Land of Oz*. The princess supposedly sent radio reports about Oz to the author.

Astronomers began listening for extraterrestrial radio signals in 1960 with Project Ozma, which monitored several nearby star systems at radio wavelengths, hoping to detect signals other than those produced naturally. The search continued in more recent years with some targeted searches of nearby stars and other projects designed to "ride piggyback" on the major radio telescopes around the world—collecting radio signals from whatever direction is being studied by astronomers for other experiments. A major new dedicated facility (Figure 84.5) is presently being built, mostly with private contributions. When completed, this array of radio telescopes will expand the SETI search capabilities dramatically.

Searching for extraterrestrial signals is not easy, because even if they exist, they are probably very weak and therefore buried in cosmic static. Improving radio technologies now allow SETI projects to monitor billions of narrow radio wavelength ranges, covering most of the wavelength range where signals from other civilizations might be detected. If any transmissions are received, they will almost certainly be so weak that they will require detailed analysis to be identified. The SETI projects therefore require huge amounts of computer time to search within the collected data for potential signals. To achieve the computer processing power needed to analyze these data, one project, called SETI@home (http://setiathome.berkeley.edu), has developed a system to transmit data from the central search telescope to millions of volunteers who allow their home computers to process the information when the computers are not otherwise in use.

Despite SETI's many clever techniques to utilize existing telescopes and computers, no signals have yet been detected. SETI researchers point out, however, that the search to date has been limited to relatively nearby stars. With a more sensitive telescope under construction, in the future it will be possible to search out to greater distances, increasing the chances of detecting other civilizations.

Even supposing many other civilizations have arisen in our Galaxy, there are other possible reasons why SETI has not yet detected any signals. The other civilizations may not be transmitting radio signals if they never developed radio communications technology, or they may rely on more carefully targeted transmission systems, such as fiber optics or laser communications. However, if other civilizations are like ours, they inadvertently broadcast a wide range of signals. Our own radio and television stations and airport radars transmit extremely powerful signals that would be detectable, with sensitive equipment, over most of our Galaxy.

There is a more disturbing possibility. The number of worlds with the ability to communicate might be reduced if intelligence does not have the survival value that we would like to believe it has. Technology also gives rise to a host of problems that are potentially dangerous to the species that invents them. Pollution, nuclear war, or some other self-inflicted disaster may cause civilizations to destroy themselves.

If technologically capable civilizations do not survive, we need to multiply N_I by another factor. We might call this the survival factor, f_S, which is the fraction of the star's lifetime that a communicating civilization lasts. After all, we have survived as an intelligent species capable of communicating with other intelligent species in our Galaxy for only about a century, that is only 0.000002% (2×10^{-8}) of the life of the Sun so far. If our descendants survive for the rest of the life of the Sun, then this fraction will rise to more than 50%. However, if most civilizations' ability to communicate typically ends after 10,000 years, then f_S might be as little as 10^{-6}. Such a small survival factor would imply that more than a million civilizations would have to have arisen for there to be much chance that another is capable of communicating at the same time as us. For example, if a civilization existed in the nearby Alpha Centauri system, but it somehow destroyed itself even as recently as 100 years ago, we could not have communicated with them by radio, and the chance has now been lost forever.

The immense distances that almost certainly separate us from any other civilizations also make it unlikely that we will be able to exchange information with them. To communicate with another civilization within a human life span, we would need to be within about 30 light-years from them. At that distance, if you send a message when you are 10 years old, and they reply immediately upon receiving it, you will get a response when you are 70! For it to be likely that another civilization is that close, our earlier calculation would have to have given us the result that there are millions of other civilizations in our Galaxy.

Humans *have* sent at least one intentional message to any aliens who might be listening. During a ceremony in 1974 celebrating the renovation of the huge Arecibo radio telescope in Puerto Rico (Unit 27), astronomers broadcast a signal toward the globular cluster M13 (Figure 84.6). This cluster is more than 20,000 light-years distant, so the message of our existence is still just beginning its voyage to the stars.

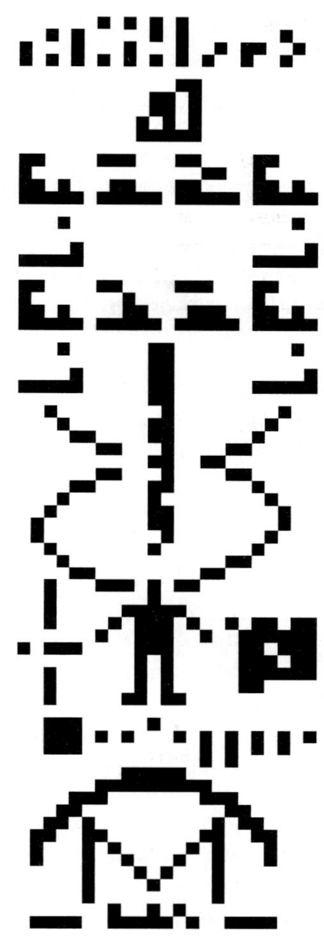

FIGURE 84.6
Message sent from Arecibo in 1974. The message communicates several fundamental ideas through this simple image. From top to bottom, it indicates: the numbers 1 through 10 in binary format; the atomic elements contained in DNA; the structure of DNA; a diagram of a human with indications of height and total population; a diagram of the Solar System, with the Earth sitting under the human figure; and a diagram of the Arecibo telescope, which also indicates its size.

Q What message would *you* send to an alien civilization? What dangers might sending a message pose for us on Earth?

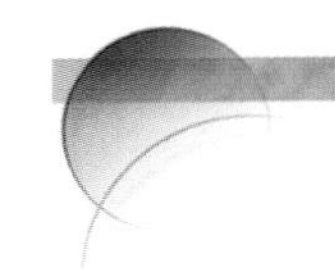

KEY TERMS

canali, 697
Drake equation, 699
exoplanet, 698
Fermi paradox, 700
loner, 699
many-worlder, 699
panspermia, 696
SETI, 702
waterhole, 702

QUESTIONS FOR REVIEW

1. What is the panspermia theory?
2. What is the significance of the meteorites from Mars?
3. How might we detect a life-sustaining planet?
4. What is the Drake equation?
5. What is Fermi's paradox?
6. What is SETI?
7. Why are radio wavelengths near 20 cm considered a likely place to find interstellar radio communications?

PROBLEMS

1. How long would it take to get to the nearest star at the Space Shuttle's speed of 30,000 kph?
2. Use a method similar to the Drake equation to estimate the number of people getting a haircut at some given moment. That is, identify the different factors involved in deciding to get a haircut, estimate the probabilities of each, then multiply the probabilities times the number of people in, for example, the United States (approximately 260 million).
3. The U.S. Department of Labor estimates that there were 651,000 barbers, hairdressers, and cosmetologists in the United States in 2002. Does this workforce level appear to be consistent with your result from problem 2?
4. Make your own estimate of the number of habitable planets within the Milky Way. What is the average distance between them?
5. After 20 years, the two *Voyager* spacecraft have reached about twice Pluto's distance from the Sun. How long will it take them to travel 4.2 ly—the distance to the nearest star?
6. Suppose that we are the only civilization in the Galaxy, and all the other civilizations have died out. For a possible 10,000 civilizations calculated by the Drake equation and a Galaxy with an age of 10 billion years, what is the typical life span of a civilization? How old is our civilization?
7. One solution suggested for our growing population in the far future is to construct a sphere surrounding the Sun with a radius of 1 AU (Earth's orbital distance). The amount of light received per square meter on the inside of this sphere would be similar to what we receive on Earth now. What is the surface area of such a sphere compared to the surface of the Earth?

TEST YOURSELF

1. The Drake equation attempts to determine
   a. the conditions under which life originated on Earth.
   b. the optimum wavelength for communicating with extraterrestrials.
   c. the age of life on Earth.
   d. the number of other technically advanced civilizations.
   e. the lifetime of our own civilization.
2. Current searches for extraterrestrial life use ____ to identify signals from intelligent beings.
   a. X-rays
   b. ultraviolet radiation
   c. visible light
   d. radio waves
3. If you wished to participate in the search for extraterrestrial life, you could
   a. volunteer your computer's downtime for SETI.
   b. set out cruelty-free traps baited with green cheese, a favorite of little green creatures.
   c. lay your speakers outside facing straight up and blast your stereo at its highest volume.
   d. get the best receiver you can afford, and tune it to receive 21- and 18-cm wavelengths.

APPENDIX

SCIENTIFIC NOTATION

Scientific notation is a shorthand method for expressing and working with very large or very small numbers. We described this method in Unit 3.2 and discuss it a little further here. We can express any number as a few digits times 10 to a power, or exponent. The power indicates the number of times that 10 is multiplied by itself.

For example, $100 = 10 \times 10 = 10^2$. Similarly, $1{,}000{,}000 = 10 \times 10 \times 10 \times 10 \times 10 \times 10 = 10^6$. Note that we do not always need to write out $10 \times \ldots$. Instead we can simply count the zeros. Thus 10,000 is 1 followed by four zeros, so it is 10^4. To express the numbers 1 and 10 in scientific notation: $1 = 10^0$ and $10 = 10^1$.

To write a number like 300, we break it into two parts: $3 \times 100 = 3 \times 10^2$. Similarly, we can write $352 = 3.52 \times 100 = 3.52 \times 10^2$. Any number can be expressed in scientific notation as a value between 1 and 10 multiplied by 10 to a power.

We can also write very small numbers (numbers less than 1) using powers of 10. For example, $0.01 = 1/100 = 1 \times 10^{-2}$. We can make this even more concise, however, by writing 1×10^{-2} as 10^{-2}. Similarly, $0.0001 = 10^{-4}$. Note that for numbers less than 1, the power is 1 more than the number of zeros after the decimal point.

We can write a number like 0.00052 as $5.2 \times 0.0001 = 5.2 \times 10^{-4}$.

Suppose we want to multiply numbers expressed in powers of 10. The rule is simple: We add the powers. Thus $10^3 \times 10^2 = 10^{3+2} = 10^5$. Similarly, $2 \times 10^8 \times 3 \times 10^7 = 2 \times 3 \times 10^8 \times 10^7 = 6 \times 10^{15}$. In general, $10^x \times 10^y = 10^{x+y}$.

Division works similarly, except that we subtract the exponents. Thus $10^5/10^3 = 10^{5-3} = 10^2$. In general, $10^x/10^y = 10^{x-y}$.

The last operations we need to consider are raising a number to a power and taking a root. In raising a power-of-ten number to a power, we multiply the powers. Thus "one thousand to the fourth power" is $(10^3)^4 = 10^{3\times 4} = 10^{12}$. Care must be used if we have a number like $(2 \times 10^4)^3$. Both the 2 and the 10^4 are raised to the third power, so the result is $2^3 \times (10^4)^3 = 8 \times 10^{4\times 3} = 8 \times 10^{12}$.

Taking a root is equivalent to raising a number to a fractional power. Thus the square root of a number is the number to the ½ power. The cube root is the number to the ⅓ power, and so forth. For example, $\sqrt{100} = 100^{1/2} = (10^2)^{1/2} = 10^1 = 10$.

SOLVING DISTANCE, VELOCITY, TIME (*d*, *V*, *t*) PROBLEMS

Many problems in this book (and in science in general) involve the motion of something. In such problems, we often know two of the three quantities distance, velocity, and time (d, V, t), and we want to know the third. For example, we have something moving at a speed V and want to know how far it will travel in a time t. Or we know that something travels with a speed V and want to find out how long it takes for the object to travel a distance d. We can usually solve such problems in our heads if the motion involves automobiles. For example, if it is 160 kilometers to a city and we travel at 80 kilometers per hour, how long does it take to get there? Or how far can we drive in 2 hours if we are traveling at 60 miles per hour?

Because we solve such problems routinely, you might find it easier to think of astronomical (d, V, t) problems in terms of cars.

Regardless of your approach, the method of solution is simple. Begin by making a sketch of what is happening. Draw an arrow to indicate the motion. Label the known quantities and put question marks beside the things you want to find. Then write out the basic relation $d = V \times t$. If you want to find d and know V and t, just multiply them for the answer. If you want to find the time and are given V and d, solve for t by dividing both sides by V to get $t = d/V$. If you want the velocity, divide d by t: $V = d/t$.

In some problems the motion may be in a circle of radius r. In that case the distance traveled will be related to the circumference of the circle, $2\pi r$. For such cases you may need to use the expression $V = 2\pi r/t$.

In most problems you will find that it is helpful to write the units of the quantities in the equation. For example, suppose you are asked how long it takes to travel 1500 km at a velocity of 30 kilometers per second. Insert the quantities so that

$$t = \frac{d}{V} = \frac{1500 \text{ km}}{30 \text{ km/sec}} = \frac{1500 \text{ km}}{30 \text{ km}} \text{ sec} = 50 \text{ sec}.$$

Note that the units of kilometers cancel out and leave us with units of seconds, as the problem requires.

APPENDIX TABLE 1 Physical and Astronomical Constants

Physical Constants		Astronomical Constants	
Velocity of light (c)	$= 2.99792458 \times 10^8$ m/sec	Astronomical unit (AU)	$= 1.495978706 \times 10^{11}$ m
Gravitational constant (G)	$= 6.67259 \times 10^{-11}$ m$^3 \cdot$kg$^{-1} \cdot$s^{-2}	Light-year (ly)	$= 9.4605 \times 10^{15}$ m $= 63{,}240$ AU
Planck's constant (h)	$= 6.62608 \times 10^{-34}$ joule · sec	Parsec (pc) = 3.26 ly	$= 3.085678 \times 10^{16}$ m $= 206{,}265$ AU
Mass of hydrogen atom (M_H)	$= 1.6735 \times 10^{-27}$ kg	Year (synodic or "tropical")	= 365.2422 days $= 3.1557 \times 10^7$ sec
Mass of electron (M_e)	$= 9.1094 \times 10^{-31}$ kg	Mass of Earth	$= 5.974 \times 10^{24}$ kg
Stefan-Boltzmann constant (σ)	$= 5.6705 \times 10^{-8}$ watts·m^{-2}·deg^{-4}	Mass of Sun ($M_\odot$)	$= 1.989 \times 10^{30}$ kg
Wien's law constant ($T \times \lambda_{max}$)	$= 2.90 \times 10^6$ K·nm	Equatorial radius of Earth	$= 6.378 \times 10^6$ m = 6378. km
		Radius of Sun ($R_\odot$)	$= 6.96 \times 10^8$ m
		Luminosity of Sun ($L_\odot$)	$= 3.83 \times 10^{26}$ watts
		Hubble's constant (H)	≈ 70 km/sec per Mpc

APPENDIX TABLE 2 Metric Prefixes

nano (n)	$= 10^{-9}$	= 1 billionth
micro (μ)	$= 10^{-6}$	= 1 millionth
milli (m)	$= 10^{-3}$	= 1 thousandth
centi (c)	$= 10^{-2}$	= 1 hundredth
kilo (k)	$= 10^3$	= 1 thousand
mega (M)	$= 10^6$	= 1 million
giga (G)	$= 10^9$	= 1 billion

APPENDIX TABLE 3 Conversion Between English and Metric Units

Length			
1 km	= 1 kilometer	= 1000 meters	= 0.6214 mile
1 m	= 1 meter	= 1.094 yards	= 39.37 inches
1 cm	= 1 centimeter	= 0.01 meter	= 0.3937 inch
1 μm	= 1 micrometer	= 10^{-6} meter	= 3.93×10^{-5} inch
1 nm	= 1 nanometer	= 10^{-9} meter	= 3.93×10^{-8} inch
1 mile	= 1.6093 km		
1 inch	= 2.5400 cm		
Mass			
1 metric ton	= 10^6 grams	= 1000 kg	= 2.2046×10^3 lb
1 kg	= 1000 grams	= 2.2046 lb	
1 g	= 1 gram	= 0.0022046 lb	= 0.0353 oz
1 lb	= 0.4536 kg		
1 oz	= 28.3495 g		

APPENDIX TABLE 4 Some Useful Formulas

Geometry

Circumference of circle = $2\pi R$

Area of circle = πR^2

Surface area of sphere = $4\pi R^2$

Volume of sphere = $\frac{4}{3}\pi R^3$

Distance Relationships

Distance–Velocity–Time: $d = V \times t$

Linear size–Angular size: $\ell = d \times \alpha/57.3°$

Distance from parallax: $d(\text{in parsecs}) = 1/p(\text{in arcsec})$

Hubble's law (for distant galaxies): $d = V/H$

Gravity

Kepler's 3rd law—orbits around Sun with semimajor axis a (in AU) and period P (in years): $P^2 = a^3$

Gravitational force between masses M and m: $F_G = G\dfrac{M \times m}{d^2}$

Gravitational potential energy: $E_G = -G\dfrac{M \times m}{d}$

Newton's modified form of Kepler's 3rd law for the total mass of two orbiting bodies: $M = \dfrac{4\pi^2}{G} \times \dfrac{d^3}{P^2}$

Mass of object for orbital speed V at distance d: $M = \dfrac{d \times V^2}{G}$

Escape velocity from a mass M at radius R: $V_{esc} = \sqrt{\dfrac{2GM}{R}}$

Light

Frequency (ν)–Wavelength (λ) relation: $\lambda \times \nu = c$

Energy of a photon: $E = h \times \nu = \dfrac{h \times c}{\lambda}$

Stefan-Boltzmann law—luminosity L of thermal source at temperature T: $L = \sigma T^4 \times$ (Surface area)

Wien's law—temperature of thermal source from wavelength of maximum emission: $T = \dfrac{2.9 \times 10^6\ \text{K} \cdot \text{nm}}{\lambda_{max}}$

Brightness (B)–Luminosity (L) relation: $B = \dfrac{L}{4\pi d^2}$

Doppler effect: Radial velocity = $V_R = c \times \dfrac{\Delta\lambda}{\lambda}$

Other Physical Relationships

Density = $\dfrac{\text{Mass}}{\text{Volume}}$: $\rho = \dfrac{M}{\mathcal{V}}$

Newton's 2nd law—acceleration a produced by force F on mass m: $a = F/m$

Kinetic energy: $E_K = ½\, m \times V^2$

Conservation of angular momentum: (Mass) × (Circular velocity) × (Radius) = Constant

Lorentz factor for special relativistic contraction at speed V: $\gamma = \dfrac{1}{\sqrt{1 - V^2/c^2}}$

APPENDIX TABLE 5 Physical Properties of the Planets

Name	Radius (Eq) (Earth units)	Radius (Eq) (km)	Mass (Earth units)	Mass (kg)	Average Density (kg/L)
Mercury	0.383	2440	0.055	3.30×10^{23}	5.43
Venus	0.949	6052	0.815	4.87×10^{24}	5.24
Earth	1.000	6378	1.000	5.97×10^{24}	5.52
Mars	0.532	3396	0.107	6.42×10^{23}	3.94
Ceres*	0.076	487	0.00016	9.43×10^{20}	2.08
Jupiter	11.21	71,492	317.9	1.90×10^{27}	1.33
Saturn	9.45	60,268	95.16	5.68×10^{26}	0.69
Uranus	4.01	25,559	14.54	8.68×10^{25}	1.27
Neptune	3.88	24,764	17.15	1.02×10^{26}	1.64
Pluto*	0.181	1160	0.0021	1.31×10^{22}	2.00
Eris*	0.188	1250	0.0028	1.67×10^{22}	2.04

*Dwarf planet

APPENDIX TABLE 6 Orbital Properties of the Planets

	Distance from Sun*		Period			
Name	(AU)	(10^6 km)	Years	Days	Orbital Inclination†	Orbital Eccentricity
Mercury	0.387	57.9	0.2409	87.97	7.00	0.206
Venus	0.723	108.2	0.6152	224.70	3.39	0.007
Earth	1.000	149.6	1.000	365.26	0.00	0.017
Mars	1.524	227.9	1.8809	686.98	1.85	0.093
Ceres‡	2.766	414.7	4.5990	1679.82	10.59	0.080
Jupiter	5.204	778.6	11.8622	4332.59	1.31	0.048
Saturn	9.582	1433.5	29.4577	10,759.22	2.49	0.056
Uranus	19.20	2872.5	84.011	30,685.4	0.77	0.046
Neptune	30.05	4495.1	164.79	60,189	1.77	0.011
Pluto‡	39.48	5906	247.68	90,465	17.15	0.248
Eris‡	67.67	10,120	557.4	203,600	44.2	0.442

*Semimajor axis of the orbit.
†With respect to the ecliptic.
‡Dwarf planet

APPENDIX TABLE 7 Satellites of the Solar System*

Planet Satellite	Radius† (km)	Distance from Planet (10^3 km)	Orbital Period (days)	Mass (10^{20} kg)	Density (kg/L)
Earth					
Moon	1738	384.4	27.322	734.9	3.34
Mars					
Phobos	13 × 11 × 9	9.38	0.319	1.3×10^{-4}	2.2
Deimos	8 × 6 × 5	23.5	1.263	1.8×10^{-5}	1.7
Jupiter‡					
Metis	20	127.96	0.295	9×10^{-4}	—
Adrastea	13 × 10 × 8	128.98	0.298	1×10^{-4}	—
Amalthea	135 × 82 × 75	181.3	0.498	8×10^{-2}	—
Thebe	55 × ? × 45	221.9	0.6745	1.4×10^{-3}	—
Io	1821	421.6	1.769	893.3	3.57
Europa	1565	670.9	3.551	479.7	2.97
Ganymede	2634	1070.0	7.155	1482	1.94
Callisto	2403	1883.0	16.689	1076	1.86
Leda	~8	11,094	238.7	4×10^{-4}	—
Himalia	92.5	11,480	250.6	8×10^{-2}	—
Lysithea	~18	11,720	259.2	6×10^{-4}	—
Elara	~38	11,737	259.7	6×10^{-3}	—
Ananke	~15	20,200	631R	4×10^{-4}	—
Carme	~20	22,600	692R	9×10^{-4}	—
Pasiphae	~25	23,500	735R	1.6×10^{-3}	—
Sinope	~18	23,700	758R	6×10^{-4}	—
Saturn‡					
Pan	10	133.58	0.574	4.2×10^{-6}	—
Atlas	19 × ? × 14	137.64	0.602	1.6×10^{-4}	—
Prometheus	74 × 50 × 34	139.35	0.613	1.4×10^{-3}	0.27
Pandora	55 × 44 × 31	141.7	0.629	1.3×10^{-3}	0.42
Epimetheus	69 × 53 × 53	151.42	0.694	5.6×10^{-3}	0.63
Janus	99 × 95 × 76	151.47	0.695	2.0×10^{-2}	0.65
Mimas	210 × 197 × 193	185.52	0.942	0.370	1.12
Enceladus	256 × 247 × 244	238.02	1.370	0.65	1.00
Tethys	536 × 528 × 526	294.66	1.888	6.17	0.98
Calypso	15 × 8 × 8	294.66	1.888	4×10^{-5}	—
Telesto	15 × 13 × 8	294.67	1.888	6×10^{-5}	—
Dione	559	377.4	2.737	10.8	1.49
Helene	18 × ? × 15	377.4	2.737	1.6×10^{-4}	—
Rhea	764	527.04	4.518	23.1	1.24
Titan	2575	1221.85	15.945	1345.5	1.88
Hyperion	185 × 140 × 112	1481.1	21.277	0.28	—
Iapetus	720	3561.3	79.331	15.9	1.0
Phoebe	115 × 110 × 105	12,952	550.4R	0.1	—
Uranus‡					
Cordelia	13	49.75	0.336	1.7×10^{-4}	—
Ophelia	16	53.77	0.377	2.6×10^{-4}	—
Bianca	22	59.16	0.435	7×10^{-4}	—
Cressida	33	61.77	0.465	2.6×10^{-3}	—
Desdemona	29	62.65	0.476	1.7×10^{-3}	—
Juliet	42	64.63	0.494	4.3×10^{-3}	—
Portia	55	66.1	0.515	1×10^{-2}	—
Rosalind	29	69.93	0.560	1.5×10^{-3}	—
Belinda	33	75.25	0.624	2.5×10^{-3}	—
Puck	77	86.00	0.764	5×10^{-3}	—

(continued)

APPENDIX TABLE 7 (continued)

Planet Satellite	Radius† (km)	Distance from Planet (10^3 km)	Orbital Period (days)	Mass (10^{20} kg)	Density (kg/L)
Uranus (*cont.*)					
Miranda	240 × 234 × 233	129.8	1.413	0.66	1.26
Ariel	581 × 578 × 578	191.2	2.520	13.5	1.65
Umbriel	584.7	266.0	4.144	11.7	1.44
Titania	788.9	435.8	8.706	35.2	1.59
Oberon	761.4	582.6	13.463	30.1	1.50
Neptune‡					
Naiad	29	48.2	0.296	1.4×10^{-3}	—
Thalassa	40	50.0	0.312	4×10^{-3}	—
Despina	74	52.5	0.333	2.1×10^{-3}	—
Galatea	79	62.0	0.396	3.1×10^{-2}	—
Larissa	104 × ? × 89	73.6	0.554	6×10^{-2}	—
Proteus	218 × 208 × 201	117.6	1.121	0.6	—
Triton	1352.6	354.59	5.875R	214	2.0
Nereid	170	5588.6	360.125	0.31	—
Eris					
Dysnomia	170	37.4	15.774	—	—
Pluto					
Charon	635	19.6	6.38718	17.7	1.83
Nix	50	48.7	24.856	—	—
Hydra	80	64.8	38.206	—	—

*Note: Authorities differ substantially on many of these values. "R" means orbit is retrograde.

†$a \times b \times c$ values for the radius are the approximate lengths of the axes for irregular moons.

‡Astronomers have recently found many new moons orbiting each of the outer planets. All are small and many are as yet unnamed. Their orbital properties and sizes are uncertain. They may be captured asteroids. A relatively up-to-date list for each planet is available at the website solarsystem.nasa.gov/planets. Click on the planet of interest and then click on the tab labeled "moons."

APPENDIX TABLE 8 Properties of Main-Sequence Stars*

Spectral Type	Luminosity ($L_\odot$)	Temperature (K)	Mass ($M_\odot$)	Radius ($R_\odot$)
O5	790,000	44,500	60	12.0
B0	52,000	30,000	17.5	7.4
B5	830	15,400	5.9	3.9
A0	54	9520	2.9	2.4
A5	14	8200	2.0	1.7
F0	6.5	7200	1.6	1.5
F5	3.2	6440	1.3	1.3
G0	1.5	6030	1.05	1.1
G5	0.79	5770	0.92	0.92
K0	0.42	5250	0.79	0.85
K5	0.15	4350	0.67	0.72
M0	0.08	3850	0.51	0.6
M5	0.01	3240	0.21	0.27
M8	0.001	2640	0.06	0.1

*Authorities differ substantially on many of these values, especially at the upper and lower mass values. Note also that the values do not always agree with the Stefan-Boltzmann law.

APPENDIX TABLE 9 The Brightest Stars

Star	Name	Apparent Visual Magnitude	Distance (ly)‡	Spectral Type	Visual Luminosity ($L_\odot$)**
—	Sun	−26.72	0.0000158	G2V	1
α CMa	Sirius	−1.44	8.6	A1V + DA2	26
α Car	Canopus	−0.74	310	F0II	14,000
α Cen	Rigel Kentaurus*	−0.27	4.3	G2V + K1V	2.3
α Boo	Arcturus	−0.05	37	K2III	220
α Lyr	Vega	0.03	25	A0V	59
α Aur	Capella*	0.08	43	G5III + G0III	91
β Ori	Rigel	0.10	860	B8Ia	50,000
α CMi	Procyon	0.40	11.5	F5IV/V + DA	7.4
α Ori	Betelgeuse	0.3–0.6†	500	M1-2 Ia-Iab	80,000
α Eri	Achernar	0.45	140	B3V	3000
β Cen	Hadar	0.61	390	B1III	74,000
α Aql	Altair	0.77	16.7	A7V	11
α Cru	Acrux*	0.78	320	B0.5IV + B1V	40,000
α Tau	Aldebaran	0.78–0.93†	67	K5III + M2V	400
α Vir	Spica	0.92–0.98†	250	B1III-IV + B2V	15,000
α Sco	Antares	1.02	550	M1Ib + B3V	60,000
β Gem	Pollux	1.16	34	K0III	46
α PsA	Fomalhaut	1.17	25	A3V	17
α Cyg	Deneb	1.25	1400	A2Ia	50,000
β Cru	Mimosa	1.25	350	B0.5IV	30,000

*A number of these stars look like individual stars to the unaided eye, but actually are in multiple-star systems, as indicated by the multiple spectral types listed. For most of these systems, the brightest star contributes the great majority of the light; however, a few would shift location in the table if seen alone: Rigel Kentaurus (Alpha Centauri) is a binary of stars with magnitude −0.01 and 1.34; Capella is a spectroscopic binary of two stars having apparent magnitudes of about 0.6 and 1.1; Acrux is a visual binary of stars having magnitudes of 1.33 and 1.73. In addition to the multiple systems noted in the table, several others have dim companions whose spectroscopic types have not been determined.

†Variable star

‡Distances from revised Hipparcos catalog.

**Luminosities corrected approximately for light emitted outside visible range.

APPENDIX TABLE 10 The Nearest Stars

Name	Distance (ly)	Spectral Type	Absolute Visual Magnitude	Visual Luminosity ($L_\odot$)	Apparent Visual Magnitude
Sun	0.0000158	G2	4.83	1.0	−26.8
Proxima Centauri	4.23	M5.5V	15.5	0.00005	11.09
α Cen A	4.32	G2V	4.4	1.7	0.01
B	4.32	K0V	5.7	0.6	1.34
Barnard's Star	5.95	M4V	13.2	0.0004	9.53
Wolf 359	7.78	M6V	16.6	0.00002	13.4
Lalande 21185	8.29	M2V	10.4	0.005	7.47
Sirius A	8.60	A1V	1.5	26.0	−1.43
B	8.60	DA2 (white dwarf)	11.3	0.003	8.44
Luyten 726-8 A	8.73	M5.5V	15.3	0.00006	12.41
B	8.73	M6V	16.1	0.00003	13.2
Ross 154	9.68	M3.5V	13.1	0.0005	10.43
Ross 248	10.3	M5.5V	14.8	0.0001	12.29
ϵ Eridani	10.5	K2V	6.2	0.3	3.73
Lacaille 9352	10.7	M1.5V	9.75	0.01	7.34
Ross 128 (FI Virginis)	10.9	M4V	13.5	0.0003	11.13
Luyten 789-6 ABC					
(EZ Aquarii)	11.3	M5V	14.7	0.0001	12.33
Procyon A	11.4	F51V-V	2.7	7.4	0.38
B	11.4	DA (white dwarf)	13.0	0.0005	10.7
61 Cygni A	11.4	K5V	7.5	0.08	5.22
B	11.4	K7V	8.3	0.04	6.03
BD + 59° 1915					
A (Struve 2398)	11.5	M3V	11.2	0.003	8.90
B	11.5	M3.5V	12.0	0.001	9.68
Groombridge 34					
A (GX Andromedae)	11.6	M1.5V	10.4	0.006	8.08
B (GQ Andromedae)	11.6	M3.5V	13.4	0.0004	11.07
ϵ Indi	11.8	K5Ve	7.0	0.13	4.68
GJ 1111 (DX Cancri)	11.8	M6.5V	17.0	0.00001	14.79
τ Ceti	11.9	G8Vp	5.8	0.4	3.50
GJ 1061	12.0	M5.5V	15.2	0.00007	13.03

Data based mainly on RECONS (Research Consortium on Nearby Stars) website.

Note: Many of the stars in the above table are known by several names. In a few instances, such alternate names are in parentheses. BD stands for Bonner Durchmusterung catalog and GJ for Gliese-Jahreiss catalog; *p* indicates a peculiar spectrum; *e* indicates that the star shows emission lines in its spectrum.

APPENDIX TABLE 11 Known and Suspected Members of the Local Group of Galaxies

Name of Galaxy	Right Ascension (hours and minutes)	Declination (degrees and minutes)	Galaxy Type	Distance (Mpc)	Diameter (kpc)	Visual Apparent Magnitude	Approximate Luminosity (millions of solar luminosities)
WLM	0 02	−15 28	Irr	1.0	3	11	50
IC10	0 20	+59 18	Irr	0.7	1	11	300
Cetus	0 26	−11 02	E	0.8	—	14	1
NGC 147	0 33	+48 30	E5	0.7	2	10	100
Andromeda III	0 35	+36 30	E5	0.8	1	15	1
NGC 185	0 39	+48 20	E3	0.7	2	9	150
NGC 205	0 40	+41 41	E5	0.8	5	8	300
Andromeda VIII	0 42	+40 37	E	0.8	14	9	150
M32	0 43	+40 52	E2	0.8	2	8	300
M31	0 43	+41 16	Sb	0.8	40	3	25,000
Andromeda I	0 46	+38 02	E3	0.8	0.5	14	4
SMC	0 53	−72 50	Irr	0.06	5	2	600
Sculptor	1 00	−33 42	E3	0.09	1	9	1
Pisces	1 04	+21 53	Irr	0.8	0.5	18	1
IC 1613	1 05	+02 07	Irr	0.7	3	9	100
Andromeda V	1 10	+47 38	E	0.8	—	15	1
Andromeda II	1 16	+33 25	E2	0.7	0.7	13	4
M33	1 34	+30 40	Sc	0.8	16	6	3000
Phoenix	1 51	−44 27	Irr	0.4	0.6	12	1
Fornax	2 40	−34 27	E3	0.14	0.7	9	15
UGCA 92	4 32	+63 36	Irr	1.44	0.6	14	5
LMC	5 23	−69 45	Irr	0.05	0.6	1	2000
Carina	6 42	−50 58	E4	0.10	0.7	16	1
Canis Major	7 35	−28 00	Irr	0.08	220	—	—
Leo A	9 59	+30 45	Irr	0.7	1	12	3
Sextans B	10 00	+05 20	Irr	1.41	1.6	12	30
NGC 3109	10 03	−26 09	Irr	1.38	5	10	100
Antlia	10 04	−27 20	E3	1.41	0.6	15	2
Leo I	10 08	+12 18	E3	0.2	0.7	10	4
Sextans A	10 11	−04 43	Irr	1.60	1.8	12	35
Sextans	10 13	−01 37	E	0.09	1	12	1
Leo II	11 13	+22 09	E0	0.2	0.7	12	1
DDO 155	12 59	+14 13	Irr	2.4	0.4	14	8
Ursa Minor	15 09	+67 13	E5	0.06	0.5	11	1
Draco	17 20	+57 55	E3	0.08	0.8	10	1
Milky Way	17 46	−29 00	SBc	0.01	40	—	20,000
Sagittarius	18 55	−30 29	Irr	0.03	—	10	30
SagDIG	19 30	−17 41	Irr	1.2	1.5	15	5
NGC 6822	19 45	−14 48	Irr	0.5	2	9	200
Aquarius	20 47	−12 51	Irr	1.0	0.6	14	2
Tucanae	22 42	−64 25	E4	0.9	0.5	15	1
UGCA 438	23 26	−32 33	Irr	1.4	0.5	14	5
Cassiopeia	23 27	+50 42	E	0.7	0.5	13	5
Pegasus	23 29	+14 45	Irr	0.8	1	13	7
Pegasus II	23 52	+24 35	E	0.8	0.9	14	3

APPENDIX TABLE 12 The Brightest Galaxies Beyond the Local Group*

Name of Galaxy	Right Ascension (hours and minutes)	Declination (degrees and minutes)	Galaxy Type	Distance (Mpc)	Diameter (kpc)	Visual Apparent Magnitude	Approximate Luminosity (millions of solar luminosities)
NGC0253	00 48	−25 17	Sc	2.6	21	7.3	17,000
NGC0300	00 55	−37 41	Scd	2.2	13	8.2	2700
NGC1068 (M77)	02 43	−00 01	Sb	13.8	28	8.7	60,000
NGC1291	03 17	−41 06	S0-a	9.2	27	8.7	25,000
IC 342	03 47	+68 06	SBc	2.5	15	8.5	51,000
NGC2403	07 37	+65 36	Sc	3.4	23	8.4	6800
NGC3031 (M81)	09 56	+69 04	Sab	3.7	23	6.9	33,000
NGC3034 (M82)	09 56	+69 41	Irr	3.7	10	8.3	15,000
NGC4258 (M106)	12 19	+47 18	Sb	7.4	41	8.4	34,000
NGC4472 (M49)	12 30	+08 00	E	16.3	43	8.3	110,000
NGC4486 (M87)	12 31	+12 24	E	15.6	32	8.6	81,000
NGC4594 (M104)	12 40	−11 37	Sa	9.2	24	8.1	58,000
NGC4736 (M94)	12 51	+41 07	Sab	5.2	11	8.1	14,000
NGC4826 (M64)	12 57	+21 41	Sab	7.4	17	8.4	30,000
NGC4945	13 05	−49 28	SBc	4.6	29	8.6	37,000
NGC5055 (M63)	13 16	+42 02	Sbc	7.7	21	8.6	25,000
NGC5128 (Cen A)	13 25	−43 01	S0	3.7	32	6.7	32,000
NGC5194 (M51)	13 30	+47 12	Sbc	8.0	26	8.3	33,000
NGC5236 (M83)	13 37	−29 52	Sc	4.6	15	7.3	28,000
NGC5457 (M101)	14 03	+54 21	Sc	7.4	47	7.8	34,000
NGC6744	19 10	−63 51	Sb	10.7	65	8.4	62,000

*Galaxies in Table 12 were selected from the HyperLeda database (http://leda.univ-lyon1.fr) based on their total visual apparent magnitude.

APPENDIX TABLE 13 The Messier Catalog*

Messier Number	Popular Name	Object Type	Right Ascension (hours and minutes)	Declination (degrees and minutes)	Visual Apparent Magnitude	Angular Size (arc minutes)	Distance (kpc)
1	Crab Nebula	Supernova remnant	05 35	+22 01	8.4	6 × 4	1.9
2		Globular cluster	21 34	−00 49	6.5	12.9	11
3		Globular cluster	13 42	+28 23	6.2	16.2	10
4		Globular cluster	16 24	−26 32	5.6	26.3	2.2
5		Globular cluster	15 19	+02 05	5.6	17.4	7.5
6	Butterfly Cluster	Open cluster	17 40	−32 13	4.2	25	0.4
7	Ptolemy's Cluster	Open cluster	17 54	−34 49	4.1	80	0.2
8	Lagoon Nebula	Diffuse nebula	18 04	−24 23	6	90 × 40	1.5
9		Globular cluster	17 19	−18 31	7.7	9.3	8.1
10		Globular cluster	16 57	−04 06	6.6	15.1	4.4
11	Wild Duck Cluster	Open cluster	18 51	−06 16	6.3	14	1.8
12		Globular cluster	16 47	−01 57	6.7	14.5	4.9
13	Hercules Cluster	Globular cluster	16 42	+36 28	5.8	16.6	7.6
14		Globular cluster	17 38	−03 15	7.6	11.7	8.8
15		Globular cluster	21 30	+12 10	6.2	12.3	10
16	Eagle Nebula	Open cluster	18 19	−13 47	6.4	7	2.1
17	Omega or Swan Nebula	Diffuse nebula	18 21	−16 11	7	11	1.5
18		Open cluster	18 20	−17 08	7.5	9	1.5
19		Globular cluster	17 03	−26 16	6.8	13.5	8.7
20	Trifid Nebula	Diffuse nebula	18 03	−23 02	9	28	1.5

(continued)

APPENDIX TABLE 13 (continued)

Messier Number	Popular Name	Object Type	Right Ascension (hours and minutes)	Declination (degrees and minutes)	Visual Apparent Magnitude	Angular Size (arc minutes)	Distance (kpc)
21		Open cluster	18 05	−22 30	6.5	13	1.3
22		Globular cluster	18 36	−23 54	5.1	24	3.1
23		Open cluster	17 57	−19 01	6.9	27	0.6
24	Sagittarius Star Cloud	Milky Way patch	18 17	−18 29	4.6	90	3.0
25		Open cluster	18 32	−19 15	4.6	32	0.6
26		Open cluster	18 45	−09 24	8	15	1.5
27	Dumbbell Nebula	Planetary nebula	19 60	+22 43	7.4	8.0 × 5.7	0.3
28		Globular cluster	18 25	−24 52	6.8	11.2	5.7
29		Open cluster	20 24	+38 32	7.1	7	1.2
30		Globular cluster	21 40	−23 11	7.2	11	8
31	Andromeda Galaxy	Spiral galaxy	00 43	+41 16	3.4	178 × 63	800
32		Elliptical galaxy	00 43	+40 52	8.1	8 × 6	800
33	Triangulum Galaxy	Spiral galaxy	01 34	+30 39	5.7	73 × 45	860
34		Open cluster	02 42	+42 47	5.5	35	0.4
35		Open cluster	06 09	+24 20	5.3	28	0.8
36		Open cluster	05 36	+34 08	6.3	12	1.2
37		Open cluster	05 52	+32 33	6.2	24	1.3
38		Open cluster	05 28	+35 50	7.4	21	1.2
39		Open cluster	21 32	+48 26	4.6	32	0.2
40		Binary star	12 22	+58 05	8.4	0.8	0.1
41		Open cluster	06 46	−20 44	4.6	38	0.7
42	The Orion Nebula	Diffuse nebula	05 35	−05 27	4	85 × 60	0.4
43		Diffuse nebula	05 36	−05 16	9	20 × 15	0.4
44	Praesepe or Beehive Nebula	Open cluster	08 40	+19 59	3.7	95	0.1
45	Pleiades or Seven Sisters	Open cluster	03 47	+24 07	1.6	110	0.1
46		Open cluster	07 42	−14 49	6	27	1.6
47		Open cluster	07 37	−14 30	5.2	30	0.4
48		Open cluster	08 14	−05 48	5.5	54	0.4
49		Elliptical galaxy	12 30	+08 00	8.3	9 × 7.5	16,000
50		Open cluster	07 03	−08 20	6.3	16	0.9
51	Whirlpool Galaxy	Spiral galaxy	13 30	+47 12	8.3	11 × 7	8000
52		Open cluster	23 24	+61 35	7.3	13	1.5
53		Globular cluster	13 13	+18 10	7.6	12.6	18
54		Globular cluster	18 55	−30 29	7.6	9.1	27
55		Globular cluster	19 40	−30 58	6.3	19	5.3
56		Globular cluster	19 17	+30 11	8.3	7.1	10
57	Ring Nebula	Planetary nebula	18 54	+33 02	8.8	1.4 × 1.0	0.7
58		Spiral galaxy	12 38	+11 49	9.7	5.5 × 4.5	17,000
59		Elliptical galaxy	12 42	+11 39	9.6	5 × 3.5	16,000
60		Elliptical galaxy	12 44	+11 33	8.8	7 × 6	16,000
61		Spiral galaxy	12 22	+04 28	9.7	6 × 5.5	15,000
62		Globular cluster	17 01	−30 07	6.5	14.1	6.9
63	Sunflower Galaxy	Spiral galaxy	13 16	+42 02	8.6	10 × 6	7700
64	Blackeye Galaxy	Spiral galaxy	12 57	+21 41	8.4	9.3 × 5.4	7400
65		Spiral galaxy	11 19	+13 05	9.3	8 × 1.5	8000
66		Spiral galaxy	11 20	+12 59	8.9	8 × 2.5	8000
67		Open cluster	08 50	+11 49	6.1	30	0.8
68		Globular cluster	12 40	−26 45	7.8	12	10
69		Globular cluster	18 31	−32 21	7.6	7.1	8.5
70		Globular cluster	18 43	−32 18	7.9	7.8	9

(continued)

APPENDIX TABLE 13 (continued)

Messier Number	Popular Name	Object Type	Right Ascension (hours and minutes)	Declination (degrees and minutes)	Visual Apparent Magnitude	Angular Size (arc minutes)	Distance (kpc)
71		Globular cluster	19 54	+18 47	8.2	7.2	3.8
72		Globular cluster	20 54	−12 32	9.3	5.9	16
73		System of four stars	20 59	−12 38	9	2.8	0.6
74		Spiral galaxy	01 37	+15 47	9.4	10.2 × 9.5	7400
75		Globular cluster	20 06	−21 55	8.5	6	18
76	Little Dumbbell Nebula	Planetary nebula	01 42	+51 34	10.1	2.7 × 1.8	1.0
77		Spiral galaxy	02 43	−00 01	8.7	7 × 6	14,000
78		Diffuse nebula	05 47	+00 03	8.3	8 × 6	0.4
79		Globular cluster	05 25	−24 33	7.7	8.7	12
80		Globular cluster	16 17	−22 59	7.3	8.9	10
81	Bode's Galaxy	Spiral galaxy	09 56	+69 04	6.9	21 × 10	3700
82	Cigar Galaxy	Irregular galaxy	09 56	+69 41	8.3	9 × 4	3700
83	Southern Pinwheel Galaxy	Spiral galaxy	13 37	−29 52	7.3	11 × 10	4600
84		Elliptical galaxy	12 25	+12 53	9.1	5	17,000
85		S0 galaxy	12 25	+18 11	9.1	7.1 × 5.2	16,000
86		Elliptical galaxy	12 26	+12 57	8.9	7.5 × 5.5	17,000
87	Virgo A	Elliptical galaxy	12 31	+12 24	8.6	7	16,000
88		Spiral galaxy	12 32	+14 25	9.6	7 × 4	17,000
89		Elliptical galaxy	12 36	+12 33	9.8	4	16,000
90		Spiral galaxy	12 37	+13 10	9.5	9.5 × 4.5	17,000
91		Spiral galaxy	12 35	+14 30	10.2	5.4 × 4.4	16,000
92		Globular cluster	17 17	+43 08	6.4	11.2	8.1
93		Open cluster	07 45	−23 52	6	22	1.1
94		Spiral galaxy	12 51	+41 07	8.1	7 × 3	5200
95		Spiral galaxy	10 44	+11 42	9.7	4.4 × 3.3	10,000
96		Spiral galaxy	10 47	+11 49	9.2	6 × 4	10,000
97	Owl Nebula	Planetary nebula	11 15	+55 01	9.9	3.4 × 3.3	0.7
98		Spiral galaxy	12 14	+14 54	10.1	9.5 × 3.2	17,000
99		Spiral galaxy	12 19	+14 25	9.9	5.4 × 4.8	17,000
100		Spiral galaxy	12 23	+15 49	9.3	7 × 6	16,000
101	Pinwheel Galaxy	Spiral galaxy	14 03	+54 21	7.8	22	7400
102	Spindle Galaxy	S0 galaxy	15 07	+55 46	9.9	5.2 × 2.3	15,000
103		Open cluster	01 33	+60 42	7.4	6	2.6
104	Sombrero Galaxy	Spiral galaxy	12 40	−11 37	8.1	9 × 4	9200
105		Elliptical galaxy	10 48	+12 35	9.3	2	11,000
106		Spiral galaxy	12 19	+47 18	8.4	19 × 8	7400
107		Globular cluster	16 33	−13 03	7.9	10	6.4
108		Spiral galaxy	11 12	+55 40	10	8 × 1	14,000
109		Spiral galaxy	11 58	+53 23	9.8	7 × 4	17,000
110		Elliptical galaxy	00 40	+41 41	8.5	17 × 10	830

*Data in Table 13 drawn primarily from the SEDS Messier Catalog Web pages. By Hartmut Frommert, Christine Kronberg, and Guy McArthur. SEDS, University of Arizona Chapter, Tucson, Arizona, 1994–2005. http://www.seds.org/messier/.

The data in the appendix tables come from many sources, including

Lang, Kenneth R. *Astrophysical Data: Planets and Stars*. New York: Springer-Verlag, 1992.

Landolt-Bornstein: *Zahlenwerte und Funktionen aus Naturwissenschaften und Technik/Numerical Data and Functional Relationships in Science and Technology.* Neue Serie/New Series. *GroupVII: Astronomy, Astrophysics, and Space Research,* vols. 1–2, Astronomy and Astrophysics. Berlin, New York: Springer-Verlag, 1965–1982.

Donald K. Yeomans of JPL for satellite dimensions and masses.

APPENDIX TABLE 14 A Cosmic Periodic Table of the Elements

Key: 321.1234 — Atomic mass of most common (or most-stable) isotope. El, 123, elementname — Atomic number.

Big Bang · Stellar Fusion · Supernovae · Radioactively unstable

1	2	3	4	5	6	7	8	9	10	11	12	13	14	15	16	17	18
1.0090 H 1 hydrogen																	4.0073 He 2 helium
7.0242 Li 3 lithium	9.0227 Be 4 beryllium											11.0221 B 5 boron	12.0140 C 6 carbon	14.0194 N 7 nitrogen	16.0135 O 8 oxygen	19.0205 F 9 fluorine	20.0157 Ne 10 neon
23.0165 Na 11 sodium	24.0129 Mg 12 magnesium											27.0129 Al 13 aluminum	28.0095 Si 14 silicon	31.0098 P 15 phosphorus	32.0093 S 16 sulfur	35.0095 Cl 17 chlorine	40.0089 Ar 18 argon
39.0090 K 19 potassium	40.0091 Ca 20 calcium	45.0082 Sc 21 scandium	48.0037 Ti 22 titanium	51.0032 V 23 vanadium	52.0009 Cr 24 chromium	55.0019 Mn 25 manganese	56.0000 Fe 26 iron	59.0017 Co 27 cobalt	58.0027 Ni 28 nickel	63.0028 Cu 29 copper	64.0035 Zn 30 zinc	69.0057 Ga 31 gallium	74.0072 Ge 32 germanium	75.0087 As 33 arsenic	80.0095 Se 34 selenium	79.0101 Br 35 bromine	84.0091 Kr 36 krypton
85.0106 Rb 37 rubidium	88.0079 Sr 38 strontium	89.0093 Y 39 yttrium	90.0093 Zr 40 zirconium	93.0144 Nb 41 niobium	98.0193 Mo 42 molybdenum	97.0191 Tc 43 technetium 14.6 Myr	102.0229 Ru 44 ruthenium	103.0252 Rh 45 rhodium	106.0267 Pd 46 palladium	107.0294 Ag 47 silver	114.0358 Cd 48 cadmium	115.0375 In 49 indium	120.0417 Sn 50 tin	121.0444 Sb 51 antimony	130.0573 Te 52 tellurium	127.0521 I 53 iodine	132.0576 Xe 54 xenon
133.0600 Cs 55 cesium	138.0656 Ba 56 barium		180.1558 Hf 72 hafnium	181.1585 Ta 73 tantalum	184.1649 W 74 tungsten	187.1732 Re 75 rhenium	192.1847 Os 76 osmium	193.1874 Ir 77 iridium	195.1915 Pt 78 platinum	197.1956 Au 79 gold	202.2055 Hg 80 mercury	205.2128 Tl 81 thallium	208.2185 Pb 82 lead	209.2234 Bi 83 bismuth	209.2255 Po 84 polonium 103 yr	210.2314 At 85 astatine 8.1 hr	211.2360 Rn 86 radon 14.6 hr
223.2791 Fr 87 francium 22 min	226.2883 Ra 88 radium 1602 yr		261.4124 Rf 104 rutherfordium 13 hr	262.4190 Db 105 dubnium 32 hr	266.4315 Sg 106 seaborgium 2.4 min	267.4383 Bh 107 bohrium 22 sec	269.4471 Hs 108 hassium 27 sec	276.4724 Mt 109 meitnerium 0.72 sec	281.4891 Ds 110 darmstadtium 9.6 sec	280.4903 Rg 111 roentgenium 3.6 sec	283.5011 112 ~5 min	284.5086 113 ~0.5 sec	288.5209 114 2.8 sec	288.5277 115 87 msec	292.5396 116 18 msec	117	118

139.0679 La 57 lanthanum	140.0682 Ce 58 cerium	141.0715 Pr 59 praseodymium	142.0728 Nd 60 neodymium	145.0813 Pm 61 promethium 17.7 yr	152.0964 Sm 62 samarium	153.0991 Eu 63 europium	158.1078 Gd 64 gadolinium	159.1102 Tb 65 terbium	164.1198 Dy 66 dysprosium	165.1221 Ho 67 holmium	166.1233 Er 68 erbium	169.1307 Tm 69 thulium	174.1412 Yb 70 ytterbium	175.1442 Lu 71 lutetium
227.2918 Ac 89 actinium 21.8 yr	232.3079 Th 90 thorium	231.3046 Pa 91 protactinium 32.8 kyr	238.3277 U 92 uranium	237.3239 Np 93 neptunium 2.14 Myr	244.3481 Pu 94 plutonium 80.8 Myr	243.3441 Am 95 americium 7370 yr	247.3577 Cm 96 curium 15.6 Myr	247.3576 Bk 97 berkelium 1380 yr	251.3716 Cf 98 californium 898 yr	252.3762 Es 99 einsteinium 1.29 yr	257.3941 Fm 100 fermium 101 days	258.3986 Md 101 mendelevium 52 days	259.4024 No 102 nobelium 58 min	262.4145 Lr 103 lawrencium ~4 hr

10,000,000,000
100,000,000
1,000,000
10,000
100
1

Cosmic Abundance Scale

The blue scale shows the relative amount of each element within the Solar System. Each full step on the scale corresponds to a factor of 10 greater abundance. Note that the scale goes two orders of magnitude higher for the two most abundant elements, hydrogen and helium.

12.3 yr

Radioactive Lifetimes

For elements with no stable isotope, the half-life of the most stable isotope is indicated. Rhenium, bismuth, uranium, and thorium are also radioactive, but their half-lives are at least several billion years, and they are found in nature.

2000 K
1500 K
1000 K
500 K
0 K

Condensation Temperature

The "thermometer" shows the temperature at which about half of each element condenses (often after first forming molecules with other elements) from a low density gas cloud such as the presolar nebula. Elements with lower condensation temperatures condense farther from the proto-Sun.

4.0073
=
(p+n).(energy)

Atomic Masses

The masses listed in the table are based on setting the mass of iron to 56.0. Iron has 26 protons and 30 neutrons, and has the least energetic nucleus of any element. On this scale, the integer part of each element's mass indicates the total number of protons+neutrons, while the decimal portion shows the mass excess that can potentially be released as energy by fission or fusion (according to $E=mc^2$).

Source: Condensation temperatures and abundance data drawn from K. Lodders (2003). *The Astrophysics Journals*, vol. 591, pp. 1220–1247.

GLOSSARY

This glossary includes definitions of the key terms and other important terms relevant to the study of astronomy. Section numbers (in parentheses) indicate where terms are introduced in the text.

Symbols

a – acceleration
a – semimajor axis of an orbit
$\mathcal{A}$ – area, such as surface area of sphere
b – baseline length, for triangulation
B – brightness, of observed light
c – speed of light
d – distance
D – diameter
e – electron
E – energy
f – fraction or percentage
F – force
g – acceleration due to gravity
G – Newton's gravitational constant
h – Planck's constant
H – Hubble's constant
ℓ – linear size
L – luminosity, or total light output
m – mass
M – mass, generally of a larger object
n – neutron
N – number, such as number of stars
p – proton
p – parallax
P – period, of rotation or of an orbit
R – radius
t – time
T – temperature
V – velocity
$\mathcal{V}$ – volume, such as volume of sphere
z – redshift, or relative shift of wavelength

Greek symbols

a – acceleration
α (alpha) – angle or angular size
β (beta) – angle
γ (gamma) – Lorentz factor
γ (gamma) – gamma ray or photon
Δ (delta) – change in…
λ (lambda) – wavelength, crest-to-crest
μ (mu) – proper or angular motion
ν (nu) – frequency, of wave oscillation
ν (nu) – neutrino
ρ (rho) – density, or mass per volume
σ (sigma) – Stefan-Boltzmann constant
τ (tau) – half-life, of a radioactive substance
Ω (omega) – density parameter of universe

A

aberration of starlight the angular shift in the apparent direction of a star caused by the orbital speed of the Earth. (52.5)

absolute magnitude the apparent magnitude a star would have if it were at a distance of 10 parsecs. (54.4)

absolute zero the lowest possible temperature, at which a material contains no extractable heat energy. Zero on the Kelvin temperature scale. (23.3)

absorption the process in which an atom or molecule intercepts light or other electromagnetic radiation and stores its energy, often by shifting an electron to a higher energy orbital. (21.4)

absorption lines dark lines superimposed on a continuous spectrum, produced when a gas absorbs specific wavelengths of light from a background source of light. (55.1)

absorption-line spectrum a spectrum in which certain wavelengths are darker than adjacent wavelengths. The missing light is absorbed by atoms or molecules between the source and the observer. (24.2)

acceleration the rate of change in an object's velocity (either its speed or its direction). This can include speeding up as well as slowing down (deceleration). (15.0)

accretion the addition of matter to a body. Examples are gas falling onto a star and asteroids colliding and sticking together. (33.3)

accretion disk a nearly flat disk of gas or other material held in orbit around a body by its gravity before it falls onto the body. (65.1, 68.3, 77.3)

achondrite a stony meteorite of rocky material from the upper layers of a differentiated body. Achondrites have no **chondrules.** (48.2)

active galactic nucleus (AGN) a small central region in a galaxy that emits abnormally large amounts of electromagnetic radiation and sometimes ejects jets of matter. Examples are radio galaxies, Seyfert galaxies, and quasars. (77.0)

A.D. abbreviation of Latin *anno domini*, "in the year of the Lord" in the Christian calendar. Equivalent to the nondenominational designation C.E. or "common era." (9.5)

adaptive optics a technique for adjusting a telescope's mirror or other optical parts to compensate for atmospheric distortions, such as **seeing,** thereby giving a sharper image. (30.2)

æther a substance that was once hypothesized to fill space and through which light waves propagated. Einstein's theory of special relativity overturned this idea. (53.1)

alpha particle a helium nucleus: two protons plus (usually) two neutrons.

alt–az mount a support structure for a telescope that allows it to rotate up and down (in altitude) and parallel to the horizon (in azimuth). (31.4)

altitude an object's angular distance above the horizon. (13.1)

amino acid a carbon-based molecule used by living organisms to build protein molecules. (83.1)

analemma the figure-8 pattern traced on the sky if observed at the same clock time each day throughout a year. (13.5)

angstrom unit a unit of length used in describing wavelengths of radiation and the sizes of atoms and molecules. One angstrom is 10^{-10} meters.

angular momentum a measure of an object's tendency to keep rotating and to maintain its orientation. Mathematically, it depends on the object's mass, *M*, radius, *R*, and rotational velocity, *V*, and is proportional to the product $M \times V \times R$. (20.3)

angular shift the change in the apparent position of an object. For example, an object will show an angular shift relative to background objects when viewed from different places. (52.2)

angular size a measure of how large an object looks to an observer. It is defined as the angle between lines drawn from the observer to opposite sides of the object. For example, the angular size of the Moon is about 1/2° as seen by an observer on Earth. (10.4)

annular eclipse an eclipse in which the Moon passes directly in front of the Sun but does not completely cover the Sun. A bright ring of the Sun's photosphere remains visible around the dark disk of the Moon. We therefore see a ring (annulus) of light around the Moon. (8.2)

Antarctic Circle the latitude line of 66.5° south, marking the position farthest from the South Pole where the Sun does not rise on the date of the summer solstice (the start of winter in the Southern Hemisphere). (6.5)

ante meridian (A.M.) before the Sun has crossed the **meridian**; before noon (7.1)

anthropic principle the principle that the properties of the universe are limited to those that permit our existence—otherwise we could not be present to observe the universe. (83.4)

antimatter a type of matter that, if brought into contact with ordinary matter, annihilates it, leaving nothing but energy. The positron is the antimatter partner of the electron. The antiproton is the antimatter partner of the proton. Antimatter is observed in cosmic rays and can be created from energy in the laboratory. (81.2)

aperture the opening in a telescope or other optical instrument that determines how much light it collects. (26.1, 27.2)

aphelion the point in an orbit where a body is farthest from the Sun.

Aphrodite a continent-like highland region on the planet Venus. (39.2)

apparent double star two stars that lie close to each other in the sky but that are not orbiting each other. (56.1)

apparent magnitude a unit for measuring the brightness of a celestial body. Also simply called *magnitude;* the word *apparent* helps to distinguish it from **absolute magnitude.** The smaller the magnitude, the brighter the star. Bright stars have magnitude 1, whereas the faintest stars visible to the unaided eye have magnitude 6. The largest telescopes can detect stars of about magnitude 30; bright objects like the Sun, Moon, and planets can have "negative" magnitudes. (54.4)

apparent noon the time at which the Sun crosses the meridian as seen from a particular location. This usually differs from noon on a clock, which represents an average within a time zone and over the year. (7.1)

arc minute (abbreviation: arcmin) an angular measure equal to 1/60th of 1 degree. (12.1, 52.1)

arc second (abbreviation: arcsec) an angular measure equal to 1/60th of 1 arc minute, or 1/3600th of 1 degree. (52.2)

Arctic Circle the latitude line of 66.5° north, marking the position farthest from the North Pole where the Sun does not rise on the date of the winter solstice (the start of winter in the Northern Hemisphere). (6.5, 7.2)

Aristarchus (about 310–230 B.C.E.) Greek astronomer who first estimated the sizes and distances of the Moon and Sun. (10.2)

Aristotle (384–322 B.C.E.) Greek philosopher who developed many early ideas of physics and astronomy. (10.1)

association see **stellar association.**

asterism an easily identified grouping of stars, sometimes a part of a larger constellation (such as the Big Dipper) or extending across several constellations (such as the Summer Triangle). (5.2)

asteroid a small, generally rocky, solid body orbiting the Sun and ranging in diameter from a few meters to hundreds of kilometers. (32.1, 41.0)

asteroid belt a region between the orbits of Mars and Jupiter in which most of the Solar System's asteroids are located. (32.1)

asteroid family a group of asteroids that appear to have originated from a single object that underwent a collision. The asteroids all have orbital parameters and spectroscopic properties that indicate they have similar compositions. (41.4)

astronomical unit (AU) a distance unit based on the average distance of the Earth from the Sun. (1.6, 52.1)

astrophysics the branch of science in which physical laws are applied to interpret astronomical observations and derive chemical and physical properties of remote objects. (54.0)

asymmetry a variation in properties where none is expected. For example, the weak force shows an asymmetry in its behavior with some particles and their antiparticles. (81.2)

atmosphere layer of gas held close to a planet or star by the force of gravity. Also a unit of pressure in which 1 atmosphere is the average pressure of Earth's atmosphere at sea level. (36.1)

atmospheric pressure the pressure produced by the weight of overlying atmosphere. (36.1)

atmospheric window a wavelength range in which our atmosphere absorbs little radiation. For example, our atmosphere allows a range of wavelengths at about 300 to 700 nm to pass through. This is our "visible" atmospheric window. The atmosphere also has a wide window at radio wavelengths and narrow windows at infrared wavelengths. (30.1)

atom a submicroscopic particle consisting of a nucleus and orbiting electrons. The smallest unit of a chemical element. (4.2)

atomic mass the mass of an atom expressed in units approximately equal to the mass of a proton or neutron. The isotope of carbon with 6 protons and 6 neutrons has an atomic mass of 12.0. (4.2)

atomic number the number of protons in the nucleus of an atom. Unless the atom is ionized, the atomic number is also the number of electrons orbiting the nucleus of the atom. (4.2, 21.4)

aurora the light emitted by atoms and molecules in the upper atmosphere. This light is a result of magnetic disturbances caused by the solar wind, and appears to us as the Northern or Southern Lights. (36.1, 43.4, 51.1)

autumnal equinox the beginning of fall in the Northern Hemisphere, on or near September 22 when the Sun crosses the celestial equator. (6.5)

averted vision looking slightly to one side of a dim object so that its light strikes slightly off-center in your field of vision. This allows you to more easily discern faint objects, although at a sacrifice of sharpness. (31.2)

azimuth a coordinate for indicating the direction of objects on the sky. The azimuth direction is measured as the angle eastward from north to the point on the horizon below the object. A star that is directly above the point due east is at azimuth 90°; south is at 180°; west at 270°; and north at 360° or 0°. (13.1)

B

Balmer lines a series of absorption or emission lines of hydrogen seen at visible wavelengths. (55.4)

barred spiral (SB) galaxy a galaxy in which the spiral arms wind out from the ends of a central bar rather than from the nucleus. (75.1)

Barringer crater an impact crater more than 1 km in diameter located in Arizona, probably produced about 50,000 years ago. (48.4)

basalt a dense, dark rock of volcanic origin. (37.1)

baseline the distance between two observing locations used for the purposes of triangulation measurements. The larger the baseline, the better the resolution attainable. (52.1)

B.C. an abbreviation used in the Christian calendar to indicate dates before the birth of Jesus ("before Christ"). Biblical scholars estimate that the actual birth year was between 4 and 8 years earlier than the date indicated by this calendar system. (9.5)

B.C.E. (before the Common Era) a nondenominational designation for referring to B.C. dates. (9.5)

belt a dark, low-pressure region in the atmosphere of Jovian planet, where gas flows downward. (43.1)

Big Bang the event that began the universe according to modern cosmological theories. It occurred about 13.7 billion years ago and generated the expanding motion that we observe today. (2.5, 79.0)

Big Crunch point of final collapse of a bound universe. (82.0)

binary star two stars in orbit around each other, held together by their mutual gravity. (56.0, 57.2)

bipolar flow the narrow columns of high-speed gas ejected by a protostar in two opposite directions. (59.2, 60.2)

blackbody an object that is an ideal radiator when hot and a perfect absorber when cool. It absorbs all radiation that falls upon it, reflecting no light; hence it appears black. Stars are approximately blackbodies. The radiation emitted by blackbodies obeys Wien's law and the Stefan-Boltzmann law. (23.1)

black dwarf a white dwarf that has cooled to the point where it emits little or no visible radiation. (64.4)

black hole an object whose gravitational attraction is so strong that its escape velocity equals the speed of light, preventing light or any radiation or material from leaving its "surface." (59.2, 68.0)

blueshift a shift in the wavelength of electromagnetic radiation to a shorter wavelength. For visible light, this implies a shift toward the blue end of the spectrum. The shift can be caused by the motion of a source of radiation toward the observer or by the motion of an observer toward the source. For example, the spectrum lines of a star moving toward the Earth exhibit a blueshift. See also **Doppler shift.** (25.1)

Bode's rule a numerical formula for predicting the approximate distances of most of the planets from the Sun. (41.1)

Bohr model first theory of the hydrogen atom to explain the observed spectral lines. In this model the electron can orbit the nucleus only at discrete radii, each orbit having a particular energy. An electron can jump between orbits only if it absorbs or emits the exact energy difference between the orbits. (24.1)

Bok globule a small, dark, interstellar cloud, often approximately spherical. Many globules are the early stages of protostars. (60.1)

B.P. a designation used primarily by archaeologists and geologists to indicate the number of years "before the present" era. (9.5)

Brahe, Tycho (1546–1601) Danish astronomer who made the finest pretelescopic measurements. His observations disproved the idea that comets and novae were atmospheric phenomena and provided the basis for Kepler's laws about the motions of the planets. (12.1)

brightness a measure of the amount of light received from a body, often measured in watts per square meter detected by a telescope or other detector. (21.2, 54.2)

brown dwarf a body intermediate between a star and a planet. A brown dwarf has a mass between about 0.017 and 0.08 solar masses—too low to fuse hydrogen in its core, but massive enough to fuse deuterium as it contracts. (34.2, 60.4)

bulge the dense, central region of a spiral galaxy. (70.4)

C

Callisto the second largest of Jupiter's satellites. (46.1)

Caloris Basin the largest impact crater region yet seen on Mercury. (38.1)

canali straight-line features on Mars observed in 1877 by Italian astronomer Giovanni Schiaparelli. He called them *canali*, by which he meant channels, but some English-speakers interpreted *canali* as canals, with the implication of intelligent beings to build them. (84.1)

carbonaceous chondrite a type of meteorite containing many tiny spheres (chondrules) of rocky or metallic material embedded in carbon-rich material. (48.2)

Cassegrain telescope a type of reflecting telescope in which incoming light hits the primary mirror and is then reflected upward toward the prime focus, where a secondary mirror reflects the light back down through a small hole in the main mirror into a detector or eyepiece. (28.2)

Cassini's division a conspicuous 1800-kilometer-wide gap between the outermost rings of Saturn. (45.3)

catadioptric telescope a hybrid telescope between a reflector and a refractor, using both a large mirror and a large lens, called a corrector plate, to focus the light entering the aperture. (31.3)

CCD charged-coupled device: an electronic device that records the intensity of light falling on it. CCDs have replaced film in most astronomical applications. (26.2)

C.E. (Common Era) a designation for the year-numbering system of the most widely used calendar system. A nondenominational equivalent of **A.D.** (9.5)

celestial equator an imaginary line on the celestial sphere lying exactly above the Earth's equator. It divides the celestial sphere into northern and southern hemispheres. (5.3)

celestial pole an imaginary point on the sky directly above the Earth's North or South Pole. (5.3)

celestial sphere an imaginary sphere surrounding the Earth. Ancient astronomers pictured the stars as all being attached to it, all at the same distance from the Earth. (5.1)

center of mass the "average" position of a collection of massive bodies weighted by their masses. (16.2, 56.3)

centripetal force a force that causes the path of a moving object to curve. (17.1)

Cepheid variables a class of yellow giant pulsating stars. Their pulsation periods range from about 1 day to about 70 days. Cepheids can be used to determine distances. See also **standard candle.** (63.2)

Ceres the first-discovered and largest of the asteroids, over 900 km in diameter. Often called a planet during the early 1800s. (41.1)

Chandrasekhar limit the maximum mass of a white dwarf stellar remnant. Approximately 1.4 solar masses—above this limit the remnant will collapse. Named for the astronomer who first calculated that such a limit exists. (65.2)

chemical energy the energy stored or released when bonds form or break between atoms and molecules. (20.1)

Chicxulub the site of a major impact crater that is the most likely explanation for the extinction of the dinosaurs about 65 million years ago. (48.5)

chondrite a meteorite containing small spherical granules called chondrules. (48.2)

chondrule a small spherical silicate body embedded in many meteorites. (48.2)

chromosphere the lower part of the Sun's outer atmosphere that lies directly above the Sun's visible surface (photosphere). (49.4)

circumpolar close enough to a celestial pole that always remains above the horizon. Circumpolar stars are closer to the celestial pole than the latitude of the observer. (5.4)

closed universe geometry that the universe as a whole would have if its gravity were great enough to curve around back on itself (positive curvature). A closed universe is finite in extent but has no edge, like the surface of a sphere. (80.4)

cluster a group of objects (for example, stars or galaxies) held together by their mutual gravitational attraction.

CNO cycle/process a reaction involving carbon, nitrogen, and oxygen (C, N, and O) that fuses hydrogen into helium and releases energy. The process begins with a hydrogen nucleus fusing with a carbon nucleus. Subsequent steps involve nitrogen and oxygen. The carbon, nitrogen, and oxygen act as catalysts and are released at the end of the process to start the cycle again. The CNO cycle is the dominant process for generating energy in main-sequence stars that are hotter and more massive than the Sun. (61.2)

coherent a relationship indicating that electromagnetic waves are in synchronization with each other. Waves that are coherent exhibit persistent interference effects. (27.1)

collecting area the total aperture over which a telescope can gather light. (27.2)

coma the gaseous atmosphere surrounding the head of a comet. (47.1)

comet a small body in orbit around the Sun, consisting of a tiny, icy core and a tail of gas and dust. The tail forms only when the comet is near the Sun. (32.1)

compact stars very dense stars whose radii are much smaller than the Sun's. These stars include **white dwarfs, neutron stars,** and **black holes.**

comparative planetology the study of bodies in the Solar System, examining and understanding the similarities and differences among worlds. (42.0)

condensation conversion of free gas atoms or molecules into a liquid or solid. A snowflake forms in our atmosphere when water vapor condenses into ice. (33.3)

cone light-sensing cell in the eye that gives us our color vision. (31.1)

conglomerate a rock made of eroded material and pebbles and later fused together by heat and pressure. (40.3)

conjunction the appearance of two astronomical objects in approximately the same direction on the sky. For example, if Mars and Jupiter happen to appear near each other on the sky, they are said to be in conjunction. If a planet is in conjunction with the Sun, sometimes it is simply said to be in conjunction without specifying the Sun. (11.3)

conservation law a theory describing how certain properties, such as electric charge, energy and angular momentum, remain constant despite interactions. (20.0)

conservation of angular momentum a principle of physics stating that the angular momentum of a rotating body remains constant unless forces act to speed it up or slow it down. Mathematically, conservation of angular momentum states that $M \times V \times R$ is a constant, where M is the mass of a body moving with a velocity V in a circle of radius R. One extremely important consequence of this principle is that if a rotating body shrinks, its rotational velocity must increase. (20.3, 67.1)

constellation an officially recognized grouping of stars on the night sky. Astronomers divide the sky into 88 constellations. (5.2)

continuous spectrum a spectrum with neither dark absorption nor bright emission lines. The intensity of the radiation in such a spectrum changes smoothly from one wavelength to the next. (24.3)

convection the rising and sinking motions in a liquid or gas that carry heat upward through the material. Convection is easily seen in a pan of heated soup on a stove. (35.3. 61.2)

convection zone the region immediately below the Sun's visible surface in which its heat is carried upward by convection. (49.3)

Copernicus, Nicolaus (1473–1543) Polish astronomer who proposed that the Earth was just one of the planets orbiting the Sun. (11.3)

core the innermost, usually hot, region of the interior of the Earth or another planet. (32.1)

core-collapse supernova a supernova of type Ib, Ic, or II. A star that begin with more than about 8 $M_{\odot}$, will fuse elements all the way up to iron in its core, which then collapses catastrophically. (66.3)

Coriolis effect the deflection of an object moving within or on the surface of a rotating body. It is caused by the conservation of angular momentum when the moving object gets closer to or farther from the rotation axis of the rotating body. The Coriolis effect makes storms on Earth spin, generates large-scale wind systems, and creates cloud belts on many planets. (36.4)

cornea the curved transparent layer covering the front part of the eye that does most of the focusing of light that we see. (31.1)

corona the outer, hottest part of the Sun's atmosphere. (49.4)

coronal hole a low-density region in the Sun's corona, probably related to the structure of the Sun's magnetic field. The solar wind may originate in these regions. (49.4)

coronal mass ejections a blast of gas moving outward through the Sun's corona and into interplanetary space following the eruption of a prominence. (51.2)

cosmic horizon the greatest distance it is possible to see out into the universe given its age and rate of expansion. The horizon lies at a **light travel time distance** in light-years equal to the age of the universe in years. (80.1)

cosmic microwave background (CMB) the radiation that was created during the Big Bang and that permeates all space. At this time, the temperature of this radiation is 2.73 K. (80.2)

cosmic rays extremely energetic subatomic particles (such as protons and electrons) traveling at nearly the speed of light. Some rays are emitted by the Sun, but most come from more distant sources, such as exploding supernovae. (50.3, 65.3)

cosmological constant a term in the equations that Einstein developed to describe the effects of gravity on the entire universe. The term has the effect of a repulsive "force" opposing gravity. (82.4)

cosmological principle the hypothesis that on average, the universe looks the same to all observers, no matter where they are located in it. (79.2)

cosmology the study of the evolution and structure of the universe. (79.0)

Crab Nebula a supernova remnant in the constellation Taurus. Astronomers in ancient China and the Far East saw the supernova explode in A.D. 1054. Astronomers discovered a **pulsar** in the middle of the nebula in the 1960s. (26.4)

crater a circular pit, generally with a raised rim and sometimes with a central peak. A crater can be formed by the impact of a solid body, such as an asteroid, with the surface of a planet, moon, or another asteroid. Crater diameters on the Moon range from centimeters to several hundred kilometers. (37.1)

crescent moon a lunar phase during which the Moon appears less than half full. (8.1)

Cretaceous period a period of time in the geologic timescale ranging from about 146 million to 65 million years B.P. (48.5)

critical density the density necessary for a closed universe. If the density of the universe exceeds the critical density, the universe will stop expanding and collapse. If the density is less than the critical density, the universe will expand forever. (82.2)

crust the solid surface of a planet, moon, or other solid body. (35.1)

curvature of space the "bending" of space or motion given to space by gravity, as described by Einstein's general theory of relativity. For example, stars curve the space around them so that the path of light traveling by them is bent from a straight line. The universe too may be curved in such a way as to make its volume finite. (68.2)

cyanobacteria a kind of photosynthesizing bacteria, formerly known as blue-green algae. Colonies of them form **stromatolites.** Thought to have been found in some microfossils as old as 3.5 billion years, but not confirmed. (83.1)

D

dark adaption the process by which the eye changes, through chemical changes in the retina and by expanding the pupil, to become more sensitive to dim light. (31.2)

dark energy a form of energy detected by its effect on the expansion of the universe, also known as the **cosmological constant.** It causes the expansion to speed up. The nature and properties of dark energy are unknown. (2.5, 82.4)

dark-line spectrum see **absorption-line spectrum.**

dark matter matter that emits no detectable radiation but whose presence can be deduced by its gravitational attraction on other bodies. (2.5, 73.1, 78.0)

dark matter halo the extended "cloud" of dark matter in which galaxies lie. The dark matter comprises most of the mass of a galaxy, but is only detected from the rapid speed of rotation of the outer parts of the galaxy. (78.1)

dark nebula a dense cloud of dust and gas in interstellar space that blocks the light from background stars. (72.1)

daylight saving time the time kept during summer months by setting the clock ahead one hour. This gives more hours of daylight after the workday. (7.4)

declination (dec) one part of a coordinate system (with **right ascension**) for locating objects in the sky. The declination indicates an object's distance north or south of the celestial equator. Declination is analogous to latitude on the Earth's surface. (5.5)

degenerate gas an extremely dense gas in which the electrons and nuclei are tightly packed. Unlike a normal gas, the pressure of a degenerate gas is not affected by its temperature. (62.3)

Deimos one of the two small moons of Mars, probably a captured asteroid. (40.5)

density the mass of a body or region divided by its volume. (3.1, 32.4)

density wave a spiral-shaped compression of the gas and dust in a spiral galaxy kept from dispersing by the gravitational attraction between all the matter in the wave. Density waves may account for the spiral arms of galaxies. According to the theory, stars and gas in one region of a galaxy may become packed more closely together (reaching a higher density). As the galaxy spins, these dense regions are pulled apart, but their gravitational influence on neighboring regions may cause them to compact in succession. This can produce "waves" of higher density that travel through the disk of a galaxy, pulling stars and interstellar gas into a spiral pattern. (73.3)

deuterium an isotope of hydrogen containing one proton and one neutron. (50.2)

differential gravitational force the difference between the gravitational forces exerted on an object at two different points. The effect of this force is to stretch the object. Such forces create **tides** and, if strong enough, may break up an astronomical object. See also **Roche limit.**

differential rotation rotation in which the rotation period of a body varies with latitude or radius. Differential rotation occurs for gaseous bodies like the Sun and Jovian planets as well as for the disks of galaxies. (73.1)

differentiation the separation of previously mixed materials inside a planet or other object. An example of differentiation is the separation that occurs when a dense material, such as iron, settles to the planet's core, leaving lighter material on the surface. (33.5)

diffraction bending of the path of light or other electromagnetic waves as they pass through an opening or around an obstacle. Diffraction limits the ability of a telescope to distinguish fine details. (29.1)

diffraction pattern a pattern that a telescope imparts to light from even something as simple as a single point of light. This is caused by diffraction around various structural elements of a telescope—such as the spikes seen surrounding the image of a stars, which are caused by the support structure for secondary mirrors. (29.1)

dirty snowball a model of the structure of a comet, suggesting it is made primarily of ices with grains of heavy elements scattered within it. (47.1)

disk the flat, round portion of a galaxy. The Sun lies in the disk of the Milky Way. (70.4)

dispersion the spreading of light or other electromagnetic radiation into a spectrum. A rainbow is an example of the dispersion of light caused by raindrops. (28.4)

DNA deoxyribonucleic acid. The complex molecule that encodes genetic information in all organisms here on Earth. (83.1)

Doppler shift the change in the observed wavelength of radiation caused by the motion of the emitting body or of the observer. The shift is an increase in the wavelength if the source and observer move apart and a decrease in the wavelength if the source and observer approach. See also **redshift** and **blueshift.** (25.0)

Doppler shift method a means of detecting planets orbiting other stars by detecting the Doppler shift caused by the star's own reflex motion in response to the orbiting planet. (34.2)

Drake equation expression that gives an estimate of the probability that intelligence exists elsewhere in the galaxy, based on a number of conditions thought to be essential for intelligent life to develop. (84.3)

dust tail a comet tail containing dust that reflects sunlight. The dust in a comet tail is expelled from the nucleus of the comet. (47.3)

dwarf a small dim star.

dwarf galaxy small galaxy containing a few million stars. (75.1, 76.1)

dwarf planet an object orbiting the Sun that is so massive that its gravity pulls it into a roughly spherical shape but, since it is not the dominant mass in the neighborhood of its orbit, it cannot be called a planet. Pluto, Eris, and Ceres are considered dwarf planets. Several dozen trans-Neptunian objects (TNOs) may soon be added to that category. (32.1, 46.5)

dynamo see **magnetic dynamo.**

E

early universe the first second or so of the universe's history after the Big Bang when the universe was composed of nothing but radiation and elementary particles. (81.2)

eccentricity how round or "stretched out" an orbital ellipse is. A circular orbit has zero eccentricity, while extremely elongated orbits have eccentricities close to one. (12.2)

eclipse the blockage of light from one astronomical body caused by the passage of another between it and the observer. The shadow of one astronomical body falling on another. For example, the passage of the Moon between the Earth and Sun can block the Sun's light and cause a solar eclipse. (8.2)

eclipsing binary a binary star pair in which one star periodically passes in front of the other, totally or partially blocking the background star from view as seen from Earth. (56.1, 57.2)

ecliptic the path followed by the Sun around the celestial sphere. The path gets its name because eclipses can occur only when the Moon crosses the ecliptic. (6.2)

Einstein, Albert (1879-1955) a German-born theoretical physicist, probably best known for his theories of special and general relativity, which describe the effects of motion and mass on space and time. (53.3)

electric charge the electrical property of objects that causes them to attract or repel one another. A charge may be either positive or negative. Unlike charges (+ and −) attract each other, but like charges (+ and + or − and −) repel each other. (4.2)

electromagnetic force the force arising between electrically charged particles, between charges and magnetic fields, and between magnets. This force holds electrons to the nuclei of atoms, makes moving charges spiral around magnetic field lines, and deflects a compass needle. (4.3)

electromagnetic radiation a general term for any kind of electromagnetic wave. (22.0)

electromagnetic spectrum the assemblage of all wavelengths of electromagnetic radiation. The spectrum includes the following wavelengths, from long to short: radio, infrared, visible light, ultraviolet, X-rays, and gamma rays. (22.0)

electromagnetic wave a wave consisting of alternating electric and magnetic energy. Ordinary visible light is an electromagnetic wave, and its wavelength determines the light's color. (21.1)

electron a low-mass, negatively charged subatomic particle. In an atom, electrons orbit the nucleus, but may at times be torn free if **ionized.** (4.2)

electron degeneracy a condition resisted by electrons because they cannot occupy the same space if they have the same energy. (65.2)

electronic transition a shift in energy level that an atom undergoes when an electron shifts from one energy level to another. (21.3)

electroweak force a combined form (or unification) of the **electromagnetic force** and **weak force** that occurs at high energies.

element a fundamental substance, such as hydrogen, carbon, or oxygen, that cannot be broken down into a simpler chemical substance. Approximately 100 elements occur in nature. (21.4)

ellipse a geometric figure related to a circle but elongated along one axis. (12.2)

elliptical (E) galaxy a galaxy in which the stars smoothly fill an ellipsoidal volume. Abbreviated as E galaxy. The stars in such systems are generally old (Pop II). (75.1)

emission the production of light or other electromagnetic radiation, by an atom or molecule when the atom drops to a lower energy level. (21.4)

emission-line spectrum a spectrum consisting of bright lines at certain wavelengths separated by darker regions in which little or no radiation is emitted. (24.2)

emission nebula a hot gas cloud in interstellar space that emits light. (72.1)

energy a quantity that measures the ability of a system to do work or cause motion. (20.1)

energy level any of the numerous levels that an electron can occupy in an atom, roughly corresponding to an electron orbital. (24.1)

epicycle a fictitious, small, circular orbit superimposed on another circular orbit. Epicycles were proposed by early astronomers to explain the retrograde motion of the planets and to make fine adjustments to the predictions of planets' positions. (11.2)

equation of time a relationship describing the offset of time measured by a sundial from the average 24-hour period from noon to noon. The offsets vary through the year because of the Earth's elliptical orbit and the tilt of its axis. (13.5)

equator the imaginary line that divides the Earth (or any other body) symmetrically into its Northern and Southern Hemispheres. The equator is perpendicular to a body's rotation axis.

equatorial mount a support structure for a telescope that allows it to rotate along celestial coordinates: north and south (in declination) and around the celestial poles (in right ascension). (31.4)

equinox the time of year when the Sun crosses the celestial equator. The number of hours of daylight and night are approximately equal, and the two dates mark the beginning of the spring and fall seasons. (6.5)

Eratosthenes (276–195 B.C.E.) Greek astronomer who first estimated the diameter of the Earth. (10.3)

Eris the largest known **trans-Neptunian object.** Eris is larger and about 27% more massive than Pluto. (46.5)

erosion the wearing down of geological features by wind, water, ice, and other phenomena of planetary weather. (36.2)

escape velocity the speed an object needs to move away from another body in order not to be pulled back by its gravitational attraction. Mathematically, the escape velocity, V, is defined as $\sqrt{2GM/R}$ where M is the body's mass, R is its radius, and G is the gravitational constant. (18.2, 68.1)

Europa the smallest of the four Galilean satellites of Jupiter. (46.1)

eukaryotes cells with nuclei that evolved after **prokaryotes.** (83.2)

Evening Star any bright planet, but most often Venus, seen low in the western sky after sunset. (13.4)

event horizon the location of the "boundary" of a black hole. An outside observer cannot see in past the event horizon. (68.2)

excited state the condition in which the electrons of an atom are not in their lowest energy level (lowest orbital). (21.4)

exclusion principle the condition that no two electrons may have the same state in an atom. Because this limitation forces electrons in extremely dense materials to occupy high energy levels, it leads to a **degenerate gas.** (62.3)

exoplanet a planet orbiting a star other than our Sun. (34.0, 84.2)

expansion redshift a lengthening of the wavelength of electromagnetic radiation caused when the waves travel through expanding space. The redshift of distant galaxies is an example of this phenomenon. (74.3)

exoplanet a planet orbiting a star other than the Sun. (34.0)

exponent a number placed to the right and above another indicating the power by which to raise the other. For example, in the expression T^4 the exponent is 4. (3.1)

extinction the dimming of starlight due to absorption and scattering by interstellar dust particles. (72.3)

extrasolar planet see **exoplanet.**

F

false-color image a depiction of an astronomical object in which the colors are not the object's real colors. Instead, they are colors arbitrarily chosen to represent other properties of the body, such as the intensity of radiation at other than visible wavelengths. (26.3)

fault in geology, a crack or break in the crust of a planet along which slippage or movement can take place. (39.2, 45.2)

Fermi paradox the question why, if there are many advanced civilizations in the galaxy, none of them are visiting us or have left indications of their existence. (84.3)

filter (for light) a material that transmits only particular wavelengths of light. (27.3)

first-quarter moon a phase of the waxing Moon when Earth-based observers see half of the Moon's illuminated hemisphere. (8.1)

fission the splitting of an atom into two or more smaller atoms.

flare an outburst of energy on the Sun. (51.2)

flat rotation curve a plot showing that the orbital speed in a galaxy remains nearly constant out to large radii. (78.1)

flat universe a universe that extends forever with no curvature. (80.4)

flatness problem astronomers' measurements show the universe is almost precisely flat, with zero curvature, although Einstein's theory of general relativity predicts that space is likely to be curved by gravity. This mystery is solved by the theory of inflation, which drives the universe toward flatness. (81.4)

fluorescence the conversion of ultraviolet light (or other short-wavelength radiation) into visible light. This occurs, for example, when an atom is **excited** into a high energy level by an ultraviolet photon, and then descends to the ground state in a series of steps, emitting lower-energy photons. (47.3)

focal length the distance between a mirror or lens and the point at which the lens or mirror brings a distant source of light into focus. (31.4)

focal plane the surface where the lenses and/or mirrors of a telescope form an image of distant object. (28.1)

focus (1) one of two points within an ellipse used to generate the elliptical shape. The Sun lies at the focus of each planet's elliptical orbit. (12.2) (2) A point in an optical system in which light rays are brought together; the location where an image forms in such systems. (26.1, 28.0)

force a push or pull; anything that can cause a body to change velocity. (14.3)

fovea a central part of the retina, where most of our color vision is concentrated. This region is not very sensitive at low light levels. (31.1)

frequency the number of times per second that a wave vibrates. (22.1)

frost line the distance from the Sun in the solar nebula beyond which it was cold enough for water to condense and therefore become a major component of planetesimals forming there. (42.4)

full moon phase of the Moon in which it appears as a completely illuminated circular disk in the sky. (8.1)

fully convective a condition seen in some stars (such as very low-mass main-sequence stars) in which convection circulates matter between the core and the surface. (61.2)

G

g the acceleration due to gravity on the surface of the Earth—about 9.8 meters/second^2. (16.3)

G Newton's universal gravitational constant, which allows us to determine the force between objects if we know their masses and separation. (16.2)

Gaia hypothesis the hypothesis that life does not merely respond to its environment but actually alters its planet's atmosphere and temperature to make the planet more hospitable. For example, by photosynthesis, plants have created an oxygen-rich atmosphere on Earth, which shields the plants from dangerous ultraviolet radiation. Gaia is pronounced *GUY-uh* in this context. (83.3)

galactic cannibalism the capture and merging of one galaxy into another. (71.3, 75.4)

Galactic plane the flat region defined by the disk of the Milky Way. (70.2)

galaxy a massive system of stars, gas, and dark matter held together by their mutual gravity. Typical galaxies have a mass between about 10^7 and 10^{13} solar masses. The Milky Way is our Galaxy, which is usually capitalized. (2.2, 70.0, 74.0)

galaxy cluster a set of hundreds or thousands of galaxies held together by their mutual gravitational attraction. The Milky Way lies on the outskirts of the Virgo Cluster. (2.4, 76.0)

galaxy group a system of from two to several dozen galaxies held together by their mutual gravitational attraction. The Milky Way belongs to the Local Group galaxy cluster. (2.4, 76.1)

Galilean relativity a method for determining the relative speeds of motion seen by observers who are moving with respect to each other. This method successfully describes motions at slow speeds, but not when speeds grow to an appreciable fraction of the speed of light. (53.1)

Galilean satellites the four moons of Jupiter discovered by Galileo: Io, Europa, Ganymede, and Callisto. (12.3, 45.1, 46.1)

Galilei, Galileo (1564–1642) Italian physicist and astronomer who studied inertia and published the first studies of the sky with a telescope. (12.3)

gamma ray electromagnetic waves that have the shortest wavelength. (22.6)

Ganymede the largest moon of Jupiter and the largest satellite in the Solar System—even larger than Mercury. (46.1)

gas giant a Jovian planet. (32.1)

general theory of relativity Einstein's theory relating how mass and energy can change the structure of space and time, "curving" them. (68.0, 79.2)

geocentric model a hypothesis that held that the Earth is at the center of the universe and all other bodies are in orbit around it.

Early astronomers thought that the Solar System was geocentric. (11.2)

giant a star of large radius and large luminosity. (62.0)

giant elliptical galaxy a large galaxy found in the crowded inner regions of **rich clusters.** These probably form by **galactic cannibalism.** (76.2)

giant molecular clouds interstellar clouds, in which atoms combine to form molecules and heavier elements combine to form interstellar dust, with masses up to about a million solar masses. (60.1)

gibbous appearance of the Moon (or a planet) when more than half (but not all) of the sunlit hemisphere is visible from Earth. (8.1)

glitch abrupt change in the pulsation period of a pulsar, perhaps the result of adjustments of its crust. (67.2)

global warming a phenomenon in which average ocean and air temperatures on Earth have increased by about 1.6 degrees Celsius over the last century. Most scientists attribute global warming to **greenhouse gases** produced by human activities such as burning of fossil fuels and deforestation. (36.3)

globular cluster a dense grouping of old stars, containing generally about 10^5 to 10^6 members. They are often found in the halos of galaxies. (69.1, 70.3)

globule see **Bok globule.**

gluon a particle that carries (or exerts) the strong force between quarks. (4.3)

grains small solid particles that condense from the heavy elements in a gas when it cools. (64.2)

grand unified theory (GUT) a theory that describes and explains the four fundamental forces as different aspects of a single force. (81.3)

granulation texture seen in the Sun's photosphere. Granulation is created by clumps of hot gas that rise to the Sun's surface. (49.3)

grating a piece of material with many fine, closely spaced parallel lines used to create a spectrum. Light may be reflected off a grating or diffracted as it passes through a transparent grating. (24.0)

gravitational force force exerted on one body by another due to the effect of gravity. The force is directly proportional to the masses of both bodies involved and is inversely proportional to the square of the distance between them. (4.2)

gravitational lens an object whose gravity curves space in a way that can focus the light from an object behind it. The light from the more distant object may be magnified, distorted, or turned into multiple images by the lens. See also **curvature of space.** (78.3)

gravitation lensing method a technique for detecting a planet orbiting a star by observing the gravitational force of the planet acting as a lens to focus the light of the star. (34.4)

gravitational potential energy the energy stored in a body subject to the gravitational attraction of another body. As the body falls, its gravitational potential energy decreases and is converted into kinetic energy. (20.1)

gravitational redshift the shift in wavelength of electromagnetic radiation (light) caused by a body's gravitational field as the radiation moves away from the body. Only extremely dense objects, such as white dwarfs or neutron stars, produce a significant gravitational redshift of their radiation. (68.4)

gravitational waves a wavelike disturbance in the curvature of space that is generated by the acceleration of massive bodies. (68.5)

graviton a particle hypothesized to carry the gravitational force. (4.3)

gravity the force of attraction between two bodies that is generated by their masses. (2.0)

Great Attractor a large concentration of mass toward which everything in our part of the universe apparently is being pulled. (76.3)

greatest elongation the position of an inner planet (Mercury or Venus) when it lies farthest from the Sun on the sky. Mercury and Venus are particularly easy to see when they are at greatest elongation. Objects may be at greatest eastern or western elongation according to whether they lie east or west of the Sun. (11.3)

Great Red Spot a reddish elliptical spot about 40,000 km by 15,000 km in size in the southern hemisphere of the atmosphere of Jupiter. The Red Spot has existed for over three centuries. (43.3)

greenhouse effect the trapping of heat by a planet's atmosphere, making the planet warmer than would otherwise be expected. Generally, the greenhouse effect operates if visible sunlight passes freely through a planet's atmosphere, but the infrared radiation produced by the warm surface cannot escape readily into space. (36.3)

greenhouse gas a molecule (such as carbon dioxide, methane, or water vapor) that efficiently absorbs infrared radiation. (39.1)

Gregorian calendar the calendar devised at the request of Pope Gregory XIII in 1582. It is the common civil calendar used throughout the world today. It improved upon previous calendars by omitting the leap year for century years not divisible evenly by 400. (9.4)

grooved terrain regions of the surface of Ganymede consisting of parallel grooves. Believed to have formed by repeated fracture of the icy crust. (46.1)

ground state the lowest energy level of an atom, generally when the electrons are in the smallest possible orbitals. (21.3)

H

half-life the time required for half of the atoms of a radioactive substance to disintegrate. (33.1)

Halley's comet a comet that reappears about every 76 years, famous because it was the first comet whose return was predicted. (47.0)

halo the approximately spherical region surrounding spiral galaxies. The halo contains mainly old stars, as are found, for example, in globular clusters. The halo also appears to contain large amounts of dark matter. (70.4)

Hawking radiation radiation that black holes are hypothesized to emit as a result of quantum effects. This radiation leads to the extremely slow loss of mass from black holes. (68.3)

heat death a scenario in which the universe dies as all available fuel is consumed, converted into heat from which no further energy can be extracted. (82.5)

heliocentric model a model of the Solar System in which Earth and the other planets orbit the Sun. (11.3)

helium flash the beginning of helium fusion in a low-mass star. The fusion begins explosively and causes a major readjustment of the star's structure. (62.3)

hertz the MKS unit of frequency: one wave or cycle per second. Named for Heinrich Hertz, who first produced radio radiation. (22.1)

HI emission 21-cm wavelength radio waves produced by atoms of un-ionized hydrogen. (72.4)

HII region a region of **ionized** hydrogen. HII regions generally have a luminous pink/red glow and often surround luminous, hot, young stars. (72.2)

highlands the older, most heavily cratered regions on the Moon. (37.1)

high-mass star a star born with a mass above about $8M_{\odot}$. High-mass stars end their lives by fusing elements up to iron in their cores and then explode as core-collapse supernovae. (59.2)

Hipparchus (about 190–120 B.C.E.) Greek astronomer who made one of the first detailed charts of the stars. (52.1)

homogeneous having a consistent and even distribution of matter that is the same everywhere. (80.1)

horizon the line separating the sky from the ground. See also **cosmic horizon** and **event horizon.** (5.1)

horizon problem the **cosmic microwave background** has almost precisely the same temperature in every direction, yet regions more than one degree apart are outside each other's **cosmic horizon,** are totally independent, and there is no obvious reason they should have the same temperature. This uniformity is explained with the theory of an inflation epoch. (81.4)

H-R diagram (or Hertzsprung-Russell diagram) a graph on which stars are located according to their temperature and luminosity. Most stars on such a plot lie along a diagonal line, called the **main sequence,** which runs from cool, dim stars in the lower right to hot, luminous stars in the upper left. (58.0)

Hubble's constant the multiplying constant H in **Hubble's law:** $V = H \times D$. The reciprocal of Hubble's constant gives the approximate age of the universe. (74.3)

Hubble's law a relation between a galaxy's distance, D, and its recession velocity, V, which states that more distant galaxies recede faster than nearby ones. Mathematically, $V = H \times D$, where H is Hubble's constant. (74.3)

Hubble time an estimate of the age of the universe obtained by taking the inverse of Hubble's constant. The estimate assumes that the universe has expanded at a constant rate. (79.4)

hydrosphere refers to the "layer" of water on the Earth consisting of oceans, lakes, rivers, ice caps, and other liquid water and ice. (36.0)

hydrostatic equilibrium the condition in which pressure and gravitational forces in a star or planet are in balance. Without such balance, bodies will either collapse or expand. (49.2, 61.1)

hyperbola the flattened curve of a trajectory at speeds higher than the **escape velocity.** (18.3)

hypernova a term used to describe a hypothesized kind of supernova explosion of a star so massive that its core collapses directly into a black hole. (66.5)

hypothesis an explanation proposed to account for some set of observations or facts. (4.1)

I

Iapetus the third largest moon of Saturn particularly noted because it has one side that is dark black while the other is bright white. (45.2)

ice giants a term sometimes used to distinguish Uranus and Neptune from the gas giants Jupiter and Saturn. Despite the name, these planets have interiors hotter than Earth's, but the name is descriptive of their likely origin from the accretion of icy planetesimals. (44.0)

ideal gas law a law relating the pressure, density, and temperature of a gas. This law states that the pressure is proportional to the density times the temperature. (49.5)

impact the collision of a small body (such as an asteroid or a comet) with a larger object (such as a planet or moon). (37.1)

inclination the angle by which an astronomical object or its orbit is tilted.

inertia the tendency of an object at rest to remain at rest and of a body in motion to continue in motion in a straight line at a constant speed. See also **mass.** (14.1)

inferior conjunction see **conjunction.**

inferior planet a planet whose orbit lies between the Earth's orbit and the Sun. Mercury and Venus are inferior planets.

inflation the enormously rapid expansion of the early universe. Between about 10^{-35} and 10^{-33} seconds after the Big Bang, the universe expanded by a factor of perhaps 10^{30}. (81.3)

infrared a wavelength of electromagnetic radiation longer than visible light but shorter than radio waves. We cannot see these wavelengths with our eyes, but we can feel many of them as heat. The infrared wavelength region runs from about 700 nm to 1 mm. (22.3)

initial mass function the relative number of stars born onto the main sequence at each mass. (69.3)

inner core the inner portion of Earth's iron–nickel core. Despite its high temperature, the inner part of the core is solid because it is under great pressure. (35.1)

inner planet a planet orbiting in the inner part of the Solar System. Usually taken to mean Mercury, Venus, Earth, or Mars. (32.1)

instability strip a region in the H-R diagram containing stars that pulsate. These stars have layers below the surface at a temperature that makes them unstable, so the outer layers expand and contract rhythmically. (63.2)

instrumentation a branch of astronomy dealing with the development of new kinds of detectors. (26.1)

interfere, interference a phenomenon in which electromagnetic waves mix together such that their crests and troughs can alternately reinforce and cancel one another. (21.1, 29.1)

interferometer a device consisting of two or more telescopes connected together to work as a single instrument. Used to obtain a high resolution, with the ability to see small-scale features. Most interferometers currently operate at radio, infrared, or visible wavelengths. (29.3)

interferometry the use of an **interferometer** to measure small angular-size features. (57.1)

international date line an imaginary line extending from the Earth's North Pole to the South Pole, running approximately down the middle of the Pacific Ocean. It marks the location on Earth at which the date changes to the next day as one travels across it from east to west. (7.3)

interstellar cloud a cloud of gas and dust in between the stars. Such clouds may be many light-years in diameter. (33.2, 59.2)

interstellar dust tiny solid grains in interstellar space, thought to consist of a core of rocklike material (silicates) or graphite surrounded by a mantle of ices. Water, methane, and ammonia are probably the most abundant ices. (60.1, 70.3)

interstellar grain microscopic solid dust particles in interstellar space. These grains absorb starlight, making distant stars appear dimmer than they truly are. (33.2)

interstellar medium (ISM) gas and dust between the stars in a galaxy. (72.0)

inverse-square law (1) any law in which some property varies inversely as the square of the distance, d (mathematically, as $1/d^2$). (2) The law stating that the apparent brightness of a body decreases inversely as the square of its distance. (21.2, 54.2)

Io the innermost of the Galilean satellites of Jupiter, noted especially for its active volcanoes. (46.1)

ionization energy the energy needed to remove an electron from an atom. (24.4)

ionize remove one or more electrons from an atom, leaving the atom with a positive electric charge. Under some circumstances, an extra electron may be attached to an atom, in which case the atom is described as negatively ionized. (21.4)

ionosphere the upper region of the Earth's atmosphere in which many of the atoms are ionized. (36.1)

ion tail a stream of ionized particles evaporated from a comet and then swept away from the Sun by the solar wind. (47.3)

iris the region of the eye surrounding the pupil that can control the amount of light entering the eye. (31.1)

iron core the endpoint of fusion in a massive star. When the iron core grows massive enough, the star undergoes a supernova explosion. (59.4)

irregular (Irr) galaxy a galaxy lacking a symmetrical structure. (75.1)

irregular variables a class of variable star that changes in brightness erratically. (63.1)

Ishtar a continent-like highland region on Venus. (39.2)

isotope an atom that has the same number of protons but a different numbers of neutrons in its nucleus. (33.1, 50.2)

J

jet narrow stream of gas ejected from any of several types of astronomical objects. Jets are seen extending from protostars and active galactic nuclei. They may be funneled into their direction of flow by magnetic fields or accretion disks surrounding the central object. (77.1)

joule the MKS unit of energy. One joule per second equals one watt. (20.1)

Jovian planet a giant gaseous planet such as Jupiter, Saturn, Uranus, or Neptune. The term *Jovian* means Jupiter-like and is based on "Jove," an alternative name for Jupiter. (32.1)

Julian calendar a 12-month calendar devised under the direction of Julius Caesar. It includes a leap year every four years. (9.3)

K

Kapteyn's Universe a model of the known universe in wide usage in the early 1900s. Based on observations of the time, it was limited to stars within the disk of the Milky Way surrounding the Sun. (70.2)

Kelvin the most commonly used temperature scale in science, defined such that absolute zero is 0 K and water freezes at 273.15 K. (23.3)

Kepler, Johannes (1571–1630) German astronomer who first discovered the elliptical orbit of Mars. (12.2)

Kepler's three laws mathematical descriptions of the motion of planets around the Sun. The first law states that planets move in elliptical orbits with the Sun off-center at a focus of the ellipse. The second law states that a line joining the planet and the Sun sweeps over equal areas in equal time intervals. The third law relates a planet's orbital period, P, to the semimajor axis of its elliptical orbit, a. Mathematically, the law states that $P^2 = a^3$, if P is measured in years and a in astronomical units. (12.2)

kiloparsec a unit of distance, equal to 1000 parsecs (pc), often used to describe distances within the Milky Way or the Local Group of galaxies. (70.2)

kinetic energy energy of motion. Kinetic energy is calculated from half the product of a body's mass and the square of its speed ($\frac{1}{2}mV^2$). (20.1)

Kirchhoff's laws a set of three "rules" that describe how continuous, bright line, and dark line spectra are produced. (24.3)

Kirkwood gaps regions in the asteroid belt with a lower-than-average number of asteroids. Some of the gaps result from the gravitational force of Jupiter removing asteroids from orbits within the gaps. (41.4)

Kuiper belt a region containing many large icy bodies and from which some comets come. The region appears to extend from the orbit of Neptune at about 30 AU, past Pluto, out to approximately 55 AU. (32.1, 46.4, 47.2)

L

laminated terrain alternating layers of ice and dust seen in the ice caps of Mars. The layers appear to reflect climate changes caused by long-term orbital variations. (40.1)

large-scale structure the structure of the universe on size scales larger than that of clusters of galaxies. (76.4)

last scattering epoch the period of time a few hundred thousand years after the Big Bang when the density and temperature of gas in the universe dropped sufficiently that photons could begin to travel freely through space. The time of the formation of the **cosmic microwave background.** (80.2)

latitude the angular distance of a point north or south of the equator of a body. The equator has a latitude of 0° while the poles are at latitudes of 90° north and south. (5.4)

law in science, generally a theory that can be expressed in a mathematical form. (4.1)

law of gravity a description of the gravitational force exerted by one body on another. The gravitational force is proportional to the product of their masses and the inverse square of the distance between them. If the masses are M and m and their separation is d, the force between them, F, is $F = G\,Mm/d^2$, where G is a physical constant. (16.2)

leap second a time adjustment added every few years to clocks around the world to adjust them to account for Earth's gradually slowing spin. (7.5)

leap year a year in which there are 366 days implemented under the Julian and Gregorian calendar systems. (9.4)

lens an optical instrument, made of glass or some other transparent material, shaped so that parallel rays of light passing through it are bent to arrive at a single focus. (28.1)

light electromagnetic energy. (21.1)

light curve a plot of the brightness of a body versus time. (57.2, 63.1, 65.3)

light pollution the illumination of the night sky by waste light from cities and outdoor lighting. Light pollution makes it difficult to observe faint objects. (30.1)

light travel time distance a measure of how far light would have traveled in the time since it was emitted if the universe were not expanding. (79.4)

light-year a unit of distance equal to the distance light travels in one year. A light-year is roughly 10^{13} km, or about 6 trillion miles. (2.3, 3.3, 54.0)

liquid core see **outer core.**

liquid metallic hydrogen a form of hot, highly compressed hydrogen that is a good electrical conductor, found in the interiors of Jupiter and Saturn. (43.2)

lobe see **radio lobe.**

Local Group the small group of three spiral galaxies and several dozen small galaxies to which the **Milky Way** belongs. (2.4, 76.1)

Local Supercluster the cluster of galaxy clusters in which the **Milky Way, Local Group,** and **Virgo Cluster** are located. (76.3)

loners those who argue that humans are probably the only intelligent beings within our galaxy. (84.3)

longitude a coordinate indicating the east–west location of a point on the Earth's surface. The line of 0° longitude runs north–south through an observatory in Greenwich, England. (5.4)

Lorentz factor a term that describes by how much time, space, and mass are altered as a result of their motion. The Lorentz factor is very close to 1 (indicating very little change) except at speeds approaching the speed of light. (53.2)

low-mass star a star born with a mass less than about 8 times that of the Sun; progenitor of a white dwarf. (59.2)

low surface brightness galaxy a galaxy with a low density of stars, making it difficult to detect in visible-wavelength surveys. (75.1)

luminosity the amount of energy radiated per second by a body. For example, the wattage of a lightbulb defines its luminosity. Stellar luminosity is usually measured in units of the Sun's luminosity (approximately 4×10^{26} watts). (23.2, 54.1)

lunar eclipse the passage of the Earth between the Sun and the Moon so that the Earth's shadow falls on the Moon. (8.2)

lunar month the time period of approximately 29½ days during which the Moon cycles from the new phase through all of its phases back to new. (8.1)

lunisolar calendar a calendar system that maintains links to both the lunar month and the solar year, such as the Chinese or Jewish calendars. Because the lunar month does not divide evenly into the year, lunisolar calendar systems must insert an extra month every few years. (9.2)

M

Magellanic Clouds two small companion galaxies of the **Milky Way.** (76.1)

magnetar a highly magnetized neutron star that emits bursts of gamma rays. (67.3)

magnetic dynamo a process whereby magnetic fields are generated by electric currents circulating inside an astronomical body. The process is thought to arise when there is electrically-conductive material flowing inside a rotating body. (35.4)

magnetic field a representation of the means by which magnetic forces are transmitted from one body to another. (35.4)

magnetogram an image of the Sun coded in a way to show the strength and direction of the magnetic field. (51.1)

magnitude see **apparent magnitude.**

main sequence the region in the H-R diagram in which most stars, including the Sun, are located. The main sequence runs diagonally across the H-R diagram from cool, dim stars to hot, luminous ones. See also **H-R diagram.** (58.1)

main-sequence lifetime the time a star remains a main-sequence star, fusing hydrogen into helium in its core. (61.3)

main-sequence star a star, fusing hydrogen to helium in its core, whose surface temperature and luminosity place it on the **main sequence** on the Hertzsprung-Russell diagram. (59.2, 61.0)

mantle the outer part of a rocky planet immediately below the crust and outside the metallic core. (35.1)

many-worlders those who argue that the development of life and intelligence are probably common occurrences within our galaxy. (84.3)

mare a vast, smooth, dark area on the Moon with a roughly circular shape, composed of basalt, a dark, congealed lava. (37.1)

maria plural of **mare.**

maser an intense radio source created when excited gas amplifies some background radiation. *Maser* stands for "microwave amplification by stimulated emission of radiation."

mass a measure of the amount of material an object contains. A quantity measuring a body's **inertia.** (14.1, 32.1)

mass excess the amount of mass—or equivalently, energy, according to Einstein's formula $E = m \times c^2$—potentially available from an element through nuclear reactions. (66.2)

massive compact halo object (MACHO) collective name for bodies of planetary or stellar mass that are hypothesized to contribute a significant amount of mass to our galaxy's halo, but that generate relatively little detectable light. (78.4)

mass–luminosity relation a relation between the mass and luminosity of stars. Higher-mass stars have higher luminosity, so that $L \approx M^{3.5}$ where L and M are measured in solar units. (58.2)

mass transfer process by which one star in a binary system transfers matter onto the other. (65.1)

matter era the times after the end of the **radiation era,** when radiation from the Big Bang cooled so that matter became the dominant constituent of the universe.

Maunder minimum the period from about A.D. 1645 to 1715 during which the Sun was relatively inactive. Few sunspots were observed during this period, and Earth's temperature was cooler than normal. (51.4)

Maxwell Montes a set of mountains on Venus, including the tallest volcano; one of the first features detected by radar imaging. (39.2)

mean solar day the standard 24-hour day. The mean solar day is based on the average day's length over a year. Using a mean is necessary because the time interval from noon to noon, or sunrise to sunrise, varies slightly throughout the year. (13.5)

megaparsec a distance unit equal to 1 million **parsecs** and abbreviated Mpc. (74.3)

merger the process in which galaxies collide and form a single larger system. (75.3)

meridian the line passing north–south from horizon to horizon and passing through an observer's zenith. The meridian is the dividing line between the eastern and western halves of the sky. (7.1)

metals astronomically, any chemical element more massive than helium. Thus carbon, oxygen, iron, and so forth are metals. (71.1)

meteor the bright trail of light created by small solid particles entering the Earth's atmosphere and burning up; a "shooting star." (47.5, 48.0)

meteor shower an event in which many meteors occur in a short space of time, all from the same general direction in the sky, typically when the Earth' s orbit intersects debris left by a comet. The most famous shower is the Perseids in mid-August. (47.5)

meteorite the solid remains of a meteor that falls to the Earth. (33.1, 48.0)

meteoroid small, solid bodies moving within the Solar System. When a meteoroid enters our atmosphere and heats up, the trail of luminous gas it leaves is called a **meteor.** When the body lands on the ground, it is called a **meteorite.** ("A meteoroid is in the void. A meteor above you soars. A meteorite is in your sight.") (48.0)

method of standard candles see **standard candle.**

metric prefix a term that can be attached to the front of a measurement unit to indicate a power of 10 times the unit. For example, the prefix *kilo*- means a thousand, so a kilometer is a thousand meters. (3.1)

metric system a set of units developed in the late 1800s to replace the confusing systems then used. See also **MKS system.** (1.1, 3.1)

microwaves a range of electromagnetic radiation lying between radio and infrared wavelengths, typically with wavelengths of from about 1 to 100 millimeters. (22.5)

mid-ocean ridge an underwater mountain range on Earth created by plate tectonic motion.

Milky Way the Galaxy to which the Sun belongs. Seen from Earth, our Galaxy is a pale, milky band in the night sky. (2.2, 70.0)

Miller–Urey experiment an experimental attempt to simulate the conditions under which life might have developed on Earth. Miller and Urey discovered that amino acids and other complex organic compounds could form from the gases that are thought to have been present in the Earth's early atmosphere, if the gases are subjected to an electric spark or ultraviolet radiation. (83.2)

millisecond pulsar a pulsar whose rotation period is about a millisecond. (67.2)

Miranda an fairly small but very unusual moon of Uranus that shows huge cliffs and canyons and dramatic differences in surface features of neighboring regions. (45.2)

MKS system the form of the **metric system** using meters, kilograms, and seconds to measure length, mass, and time, respectively. (3.1)

model a theoretical representation of some object or system. (4.1)

molecule two or more atoms bonded into a single particle, such as water (H_2O—two hydrogen atoms bonded to one oxygen) or carbon dioxide (CO_2—one carbon atom bonded to two oxygen atoms).

molecular cloud relatively dense, cool interstellar cloud in which molecules are common. (60.1, 72.4)

month an interval of time loosely resembling the lunar month in some calendar systems, but closely tied to the cycle of moon phases in others. (8.0)

Moon illusion the optical illusion that the Moon appears larger when nearer the horizon than when seen high in the sky. (8.4)

Morning Star the planet Venus when it is seen in the eastern sky. (13.4)

mountain ranges long parallel ridges of mountains formed by the buckling surface of colliding plates. See also **plate tectonics.** (35.3)

N

nadir the point on the celestial sphere directly below the observer and opposite the **zenith.** (5.2)

nanometer a unit of length equal to 1 billionth of a meter (10^{-9} meters) and abbreviated nm. Wavelengths of visible light are several hundred nanometers. The diameter of a hydrogen atom is roughly 0.1 nm.

neap tide a weak tide occurring when the Moon is at first or third quarter, so the Sun's and Moon's gravitational effects on the ocean partially offset each other. (19.3)

near-Earth object an asteroid with an orbit that crosses or comes close to the Earth's orbit. (41.4)

nebula a cloud in interstellar space, usually consisting of gas and dust.

negative curvature a form of curved space sometimes described as being "open" because it has no boundary. Negative curvature is analogous to a saddle shape. (80.4)

Nereid a moderately large satellite of Neptune with an irregular orbit. (46.3)

neutrino tiny neutral particle with little or no mass and immense penetrating power. These particles are produced by stars when they fuse hydrogen into helium. They are also created in the early universe and during a supernova when a star's core collapses to form a neutron star. (4.4, 50.0)

neutron a subatomic particle of nearly the same mass as the proton but with no electric charge. Neutrons and protons comprise the nucleus of the atom. (4.2, 50.1)

neutron star a very dense, compact star composed primarily of neutrons. Protons and electrons combine to form neutrons because of the enormous gravitational pressure squeezing the iron core of a massive star late in its evolution. (59.2, 67.0)

new moon phase of the Moon during which none of the sunlit side is visible so it appears completely dark. (8.1)

Newton, Isaac (1642–1722) English physicist, mathematician and astronomer who pioneered the modern studies of motion, optics, gravity, and calculus, and showed how to apply them to the universe. (14.3)

Newtonian telescope a reflecting telescope designed so that the focused light is reflected by a small secondary mirror out to the side of the telescope, where it can be viewed. (28.2)

newton the standard unit of force in the metric (MKS) system. (15.2, 16.3)

Newton's first law of motion a body continues in a state of rest or in uniform motion in a straight line unless acted upon by an external force. See also **inertia.** (14.2)

Newton's second law of motion $F = ma$. The amount of acceleration, a, that a force, F, produces depends inversely on the mass, m, of the object being accelerated. (15.2)

Newton's third law of motion when two bodies interact, they exert equal and opposite forces on each other. (15.3)

Newton's version of Kepler's third law a generalized version of the law ($P^2 = a^3$) relating the periods and semimajor axes of orbits for systems in which the mass is not dominated by the Sun. The modified form can be expressed as $(M_1 + M_2)\,P^2 = a^3$ where the sum of the masses is measured in solar masses. See also **Kepler's three laws.** (17.2)

nonthermal radiation radiation emitted by charged particles moving at high speed in a magnetic field. The radio emission from pulsars and radio galaxies is nonthermal emission. More generally, *nonthermal* means "not due to high temperature."

north celestial pole the point on the **celestial sphere** directly above the Earth's North Pole. The star Polaris happens to lie close to this point, and other objects in the northern sky appear to circle around this point.

North Star any star that happens to lie very close to the north celestial pole. Polaris has been the North Star for about 1000 years, and it will continue as such for about another 1000 years, at which time a star in Cepheus will be nearer the north celestial pole. (6.6)

nova an explosion of a surface layer of hydrogen that builds up on a white dwarf and then fuses rapidly into helium, causing the white dwarf to shine brightly for several weeks. Nova explosions may be recurrent. (63.1, 65.1)

nuclear fusion the binding of two lighter nuclei to form a heavier nucleus, with some nuclear mass converted to energy—for example, the fusion of hydrogen into helium. This process supplies the energy of most stars and is commonly called "burning," but burning is a chemical process that does not alter atoms' nuclei. (50.1)

nucleosynthesis the formation of elements, generally by the fusion of lighter elements into heavier ones—for example, the formation of carbon by the fusion of three helium nuclei. (66.2)

nucleus the core of an atom around which its electrons orbit. The nucleus has a positive electric charge and comprises most of an atom's mass. (4.2)

nucleus (of a comet) the core, typically a few kilometers across, of frozen gases and dust that make up the solid part of a comet. (47.1)

nucleus (of a galaxy) the central region of a galaxy where there is a supermassive black hole and energetic activity. (70.4)

O-B-A-F-G-K-M-L-T the sequence of stellar-spectral classifications from hottest to coolest stars. (55.3)

Occam's razor the principle of choosing the simplest scientific hypothesis that correctly explains any phenomenon. (11.2)

Olbers' paradox an argument that the sky should be bright at night because of the combined light from all the distant stars and galaxies. (80.1)

Olympus Mons a volcano on Mars, the largest volcanic peak in the Solar System. (40.1)

Omega (Ω) the ratio of the average density of matter or energy in the universe to the **critical density.** In a **flat universe,** $\Omega = 1$. (82.2)

Oort cloud a vast region in which comet nuclei orbit. This cloud lies far beyond the orbit of Pluto and may extend halfway to the next nearest star. (32.1, 47.2)

opacity the blockage of light or other electromagnetic radiation by matter.

open cluster a loose cluster of stars, generally containing a few hundred members. (69.1)

opposition the configuration of a planet when it is opposite the Sun in the sky. If a planet is in opposition, it rises when the Sun sets and sets when the Sun rises. (11.3)

orbit the path in space traveled by a celestial body.

orbital a three-dimensional electron wave in an atom. The closest analog at larger scales is an orbit, however because of the wave nature of an electron at microscopic scales it does not orbit the atoms nucleus like a planet. Each orbital has a specific **energy level.** (21.3)

orbital plane the flat plane defined by the path of an orbiting body. In the Solar System, most of the bodies orbiting the Sun have similar orbital planes. (32.2)

orbital velocity the speed needed to maintain a (generally) circular orbit. (18.1)

organic compound a compound containing carbon, especially a complex carbon compound. Organic compounds can be produced by chemical processes, so they do not necessarily indicate the presence of living organisms. (48.2)

outer core the molten portion of the Earth's iron-nickel core surrounding the solid inner core. (35.1)

outer planet a planet whose orbit lies in the outer part of the Solar System. Jupiter, Saturn, Uranus, and Neptune are outer planets. (32.1)

ozone a molecule consisting of three oxygen atoms bonded together. Its chemical symbol is O_3. Because it absorbs ultraviolet radiation, ozone in our planet's stratosphere shields us from the Sun's harmful ultraviolet radiation. Too much ozone in the lower atmosphere, however, is harmful to human respiration. (36.3)

P

pancake dome an unusual surface feature on Venus, they appear as "blisters" of uplifted rock that may be produced by volcanic activity. (39.2)

panspermia a theory that life originated elsewhere than on Earth and came here across interstellar space either accidentally or deliberately. (84.0)

parabola the mathematical curve of a trajectory at **escape velocity.** (18.3)

parallax the shift in an object's perceived position caused by the observer's motion. Also a method for finding distance based on that shift. (10.2, 52.2)

parsec a unit of distance equal to about 3.26 light-years (3.09×10^{13} km). It is the distance at which an object would have a parallax of one arc second. (3.3, 52.3)

partial eclipse an alignment of two celestial bodies during which only a part of the light from one body is blocked by the other. (8.2)

pattern speed the speed at which a spiral arm orbits the center of a galaxy, which differs from the speed of any individual star. (73.3)

perfect gas law see **ideal gas law.**

period–luminosity relation the relationship between the period of brightness variation and the luminosity of a variable star. Cepheid variables with longer periods are more luminous. (63.3)

penumbra the outer part of the shadow of a body where sunlight is only partially blocked by the body. (8.2)

penumbral eclipse a lunar eclipse in which the Moon passes only through the Earth's **penumbra.** (8.2)

perihelion the point in a planet's orbit where it is closest to the Sun. (38.2)

period the time required for a repetitive process to repeat. For example, *orbital period* is the time it takes a planet or star to complete an orbit. *Pulsation period* is the time it takes a star to expand and then contract back to its original radius. (12.2, 63.1)

periodic table a table displaying the properties of each type of atom, organized in order of the number of protons, or the atomic number. (4.2)

phase the changing illumination of the Moon or other body that causes its apparent shape to change. The following is the cycle of lunar phases: new, crescent, first quarter, gibbous, full, gibbous, third quarter, crescent, new. (8.1)

Phobos one of the two small moons of Mars, thought to be a captured asteroid. Its close orbit will cause it eventually to spiral into the planet. (40.5)

photo dissociation the breaking apart of a molecule by intense radiation.

photoelectric effect emission of electrons from a material when light of a high-enough frequency strikes it. Regardless of the brightness of the light, no electrons are emitted unless the photons' energy is greater than a value that depends on the material. (26.2)

photometer a device that measures the total amount of light received from a celestial object over a chosen range of wavelengths. (54.2)

photon a particle of visible light or other electromagnetic radiation. (4.2, 21.1)

photopigment a chemical that undergoes a chemical or physical change when light shines on it. Vision is possible because of cells in the eye that have photopigments, and the chlorophyll in plants is a photopigment. (31.1)

photosphere the layer in the Sun or any other star where photons can escape into space. This appears to be a surface when we look at the Sun, but there is higher-density gas below and lower-density gas above the photosphere. (49.2, 55.1)

pixel a "picture element," consisting of an individual detector in an array of detectors used to collect light to construct an image. (26.2)

Planck time the brief interval of time, about 10^{-43} second immediately after the Big Bang, when quantum fluctuations are so large that current theories of gravity can no longer describe space and time. (81.3)

planet a body in orbit around a star. (1.1, 11.0)

planetary nebula a shell of gas ejected by a low-mass star late in its evolutionary lifetime. Planetary nebulae typically appear as a glowing gas ring around a central star. (59.2, 64.0)

planetesimal one of the numerous small, solid bodies that, when gathered together by gravity, form a planet. (33.3)

plasma a fully or partially ionized gas. (49.1)

plate tectonics a model of the movement of the Earth's crust, caused by slow circulation in the mantle. The Earth's surface is divided into large regions (plates) that move very slowly over the planet's surface. Interaction between plates at their boundaries creates mountains and activity such as volcanoes and earthquakes. (35.3)

Pleiades a nearby star cluster in the constellation Taurus. (69.1)

plutino a trans-Neptunian object whose orbital period (like that of Pluto) is in a 2:3 resonance with the orbit of Neptune. (46.5)

Pluto a small icy "planet" discovered in 1930. Recent discoveries of similar and larger bodies in the outer Solar System have resulted in Pluto being recategorized as a **dwarf planet** or **plutoid.** (46.4)

plutoid a **trans-Neptunian object (TNO)** that is also a **dwarf planet.** (46.5)

Polaris a moderately bright star in the constellation Ursa Minor. Polaris is also known as the "North Star" because it currently lies near the north celestial pole, but this steadily changes because of the **precession** of Earth's axis. (5.3)

poor cluster a galaxy cluster with few members. The galaxy cluster to which the Milky Way belongs, the Local Group, is a poor cluster. (76.2)

Population (Pop) I the younger stars, some of which are blue, that populate a galaxy's disk, especially its spiral arms. (71.1)

Population (Pop) II the older, redder stars that populate a galaxy's halo and bulge. (71.1)

Population (Pop) III a hypothetical stellar population consisting of the first stars that formed, composed of only hydrogen and helium. (71.2)

positive curvature bending of space leading to a finite volume. A **closed universe** with positive curvature is analogous to the surface of a sphere. (80.4)

positron a subatomic particle of antimatter with the same mass as an electron but a positive electric charge. It is the electron's antiparticle. (50.2, 81.2)

post meridian (P.M.) after the Sun has crossed the **meridian;** after noon (7.1)

potential energy the energy stored in an object as a result of its position or arrangement. For example, a mass lifted to a greater height gains gravitational potential energy. (20.1)

powers of 10 see **scientific notation.**

precession the slow change in the direction of the pole (rotation axis) of a spinning body. (6.6)

pressure the force exerted by a substance such as a gas on an area, divided by that area. That is, pressure = force/area. (49.2)

primary mirror the large, concave, light-gathering mirror in a reflecting telescope. (28.2)

prism a wedge-shaped piece of glass that is used to disperse white light into a spectrum. (28.4)

prokaryotes cells without nuclei. The first life forms on Earth were probably prokaryotes. (83.2)

prominence a cloud of hot gas in the Sun's outer atmosphere. This cloud is often shaped like an arch and is supported by the Sun's magnetic field. (51.2)

proper motion the position shift of a star on the celestial sphere caused by the star's own motion relative to the Sun. (52.4)

proper motion method a technique for detecting a planet orbiting another star by observing the side-to-side "wobble" of the star on the sky caused by the pull of the orbiting planet. (34.4)

proteinoids strings of amino acids that can form through chemical and physical interactions. A possible precursor to the origin of life. (83.2)

proton a positively charged subatomic particle. The constituent of an atom's nucleus that determines the type of the atom—any atom with six protons is a carbon atom, for example. (4.2, 50.1)

proton decay a prediction by **grand unified theories** that protons, long thought to be stable particles, might have pathways to break down into lighter particles. (82.5)

proton–proton chain the nuclear fusion process that converts hydrogen into helium in stars like the Sun and thereby generates their energy. This is the dominant energy generation mechanism in main-sequence stars with masses smaller than the Sun's. (50.2, 61.2)

protoplanetary disk a disk of material encircling a protostar or a newborn star. (34.1)

protostar a star still in its formation stage. (59.2, 60.2)

Proxima Centauri the nearest star to the Solar System at a distance of 1.30 parsecs. Despite its proximity, this M-type main-sequence star is too dim to be seen without a small telescope. (52.3)

Ptolemy, Claudius (about 100–170 C.E.) Roman astronomer who developed a geocentric model of the Solar System that could predict most of the observable motions of the planets. (11.2)

pupil the aperture of the eye, which can be adjusted in size to allow more or less light into the eye. (31.1)

pulsar a spinning **neutron star** that emits beams of radiation each time the star spins.

These beams flow out along the directions of the poles of the strong magnetic fields of these stars. If the beam happens to point in the Earth's direction, we observe the radiation as regularly spaced pulses as the star spins. (67.1)

pulsating variable a variable star that expands and contracts in size, producing a repetitive pattern of variations in its luminosity and surface temperature. (63.1)

P wave a "primary" or "pressure" wave. P waves form as matter in one place vibrates against the adjacent matter ahead of it, producing compression and decompression in an oscillating fashion. Unlike **S waves,** P waves can travel through liquids. (35.1)

Pythagoras (about 560–480 B.C.E.) ancient Greek mathematician who developed early ideas about the nature of the Earth and sky and geometrical techniques for studying them. (10.1)

Q

quadrature the points in the orbit of an outer planet when it appears (from Earth) to be at a 90° angle with respect to the Sun. From the outer planet at the same times, the Earth would appear to be at **greatest elongation.** (11.3)

quantized the property of a system that allows it to have only discrete values. Subatomic matter is generally quantized in nearly all of its properties. (21.1)

quantum fluctuation a temporary random change in the amount of energy at a point in space. These tiny subatomic variations are constantly occurring everywhere in space. (81.3)

quark a fundamental particle of matter that interacts via the strong force; basic constituent of protons and neutrons. (4.4, 81.2)

quarter moon a phase of the Moon when it is located 90° from the Sun in the sky. (8.1)

quasar a highly energetic source in the nucleus of a galaxy generally seen at a large redshift. Quasars are among the most luminous and most distant objects known to astronomers. (76.5, 77.2)

R

radial velocity the velocity of a body along one's line of sight. That is, the part of a body's motion directly toward or away from the observer. (25.1, 52.4)

radiant the point in the sky from which meteors in showers appear to originate. See also **meteor shower.** (47.5)

radiation era the period of time up until about 50,000 years after the Big Bang, when radiation rather than matter was the dominant constituent of the universe. (81.0)

radiation pressure the force exerted by photons when they strike matter. (47.3)

radiative zone the region inside a star where its energy is carried outward by radiation (that is, photons). In the Sun, this region extends about two-thirds of the way out from the core. (49.3)

radioactive decay the spontaneous breakdown of an atomic nucleus into smaller fragments, often involving the emission of other subatomic particles as well. (33.1, 35.2)

radioactive element an element that undergoes **radioactive decay** and spontaneously breaks down into lighter elements. (35.2)

radio galaxy a galaxy, generally an elliptical system, that emits abnormally large amounts of radio energy. (77.1)

radio lobe a region lying outside the body of a radio galaxy where much of its radio emission comes from. Radio lobes are usually located on opposite sides of the galaxy and contain hot gas ejected from an **active galactic nucleus** into intergalactic space through **jets.** (77.1)

radio waves long-wavelength electromagnetic radiation. (22.5)

rays long, narrow, light-colored markings on the Moon or other bodies that radiate from young craters. Rays are debris "splashed" out of the crater by the impact that formed it. (37.1)

recession velocity the apparent speed of a galaxy (or other distant object) moving away from us, caused by the expansion of the universe. (74.3)

recurrent nova a white dwarf in a binary system which undergoes repeated **nova** outbursts. (65.1)

reddening the alteration in a star's color seen from Earth as the star's light passes through an intervening interstellar dust cloud. The dust preferentially scatters the star's blue light, leaving the remaining light redder. (72.3)

red giant a cool, luminous star with a radius much larger than the Sun's. Red giants are found in the upper right portion of the H-R diagram. (58.1, 59.2)

red giant branch the section of the evolutionary track of a star corresponding to intense hydrogen shell burning, which drives a steady expansion and cooling of the outer envelope of the star. As the star gets larger in radius and its surface temperature cools, it becomes a red giant. (62.1)

redshift a shift in the wavelength of electromagnetic radiation to a longer wavelength. For visible light, this implies a shift toward the red end of the spectrum. The shift can be caused by a source of radiation moving away from the observer or by the observer moving away from the source (see **Doppler shift**). Light traveling through expanding space undergoes an **expansion redshift,** and light traveling away from a strong gravitational source undergoes a **gravitational redshift.** (25.1, 74.3)

reflection nebula an interstellar cloud in which the dust particles reflect starlight, making the cloud visible. (72.1)

reflector a telescope that uses a mirror to collect and focus light. (28.2)

refraction the bending of light when it passes from one substance and enters another, generally with a different density. (28.1)

refractor a telescope that uses a lens to collect and focus light. (28.1)

regolith the surface rubble of broken rock on the Moon or other solid body. (37.2)

regular orbit an orbit of a satellite that is nearly circular, in the same direction as the planet spins, and close to the plane of the planet's equator. (45.1)

resolution the ability of a telescope or other optical instrument to discern fine details of an image. (26.1, 29.0)

resonance a condition in which the repetitive motion of one body interacts with the repetitive motion of another to reinforce the motion. For example, planets or satellites orbiting with orbital periods that have a simple fractional ratio (2:1, 3:2, etc.) are in resonance and can have a strong influence on each other's orbit. (41.4, 45.3)

rest frame a system of coordinates that appear to be at rest with respect to the observer. (53.1)

retina the back interior surface of the eye onto which light is focused. (31.1)

retrograde motion the westward shift of a planet against the background stars. Planets usually shift eastward because of their orbital motion, but they appear to reverse direction from our perspective when the Earth overtakes and passes them (or when an inferior planet overtakes and passes the Earth). (11.1)

retrograde spin a spin backward from the usual orbital direction. For example, seen from above their north poles, most of the planets orbit and spin in a counterclockwise direction; however, a few have a retrograde spin. (39.3)

revolve to orbit around another body. (6.1)

rich cluster a galaxy cluster containing hundreds to thousands of member galaxies. (76.2)

rifting the pulling apart of a geological plate by currents in the mantle beneath it. (35.3)

right ascension (RA) a coordinate for locating objects on the sky, analogous to longitude on the Earth's surface. Usually measured in hours and minutes of time. Together with **declination**, this defines the position of an object on the celestial sphere. (5.5)

rille narrow canyon on the Moon or other body. (37.1)

ring (planetary) numerous small particles orbiting a planet within its Roche limit. (45.0)

ring galaxy a galaxy in which the central region has an abnormally small number of stars, causing the galaxy to look like a ring. Caused by the collision of two galaxies. (75.4)

ring system particles organized into thin, flat rings encircling a giant planet, such as Saturn. (45.3)

ringlet any one of numerous, closely spaced, thin bands of particles in Saturn's ring system. (45.3)

Roche limit the distance from an astronomical body at which its gravitational force breaks up another astronomical body. (45.4)

Roche lobe the region around a star in a binary system in which the gravity of that star dominates. Matter pushed outside the Roche lobe (which may occur during a star's red giant phase) may be captured by the companion star. (65.1)

rod light-sensing cell in the eye that gives us our black-and-white vision. (31.1)

rotate to spin on an axis. The distinction between revolving and rotating can be confusing—a galaxy may rotate, but the stars in it revolve around the center. (6.3)

rotation axis an imaginary line through the center of a body about which the body spins. Analogous to the handle and point of a spinning top. (6.3)

rotation curve a plot of the orbital velocity of the stars or gas in a galaxy at different distances from its center. (73.1, 78.1)

round off to express a number with fewer digits by replacing the trailing digits with the nearest decimal value. For example, 12.34567 can be rounded off to 12.346, to 12.3, or to 12 depending on the precision needed. (3.4)

RR Lyrae variables a type of pulsating variable star with a period of about one day or less. They are named for their prototype star in the constellation Lyra, RR Lyrae. (63.2)

runaway greenhouse effect an uncontrollable process in which the heating of a planet leads to an increase in its atmospheric greenhouse effect and thus to further heating. The process significantly alters the composition of the planet's atmosphere and the temperature of its surface. (39.1)

S

S0 galaxy galaxy that shows evidence of a disk and a bulge, but that has no spiral arms and contains little or no gas. (75.1)

Sagittarius A* (Sgr A*) the powerful radio source located at the core of the Milky Way Galaxy. (73.2)

satellite a body orbiting a planet. (1.2, 32.1)

scarp a cliff produced by vertical movement of a section of the crust of a planet or satellite. (38.1)

scattering the random redirection of a light wave or photon as it interacts with atoms or dust particles in its path. (72.3)

Schwarzschild radius the distance from the center of a black hole to its **event horizon.** No light (or anything else) can escape from within the Schwarzschild radius. (68.1)

scientific method the process of observing phenomena, proposing hypotheses to explain them, and testing the hypotheses. Scientifically valid hypothesis must offer a means of being tested and rejected. (4.1)

scientific notation a shorthand way to write numbers using 10 to a power. For example, $1{,}000{,}000 = 10^6$. Numbers that are not simple powers of 10 are written as a number between 1 and 10 and 10 to a power—for example, $123.45 = 1.2345 \times 10^2$. (3.2)

scintillation the twinkling of stars, caused as their light passes through moving regions of different density in the Earth's atmosphere. (30.2)

secondary mirror in a reflecting telescope, a mirror that directs the light from the primary mirror to a focal position. (28.2)

Sedna a large body (about two-thirds the diameter of Pluto) orbiting far beyond Pluto's orbit. (46.5)

seeing a measure of the steadiness of the atmosphere during astronomical observations. Bad seeing results from **scintillation**

and makes fine details difficult to observe. (30.2)

seismic waves waves generated in the Earth's interior by earthquakes. Similar waves occur in other bodies. Two of the more important varieties are S and P waves. S waves can travel only through solid material; P waves can travel through either solid or liquid material. (35.1)

selection effect an unintentional selection process that omits some set of the objects being studied and leads to invalid conclusions about the objects. For example, selecting the brightest stars in the night sky does not provide a representative sample of all star types. (69.3)

self-propagating star formation a model that explains a galaxy's spiral arms as arising from the explosion of massive stars, triggering the birth of other stars around them. The resulting pattern is then drawn out into a spiral by the galaxy's differential rotation. (73.3)

semimajor axis half the long dimension of an ellipse. (12.2)

SETI Search for Extra Terrestrial Intelligence. Some SETI searches involve automated "listening" to millions of radio frequencies for signals that might be from other civilizations. (84.4)

Seyfert galaxy a type of galaxy with a small, abnormally bright **active galactic nucleus.** Named for the astronomer Carl Seyfert, who first drew attention to these objects. (77.1)

shell fusion nuclear energy generation in a region surrounding the core of a star rather than in the core itself. This occurs during the red giant phase of stellar evolution. (62.1)

shepherding satellite a satellite that by its gravitational attraction prevents particles in a planet's rings from spreading out and dispersing. Saturn's F-ring is held together by shepherding satellites. (45.3)

shielding the process in which dust particles in an interstellar cloud block light from entering the interior of the cloud, thereby allowing it to cool. (72.2)

short-period comet a comet whose orbital period is shorter than 200 years. For example, Halley's comet has a period of 76 years. (47.4)

sidereal day the length of time from the rising of a star until it next rises. The length of the Earth's sidereal day is 23 hours 56 minutes. (7.1)

sidereal month the length of time (about 27.3 days) required for the Moon to return to the same apparent position among the stars. Compare with **lunar month.** (8.1)

sidereal period the time it takes a body to turn once on its rotation axis or to revolve once around a central body, as measured with respect to the stars.

sidereal time a system of time measurement referenced to the motion of stars across the sky rather than of the Sun. (7.1)

significant digits all the leftmost digits in a number (excluding any leading zeros) whose values are well determined by measurement or calculation. For example, someone's height might be measured as 1.93524 meters, but if the accuracy of the measurement is only about a millimeter (0.001 meters), the significant digits are 1.935. (3.4)

silicate mineral composed primarily of silicon and oxygen. Most ordinary rocks are silicates. For example, quartz is silicon dioxide, and a wide variety of minerals are made of silicates containing an additional element. (35.1)

singularity a theoretical point of zero volume and infinite density to which any object that becomes a black hole must collapse, according to the **general theory of relativity.** (68.1)

solar activity phenomena that cause changes in the appearance or energy output of the solar atmosphere, such as sunspots and flares. (51.0)

solar cycle the cyclic change in solar activity, such as that of sunspots and of solar flares, that peaks once every 11 years, approximately. (51.3)

solar day the time interval from one sunrise to the next sunrise or from one noon to the next noon. That interval is not always exactly 24 hours but varies throughout the year. For that reason, we use the **mean solar day** (which, by definition, is 24 hours) to keep time. (7.1)

solar eclipse the passage of the Moon between the Earth and the Sun so that our view of the Sun's photosphere is partially or totally blocked. During a **total eclipse** of the Sun we can see the Sun's chromosphere and corona. (8.2)

solar nebula the rotating disk of gas and dust from which the Sun and planets formed. (33.2)

solar nebula theory the theory that the Sun and planets formed all at approximately the same time from a rotating cloud of gas and dust. (33.2)

solar seismology the study of pulsations or oscillations of the Sun to determine the characteristics of the solar interior. (49.5)

Solar System the Sun, planets, their moons, and other bodies, such as meteors and comets, that orbit the Sun. (1.5, 32.0)

solar wind the outflow of low-density, hot gas from the Sun's upper atmosphere. It is partially this wind that creates the tail of a comet by pushing a comet's gases away from the Sun. (47.3, 49.4)

solid core see **inner core.**

solstice (winter and summer) the beginning of winter and summer. The solstice occurs when the Sun is at its greatest distance north (about June 21) or south (about December 21) of the celestial equator. (6.5)

south celestial pole the imaginary point on the **celestial sphere** directly over the Earth's South Pole. Objects on the sky appear to circle around this point as viewed from Earth's Southern Hemisphere.

space velocity the overall velocity of a star, determined by vector addition of the star's **radial velocity** and **transverse velocity.** (52.4)

special relativity a theory developed by Einstein to explain why light is always measured to travel through space at the same speed—regardless of the motion of the source of light or the observer. This very well-established theory shows that the measurements of lengths, times, masses, and other quantities change depending on the speed of the observer relative to the object being measured. (53.0)

speckle interferometry a technique to overcome the limitations imposed by the effects of scintillation in our atmosphere by analyzing a set of images collected over very short time intervals. (57.1)

spectral type an indicator of a star's temperature. A star's spectral type is based on the appearance of its spectral lines, with different strengths and weaknesses indicating stellar composition and temperature. The fundamental types are, from hot to cool, O, B, A, F, G, K, M, L and T. (55.3)

spectrograph a device for making a spectrum by spreading light out into its component wavelengths.

spectroscopic binary a type of binary star in which the spectral lines exhibit alternating red and blue Doppler shifts. This is caused by the orbital motion of one star around the other, causing the stars to alternately move away from us and back toward us. (56.1)

spectroscopy the study and analysis of spectra. (24.0)

spectrum electromagnetic radiation (for example, visible light) spread into its component wavelengths or colors. A rainbow is a spectrum produced naturally by water droplets in our atmosphere. (55.0)

spicule a thin column of hot gas in the Sun's chromosphere. (49.4)

spindown the slowing down of a body's rotation, often referring to pulsars. (67.2)

spiral arm a long, narrow region containing young stars and interstellar matter that winds outward in the disk of spiral galaxies. (70.4)

spiral density wave a mechanism for the generation of spiral structure in galaxies by the gravitational influence of a denser concentration of material in a spiral arm to influence the orbits of other material in the galaxy to reinforce the spiral structure. A density wave can interact with interstellar matter and trigger the formation of stars. Spiral density waves are also seen in the rings of Saturn. (45.3)

spiral (S) galaxy a galaxy with a disk in which its brighter stars form a spiral pattern of two or more **spiral arms.** (75.1)

spring tide a strong tide that occurs at new and full moon, when the Moon's and Sun's tidal forces work in the same direction to make the tide more extreme. (19.3)

standard candle a type of star or other astronomical light source in which the luminosity has a known value, allowing its distance to be determined by measuring its apparent brightness and applying the **inverse-square law.** Good examples include Cepheid variable stars and Type Ia supernovae. (54.3)

standard model the current theoretical model that describes the fundamental particles and forces in nature. (4.4)

standard time a common time kept within a given region so that all clocks in that region agree. (7.3)

star a massive, gaseous body held together by gravity and generally emitting light. Stars generate energy by nuclear reactions in their interiors. (1.4)

starburst galaxy a galaxy in which a very large number of stars have formed recently, making them very luminous. (75.4)

star cluster a group of stars numbering from hundreds to millions held together by their mutual gravity. (69.0)

steady-state theory a model of the universe in which the average properties of the universe do not change with time. An early competitor of the **Big Bang** model.

Stefan-Boltzmann constant constant that appears in the laws of thermal radiation (23.5)

Stefan-Boltzmann law the amount of energy radiated by one square meter in one second by a hot, dense material depends on the temperature T raised to the fourth power (σT^4). (23.5, 57.3)

stellar association an extended, loose grouping of young stars and interstellar matter. (69.1)

stellar evolution the gravity-driven changes in stars as they are born, age, and finally run out of fuel. (2.1, 59.0)

stellar model the result of a theoretical calculation of the physical conditions in the different layers of a star's interior. (59.1)

stellar remnant the body remaining after a star ceases supporting itself with the heat and pressure generated by nuclear fusion in its interior, such as a white dwarf, neutron star, or black hole. (59.1)

stellar spectroscopy the study of the properties of stars based on information that can be learned from their spectra. (55.3)

stellar wind an outflow of gas from a star, sometimes at speeds reaching hundreds of kilometers per second. (59.2, 64.2)

stratosphere the region of the Earth's atmosphere extending from about 12 to 50 km above the surface. A protective layer of **ozone** is located in the stratosphere. (36.1)

stromatolites layered fossil formations caused by ancient mats of algae or bacteria, which build up mineral deposits season after season. (83.1)

strong force the force that binds quarks together and holds protons and neutrons in an atomic nucleus. Sometimes called the *strong nuclear force*. (4.3, 50.1)

subatomic particles particles making up an atom, such as electrons, neutrons, and protons, as well as other particles of similar small size.

subduction the sinking of one crustal plate under another where they are driven together by plate tectonics. (35.3)

sublimate to change directly from a solid into a gas without passing through a liquid phase. (47.2)

sunspot a dark, cooler region on the Sun's visible surface created by intense magnetic fields. (12.3, 51.1)

supercluster a cluster of galaxy groups and clusters. Our Milky Way belongs to the Local Group, which is but one of many galaxy groups making up the Local Supercluster. (2.4, 76.0)

supergiant a very large-diameter and luminous star, typically at least 10,000 times the Sun's luminosity. (58.3)

superior conjunction see **conjunction.**

superior planet a planet orbiting farther from the Sun than the Earth. Mars, Jupiter, Saturn, Uranus, and Neptune are superior planets. (32.1)

supermassive black hole a huge black hole with a mass millions to billions times larger the mass of the Sun. Observations suggest that they may be present in the nuclei of all large galaxies. **(77.3)**

supernova an explosion marking the destruction of some white dwarfs that reside in binary systems (**Type Ia supernova**), as well as the end of most massive stars' evolution (**core-collapse supernova**). (59.2, 63.1)

supernova remnant the debris ejected from a star when it explodes as a supernova. Typically this material is hot gas, expanding away from the explosion at thousands of kilometers per second. (65.3, 66.4)

surface brightness the amount of light emitted by a source divided by the area from which the light is emitted. For example, a galaxy with a high density of stars will have a high surface brightness, but one with the same number of stars more widely spread out will have a lower surface brightness. (31.6, 74.1)

surface gravity the acceleration an object will experience near the surface of a planet (or other body) because of the planet's gravitational pull. (16.3)

S wave a "secondary" or "shear" wave. A type of seismic wave produced by side-to-side motion. S waves can travel only through solid material and move more slowly than **P waves.** (35.1)

symbiosis the mutually beneficial relationships of organisms with other organisms. For example, the human body needs the bacteria in its intestines to break down its food, and the bacteria are also fed in the human body. Neither would survive without the other. (83.1)

synchronous rotation the condition that a body's rotation period is the same as its orbital period. The Moon rotates synchronously. (37.4)

synchrotron radiation a form of nonthermal radiation emitted by charged particles spiraling at nearly the speed of light in a magnetic field. Pulsars and radio galaxies emit synchrotron radiation. The radiation gets its name because it was first seen in synchrotrons, a type of atomic accelerator. (67.2, 77.1)

synodic period the time between successive configurations of a planet or moon. For example, the time between oppositions of a planet or between full moons. (13.4)

T

tail the plumes of gas and dust from a comet. These are produced by the solar wind and radiation pressure acting on gas and dust evaporated from the comet's nucleus. The **dust tail** and **ion tail** point away from the Sun and get longer as the comet approaches perihelion.

telescope an instrument for gathering and focusing light and magnifying the resulting image of remote objects. (26.1)

terrestrial planet a rocky planet similar to the Earth in size and structure. The terrestrial planets are Mercury, Venus, Earth, and Mars. (32.1)

Tharsis bulge a large volcanic region on Mars rising about 10 km above surrounding regions. (40.1)

theory a hypothesis or set of hypotheses that have become well-established through repeated and diverse testing. (4.1)

thermal energy the kinetic energy associated with the motions of the molecules or atoms in a substance. (20.1, 23.3)

thermal radiation the continuous spectrum of electromagnetic radiation emitted as a result of the thermal energy in any relatively dense material. Also see **blackbody.** (23.0)

thermonuclear runaway a condition in which a nuclear reaction generates enough heat to cause an increasing rate of reaction, rapidly accelerating into an explosion. (65.1)

third-quarter moon a phase of the waning Moon when Earth-based observers see half of the Moon's illuminated hemisphere. This occurs three-quarters of the way through the lunar month. (8.1)

tidal braking the slowing of one body's rotation as a result of gravitational forces exerted on it by another body. (7.5, 19.4)

tidal bulge a bulge on one body created by the gravitational attraction on it by another. Two tidal bulges form, one on the side near the attracting body and one on the opposite side. (19.1)

tidal force relative forces arising between the parts of a body as a result of differences in the strength of the gravitational attraction by another body. (19.1)

tidal lock circumstance in which tidal forces have caused a body to rotate at a rate closely tied to its orbital period. For example, Mercury rotates 3 times in each 2 revolutions around the Sun. The Moon always keeps the same face turned toward the Earth, in **synchronous rotation,** where it rotates once for each revolution. (38.4)

tides the rise and fall of the Earth's oceans created by the gravitational attraction of the Moon. Tides also occur in the solid crust of a body and its atmosphere. (19.0)

time zone one of 24 divisions of the globe at every 15 degrees of longitude. In each zone, a single standard time is kept. Most zones have irregular boundaries, usually because they follow geopolitical borders. See also **universal time.** (7.3)

Titan the largest moon of Saturn, even larger than Mercury. Titan has a substantial atmosphere and "weather," but its surface temperature is far colder than the terrestrial planets. (46.2)

torque a twisting force that can change an object's angular momentum. (20.3)

total eclipse an eclipse in which the eclipsing body completely covers the other body. (8.2)

transit the passage of a planet directly between the observer and the Sun. At a transit, we see the planet as a dark spot against the Sun's bright disk. From Earth, only Mercury and Venus can transit the Sun. (52.1)

transit method a method for detecting planets orbiting other stars by detecting the slight dimming if the planet's orbit causes it to cross in front of the star. (34.4)

trans-Neptunian object (TNO) an object orbiting in the **Kuiper belt** or outer Solar System with a semimajor axis larger than Neptune's. These include Pluto and many other objects ranging up to sizes even larger than Pluto. (32.1, 46.5)

transverse velocity the actual speed of a star across, or transverse to, our line of sight, determined from the **proper motion** and distance of the star. (52.4)

triangulation a method for measuring distances. This method is based on constructing a triangle, one side of which is the distance to be determined. That side is then calculated by measuring another side (the baseline) and the two angles at either end of the baseline. (52.1)

triple alpha process the fusion of three helium nuclei (**alpha particles**) into a carbon nucleus. This process is sometimes called *helium burning,* and it occurs in many old stars. (62.2)

Triton the largest moon of Neptune, somewhat larger than and probably fairly similar to Pluto, possibly a **TNO** captured by Neptune. Triton shows evidence of a thin atmosphere and some volcanic activity. (46.3)

Tropic of Cancer the latitude line of 23.5° north, marking the distance farthest north where the Sun can pass directly overhead (on the summer solstice). (6.5)

Tropic of Capricorn the latitude line of 23.5° south, marking the distance farthest south where the Sun can pass directly overhead (on the winter solstice). (6.5)

troposphere the lowest layer of the Earth's atmosphere, extending up to an elevation of about 12 km, within which convection produces weather. (36.1)

true binary a pair of stars that actually orbit each other as opposed to just appearing to be close to each other on the sky (see **apparent double star**). (56.1)

T Tauri star a type of extremely young star or protostar that varies erratically in its light output as it settles toward the main-sequence phase. (60.2)

Tunguska event a large aerial explosion that occurred over Siberia in 1908, probably when a large meteor entered the atmosphere and heated up and exploded before reaching the ground. (48.4)

turnoff point the location on the main sequence where a star's evolution causes it to move away from the main sequence toward the red giant region. The location of the turnoff point can be used to deduce the age of a star cluster. (69.2)

21-centimeter radiation radio emission from a hydrogen atom caused by the flip of the electron's spin orientation. (72.4)

twin paradox a supposed paradox in special relativity arising from the differences in time measurement for two observers moving at speeds relative to each other. The paradox is usually phrased in terms of the relative aging of one twin taking a trip in a spacecraft while the other remains on Earth. Special relativity (and experiment) shows that there is no paradox—the twin on the spacecraft ages less. (53.5)

Type Ia supernova an extremely energetic explosion produced by the abrupt fusion of carbon and oxygen in the interior of a collapsing white dwarf star. See **supernova.** (65.3)

Type II supernova A type of supernova explosion which shows spectral lines of hydrogen, as opposed to type I supernovae, which do not exhibit hydrogen lines. Also see **core-collapse supernova.** (66.3)

U

ultraviolet a portion of the electromagnetic spectrum with wavelengths shorter than those of visible light but longer than those of X-rays. By convention, the ultraviolet region extends from about 10 to 300 nm. (22.3)

umbra The inner portion of the shadow of a body, within which sunlight is completely blocked. (8.2)

uncompressed density the density a planet would have if its gravity did not compress it. (42.3)

unit a quantity used for reporting measurements. (1.6, 3.0)

universality the assumption that the physical laws observed on Earth apply everywhere in the universe. (4.2)

universal time (UT) the time kept at Greenwich, England. Universal time is the same as Greenwich mean time. This is the starting point from which the Earth's 24 **time zones** are calculated. (7.3)

universe the largest astronomical structure we know of. The universe contains all matter and radiation and encompasses all known space. (2.4)

Ursa Major group the nearest star cluster to the Solar System, including the stars of the Big Dipper and many other bright stars over a large part of our sky. (69.1)

V

Valles Marineris a huge canyon feature on Mars stretching thousands of kilometers. (40.1)

valve mechanism a process causing some kinds of stars to pulsate. Radiation becomes trapped (like a closed valve), making the atmosphere heat and expand. The radiation is no longer trapped when the atmosphere is expanded enough (open valve), and the star shrinks back down in size. (63.2)

Van Allen radiation belts doughnut-shaped regions surrounding the Earth containing charged particles trapped by the Earth's magnetic field. (36.1)

variable star a star whose luminosity changes over time. (31.2, 59.2, 63.0)

velocity a physical quantity that indicates both the speed of a body and the direction in which it is moving. (14.2)

vernal equinox or spring **equinox.** Spring in the Northern Hemisphere begins on the vernal equinox, which is on or near March 20. (6.5)

Virgo Cluster the nearest large galaxy cluster. The gravity of the thousands of galaxies in this cluster is predicted to eventually pull in the Milky Way and Local Group. (2.4)

virtual particles a particle and its antiparticle, which are created simultaneously in pairs and which quickly disappear. Virtual particles are created from **quantum fluctuations.** (68.3)

visible spectrum the part of the electromagnetic spectrum that we can see with our eyes. It consists of the familiar colors red, orange, yellow, green, blue, and violet. (22.1)

visible universe the portion of the universe in which light has had time to reach us within the history of the universe. This region is limited, therefore, to distances for which the travel time of the light is less than about 13.7 billion years. (80.1)

visual binary a pair of stars, held together by their mutual gravity and in orbit about each other, that can be seen with a telescope as separate objects. (56.1)

void a region between clusters and superclusters of galaxies that is relatively empty of galaxies. (76.4)

volatile element or compound that vaporizes at low temperature. Water and carbon dioxide are examples of volatile substances. (32.3)

vortex a strong spinning flow within a gas or liquid, such as the Great Red Spot on Jupiter. (43.3)

W

wane to gradually decrease, as in the "waning crescent Moon" or the "waning gibbous Moon." (8.1)

watt the MKS unit of power. It represents a rate of energy generation or consumption of one joule per second.

waterhole the interval of the radio spectrum between the 21-cm hydrogen radiation and the 18-cm OH radiation. Proposed as likely wavelengths to use in the search for extraterrestrial life. (84.4)

wavelength the distance between wave crests. Wavelength determines the color of visible light and is generally denoted by the Greek letter lambda (λ). (22.1)

wave–particle duality the theory that electromagnetic radiation may be treated as either a particle (photon) or an electromagnetic wave. All subatomic particles appear to have both wavelike and particlelike properties. (21.1)

wax to gradually increase, as in the "waxing crescent Moon" or the "waxing gibbous Moon." (8.1)

weak force the force responsible for radioactive decay of atoms. Now known to be linked to electric and magnetic forces in what is called the **electroweak force.** (4.3, 50.4)

weakly interacting massive particle (WIMP) a hypothetical class of subatomic particles that interacts only through the weak force, making it difficult to detect except by gravitational interactions. A possible dark matter candidate. (78.4)

weight the gravitational force exerted on a body by the Earth (or another astronomical object). Sometimes more broadly used to indicate the net force on a body after other forces (centripetal forces, buoyancy, and so on) and effects of the motion of the surrounding rest frame are accounted for. Thus, astronauts in an orbiting spacecraft are "weightless" because their spacecraft is in orbit; however the gravitational force on them is only slightly less than on the Earth's surface. (14.1)

white dwarf a dense star whose radius is approximately the same as the Earth's but whose mass is comparable to the Sun's. White dwarfs burn no nuclear fuel and shine by residual heat. They are the end stage of stellar evolution for low-mass stars like the Sun. (58.1, 59.2, 64.0, 65.0)

white light visible light exhibiting no color of its own but composed of a mix of all colors. Our eyes adapt to color differences, making many artificial light sources appear "white." (22.2)

Wien's law a relation between a body's temperature and the wavelength at which it emits radiation most intensely. Hotter bodies radiate more intensely at shorter wavelengths. Mathematically, the law states that $\lambda_{max} = 2.9 \times 10^6/T$, where λ_{max} is the wavelength of maximum emission in nanometers, and T is the body's temperature on the Kelvin scale. (23.2)

winding problem because a galaxy's differential rotation, its spiral arms should become tightly "wound up" as the inner parts wrap around the galaxy several times for every single orbit of the outer part of the arm. This seeming problem is explained by **spiral density waves.** (73.3)

X

X-ray the part of the electromagnetic spectrum with wavelengths longer than gamma rays but shorter than ultraviolet. (22.6)

X-ray pulsar a neutron star from which periodic bursts of X-rays are observed. These are thought to consist of a neutron star and a normal star in a close binary system. Mass from the normal star spills onto the neutron star, where it slowly accumulates and then undergoes a burst of fusion. The hot region of the neutron star where the burst occurred spins causing more fusion and leading to an explosion that we observe as the X-ray burst. (67.3)

Y

year the time that it takes the Earth to complete its orbit around the Sun; that is, the **period** of the Earth's orbit.

yellow giant a phase stars pass through after moving off the main sequence, and sometimes again after starting helium fusion in their cores. Many yellow giants become unstable and pulsate as they cross the **instability strip** in the H-R diagram. (63.2)

yellow supergiant a yellow giant star with a luminosity greater than about 10,000 solar luminosities. (63.2)

Z

Zeeman effect the splitting of a single spectral line into two or three lines by a magnetic field. The Zeeman effect allows astronomers to detect magnetic fields in objects from the appearance of their spectral lines. (51.1)

zenith the point on the celestial sphere that lies directly above the observer's location. (5.2)

zero-age main sequence main sequence on the H-R diagram for a system of stars that have completed their contraction from interstellar matter and are just beginning to derive all their energy from hydrogen-to-helium fusion in their core. (60.3, 61.4)

zodiac a set of 12 constellations along the ecliptic, in a band around the celestial sphere. The Sun, Moon, and planets move through the constellations of the zodiac from our vantage point on Earth. (6.2)

zone a bright, high-pressure region in the atmosphere of a Jovian planet, where gas flows upward. (43.1)

zone of avoidance a band running around the sky in which few galaxies are visible. It coincides with the plane of the Milky Way and is caused by dust that is within our Galaxy. This dust blocks the light from distant galaxies. (74.1)

ANSWERS TO TEST YOURSELF QUESTIONS

Part One

Unit 1: 1–e; 2–e; 3–d.
Unit 2: 1–c; 2–e; 3–b.
Unit 3: 1–a; 2–d; 3–b.
Unit 4: 1–b; 2–d; 3–a.
Unit 5: 1–b; 2–c; 3–d.
Unit 6: 1–e; 2–a; 3–c.
Unit 7: 1–d; 2–a; 3–c.
Unit 8: 1–b; 2–b; 3–b.
Unit 9: 1–b; 2–a; 3–a.
Unit 10: 1–a; 2–d; 3–a.
Unit 11: 1–b; 2–a; 3–a.
Unit 12: 1–e; 2–a; 3–b.
Unit 13: 1–b; 2–d; 3–e.

Part Two

Unit 14: 1–e; 2–d; 3–d.
Unit 15: 1–c; 2–d; 3–b.
Unit 16: 1–c; 2–d; 3–d.
Unit 17: 1–e; 2–c; 3–d.
Unit 18: 1–b; 2–c; 3–e.
Unit 19: 1–a; 2–e; 3–c.
Unit 20: 1–b; 2–c; 3–b.
Unit 21: 1–b; 2–c; 3–b.
Unit 22: 1–e; 2–c; 3–a.
Unit 23: 1–c; 2–d; 3–e.
Unit 24: 1–d; 2–d; 3–e.
Unit 25: 1–a; 2–d; 3–b.
Unit 26: 1–d; 2–b; 3–c.
Unit 27: 1–b; 2–a; 3–c.
Unit 28: 1–b; 2–c; 3–b.
Unit 29: 1–c; 2–d; 3–a.
Unit 30: 1–a; 2–c; 3–d.
Unit 31: 1–b; 2–c; 3–a.

Part Three

Unit 32: 1–c; 2–a; 3–b.
Unit 33: 1–c; 2–e; 3–d.
Unit 34: 1–b; 2–d; 3–a.
Unit 35: 1–a; 2–e; 3–d; 4–d.
Unit 36: 1–b; 2–a; 3–b.
Unit 37: 1–b; 2–a; 3–d.
Unit 38: 1–d; 2–a; 3–c.
Unit 39: 1–c; 2–a; 3–b.
Unit 40: 1–c; 2–b; 3–a.
Unit 41: 1–d; 2–d; 3–a.
Unit 42: 1–b; 2–d; 3–d.
Unit 43: 1–d; 2–c; 3–a; 4–b
Unit 44: 1–b; 2–c; 3–c.
Unit 45: 1–a; 2–a; 3–d.
Unit 46: 1–c; 2–a; 3–d.
Unit 47: 1–e; 2–b; 3–d.
Unit 48: 1–d; 2–c; 3–c.

Part Four

Unit 49: 1–d; 2–a; 3–b.
Unit 50: 1–d; 2–d; 3–e.
Unit 51: 1–c; 2–a; 3–b.
Unit 52: 1–d; 2–a; 3–d.
Unit 53: 1–e; 2–c; 3–b.
Unit 54: 1–a; 2–e; 3–d.
Unit 55: 1–e; 2–e; 3–a.
Unit 56: 1–b; 2–e; 3–a.
Unit 57: 1–e; 2–c; 3–c.
Unit 58: 1–a; 2–b; 3–e.
Unit 59: 1–e; 2–a; 3–c.
Unit 60: 1–e; 2–b; 3–d.
Unit 61: 1–d; 2–a; 3–d.
Unit 62: 1–b; 2–d; 3–e.
Unit 63: 1–d; 2–e; 3–e.
Unit 64: 1–e; 2–c; 3–a.
Unit 65: 1–a; 2–b; 3–c.
Unit 66: 1–b; 2–c; 3–c.
Unit 67: 1–d; 2–b; 3–c.
Unit 68: 1–c; 2–e; 3–d.
Unit 69: 1–a; 2–d; 3–b.

Part Five

Unit 70: 1–b; 2–b; 3–b.
Unit 71: 1–a; 2–e; 3–b.
Unit 72: 1–e; 2–b; 3–c.
Unit 73: 1–e; 2–a; 3–d.
Unit 74: 1–b; 2–c; 3–c.
Unit 75: 1–d; 2–b; 3–d.
Unit 76: 1–e; 2–e; 3–a.
Unit 77: 1–b; 2–a; 3–e.
Unit 78: 1–e; 2–d; 3–e.
Unit 79: 1–b; 2–b; 3–a.
Unit 80: 1–e; 2–a; 3–b.
Unit 81: 1–c; 2–c; 3–a.
Unit 82: 1–e; 2–b; 3–a.
Unit 83: 1–a; 2–c; 3–d.
Unit 84: 1–d; 2–d; 3–a.

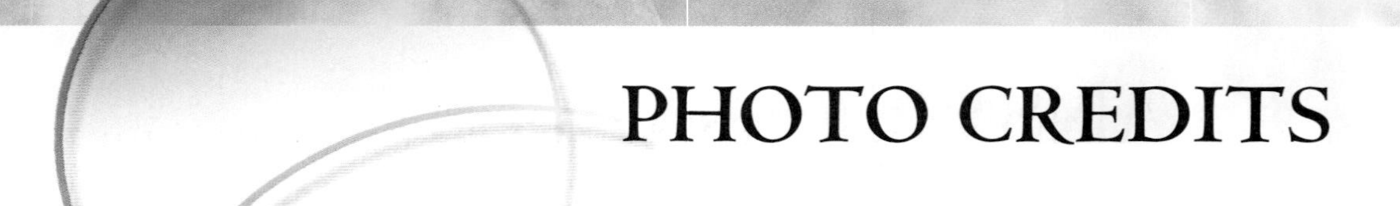

PHOTO CREDITS

Photographs

Looking Up #1 Northern Circumpolar Constellations
Background (both): © Akira Fujii/DMI; **M52:** Hartmut Frommelt; **M81 and M82:** © Robert Gendler; **M101:** Adam Block/NOAO/AURA/NSF

Looking Up #2 Ursa Major
Background: © Akira Fujii/DMI; **M97:** Gary White and Verlenne Monroe/Adam Block/NOAO/AURA/NSF; **Mizar and Alcor:** Courtesy of DSS/Processing by Coelum (www.coelum.com); **M51:** © Tony and Daphne Hallas

Looking Up #3 M31 & Perseus
All photos: © Akira Fujii/DMI

Looking Up #4 Summer Triangle
Background: © Akira Fujii/DMI; **M57:** Courtesy of H. Bond et al., Hubble Heritage Team (STScI/AURA), NASA; **M27:** © IAC/RGO/Malin; **Alberio:** Courtesy of Randy Brewer

Looking Up #5 Taurus
Background: © Akira Fujii/DMI; **M1:** Courtesy of R. Wainscoat; **M45:** Courtesy of Anglo-Australian Observatory, photographs by David Malin; **Hyades;** © Akira Fujii/DMI

Looking Up #6 Orion
Background: © Akira Fujii/DMI; **Horsehead Nebula:** Courtesy of Anglo-Australian Observatory, photograph by David Malin; **Close-up of Horsehead Nebula:** Courtesy NOAO/AURA/NSF; **Betelgeuse:** Courtesy A. Dupree (CFA), NASA, ESA; **Pink Orion Nebula:** Courtesy of Carol B. Ivers and Gary Oleski; **Orion Nebula with Dust and Gas:** Courtesy Gary Bernstein (U. Pennsylvania), copyright U. Michigan, Lucent; **Protoplanetary Disk;** Courtesy of A.M. Lagrange, D. Mouillet, and J.L. Beuzit, Grenoble Observatory

Looking Up #7 Sagittarius
Background: Courtesy of Till Credner, AlltheSky.com; **M16 Close-up:** Courtesy of NASA, HST, J. Hester & P. Scowen (ASU); **M16 with Gas Cloud:** Courtesy of Bill Schoening/NOAO/AURA/NSF; **M20:** Courtesy of Jason Ware; **M22:** Courtesy of N.A. Sharp, REU program/NOAO/AURA/NSF

Looking Up #8 Centaurus and Crux, The Southern Cross
Background: Anglo-Australian Observatory/David Malin Images; **Centaurus A:** Courtesy Peter Ward, 2004; **Omega Centauri:** Al Kelly; **Eta Carinae:** J. Hester/Arizona State University, NASA; **The Jewel Box:** Research School of Astronomy and Astrophysics, the Australian National University

Looking Up #9 Southern Circumpolar Constellations
Background: Christopher J. Picking; **Hourglass Nebula:** Raghvendra Sahai and John Trauger (JPL), the WFPC2 science team, and NASA; **Thumbprint Nebula:** Courtesy of STScI; **Tarantula Nebula:** WFI/2.2-m/ESO

Part 1(Unit 2): Image adapted from NASA; (Unit 9): © Bettmann/Corbis

Unit 1 Fig. 1.1: NASA; 1.2a: © Vol. 74 PhotoDisc/Getty; 1.2b: Dr. F.A. Ringwald; 1.3(all) Pictures of planets Courtesy of NASA/JPL; 1.4 (Jupiter): Courtesy of NASA/JPL/University of Arizona; (Earth): Courtesy of NASA; (Sun): Courtesy of SOHO, ESA & NASA

Unit 2 Opener: Image adapted from NASA; Fig. 2.1a: Atlas Image courtesy of 2MASS/UMass/IPAC-Caltech/NASA/NSF; 2.1b: Courtesy Sloan Digital Sky Survey; 2.2a: Courtesy of Anglo-Australian Observatory, photograph by David Malin; 2.2b: Courtesy Anglo-Australian Telescope Board; 2.3: George Greaney; 2.5: Robert Gendler; 2.6: NASA

Unit 3 Fig. 3.1: © The Bridgeman Art Library/Getty

Unit 5 Fig. 5.2a (left): © Roger Ressmeyer/Corbis; (right): Roger Ressmeyer/Corbis, digitally enhanced by Jon Alpert; 5.2b (both): Courtesy of Eugene Lauria

Unit 6 Fig. 6.1: jupiterimages; 6.7c: © English Heritage Library; 6.8: NASA

Unit 8 Fig. 8.1: NASA; 8.4: © Roger Ressmeyer/Corbis; 8.5: John Walker; 8.7b: Stefan Seip-www.astromeeting.de; 8.8b: Courtesy of JPL/California Institute of Technology/NASA; 8.9: © Roger Ressmeyer/Corbis

Unit 9 Opener: © Bettmann/Corbis; Fig. 9.1: © Araldo de Luca/Corbis; 9.2: Steve Schneider; 9.3: Chinese Rare Book Collection/Asian Division/Library of Congress; 9.4: © Bettmann/ Corbis

Unit 10 Fig. 10.1b: Rod Somerville

Unit 11 Fig. 11.3: Courtesy Tenc Tezel; 11.6: © Art Resource/Erich Lessing

Unit 12 Fig. 12.1: © Bettmann/Corbis; 12.2: © Stapleton Collection/Corbis; 12.3: © Art Resource/Erich Lessing; 12.6: Courtesy of the Bridgeman Art Library/Getty

Unit 13 Fig. 13.5: © Hermann Eisenbaiss/Photo Researchers, Inc.; 13.6 (both): Steve Schneider; 13.7: Laurent Laveder-PixHeaven.net; 13.8a: Copyright 2008 Anthony Ayiomamitis

Part 2 (Unit 29): R. Thompson, (University of Arizona) and NASA

Unit 14 Fig. 14.2: © Bettmann/Corbis

Unit 15 Fig. 15.4: © Stone/Getty Images

Unit 16 Fig. 16.4: NASA Johnson Space Center (NASA-JSC)

Unit 19 Fig. 19.2b (both): Tom Arny; 19.3: NASA

Unit 21 Fig. 21.5a,b: © McGraw-Hill Companies/photographer Joe Franek

Unit 22 Fig. 22.1(left): © Ewing Galloway, Inc.; 22.4(Pulsar): Courtesy of NASA/CXC/SAO; (Sun): SOHO (ESA & NASA); (other stars): Courtesy of NOAO and Anglo-Australian Observatory; (Cold Interstellar cloud): Courtesy of Anglo-Australian Telescope Board, photo by David Malin; (Cosmic microwave background): Courtesy of NASA/WMAP; (Active galaxy): NRAO/AUI/NSF

Unit 23 Fig. 23.1: Photo Researchers, Inc.; 23.4(both): Spitzer Science Center and the Infrared Processing and Analysis Center; 23.5: Courtesy of SOHO/MDI consortium. SOHO is a project of international cooperation between ESA and NASA

Unit 24 Fig. 24.7a: © Courtesy of Mees Solar Observatory, University of Hawaii

Unit 26 Fig. 26.1: © Roger Ressmeyer/Corbis; 26.4a,b: Courtesy of NRAO/AUI; 26.5b: Courtesy of Richard Wainscoat; 26.5c: Courtesy of NRAO; 26.5d: NASA/CXC/ASU/J. Hester et al

Unit 27 Fig. 27.1: Photograph courtesy of Gara Mora, IAC; 27.2: © European Southern Observatory; 27.3: Courtesy of the NAIC-Arecibo Observatory, a facility of the NSF

Unit 28 Fig. 28.2: Tom Arny; 28.4: Photo courtesy of Yerkes Observatory; 28.8a: © Roger Ressmeyer/Corbis; 28.8b: US Gemini Project/AURA/NOAO/NSF; 28.9: Image courtesy of NRAO/AUI; 28.10: The Video Encyclopedia of Physics Demonstrations © The Education Group

Unit 29 Opener: R. Thompson, (University of Arizona) and NASA; Fig. 29.1a: © E.R. Degginger; 29.1b: R. Thompson, (University of Arizona) and NASA; 29.2: California Institute of Technology; 29.3: Courtesy of Steve Criswell; 29.4a,b: Courtesy of Andrea Gehz, UCLA

Unit 30 Fig. 30.2a: Courtesy of Kitt Peak National Observatory; 30.2b: Courtesy Astronomical Society of the Pacific; 30.2c: W.T. Sullivan, © 1993; 30.4: Courtesy of USAF; 30.5c: Photo courtesy of Patrick Watson; 30.6 (HST): Courtesy of NASA/JPL; (EUVE): Courtesy of NASA, ESA, and Max Planck Institute for Extraterrestrial Physics; (Chandra): CXC/TRW: 30.7(all): Courtesy of NASA and The Hubble Heritage Team

Unit 31 Fig. 31.5: Courtesy of Alan Dyer; 31.6: Joe Orman; 31.7a: © Roger Ressmeyer/Corbis; 31.7b: © Carol B. Ivers and Gary Oleski

Part 3 (Unit 34): © ESO; (Unit 35): Photo courtesy of NASA; (Unit 36): SVS/TOMS/NASA; (Units 37-39): Courtesy of NASA; (Unit 40): ESA/DLR/FU Berlin (G. Neukum); (Unit 41): NASA Jet Propulsion Laboratory (NASA-JPL); (Unit 43): NASA/JPL/Space Science Institute; (Unit 44): Courtesy of STScI; (Unit 45) (Tethys – Oberon): NASA Jet Propulsion Laboratory (NASA-JPL); (Larissa): NASA Goddard Space Flight Center; (Unit 46): NASA Jet Propulsion Laboratory (NASA-JPL); (Unit 47): Courtesy of Mike Skrutskie, University of Virginia; (Unit 48): Courtesy of John A. Wood, Harvard-Smithsonian Center for Astrophysics

Unit 32 Fig. 32.1(planets): Courtesy of NASA/JPL; (sun): Courtesy SOHO/ESA/NASA

Unit 33 Fig. 33.2: Courtesy of Anglo-Australian Observatory, photograph by David Malin; 33.4: © Fundamental Photographs/ Diane Schiumo; 33.8a: Tom Pfeiffer/Volcano Discovery

Unit 34 Opener: © ESO; Fig. 34.1: Mark McCaughrean (Max-Planck-Institute for Astronomy), C. Robert O'Dell (Rice University), and NASA; 34.2a: Smith (University of Hawaii), G. Schneider (University of Arizona), E. Becklin and A. Weinberger (UCLA) and NASA; 34.2b: NASA, ESA, D. Golimowski (Johns Hopkins University), D. Ardila (IPAC), J. Krist (JPL), M. Clampin (GSFC), H. Ford (JHU), and G. Illingworth (UCO/Lick) and the ACS Science Team; 34.6: © ESO

Unit 35 Opener: Photo courtesy of NASA; Fig. 35.1: Copyright © 2005 Planetary Visions Limited; 35.2: © Taxi/Getty Images; 35.10a: Copyright by Marie Tharp 1977/2003. Reproduced by permission of Marie Tharp Oceanographic Cartographer, One Washington Ave., South Nyack, New York 10960; 35.12: Tom Arny

Unit 36 Opener: SVS/TOMS/NASA; Fig. 36.1: Courtesy of NASA; 36.2: Steve Schneider; 36.5a: Courtesy Paul Schneider; 36.5b: Courtesy of NASA; 36.6: Sylvain Grandadam/Photo Researchers, Inc.; 36.7(top): © Maso Hayashi/Photo Researchers; (bottom): © Douglas/Photo Researchers; 36.8: SVS/TOMS/NASA; 36.12: Courtesy of NOAA

Unit 37 Opener: Courtesy of NASA; Fig. 37.1: NASA/JPL/USGS; 37.2a,b: UC Regents/Lick Observatory. Unauthorized use prohibited; 37.3: Courtesy of NASA; 37.4a: Courtesy of NASA; 37.5b: © John Gillmoure/Corbis; 37.5: NASA; 37.6: Astrogeology Team, U.S. Geological Survey, Flagstaff, Arizona; 37.7: © UC Regents, UCO/Lick Observatory image; 37.10: Courtesy of NASA; 37.12b: Courtesy of Robin Conup, Southwest Research Institute

Unit 38 Opener: Courtesy of NASA; Fig. 38.1: NASA/Johns Hopkins University Applied Physics Lab/Carnegie Institution of Washington; 38.2: Courtesy of NASA; 38.3a: Courtesy of NASA; 38.3b: NASA/ Johns Hopkins University Applied Physics Laboratory/Carnegie Institution of Washington; 38.4: Courtesy of NASA; 38.5: NASA/ Johns Hopkins University Applied Physics Laboratory/Carnegie Institution of Washington

Unit 39 Opener & Fig. 39.1: Courtesy of NASA; 39.2: National Space Science Data Center; 39.3-39.5: Courtesy of NASA/JPL

Unit 40 Opener: ESA/DLR/FU Berlin (G. Neukum); Fig. 40.1: Courtesy of HST; 40.2: Courtesy of A.S. McEwen, USGS; 40.3: MOLA Science Team, MGS, NASA; 40.4a: Courtesy of A.S. McEwen, USGS; 40.5: Courtesy of NASA; 40.6a: Courtesy of NASA/JPL; 40.6b: Courtesy of A.S. McEwen, USGS; 40.6c: NASA/JPL-Caltech/University of Arizona; 40.7a: Courtesy of NASA; 40.7b: NASA/Calvin J. Hamilton; 40.8a: Courtesy of NASA/JPL; 40.8b: Courtesy of NASA/JPL/Malin Space Science Systems; 40.8c: ESA/DLR/FU Berlin (G. Neukum); 40.9a: Courtesy of NASA; 40.9b: NASA/JPL/Malin Space Science Systems; 40.10: Courtesy of NASA/JPL; 40.11: Courtesy of Mars Exploration River Mission, JPL , NASA; 40.12 & 40.13: Courtesy of NASA/JPL/Cornell; 40.14a,b & 40.15: Courtesy of NASA/JPL

Unit 41 Opener: NASA Jet Propulsion Laboratory (NASA-JPL); Fig. 41.2 (Vesta): NASA Goddard Space Flight Center (NASA-GSFC); (Eros): NEAR Shoemaker (JHU/APL); (Ceres): NASA, ESA, J. Parker (Southwest Research Institute), P. Thomas (Cornell University), and

L. McFadden (University of Maryland, College Park); 41.3a: NASA Jet Propulsion Laboratory (NASA-JPL); 41.3b: ISAS/JAXA; 41.4: Courtesy of NASA/JPL

Unit 42 Fig. 42.1a (Mercury): NASA Headquarters – Greatest Images of NASA (NASA-HQ-GRIN); 42.1b-j: Courtesy of NASA/JPL/USGS; 42.1k(swirling clouds): NASA Johnson Space Center – Earth Sciences and Image Analysis (NASA-JSC-ES&IA); 42.1l-v: Courtesy of NASA/JPL/USGS; 42.3: Courtesy of NASA/JPL; 42.5 (Mercury): Courtesy NASA's Planetary Geology and Geophysics Program/Arizona State University; (Mars, Manicougan Crater on Earth, Dione, Tethys): Courtesy of NASA

Unit 43 Opener: NASA/JPL/Space Science Institute; Fig. 43.1(Jupiter): Courtesy of NASA/JPL; (Earth): Courtesy of NASA; (Saturn): Courtesy of Cassini Imaging Team/SSI/JPL/ESA/NASA; 43.2: Courtesy of NASA; 43.5a: Courtesy of NASA/JPL; 43.6: Lawrence Sromovsky, University of Wisconsin-Madison/W.M. Keck Observatory; 43.7a: Courtesy of NASA/JPL; 43.7b: J.T. Trauger (Jet Propulsion Laboratory) and NASA; 43.7c: Courtesy of NASA/JPL

Unit 44 Opener: Courtesy of STScI; Figs. 44.1a,b & 44.4: Courtesy of NASA/JPL; 44.5: Courtesy of STScI; 44.7: Lawrence Sromovsky, University of Wisconsin-Madison/W.M. Keck Observatory

Unit 45 Opener (Tethys – Oberon): NASA Jet Propulsion Laboratory (NASA-JPL); (Larissa): NASA Goddard Space Flight Center; Fig. 45.1: Courtesy Don Wheeler and Brian Dunn/Louisiana Delta Community College; 45.3 (lo): NASA Jet Propulsion Laboratory (NASA-JPL); (Europa): NASA; (Ganymede, Callisto, Titan): NASA Jet Propulsion Laboratory (NASA-JPL); (Triton): National Space Science Data Center; 45.4 (Amalthea & Mimas): © Calvin J. Hamilton; 45.4 (Enceladus – Oberon): NASA Jet Propulsion Laboratory (NASA-JPL); 45.4 (Larissa & Proteus): NASA Goddard Space Flight Center; 45.4 (Nereid): NASA Jet Propulsion Laboratory (NASA-JPL); 45.5a: NASA, Voyager 1, © Calvin J. Hamilton; 45.5b: NASA/JPL/ Space Science Institute; 45.6: NASA Jet Propulsion Laboratory (NASA-JPL); 45.7, 45.8 & 45.9: Courtesy of NASA/JPL; 45.10(Uranus): NASA and Erich Karkoschka, University of Arizona; 45.10 (Jupiter, Neptune): NASA Jet Propulsion Laboratory (NASA-JPL); 45.11: Courtesy of NASA/JPL; 45.12: Bjorn Jonsson; 45.13: Courtesy of STSci/HST

Unit 46 Opener: NASA Jet Propulsion Laboratory (NASA-JPL); Fig. 46.1(Io, Europa, Ganymede, Calliso, Titan, 2003UB313, Orcus, Sedna): NASA Jet Propulsion Laboratory (NASA-JPL); 46.1(Triton): National Space Science Data Center; 46.1(Pluto & Charon): Images courtesy of Marc W. Buie/Lowell Observatory; 46.1(Moon): © Digital Vision RF/Punchstock; 46.1 (Earth): © Stocktrek/Getty Images RF; 46.2a-c & 46.3a,b: NASA Jet Propulsion Laboratory (NASA-JPL); 46.4a(top): NASA Jet Propulsion Laboratory (NASA-JPL); 46.4a(bottom): NASA Goddard Space Flight Center; 46.4b,c (all): NASA Jet Propulsion Laboratory (NASA-JPL); 46.5a: NASA Jet Propulsion Laboratory (NASA-JPL); 46.5b: Courtesy of NASA/JPL/University of Arizona; 46.5c: NASA/JPL; 46.6a-c & 46.7: NASA Jet Propulsion Laboratory (NASA-JPL); 46.8a: © UC Regents; UCO/Lick Observatory image; 46.8b: Courtesy of NASA; 46.8c: Courtesy of STScI

Unit 47 Opener: Courtesy of Mike Skrutskie, University of Virginia; Fig. 47.1: © Robert McNaught/Photo Researchers, Inc.; 47.3a,b: NASA Jet Propulsion Laboratory (NASA-JPL); 47.4: NASA/JPL-Caltech/UMD; 47.7: Courtesy of Mike Skrutskie, University of Virginia; 47.10: Courtesy Juan Carlos Casado and Isabel Graboleda

Unit 48 Opener: Courtesy of John A. Wood, Harvard-Smithsonian Center for Astrophysics; Fig. 48.1: Courtesy of Ronald A. Oriti, Santa Rosa Junior College, Santa Rosa, Calif.; 48.3 & 48.4: Steve Schneider; 48.5: NASA Jet Propulsion Laboratory (NASA-JPL); 48.6a: Courtesy of John A. Wood, Harvard-Smithsonian Center for Astrophysics; 48.8: Courtesy of Landsat/EOS; 48.9: NASA Jet Propulsion Laboratory (NASA-JPL); 48.10b: A. Hildebrand, M. Pilkington, and M. Connors

Part 4 (Unit 51): Courtesy of NOAO/AURA/NSF; (Unit 55): NOAO/AURA/NSF

Unit 49 Fig. 49.1a: SOHO/MDI project; 49.1b: Courtesy of SOHO/Extreme Ultraviolet Imaging Telescope (EIT) consortium; 49.6: Prof. Goran Scharmer/Dr. Mats G. Lofdahl/Institute for Solar Physics of the Royal Swedish Academy of Sciences; Courtesy of Jacques Guertin, Ph.D.; 49.8: Courtesy NOAO/AURA/NSF; 49.9: 1988 eclipse image courtesy High Altitude Observatory (HAO), University Corporation for Atmospheric Research (UCAR), Boulder, Colorado. UCAR is sponsored by the National Science Foundation; 49.10: Courtesy of LMSAL and NASA; 49.12: Courtesy NOAO/AURA/NSF

Unit 50 Fig. 50.4: Courtesy of Ernest Orlando, Lawrence Berkeley National Laboratory; 50.5: R. Svoboda and K. Gordan (LSU)

Unit 51 Opener: Courtesy of NOAO/AURA/ NSF; Fig. 51.1: Courtesy of Royal Swedish Academy of Sciences; 51.2a(left): NSO/AURA/NSF; (right): Courtesy of NOAO/AURA/NSF; 51.2b: Courtesy of NOAO/AURA/NSF; 51.4: NASA/ESA; 51.5b-d: Courtesy of NOAO/AURA/NSF; 51.6a: Courtesy of SOHO-EIT Consortium, ESA, NASA; 51.6b: Courtesy of NOAO/AURA/NSF; 51.7: Courtesy of Eugene Lauria

Unit 52 Fig. 52.4: Courtesy of Stefan Seip/www.astromeeting.de

Unit 53 Figs. 53.3 & 53.5: © Hulton-Deutsch Collection/Corbis

Unit 54 Fig. 54.1: © Edward P. Flaspoehler, Jr.; 54.3: © Getty Images; 54.4: © 2001 Bert Katzung; 54.5: Courtesy of Dean Ketelsen

Unit 55 Opener: NOAO/AURA/NSF; Fig. 55.2: © Akira Fujii/DMI; 55.4: Courtesy of Dean Ketelsen; 55.5: Courtesy of Harvard College Observatory; 55.6 NOAO/AURA/NSF; 55.7: Courtesy of Katherine Haramundanis; 55.8: NOAO/AURA/NSF

Unit 57 Fig. 57.1: Astrophysics Group, University of Cambridge, Department of Physics

Unit 58 Fig. 58.6: NOAO/AURA/NSF

Unit 60 Fig. 60.1: Five College Radio Astronomy Observatory (FCRAO); 60.2: NASA and The Hubble Heritage Team (STScI/AURA); 60.3: NASA, ESA, M. Robberto (Space Telescope Science Institute/ESA) and the Hubble Space Telescope Orion Treasury Project Team; 60.4(inset): © European Southern Observatory; 60.5a: Courtesy of STScI; 60.5b: © European Southern Observatory; 60.6a: NASA, ESA, N. Smith (University of California Berkeley), and The Hubble Heritage Team (STScI/AURA); 60.6b: J. Alves (ESO), E. Tostoy (Groningen), R. Fobury (ST-ECF), & R. Hook (ST-ECF), VLT; 60.7a: Courtesy of Anglo-Australian Observatory, photograph of David Malin

Unit 61 Fig. 61.4: Courtesy of Anglo-Australian Observatory; photograph of David Malin

Unit 63 Fig. 63.1: © Akira Fujii/DMI

Unit 64 Fig. 64.1: Courtesy of NASA, ESA, and Valentin Bujarrabal-Observatorio Astronomico National, Spain; 64.5: NASA/JPL-Caltech; 64.6a: Courtesy of Hubble Heritage Team (AURA, STScI/NASA); 64.6b: Courtesy of Anglo-Australian Observatory; photograph by David Malin; 64.6c: Raghvendra Sahai and John Trauger (JPL), the WFPC2 Science Team, and NASA; 64.6d: Bruce Balick, University of Washington; Vincent Icke, Leiden University, Netherlands; Garrelt Mellema, Stockholm University/NASA

Unit 65 Fig. 65.3: Courtesy of Paresce, R. Jedrzejewski (STScI), NASA, ESA; 65.4: NASA, ESA, and Z. Levay (STScI); 65.6: Pablo Candia; 65.8a,b: NASA/CXC/Rutgers/J.Warren & J.Hughes et al

Unit 66 Fig. 66.5 (both): Courtesy of Anglo-Australian Observatory, photograph by David Malin; 66.6a: NASA/CXC/MIT/UMass Amherst/M.D.Stage et al.; 66.6b: Courtesy of R. Wainscoat; 66.6c: NASA and The Hubble Heritage Team (STScI/AURA)

Unit 67 Fig. 67.3b: Photos courtesy of NOAO; 67.6: © European Southern Observatory

Unit 69 Fig. 69.1a: NASA, ESA and AURA/Caltech; 69.1b,c: Courtesy of Anglo-Australian Observatory, photograph by David Malin; 69.2a: Max Kilmister; 69.2b: Robert Lupton and the Sloan Digital Sky Survey Consortium

Part 5 (Unit 71): David Law/University of Virginia; (Unit 73): Courtesy of UCLA Galactic Center Group; (Unit 76): Courtesy of NOAO; (Unit 83): Courtesy of Ken Eward/BioGrafx; (Unit 84): National Astronomy & Ionosphere Center (NAIC)

Unit 70 Fig. 70.1: Courtesy of Steward Observatory and NOAO; 70.2: Courtesy of Anglo-Australian Observatory, photograph by David Malin; 70.6(top): Courtesy of Hubble Heritage Team (AURA/STScI/NASA; 70.7: Courtesy Sloan Digital Sky Survey

Unit 71 Opener & Fig. 71.4: David Law/University of Virginia; 71.5a-c: Courtesy of Neal Katz, University of Massachusetts, and James E. Gunn, Princeton University; 71.5d: David Law/University of Virginia; 71.5e: Courtesy of Neal Katz, University of Massachusetts, and James E. Gunn, Princeton University

Unit 72 Fig. 72.1: Courtesy of NASA and The Hubble Heritage Team (STScI/AURA); 72.2a-c, 72.3b & 72.4: Courtesy of Anglo-Australian Observatory, photograph by David Malin; 72.7a: Canada-France-Hawaii Telescope/J.C. Cuillandre/Coelum; 72.8: Atlas Image courtesy of 2MASS/UMASS/IPAC-Caltech/NASA/NSF

Unit 73 Opener: Courtesy of UCLA Galactic Center Group; Fig. 73.3: Courtesy of NASA, UMass, D. Wang et al; 73.4: Courtesy of Rainer Scholdel (MPE) et al., NAOS-CONICA, ESO; 73.5: Courtesy of UCLA Galactic Center Group; 73.7a: Stefan Seip/Adam Block/NOAO/AURA/NSF; 73.7b: Bruce Hugo and Leslie Gaul/Adam Block/NOAO/AURA/NSF

Unit 74 Fig 74.1: Courtesy of George Greany; 74.2: Courtesy of William C. Keel; 74.4b: Dr. Wendy L. Freedman, Observatories of the Carnegie Institution of Washington, and NASA

Unit 75 Fig. 75.1a: Courtesy of Bruce Hugo and Leslie Gaul, Adam Block, NOAO, AURA, ASF; 75.1b: Courtesy of Hubble Heritage Team, AURA/STScI/NASA; 75.1c: © European Southern Observatory; 75.2a: California Institute of Technology; 75.2b: NOAO; 75.2c, 75.3 & 75.4: Courtesy of Anglo-Australian Observatory, photograph by David Malin; 75.5 & 75.6: Courtesy Sloan Digital Sky Survey; 75.7a: Courtesy of Anglo-Australian Observatory, photograph by David Malin; 75.7b: Steve Schneider; 75.8: Courtesy Sloan Digital Sky Survey; 75.9: Roff/Lowell Observatory; 75.10a,b: Images courtesy of STScI; 75.11: Courtesy of F. Schwizer, Carnegie Institution of Washington; 75.13: Courtesy of Joshua Barnes, University of Hawaii; 75.14a: NASA, ESA, and The Hubble Heritage Team (STScI/AURA); 75.14b: Courtesy of STScI and NOAO; 75.14c: Courtesy of NASA, H. Ford (JHU), G. Illingworth (USCS/LO), M. Clampin (STScI), G. Hartig (STScI), the ACS Science Team, and ESA; 75.15: Courtesy of Mary E. Putman, University of Michigan; Lister Staveley-Smith, Australia Telescope National Facility; Kenneth C. Freeman, Australian National University; and Brad K. Gibson and David G. Barnes, both Swinburne University of Technology. The Parkes Telescope is operated by CSIRO

Unit 76 Opener: Courtesy of NOAO; Fig. 76.1: Two Miron All Sky Survey, a joint project of the University of MA and the Infrared Processing and Analysis Center/CA Institute of Technology, funded by NASA/National Science Foundation/T.H. Jarrett, J. Carpenter, & R. Hurt. Perha; 76.3: Martin & Rodrigo Ibata, Observatoire de Strasbourg, 2003; 76.4: Courtesy of NOAO; 76.5: Robert Gendler; 76.6a: Courtesy of A. Fabian (IoA Cambridge) et al., NASA; 76.6b: NASA/CXC/ESO/P. Rosati et al; 76.9: Numerical simulations performed at the National Center for Supercomputing Applications (NCSA, Urbana-Champaign, Illinois) by Andrey Kravtsov (The University of Chicago) and Anatoly Klypin (New Mexico State University)

Unit 77 Fig. 77.1: AURA/STSci/NASA; 77.2a-c: Images courtesy of NRAO/AUI; 77.3: Courtesy of NOAO; 77.4: Courtesy of John Bahcall, Institute for Advanced Study; M. Disney, University of Wales; and NASA; 77.7: Courtesy of W. Jaffe, Leiden Observatory, and H. Ford, Johns Hopkins University, Space Telescope Science Institute; and NASA; 77.7(inset): Courtesy of NRAO and California Institute of Technology; 77.10: NASA/CXC/CfA/INAF/Risaliti Optical: ESO/VLT

Unit 78 Fig. 78.5a: Courtesy of NASA; 78.6: Courtesy of STScI; 78.7: NASA/STScI; Magellan/U. Arizona/D. Clowe et al

Unit 80 Fig. 80.1a: Courtesy of Seldner, M., Siebers. B: Groth, E.J., and Peebles, P.J.E., A.J. 82, 4, "New Reduction of the lick Catalog of Galaxies"; 80.6: Courtesy S. Beckwith and the HUDF Working Group (STScI), HST, ESA, NASA; 80.7: Courtesy NASA/WMAP Science Team

Unit 81 Fig. 81.7a,b: Courtesy NASA/WMAP Science Team

Unit 83 Opener: Courtesy of Ken Eward/BioGrafx; Fig. 83.1a: Reproduced with permission from Dr. J. William Schopf, 1992-1993, CSEOL/UCLA; 83.1b: © Dwight Kuhn; 83.2: © Paul Hoffman; 83.2(inset): Courtesy of Donald R. Lowe; 83.4(whale): © Brandon Cole/Visuals Unlimited; (butterfly): © Corbis RF; (giraffe): © DV RF/Getty; 83.6: Courtesy of Ken Eward/BioGrafx; 83.8: Dr. Sidney Fox; 83.9a,b: © Dudley Foster/Woods Hole Oceanographic Institution; 83.10: Andrew H. Knoll, Harvard University; 83.11: © R. L. Batten

Unit 84 Opener: National Astronomy & Ionosphere Center (NAIC); Fig. 84.1: Courtesy of NASA/JPL/Cornell; 84.2: NASA; 84.5: Seth Shostak; 84.6: National Astronomy & Ionosphere Center (NAIC)

INDEX

C

D

E

F

G

H

I

J

K

L

M

P

Q

R

S

T

U

V

W

X

Y

Z

SPRING NIGHT SKY

When to Use This Star Map

Early March:	2 a.m.
Late March:	1 a.m.
Early April:	midnight
Late April:	11 p.m.
Early May:	10 p.m.
Late May:	Dusk

This star chart is most accurate if used within an hour or so of the times listed and is plotted for observers located between 30° and 50° north latitude. All times are standard time; if daylight-saving time is in effect, add one hour.

To use this chart, hold it in front of you and rotate it so that the yellow label corresponding to the direction you are facing is positioned at the bottom, right-side up. The stars in the sky should match those depicted on the chart. The center of the chart is the zenith, the point in the sky directly overhead.

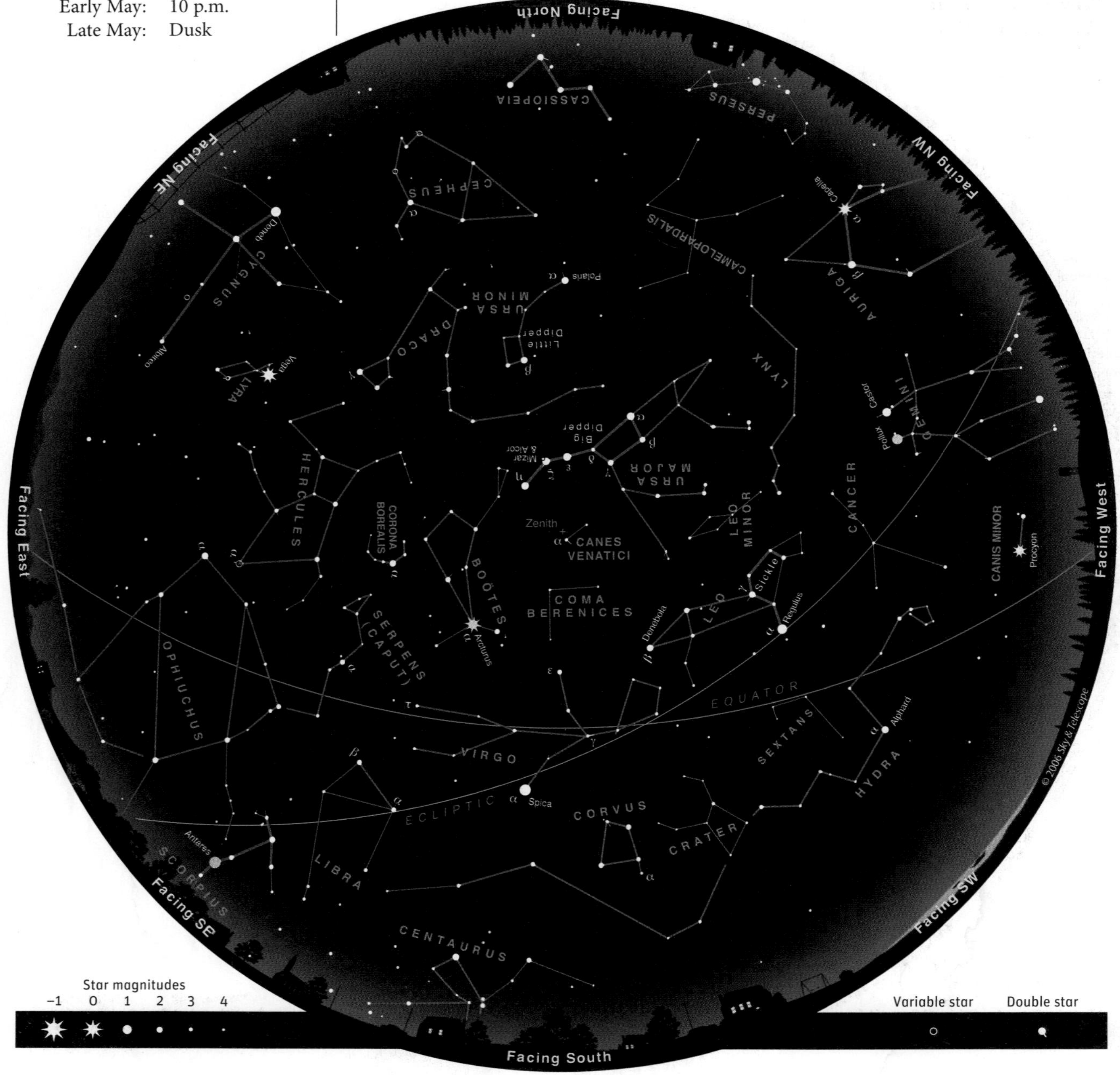

SUMMER NIGHT SKY

When to Use This Star Map

Early June:	1 a.m.
Late June:	midnight
Early July:	11 p.m.
Late July:	10 p.m.
Early August:	9 p.m.
Late August:	Dusk

This star chart is most accurate if used within an hour or so of the times listed and is plotted for observers located between 30° and 50° north latitude. All times are standard time; if daylight-saving time is in effect, add one hour.

To use this chart, hold it in front of you and rotate it so that the yellow label corresponding to the direction you are facing is positioned at the bottom, right-side up. The stars in the sky should match those depicted on the chart. Ignore all the parts of the map above horizons you are not facing.

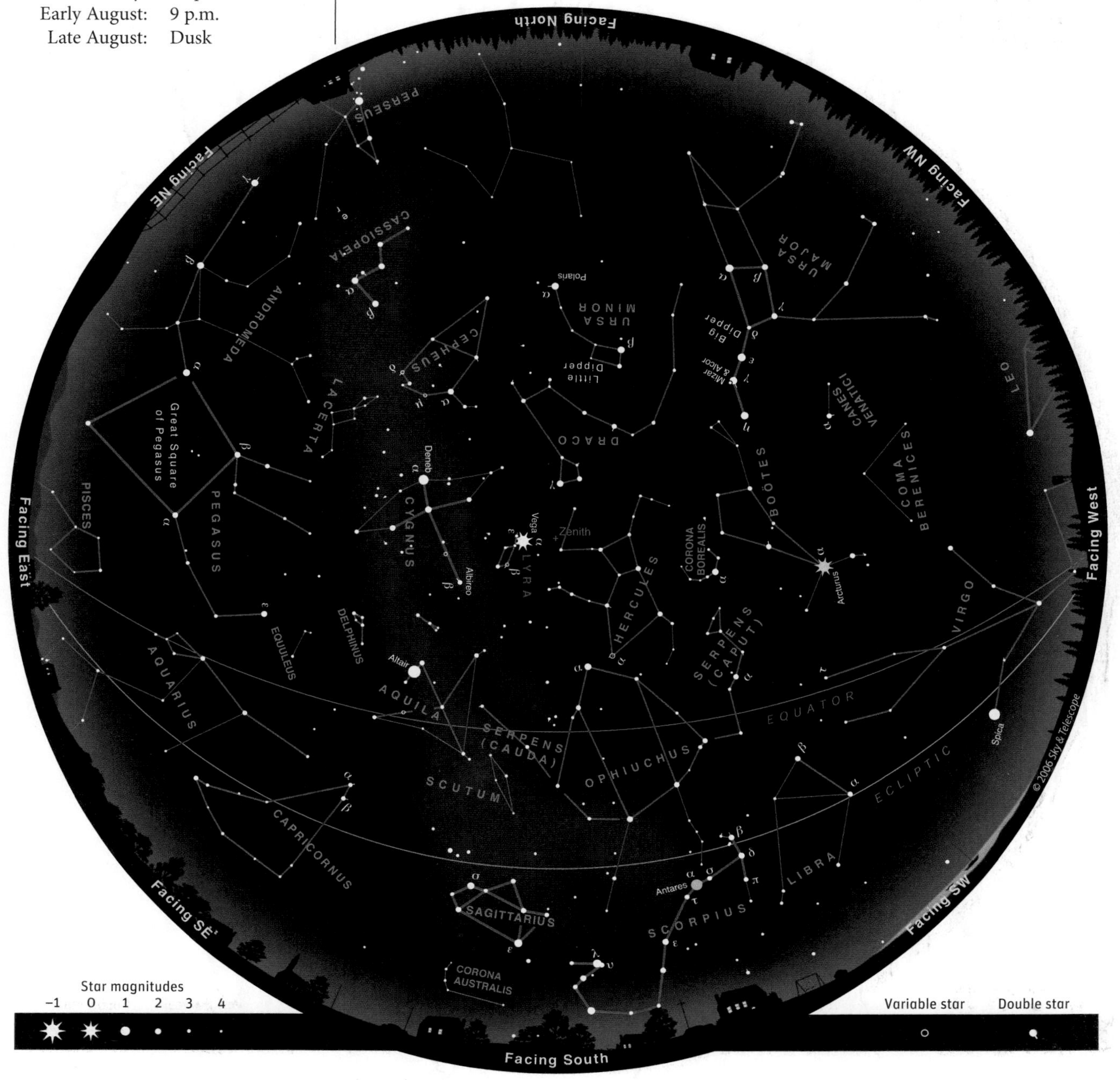

AUTUMN NIGHT SKY

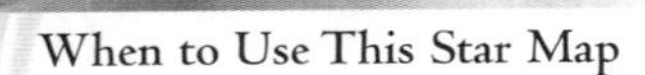

When to Use This Star Map

Early September:	midnight
Late September:	11 p.m.
Early October:	10 p.m.
Late October:	9 p.m.
Early November:	8 p.m.
Late November:	7 p.m.

This star chart is most accurate if used within an hour or so of the times listed and is plotted for observers located between 30° and 50° north latitude. All times are standard time: if daylight-saving time is in effect, add one hour.

To use this chart, hold it in front of you and rotate it so that the yellow label corresponding to the direction you are facing is positioned at the bottom, right-side up. The stars in the sky should match those depicted on the chart. The farther up from the map's edge they appear, the higher they'll be shining in your sky.

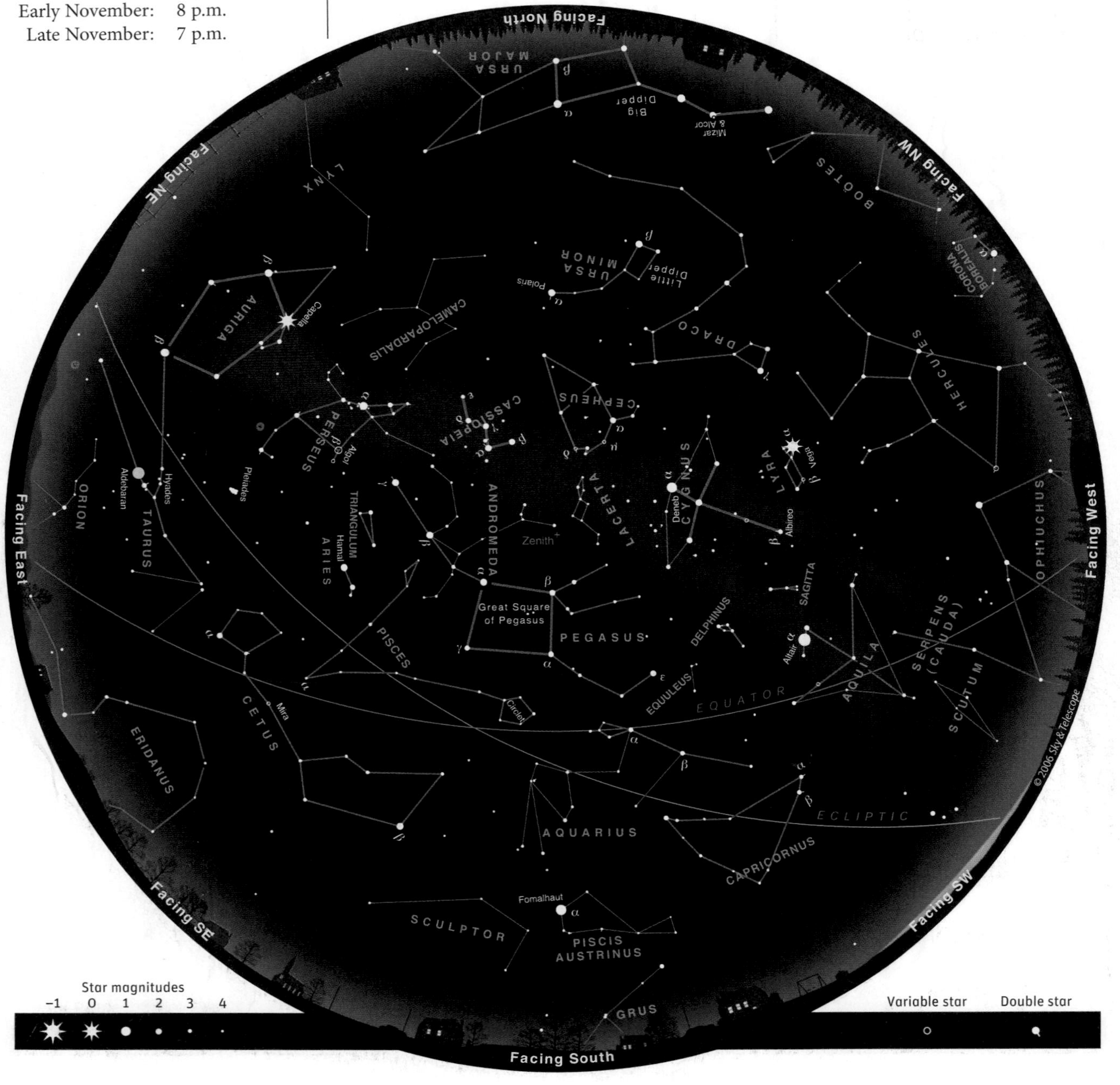

WINTER NIGHT SKY

When to Use This Star Map

Early December:	midnight
Late December:	11 p.m.
Early January:	10 p.m.
Late January:	9 p.m.
Early February:	8 p.m.
Late February:	Dusk

This star chart is most accurate if used within an hour or so of the times listed and is plotted for observers located between 30° and 50° north latitude. All times are standard time.

To use this chart, hold it in front of you and rotate it so that the yellow label corresponding to the direction you are facing is positioned at the bottom, right-side up. The stars in the sky should match those depicted on the chart. Remember that star patterns will look much larger in the sky than they do here on paper.

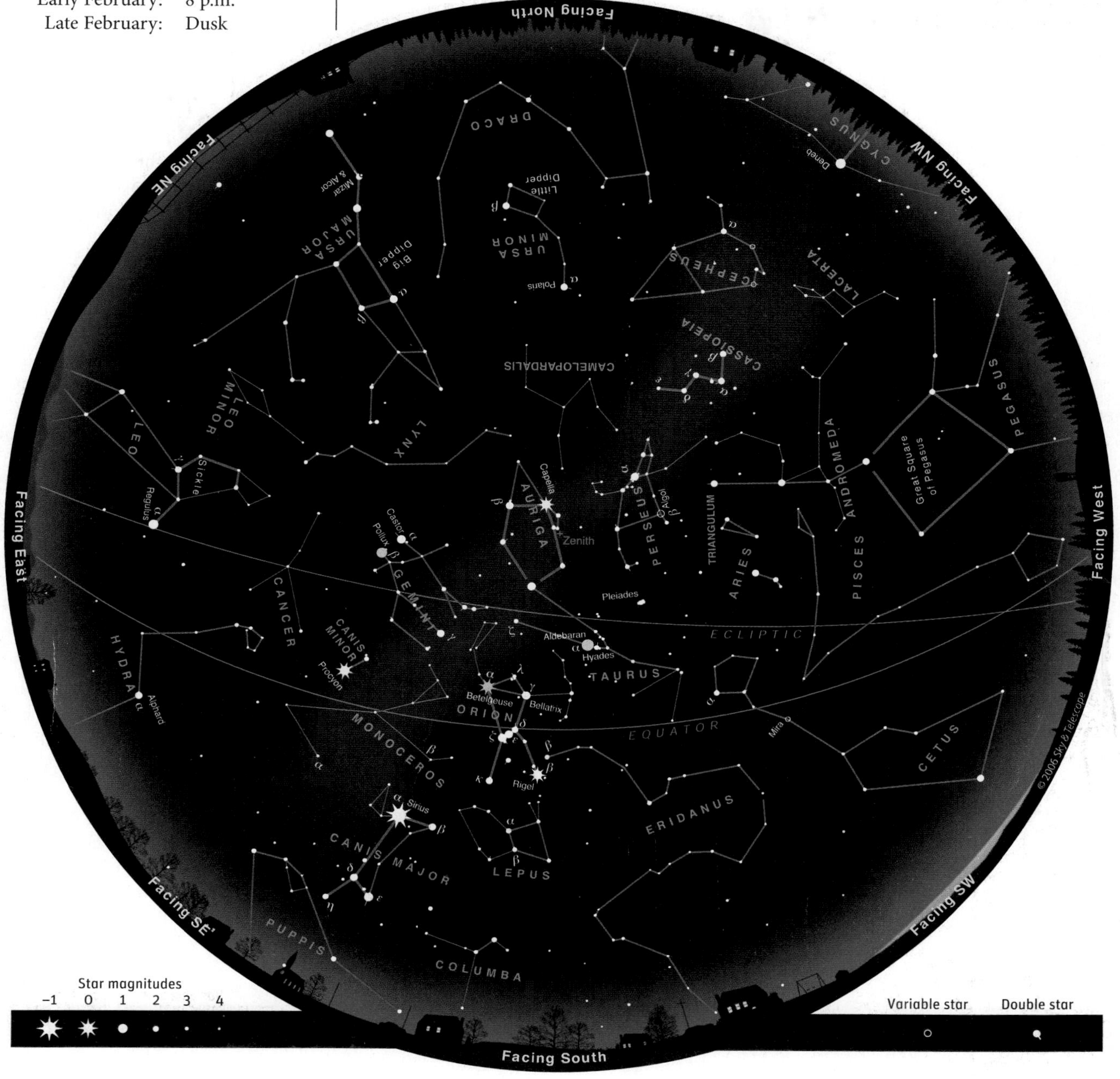

Using the Foldout Star Charts➤

These star charts cover the entire sky and can be used to identify stars and constellations from any location at any time. They were produced at Stephen F. Austin State University by Professor Dan Bruton and his students, with additions by the publisher. The large rectangular chart covers a wide region around the celestial equator, and the two circular charts are centered on the northern and southern celestial poles. The charts are labeled with Right Ascension and Declination coordinates, which are like longitude and latitude on the celestial sphere (see Unit 5). Stars visible to the naked eye have sizes on the charts that depend on their brightness (or apparent magnitude—see Unit 54). Lines connecting the stars help to identify the pattern and primary stars in each constellation. The pale blue region shows the approximate location of the Milky Way.

For viewers in the northern hemisphere, to observe the northern half of the sky, use the **Northern Region** chart. If you are looking at the sky around 8 P.M., you should locate the closest date along the outer perimeter of the northern region chart. Rotate the chart so that the correct date is at the top. For each hour earlier/later that you are outside, you should rotate the map so that a Right Ascension an hour earlier/later is at the top. For example, if it is 8 P.M., on September 18, you would rotate the map so that "Sep 20" (Right Ascension of 20^h) is at the top. On the same date at 9 P.M., you would rotate the map so that a Right Ascension of 21^h is at the top. Held with this orientation, the chart should match what you see over the northern sky. The stars that are due north will match what you see at the center of the chart. Stars at the bottom of the chart will be below the horizon and stars near the top of the chart will be overhead.

Looking southward from the northern hemisphere, use the **Equatorial Region** chart. Find the date along the top of the chart, and adjust to an earlier or later Right Ascension according to how much earlier or later than 8 P.M. you are observing the sky as explained for the Northern Region chart. Stars located at the Right Ascension you have identified will lie straight up from the point on the horizon that is due south of you, along the "meridian" (see Unit 7). For observers north of latitude 30°, some stars along the bottom of the chart will never rise above the horizon.

The curving "sine wave" line running through the middle of the Equatorial Region chart is the ecliptic (Unit 6), which marks the path of the Sun among the stars. The dates along this line indicate where the Sun is located throughout the year. The Moon and planets also remain close to this line as they move around the celestial sphere.

The star charts can be used from the southern hemisphere using the same instructions but swapping the words "north" and "south" wherever they occur. Looking northward from the southern hemisphere, you will also need to hold the Equatorial Region chart upside down.

Using the Moon and Planet Finder Tables

Along the bottom of the foldout chart, information is provided to help you find where the Moon and five bright planets will be located each month. The Moon will be in its new phase and will lie close to the ecliptic at the Sun's position on the date given in the table. It then shifts about 13° to the east each subsequent day. If the date of the new moon is in a black circle, there will be a total solar eclipse on that date. The dates of total lunar eclipses of the *full moon* are also provided, listed in red circles.

The locations of Venus, Mars, Jupiter, and Saturn each month are indicated by the abbreviations for the constellations they are in. The abbreviation is listed in green when the planet is primarily in the morning sky and listed in blue for the evening sky. The listing is in black when the planet is in opposition to the Sun, indicating that it rises at sunset and sets at sunrise. The word *Sun* is listed when the planet is nearly in conjunction and therefore difficult to see. For Venus, the month when it is at its greatest elongation from the Sun is marked by an asterisk.

Mercury is always fairly close to the Sun and can be seen only shortly after sunset or shortly before sunrise. The date of its greatest elongation is given in green when the planet is in the morning sky and in blue when it is in the evening sky. The best opportunity for seeing Mercury is generally within about one week of this date.